Second Edition

BROGDON'S FORENSIC RADIOLOGY

Second Edition

BROGDON'S
FORENSIC RADIOLOGY

Edited by Michael J. Thali, M.D., Mark D. Viner, B. G. Brogdon

CRC Press
Taylor & Francis Group
Boca Raton London New York

CRC Press is an imprint of the
Taylor & Francis Group, an **informa** business

First published in paperback 2024

First published 2011 by CRC Press
2385 NW Executive Center Drive, Suite 320, Boca Raton FL 33431

and by CRC Press
4 Park Square, Milton Park, Abingdon, Oxon, OX14 4RN

CRC Press is an imprint of Taylor & Francis Group, LLC

Publisher's Note
The publisher has gone to great lengths to ensure the quality of this reprint but points out that some imperfections in the original copies may be apparent.

Library of Congress Cataloging-in-Publication Data

Brogdon's forensic radiology / edited by Michael J. Thali, Mark D. Viner, B.G. Brogdon. -- 2nd ed.
p. ; cm.
Forensic radiology
Rev. ed. of: Forensic radiology / B.G. Brogdon. c1998.
Includes bibliographical references and index.
Summary: "Brogdon's Forensic Radiology demonstrates the application of radiology in cases involving identifying remains, evaluating causes of death, and establishing evidence. This second edition features new material on the large scale investigation of war crimes and human rights abuses; recent technological changes and the role of virtual autopsies; the use of sectional imaging for dental identification; reconstruction of gunshot injuries; new techniques for portable x-ray, computerized data recording, and image analysis post-processing; as well as increased coverage of applications in non-biological materials. Case studies feature tsunamis and the London bombings"--Provided by publisher.
ISBN 978-1-4200-7562-5 (hardcover : alk. paper)
1. Forensic radiography. I. Thali, Michael J. II. Viner, Mark D. III. Brogdon, B. G. (Byron Gilliam) IV. Brogdon, B. G. (Byron Gilliam). Forensic radiology. V. Title: Forensic radiology.
[DNLM: 1. Forensic Medicine. 2. Radiology. W 700]

RA1058.5.B76 2011
614'.1--dc22 2010037028

ISBN: 978-1-4200-7562-5 (hbk)
ISBN: 978-1-03-291793-1 (pbk)
ISBN: 978-0-367-80570-8 (ebk)

DOI: 10.4324/9780367805708

Visit the Taylor & Francis Web site at
http://www.taylorandfrancis.com

and the CRC Press Web site at
http://www.crcpress.com

The "feu sacre" is the "permetuum mobile" which makes the scientific world goes around—I dedicate this book to all the open minded and cutting edge human beings who were and are supporting the positive development of this process, especially my family.

Michael J. Thali

"Imagination is more important than knowledge"—Albert Einstein

Dedicated to Reginald Blackall, Ernest Harnack, Harold Suggars and Ernest Wilson, pioneer radiographers at the Royal London Hospital from 1896, who gave their lives to the development of our science. And to all others, both past and present, through whose imagination and tenacity, forensic radiology has advanced in the pursuit of truth.

Mark D. Viner

Again, with enduring love to my precious Babs, who can no longer appreciate how much I appreciate the love and happiness she has given me throughout all of our years together.

B.G. Brogdon

Contents

SECTION VII Virtual Imaging

SECTION VIII Facilities, Equipment, and Organization

Foreword

Forensic radiology is the medical specialty of radiology applied to help answer issues that arise for the law. My own involvement has been in its intersection with the medico-legal death investigation (MLDI), at the heart of which is the intellectual discipline of forensic pathology.

The public would be surprised to learn that forensic pathology has not been a "high tech" medical specialty like other branches of pathology, let alone surgery and other forms of investigative and procedural medicine and forensic science. Observation and dissection have been its mainstays for the last 100 years, with photography and histology increasingly important in the last 60-70 years. But things have progressed. My mentor at Guys Hospital Medical School, Professor Keith Mant, undertook his own scientific analyses (eg hairs, fibres and toxicology) in his early days as a forensic pathologist in the 1940's and 50's. As time passed, these forensic specialties, and others, developed and established themselves separately. Forensic radiology has come about slightly differently because of its dependence on expensive equipment, fixed and not easily available to the pathologist. Despite this, it has been helping answer some questions that confront the forensic pathologist since Roengten first discovered x-rays in 1895.

The incorporation of the CT scanner into daily operational forensic pathology practice, however, promises to strengthen our discipline enormously, and perhaps, in time, even to redefine it. But this will not happen in the way it was first thought. It will not replace the autopsy in those cases of concern to forensic pathology. What it will do, and is already doing in those places fortunate enough to be using the technology routinely, is two things: first, in some cases, provide sufficient information to satisfy the decision maker about the necessity for an autopsy; and second, it adds great strength in those cases where there has been an autopsy. In relation to the first, there needs to be real understanding of what is required of the medico-legal death investigation and what contribution the autopsy will make.

"Forensic Radiology" then is at the forefront, guiding forensic pathologists (and those clinically based radiologists working with forensic pathologists) through new terrain which is familiar in one sense, but quite foreign in another. Having become, as I have, much more familiar with radiologists than previously, I have been quite struck with how radiologists talk about what they are seeing: it sounds very similar to how forensic and anatomical pathologists talk. After all, radiologists are simply viewing anatomy and pathology in a different way. In this sense, and contributing in this way, radiologists fit very comfortably, merge almost seamlessly, into a forensic pathology setting. As a consequence, we pathologists are becoming comfortable with the new ways of visualising our autopsy findings, and increasingly impressed with seeing things we did not see when we relied only upon the autopsy.

What distinguishes forensic pathologists from their clinical pathology colleagues (and clinical radiologists) is a focus on the end point of the forensic investigation, which is a judicial process of some form, often a trial. This is quite different to our clinical colleagues who are focussed on providing useful diagnostic advice to treating doctors. This difference is manifest in a number of ways including:

- Understanding, training and experience in the forensic application of pathology: formulating opinions and conclusions relevant to the legal process and communicating them
- Interacting with police in particular cases
- Understanding the legal investigative process
- Understanding the court based processes of inquests and trials
- Understanding the legal principles underpinning these processes so that the forensic pathology contributions can be framed to meet their needs
- The skill of fashioning reports in this context
- The special knowledge of injuries and disease, their interaction and relationship to the alleged responsibility of a person for causing them

This is the terrain of forensic pathology, and it is for this reason that at the Victorian Institute of Forensic Medicine all the post mortem medical and scientific contributions to the MLDI are channelled through the forensic pathologist and form attachments to the MLDI report. This includes toxicology, neuropathology, microbiology and, in more recent times, the CT reports being relied upon by the forensic pathologist responsible for the overall MLDI.

I am deeply honored to have been asked by Michael Thali and Gil Brogdon to write this foreword. Over many years between them, they have had the vision, initiative and energy to be building the knowledge base for the contributions of radiology, including CT, to forensic medicine. We are benefiting from this on a daily basis in our practice in Victoria, Australia, and as time passes, so will many many more. We owe them an enormous debt.

Stephen Cordner
Professor of Forensic Medicine, Monash University
Director of the Victorian Institute of Forensic Medicine

Preface to the Second Edition

Gil Brogdon published *Forensic Radiology* in 1998. It was and remains the first and only book of its kind that aimed to cover the entire scope of radiological applications in the forensic sciences. Over ten years following its publication, it is seen as a milestone in the forensic world and this is proven by the fact that it is a forensic "best seller" not only in the United States but throughout the world and has become a standard reference text for practitioners and educators in Europe and Scandinavia, the Middle East, the Far East including Japan, India, Singapore and Malaysia, Latin America and Australasia.

In this influential text, Gil Brogdon described the history and development of forensic radiology and the application of both traditional and modern radiological techniques in forensic investigation in pursuit of the truth.

On the "Problems and Promise of Interdisciplinary Effort" he wrote:

> A rather sad commentary on the research examples cited and, indeed, on most forensic research employing radiologic methods and modalities, is that the overwhelming majority of investigators have been nonradiologists. Often they must struggle with substandard equipment and in ignorance of well-known radiologic tenets not published in their literature. On the other hand, most radiologists have little connection with the forensic sciences and are unaware of the research possibilities in that field, or of the problems that need solution. It is believed that forensic scientists in other disciplines would find radiologists in their area interested in cooperative efforts. Sharing of interdisciplinary skills and knowledge would improve the economy and effectiveness of investigative efforts, prevent some false starts and/or reinventions of well-worn wheels, and most important, expand scientific horizons.

This single statement has inspired many to take up the challenge to research, develop, and share knowledge and best practice in the field of forensic radiology. It represents both the founding and guiding principles by which the radiological portion of the Virtopsy project in Bern, Switzerland was conceived and is directed, and it also inspired the creation of the International Association of Forensic Radiographers dedicated to the development and sharing of skills between all those involved in forensic imaging and investigation.

With just 477 pages, *Forensic Radiology*, in comparison with the other forensic literature, was a small step forward, but marked an enormous milestone for the forensic community. At a time in his life when many of his contemporaries were content to slow down just a little, Gil Brogdon has continued to be both active and influential in the radiological and forensic fields, both in his national societies and internationally. Thus, in 2003, he published *A Radiological Atlas of Abuse, Torture, Terrorism and Inflicted Trauma* (CRC Press).

The fields of both forensic science and radiology have developed considerably since the publication of the first edition in 1998. Cognizant of this and the increased interest in forensic radiology amongst both practitioners and researchers alike, Gil approached us to coedit a second edition of *Forensic Radiology*.

We were both greatly honored to be asked to assist Gil with this important work. However, we received his request to edit such a large opus for the new edition with some trepidation as we were of the opinion that the work of putting together a new edition belonged to him as first editor and only he would be able to carry this out successfully. Only after he ardently begged us to support him, we accepted his offer gratefully and we would like to record here once again our thanks to him for the trust he has imbued us with.

In this second edition, existing chapters from the 1998 edition have been reviewed and significantly supplemented by Gil Brogdon. Virtopsy methods have been added in Sections III, IV, VI, and VII by Michael Thali and the Virtopsy Team, whilst the radiography chapters from the first edition have been significantly revised and expanded in Sections VIII, IX, and X by Mark Viner together with both the original and new contributors.

We would like to record our thanks to all of those who contributed to the first book and laid such a solid foundation for this new edition. We are exceedingly fortunate and grateful that so many outstanding contributors have agreed to assist with this new edition by imparting their knowledge and expertise in the revision of existing chapters and writing new material. This book would not have been possible without their time-consuming efforts.

We would like to thank all of those who have assisted with this project. In particular Gil would like to record his thanks to his secretary, Alecia Mackie, and also to Tolley Tollefson of the photography section of the radiology department at the University of South Alabama who contributed to many of the old and new chapters. Mark would like to thank all of his co-contributors and in particular Jerry Conlogue, not only for contributing to several chapters, but also for his technical advice and encouragement. He would also like to thank his wife, Kim, for the many hours of proofreading and correcting the errors and omissions that would otherwise have gone unnoticed. Michael would like to thank all the Virtopsy Team members (www.virtopsy.com), who supported the project over the last years and specially his family, who missed him over several days and nights at home.

Finally, despite some inevitable delays, we are pleased that we, together with Gil, are able to make this book available in 2010. In honor of Gil Brogdon's lifelong outstanding contributions to the field of forensic radiology through his activities and publications, Michael and Mark have decided that the title of this book and any future editions will be *Brogdon's Forensic Radiology.*

Michael J. Thali
Mark D. Viner

Preface to the First Edition

There are about 25,000 physicians in the United States engaged in the practice of diagnostic radiology. No more than three dozen of them have had enough experience or interest in forensic radiology to have published in the field. Just over a handful have had sufficient involvement to have become members of the American Academy of Forensic Sciences (AAFS), and only three of those have satisfied the requirements for fellowships in that organization.

There is no set definition or standard for a "forensic radiologist." There is no specialized training or fellowship available in that field; there is no separate society for forensic radiology, nor is there certification for a subspecialty or added qualification in forensic radiology. It is doubtful whether any radiologist in North America, regardless of his level of interest, devotes as much as 10% of his active practice to forensic radiology.

Those few of us who do maintain some continuing, albeit sporadic, activity in forensic radiology mostly became involved by happenstance, circumstance, curiosity, or just plain good luck. My own introduction to the fascination of forensic problems came in the mid-1960s when I was Radiologist-in-Charge of the Division of Diagnostic Radiology at Johns Hopkins. The famous Dr. Russell S. Fisher (Figure 1), longtime medical examiner of the State of Maryland and AAFS President in 1960–1961, first asked for my help in sorting out the commingled body parts of two children (only 15 months apart in age) who were victims of a light aircraft accident. Later, he allowed me to help in other cases. I was hooked for life.

When I left Johns Hopkins to accept the Chair of Radiology at the fledgling School of Medicine at the University of New Mexico, good fortune followed me to Albuquerque. I had several years there of exhilarating experience working with two outstanding forensic scientists, Jim Weston and Homer Campbell (Figure 2). When Jim came to New Mexico he designed and built both a model state-wide medical examiners system and a state-of-the-art forensic sciences building on the medical center campus. He already was a world figure in forensic pathology and served as president of AAFS in 1976–1977. His untimely death during one of his daily morning runs on the UNM golf course was a great loss to the forensic sciences. Homer Campbell at that time was in the practice of general dentistry and pursued an interest in forensic odontology as a part-time avocation. Homer was, in fact, *my* dentist and cleaned my teeth twice a year. (Fortunately, I never needed Jim Weston's professional services!) Later, Homer left his practice in favor of full-time forensic work and became president of the AAFS in 1991–1992.

Since coming to the University of South Alabama in 1978, I have enjoyed an entirely pleasant professional and personal relationship with LeRoy Riddick, a trainee of the legendary Milton Helpern, and now Professor of Pathology and Alabama State Medical Examiner (Figure 3). Roy and his associates and fellows (several of whom have gone on to responsible positions of their own) have invariably treated me with a most gratifying and warm collegiality as their Consultant in forensic radiology.

As my participation in forensic activities and meetings increased, I was astounded by the volume of radiological images presented and/or published by (of necessity) nonradiologists—dentists, pathologists, physical anthropologists, and other forensic scientists. I have been astonished at the good

FIGURE 1 Russell S. Fisher, MD.

FIGURE 2 Left to right: James T. Weston, MD, Homer R. Campell, Jr., DDS, and B.G. Brogdon, MD on the occasion of a Science Writers' Forum on Forensic Radiology in New York City on April 13, 1977.

FIGURE 3 LeRoy Riddick, MD.

quality of some of those images and saddened by how awfully bad others were. I have seen colleagues struggle with the interpretation of conditions that are well known and easily recognized by almost any radiologist. I have watched others agonize over problems that could be approached rather easily by the radiological method. I have reviewed elaborate and costly research projects based on null hypotheses both proved and disproved years ago in our discipline. On the other hand, I have become acquainted with applications of my own technology in forensic sciences that are largely unknown to me and my fellow radiologists. None of this is totally surprising when one recognizes that there is essentially no common fund of radiological knowledge, comprehensive radiological literature, or data bank of radiological information readily available to the many from disparate disciplines who may wish or need to use the x-ray or other radiological modalities in their individual forensic pursuits.

The sad truth is that a century after the first x-ray was introduced as evidence in a court of law, there is no general appreciation of the extent of the radiological potential in the forensic sciences. This was brought most forcefully to my attention when at a recent meeting no less an exemplar than the renowned criminalist, Dr. Henry C. Lee, showed a slide depicting the 19 disciplines or specialties comprising the body of forensic sciences; radiology was not among them!

The generally favorable response to my own efforts, and those of a few others, to spread the gospel of forensic radiology through talks, refresher courses, and publications directed to both forensic scientists and fellow radiologists finally has prompted me to undertake this book on forensic radiology. It represents an attempt to illustrate the applicability of diagnostic radiology to the broad range of the forensic sciences in the hope that it will both assist and stimulate others in the utilization of this useful tool in the solution of their problems. It is not an encyclopedia. It is not, for the most part, a how-to-do-it book. However, the three chapters in Section VIII will serve as a practical technical primer devoted to simplification and solution of technical problems which may mystify or bedevil the relatively untrained person who is forced to undertake unfamiliar tasks.

There is the risk that this volume, in trying to reach a large multidisciplinary readership, will be too difficult for some and too simple for others. As I try to leap unscathed between the horns of this dilemma, I hope the reader will view the attempt with tolerance and understanding.

The goal of this book is to cover the entire scope of radiological applications in the forensic sciences; the range of radiological applicability may be surprising to some readers. To make this volume timely and up to date, both current and anticipated uses of the exciting new modalities and techniques that have graced our discipline in recent years have been included. I am exceedingly fortunate and most grateful that an outstanding group of contributors has agreed to impart a unique body of knowledge and expertise in several chapters. This book would not have been possible without the time-consuming efforts of those extremely busy colleagues who herein share unstintingly the essence of their own forensic investigations and experience.

I have leaned heavily on other resources in this endeavor. Much information and a number of illustrations have been gleaned from the previously published material of others, and I appreciate the generosity of many authors and publishers in releasing their material for my use.

Sources of material have been acknowledged appropriately whenever possible. Some of the figures were produced from radiographs and slides that have resided in my files for many years; the origin of some of these is lost to both record and memory. My apology is extended for any oversight where credit is due, with the hope that it will be accepted in the same spirit of magnanimity characterized by the original offering.

I cannot apologize for the quality of the illustrations, some of which admittedly are suboptimal. I wish it were not so. But many postmortem roentgenograms are obtained under the most difficult conditions, in the morgue or in the field, by minimally trained or self-taught personnel. Often there is no opportunity for a repeat effort. Furthermore, there is no way to influence the quality of antemortem roentgenographs obtained from outside sources. Whatever one gets is considered a benison when the resolution of a case is at stake.

Tolley Tollefson and Teri Deese of the University of South Alabama Department of Radiology's photography section cheerfully bore the added burden of this book even though they were already heavily laden with the routine work of a busy academic department. The illustrations in this book reflect their high degree of professionalism since, unfortunately, some of the original material with which they worked was of inferior quality, for reasons already given.

Vanessa Brown cheerfully and efficiently typed the several versions of my portion of this manuscript (as well as those of several of the contributors). She has coped with my hillbilly accent on dictated material, interpreted my nearly illegible cursive script and, in all cases, corrected my abominable spelling. This book was added to the exceptional demands she daily faces as Departmental Secretary, with direct responsibility as amanuensis to two senior faculty members and overseer of all other secretarial activities in the department. I thank her for a prodigious effort accomplished with patience and good humor.

My wife, Babs, called upon her experience (in a previous life) as an editor of a biomedical journal to compensate for my well-documented failure as a copyeditor and proofreader. Her influence on this book, and my life, are acknowledged with thanks here and elsewhere.

Steven K. Teplick, MD, now professor and chairman of the Department of Radiology, University of South Alabama, has graciously supported this work by making time and departmental facilities and resources available to me, for which I thank him.

The staff of the Biomedical Library of the University of South Alabama has been unfailingly kind, courteous, and professionally competent in meeting all requests for help in finding and acquiring reference materials.

My editors, first George V. Novotny and later Bob Stern of CRC Press, have provided useful assistance and advice, continuous encouragement, and gentle prodding. They have borne my literary shortcomings, missed deadlines, and other problems with patience and aplomb. I hope the result was worth their effort.

Judge Haskell M. Pitluck, Past President of the American Academy of Forensic Sciences and valued friend, was kind enough to critically review my effort on "Coping with the Courts"; however, any residual errors or inaccuracies are solely my responsibility.

Finally, I must thank an unnamed multitude of friends, acquaintances, associates, and colleagues in Radiology, and a similar horde in the Forensic Sciences, who have been openhanded with their friendship, assistance, case material, and the lore of their respective disciplines. I have gained far more from those associations than have they and, especially, have learned far more than I have taught. I hope they will accept this blanket acknowledgment and my thanks for their contributions—in uncounted ways—to this opus.

Gil Brogdon
Mobile, Alabama

Editors

Michael J. Thali, MD, Executive MBA, HSG, has been working since 1995 in the field of forensic medicine. He had a two-year fellowship in clinical radiology with Peter Vock and Gerhard Schroth, MD and long experience in wound ballistic by Dr. sc. forens. Beat Kneubuehl.

In 2001–2002, he was a fellow at the office of the Armed Forces Medical Examiners at the Armed Forces Institute of Pathology (AFIP) in Washington, DC. He has written many virtual autopsy papers (see www.virtopsy.com), and he is editor of the book *The Virtopsy Approach*. He has been full professor of forensic medicine at the University of Bern, Switzerland since 2006 and is director of the Forensic Institute at the University of Bern which introduced the world's first production stream of consecutive forensic examinations including 3D optical surface and radiology scanning using multislice CT, TIM–MRI, and postmortem angiography and biopsy. He is well integrated in the global society of forensic sciences. Together with Stephan Bolliger, Steffen Ross, Lars Oesterhelweg, and Beat Kneubuehl he was the 2009 IgNobel prize winner for peace. Working in forensics is for him serving the society for a better world. He will be the chairman/director of the University Forensic Institute in Zurich in 2011.

Mark D. Viner is a fellow of Cranfield University Forensic Institute, chief executive of the Inforce Foundation and a senior manager at St. Bartholomew's and The Royal London Hospitals, London, United Kingdom. He has a long held interest in forensic imaging and emergency planning, developed during almost 30 years experience as radiographer and radiology manager in the United Kingdom. During this time, he was involved in the clinical and forensic response to several major incidents including the IRA terrorist attacks in London from 1990 to 1997 and also led the forensic radiography response to the terrorist bombings of July 7, 2005.

From 1996 to 2001, he coordinated the forensic radiography team for the United Nations International Criminal Tribunal for the Former Yugoslavia (ICTY) and has worked as forensic radiography consultant in Bosnia, Kosovo and Croatia, Sierra Leone, the Irish Republic, and for several high-profile cases in the United Kingdom. He continues to be regularly involved in forensic casework in the United Kingdom and through his charitable work with the Inforce foundation is actively involved with the education and training of forensic practitioners in the forensic investigation of mass fatalities and atrocity crimes on a worldwide basis.

He is a founding member and first chair of the International Association of Forensic Radiographers, and advisor to the UK Home Office, The Government Office for London and to the International Criminal Court in The Hague. He has lectured widely at international conferences and contributed a number of radiology chapters to forensic textbooks. He has been an active member of the College of Radiographers, holding office at local, regional, and national levels. He is also a member of the Forensic Science Society and the American Academy of Forensic Sciences and was awarded a Winston Churchill Travelling Fellowship and the Fellowship of the College of Radiographers in 2005. In 2010, he was awarded the Gold Medal of the Society & College of Radiographers.

B.G. Brogdon, MD—Gil Brogdon's interest and experience in forensic radiology goes back more than 40 years to a time when he was radiologist-in-charge, Division of Diagnostic Radiology at Johns Hopkins. His involvement in the field continued throughout his tenures as professor and chair of the Departments of Radiology at the University of New Mexico and, later, at the University of South Alabama where he is now University Distinguished Professor Emeritus. In 1998, the first edition of *Forensic Radiology* was published, followed in 2002 by the award-winning *A Radiological Atlas of Abuse, Torture, Terrorism, and Inflicted Trauma* with Vogel and McDowell. His 300 plus publications span more than 50 years and most of the development of diagnostic radiology to its present imaging capabilities and applications.

Brogdon is a fellow, past-president, and gold medalist of the American College of Radiology; past-president and gold medalist of the Association of University Radiologists; gold medalist of the American Roentgen Ray Society; silver medalist of the International Skeletal Society; and Schinz medalist of the Swiss Society of Medical Radiology. He has received the Medal of Honor of Leopold-Franzens University, Medals of the University of Leuven and the City of Brescia, and the Austrian Cross of Honor for Science and Art, First Class, among other offices and honors. He is a distinguished fellow and Hunt awardee of the American Academy of Forensic Sciences and patron of the International Association of Forensic Radiographers.

Contributors

Nancy Adams
Department of Radiologic Sciences
Itawamba Community College
Fulton, Mississippi

P.R. Algra
Radiology
Medical Center Alkmaar
Alkmaar, The Netherlands

Garyfalia Ampanozi
Center for Forensic Imaging and Virtopsy
Institute of Forensic Medicine
University of Bern
Bern, Switzerland

Leonard Berlin
Department of Radiology
Rush University
Chicago, Illinois

and

NorthShore University
HealthSystem Skokie Hospital
Skokie, Illinois

Tania Blyth
Diagnostic Imaging Program
Quinnipiac University
Hamden, Connecticut

Michael Bolliger
Center for Forensic Imaging and Virtopsy
Institute of Forensic Medicine
University of Bern
Bern, Switzerland

Stephan A. Bolliger
Center for Forensic Imaging and Virtopsy
Institute of Forensic Medicine
University of Bern
Bern, Switzerland

B. G. Brogdon
Department of Radiology
University of South Alabama Medical Center
Mobile, Alabama

Ursula Buck
Center for Forensic Imaging and Virtopsy
Institute of Forensic Medicine
University of Bern
Bern, Switzerland

Andreas Christe
Department of Interventional, Pediatric and Diagnostic Radiology
University Hospital
Bern, Switzerland

Gerald Conlogue
Bio-Archaeology Research Institute
Quinnipiac University
Hamden, Connecticut

Eric Crubézy
Laboratoire d'Anthropobiologie
AMIS – FRE
Toulouse, France

Fabrice Dedouit
Service de Médecine Légale

and

Service de Radiologie
CHU Toulouse-Rangueil

and

Laboratoire d'Anthropobiologie
AMIS – FRE
Toulouse, France

Richard Dirnhofer
Institute of Forensic Medicine
University of Bern
Bern, Switzerland

Lars C. Ebert
Center for Forensic Imaging and Virtopsy
Institute of Forensic Medicine

and

Institute for Surgical Technology and Biomechanics
University of Bern
Bern, Switzerland

Emily Faircloth
International Association of Forensic Radiographers
Plymouth, United Kingdom

Laura Filograna
Center for Forensic Imaging and Virtopsy
Institute of Forensic Medicine
University of Bern
Bern, Switzerland

Patricia M. Flach
Centre for Forensic Imaging and Virtopsy
Institute of Forensic Medicine
University of Bern

and

Department of Diagnostic and Interventional Neuroradiology
University Hospital
Bern, Switzerland

Tanja Germerott
Center for Forensic Imaging and Virtopsy
Institute of Forensic Medicine
University of Bern
Bern, Switzerland

Howard T. Harcke
Department of Radiologic Pathology
Armed Forces Institute of Pathology
Washington, DC

Gary Hatch
Center for Forensic Imaging and Virtopsy
Institute of Forensic Medicine
University of Bern
Bern, Switzerland

Chucri M. Jalkh
Departrment of Radiological Sciences
University of South Alabama
Mobile, Alabama

A. Everette James
Department of Radiology & Radiological Sciences
Vanderbuilt University School of Medicine
Nashville, Tennessee

and

Department of Radiology
University of North Carolina School of Medicine
Chapel Hill, North Carolina

Francis Joffre
Service de Radiologie
CHU Toulouse-Rangueil
Toulouse, France

Beat P. Kneubuehl
Centre for Forensic Physics and Ballistics
Institute of Forensic Medicine
University of Bern
Bern, Switzerland

Paul F. Laudicina
Radiology Technology Services, Inc.
Du Page County, Illinois

Angela D. Levy
Department of Radiology
Georgetown University Hospital
Washington, DC

Joel E. Lichtenstein
Department of Radiology
University of Washington
Seattle, Washington

Gregory Mahoney
Boston Police Department
Office of Multi-Media Production
Boston, Massachusetts

and

Centre for Anatomy and Human Identification (CAHId)
University of Dundee
Dundee, Scotland

Sandra Mathier
Center for Forensic Imaging and Virtopsy
Institute of Forensic Medicine
University of Bern
Bern, Switzerland

John D. McDowell
Department of Oral Medicine and Forensic Science
University of Colorado School of Dental Medicine

and

Department of Family Medicine
University of Colorado School of Medicine
Aurora, Colorado

James M. Messmer
Department of Radiology
Virginia Commonwealth University School of Medicine
Richmond, Virginia

Silvio Näther
Center for Forensic Imaging and Virtopsy
Institute of Forensic Medicine
University of Bern
Bern, Switzerland

Charles W. Newell
Department of Radiological Sciences
University of South Alabama
Mobile, Alabama

Haskell M. Pitluck
Retired Circuit Court Judge
State of Illinois,
Crystal Lake, Illinois

Ulrich Preiss
Center for Forensic Imaging and Virtopsy
Institute of Forensic Medicine
University of Bern
Bern, Switzerland

LeRoy Riddick
Department of Pathology
The University of South Alabama Medical Center
Mobile, Alabama

Steffen G. Ross
Center for Forensic Imaging and Virtopsy
Institute of Forensic Medicine
University of Bern

and

Department of Diagnostic Radiology
University Hospital Bern
Bern, Switzerland

Daniel Rougé
Service de Médecine Légale
CHU Toulouse-Rangueil

and

Laboratoire d'Anthropobiologie
AMIS – FRE
Toulouse, France

Hervé Rousseau
Service de Radiologie
CHU Toulouse-Rangueil
Toulouse, France

Thomas Ruder
Center for Forensic Imaging and Virtopsy
Institute of Forensic Medicine
University of Bern
Bern, Switzerland

Christopher Rynn
Centre for Anatomy and Human Identification (CAHId)
University of Dundee
Dundee, Scotland

Norbert Telmon
Service de Médecine Légale
CHU Toulouse-Rangueil

and

Laboratoire d'Anthropobiologie
AMIS – FRE
Toulouse, France

Michael J. Thali
Center for Forensic Imaging and Virtopsy
Institute of Forensic Medicine
University of Bern
Bern, Switzerland

Mark D. Viner
Inforce Foundation
Cranfield Forensic Institute
Cranfield University
Shrivenham, United Kingdom

and

St. Bartholomew's and The Royal London Hospitals
London, United Kingdom

H. Vogel
Forensic Radiologist
Institute for Legal Medicine of the University Hospital
Hamburg, Germany

Richard A. Weems
Department of Diagnostic Sciences
School of Dentistry
University of Alabama at Birmingham

and

Dental Consultant
Department of Forensic Sciences and Jefferson County
Office of the Chief Medical Examiner
Birmingham, Alabama

Section I

Introduction to Forensic Radiology

1 Definitions in Forensics and Radiology

B.G. Brogdon

CONTENTS

Forensic is derived from the Latin *forens(is)*: of or belonging to the forum, public, equivalent to *for(um)* forum + *ens*—of, belonging to + *ic*. By extension it came to also mean *disputative, argumentative, rhetorical, belonging to debate or discussion*. From there it is but a small step to the modern definition of forensic as *pertaining to, connected with, or used in courts of judicature or public discussion and debate*. Thus the forensic sciences encompass the application of specialized scientific and/or technical knowledge to questions of civil and criminal law, especially in court proceedings.

Forensic Medicine has come to be recognized as a special science or discipline that deals with relationships and applications of medical facts and knowledge to legal problems. Some prefer to call it *legal medicine* or *medical jurisprudence*.

Forensic medicine is often considered to be synonymous with forensic pathology because full-time involvement of a physician with forensic activity is almost exclusively the province of that specialty.[1] The forensic pathologist is concerned principally with the post-mortem examination and, hence, deals mostly with the dead. In acknowledgment of this, Milton Helpern, the third Chief Medical Examiner of the City of New York, caused to be inscribed upon the lobby wall of his new office building in 1961 the Latin admonition, *TACEAT COLLOQUIA. EFFUGIAT RISUS. HIC LOCUS EST UBI MORS GAUDET SUCCURRERE VITAE*, (Let conversation cease. Let laughter flee. This is the place where death delights to help the living).[2] While other medical specialists may consult with the pathologist in the evaluation of death, virtually all of their other professional activities may have medicolegal ramifications and hazards involving both the living and the dead. These can embrace a large body of legal issues (e.g., age determination, assault, civil rights violations, inheritance, larceny, malpractice, parentage, personal injury, product liability, sexual offenses, smuggling, virginity, and wrongful birth or death).

Thus, in Gradwohl's[3] the definition of legal medicine was expanded to include "the application of medical knowledge to the administration of law and to the furthering of justice and, in addition, the legal relations of the medical man."

Evidence of the origin of legal or forensic medicine can be found in records of ancient people some thousands of years ago, when occasionally a law appears to influence medicine or medicine is found to influence or modify a law.[3,4] The Egyptian, Imhotep, may have been the first to apply both the law and medicine to his surroundings. Hammurabi codified medical law circa 2200 BC, and medicolegal issues were covered in early Jewish law. Later, other civilizations—the Greeks, ancient India, the Roman Empire—evolved jurisprudential standards involving medical fact or opinion.

Early cultures recognized the desirability of controlling the organization, duties, and liabilities of the medical profession. They also were acquainted with the importance of the knowledge and opinion of the medical person in the legal consideration of issues of great moment such as the use of drugs or poisons, the duration of pregnancy, virginity, superfetation, the prognosis of wounds in different body locations (a physician determined that only one of Caesar's 23 stab wounds was fatal), sterility and impotence, sexual deviation, and suspicious death.

Early in the sixteenth century a separate discipline of forensic medicine began to emerge. New codes of law required expert medical testimony in trials of certain types of crime or civil action. The first medicolegal books appeared in the late sixteenth and early seventeenth centuries and, after 1650, lectures on legal medicine were given in Germany and France. The first book on medical jurisprudence in the English language appeared in 1788, and 19 years later the first Regius Chair in Forensic Medicine was recognized by the Crown at the University of Edinburgh. The English coroner's system was imported to the Colonies in North America in 1607, and it was not until 1871 that Massachusetts, later followed by New York and other jurisdictions, established a medical examiner system. Upon this base of professionalism in death investigation, supported by the framework of solid scientific and technical advances during the twentieth century, was erected the modern structure of forensic medicine, which covers a heterogeneous, sometimes loosely related, family of numerous disciplines or subspecialties sharing a common interest.

Among those, Forensic Radiology usually comprises the performance, interpretation, and reportage of those radiological examinations and procedures that have to do with the courts and/or the law. Until but a few decades ago, Radiology could be defined as that special branch of medicine employing ionizing radiant energy in the diagnosis and treatment of disease. Now, the specialty of radiology is divided into two quite distinct disciplines sharing only a common historical origin and a single certifying body, the American Board of Radiology. One of those two branches is Radiation Oncology, which utilizes high-energy ionizing radiant energy wavelengths and particles in the treatment of (almost exclusively malignant neoplastic) disease.

It is the other major branch of the specialty, Diagnostic Radiology, with which this book will be concerned. Diagnostic radiology is devoted primarily to the study of images of the internal structures of the human body. Perhaps Harry Z. Mellins, a superbly talented radiologist and teacher, best captured the eidolon of the diagnostic radiologist more than 40 years ago when he wrote,[5]

> The [diagnostic] radiologist perceives a shadow, sees a lesion, and imagines a man. The bedside physician sees the man, perceives the signs and images the lesion. They practice from the outside in and we from the inside out.

Nowadays, images are acquired by an array of modalities and techniques:

The *x-ray* or *roentgen ray* is an energy form of ionizing radiation from which may be produced fluorescent or photographic images. The latter are sometimes also called "x-rays" but are correctly termed *roentgenograms*, less accurately *radiographs*, and vulgarly as *films* (e.g., "chest film") (Figure 1.1).

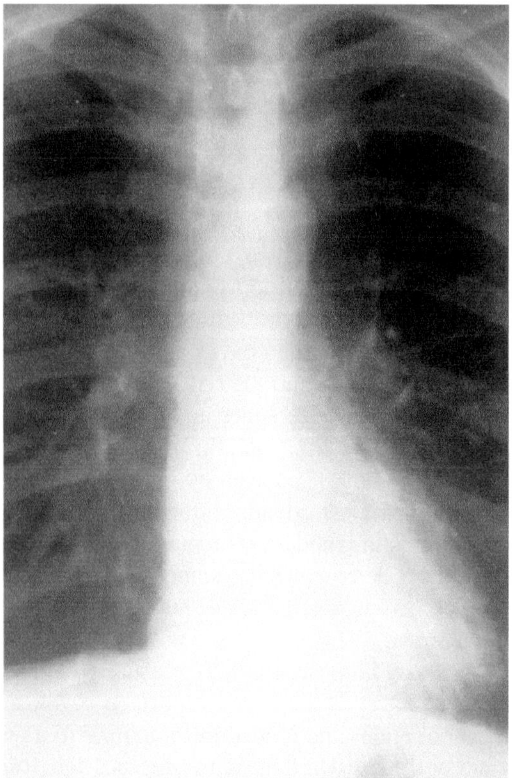

FIGURE 1.1 Chest roentgenogram, radiograph, or "film." This patient has an intracardiac tumor but it cannot be distinguished from the heart muscles or the blood inside the heart since all have approximately the same ability to absorb x-rays.

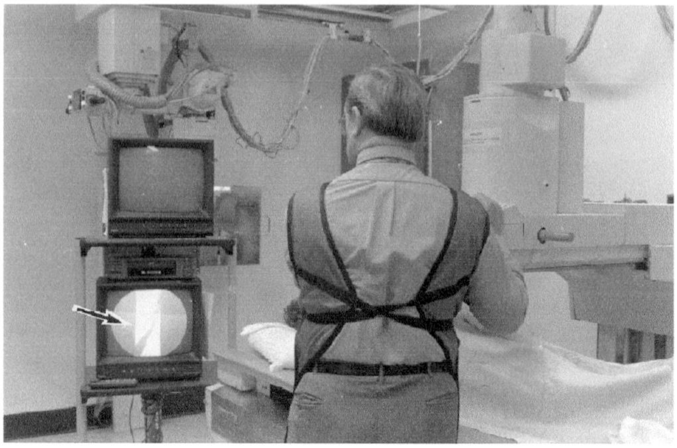

FIGURE 1.2 A modern fluoroscope with an image intensifier connected to a television camera. The televised image (arrow) can be seen without darkening the room.

The *fluorescent image* can be electronically enhanced and directly visualized in real-time motion, cine-photographed, videotaped, or digitized and stored on magnetic tape or disks for replay (Figure 1.2).

In the subspecialty of *nuclear medicine* or *nuclear radiology*, radioactive materials or isotopes can be directed to internal target organs or tissues by injection, inhalation, or ingestion; the radiant energy escaping from inside the body can be collected on sensitive films or phosphors to create images, *scintiscans*, of the internal targets (Figure 1.3).

Sound waves generated outside the body by transponders are reflected back from internal structural interfaces to be recaptured and converted into real-time or static images.

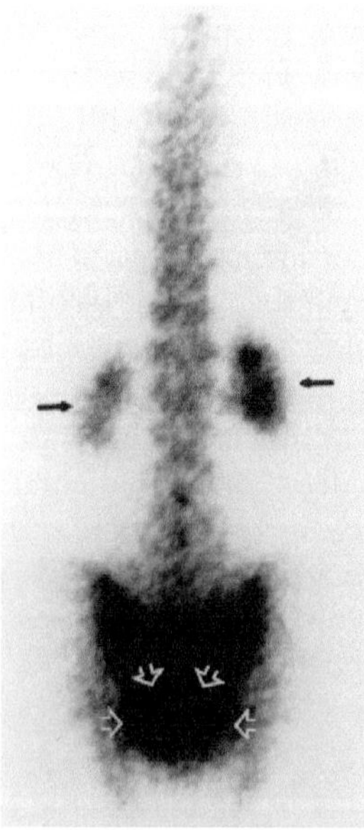

FIGURE 1.3 This is a nuclear scan using a bone-seeking isotope so the skeleton is imaged. Some of the isotope is taken up by the kidneys (arrows) and excreted into the bladder (open arrows) which should have been emptied before the scan was done. Unfortunately the full bladder obscures a tumor (chordoma) in the sacrum. (See Figure 1.7b–e, same case.)

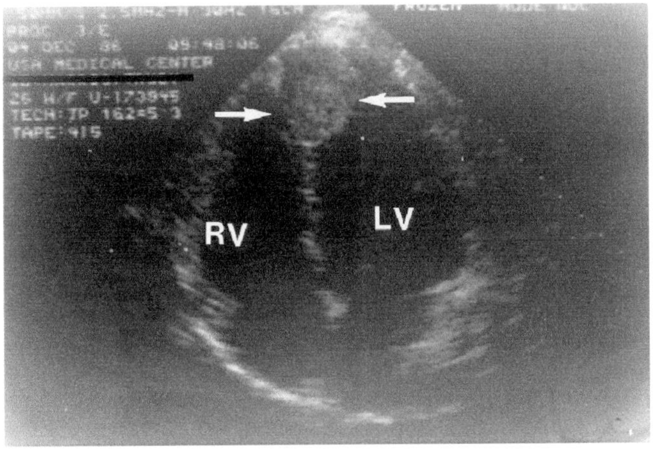

FIGURE 1.4 The ultrasound image of the same heart shown in Figure 1.1 is displayed as a cross-sectional image of the heart. The tumor is visualized in the interventrical septum (arrows) between the right ventricles (RV) and left ventricles (LV).

The modality is called *ultrasound* or *ultrasonography*. The image is a *sonogram* (Figure 1.4).

With special equipment, a roentgenogram of a thin section or slice of the body or body part can be acquired in the sagittal, coronal, or oblique plains (Figure 1.5). This technique is known as *tomography* and the processed image is a *tomogram* (now qualified as a *conventional tomogram* to distinguish it from a *computed tomogram*). Conventional tomography has essentially disappeared from medical imaging because of the superior sectional images produced by computed tomography (CT).

Conventional roentgenograms result when x-ray photons (or light rays from intensifying screen phosphors excited by x-rays) "expose" a sensitive silver halide emulsion which, when processed, leaves a grain of reduced silver where the light-ray or x-ray photon struck. These microscopic black "dots" coalesce where the most x-rays were transmitted to the film; they are sparse to absent where the interposed body absorbs the radiation and are intermediately dispersed in a gray scale according to the atomic number of the interposed tissue. This is an analog image. With proper instrumentation or equipment an *analog image* can be converted into a *digital image*. Most newer radiographic equipment modalities (i.e., digital fluoroscopy, digital subtraction angiography, some ultrasonography, CT, and magnetic resonance imaging (MRI) produce a digital image directly. The advantage of a digital image is that it can be manipulated by such techniques as contrast shift, density range adjustment, back/white reversal, and edge enhancement (Figure 1.6). Although the analog image has better spatial resolution than the digital image, maximal contrast resolution is required to exploit the spatial resolution, and shades of gray are not easily separated. Thus the analog roentgenogram may display muscle, water, blood, or brain in a gray scale in which they are indistinguishable one from another (Figure 1.7). The digital image can be computer processed or manipulated to separate shades of gray,

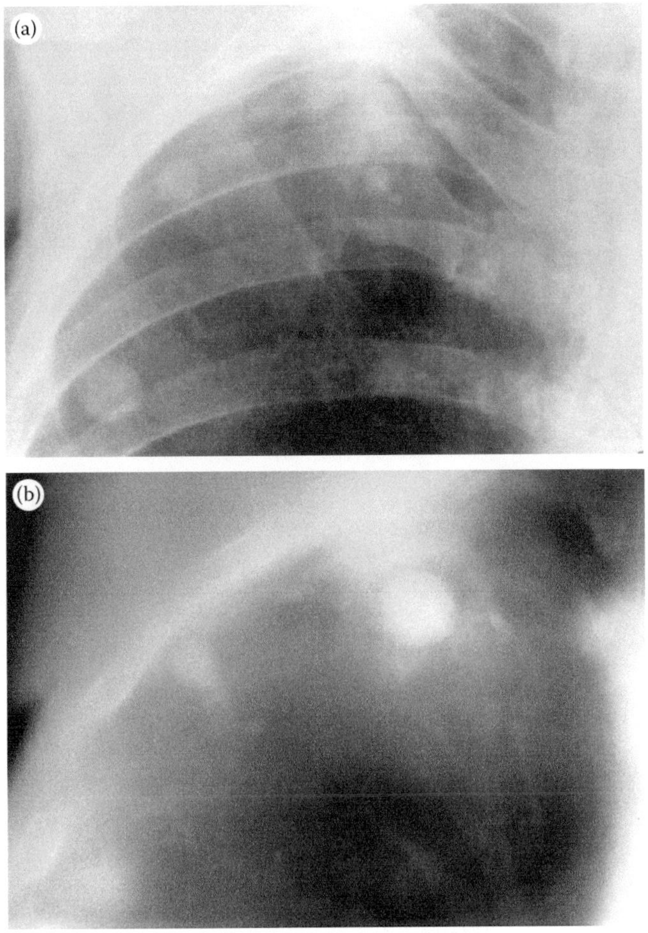

FIGURE 1.5 Example of conventional tomography. (a) Chest roentgenogram: close-up of upper lung showing a large plaque of thickened pleura which obscures the underlying lung. (b) Conventional tomogram: slice through the lung shows a nodular lesion that was hidden by the overlying pleural thickening.

thus permitting, for instance, visual separation of the brain substance from its interventricular fluid or the gallbladder wall from the bile it contains.

Computed axial tomography caught public fancy as the "CAT Scan," but is now more commonly and properly referred to as *computed tomography* or simply CT, and the product is also called a CT or, preferably, a CT *scan*. This device uses an array of photoreceptors to detect slight differences in attenuation of roentgen rays emitted by a rotating x-ray tube as they pass through the body or body part over multiple diametric pathways in the axial or cross-sectional plane. The computer-processed result is a series of images of cross-axial sectional *slices* of the body or part, allowing a much higher differentiation between body tissues than can be achieved from a conventional x-ray machine (Figure 1.8). The electronically acquired and computer-processed image can be manipulated in a variety of ways; for instance, to reconstruct images in coronal or sagittal planes or even two-dimensional (2-D) displays of three-dimensional (3-D) reconstructions. Now, contemporary CT units can image

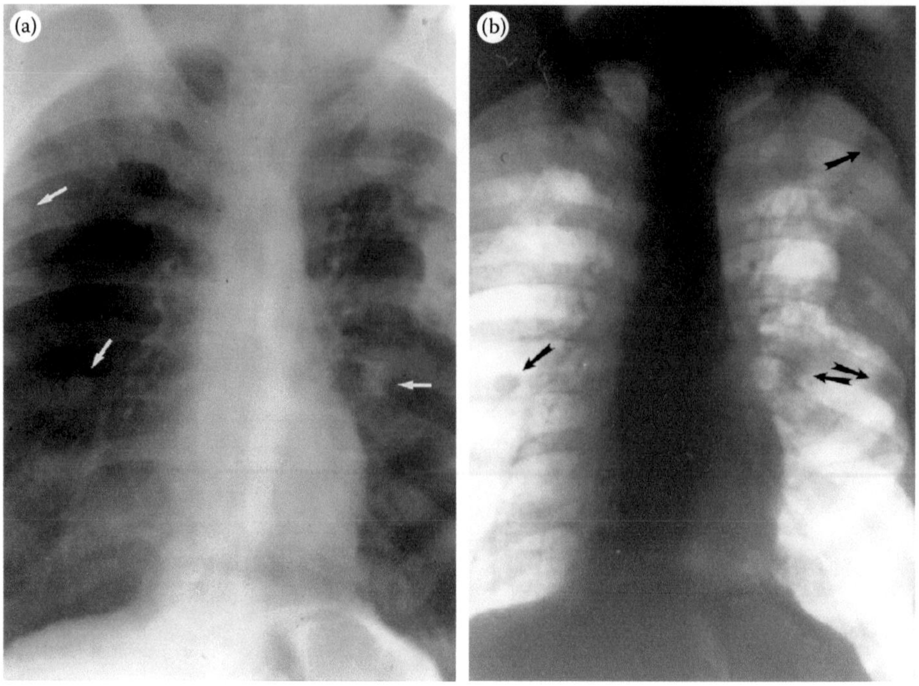

FIGURE 1.6 (a) and (b) are digital images of a chest containing multiple pulmonary nodules (arrows). The visual effect of black/white reversal image manipulation is demonstrated.

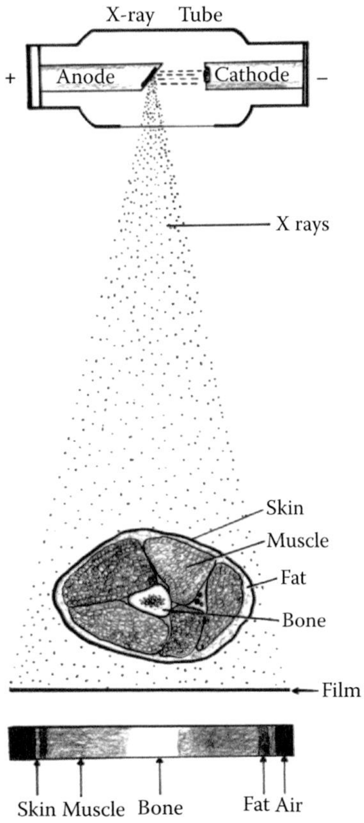

FIGURE 1.7 Schematic drawing showing differential absorption of x-rays by the tissues of the body in conventional radiography. From the most to least radiodense, these are (1) bone or calcium; (2) all soft tissues and liquids (muscle, blood, brain, heart, liver, urine, etc.) except (3) fat; (4) air or other gases.

16–64 slices of tissue simultaneously, allowing for faster throughput, thinner slices, finer resolution, and virtually seamless reconstruction.

Magnetic resonance imaging (MRI) utilizes strong magnetic fields to generate electromagnetic signals from elements and compounds found in body fluids and tissues (Figure 1.9). With computer manipulation one can obtain multiplanar, multidirectional, sectional images or *slices (MR scans)*. The signal strengths for individual tissues can be identified or manipulated by varying computer protocols.

The subspecialty of *interventional radiology* is almost a separate discipline. A complex armamentarium of tools, devices, and instruments (e.g., catheters, balloons, lasers, needles, drains, stents, "wooly worms," macrospheres, enzymes, etc.) are used to invade the body—not only to create images and enhance diagnoses, but also to intervene therapeutically in disease processes or anatomic abnormalities. But CT and MR imaging of the cardiovascular system has replaced many nontherapeutic interventional procedures.

Heretofore, forensic radiology has depended almost exclusively on the x-ray and the static image captured on the roentgenogram. The newer imaging methods have revolutionized the field of diagnostic radiology in a time span so short that it falls within the career experience of radiologists still engaged in active practice. Now these advances are being incorporated into forensic studies. If problems of accessibility and cost can be resolved, the newer radiological techniques and modalities may be widely appropriated by the forensic sciences.

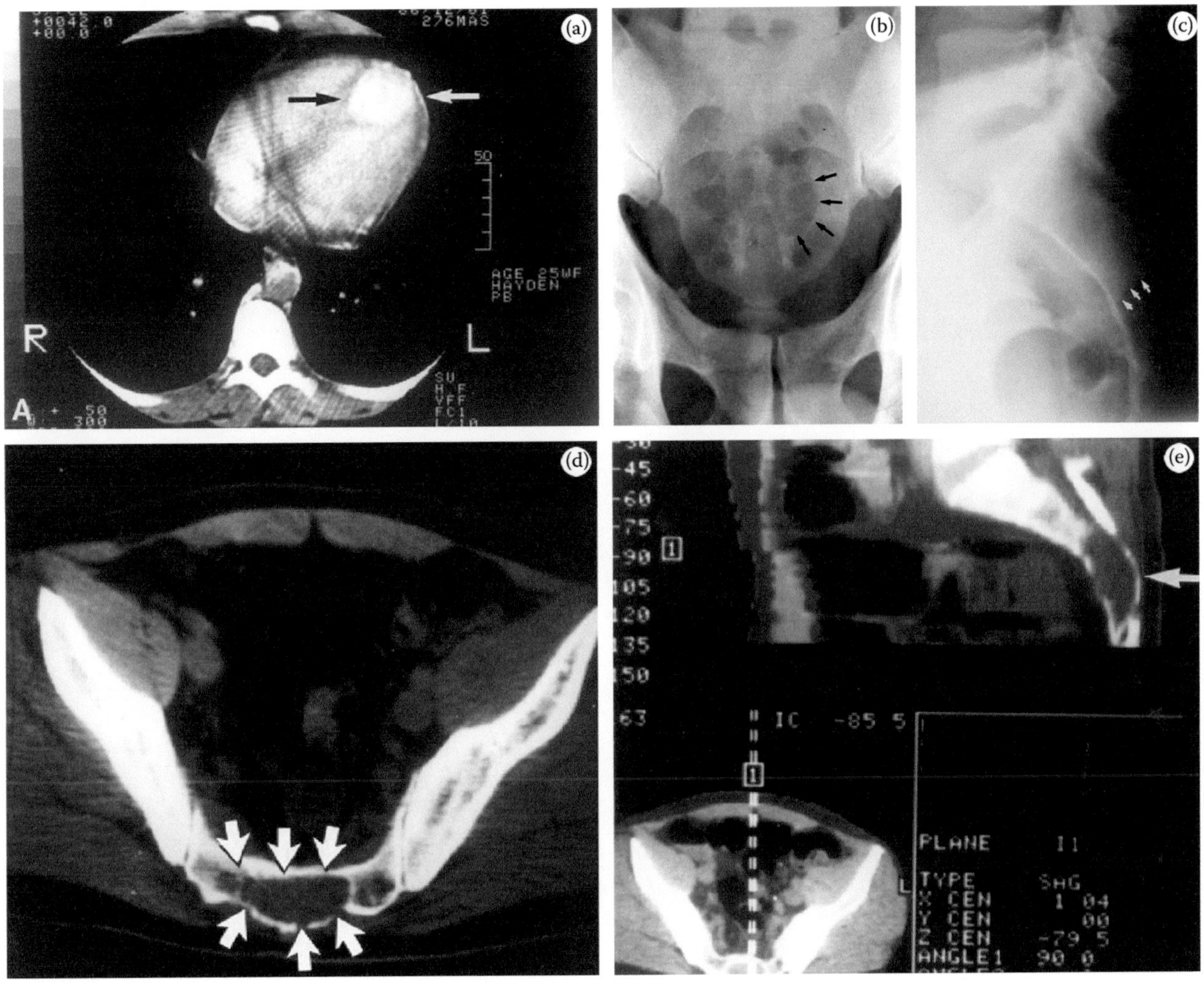

FIGURE 1.8 Examples of CT (a) CT slice showing tumor (arrows) in the interventricular septum (the same case as in Figures 1.1 and 1.4). (b) and (c) Frontal and lateral roentgenograms of a sacrum being partially destroyed by a tumor (arrows). (e) CT reconstruction of saggital (lateral) view from cross-sectional image (d). Note the similarity to the lateral view in (c).

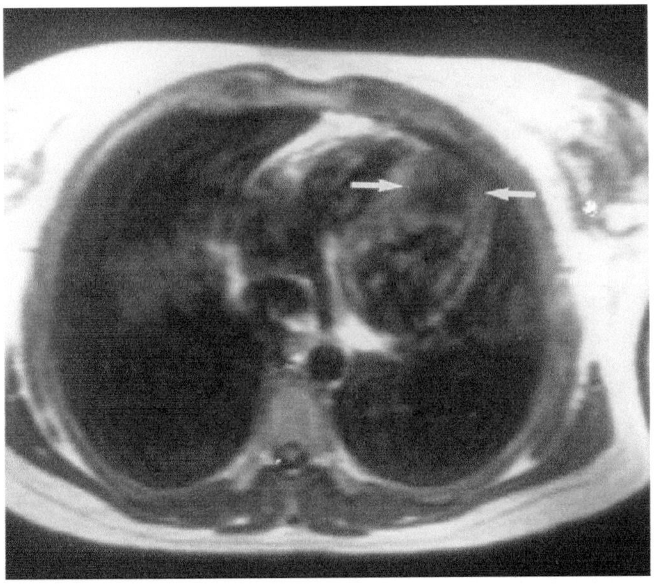

FIGURE 1.9 An MRI of the cardiac tumor (arrows) shown in Figures 1.1, 1.4, and 1.8A.

After all, it took six or seven thousand years of growth and development for forensic medicine to reach its present level of fruition. The flowering of forensic radiology had to wait on Professor Röntgen.

REFERENCES

1. Knight, B., How radiography aids forensic medicine, *Radiography*, 50, 5, 1984.
2. Helpern, M. and Knight, B., *Autopsy*, New American Library, New York, 1979, 184.
3. Camps, F. E., Ed., *Gradwohl's Legal Medicine*, 3rd ed., Yearbook, Chicago, 1976, chap. 1
4. Wecht, C. H., Forensic use of medical information, in *Legal Medicine: Legal Dynamics of Medical Encounter*, 2nd ed., Am. Coll. Legal Med., Mosby-Yearbook, St. Louis, 1991, chap. 47.
5. Mellins, H. Z., Personal communication, 1963.

2 Forensic Radiology in Historical Perspective

B.G. Brogdon and Joel E. Lichtenstein

CONTENTS

RÖNTGEN AND HIS RAYS

The discovery of "a new kind of ray" while working alone in the laboratory on an autumn afternoon firmly established the place of Wilhelm Conrad Röntgen among the great investigative scientists of all time. But the mantle of greatness did not descend upon obscurity—Röntgen already enjoyed a solid scientific reputation built on the foundation of academic excellence and painstaking research.

Wilhelm Conrad Röntgen (Figure 2.1) was born on March 27, 1845, in Linnet, a provincial town on the edge of the Ruhr Valley. The family moved to Apeldoorn in Holland when he was three years old. There he enjoyed a pleasant childhood as the only son of a prosperous textile merchant. His schooling was erratic; it was only after considerable difficulty that the 23-year-old Röntgen finally won a diploma in Mechanical Engineering from the Polytechnical School in Zürich. He stayed on in Zürich for some additional courses in mathematics and physics and there found his life-interest in the physics laboratory. He obtained a PhD. at the University of Zürich in 1869. There he also met his future wife, Bertha (Figure 2.2), and reached the decision to follow a university career in physics.

Except for some initial delays, the result of his rather informal education, the next quarter of a century brought steady progress in Röntgen's career. At age 43, he received an appointment as professor of Physics and director of the Physical Institute at the University of Würzburg; only 6 years later he was elected as Rector of the University, largely as an acknowledgment of his research contributions. Thus, by his fiftieth birthday in 1895, Röntgen could justifiably look upon his achievements in life with complacent satisfaction. He held the highest office bestowed by the university, and his 48 published communications had brought widespread recognition and admiration from his peers.

The Röntgens enjoyed comfortable living quarters on the top floor of the Physical Institute. His laboratory was readily accessible on the floor below (Figure 2.3). Thus, unlike many present-day researchers whose accomplishments in the laboratory are "rewarded" by appointment to high academic and administrative positions, Röntgen was able to continue to devote a large portion of his time to independent investigation. In an address upon assuming the rectorship of the university, Röntgen epitomized a strong scientific credo when he said[1]:

> The university is a place of scientific research and mental education ... the experiment is the most powerful and most reliable lever enabling us to extract secrets from nature and ... experiment must constitute the final judgment as to whether a hypothesis should be retained or be discarded. ... If the result does not agree with reality it must necessarily be wrong.

In the autumn of 1895, Dr. Röntgen became interested in the cathode ray experiments of Hittorf, Crookes, and others. On November 8, 1895, he was surprised and immensely excited to observe the fluorescence of a screen coated with barium platinocyanide, which was lying on a bench at some distance from an energized Hittorf tube which had been completely light-proofed with black cardboard. Repetition of the experiment produced the same result even when the barium platinocyanide screen was moved at greater and greater distances from the tube—even up to 2 m. Since no visible light could escape from the heavily shielded tube, and since cathode rays were known to penetrate only a few centimeters of air, the fluorescence must be caused by a completely new kind of emanation from the tube![2] Much later, a correspondent from McClure's magazine reported an interview with Röntgen concerning his discovery. He asked the great physicist what his thoughts were upon perceiving that first fluorescent glow in the darkened laboratory. "I did not think," answered Röntgen, "I investigated."[3]

And so he did!

Röntgen embarked upon 50 days of intensive research culminating in the submission of his initial manuscript, "Ueber eine neue Arte von Strahlen" (On a new kind of ray), on December 28, 1895.[4] He worked day and night, taking his meals in the laboratory and often sleeping there.

He had found by accident that the rays penetrated black paper. He then determined that the rays were not stopped by cardboard, cloth, a thick plank, or a book of 2000 pages. Metals such as copper, lead, gold, platinum, and silver were less penetrable; the densest of them were almost opaque.[5]

FIGURE 2.1 Wilhelm Conrad Röntgen, 1845–1923. Taken at the time of his discovery of the x-rays. (From Glasser, O., *Am. J. Roentgenol.*, 25, 437, 1931. With permission.)

FIGURE 2.3 The Physical Institute as it stands today on the Röntgenring in Würzburg. The Röntgens' living quarters were on the top floor (arrow) just above the laboratory where Röntgen discovered his rays. The inscription on the front exterior wall of the laboratory reads, "In diesum hause entdeckte W.C. Röntgen im jabre 1895 die nach ihm benannten strahlen."

Glasser[6] describes Röntgen's most startling observation,

> To test further the ability of lead to stop the rays, he selected a small lead piece, and in bringing it into position observed to his amazement not only that this round shadow of the disk appeared on the screen, but that he suddenly could distinguish the outline of his thumb and finger, within which appeared darker shadows—the bones of his hand.

The professor had just witnessed the first fluoroscopic image of the internal structures of a human body!

To better document his findings, Röntgen replaced the fluorescent screen with photographic plates and preserved the images of metal weights in a wooden box, a compass, a piece of welded metal, and a piece of wood wrapped in wire[5,7] (Figure 2.4).

Bertha Röntgen was somewhat mystified, and more than a little put off by her husband's obsessive and solitary activity: catching naps in the laboratory, being late for meals, and being generally absent-minded and inattentive to ordinary family affairs. Finally, one night she became very angry when he failed to comment upon the excellent dinner she had prepared. Perhaps to mollify her, he led his wife downstairs to the laboratory for a rather disturbing revelation of the

FIGURE 2.2 Bertha Röntgen, née Ludwig.

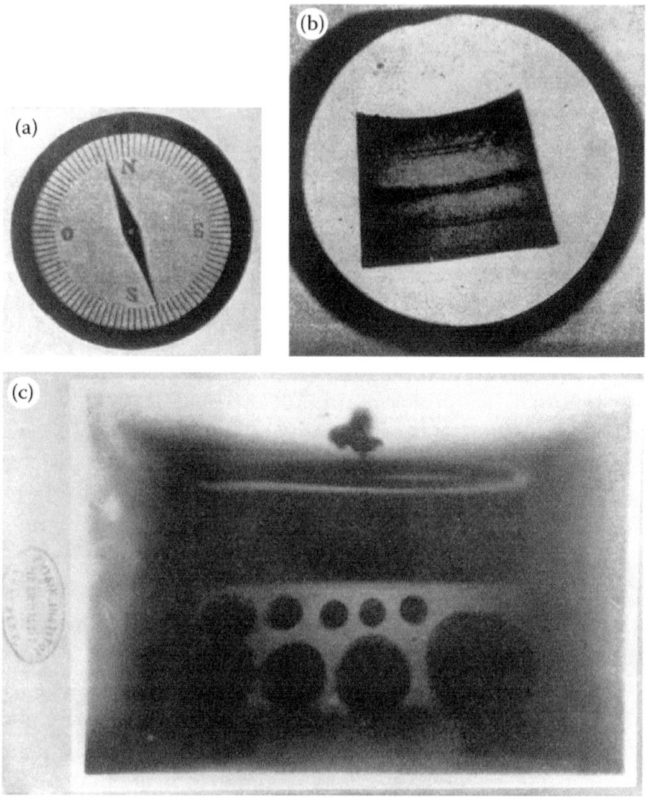

FIGURE 2.4 Photographic prints of articles Röntgen exposed to the rays during his discovery period. (a) a compass; (b) a welded piece of metal; (c) weights in a wooden box inside a purse.

reason for his excitement and single-mindedness. Let Glasser tell it:

> At his instruction she placed her hand on a cassette loaded with a photographic plate, upon which he directed rays from his tube for 15 minutes. On the developed plate the bones of her hand appeared light within the darker shadow of the surrounding flesh; two rings on her finger had almost stopped the rays and were clearly visible. When he showed the picture to her, she could hardly believe that this bony hand was her own and shuddered at the thought of seeing her skeleton. To Frau Röntgen, as to many others, this experience gave a vague premonition of death.[6]

This one radiograph, when widely circulated, probably did more to change the face of medicine than any other single item in history (Figure 2.5).

Working with his screens and plates, Röntgen made all of the fundamental observations that were the basis for his first two papers on the "x"-rays: so named because "x" was the symbol for the unknown. Röntgen's experimental design was so thorough and his reportage so meticulous and detailed that it was years before other investigators could add anything new to the subject. He showed that the rays were propagated in straight lines and were not influenced by magnetic fields, that the beam could be "hardened" by passing through absorptive material, that secondary radiation could be produced from certain targeted materials, and noted many other basic properties of the strange new ray.[7]

His manuscript was immediately accepted for publication in the December 1895 issue of the *Annals of the Würzburg Physical-Medicine Society* but, since no meetings or lectures were given in German universities during Christmas vacation, the formal presentation of Röntgen's seminal preliminary communication[4] had to await the January 1896 meeting of the society.

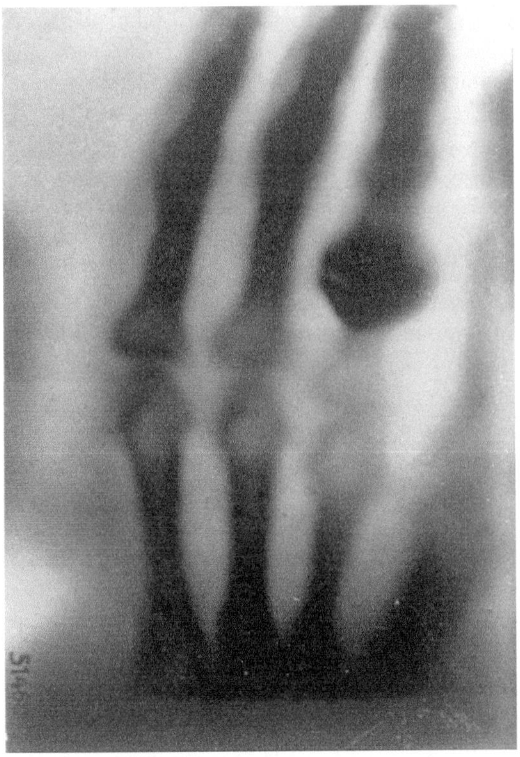

FIGURE 2.5 The first roentgenogram of a human, Bertha Röntgen's hand, exposed in the laboratory during November 1895.

The alacrity with which Röntgen's offering was accepted is astonishing by today's sluggish standards when months (or years) elapse between the submission, acceptance and publication of a scientific work. Nonetheless, then, as now, it seems there was sufficient impatience with the system to stimulate "leakage" of an important scientific breakthrough to the popular press.

On New Year's Day, 1896, Röntgen sent preprints of "Uber eine neue Arte von Strahlen" and a few of his first x-ray pictures to friends in Bavaria and Austria. Not surprisingly, the news of his discovery ran in *Vienna Presse* on January 5, 1896, and was flashed around the world, hitting the London and New York papers on the following day.

World reaction to this news, often garbled and inaccurate, was as astonishing and sensational as the news itself.[8,9] Lead-lined underclothes for modest young ladies were advertised by a London entrepreneur. A New Jersey assemblyman proposed a bill prohibiting x-ray opera glasses. In New York, a newspaper stated that at the College of Physicians and Surgeons the roentgen rays were used "to reflect anatomic diagrams directly into the brains of the students, making a much more enduring impression than the ordinary methods of learning anatomical details."[10,11] A young man in Jefferson County, Iowa, claimed that with x-rays he was able to transmute 13¢ worth of base metal into pure gold worth $153.[8] The x-ray was proposed as a solution for the issues of vivisection, spiritualism, soul photography, the temperance movement, and telepathy.[7] A private detective, Mr. Henry Slater, offered to introduce the "New Photography … in divorce matters free of charge," presumably to discover the skeleton that every closet is said to contain.[8,11]

For the most part, however, physicians and scientists, as well as responsible jurists and journalists, quickly understood that Röntgen's rays were a boon of great potential in the fields of medicine, biological and physical sciences, industry, the arts, and law:

> The surgeon … could … determine the extent of a complicated fracture without the manual examination … so painful to the patient [or] find the position of a foreign body, such as a bullet or piece of shell … without any painful examinations with a probe. (*Frankfurt Zeitlung*, January 7, 1896)

> Knowing the existence of a fracture in a person, who has been burned or mutilated beyond recognition, we can hope to identify him by the x-ray …. (Dr. Fovau d'Courmelles, *Am. X-Ray J.*, October 1898)

It is no wonder then, that on January 23, 1896, a large crowd of representative scientists and members of the Society, university faculty and students, city officials, and representatives from the army filled the auditorium of the Physical Institute for Röntgen's first and only lecture on the x-ray (Figure 2.6).

FIGURE 2.6 Röntgen's lecture at the January 23, 1896, meeting of the Würzburg Physical Medicine Society as portrayed by Robert A. Thom in a painting commissioned in 1962 by Park, Davis & Co., now a division of Pfizer. The noted anatomist, von Kolliker (arrow), was a willing subject for the demonstration of "a new kind of ray." (From Collection of the University of Michigan Health System. Gift of Pfizer Inc. With permission.)

He was modest in demeanor, generous in praise of his scientific forebears, and precise in his description of his "*arbeit.*" He showed x-ray pictures of various test objects and actually created an x-ray image of the hand of the famous anatomist, von Kolliker, during the meeting (Figure 2.7). Von Kolliker led three cheers, and amid general applause proposed that the rays be called Röntgen's rays in honor of the great discoverer.[5]

Never has a new scientific or technological breakthrough been so quickly and universally adopted by the medical/scientific community. The usefulness of the x-ray in noninvasive examinations of persons and objects was self-evident. The relatively simple tools and devices necessary to assemble an x-ray generator were readily available in most Western countries. Finally, thanks to a popular press thriving on sensationalism and a scientific press more rapidly responsive than can be even imagined today, the "how-to-do-it" instructions were handy to all interested parties.

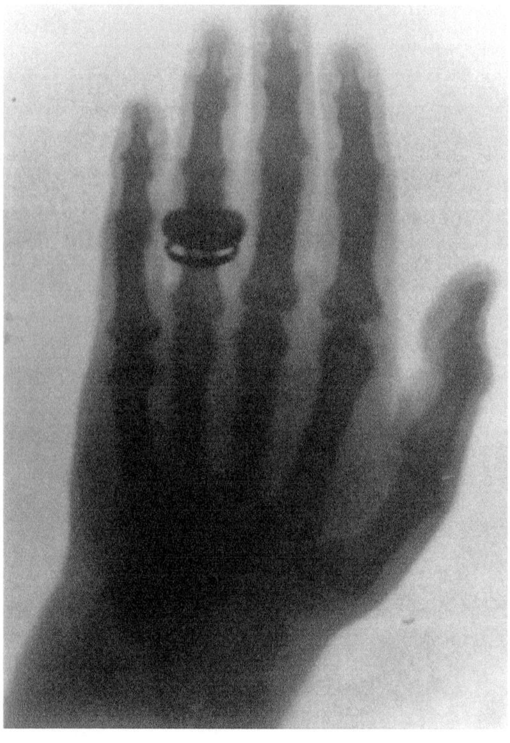

FIGURE 2.7 Photographic print of the roentgenogram of Professor von Kolliker's hand during the January 23, 1896, presentation by Röntgen.

THE FLOWERING OF FORENSIC RADIOLOGY

By its very nature, the science of radiology solves mysteries as it reveals deep within the body hidden secrets that are otherwise inaccessible to exposure. This potential was obvious from the first few images Röntgen produced in those first fateful 50 days in Würzburg. It seems unsurprising and only natural that the application of x-rays to the solution of problems and mysteries that fall within the purview of law and the courts should have been so quickly realized. Indeed, the first news of Röntgen's discovery in the United States in the *New York Sun* on January 6, 1895, reported erroneously—but prophetically—that, "The professor is already using his discovery to photograph broken limbs and bullets in human bodies."[12]

Some credit Professor A. W. Wright of Yale University with being the first American to produce an x-ray image. Wright wrote,

> A rabbit, purchased in one of the markets of the city, after an exposure of one hour to the rays, left upon the plate a complete representation. ... Particularly interesting in this photograph were several small round spots which appeared dark on the positive print. These were surmised to be shot[s] ... were readily found and extracted. The mode of death of the animal was not previously known.[13]

So this one small animal experiment was a precursor of the familiar forensic activities of localization and extraction of missiles and establishment of the cause and manner of death, in this case: "multiple gunshot wounds of vital organs" and, since the rabbit did not commit suicide and was shot intentionally, "canaliculacide" respectively.

Actually, the first court case involving the x-rays in North America commenced on Christmas Eve, 1895 (three days before Röntgen submitted his first communication to the Physical-Medicine Society of Würzburg. In Montreal, a Mr. George Holder shattered the peace of that wintry evening by shooting in the leg a Mr. Tolson Cunning. Attempts to locate the bullet by probing failed; the wound healed but remained symptomatic. A professor of physics at McGill University, John Cox, was requested by Cunning's surgeon, Dr. R. C. Kirkpatrick, to make an x-ray photograph of the wounded extremity. In the Physics Lecture Theater appropriate equipment was assembled and, with a 45-min exposure, a plate was obtained which showed the flattened bullet lying between the tibia and fibula (Figure 2.8). Dr. Kirkpatrick removed the bullet, and Mr. Cunning was discharged 10 days later. The x-ray plate was submitted to the court during the

FIGURE 2.8 Tolson Cunning's leg was exposed to x-rays in order to locate the bullet fired by George Holder on Christmas Eve, 1895. The examination took place in the Physics Laboratory at McGill University, Montreal on February 7, 1896. The result was the first x-ray plate to be admitted to a court in North America. (Reprinted from Center for the American History of Radiology. No other representation of this material is authorized without expressed, written permission from the American College of Radiology. With permission.)

trial, with the subsequent conviction of Mr. Holder for attempted murder. He was sentenced to 14 years in the penitentiary.[14–16]

The first instance in which a roentgenogram was brought to court in England was a personal injury case tried by Mr. Justice Hawkins and a special jury in Nottingham.[17] Early in September 1895, a Miss Ffolliott, a burlesque and comedy actress, suffered an injury while carrying out an engagement at a local theater. Hurrying to change costumes between acts, Miss Ffolliott fell on the staircase leading to her dressing room and severely injured her left foot. Even after a month of bed rest, she was unable to resume her professional activities. Finally, early in 1896, a Dr. Frankish sent her to University College Hospital where both her feet were examined by x-rays (the first comparison views?). There, clearly, was demonstrated a displacement of the cuboid bone in Miss Ffolliott's left foot, and when the two negatives were taken into court, both judge and jury could see the difference between the injured and uninjured members. Neither side could further argue the point; the only defense was a charge of contributory negligence against Miss Ffolliott.

Murder was the crime in one of the earliest forensic applications of Röntgen's discovery.[18,19] The murder took place on April 23, 1896, in a Lancashire textile town when Hargreaves Hartley fired four shots from a pistol into the head of his wife, Elizabeth Ann, then killed himself by jumping into the Leeds-Liverpool Canal and drowning. Mrs. Hartley survived the attack and was conscious.

Her physician, Dr. William Little, had learned of the potential of x-rays from a letter calling attention to Röntgen's work in the *Manchester Guardian* from Arthur Schuster, Professor of Physics at Owens College, Manchester. After a local photographer from Burnley tried unsuccessfully to x-ray the head of the victim, Dr. Little called upon Professor Schuster for assistance in locating the bullets. Unfortunately, Professor Schuster was ill and sent two assistants, C. H. Lees and A. Stanton, to the small home at 20 North Street, Nelson, on April 29. The tiny bedroom in this typical working-class terraced house was soon crowded with equipment: three Crookes tubes (two as spares), a high-tension coil, glass photographic plates, and storage batteries (there was no electricity in the house).

The first two radiographs, the results of 60- and 70-minute exposures, were developed by Professor Schuster and showed three of the bullets. On May 2, Professor Schuster himself

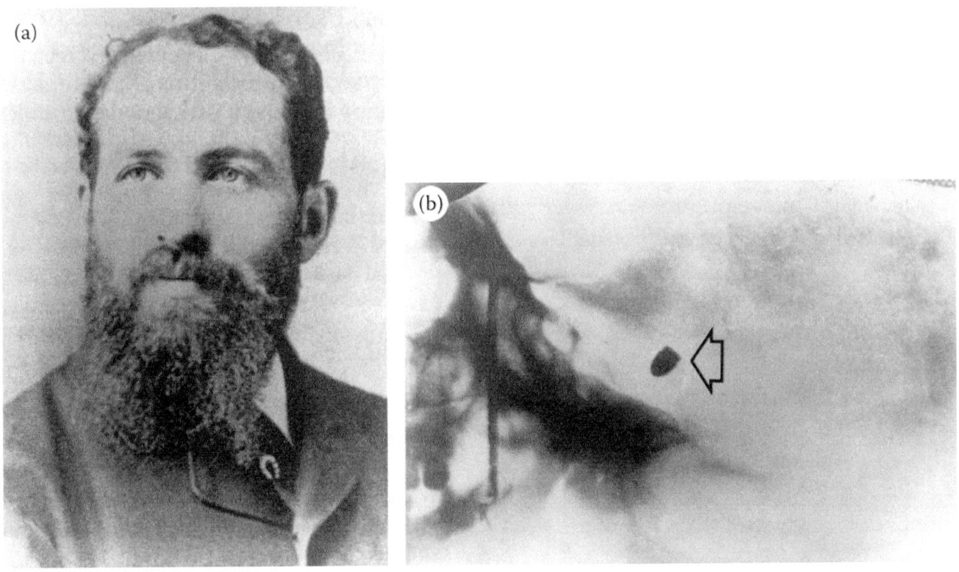

FIGURE 2.9 Murder in Lancashire. (Adapted from Eckert, W. C. and Garland, N., *Am. J. Forensic Med. Pathol.*, 5, 53, 1984.) (a) Professor Arthur Schuster, who found the fourth bullet. The excitement and stress of the event caused one of Schuster's associates to have a nervous breakdown. (b) Is this the "fourth bullet" found by Professor Schuster on the second examination of Mary Elizabeth Hartley with Röntgen's ray? No!

went to Nelson and was able to find the fourth bullet (Figure 2.9a). All were beyond the reach of a probe. Because of Mrs. Hartley's weakened condition, no attempt was made to extract the bullets, no treatment was undertaken, and the patient died on May 9. (This case can be considered an early manifestation of our tendency to use elaborate procedures and the newest technology, whether or not it will influence the outcome. At best, the radiographs probably replaced an autopsy in this case.)

The entire exercise was quite a spectacle for the town, with the mayor, town clerk, and others crowded into the room where the "experiment" took place, and it was reported extensively in the local press. The excitement and tension associated with the effort was blamed for the nervous breakdown of one of the associates, Mr. Stanton, some time later.

There is a peculiar twist to this story. Eckert and Garland's paper[19] on this historic episode includes a photograph of Professor Schuster (Figure 2.9a), accompanied by a radiograph (Figure 2.9b), with the following caption: "A picture of an x-ray from one of Schuster's cases, showing [the] location of one of the bullets, which had been shot into a victim's head by her husband." One might infer that this represents the fourth bullet that Professor Schuster found. Not true! In the first place, only one bullet is displayed. Secondly, Lichtenstein pointed out the long shadow of a coil spring typically found in a demonstration skull with an articulated mandible. Further, Evans and coauthors[18] in their earlier and more detailed descriptions of the event, based on a contemporary newspaper account, wrote, "There is neither a record of a post-mortem or any photograph with an articulated mandible, but it is known that the bullets were beyond the reach of a surgical probe." Eckert and Garland give no clue to the provenance or purpose of the radiological contrivance found in their paper.

The first civil case in which x-ray evidence was accepted in a U.S. court took place in Denver, Colorado.[16,20–22] It all began on June 15, 1895, when James Smith, a poor clerk who was "reading law" and doing odd jobs to meet expenses, fell from a ladder while pruning a tree. He procrastinated (perhaps because of financial considerations) for almost a month before calling upon Dr. W. W. Grant for professional services. Dr. Grant was well known as a distinguished surgeon, one of the founders of the American College of Surgeons, who had performed the first appendectomy in the country on a 22-year-old female, Mary Gartside.[23]

Dr. Grant did a thorough examination and found no evidence of fracture; he did not restrict Mr. Smith's activity but asked him to return for an office visit in one week. The second examination too disclosed no fracture. Dr. Grant neither saw nor heard from the patient again until April 1896 when Mr. Smith brought a $10,000 suit for malpractice against Dr. Grant, claiming limb shortening and disability due to a misdiagnosis of an impacted fracture of the left femoral neck (Figure 2.10).

Since James Smith had been employed by several lawyers to take dictation and type up legal instruments and was himself studying the law, it should come as no surprise that he appreciated the potential for legal redress for his incapacity

FIGURE 2.10 Illustration in the *Daily News*, December 3, 1896, of some of the principals and witnesses in *Smith v. Grant*. (From Collins, V. P., *Classic Descriptions in Diagnostic Radiology*, Vol. 2, Bruwer, A. J., Ed., Charles C. Thomas, Springfield, IL, 1964, 1578. With permission.)

and could engage two of the brightest young minds in the legal community to prosecute his case. One, Ben B. Lindsey, later became the founder of the Denver Court of Domestic Relations or the Juvenile Court. The other, Fred W. Parks, became the youngest senator from Colorado.

In the interval between the fall from the ladder and the filing of the lawsuit, radiology had come to Denver. Dr. Chauncey Tennant Jr. of the Denver Homeopathic College, Mr. Harry H. Buckwalter, a local photojournalist, and a Mr. Hall, who was President of the Diamond Lamp Company (and thus could manufacture tubes) had been experimenting with the new rays for some time and had given an exhibition before the Denver and Arapahoe Medical Society. There was speculation at that meeting whether this new science might play a role in the malpractice suit filed against Dr. Grant earlier that same month.

Smith's young attorneys were quick to grasp at new opportunities and, as the trial date drew near, approached Mr. Buckwalter, who agreed to x-ray their client. On November 7, 11, 21, and 29 of 1896 Mr. Buckwalter and Dr. Tennant made several plates of Mr. Smith's hip with exposures ranging up to 80 min, the last of which showed the outline of an impacted fracture of the proximal femur. The attempt to admit these roentgenograms in evidence prompted lengthy argument.

There were two issues here: one more medical than legal, the other more legal than medical. The first had to do with whether the shadows on the roentgen plate could accurately represent the condition of the proximal femur, with its complex bends and angles, lying at some distance from both the x-ray tube and the photographic plate. This, basically, was simply a question of proper positioning and alignment of patient, tube, and plate, and was resolved by a courtroom demonstration.

The other issue was more fundamental. Photographs had long been accepted as "secondary" evidence; a witness could testify that the photograph was a true representation of what had actually been seen. The roentgenogram, on the other hand, purported to show the true nature of a condition totally hidden from direct observation. For that reason, judges in the eastern United States had refused to admit roentgenograms as evidence saying, "It is like offering the photograph of a ghost."

The arguments raged all day before Judge Owen E. LeFevre in the District Court of Arapahoe County. Judge LeFevre was a well-known and colorful character in Colorado. He was a Civil War veteran, University of Michigan graduate, and a respected lawyer and jurist. He had made a fortune in gold mining, bred horses, farmed, had broad social interests, and owned the finest collection of contemporary French paintings in the state. He decided to sleep on the matter, and on the morning of December 2 or 3, 1896 (the record is not clear), handed down his landmark decision in the elegant language of those days:

> We ... have been presented with a photograph taken by means of a new
> scientific discovery .... It knocks for admission at the temple of learning;
> what shall we do or say?
> Close fast the door or open wide the portals?
> These photographs are offered in evidence to show the present condition of the head and neck of the femur bone, which is entirely hidden from the eye of the surgeon .... Modern science has made it possible to look beneath the tissues of the human body, and has aided surgery in telling of the hidden mysteries. We believe it is our duty to be the first ... in admitting in evidence a process known and acknowledged as a determinate science. The exhibits will be admitted in evidence."

X-rays already had been in court before Judge LeFevre's decision, but not as images. This involved another malpractice case but with x-rays as the cause of the injury. In Chicago, Frank Bolling was thrown from his buggy and suffered a fracture of his right ankle on September 2, 1895. By May, 1896, he was able to return to work but still had symptoms. On September 10, 1896, three x-ray photographs were made under the supervision of Dr. Otto L. Smith and Professor W. C. Fuchs, using exposure times of 35 to 40 min with the tube only 6 in from the ankle. The resultant radiation damage eventuated in amputation of the foot and ankle. The jury awarded $10,000 in damages to Mr. Bolling.[16,21]

The first criminal case in the United States involving x-rays was the October 1897 Haynes murder trial in Watertown, New York.[16,21] The victim was shot in the jaw with a .32 caliber bullet. Another foreign object was discovered lodged in the back of the head. Was this a second bullet or a fragment from the first? Dr. Gilbert Cannon gave testimony on the findings of the roentgenogram (not a second bullet), which subsequently was accepted as evidence by Judge Wright.

Acceptance of radiology—and the radiologist—in court was not immediate, universal, or standardized. As late as 1919, a court in Iowa accepted x-ray photographs as the best evidence, but disallowed expert testimony to explain what they showed. The physician witness might describe the expected roentgenographic manifestations of an injury, then the jury would view the images and draw their own conclusions.[21]

A more sensible approach was that of a Judge Pound in 1915, who denied that the doctrine—a photograph is the best evidence of what it contains—should be extended to the radiograph. Rather, he said, "... the x-ray picture is not ... the best evidence to laymen of what they contain. The opinion of the expert is the best evidence of what they contain—the only evidence."[21]

Dr. Frank W. Ross pointed out that there might be different standards for evidence from a physician who was an x-ray expert and that from his counterpart in general clinical practice. He wrote in 1899, "... evidence which ... may be sufficient for us ... may count for naught in a court of law .... Best intentions are often looked upon with suspicion by the court and the jury. Especially is this true in regard to new discoveries, which are viewed rather in the light of experiments."[16] (Ross could have been talking about any subsequent technological breakthrough in radiology or in other fields—DNA for instance.)

Other forensic applications of Röntgen's rays were quickly proposed as the news of his experiments spread and others repeated them. The May 30, 1896, issue of *JAMA* mentions an article by Dr. T. Bordas of the Faculty of Medicine of Paris in the May issue of the *Annales d'Hygiene Publique et le Medicine Legale*.[21] Monsieur le Docteur suggested that x-ray be used not only for identification through the visualization of old fractures, bullets, or other known peculiarities, but also recommended its use on suspicious packages suggestive of being infernal machines. Unfortunately, his advice is still timely today (Figure 2.11).

A somewhat logical extension of the concept of examining packages for infernal machines was the use of an x-ray device in customs houses (Figure 2.12). The Bureaux de Douanes in 1897 used fluoroscopes to examine passengers' luggage, purses, hats, hair, and so on. for contraband at the Pavillion de Rohan and at the Gare du Nord.[21] (A modern version of this initiative is discussed in Chapter 19.)

d'Courmelles' expectations for identification of persons by radiographic comparison (see page 11) seem not to have been realized before the turn of the century. The popular scientific method of human identification at the time of d'Courmelles' prediction had been devised by Alphonse Bertillon, a Parisian anthropologist.[24] In 1879 he founded an anthropometric department at the prefecture of police. There he introduced an anthropometric system designed to identify

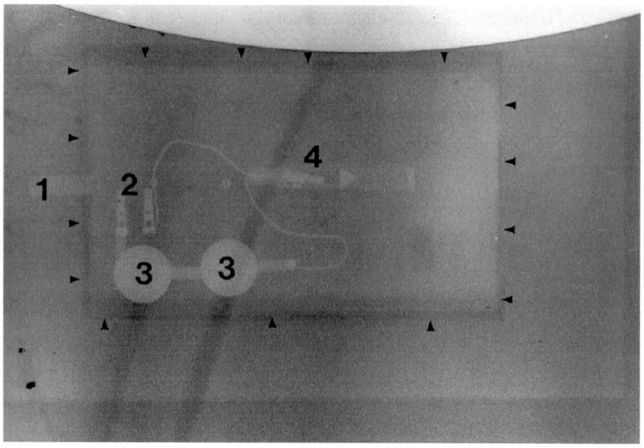

FIGURE 2.11 Radiograph of a modern letter bomb. Arrowheads indicate plastic explosive inside envelope. The triggering string (1) closes the contact (2) so that current from the batteries (3) energizes the detonator (4). (Courtesy Dr. Rafic E. Melhem. See *Radiology*, 151, 606, 1984.)

an individual throughout his life—this on a classification (small, medium, and large) of 12 precise measurements of the length and breadth of the head; the length of the radius, foot, and left elbow; the height of the trunk and bust; and also color of the iris of the right eye. This description was supplemented by special visible features and front and side photographs. Bertillon's system was accepted widely and used internationally by 1892.[21] Even Bertillon recognized inherent problems in the method—variations in measurements made by different investigators, and changes in individual measurements with variations in body weight or as the result of pathologic processes.

A Berliner, Levinsohn, suggested that direct measurements of the skeleton through x-ray photography would be more accurate.[25] His paper attracted some notice, but apparently his method was not widely employed.

FIGURE 2.12 French custom officer (on right) using a hand-held fluoroscope to examine a piece of luggage. The glass x-ray tube (arrow) is on the table between him and the other official with the chevron on his sleeve. (From Angus, W.M., *Radiographics*, 9, 1225, 1989. With permission.)

A radiologist and compatriot of Bertillon was the Parisian, Henri Béclère.[26,27] Béclère advocated *dactylography* in which the skin of the fingers is lightly coated with powered lead tetroxide and exposed to soft x-rays. The resultant roentgenogram produced fine fingerprints (Figure 2.13). Béclère also made much of the configuration of the nails; but it is ironic that he paid no attention to the equally unique configuration and trabecular pattern of the underlying bones commenting only that the skeletal shadows do not compromise the distinctness or clarity of the ungula furrows.

Béclère may not have known of the work of Dr. David Walsh who, 21 years earlier, also produced images of knuckle folds, palmar lines, and fingerprint furrows on roentgenograms of hands impregnated with bismuth subnitrate.[28] The question of primacy is of no practical importance since both systems quickly were relegated to the status of historical curiosities by the cheap and rapid ink print.

The x-ray enjoyed a modest resurgence of activity in fingerprint work in the recovery of latent prints from difficult surfaces (i.e., multicolored documents, cloth, polythene, wax, cardboard, hardboard, varnished and untreated wood, rubber, pigskin, and the skin of human corpses or nonvital separated body parts). Thus, by this fingerprint method the corpse

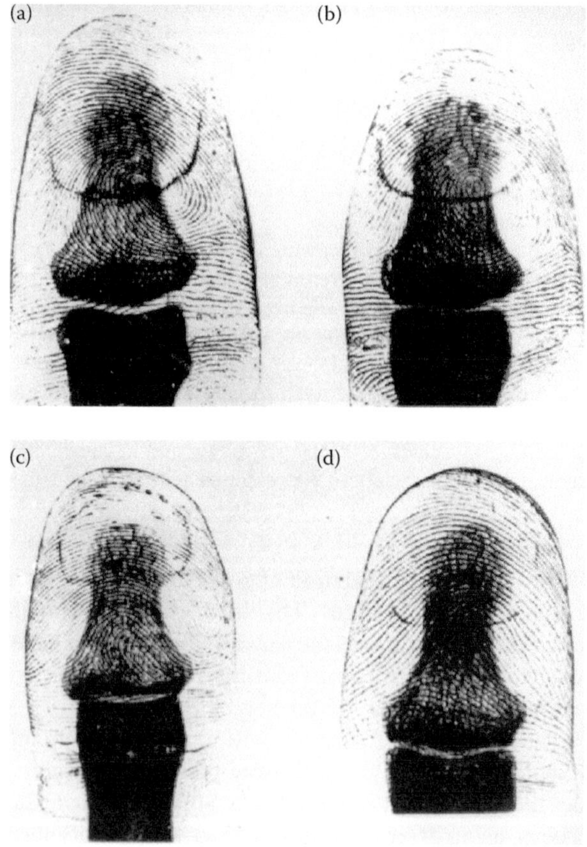

FIGURE 2.13 Roentgenographic fingerprints obtained by H. Béclère by coating the fingers with lead tetroxide before exposure to soft x-rays. (From Collins, V. P., Origins of medico-legal and forensic roentgenology, in *Classic Descriptions in Diagnostic Radiology*, Vol. 2, Bruwer, A. J., Ed., Charles C. Thomas, Springfield, IL, 1964, 1578. With permission.)

may be able to identify his or her assailant. Introduced by Graham[29] and followed up by Winstanley,[30] the method is known as x-ray fluorescence radiography (XFR) or backscatter radiography. It relies on the emission of low-energy x-rays (secondary or Compton scatter) from heavy metal particles (lead dust) when bombarded with suitable high-energy x-rays. Special equipment is required and the lead dust represents a potential health hazard.

As an aid to identification, a prescient Dr. Angerer in Munich suggested in 1896 the use of wrist-bone development as a measure of bone age.[31]

Forensic radiology of celebrities, whether famous or notorious, has a special fascination and has contributed to the public awareness of the field. His Excellency A. von Kolliker, the famous anatomist whose hand Röntgen exposed at the lecture on January 23, 1896, certainly was a scientific celebrity and the resultant image was widely circulated. Various early participants in court cases involving x-rays enjoyed fleeting celebrity status in the popular press because of the novelty of the method. Perhaps the first American of national or international stature to undergo x-ray examination in a situation with forensic connotations was Theodore Roosevelt. Mr. Roosevelt was the fifth of ten presidents to become the target of an assassination attempt (Jackson, Lincoln, Garfield, McKinley, both Roosevelts, Truman, Kennedy, Ford twice, and Reagan—all but Jackson and T. Roosevelt while in office).

President McKinley was shot by an anarchist while viewing the 1901 Pan-American Exposition in Buffalo, the first such outrage to occur since Röntgen's discovery. Although x-ray equipment was on display at the fair, and many of the buildings had electric lights, such was not the case in the Exposition "hospital." Lamps, lanterns, and candles were excluded for fear of ether explosion. Consequently, baker's pans were employed in an attempt to reflect daylight into the room where several physicians were examining and probing the fallen President, perhaps inoculating him with the infection which killed him eight days later.

Theodore Roosevelt succeeded McKinley as President, but was shot after leaving office while running as the Bull Moose Party candidate for a third term. The date was October 14, 1912. As Roosevelt was leaving his hotel in Milwaukee to speak at a rally, "a half-crazed fanatic" stepped close to his automobile and shot him in the chest. The bullet, which ordinarily would have been fatal, was decelerated and deflected by having to pass through Mr. Roosevelt's overcoat, suit, shirt, and underclothes. Along the way it penetrated the folded 50-page manuscript of his speech and a steel eyeglass case, before entering his heavily muscled chest. Mr. Roosevelt remained upright. An experienced hunter, he correctly assumed that his lung had not been penetrated since he was not spitting blood. Mr. Roosevelt insisted on proceeding. He asked the crowd to please be quiet as "... I have just been shot, but it takes more than that to kill a Bull Moose !". With blood seeping onto his shirt, he addressed the audience for 90 min, then shook hands with some of the crowd. (He did admit that this last activity was somewhat painful.)

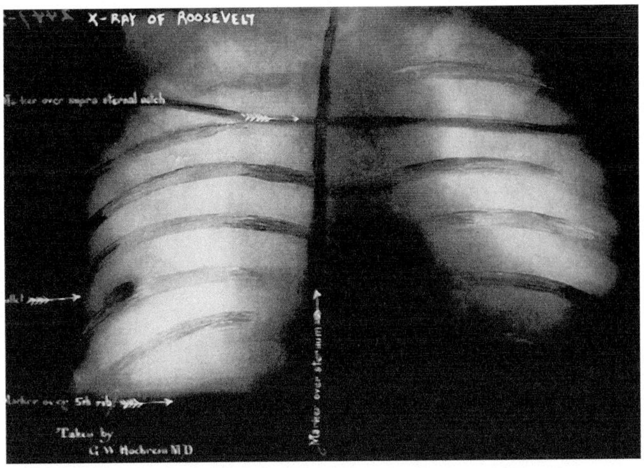

FIGURE 2.14 Radiograph of Theodore Roosevelt's chest obtained a few hours after the assassination attempt by G.W. Hochrein, M.D., in Chicago. The bullet is embedded in an anterior rib (arrow).

Later that evening, Mr. Roosevelt went by special train to Chicago, walked to a waiting ambulance, sat up in it en route to the hospital, and walked in for an examination of his wound with x-rays. The bullet had entered his right chest, medial and inferior to the nipple, and was embedded in a rib, splintering its internal surface so that the pleura was compromised. The bullet was not extracted for fear of massive pneumothorax and/or subsequent empyema. He carried the bullet with him to the grave[32] (Figure 2.14).

The assassination attempt on Franklin Roosevelt missed him, but killed the mayor of Chicago. The Puerto Rican extremists who assaulted Truman in Blair House never got close to him. But few events in history have had the world impact, or generated so much enduring controversy, as the November 22, 1963, assassination of President John F. Kennedy in plain view of a large crowd and a worldwide television audience. The chronology of that Dallas afternoon, or its aftermath, needs no reiteration here.

The postmortem radiographs of Mr. Kennedy have added some fuel to the firestorm of debate which burns so fiercely even today, yet seems to cast so little light upon the truth of the matter. The radiographs are central to the controversy about the number of shots, direction and trajectory of the shots, and the number and the location of the shooter(s). The x-ray set, a frontal and two lateral views of the skull, are of poor quality, being somewhat overexposed. The location of the bullet wound, trajectory, fracture patterns, and conflicting expert opinion have been used paradoxically in support of opposing theories.[33–39] It is not within the interests or province of this book to enter into those arguments.

In February, 1896, W. Koenig was taking intraoral films of the teeth showing restorations (Figure 2.15), thus leading the way for the science of Forensic Odontology which has flourished only since the 1940s.[11,40] The case of Adolph Hitler dates from that decade. Following the unsuccessful assassination attempt with a bomb in the Wolf's Lair bunker on July 20, 1944, Hitler had many residual physical symptoms and

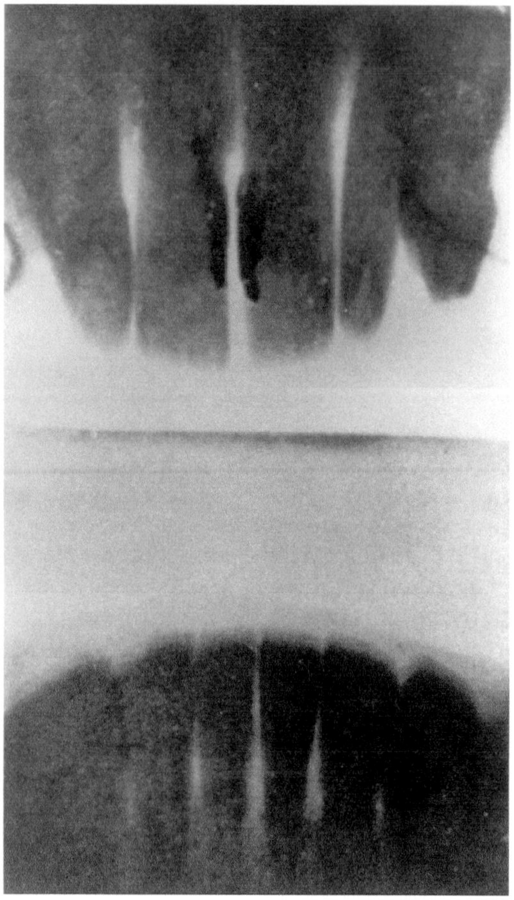

FIGURE 2.15 Dental x-rays by W. Koenig in 1896. Restorations are seen in the maxillary central incisors. (new)

disabilities. Persistent headaches finally forced him to follow the advice of his otolaryngologist, Dr. Paul Giesler.[41] On September 19 he was driven to an army field hospital at Rastenburg where three roentgenograms of his skull were obtained[42] (Figure 2.16). Those films survived the war; Hitler, of course, did not—although speculation and rumors abounded that he had somehow escaped. The dental work displayed on Hitler's roentgenograms was quite distinctive, however, and the Russians were able to make a comparison with the burned remains found in the ruins of the chancellery garden. Although positive identification was made by this dental comparison, the Russians kept secret this information for more than two decades.[43]

One of the worst of Hitler's villains escaped retribution. Dr. Joseph Mengele, the Doctor of Death at Auschwitz, left the camp before its liberation. He lived for some time near Günzberg protected by family influence and local officials. He then went to South America where he practiced medicine in various places under false names; advised General Stroessner, the dictator of Paraguay, on how to annihilate the Indian population; disappeared from Argentina just in time to escape extradition to West Germany; perhaps outsmarted and murdered a female Israeli spy who tried to entrap him; and perhaps was involved in the drug trade. Mengele finally died of a heart attack while swimming in 1979.[44]

In 1985, rumors spread outside South America of a body in Brazil believed to be that of Mengele. A vast team of forensic scientists, representing several disciplines, were employed in the complex identification process.[45] Skeletal abnormalities were found consistent with accidents and illnesses documented in Mengele's history, but there were no antemortem radiographs to be found for comparison. Finally, the reluctant family made Mengele's diary available, and in it were references to root canal work. This led to recovery of antemortem dental radiographs. These, along with superimposition techniques using the skull and known photographic likenesses of Mengele, and an oral-antral fistula, finalized an inarguable, positive identification,[46–48] eventually confirmed by DNA typing.

Any reflection on the historical genesis of forensic radiology ultimately must include consideration of early "practical use of roentgen rays for non-medical purposes" as summarized by Glasser in his classic biography.[49] One of Röntgen's first test objects was a piece of welded metal (Figure 2.4). Röntgen was an avid hunter and outdoorsman so it is appropriate that one of his best x-ray pictures, made a bit later, was of his shotgun (Figure 2.17). Röntgen was delighted that not only were the "bullets" there for everyone to see, but also small irregularities in the steel could be discerned. Consequently, early in 1896, the War Ministries of Germany and of Austria proposed using Röntgen's method to detect defects in guns and armor. U.S. ordnance officials were similarly influenced by demonstrations of invisible welds in metal by Professor A. W. Wright of Yale, previously mentioned for his dead- rabbit investigation. That same year, the Carnegie Steel Works in Pittsburgh also employed x-rays for nondestructive testing. Industrial radiography is not within the scope of this book, but is still a useful tool for our colleagues in forensic engineering. Apart from mail-bomb searches already mentioned, x-rays were demonstrated to British postal authorities as a means of finding coins in newspapers, embedded in sealing wax, and otherwise posted in violation of existing regulations.

A Mr. B. Hicks made an early excursion into the field of "Questioned Documents" when he used a roentgenogram of a document to show the court how the parchment was extremely thin in one area "as if some names had been erased in order to be replaced by others."

Adulteration of foodstuffs was proven by x-ray in 1896. In the same year roentgenograms of mummies were first obtained. The roentgenogram and its sophisticated progeny, computed tomography (CT), still are used to nondestructively evaluate mummies for content, age, sex, embalming methods, hidden valuables, injuries, and disease. Also contemporaneously, roentgen rays were employed to detect fake jewels and alterations in paintings (see Chapters 21 and 22).

Thus, we see that the foundation for most of modern forensic radiology had been laid down before the end of the first decade after Röntgen's discovery. Some seminal work lay ignored for years only to be rediscovered or renovated at a later date. Some building blocks were missing, more from social scotomata than scientific oversight—our blindness to abuse in all of its ugly forms, for instance.

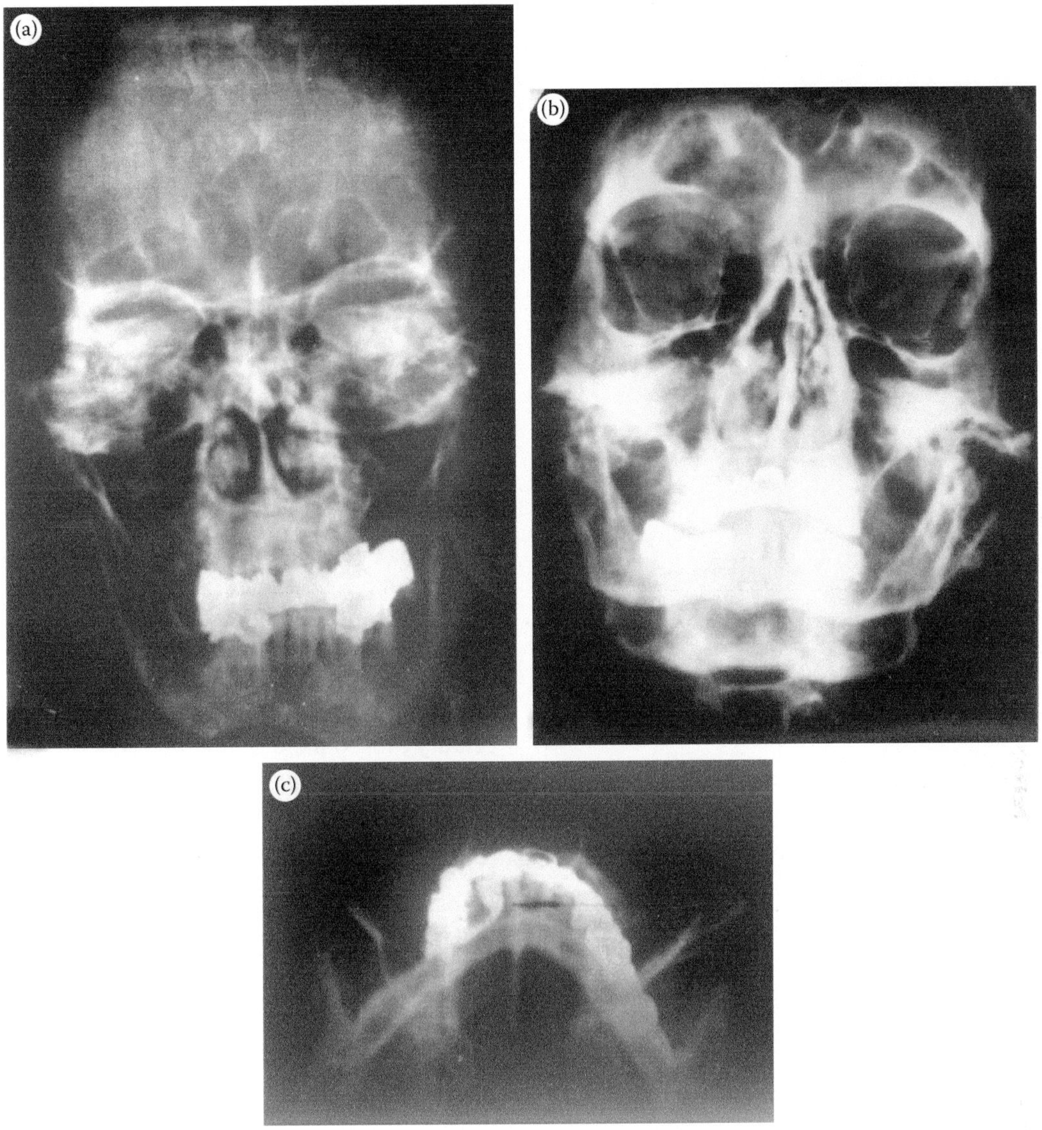

FIGURE 2.16 (a, b, c): Antemortem roentgenograms of A. Hitler from September 19, 1944. Note the extensive and unusual dental work which confirmed identification of his remains.

Professor Röntgen furnished the tool. His contemporaries showed us how to use it. Realization of the full scope of forensic radiology was to depend upon the imagination and industry of modern scientists and the indulgence or approval of the courts.

But the field of forensic radiology did not immediately flower; rather it remained relatively stagnant for several decades, while medical radiology, in general, flourished. Improved generators allowed exposure times short enough to stop respiratory, pulsatory and peristaltic motion—not very important in the morgue! Image intensification took fluoroscopy out of its dark age—immaterial in forensic work. Endovascular and cardiovascular procedures were introduced, as were radioisotope studies and sonography. None had much of an impact on forensic imaging. However, some of these technical advances were embraced and modified for forensic application. Examples would include Venezi's contrast techniques in studying vertebral arterial systems in cadavers[50]; the use of solidifying silicone rubber with lead oxide in autopsy studies of vascular structures[51]; and similar applications for the demonstration of esophageal, tracheal, and aortic fistulae.[52] Beginning with H. C. Andrews' 1919 book on aircraft accidents,[53] there were accident investigations and research using radiology. In 1944, Haas used radiography extensively to study the relationships between force, injuries, and aircraft structures[54]; Simson showed the value of the aircraft accident at necropsy almost 40 years ago in a clever reconstruction of a fatal crash from radiographic evidence.[55] He also showed that cervical spine fracture patterns were related to impact velocities. An innovative prospective approach was the work done by Jones and associates using fresh cadavers and the then-famous rocket sled facility at Holloman Air Force Base,[56] again relating fracture and dislocation patterns to impact forces. One of the earlier accident investigation studies with forensic applications in the radiological literature involved postmortem radiology of head and

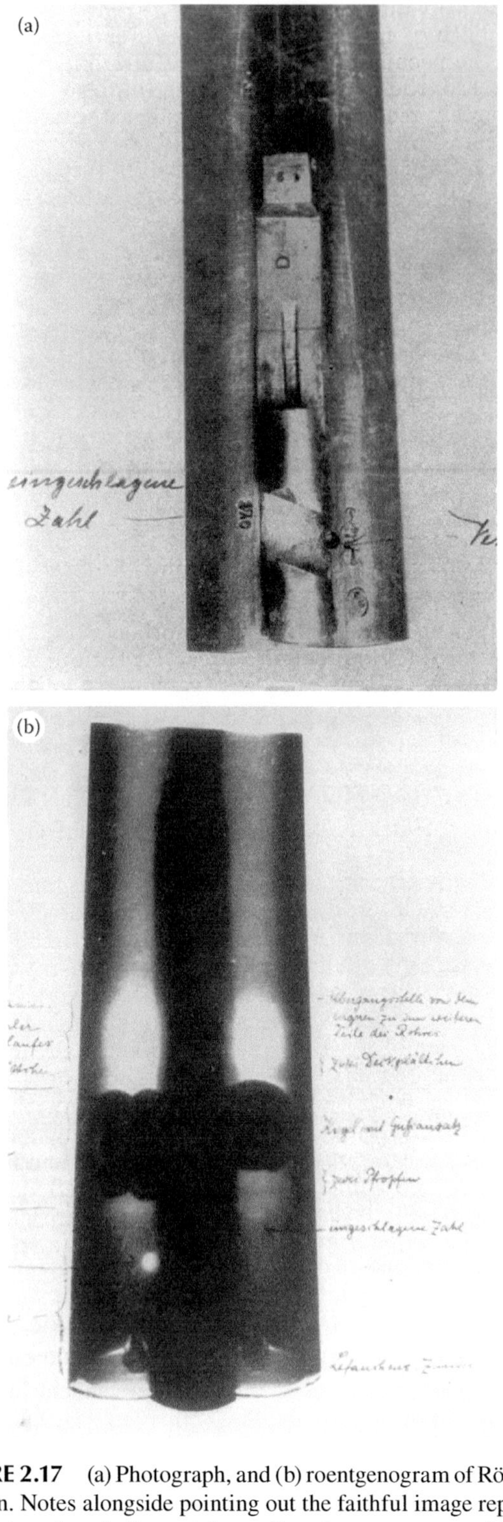

FIGURE 2.17 (a) Photograph, and (b) roentgenogram of Röntgen's shotgun. Notes alongside pointing out the faithful image reproduction of stampings in the metal as well as the components of the load are in Röntgen's handwriting.

neck injuries, which revealed a high incidence of cervical spine injuries and both intravascular and intracardiac air which might be missed by the unforewarned autopsy surgeon (Table 2.1) (Figure 2.18).[57]

In 1946, Caffey published his seminal work leading to what is perhaps radiology's most important contribution to

TABLE 2.1
Incidence of Head and Neck Injuries in 146 Victims of Fatal Traffic Accidents

Craniocervical trauma			82 (56.2%)
Skull fractures only	51 (34.9%)		
Skull and cervical injury	10 (6.8%)	61 (41.7%)	
		31 (21.2%)	
Cervical injury only	21 (14.3%)		
No craniocervical trauma			64 (43.8%)
Total			146 (100%)

Source: From Akers, G.T. Jr. et al., Radiology, 14, 611, 1975. With permission.

the forensic sciences, the conceptualization and diagnosis of intentional physical trauma to children by their caregivers.[58]

In 1949, Singleton became the father of mass casualty radiology with a success rate rarely, if ever, matched since then.[59]

The widespread adoption of CT in the eighth decade of the century evoked more dismay and concern than enthusiasm in forensic circles originally. How could one compare or match antemortem CT images with conventional postmortem radiographs? Although the cost, availability, and accessibility of CT scans was substantial, physical anthropologists and forensic pathologists were quick to see the potential usefulness of CT as it burgeoned in clinical use. Examples are Haglund's use of a toposcan for identification by comparison of frontal sinuses,[60] and identification of comparative CT studies of the lumbar spine in Germany (Figure 2.19).[61] Rougé et al. compared CT images to exclude a possible identification.[62] Oliver

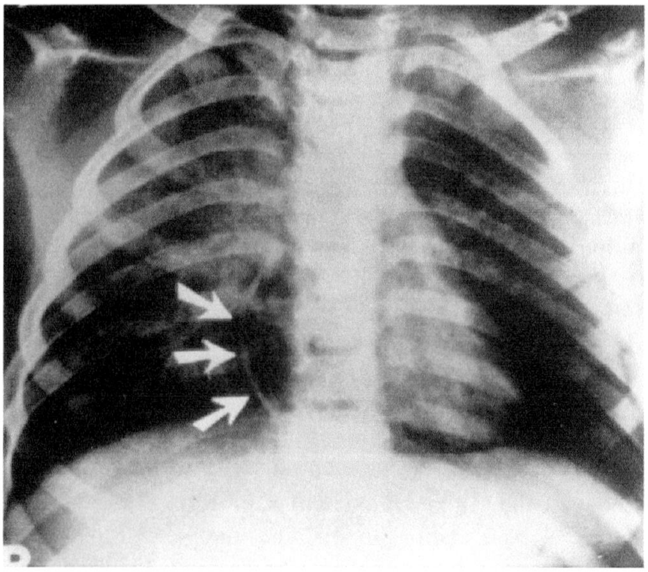

FIGURE 2.18 Example of air in the heart of a person suffering fatal massive injuries in a vehicular accident. (From Akers, G.T. Jr. et al., Radiology, 14, 611, 1975. With permission.)

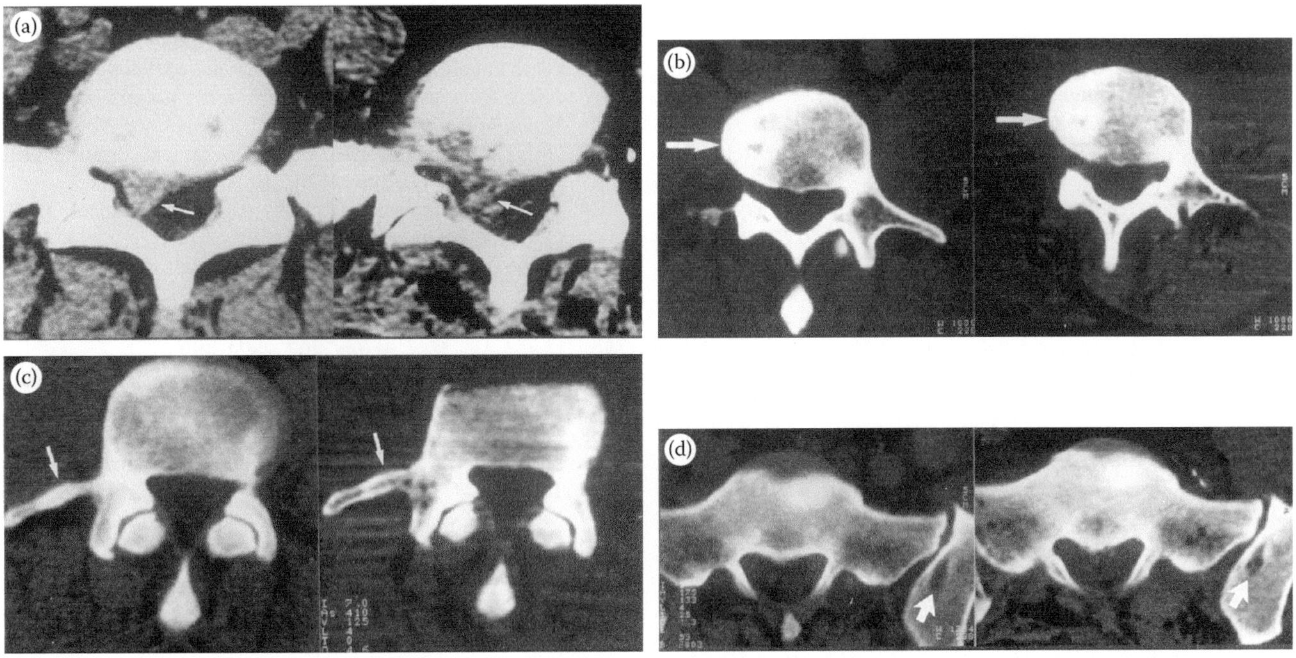

FIGURE 2.19 Identification through comparison of antemortem and postmortem CT. The antemortem study is on the viewer's left in each instance. (a) shows a posterolateral disk herniation at L5–S1. (b) shows a Schmorl's node in the inferior endplate of L4. (c) shows a peculiar thickening of the right transverse process of L4. (d) shows identical lucencies in the left ilium at the sacroiliac joint. (Original images courtesy the authors, copyright ASTM, reprinted. With permission.)

et al. used an innovative combination of CT images and radiographs taken at autopsy to generate a three-dimensional (3-) reconstruction of a bullet path—a powerful exhibit for the courtroom or the laboratory.[63]

A number of factors inhibited the adoption and application of these remarkable advances in radiological imaging in routine forensic activities and research. The equipment was costly and expensive to operate and maintain. It resided almost totally in clinical facilities devoted to the care of live patients. The overwhelming majority of forensic investigators are nonradiologists who usually struggle with substandard equipment and in ignorance of well-known radiological tenets not published in their literature. On the other hand, most radiologists have little connection with the forensic sciences and are unaware of the research possibilities in that field or of the problems that need solution. Sharing of interdisciplinary facilities, skills, and knowledge was impacted by conflicting clinical responsibilities, manpower shortages, financial considerations, and organizational patterns. The possibilities for interdisciplinary interaction at the academic level were much larger in Europe where standalone institutes or departments of forensic or legal medicine are common, whereas only two such entities are to be found in U.S. medical schools.

Still, limited forensic research investigation with CT or magnetic resonance imaging (MRI) began in the United States as early as 1990.[64] But the first forensic CT scan is attributed to Schimacher and associates to describe a gunshot injury to the head in 1977.[65] In Japan, which has only 2% of the world's population, one-third of the world's CT units but only five cities with a Medical Examiners System (85% of the population live outside a Medical Examiner System), postmortem CTs have been done since 1985 or earlier.[66] In this decade, large postmortem CT series, some with MRIs, have been collected in Western Europe, Australia, and Scandinavia. A model interdisciplinary research effort was fashioned in Bern by Professor Dirnhofer of the Institute of Forensic Medicine and Professor Vock of the Institute of Diagnostic Radiology. The remarkable results of that liaison will be found elsewhere in this book.

The success of virtual autopsy has caught the attention of the forensic sciences, but no less important are the advantages CT and MRI have brought to forensic evaluation of the living. Great contributions have been made in the areas of abuse, sharp and blunt trauma, gunshot wounds, and accident investigation.

REFERENCES

1. Glasser, O., *Wilhelm Conrad Röntgen and the Early History of the Roentgen Rays*, Charles C. Thomas, Springfield, IL, 1934, chap. 5.
2. Glasser, O., *Wilhelm Conrad Röntgen and the Early History of the Roentgen Rays*, Charles C. Thomas, Springfield, IL, 1934, chap. 2.
3. Glasser, O., *Wilhelm Conrad Röntgen and the Early History of the Roentgen Rays*, Charles C. Thomas, Springfield, IL, 1934, chap. 1.
4. Röntgen, W. C., Ueber eine neue art von strahlen, Sitzgber. *Physik-Med. Ges*. Würzberg, December, 132, 1895.
5. Glasser, O., *Wilhelm Conrad Röntgen and the Early History of the Roentgen Rays*, Charles C. Thomas, Springfield, IL, 1934, chap. 4.

6. Glasser, O., *Dr. W. C. Röntgen*, 2nd ed., Charles C. Thomas, Springfield, IL, 1958, chap. 3.
7. Glasser, O., W. C. Roentgen and the discovery of the roentgen rays, *Am. J. Roentgenol.*, 25, 437, 1931.
8. Glasser, O., *Wilhelm Conrad Röntgen and the Early History of the Roentgen Rays*, Charles C. Thomas, Springfield, IL, 1934, chap. 6.
9. Linton, O. W., News of x-ray reaches America days after announcement of Roentgen's discovery, *Am. J. Roentgenol.*, 165, 471, 1995.
10. Glasser, O., *Dr. W. C. Röntgen*, 2nd ed., Charles C. Thomas, Springfield, IL, 1958, chap. 7.
11. Dewing, S. B., *Modern Radiology in Historical Perspective*, Charles C. Thomas, Springfield, IL, 1962, p. 36.
12. Brecher, R. and Brecher, E., *The Rays: A History of Radiology in the United States and Canada*, Williams & Wilkins, Baltimore, 1969, 9.
13. Brecher, R. and Brecher, E., *The Rays: A History of Radiology in the United States and Canada*, Williams & Wilkins, Baltimore, 1969, 14.
14. Brecher, R. and Brecher, E., *The Rays: A History of Radiology in the United States and Canada*, Williams & Wilkins, Baltimore, 1969, 18.
15. Halperin, E. C., X-rays at the bar, 1896–1910, *Invest. Radiol.*, 23, 639, 1988.
16. Cox, J. and Kirkpatrick, R. C., The new photography with report of a case in which a bullet was photographed in the leg, *Montreal Med. J.*, 24, 661, 1896.
17. Glasser, O., First roentgen evidence, *Radiology*, 17, 789, 1931.
18. Evans, K. T., Knight, B., and Whittaker, D. K., *Forensic Radiology*, Blackwell Scientific, Oxford, 1981, chap. 1.
19. Eckert, W. C. and Garland, N., The history of the forensic applications in radiology, *Am. J. Forensic Med. Pathol.*, 5, 53, 1984.
20. Withers, S., The story of the first roentgen evidence, *Radiology*, 17, 99, 1931.
21. Collins, V. P., Origins of medico-legal and forensic roentgenology, in *Classic Descriptions in Diagnostic Radiology*, Vol. 2, Bruwer, A. J., Ed., Charles C. Thomas, Springfield, IL, 1964, 1578.
22. Pear, B. L., 1896: The first year of x-rays in Colorado, *Am. J. Roentgenol.*, 165, 1075, 1995.
23. Anon., *Associated Press release*, Mobile Register, January 4, 1989, 1D.
24. Saban, R., Salmar, A., and Potier, M., Biographical notes in: Musees d'Anatom'e Delmar- Orfila-Rouvìere, *Surg. Radiol. Anat.*, 17 (Suppl. 1), 5129, 1995.
25. Levinsohn, Beitraz zur feststellung der identität, *Arch. Krim.-Anthrop. Leipzig*, 2, 211, 1899.
26. Béclère, H., La radiographe anthropométric du pouce (superposition des empreites digitales, du sequelette et de l'ongle), *C. R. Acad. Sci.*, 167, 499, 1918.
27. Béclère, H., La radiographe cutanée, *J. Radiol. Electrol.*, 4, 145, 1920.
28. Walsh, D., Skin pictures with x-rays, *Br. Med. J.*, March 27, 787, 1897.
29. Graham, D., *The Use of X-ray Techniques in Forensic Investigations*, Churchill Livingstone, Edinburgh, 1973, 16–22.
30. Winstanley, R., Recovery of latent fingerprints from difficult surfaces by an x-ray method, *J. Forensic Sci. Soc.*, 17, 121, 1977.
31. Goodman, P. C., The new light: Discovery and introduction of the x-ray, *Am. J. Roentgenol.*, 165, 1041, 1995.
32. Bishop, J. B., *Theodore Roosevelt and His Time, Shown in His Own Letters*, Vol. II, Schriber's, New York, 1920, 337.
33. Clark Panel Report, *1968 Panel Review of Photographs, X-ray Films, Documents and Other Evidence Pertaining to the Fatal Wounding of President John F. Kennedy on November 22, 1963 in Dallas, Texas*, U.S. Government Printing Office, Washington, DC, 1969, 12.
34. Warren Commission Report, U.S. Government Printing Office, Washington, DC, 1964.
35. Breo, D. L., JFK's death—the plain truth from the MDs who did the autopsy, *J. Am. Med. Assoc.*, 267, 2794, 1992.
36. Breo, D. L., JFK's death. II. Dallas MDs recall their memories, *J. Am. Med. Assoc.*, 267, 2804, 1992.
37. Breo, D. L., JFK's death. III. Dr. Finch speaks out: 'two bullets, from the rear', *J. Am. Med. Assoc.*, 268, 1748, 1992.
38. Artwohl, R. R., JFK's assassination: Conspiracy, forensic science and common sense, *J. Am. Med. Assoc.*, 269, 1540, 1993.
39. Morgan, R. S., Unpublished data, presentation before The Rocky Mountain Radiological Society, Denver, August 19, 1972.
40. Ford, M. A. and Ashley, K. F., The role of the forensic dentist in aircraft accidents, in *Aerospace Pathology*, Mason, J. K. and Reals, W. J., Eds., College of American Pathologists Foundation, Chicago, 1973, chap. 8.
41. Toland, J., *Adolph Hitler*, Vol. 2, Doubleday, Garden City, NJ, 1976, chap. 28.
42. Toland, J., *Adolph Hitler*, Vol. 2, Doubleday, Garden City, NJ, 1976, chap. 29.
43. Sognnaes, R. F., Dental evidence in the postmortem identification of Adolph Hitler, Eva Braun, and Martin Borman, *Legal Medicine Annual*, 1977, 173.
44. Lifton, R. J., *The Nazi Doctors: Medical Killings and the Psychology of Genocide, Basic Books*, New York, 1986, chap. 17.
45. Joyce, C. and Stover, E., *Witnesses from the Grave*, Little, Brown, Boston, 1991, chap. 9.
46. Joyce, C. and Stover, E., *Witnesses from the Grave*, Little, Brown, Boston, 1991, chap. 10.
47. Curran, W. J., The forensic investigation of the death of Joseph Mengele, *N. Engl. J. Med.*, 315, 1071, 1986.
48. Teixeira, W. R. G., The Mengele report, *Am. J. Forensic Med. Pathol.*, 6, 279, 1985.
49. Glasser, O., *Wilhelm Conrad Röntgen and the Early History of the Roentgen Rays*, Charles C. Thomas, Springfield, IL, 1934, chap. 18.
50. Vanezi, P., Techniques used in the evaluation of vertebral artery trauma at postmortem, *Forensic Sci. Int.*, 13, 159, 1979.
51. Karhunen, P.J., Mannikko, A., Penttilä, A., and Kiastro, K., Diagnostic angiography in postoperative autopsies, *Am. J. Forensic Med. Pathol.*, 14, 303, 1989.
52. Karhunen, P.J. and Lalu, K., Radiographic demonstration of esophageal and tracheal fistulas at autopsy using a contrasting medium that vulcanizes at room temperatures, *J. Forensic Sci.*, 36, 1129,1991.
53. Andrews, H.G., *The Medical and Surgical Aspects of Aviation*, Oxford University Press, London, 1919.
54. Haas, G. M., Relations between force, major injuries and aircraft structure in design of aircraft, *Aviat. Med.*, 15, 395, 1944.
55. Simson, L.R. Jr., Roentgenography of the human factors of fatal aviation accidents. *Aerosp. Med.*, 43, 81, 1972.
56. Jones, A.M., Bean, S.P., and Sweeney, B.G., Injuries to cadavers resulting from experimental rear impact, *J. Forensic Sci.*, 23, 730, 1978.

57. Akers, G.T. Jr., Oh, Y.S., Leslie, E.V., Lehotay, J., Panaro, V.A., and Eschner, E.G., Postmortem radiology of head and neck injuries in fatal traffic accidents, *Radiology*, 114, 611, 1975.
58. Caffey, J., Multiple fractures in long bones of children suffering from chronic subdural hematomas, *Am. J. Roentgenol.*, 56, 163, 1946.
59. Singleton, A.C., The roentgenological identification of victims of the "Noronic" disaster, *Am. J. Roentgenol.*, 66, 375, 1951.
60. Hagland, W.D. and Fligner, C.L., Confirmation of human identification using computerized tomography, J. *Forensic Sci.*, 38, 708, 1993.
61. Riepert, T., Rittner, C., Ulmcke, D., Oghuihi, S., and Scheveden, E., Identification of a unknown corpse by means of computed tomography (CT) of the lumbar spine, *J. Forensic Sci.*, 40, 126, 1995.
62. Rougé, D., Telmon, N., Arrue, P., Larrouy, G., and Arbur, L., Radiographic identification of human remains through deformities and anomalies of post-cranial bones: A report of two cases, *J. Forensic Sci.*, 38, 997, 1993.
63. Oliver, W.R., Chancellor, A.S., Soltys, M. et al., Three-dimensional reconstruction of a bullet path: Validation by computed tomography, *J. Forensic Sci.*, 40, 321, 1995.
64. Ros, P.R., Li, K.C., Vo, P., Baer, H., and Staalo, E.V., Preautopsy magnetic resonance imaging: comparison with findings at autopsy, *Am. J. Roent.*, 168, 41, 1990.
65. Wullenweber, R., Schneider, V., and Gremme T., A computer-tomographical examination of cranial bullet wounds [in German], *Z. Rechtsmed.*, 80, 227, 1997.
66. Seiji Shiotani, M.D., Personal communication.
67. Angus, W.M., A commentary on the development of diagnostic imaging technology, *Radiographics*, 9, 1225, 1989.

3 Scope of Forensic Radiology

B.G. Brogdon

CONTENTS

INTRODUCTION

Forensic radiology rests, as do all other academic and scientific disciplines, on the sometimes unsteady four-legged stool of service, education, research, and administration. The scope of forensic applications of diagnostic medical radiology as currently understood and practiced is summarized in Table 3.1. As the field of diagnostic radiology has undergone rapid expansion in technology and utilization in the past four decades, so will the range of forensic applications burgeon in the near future.

DETERMINATION OF IDENTITY

Radiological identification depends in the early stages on general biomedical knowledge and utilization of various standards and tables in establishing the basic issues such as animal vs. human remains, commingling, age, sex, stature, and ethnicity. Radiological determination of individual identity may be presumptive upon demonstration of preexisting injuries, illness, or congenital and/or developmental peculiarities (Figure 3.1). Positive radiological identification requires direct comparison of antemortem and postmortem images of the body or its parts (Figure 3.2). The emergence of multiple radiological imaging modalities in recent years initially complicated rather than simplified this comparison. This is because, for the most part, the newer modalities display body parts in sectional and planar images quite different from the routine frontal and lateral views typifying most conventional roentgenograms. Identification of human remains by radiological methods is covered extensively in Section III.

EVALUATION OF INJURY OR DEATH

Evaluation of injury or death by radiological methods is greatly enhanced by historical information, physical findings, and appropriate laboratory data when available. Such evaluation frequently requires elements of detection, pattern recognition, interpretation, and comparison, all solidly based on radiological training and experience with normal and abnormal findings in patients of both sexes and all age groups. The osseous skeleton is the prime target of forensic radiological evaluation, but in many situations the soft tissues of the musculoskeletal framework and the abdominal and thoracic viscera may offer key findings (Figure 3.3).

Osseous injuries are best detected and studied in the postmortem state if the body parts can be manipulated to replicate the standard positions for those parts normally used for roentgenography in clinical practice (see Chapter 39). Typically, in medical radiology, the patient or his body part is examined in at least two positions or projections (usually at right angles to one another). As a result, objects or parts obscured in one view may be visible in the other (Figure 3.4 is a rather blatant example of this). Also, even though a roentgenogram depicts findings in only two planes or dimensions, the trained observer can use two radiographs taken from different perspectives to conceptualize the third dimension of depth (Figure 3.5).

The location and type of fracture, considered with reference to the history, age, and sex of the victim and the expected level of activity of the individual, may be highly suggestive of whether the injury is accidental or inflicted. Certain fractures, dislocations, and epiphyseal separations are relatively common in the course of "normal" activities in certain age ranges; others are virtually impossible to sustain accidentally within those parameters (Figure 3.6).

The configuration and direction of fractures in the skull may locate the impact point and direction of impact, indicate the sequence of repetitive blows (Figure 3.7) and, sometimes, the shape of the wounding object or weapon. The appearance and location of some skeletal injuries may be clues to their

TABLE 3.1
Scope of Forensic Radiology

I. Service
  A. Determination of Identity
  B. Evaluation of Injury and Death
    1. Accidental
    2. Nonaccidental
      a. Osseous injury
      b. Missiles and foreign bodies
      c. Other trauma
      d. Other causes
  C. Criminal Litigation
    1. Fatal
    2. Nonfatal
  D. Civil Litigation
    1. Fatal
    2. Nonfatal
  E. Administrative Proceedings
II. Education
III. Research
IV. Administration

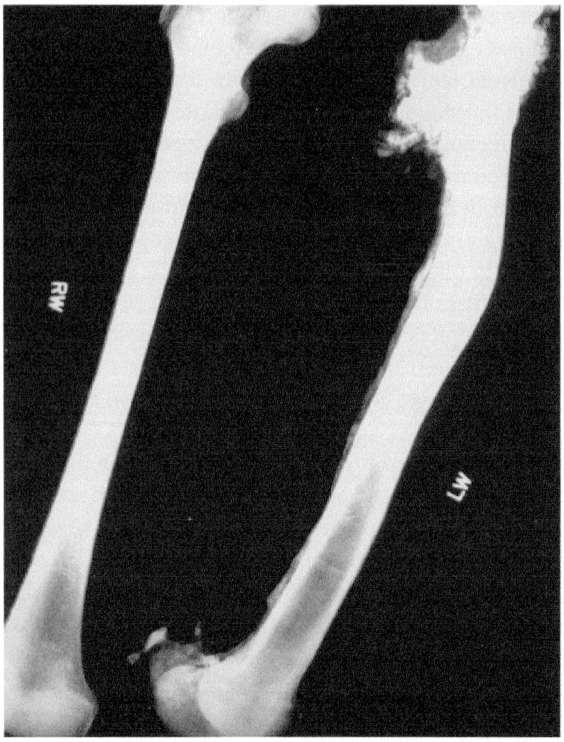

FIGURE 3.1 Skeletonized remains from the desert show an old healed fracture of the left proximal femoral shaft with residual bowing and shortening. This correlated with the history of a suspected decedent, but a positive identification was not possible because the antemortem roentgenograms were no longer available.

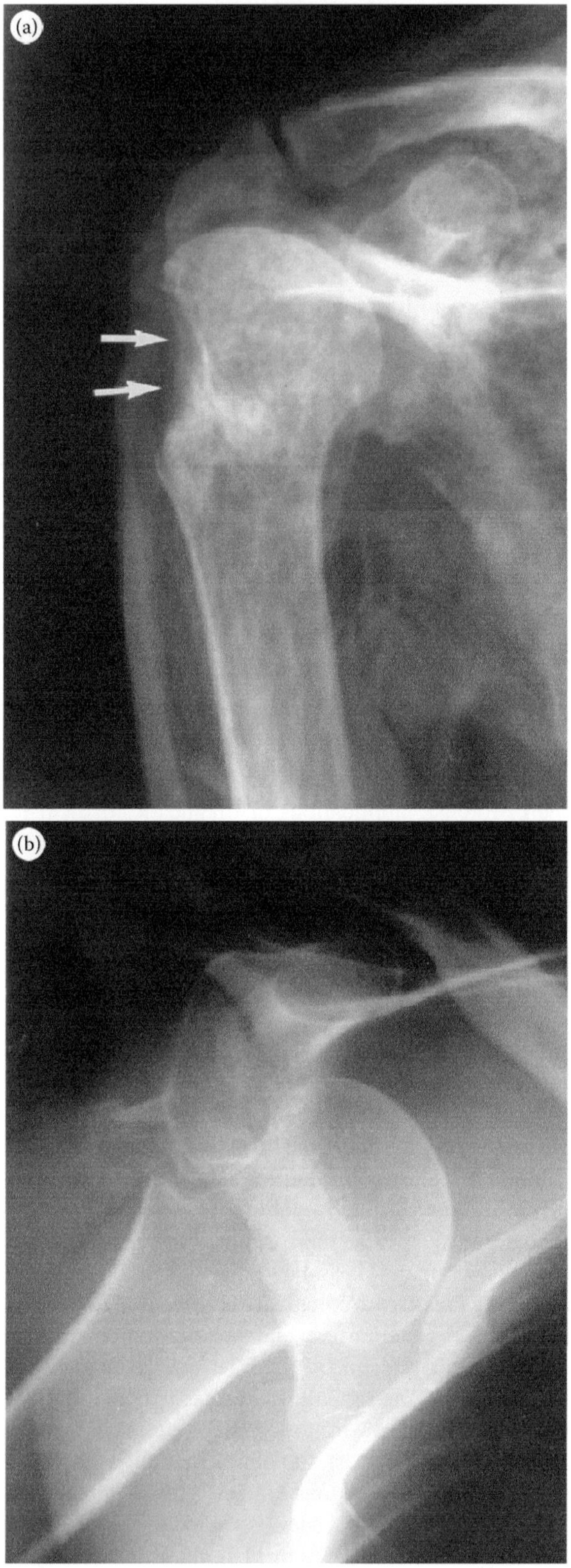

FIGURE 3.2 (a) A truck driver who was burned beyond recognition after a wreck had a history of dislocation of the right shoulder. The postmortem roentgenogram shows a typical Hill–Sachs deformity, an impaction fracture of the humeral head associated with anterior dislocation of the shoulder (arrows). (b) Antemortem examination shows the dislocation and Hill-Sachs deformity, a positive identification.

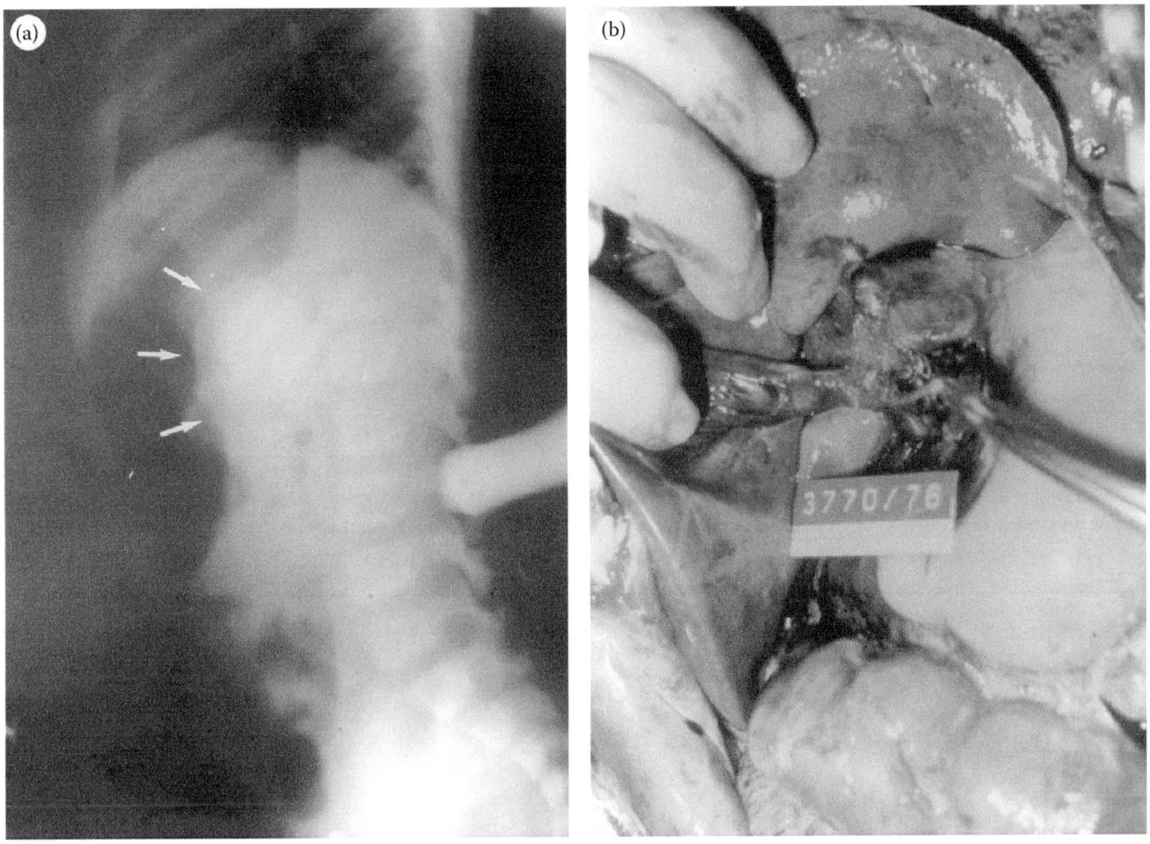

FIGURE 3.3 (a) A battered child was moribund on arrival in the Emergency Department. Cross-table lateral view of the abdomen shows a large retroperitoneal mass (arrows) which at autopsy (b) was shown to be a collection of blood and other body fluids from laceration of the liver and pancreas and transection of the bile duct. (Dr. Weston made a cast of the stepmother's fist, which exactly matched a bruise of the skin over the epigastrium and led to a conviction.)

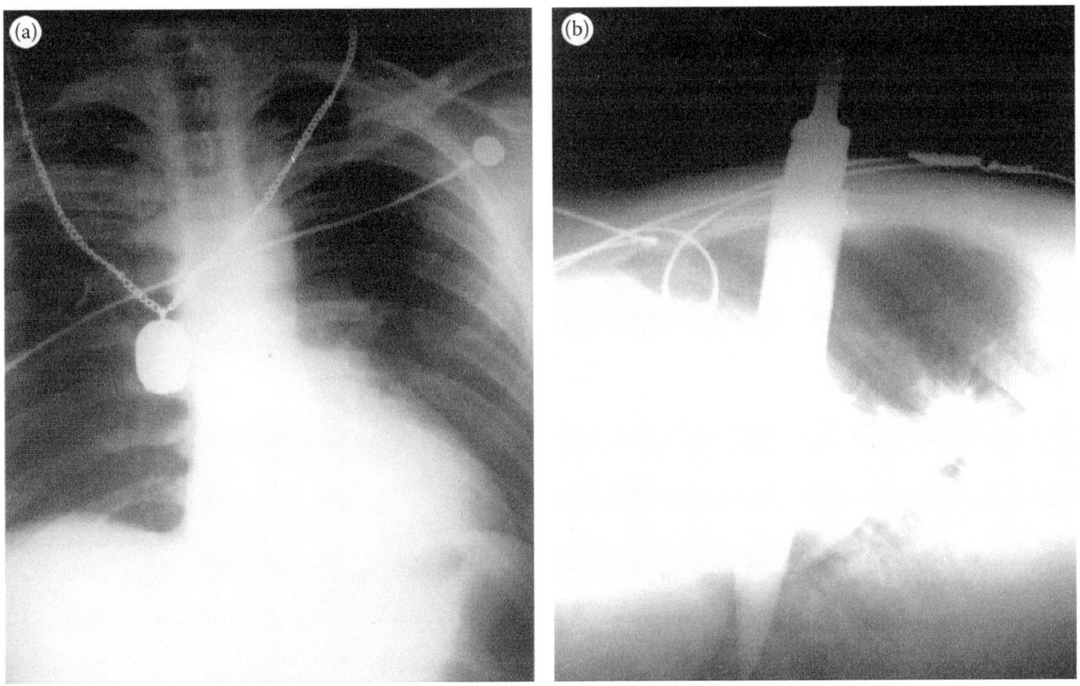

FIGURE 3.4 (a) Frontal roentgenogram of a body does not reveal a cause of death quite obvious to the autopsy surgeon. (b) The lateral view, while not necessary for the edification of the medical examiner, makes a powerful display for the jury.

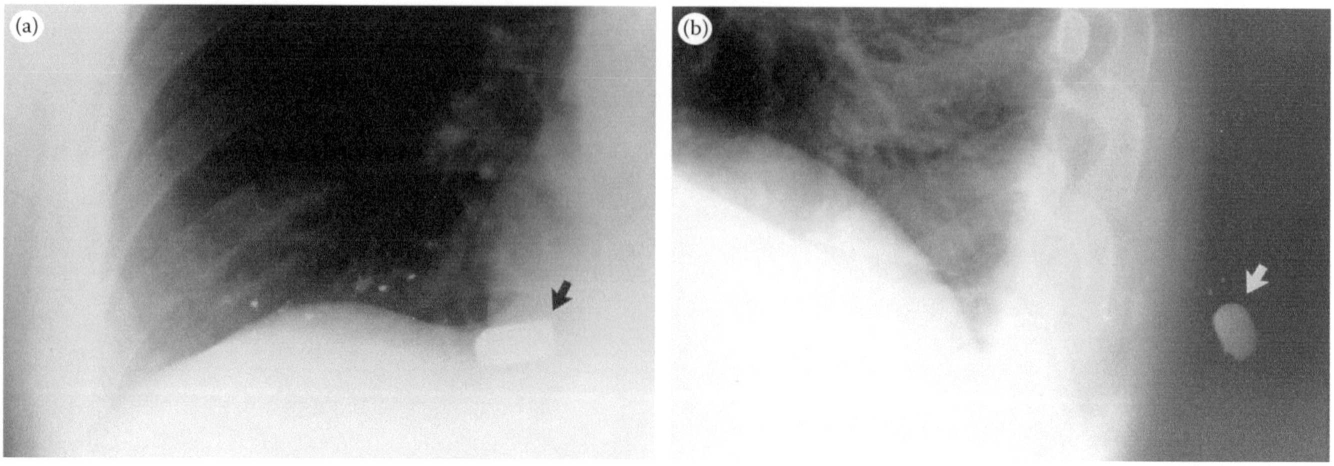

FIGURE 3.5 A foot patrolman, seeing a man inside a closed and darkened store, called for him to come out. He came out running and fled down the street while the officer fired at him with his service revolver. Sometime later a man was admitted to a local hospital with a bullet wound. (a) The frontal view of the chest showed a track of lead fragments leading from the midportion of a rib to the major bullet fragment (arrow) overlying the right margin of the heart. It gave no clue as to depth of the missile within the body. (b) A lateral view showed that the slug (arrow) had simply bounced off the rib and came to rest just beneath the skin of the back. With a court order, the slug was popped out through a small incision and matched by ballistics with the officer's weapon.

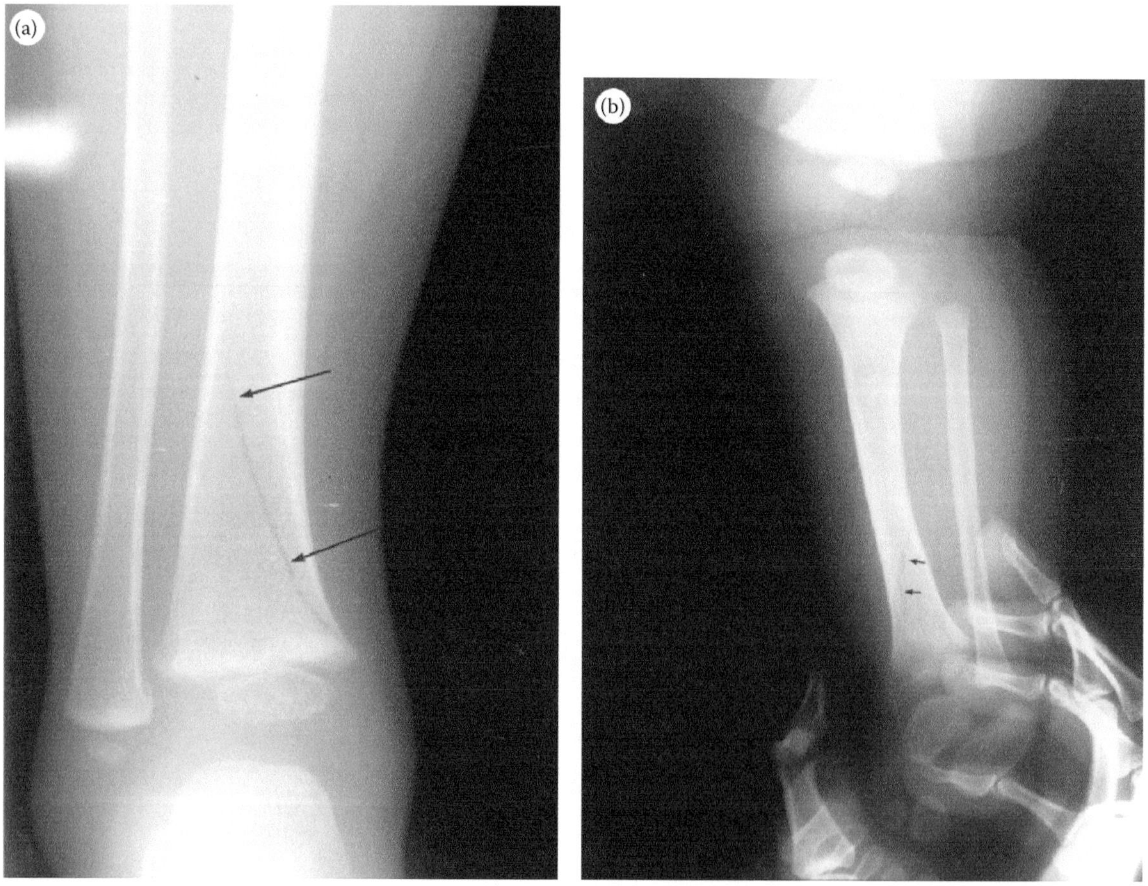

FIGURE 3.6 (a) This so-called "toddler's fracture" is commonly seen in young children in the early years as they begin to walk and run, but are not yet very steady or coordinated. It is a "normal" fracture for this age group. (b) This nonambulatory infant has a similar-looking fracture, but one impossible to acquire naturally in the course of infantile movement. Rather, this fracture was caused by a twisting force or torsion at the hands of an adult caregiver.

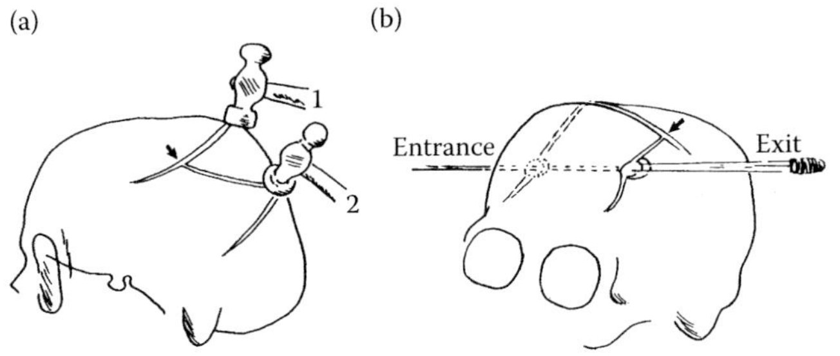

FIGURE 3.7 Schematic drawings indicate that (a) a linear fracture from an earlier blow will stop propagation of a fracture from a second blow. (b) a linear fracture from an entrance wound travels faster than the bullet causing it; a fracture from the exit wound will terminate on meeting the preexisting fracture. (This helps in deciding between entrance and exit wounds when the skull defect is in thin bone, i.e., temporal bone, or when beveling is obscured by fire or decomposition.) The trajectory or direction of fire is suggested by the angle of eccentric beveling. (Adapted from Dixon, D. S., *J. Forensic Sci.*, 29, 651, 1984.)

origin (Figure 3.8). Some skeletal injuries are typically defensive in nature (Figure 3.9). The time span since the original injury often can be estimated with some degree of accuracy and may be important. Of course, multiplicity of injuries and injuries in various stages of healing are highly significant findings. Fractures of the hyoid bone or thyroid cornua usually suggest strangulation (Figure 3.10). In vehicular injuries, certain fracture/dislocations may actually suggest the velocity of impact or deceleration. We have seen a pattern-like series of craniocervical distractions and/or dislocations in motorcycle riders, at least some of whom are known to

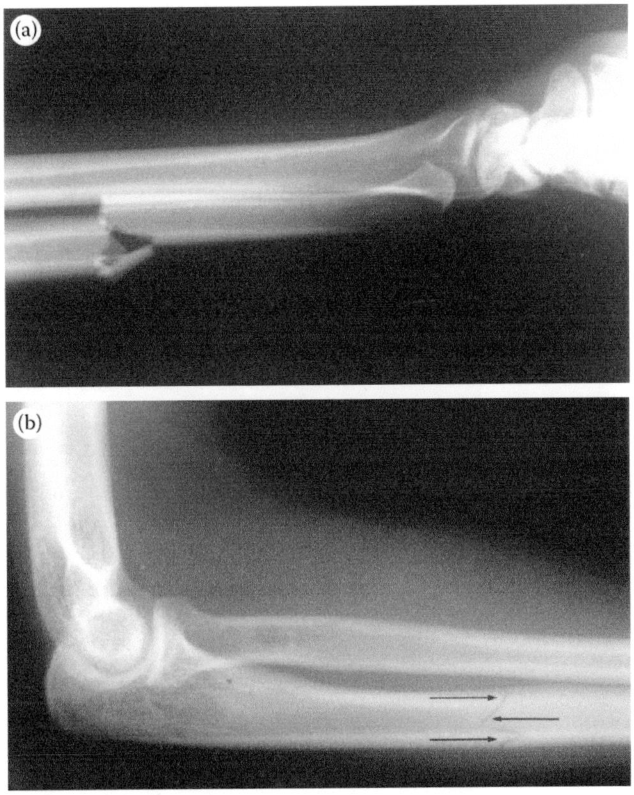

FIGURE 3.9 (a) A "fending fracture" of the ulna—the result of trying to ward off a blow by blocking it with the upraised forearm. These have also been called "night-stick" or "pool-cue" fractures. (b) A subtle, undisplaced fending fracture.

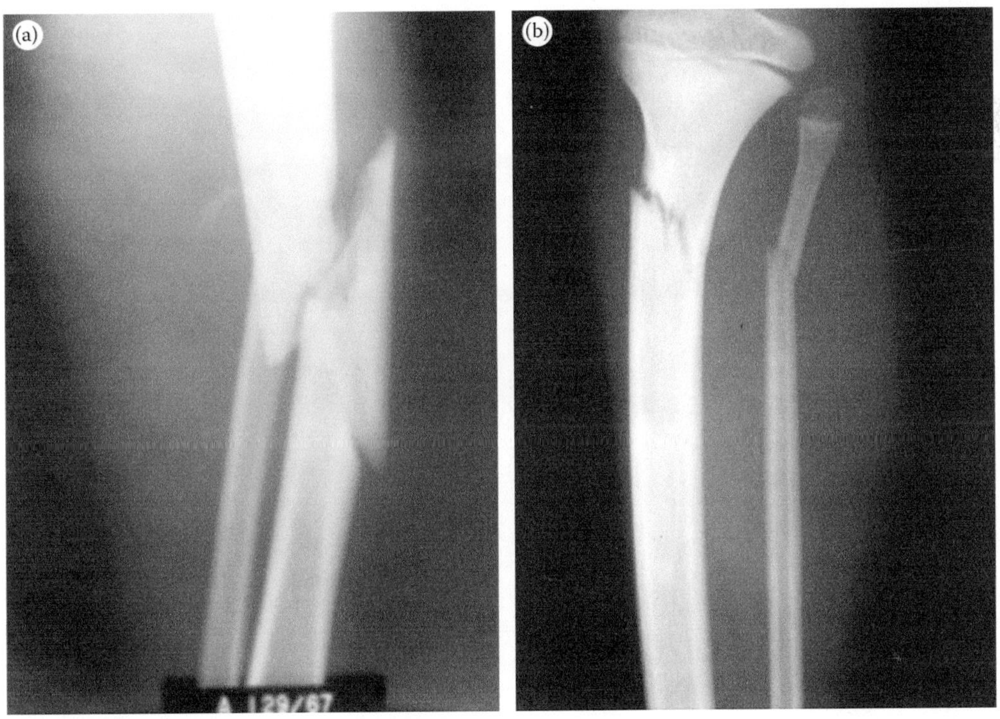

FIGURE 3.8 (a) A typical "bumper fracture" in an adult pedestrian hit from the left. The apex of the triangular or butterfly fragment points away from the impact. (b) "Bumper fracture" in a child hit from the left is higher in the leg. Because of the increased elasticity and plasticity of young bones, the impact produced an incomplete or "green-stick" fracture.

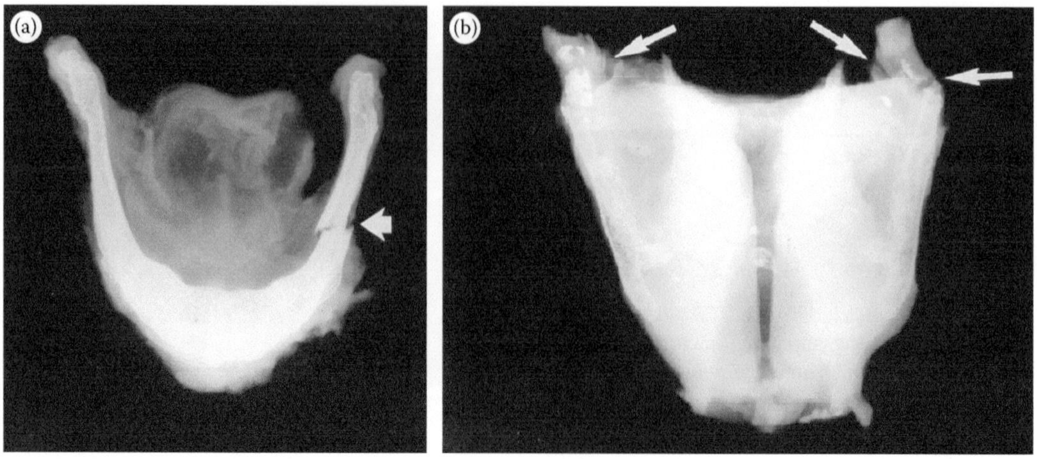

FIGURE 3.10 (a) A fracture of the hyoid bone (arrow) from strangulation. (b) Fractures of the superior cornua of the thyroid cartilage (arrows) from strangulation. (Both figures courtesy of Dr. J.C.U. Downs.)

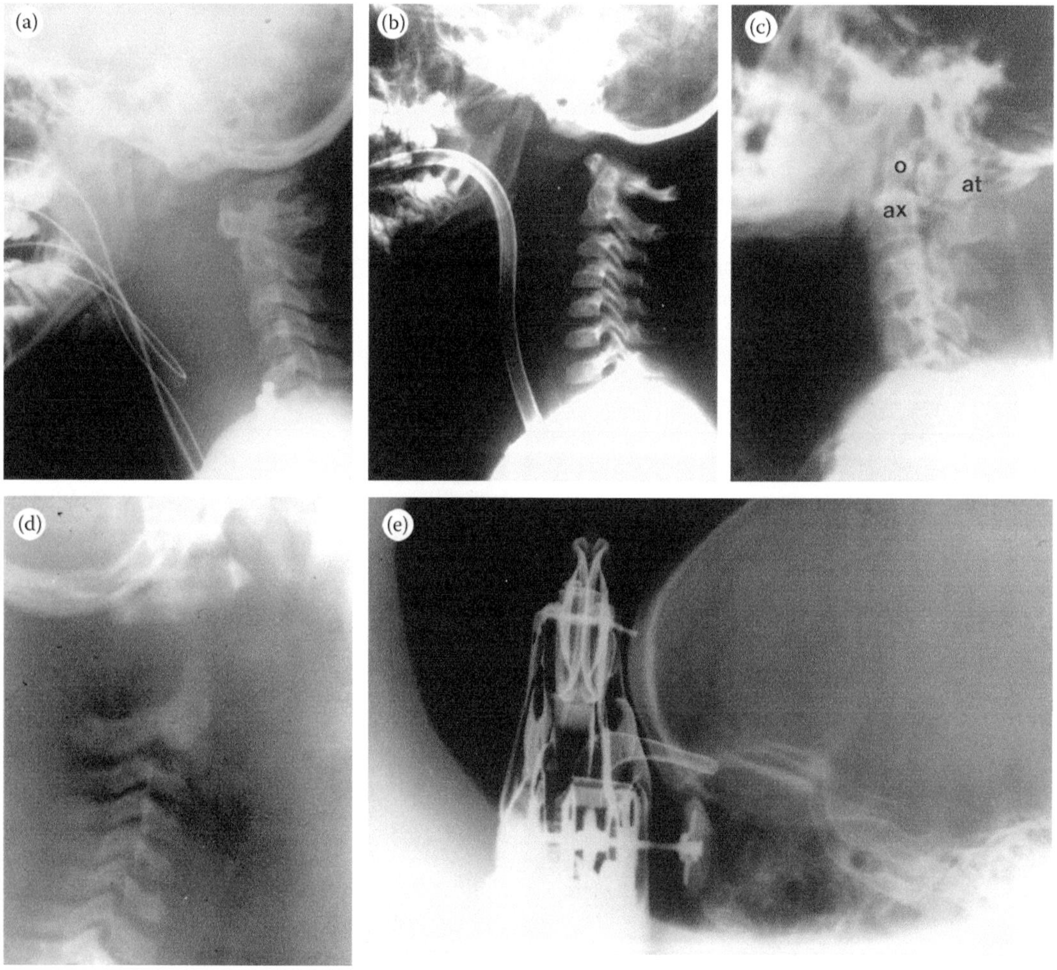

FIGURE 3.11 Biker injuries. (a) Separation or distraction of the base of the skull from the cervical spinal column. (b) Another craniocervical separation. (c) The atlas or first cervical vertebra (at), still articulating with the base of the skull, has jumped posteriorly entirely over the odontoid process (o) of the axis (ax) or second cervical vertebra. (d) Wide distraction and separation of the axis and atlas in a child who was riding a motor scooter. (e) The only "good" motorcycle injury the author has ever seen. This three-year-old was brought to the emergency room with a toy motorcycle stuck to his face. Radiography showed that the axle of the front wheel had barely penetrated the outer table of the skull of the forehead. The toy was simply lifted off, leaving only a tiny puncture wound.

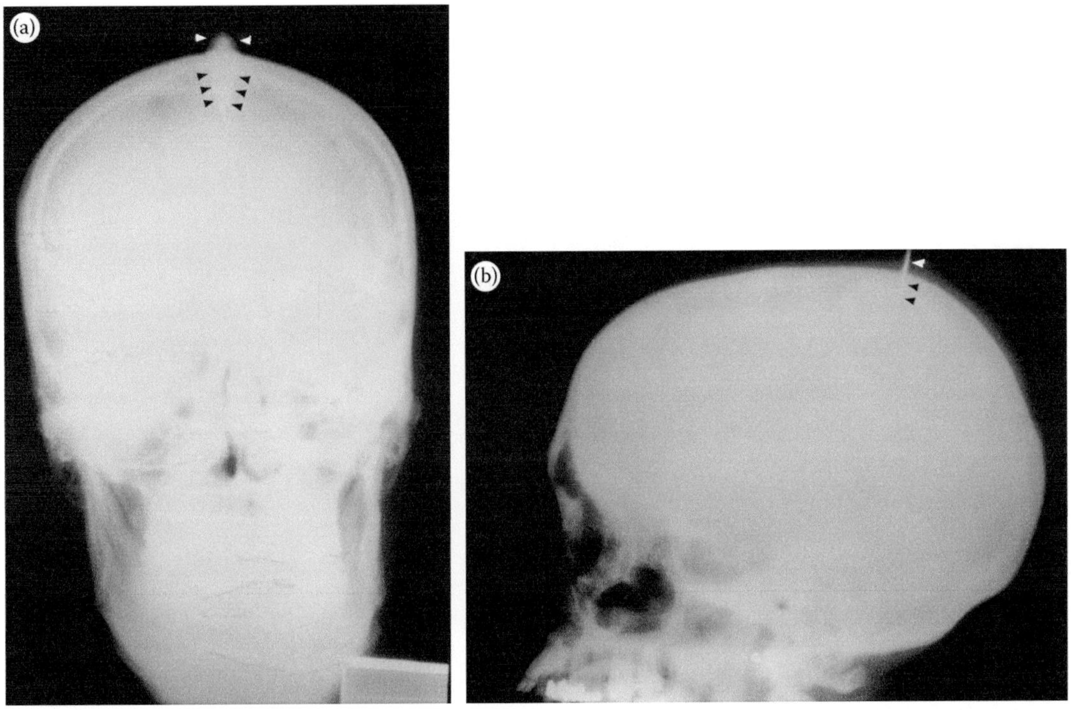

FIGURE 3.12 This convenience store cashier was stabbed in the head by a would-be-robber. He drove the thief off, then walked into the emergency room for extraction of the knife (arrows). (a) Frontal view. The knife blade is optically widened by scatter in the adjacent soft tissues. (b) Lateral view (B.G.B., Mobile, AL).

have ridden at speed into a head-high obstacle such as a traffic sign or truck bed (Figure 3.11).

Forensic pathologists like to categorize trauma as "sharp" or "blunt" which really indicates the causative agent. Still, it can be a useful classification, especially if the type of weapon is unknown initially. Those producing sharp trauma usually are either edged or pointed, although the edge may be dull and the point blunted or truncated.

Examination of sharp trauma to the head is often quite informative since the predominant bony structure tends to entrap or retain the weapon (Figures 3.12 and 3.13). However,

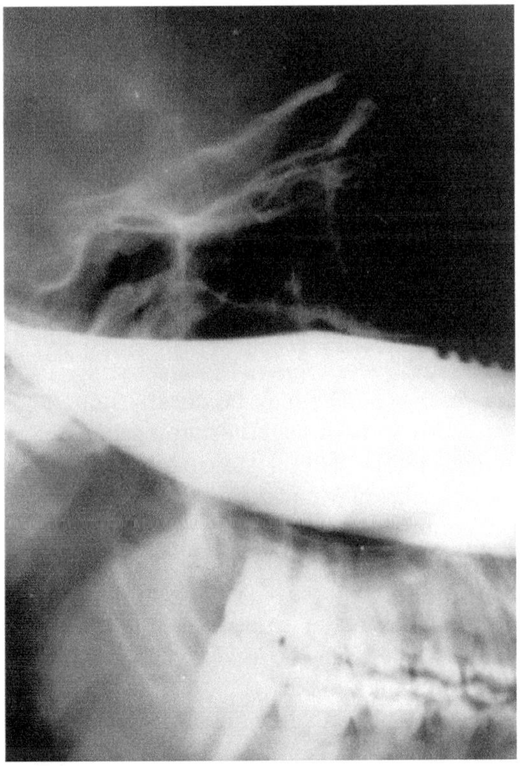

FIGURE 3.13 Knife piercing the maxillary antrum and orbital floor extends into the temporal lobe, the result of social fighting after payday or during a sporting event (South Africa). (Courtesy of Professor Dr. H. Vogel.)

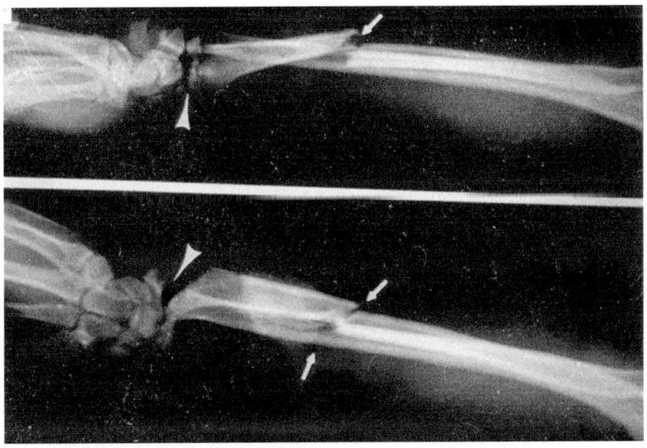

FIGURE 3.14 Two views of the forearm and wrist show results of a cane-worker's machete fight. There are fractures of both bones of the forearm (arrows) and through the wrist (arrowheads) (Columbia). (Courtesy of Professor Dr. H. Vogel.)

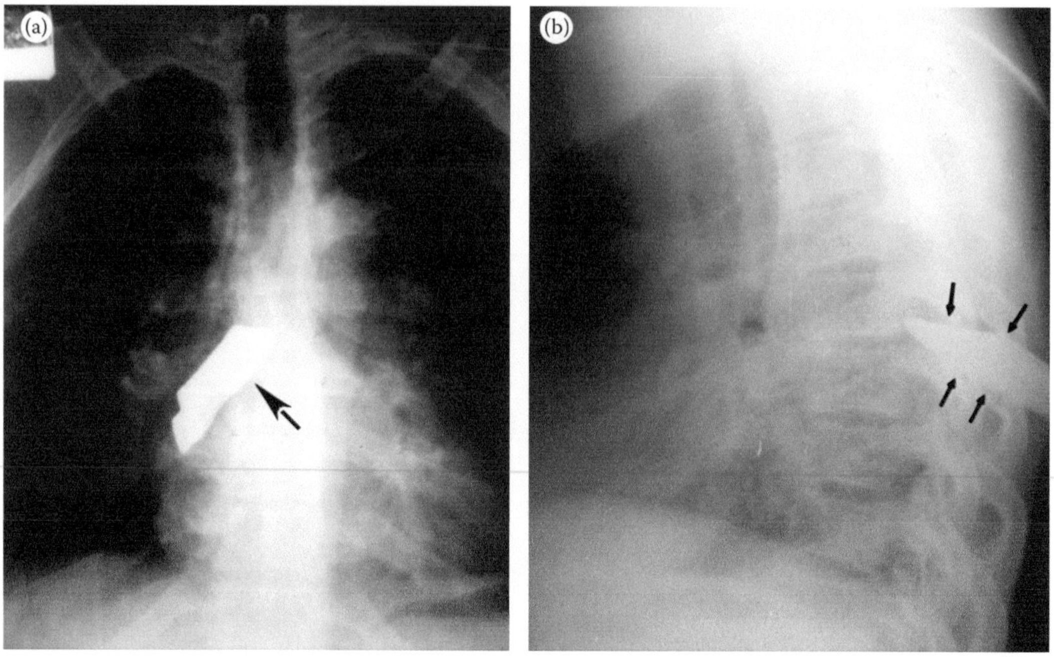

FIGURE 3.15 This knife was stabbed into the chest and stuck there between the ribs (arrows). (a) Frontal view. (b) Lateral view (Zimbabwe). (Courtesy of Professor Dr. H. Vogel.)

sharp objects are more likely to slice into and out of more fleshy parts (Figure 3.14). Blades are rarely found within the chest, abdomen, or extremities; they occasionally become entangled in ribs or vertebrae (Figures 3.15 through 3.17).

Blunt trauma (Figures 3.18 through 3.22) results from weapons with a broader impact footprint ranging from fists to clubs, from the throttling thumb to the more focused ligature. There often is extensive bruising of underlying soft tissues, fractures of underlying bones, and damage to internal organs. Sectional imaging, CT and MRI, has expanded greatly our ability to appreciate the multilayered extent of blunt trauma injury.

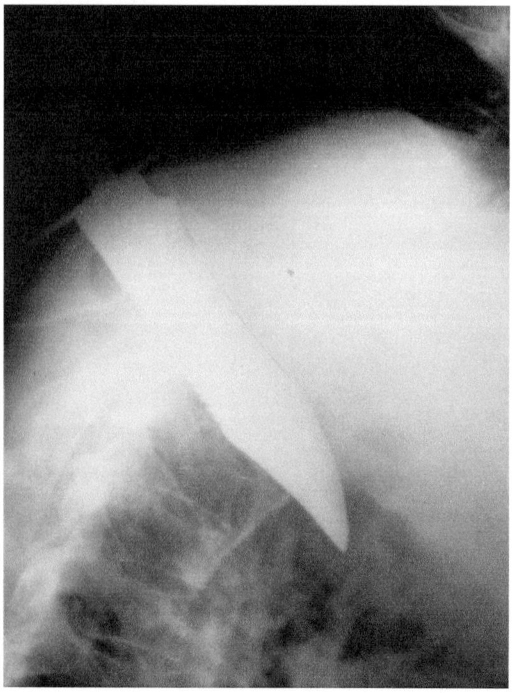

FIGURE 3.16 Stab wound in the back of the chest. The victim could not remove the knife (South Africa). (Courtesy of Professor H. Vogel.)

Finally, many diseases leave their "mark" on the skeleton; some of them actually are quite legible "signatures" allowing positive identification of the antecedent process. Those distinctive affectations of the skeleton are too numerous to list here (and are the subject of many volumes of radiological textbooks and at least one excellent anthropological text[1]) but include infections, infestations, metabolic processes, dietary abnormalities, tumors, and even poisons (Figure 3.23).

Missiles and other foreign bodies are the object of many forensic radiological examinations. Gunshot wounds are the subject of Section IV to follow, and their radiological evaluation may provide important information in a variety of ways.

Other foreign bodies may be equally important in any given case. All sorts of materials are opaque to x-rays and may be detected by careful radiographic examination and visual search. One may find the snapped-off point of a knife; fragments of broken glass (Figure 3.24); bomb fragments or shrapnel, parts of the automobile or aircraft in which the victim was riding (Figures 3.25 and 3.26); and animal, mineral, or vegetable matter embedded, aspirated, or injected (Figure 3.27). There may be aspirated dirt from a cave-in or sand from drowning in surf (Figure 3.28). Opaque poisons may be seen in soft tissues or the gut. Foreign bodies in the vagina, rectum, bladder, or other tissues can indicate sexual abuse, autoeroticism, or psychosis (Figure 3.29).

Other trauma can include such findings as intracranial hemorrhage from shaking, penetrating wounds, which can be demonstrated with injected contrast media, as so can vascular

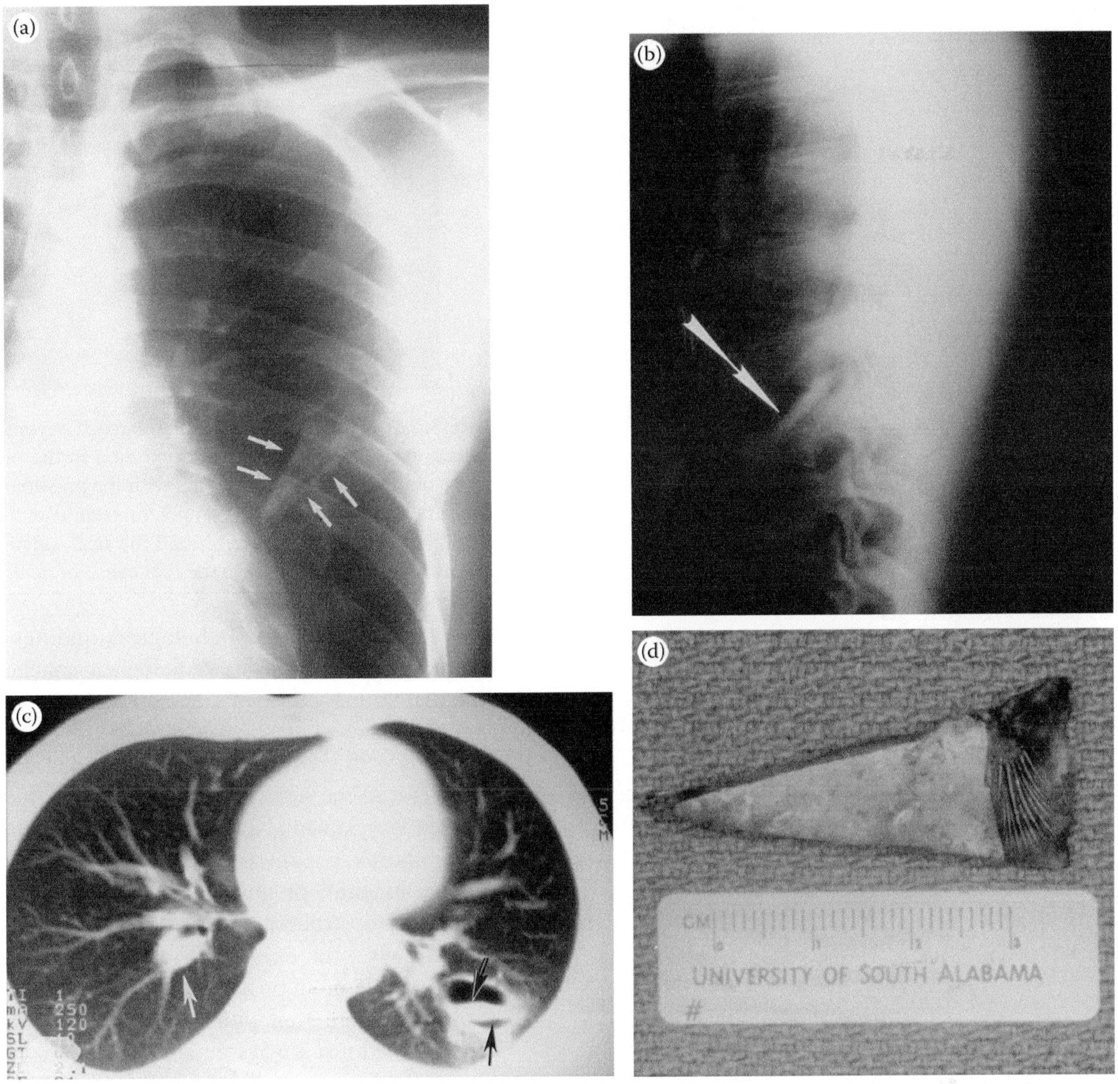

FIGURE 3.17 (a) Shard of glass (arrows) broken off in the left lung after the victim was repeatedly stabbed in the back with a broken beer bottle. (b) Lateral view of the chest shows the shard (arrow) lies in the lung after having slipped through between the ribs. (c) A CT scan shows the shard of glass (black arrows) surrounded by a pneumatocele in the left lower lobe. On the right is another pneumatocele (arrow) from an earlier stab wound before the point of the broken beer bottle broke away. This pneumatocele is partially filled with fluid, probably blood. (d) Photograph of the shard of beer bottle glass removed from the left lung at surgery.

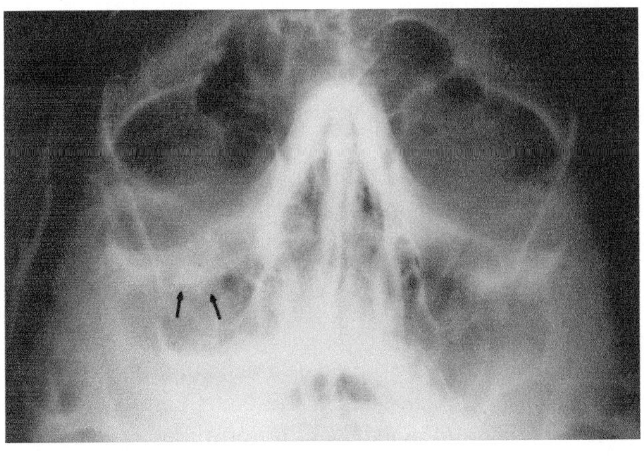

FIGURE 3.18 Blowout fracture of the right orbit. The orbital floor is depressed. There is herniation of orbital content (arrows) into the right maxillary sinus, which is partially filled with blood.

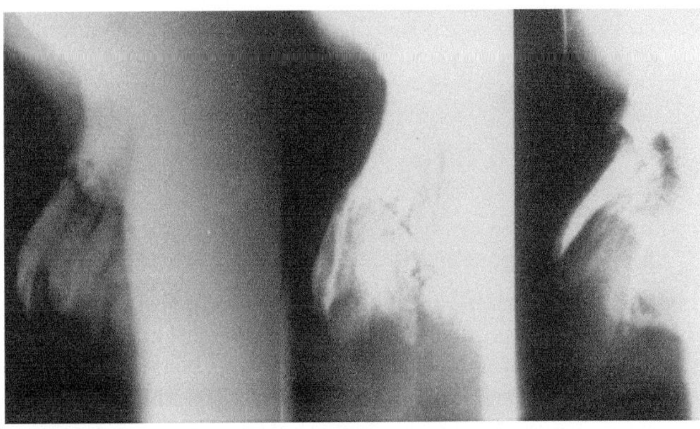

FIGURE 3.19 Three different victims with nasal bone fractures.

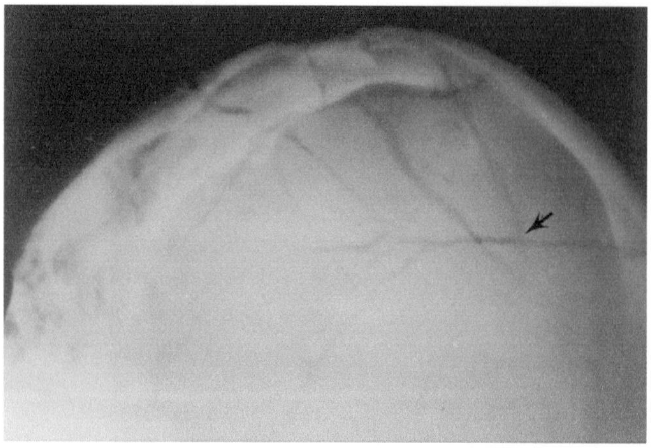

FIGURE 3.20 Multifragmented fissural fractures through the top of the skull (lateral view) with a hammer club. The patient was conscious. We are seeing both sides of the skull, so the law of intersecting fractures is not abrogated. An arrow points to an example where one fracture did not propagate across a preexisting fracture. (Courtesy of Professor Dr. H. Vogel.)

tears or avulsions be revealed. Fractures of the laryngeal soft tissues have been seen with hanging (Figure 3.30) and massive soft tissue contusions may follow beatings.

Other causes of injury or death may have radiological implications. These would include drowning or near drowning,

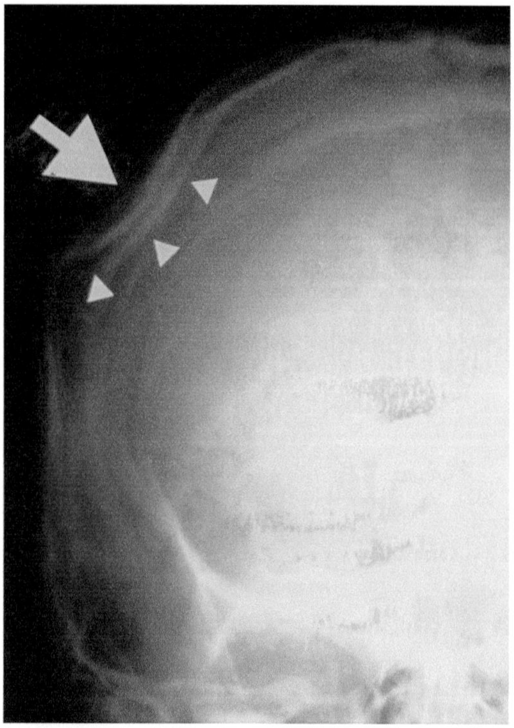

FIGURE 3.21 Massive impression fracture of the parietal bone conforming to the size and shape of the weapon. This type of fracture is typical for Africa south of the Sahara. In some areas, it could be due to a coconut. In this case, it exactly corresponds to the wooden club used in the assault (South Africa). (Courtesy of Professor Dr. H. Vogel.)

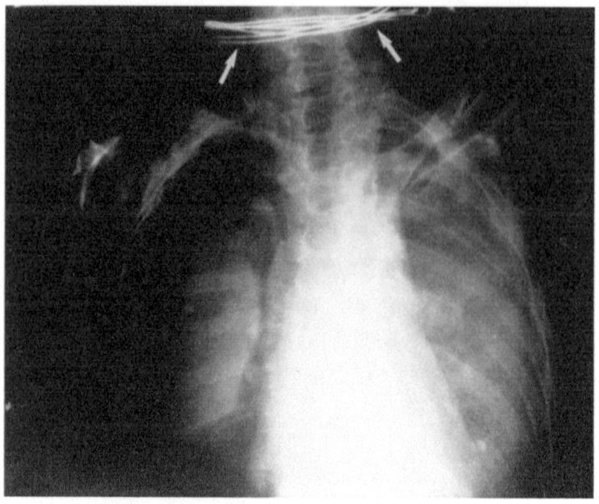

FIGURE 3.22 This woman was found burned beyond recognition in a house fire. It was assumed she had died in the blaze. A chest x-ray was obtained to try for a match with the presumed decedent's antemortem chest radiograph. The postmortem film disclosed several coils of wire as a ligature around the neck (arrows). The fire was set in an attempt to obscure the real cause of death.

wherein there usually are radiological findings. (However, one cannot depend upon the radiological appearance of the chest to distinguish between saltwater and freshwater drowning, as once believed.) Poisoning already has been mentioned. Radiography is the best and earliest method of demonstrating air embolism to the heart, brain, or vascular tree. Similarly, the autopsy surgeon can be alerted in advance to the presence of pneumothorax, pneumopericardium, pneumomediastinum, pneumoperitoneum, or abnormal air collections associated with abscess, obstruction, or paralytic dysfunction (Figure 3.31). Otherwise, the exact location and confines of the air may be difficult or impossible to discern with an ordinary operative approach, even with submersion techniques.

Finally, one must always remember that arson often hides other crimes. It is essential that all bodies burned beyond recognition be investigated radiographically (Figures 3.22 and 3.32).

CRIMINAL LITIGATION

The usefulness of appropriate radiological images in cases of murder, suicide, attempted murder, manslaughter, mayhem, assault, battery, abuse, terrorism, or any other type of criminal activity directed to the person is self-evident and widely known. Less well-appreciated is the contribution the radiological method can make in other nonviolent crimes such as smuggling, larceny and fraud, faking, or counterfeiting. Further comments pertinent to nonviolent crimes will be found in Section VI.

CIVIL LITIGATION

The radiologist may be called as a defendant, plaintiff, ordinary witness, or expert witness in court cases dealing with liability, be it professional liability or malpractice, personal liability, property liability, or product liability. Expert testimony may be required in civil actions in cases of wrongful death or birth, civil rights violation or, quite commonly,

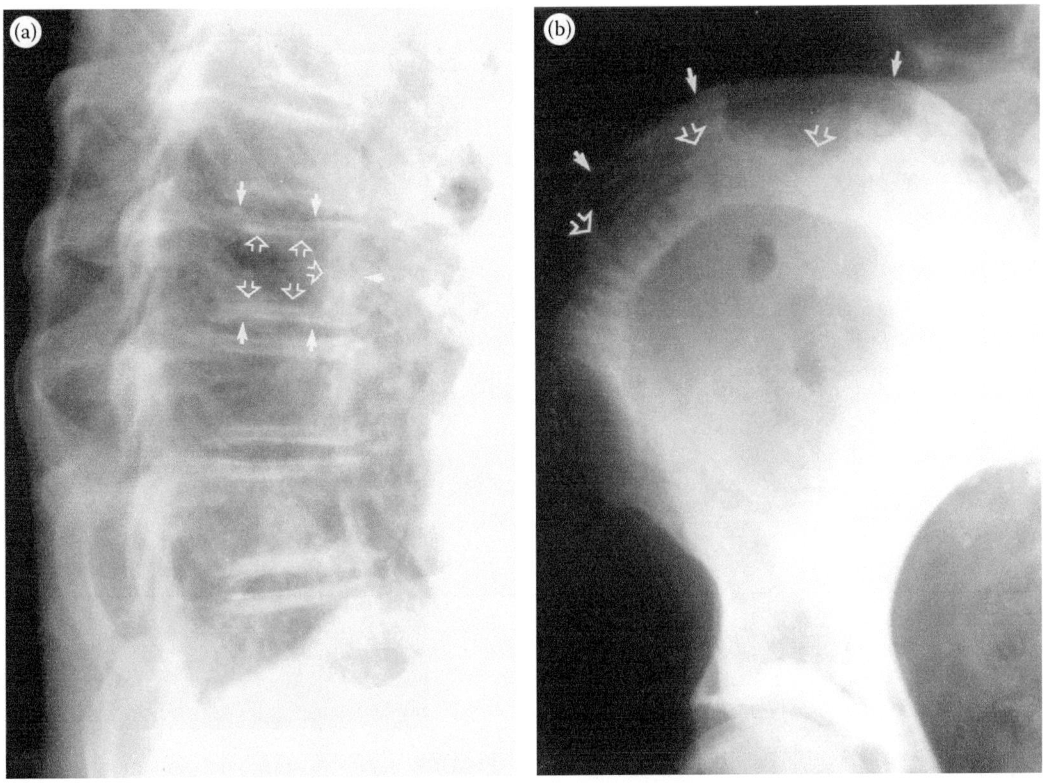

FIGURE 3.23 Bone-in-a-bone. Note that in the lateral view of the spine (a) there is the outline of a smaller vertebra (open arrows) residing inside the adult vertebral margin (closed arrows). Similarly, in the pelvis (b) a small pelvic outline (open arrows) is within the adult pelvis (closed arrows). This woman, now 66, had a habit of chewing matches as a teenager. Phosphorus was laid down in the growing margin of the bone at that time. (This finding is not specific to match chewers but can be seen with lead ingestion, severe childhood illness with cessation of growth for several weeks, and with some diseases such as osteopetroses or treated Langerhans-cell histiocytosis.)

personal injury. These cases may demand a high level of professional knowledge and expertise, but these issues are not within the scope of this treatise. The section on Coping with the Courts (Section II) contains advice on general courtroom procedure, conduct, and demeanor.

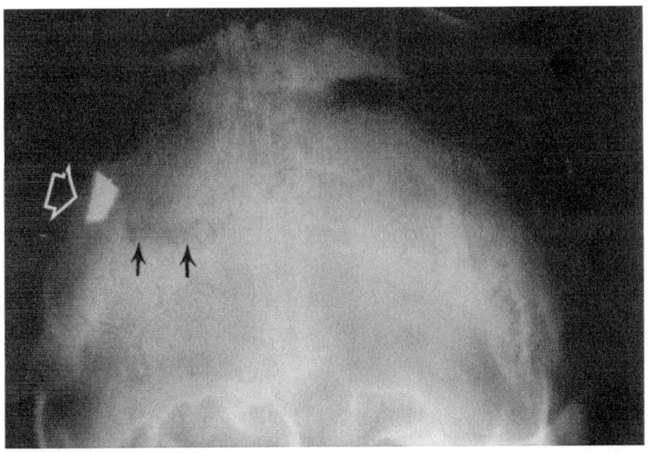

FIGURE 3.24 A beer bottle was smashed against the right frontal area of this victim, producing a fracture (arrows) and shattering the bottle as well. A piece of the bottle penetrated the scalp and stuck in the outer table of the skull (open arrow). (This is an example of both sharp and blunt trauma.)

ADMINISTRATIVE PROCEEDINGS

The radiologist may be called to testify in an administrative legal proceeding that is not, strictly speaking, in a court of law. This would include such actions as a Workman's Compensation Hearing or a Military Board to determine "Line of Duty" status of an injury or illness as well as the extent of any resultant disability and eligibility for a pension or award.

EDUCATION

All physicians have an educational obligation dating back to the Hippocratic Oath "to impart a knowledge by precept, by lecture, and by every other mode of instruction to my sons, to the sons of my teacher, and to pupils who are bound by stipulation and oath, according to the law of medicine." However, there is no formal curriculum in forensic radiology nor, to our knowledge, is there any regular course of instruction in forensic radiology in this country. There has been a smattering of "refresher courses" on the subject lasting from one and one-half to six hours presented at random intervals during the annual meetings of the American Academy of Forensic Sciences, the American Roentgen Ray Society, and the Radiological Society of North America. Most instructions, and learning, in forensic radiology is by the individual case

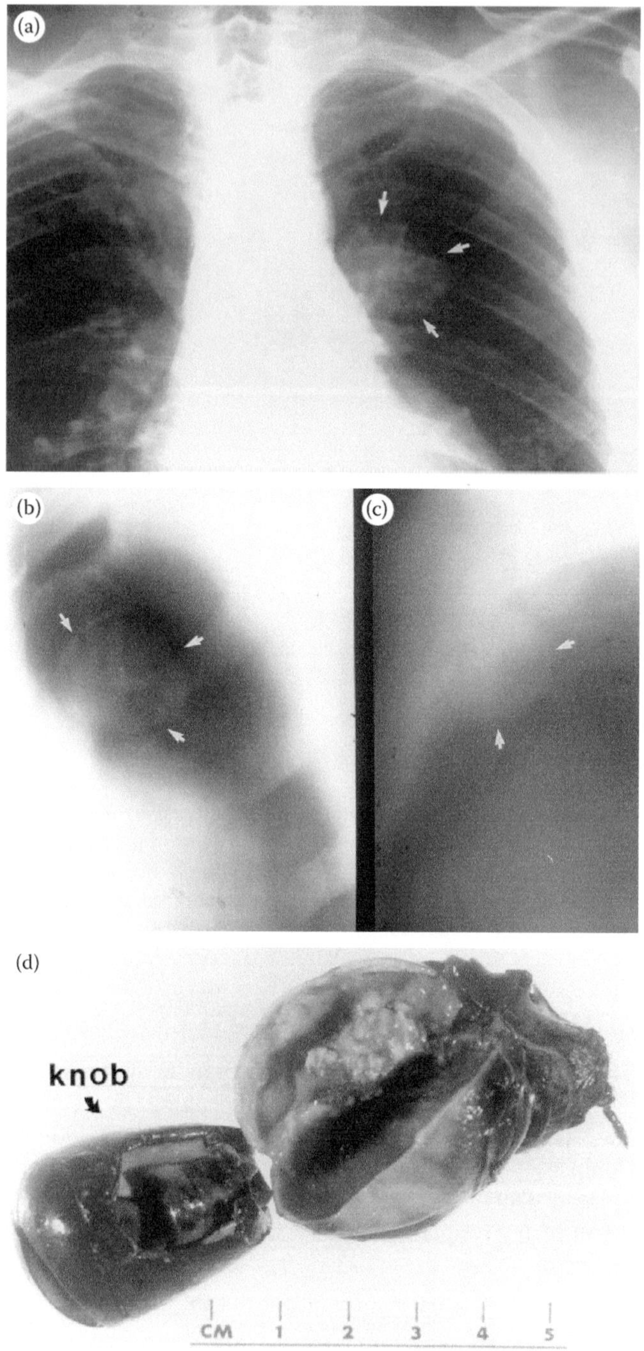

FIGURE 3.25 This middle-aged man was sent for a chest film (a) because of suspected heart disease. A round mass in the left lung prompted a tomogram (b and c) which defined the mass in frontal and lateral projections. At surgery (d) a gearshift knob encapsulated in fibrous scar was removed. The man had been in an automobile accident 22 years earlier but did not recall any penetrating injury at the time!

study method, hands-on experience, or published reports in a variety of scientific journals. Similarly, there are scattered talks related to the utilization of radiology and forensic matters at meetings of a wide spectrum of scientific disciplines. The quality and quantity of these offerings is as varied as the audience. Many come from nonradiologists. Efforts to develop an organization of individuals interested in the field, or a section in forensic radiology under the aegis of the AAFS, have been unsuccessful and discouraging so far.

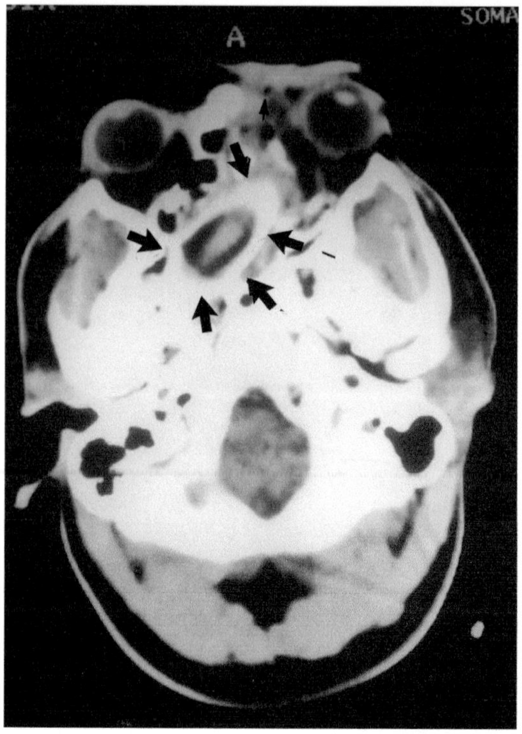

FIGURE 3.26 A shift knob mounted on the steering column pierced the skull of the front-seat passenger (arrows) during an automobile accident.

Because of the rapid and recent technological developments in the field of diagnostic radiology *per se*, there is a compelling need for better-planned, better-organized, and systematic educational efforts in forensic radiology. The newer modalities have both great promise and great problems in application to

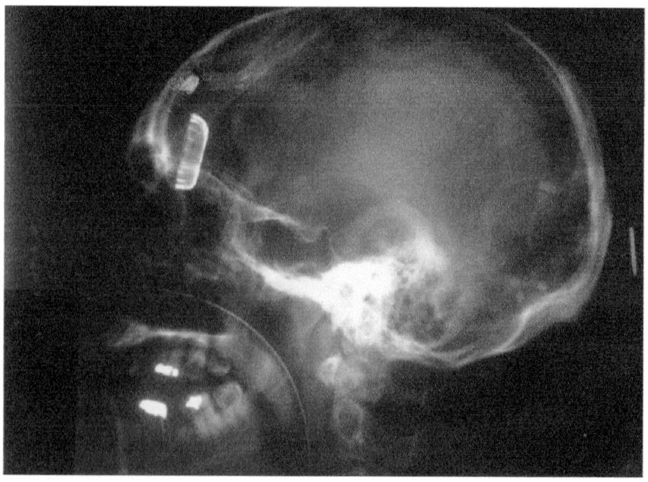

FIGURE 3.27 A bottle was driven into the victim's face. The cap stayed behind as the bottle was withdrawn.

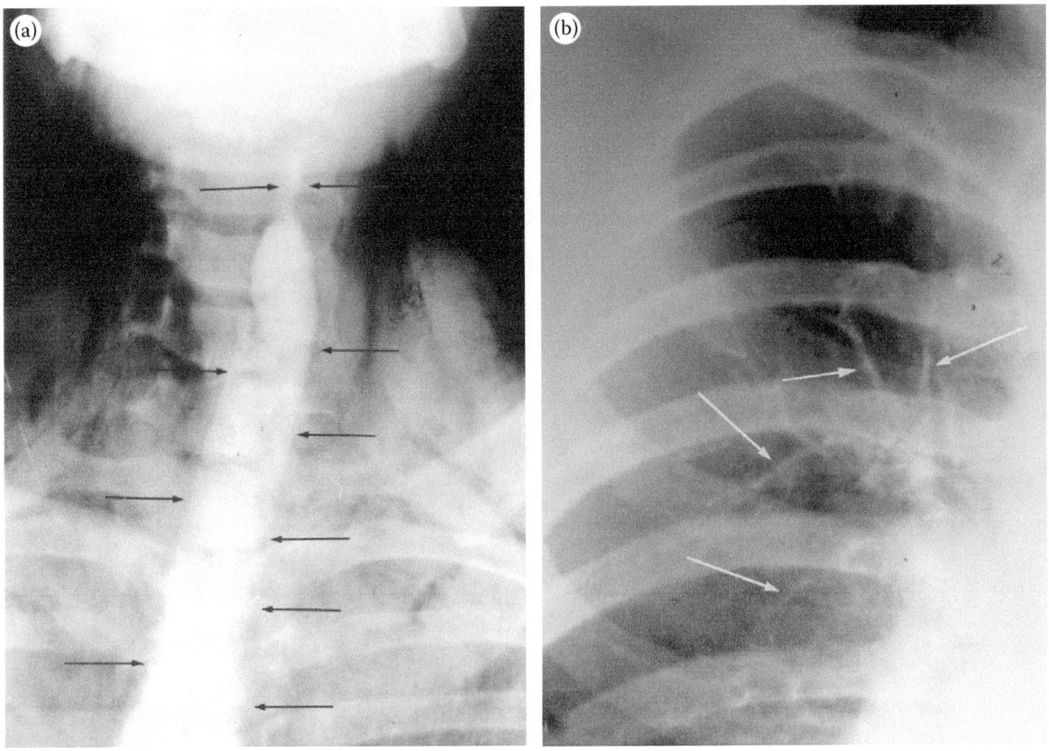

FIGURE 3.28 (a) Sand completely packs the tracheobronchial tree of a man who drowned in high surf. (b) A man who recovered from near drowning in surf was found to have sand impactions in upper lobe bronchi. (From Bonilla-Santiago, J. and Fill, W.L., *Radiology*, 128, 301, 1978. With permission.)

forensic problems. More widespread understanding of, and familiarity with, the discipline will expedite the assimilation of these tools into the forensics armamentarium.

The forensic sciences have another educational obligation to the public, that is, to alert them to situations discovered to be hazardous to public health or public safety. Examples would be research leading to bicycle helmet laws, warning of the danger of larger water-filled buckets to infants and children (who have a high center of gravity, and cannot extricate themselves after tumbling in), or publicity alerting citizens to the peril of urban hyper- or hypothermia during spells of severe weather. Perhaps the finest example of public

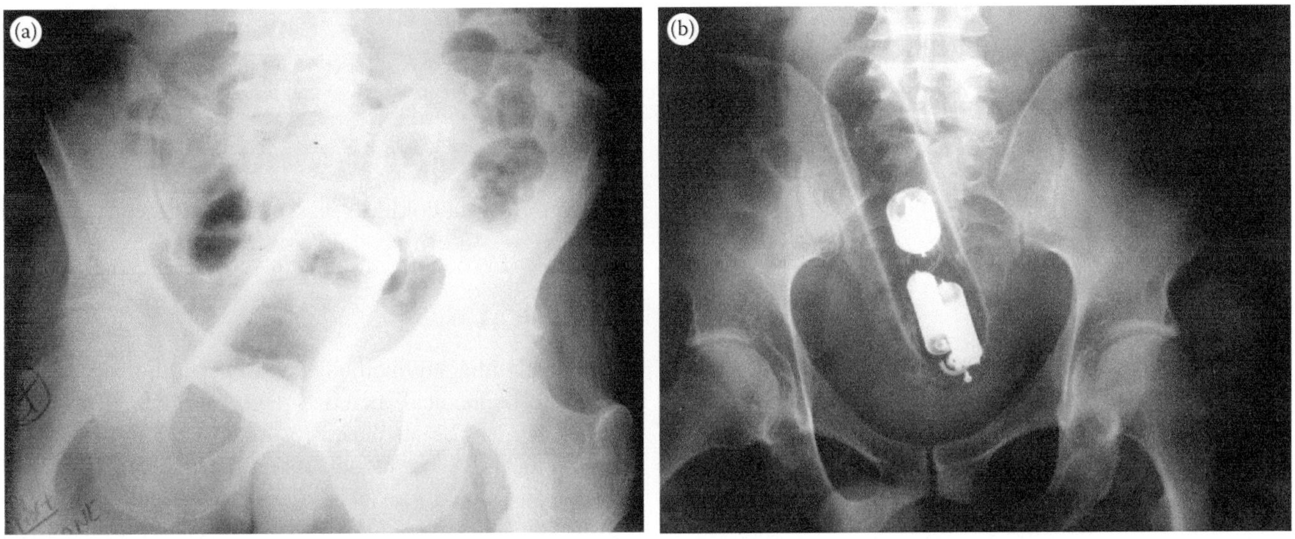

FIGURE 3.29 (a) Water glass in the rectum. (b) plastic vibrator lost in the rectum. (c and d) A disturbed youth who obviously is left-handed inserts wires—straightened paper clips—(arrows) beneath his skin. One has traveled through the venous system to the right ventricle (open arrow).

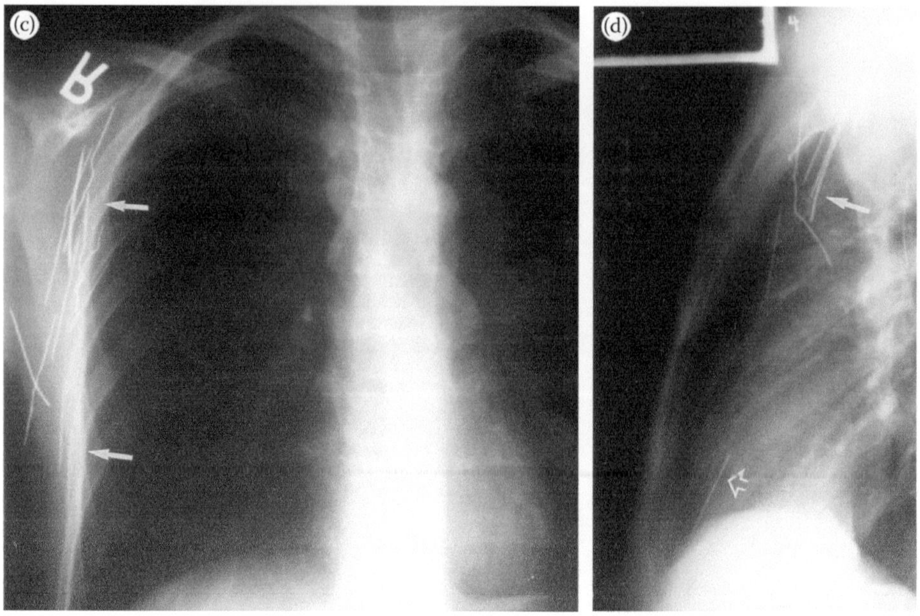

FIGURE 3.29 (*Continued*)

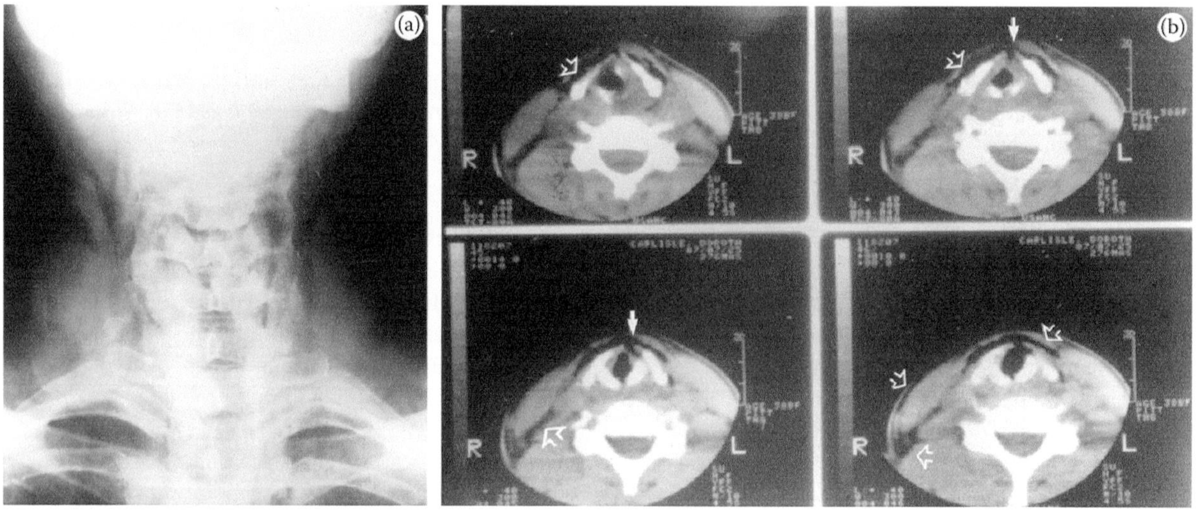

FIGURE 3.30 Attempted suicide by hanging. (a) Cervical spine examination shows massive dissection of air in the soft tissue planes of the neck. (b) The CT scan demonstrates the fracture of the anterior commissure of the larynx (arrow) and air dissection in the soft tissues of the neck (open arrows).

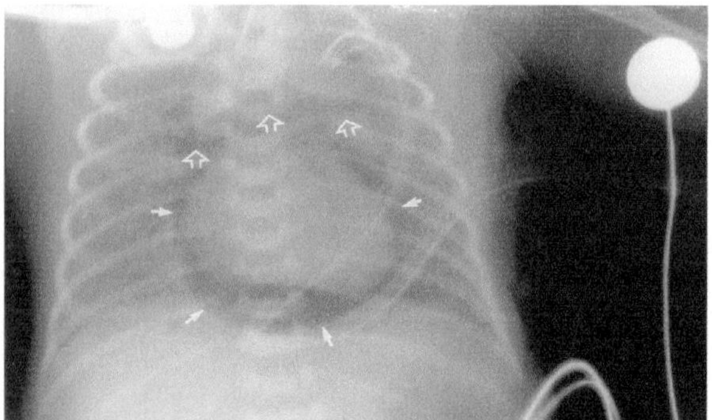

FIGURE 3.31 Pneumopericardium represented by the dark halo of air surrounding the heart (arrows). There also is pneumomediastinum outlining the inferior border of the thymus (open arrows).

education by radiologists has been in the early awareness campaigns concerning child abuse.

RESEARCH

Diagnostic imaging methods are, in general, underutilized in forensic biomedical practice and research. Part of this deficiency can be attributed to the cost of, and difficulty of access to, the newer modalities and techniques. These problems may gradually abate as equipment becomes more widely distributed, more commonplace, and, consequently, cheaper and more accessible. Some very promising applications of modern radiological methods in the courts have come about already through individual effort, innovation, and inspiration.

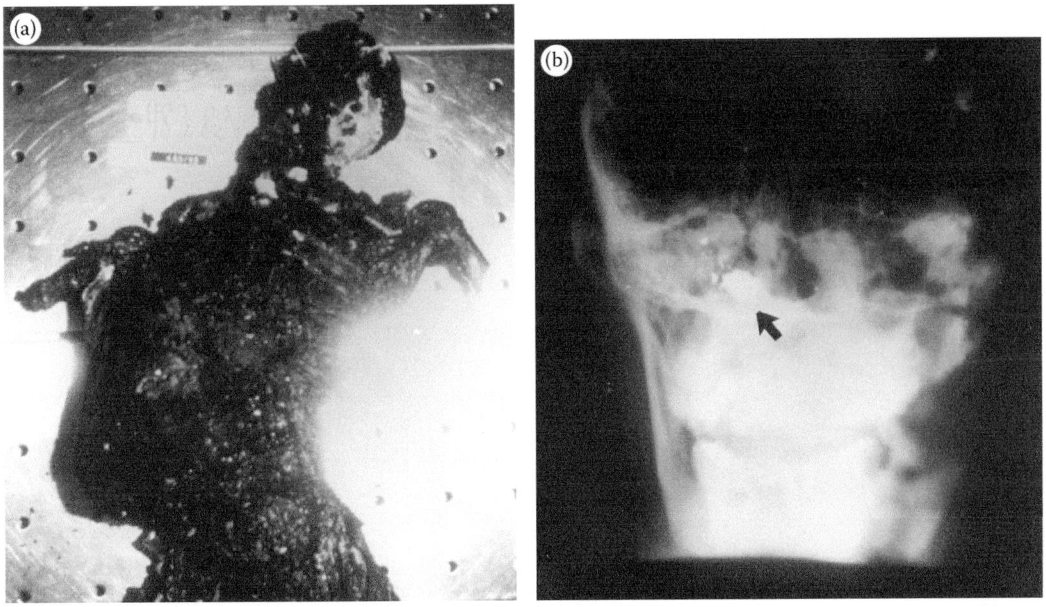

FIGURE 3.32 (a) An almost totally incinerated body with extremities and half of the head missing. (b) Large caliber bullet (arrow) and smaller fragments found in the remaining part of the skull.

ADMINISTRATION

Administratively, the forensic radiologist usually functions as a consultant to the Office of the Medical Examiner or its local equivalent. He or she may have an official appointment to the Office and serve as a regular member of a team that in large offices may include the pathologist, anthropologist, toxicologist, criminalist, ballistics expert, and others. The radiologist is rarely a sworn officer. In the forensic organization, his or her administrative duties at best are related to selection and maintenance of equipment, development, and ensuring conformity with diagnostic protocols, radiation protection, instruction and supervision of technical personnel responsible for image production, proper identification and storage of images, and proper recording of findings.

In his or her larger activities as a practicing clinical diagnostic radiologist, there may be substantial administrative duties depending on the organization and hierarchy of the parent Department of Radiology. Those administrative responsibilities will not overlap unless local conditions require radiography of forensic specimens and bodies in the clinical facility. In such cases, great administrative and diplomatic skills may be required to ensure optimal handling of competing and conflicting demands for space, equipment, time, and personnel in the provision of both clinical and forensic examinations in shared facilities.

REFERENCES

1. Ortner, D. S. and Putschar, W. G. J., *Identification of Pathological Conditions in Human Skeletal Remains*, reprint ed., Smithsonian Institution Press, Washington, DC, 1985.
2. Bonilla-Santiago, J. and Fill, W.L., Sand aspiration in drowning and near drowning, *Radiology*, 128, 301, 1978.

CREDITS

From Brogdon, B.G., Vogel, H. and McDowell, J.D., Eds., A Radiological Atlas of Abuse, Torture, Terrorism and Inflicted Trauma. CRC Press, Boca Raton, FL, 2002. Figures 3.12–16, 3.18–22, 3.24.

Section II

Coping with the Courts

PREFATORY REMARKS

So far, this treatise has dealt with history, science, medicine, and social problems in establishing a basis for forensic radiology as a field of special expertise within the forensic sciences. Now we approach the problem of interfacing this medically based forensic science with the law, and this is not always an easy fit. Perhaps we cannot expect a seamless junction between radiology, the legal profession, and the judiciary, but knowledge of the system and the rules and techniques for dealing with it can smooth out some of the rough spots as they come together. That is the purpose of this section: to relieve anxiety and to instill confidence in the forensic radiologist (especially the tyro or relatively inexperienced) as he enters our great adversarial system of justice.

B.G. Brogdon

4 The Radiological Expert

B.G. Brogdon

CONTENTS

INTRODUCTION

The author makes no claim of legal training, a law degree, or special knowledge of the law. This section is written from the standpoint of an experienced forensic radiologist who reads a lot, listens attentively, and leans heavily on outside sources and experts. The effort here is to provide introductory background information for the colleague undertaking to act as an expert in this field. If the reader is looking for sound legal advice, he should stop right here and go call his personal attorney, or find one.

The reader who subsequently becomes involved in a case in litigation must depend heavily upon the attorney engaging his assistance and who eventually will sponsor the expert in court. If not entirely comfortable with that dependency on the hiring attorney, the expert should decline the case and the association.

A final disclaimer: it is very difficult, and somewhat awkward, to write gender-neutral text. The masculine pronoun has been used throughout this section for purposes of convenience, *not* because of chauvinism or bias. Please read every "he," "him," and "his" as "she," "her," and "hers"; the words are totally interchangeable in this text and context.

The forensic scientist who deals with images or other data derived by radiological methods is, by definition, likely to become involved in court proceedings as a witness. At first, this experience may be anticipated with dread and anxiety. An understanding of the process and the Rules of Procedure, Law, and Evidence which establishes its parameters can help to alleviate this apprehension.

A radiological scientist, like any other person, may be involved in a legal process as a "party" (i.e., as a plaintiff or as a defendant) to a lawsuit, or as a "witness" (by virtue of his five senses) to events relevant to proof of the facts at issue. In either capacity, he may be *required* to give testimony, under oath or affirmation, as an ordinary lay or fact witness—except during Fifth Amendment protection against self-incrimination.

The fact witness can testify on the basis of personal knowledge of matters perceived through the use of his five senses. The only opinions or inferences admissible in court are those rationally based on the perception of the witness and helpful in clarifying his testimony or the determination of a fact at issue.

Testimony is the verbal statement of a witness under oath or affirmation to the trier of fact, that is, the judge or jury. Such testimony may be given orally from the witness stand, or in writing or videotaping in evidence deposition.

Evidence is any and all data presented to the judge or jury in proof of the facts at issue. It includes not only testimony from witnesses, but also records, documents, or objects.

Issues in litigation have become increasingly complex. In many cases, the matter of inquiry is of such complexity that the average trier of fact is unable to understand or come to a correct judgment upon it since it falls beyond the range of common experience or knowledge. In those cases an "expert" with special experience or knowledge may be called upon to assist the trier of fact in understanding evidence or determining a fact at issue.

This *expert witness* differs from the lay witness primarily and most importantly in that he can testify not only on the basis of personal knowledge, but also in the form of opinion. This unique privilege, and responsibility, is afforded to the expert witness in support of his *raison d'etre*, his ability to translate complex scientific or technical issues into language understandable to the trier of fact who is not knowledgeable or experienced in those matters. In doing so, the expert must collect, test, and evaluate evidence, form an opinion as to the evidence, then ably communicate that opinion—and the bases from which it was derived—to the trier of fact.[1]

ADMISSIBILITY OF SCIENTIFIC EVIDENCE: THE FRYE TEST

For many years, the standard for admitting "scientific" testimony as evidence was the so-called "Frye Test" derived from a decision handed down by the Court of Appeals of the District of Columbia in 1923.[2] The case had to do with "the systolic blood pressure deception test," the precursor of the modern polygraph test. The lower court's refusal to admit an expert witness to testify as to the results of this test was sustained on appeal.

Counsel for the defendant argued on this admissibility:

> The rule is that the opinions of experts or skilled witnesses are admissible in evidence in those cases in which the matter of inquiry is such that inexperienced persons are unlikely to prove capable of forming a correct judgement upon it, for the reason that the subject-matter so far partakes of a science, art, or trade as to require a previous habit or experience or study in it, in order to acquire a knowledge of it. When the question involved does not lie within the range of common experience or common knowledge, but requires special experience or special knowledge, then the opinions of witnesses skilled in that particular science, art, or trade to which the question relates are admissible in evidence.

In response to which Associate Justice Van Orsdel stated in his opinion:

> Numerous cases are cited in support of this rule. Just when a scientific principle or discovery crosses the line between the experimental and demonstrable stages is difficult to define. Somewhere in this twilight zone the evidential force of the principle must be recognized, and while courts will go a long way in admitting expert testimony deduced from a well-recognized scientific principle or discovery, *the thing from which the deduction is made must be sufficiently established to have gained general acceptance in the particular field in which it belongs.* (Emphasis added.)

This rule of "general acceptance" was adopted in all of the Federal Circuits and most of the individual states.[3]

FEDERAL RULES OF EVIDENCE

In 1975, the Federal Rules of Evidence (FRE) were adopted, and most states have adopted their own rules of evidence modeled on the Federal Rules.[3] Several of them are relevant to scientific evidence and the expert (vs. the lay) witness:

THE FEDERAL RULES OF EVIDENCE

ARTICLE VII

Opinions and Expert Testimony

Rule 701
Opinion Testimony by Lay Witnesses

If the witness is not testifying as an expert, the witness' testimony in the form of opinions or inferences is limited to those opinions or inferences which are (a) rationally based on the perception of the witness and (b) helpful to a clear understanding of the witness' testimony of the determination of a fact in issue.

Rule 702
Testimony by Experts

If scientific, technical, or other specialized knowledge will assist the trier of fact to understand the evidence or to determine a fact in issue, a witness qualified as an expert by knowledge, skill, experience, training, or education, may testify thereto in the form of an opinion or otherwise.

Rule 703
Bases of Opinion Testimony by Experts

The facts or data in the particular case upon which an expert bases an opinion or inference may be those perceived by or made known to the expert at or before the hearing. If of a type reasonably relied upon by experts in the particular field in forming opinions or inferences upon the subject, the facts or data need not be admissible in evidence.

Rule 704
Opinion on Ultimate Issue

a. Except as provided in subdivision (b), testimony in the form of an opinion or inference otherwise admissible is not objectionable because it embraces an ultimate issue to be decided by the trier of fact.

b. No expert witness testifying with respect to the mental state or condition of a defendant in a criminal case may state an opinion or inference as to whether the defendant did or did not have the mental state or condition constituting an element of the crime charged or of a defense thereto. Such ultimate issues are matters for the trier of fact alone. (As amended October 12, 1984.)

Rule 705
Disclosure of Facts or Data Underlying Expert Opinion

The expert may testify in terms of opinion or inference and give reasons therefor without first testifying to the underlying facts or data, unless the court requires otherwise. The expert may in any event be required to disclose the underlying facts or data on cross-examination.

Rule 706
Court Appointed Experts

a. *Appointment.* The court may on its own motion or on the motion of any party enter an order to show cause why expert witnesses should not be appointed, and may request the parties to submit nominations. The court may appoint any expert witnesses agreed upon by the parties, and may appoint expert witnesses of its own selection. An expert witness shall not be appointed by the court unless the witness consents to act. A witness so appointed shall be informed of the witness' duties by the court in writing, a copy of which shall be filed with the clerk, or at a conference in which the parties shall have opportunity to participate. A witness so appointed shall advise the parties of the witness' findings, if any; the witness' deposition may be taken by any party; and the witness may be called to testify by the court or any party. The witness shall be subject to cross-examination by each party, including a party calling the witnesses.

b. *Compensation.* Expert witnesses so appointed are entitled to reasonable compensation in whatever sum the court may allow. The compensation thus fixed is payable from funds which may be provided by law in criminal cases and civil actions and proceedings involving just compensation under the fifth amendment. In other civil actions and proceedings the compensation shall be paid by the parties in such proportion and at such time as the court directs, and thereafter charged in like manner as other costs.

c. *Disclosure of appointment.* In the exercise of its discretion, the court may authorize disclosure to the jury of the fact that the court appointed the expert witness.

d. *Parties' experts of own selection.* Nothing in this rule limits the parties in calling expert witnesses of their own selection.

These Rules from Article VII need to be read in conjunction with the Rules on relevance:

Rule 401
Definition of "Relevant Evidence"

Relevant evidence" means evidence having any tendency to make the existence of any fact that is of consequence to the determination of the action more probable or less probable than it would be without the evidence.

Rule 402
Relevant Evidence Generally Admissible: Irrelevant Evidence Inadmissible

All relevant evidence is admissible, except as otherwise proscribed by the Supreme Court pursuant to statutory authority. Evidence which is not relevant is not admissible.

Rule 403
Exclusion of Relevant Evidence on Grounds of Prejudice, Confusion, or Waste of Time

Although relevant, evidence may be excluded if its probative value is substantially outweighed by the danger of unfair prejudice, confusion of the issues, or misleading the jury, or by considerations of undue delay, waste of time, or needless presentation of cumulative evidence.

Over the years, there has been argument that FRE 702 superseded *Frye* in defining the basis for admission of scientific evidence. The U.S. Supreme Court essentially accepted this argument in what has become known as the *Daubert* rule or standard,[3] which applies to the Federal Courts in those states that follow the Federal Rules of Evidence.

DAUBERT

In *Daubert v. Merrell Dow Pharmaceuticals*, it was claimed that a drug, Bendectin, marketed by Merrell and prescribed for "morning sickness" caused birth defects in children born to women who had taken it. The case went all the way to the Supreme Court on appeal, hinging on the admissibility of a Dr. Gross and his "reanalysis" of an earlier epidemiological study. Writing for the 7 to 2 majority, Justice Blackmun considered that FRE 702 superseded the too-demanding "general acceptance" standards of *Frye*. The decision assigned a "gatekeeping" role to the trial judge in ascertaining whether scientific evidence and testimony offered in the court was obtained by the "scientific method" and was relevant. Suggested factors or criteria for consideration in evaluating such "science" include, but are not limited to the following: (1) the technique or theory can be, and has been, tested; (2) peer review and publication; (3) known or potential error rate; (4) existence and maintenance of control standards for performance or operation; (5) widespread acceptance within a relevant scientific community or discipline; (6) research used as a basis of testimony must have been conducted independent of the litigation at hand.[3–5]

Daubert engendered extensive contention and confusion in the legal profession, the judiciary, and the forensic sciences regarding standards for scientific evidence and expert testimony. A survey published in 1995[3] revealed that one year following *Daubert* there was great variation in admissibility standards in state law and state judicial rules governing most trials in the United States. According to that author, the standards were clear or fairly clear in 39 states, fairly unclear in 5 states, and completely unclear in 6 states.

All of the above notwithstanding, there would appear to be no problem in admission of scientific evidence in the form of radiological images or data collected in the course of diagnostic radiological procedures. Further, of those "expert" witnesses proffered to explain or translate that evidence to the trier of fact, few, if any, are likely to be rejected by the court, if properly qualified as experts.

THE EXPERT WITNESS

> If the law has made you a witness, remain a man of science. You have no victim to avenge, no guilty or innocent person to convict or save—you must bear testimony within the limits of science.
>
> **Dr. P. C. H. Brouardel**
> *(late nineteenth-century French medico-legalist[6])*

ENGAGEMENT OF THE EXPERT WITNESS

A person with scientific, technical, or other specialized knowledge which may help the trier of fact, judge or jury, to better understand the evidence, or to better understand and determine a fact at issue in a lawsuit, can be approached by either side—plaintiff or defendant, prosecutor or defense—to serve the court as an expert witness. An expert witness can be appointed by the court (judge) with or without the consent of the contending parties. However, the expert witness' consent is required. He cannot be forced to serve by subpoena.

Here, for physicians and dentists and other health-care providers who may have special knowledge that ordinarily would qualify them as an expert, a special exception must be noted. If that person has provided professional services to one of the parties in a lawsuit, he may be subpoenaed and required to testify as an ordinary witness. Examples would be a radiologist who performed and interpreted radiological examinations on a patient who later becomes a party in a personal injury case, or in a corporate liability action, or a malpractice suit. In such instances, the health provider may be required to testify without compensation, but ordinarily the party calling the witness will offer compensation in order to avoid dealing with a "hostile" witness.

Parties on either side of the lawsuit can call expert witnesses of their own selection. Indeed the American Bar Association's (ABA) *Model Rules of Professional Conduct* requires an attorney to seek expert services if the client needs them. Failure to do so may be considered "ineffective assistance of counsel" or may cause counsel to be found liable for malpractice.[7]

There are further standards concerning the relations between the hiring attorney and the expert witness. Counsel must respect the expert's independence, cannot dictate the expert's opinion, and must explain to the expert his impartial role as an assistant to the trier of fact. Also, counsel is warned against paying excessive or contingency fees in order to influence expert opinion or testimony. Fee splitting, or sharing of fees between counsel and expert, is explicitly forbidden.[8]

AMERICAN BAR ASSOCIATION STANDARDS RELATING TO THE ADMINISTRATION OF CRIMINAL JUSTICE

Standard 3-3.3 Relations with Expert Witnesses

a. A prosecutor who engages an expert for an opinion should respect the independence of the expert and should not seek to dictate the formation of the expert's opinion on the subject. To the extent necessary, the prosecutor should explain to the expert his or her role in the trial as an impartial expert called to aid the fact finders and the manner in which the examination of witnesses is conducted.
b. A prosecutor should not pay an excessive fee for the purpose of influencing the expert's testimony or to fix the amount of the fee contingent upon the testimony the expert will give or the result in the case.

Standard 4-4.4 Relations with Expert Witnesses

a. Defense counsel who engages an expert for an opinion should respect the independence of the expert and should not seek to dictat e the formation of the expert's opinion on the subject. To the extent necessary, defense counsel should explain to the expert his or her role in the trial as an impartial witness called to aid the fact finders, and the manner in which the examination of witnesses is conducted.
b. Defense counsel should not pay an excessive fee for the purpose of influencing an expert's testimony or fix the amount of the fee contingent upon the testimony an expert will give or the result in the case.

CONTACT AND AGREEMENTS BETWEEN COUNSEL AND EXPERT

The initial contact between the counsel representing a party anticipating or already in litigation and an expert is usually by telephone or e-mail. The expert may be approached for one or more of several reasons: the attorney may have worked with him before; the attorney may know of the expert's performance in other cases; the expert is widely known as an authority of excellent reputation in his field; the expert may have rare or unique expertise; the expert may have been recommended by other law firms; or the expert may have been selected from published lists of forensic scientists willing to serve as expert witnesses.

After the usual introductions and pleasantries, the counsel will ascertain if the expert is willing to consult on a case which may or may not be summarized briefly at that time.

In the field of forensic radiology particularly, many experts—and many attorneys—prefer that the image(s) first be seen and evaluated "cold," that is, without any introduction of bias excepting the inevitably heightened search expectations accompanying the knowledge that this image(s) ultimately may become evidence in a lawsuit.

During the initial contact the professional *bona fides* of both parties should be established to the extent that they can be independently verified (by looking through listings in Dun and Bradstreet, professional directories, Who's Who, state and national licensing boards, professional associations, etc.). The expert can ask for an estimate of the maximum involvement that might be required of him. This is the time to bring up the subject of fee, quote fee schedules if they are set, and establish responsibility for prompt payment of those fees as service is provided. There is case law holding a law firm hiring an expert witness on behalf of a client liable for expert witness fees after the client refused to pay.[9] There is a modern trend to hold attorneys liable for expert or other litigation provider service fees in the absence of an express disclaimer of responsibility.[7] If the caller seems taken aback by your fee schedule or says he has to check back with his firm or his client, watch out! He probably is underfunded.[10]

If the proposition is appealing to both the expert and the attorney, then a date can be set for the initial consultation. This should be scheduled no sooner than two working days in order to allow the expert an opportunity to check out the attorney and/or his firm. Of course, *no opinion should be expressed during the initial telephone call.*

Immediately after the initial contact, it is a good idea for the expert to summarize his understanding of the requirement for his services by mail, fax, or express letter to the attorney, including a fee schedule and a *curriculum vitae* (CV), and reconfirming the date and time for the first consultation. This letter should be sufficiently clear and detailed that it disallows room for misunderstanding and may also serve as a sort of contract should problems later arise regarding fees, and so on. It has been suggested that an advance retainer or fee be obtained for the initial consultation or opinion.[11] (The author has not found that necessary and believes a better relationship is established, with small risk, if a bill for the first service is sent immediately after that service is provided. Usually it becomes obvious early on whether or not the attorney's firm is adequately capitalized to carry the cost of long-term litigation, which may be on a contingency basis. Incremental billing by the expert as he provides services also reduces the risk of loss. Of course, if there is difficulty or delay in collecting expert fees from the hiring attorney, several remedies are available through your own attorney or the Bar Association.)

FEES

The fees charged for professional services in a forensic setting are entirely personal and not bound by any limits other than the reference to those sufficiently "excessive" to skew honesty and influence opinion (cited earlier in the ABA Standards). Indeed, there is more independence and leeway in setting these fees than there is nowadays in the establishment of professional fees for patient care. Consultation with colleagues of similar stature and expertise can help in determining reasonable and customary ranges of fees for expert forensic services.

It is convenient and commonplace to charge according to time and level of activity. The time component easily is broken down into hours, quarter-hours, or even minutes. (One may wish to follow the lead of many lawyers by establishing a "least billable segment" of 5–15 min, thus allowing "rounding off" and relieving the burden of time-keeping to the exact minute.) Time charges related to activity devoted to the case at hand include, but are not limited to, library or bench research, consultation face to face or by "wire," correspondence and/or report preparation, travel, preparation of exhibits, review of contributing materials, and so on.

The fee schedule by unit of time quite properly can vary by level and location of activity. The lowest unit charge would be for work accomplished in- house, that is, at the place selected by the expert. The next higher level of activity charged would be for deposition and might vary according to where the deposition is taken. The highest level of activity, of course, is appearance in court. Here travel and travel expenses must be considered and the time scale may change to half-day or daily increments.

It must be remembered that determination of fees by time and activity is simply a convenience, no more. The expert is not being recompensed just for his time. Rather the expert's fee is a recognition of a totality of qualification including his education, experience, research activity, scholarly productivity, rare or unique talents, communicative skills, overall reputation, the esteem of his peers as evidenced by offices and awards, personal character, professional ethics, and above all, his unimpeachable honesty and impartiality.

Finally, established or quoted fees can obviously be modified from case to case.

THE FIRST CONSULTATION

After a mutually agreeable initial contact and the establishment of *bona fides* comes the first consultation involving the expert and the attorney in the particulars of a specific case or question. Often, in the field of forensic radiology, this eventuates as the arrival of one or more radiographs (originals or good first copies) and/or other images either in hard copy or on disks, with or without accompanying documents or explanatory material. At other times, the attorney or his paralegal will bring the images and other material to the expert's office. In either event, the expert must not rush to an opinion. He must review and interpret the radiological images, consider any other pertinent material furnished, and formulate a preliminary opinion on the basis of all the information available to that point in time. That opinion is then conveyed to the attorney in person, through his paralegal, by telephone, or by hard-copy correspondence. Except for the latter instance, it may be best for the expert to avoid making any notes or permanent record of his deliberations up to that time, as these may be subject to discovery by the opposing party. After receiving the expert's verbal opinion, the attorney may or may not ask for a written opinion or affidavit. Also, it is wise to submit a bill for services to date at this time. Very often this first consultation will also be the last one because the vast majority of lawsuits will be settled after some expert input is obtained.

The first consultation also affords an excellent opportunity to formulate an opinion of the attorney. Does he have a firm grasp of the case? Is he experienced in this type of litigation? Is he organized? Is he an effective communicator and time manager? Is he on a fishing trip to learn whether a case has merit? Does he try to rush or shade your opinion? Can you work together with mutual trust and respect? As the case evolves, will the attorney keep you advised of progress or stagnation, new information or changes, opposing opinion, and so on.? Are there money problems?[10]

If the litigation continues there may be additional consultations face-to-face or by wire and sometimes extensive in-house preparation, review, and research by the expert. The mutually agreed upon arrangement for incremental billing for these activities should be worked out in advance. As the case matures, and no settlement intervenes, the likelihood of deposition increases.

DISCOVERY DEPOSITION

Discovery has been defined as the ascertainment of that which was previously unknown.[1] Thorough pretrial preparation for cross-examination requires advance knowledge of expert testimony to be presented. Discovery, in effect, is the compulsory disclosure of the expert's opinion and material associated with the formulation of that opinion. Although rules of disclosure vary somewhat among jurisdictions and between civil and criminal procedure, the expert is best advised to assume that every image, report, document, note or other material related to a case is subject to discovery by opposing counsel. Federal Rule 26 (a)(2)(B) which requires an expert to furnish the "data or other information considered by the witness in forming the opinions" could be interpreted to include almost all interaction between the expert and his attorney, including conversation.[12] A list of cases in which the expert has testified by deposition or in court during the past 5 years may be requested and will be required if the case is going to the Federal Court. Apart from requiring copies of all films, reports, correspondence, and so on. related to the case, the principal method of discovery by opposing counsel is by deposition.

Discovery deposition is a method of taking testimony under oath but not before the court.[1] One must prepare for a deposition with the same thoroughness and seriousness as for a court appearance. Consequently, the deposition should be scheduled far enough in advance to allow for this preparation, including adequate communication between the expert and his hiring attorney. The discovery deposition will be requested by the opposing counsel. The time and location of deposition is reached by mutual agreement of the minimal number of participants required: the expert, the two opposing attorneys, and the court reporter who will administer the oath and make a legal record of the proceedings. The expert can expect exhaustive (and sometimes hostile) questions about his entry into the case, his knowledge of it, materials reviewed, and other preparation incidental to the formulation of an opinion. All questions must be answered, truthfully, unless they are unanswerable as posed, or upon the expert's attorney's advice not to answer because of his objection to the question. Since there is no judge to rule on objections, such interaction between opposing counsel is recorded and the expert is not at risk for following advice. Because the opposing side has requested the deposition and has posed the questions, there is no "cross-examination," as such, on discovery deposition. However, the attorney retaining the expert can requisition him to clarify issues brought into the deposition by the opposing counsel.

The expert may be asked if he is willing to "waive signature" upon completion of a deposition. Decline! Rather, the expert must make certain that the deposition scheduled allows adequate time for him to receive, review, correct, and approve the transcript of his deposed testimony well before the case goes to trial. Any inconsistency between testimony at deposition and in court may be used to impeach the witness or weaken his impact on the trier of fact.

Fees for discovery deposition may be billed to the hiring attorney, or directly to the opposing counsel. It is a good practice to bring and exchange business cards with both the opposing attorney(s) and the court reporter at the beginning of this session while determining to which attorney the invoice will be sent. When the government takes a deposition, or when the deposing party is indigent, or when the expert is court appointed, expenses and/or fees of the witness may be paid by the court.[1]

An *evidence deposition* can be requested (usually by the hiring counsel) when it is known in advance that the expert cannot be physically present in court at the trial. The procedure for this deposition is much the same as for a discovery deposition except for one major difference: the expert is subject to cross-examination by the opposing counsel, because the party calling the witness questions first. A minor difference is that this type of deposition is more likely to be videotaped so that the trier of fact can see as well as hear the witness if the deposed testimony is admitted at trial.

EXPERT TESTIMONY IN COURT

As previously stated, the vast majority of lawsuits in which expert opinion in the field of radiology is solicited will be settled or plea-bargained during or after the discovery phase. Only a small proportion will come to trial, requiring expert testimony in court. Even so, every case accepted for consultation must be treated as if it will go to trial, with equally serious attention and preparation. The forensic radiologist is most likely to actually testify in court in cases of malpractice, personal injury, abuse, establishing identity of human remains, or criminal cases not resolved by a plea of guilty of the crime or a lesser charge or by reason of insanity.

Of course, the expert must be thoroughly prepared for a court appearance and take with him all notes, radiographs, or other images, and any other material needed to support the testimony—sometimes including viewboxes, since one may not be available in the courthouse. To avoid embarrassment, one's briefcase should be purged before entering the courtroom. Anything brought to the stand is subject to examination, including your traffic tickets, overdue bills, bank statement, or a day-old tuna fish sandwich.[11]

If visual aids such as drawings or photographic enlargements of radiological images are to be used, they must be reviewed for adequacy and accuracy.

Lines of anticipated questioning by the counsel from both sides must have been reviewed with the sponsoring attorney. The expert may be required to disclose the underlying facts or data leading to his opinion or inference and must be prepared to do so (see Model Rule 705 above).

For the expert witness, the courtroom experience most often can be likened to seeing only one scene of a three-act play. Virtually never is he in attendance for jury selection, opening statements, summations, or jury instructions. Rule 615 of the Federal Rules of Evidence provides for the sequestration (separation) of witnesses at the request of either party, or the court may so order on its own motion, so that the expert's testimony is not tailored in response to the testimony of other witnesses. Other jurisdictions have a similar Rule although the exact wording may vary. The judge has discretionary power to exclude an expert witness from the Rule How ever, for the most part, witnesses are not allowed in the courtroom during testimony by others. Consequently, the expert walks into court with no direct knowledge of previous testimony, testifies, then walks back out and is dependent upon the press for events following his appearance. This is somewhat frustrating, and incomprehensible, to the witness since, particularly in civil trials, it is already well established what experts on either side will opine, as a result of access to affidavits filed, the required Designation of Expert Witness (including there a summary of the expert's expected testimony), and previously available discovery depositions from which one may not stray. It is far more entertaining and instructive to attend parts of the trial than to sit out on the typically hard benches in a courthouse hallway waiting to be called.[1]

To reduce waiting time outside the courtroom (and fees) the hiring counsel will try to closely estimate just when the expert will be called. But this is guesswork. Trials tend to develop their own pace, and it is wise to arrive with professional or recreational reading material to while away the hours.

When finally called to testify, the expert forensic scientist should enter the court looking like one—conservatively dressed appropriate to the region and venue, neat and trimmed, with a pleasant yet dignified mien. While walking to the stand it is good to make eye contact with the judge and jury. The expert will take the stand and be sworn.

THE OATH

Every witness must by oath or affirmation declare that he will testify truthfully. The form of the oath is designed to stimulate the conscience of the witness and inspire his sense of duty and responsibility to the truth.[13] However, the language of the oath requiring "the whole truth" must be interpreted within the confines of the adversary system. The witness must supply the whole truth only to the extent required by the question asked of him. The witness has no duty to go beyond the question or to answer questions unasked. Indeed, to do so may result in the answer being stricken from the record. When necessary, the witness can usually indicate subtly a wish to go further with an answer or to make additional comments. This will be picked up quickly by a perceptive attorney or by the trial judge who, of course, can question any witness.

Testifying falsely under oath can lead to perjury charges and also malpractice action against the expert.[4,5,13]

QUALIFICATION OF THE EXPERT WITNESS

After being sworn and identified the expert witness is "qualified" through a series of questions by the attorney who engaged him. This qualification will be based on the witnesses' CV and experience. If the CV is extensive, the expert should have, in advance, helped the counsel winnow out the significant parts in conversation or by highlighting or abridging the CV. It has been suggested that the witness furnish sample questions to facilitate the qualification process.[14]

The expert witness does not have to be the greatest star in his field. He simply is required to have sufficient expertise to assist the trier of fact. However, the impact of testimony on the judge or jury is probably influenced by the expert's level of expertise and reputation in his field. Consequently, the attorney presenting the expert witness will lead him through a long litany of education, degrees, academic appointments, professional experiences, publications and other scholarly productivity, professional honors and offices, previous jurisdictions in which testimony has been admitted, and any other material reflecting favorably on the witness and by any stretch pertinent to the case. The opposing counsel may try to interrupt and truncate this exposition with an offer to "stipulate" that the witness is an expert, that is, agree to his qualifications forthwith. However, presenting counsel usually will insist on continuing the inventory of his witness's achievements.

When the presentation of the expert qualification is completed, the opposing lawyer has the opportunity to challenge the witness in the "*voir dire*" (from the French, meaning "to speak the truth") by bringing out matters that might prevent his qualification.[1] Following this, the judge decides on the admissibility of the witness.

It is improbable that anyone proposed as an expert involving testimony related to diagnostic radiology, or a product thereof, would fail to qualify. Although in 1919 an Iowa court ruled that the x-ray speaks for itself (like a photograph) and that the evidence reposed within it could be interpreted by the jury,[15] an earlier and more sensible view has prevailed:

> The evidence shows that nobody but an x-ray expert could tell anything from the plate and if they had been produced they would have done the court, the jury or the defendant's ordinary physician no good. I do not think that the doctrine that an ordinary photograph is the best evidence of what it contains should be applied to x-ray pictures. They constitute an exception to the rule concerning ordinary documents and photographs, for the x-ray pictures are not in fact the best evidence to laymen of what they contain. The opinion of the expert is the best evidence of what they contain—the only evidence.[16]

It is commonly held that almost any physician is qualified to "read" x-rays, and that any dentist can evaluate dental x-rays. Still, there is an opinion that simply being a physician does not in and of itself qualify one as a competent interpreter of a diagnostic imaging study.[17] Dr. F.J. Baetjer, the first radiologist at Johns Hopkins, early emphasized:

> There is no such thing as an x-ray picture. A roentgenogram is a projection upon a photographic plate of a series of shadows of varying density representing the various structures through which the rays have passed. The correctness of the diagnosis depends entirely upon the skill with which these various shadows are separated and interrupted. To interpret these shadows correctly one must know not only the appearance of the normal structure, but also the alterations that take place when there is a pathological process present .... Roentgenography is ... a medical procedure based upon careful analysis and logical deductions from the shadows observed upon an x-ray plate, and translated into pathological terms. This means—and it cannot be too strongly emphasized—that the skill of a roentgenologist will vary directly with his medical knowledge: the value of the roentgenologist to the medical profession (and patient) will be based upon this fact and not upon his technical ability.[18]

The more highly qualified and specialized the expert, the more weight his opinion will carry. Most juries will appreciate that the diagnostic images must be interpreted and explained to them by a competent expert familiar with both medical and technical factors applicable to the evidence; therefore, an expert in the field of radiology will be preferred.[19] Hence, contending parties are apt to want a qualified radiologist, even a radiological subspecialist, to testify on radiological images especially if produced by one of the newer, highly technical modalities. A dentist or oral surgeon well experienced in dental radiology, or an orthopedic surgeon well versed in radiology of bones and joints, or a neurologist or neurosurgeon familiar with neuroradiological procedures, will be more easily and effectively qualified than a generalist. A physical anthropologist may qualify as an expert to testify on radiographs of the skeleton, but probably will be required to explain or document special training or experience in the use of this tool. Rarely, a technologist will be asked to testify on the production, identification, or chain of custody of images.

DIRECT EXAMINATION

Once qualified and admitted, the expert witness begins his testimony on direct examination by the attorney bringing him to the stand. Direct examination allows an opportunity for the expert to use his training, knowledge, experience, and special skills to present his evidence, describe it, and demonstrate how his opinion is reached on the basis of certain facts.[1] The expert is led through his testimony with a series of well-rehearsed questions designed to bring out the best in the witness and his evidence by a friendly, considerate, and understanding lawyer. This is the first opportunity for the trier of fact, judge or jury, to see and appreciate the appearance, demeanor, and communicative skills of the witness. Questions should be answered slowly and distinctly in a pleasant, confident conversational rhythm. Answers should be directed to the jury when present. One should speak loudly enough to reach a hearing-impaired juror, judge, and counsel at both tables. Although the expert witness is cast in the role of teacher to the jury, he must not be pedantic, aloof, overbearing, or smug.

> There is nothing worse than a pompous expert.
>
> **Judge Haskell M. Pitluck**

The witness should exude an air of friendly yet dignified eagerness to be helpful. Direct eye contact with individual jurors is most desirable and effective. Levity and flippancy must be avoided at all costs.

One must expect to demonstrate images to the jury. Bring a pointer. Make sure the entire jury can see the image(s). Many courtrooms now are equipped, or can be arranged, to permit demonstration of images by PowerPoint projection. Remember, viewboxes, some projection screens, and television monitors are not friendly to laser pointers. It is wise to be equipped with both kinds. It is acceptable, sometimes desirable, to use scientific nomenclature if it is immediately translated to lay terminology and/or thoroughly explained. The jury should be carried along in the process from interpretation of findings on the image(s) through consideration of other influencing knowledge, facts, or experience to the formulation and clear statement of the expert's opinion. Questions on direct examination tend to be open-ended, allowing the expert considerable freedom for unfettered discourse to that end, insofar as allowed by the patience of the judge and the unsustained objections of the opposing counsel.

Finally, testimony on the witness stand must be in absolute consonance with previous deposition, or else a very good explanation for any discordance must be firmly in hand.

ADMISSIBILITY OF RADIOLOGICAL IMAGES AND RESULTS

As already noted in Chapter 2, roentgenograms were admitted as evidence in courts in England, Canada, and the United States in 1896, just months after Röntgen's discovery. The admissibility of the product of a radiological examination is unlikely to be questioned in a modern courtroom. There may be a requirement to show that it was obtained by an accurate and generally recognized methodology and accurately represents the object investigated. For newer modalities such as ultrasonography, CT, MRI, and nuclear studies, the expert should be prepared to explain and defend the procedures in some detail.[19]

CROSS-EXAMINATION

The relative comfort during direct examination with a considerate counsel coddling the witness evaporates quickly at cross-examination.

The expert has been put on the stand because his opinion is seen as favorable to the side sponsoring him, and the direct examination is designed to present that opinion in the most effective and positive manner. In our adversarial system of justice, it is the duty of the opposing counsel to tear down that testimony if possible. If direct testimony has been so good as to be virtually incontestable, or so ineffective that it poses no problem, the opposing counsel may pass the witness. That is rare.

Usually, the opposing counsel will use cross-examination to weaken or impeach prior testimony or the witness himself. The witness must not react to this new setting with any change in his demeanor or attitude. The voice and rhythm of his responses should remain unaltered. The expert is empowered by the very expertise that brought him to this confrontation. No matter how hard the attorney has crammed the science and studied the specifics of the case, he will be no match for the well-prepared expert and may well be a little in awe of him.[1] On the other hand, it must be remembered that the attorney is at home in the courtroom. The expert is the visitor. And the playing field is not level, but scattered with pits and traps into which the expert must not stumble. The well-prepared expert witness will use his common sense to avoid the pitfalls of cross-examination, some of which are catalogued below, along with suggested avoidance tactics.

THE RACEHORSE[1,11,20,21]

The cross-examiner frequently will try to rattle the witness by changing the tempo of questioning, usually by stepping up the cadence. Watch out! Don't let staccato questions tempt rapid-fire responses. Pause. Think. Answer at your established pace.

Don't let the questioner step on (cut off) your answers. Insist on being allowed to complete your reply, but pleasantly, so the jury will notice what a nice fellow you are.

Don't be lulled by a series of rapid sequential answers: "Yes, yes, yes, yes . . . Uh Oh!" Keep it slow.

Don't be confused by lawyer-talk or terminology at variance with scientific terms. Ask for clarification or rephrasing and qualify your answer.

Don't be bullied by "Answer 'Yes' or 'No'!" questions if they cannot be answered without some further qualification. Say so, pleasantly. If necessary, appeal to the judge.

If the attorney opens with, "Now you previously testified (or wrote) . . ., " don't buy it. Ask that the reporter read back the cited questions and answers, or that the written statement be presented for your review. Never hesitate to have a question repeated in full.

If in his hurry, or ignorance, or cleverness, the cross-examiner mis-states the facts as the expert understands them, or mistakes scientific principle, don't answer the question but, rather, correct his error.

Don't let the insistence of rapid questioning distract your attention from the jury to the attorney. Try to respond to the jury on all questions.

Don't ride with a racehorse attorney. You can dismount at will and proceed at your own pace.

DON'T OVERREACH[5,13,14,21]

Do not overblow your qualifications. Make sure your C.V. is up-to-date and error free. Exaggerations will not be well received by the trier of fact. Falsification can lead to professional disaster. You can be sure that opposing counsel has studied your CV with a jaundiced eye toward that end.

Don't be lured outside your scope of expertise. If you are describing cerebral edema and intracranial bleeding in a child abuse case, don't get trapped into discussing autism or head-bangers. Insist firmly and unashamedly that your special qualification is in diagnostic imaging—not pediatrics or neurology. Know your limits and do not exceed them. The jury will not think less of you.

Don't volunteer. Do not extend answers beyond the limitation of the question. Such answers may be judged irrevelant and stricken. Oration or argument may appear as partisanship to the jury. Don't overstate. The expert who tries to make more of his evidence than it is worth is unlikely to survive competent cross-examination.

JUST SAY "NO"[11,21]

Do not hesitate to say, "No," or "I don't know," to a question if honesty demands it. A frank answer admitting a specific lack of knowledge or expertise is unlikely to do much harm and may enhance your aura of sincerity. No one is expected to know everything. A bluff successfully called by opposing counsel may cause irreparable damage.

SENSITIVE ISSUES[1,5,11]

Everyone has a few chinks in his armor—a few sensitive areas he would rather not have probed in public. One must assume that the opposing counsel knows of them. They may be trivial, or ancient history, but still may be used to dull the luster of your testimony or to imply bias. How did you get involved in this case? How much are you being paid for the testimony? Do you advertise for work as an expert? Do you always, or nearly always, testify only for the plaintiff (or the defendant)? Have you been sued for malpractice? Bad debts? Did you fail licensure/specialty board examinations? Are you on retainer to this law firm? Have you practiced your testimony with your lawyer? Did you have a different opinion on a similar issue two years ago?

There are answers for all of the questions; some are easy, others require preparation. All answers should reinforce your integrity, honesty, objectivity, and the distance between the scientist and the other parties to the case.

Sometimes it is advisable to have one's own attorney bring up a sensitive issue on direct examination in order to cast the best possible light upon it. Remember that fees require no apology or explanation. They represent fair recompense for a totality of qualification. You should discuss this issue prior to trial with your retaining counsel. It should be emphasized that expert fees are not contingent on the outcome of the case.

DON'T PLAY THE NUMBERS[11,14]

Cross-examiners love to try to pin the expert down to numbers on issues of certainty, likelihood, ratios, chances of, incidence, and so on. Usually these "numbers" questions are brought on matters with no propensity for precision. Do the best you can with them; if the question is impossible to answer, say so, and appeal to the judge if necessary. If still required to answer, try to use ranges rather than precise numbers and qualify your answer as best you can. At the same time, try not to appear wimpish or sound evasive. Watch out for "weasel words:" perhaps, maybe, could be, probably, and so on.

The law likes to deal with terms such as "reasonable medical certainty," "beyond reasonable doubt," or "reasonable medical probability." Unfortunately, these do not seem to have a precise numerical basis either. If you use them, be prepared to provide your own definition.

HYPOTHETICAL QUESTIONS[1,11,19]

A common courthouse tactic is to pose a question to the expert witness based on a hypothetical situation. This offers great latitude in presenting (or eroding) the opinion of the witness. In most jurisdictions, especially on direct examination, hypothetical questions must be based only on facts in evidence. Those restrictions do not apply to hypothetical questions posed on cross-examination, which can sometimes be quite fanciful and labyrinthine.

The expert must be certain he understands the question, that there are enough facts upon which to base an opinion, and that the "facts" proposed in the premise are not scientifically impossible. Do not hesitate to ask for repetition or explanation of the question. Take time in answering. Watch out for traps. On cross-examination the hypothetical question most likely will be antithetical to your previous testimony.

A ubiquitous phrase loved by attorneys during cross-examination (and sometimes hiding an unannounced hypothetical) goes like this, "Now, Doctor, would you not agree that..." Watch out! Take a few deep breaths while trying to parse this confusing question. Try to avoid a yes or no answer and state your position on the issue in as simple a declarative sentence as you can construct.

LEARNED TREATISES AND TEXTBOOKS[1,11,20,22]

A specific method of impeaching an expert witness is through the use of learned treatises or textbooks which are in disagreement with the expert's testimony and opinions. To be used as an instrument of impeachment, the treatise or text must be *relied* upon by the witness in reaching his opinion, or *recognized* as authoritative by the witness. Rarely, the treatise or text will be *established by other means* including the testimony of other experts or by judicial notice. The text or treatise can be read or quoted in court only when the witness is on the stand, and cannot be entered as evidence where it might be misunderstood or misused in the jury room.

The presentation of a treatise or text to the expert witness may be somewhat veiled or insidious. The witness may be asked if he knows a certain person eminent in this field. Is that person reputable and authoritative? Is the witness familiar with his writings or book? Are they authoritative?

While the expert witness must be familiar with the literature and authoritative opinion in his field, he must not give a blanket endorsement to any individual or his work. The expert witness must leap between the horns of this dilemma by admitting of knowledge, and even admiration, of the person or treatise proposed, while declining total approval or agreement. The expert can label the work as out-of-date, uneven in content, controversial, or erroneous in part. The expert must ask to read for himself (not aloud) quotations, sometimes incomplete or out of context, offered by opposing counsel. The expert witness should not hesitate to disagree with quoted "authority" within the security of his own knowledge and experience. After all, the expert witness is the only expert so far qualified as such by an adversarial procedure in court. He must not give away equal billing to an absent "authority" represented only by excerpts of written work.

TAKING ABUSE[1,4,5,20,21]

Rarely, and usually in desperation, the cross-examiner will try to tear down an expert and his prior qualification and testimony by abusive attack. He may speak harshly, omit titles, show disdain or anger or exasperation, use aggressive body language, gesture too closely or "get in the face" of the witness. He may wisecrack, be flippant, even insulting, in order to upset or goad the witness. It is risky business, but more so for the examiner than for the witness *if* the expert stays cool.

The expert witness, already empowered by qualification and acceptance as an authority figure, must maintain the persona and power already established; he must remain wrapped in a cloak of dignified professionalism.

The judge and jury may harbor protective emotions toward the witness under assault and reproach for the aggressor. Well-placed objections by sponsoring counsel will further tip the scales against the imprudent assailant. Attorneys have been sanctioned at the bar for abuse on cross-examination.[7]

However, if the witness reacts to the abuse, takes the bait, hits back, becomes an adversary, then he will lose his standing of impartiality with the jury, becoming just another fighter in an ugly fray. In this courtroom setting, meeting force with equal force will only put the attorney in control.

> Don't get into a mud-fight with a pig. You'll get mud all over you, and the pig will like it!
>
> **Judge Haskell M. Pitluck**

If cross-examination has been harmful to the expert witness and his client's case, there is one more opportunity for damage control, the *redirect examination* of the witness by the party calling him, after cross-examination. This affords an opportunity to correct possible misunderstandings by the jury, and allows reaffirmation of points that may have been blunted by opposing counsel. This will conclude the testamentary experience for the expert in this trial. He will be excused by the judge and will leave the courtroom.

JURY CHARGE

The expert witness's appearance and testimony can be effectively voided by the court. Standard jury instructions in various jurisdictions may go something like this:

> Expert witnesses are like any other witnesses, with one exception—the law permits an expert witness to give his opinion. However, an expert's opinion is only reliable when given on a subject about which you believe him to be an expert. Like other witnesses, you may believe or disbelieve all or any part of an expert's testimony.[4]

An expert witness's testimony carries great weight and influence in the jury room.[7] At the discretion of the court, relevant evidence can be excluded under the Federal Rules of Evidence, Rule 403 "If its probative value is substantially outweighed by the danger of unfair prejudice, confusion of issues, or misleading the jury ...."[13] Rule 403 has general application to all testimony, including expert testimony. "Expert testimony, like any other testimony, may be excluded if, compared to its probative worth, it would create a substantial danger of undue prejudice or confusion."[23]

THE VERDICT

In the purest sense in a perfect world, the verdict in the case just completed should be of no more interest than that of a trial scanned briefly in a newspaper or glimpsed on the evening news. After all, the expert is expected to be a scientist first and a forensic scientist second.[24] He is expected to be honest, competent, reliable, and totally objective and nonpartisan.[25] However, the adversarial system works against impartiality.

The forensic scientist who gets so far into a case that he becomes a witness, surely has opinions, however objective, that support the contentions of the side bringing him rather than those of the other party. The jury probably assumes he is a partisan of the side that brings him, and the opposing counsel may try to exploit this.[19]

If the expert expresses an opinion contrary to the cause of the party first contacting him, then his first consultation will be his last and he is out of the case. (He must not try to sell himself to the other side at this point.[11])

Even if brought as an expert witness by the court rather than by one or the other party, the forensic scientist is likely to lean more toward one side than the other as the case unfolds.

The difficulties of evenhandedness and ethical behavior in an adversarial legal system are well recognized but not readily solved.[13,24–27] The "'search for truth' in the context of the law is simply part of the process by which the goal, 'justice', is strived for. In science, on the other hand, truth is the goal."[25] The forensic radiologist is fortunate in that the evidence he works from, radiological image(s), lends itself to objectivity. The forensic radiologist then must exert every effort to remain a man of science, resolutely preserving scientific objectivity, regardless of unbidden emotion seeking to lead him astray.

> I will bear in mind always that I am a truth-seeker, not a case-maker; that it is more important to protect the innocent than to convict the guilty.[28]
>
> **Anonymous**

REFERENCES

1. Kuzmack, N. T., Legal aspects of forensic science, in *Forensic Science Handbook*, Saferstein, R., Ed., Prentice-Hall, Englewood Cliffs, NJ, 1982, chap. 1.
2. *Frye v United States*, 293 F. 1073 (D.C. Cir. 1923).
3. Bohan, T. L. and Heels, E. J., The case against *Daubert*: The new scientific evidence "standard" and the standards of the several States, *J. Forensic Sci.*, 40, 1030, 1955.
4. Henderson, C., *Jurisprudence: Science in court, presented (with handout material) at a Multidisciplinary Symp. Uses of Forensic Sci.*, American Academy of Forensic Sciences, Nashville, February 20, 1996.
5. Henderson, C., *Jurisprudence: Science in court, presented (with handout material) at a Multidisciplinary Symp. Uses of Forensic Sci.*, American Academy of Forensic Sciences, Seattle, February 14, 1995.
6. Helpern, M., *Autopsy*, Signet, New York, 1979, p. 66.
7. Garcia, C. H., Legal and ethical considerations in using expert witnesses in litigation, *Shepard's Expert and Scientific Evidence Quarterly*, 1, 717, 1994.
8. American Bar Association, ABA Model Rules of Professional Conduct, Rule 5.4 (1) 1993.
9. *Copp v Breskin*, 782 P. 2nd 1104 (Wash. Ct. App. 1989).
10. Anon., Quoting Lewis, P. R., Experts who recognize red flags avoid cases with traps and pitfalls, *Testifying Expert*, 3, 1, 1995.

11. Stimson, P. G., Rules of evidence and document preparation for expert testimony pretrial information; Expert testimony: Preparation, delivery and cautions, presented (with handout material) at Forensic Investigation of Abuse and Violence, 23rd Annu. FL Med. Examiners and 3rd Annu. Invest. Identification Combined Educational Conf., Pensacola, September 20, 1995.
12. Anon., Quoting Rothstein, P., Experts may have to disclose phone calls with case attorney, *Testifying Expert*, 3(5), 7, 1995.
13. Giannelli, P. C., Evidentiary and procedural rules governing expert testimony, *J. Forensic Sci.*, 34, 730, 1989.
14. Pitluck, H. M., *A bench-eye view of expert testimony, Forensic Science Update: Crime Scene to Verdict, 4th Annu. Invest. Identification Educational Conf.*, Pensacola, November 15, 1996.
15. *Lang v Marshalltown*, 185 Iowa 940, 170 NW 463, 1919.
16. *Marion v Coon Construction Co.*, 216 NY 178, 110 NE 444, 1915.
17. *Raleigh v Donoho*, 238 Ky. 480, 38 S.W. 2nd 227, 1931.
18. Baetjer, F. J. and Waters, C. A., *Injuries and Diseases of the Bones and Joints*, Paul B. Hoeber, New York, 1922; quoted in: Garland, L. H., Forensic skiagraphy, *Calif. Med.*, 87, 295, 1957.
19. James, A. E. and Hall, D. J., The law of evidence and diagnostic imaging techniques, in *Legal Medicine with Special Reference to Diagnostic Imaging*, James, A. E., Ed., Urban & Schwarzenberg, Baltimore, 1980, chap. 5.
20. Anon., Quoting Whitaker, E., The O.J. Simpson trial: The cross-examination trap, dealing with the aggressive lawyer, *Testifying Expert*, 3(6), 7, 1995.
21. Donaldson, S. W., *The Roentgenologist in Court*, 2nd ed., Charles C Thomas, Springfield, IL, 1954, p. 255.
22. Donaldson, S. W., *The Roentgenologist in Court*, 2nd ed., Charles C Thomas, Springfield, IL, 1954, p.153.
23. *United States v Schmidt*, 711 F. 2nd 595, 599 (5th Cir. 1983), *cert. denied*, 464 U.S. 1041 (1984).
24. Lucas, D. M., The ethical responsibilities of the forensic scientist: Exploring the limits, *J. Forensic Sci.*, 34, 719, 1989.
25. Petersen, J. L., Symposium: Ethical conflicts on the forensic sciences, *J. Forensic Sci.*, 34, 717, 1989.
26. Peterson, J. L., Forensic science ethics: Developing an integrated system of support and enforcement, *J. Forensic Sci.* 34, 749, 1989.
27. Frankel, M. S., Ethics and the forensic sciences: Professional autonomy in the criminal justice system, *J. Forensic Sci.* 34, 763, 1989.
28. Anon., Quotation on the wall of the Office of the Medical Examiner, State of New Mexico.

5 The Expert Witness as Viewed from the Bench

Haskell M. Pitluck

There are thousands of courtrooms in the U.S. judicial system and each one of them is presided over by a judge. While the basic principles of law are similar in all courtrooms, the business of justice is conducted in a manner determined by the judge presiding in that particular courtroom. The views from the bench of everyone, including experts, as seen from the eyes of the judge presiding will vary from judge to judge. It is important to understand that any observations or statements that I make in this chapter, I make *personally* and are mine alone, and that I can not speak for any other judge. However, in my many years of experience as an attorney and a judge, I have made some observations that I believe can be helpful to anyone appearing in court. Remember that these are my perceptions and should not be attributed to any other judge.

One of the many traits that judges have in common, however, is a desire to conduct the business of their courtroom in an efficient, expedient and fair manner insuring that justice is dealt fairly and equally to all. The medical profession too would concur that their goal is to help all patients in the best manner possible, and in that sense, the goals of both professions, medical and legal, are similar. However, when the two professions meet in court, conflict can result. Forensic science is the application of science to law, but the law's desire for exactness can sometimes conflict with the inexactness of science.

No matter what the case, the better prepared the litigants, the better the courtroom will run. A well-prepared expert can greatly assist the process. There is an old joke that the definition of an expert is someone who is over 50 miles from home and uses charts and graphs. Today you would have to add video and PowerPoint presentations to that definition.

The first step in litigation, in relation to the expert witness, is to determine whether an expert, whose job it is to assist the trier of fact in viewing and evaluating the evidence, is needed at all in court. Trier of fact refers to the jury in a jury trial or the judge in a bench trail. A bench trial takes place in front of a judge only, without a jury. If the reason the evidence is being introduced is to prove something that is easily apparent and understandable by the trier of fact, there is no need for an expert. Many times an attorney will try and use a prestigious or well-known expert to give more credibility to his case. Judges are usually alert to this type of behavior, and will insure that everyone plays by the rules. I once had an expert tell me "any idiot would know that." I told him if any idiot would know that, I did not need his testimony and I barred him from testifying in the case. He wasn't needed.

It is probably safe to assume that any case with important medical testimony would require that an expert, or experts, be involved.

A trial is much like a puzzle. All the witnesses have a piece to add to that puzzle until there is a complete picture. The fewer witnesses it takes to finish the picture, the easier it is for the trier of fact to sort out the issues.

If an expert is needed, and is called to testify, he has to be qualified as an expert. The judge is the one who determines if a witness is qualified to testify as an expert. There is nothing magical about it. Before a judge can make that determination he has to hear the expert's qualifications, or in some cases, lack of qualifications. This is called *voir dire* and is done by the attorneys asking questions of the expert who will be sworn under oath to respond truthfully. This will probably be the first time the judge has seen the expert in person, and depending on the circumstances, he too may ask the expert questions. I know of some experts who have given questions and answers highlighting their qualifications and experience before hand to the attorney who is calling them to testify in order to aid in the qualification process. This is especially helpful if the expert's curriculum vitae (CV) is voluminous.

Depending on the jurisdiction, the *voir dire* may be with or without the jury present in the courtroom. The reason the jury may not be present is to keep them from hearing possible improper material having nothing to do with what they must decide in the case. Attorneys for all parties may question the expert as to their qualifications. This is done before the expert gives any testimony or opinions about the case itself. If the judge feels the witness is qualified to testify as an expert and the testimony will assist the trier of fact in evaluating the evidence, the judge so rules and the witness may testify as an expert. This is not to be confused with pretrial hearings such as a *motion in limine* to bar the testimony for some legal reason. If the witness is not allowed to testify as an expert, the trier of fact would not get to hear the testimony and would never reach the question of the admissibility of the testimony or the weight to be given to the testimony if admitted.

If the *voir dire* was done outside the presence of the jury, and the witness was allowed to testify as an expert at trial, the jury will probably get to hear all the same questions and answers that they missed. The attorney will want the jury to hear the testimony of the expert's qualifications as it goes to their acceptance of the expert and his opinion.

The manner in which the expert handles the *voir dire* and gives testimony to the trier of fact could make or break the case. Most definitely, the style of the expert's testimony will make or break the perception and acceptance of the expert by the trier of fact.

Occasionally, an attorney may offer to stipulate to an expert's qualifications. This can be a two-edged sword. If an attorney stipulates to an expert's qualifications, it would be hard to argue that the expert is not qualified later in the trial. But when the attorney stipulates to the expert's qualifications, the trier of fact will not hear the extensive qualifications of the expert. For this reason, many times the attorney offering the witness will not accept the stipulations and will proceed with testimony outlining the credentials of the witness, knowing that the opposition has already indicated that they know the witness is qualified. This depends on the case and the reason for which the expert is there.

Obviously, for this process to be smooth and efficient, the expert witness and the attorney must be on the same page. The way to accomplish this is by thorough preparation. Despite the reputation of the attorneys and/or the experts, if they are not well prepared, the trial can go downhill rather quickly. The trial may then not proceed in a manner advantageous to their position.

Often times, attorneys will ask the judge for a sidebar, which is a discussion between the attorneys and the judge at the side bar of the bench. A side bar is a conversation that is not to be heard by the members of the jury. It usually concerns something the witness has said or is about to say. The courtroom is usually arranged so the witness cannot help but hear the conversation, and I am sure members of the jury try hard to hear the sidebars too! The witness should not react to the sidebar conversation, even though, at times, it may be hard not to do so. When sitting as a trial judge, I tried to keep sidebars to a minimum on minor matters. If something was important enough to be heard outside the presence of the jury, I would take the attorneys and the court reporter to make a record outside the court room so no one else could hear.

In this day and age, there is very little trail by ambush. Broad discover rules in most jurisdictions insure that, in almost all cases, all sides will know who the witnesses are and what their testimony will be prior to trial. The attorneys will be provided with a CV of the experts. Many experts have voluminous CVs that are often better reading than much of the case.

An expert should update and keep current and date his CV on a regular basis. A date of the last update lets everyone know how current the CV is. In conjunction with this, it should go without saying that an expert should know his CV. Oftentimes there may be conflicting or incorrect information on a CV which has to be explained before pertinent testimony can even begin. Not only would erroneous information be embarrassing to the expert as well as the attorneys, it can also diminish an expert's credibility before he starts his testimony.

One of my pet peeves is when an expert has many publications on his CV and Counsel will ask "Do any of these publications have anything to do with this case?" I have seen experts spend long and painful minutes reading over the list of publications in their own CV to see if any apply. It is bad enough when the expert finds some that do apply, but even worse when they find none. The trier of fact may start to wonder if the witness is really an expert or is qualified. A little preparation would save a great deal of time and embarrassment for the witness. With proper preparation, the attorney and the expert could be ready for the question and have an answer prepared. For example, "None of the listed publications have anything to do with this case" or "Yes, so-and-so publications involved the same subject matter as this case" are good responses. Either way, it indicates the expert knows his CV, and more importantly, knows the case—or at least appears to do so. With proper preparation, the witness can also advise the attorney of any pertinent publications, or if none, why that would not affect his testimony and opinion in the case.

An expert, in most cases, while assisting the trier of fact in viewing and evaluating the evidence, will also be called upon to give an opinion—an expert opinion—on their view of the evidence. This opinion should be based upon where the science leads the expert in making that opinion. If the opinion is based upon good science, and the attorney cannot attack the science, the attorney will have to attack the person giving the opinion—the scientist expert witness.

It is also important for the witness to recall if he has ever testified or written contrary to his opinion in this case and if he has changed their opinion, why he has done so.

The truth, credibility and any bias or interest of the expert is material and is always fair game for cross-examination. If the attorneys prepare properly, they will know everything about the expert and probably even have prior testimony and depositions given in other cases. In today's electronic age, it is easy for attorneys to obtain and use this information.

In conjunction with this, I have never been able to understand why an expert witness waives signature at a discovery deposition. I know why they do it but I do not understand why they do it. Waiving signature means that the expert will waive the right to read the transcript of the deposition and make any corrections before he signs off on it. Everyone is busy and there is no question that reading the transcript will take time, but it is time well spent. Court reporters are human and make mistakes in transcription, even with electronic recording. Furthermore, statements may be taken or said out of context and, when read by the expert, easily seen and explained. Finally, it is good practice. The more transcripts an expert, reads, the more he learns. He learns that people do not always speak in complete sentences or he will learn how to better express himself when testifying. With the statement "We will waive signature," the appearance is that the attorney and expert are being gracious and saving everyone time. However, if you, the reader, are an expert, I do not care how busy you are, do not do it. If you do not waive signature, you not only get to read your testimony in the deposition, you also get to make corrections to insure that your answers are properly conveyed. This only applies to discovery depositions. An evidence deposition is one used at trial to present the witness's testimony. It is used when a witness will not be available at the time of the trial and the attorneys want to use his testimony. A witness in an evidence deposition will be subject to cross examination and objections, just as if he were testifying in person at the trial.

The ultimate use of an expert after he has assisted the trier of fact to evaluate the evidence is to give an opinion as to identification or condition. A person can be the greatest scientist or expert in their field, but if he cannot convey that opinion to the trier of fact in a meaningful way that can be understood, the findings are not worth anything and will not add much to the case. He has to say what he means and mean what he says.

Experts convey their opinions in their own style and the opinions can be all over the board. Some examples of opinions used by witnesses include: match, possible, probably, cannot exclude, on the money, are the same, no question about it, no way, one out of (whatever number), including the qualifying buzz words "to a reasonable degree of medical or scientific certainty" (whatever that is)! Another concern may be that one expert's probable may be another expert's possible; another's 'match' may be 'consistent with' by another's determination. You get the idea. There is no standard for the result of the expert's opinion. Even though some professional organizations are trying to get their members to be more consistent in their opinions, it is difficult to tell experts how to testify in a uniform way.

Whatever their opinion, they will probably be cross-examined extensively on what caused them to reach their opinion. Or, as I like to think about it, what brought them to that point of personal conversion when they decided what their opinion was based on the scientific evidence. It is a little like the Supreme Court justice who once stated that he could not define pornography, but he knew it when he saw it.

There are experts and there are experts. Experts can be plotted on a Bell curve. Most of them will be in the middle. They are well trained, competent and reputable—even though they may, at times, have differences of opinion. The ones an attorney has to be careful of are the ones who are at either end or on the fringes of the Bell curve. There are always some who can see what no others can see and have theories to which there are no other subscribers. They are the true hired guns who often border close to the line of junk science.

I am always leery of the expert who has all the answers. I will give more credence to the expert who sometimes does not know the answer or cannot recall something. I also have a small rule of thumb that the more letters a person has after his name, the closer I will scrutinize his qualifications and testimony.

It is important for experts to remember that even though they have been called to testify by one side or another, their job is to assist the trier of fact in understanding and evaluating the evidence and for the truth. I know this may be difficult at times as it is human nature to become involved or sympathetic to the side who hired them and where they have spent the most time, effort and energy. The expert has to explain to the party who hired him that he will go where the science takes him and that they may not like the opinion. Of course, in that case, he may not be called to testify, but he will have maintained his honesty, ethics and integrity and will live to work another day. There is no substitute for honesty, ethics and integrity and as a professional and a person, the expert should not compromise them. It bothers me when I hear experts say "We won that case" or "We got that person off." They were there for the truth and the trier of fact made the decision.

I have stressed that there is no substitute for good preparation. If you are testifying as an expert, there are some things it is important to remember. The old saying, "When in Rome, do as the Romans do," can also apply to expert testimony. Check with the attorney who hired you regarding proper dress. Different parts of the country, and even different judges, may have different customs. Being well groomed and wearing conservative business attire is usually acceptable. This is especially important because you are what you say and how you look to the trier of fact. If they do not understand what you say or how you say it, how you look can become even more important to the acceptance of your testimony.

You should also check with the attorney as to things you should not say or get into. I do not mean that anyone should tell you what to say, but, for example, in a criminal case, if a defendant is advised of his Miranda warnings and exercises the right to remain silent, that cannot be mentioned before the jury. It would be grounds for a mistrial and the whole trial would have to start over. Likewise, in a civil case, in most jurisdictions, you cannot mention insurance. This would also be grounds for a mistrial, even though I am sure it is one of the first things jurors talk about during deliberations on the case.

Do not talk down to the judge or jury. Use technical terms for the record, but explain them in plain English to the trier of fact. It is very helpful if an expert has a list of correctly spelled technical or medical terms used in their testimony to provide to the attorneys, the judge, and more importantly, the Court reporter. The expert can provide this list to them in the courtroom just prior to testimony. This list will win you points with all concerned and make you appear as a true professional to the trier of fact. Be human. If something is funny, do not be afraid to smile or even laugh. If you do not know the answer, say you do not know. If you cannot recall, say you cannot recall. If you need to refresh your memory, say so, and the attorney will know what to do.

After you finish your direct testimony, you will be cross-examined. This is when the opposing attorney will try to negate or impeach your testimony. There is nothing magic about impeachment. No lights flash or sirens blow when it happens. But you can be sure the attorney will make much of it in final summation or argument to the trier of fact. As stated earlier, almost no holds are barred in cross examination of your testimony. Any interest, bias or prejudice you may have as well as the manner in which your opinion is stated are all open to question during cross-examination.

In addition, leading questions are allowed. Do not lose your composure. Many times an attorney will try to make you answer yes or no to a question that cannot be answered with a simple yes or no. Do not be afraid to tell the judge that you cannot answer yes or no. Most of the time, the judge will understand, but unfortunately, sometimes that is not the case. The attorney who hired you should also be objecting at the same time. An example of such a question is: "Is it true or not true that you did A to B?" There is no way that question can be satisfactorily answered yes or no. Do not let anyone put words in your mouth. Do not exceed your expertise. If you made a mistake, say so. Hopefully, you can explain it.

There are two telling questions on cross examination that are good indicators of your expertise and your preparedness. First, "Did you talk to the attorney before you testified here today?" If you hesitate in answering that question, I know you have not properly prepared. Of course you spoke with the attorney. If you have not, you have not done your job. Who would want to call a witness without knowing what they are going to say. It is perfectly all right to talk to the attorney and in fact, it is mandatory that you should. The important thing to get across to everyone is that the attorney is not telling you what to say. Just tell the truth. If your answer is yes right away, the cross-examining attorney will probably drop that line of questioning as he knows you are prepared. If you hesitate or say no, the cross examiner will keep going. Most likely, he will embarrass you and affect your credibility to the trier of fact.

The second question is "How much are you being paid for your opinion?" You are NOT being paid for your opinion. You are being paid for your time to be in court to testify, your knowledge, your experience and your reputation. Your opinion is part of your testimony. Try to make it clear that no matter who is paying you, it does not affect your testimony or your opinion. If money has an effect on your testimony, you do not belong in the business. After stating that you are not being paid for your opinion, the cross examiner may move on. But he may continue inquiry into your compensation to see if it contributes to bias, interest or prejudice. Whatever you do regarding compensation, *do not ever* under any circumstances, take a contingent fee based upon the outcome of the case for your compensation.

Depending on the situation, you may have to identify in court a party whom you have examined. If would be nice if you could identify them. This may not be easy. Sometimes years may pass between your initial involvement in the case and when you are called to testify. If you do not have a photograph from your meeting with the individual, you should check with the attorney to help you refresh your memory to make an identification. You can say you have many cases and you cannot remember everyone. Explain why you can or cannot indentify the individual. If you cannot, there had better be something to tie you to the individual and the case.

At times the opposing counsel may show you something in court that you have never seen before and ask you to look at it for an opinion. Do not make snap judgments and do not get caught in that trap. If fact, it is a perfect opportunity for you to score points. You can respond with, "Counsel, I cannot make any judgments or form an opinion by looking at this in Court. I have never seen this material previously. I spent hours going over the evidence in this case in my office (or laboratory) and would have been happy to do the same with this material if you had given it to me in advance." The trier of fact can sympathize with this and will wonder why if it was so important you were not given it in advance. Such a scenario makes you, the expert, look good as the complete professional and the attorney who asked the question appears to be grandstanding and unprofessional.

When you are testifying, do not talk too much. Listen carefully to the question asked and answer just that question. Depending on the jurisdiction, you most likely will not be in the courtroom when other experts testify as some jurisdictions sequester witnesses and only allow them in court when they are testifying. Do not assume that someone is an expert just because they say they are. Sequestration is an old rule intended to keep witnesses from constructing their testimony based upon what another witness says. With today's liberal discovery, all sides usually have a good idea what the testimony of the witnesses will be. So the old rule might be outdated. However, you still get some experts with different stories who are caught in inconsistencies. I once had an attorney who was watching a trial in my court room tell me that with all the "testiphony" (sic) I was going to have to decide the case on a preponderance of the perjury!

In addition to what I have said regarding your courtroom experiences, there are some things that should be so obvious they barely deserve mention, yet they have a great deal to do with how you are perceived and accepted. Do not chew gum or candy. Turn off all cell phones, pagers, and electronic devices. Do not be argumentative, disrespectful, or pompous. Speak up so everyone can hear you.

There are two final concerns of which you should be aware. The first is the CSI effect. Television shows, such as CSI, while they may be good entertainment to some, have also given the public some unreasonable expectations of what forensic science can and cannot do, including solving a case in less than an hour. The public, from whom juries are chosen, may wonder why certain tests were not conducted and why other investigative steps were not taken. Attorneys and expert witnesses have to be aware of some of these concerns and deal with them in their presentation of the case so the trier of fact decides the case on the evidence presented, and not what they think ought to have been presented.

Last, but not least, expert witnesses used to have almost absolute immunity in regard to actions upon their work and testimony. This is changing. If a witness intentionally misrepresents his reports or testimony, he may be sued or charged criminally. In some cases, he may face actions for negligent behavior. It depends on the jurisdictions again, but the legal climate of absolute immunity for experts is changing, as it should. Experts should be held responsible for improper behavior. Experts should take every case seriously so they can do the best job possible for all parties to receive a fair and impartial trial. Remember that we seldom get a perfect trial, but we should all strive to have the best trial possible.

6 The Radiologist in the Courtroom Witness Stand
Good, Bad, and Indifferent

Leonard Berlin

CONTENTS

Man, proud man,
Dressed in a little brief authority,
Most ignorant of what he is most assured.

Shakespeare, "Measure for Measure"[1]

INTRODUCTION

Radiologists are called to the witness stand to testify in a courtroom trial for one of three possible reasons. Radiologists may be defendants, in which case they are usually called to testify involuntarily. Radiologists may be fact witnesses, in which case they are usually called involuntarily via a court order or subpoena. Or, radiologists may be called to testify voluntarily as expert witnesses. This chapter will focus on the conduct of radiologists while testifying on the witness stand, giving examples of testimony that can be characterized as desirable ("good"), undesirable ("bad"), or neutral ("indifferent"). Although when testifying radiologists should adhere to a certain basic standard of conduct regardless of whether they are defendants, fact witnesses, or expert witnesses, nonetheless, witness expectations can differ depending on whether the radiologist witness is there to defend himself, simply to recite facts and incidents, or to offer opinions regarding breaches of the standard of care.

THE DEFENDANT WITNESS

The radiologist-defendant is almost always required to be in the courtroom and to testify on his own behalf. He is usually questioned first by his defense attorney, whose purpose is to support and bolster the position of the defendant. The radiologist is then cross-examined by the opposing plaintiff's lawyer, whose purpose is to make the defendant appear culpable for the alleged negligent conduct with which he has been charged in the lawsuit. The defendant-radiologist has one goal in mind: to attempt to convince the jury that he is innocent of wrongdoing.

THE FACT WITNESS

The radiologist who is subpoenaed by the court to testify regarding medical information known to him, usually consisting of radiological reports he has authored, is called to the witness stand to simply and honestly recite what he did and what he said or reported. The radiologist is not there to attempt to convince the jury that his testimony is necessarily accurate, nor is he there to offer any opinions about whether a defendant did or did not adhere to the standard of care. As a California Appellate Court has stated, "A doctor who has treated or examined a patient may be compelled to testify as an ordinary witness and answer pertinent questions concerning facts relating to the patient's condition, knowledge of which he acquired through his examination and treatment of the patient."[2] Thus, if a radiologist is subpoenaed to testify as a fact witness, he must comply and appear. Failure to do so could subject him to a contempt of court citation.

THE EXPERT WITNESS

The radiologist who agrees to serve as an expert witness does so voluntarily: "There is no duty on a doctor's part to agree to serve as an expert witness for one with whom he has no pre-existing contractual relationship."[2]

There can be situations when the distinction between a fact witness and an expert witness is not clear, particularly if there is a conflict regarding a witness fee which is ordinarily paid to an expert witness but not to a fact witness. Such was the case in a New York lawsuit, where a physician-witness claimed he was an expert and demanded payment of a fee, but the lawyer who retained him asserted the physician was merely a fact witness.

The New York Court ruled as follows:

> The distinction between a lay witness' and an expert witness' opinion is that a lay witness provides an opinion resulting from a process of reasoning familiar with everyday life, while an expert testimony results from a process of reasoning which can be mastered only by specialists in the field. If the opinion provided is based upon scientific, technical, and other specialized knowledge, then the witness is deemed to be an expert.[3]
>
> When the doctor's opinion testimony extends beyond the facts disclosed during care and treatment of the patient and the doctor is specifically retained to develop opinion testimony, he or she is considered an expert witness. Whether a report must be prepared or whether the expert is designated by the attorney as an expert witness does not control whether a witness is an expert. Rather, it is the substance of the testimony that controls whether it considered expert or lay testimony. Upon review of the transcript submitted here, there are questions asked by the attorneys that could reasonably be interpreted as calling for opinion testimony. Further, such opinion testimony would require specialized knowledge, skill, experience, training and/or education. As such, parts of the testimony can fairly be characterized as falling within the peer view of expert testimony and accordingly, attorneys are obligated to pay a reasonable fee for the expert's time spent responding to discovery.

SHOULD RADIOLOGISTS VOLUNTEER TO BE EXPERT WITNESSES?

Although radiologists and other physicians are not compelled to testify as an expert witness, nonetheless physicians are *encouraged* to do so. *The Code of Medical Ethics* of the Council on Ethical and Judicial Affairs of The American Medical Association (AMA) states that because the physician is a professional with special training and experience, the physician has "an ethical obligation to assist in the administration of justice."[4] In a similar fashion, the American College of Radiology (ACR), in its *Digest of Council Actions* 1999–2008, points out that "[i]t is in the public interest that medical expert testimony [by radiologists] be readily available."[5]

Once a radiologist has agreed to assume the role of an expert witness, guidelines are available that will assist the radiologist in comporting himself in a professional and ethical manner while on the witness stand. We shall discuss these briefly.

GUIDELINES FOR EXPERT WITNESSES

The AMA

The ethical codes published by the AMA and the ACR go into some detail regarding how physicians should conduct themselves when offering expert testimony. The AMA states that medical experts "should have recent and substantive experience in the area in which they testify ... and should testify honestly and truthfully to the best of their medical knowledge .... The medical witness must not become an advocate or a partisan in the legal proceeding .... The attorney for the party who called the physician as a witness should be informed of all favorable and unfavorable information developed by the physician's evaluation .... It is unethical for a physician to accept compensation that is contingent upon the outcome of litigation."[4]

The ACR

The ACR not only affirms those guidelines but adds even more specific language: "The radiologist expert witness must maintain an active practice of radiology and should be certified ... by the American Board of Radiology [or comparable organization]."

The ACR also admonishes that radiology expert witnesses "should be familiar with ... and actively involved in the clinical practice of the subject matter of the case for three of the previous five years at the time of the testimony." According to the ACR, the radiology expert witness should review the standards of practice prevailing at the time of occurrence and also should be aware that transcripts of depositions and courtroom testimony are public records and thus subject to independent peer review. As a final caveat, the ACR also reminds its members that "[a]n individual holding an official capacity with the College who gives evidence for use in litigation must exercise great care to distinguish between his or her personal opinion on the merits of the matter at issue and the policy positions of the College ... and must not state or imply in a written opinion or deposition or trial testimony that he or she is speaking as a representative of the College or is testifying to the views of the College on the merits of a particular case."[5]

Still another ACR document, the *ACR Practice Guideline on the Expert Witness in Radiology and Radiation Oncology* reiterates these precepts.[6] All radiologists should acquaint themselves with these documents. The ACR also refers to expert witness testimony in the *Code of Ethics* section of its 2008–2009 bylaws: "In providing expert medical testimony, members should exercise extreme caution to ensure that the testimony provided is non-partisan, scientifically correct, and clinically accurate."[7]

The Court's Perspective

As already mentioned, to render standard-of-care testimony on behalf of or against a medical practitioner, the expert must be scientifically or medically qualified. An Illinois Appellate Court has expanded on this as follows:

> [For an expert to be medically qualified], a two-prong showing must be made. First, the expert must be a licensed member of the school of medicine about which he or she proposes to opine—the "licensure" prong. Second, the expert must be familiar with the methods, procedures, and treatments that similarly situated physicians as the defendant would ordinarily observe—the "familiarity" prong. These are the "foundational" requirements and form a threshold determination. If this threshold determination is not met, the analysis ends and the trial court must disallow the expert's testimony .... Once the foundational requirements have been met, the trial court has discretion to determine whether a physician is qualified and competent to state his opinion as an expert regarding the standard of care .... It is insufficient for a plaintiff to merely present that "another physician would have acted differently from defendant," since medicine is not an exact science. Differences in opinion are consistent with the exercise of due care.[8]

STANDARD OF CARE AND BREACHES THEREOF

Determining what the standard of care is in a specific case is not an easy task. Most people presume that physicians are capable of determining with pinpoint accuracy whether in a given case a defendant colleague met the standard of care, but this presumption has little basis in fact. Studies show that there is very poor agreement and reliability among physicians in identifying adverse events when receiving medical records, much less determining whether those adverse events did or did not constitute negligence.[9,10]

As for the obligation of an expert witness to opine whether there was a breach of the standard of care committed by a given physician, not only is disagreement among medical expert witnesses themselves a well-accepted fact within the medical community, but disagreement among witnesses is accepted by the courts as well. An Illinois Appellate Court stated the following:

> People qualified in their field stated their views and gave their reasons for these opinions. Not so surprisingly, the plaintiffs' expert did not agree with the defendants' experts—not an unusual situation. It was the jury's job to listen to the conflicting evidence and use its best judgment about where the truth could be found. This is what juries do best.[11]

Another Illinois Appellate Court decision expanded on this point:

> What we have here is a case of dueling experts. When the experts conflict, the jury is expected to resolve the conflict by evaluating the relative merits of the experts and their opinions, and decide the weight to be given to their testimony.[12]

Added another court:

> Where the parties offer conflicting medical testimony regarding the applicable standard of care and the defendant's breach of that standard, the jury is uniquely qualified to resolve the conflict. It is the function of the jury to weigh contradictory evidence, judge the credibility of the witnesses, and draw the ultimate conclusions as to the facts of the case.[13]

THEORY VS. REALITY OF WHAT IS EXPECTED OF THE EXPERT

Notwithstanding the "theoretical expectations" of expert witnesses called for by the AMA and the ACR, the "realistic expectations" called for by the plaintiff's and defense attorneys "in the real world" can be somewhat different. This point is illustrated in the following "transcript" of a telephone call:

> Hello, Doctor. My name is Sue Barrister, and I am an attorney who is seeking your assistance in a medical-legal matter. I represent one of the parties in a medical malpractice lawsuit in which a radiologist is accused of missing a lung lesion on a chest x-ray, resulting in a 1-year delay in diagnosis of lung cancer. The patient is still alive but has metastatic disease.

Let me explain what I would like you to do. I will send you the chest x-rays with the corresponding interpretation rendered by the defendant-radiologist and pertinent medical records. I ask that you review this material in an objective and unbiased manner, and then telephone me with your opinion as to whether the defendant-radiologist breached the standard of care by missing the chest pathology. Do not put anything in writing. If your opinion does not support the position of my client in this case, you may send me a bill for your time but I will not be requiring any further assistance from you. On the other hand, if your opinion does support the position of my client, I will need you to appear for a deposition and also at a trial if one is to be held. For your deposition, the opposing attorney and I, along with a court reporter, will come to your office. If there is to be a trial, you are expected to come to court to testify. We will, of course, compensate you for your time and pay for your expenses. In my experience, radiology expert witnesses charge anywhere from $300 to $700 per hour for reviewing cases and depositions, and from $5,000 to $12,000 per day for court testimony. Of course, under no circumstances should potential remuneration sway you in any way or influence your opinions. However, if I find that you are an effective witness and have been helpful in convincing the opposing attorney and the jurors to reach a resolution of this case favorable to my client, you can expect that I will be calling upon you again in future cases.

One more thing. I want to explain to you that testifying about your professional opinions in this case during a deposition or trial is quite unlike your everyday practice of discussing radiological findings with your referring physicians, radiological colleagues, or residents. In the latter scenarios, you tend to discuss differential diagnoses and probabilities, frequently using such words as "maybe," "possibly," "somewhat more or less likely," "perhaps," and "I'm not sure." These terms are not to be used in a legal proceeding if you are an expert witness. In such proceedings you must be definite in your opinion, and cannot be "wishy-washy" or "hem and haw." Furthermore, let me remind you that when you are talking to medical colleagues, they may gently debate with you about your radiological opinions and may even offer some polite counter-arguments. In the courtroom, however, be prepared for the likelihood that the opposing lawyer may not only question the validity of your opinions, but may harass you, intimidate you, and/or attempt to discredit you. If this occurs, I expect you to hold firm to your opinions. Keep in mind that during direct and cross-examination by attorneys, there is little value in expert witnesses expressing "gray" opinions; opinions must be "black" or "white."

So, Doctor, would you care to assist me in this matter?

The above telephone monologue never occurred. Although it is hypothetical, in all likelihood variations of it have taken place many times.

EXPERT WITNESSES AND ETHICAL ISSUES

It should now be easy to recognize how radiologists who desire to become expert witnesses can quickly be confronted with ethical issues. Remaining completely impartial and unbiased, which is required in theory, can in the reality of the adversarial environment of a courtroom trial present an ethical dilemma for many physician expert witnesses. As the then-editor, George Lundberg, of the *Journal of the American Medical Association* (JAMA) editorialized a decade ago, our legal system "flourishes as an expression of the aggressive nature of our society."[14] In theory, all physician-expert witnesses "seek truth and, while under oath in court, must speak truth," pointed out Lundberg, but he then observed, "It is very difficult for any person who has grown up on a system of 'sides' on the football field or in the debating hall not to share an advocacy position when his or her services have been secured by counsel committed to such advocacy."

In a later JAMA article, two medical legal commentators agreed that in the real world, complying with the ideals put forth by the AMA and the ACR and achieving their "lofty goals" are difficult. The commentators pointed out that the everyday practice of medicine is considerably different from the giving of medical testimony in court.[15]

> Unlike the cooperative setting in medicine where all parties are seeking an accurate diagnosis and appropriate treatment, medical testimony in the courtroom exists in an adversarial setting in which each party seeks to present its strongest case. By its very nature, this clash invariably leads to conflicts among experts representing both sides.

It is thus unavoidable that in the midst of a highly charged and hotly contested legal adversary proceeding, the physician expert witnesses retained by the plaintiff's attorney and the defense attorney will disagree—sometimes forcefully, if not bitterly.

EXPERT WITNESSES ARE APPRISED OF THE "FACTS OF COURTROOM LIFE"

Explaining to potential expert witnesses what they may face when they sit in the courtroom witness stand was portrayed very well in two unrelated nonfiction books. In the first book, *A Civil Action*, author Jonathan Harr focused on the book's protagonist, a plaintiff's attorney named Schlichtmann who was involved in litigation against an industrial firm for allegedly polluting the environment. Portraying the education of a potential medical expert witness by the attorney who retained him, Harr wrote[16]:

> Schlichtmann liked inexperienced witnesses. They were generally impressionable enough to follow his advice and listen to his warnings. They might get frightened under cross-examination, but Schlichtmann felt this was no handicap. The honest, scared witness is best, he once explained. They exude honesty when they are nervous. Judges and juries love them .... The witness was getting a final lesson in witness-stand etiquette. "You will never, ever, put your hands to your face. Just keep your hands clasped in front of you. Don't leave them free. Do not cross your arms. It looks hostile and withdrawn. Do not slouch. Sit on the edge of the chair. It keeps you alert. And keep your eyes on who ever is talking. When you are being cross-examined, you never look at me .... They are going to ask, 'Are you being paid for your testimony?' You answer; 'You are being compensated

> for your time.' Do not wear a flashy coat or tie. I want you to wear a conservative suit." ... Schlichtmann had to convince the medical experts that the rules of evidence were different from the rules of science, that the witness stand was no place to express doubt or uncertainty. They would have to understand that the other side would try to make them say things that could be used against them later in the courtroom.

The second book is *Damages,* and deals with a lawsuit alleging obstetrical malpractice resulting in severe birth injuries to a newborn. In it, author Barry Werth described the frustration and exasperation experienced by a pathology expert witness, named Benirschke, who specialized in placental pathology. Werth wrote[17]:

> Benirschke quickly discovered that law wasn't medicine. Law was based on impression, not data; lawyers are prosecutorial, not collegial; conferred more value on questions than on answers. It was debating, not science. No matter how certain he was of the pathology in a given case, he was regularly dismayed by how the lawyers drew their own conclusions, as if they hadn't heard him .... He was disgusted. All he'd wanted to do was to say what, medically speaking, probably had occurred. But the lawyers don't really want the truth; they just want to bend the truth to their party's interest. They were conspirators against the truth, asking narrow, partial indirect questions from which they had hoped certain assumptions would be inferred, and then objected when the other side made the least effort to inquire further. They never hear what you want to say, that's what hurts so much .... The jury would never hear Benirschke's opinions. In a courtroom, unlike a morgue or a lab, truth emerges not from the accretion of carefully collected information but from a paradox: the only questions answered are those that are asked, yet the only ones asked are those to which the answers are already known. Benirschke had hoped to tell the jurors what he thought about the issue of blame. No one's views were more authoritative or crucial to their ability to reach a proper reckoning, he thought. But the lawyers joining preemptively to edit him, combined to bar him from addressing the subject.

EXPERT WITNESS TESTIMONY: THE GOOD, THE BAD, THE INDIFFERENT

At this point let us look at various excerpts taken from actual transcripts of expert witness testimony. I shall arbitrarily categorize them as "bad, indifferent, or good," in that order.

THE "BAD" TESTIMONY: MAMMOGRAPHY

The following is an excerpt from testimony given by a radiologist presented by an attorney as an expert witness in a lawsuit filed against a radiologist and a hospital, alleging that a patient's breast was injured while undergoing mammography. The reader can ponder whether the "expert" was truly an expert.

Q: What do you, as a board-certified radiologist, do when mammography is being performed on patients?

A: We make sure the tech is performing the procedure properly, we do spot-checks and audits on the tech, make sure positioning is right, the films are placed properly, the film plates are clean, and the developer is functioning normally.

Q: This is your usual activity as a board-certified radiologist?

A: If my daily rotation is mammography, I read films, and in between reading them I periodically go into the room and look at the machinery, making sure that the patient is being positioned properly and the machinery is functioning properly.

Q: You are trained [in] how the machinery works?

A: Yes, in my residency we actually did several mammograms on patients to learn how it's done, what is appropriate for the patient, and what all the bells and whistles do.

Q: You're aware, Doctor, that among the issues in case is whether or not the technologist while performing the mammogram injured the patient by deviating from the standard of care. As part of your training as a radiologist, have you ever been certified to perform a mammogram?

A: Yes. As part of my MQSA [Mammography Quality Standards Act]; certification I'm certified to do mammograms.

Q: But, your MQSA certification has to do with interpretation of mammographic films, does it not?

A: That's one element of it. The MQSA, the ACR, and the FDA look at all aspects of mammography. One of the elements is that you meet certain reading. requirements, but the radiologist also has to demonstrate a certain level of ability to perform the technical functions of mammography.

Q: But, when you were deposed, you indicated that you [had] never performed a mammogram, true?

A: Correct, not by myself. But I did as a resident. There was a one-month period where I helped perform mammograms with the technologist.

Q: So, if I understand your testimony, you are saying that as a licensed diagnostic radiologist, you perform mammograms.

A: I don't perform them now; I mean I have done it in the past.

Q: Doctor, I understand you have a medical opinion that a breach in the standard of care during the taking of the mammogram on the plaintiff caused injury to her breast. Could you explain?

A: Well, the swelling of the breast indicates that the lymph channels and veins, even small veins, were likely damaged during the exam and as a result the breast can't drain the fluids properly. The pain may be related to underlying nerve damage, because there are small nerves also throughout the breast. And there were likely some muscular contusions also explaining the pain.

Q: Doctor, have you had discussions in your radiology practice with your peers about whether patients can sustain lymphatic injury within a breast by mammogram compression?

A: Yes, it's been discussed at mammogram meetings and seminars.

Q: Doctor, in your practice now, do you perform or interpret mammography?

A: I do not.

Q: Are you MQSA certified?

A: Currently, No.

Q: Doctor, do you consider yourself qualified to offer opinions with respect to chest wall nerve injuries?

A: No I'm not a neurologist or a neurosurgeon or a chest surgeon. But as a radiologist, I understand the anatomy and what happens when you get a mammogram and a potential injury that can result from it. I know nerves are there.

Q: Now, you have previously testified that it's your opinion that the patient's injury occurred because her breast was crushed, isn't that what you said?

A: I used that word, yep.

Q: If her breast had been crushed during the performance of the mammogram to cause the kind of injury you testified you believed it caused, would you have expected her breast to at least have redness?

A: Yeah, there should be some, I guess.

Q: Wouldn't you also expect with this kind of injury there would be some bruising of her breast?

A: It depends. A bruise results when the external part of the skin and the blood vessels in that area are damaged and then what you are seeing is the blood underneath the skin. If somehow the blood never made it to the surface or was somewhere else, you may not see it.

Q: Doctor, you've testified that the patient suffered lymphedema as a result of the mammogram. If someone were to develop lymphedema as a result of a crushing pressure on their breast, in what timeframe do you believe the swelling of the breast would occur?

A: Hard to tell. Say, for example, an air conditioner or fan was not working properly and it just gradually kept getting worse. The air conditioner may work for a while and then over time it just keeps getting worse and worse. So it's this kind of thing. It's hard to predict really. I mean it could function for a little while and then gradually get worse and worse.

Q: The complaint here, Doctor, is not about air conditioners, it's about a woman's breast. Can you give me an answer with respect to your opinion concerning when the breast would begin to exhibit observable changes that you would attribute to lymphedema?

A: That's why I say, I can't predict. I mean it could be one month, six months, eight months, I don't know.

Q: Doctor, how many women have you treated during your medical career for lymphedema?

A: Well, I'm a radiologist so I don't treat any.

Q: When did you gain any knowledge about lymphedema and the process that causes that condition?

A: Medical school. I don't really know exactly when. I mean, you know, you learn about it a lot. Lymph is all over the body, you know, so just about every disease has some interplay with the lymph tissue. So I, you know, deal with it almost on a routine basis.

Q: But you've never treated anybody for it, true?

A: Right.

Q: Where did you obtain knowledge that lymphedema can result from the performance of a mammogram?

A: In my training.

Q: Do you have any recollection as to who taught you that lymphedema can result from performance of a mammogram?

A: Well, you know, we had a lot of people who lectured to us, and I don't remember exactly who.

Q: But it would be fair to say that this is something that you recollect from your residency training years ago. Correct?

A: Definitely.

THE "BAD" TESTIMONY: ANAPHYLACTIC SHOCK

The following is an excerpt from testimony given by a radiology expert witness in a lawsuit filed against a radiologist and an imaging center alleging negligence for failing to treat a fatal anaphylactic reaction following administration of contrast media.

Q: Doctor, let me ask you about the criticisms you have already expressed in writing. Your first one states that "[i]t was below the standard of care not to give the patient a choice between ionic and non-ionic contrast material."

A: That is correct, the choice was not given to the patient.

Q: Have you ever read any of the studies in the literature about the difference in severe adverse reactions such as death between ionic and non-ionic material?

A: Not that I recall.

Q: Have you ever read any data on contrast media that indicate that the rate of death from non-ionic contrast is higher than the rate of death from ionic contrast?

A: No, but my long term experience with both contrast media is that we see much less serious reactions with non-ionic media.

Q: Is it your position that the medical literature supports that opinion?

A: The literature I quote does. The largest literature, mainly from Australia and Japan, supports my opinion.

Q: Do you have the names of any of the authors of those studies?

A: I don't recall.

Q: What should people undergoing contrast injections be advised?
A: I think patients should be told there are newer contrast media available that are more expensive but with less incidence of reactions. Patients should be offered an alternative.
Q: So your first criticism of the defendant radiologist is failure to give the patient a choice?
A: Yes.
Q: The next criticism is, "A patient should have been monitored closely and intubated with an endotracheal tube much sooner than he was intubated." Do you have any training on intubation?
A: Yes.
Q: Where did you receive your training?
A: I received my training over many, many years throughout my entire practice. Initially as a medical student and then as an intern, I was taught about resuscitation.
Q: Is there anything in the records that indicates that the defendant-radiologist fell below the standard of care in exercising the judgment not to intubate this patient on the site?
A: My only information is that the patient did not survive, and that raises the question of whether the judgment was correct.
Q: Your next criticism states, "The standard of care for responding to adverse reactions to contrast media includes the intravenous administration of three medications, which should include an antihistamine, an h2 blocker such as Tagament, and a corticosteroid." Why is that the appropriate way to respond to this adverse reaction [to] contrast material?
A: If the reaction is a histamine reaction, which is probably the most common reaction, it requires two agents to block it. The antihistamine is one. In other words, an h1 blocker. The h2 blocker such as Tagament or Pepcid blocks the h2 site, and to stop a histamine reaction you have to block both sites. The corticosteroid is the drug that blocks the cascade which follows the reaction and stops it. That is standard treatment.
Q: What is in the record that indicates to you that this patient had a histamine reaction?
A: The very fact that they gave Benadryl would indicate that they must have thought the patient was having a histamine reaction, but yet they did not do an injection of an h2 blocker, which therefore would be an incomplete treatment for what they were seeking to block.
Q: Is there anything inappropriate in giving epinephrine to the patient?
A: Yes there is. Epinephrine is very dangerous drug in a very elderly patient such as this 70 year old. It produces a very rapid heart rate and produces constriction of the blood vessels throughout the body, increasing the stress on the heart, and [it] will also constrict the coronary arteries. A patient who has heart disease and is elderly can be killed with epinephrine. Epinephrine is a drug that you use only as a last choice, after you've tried the other drugs, in order to prevent the demise of the patient,
Q: If you have a patient who is not having established blood flow, is it appropriate to give him epinephrine to try to establish the blood flow?
A: Not as the first drug because you are going to produce cardiac arrest and arrhythmias by doing so.
Q: Any other criticism regarding the response to the adverse reaction to the contrast media other than what you have already outlined?
A: The crash cart must have the proper equipment available immediately. It is my understanding that the suction device did not work due to the fact that the personnel who had cleaned the suction device had not reassembled it properly. And so, we have a problem with the personnel not being properly trained and properly reassembling it. So we have a deficiency, in that somebody who was inspecting that crash cart should be testing the suction device and make sure it is working.
Q: Are you able to state that had the suction equipment been operating, that this patient would have survived?
A: I don't know that he would have survived, but he would have had a chance if they had properly suctioned his airway.
Q: Have you conducted any kind of a literature search in this case in order to arrive at your opinion?
A: No.
Q: Have you consulted any other colleagues?
A: No
Q: In these studies that you talked about in Australia and Japan—when is the last time that you looked at those studies?
A: It's been about 10 years ago.
Q: Have you looked at the statistics of any other major studies on ionic versus non-ionic contrast outside of those two major studies?
A: I have, but not that I can recall at the present time.

THE "BAD" TESTIMONY: INTERVENTIONAL NEURORADIOLOGY

The following is an excerpt from the testimony of a neuroradiology expert witness who testified in a lawsuit filed against an interventional neuroradiologist for complications arising from embolization of a brain tumor.

Q: I understand that you have been retained in this case as an expert witness.
A: Yes, as a board certified neuroradiologist to give the truth in this case.
Q: Are you a board certified neuroradiologist?

A: I am not. There is no such thing as a board certified neuroradiologist. There is a certificate of added qualification for neuroradiology.

Q: Do you hold such a certificate?

A: I do not. And the reason is, I don't because I have been practicing for 30 years and I really don't think it is necessary for me to take this little test. But I am board certified in radiology and I have done a fellowship in neuroradiology so I know neuroradiology.

Q: You have been a member in the American College of Radiology for the past 25 years?

A: I'm not exactly sure. My wives have taken care of that and sometimes I wouldn't bet that my first wife took care of it consistently, so it is possible that my membership has lapsed.

Q: Let's talk about your other affiliations such as Radiological Society of North America. What is that?

A: That is a society where you send money and you get a journal and a monthly magazine and then you go to the convention once a year.

Q: Have you been a member of the Radiological Society of North America for the past 25 years?

A: Probably not, probably it has lapsed and then renewed and lapsed, so it has been irregular. I know for sure I'm a member of the American Society of Neuroradiology, but I'm not sure if I am currently a member of the AMA.

Q: Those are three societies that you have got listed here on your CV, and you're telling me that despite what is on your CV, you don't know whether or not you are currently a member of the Radiological Society of North America?

A: That's correct.

Q: But you are a member of the American Society of Neuroradiology?

A: Again, I'm not sure if it's lapsed, okay?

Q: In the report that you have issued in this case, you made a statement that you were a member of the American Society of Neuroradiology. But you may not be a member?

A: You know, I wasn't under oath when I wrote the report. I am here to tell 100% truth, but after thinking about it in depth, there may be periods when my membership might have lapsed.

Q: But your testimony here today is that you are currently a member of the American Society of Neuroradiology: And you state that you were a member when you issued your first report in this case.

A: I believe so, but the way you are asking these questions, it is sort of making me wonder whether or not I was at the time. So I would have to check it. I know for sure I'm a member of the American College of Radiology because a plaque hangs on my wall in my office.

Q: But you made those representations when you signed the report.

A: I believed at the time I thought I was, but you know there's a lot of societies, sort of like, I'm not sure if I subscribe to the *New York Times* now. It might have lapsed. I am not 100% sure.

Q: But you didn't talk about whether you subscribed to the New York Times when you wrote your report to advise the court and the parties of why you were an expert witness in this case, did you?

A: No. I did not.

Q: In fact, what you put on your CV, and what you told these lawyers is that you'd better check out the veracity of what you've got on your CV because you are not representing that the CV is accurate. I will read onto the record what you have stamped on the face of your CV on page 1. It states, "It is the responsibility of the person requesting the expert's service to verify the contents of this resume." What does that mean to you?

A: I don't know what it means.

Q: What types of memberships exist in the American Society of Neuroradiology?

A: There is junior membership and there is senior membership.

Q: Do you know whether you have ever attained either junior or senior membership in that society?

A: I don't. I know I was for sure a junior member when I finished my fellowship. I don't know what I am today.

Q: Does the American Society of Neuroradiology publish any kind of journal?

A: Yes, but I don't know the exact name.

Q: Do you get the journal?

A: The office where I work gets the journal.

Q: Do you find that the journal is a good journal or bad journal? In other words, do you consider it an authoritative journal?

A: No, I don't consider anything authoritative.

Q: Do you know of any other medical societies that exist in this country that you are involved with?

A: There is the Interventional Society of some sort. I don't know what the name of it is. I'm not sure there is a neuroradiological interventional society.

Q: Would you be surprised to know that there is a Society of Interventional Neuroradiologists?

A: No, I would not.

Q: The last thing you've got noted in terms of a society in your CV is the AMA Radiological Society of North America, what is that?

A: It is a division of the AMA.

Q: Are you a member of the AMA?

A: I don't know, I'd have to check it. My wife takes care of those things.

Q: We're back to you don't know whether what you've got on your CV is accurate or not?

A: Well I don't know what societies are current, other than for sure the American College of Radiology because the plaque is displayed in my office, and I thought I was but I may not be.

Q: Have you ever engaged yourself with an expert witness service to act as an expert witness?

A: Yes. TASA.

Q: What is TASA?

A: It is an expert witness service.

Q: Are you still listed through TASA?

A: I'm not sure.

Q: Your CV states that you are a doctor licensed to practice medicine in the United States. I didn't know that the United States actually licenses doctors, do they?

A: No.

Q: Have you done any interventional radiology of any kind in the past 15 years?

A: No.

Q: So as of the date when you issued your report on this case, you had not performed an interventional radiological procedure for 14 years?

A: Correct.

Q: Have you ever embolized a meningioma in the brain?

A: Yes, many times.

Q: The last time you did that was 15 years ago?

A: Yes.

Q: Do you know what advances have been made in interventional radiology since that time?

A: Not all of them, but I've tried to keep abreast of them.

Q: What would you look at in terms of trying to identify a standard of care that would exist for the treatment of the tumor of the brain such as was present in this case?

A: I'd look in the mirror.

Q: What do you mean, "I'd look in the mirror to determine the standard of care?"

A: What I mean by that is, I [would] look in the mirror and ask myself, "What does the typical neuroradiologist do in a certain situation, in other words, what is the prudent, wise, and intelligent thing to do?

Q: Would you look at any literature that would be published by the Society of Interventional Neuroradiologists about the criteria for when one should consider the embolization of a brain tumor?

A: I deal in neuroradiology on a daily basis. I interface weekly with interventional neuroradiologists. I know which cases are appropriate to be embolized and which are not. I don't do these cases today. But I know what is appropriate and what is not.

Q: Insofar as to what the standard of care requires for a reasonably prudent interventional neuroradiologist, would you agree that the criteria that existed and that had been published by the Society of Interventional Neuroradiology for embolization for head and neck tumors would indicate what the standard of care is?

A: The standard of care is what a physician of reasonable or ordinary prudence would do in a case like this, with a meningioma fed by the internal carotid artery via the right anterior cerebral artery, and that it is malpractice to embolize that patient. To attempt to embolize that patient was a tragedy.

Q: Do you know the criteria that were published by the Society of Neuroradiology at the time of the event in this case?

A: No I don't, and I don't need to know.

Q: Did you know that the *American Journal of Neuroradiology* published in September of 2001 "The Standards of Practice of the American Society of Interventional on Therapeutic Neuroradiology?"

A: No.

Q: Did you know that the organization that you believe you've been a member of published the standards in its journal?

A: I was not aware of that.

Q: Do you know what the Society's position was in regard to the Standards of Practice and criteria for doing embolization of the type of tumor that the patient in this case had?

A: No, I don't know what they state the standard of care is, but I do know what the standard of care is.

Q: Are their Standards not relevant to you?

A: Well they're of relevance, but I know what a prudent and reasonable treatment is for these kinds of tumors under these circumstances by any practicing neuroradiologist and interventional neuroradiologist.

Q: How many meningiomas do you believe that you have ever done angiography on, and how many have you embolized?

A: I've done a lot of angiograms. I can't give you an exact number; I would say 50 to 100, in that range.

Q: Do you know which manufacturers make the catheters that are used for embolizations today?

A: Who exactly are the manufacturers today, I don't know. I don't know the specifics about the equipment, but I understand exactly what is necessary.

Q: What is your opinion regarding what happened to this patient?

A: Well, the brain was infarcted by the radiologist and seizures probably were caused by the infarction.

Q: Are you qualified to give an opinion as to the cause of the seizures?

A: I think I am. I have been a neuroradiologist for almost 30 years, and I've dealt with many cases and it just makes sense that the seizure was caused by the subarachnoid hemorrhage which was caused by the radiologist. That makes sense to me. The tumor should never have been embolized. There should never have been any further imaging. The case should have been terminated at that time and the patient should have gotten a gamma knife treatment. It's a gamma knife case. It's a radiation treatment case.

Q: Do you remember one of the neurosurgical experts in this case testifying that he didn't believe that radiation therapy would be proper treatment in this case? Are you in a position to discuss surgical treatment versus radiation in this case?

A: I'm not a neurosurgeon, but I've seen cases like this before, and I know what the treatment of choice is. These are my opinions because I work with neurosurgeons on a daily basis. This is not an embolization case. This is not a surgical case. This is a gamma knife case.

Q: Do you have any medical literature that would indicate when a gamma knife procedure is indicated in a case of this nature?

A: No.

Q: Do you know of any medical literature that exists that would indicate that it's inappropriate to embolize a brain tumor of this type?

A: I don't need any literature, Counsel. It's common sense. You don't embolize these cases.

Q: Do you know of any medical literature that supports your opinion?

A: Not at this time. I have not yet done a search on this.

Q: Do you know whether or not other interventional neuroradiologists throughout the country embolize meningiomas that have a primary blood supply from the right anterior cerebral artery?

A: I don't believe they do and if they do, as far as I'm concerned, they're taking an extreme risk and they're creating negligence and malpractice.

Q: Do you know of any interventional neuroradiologists that refuse to do embolization of meningiomas that have as their primary blood supply the right anterior cerebral artery.

A: I don't know of any interventional neuroradiologist who would embolize such a case.

Q: Do you know of any who would have refused to do so?

A: Every one of them.

Q: Tell me one by name.

A: I'll have that for you sometime in the future.

Q: Doctor, let me summarize what you have testified to, namely that there is a certain select group of neuroradiologists who perform certain procedures that they call "interventional procedures," and some of these have joined a society and now call themselves "Interventional Radiologists," and unless you are a member of that society and have paid your dues in that group, you don't know anything about interventional procedures, and thus cannot pass judgment on anything that these people in that society do. Is that what you are saying?

A: Yes.

Q: In other words, if you have a group of doctors who are doing a particular procedure, and they say, "Well, let's form a Society; we will call ourselves a new name, and after that, nobody can criticize our conduct unless they are in our Society, and furthermore, we are not going to criticize each other." Is that correct?

A: Yes, what they say is a bunch of hogwash.

Author's Note: The Appellate Court in the state in which this testimony was given disallowed the opinions expressed by the radiology expert witness, ruling that it found "no basis for concluding that the radiology expert's training or experience qualified him to offer an expert opinion on the standard of care for the performance of embolization of brain tumors."

THE "BAD" TESTIMONY: MISCELLANEOUS EXCERPTS

The following five short excerpts are the testimonies of radiology expert witnesses given in various malpractice lawsuits alleging negligence by a defendant-radiologist.

Q: Doctor, what is the standard of care for a radiologist?

A: To make the correct diagnosis on an x-ray.

Q: Doctor, you have told us you are an examiner for the American Board of Radiology. If you were testing a resident for board certification, and there was this abnormality on the chest radiograph, and that resident didn't describe the abnormality on the report, are you saying that the resident would fail the examination?

A: Absolutely.

Q: Doctor, tell us again what position you hold in the American College of Radiology and how that helps you form an opinion on this case.

A: I have been elected to the board of The American College of Radiology and my assignment is to be in charge of Quality Assurance for all of the United States. As such, I will tell you that this radiologist's report is inadequate. Furthermore, it would not suffice if this man were in my radiology group.

Q: Doctor, tell us why you feel that the missing of this radiographic abnormality amounts to negligence.

A: I looked at this film as a neutral and with complete disinterest, and saw the lesion immediately, cold, without a history or clinical information.

Q: Doctor, do you think the missed lesion is subtle?

A: No, even a student could see a lesion on these films.

Q: Doctor, would you call this lesion obvious?

A: Yes, the abnormality was so obvious my secretary saw it from across the room.

Q: So, Doctor, unless a radiologist is certified by the American Board of Radiology, that radiologist would not be qualified to interpret x-rays?

A: Correct.

Q: And if a radiologist who is not board certified was to review x-rays, in your opinion that physician is not qualified to do so, and would fall below the standard of care?

A: Correct.

Q: Doctor, in your experience, do you know if you have ever had occasion to miss a nodule on a chest x-ray?

A: I don't know, unfortunately; I see the ones other people miss, and they see the ones that I miss.

Q: Do you have any estimate?

A: Very rarely.

Q: What does that mean?

A: Maybe once or twice in my life.

The "Indifferent" Testimony: General Radiology

The following two excerpts are from the testimonies of radiology expert witnesses given in two lawsuits involving alleged negligence for failure to diagnose findings on a plain film and a head CT, respectively.

Q: Do you think that the American College of Radiology would agree that what you're saying is mainstream medical thought?

A: I'd like you to send it to them but they'll probably throw it out. You know the ACR is an office with some secretaries in it and some people doing some tests, making notations and doing papers, and collecting information from other places. They don't set standards, they recommend courses of action. They're a college. They represent us in certain medical boards and certain medical meetings.

Q: If I sent your testimony to the ACR Ethics Committee, aren't there men and women sitting on that committee who would review it?

INTERRUPTION BY THE WITNESS'S ATTORNEY: Don't answer that question. It is oppressive, burdensome, harassing, and annoying. My client will not be intimidated by that type of questioning which is designed not to answer questions but to intimidate and harass the witness.

Q: All I want to know is if this is mainstream medical thought. Do you know if you have any citation to any learned treatise or any document where I can go to verify that what you are testifying to is indeed mainstream thought?

A: It's common knowledge in the medical business. You don't have to go look at some journal article. You can see it everyday at work, it's very common, and it wouldn't make it into the journal, because it's too simple and straightforward and easy to observe and be seen everyday.

Q: What is the standard of care, Doctor? How do you define standard of care?

A: I'm not a lawyer. I don't know.

Q: You don't know how to define standard of care?

A: I don't define it. Lawyers define it.

Q: Well, you are here telling us that these doctors did not comply with the standard of care.

A: I don't set the standard of care. In my professional opinion, I have an opinion about what I think is the appropriate level of reading of a film. I think that this would be a good case to show at the American Board of Radiology examination.

Q: Why is that?

A: I think that this is something that somebody coming out of residency should be able to identify. If I showed this to somebody at the American Board of Radiology exam and they passed over it and did not say, "Gee, I think this could be a dural sinus thrombosis, I would give that person a low score." That's the best reference standard I can give.

Q: Are you going to opine in this case that the radiologist deviated from the standard?

A: Yes.

Q: Why did the radiologist deviate from the standard of care?

A: Because the films indicate the presence of a dural sinus thrombosis.

Q: Doctor, just because somebody misses something on a film that does not necessarily translate into negligence, correct?

A: That is absolutely correct.

Q: So you can miss a finding and still not be negligent, correct?

A: Yes, but a miss plus a severe outcome is negligence.

A: So a miss if the outcome is not severe is not negligence, correct?

A: Yes.

Q: Let me ask you this. What if you miss a very subtle finding and the patient dies, is that negligence?

A: I think that's a very broad question. I'm not sure. Let's put it this way. Let's use again the board examination. If I showed a fourth year resident a subtle finding and that person missed it, I would say, yeah, you simply missed this thing. I don't feel that way about the finding in this lawsuit. I feel that if I showed it to most people studying for the Board examination, they would put their finger on it and make the diagnosis. I completely agree that it's a subtle finding, but there was a severe outcome.

Author's Note: It is obvious that the expert agreed that the finding that was missed was subtle, but because there was a "severe outcome," he incorrectly considered the "miss" to be an act of negligence.

THE "INDIFFERENT" TESTIMONY: MAMMOGRAPHY

Q: Do you think there is a different level of competence between general radiologists who do not spend the majority of their time interpreting mammograms versus somebody like yourself who spends the vast majority of their time interpreting mammograms?

A: I believe the general radiologist can observe the same thing that a specialized breast imager can observe but the difference may be in the interpretation of the findings. For example, if I've seen 3000 cases of fat necrosis and 15,000 cases of breast cancer, my ability to discern differences and my ability to make a decision concerning what is fat necrosis and what is cancer is probably a better educated decision than someone who has seen five cases of each. But, we should all see calcifications.

Q: What should the defendant-radiologist have seen on this mammogram?

A: The radiologist should have recognized the presence of calcifications.

Q: Is it surprising to you that in this case I've shown the films to five different radiologists blind and none of them saw the calcifications?

A: Yes, I think that is a surprise to me.

Q: Doctor, in radiology specifically involving mammography, is the standard of care perfection?

A: The standard of care is based on the minimum level of attainment that you can expect will be achieved by all radiologists. It is not a pinnacle, but it is something that everyone should be able to adhere to, general radiologists, specialists. It's not asking for the impossible.

Q: The real question is, can radiologists retrospectively have missed an abnormality on a mammogram but still have been within the standard of care?

A: No.

Q: Just so I'm clear then, anytime even retrospectively an abnormality is missed in the field of radiology, in your opinion it is a deviation from the standard of care?

A: If a radiologist missed it, it is not the standard of care. It's a mistake, but it is not the standard of care. You cannot make a mistake and say it is the standard of care.

Q: Are you aware that there are studies that have taken mammograms that are known to possess cancer and they are presented to mammographers and as many as one-third of the cancers are not called out by those mammographers? So what you are telling me is in those studies that a third of mammographers are acting below the standard of care?

A: You didn't tell me in these studies where a third of the cancers were missed, what the density of the breast tissue was, and whether the misses were masses or calcifications. Calcifications should not be missed. They are missed but they shouldn't be.

Q: In other words, if microcalcifications turn out retrospectively to be a cancer, your opinion is that hundred percent of the time, when a radiologist does not call them out initially, that that is a deviation of the standard of care.

A: The answer is yes.

Q: Would you expect other mammographers to agree with you that it is a reasonable standard to hold all radiologists negligent for missing calcifications?

A: Yes, I would expect that.

Q: Do you have a miss rate?

A: Maybe one or two percent.

Q: Do you consider that could be acceptable, one or two percent?

A: For mistakes, one or two percent, yes.

Q: Do you know what the definition of standard of care is?

A: I do know what I use as the definition of standard of care.

Q: Did you ever miss a cancer?

A: I think that in some instances there may have been a delay and looking retrospectively probably there were some that I might have called on earlier mammograms, yes.

Q: Did you feel you committed medical malpractice in those cases?

A: No. I felt I met the standard of care.

Q: So you can have a miss, not see a cancer, and still be within the standard of care, as has happened to you?

A: I didn't say I missed it. We're talking about sensitivity levels. There's a difference between a sensitivity and a miss. A miss is a mistake. A miss is malpractice if it has detrimental effect. Have I ever missed something? Not that I'm aware of. You can look back at the previous films when you're dealing with a newly diagnosed cancer to see your sensitivity level was at a lower bar. But if you see something looking back and say to yourself it was a cancer, but I did not diagnose it at the time, then you've got a miss.

Q: That has never occurred to you?

A: No. If you look back at a film and you ask yourself could I have called it, do I see it—is there something in retrospect, might it be there, then that is not a miss. That is a sensitivity.

Q: That latter description has occurred to you?

A: It has occurred to everybody.

Q: That latter description can occur and that is not malpractice.

A: No it is not.

Q: Doctor, could you explain to the jury what the duty of care is for a radiologist interpreting a screening mammogram?

A: The duty of the radiologist who's screening is to detect potential abnormalities. The radiologist then has a further obligation. Federal law, MQSA, mandates that there should be a recommendation for suspicious abnormalities or biopsy for an indeterminate abnormality.

Q: Do you have an opinion whether or not the defendant-radiologist deviated from the accepted standard of care in his interpretation of the mammograms in this case?

A: Yes. The deviation is with respect to what the obligations are when one is doing mammography, which is that if an abnormality is seen or a potential abnormality, the obligation is to have the patient brought back for additional studies. The radiologist should have seen calcifications and recalled the patient for additional views.

Q: Doctor, it is your opinion that if calcifications retrospectively turn out to be cancer, hundred percent of the time radiologists who do not call them out originally are acting below the standard of care?

A: The standard of care requires that if you see microcalcifications on a film you need to evaluate them further. That is the standard of care.

Q: But Doctor, isn't it true that in the field of radiology, often times radiologists make a diagnosis of cancer and they look back at prior mammograms and they may see signs of that cancer earlier that they did not call out initially?

A: Yes, that is true.

Q: And again it is your opinion that hundred percent of the time any radiologist who did not call out those calcifications initially deviated from the standard of care?

A: Yes, they have deviated from the standard of care.

Q: And Doctor, you yourself have never missed a cancer have you?

A: No. I don't think I have, but I don't know. You need to have a tumor registry to be able to follow up your negatives.

Q: And Doctor, have you never had occasion to look back and say to yourself, it was a cancer on the previous mammogram but I didn't diagnose it until a later mammogram was obtained?

A: It has happened to me and it has happened to everyone else who does mammography, which is why you try to see how you can modify your threshold for sensitivity.

Author's Note: In a personal communication, the head of the section on mammography at a large academic institution was asked to audit mammographic interpretations in his department. The investigation revealed that in the normal course of reading, experienced mammographers missed calcifications that turned out to be malignant in 41% of the cases.

THE DEFENDANT-RADIOLOGIST'S TESTIMONY: "INDIFFERENT"

Before leaving the excerpts illustrating what I consider to be "bad" or "indifferent" testimony by expert witnesses, I shall present an excerpt of the testimony of a defendant-radiologist who was accused of failing to communicate an abnormal radiological finding directly to the referring physician. I consider the testimony to be "indifferent."

Q: Were you aware of any standards in place within your area of specialty at that point in time that called upon radiologists to make follow up recommendations and incorporate them into the x-ray report?

A: No, I am not aware of a standard like that.

Q: To your knowledge, were there any standards within your field of radiology that you were familiar with that called for and defined for radiologists what type of findings should be considered as significant?

A: Well, the ACR does promulgate a set of standards, but I'm not sure it meets the criteria you are talking about.

Q: Were you familiar with those Standards at the time you read the x-rays in question?

A: The ACR Standards is a very thick book. It's like a reference tool that is available to radiologists when they have specific questions or concerns about a particular standard. It for the most part applies to performance of examinations.

Q: Would that particular set of Standards from the ACR define those types of findings considered critical findings?

A: No, I do not believe it has that kind of detail.

Q: Do the Standards of the ACR define what type of findings are urgent findings?

A: No.

Q: Do the Standards of the ACR provide you with any guidelines on how and when to have direct communication with a referring physician?

A: There is a guideline concerning that in the ACR Standards.

Q: Were you familiar with that guideline when you read these x-rays?

A: No.

Q: What's your understanding of the Standard that pertains to under what circumstances a radiologist should have direct communication with a referring physician?

A: Well, I've read the ACR Standard on Communication. Do you want me to tell you what I remember of what I read in that Standard?

Q: Yes, I do.

A: I was unaware of the existence of that particular Standard, so I looked in the ACR Guidelines to see if they had a Standard on communication, and they do have one. Now, the ACR Standards are a set of guidelines. They don't create a legal standard. They don't even create a community standard. They are revised from time to time, and they don't represent a set of standards that someone can sit down and read. They represent, you know, more of a reference text, and my recollection on precisely what they say for direct communication or even what their definition is for direct communication, I can't recall specifically.

Q: Do you feel that at the time you read these x-rays it was important for you to follow the Standards promulgated by the ACR?

A: To the best of my knowledge, I believe I have always followed as close as I could the Standards as set forth by all of the important bodies in the field of radiology, including the ACR and the RSNA, and I certainly felt that I was meeting the Standard set forth by them in respect to this case.

Q: And you have attempted to follow these Standards because these Standards help to define the standard of care that radiologists should follow?

A: No. These are guidelines that are set up by men just like me getting together and trying to promulgate what they think might be the right way to handle cases to promote patient care. These are not things that I consider authoritative. They are not things that are backed up by any documentary evidence. They are things that are revised from time to time. These are simple guidelines. They do not establish a legal standard. They do not establish the community standard. They certainly don't establish the community standard anywhere I have practiced.

Q: When you went back and consulted those Standards, did you find that the Standards called for direct communication with a referring physician when significant findings were reported like those contained in your report on these films?

A: I'm not sure that I would characterize it that way, and I don't think they use the term significant finding, but I'm not certain. I don't have that in front of me right now. I would have to have the document to review in order to answer your question.

Q: Well, do you believe there was any responsibility on your part to attempt to have direct communication with the referring doctor, if indeed you were attempting to follow the guidelines of the ACR?

A: First of all, even though I was aware of and had used the ACR College Guidelines in instances prior to this case, I was not aware of any specific ACR Guideline dealing with communication. The majority of the guidelines deal with how to perform examinations.

Q: I understand you weren't aware of the particular guideline from the ACR as it relates to direct communication with the referring physician, but what I'm asking now is do you believe that you had a responsibility to have direct communication with the referring doctor, if you were, in fact, attempting to follow the ACR Guidelines?

A: No.

Q: And the basis for your opinion is that you had no responsibility to contact the referring physician?

A: The basis is that I reviewed an examination, I found some findings, some important findings, and I reported them and I generated a report directly for the referring physician, outlining my findings. And furthermore, my findings were not unexpected.

Q: This gentleman's history that was presented to you on a prescription was pneumonia. Is a right hilar mass an expected finding associated with pneumonia?

A: It certainly can be.

Q: Did anything about this gentleman's requisition form indicate to you knowledge or concern on the part of the referring physician that this patient might have a clinical diagnosis of lung cancer?

A: No. I might add something to that, though. The occurrence of pneumonia in a relatively young, healthy person, and especially recurrent pneumonias in that type of person, frequently is caused by lung cancer. That's a common finding, and a hilar mass can represent a lung cancer, although it can also represent enlarged lymph nodes from an infection.

Q: You've indicated that based upon your routine practice that year, you would not have had direct communication with the referring physician unless your findings indicated to you that it was an immediately life-threatening process. What would you have had to see on this study, in your opinion, that would have required you, following the standard of care in this community, to have direct communication with the referring physician?

A: Several things could fall into that category of immediate life-threatening processes; a pneumothorax where the lung was deflating, a dissecting aneurysm, acute congestive heart failure, or a new bout of pneumonia.

Author's Note: I leave it to the reader to decide whether this testimony would likely convince the jury to find in favor of the defendant-radiologist, or in favor of the plaintiff-widow of the patient who died of lung cancer.

THE "GOOD" TESTIMONY: A MISSED LUNG CANCER

All of the excerpts presented thus far have in my opinion been examples of "bad" or "indifferent" testimony. The following is an excerpt of testimony given by an expert radiology witness which I categorize as "good."

Q: Doctor, have you formed an opinion as to the professional conduct of the defendant-radiologist in this lawsuit?

A: Yes, I have.

Q: What is your opinion?

A: It is my opinion that the defendant-radiologist breached the standard of care when he rendered his interpretation of the chest radiographs.

Q: On what basis did you arrive at that opinion?

A: I arrived at that conclusion based on my many years of radiologic practice, and on what I believe constitutes the standard of radiologic care.

Q: How do you define the standard of radiologic care?

A: The standard of radiologic care calls for radiologists to conduct themselves in a reasonable and ordinary manner, similar to the manner of conduct of other radiologists in the same or similar situations.

Q: Doctor, do you feel that the defendant-radiologist made a mistake when he interpreted the chest films that are the focus of this lawsuit?

A: Yes, I believe his failure to observe the lesion in the patient's left lung was a mistake.

Q: Do you believe all mistakes constitute medical malpractice?

A: No, certainly not. Many mistakes that are made do not, in my opinion, equate to a breach in the standard of care.

Q: I don't quite understand what you are saying, Doctor. Please elucidate.

A: Well, as I have just said, I believe the standard of care calls for reasonable conduct. In other words, as it relates to this case, a reasonable interpretation of the chest radiographs. Whether the interpretation turns out to be correct or incorrect is in my opinion not the issue; the issue is whether a radiologist practicing in a reasonable manner would render a similar interpretation, no matter whether that interpretation is right or wrong.

Q: Well, Doctor, you seem to be saying that some errors or mistakes do constitute a breach in the standard, and others do not, correct?

A: Yes.

Q: Well, how do you decide which mistakes still conform to the standard of care and which do not?

A: In dealing with radiologic misses, as we are in this case, I believe the answer has to do with evaluating the conspicuity of the lesion, in other words to what degree it stands out compared to the surrounding tissue, along with the location of the abnormality. In other words, the degree to which it is obvious versus the degree to which it is very subtle.

Q: Would you consider this lesion obvious or subtle?

A: In my opinion, the lesion is obvious.

Q: So that is why you are saying that missing it is a deviation from the standard of care, in other words, negligence?

A: Yes.

Q: Do you think that other radiologists would agree with you?

A: I really have no idea. My opinion is based on my own evaluation of the films, not the evaluation of other radiologists.

Q: Have you shown these films to other radiologists or discussed them with other radiologists?

A: No.

Q: Doctor, I am sure you are aware that there is another expert radiology witness testifying in this case on behalf of the defense who is going to be saying that in his opinion the defendant-radiologist's missing of the lesion did not constitute a deviation of the standard of care. What do you think of that expert's opinion?

A: I respect his opinion, but I just happen to disagree with it.

Q: Do you believe that this other radiology expert witness is well qualified and a knowledgeable radiologist?

A: I do not know the radiology expert witness personally, but I am aware of his name and reputation, and have no reason to doubt that he is highly credentialed and knowledgeable.

Q: Well, then how is it that you, who claim to be a highly credentialed and knowledgeable radiologist, and this other witness who also claims to be highly credentialed and knowledgeable, can disagree as to whether the missing of the abnormality on this chest film does or does not constitute negligence?

A: I see nothing unusual in that. On most occasions, radiologists who review the same radiographs come to essentially similar conclusions, but in some cases they do not. Assuming that radiologists are rendering opinions in the best of faith and consistent with adequate knowledge, radiologists will occasionally disagree. In this particular case, obviously the expert witness for the defense and myself disagree. I respect the other radiologist's opinion, and I hope that he respects mine as well. But we simply disagree.

Q: Well, Doctor, how do you plan on convincing the jury that your opinion that the defendant-radiologist breached the standard of care should supersede the opinion of the defense's expert radiology witness who is testifying that the defendant-radiologist did not breach the standard of care?

A: Sir, I do not see my role as trying to convince the jury of anything. I am here to give my opinion based on my own experience and knowledge. I realize that one or more witnesses in this case will give a contrary opinion. The jury will hear my opinion as well as the opinions of others. As I understand the legal system, which I respect, it's up to the jury, not the other witness or myself, to eventually decide whether the defendant-radiologist's actions did or not breach the standard of care.

Q: Thank you, Doctor. That is all I have.

When comparing the testimony of this radiology expert witness to others that have been documented in this chapter, it appears that this last excerpt illustrates what I consider to be "good" expert testimony. It is objective rather than subjective or emotional. The witness is firm in his opinion but nonetheless respectful of contrary opinions. The witness defines the standard of care correctly, acknowledging that the standard calls for reasonable interpretations, whether they be correct or incorrect; and the witness recognizes that his primary role as an expert is more to *educate* and *explain* medical facts to the jurors, rather than to attempt to *persuade* them. Furthermore, the expert witness does not appear to be self-aggrandizing, insulting, or contemptuous of others; he is well-prepared, matter-of-fact, seemingly low-keyed, and apparently professional in his responses.

EXPERT TESTIMONY: THE GOOD, THE BAD, THE INDIFFERENT

I leave it to the reader to decide whether the numerous excerpts shown above represent expert testimony that is good, bad, or indifferent. Notwithstanding that expert witnesses are expected to be sufficiently knowledgeable to proffer realistic assessments of the appropriate standard of medical care and whether a defendant physician complied with or breached it, many in the legal and medical communities have often expressed doubts that a number of such experts are competent to do so. A perusal of the excerpts shown above not only fails to dispel such doubts but perhaps may even connote a meaning totally different from and never previously envisioned to Shakespeare's words that are quoted at the beginning of this chapter.[1]

CREDIBILITY OF MEDICAL EXPERT TESTIMONY

Concern about the credibility of medical expert testimony in malpractice litigation is nothing new. An editorial in JAMA as far back as 1892 lashed out at what it called "the disgraceful exhibition of medical experts who are hired [to give] opinions:

> The lawyers, acting as generals, lead the experts up to conflict, enthused with the idea that the truth is the great object of the struggle. In reality, both sides care nothing for the truth; winning the case is paramount to every other object. The expert physician is seductively ... made to give a jumbled, confused mass of half truths and facts open to question .... Both sides avoid informing the jury, and are always eager to deceive them.[18]

An article published in an 1897 issue of the *Harvard Law Review* also made reference to the low esteem in which expert medical testimony was held by quoting an attorney's statement to a jury:

> Gentlemen of the Jury, there are three kinds of liars: the common liar, the damned liar, and the scientific expert.[19]

Today, 114 years later, concern surrounding the quality of medical expert testimony has diminished greatly, but to a lesser extent still remains. An insightful perspective of the role played by expert witnesses in today's environment was recently rendered by an Ohio court:

> The common perception exists that the recent proliferation of medical malpractice cases somehow is due to the onerous efforts of lawyers. Without being drawn into that argument, it has been the experience and observation of this court that in all the medical malpractice trials over which it has presided, the ultimate beneficiaries in an economic sense are truly the physicians who demand and usually obtain exorbitant compensation for their testimony as expert witnesses. Ordinary checks and balances are non-existent in medical malpractice cases and the standard appears to be to get whatever the traffic will bear. In too many medical malpractice cases, unfortunately, the Hippocratic Oath has been supplanted by opportunism and greed by those who participate as medical expert witnesses.[20]

The Illinois Supreme Court has echoed similar sentiments:

> Many experts today spend so much of their time testifying throughout the country that they might be deemed not only experts in their field but also experts in the art of being a persuasive witness and in the art of handling cross-examination. Little has the nonlitigating public (including the jury) realized the true rhetorical masterpieces that come from the lips of medical experts .... There is [sic] a growing number of experts whose livelihood is dependent in large part upon the litigation process. Such experts with their vast amount of litigation experience become exceptionally proficient in the art of expert witness advocacy .... The financial advantage which accrues to an expert witness in a particular case can extend beyond the remuneration he receives for testifying in that case. A favorable verdict may well help him establish a "track record" which, to a professional witness, can be all-important in determining not only the frequency with which he is asked to testify but also the price he can demand for such testimony.[21]

SUMMARY

As the title of this chapter indicates, the testimony of expert witnesses can be categorized as "bad," "good," and "indifferent." Examples of each of these categories of testimony have been given. Most radiologists and other physicians who step forward to assume the role of expert witness for either the plaintiff or the defendant in a medical malpractice lawsuit are sincere, well-trained, and honorable individuals, intent on doing the right thing. Undoubtedly, the overwhelming majority of potential expert witnesses desire to offer testimony that is considered "good."

"Good" potential witnesses should keep in mind that they will be asked to be seated in a courtroom witness chair, not asked to stand on a soapbox or behind a podium in an auditorium. They will be required to answer as objectively and dispassionately as possible questions posed by attorneys, under direct and cross-examination, pertaining to their credentials, experience, and other qualifications; they will not be required to inflate their credentials or qualifications for the purpose of diminishing the performance of, or demeaning in any other way, the defendant-radiologist. They will be required to testify as to what they believe to be the appropriate standard of care applicable under the circumstances of the specific case being adjudicated. In other words, what a reasonable practicing radiologist should have done or not done. They will be required to state their opinion as to whether a defendant-radiologist complied with, or deviated from, that standard of care; they will not be required to expound, lecture, philosophize, professorialize, or pontificate about political, social, economic, or other matters or personal pet peeves not germane to the case at hand. They will be required to offer criticism when warranted, but they will not be required to offer inflammatory or other pejorative comments. They will be knowledgeable of the specialty, standards, and facts of the case, honest and truthful, of firm opinion, objective, and be a good communicator, using easy to understand language; they will not be apologetic, accusatory, inflammatory, self-aggrandizing, or ego-gratifying.

In the final analysis, the potential expert witness whose goal is to provide the court and the jury with "good" testimony will do so by adhering to the Biblical Commandment, "Thou shalt not bear false witness against thy neighbour."[22]

REFERENCES

1. Measure for measure, act 2, scene 2, lines 117–118, in *The complete works of William Shakespeare*, Rex Library, London, 1973, 800.
2. Agnew v Parks, 343 P2d 118 (Cal App 1959).
3. Lamere v New York State Office For The Aging, Lexis 13281 (US Dist. 2004).
4. American Medical Association Council on Ethical and Judicial Affairs, 2008–2009 Ed., §9.07, Medical testimony, in *Code of Medical Ethics*, American Medical Association, Chicago, 2008, 314–315.
5. American College of Radiology, *2009 Digest of Council Actions, Section II, K 1–3, testimony*, American College of Radiology, Reston, VA, 2009, 87–88.
6. American College of Radiology, ACR practice guideline on the expert witness in radiology and radiation oncology. IN: 2008 Practice Guidelines & Technical Standards, American College of Radiology, Reston, VA, 9–12, 2008.
7. American College of Radiology, 2008–2009 Bylaws, article XIII, code of ethics: Section 2, rules of ethics, American College of Radiology, Reston, VA, 2008.
8. McWilliams v. Dettore, 1-07-0678, (IL App 2009).
9. Thomas E. J., Studdert D. M., Brennan T. A., The reliability of medical record review for estimating adverse event rates, *Ann Intern Med*, 136, 812–816, 2002.
10. Brennan T. A., Leape L. L., Laird N. M., et al., Incidence of adverse events and negligence in hospitalized patients: Results of the Harvard Medical Practice Study 1, *N Engl J Med*, 324, 370–376, 1991.
11. Hallowell v University of Chicago Hospital, 777 NE2d 435 (Ill App 2002).
12. LaSalle Bank v C/HCA Development Corp. 1-06-1859 (Ill App 2008).
13. Bosco v Janowitz. 1-07-0617 (Ill App 2009).
14. Berlin L. Lundberg G. D., Expert witness for whom? *JAMA*, 252, 251, 1984.
15. Kassirer J. P., Cecil J. S., Inconsistency in evidentiary standards for medical testimony, *JAMA*, 288, 1382–1387, 2002.
16. Harr J. *A Civil Action*, Vintage Books, New York, 219, 253, 298–299, 1995.
17. Werth B, *Damages*, Simon & Schuster, New York, 216, 308–309, 1998.
18. Expert medical testimony in jury trials, *JAMA*, 18, 304, 1892.
19. Murphy, J. P., Expert witnesses at trial: Where are the ethics? *Georgetown Journal of Legal Ethics*, 14, 217–240, 2000.
20. Kirby v Ahmad, 635 NE2d 98 (Ohio Misc 1994).
21. Trower v Jones, 520 NE2d 297 (Ill 1988).
22. *The Holy Bible: Authorized King James version*, Exodus 20:16, Christian Science Publishing Society, Boston, 1987.

Section III

Identification

7 Identification of the Dead

LeRoy Riddick

CONTENTS

THE NEED TO IDENTIFY

There is no greater challenge and no heavier responsibility than that facing the professionals charged with identifying the dead. Something deep within the human psyche abhors loss of identity, even for the deceased. The reverence paid to the Tombs of Unknown Soldiers throughout the world attests to the pathos and disquiet we feel about those who have lost their identity in death. Every unidentified body brought to the medical examiner or coroner represents a person missing to friends and family. Without proper identification of these bodies, loved ones bear the heavy burden of a continuing search for that missing person; they are denied the closure that can come with mourning. The religious ceremonies and the social mores concerning the dead bear witness to this need for the living to put the dead to rest. Without proper identification, this is impossible. As a corollary of these intense emotional issues, few activities satisfy medicolegal professionals so much as making a proper identification of a body originally found unidentified.

The solution of legal problems associated with death also requires proper identification of the deceased. Without a death certificate, which demands an identifiable decedent (except in cases of fraud), families cannot probate wills, receive death benefits, enter a safety deposit box, and so on; a spouse cannot remarry until the missing person can legally be declared dead. In homicides and suspicious cases, detectives have almost no leads upon which to start an investigation when the body remains unidentified. Moreover, if the remains are decomposed or severely mutilated by fire, investigators may not even have an adequate cause of death, which might be determined from the victim's history if his identify were known. Even in those rare cases in which a suspect is associated with a missing person and all of the circumstantial evidence points to a homicide, without the body and proper identification the case is difficult to adjudicate. And without justice, the loved ones of the decedent are also denied closure.

METHODS

In general, the condition of the body, whether the decedent was known in the community in which the death occurred, the number of victims, and the capabilities of the medicolegal professionals will dictate the methods used. In practice, two methods are employed to identify the decedent: (1) the least reliable, that is the least reproducible, method involves a visual review of the remains, photographs of the remains, details such as tattoos on the remains, or personal effects found either on or about the decedent by family or friends; and (2) the more reliable method uses documentation on or in the body of certain anatomic characteristics, such as fingerprints, dental restorations, healed features, surgical sutures, and so on that can be compared with similar documentation— fingerprints, photographs, radiographs, or dental records—prepared prior to death. The condition of the body, the age and history of the victim, and the resources of the authorities set stringent parameters on what can be done to make positive, reproducible identification of the dead. The caveat is that some remains cannot be positively identified.

Almost without exception two generalities apply to whatever method is used to make or confirm an identification. First, and almost fatuously, all techniques from visual observation of the remains to sophisticated DNA profiling rely on pattern recognition. The more intact the pattern—whether fingertip pads or teeth—and the more conversant the examiner with analyses of the patterns, the more likely a positive, reliable identification. Secondly, the process of identification requires a team approach. Even in cases where visual inspection of the body is used, a law enforcement officer will most often notify the family or friends of missing persons of the death of someone who has to be identified and the medical examiner will witness the observation and confirm the identity. In complex cases requiring a variety of professionals ranging from fingerprint examiners to anthropologists to radiologists, the team can be quite large. All members of the team should remember that it is the medical examiner certifying the death who bears the ultimate responsibility for assuring the identity of the body and is, thus, the *de facto* leader of the team.

Most deaths occur in the community in which the decedent lived and in most instances family members and friends are either in attendance at the death or, in the case of deaths in the hospital or nursing homes, have visited the victim recently. In such cases the health care personnel notify the next of kin about the death and, in many instances, prepare the body to be viewed by the loved ones. Except in medicolegal cases, decedents are most often released directly to a funeral home and the death certificate completed on the basis of visual inspection and/or history provided by caregivers.

In cases of easily recognizable victims dying at home, in the hospital, or in a location in the community in which they lived, but whose death comes under the coroner's or medical examiner's jurisdiction, the method of identification will depend upon the resources and the philosophy of the medicolegal authority. In small communities where most people are known in life and readily identified in death, the hospital identification band and/or the police report may suffice. In large urban centers where the living and the dead are often anonymous even to next-door neighbors, the authorities may require visual identification of the body or of an image of the body on video or photograph by someone who knew the decedent in life. One caveat about accepting hospital identification must be mentioned: victims who die in transit, or in the emergency department or elsewhere in the hospital, must be properly identified by relatives or friends and not by documents in or around their bodies at the scene and transported with them to the hospital. These documents can easily have been forged, borrowed, or stolen.

When the body has decomposed or has been mutilated by fire or excessive forces precluding visual identification of the remains, the investigators know from the start that scientific methods must be used and should begin collecting dental and medical records, antemortem radiographs, fingerprints, and so on to expedite the process. In cases of badly decomposed but nonskeletonized remains, the skin of the hands can sometimes be carefully peeled off and used as a glove on an investigator's hand to obtain adequate fingerprints. If that skin has sloughed off and is not usable, prints of the underlying dermal ridges can be used. If the fingers are mummified, thus distorting the dermal ridges, the fingerprint analyst may obtain adequate fingerprints after injecting glycerine and saline subcutaneously to rehydrate and expand the tissues. Tattoos and other scars can be delineated by scraping away the discolored dermis. Dental records and radiographs are the most productive tool in identifying such mutilated remains. Radiographs of a variety of anatomic regions and body parts, ranging from those of extremities with healed fractures to scout- films of the abdomen with distinctive phleboliths, may prove invaluable.

EXAMPLE CASES

Two cases illustrate the power of radiographs to establish both the identity and the cause of death in severely mutilated remains.

Case 4-1 (Figure 7.1) involved the badly decomposed, almost skeletonized, remains of a female. The body was found in the woods near a neighborhood where a 46-year-old mentally retarded female had disappeared 3 months earlier. A search for records revealed a dental x-ray taken some years prior to her death; this was compared with the postmortem radiograph of the mandible which contained only one tooth (Figure 7.2). Fortuitously, that one tooth had a restoration and next to the tooth in the mandible was the broken bit of a drill; these two findings established the identity beyond any reasonable doubt.

FIGURE 7.1 Partially skeletonized, badly decomposed remains of a female body.

Some days investigators get lucky. *Case 4-2* involved a dismembered torso of a female, which was found in a sewer (Figure 7.3). A radiograph of the torso disclosed a peculiar beak-like configuration of the first ribs at the costochondral junctions (Figure 7.4a), which matched those in an antemortem chest x-ray of the suspected victim (Figure 7.4b). (We gratefully acknowledge Dr. Brian D. Blackbourne, Chief Medical Examiner of San Diego County, CA for this case.)

In cases of badly decomposed (skeletonized) and severely mutilated bodies for which there are no antemortem dental records (assuming the body was not edentulous) or radiographs of the remaining body parts available, the investigators must use more esoteric methods. A physical anthropologist may be able to provide a reasonable evaluation of the age, sex, race, and stature of the decedent and thus assist in narrowing the identification process. If the skull is present and reasonably intact, an anthropologist skilled in reconstruction may be able to render a likeness that family or acquaintances can recognize enough either to identify or to point to other documents that can lead to a positive identification. Similarly, portrait artists (especially those trained in reconstruction) may be able to create a recognizable portrait. With scanners and the hardware and software necessary for digital imaging, a skilled computer guru may be able to superimpose images from suitable photographs of

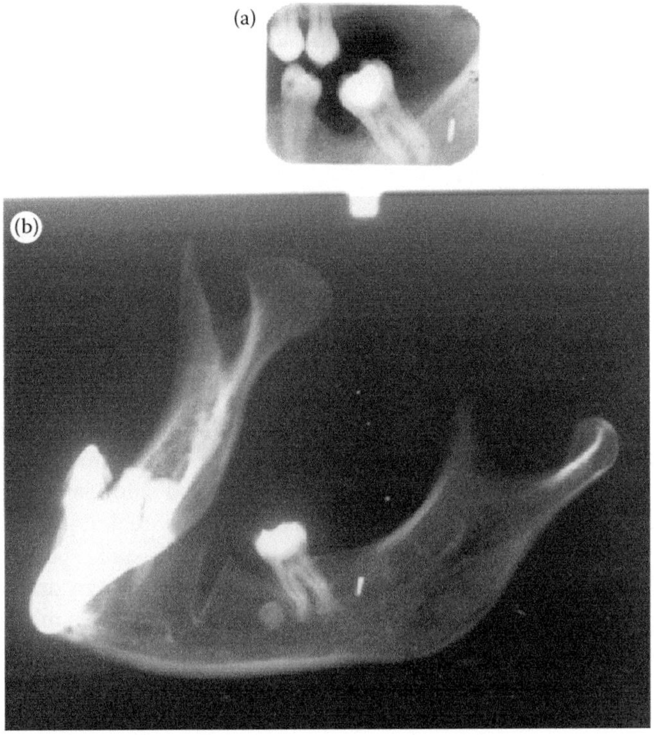

FIGURE 7.2 Comparison of (a) antemortem dental radiograph, with (b) postmortem radiograph of disarticulated mandible. There is a perfect match of both the restoration in the molar and the broken-off drill bit tip.

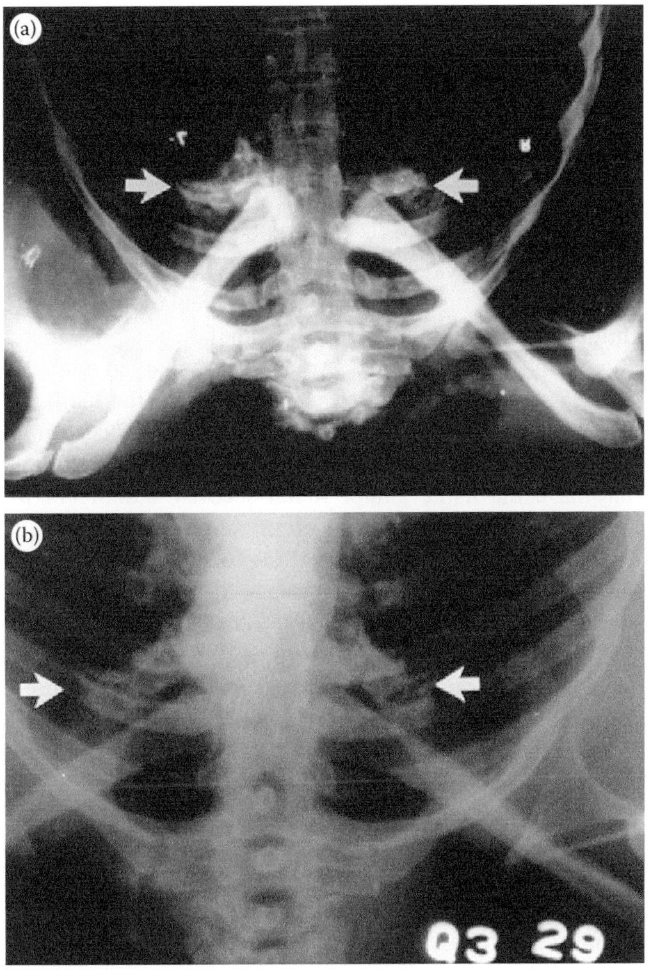

FIGURE 7.4 (a) Postmortem radiograph shows peculiar beak-like calcification of the 1st costochondral junction bilaterally (arrows); (b) antemortem chest with identical calcific configuration (arrow).

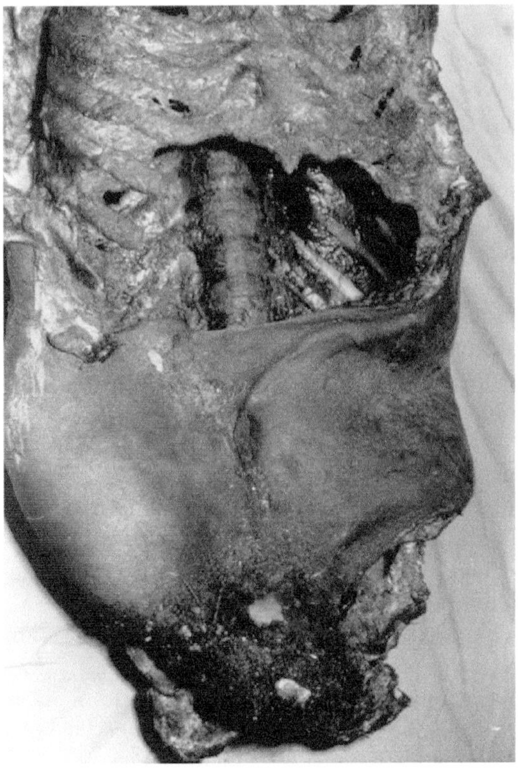

FIGURE 7.3 Dismembered partial torso of a female found in a sewer.

the missing person onto images of the skull and render a reasonable opinion as to identity.

Biochemical genetic markers may also prove useful in badly mutilated or decomposed bodies. In those cases in which blood is available (and often enough can be milked out of a limb or other body part to suffice), traditional serological methods used for typing blood for transfusions may be sufficient. In other cases where cells of one type or another (including osteocytes) containing DNA are still present, analysts proficient in the techniques and interpretation of DNA analysis may be able to determine the identity through a comparison with DNA analyses of relatives. This method may be the only reliable means of positively identifying infants and children for whom no antemortem dental records or radiographs exist.

When the victim is not known in the community and does not fit the description of anyone missing, that is, the decedent is a John or Jane Doe, the process of identification takes on other dimensions. If the remains have readily identifiable features and if presumptive documents of identification are found on or around the body, the major problem becomes one of searching for the next of kin or someone knowledgeable about the victim. The authorities in locations listed on the documents and the media may prove helpful. The next step is having all the pertinent law enforcement agencies search their fingerprint files for a match with the victim (all dead bodies brought to the medical examiner's office should be fingerprinted). Those agencies connected with the Automated Fingerprint Identification System (AFIS) can broaden the search to a nationwide data bank. The problem with fingerprint identification is that so few persons, particularly women and children, have their fingerprints on file. If the search to match fingerprints proves fruitless, then those charged with the identification must document and radiograph the teeth and take full body radiographs. Dental and other prostheses may have unique characteristics and, as in the case of dentures, may have the name and/or social security number of the decedent. The data from those examinations coupled with all the observations made during the external examination (clothing, eyeglasses, jewelry, scars, dental or orthopedic prostheses, etc.) and the internal (postmortem) examinations (congenital anomalies, missing organs, medical devices such as pacemakers, etc.) must be entered into the Federal Bureau of Investigation National Crime Information Center (NCIC) Unidentified Person (UP) and Missing Person (MP) computer data banks for a possible match with data of other missing persons entered into the same system. Contacting the local media as well as nationwide TV programs such as "Unsolved Mysteries" may assist in the investigation and identification. The success of such searches depends not only upon the quantity and quality of the information documented and encoded by professionals examining the dead body, but also upon the quantity and quality of the information available to and encoded by the investigators searching for a missing person. Moreover, someone must first report an individual missing for a search to be initiated. In this land of continual movement, families who have lost contact with members may have no idea that a relative is missing and, thus, data are never entered into the system.

In mass disasters, which in practical terms means having more bodies than can be handled by the facility available to those charged with making the identifications, all the techniques discussed previously may need to be employed. More pertinent, the organization and management of the personnel involved with the recovery of the remains and their identification are among the most important aspects of processing multiple dead bodies. All professionals with expertise in the various aspects of scene investigation and body identification must work as a team. Plans, practice, and patience are necessary ingredients in assuring that these horrendous tasks are accomplished efficiently. With multiple deaths, extreme care and caution must be exercised at the scene by those charged with the recovery. In airplane crashes and/or bombings, where great forces not only shred the remains into small pieces but also scatter them over broad areas, the investigators must meticulously chart which remains were found where and carefully label them accordingly, both to assist in the identification and also to assist in reconstructing the event. Where body parts are commingled, the assistance of the anthropologist becomes all important.

In transportation-related disasters, manifest lists, seating charts, and other documents will prove indispensable in identifying the bodies. As in individual deaths, the bodies may be, at least at first glance, readily recognizable either directly or through photographs. We recommend however that, where possible, all remains be identified scientifically with conventional means such as dental records or radiographs. In such circumstances relatives and friends may not have seen the victim for years, making visual identification tricky at best. Moreover, comparison of radiographs, dental records, and fingerprints by professionals reduces the involvement of loved ones at an extremely stressful time for everyone. In such trying circumstances as when the bodies are mutilated by fire or dismemberment, the patience and stamina of the investigators will prove as valuable as the techniques of identification at their disposal. An articulate spokesperson from the Emergency Management Agency who can deftly deal with the media may prove to be the most valuable member of the identification team.

Whether the remains are those of an individual or a host of persons, the media mavens will hover like vultures around the case, pecking at every tidbit of information whether verifiable or a rumor. One expert who is adroit at dealing with these persons should be appointed to do so, enabling the professionals to get the job done.

As we move further into the twenty-first century, the problems of a proper scientific identification of the dead have increased. The tragedy of the destruction of the World Trade Center Towers in New York City on September 11, 2001, illustrates this vividly. The number of decedents was only one of many problems. The combination of explosive, incendiary and gravitational forces literally tore or burned many of the victims into small charred pieces. Identifying literally millions of fragments through sophisticated DNA analysis proved to be a monumental task, competently discharged by the Office of the Chief Medical Examiner of New York City. That many of the decedents were foreign nationals further complicated the matter through the lack of readily available antemortem documents and specimens. This issue of available medical or other records for identification of foreign travelers will continue to be an issue. Such terrorist attacks as occurred in Mumbai took the lives of many foreigners to India. Tsunamis, earth quakes, avalanches and other natural disasters around the world have and will kill the tourist far from home. Thousands if not more of the victims of such wars that occurred in Bosnia and are occurring now in the Sudan and the Republic of the Congo will remain unknown. Then there are illegal immigrants. Many of the Mexicans who cross the border into the desert lands of southwestern United States die from hyperthermia or dehydration and quickly become skeletonized. Having little or no personal identification, the process of finding out who the remains are and returning them home is an ongoing task.

Just as the economic crisis of 2008–2010 is a worldwide event so too in this global village is the issue of proper identification of the dead. We would hope that governments and their forensic institutions and scientists will continue to develop the systems and means to scientifically identify the dead persons in their jurisdiction.

Finally, all those persons having the responsibility of interviewing and communicating with relatives and other loved ones must keep in mind the emotions involved at this most difficult time. The sudden, unexpected loss of a loved one, particularly if the death is violent, far from home stresses even the strongest person who then may lash out from frustration at those who are trying their best to help the bereaved. At these moments, Sir William Osler's essay on *Equanimity* should sustain the investigators in their task to positively identify the dead.

BIBLIOGRAPHY

1. Knight, B., *Forensic Pathology*, Oxford University Press, New York, 1991, chap. 3.
2. Osler, W., *Aequanimitas: With Other Addresses to Medical Students, Nurses and Practitioners of Medicine*, 3rd ed., Blakiston Co., Philadelphia, 1932, chap. 1.
3. Stahl, J., III and Fierro, F., Identification, in *Handbook of Forensic Pathology*, Froede, R. C., Ed., College of American Pathologists, Northfield, IL, 1990.

8 Radiological Identification
Anthropological Parameters

B.G. Brogdon

CONTENTS

INTRODUCTION

Common methods of identifying human remains—facial features, scars, birthmarks, tattoos, fingerprints, palmprints, and footprints—depend on preservation of the soft tissue components of the body in question. These methods are thwarted when the remains are so decomposed, burned, mutilated, skeletonized, or fragmented that the surface topography is unrecognizable or featureless. It is then that medical and dental radiological methods may be required.

Sometimes, before any attempt at individual identification of such remains can be undertaken, certain anthropological parameters need be established. Are the remains human or animal? If human, can the sex, age at death, population ancestry, and stature of the individual be determined or estimated within reasonable limits? Are the remains of more than one body commingled?

The radiological evaluation of these anthropological parameters, sometimes called the biological profile, is particularly applicable when the skeletal components are still partially or totally sheathed in soft tissues—however burned, macerated, decomposed, or mutilated. If the remains are completely skeletonized, or there are both facilities and time to deflesh the specimen, then the forensic anthropologist can do the job with equal, or in some cases, superior accuracy. Still, it is useful to x-ray the bones for possible comparative matching with antemortem radiological images in the future.

ANIMAL OR HUMAN

It is not uncommon for animal parts or bones to be brought to the attention of law enforcement agencies or forensic investigators. Ordinarily, the trained forensic scientist can easily determine the nonhuman nature of the specimen. At times, the true nature of the item may be obscured by its condition. A bear's paw, with claws and terminal phalanges torn away for souvenirs, and with hide and fur removed by the skinning knife or by decomposition, may closely resemble a human hand to the lay hunter or hiker who discovers it. The radiograph quickly reveals the difference (Figure 8.1).

Partially fleshed or skeletonized remains of large mammals may be confusing to the untrained finder. Again, someone trained in human anatomy or osteology—physician, dentist, physical anthropologist—will have no difficulty in detecting nonhuman characteristics of size, architecture, and configuration of intact animal bones (Figure 8.2). However, the most distinctive parts of both animal and human bones, the ends and/or articular surfaces, may be missing because of carnivoral activity, decomposition, or (especially in the case of immature bones with unfused epiphyses) scattering (Figure 8.3).

If only fragments of the shaft or diaphysis of bones remain, roentgenography can be most helpful. Bony ridges, processes, and excrescences related to muscle organs and insertions will be different in four-legged animals. Chilvarquer et al. point

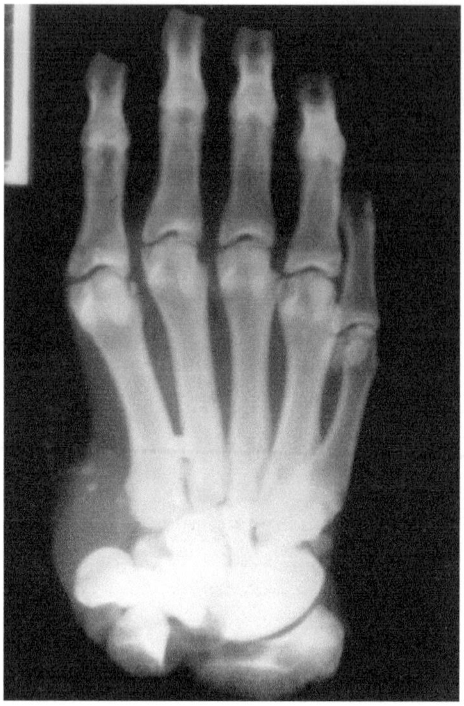

FIGURE 8.1 Bear paw mistaken for a human hand.

out differences in the roentgenographic appearance of the midshafts of human vs. animal bones.[1] The pattern of the human spongiosa and medullary canal is fairly regular, with rounded or ovoid "spaces" between coarse primary trabeculae and finer secondary trabeculae. The zone of transition between the dense cortex and less dense medulla may be 1–3 mm wide. (Disease states or aging causing osteoporosis sharpens the transition; conditions producing osteomalacia diffuses the corticomedullary junction.)

In animals the corticomedullary junction will be sharp. The spongiosa is less patterned and is more homogeneous or granular in appearance. Cortical spicules or invaginations may extend into the medullary canal from the cortical endosteum.

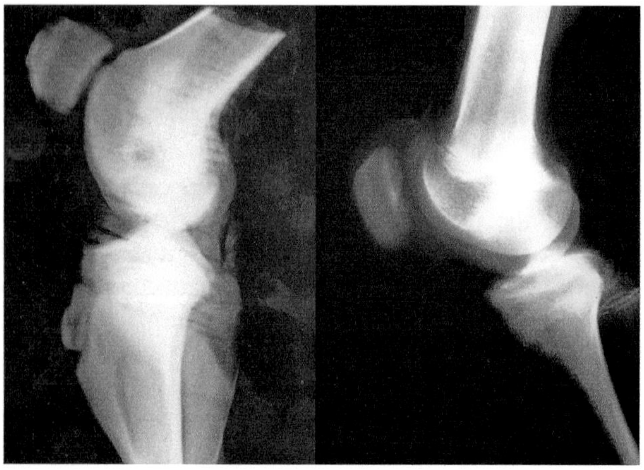

FIGURE 8.2 Left: lateral view of the knee of an immature pig (cured ham). Compare with Right: knee of adolescent human.

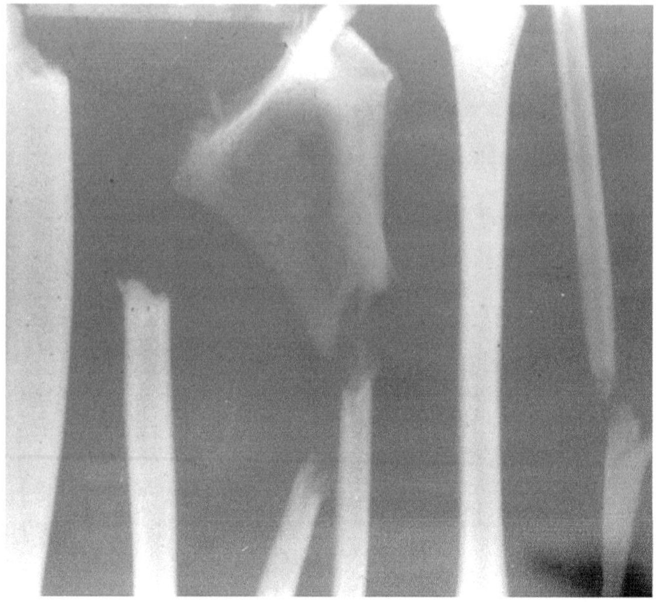

FIGURE 8.3 Radiograph of human skeletal remains recovered from coastal wetlands. The ends of most of the bones were destroyed or missing.

AGE DETERMINATION

Determination of age at time of death is an important step toward identification of unknown remains. Age can be established with considerable accuracy by roentgenography of the skeleton from the time of its appearance about the 20th week of gestation until early adulthood. This is possible due to the complex but dependable system by which the osseous framework of the body develops, grows, and matures. Most of the 206 bones of the human adult skeleton develop in cartilage precursors or anlagen from one or more primary centers of ossification (which make up the shaft or diaphysis of a long bone, the centrum of an axial or round bone) and secondary centers which develop the articular ends of the bones (epiphyses) or nonarticular processes (apophyses) for attachment of muscles, ligaments, and tendons (Figure 8.4). The appearance of these centers, and the fusion of secondary centers with the primary, follows a timetable, allowing rather precise aging if appropriate skeletal parts are available for evaluation.

Fetal age can be measured by crown-rump measurements, fetal length, femoral length, biparietal diameter, or skeletal maturation[2] (Figure 8.5). Fetal parts and soft tissues, if extrauterine, are small enough that radiological magnification will not be a major problem in view of the rather wide range of standard deviations for the various fetal measurements, most of which nowadays, are based on real-time intrauterine measurements by ultrasonography. Intrauterine fetuses imaged roentgenographically will be magnified (Figures 8.6 and 8.7 show systems of correcting for magnification).

Under ideal conditions, the intrauterine fetal skeleton may be seen as early as the 10th week of gestation,[3] but in practice it is not often visualized before the 18th or 20th week. The ossification centers that appear in the posterior elements of

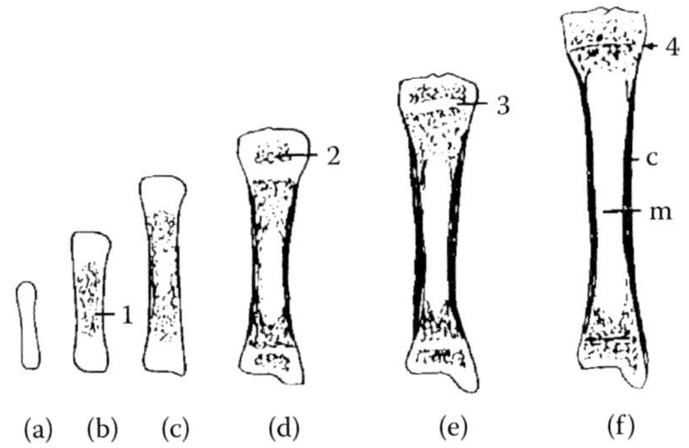

FIGURE 8.4 Development and ossification of a long bone. (a) cartilaginous anlage; (b) (1) Appearance of primary ossification center in shaft or diaphysis; (c) Primary center begins reorganization into cortex, medulla, and (with perichondral ossification) the metaphysis or growing end of the shaft; (d) Appearance of (2) secondary ossification center or epiphysis, the diaphysis and metapyhses are continuing to develop and mature; (e) The differential between cortex and medulla is well established, epiphyseal development continues with growth in length taking place at (3) the epiphyseal cartilaginous plate or physis; (f) The epiphyseal plates (4) have closed by ossification, growth has ceased; the bone has been modeled into its adult shape and form with well-defined cortex (c) and medullary canal (m).

the spine often are the first skeletal components seen radiologically as a chain of densities or "string of beads" sometimes accompanied by tiny rib shadows. The base of the skull and long bones may be visualized between 20 and 25 weeks if not obscured by maternal gas, bones, or other tissues (Figure 8.8).

The ossification center for the calcaneus appears between 24 and 26 weeks of gestation, followed by the center for the talus (Figure 8.9) in 2 weeks. In the live fetus, intrauterine movement often obliterates the image of these small parts. Between 36 and 40 weeks, the distal femoral epiphysis, followed by the proximal tibial epiphysis, will appear (Figure 8.10).

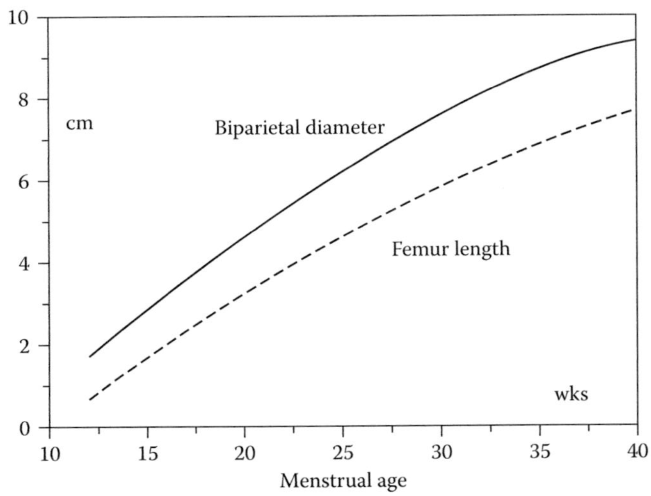

FIGURE 8.5 Graph presenting average biparietal diameter and femoral lengths in centimeters plotted against menstrual age of fetus. (Plotted from data in Reference 2 by M.D. Harpen, PhD.)

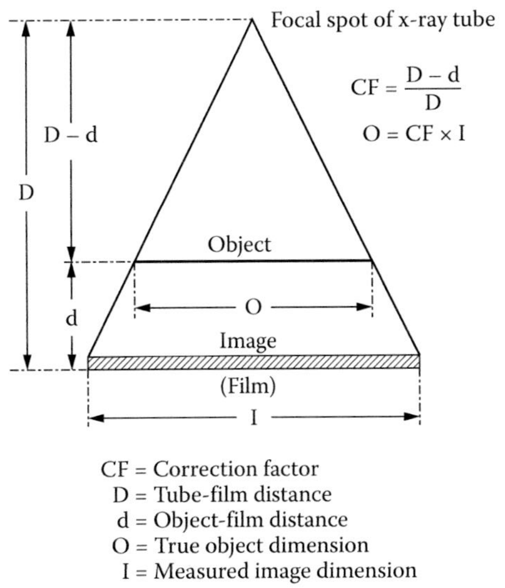

FIGURE 8.6 Diagrammatic representation of factors involved in roentgenographic image magnification and formula for magnification correction.

Before obstetrical ultrasonography, this was the best method of determining fetal maturity. The distal femoral epiphysis will be found in 90% or more of term fetuses, the proximal tibial epiphysis in 85% or more.

At birth the primary ossification centers (diaphyses) of the long bones of the extremities, including the hands and feet, are present. The vertebral bodies and posterior elements have begun their process of ossification, as have the scapulae, pelvic, clavicles, base of the skull, calvaria, and facial bones.[4]

For the next two decades, radiological determination of age is based on the appearance and eventual fusion of the secondary ossification centers (Figure 8.11) (Table 8.1). There are many standards for radiological bone age determination.[4–9] In general, female skeletal maturation precedes that of the male after the first few months of life. The final epiphysis to close is at the medial end of the clavicle during the third decade of life.[4]

The range of standard deviations for skeletal maturation of various ages is quite broad. Most of the radiological standards are based on Caucasian population ancestry, and even greater standard deviations may be found in children of different population ancestries.

For precise age determination in mass casualty studies and sorting out commingled children's remains, we tend to rely heavily on standards of development of the knee, foot, ankle, and especially, the hand and wrist[10–12] (Figures 8.12 through 8.14) (Table 8.2a and b). These standards must be used with reservations, particularly in Black and Hispanic girls and in Asian and Hispanic boys in late childhood and adolescence when their bone age may exceed the chronological age by 9 to 1/2 months.[13] It must be pointed out that, for subadults, dental age estimation, when feasible, is a more reliable estimate of chronological age than is bone age evaluation.[14]

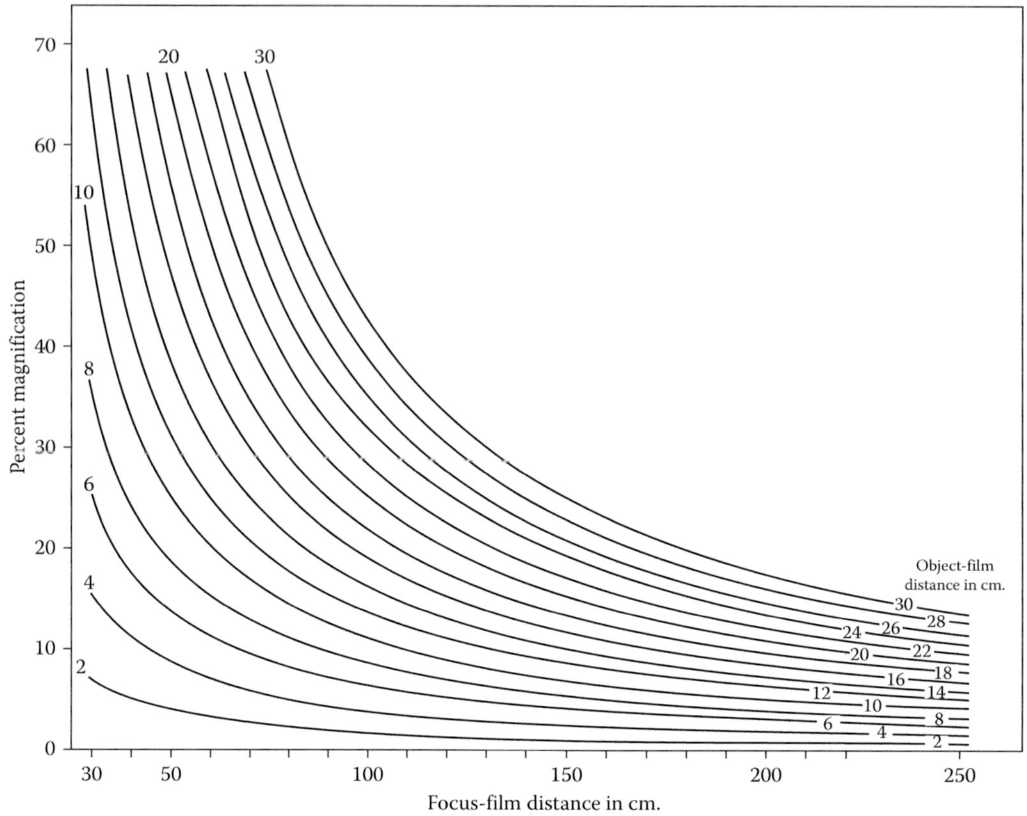

FIGURE 8.7 Graph constructed by T. H. Oddie, D.Sc. allowing calculation of percent magnification of an object when target-film and object-film distances (in centimeters) are known. (Reproduced from Meschan, I, *An Atlas of Anatomy Basic to Radiology*, W. B. Saunders, Philadelphia, 1975, p. 1078. With permission.)

From the age range of about 25–50, anthropologists rely heavily on the external appearance of the skull sutures, the pubic symphysis, and rib ends, and these criteria do not lend themselves readily to routine radiological evaluation.

Beginning at about age 40, wear-and-tear degenerative changes start to appear at the margins of the articular surfaces of major joints, especially at the margins of vertebral end-plates (Figure 8.15). These changes of spurring and lipping, sometimes called deforming spondylosis in the spine, are quite variable and modified by occupation, level and kinds of activity, disease, and heredity. These degenerative changes progress with age and are usually more prominent in males. In females vertebral changes of osteoporosis, deformity, and collapse began about the time of menopause and progress. Here again, variability depends on many factors including exercise, diet, and hormonal balance. Ordinarily, an experienced radiologist can estimate adult age from

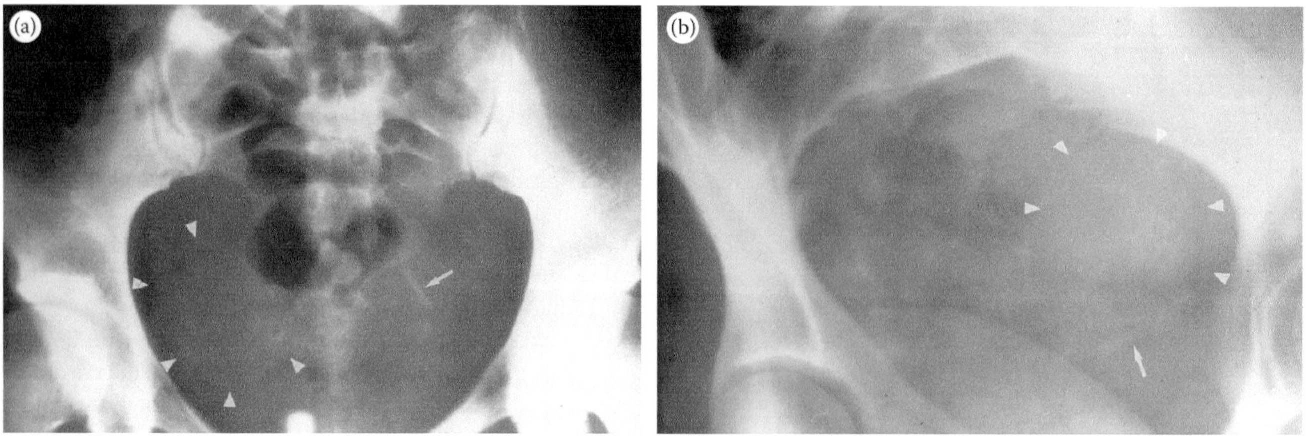

FIGURE 8.8 (a) Frontal and (b) oblique roentgenogram of a 20-week fetus *in utero*. The skull (triangles) and long bones (arrows) are seen faintly within the maternal pelvis.

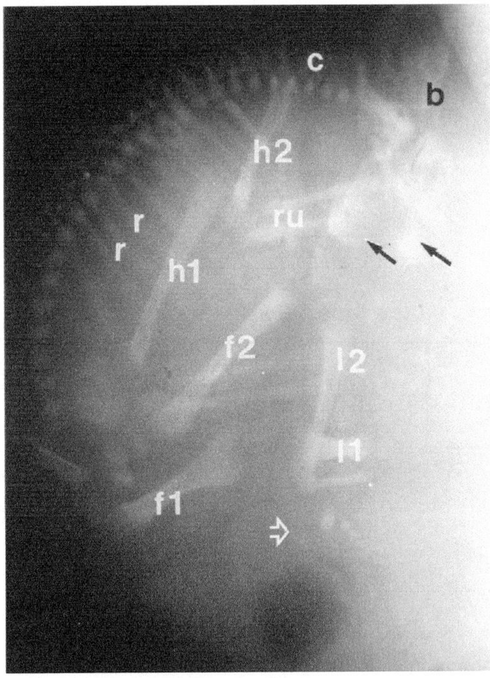

FIGURE 8.9 A 28-week fetus *in utero* (mother lying on her side). The upside extremities are more magnified than the downside extremities (which are closer to the film). Some bones are parallel to the film; others are angled away from the film and appear distorted and foreshortened. Code: (b) base of skull; (c) cervical vertebrae; (h1) upside humerus; (h2) downside humerus; (ru) downside forearm (radius and ulna) with faint metacarpals at end; overlapping mandible and maxilla (arrows); (r) ribcage; (f1) upside femur, foreshortened; (F2) downside femur; (l1) upside leg (tibia and fibula, foreshortened; (l2) downside leg (tibia and fibula overlapped) with foot at end (open arrow) containing calcaneus, talus, and metatarsal bones.

skeletal findings in a range of ±5–10 years, the range widening with age.

The costal cartilages tend to become mineralized with age and may be visualized radiographically on prepared "chest plate" specimens as early as age 15 years. (The "chest plate" in anthropological terminology consists of the terminal 4–6 cm of rib ends, the costal cartilages, the sternum and, sometimes. the medial ends of the clavicles removed from the body for fine detail radiography.) In standard chest radiography, such ossification is rarely noted before 30 years of age and is not prominent before age 50. Again, there is great variability between individuals. Attempts to establish chronological age or age at death by evaluation of costal cartilage mineralization is imprecise. McCormick[15] found that similar degrees of ossification "over a wide age span during middle years seriously limits the value of this method," and its best recommendations for use were "ease, rapidity and relative inexpensiveness." Barrès and coworkers[16] found that the accuracy of age estimation by radiological evaluation of "chest plates" had a standard error of 8.5 years, which translates to a 95% confidence interval of ±17 years. Stewart and McCormick[17] have described a particular pattern of costal cartilage ossification (which they term Type A) specific to females of advanced age. Type A ossification usually appears no earlier than the mid-50s and becomes well-developed and relatively dense only after age 65 years. In mature adults, the physical anthropologist working with defleshed bones may achieve a higher range of accuracy in determining age at death.

So far, assessment of age has been discussed in the context of identification of remains. There is another medicolegal

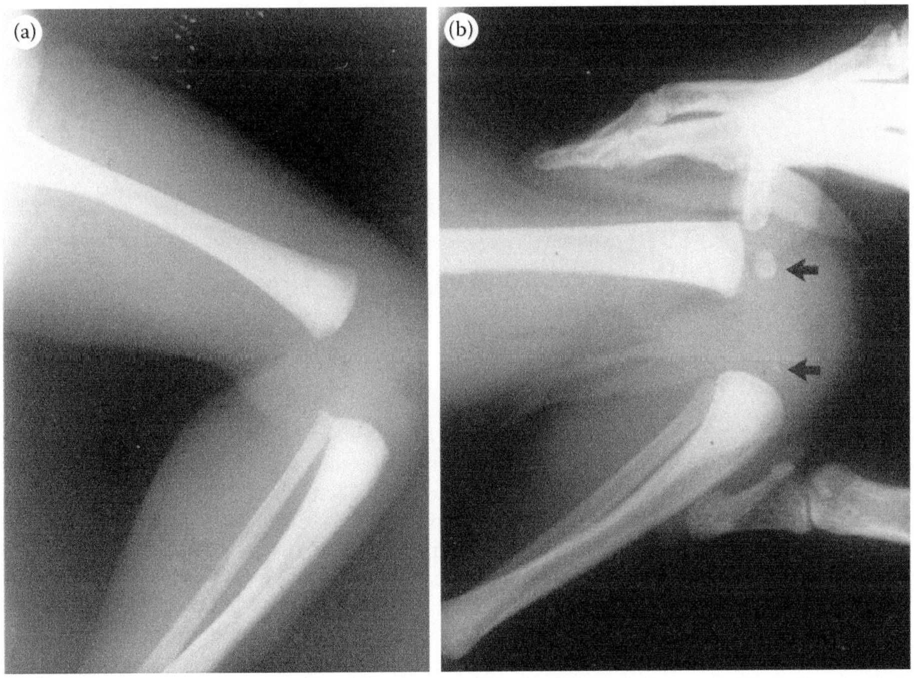

FIGURE 8.10 (a) Knee of premature newborn. The distal femoral and proximal tibial epiphyses are not ossified. (b) Term newborn with knee epiphyses (arrows) present.

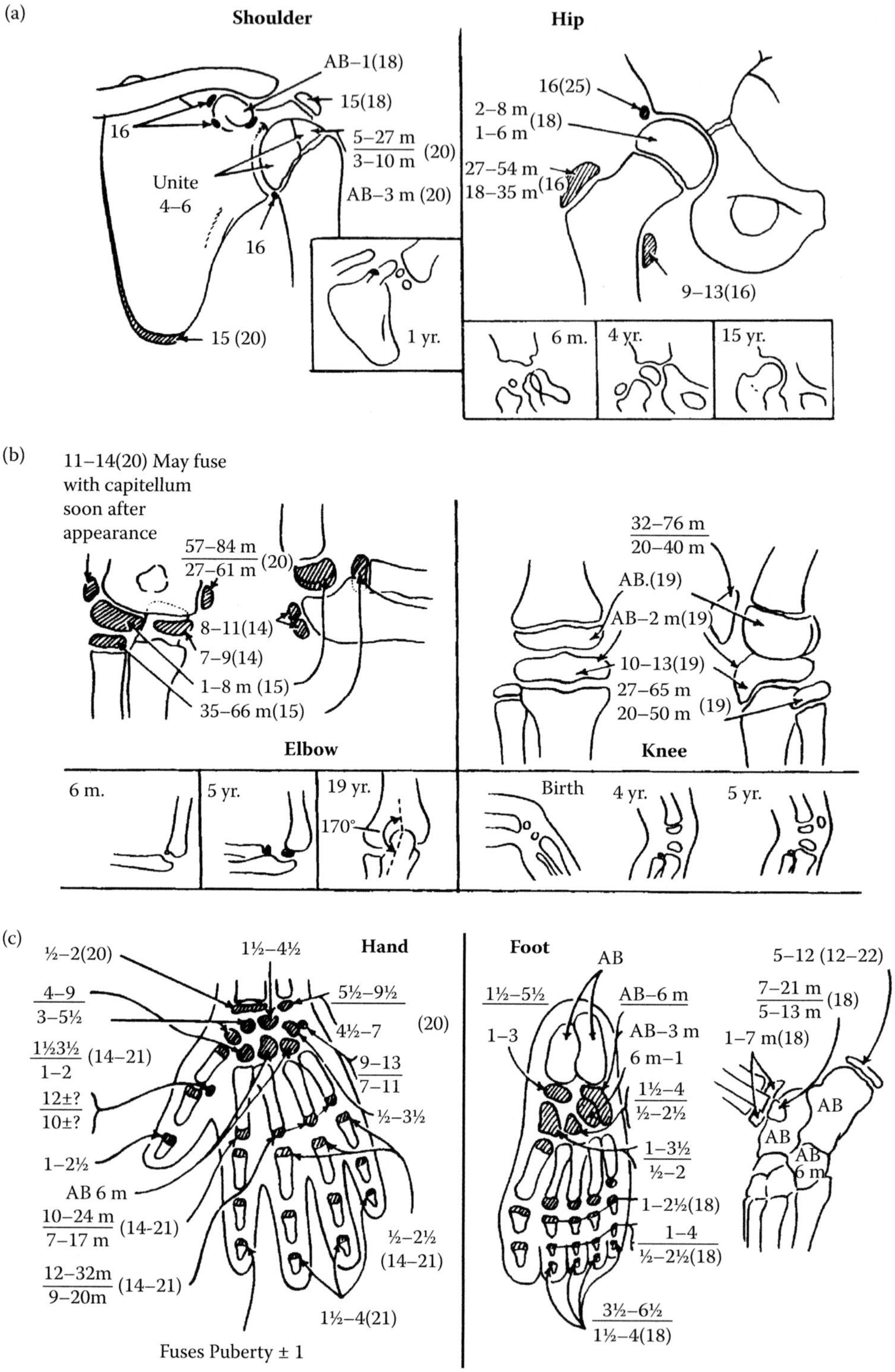

FIGURE 8.11 Skeletal maturation chart from Girdany and Golden. (a) Shoulder and hip; (b) elbow and knee; (c) hand and foot; (d) vertebra; (e) sacrum and coccyx; (f) rib and clavicle. Note: The numbers in figures indicates the range of appearance time for centers of ossification from the 10th to the 90th percentile. Numbers followed by an "m" mean months, all others are in years. When two sets of numbers are given for a single ossification center, the upper one refers to males, the lower one to females. A single get of figures applies to both sexes. AB indicates the center is present at birth. Numbers in parantheses give the appromate time of fusion. (From Girdany, B. R. and Golden, R., *Am. J Roentgenol.*, 68, 922, 1952. With permission.)

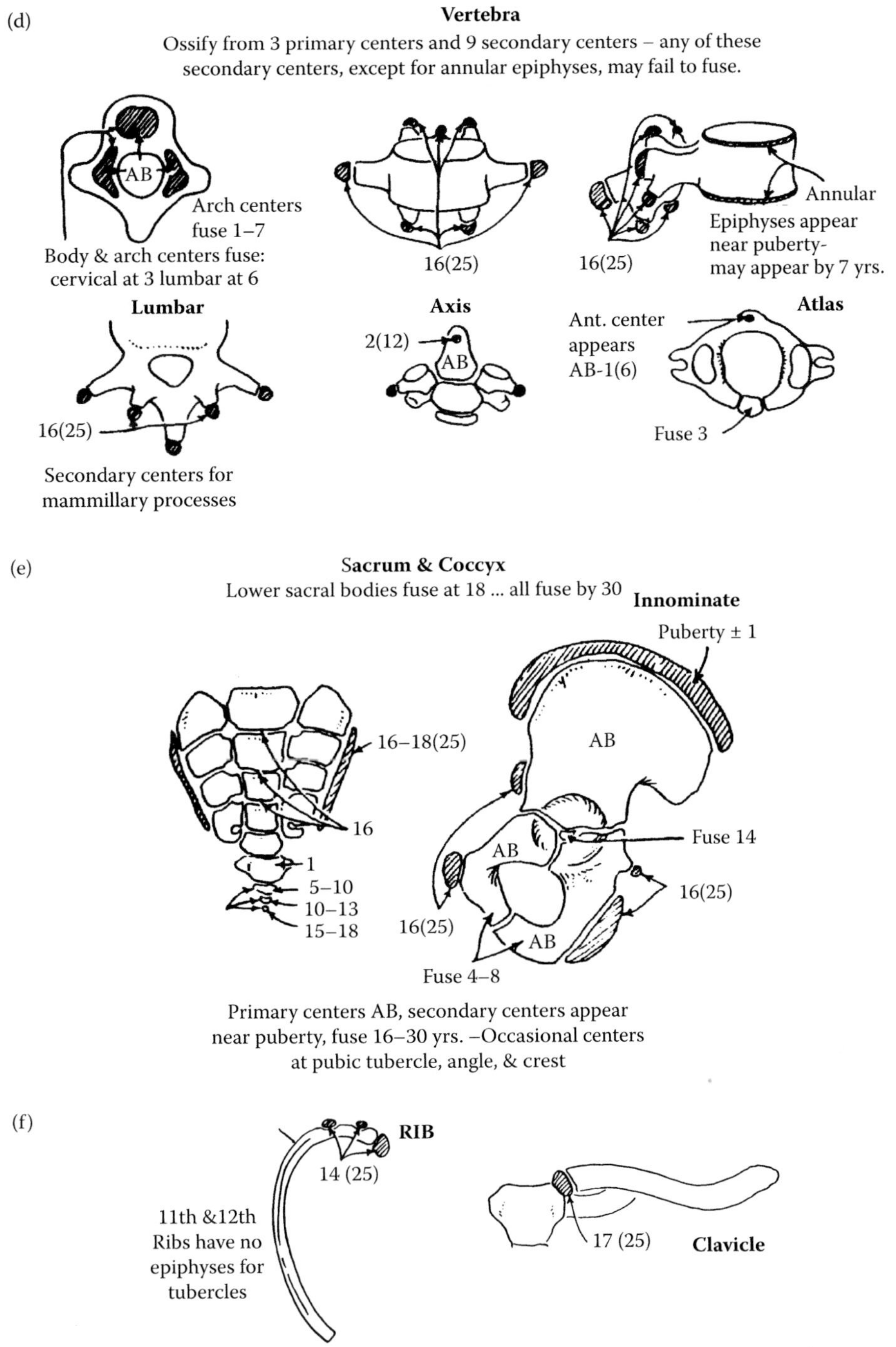

FIGURE 8.11 *(Continued)*

indication for radiological age estimation in the living. When the age of a defendant is unknown (usually in primitive cultures) bone age may determine whether he is tried as a juvenile or an adult, or may determine the severity of the punishment. According to newspaper accounts, the young African native who murdered Joy Adamson (of Elsa the lion and *Born Free* fame) was saved from the hangman's rope by a justice who ruled, contrary to a radiologist's expert opinion, that the perpetrator was below age 18 when the crime was committed.

SEX DETERMINATION

It has been pointed out already that skeletal development maturation in females is accelerated over that of males after the third or fourth year of life. However, differentiation of sexes by skeletal radiology is unreliable until after puberty.[4] It is then that the sexual characteristics discernible by radiography begin to appear.

In general, the male skeleton is more robust and heavier, with more prominent attachment for muscles and tendons.

TABLE 8.1
Age–at–Appearance (Years–Months) Percentiles for Selected Ossification Centers

		Boys			Girls		
	Centers	**5th**	**50th**	**95th**	**5th**	**50th**	**95th**
1.	Humerus, head	—	0–0	0–4	—	0–0	0–4
2.	Tibia,proximal	—	0–0	0–1	—	0–0	0–0
3.	Coracoid process of scapula	—	0–0	0–4	—	0–0	0–5
4.	Cuboid	—	0–1	0–4	—	0–1	0–2
5.	Capitate	—	0–3	0–7	—	0–2	0–7
6.	Hamate	0–0	0–4	0–10	—	0–2	0–7
7.	*Capitellum of humerus*	0–1	0–4	1–1	0–1	0–3	0–9
8.	Femur, head	0–1	0–4	0–8	0–0	0–4	0–7
9.	Cuneiform 3	0–1	0–6	1–7	—	0–3	1–3
10.	Humerus, greater tuberosity	0–3	0–10	2–4	0–2	0–6	1–2
11.	Toe phalanx 5 M	—	1–0	3–10	—	0–9	2–1
12.	Radius, distal	0–6	1–1	2–4	0–5	0–10	1–8
13.	Toe phalanx 1 D	0–9	1–3	2–1	0–5	0–9	1–8
14.	Toe phalanx 4 M	0–5	1–3	2–11	0–5	0–11	3–0
15.	Finger phalanx 3 P	0–9	1–4	2–2	0–5	0–10	1–7
16.	Toe phalanx 3 M	0–5	1–5	4–3	0–3	1–0	2–6
17.	Finger phalanx 2 P	0–9	1–5	2–2	0–5	0–10	1–8
18.	Finger phalanx 4 P	0–10	1–6	2–5	0–5	0–11	1–8
19.	Finger phalanx 1 D	0–9	1–6	2–8	0–5	1–0	1–9
20.	Toe phalanx 3 P	0–11	1–7	2–6	0–6	1–1	1–11
21.	Metacarpal 2	0–11	1–7	2–10	0–8	1–1	1–8
22.	Toe phalanx 4 P	0–11	1–8	2–8	0–7	1–3	2–1
23.	Toe phalanx 2 P	1–0	1–9	2–8	0–8	1–2	2–1
24.	Metacarpal 3	0–11	1–9	3–0	0–8	1–2	1–11
25.	Finger phalanx 5 P	1–0	1–10	2–10	0–8	1–2	2–1
26.	Finger phalanx 3 M	1–0	2–0	3–4	0–8	1–3	2–4
27.	Metacarpal 4	1–1	2–0	3–7	0–9	1–3	2–2
28.	Toe phalanx 2 M	0–11	2–0	4–1	0–6	1–2	2–3
29.	Finger phalanx 4 M	1–0	2–1	3–3	0–8	1–3	2–5
30.	Metacarpal 5	1–3	2–2	3–10	0–10	1–4	2–4
31.	Cuneiform 1	0–11	2–2	3–9	0–6	1–5	2–10
32.	Metatarsal 1	1–5	2–2	3–1	1–0	1–7	2–3
33.	Finger phalanx 2 M	1–4	2–2	3–4	0–8	1–4	2–6
34.	Toe phalanx 1 P	1–5	2–4	3–4	0–11	1–7	2–6
35.	Finger phalanx 3 D	1–4	2–5	3–9	0–9	1–6	2–8
36.	Triquetrum	0–6	2–5	5–6	0–3	1–8	3–9
37.	Finger phalanx 4 D	1–4	2–5	3–9	0–9	1–6	2–10
38.	Toe phalanx 5 P	1–6	2–5	3–8	1–0	1–9	2–8
39.	Metacarpal 1	1–5	2–7	4–4	0–11	1–7	2–8
40.	*Cuneiform 2*	1–2	2–8	4–3	0–10	1–10	3–0
41.	Metatarsal 2	1–11	2–10	4–4	1–3	2–2	3–5
42.	Femur, greater trochanter	1–11	3–0	4–4	1–0	1–10	3–0
43.	Finger phalanx 1 P	1–10	3–0	4–7	0–11	1–9	2–10
44.	Navicular of foot	1–1	3–0	5–5	0–9	1–11	3–7
45.	Finger phalanx 2 D	1–10	3–2	5–0	1–1	2–6	3–3
46.	Finger phalanx 5 D	2–1	3–3	5–0	1–0	2–0	3–5
47.	Finger phalanx 5 M	1–11	3–5	5–10	0–11	2–0	3–6
48.	Fibula, proximal	1–10	3–6	5–3	1–4	2–7	3–11
49.	Metatarsal 3	2–4	3–6	5–0	1–5	2–6	3–8

TABLE 8.1 (continued)
Age–at–Appearance (Years–Months) Percentiles for Selected Ossification Centers

		Boys			Girls		
	Centers	5th	50th	95th	5th	50th	95th
50.	Toe phalanx 5 D	2–4	3–11	6–4	1–2	2–4	4–1
51.	Patella	2–7	4–0	6–0	1–6	2–6	4–0
52.	Metatarsal 4	2–11	4–0	5–9	1–9	2–10	4–1
53.	Lunate	1–6	4–1	6–9	1–1	2–7	5–8
54.	Toe phalanx 3 D	3–0	4–4	6–2	1–4	2–9	4–1
55.	Metatarsal 5	3–1	4–4	6–4	2–1	3–3	4–11
56.	Toe phalanx 4 D	2–11	4–5	6–5	1–4	2–7	4–1
57.	Toe phalanx 2 D	3–3	4–8	6–9	1–6	2–11	4–6
58.	Radius, head	3–0	5–3	8–0	2–3	3–10	6–3
59.	Navicular of wrist	3–7	5–8	7–10	2–4	4–1	6–0
60.	Greater multangular	3–6	5–10	9–0	1–11	4–1	6–4
61.	*Lesser multangular*	3–1	6–3	8–6	2–5	4–2	6–0
62.	Medial epicondyle of humerus	4–3	6–3	8–5	2–1	3–5	5–1
63.	Ulna, distal	5–3	7–1	9–1	3–3	5–4	7–8
64.	Calcaneal apophysis	5–2	7–7	9–7	3–6	5–4	7–4
65.	Olecranon of ulna	7–9	9–8	11–11	5–7	8–0	9–11
66.	Lateral epicondyle of humerus	9–3	11–3	13–8	7–2	9–3	11–3
67.	*Tibial tubercle*	9–11	11–10	13–5	7–11	10–3	11–10
68.	Adductor sesamoid of thumb	11–0	12–9	14–7	8–8	10–9	12–8
69.	Os acetabulum	11–11	13–6	15–4	9–7	11–6	13–5
70.	Acromion	12–2	13–9	15–6	10–4	11–11	13–9
71.	Iliac crest	12–0	14–0	15–11	10–10	12–9	15–4
72.	Coracoid apophysis	12–9	14–4	16–4	10–4	12–3	14–4
73.	Ischial tuberosity	13–7	15–3	17–1	11–9	13–11	16–0

Source: From Graham, C. B., *Radiol. Clin. N. Am.*, 10, 185, 1972. With permission.
Note: P = proximal, M = middle, D = distal. Important "happenings" at various ages are in bold face type.

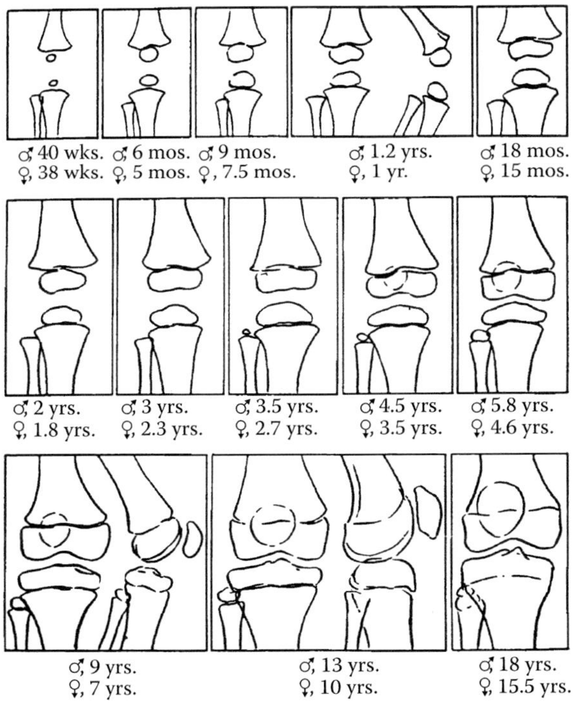

FIGURE 8.12 Tracings (From Meschan, I, *An Atlas of Anatomy Basic to Radiology*, W. B. Saunders, Philadelphia, 1975, p. 56. With permission) of standard roentgenograms of the maturing knee. (From Pyle, S. I. and Hoerr, N. L., *Atlas of Skeletal Development of the Knee*, Charles C. Thomas, Springfield, IL, 1955. With permission.)

With aging, there is a tendency for more degenerative and hyperostotic changes in the male skeleton. Male long bones are about 110% the length of female long bones. The male femoral head is larger in all dimensions. All of these general findings are helpful but not definitive in establishing the sex of unidentified human remains. There are certain skeletal components, and both skeletal and extraskeletal findings, which are more useful in determining sex.

PELVIS

The bony pelvis often survives the onslaught of factors which diminish or destroy the usefulness of other body parts. This is fortunate since the pelvis offers the most definitive traits of sexual differentiation[18,19] (Figure 8.16).

The *subpubic arch* (subpubic concavity) is narrow and triangular with an inverted V-shape in the male and broad and with an inverted U-shape in the female.

The *pubic bone* tends to be long and narrow in the male and broad and rectangular in the female.

The *sciatic notch* is deep and narrow in the male, and is wide and shallow in the female.

The *preauricular sulcus* (paraglenoid sulcus) when present is one of the most dependable indicators of femaleness. This variable groove in the ilium at the inferior end of the sacroiliac joint is missing or manifest very rarely as a thin groove in the male,[19] and a deep groove is found only in females[20] (Figure 8.17). The groove or sulcus is believed to represent resorption of bone at the insertion of the anterior sacroiliac ligament, much as the costoclavicular ligament produces the rhomboid fossa in the anteroinferior end of the clavicle. Deep grooves are found only in subjects in the mid-fourth decade of life or beyond and only in parous women. The depth of the groove is influenced by multiple pregnancies, genetic differences, level of physical activity, and perhaps, the degree of lumbar lordosis.[20] While a deep groove denotes a female, not all women have a preauricular sulcus. Among the bodies recovered from the Air India disaster, a small notch was found in a 14-year-old female pelvis, and the oldest female without a periglenoid sulcus was 51 at the time of death.

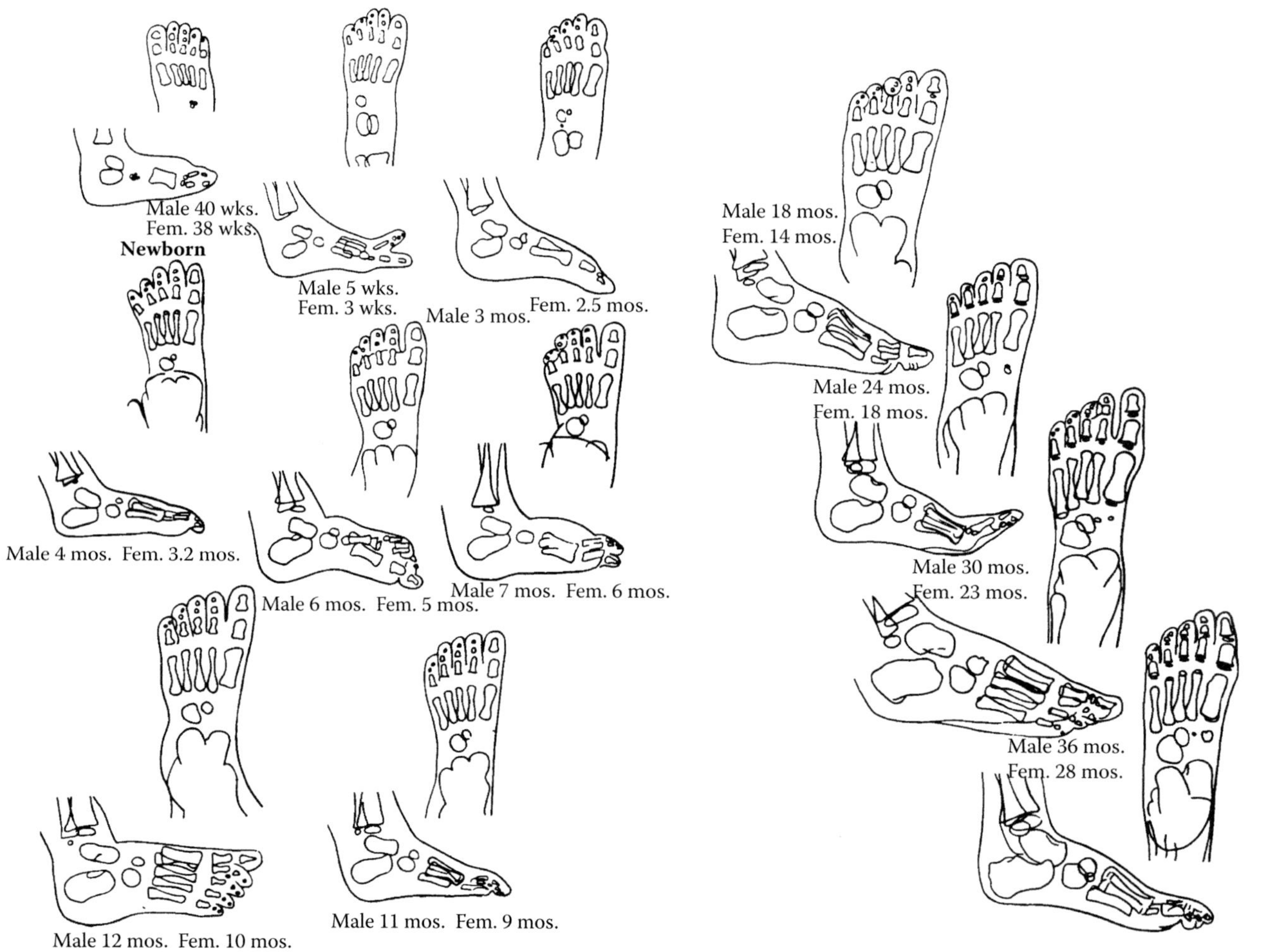

FIGURE 8.13 Tracings (From Meschan, I., *An Atlas of Anatomy Basic to Radiology*, W. B. Saunders, Philadelphia, 1975, pp. 54–55. With permission) of standard roentgenograms of the maturing foot and ankle. (From Hoerr, N. L., Pyle, S. I., and Francis, C. C., *Radiologic Atlas of the Foot and Ankle*, Charles C. Thomas, Springfield, IL, 1962. With permission.)

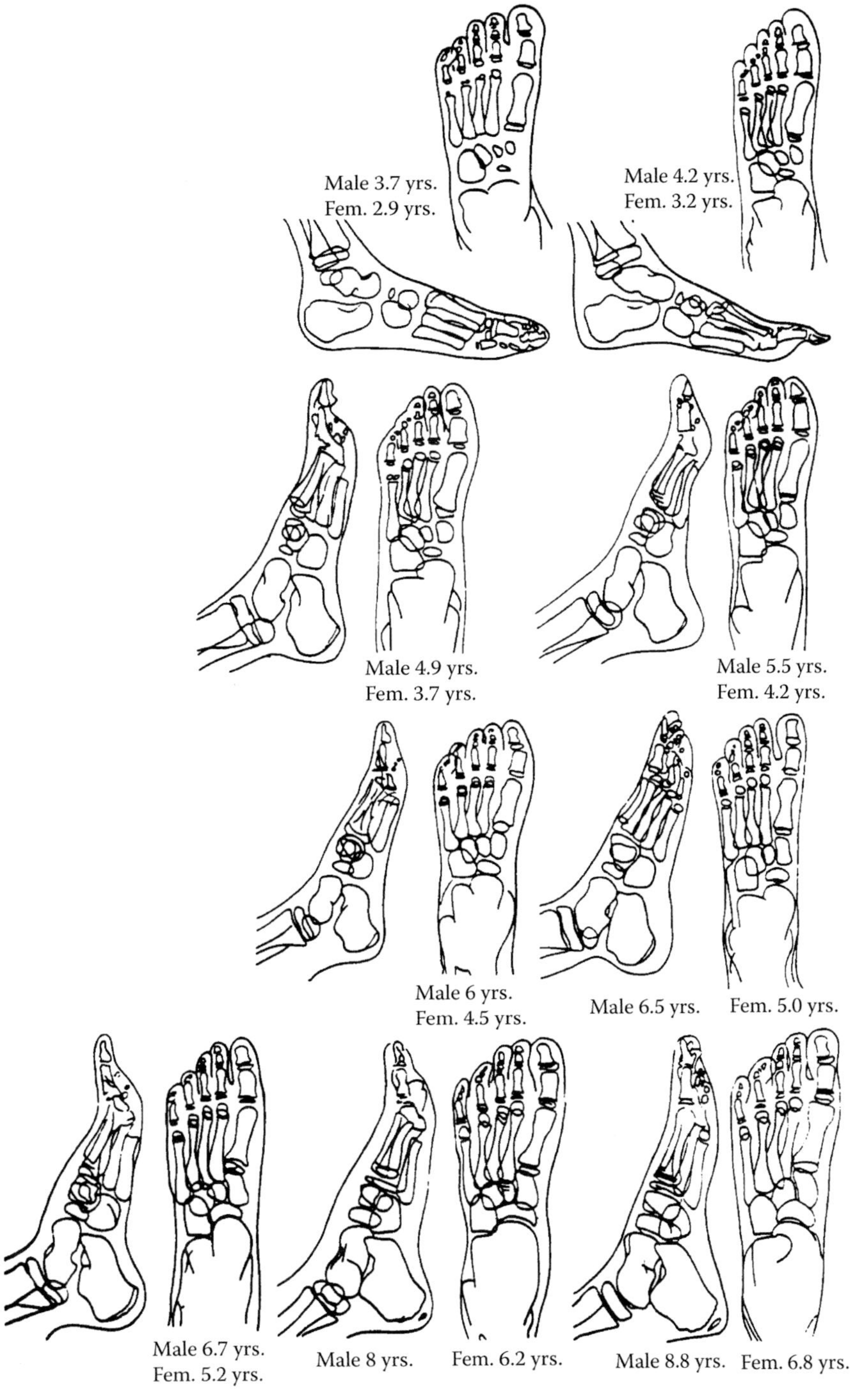

FIGURE 8.13 (*Continued*)

The *obturator foramina* are large and oval or round in the male, but small and triangular in the female.

The *acetabular fossae* are large and directed laterally in the male, and are smaller and directed anterolaterally in the female.

The *ilial alae* are high and vertical in the male, and broad and laterally divergent in the female.

The *sacrum* in the male is narrow, has a relatively flattened curve and has five or more segments. The female sacrum is broad and short with five segments and an anterior concavity.

The *pelvic inlet* is triangular or heart-shaped in the male, ovoid in the female.

The *muscle markings* are more prominent and rugged in the male. The female bony pelvis tends to be smooth and gracile.

Osteitis condensans ilii is a triangular area of increased bony density on the ilial side of the sacroiliac joint, usually bilateral, found almost exclusively in parous women in the child-bearing years (Figure 8.18). The joint, per se, is unaffected. It may be caused by the stress of pregnancy and childbirth. There

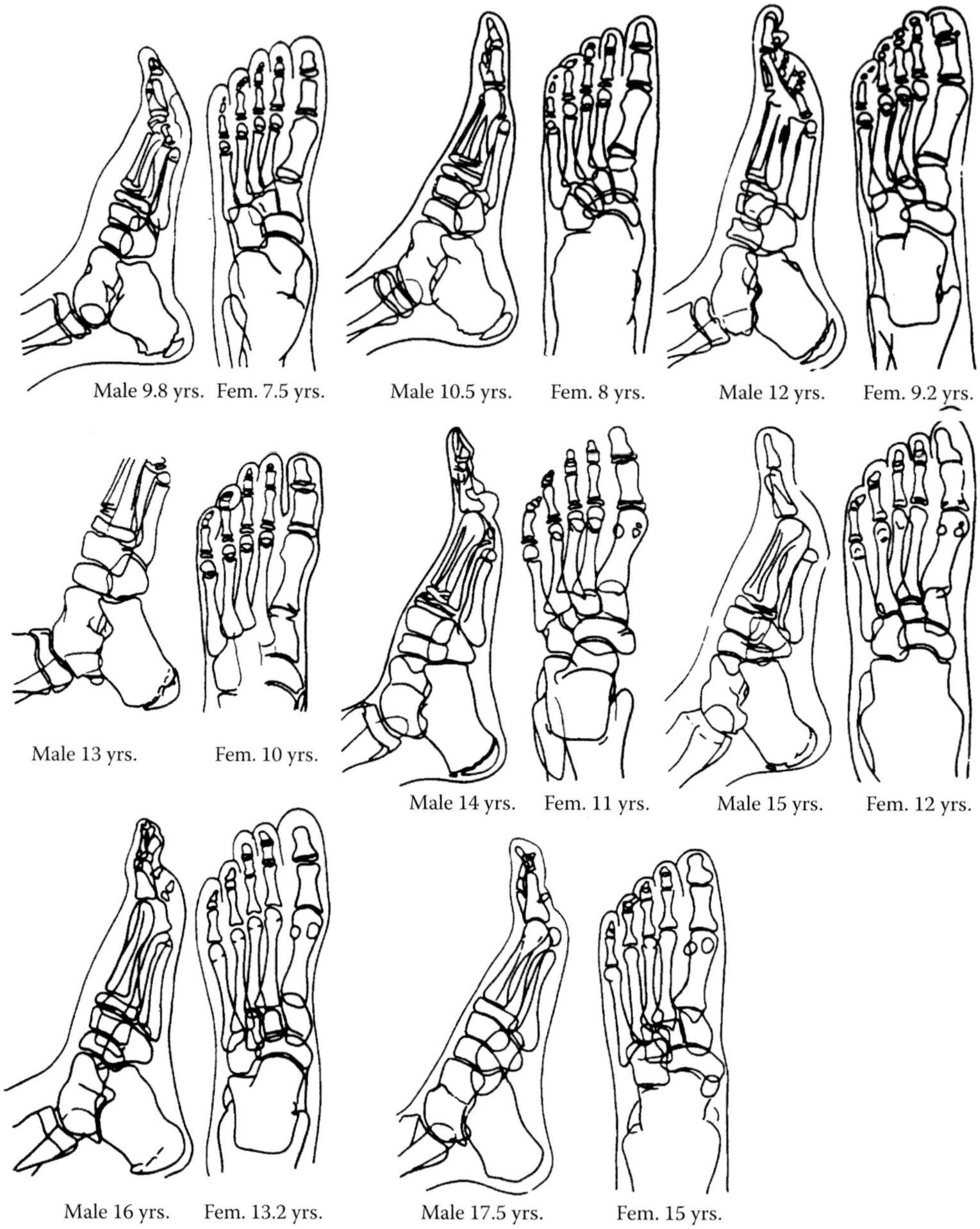

FIGURE 8.13 (*Continued*)

is correlation with the presence of deep preauricular sulci. The condition is self-limited and disappears; consequently, it is not to be found in the elderly female.[20–22]

SKULL AND MANDIBLE

Bony characteristics of the skull and mandible may be useful in assigning sexual identification to unknown remains[23–25] (Figure 8.19). The male skull tends to range from mesocephalic to dolichocephalic; the female skull is more likely to be mesocephalic to brachycephalic. The male has a larger brow or supraorbital ridge and a more sloping forehead. The male zygomatic arch is wider and heavier. The male inion or nuchal crest is prominent. The male mastoid process is larger and heavier. The male mandible is larger and more rugged with a wide ascending ramus. Male orbits tend to be larger and higher. The inferior nasal spine is longer in the male.

Hyperostosis interna frontalis is an overgrowth of the inner table of the frontal bone, often florid, found almost exclusively in middle-aged or older females and is a valuable characteristic for sex determination (Figure 8.20).

Parietal thinning[21,26] is a condition of postmenopausal females in which profound osteoporosis causes symmetrical resorption and virtual disappearance of the outer table and diploë of the parietal bones (Figure 8.21).

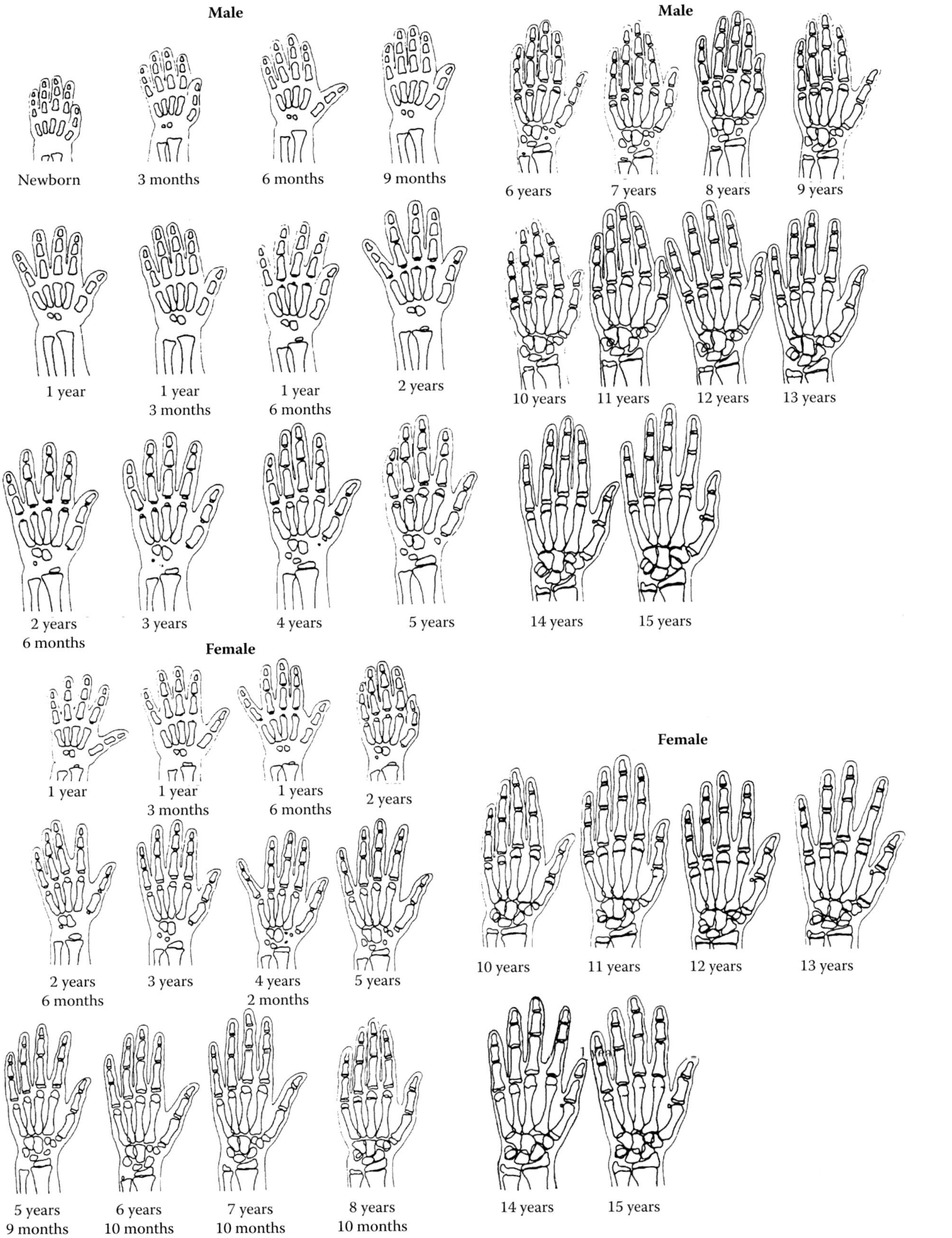

FIGURE 8.14 Tracings of male and female standards of skeletal development of the hand and wrist. There is no difference between sexes for the first year of life; then development accelerates in females. (From Greulich, W. W. and Pyle, S. I., *Radiographic Atlas of Skeletal Development of the Hand and Wrist*, 2nd ed., Stanford University Press, Stanford, CA, 1959. With permission.) (Tracings from Keats, T. E, *Atlas of Roentgenographic Measurement*, 6th Mosby Year Book, St. Louis, 1990, chap. 4B. With permission.)

TABLE 8.2a
Means and Standard Deviations for Skeletal Age (Hand)—Boys

		Skeletal Age (in Months)				Skeletal Age (in Months)	
Chronological Age	Number of Hand-Films	Mean	Standard Deviation	Chronological Age	Number of Hand-Films	Mean	Standard Deviation
12 months	66	12.7	2.1	7 years	66	80.6	10.1
18 months	67	17.5	2.7	8 years	63	92.5	10.8
2 years	67	22.6	4.0	9 years	63	104.9	11.0
1/2 years	67	28.1	5.4	10 years	63	118.0	11.4
3 years	67	33.8	6.0	11 years	65	132.1	10.5
1/2 years	67	39.5	6.6	12 years	64	144.5	10.4
4 years	65	44.8	7.0	13 years	66	156.4	11.1
1/2 years	64	50.3	7.8	14 years	65	168.5	12.0
5 years	64	56.2	8.4	15 years	65	180.7	14.2
1/2 years	64	62.4	9.1	16 years	65	193.0	15.1
6 years	66	68.4	9.3	17 years	60	206.0	15.4

Source: From Greulich, W. W. and Pyle, S. I., *Radiographic Atlas of Skeletal Development of the Hand and Wrist*, 2nd ed., Stanford University Press, Palo Alto, CA, 1959. With permission.

TABLE 8.2b
Means and Standard Deviations for Skeletal Age (Hand)—Girls

		Skeletal Age (in Months)				Skeletal Age (in Months)	
Chronological Age	Number of Hand-Films	Mean	Standard Deviation	Chronological Age	Number of Hand-Films	Mean	Standard Deviation
12 months	65	12.7	2.7	6 years	67	70.4	9.0
18 months	66	18.4	3.4	7 years	67	82.0	8.3
2 years	66	23.7	4.0	8 years	67	94.0	8.8
1/2 years	65	29.0	4.8	9 years	67	105.9	9.3
3 years	66	34.5	5.6	10 years	66	119.0	10.8
1/2 years	66	40.6	6.5	11 years	66	132.9	12.3
4 years	67	46.4	7.2	12 years	66	147.2	14.0
1/2 years	67	52.3	8.0	13 years	66	160.3	14.6
5 years	67	58.1	8.6	14 years	63	172.4	12.6
1/2 years	67	63.9	8.9	15 years	61	184.3	11.2

Source: From Greulich, W. W. and Pyle, S. I., *Radiographic Atlas of Skeletal Development of the Hand and Wrist*, 2nd ed., Stanford University Press, Stanford, CA, 1959. With permission.
CC: Mongoloid (see Table 8.2b).

STERNUM

The gender-predictive value of sternal length is not often used radiographically because it requires cross-table radiographs of the chest with a partially radiopaque ruler in place. With "chest plate" preparations direct measurements can be obtained. A combined length of manubrium and gladiolus of 17.3 cm includes only males; a combined length of manubrium and glandulous of less than 12.1 cm includes only females. Sternal lengths of 14.3 to 15.7 cm were indeterminate.[27]

OTHER AREAS

Bi partite patella is a common anatomical variant in adolescents, and is nine times more common in boys than girls. It is seen as a separate ossicle (or ossicles) occupying the upper outer quadrant of the patella. It occurs in approximately 2% of the population and is bilateral in 40–80% of cases.[28]

Krogman states that the chance of correctly sexing bones is 100% if the entire skeleton is available, 95% with the skull and pelvis or with the long bone and pelvis, and 90% with the skull alone or with the long bones and skull.[25]

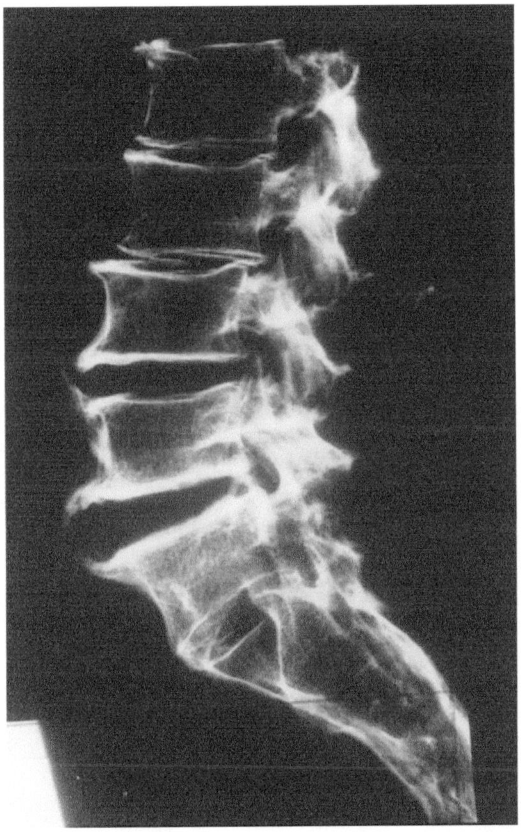

FIGURE 8.15 Degenerative hypertrophic changes (spurring, lipping, osteophyte production) at the margins of vertebral endplates.

Costal cartilage mineralization patterns as a distinctive finding between sexes was first reported by Sanders[29] (Figure 8.22). He noted that the typical male pattern is that of continuous parallel ossification of the upper and lower borders of the cartilage as it extends from the rib end (Figures 8.22A and 8.23). The typical female pattern is a tongue like or triangular mineralization extending from the rib end into the centrum of the cartilage (Figures 8.22C and 8.24). An uncommon pattern, more common in females, is that of two parallel lines extending from the center of the rib and into the adjacent

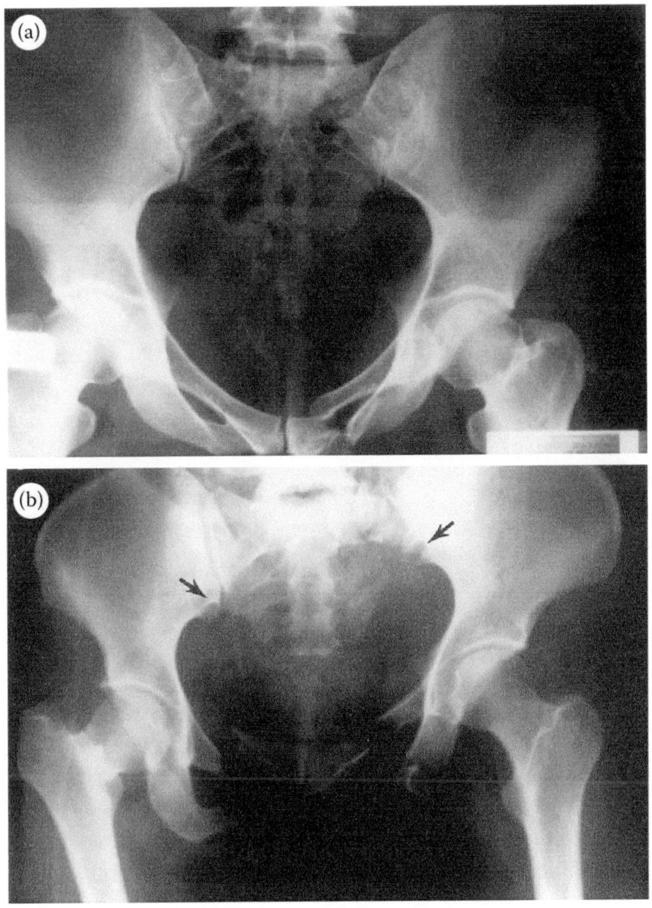

FIGURE 8.17 (a) Male pelvis and (b) female pelvis, victims of the Air India disaster. Note the differences in configuration and, especially, the preauricular sulcus (arrows) in the female.

cartilage (Figure 8.22D). A pitfall is that the male pattern tends to first appear on the inferior images of the costal cartilage (Figures 8.22B and 8.25) and may be mistaken for the female "tongue," which is always central to the cartilage. Finally, the pattern of a central rounded mineral collection in costal cartilage, sometimes with a more lucent center, is believed to be specific for elderly females[17] (Figures 8.22E). Navani et al.[30] believe that the predictive value of the parallel

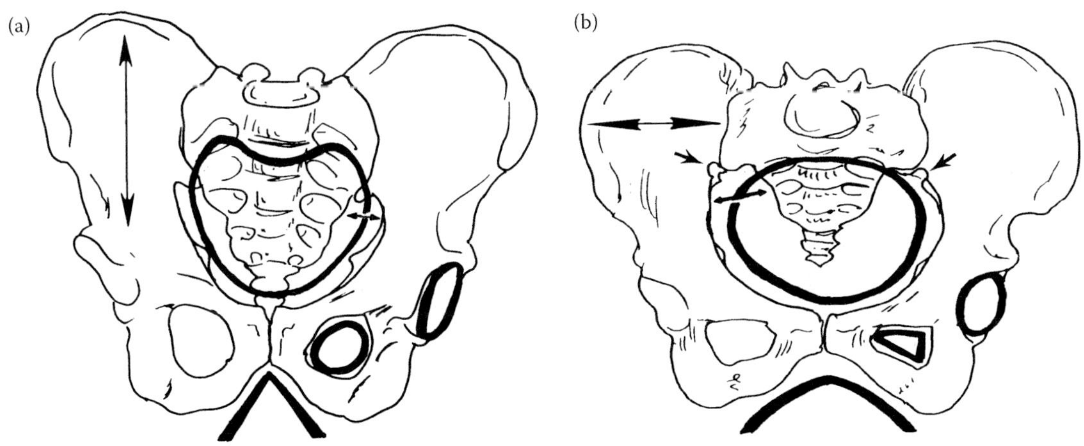

FIGURE 8.16 (a) Male pelvis. (b) Female pelvis. Differential characteristics are emphasized with bold lines and arrows (see text).

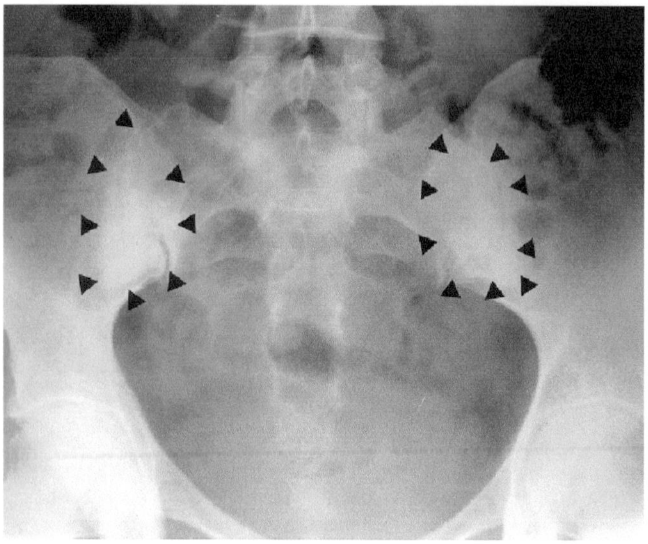

FIGURE 8.18 Osteitis condensans ilii. A triangular area of increased density (arrowheads) on the ilial side of the sacroiliac joint in women during the child-bearing years.

marginal male pattern to be 95%, the predictive value of the tongue-like central female-type mineralization to be 93%, and mixtures or combinations of those two patterns to be more likely found in females (predictive value 57%).

Calcification of tracheobronchial cartilage is found in only a small percentage of the people, overwhelming female.[21]

Calcification or ossification of the thyroid cartilage anteriorly is more common in males while *arytenoid cartilage* calcification is four times more common in women.[21]

Perhaps the only *absolute roentgenographic indicator of sex* was present in one of the victims of the Air India catastrophe (Figure 8.26). Many of the recovered bodies had abdominal viscera displaced into the chest. One young female chest contained a fetal skeleton estimated at 18–22 weeks.

DETERMINATION OF RACE OR POPULATION ANCESTRY

Physical anthropologists have many elaborate methods of evaluating race or population ancestry if bare bones are available. Some fleshed or partially fleshed remains can be evaluated radiographically.

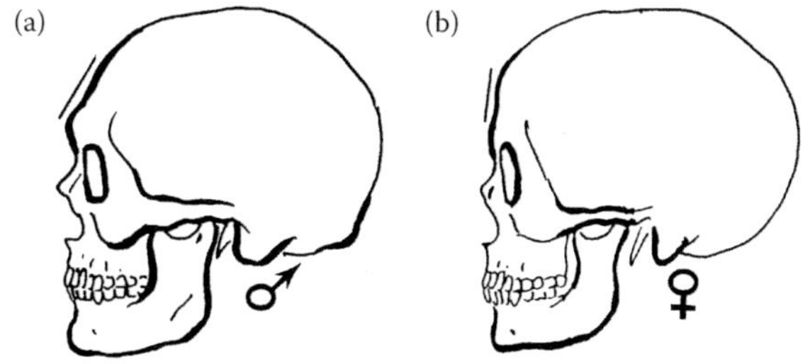

FIGURE 8.19 (a) male and (b) female characteristics in the skull. Differential areas are emphasized by bold lines (see text).

SKULL

The skull reflects certain characteristics of population ancestry that are reasonably dependable,[23,24] but may be confusing on account of racial mixing. Figure 8.27 summarizes differential features in frontal and lateral projections.

INTERCONDYLAR SHELF ANGLE

A method of determining race from the intercondylar shelf angle can be used with skeletal or fleshed remains[31]

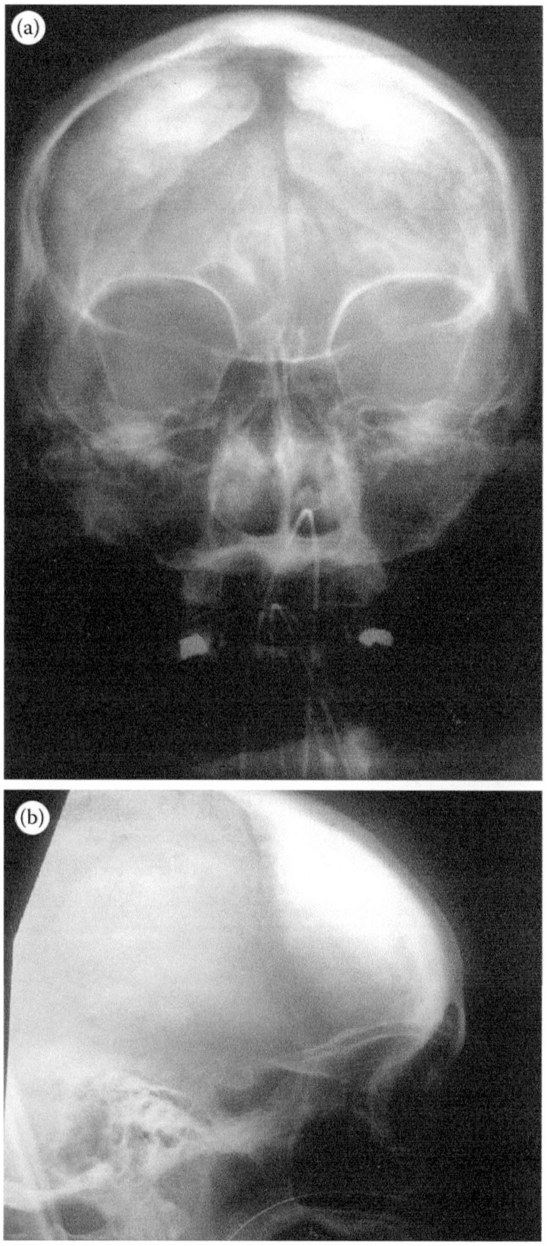

FIGURE 8.20 Hyperostosis interna frontalis. Dense bony thickening of the inner table of the frontal bone seen in (a) frontal and (b) lateral roentgenograms, and overleaf (c) on a CT scan of the skull using a "bone window." (d) The "topogram" preliminary to the CT scan, is sufficiently detailed that it could be used for purposes of identification comparison.

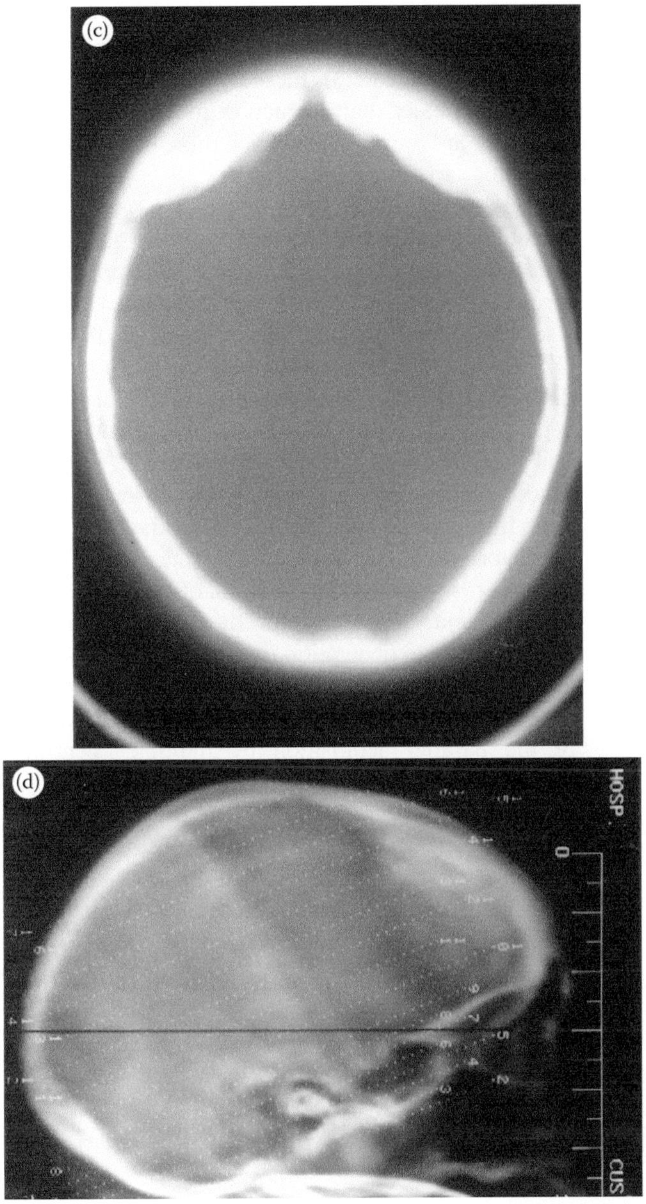

FIGURE 8.20 *(continued)*

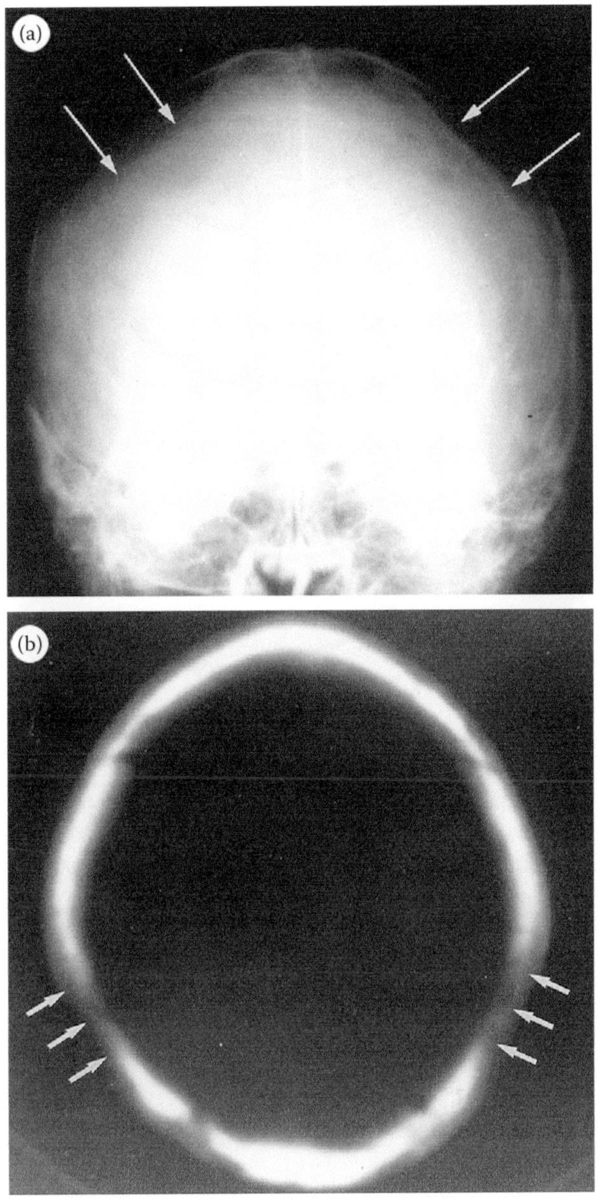

FIGURE 8.21 Parietal thinning. The resorption of the outer table of the skull and diploë is striking in both the frontal radiograph (a) and the CT scan (b).

(Figure 8.28). It requires only true lateral positioning of the distal femur. The measurement of the angle between the roof of the intercondylar notch (or intercondylar shelf) and the long axis of the femoral shaft is independent of magnification. Figure 8.29 shows the bimodal nature of the racial curves with a fairly narrow overlap range of only 18% between sectioning points.

LONG BONES

In Blacks, the tibia is long relative to the femur and the radius is long relative to the humerus, but the ratios are variable and overlap in the U.S. population, probably due to racial mixing.[4] Compared to Blacks, the femoral shafts are bowed anteriorly in Whites and Mongoloid populations. Again there is considerable variability, but a markedly bowed femur is unlikely to belong to a Black.[4]

STATURE

Estimations of stature from measurements of long bones have been the province of anatomists and physical anthropologists for many years. Most have been based on extensive research on World War II and Korean War casualties.[32,33] The length of the femur is the most reliable basis for calculation of stature.[25] Controversy has arisen recently concerning the accuracy of tibial measurements in Trotter and Gleser's data.[34]

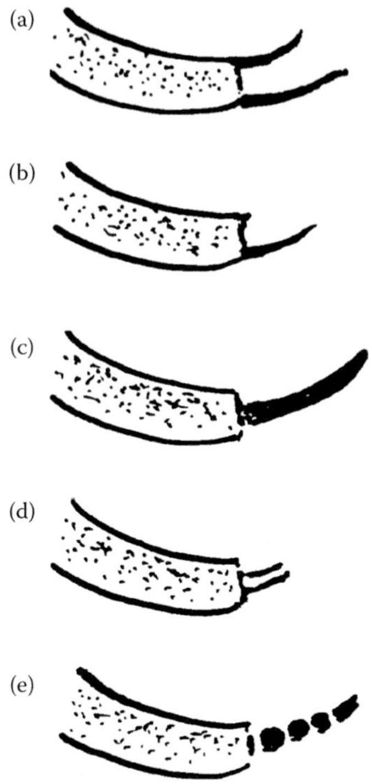

FIGURE 8.22 Schematic drawing of patterns of costal cartilage ossification useful in determining the sex of human remains. (a) Typical male pattern; (b) Male pattern initially involving only the inferior margin of the costal cartilage; (c) Typical female pattern; (d) Uncommon pattern more often found in females than males; (e) String of rounded ossifications believed specific for elderly females.

The equations furnished for stature estimation from long bone measurements are based on direct measurement of the unfleshed bones[35] (Table 8.3). Fleshed remains can be measured radiographically, but correction for magnification is essential. There are four ways in which this correction can be accomplished:

1. Formula to determine correction factor for magnification (see Figure 8.6 and accompanying formula).[36]
2. Use of an excessively long tube-to-specimen distance (72 in. or more) with minimal specimen-to-film distance (table-top grid cassette) will essentially negate magnification; this was used in Maresh's seminal studies on long bone growth in children.[37]
3. Since a plain radiograph or photograph views a long bone from a point source, the ends of the bone will be distorted by the divergence of the beam. This may confuse precise identification of landmarks for measurement, especially for the tibia. A reconstructed caronal or saggital image from a multislice computed tomography (MSCT) will be an undistorted representation of the defleshed bone because of the parallel acquisition by multiple "central rays"[38] (Figure 8.30).
4. Lacking CT, one can adapt an old technique for accurately measuring long bones to determine correct sizes of intramedullary nails for trauma cases or to measure leg length discrepancies. This technique depends on carefully collimated x-ray beams

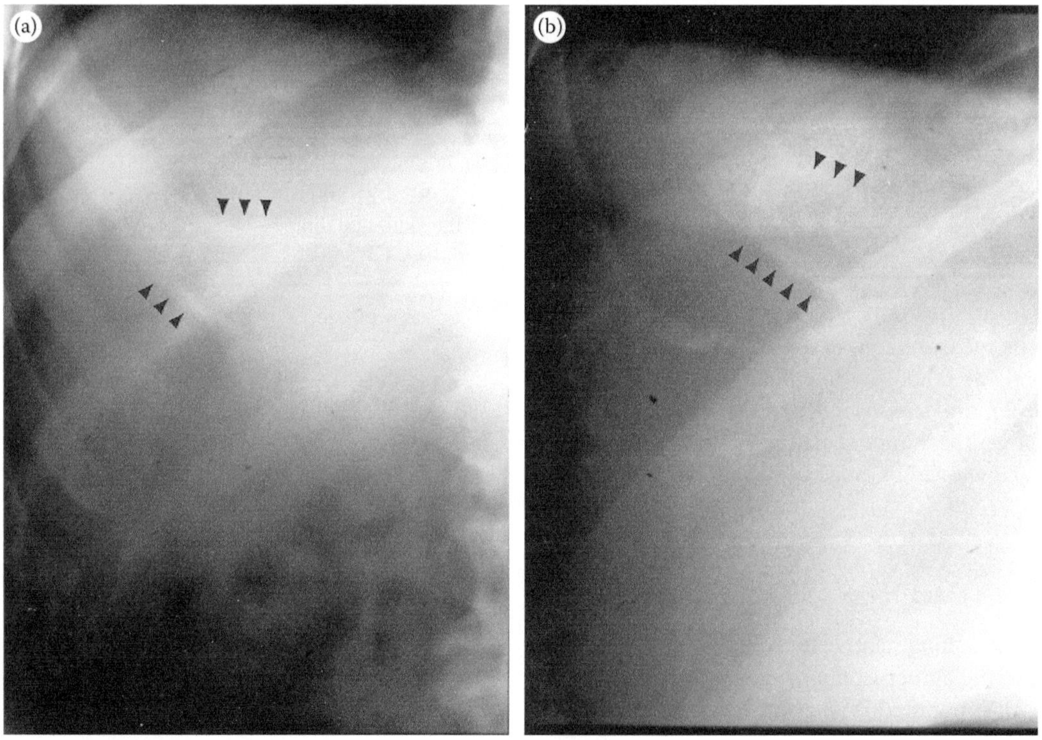

FIGURE 8.23 (a) and (b) Typical male pattern costal cartilage ossification.

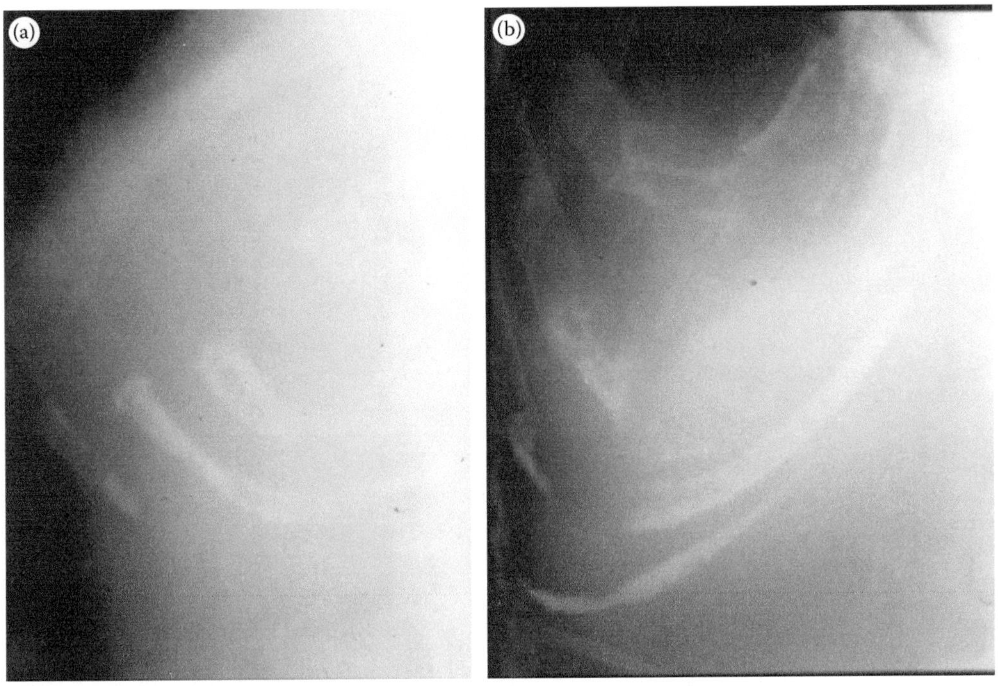

FIGURE 8.24 (a) and (b) Typical female pattern costal cartilage ossification.

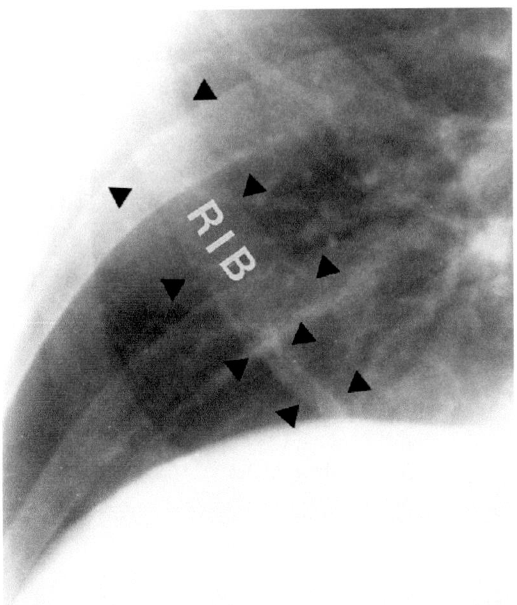

FIGURE 8.25 Male pattern ossification involving only the inferior border of the costal cartilage, easily confused with female pattern if not traced back to the rib end.

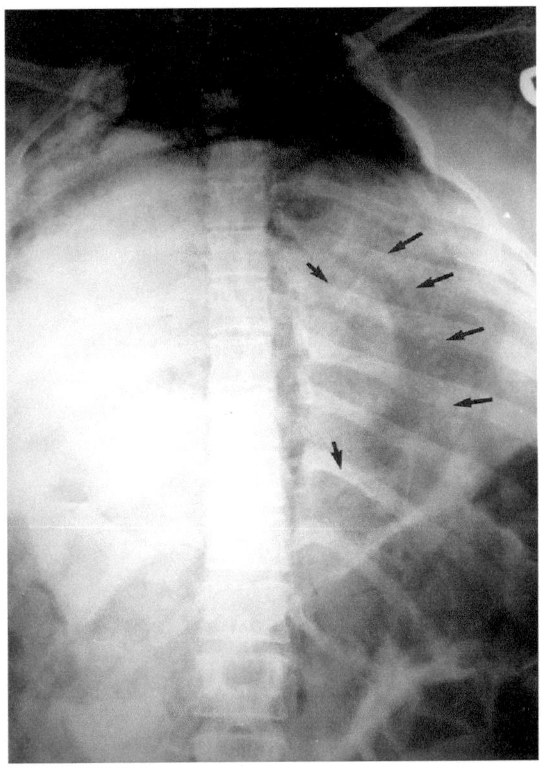

FIGURE 8.26 Fetal skeleton displayed into the thorax of an Air India crash victim. The chain of ossified vertebral body centers ("string of beads") is marked by long arrows. Short arrows indicate long bones in extremities. The base of the skull is seen at the upper end of the spine.

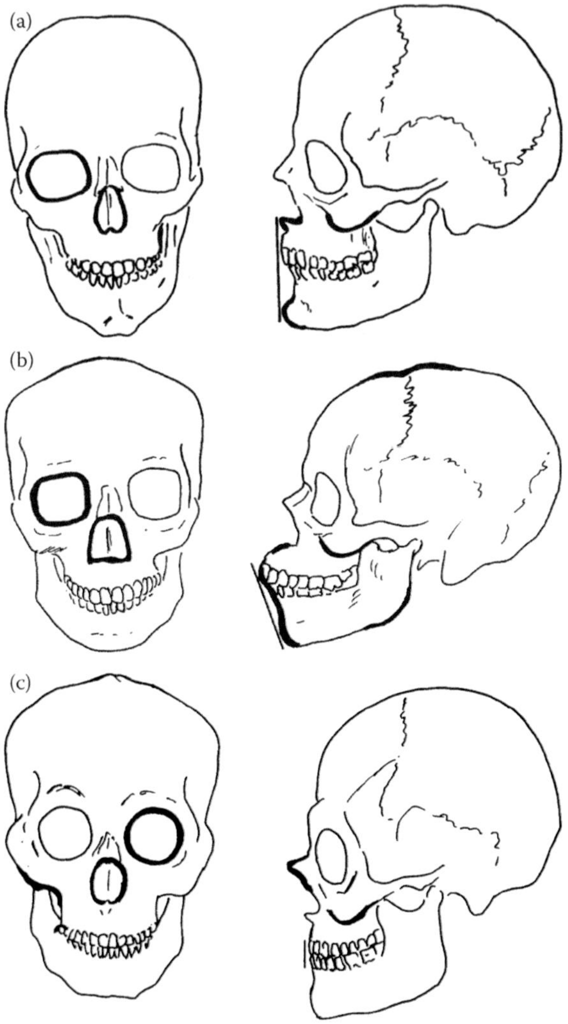

FIGURE 8.27 Drawings emphasizing characteristic differences in the skull between humans of diverse population ancestry; in frontal and lateral views. (a) Caucasoid; (b) Negroid; (c) Mongoloid (see Table 8.2b).

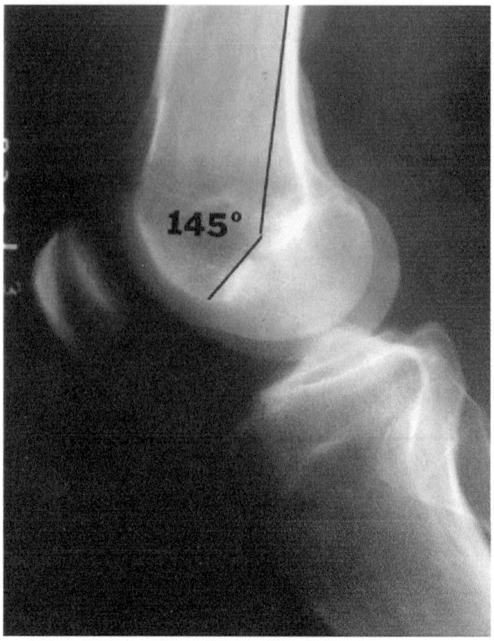

FIGURE 8.28 Lateral roentgenogram of the knee illustrating method of measuring the intercondylar shelf angle.

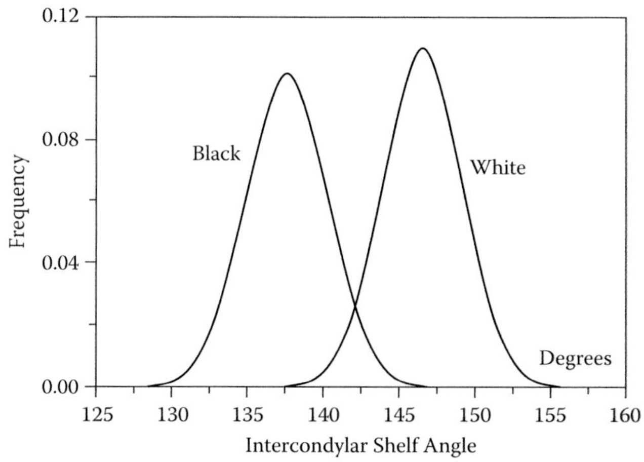

FIGURE 8.29 Graphic representation of racial distribution of intercondylar shelf angles. (Courtesy of Michael D. Harpen., from data presented in Reference 31.)

TABLE 8.3
Equations to Estimate Living Stature (cm)—with Standard Errors—from the Long Bones of American Whites and Negroes between 18 and 30 Years of Age[a]

White Males				Negro Males			
3.08 Hum	+	70.45	±4.05	3.26 Hum	+	62.10	±4.43
3.78 Rad	+	79.01	±4.32	3.42 Rad	+	81.56	±4.30
3.70 Ulna	+	74.05	±4.32	3.26 Ulna	+	79.29	±4.42
2.38 Fem	+	61.41	±3.27	2.11 Fem	+	70.35	±3.94
2.52 Tib	+	78.62	±3.37	2.19 Tib	+	86.02	±3.78
2.68 Fib	+	71.78	±3.29	2.19 Fib	+	85.65	±4.08
1.30 (Fem + Tib) + 63.29		±2.99		1.15 (Fem + Tib) + 71.04			±3.53

TABLE 8.3 (continued)
Equations to Estimate Living Stature (cm)—with Standard Errors—from the Long Bones of American Whites and Negroes between 18 and 30 Years of Age[a]

White Females				Negro Females			
3.36 Hum	+	57.97	±4.45	3.08 Hum	+	64.67	±4.25
4.74 Rad	+	54.93	±4.24	2.75 Rad	+	94.51	±5.05
4.27 Ulna	+	57.76	±4.30	3.31 Ulna	+	75.38	±4.83
2.47 Fem	+	54.10	±3.72	2.28 Fem	+	59.76	±3.41
2.90 Tib	+	61.53	±3.66	2.45 Tib	+	72.65	±3.70
2.93 Fib	+	59.61	±3.57	2.49 Fib	+	70.90	±3.80
1.39 (Fem + Tib) + 53.20			±3.55	1.26 (Fem + Tib) + 59.72			±3.28
Mongoloid Males				**Mexican Males**			
2.68 Hum	+	83.19	±4.25	2.92 Hum	+	73.94	±4.24
3.54 Rad	+	82.00	±4.60	3.55 Rad	+	80.71	±4.04
3.48 Ulna	+	77.45	±4.66	3.56 Ulna	+	74.56	±4.05
2.15 Fem	+	72.57	±3.80	2.44 Fem	+	58.67	±2.99
2.39 Tib	+	81.45	±3.27	2.36 Tib	+	80.62	±3.73
2.40 Fib	+	80.56	±3.24	2.50 Fib	+	75.44	±3.52
1.22 (Fem + Tib) + 70.37			±3.24				

[a] To estimate stature of older individuals subtract 0.06 (age in years—30) cm; to estimate cadaver stature, add 2.5 cm.

Source: From Trotter, M., *Personal Identification in Mass Disasters*, Stewart, T. D., Ed., National Museum of Natural History, Smithsonian Institution, Washington, DC, 77, 1970. With permission.

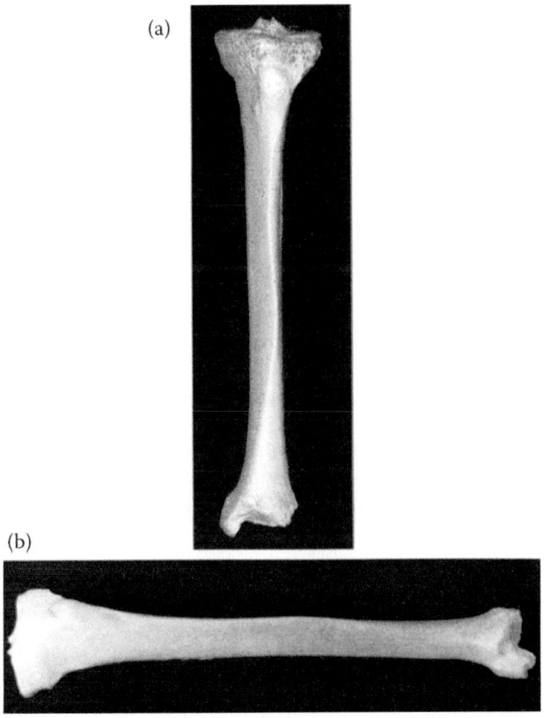

FIGURE 8.30 (a) This is a virtual image of the tibia created from CT imaging data. (b) A photograph of the same bone after defleshing to demonstrate the differing appearance of the ends of the bone resulting from the two methods. The distortion of the bone ends in the photograph is caused by beam divergence from the point source of the x-ray, while the ends are undistorted by the perpendicular scanning beam of the CT unit. (From Robinson, C., et al. *J. Forensic Sci.*, 53, 1289, 2008. With permission.)

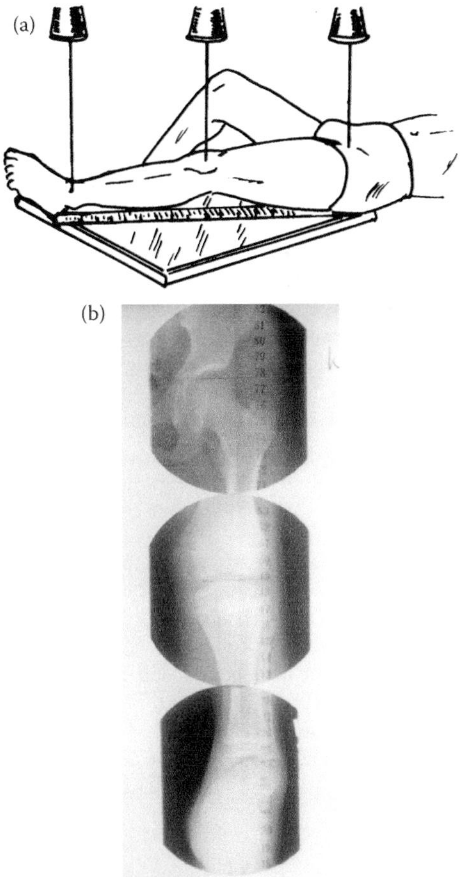

FIGURE 8.31 (a) Method of obtaining magnification-free measurements of long bone length using collimated, nondivergent x-rays over the bone ends with a partially radiopaque ruler in the field of exposure. (b) Example of image obtained by above method.

at conventional tube—object distances, so that only the central nondivergent x-rays are used to expose the ends of the bones. A partially radiopaque ruler included in the field of exposure can be used for direct measurement (Figure 8.31).

REFERENCES

1. Chilvarquer, I., Katz, J. O., Glassman, D. M., Prihoda, T. J., and Cattone, J. A., Comparative radiographic study of human and animal long bone patterns, *J. Forensic Sci.*, 32, 1645, 1987.
2. Hadlock, F. P., Deter, R. L., Harrist, D. B., and Park, S. K., Estimating fetal age: Computer-assisted analysis of multiple fetal growth parameters, *Radiology*, 152, 497, 1984.
3. Hartley, J. B., Radiological estimation of foetal maturity, *Br. J. Radiol.*, 30, 561, 1957.
4. Krogman, W. M. and Iscan, M. Y., *The Human Skeleton in Forensic Medicine*, 2nd ed., Charles C. Thomas, Springfield, IL, 1986, chap. 13.
5. Graham, C. B., Assessment of bone maturation—methods and pitfalls, *Radiol. Clin. N. Am.*, 10, 185, 1972.
6. Girdany, B. R. and Golden, R., Centers of ossification of the skeleton, *Am. J. Roentgenol.*, 68, 922, 1952.
7. Keats, T. E, *Atlas of Roentgenographic Measurement*, 6th Mosby Year Book, St. Louis, 1990, chap. 4B.
8. Meschan, I., *Roentgen Signs in Clinical Practice*, Vol. I, W.B. Saunders, Philadelphia, 1966, chap. 4.
9. Sontag, I. W., Snell, D., and Anderson, M., Rate of appearance of ossification centers from birth to the age of five years, *Am. J. Dis. Child.*, 58, 949, 1939.
10. Pyle, S. I. and Hoerr, N. L., *Atlas of Skeletal Development of the Knee*, Charles C. Thomas, Springfield, IL, 1955.
11. Hoerr, N. L., Pyle, S. I., and Francis, C. C., *Radiologic Atlas of the Foot and Ankle*, Charles C. Thomas, Springfield, IL, 1962.
12. Greulich, W. W. and Pyle, S. I., *Radiographic Atlas of Skeletal Development of the Hand and Wrist*, 2nd ed., Stanford University Press, Palo Alto, CA, 1959.
13. Ontell, F. K., Ivanovic, M., Ablin, D. S., and Barlow, T. W., Bone age in children of diverse ethnicity, *Am. J. Roentgenol.*, 167, 1395, 1996.
14. Sundick, R. I., Age and sex determination of sub adult skeletons, *J. Forensic Sci.*, 22, 141, 1977.
15. McCormick, W. F., Mineralization of costal cartilages as an indicator of age: Preliminary observations, *J. Forensic Sci.*, 25, 736, 1980.
16. Barres, D. R., Durigon, M., and Paraire, F., Age estimation from quantitation of features of "chest plate" x-rays, *J. Forensic Sci.*, 34, 228, 1989.
17. Stewart, J. H. and McCormick, W. F., A sex and age-limited ossification pattern in human costal cartilages, *Am. J. Clin. Pathol.*, 81, 765, 1984.
18. Sutherland, L. D. and Suchey, J. M., Use of the ventral arc in pubic sex determination, *J. Forensic Sci.*, 36, 501, 1991.
19. Rogers, T. and Saunders, S., Accuracy of sex determination using morphological traits of the human pelvis, *J. Forensic Sci.*, 39, 1047, 1994.
20. Schemmer, D., White, P. G., and Friedman. L., Radiology of the paraglenoid sulcus, *Skeletal Radiol.*, 24, 205, 1995.
21. Kurihara, Y., Kurihara, Y., Ohashi, K., Kitagawa, A., Miyasaki, M., Okamoto, E., and Ishikawa, T., Radiologic evidence of sex differences: Is the patient a woman or a man? *Am. J. Roentgenol.*, 167, 1037, 1996.
22. Wells, J., Osteities condensans ilii, *Am. J. Roentgenol.*, 76, 1141, 1956.
23. Bass, W. M., III, Forensic anthropology, in *CAP Handbook for Postmortem Examination of Unidentified Remains*, Fierro, M. F., Ed., College of American Pathologists, Skokie, IL, 1990, chap. 8.
24. Heglund, W. D., How can the forensic anthropologist help? Handout presented at the American Academy of Forensic Science, Seattle, February 14, 1995.
25. Krogman, W. M., Will Mr. X please come forward? *Del. Med. J.*, 51, 399, 1979.
26. Steinbach, H. L., The significance of thinning of the parietal bones, *Am. J. Roentgenol.*, 78, 39, 1957.
27. Stewart, J. H. and McCormick, W. F., The gender predictive value of sternal length, *Am. J. Forensic Med. Pathol.*, 4, 217, 1983.
28. Lawson, J. P., Clinically significant radiologic anatomic variants of the skeleton, *Am. J. Roentgenol.*, 163, 249, 1994.
29. Sanders, C. F., Correspondence, *Br. J. Radiol.*, 39, 233, 1966.
30. Navani, S., Shak, J. R., and Levy, P. S., Determination of sex by costal cartilage calcification, *Am. J. Roentgenol.*, 108, 771, 1970.
31. Craig, E. A., Intercondylar shelf angle: A new method to determine race from the distal femur, *J. Forensic Sci.*, 40, 777, 1995.
32. Trotter, M. and Gleser, G. C., Estimation of stature from long bones of American Whites and Negroes, *Am. J. Phys. Anthropol.*, 10, 463, 1952.
33. Trotter, M. and Gleser, G. C., A re-evaluation of estimation of stature based on measurements of stature taken during life and of long bones after death, *Am. J. Phys. Anthropol.*, 15, 79, 1958.
34. Jantz, R. L., Hunt, D. R., and Meadows, L., The measure and mismeasure of the tibia: Implications for stature estimation, *J. Forensic Sci.*, 40, 758, 1995.
35. Trotter, M., Estimation of stature from intact long limb bones, in *Personal Identification in Mass Disasters*, Stewart, T. D., Ed., National Museum of Natural History, Smithsonian Institution, Washington, DC, 77, 1970.
36. Brown, G. H., Jr., Automatic compensation in roentgenographic polycephalometry, *Am. J. Roentgenol.*, 78, 1063, 1957.
37. Maresh, M. M., Growth of major long bones in healthy children, *Am. J. Dis. Child.*, 66, 227, 1943.
38. Robinson, C., Roos, E., Morgan, B., et al. Anthropological measurement of lower limb and foot bones using multidetector computed tomography, *J. Forensic Sci.*, 53, 1289, 2008.

9 Modern Cross-Sectional Imaging in Anthropology

Fabrice Dedouit, Norbert Telmon, Hervé Rousseau, Eric Crubézy, Francis Joffre, and Daniel Rougé

CONTENTS

INTRODUCTION

This chapter concerns the diagnostic possibilities offered by modern cross-sectional imaging in anthropology. Its application in this field is generally considered as a recent development. However, the first computed tomography (CT) scan for anthropological purposes was performed in 1976 by Lewin and Harwood-Nash.[1,2] The mummified brain of a 14-year-old child who died in ancient Egypt 3200 years previously was studied only four years after Hounsfield developed this technique.[3] From that time onwards, the various technological advances of scanning by CT found many applications in anthropology. The CT systems used for examination of the mummy were in phase with the most advanced technology of the time: sequential acquisition was followed by monoslice helical scanning, which was itself followed by multislice scanning technology. Over the last 20 years, advances in CT scanning, and especially multislice acquisition, have resulted in significant improvement in spatial definition and isotropic image quality. Progress in computer technology led to the development of postprocessing, yielding three-dimensional (3D) reconstructions. Initially, only the surface-shaded display (SSD) technique was available, but nowadays most 3D reconstructions use the volume rendering technique (VRT) and maximum intensity projection (MIP) modes, which offer 2- and 3D reconstructions of high quality. Two-dimensional reconstructions are also easily obtained not only for the orthogonal planes used in clinical radiology (axial, frontal, and sagittal), but also for oblique and curved slices.

Two types of investigation are possible: multislice computed tomography (MSCT) and magnetic resonance imaging (MRI).

With regard to MRI, there are a few publications on its successful use in anthropological imaging.[4–6] The major limitation of this technique in anthropology is the need for water and protons in order to obtain an efficient signal and, consequently, an image. This is an important drawback for the study of dehydrated tissues such as mummies.[4,7] However, MRI has already been used for forensic determination of the skeletal age of living persons.[8–11]

Nowadays, the imaging modality most frequently used in anthropology is undeniably MSCT. The main published studies on the use of MSCT concern mummies. The applications of modern slice imaging in anthropology, paleoanthropology, bioarcheology and paleopathology are grouped under the term paleoradiology.[12] In recent years, a new approach based on CT studies has been developed called virtual anthropology.[13] This concept rejoins that of virtual autopsy, or postmortem imaging investigations.[14] In some cases where identification of bodies becomes necessary, both these investigations can be carried out simultaneously through a single MSCT acquisition.[15,16] The first analysis identifies causes of death, possible pre-existing disease conditions and nonlethal abnormalities, while the second detects criteria potentially useful for identification of the deceased. As with plain x-rays, identification may be based on detection of surgical material, variants of normal radiological appearances, and pre-existing abnormalities, whether congenital or acquired, and antemortem MSCT images can be compared with postmortem images.[17–19] As an increasing number of investigations are performed in clinical and surgical situations in an increasing number of individuals, the opportunity for such comparisons becomes more and more frequent.

The anthropological applications of MSCT are based on two main approaches: the use of criteria and characteristics specific to the technique, or the transposition of techniques used on dry bones in physical anthropology. Consequently, the radiologist who performs such investigations must be familiar with the techniques of physical anthropology. Identification can, in fact, be based on assessment and determination of some important anthropological parameters: racial phenotype, age at death, sex, stature.

For current anthropological purposes, MSCT has many advantages over dry bone analysis. One of the major assets in its use in forensic anthropology is the elimination of lengthy bone preparation, which may sometimes cause anatomical damage, especially when bone is already fragile. This can be particularly useful when bones are very burned or charred.[15] Documentation by radiological imaging is classically described as observer-independent, objective, and noninvasive. Digitally stored data may be recalled at will and provide fresh, intact reconstructions. A new approach to quality control and expert supervision becomes possible, as well as image transmission and use in forensic telemedicine. Additionally, MSCT can be performed in the country where the bones or body were discovered, while further work on image analysis and reconstruction can be performed in another country. Image and data processing and the high spatial resolution of MSCT offer objective visualization and recapitulation of forensic results. This provides an amazing opportunity for creating a virtual skeletal collection or museum, where bones can be studied by scientists on their computer screen, thousands of miles or kilometers from the actual bones. MSCT presents numerous inherent advantages over plain x-rays, in particular the ability to provide 3D information. Postprocessing allows segmentation of an individual bone, which can be highly useful for its analysis.

Unfortunately, the main drawbacks to routine use of MSCT in anthropology are the limited accessibility of such systems, the cost of the investigations, and the real need for the radiologist to be familiar with anthropological techniques.

RACIAL PHENOTYPE OR POPULATIONAL ANCESTRY DETERMINATION

It is nearly impossible to establish the identity of skeletal remains without determining populational ancestry. Forensic examiners must be extremely cautious with regard to this aspect of the identification of unknown remains. The most interesting and accurate anatomical part for populational ancestry assessment is indisputably the skull.[20] Craniofacial detail has long been recognized as differentiating populations.[21] In the postcranial skeleton, few criteria can reliably determine race. Indices such as the ratio of tibia-to-femur length and the radius-to-humerus length, as well as the anterior curvature of the femoral shaft and the intercondylar angle, have been suggested as good indicators of racial affinity.[22] Racial phenotype can be schematically assessed from the skeleton using morphological (osteoscopic) and metric (osteometric) traits.

Morphological traits present shape and size differences that can be evaluated osteoscopically. Frontal and profile views of Caucasoid, Negroid and Mongoloid skulls are discussed in Chapter 8 of this book. It is important for the forensic radiologist to be aware of the limitations of these evaluations, which must be made with great caution. Currently, the assessment of populational ancestry is complicated by every kind of admixture as well as by the variability that exists within races at the population level.[20] Furthermore, in most cases, dry bone study is possible and so radiological investigation is superfluous.

When visual morphological examination is not possible, for example in mummies with dehydrated cephalic soft tissues that must be preserved, numerical analysis of crania can provide assistance through a statistical comparative population approach.[21] In such cases, a major advantage of MSCT is the possibility of metric studies, of the skull in particular. These can be done using the existing methods developed on dry bones. Metric determination of racial phenotype is based on the use of selected measurements that show statistically significant interpopulation differences.[20] This technique necessitates a thorough knowledge of skeletal landmarks, proper equipment, and precise measuring skills. There can be complex combinations of shape and size differences, which are not morphologically obvious, between populations, and these may be quantified and evaluated by using a set of measurements. The statistical technique most commonly used is discriminant function analysis, which assumes that human variation spans a continuum across space and populations, but concentration of people with similar features can be found toward the centers, whereas at the peripheries there may be an overlap with neighboring groups. In this context, a recent study validated the accuracy and exactness of craniometric measurements performed on skull CT reconstructions, with results identical to those obtained on the same dry skulls.[23]

One of the earlier discriminant functions used to assess racial ancestry from the skull is that reported by Giles and Elliot who worked on white, black, and Indian American populations.[24] Krogman and Iscan studied white, black, and Japanese populations.[25] Using multivariate stepwise discriminant function analysis, the most discriminant measures yielding maximum differentiation between the three samples were determined. Racial phenotype can be calculated for two functions (1 and 2) by multiplying each dimension by its coefficient and adding them up along with the constant. The results of both functions can be placed on a figure showing the territorial placement of the three samples according to gender or compared with the group centroids of each population. However, it has been questioned whether discriminant function statistics can be applied to populations of similar racial origins. It is unlikely that standards based on American Blacks will be applicable to their African counterparts.[20] However,

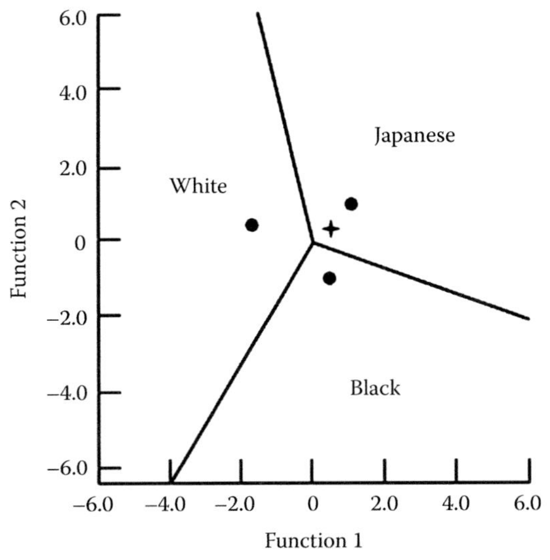

FIGURE 9.1 Racial phenotype determination of a natural female mummy from Yakutiya (Siberia). The asterisk indicates the position of the individual among American White and Black and Japanese female populations determined using cranial dimensions. (From Iscan, M. Y., Loth, S. R., and Steyn, M., Determination of racial affinity, in *Encyclopaedia of Forensic Sciences*, Siegel, J., Knupfer, G. and Saukko, P., Eds., Academic Press, 227–35, 2000. With permission.)

this technique of populational ancestry was successfully applied to a natural Siberian mummy.[26] The Yakut female mummy was correctly assigned to the Japanese territorial placement (mongoloid subject) (Figure 9.1).

Also useful for populational ancestry assessment by craniometric study are Howells' data.[27] Howells collected craniometric data of 28 populations based on 77 cranial measurements. These data are freely accessible through the Internet. We successfully determined the populational ancestry of the natural female mummy from Yakutiya, Siberia. We measured only 9 craniometric lengths using a landmark system (Figure 9.2a) and these were introduced in a principal analysis component of Howells' data for females by means of R software. The result was interesting because the craniometric characteristics of the mummy were similar to those of the Buryat population, which is geographically the closest population to Yakutiya available in Howells' data (Figure 9.2b).[26]

Another possibility is provided by Fordisc®, a computer program in which the user can enter up to 21 cranial measurements that are compared with the 28 populations measured by Howells.[28] The user enters the available measurements of the skull to be studied and the program calculates the variance–covariance matrices, followed by linear discriminant analysis, to estimate the probability that the skull belongs to, or at least resembles, one or other of the populations in Howells' data. Users can select the Howells populations with which to compare the skull studied. The results obtained do not give a definitive diagnosis concerning the population to which the skull belongs, but statistical classification within groups in the database is achieved by typical probabilities that allow the user to draw conclusions. The program is qualified through selection of reasonable populations for comparative analysis.[21] Another morphometric analysis software package, Cranid®, filled the gap in Howells' database by including data of South Asian individuals. It has already been successfully used for morphological analysis of the cranium of an ancient Egyptian mummy.[29]

SEX DETERMINATION

Because the major features of sexual dimorphism (differences in shape between males and females) develop during puberty, most forensic anthropologists agree that sex determination from the skeleton is only practical for late teens or adults.[22] Although sex differences have been quantified in fetal and child skeletons, they are subtle and highly variable until the secondary sex characteristics develop during the juvenile period. Race and sex differences must always be considered in age-estimation standards and decisions.[30] When estimating age from the skeleton, one must remember that what has been analyzed in research is the skeletal age rather than birth certificate-based chronological age. The physiological processes that underlie aging are dependent on many internal and external influences including genetic make-up, health and nutritional status, and substance abuse. During growth, the situation becomes especially complex when age estimation is linked to bone lengths in children because no matter how much variation is incorporated, the potential is there to make errors. The same applies to remodeling in adulthood. That is why it is so important to provide age ranges so that extremes are not omitted from the police list of missing victims and to provide a meaningful range in which there is a high probability (95%) that the true civil age will fall. Sex determination from the skeleton can be schematically carried out using morphological (osteoscopic) and metric (osteometric) traits.

Morphological Determination

Observations are generally made of the cranium or the bones of the postcranial skeleton, in particular the pelvis and os coxae (Table 9.1).[31] Generally, male skeletons are larger and more robust than female skeletons.[25] By studying features of the skull and especially the hip bones, experienced osteologists should be able to identify sex with more than 95% accuracy.[32] Three features of the pubic regions are particularly useful: the ventral arc, the subpubic concavity, and the shape of the ischiopubic ramus. General sexually dimorphic features of the skull include the brow ridges, mastoid processes, and external occipital protuberance, all of which are usually more pronounced or larger in males. The chin portion of the mandible tends to be squarer in males and more pronounced or 'V' shaped in females.[32]

These morphological differences have already been used to identify unknown bodies.[15,16,33] Figures 9.3 through 9.10 illustrate the possibilities offered by 3D MSCT reconstructions for pelvic and skull sexing.

(a)

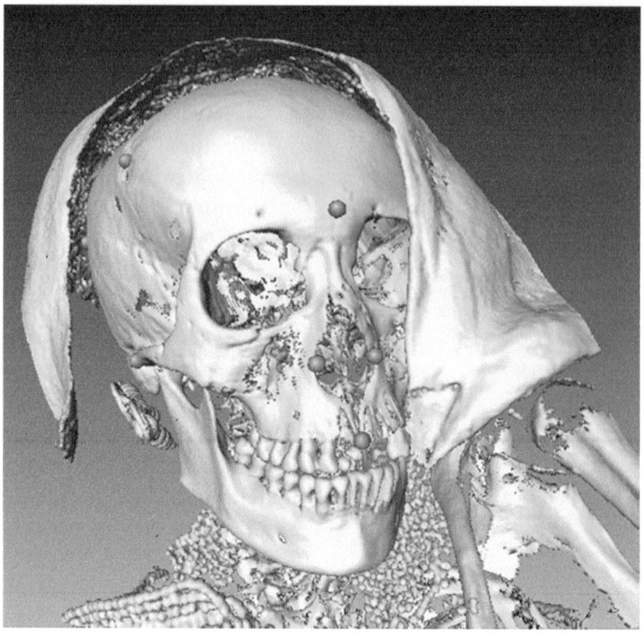

(b)

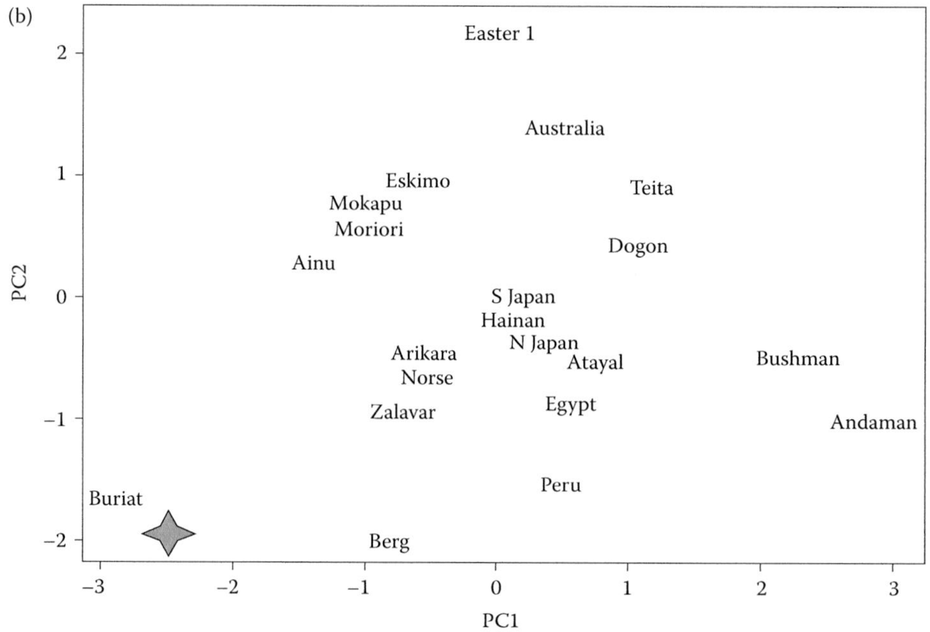

FIGURE 9.2 Racial phenotype determination of a natural female mummy from Yakutiya (Siberia) based on Howell's data for females. (a) 3D VRT craniofacial reconstruction with positioning of different landmarks (circles) representing craniometric points. (b) Position of the individual among the different populations studied by Howells (asterisk).

Metric Analysis

Metric analysis can easily be done on multiplanar reconstructions (MPR) and MIP MSCT reconstructions.

As with racial phenotype determination, craniometric study is possible with a reported accuracy of between 82% and 89%.[24]

In long bones, it has been observed that epiphyseal measurements are better indicators of sex than length or diaphyseal dimensions.[32] It has also been shown that measurements of sexual dimorphism in the long bones are more diagnostic than in the skull and necessitate the use of fewer dimensions. In all samples, males are significantly larger than females. In general, the limb bones of females are more gracile and less clearly marked by muscle attachments than those of males. The articular ends of the bones are smaller and the shafts less robust.[31] However, metrics are group-specific and populations differ from each other in the degree and range of dimorphism exhibited by various dimensions.[32] Size differences between population groups mean that the sex determination from the limb bones is population-specific. Some measurements used to estimate sex in postcranial remains are easily accessible with MSCT reconstructions (Table 9.2). More

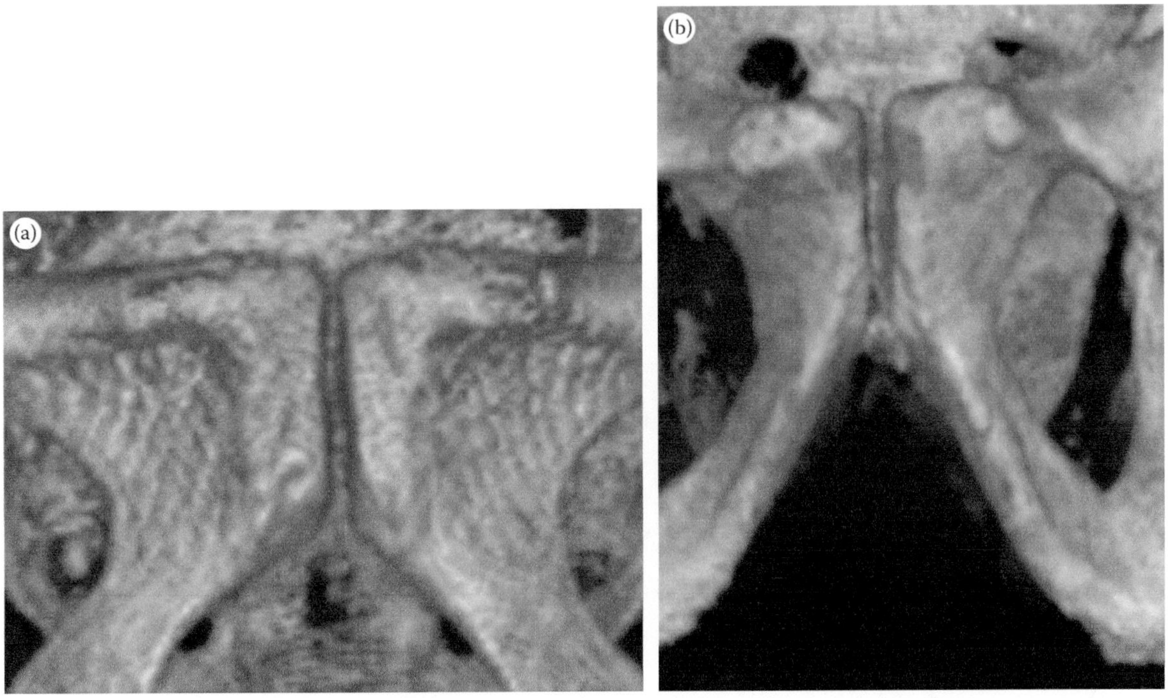

FIGURE 9.3 Appearance of the ventral arc. (a) Female subject, 3D reconstructions, VRT mode, front view: the ventral arc is a curved ridge of bone on the anterior surface of the pubic joint. (b) Male subject, 3D reconstructions, VRT mode, front view: the ventral arc is absent. (From Dedouit, F., et al. *Forensic Sci Int* 173, 182–7, 2007. With permission.)

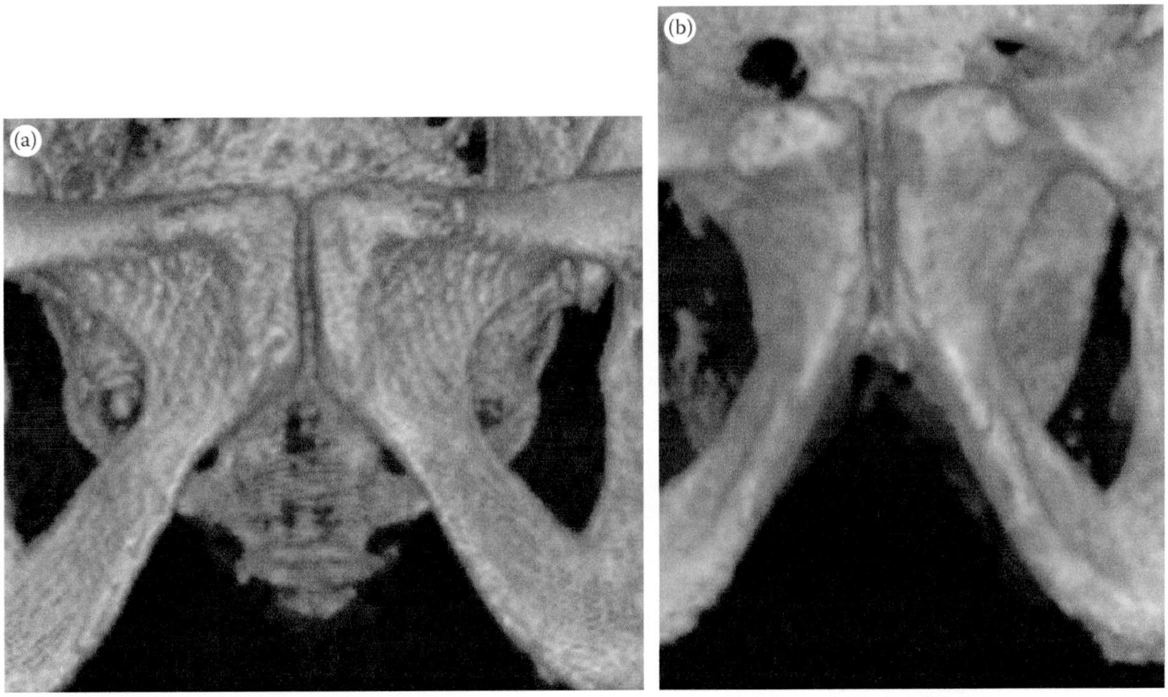

FIGURE 9.4 Appearance of the subpubic concavity and subpubic angle. (a) Female subject, 3D reconstructions, VRT mode, front view: the medial edge of the bone that extends laterally and inferiorly from the joint between the pubic bones is concave. The subpubic angle is wide. (b) Male subject, 3D reconstructions, VRT mode, front view: the medial edge of the bone that extends laterally and inferiorly from the joint between the pubic bones is straight. The subpubic angle is narrow. (From Dedouit, F., et al. *Forensic Sci Int*, 173, 182–7, 2007. With permission.)

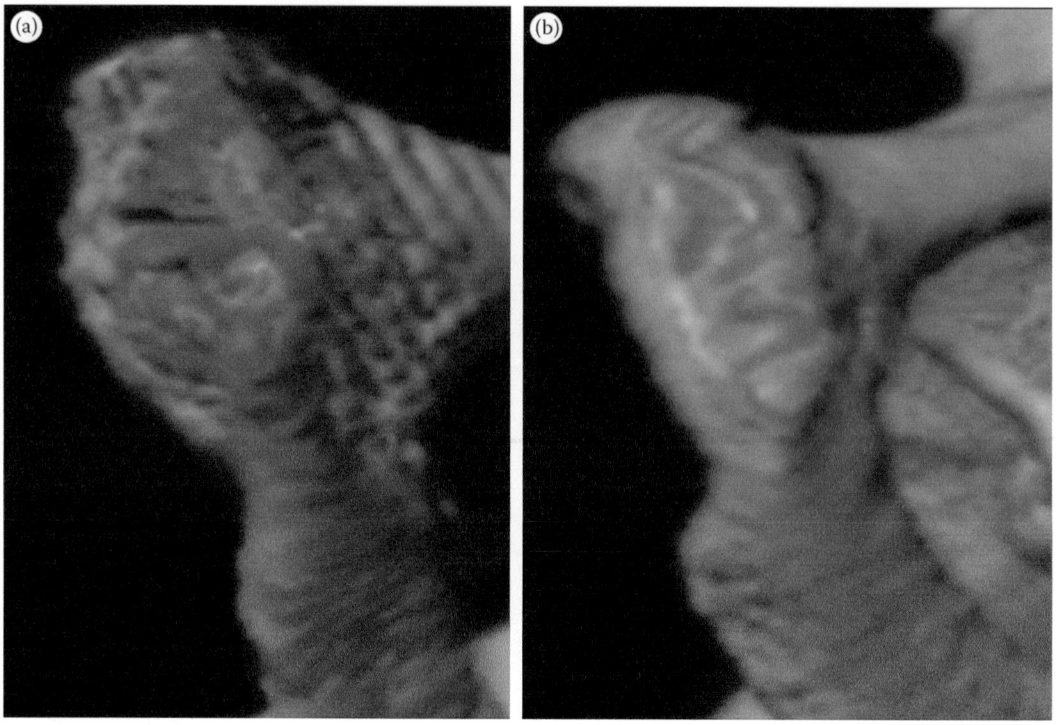

FIGURE 9.5 Appearance of the ischiopubic ramus. (a) Female subject, 3D reconstructions, VRT mode, inner view: the medial edge of the bone connecting the pubic bone and the ischium is a narrow bridge. (b) Male subject, 3D reconstructions, VRT mode, inner view: the medial edge of the bone connecting the pubic bone and the ischium is wide and flat.

details, including ranges and standard deviations, are presented in manuals of human osteology.

Some specific MSCT studies with metric analysis to determine sex have already been conducted for identification purposes.[15] Measurements based on foramen magnum dimensions have been performed on 3D reconstructions. The length and width of the right occipital condyle and the foramen magnum diameters were found to be statistically different in each sex with 81% accuracy.[34] Measurements based on laryngeal MSCT have also revealed sexual dimorphism.[35] The angles of both laryngeal laminae of the thyroid cartilage were estimated at 102° for females (standard deviation = 16°) and 67° for males (standard deviation = 15°).

AGE ASSESSMENT

We prefer to use the term age assessment rather than age determination. Unfortunately, radiological methods allow only an approximate approach to civil age; they are merely the reflection of biological age, which is subject to interindividual variations.[36] Estimation of age is based on biological changes that take place throughout life. There is a high statistical correlation between the chronological age of a person and the biological stage of growth and development. The assessment of biological age is usually most accurate in the early phases of development and decreases as the individual gets older. For each method used, the forensic radiologist has to reason in terms of accuracy and sensitivity. For age-at-death evaluation of adult or nonadult subjects, many criteria are available. This chapter cannot address this subject in detail and it is dealt with exhaustively elsewhere.[37] As previously indicated for plain x-rays, in the case of children or young immature adults, analysis of the secondary ossification centers is an accurate method of age assessment that is often applied in current anthropological radiology.

Measurement of bone lengths in fetuses is also helpful. These measures are also possible on MSCT MPR reconstructions for long or cranial bones (Figure 9.11).[38]

Some authors have shown interest in forensic age estimation in living adolescents or young adults using MSCT. The bone most often studied in this forensic context is undeniably the medial clavicular extremity[39–41]. This anatomical structure is particularly interesting because it is the last secondary ossification center in the body to fuse with its adjacent metaphysis. Fusion occurs around the age of 21 years, which corresponds to penal liability in some European countries. Analysis of the morphology of the medial clavicular extremity is therefore helpful to determine whether the adult or juvenile penal system is applicable. The authors described different CT appearances and the corresponding time of fusion (Table 9.3; Figure 9.12).[39] Evaluation of metaphyseal and epiphyseal maturation of the medial clavicular extremities has already been used for age evaluation in mummies.[26,42]

In forensic anthropology, many anatomical sites have been studied to assess age at death of adult subjects. Age estimation is based on degenerative skeletal changes.[22] In current anthropological use, the preferred anatomical sites are the auricular surface, the pubis symphysis and the sternal extremity of the right fourth rib. The gradual closure of the cranial sutures

TABLE 9.1
Morphological and Sex Differences in Skull and Pelvis

Trait	Male	Female
	Skull	
General appearance	Rugged	Smooth
Supraorbital ridges	Medium to large	Small to absent
Mastoid processes	Medium to large	Small to medium
Occipital area	Muscle lines and protuberance marked or hooked	Muscle lines and protuberance not marked
Frontal eminences	Small	Large
Parietal eminences	Small	Large
Orbits	Rectangular	Rounded
	Small relative to face	Large relative to face
	Rounded margins	Sharp margins
Forehead	Sloped	Vertical
Mandibular ramus flexure	Ramus flexure	Straight ramus
Palate	Larger, broader, tends to U-shape	Small, tends to parabola
Teeth	Large, lower M1 more often 5- cusped	Small, molars often 4- cusped
	Pelvis	
Pubic symphysis	Higher	Lower
Pubic bone	Triangular	Lower
Subpubic angle	Narrow V-shaped, acute angle	Wide U-shaped, obtuse angle
Acetabulum	Large, tends to be directed laterally	Small, tends to be directed anterolaterally
Greater sciatic notch	Narrow, deep	Wide, shallow
Ischiopubic rami	Rough everted margin	Gracile, narrow near symphysis
Sacroiliac joint	Large	Small, oblique
Postauricular space	Narrow	Wide
Preauricular sulcus	Rarely present	Often present, well developed
Iliac tuberosity	Large, not pointed	Small or absent, pointed or varied
Sacrum	Longer, narrower with more evenly distributed curvature	Shorter, broader, with tendency to marked curvature at S1–S2, S2–S5
	Often 5 or more segments	5 segments is the rule
Pelvic inlet	Heart-shaped	Circular, elliptical

Source: Adapted from Loth, S. R. and Iscan, M. Y., Morphological age estimation, in *Encyclopaedia of Forensic Sciences*, Siegel, J., Knupfer, G., and Saukko, P., Eds., Academic Press, 242–52, 2000. With permission.

during adulthood is an unreliable procedure when used alone, but can be helpful when considered together with other age indicators.[30] There is no precise age indicator in the skull, and one can only give a very rough estimate such as young adult, middle-aged, or elderly. In the skull, the cranial suture closure pattern and its relation to age have been rigorously investigated for nearly a century. Variations of this marker are so large that it has been practically abandoned as an age marker.

In forensic radiology some indicators have already been studied and are detailed below. Of course, these are only examples, but they are derived from robust, universally used methods.

Right Fourth Rib

The aging process in the rib has been assessed by studying changes at the costochondral junction of the fourth rib (Figure 9.13).[30] The metamorphosis of the rib begins after the completion of growth at about the age of 14 in white females and 17 in white males. It is characterized by the disappearance of the epiphyseal line and the beginning of pit formation at the nearly flat, billowy surface of the sternal end of the rib. Within a few years, the pit has taken a definite V-shape and scallops appear on the rim of the bone. The pit gradually deepens and widens to a wide V in females or U-shape for both genders. The rounded edges of youth begin to thin and sharpen by the mid-30s. With increasing age, the rim becomes irregular, the interior of the pit becomes porous, and the bone quality deteriorates until the ribs of most individuals over 70 years are thin and irregular with bony projection at the costochondral junction. The observer must be aware that there are considerable race and sex differences in bone density with age. Although standards are based on the right fourth rib, there are no significant differences between rib sides and no significant intercostal variation between ribs 2–7, or between ribs 3, 4, and 5 in particular. In Iscan's studies, observations were made at the costochondral junction with special attention to pit formation (its shape and depth),

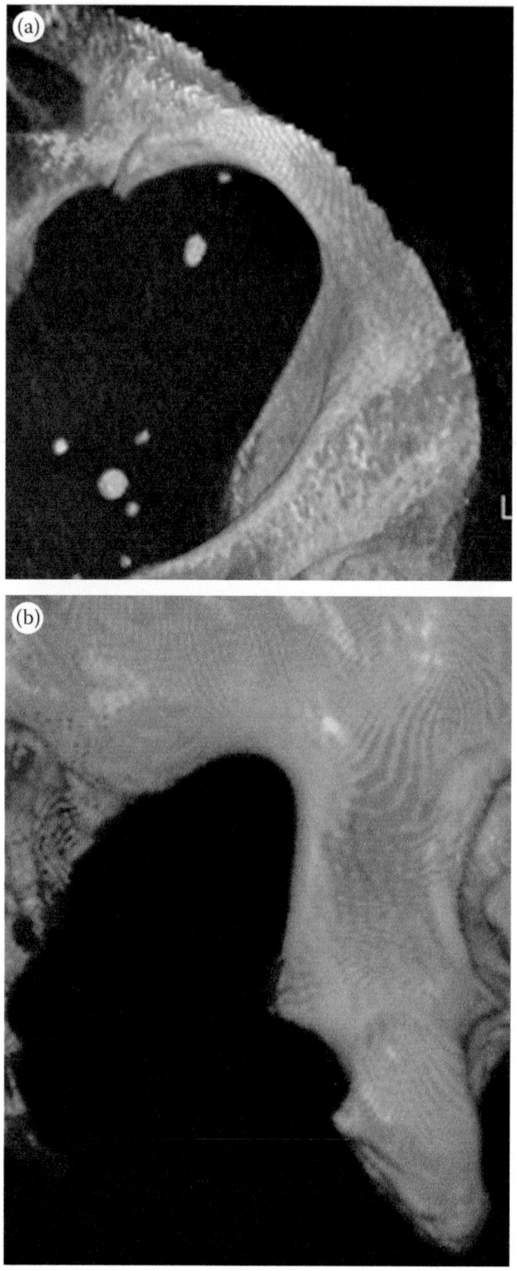

FIGURE 9.6 Appearance of the sciatic notch. (a) Female subject, 3D reconstructions, VRT mode, lateral view: the sciatic notch is large and "U" shaped. (b) Male subject, 3D reconstructions, VRT mode, lateral view: the sciatic notch is narrow and "V" shaped.

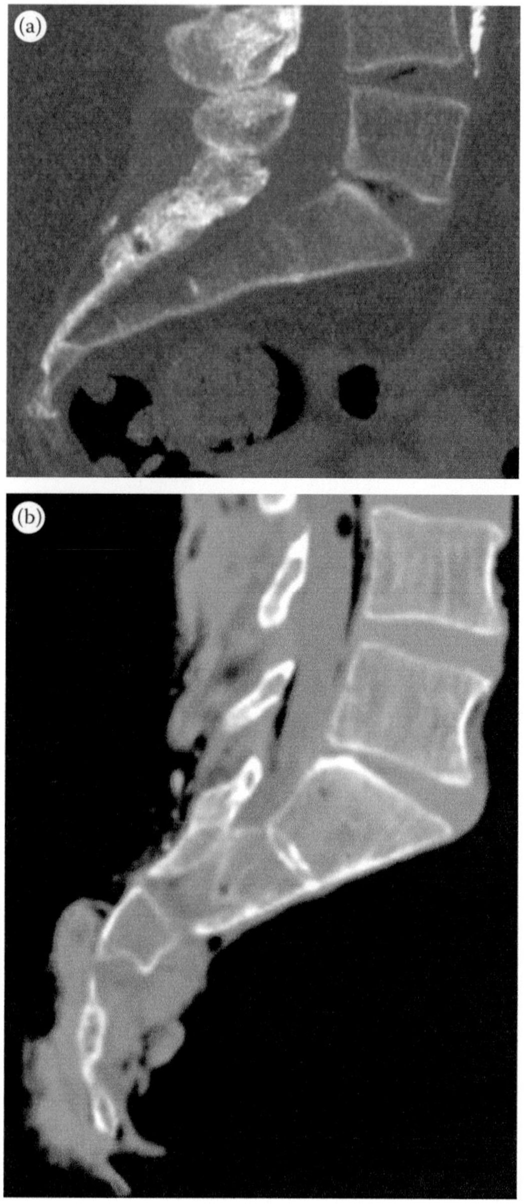

FIGURE 9.7 Appearance of the sacrum. (a) Female subject, 2D reconstructions, MPR mode, sagittal slice: flattened anterior contour from the first to the third sacral vertebra. (b) Male subject, 2D reconstructions, MPR mode, sagittal slice: regular curve anteriorly from the first to the fifth sacral vertebra.

changes in the walls and surrounding rim, and overall bone density and texture.[43,44] Based on changes in these areas, the ribs were separated into 9 phases (0 through 8) of progression spanning 7 decades from the teens through the 70s. The major morphological features indicating changes in the rib were summarized and illustrated with photographs. Age is estimated from the mean age and the 95% confidence interval of the mean for each phase and each sex. These data are readily accessible in human osteology manuals.

The MSCT transposition of Iscan's classification of dry bones has already been studied.[45] With both these methods, intraobserver variability was found to be the same and there was a tendency to overestimate the phases at the second estimation. Concerning interobserver variability, less variability was noted with MSCT reconstructions than with dry bones: divergence was less and concordance better. Intermethod error varied according to the anthropological experience of the observer. However, phase estimations seldom showed complete agreement between the two methods, varying from 23% to 44% depending upon the observer. The percentage of estimations which differed by one phase (above or below) varied from 64% to 81% depending upon the observer. There was a general tendency for the various observers to underestimate phases 5 and after and to overestimate younger

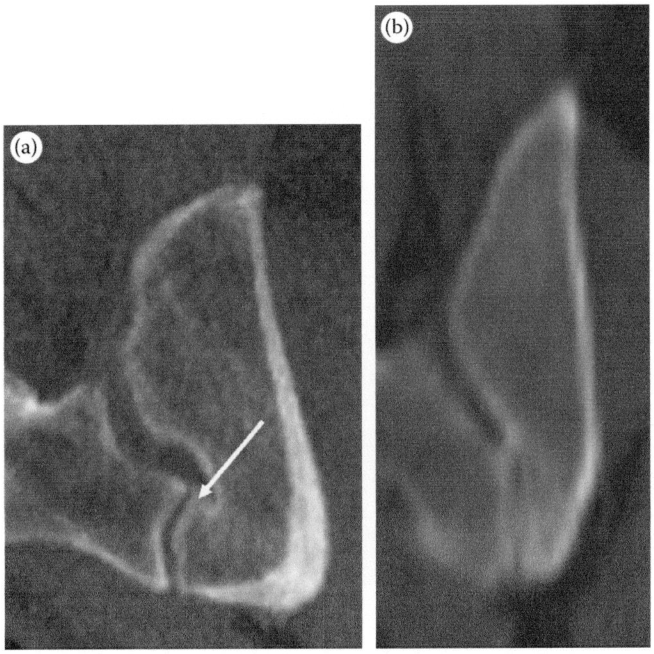

FIGURE 9.8 Appearance of the postauricular sulcus. (a) Female subject, 2D reconstructions, MPR mode, frontal slice: present and marked (arrow), elevated sacro-iliac articulation. (b) Male subject, 2D reconstructions, MPR mode, frontal slice: flat sacro-iliac articulation.

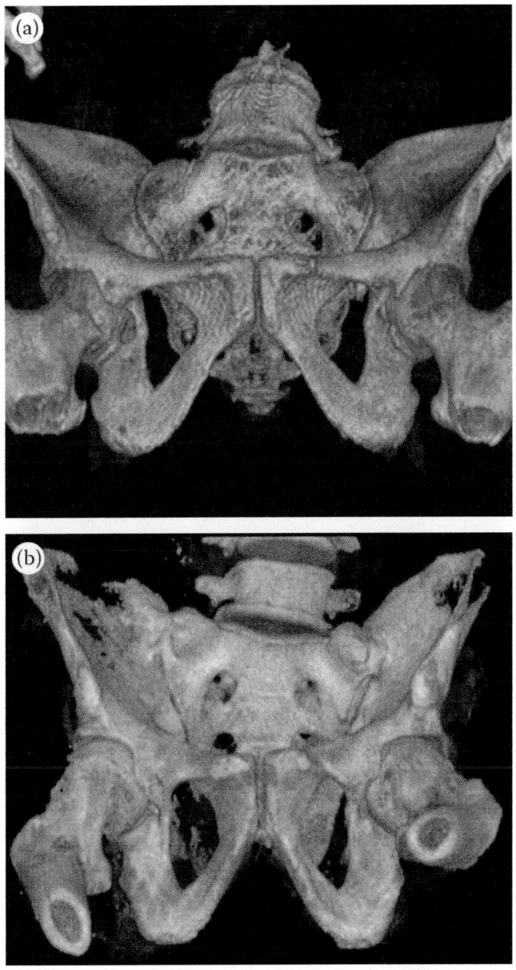

FIGURE 9.9 Shape of the pelvis. (a) Female subject, 3D reconstructions, VRT mode, front view: wide pelvic outlet. (b) Male subject, 3D reconstructions, VRT mode, front view: narrower pelvic outlet. (From Dedouit, F., et al. *Forensic Sci Int*, 173, 182–7, 2007. With permission.)

phases. Considerable variability was also noted for phase five determination. One of the major problems of the MSCT method concerns threshold determination. If it is set too high, some parts of fragile and damaged bone are erased and it is impossible to analyze the edges correctly. If set too low, superimpositions with the chondrosternal cartilage hinder analysis of the edges on MSCT reconstructions. Furthermore, from phase 4 and older, diminishing wall thickness was best evaluated on 2D images. In this study, civil age calculated according to the Iscan method was based on dry bone analysis in 58.3% of cases and on MSCT reconstructions in 63.9%. One major advantage of MSCT reconstruction is that fragile osteophytes and calcifications can be examined with no risk of damage.

Pubis

Estimation of age of adult skeletal remains can also be performed on the pelvis. The surface of the pubic symphysis shows degenerative changes during life. An understanding of the systematic osteological changes that the normal symphysis undergoes enables an estimation to be made of age at death.[36] In young adults, the symphyseal surface is rugged in appearance and is characterized by horizontal ridges. By age 35, the sharp features of the surface are less defined, and a rim forms around the edge. The following years are characterized by progressive degenerative changes, including formation of a rough symphyseal surface, and irregular borders. Until recently, the most commonly used method for estimating age using the architecture of the pubic symphysis was that of Todd, who described 10 stages of pubic symphysis change during aging.[46]

More recently, researchers Suchey and Brooks described six phases of degenerative changes in the pubic symphysis (Figure 9.14).[47,48] The major limitation of the original Todd method was that it overestimated age at death for most individuals, particularly those under the age of 40; it did not account for individual variation and did not accurately age older individuals. Many authors found that the modified six-phase stage was a more accurate estimator than the original Todd system. Consequently, the Suchey–Brooks system is currently the universal standard for estimating age from the pubic symphysis. This six-phase graduation is well described but still suffers the limitation of relatively arbitrary classification of a continuum of change in a biological structure. Also, these methods require long and tedious preparation of the bone specimens and it is often difficult to avoid anatomic damage. Moreover, they cannot be applied to living individuals. The Suchey–Brooks method has already been studied with CT.[49] Intermethod agreement was better for an experienced

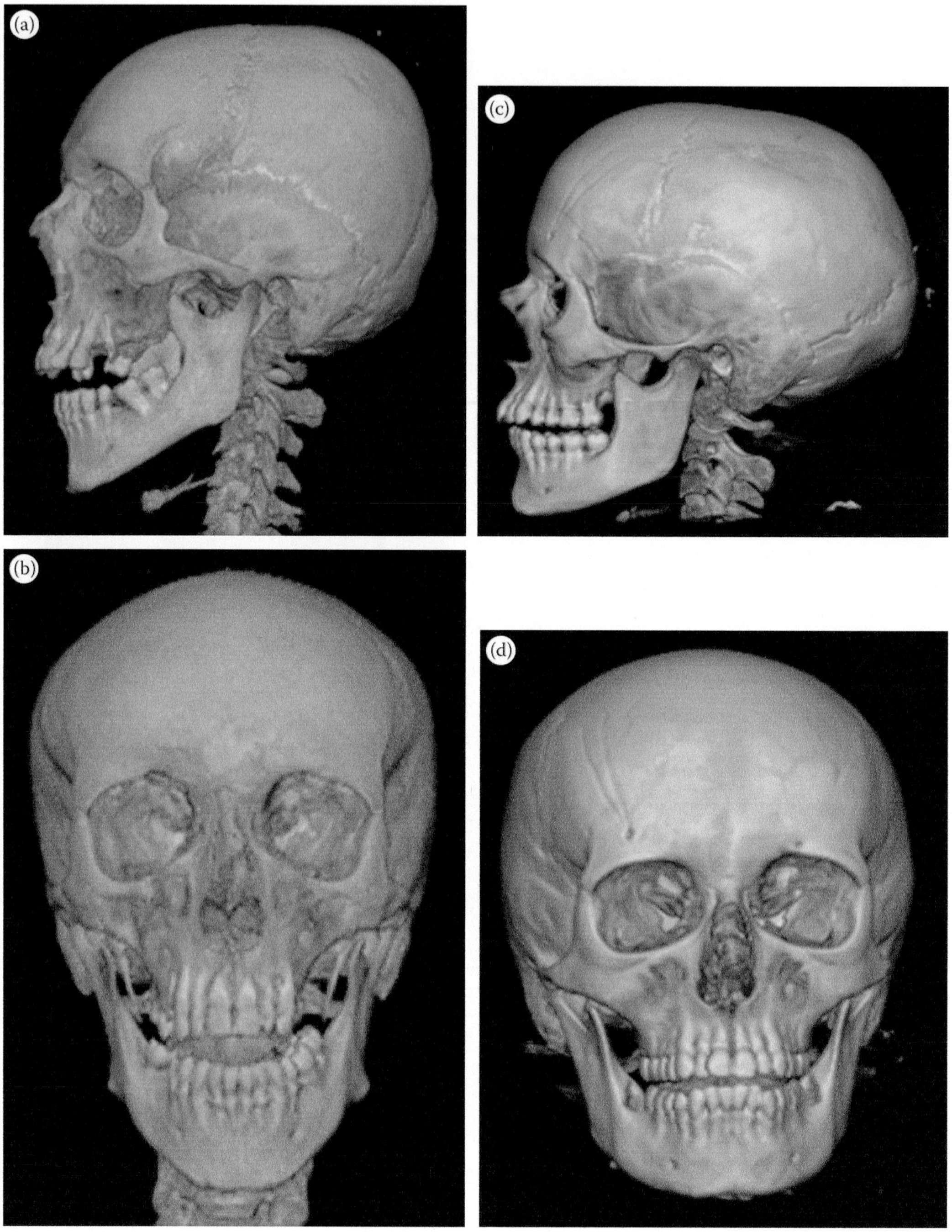

FIGURE 9.10 Sex differences in the skull, 3D VRT reconstructions. (a, b) Male features: sloping forehead and rounded skull contour; large brow ridge; robust occiput; ramus flexure. (c, d) Female features: vertical forehead; rounded occiput; straight ramus.

observer than an inexperienced observer (agreement in 81% of cases vs. 71%, respectively). Errors consisted of overestimation or underestimation of a single phase and concerned phases III/IV, IV/V, and V/VI. An experienced observer compared the different features of bones and CT reconstructions, and the following agreement was found: 100% for ridges and delimitation of extremities, 95% for ligamentous outgrowths, 90% for bone texture and face depression, 86% for the rim, and 81% for the ventral rampart. Consequently, 3D CT imaging seems to be applicable to study the pubic symphysis for age estimation. The various Suchey–Brooks features can be observed on the CT reconstructions. The two approaches, using CT images and dry bones, appeared to yield almost totally concordant results for phase estimation and feature analysis, with nevertheless a few differences.

Other Areas of Interest

Another pelvic site which can be examined is the auricular surface of the ilium. However, this method is very difficult to

TABLE 9.2
Measures of Some Useful Postcranial Bones

Anatomical Segment	Female (mm)	Indeterminate (mm)	Male (mm)
Capula length	<129	140–159	>160
Glenoid cavity length	<34	34–36	>37
Mean clavicular length	140		158
Vertical humeral head diameter	<43	44–46	>47
Mean humeral length	305		340
Mean humeral epicondylar breadth	57		64
Diameter of radial head	<21		>22
Vertical diameter of the femoral head	<41.5	43.5–44.5	>45.5
Femoral length	439		477
Femoral bicondylar width	<72	74–76	>78

Source: From Briggs, C. A., Sex determination, in *Encyclopaedia of Forensic and Legal Medicine*, Payne-James, J., Byard, R. W., Corey, T. S., and Henderson, C., Eds., Elsevier Academic Press, 129–37, 2005. With permission.

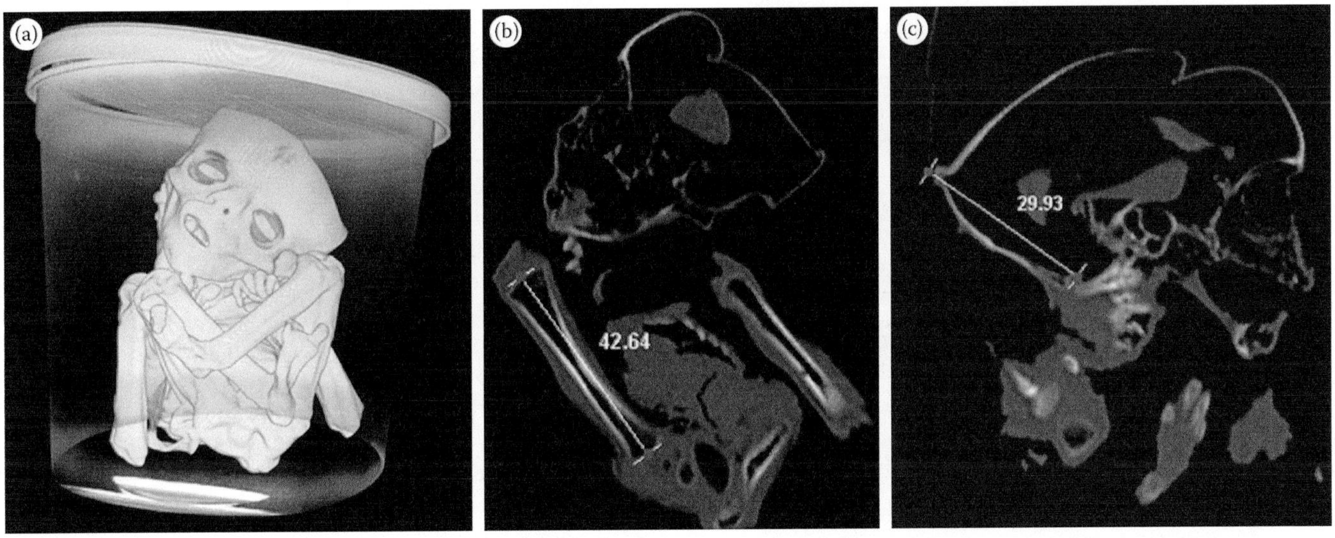

FIGURE 9.11 Age assessment of a fetus. (a) General front view of the fetus, VRT mode. (b) Measure of the right femoral diaphyseal length in an MPR reconstruction. (c) Mid-sagittal measure of the occipital pars squama in an MPR reconstruction.

TABLE 9.3
Statistical Parameters, in Years and by Sex, for the Ossification of the Medial Clavicular Extremity from Stages 2 to 5

Stage	Female		Male	
	Mean ± SD [min, max]	Median, Lower Quartile, Upper Quartile	Mean ± SD [min, max]	Median, Lower Quartile, Upper Quartile
2	18.2 ± 1.6 [15.0, 21.6]	18.5, 16.9, 19.4	18.9 ± 1.7 [15.2, 23.9]	18.9, 16.9, 20.0
3	20.5 ± 2.7 [16.6, 28.6]	20.0, 18.4, 22.0	20.9 ± 1.9 [17.5, 27.2]	20.7, 19.4, 21.9
4	25.1 ± 2.8 [21.5, 29.9]	24.3, 22.8, 27.8	25.2 ± 2.7 [21.2, 30.4]	24.7, 23.1, 27.4
5	27.4 ± 2.3 [21.9, 30.9]	27.9, 25.7, 29.5	27.6 ± 2.3 [22.4, 30.9]	27.8, 26.0, 29.7

Source: From Schulz, R., et al. *Int J Legal Med*, 119, 142–5, 2005. With permission.
Note: SD, standard deviation; min, minimum age; max, maximum age.

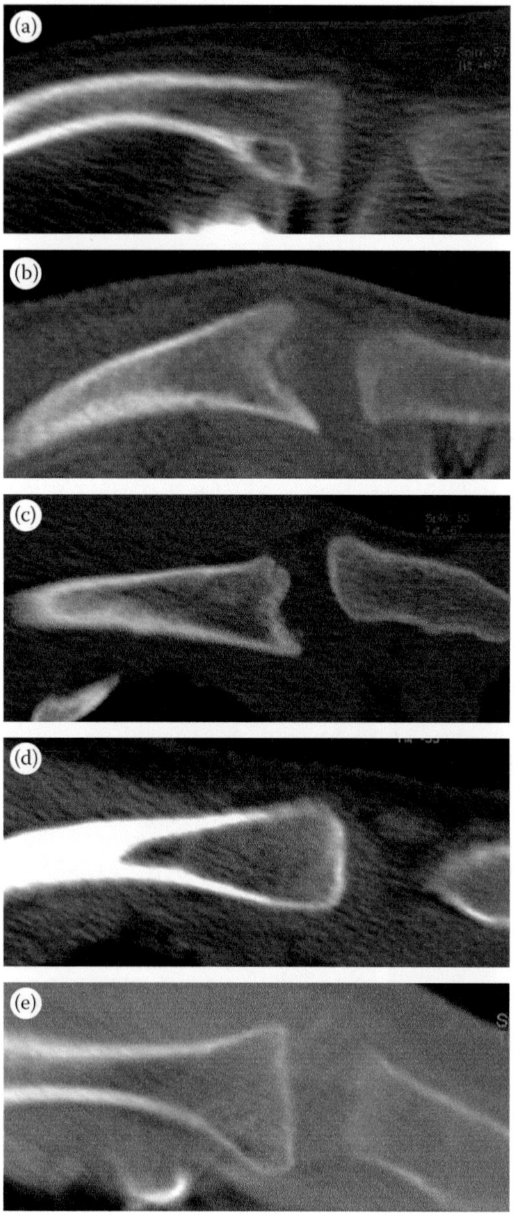

FIGURE 9.12 MPR reconstructions of the medial clavicular extremities. (Adapted from Schulz, R., et al., *Int J Legal Med*, 119, 142–5, 2005. With permission.) (a) Stage 1, no signs of epiphyseal ossification centers. (b) Stage 2, epiphyseal ossification centers without signs of bridging. (c) Stage 3, epiphyseal ossification with bridging. (d) Stage 4, complete bony fusion of epi- and metaphysis with an epiphyseal scar visible. (e) Stage 5, epiphyseal scar no longer visible.

apply and independent testing indicated that this site was not suitable for use in individual forensic cases.[30]

CT scanning has also been applied to study the ossified volume of the thyroid cartilage and its morphological progression. Unfortunately, the results were not sufficiently accurate for individual forensic use.[50]

The complex method of Nemeskeri and co-workers studying endocranial sutures and the trabecular structure of the humeri, femora, and pubic bone has been tested in one study.[33] Features of secondary interest examined were ectocranial sutures, degenerative joint disease, degree of epiphyseal fusion, presence of osteophytes, ossification of the rib and laryngeal cartilage, degeneration of the spine, texture of the clavicle, and state of the dentition. Correct age-at-death estimations varied from 40.9% to 81.8%, depending on the observer.

STATURE DETERMINATION

With regard to dry bones, tables exist and can be applied to MSCT reconstructions.[15] Anthropologists usually use one of two methods to estimate living height from the skeleton: the anatomical method or the mathematical method based on limb bone proportions.[22]

The *anatomical method* involves measuring all of the skeletal elements that contribute to height and then adding a constant.[51] Skeletal height is summed from measurements of the basion-bregma height of the cranium, maximum vertebral body heights of C2–L5 measured separately, anterior height of the first sacral segment, oblique length of the femur, maximum length of the tibia, and the articulated height of the talus and calcaneus. For the femur, tibia, and talus and calcaneus, the average value of the right and left measurements should be used in calculating skeletal height. It is proposed that the following correction factors are to be added to calculated skeletal height in order to obtain a final estimation of living stature: when the skeletal height is below or equal to 153.5 cm, 10 cm must be added; between 153.6 and 165.4 cm, 10.5 cm; and equal to or above 165.5 cm, 11.5 cm. Recently, Raxter modified this equation.[52] When age is known: living stature $= 1.009 \times$ skeletal height $- 0.0426 \times$ age $= 12.1$. When age is unknown: living stature $= 0.996 \times$ skeletal height $+ 11.7$. This anatomical method of stature estimation is described as independent of ethnic affiliation as well as of gender.

We applied this method to a natural female mummy because the skeleton was complete, with no missing bones.[26] The various bone measurements made according to Raxter et al. showed that the young woman was 146 cm tall. During autopsy, anatomical measurement of the mummy's stature was 145 cm, probably slightly underestimated because of soft-tissue dehydration.

The *mathematical method* of stature estimation is based on the correlation between discrete bones and body parts and stature.[22] Regression formulae based on measurements of single bones and on combinations of various bones specific for sex and population are used. The best bones from which to reconstruct living stature are the long bones of the lower limb, since they are the most important components of height.

The limb bone proportion method relies on well-studied relationships between the length of limb bones and living height. Because of the variation in long bone-to-height ratio among the world's populations, different formulas are provided for different groups. Numerous tables exist for gender and different racial phenotypes and are available in many osteology manuals. With the use of postprocessing imaging (MPR and MIP), the lengths of certain long bones can be

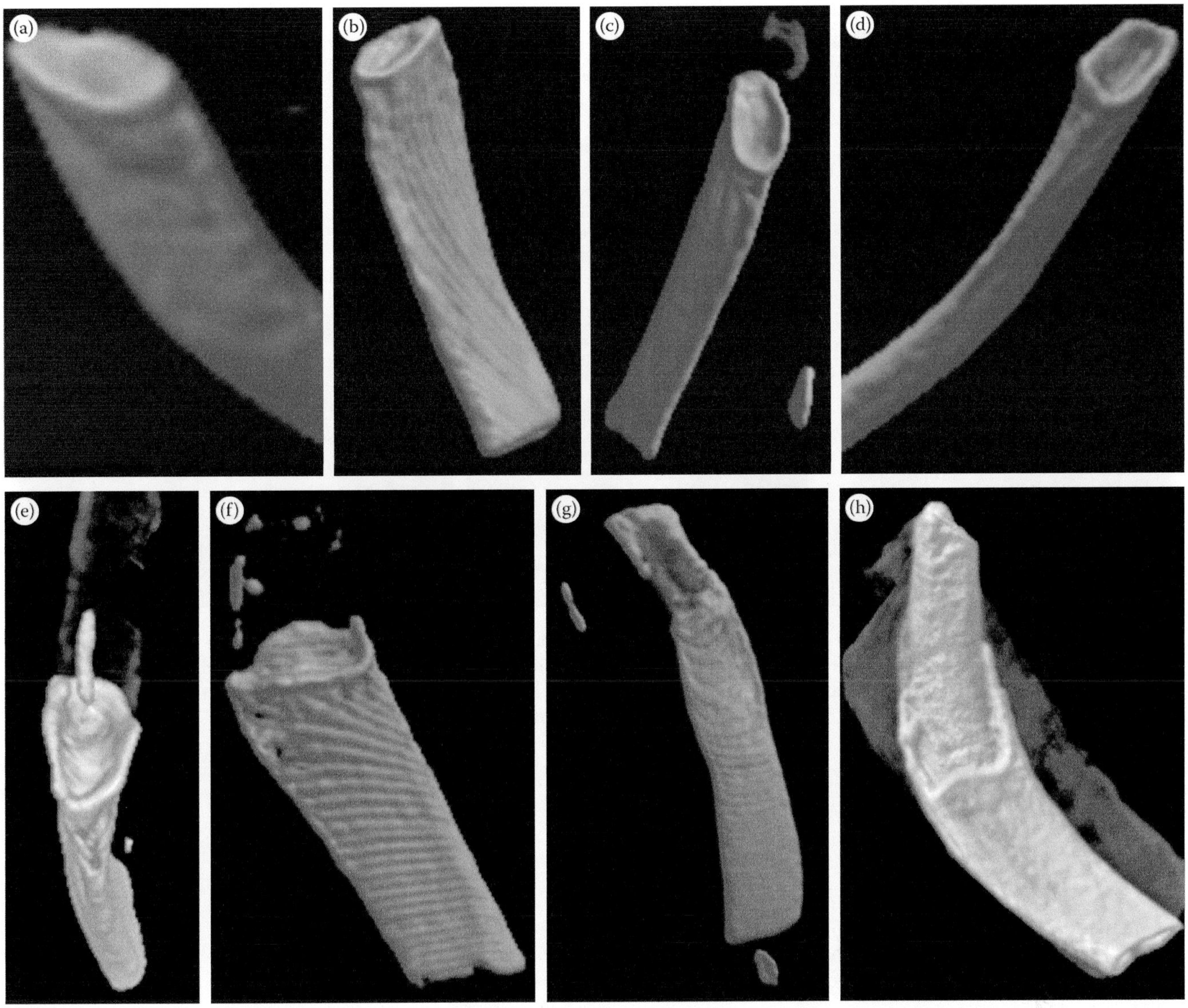

FIGURE 9.13 Iscan classification transposed on VRT reconstructions of the sternal extremity of the right fourth rib, with the classic description of the appearance of each stage. (Adapted from Dedouit, F., et al., *J Forensic Sci*, 53, 288–95, 2008. With permission.) (a) Phase 1: there is a beginning of an amorphous indentation in the articular surface, but billowing may also still be present. The rim is rounded and regular. In some cases, scallops may start to appear at the edges. The bone is still firm, smooth and solid. (b) Phase 2: the pit is now deeper and has assumed a V-shaped appearance formed by the anterior and posterior walls. The walls are thick and smooth with a scalloped or slightly wavy rim with rounded edges. The bone is firm and solid. (c) Phase 3: the deepening pit has taken on a narrow to moderately U-shape. Walls are still fairly thick with rounded edges. Some scalloping may still be present but the rim is becoming more irregular. The bone is still quite firm and solid. (d) Phase 4: pit depth is increasing, but the shape is still a narrow to moderately wide U. The walls are thinner, but the edges remain rounded. The rim is more irregular with no uniform scalloping pattern remaining. There is some decrease in the weight and firmness of the bone; however, the overall quality of the bone is still good. (e) Phase 5: there is a little change in pit depth, but the shape in this phase is predominantly a moderately wide U. Walls show further thinning and the edges are becoming sharp. Irregularity is increasing in the rim. The scalloping pattern is completely gone and has been replaced with irregular bony projections. The condition of the borne is fairly good; however, there are some signs of deterioration with evidence of porosity and loss of density. (f) Phase 6: the pit is noticeably deep with a wide U-shape. The walls are thin with sharp edges. The rim is irregular and exhibits some rather long bony projections that are frequently more pronounced at the superior and inferior borders. The bone is noticeably lighter in weight, thinner and more porous, especially inside the pit. (g) Phase 7: the pit is deep with a wide to very wide U-shape. The walls are thin and fragile with sharp, irregular edges and bony projections. The bone is light in weight and brittle with significant deterioration in quality and obvious porosity. (h) Phase 8: in this final phase the pit is very deep and widely U-shaped. In some cases the floor of the pit is absent or filled with bony projections. The walls are extremely thin, fragile, and brittle, with sharp, highly irregular edges and bony projections. The bone is very lightweight, thin, brittle, friable, and porous. "Window" formation is sometimes seen in the walls.

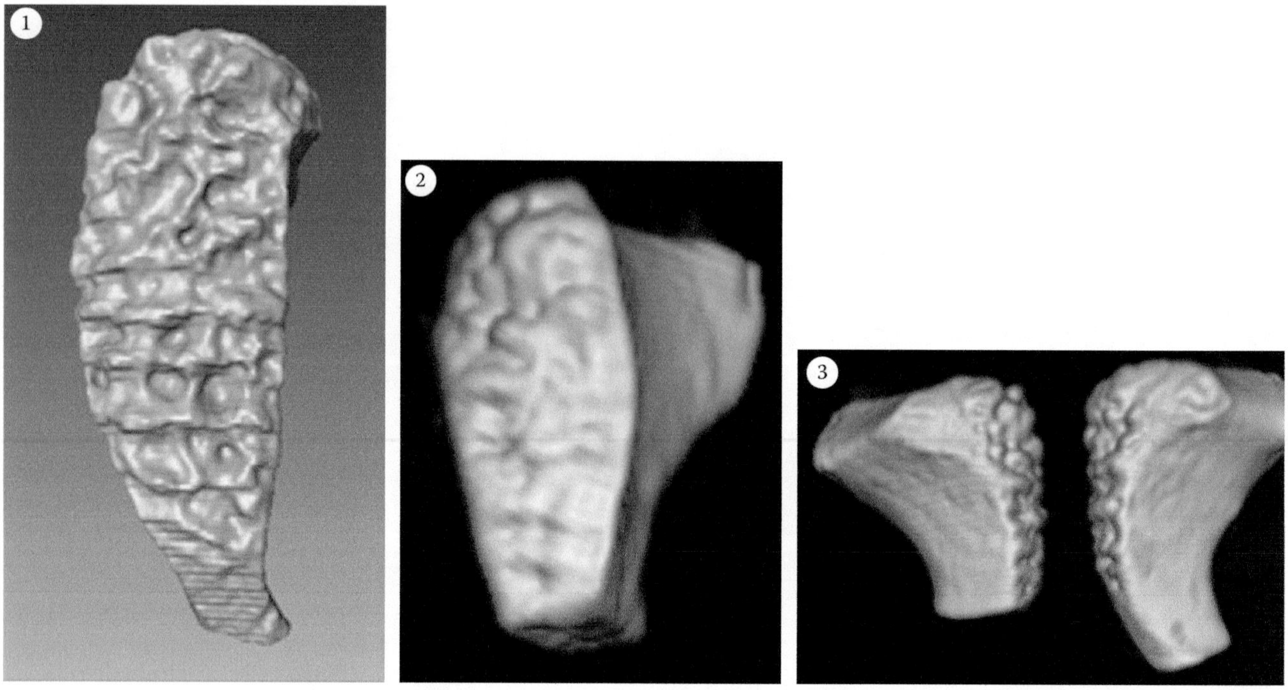

FIGURE 9.14a 3D VRT reconstruction of the medial surface of the symphysis pubis of a female subject. Phase 1: the symphyseal face has a billowing surface composed of ridges and furrows which includes the pubic tubercle; the horizontal ridges are well-marked; ventral beveling may be commencing; lack of delimitation for either extremity (upper or lower). (1) 3D reconstruction based on the CT scan of a cast. (2) 3D VRT reconstruction of fleshed bones, medial view. (3) 3D VRT reconstruction of fleshed bones, front view.

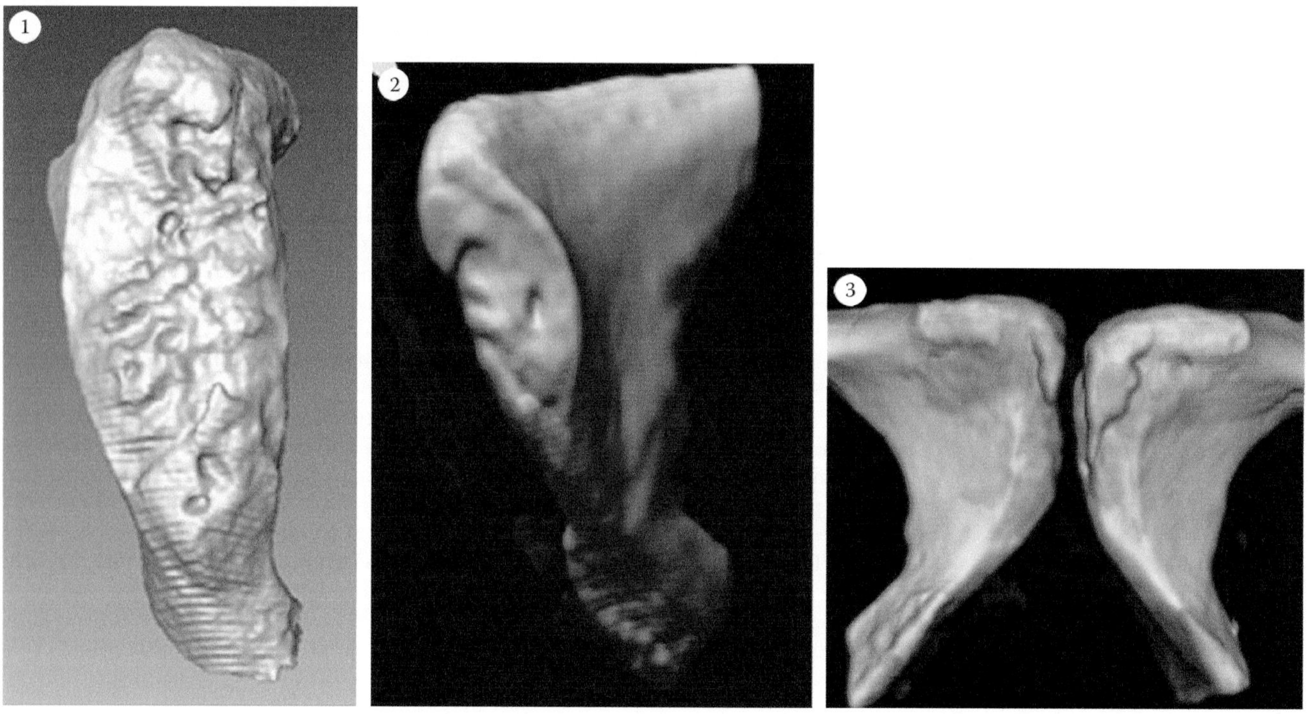

FIGURE 9.14b 3D VRT reconstruction of the medial surface of the symphysis pubis of a female subject. Phase 2: symphyseal face may still show ridge development; lower and upper extremities show early stages of delimitation, with or without ossific nodules; ventral rampart may begin formation as extension from either or both extremities. (1) 3D reconstruction based on CT scan of a cast. (2) 3D VRT reconstruction of fleshed bones, medial view. (3) 3D VRT reconstruction of fleshed bones, front view.

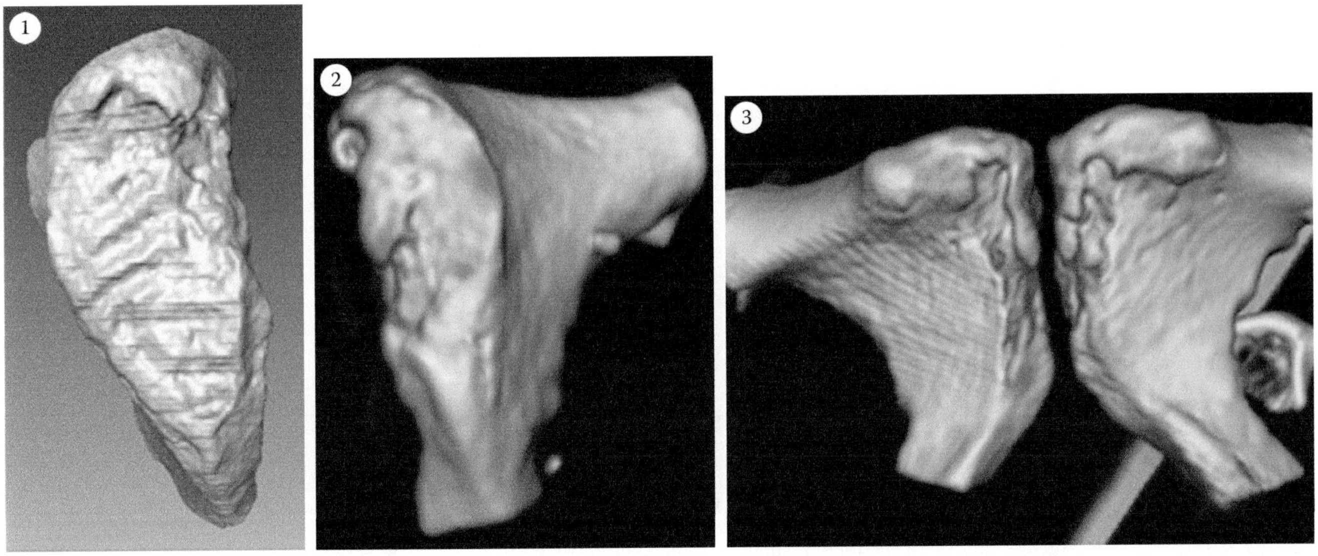

FIGURE 9.14c 3D VRT reconstruction of medial surface of the symphysis pubis of a female subject. Phase 3: symphyseal face shows lower extremity and ventral rampart in process of completion; fusing ossific nodules may form upper extremity and extend along ventral border; symphyseal face may either be smooth or retain distinct ridges; dorsal plateau is complete; no lipping of symphyseal dorsal margin or bony ligamentous outgrowths.(1) 3D reconstruction based on the CT scan of a cast. (2) 3D VRT reconstruction of fleshed bones, medial view; (3) 3D VRT reconstruction of fleshed bones, front view.

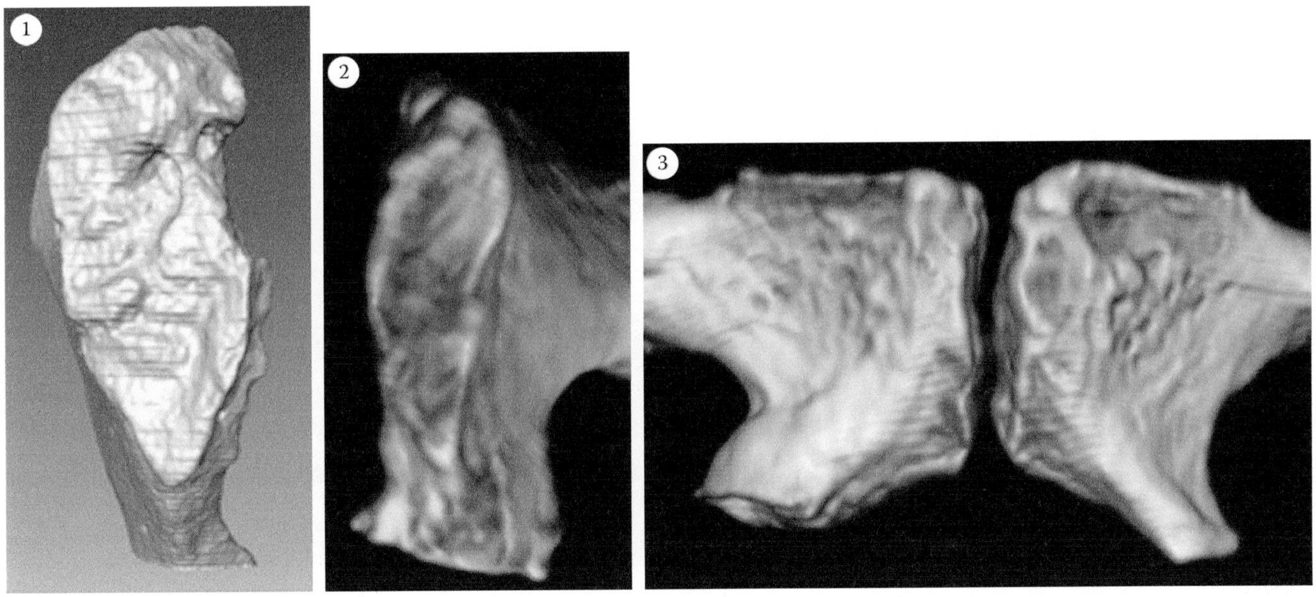

FIGURE 9.14d 3D VRT reconstruction of the medial surface of the symphysis pubis of a female subject. Phase 4: symphyseal face generally fine-grained, although remnants of ridge and furrow system may remain, may have a distinct rim; oval outline usually complete at this stage, though a hiatus may occur on the upper aspect of the ventral circumference; ventrally, bony ligamentous outgrowths may occur in the inferior portion of the pubic bone adjacent to the symphyseal face; slight lipping may appear on the dorsal border. (1) 3D reconstruction based on the CT scan of a cast. (2) 3D VRT reconstruction of fleshed bones, medial view. (3) 3D VRT reconstruction of fleshed bones, front view.

directly measured on the reconstructions. Based on these lengths and the use of various formulas, stature can be estimated with relative ease.[42,53] In mummies lying in the fetal position, stature can be estimated by measuring the length of the long bones (femur and tibia) and using the Trotter and Gleser tables.[54,55]

FORENSIC TAPHONOMY

In addition to the previously described classic areas of enquiry in physical anthropology (racial phenotype, sex determination, age assessment, and stature determination), another field is accessible to MSCT investigation: taphonomy.

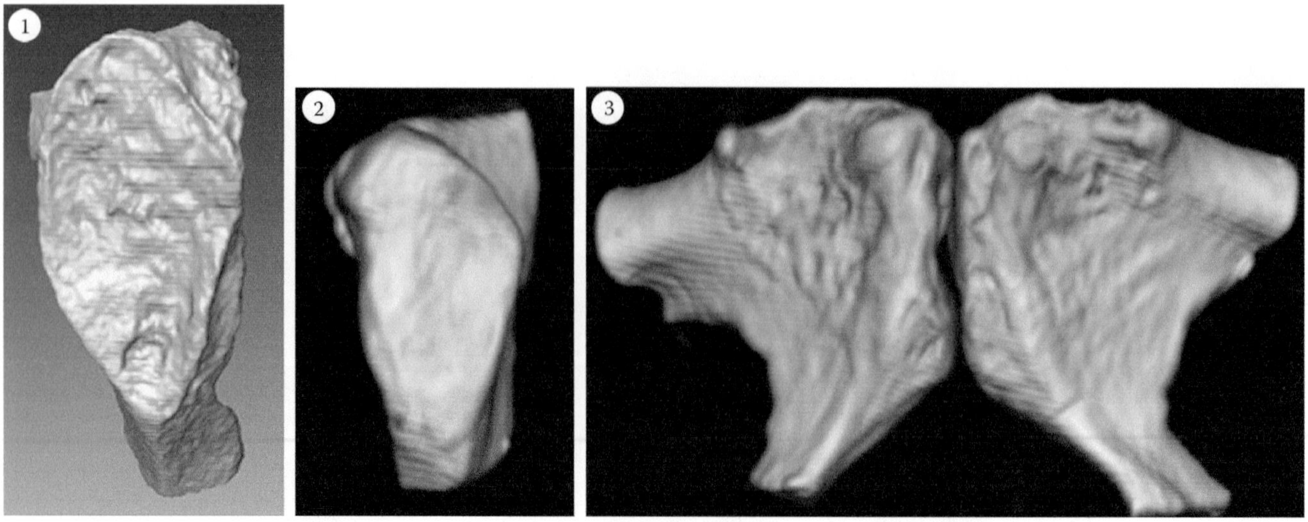

FIGURE 9.14e 3D VRT reconstruction of the medial surface of the symphysis pubis of a female subject. Phase 5: slight depression of the face relative to a completed rim; moderate lipping is usually found on the dorsal border with prominent ligamentous outgrowths on the ventral border; little or no rim erosion, though breakdown is possible on the superior aspect of the ventral border. (1) 3D reconstruction based on the CT scan of a cast. (2) 3D VRT reconstruction of fleshed bones, medial view. (3) 3D VRT reconstruction of fleshed bones, front view.

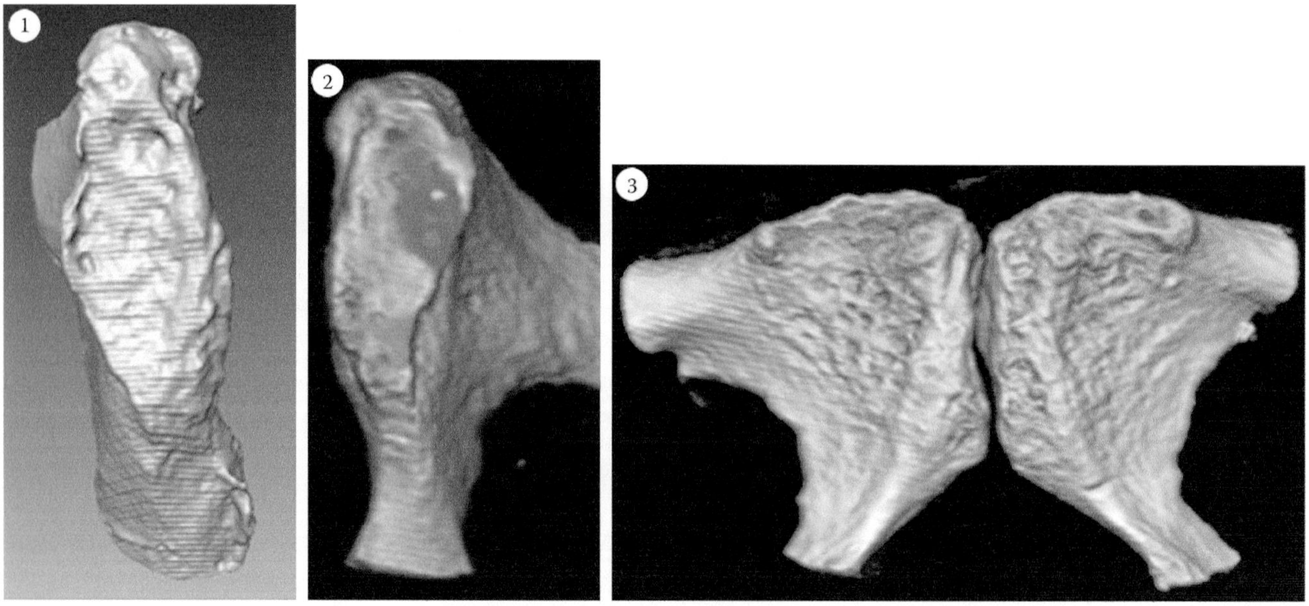

FIGURE 9.14f 3D VRT reconstruction of medial surface of the symphysis pubis of a female subject. Phase 6: symphyseal face shows ongoing depression as rim erodes; ventral ligamentous attachments are marked; pubic tubercle may appear as a separate bony knob. (1) 3D reconstruction based on the CT scan of a cast. (2) 3D VRT reconstruction of fleshed bones, medial view. (3) 3D VRT reconstruction of fleshed bones, front view.

Taphonomy is the study of an organism from death through resorption.[56] Forensic taphonomy is the interdisciplinary study and interpretation of postmortem processes of human remains in their depositional context.[57] Taphonomic details are critically important for estimating time since death and for differentiating injuries from postmortem changes.

The radiologist who performs postmortem MSCT must be familiar with late postmortem changes to avoid misinterpreting normal changes as traumatic injuries.[58] Such confusion, especially when exhumed bodies are concerned, could potentially have significant judicial consequences.

The post-mortem changes of the ossicular chain of the middle ear have been illustrated in a study of six unprepared, naturally skeletonized skulls.[58] The post-mortem MSCT disposition of the ossicular bones was different from the normal aspect seen in living subjects. No ossicular chain was intact.

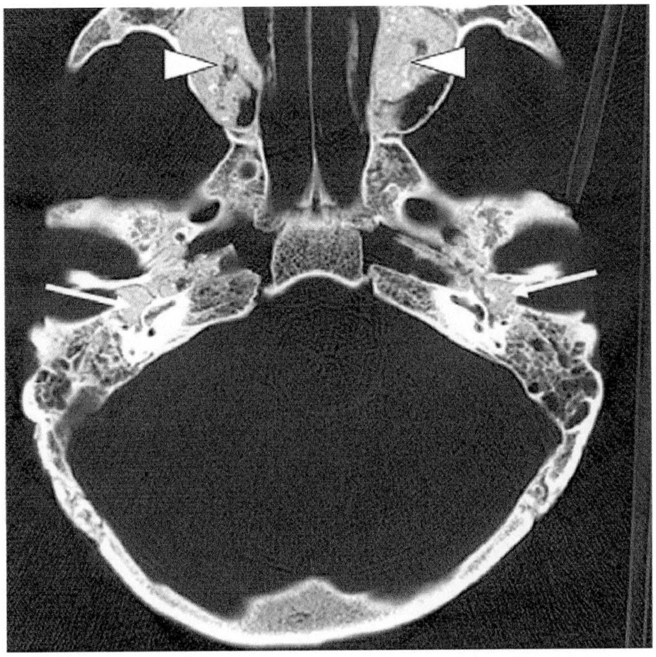

FIGURE 9.15 Axial MSCT image of an unprepared dry skull. Hyperdense concretions filling both tympanic cavities (arrows); both maxillary sinuses are filled with the same material (arrowheads). (From Dedouit, F., et al. *Forensic Sci Int*, 175, 149–54, 2008. With permission.)

In all skulls, it was disrupted or the ossicular bones were missing. The most common change was dislocation of the malleoincudal complex, present in 7 of the 12 middle ears. Other changes observed were absence of ossicular bones, migration of the incus and/or malleus into the aditus ad antrum, and internal stapediovestibular dislocation. The inner ear, including the otic capsule and cochlea, vestibule and semicircular canals, was normal in all skulls. It is important to note that concretions may be present within the external auditory meatus and the tympanic cavities if the body has been buried in soil (Figure 9.15). These concretions have a spontaneous hyperdense appearance and must not be confused with radio-opaque foreign bodies such as metal projectiles. Like other appendicular joints, the ossicular chains undergo postmortem changes and may have a pseudotraumatic appearance. The radiologist who performs postmortem imaging must be aware of the potential changes secondary to taphonomic processes, especially in exhumed bodies, where ossicular disruptions must not be misinterpreted as antemortem or perimortem traumatic injuries. Any confusion could obviously have significant judicial consequences.

Other classic postmortem changes visible on mummified, putrefied, or exhumed subjects are axial and appendicular joint disarticulations. Because of the loss of the soft tissues and the costal cartilages, the ribs, sternum, and clavicles may collapse into the chest cavity (Figure 9.16). The hyoid bone may also fall near the spine. Especially in cases of putrefied, partially skeletonized, or exhumed bodies, MSCT is a precious tool for identifying the different laryngeal components,

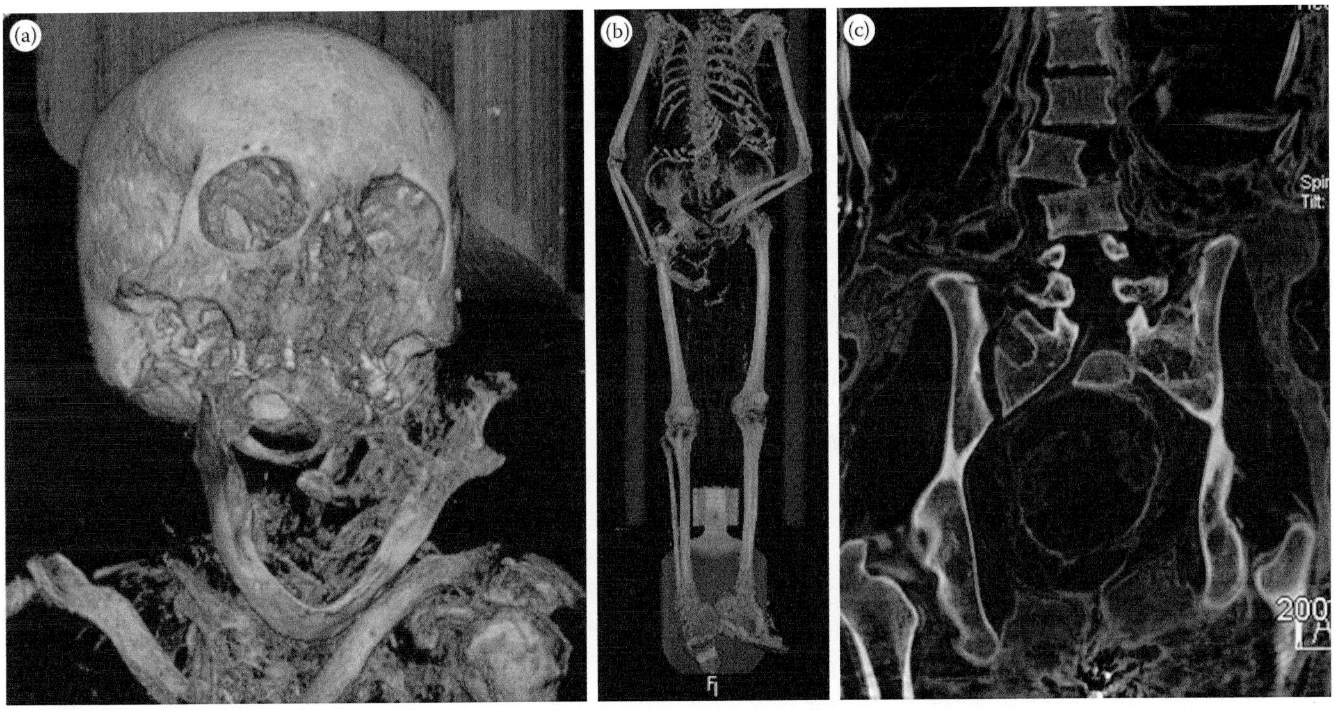

FIGURE 9.16 Appendicular and axial joint disarticulation in an exhumation case. (a) 3D VRT reconstruction of the cephalic extremity, front view. Mandible completely disarticulated. (b) 3D VRT reconstruction of the body, front view. Disarticulation of numerous joints (sternocostal, wrists, carpals, and fingers). (c) Frontal MPR reconstruction of the lumbar spine. Vertebral bodies' subluxation.

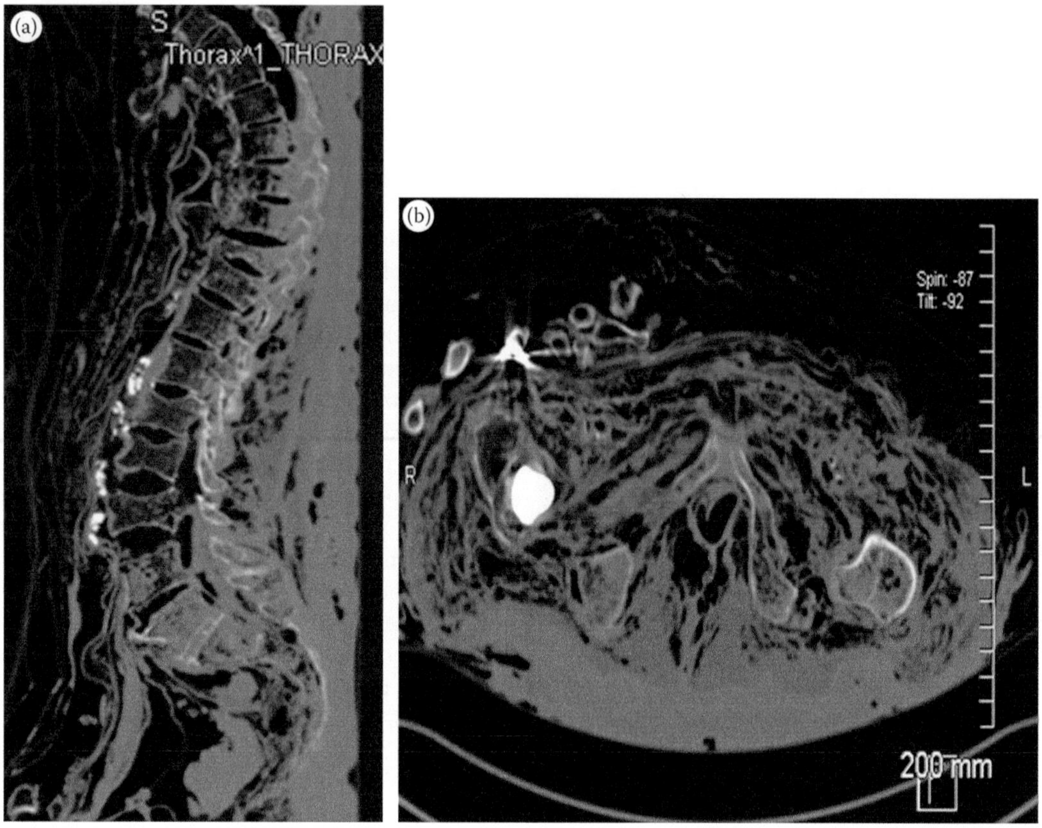

FIGURE 9.17 Antemortem fractures in an exhumation case. (a) Frontal MPR reconstruction of the spine. Note the presence of multiple vertebral body fractures. (b) Axial MSCT image of the pelvis in an exhumation case. Healed fracture with hard callus of the left ischiopubic ramus.

namely the hyoid bone and the thyroid cartilage. The forensic radiologist must be aware that for forensic pathologists, recovery of this anatomical part may be particularly difficult because it often decomposes into dark, semi-liquid putrefied tissue. However, its recovery is always essential because it may enable detection of fractures of the thyroid cartilage or the hyoid bone, stigmata of strangulation, or hanging.

The mandible may also be disarticulated and some teeth may be absent because they have fallen postmortem. This must not be confused with post-traumatic lesions.

Like pseudotraumatic lesions, true post-traumatic lesions can of course be diagnosed, in particular healed antemortem fractures (Figure 9.17). These are important findings that may concur with identification by comparison of ante- and postmortem radiographs or CT investigations. Occult lesions that are revealed on imaging may raise difficulties for identification if previously unknown.[17]

IN CURRENT PRACTICE

First of all, indications of radiological examinations and investigations have to be strictly selected. When direct visual dry bone study is possible or accessible and provides adequate answers to the questions raised, why make additional costly investigations? Consequently, such investigations need the agreement of the anthropologists or forensic anthropologists concerned. As in clinical situations, the radiologist must be informed of their purpose. Furthermore, in some identification cases, the definitive answer will not be radiological but biological: it will be provided by DNA analysis. The radiologist must therefore be aware of the limitations of the techniques employed and of the possibilities of obtaining an accurate answer in terms of percentage of efficiency.

REFERENCES

1. Lewin, P. K. and Harwood-Nash, D. C., X-ray computed axial tomography of an ancient Egyptian brain, *ICRS Med Sci*, 5, 78, 1977.
2. Lewin, P. K., and Harwood-Nash, D. C., Computerized axial tomography in medical archaeology, *Paleopathol Newsl*, 17, 8–9, 1977.
3. Hounsfield, G. N., Computerized transverse axial scanning (tomography): Part 1. Description of system, *Br J Radiol*, 46, 1016–22, 1973.
4. Piepenbrink, H., Frahm, J., Haase, A., and Matthaei, D., Nuclear magnetic resonance imaging of mummified corpses, *Am J Phys Anthropol*, 70, 27–8, 1986.
5. Karlik, S. J., Bartha, R., Kennedy, K., and Chhem, R. K., MRI and multinuclear MR spectroscopy of 3,200-year-old Egyptian mummy brain, *Am J Roentgenol*, 189, W105–10, 2007.
6. Rühli, F. J., von Waldburg, H., Nielles-Vallespin, S., Böni, T., and Speier, P., Clinical magnetic resonance imaging of ancient dry human mummies without rehydration, *JAMA*, 298, 2618–20, 2007.

7. Notman, N. H., Tashjian, J., Aufderheide, A. C., Cass, O. W., Shane, O. C. III, Berquist, T. H., and Gedgaudas, E., Modern imaging and endoscopic biopsy techniques in Egyptian mummies, *Am J Roentgenol*, 146, 93–6, 1986.
8. Dvorak, J., George J., Junge A., and Hodler, J., Age determination by magnetic resonance imaging of the wrist in adolescent male football players, *Br J Sports Med*, 41, 45–52, 2007.
9. Dvorak, J., George J., Junge A., and Hodler, J., Application of MRI of the wrist for age determination in international U-17 soccer competitions, *Br J Sports Med*, 41, 497–500, 2007b.
10. Schmidt, S., Mühler, M., Schmeling, A., Reisinger, W., and Schulz, R., Magnetic resonance imaging of the clavicular ossification, *Int J Legal Med*, 121, 321–4, 2007.
11. Dedouit, F., Auriol, J., Braga, J., et al., *Age Determination by Magnetic Resonance Imaging of the Knee: A Preliminary Study, Presented at the American Association of Physical Anthropologists*, Chicago, April, 2009.
12. Chhem, R. K. and Brothwell, D. R., *Paleoradiology: Imaging Mummies and Fossils.* Springer, Heidelberg, 2008.
13. Weber, G. W., Virtual anthropology (VA): A call for glasnost in paleoanthropology, *Anat Rec*, 265, 193–201, 2001.
14. Thali, M. J., Yen, K., Schweitzer, W., et al., Virtopsy, a new imaging horizon in forensic pathology: Virtual autopsy by postmortem multislice computed tomography (MSCT) and magnetic resonance imaging (MRI)—a feasibility study, *J Forensic Sci*, 48, 386–403, 2003.
15. Dedouit, F., Costagliola, R., Telmon, N., Otal, P., Joffre, F., and Rougé, D., Virtual anthropology and forensic identification: Report of one case, *Forensic Sci Int*, 173, 182–7, 2007a.
16. Dedouit, F., Guilbeau-Frugier, C., Telmon, N., et al., Virtual autopsy and forensic identification—practical application: A report of one case, *J Forensic Sci*, 52, 960–4, 2007b.
17. Dedouit, F., Telmon, N., Costagliola, R., et al., New identification possibilities with post-mortem multislice computed tomography, *Int J Legal Med*, 121, 507–10, 2007c.
18. Pfaeffli, M., Vock, P., Dirnhofer, R., Braun, M., Bolliger, S. A., and Thali, M. J., Post-mortem radiological CT identification based on classical ante-mortem x-ray examinations, *Forensic Sci Int*, 171, 111–7, 2007.
19. Sidler, M., Jackowski, C., Dirnhofer, R., Vock, P., and Thali, M., Use of multislice computed tomography in disaster victim identification—advantages and limitations, *Forensic Sci Int*, 169, 118–28, 2007.
20. Iscan, M. Y., Loth, S. R., and Steyn, M., Determination of racial affinity, in *Encyclopaedia of Forensic Sciences*, Siegel, J., Knupfer, G., and Saukko, P., Eds., Academic Press, London, 227–35, 2000.
21. Marks, M. K., and Synstelien, J. A., Determination of racial affinity, in *Encyclopaedia of Forensic and Legal Medicine*, Payne-James, J., Byard, R. W., Corey, T. S., and Henderson, C., Eds., Elsevier Academic Press, London, 137–42, 2005.
22. Kahana, T., Anthropology—Overview, in *Encyclopaedia of Forensic and Legal Medicine*, Payne-James, J., Byard, R. W., Corey, T. S., and Henderson, C., Eds., Elsevier Academic Press, London, 80–89, 2005.
23. Verhoff, M. A., Ramsthaler, F., Krähahn, J., Deml, U., Gille, R. J., Grabherr, S., Thali, M. J., and Kreutz, K., Digital forensic osteology–possibilities in cooperation with the Virtopsy project, *Forensic Sci Int*, 174, 152–6, 2008.
24. Giles, E. and Elliot, O., Sex determination by discriminant function analysis of crania, *Am J Phys Anthropol*, 21, 53–68, 1962.
25. Krogman, W. M. and Iscan, N. Y., *The Human Skeleton in Forensic Medicine*, Thomas, Springfield, IL, 1986.
26. Dedouit, F., Geraut, A., Baranov, V., et al., Virtual and macroscopical studies of mummies—Differences or complementarity? Report of a natural frozen Siberian mummy, *Forensic Sci Int*, in press, 2010.
27. Howells, W. W., Howells' craniometric data on the Internet, *Am J Phys Anthropol*, 101, 441–2, 1996.
28. Ousley, S. D., and Jantz, R. L. *Fordisc 2.0*, Forensic Anthropology Center, University of Tennessee, Knoxville, TN, 1996.
29. Hughes, S., Wright, R., and Barry, M., Virtual reconstruction and morphological analysis of the cranium of an ancient Egyptian mummy, *Australas Phys Eng Sci Med*, 28, 122–7, 2005.
30. Loth, S. R., and Iscan, M. Y., Morphological age estimation, in *Encyclopaedia of Forensic Sciences*, Siegel, J., Knupfer, G., and Saukko, P., Eds., Academic Press, London, 242–52, 2000.
31. Briggs, C. A. Sex determination, in *Encyclopaedia of Forensic and Legal Medicine*, Payne-James, J., Byard, R. W., Corey, T. S., and Henderson, C., Eds., Elsevier Academic Press, London, 129–37, 2005.
32. Loth, S. R., and Iscan, M. Y., Sex determination, in *Encyclopaedia of Forensic Sciences*, Siegel, J., Knupfer, G., and Saukko, P., Eds., Academic Press, London, 252–60, 2000.
33. Grabherr, S., Cooper, C., Ulrich-Bochsler, S., et al., Estimation of sex and age of "virtual skeletons"—a feasibility study, *Eur Radiol*, 19, 419–29, 2009.
34. Uysal, S., Gokharman, D., Kacar, M., Tuncbilek, I., and Kosa, U., Estimation of sex by 3D CT measurements of the foramen magnum, *J Forensic Sci*, 50, 1310–4, 2005.
35. Dang-Tran, K. D., Dedouit, F., Telmon, N., Joffre, F., Rougé, D., and Rousseau, H., Apport de l'étude tomodensitométrique du cartilage thyroïde en anthropologie médico-légale, presented at the Journées Françaises de Radiologie, Paris, October 27, 2008.
36. Simpson, E. K., Morphological age estimation. in *Encyclopaedia of Forensic and Legal Medicine*, Payne-James, J., Byard, R. W., Corey, T. S., and Henderson, C., Eds., Elsevier Academic Press, London, 119–23, 2005.
37. Scheuer, L. and Black, S., *Developmental Juvenile Osteology.* Academic Press, San Diego, 2000.
38. Dedouit, F., Guilbeau-Frugier, C., Loubes-Lacroix, F., et al., Virtual autopsy and forensic anthropology of a mummified fetus: A report of one case, *J Forensic Sci*, 53, 208–12, 2008.
39. Kreitner, K. F., Schweden, F. J., Riepert, T., Nafe, B., and Thelen, M., Bone age determination based on the study of the medial extremity of the clavicle, *Eur Radiol*, 8, 1116–22, 1998.
40. Schulz, R., Mühler, M., Mutze, S., Schmidt, S., Reisinger, W., and Schmeling, A., Studies on the time frame for ossification of the medial epiphysis of the clavicle as revealed by CT scans, *Int J Legal Med*, 119, 142–5, 2005.
41. Schulze, D., Rother, U., Fuhrmann, A., Richel, S., Faulmann, G., and Heiland, M., Correlation of age and ossification of the medial clavicular epiphysis using computed tomography, *Forensic Sci Int*, 158, 184–9, 2006.
42. Chan, S. S., Elias, J. P., Hyssel, M. E., and Hallowell, M. J., CT of a Ptolemaic period mummy from the ancient Egyptian city of Akhmin, *Radiographics*, 28, 2023–32, 2008.
43. Iscan, M., Loth, S. R., and Wright, R. K., Age estimation from the rib by phase analysis: White males, *J Forensic Sci*, 29, 1094–1104, 1984.
44. Iscan, M., Loth, S. R., and Wright, R. K., Age estimation from the rib by phase analysis: White females, *J Forensic Sci*, 30, 853–863, 1985.

45. Dedouit, F., Bindel, S., Gainza, D., et al., Application of the Iscan method to two- and three-dimensional imaging of the sternal end of the right fourth rib, *J Forensic Sci*, 53, 288–95, 2008.
46. Todd, T. W., Age changes in the pubic bone: I. The white male pubis, *Am J Phys Anthropol*, 3, 285–334, 1920–21.
47. Suchey, J. M., *Male Pubis Age Determination-instructional Casts. Typescript Materials Distributed with the Suchey-Brooks Male Instructional Casts*, France Casting, Bellvue, CO, 1987.
48. Suchey, J. M., Brooks, S. T., and Katz, D., Instructional materials accompanying female pubic symphyseal models of the Suchey-Brooks system, distributed by France Casting: Diane France, 2190 West Drake Road, Suite 259, Fort Collins, CO 80526, 1988.
49. Telmon, N., Gaston, A., Chemla, P., Blanc, A., Joffre, F., and Rougé, D., Application of the Suchey Brooks method to three-dimensional imaging of the pubic symphysis, *J Forensic Sci*, 50, 507–12, 2005.
50. Dang-Tran, K. D., Dedouit, F., Joffre, F., Rougé, D., Rousseau, H., and Telmon, N. Thyroid cartilage ossification and multislice computed tomography exploration: A useful tool for age assessment? *J Forensic Sci*, 55, 677–83, 2010.
51. Fully G., Une nouvelle méthode de détermination de la taille, *Ann Med Leg*, 36, 266–273, 1956.
52. Raxter, M. H., Auerbach, B. M., and Ruff, C.B., Revision of the fully technique for estimating statures, *Am J Phys Anthropol*, 130, 374–84, 2006.
53. Gardner, J. C., Garvin, G., Nelson, A. J., Vascatto, G., and Conlogue, G., Paleoradiology in mummy studies: The Sulman mummy project, *Can Assoc Radiol J*, 55, 228–34, 2004.
54. Trotter, M. and Gleser, G. C., A re-evaluation of estimation of stature based on measurements of stature taken during life and of long bones after death, *Am J Phys Anthropol*, 16, 79–123, 1958.
55. Cesarani, F., Martina, M. C., Ferraris, A., et al., Whole-body three-dimensional multidetector CT of 13 Egyptian human mummies, *AJR Am J Roentgenol*, 180, 597–606, 2003.
56. Matshes, E. and Lew, E., Forensic osteology, in *Forensic Pathology: Principles and Practices*, Dolinak, D., Matshes, E., and Lew, E., Eds., Elsevier Academic Press, London, 563–603, 2005.
57. Haglund, W. D. and Sorg, M. H., Taphonomy, in *Encyclopaedia of Forensic and Legal Medicine*, Payne-James, J., Byard, R. W., Corey, T. S., and Henderson, C., Eds., Elsevier Academic Press, London, 94–100, 2005.
58. Dedouit, F., Loubes-Lacroix, F., Costagliola, R., et al., Postmortem changes of the middle ear: Multislice computed tomography study, *Forensic Sci Int*, 175, 149–54, 2008.

10 Radiographic Applications in Forensic Dental Identification

Richard A. Weems

CONTENTS

ORIGINS OF HUMAN IDENTIFICATION VIA DENTAL COMPARISON

It has been proposed that a human was definitively identified through dental comparison in ancient history.[1] This occurred during the first century CE, when Agrippina, the mother of Roman Emperor Nero, made a clandestine financial contract to murder Lollia Paulina. To ensure that the contract had actually been completed, Agrippina instructed that Paulina's head be brought to her. The confirmation of identification was made based on dental anomalies and peculiar alignments of Paulina's teeth. One of the first recorded dental identifications in the United States is attributed to Paul Revere.[2,3] In Boston in 1776, Dr. Joseph Warren was killed in the Revolutionary

War in the Battle of Breed's Hill. Fatal trauma induced to the face of Dr. Warren made him unrecognizable, which hindered a proper burial. Paul Revere, a local silversmith, identified the remains of Dr. Warren by the small dental appliance that he had fabricated for him at some point in the past. The appliance was fashioned from ivory and was anchored by silver wires. The identification made it possible to bury Dr. Warren with full military honors on April 8, 1776. Dr. Oscar Amoedo is considered by many to be the Father of Forensic Odontology.[4] He was sent as a delegate to the International Dental Congress in Paris in 1889. He became a dental instructor and teacher at the Ecole Odontotechnique de Paris in 1890 and rose to the rank of Professor. A tragic fire at the Bazaar de la Charitè stimulated his interest in the field of forensic odontology. His thesis "L' Art Dentaire en Mèdicine Lègle" served as the basis for his book by the same name; the first comprehensive text on forensic odontology.

Radiographic images of human dentition were made within days of Wilhelm Roentgen's discovery of x-rays in 1895, but the first practical and routine use of dental radiography in a dental practice is attributed to C. E. Kells of New Orleans.[5] Early users of dental x-ray equipment were unaware of the life- threatening qualities involved, and Kells was eventually diagnosed with the fatal effects of radiation that he had received over the years and he eventually took his life. It is probable that the first comparison of dental features seen in radiographs most likely occurred on the first occasion that Dr. Kells inadvertently mixed radiographs from two different patients and had to sort them out properly. The first published use of dental radiography as a means of achieving human identification was in 1943.[6] Today dental radiography plays a major role in the identification of human remains both on an individual basis and following incidents of mass fatality.

Forensic odontologists commonly provide analyses and legal testimony concerning human identifications, bite mark comparisons, domestic violence, and age estimation, and render opinion in cases of civil litigation addressing dental injury and negligence. However, this chapter addresses concerned with the identification of dental remains with particular emphasis on the techniques of radiographic comparison.

THE NEED FOR HUMAN IDENTIFICATION

One reason for establishing a scientific identification of human remains involves the welfare of the families of missing loved ones. One can only hope to never be in a position of the parent or spouse who has lost a loved one "without a trace." Most experts in the field believe that grief has distinct progressive stages. One of the most important stages involves the all-important acceptance that the person is indeed deceased. This acceptance is extremely difficult if the disposition of the missing loved one is truly unknown. Thus, victims' families may never get over the deep loss that they experience.

Second, human identification is necessary to resolve several medicolegal issues including issue of death certificates, remarriage, estate settlements, and the filing of insurance claims. Also, when no identity is available and foul play is suspected, the investigation that should follow cannot even begin since suspects are usually derived from those most closely associated with a victim. Murder trials without scientifically identified remains are extremely rare due to the difficulty of establishing *corpus delicti*.

Currently accepted forms of scientific human identification include fingerprints, dental comparison, and DNA analysis.

WHY TEETH ARE SUITABLE FOR SCIENTIFIC IDENTIFICATION

Durability of Teeth

Most scientific analyses performed in the forensic sciences are essentially a critical comparison of a known sample with an unknown in order to establish if they are the same. Dental identifications are accomplished by comparing dental findings from a known individual (antemortem dental records) to similar records compiled by the forensic odontologist (postmortem dental records) of the victim in question. It is ironic that during life the teeth are subject to decay and diseases of the periodontium and have no ability to repair themselves once damage has occurred. Dental procedures must be conducted to mechanically replace lost structures. However, in death, caries and periodontal disease cease and teeth become almost indestructible, often lasting hundreds of years, and tooth pulp chambers are often a good source of DNA material. Pulp material should be retrieved by snapping crowns at the neck to avoid heat desiccation, which can result from high-speed drills. Saliva and check swabs of fresh oral tissues normally contain sloughed mucosal cells rich in cellular DNA.

Teeth can also withstand a great deal of trauma, including insult from a common cause for the need of a scientific identification—accidental or intentional exposure to fire. Teeth can normally withstand temperatures resulting from a typical house (650°C) with some dental restorative materials easily withstanding cremation temperatures (Table 10.1). It should be noted, however, that the resulting reactions to extreme temperatures are greatly affected by the duration of the event.[7]

Also, the damage resulting from exposure to heat will greatly depend on the location of the tooth within the dental arches. It is a common finding to see almost complete

TABLE 10.1
Tooth and Restorative Materials' Reaction to Temperature

Unprotected teeth	540–650°C
Amalgam	500–1000°C
Composite	Inorganic (filler) hardly altered
Gold	870–1070°C
Porcelain	1100°C
Porcelain fused to metal	1150–1450°C
Removable partial	1275–1500°C

carbonization and destruction of the crowns of the anterior teeth after a fire, and yet discover that the posterior teeth are in pristine condition. This is due to the insulating effect of the soft tissues surrounding the dental arches including the tongue, soft palate, cheeks, and the muscles of mastication.

Care should be taken when removing the jaws from victims of fire, as it can be expected that the posterior teeth will still be rather intact and they can be used to contribute to a solid dental identification. The more anterior teeth will likely be in a carbonized condition, with the teeth being in a very fragile, chalky state.

Uniqueness of Teeth and Dental Restorations

Most dentists agree in principle that an individual's dentition is unique when taken as a whole and assuming all or most of the teeth remain.[8] Any variation from the norm is helpful when attempting to make a dental identification, but changes seen resulting from dental restoration of the structures is by far preferred over normal anatomical landmarks in making dental matches. There are 32 permanent teeth with 5 surfaces per tooth. There are numerous ways to restore each of the 5 tooth surfaces or replace missing teeth with a great variety of dental materials: amalgam (silver), composite resins, cements, root canals, crowns, bridges, removable partial dentures, and dental implants. As these surfaces are assessed one by one, the weight of the findings mounts statistically. In addition to the restorative material, radiographs taken of the teeth with fillings in place will show the internal forms or shapes of restorations. Most forensic odontologists would rather have one tooth with a large filling than 32 with none. Indeed, dental identifications are occasionally made through the appearance of a single restored tooth.

In addition to the crowns of the teeth there are other tooth components that can be compared, such as roots, pulp chambers, and root canals. Even within the maxillary sinus, anatomical features as well as bony trabeculation can be of use. However, care must be taken in such comparisons, as the normal dental structures of separate individuals may look extremely similar.

Teeth Form and Erupt in a Predictable Manner

The teeth are slowly formed from ectodermal embryonic tissues with mineralization, starting with the cusp tips and continuing to the apex of the root beginning at birth and continuing until approximately age 24. The actual number of teeth a person eventually possesses and the ages at which the different teeth erupt and are lost varies greatly from one individual to another.[9] Tooth formation also may vary greatly according to race and gender.[10] However, there is a statistical prediction in which teeth typically form and/or erupt that can be used to "age" a victim. However, eruption is less accurate than formation. After the victim's third molars erupt and the roots are complete, dental age estimates become much less reliable, although there are further tests that can be conducted if the tooth can be sacrificed for a more detailed analysis. One such technique is the Gustafson technique, which includes occlusal attrition, gingival attachment, secondary dentin, root transparency, and apical resorption.

Teeth React in Known Ways to One's Genetics and Life History

There are anomalies in teeth that are extremely useful in establishing the identity of human remains. Some of these occur randomly while others manifest due to trauma or insult during the tooth's life or formation. For example, young individuals taking certain medications (Tetracycline) as the tooth enamel is being formed will possess discolored striations in the clinical crown that are irreversible. If the missing individual has a history of taking such medications at that specific age then that knowledge can contribute to the dental comparison. Also, a traumatic blow to a tooth may result in an incomplete, blunted root formation and may again be compared to the individual's dental history. Blunted root apicies may indicate that the individual underwent orthodontic treatment. Even habits such as sucking lemons or pipe smoking may provide useful information. Finally, there are numerous genetic syndromes that alter the victims' dentition. One such syndrome, for example, is cleidocranial dysplasia in which the patient's clavicles do not form, but have numerous supernumerary (extra) teeth.

Dental Data Availability

Most individuals have at one time or another visited a dental office. Therefore comparative dental records, including dental x-rays, can almost always be found on the missing person if there is a putative victim. In the absence of dental radiographs, patients' diagnostic photographs or working plaster models may also be of use in some cases.

Also, dental information may be obtained via the FBI's National Crime Information Records on Missing persons (NCIC). In 2000, the dental findings portion of the NCIC searchable database was separated from the other physical identifiers portion, which should improve the effectiveness of the system. However, dental information is all too often not input into the missing persons section of the system.

The Department of Justice recently established the National Missing and Unidentified Persons System (NamUs) to provide the nation's medical examiners, coroners, victim advocates, law enforcement agencies, and the general public with the ability to simultaneously search the records of missing persons and unidentified human remains online in an effort to solve missing persons cases. The Unidentified Decedents Database records are entered by medical examiners and coroners, and it allows searches based on physical characteristics such as demographics, anthropologic analysis, dental information, and distinct body features. The

Missing Persons Database provides a national online repository of similar data related to missing persons cases. New cases can be added by law enforcement officers and also by the general public. The two NamUs Databases will be fully integrated to allow simultaneous searching of the Missing Persons records against cases in the Unidentified Decedents Database.[11]

Dental and craniofacial structures may possibly be seen in radiographs from hospitals, chiropractors, prisons, and insurance companies. Also, military dental records are retained on all veterans indefinitely at the National Military Records Repository in St. Louis, Missouri.

DENTAL RADIOGRAPHIC SERIES[12]

Panoramic Radiography

Panoramic radiographs (Figure 10.1) are created using extraoral "green sensitive" film encased in a cassette containing two rare earth scintillation screens which rotate about the subject to create a pseudotomograph of the oral cavity. The film is normally 30 cm × 15 cm in size, possesses emulsion on both sides and is extremely light-sensitive. Modern panoramic systems require a Kodak GBX II for safe lighting in the darkroom to avoid film fog. Panoramic's strength, as its name indicates, is that it provides a "panorama" view of all of the teeth, tooth bearing structures, sinuses, and craniofacial structures. It is also a fast and relatively simple technique. Its weaknesses include a lack of sharpness to the point that some detailed features of teeth and bone cannot be completely evaluated and a very narrow zone of sharpness in the anterior region, which can cause significant blurring of the anterior teeth. Regardless of its flaws, panoramic radiography is extremely useful when performing a dental comparison due to the vast amount of tissues revealed. Excised mandibles, maxillas, and even entire skulls may be placed in a panoramic unit for postmortem exposure if circumstances so dictate.

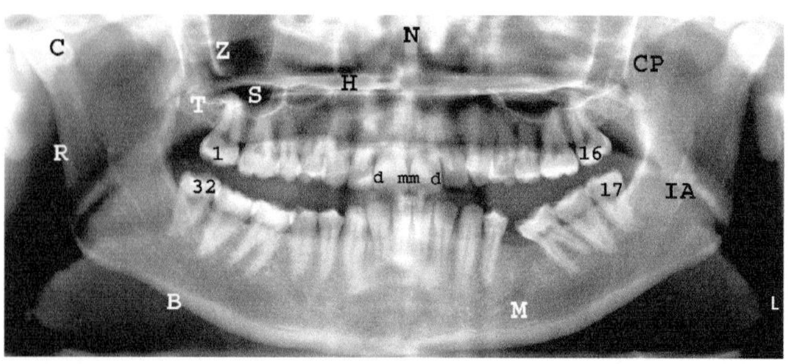

FIGURE 10.1 Panoramic radiograph with anatomic structures: C—condyle, T—maxillary tuberosity, Z—zygomatic arch, S—sinus cavity, H—hard palate, N—nasal septum, CP—coronoid process, IA—inferior alveolar nerve canal, M—mental foramen, B—body of the mandible, R—ramus, Teeth nos.: 1,16,17,32, m—mesial surfaces, d—distal surfaces.

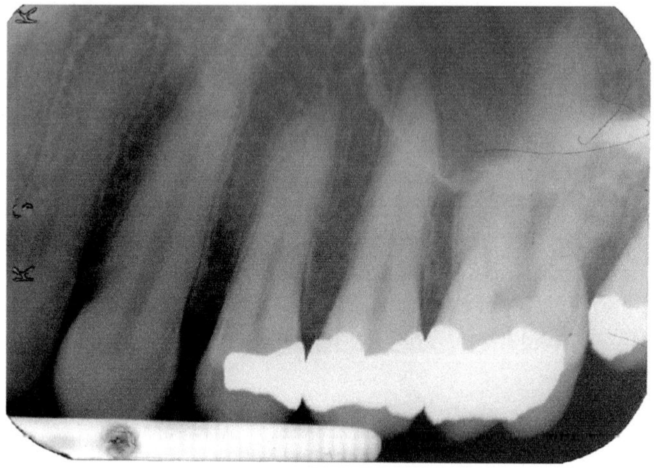

FIGURE 10.2 Periapical radiograph.

Intraoral Dental Radiography

Intraoral radiography is performed using small nonscreen films placed within the oral cavity during exposure. There are four sizes of film: 0-pediatric, 1-anterior adult, 2-posterior adult, 3-posterior wide, and 4-occlusal. The film is composed of an acetate layer covered on both sides with an emulsion of silver halide. There are two currently available film speeds; film speed D and film speed F. F-speed is approximately twice the speed of D- speed. Most dental x-ray tubeheads operate within a range of approximately 8 mA and 70 kVp. Density of the image should always be controlled with exposure time or milliampere since increasing kilovolt peak will result in a long scale of contrast (loss of visual contrast).

Intraoral radiographs are typically exposed using one of two standard projections. The periapical projection (Figure 10.2) displays the crown, root, apex, and surrounding bone. The bitewing projection (Figure 10.3) displays the crowns of both arches with the teeth in occlusion. Many dentists take

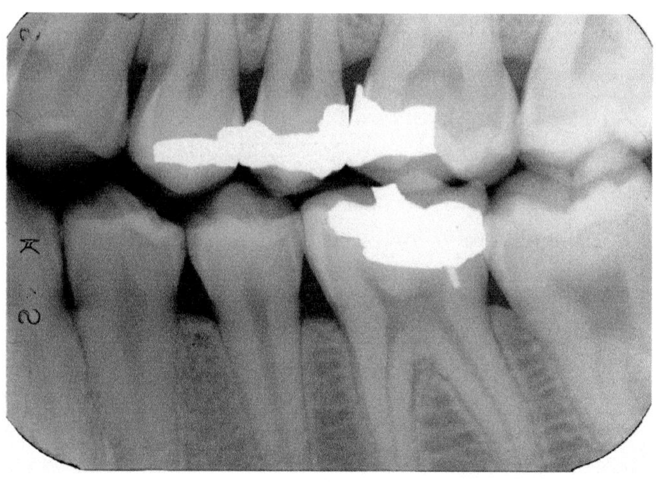

FIGURE 10.3 Bitewing radiograph.

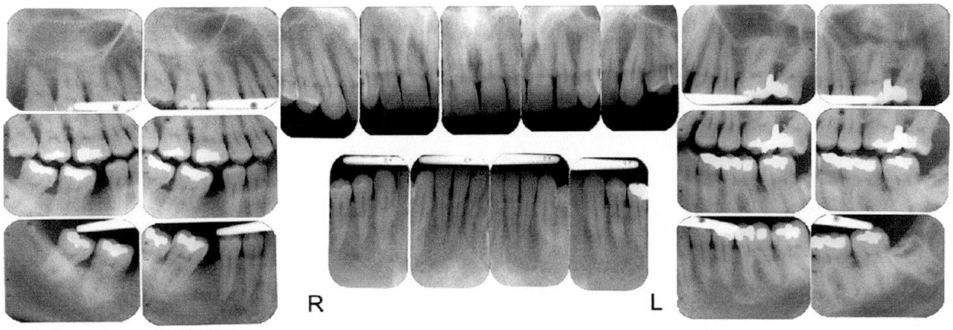

FIGURE 10.4 Complete mouth radiographic series.

approximately 20 periapical and bitewing radiographs of the dental arches, which are collectively known as a Complete Mouth Radiographic Series (Figure 10.4). Intraoral radiographs possess a very high degree of sharpness and resolution and are far superior to that seen in panoramic radiography. However, the intraoral technique takes much more time and requires a much higher degree of skill. The standard projection geometry of the periapical and bitewing radiographs is shown in which the film is typically parallel to the tooth and the beam is perpendicular to both (Figures 10.5 and 10.6). A complete mouth radiographic series should clearly show all tooth-bearing areas, all of the apicies and crowns of the teeth, and all open interproximal tooth contacts without overlap.

Most postmortem series of radiographs consist of a combination of periapicals and bitewings exposed on the specimen. Bitewings are the most common images taken in the dental office. Therefore, it is advisable to be sure to include them even in the presence of periapicals, because it is critical to recreate the antemortem projection angle or the restoration shapes will not coincide.

Film Orientation

Intraoral dental film packets have an exposure side and a nonexposure side. Once the image is developed, there is a bump in one corner of the film to help in the orientation process (left or right is not labeled as in medical films).

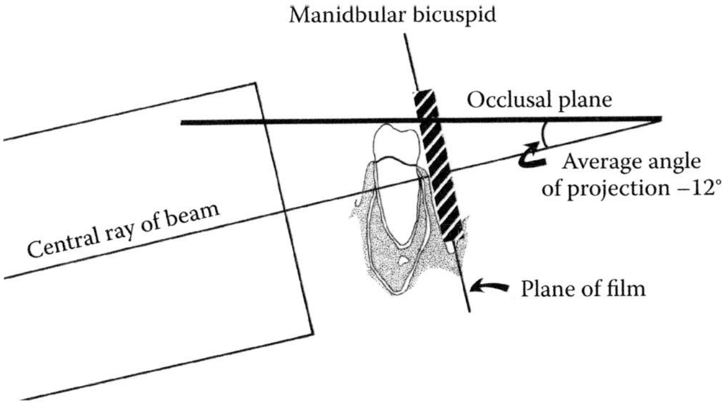

FIGURE 10.5 Projection geometry for periapical radiographs.

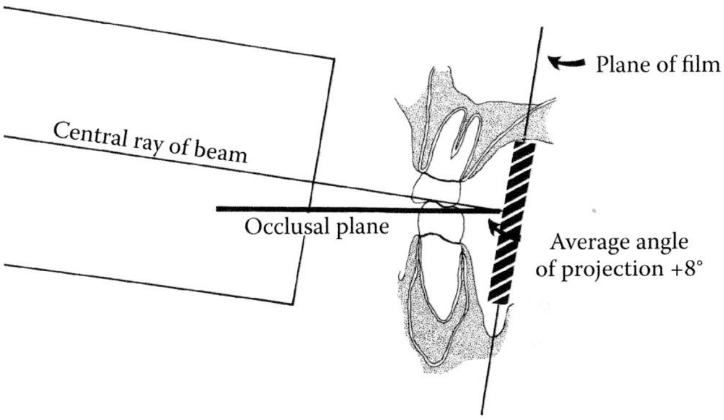

FIGURE 10.6 Projection geometry for bitewing radiographs.

When the film is viewed with the bump facing the viewer, it is oriented as if the viewer were facing the patient (the patient's right is on the viewer's left). If the bump points away, it is as if the viewer is observing the teeth from the perspective of the tongue. Should the film be placed backwards during exposure, the beam must pass through a lead layer, which creates a repeating geometric pattern that will be visible in the image (Figure 10.7). This film should be

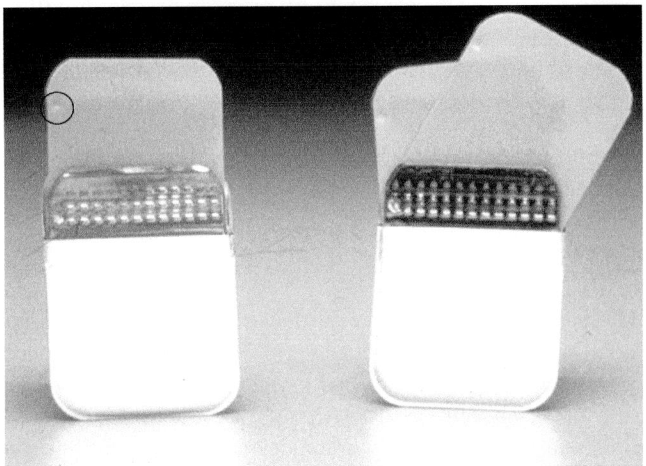

FIGURE 10.7 Intraoral film packet showing orientation bump and patterned lead backing.

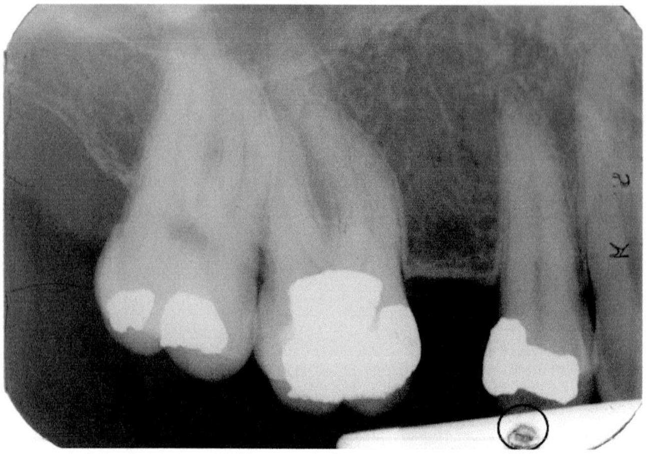

FIGURE 10.8 Duplicate of a periapical radiograph with bump in the lower right corner. This shows that the bump was oriented "up" when the duplicate was made.

immediately discarded because it was taken backwards and will be misleading as to "right" and "left" and can result in serious errors.

Another very common pitfall occurs when exposing small fragments of human dental remains. This involves placing the film buccal to the surface of the fragment with the beam originating from the lingual (tongue). This alignment cannot possibly occur in antemortem image acquisition and can be quite misleading. It results in an image that appears to have been taken on the opposite side of the arch and might lead to the exclusion of that suspect. Care must always be taken to simulate only what is possible *in vivo*. Also, teeth that are no longer in sockets should only be replaced if the event is witnessed. It is better to take individual radiographs of the teeth outside of the sockets rather than to have erroneously placed them in the incorrect locations.

Another common problem is to receive antemortem duplicated radiographs in which it is unknown in which direction the bump of the original film faced during duplication. If the bump was in the proper bump-up orientation, then it will be seen in the lower right or upper left corner when the film is oriented horizontally. If it is in the opposite corners, then the bump was down or away when duplicated (Figure 10.8).

TOOTH ANATOMY, NUMBERING, AND NOMENCLATURE

Tooth Components

Humans typically possess 20 deciduous teeth which erupt after birth and are lost during childhood and adolescence. These teeth are then replaced by 32 permanent teeth which may or may not remain for a lifetime. As the deciduous teeth are being lost and the permanent teeth are erupting, the patient is said to have "mixed" dentition.

Each tooth consists of a layer of inert enamel covering a segment of tubular vital material known as dentin (Figure 10.9). Tooth enamel is extremely opaque on radiographs with dentin also being opaque but to a lesser degree. Within the dentin is the pulp chamber filled with nerve connections and vascular components. The pulp chamber and nerve canals are radiolucent. These three entities collectively form the crown. Below the crown is the root structure composed of cementum tissue surrounding the nerve canal, which continues from the pulp chamber to the root apex. The tooth root is surrounded with a tissue that provides cushioning and is known as the periodontal ligament. Finally, there is a solid or corticated wall surrounding the entire complex known as the lamina dura which connects to the cancellous. Each of these tissue components produces different relative radiographic densities, with enamel being the most radiopaque and the pulp chamber and canal being the most radiolucent.

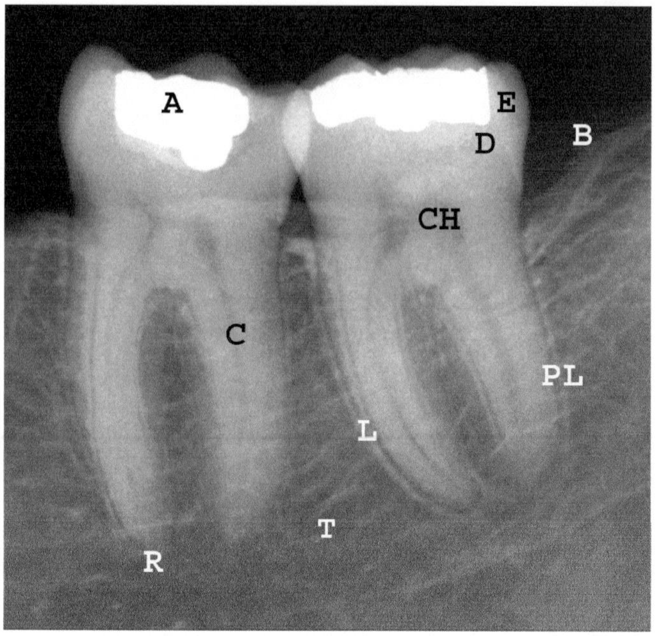

FIGURE 10.9 Tooth components: E—enamel, D—dentin, CH—pulp chamber, C—root canal, PL—periodontal ligament space (lucent), L—lamina dura (opaque), R—root apex, A—amalgam, B—bone crest, T—medullary trabeculation.

There are four basic types of permanent teeth: incisors, cuspids, bicuspids, and molars. The most logical way to consider human dentition is to segment the two dental arches into four mirrored quadrants; two maxillary and two mandibular. In each quadrant of the adult oral cavity there is (starting at the anterior midline) a central incisor, a lateral incisor, a cuspid, a first bicuspid, a second bicuspid, and a first, second, and third molar.

Tooth Numbering

Dentists worldwide use numerous different numbering or notation systems to indicate specific individual teeth but the two most common systems are the Universal System and the FDI System.[13] The Universal System is used primarily in the United States. The FDI system was developed by the

World Dental Federation and is widely used by dentists internationally.

Each tooth is serially numbered in the dental arches when using the Universal System (Figure 10.10). The numbering starts with the patient's maxillary right third molar being tooth 1 and continues across the maxilla to tooth 16 which is the maxillary left third molar. The mandibular left third molar is tooth 17 and the sequence continues to the mandibular right third molar which is tooth 32. Thus, a mandibular first molar on the patient's right side would be tooth 30. Deciduous teeth are specified in a similar manner and pattern except that the alphabet is used instead of numbers (i.e., teeth "A" through "T").

The FDI System uses two numerals with the first number representing the quadrant indicated and the second number indicating the type of tooth with the tooth numbers starting with '1' at the midline. The quadrants are numbered beginning with the maxillary right quadrant being designated as 1 and continuing in a clockwise manner to quadrants 2, 3, and 4. The teeth are numbered such that 1 indicates a central incisor and 8 is the third molar. Thus, the mandibular first molar on the patient's right side would be tooth 46. The deciduous teeth are numbered similarly except that the quadrant numbers are 5–8. Similar to the FDI System is the Palmer Notation which is used by many pediatric dentists and orthodontists. The teeth are numbered from the midline like the FDI so that each patient's "sixes" refer to their first molars. However, rather than number the quadrants, brackets are drawn adjacent to the tooth number to indicate the quadrant in question.

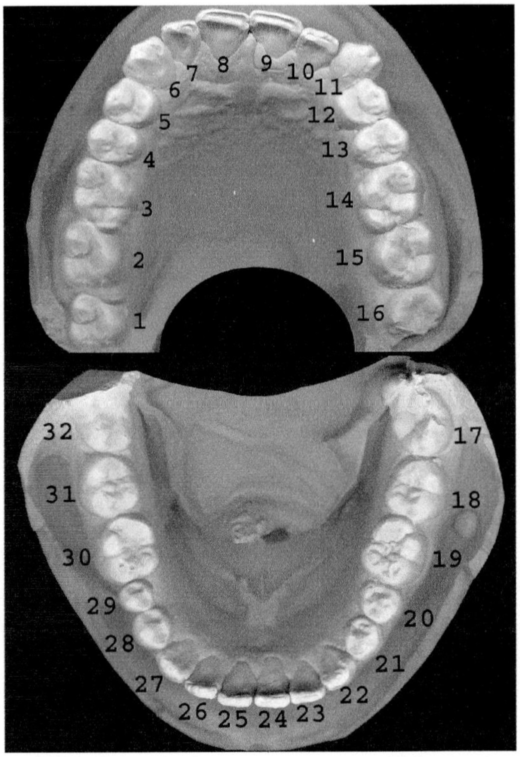

FIGURE 10.10 Universal tooth numbering system.

Tooth Surfaces

As mentioned previously, each tooth consists of five surfaces. The chewing surfaces of the posterior teeth are known as the *occlusal (O)* surfaces. In the anterior teeth (incisors and cuspids), this surface or edge is known as the *incisal (I)* surface. The surfaces of all of the teeth facing the tongue are known as the *lingual (L)* surfaces. Surfaces facing the cheeks or lips are known as the *buccal (B)* surfaces in the posterior and *facial (F)* surfaces in the anterior region including incisors and cuspids. Tooth surfaces which touch or "approximate" each other are known as *mesial (M)* or *distal (D)*; the *mesial* surfaces are toward the anterior midline while the *distal* surfaces face the posterior of the quadrant. (Figure 10.11)

It is typical for the surface abbreviations to be included when describing a dental restoration. For example, an amalgam restoration on tooth 3 which involves the mesial, occlusal, and distal surfaces is noted as an MOD amalgam. An anterior resin restoration involving the distal and lingual of a tooth would be noted as a DL composite resin. It is not uncommon for an amalgam restoration to involve ALL tooth surfaces with a notation of MODBL amalgam.

Dental Disease

Dental disease can be easily seen in radiographs but rarely provide the specificity to solely establish a dental identification. However, in combination with enough dental features, disease may carry some weight .

Dental decay or caries is classified according to the type of tooth surfaces involved. Class I caries involves the occlusal surface of the tooth only. Class II caries involved the proximal surface where two teeth touch in the posterior region of the quadrant. Proximal caries in the anterior region (incisors and cuspids) is classified as Class III caries. Class IV caries involves the proximal and incisal edge of the anterior teeth. Finally, caries forming at the gingival margin, whether buccal/facial or lingual is classified as Class V caries. Note that it is impossible to determine if Class V caries, or restorations, are facial or lingual on a single radiograph. This is because a radiograph is a 2D representation of a 3D object.

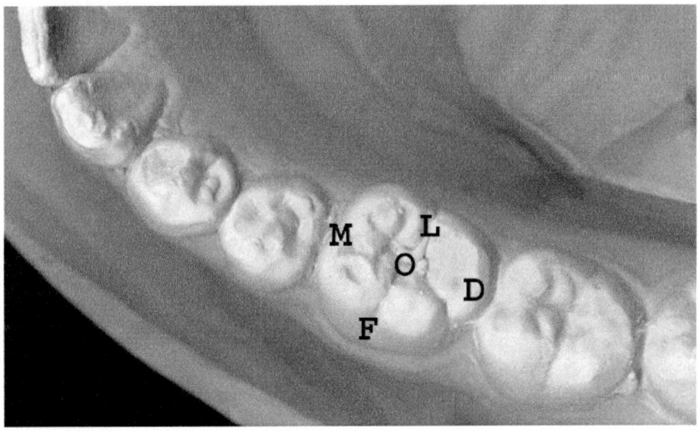

FIGURE 10.11 Tooth surface nomenclature using WinID codes: M–mesial, O–occlusal, D–distal, F–facial, L–lingual.

Other findings may include periodontal bone loss, periapical abscesses and cysts, impacted teeth, rotated teeth, extra teeth, and various neoplasms and cancers.

DENTAL RESTORATIVE MATERIALS

As stated previously, dental restorations of all types are much more suitable in attaining an identification compared to typical dental structures and anatomy. The availability and variety of high quality dental restorative techniques and materials has increased dramatically over the past 50 years. The following section will address the radiographic appearance of the materials that are likely to be seen when making a dental comparison in order to establish an identity.

The first category of materials is the most commonly seen materials in the mouth and are termed "fillings" in the sense that a cavity preparation is made via tooth and caries removal followed by filling the preparation with materials which eventually harden. A silver or "amalgam" filling (Figure 10.12) is made from an amalgamation of silver with several other metals combined with mercury and is one of the most traditional and time-tested restorations used in dentistry. The amalgam may replace one tooth surface or it may include all five. The amalgam will be seen within the tooth radiographically and will be highly radiopaque. The other common filling material is composite resin which contains acrylic with various filler materials including glass ionomers. These restorations are commonly called "composites" and have been used traditionally in the anterior teeth for aesthetic reasons as these materials are made to match the tooth in color and texture. At one time, composites were cured chemically but now are cured by exposing the material to UV light. They will also be seen radiographically within the tooth, but may appear radiolucent in older restorations and slightly radiopaque in newer restorations. Since composites often match the tooth so well, that they are difficult to observe clinically but will fluoresce less that tooth enamel when exposed to a UV light source. A portable source is now very affordable and is essential to have in the morgue to detect today's exquisite tooth-matching technology.

The second grouping of restorative materials involves the fixed prosthodontic restorations which include crowns and fixed bridges. These restorations are fashioned or cast after the outer tooth material has been removed and impressions of the remaining tooth structures have been made. The two primary materials are gold and (Figure 10.13). Single crowns or facial veneers used in the anterior segment for esthetic purposes may also be made from porcelain alone. Radiographically, the restorative material will cover the tooth and will be totally radiopaque when gold is used. A PFM crown or bridge will display two different opacities, with the outer covering of porcelain being less opaque than the underlying metal coping. A fixed bridge typically replaces one or more missing teeth by connecting two crowns to a solid false tooth known as a pontic. Fixed bridges may be made from the same two materials previously mentioned; gold or PFM (Figure 10.14).

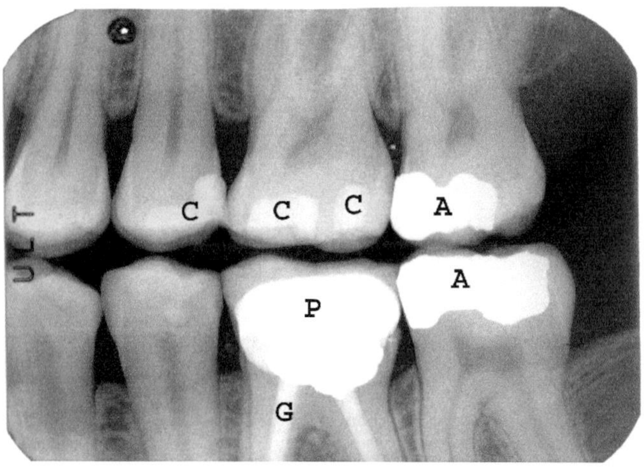

FIGURE 10.12 Restorative materials: C–composite resins, A–amalgam restorations, P–PFM crown, G–gutta percha root canal filling.

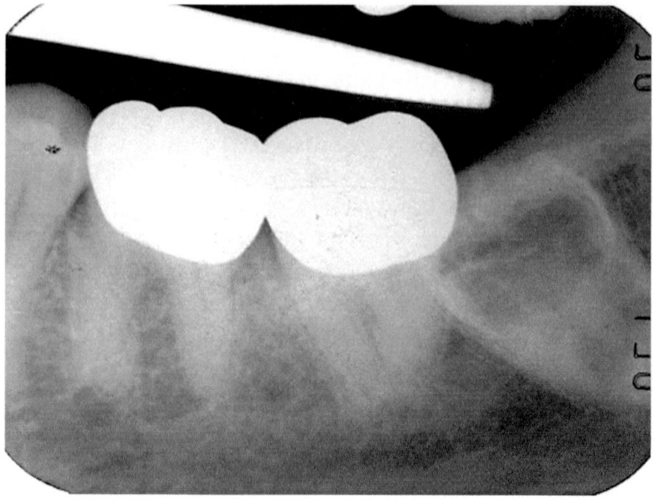

FIGURE 10.13 Full gold crown restorations.

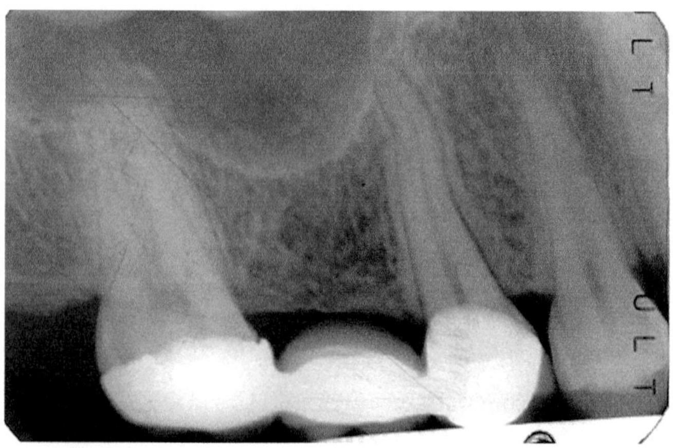

FIGURE 10.14 Three-unit PFM fixed bridge with pontic replacing tooth 3.

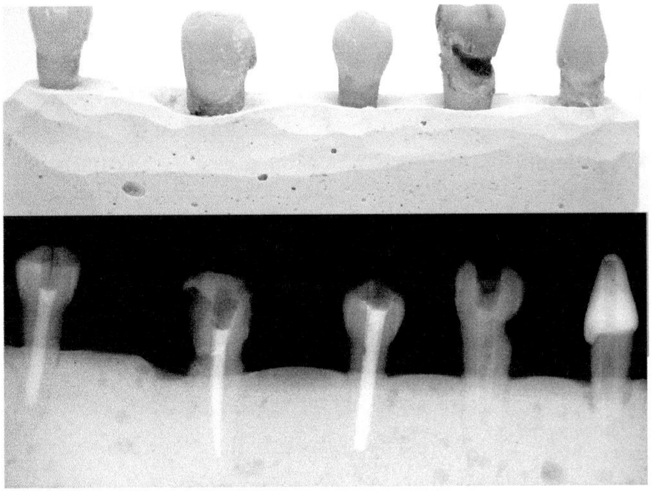

FIGURE 10.15 Radiograph test of five manufacturers' composite endodontic posts. Note that three of the posts are radiopaque but two are radiolucent and impossible to detect. Courtesy of Dr. Robert E. Wood.

Endodontically treated teeth have had the central portion of the tooth mechanically removed including the pulp chamber and nerve canal tissues. The central portion is then obturated or filled with a gum-type material known as gutta percha. It is also very common to see a metal post within the root canal placed for mechanical support of a weakened tooth, particularly if it is to receive a crown after the root canal is completed. These materials will be seen radiographically within the canal, with the post being normally radiopaque and the gutta percha being less opaque compared to the metal post. Unfortunately, relatively new prefabricated composite fiber endodontic posts are becoming popular that do not project a radiopaque profile, thus making it very difficult to detect them radiographically (Figure 10.15).[14]

Removable prosthodontic devices include full and partial dentures. Full dentures are typically fabricated from acrylic with either porcelain or acrylic teeth. Partial dentures normally are partly acrylic but have metallic frameworks which clasp onto the teeth. It should be noted that dentists often record the exact type of prosthodontic tooth used and its shade number in the dental record. Many patients request decorative gold teeth placed within their denture when fabricated and, in some cases, verge on the side of the bizarre (Figure 10.16). These special features may, however, lead to the findings of an identity for that individual when compared to facial photographs of the victim. Also, many dentists today include the patient's name on the tissue side of the denture which is extremely helpful when making identifications. Indeed, this laboratory procedure is mandatory in some states.

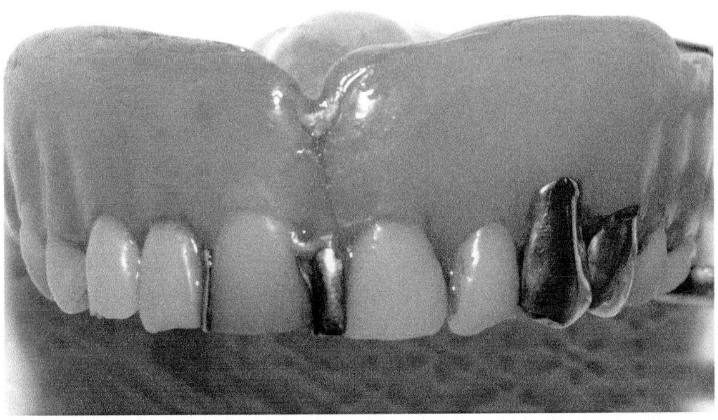

FIGURE 10.16 Maxillary full denture with decorative gold adornment.

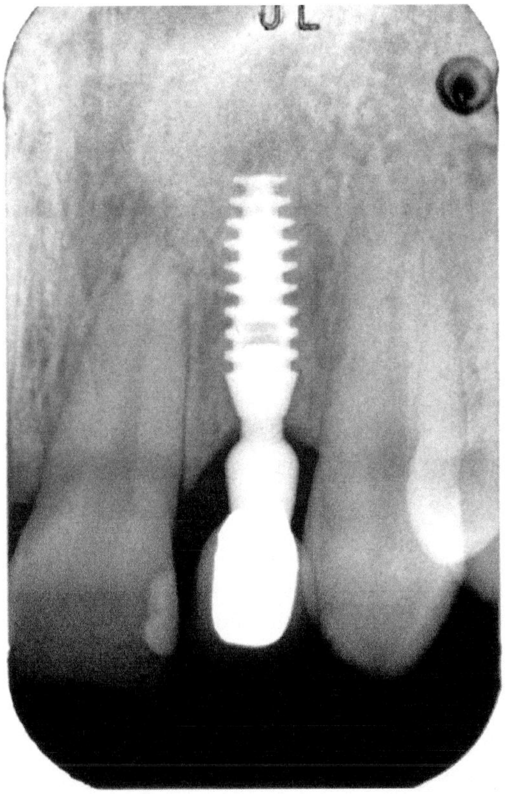

FIGURE 10.17 Intraosseous dental implant with crown.

Finally, the replacement of teeth through the surgical placement of dental intraosseous implants is growing in popularity and will most likely revolutionize restorative and endodontic dentistry. There are numerous advantages to replacing missing teeth with dental implants but the most beneficial character is that the implant retains the alveolar bone which is normally lost beginning immediately after a tooth is extracted and continuing over time. There is currently research in the field to categorize unique implant characteristics from various different manufacturers which could be another aid in forming a dental identification (Figure 10.17).

CHARTING EXISTING FINDINGS: ANTEMORTEM RECORDS

Record of Treatment

Equally important to comparing dental radiographs is the assessment of the written dental record. Typically dentists make a series of radiographic surveys on new and follow-up patients. A treatment plan of procedures to be accomplished is then formulated. As the treatment is performed, the procedures are noted in the "record of treatment," without radiographs of the new restorations being taken at that time. Thus, the record of treatment fills in the blanks and tells the odontologist what teeth have been treated, when, and how. For example, tooth 30 may be missing when comparing antemortem and postmortem radiographs creating a disparity. This disparity may be easily explained if the record of treatment indicates that the tooth was extracted subsequent to the date of the initial radiographs. The written dental record *must* be included in the data provided to the forensic odontologist by those tasked with recovering the postmortem dental evidence before an analysis and comparison is attempted.

The Odontogram

Dental findings charted on a graphic "tooth chart" is known as an odontogram. Some dentists chart an odontogram of existing restorations on new patients as well as charting a second odontogram of dental procedures to be done, and this is most helpful in establishing an identity. Unfortunately, the charting is normally only a graphic representation of what is to be completed. To make a dental comparison, two odontograms are completed by the forensic odontologist. The first is made from all of the findings of the postmortem examination both clinical and radiographic in nature. This process is usually rather straight forward if all of the dentition is in tact.

A second odontogram is created from the antemortem findings that are illustrated via a careful and systematic review of the suspected victim's written records and radiographs. This can be a tedious and difficult process. The objective is to produce an odontogram that is a "snap shot" of the dentition of the individual accurate up to the day of disappearance. Part of the difficulty in this process is to decipher the dentist's notations in the record of treatment. Most dentists expect their written records to be scrutinized only by themselves. This leads to shorthand notations and unique abbreviations that are foreign to others, not to mention handwriting that might require the services of a documents expert. Additionally, dental procedures are, by nature, dynamic over time. That is, a tooth may first be restored with a small amalgam restoration initially but may require further treatment such as a full crown at some point in time. Eventually the tooth may even require extraction. Scrutinizing a voluminous record with numerous series of radiographs over a 20-year time frame can be a daunting task. One short cut often used is to evaluate the record of treatment and radiographic images in reverse order of time. Regardless, the task must be completed in as precise a manner as possible whether there is a massive amount of data or, even worse, too little due to inadequate record keeping.

Once the two odontograms are completed, dental comparisons may begin. Obviously the latest antemortem radiographs are most likely to be comparable to those taken after death.

Odontogram Primer

The following odontogram (Figure 10.18) has been created as a primer and exercise illustrating the charting via an odontogram of the more common dental findings. It is especially important to note that entering no information is better than assumed and possibly incorrect information, especially concerning antemortem data. Nothing should ever be "assumed."

ESTABLISHING CONCLUSIONS IN DENTAL IDENTIFICATIONS

Complexity and Expertise

It is not highly unusual for a nondentist to attempt to establish an unknown victim's identity through dental findings. However, it is not recommended, particularly in cases with complex or scant dental findings or when the dental findings form the sole source of scientific identification. With the recent climate of questioning the "science" in forensic sciences, acceptance of all forensic expert testimony in the courts will likely be under more stringent scrutiny in the near future.[15]

The rate of newly released restorative materials and techniques alone make it difficult even for dentists to keep abreast of dental technology and materials. Also, there is no substitute for experience in determining what findings are unusual or even unique.

Dental identifications should be made by an experienced dentist with additional training and experience in forensic odontology, which is normally absent in dental school curricula. Such training is available through several courses given annually throughout the United States and Canada and by attending annual meetings of forensic odontology groups. These meetings include excellent scientific sessions in which many highly experienced individuals present state-of-the-art information and case studies. If at all possible, medical examiners should attempt to form working relationships with forensic odontologists who have been certified by the American Board of Forensic Odontology (ABFO). Diplomates of the ABFO are initially required to demonstrate a substantial body of work and undertake a challenging and extensive examination with written, practical, and oral components. Diplomates must also be recertified every five years. The most recent checklist for accreditation by the National Association of Medical Examiners (NAME) includes the question as to an affiliation with an ABFO Diplomate.[16]

Examining and Radiographing Postmortem Dental Remains

Dental radiographs are easiest to expose on skeletal remains. *In situ* radiographs on intact bodies are often very difficult to obtain. Jaw resections or facial tissue flaps are often performed to facilitate radiography on decomposing, charred,

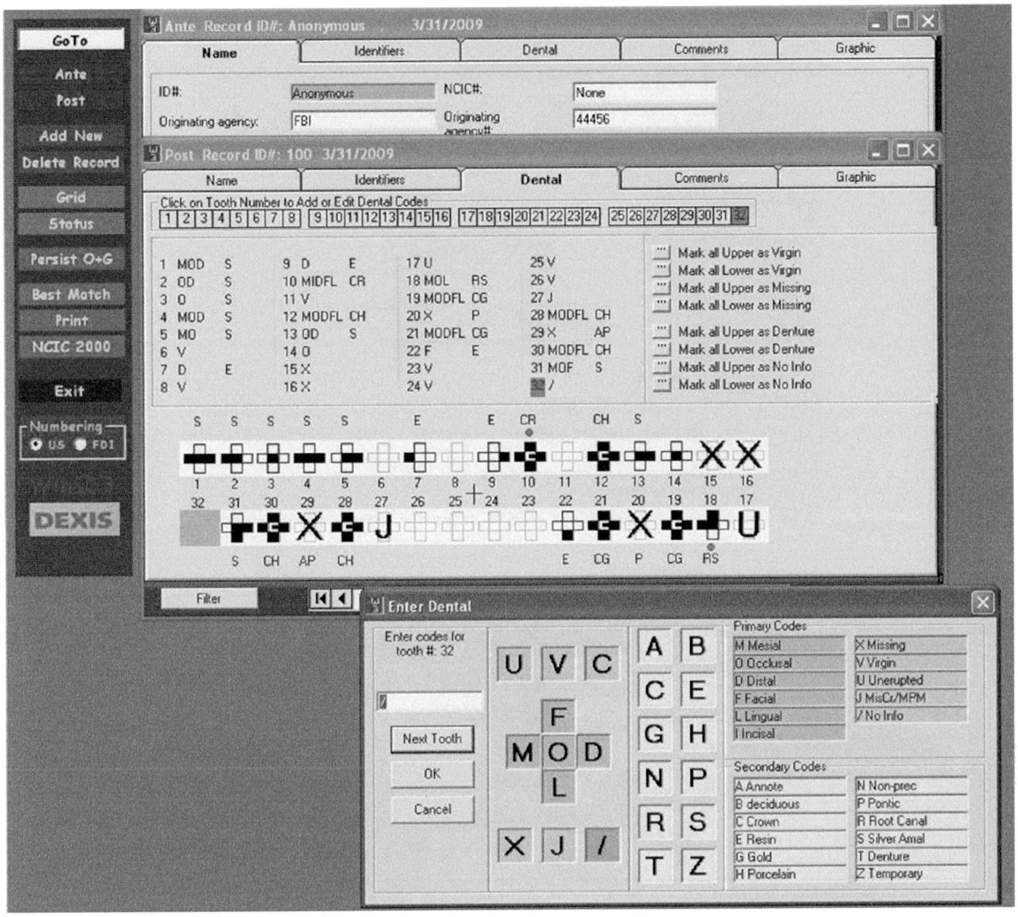

FIGURE 10.18 Odontogram primer. 1-MOD Am, 2-DO Am, 3-OAm, 4-MOD Am, 5-MO Am, 6, 8, 11, 23, 24, 25 & 26 are all Virgin. 7-D Comp resin, 9-D Comp resin, 10-Crown w/Root canal, 12-Crown Porcelain, 13-DO Am, 14-O Am, 15 & 16 are Missing (healed). 17-Unerupted. 18-MOL Am w/Root canal, 19-21-Gold bridge replacing 20(pontic), 22-F Comp resin, 27-Missing post mortem, 28-30-Porcelain bridge replacing 29(pontic), 31-MOF Am, 32-No information. Odontogram graphics courtesy of Dr. James McGivney.

and traumatized remains, but only with the clear approval of the medical examiner in charge of the case (Figure 10.19).

Carbonized (burned) teeth and bone are friable and can crumble with the least disturbance. Teeth might disintegrate when the jaws are removed but can be made more stabile by applying cyanoacrylate or spray acrylic. Even with stabilization, photographs and as many radiographs as possible should be made before removal. One successful technique is to remove the mandible with tree and shrub trimming shears just below the condyles. This produces less vibration than vibrating saws and gains easy access to the maxilla without disturbing the more fragile structures. If the maxilla is to be removed, care should be taken to ensure that the section is made well above the apicies of the third molars since the formation of the roots may be of importance in estimating the victim's age. Also, the third molar is usually located higher in the tuberosity than the first and second molars.

The x-ray energy should be lowered by approximately 30% when exposing skeletal remains and by as much as 50% when exposing severely charred remains as the tissues will be desiccated. This is normally accomplished by lowering the exposure time of the tubehead.

Categories and Terminology used in Dental Identifications

Once all possible postmortem and antemortem dental evidence has been recovered and carefully evaluated, an attempt to reach a conclusion about the possible identity of the victim should be made. It must be noted that the forensic odontologist provides only an opinion regarding the strength and putative weight of the dental comparison to the Medical Examiner who solely makes the decision toward or against a positive identification based on all of the available information. There is no requisite number of points needed to establish a dental identification. When sufficient concordance and specificity exists in the absence of unexplainable disparity, identity can be confirmed. A single unique feature may be specific enough, or a list of less unique features may combine to establish uniqueness.

Any unexplainable disparity eliminates the putative identity being compared. This typically involves a tooth that contained a restoration at an earlier point in time, but is observed at a later time to be unrestored. Another example is a tooth that has been extracted at an earlier date but in the victim

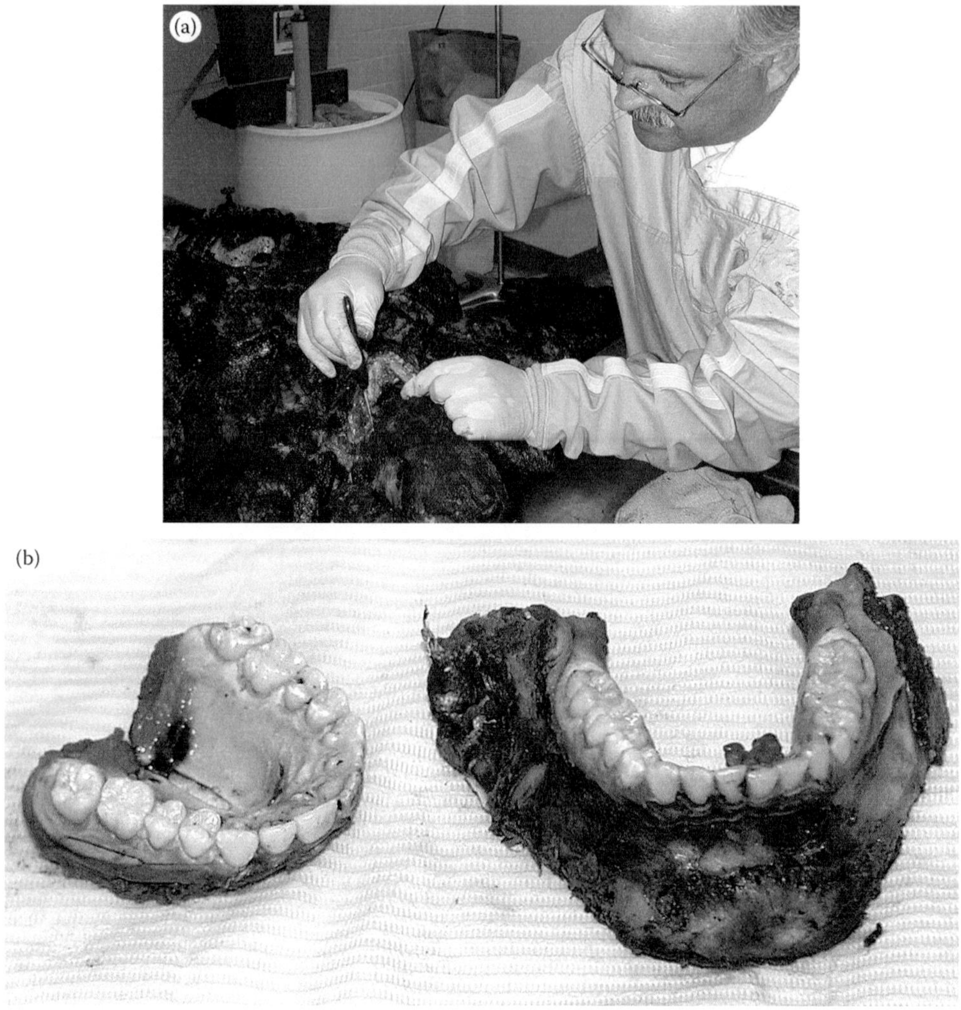

FIGURE 10.19 Resection of mandible involving (a) charred remains. (b) resected maxilla and mandible.

appears to have "reappeared." Since these disparities unequivocally preclude a positive identity being made care must be taken to view records and radiographs in the proper time progression to avoid common temporal errors.

The four categories of dental identification recommended by the ABFO are as follows[17]:

1. *Positive Identification:* The antemortem and postmortem data match in sufficient detail to establish that they are from the same individual. In addition, there are no irreconcilable or unexplainable disparities.
2. *Possible Identification:* The antemortem and postmortem data have consistent features but, due to the quality of either the postmortem remains or the antemortem evidence, it is not possible to positively establish a dental identification.
3. *Insufficient Evidence:* The available information is insufficient to form any basis for a conclusion.
4. *Exclusion:* The antemortem and postmortem data are clearly inconsistent. However, it should be understood that identification by exclusion is a valid technique in certain circumstances.

While each of the above categories is of use, there are similarities and some ambiguity between categories "2" and "3." It is very subjective to claim that there were a sufficient number of consistent features to claim a finding of "possible" versus "insufficient." A categorization of "exclusion," however, may be as powerful as a "positive identification," particularly in a case of multiple fatalities in a closed population.

Sample Comparisons

The objective of exposing postmortem dental radiographs is to ultimately subject those radiographic images to a point-to-point comparison with antemortem images of a known individual. The visual "pattern matching" comparison of antemortem and postmortem dental radiographs is by far the "gold standard" when compared with the written record, as radiographs are objective and the handwritten record is not and may contain errors. In fact, it is not unusual to have a situation in which the written record opposes what is clearly seen on the radiographs. Once the ownership of the radiographs is confirmed, the radiographic images override errors in the written record. The reliability of findings that support

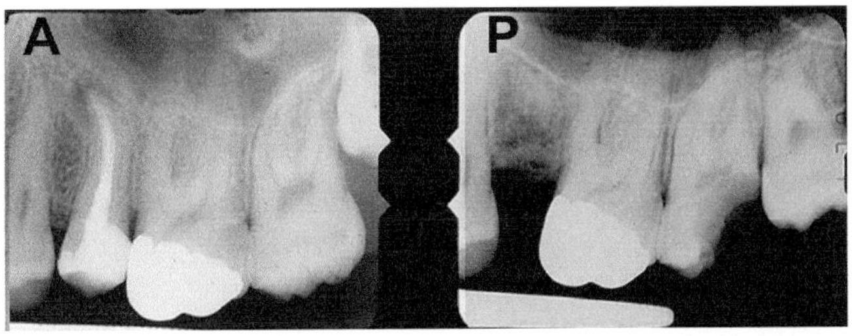

FIGURE 10.20 Sample case 1. Antemortem and postmortem radiographs.

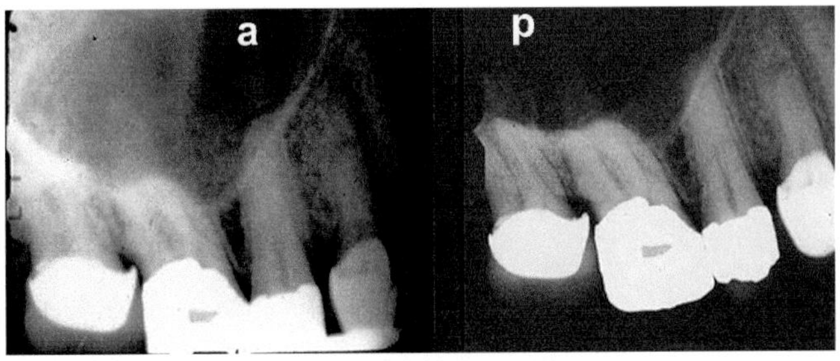

FIGURE 10.21 Sample case 2. Antemortem and postmortem radiographs.

the conclusions is: radiography over written records and manmade restorations over common anatomical structures. Dentistry does not have a minimal numerical threshold of comparative points as does fingerprint analysis. Thirty-two virgin teeth may be insufficient while one tooth with a bizarre restoration or anatomical finding may be sufficient to regard a comparison as "positive."

Sample case 1 (Figure 10.20). This case involved a low impact aircraft crash followed by a severe fire. The plane was descending in a severe summer storm and crashed into a home. As the plane and the front of the house disintegrated on impact, some of the 13 victims were thrown into the house and suffered severe charring of the tissues of the maxillofacial region. The antemortem radiograph of one victim showed that the maxillary bicuspid had been treated with root canal therapy. A large amalgam filling had been placed on the maxillary first molar and the roots of the maxillary second molar curved severely to the distal. When compared with the postmortem image of a dental fragment, disparity could be seen, including the fact that the maxillary second bicuspid was missing and the maxillary second molar contained a large carious lesion. The third molar was also more erupted.

However, the date of the antemortem image was five years previous to the accident. Therefore, it was not impossible for the caries to have reached that size and for the third molar to have erupted in the interim. Also, the victim's written record indicated that the maxillary second bicuspid had been extracted due to endodontic failure within the stated interval of time. The exact shape of the irregular underside of the large amalgam and the severely curved molar roots were sufficient to make a positive identification in this closed population of victims.

Sample case 2 (Figure 10.21). The second comparison resulted from a motor vehicle accident involving an automobile and a speeding dump truck carrying hot asphalt colliding head-on. The victim sustained blunt force trauma and severe thermal damage. Only the segment of bone that represented the maxillary right posterior segment survived the event. However, when the postmortem image was compared with the antemortem image of the suspected victim, a wealth of concordant dental restorative findings were apparent. The maxillary bicuspids in both images had been restored with a PFM crown and an MOD amalgam with a space between the two. The maxillary first molar had been restored with an amalgam in a manner that encircled the crown. However, the most unique finding was the crown on the maxillary second molar. This PFM crown very poorly fit the margins of the tooth structure with excess metal extending into the embrasure. This crown would normally have been replaced by any prudent practitioner as such an ill fitting crown will surely lead to a food trap resulting in periodontal disease and dental caries of the root surfaces. This was determined to be a positive dental identification.

Sample case 3 (Figure 10.22). This case involved a completely decomposed skeleton recovered by hunters in a dense

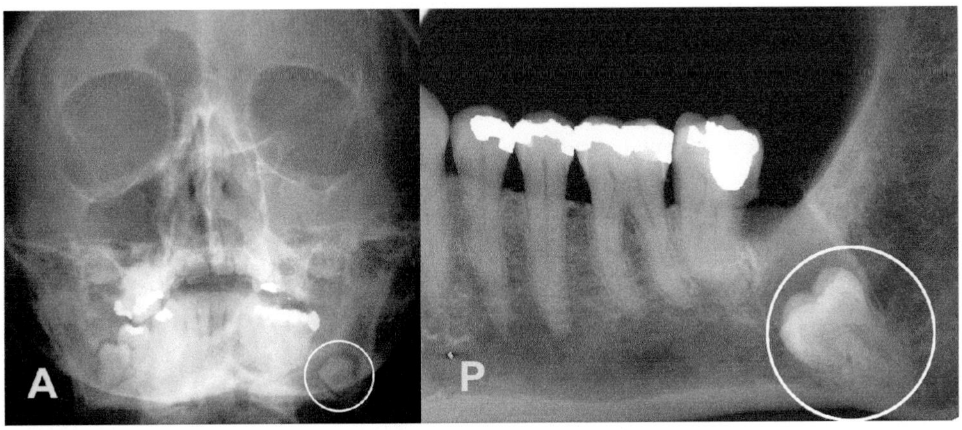

FIGURE 10.22 Sample case 3. Antemortem and postmortem radiographs.

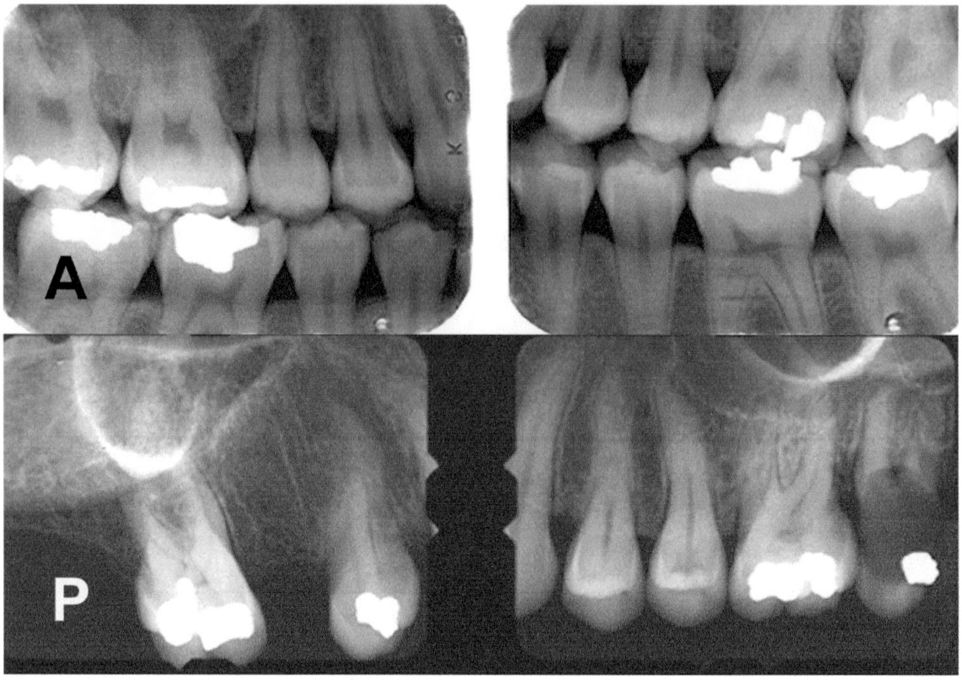

FIGURE 10.23 Antemortem and postmortem radiographs.

forest. The skeletal remains were anthropologically similar in the sex, age, and stature with an individual who had been missing from the area for two years, and evidence of foul play was discovered at the putative victim's home. A thorough search for dental records was unsuccessful. However, medical skull radiographs had been taken on the victim a few years earlier following a vehicle accident. The posterior–anterior skull image showed an extremely unusual case of both mandibular third molars being impacted at the absolute lower border of the mandibular cortical plate. Postmortem dental radiographs of the same areas clearly showed the same unusual tooth impactions. This illustrates how a rare anatomical anomaly may be a powerful indicator of identity. The experience of the forensic odontologist aids in determining how unique the anomaly is in the human population. In this case it was considered unique enough to form the basis of a positive dental identification.

Sample case 4 (Figure 10.23). This case involved the comparison of the dental records of five individuals, who had been missing for several years, with a single dry skull. Such comparisons are always difficult because dental conditions could have changed dramatically in the interim period. Four of the five putative victims could be excluded based on unexplainable disparities. The last individual showed several dental disparities, but all of the changes could have occurred over time. For example, the maxillary right second molar and bicuspid could have been extracted since the antemortem image was made. Also, the amalgams in teeth 3, 5, and 14 may have been replaced and the tooth 15 may have become decayed. Therefore, all of the disparities were explainable but none of the entities within the image clearly matched. These findings could only fit in the dental identification categories of possible or insufficient evidence.

Special consideration must be given when dry or burned specimens are evaluated when considering missing teeth. Once the periodontal ligament surrounding the tooth is degraded, teeth may fall from their sockets, especially in the anterior teeth, which have conically shaped roots. These sockets, when radiographed, show the presence of the opaque lamina dura that still outlines the root structure (Figure 10.24). Also, the crest of the bone between these sockets still remains "squared" in appearance. These teeth should be indicated as "missing-postmortem" and not missing from being extracted. Within several months of healing after an extraction, the socket outline and squared crestal bone will not appear as sharply. If this question cannot be resolved, it is often prudent to report the condition of that tooth as "no information."

NEW TECHNOLOGY IN DENTAL FORENSIC IDENTIFICATIONS

Handheld Battery-Operated X-ray Generators

New handheld battery-operated x-ray generators are extremely useful in forensic dentistry, particularly in mass fatality morgues. The Arebex™ Nomad™ is now considered state-of-the-art by most mass fatality forensic dentists and is the x-ray tubehead currently used by the Disaster Mortuary Operational Response Team (DMORT). Receiving FDA approval in 2005, the Nomad weighs approximately 4 kg and is powered by a small battery that is

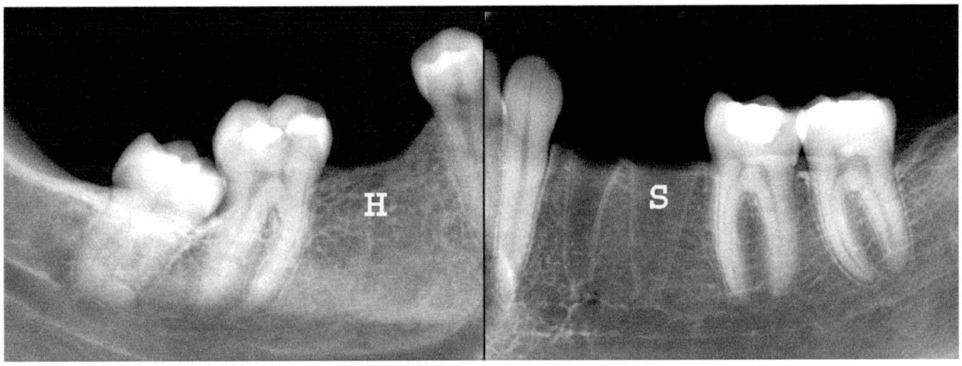

FIGURE 10.24 Healing bone (H) from previous extractions compared to images of sockets (S) where healing has not begun and lamina dura is observed. These teeth were lost postmortem.

rechargeable. During morgue operations involved in the DMORT response to Hurricane Katrina, personal experience was that more than one hundred exposures are possible before a recharge is necessary. The position indicating device (cone) includes a lightweight circular acrylic shield that provides operators with a "zone of safety" relative to scatter radiation (Figure 10.25).

The Nomad was considered a safe product initially, on the basis of a dose study conducted by the manufacturer[18] and the device's approval by the FDA. The study demonstrated extremely small tubehead leakage and backscatter to the operator. However, some state's radiation safety regulations still prohibit *any* handheld x-ray devices. The number of states approving the NOMAD is increasing at a rapid rate, but those considering taking advantage of this device should check with their local and state authorities.

Recent research simulating morgue usage in a mass fatality setting has proven the safety of those working within the dental operating area. A study simulating doses to that received by a dental team over a 2-week disaster deployment (5760 exposures) showed that the team member receiving the highest dose was at a position 60° to the side of the emanating beam, and received an exposure of 0.253 mSv.[19] This dose corresponds to 1/200th of the annual occupational maximum permissible dose (MPD) of 50 mSv and approximately 3.5 weeks of the U.S. average background radiation.[20] Another study simulated the dose received at numerous operator body locations during 915 exposures. Extrapolating the data as an expression of average annual operator exposure resulted in a whole body dose of 0.4536 or 0.9% of the annual MPD.[21]

The studies above were based on the operator strictly following the manufacturer's guidelines, which maximize

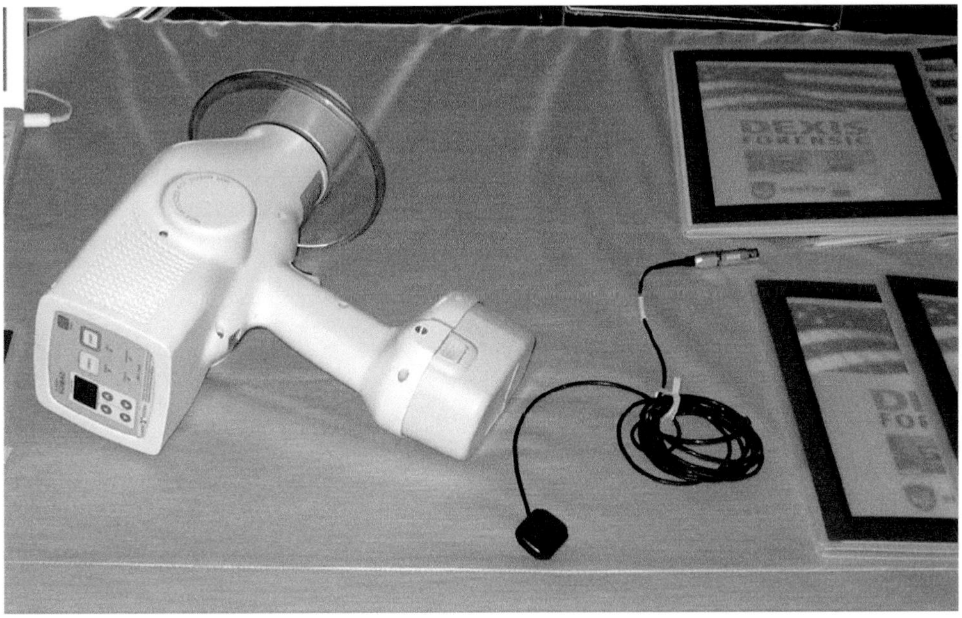

FIGURE 10.25 Aribex NOMAD™ portable x-ray tubehead and Dexis™ x-ray sensor. Note that the shield is not at the safe, recommended position at the end of the position indicating device (PID) (also called the cone).

protection via a circular acrylic shield. If the tubehead and shield are not always as ideally aligned as recommended by the manufacturer, exposures may be slightly higher than the studies above indicate. It would be prudent, therefore, for the primary operator using the Nomad in a morgue setting to wear a lead apron during exposures. Also, the NOMAD shield remain at the far end of the position-indicating device (cone) to provide maximum protection and a larger zone of safety.

Digital Dental Radiography

Digital dental radiography has become a mainstream radiographic technique in the private practice of dentistry over the last 25 years.[22] It is particularly well suited for forensic odontology, especially in mass fatality incidences where film, darkrooms, dental film processors, and processing chemicals are totally eliminated. Digital images, both radiographic and photographic, are also much easier to manipulate, enhance, categorize, store, and recall compared to film (Figure 10.26). The oral digital receiver is typically a direct solid state digital charge-coupled device (CCD) or complementary metal-oxide semiconductor (CMOS) detector. Photostimulable phosphor plates are another commonly used dental digital x-ray system but require a laser scan device to produce the resultant image and do not provide an "immediate" x-ray image. Contamination of the plates and scanner are also a definite concern in a morgue setting.

Both CCD and CMOS sensors provide an immediate image, which is extremely desirable for the forensic odontologist's task. It dramatically reduces the time necessary to examine and radiograph victims in a mass fatality setting by reducing the time previously required to develop a series of dental x-ray films (approximately 5–15 min) and the subsequent "retakes" for unacceptable images. Any necessary retake is displayed on the monitor immediately and the exposure can be retaken in a matter of seconds. Also, a comparative study showed that digital images are of equal quality to film when evaluating interproximal caries.[23] The task of the forensic odontologist is typically less demanding than what is required in the routine practice of dentistry.

However, digital sensors are delicate, expensive and, in some cases, difficult to place. The most useful feature of digital systems is the ability to import the images into third-party information management software programs such as WinID™, which is described in detail later in this text. To ensure interoperability with third-party software, only digital devices that are compatible with the Digital Imaging and Communications in Medicine (DICOM) standards, in this case DICOM3 compatible, should be used.

Digital radiography requires approximately one-fourth or less of the exposure energy required by film. This undoubtedly lowers the amount of scattered x-ray exposure, particularly to multiple operators in a mass fatality morgue facility. However, the ultra-short exposure times for digital imaging requires that x-ray tubeheads be capable of producing accurate exposures in the extremely short timer range. This often excludes the use of older x-ray equipment.

If digital radiography is not available, all antemortem and postmortem radiographs should be digitally scanned on a

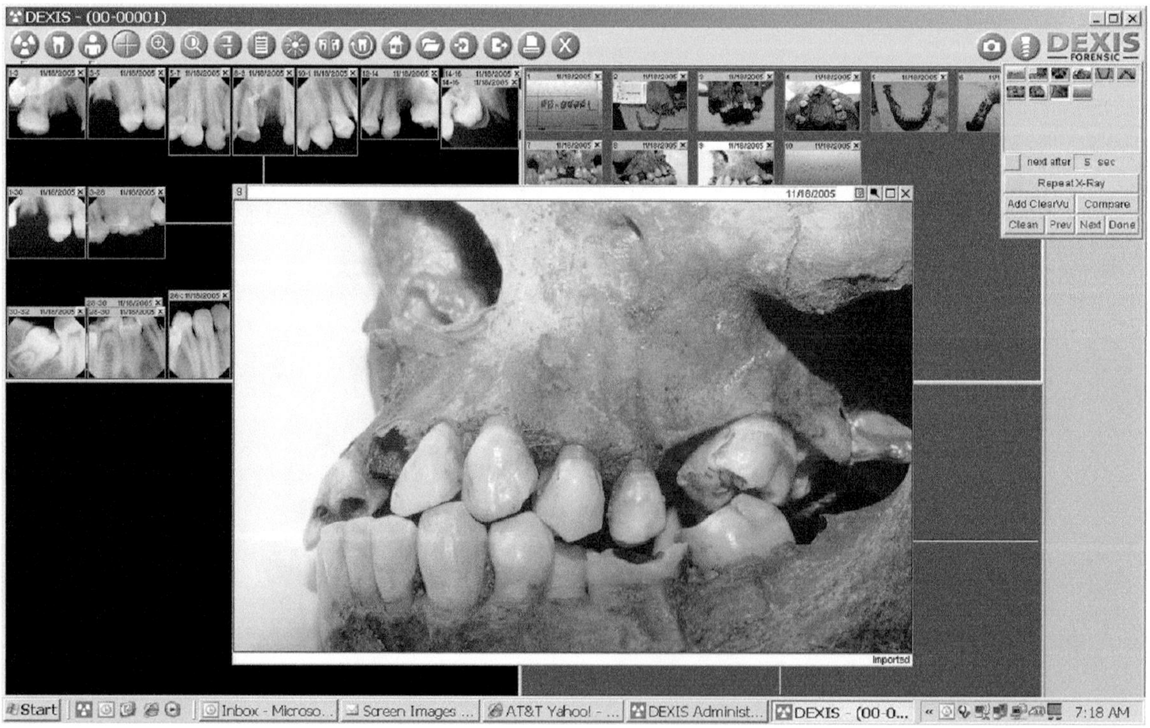

FIGURE 10.26 Screen image of complete mouth radiographic series and imported oral photographs in the Dexis™ digital x-ray system.

flat-bed scanner for archival purposes. The scanner must have transparency/direct positive capabilities. Once digitized, these images can be integrated with any software system used in forensic dentistry; they are similar and equal to images taken by direct digital radiography.

WinID3©: Dental Data Management and Matching via Computers

WinID3 is the most widely used dental identification/data management system used for timely dental identification involving large numbers of individuals, particularly in mass fatality incidences. WinID was created by Dr. James McGivney and initially used Microsoft Access as a database configured with an electronic odontogram using a very basic and simplified set of dental charting codes (primary and secondary). WinID3 was released in 2001 for use in the World Trade Center Disaster. The new version uses Microsoft ActiveX Data Objects, which improved the data management tools and increased graphics capabilities. Antemortem and postmortem victim identifiers and radiographic images are entered into the database via data fields and the odontogram. The program has many features including the ability to identify the most likely ordinal dental matches, specific multiple search filters, and the ability to store and view antemortem and postmortem radiographic series side-by-side on the computer display at the touch of the keyboard.

Likely dental "matches" are listed in order based on the number of common "hits" that represent the presence or absence of a missing tooth, or a restoration pattern on the same tooth as being identical when all of the antemortem and postmortem cases are compared. The operator may then scroll through the most likely matching odontograms or compare the stored radiographic series to establish a victim's identity (Figure 10.27). The amount of time saved when making point-by-point comparisons involving hundreds or thousands of sets of dental records and radiographs is enormous. This was seen in the 9/11 disaster in New York. While WinID3 made its name as an indispensable tool for mass fatalities, many forensic dentists enter dental and radiographic data from their local individual cases because of the streamlined image storage and data management provided by the system. A new version, "WinID-on-the-web," is now in the testing stages, and it will allow the use of the system online from any computer in the world.

Cone Beam Computed Tomography

Dental Cone Beam Computed Tomography (CBCT) is a relatively new digital x-ray technology that has come to be highly accepted in the dental community.[24] The essence of the system is that it gathers 360° of radiographic images of the craniofacial structures and provides the practitioner with the third dimension, depth. Various standard plain film projections and tomographic representations can also be created from a single scan (Figure 10.28). The resulting images are without the usual inaccuracies of magnification, blurring, and superimposed contralateral structures encountered in plain film radiography. Also, numerous software applications are now available which allow 3D bony and facial reconstruction images from the raw DICOM3 scan data.

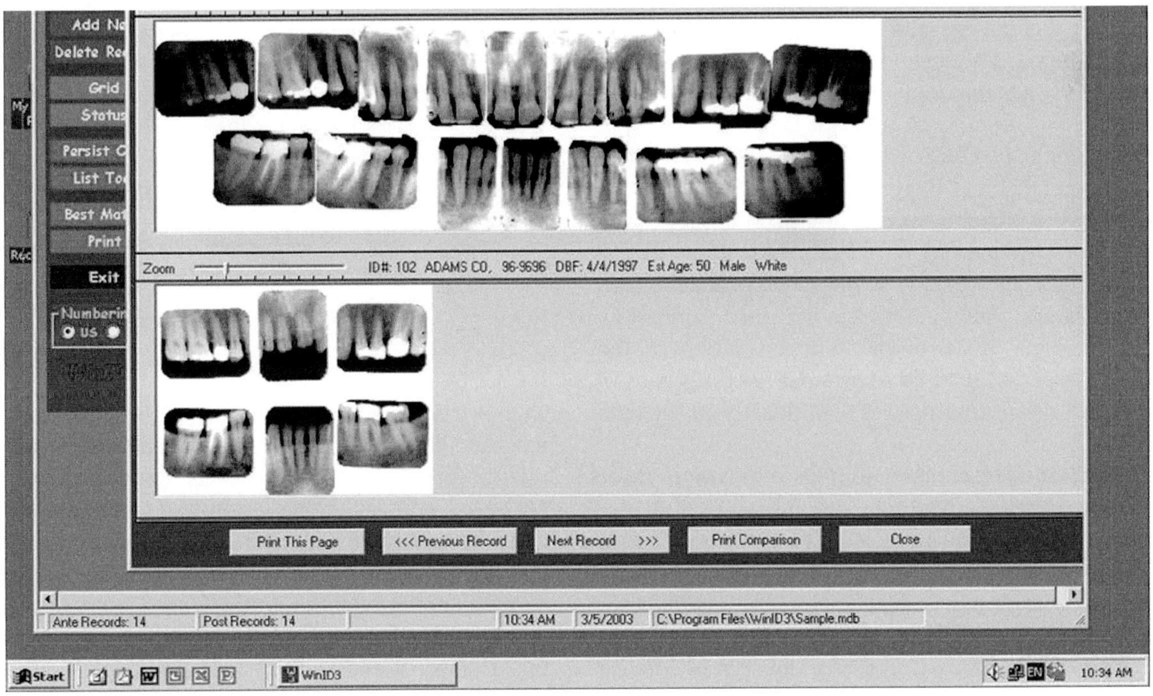

FIGURE 10.27 Screen image comparing postmortem radiographs to possibly matching antemortem radiographs that were identified by the "best match" feature of WinID3©. Courtesy of Dr. James McGivney.

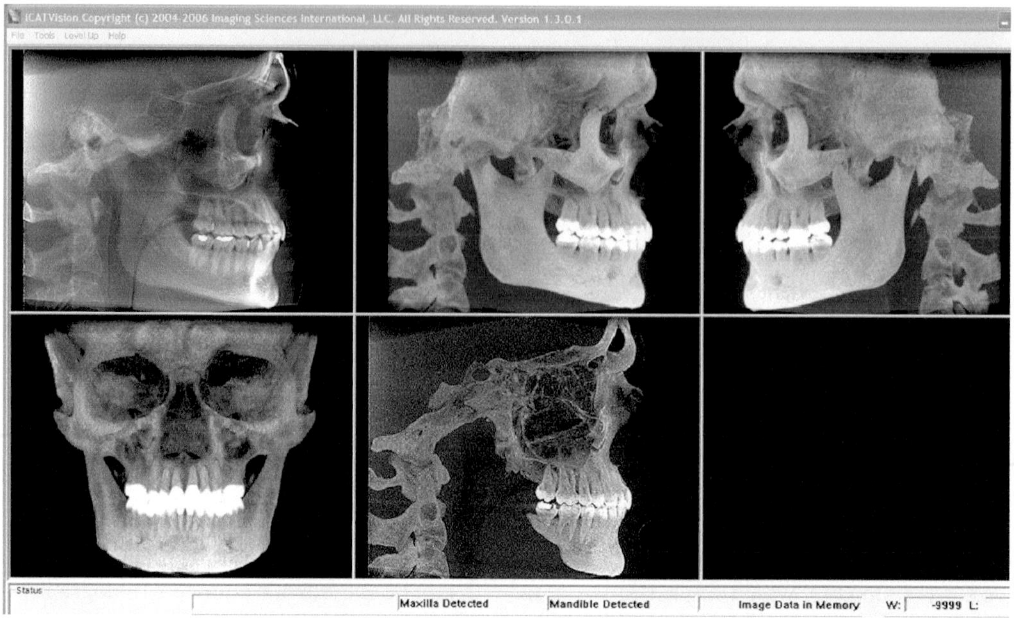

FIGURE 10.28 Multiple PA and lateral skull images produced by a single scan using Imaging Sciences International's i-CAT® Cone Beam CT unit and iCATVision© software.

CBCT units create images by acquiring multiple (300 or more) captures of the variably selected tissue field. The system is termed cone beam because the full-sized, cone-shaped beam is not thin or fan-shaped such as that used in typical medical CT units. The transmitted energy from the tube is received by either an image intensifying tube married to a CCD or a flat panel receiver composed of amorphous silicon. After the captured images are acquired they are reconstructed by the system's computer algorithm and visualized on a computer display. The various CBCT units vary in size, shape, energy receiver technology, and the size of the tissue field that may be captured. The scan field can vary from a volume 2 × 3 cm vertically and horizontally to a volume of 8 × 22 cm. The units utilize positioning of the patient in a sitting or, in some cases, supine position, which would be ideal for a morgue setting. The CBCT image comprises 3D image units known as voxels. A voxel is similar to a pixel except that it includes the third dimension. Where a pixel is a lateral and vertical image unit, the voxel is in the shape of a cube (i.e., 0.2 × 0.2 × 0.2 mm), which allows for the third dimension. Thus, CBCT voxels are isotropic. The resolution of the images can range from 0.4 mm to as small as 0.125 mm in some units, and the patient dose is significantly lower than medical CT.

The initial image presentation after a scan involves three geometrically related tomographs providing multiplanar slices: sagittal, coronal, and axial. Any intraoral or extraoral plain film image view can then be reconstructed from one scan, with the operator being able to select the desired slice location, thickness, and orientation. For most dental purposes a panoramic image is created for the area of interest with no superimposed anatomical structures to obscure structures. Also, computer algorithms correct for the geometric distortions that are present in all 2D images. Therefore, there is no magnification and the images, when printed, are 1:1 representations.

Eventually, identifications of human remains will be solved using the image data directly from antemortem and postmortem CBCT scans. Also, any single CBCT postmortem scan of the victim could later be compared with any possible variety of submitted antemortem plane film images.

Also, it is not unusual for skeletal remains to be missing both the mandible and maxillary arch, because of animal scavenging. CBCT multiplanar views demonstrate craniofacial structures that have not been discernable previously. For example, the paranasal sinuses are clearly seen in CBCT scans (Figure 10.29) and early studies show that they provide enough specificity for human identifications.[25] Stereolithic reconstructions and images of the soft tissue of the face may also be recreated from scan data with supporting 3D software.

SPECIAL CONSIDERATIONS IN MASS FATALITY INCIDENCES

Dental members of DMORT have been deployed in several mass fatality incidences to aid local authorities in identifying the dead and replacing caskets disinterred by flooding. Massive amounts of antemortem and postmortem dental data must be collected and subsequently compared when the number of those who have perished is in the thousands. This was the case in both the World Trade Center disaster (WTC) and with Hurricanes Katrina and Rita. Compared with making a single identification with a putative victim, disaster operations are exponentially more difficult. The answer to the complexity is an organized and strict protocol for morgue operations and taking advantage of the new technologies

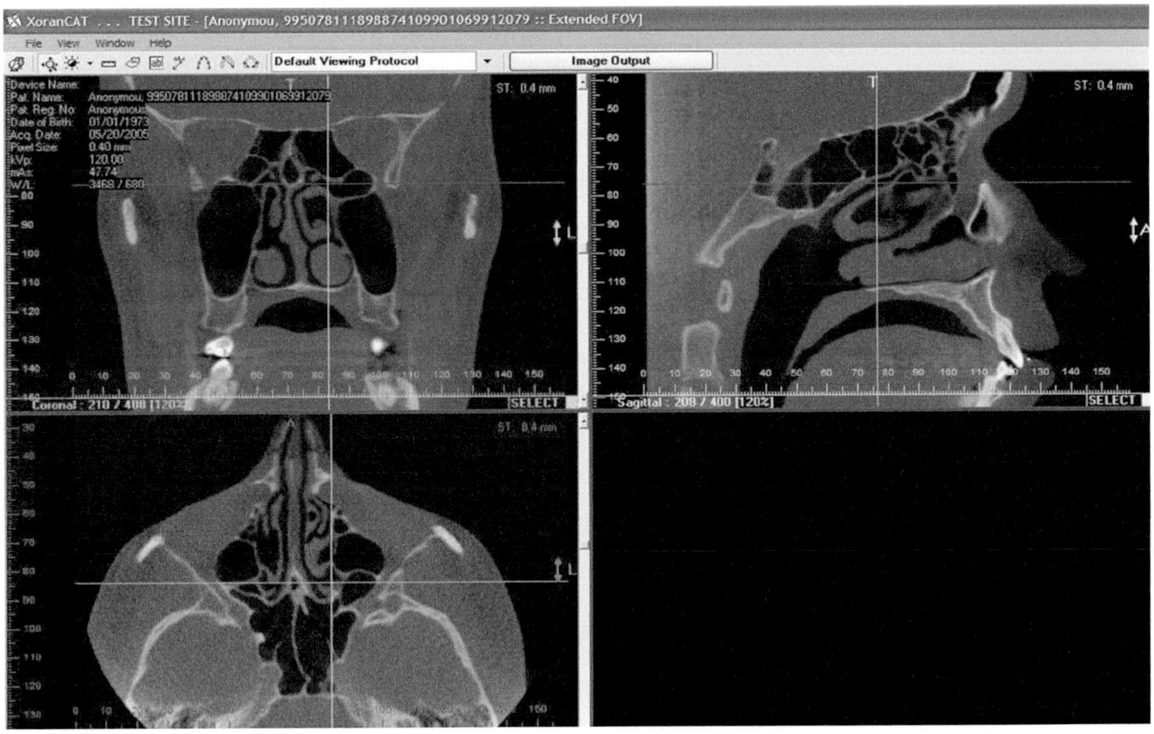

FIGURE 10.29 Multiplanar CBCT views of paranasal sinuses: coronal, sagittal, and axial. These may be specific enough features to allow scientific human identification.

mentioned above. Failure to follow protocols will inevitably lead to delays, confusion, errors, possible loss of data, and the need to repeat difficult work. If a misidentification occurs in a Mass Fatality Incident (MFI), then a second victim will also be misidentified or not identified at all.

In response to hurricanes Katrina and Rita, DMORT established two morgue operations, one in Gulfport, Mississippi, and one in St. Gabriel, Louisiana. The dental identification process for both morgues was "film-less" and "paper-less." The dental chartings were created directly in

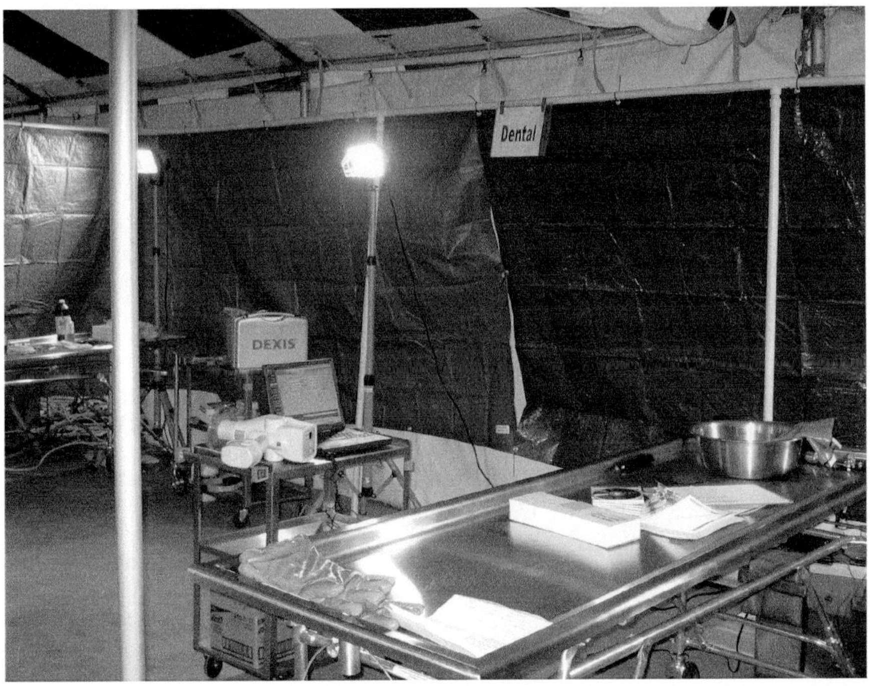

FIGURE 10.30 Dental examination station in the DMORT deployable portable morgue unit (DPMU) in response to Hurricane Katrina in Gulfport, Mississippi.

WinID3 at the dental station and all of the postmortem radiographs were captured using direct digital radiography (Figure 10.30). As antemortem dental radiographs were received, they were scanned into WinID3 using a digital flatbed scanner. All photographs of the victim's dental conditions were captured and stored digitally. Radiographic exposures were made using the Nomad handheld battery-operated x-ray device.

Dexis™ Corporation showed a true commitment to forensic dentistry in sharing source codes for their direct digital x-ray sensor software so that the images taken directly or scanned could be sent directly to WinID3. Dr. McGivney equally contributed to this effort by opening and revising his WinID3 code. This allowed total interoperability as the operator could move seamlessly between WinID3 and the Dexis imaging software. The Dexis "forensic version" was also used to scan the antemortem radiographs and store all digital camera images taken by the dental team. Many state and local disaster teams are now incorporating this equipment into their systems and training. These methods and technology are, collectively, the paradigm for the future and have been proven in the field during major mass fatality events.

SUMMARY

Forensic odontologists rely on the principles and science of dentistry and dental radiography in establishing scientific victim identifications. Research continues in many areas, including new radiographic imaging systems and techniques and digital automated x-ray comparison systems, to establish more scientifically the uniqueness of all oral conditions and anomalies. True science in forensic sciences must be a common goal. Those interested in contributing in some way to forensic odontology and advancing the science should join one or all of several current organizations: the American Society of Forensic Odontology; the Odontology Section of the American Academy of Forensic Sciences; and the American Board of Forensic Odontology. All three organizations support our standards and also provide grant funding for those wishing to gain financial support for worthy research projects.

Dentists are taught during their training never to release their original dental records and radiographs. Many dentists are also unaware of the details and releases from the Health Insurance Portability and Accountability Act of 1996 (HIPAA).[26] Section 45 CFR 164.512(g) states, in part, that "[a] covered entity may disclose protected health information to a coroner or medical examiner for the purpose of identifying a deceased person, determining a cause of death, or other duties as authorized by law." Presenting dental offices with an official copy of this section and release form as a matter of routine will allay concerns and help expedite the transfer of high-quality antemortem dental information to the forensic odontologist in a timely manner.

REFERENCES

1. Cassius Dio, C., Baldwin, E. F., *Herbert Dio's Roman History*, The Loeb Classical Library, W. Heinemann; The Macmillan Co., London, 1914.
2. Forbes, E., *Paul Revere and the World He Lived in*, Houghton Mifflin Co., Boston,, 1943.
3. Ring, M. E., Paul Revere—dentist, and our country's symbol of freedom, *N Y State Dent J.*, 42, 598–601, 1976.
4. Amoëdo, O., *L'Art Dentaire en Méicine Légale*, Masson et Cie, Paris, 1898.
5. Langland, O. E., Sippy, F. H., Langlais, R. P., *Textbook of Dental Radiology*, 2nd ed., Charles C. Thomas, Springfield, IL, 1984.
6. Fry, W. K. The Baptist Church Cellar Case, *Br Dent J*, 75, 154, 1943.
7. Herschaft, E., Alder, M., Ord, D., Rawson, R., Smith, S., *Manual of Forensic Odontology*, 4th ed., American Society of Forensic Odontology. Impress Printing & Graphics, Inc., Albany, NY, 2007.
8. Stimson, P. G., Mertz, C. A. *Forensic Dentistry*, CRC Press, Boca Raton, FL, 1997.
9. Mincer, H. H., Harris, E. F., Berryman, H. E., The A.B.F.O. study of third molar development and its use as an estimator of chronologic age, *J Forensic Sci*, 38, 379, 1993.
10. Harris, E. F., McKee, J. H., Tooth mineralization standards for blacks and whites from the middle southern United States, *J Forensic Sci*, 35, 859, 1990.
11. U.S. Department of Justice, Office of Justice Programs. National Missing and Unidentified Persons System. Available online at http://www.namus.org
12. Goaz, P., White, S., *Oral Radiology: Principles and Interpretation*, 3rd ed., Mosby, St. Louis, 1994.
13. Fixott R., Forensic odontology, in *Dental Clinics of North America*, W.B. Saunders Co., Philadelphia, 2001.
14. Weems, R., Broome, J., Heaven, T., Yarbrough, R., Radiopacity of endodontic posts in dental identifications. *Proceedings of the Annual Meeting of the American Academy of Forensic Sciences*, Dallas, Texas, 2004.
15. National Research Council of the National Academies, *Strengthening Forensic Sciences in the United States: A Path Forward*, The National Academies Press, Washington, DC, 2009.
16. National Association of Medical Examiners. Available online at http://www.thename.org
17. American Board of Forensic Odontology, I.D. and Bitemark Guidelines. Available online at http://abfo.org
18. Turner, D. C., Kloos, D. K., Morton, R. *Radiation Safety Characteristics of the NOMAD™ Portable X-ray System*, Aribex, Inc., Orem, UT, 2005.
19. Hermsen, K., Stanley, J., Jaeger, M., Radiation safety for the Nomad™ portable X-ray system in a temporary morgue setting, *J Forensic Sci*, 53, 917–19, 2007.
20. NCRP Report No. 145, *Radiation Protection in Dentistry*, National Council on Radiation Protection and Measurements, Bethesda, MD, 2004.
21. Danforth, R., Herschaft, E., Leonowich, J., Operator exposure to scatter radiation from a portable hand-held dental radiation emitting device (Aribex™ NOMAD™) while making 915 intraoral dental radiographs, *J Forensic Sci*, 54, 415–421, 2009.
22. Dunn, S., Kantor, M., Digital radiography. Facts and fiction, *J Am Dent Assoc*, 124, 38–47, 1993.

23. Duncan, R., Heaven, T., Weems, R., Firestone, A., Greer, D., Patel, R., Using computers to diagnose and plan treatment of approximal caries detected in radiographs, *J Am Dent Assoc*, 126, 873–882, 1995.
24. Scarfe, W., Farman, A. G., Sukovic, P., Clinical applications of cone-beam computed tomography in dental practice, *J Can Dent Assoc*, 72, 75–80, 2006.
25. Weems, R., Cone beam CT radiography for dental identifications. Paper presented at the annual meeting of the American Academy of Forensic Sciences, Washington, DC, 2008.
26. U.S. Department of Health and Human Services, Health Information Privacy Act. Available online at http://www.hhs.gov

11 Virtopsy
Dental Scan and Body CT Imaging as a Screening Tool for Identification

Michael J. Thali, Ulrich Preiss, and Stephan A. Bolliger

CONTENTS

Radiology plays an important role in forensic identification. Dental identification uses the teeth, jaws, and orofacial characteristics in general as well as the specific features of dental work with metallic or composite fillings, crowns, bridges, and removable prostheses. It also includes the distinctive configuration of the bony structures of the jaw (mandible and maxilla), the presence and shape of teeth including the roots, the configuration of maxillary sinuses, and longstanding pathology, such as prior fractures and orthopedic procedures.

Cross-section imaging plays an increasingly important role in mass disaster management and in the forensic identification process. In a Dentalscan program, a curved line is produced by the user on an oblique cross-section, the direction of which has been already adapted to the line of the jaw. Along this line, the software creates panoramic images in user-defined thickness and number.

A reformatted panoramic overview created by Dentalscan delivers in a noninvasive way an overview of the jaws, showing the basic components of teeth (enamel, dentin, and pulp); the anatomic structure of the alveolar bone (with mandibular and maxillary landmarks such as, e.g., the mandibular nerve canal or the floor of the nasal cavity and the maxillary sinuses); pathology (caries, radiolucencies, radiopacities, or the position of the third molars); and restorations.

The most important advantage of Dentalscan programs in contrast to classical methods is that documentation can be made in a noninvasive and digital way without jaw resection, which is often performed to facilitate classical radiological documentation on decomposed, charred, and mutilated

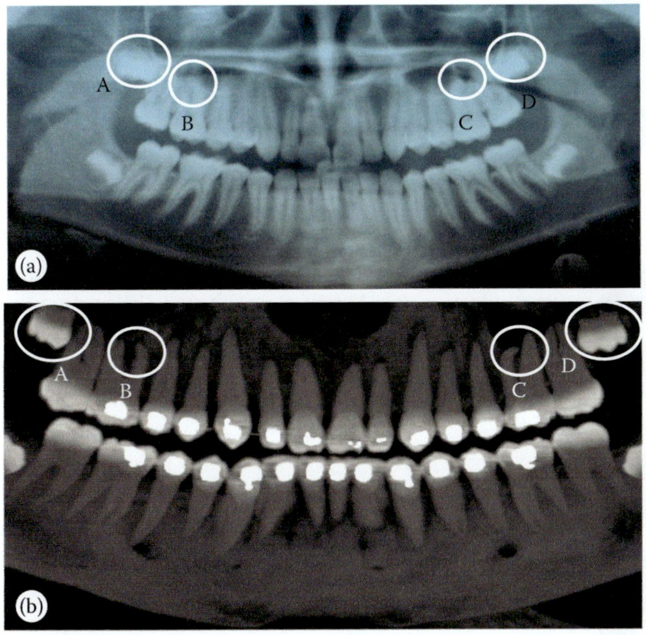

FIGURE 11.1 Dental ID. A 17-year-old male of ambiguous identity. (a) Antemortem images (b) Postmortem images. Since he had not undergone any dental surgery, the shape of the roots and the position of the retained teeth had to be compared.

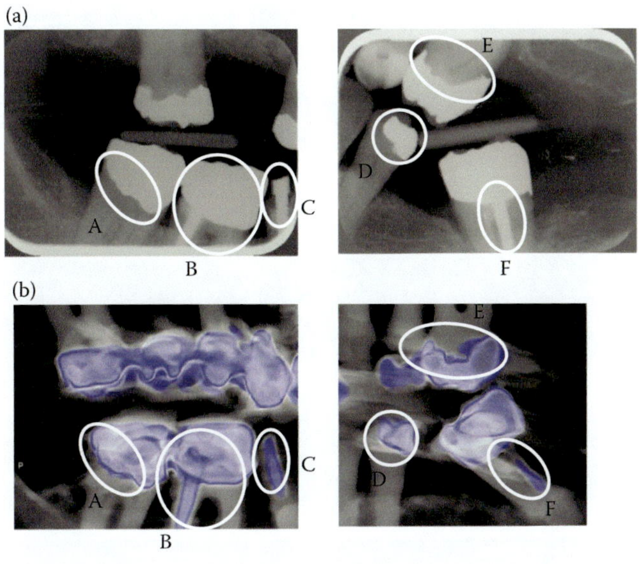

FIGURE 11.2 Dental ID. A 62-year-old female of ambiguous identity. Antemortem images (a) were compared to postmortem images (b). Although the person had obviously undergone dental surgery between the antemortem and the postmortem imaging, it was possible to match three other distinctive features and the suspected identity could be confirmed.

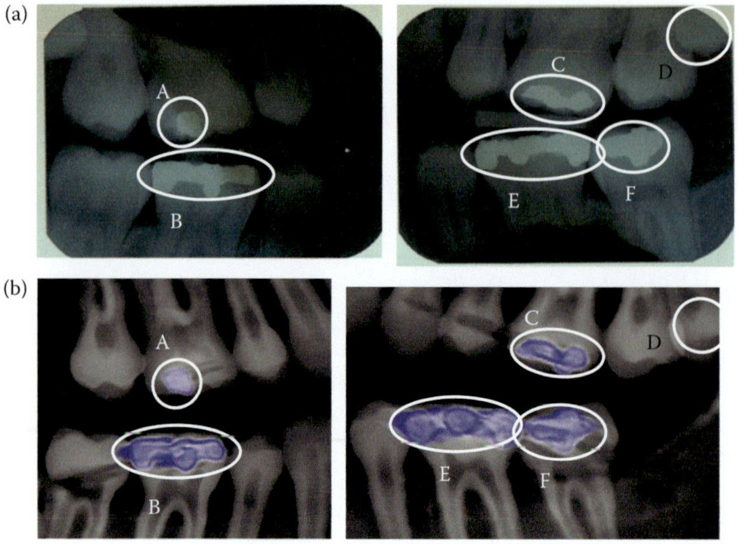

FIGURE 11.3 Dental ID. A 37-year-old male of ambiguous identity. (a) Antemortem images (b) postmortem images. Because of advanced decomposition, the identification of this deceased needed to be confirmed through dental images. There is a match of more than 3 features between the antemortem and the postmortem images.

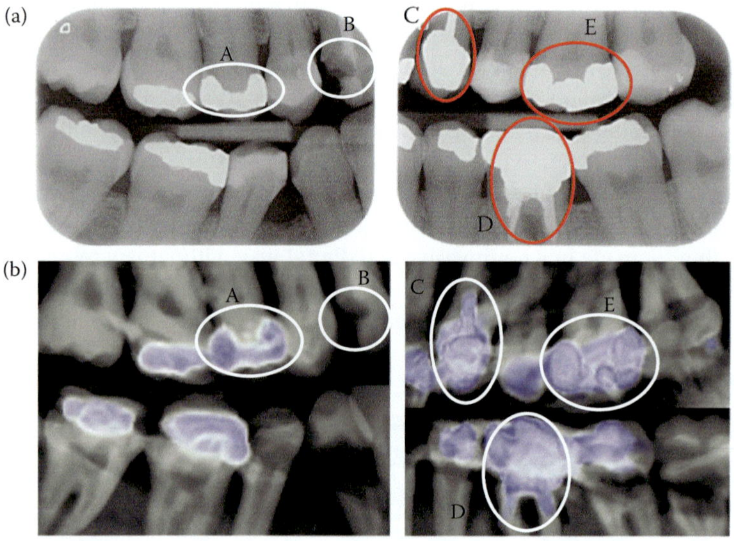

FIGURE 11.4 Dental ID. A 55-year-old female of ambiguous identity. (a) Antemortem images (b) postmortem images. The corpse was severely damaged after her death in a railway accident. Parts of the remaining teeth have been rearranged digitally and her identity could thus be established.

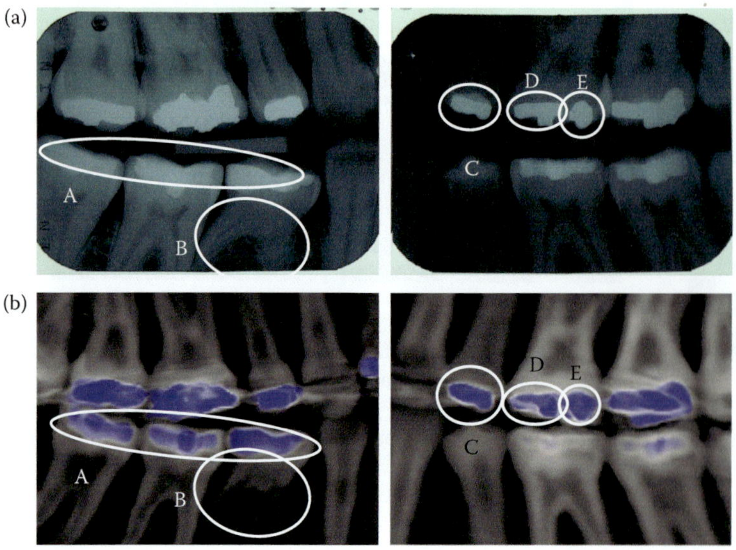

FIGURE 11.5 Dental ID. A 40-year-old female of ambiguous identity. (a) Antemortem images (b) postmortem images. A plethora of matching distinguishing characteristics was identified in this case and identity could be established.

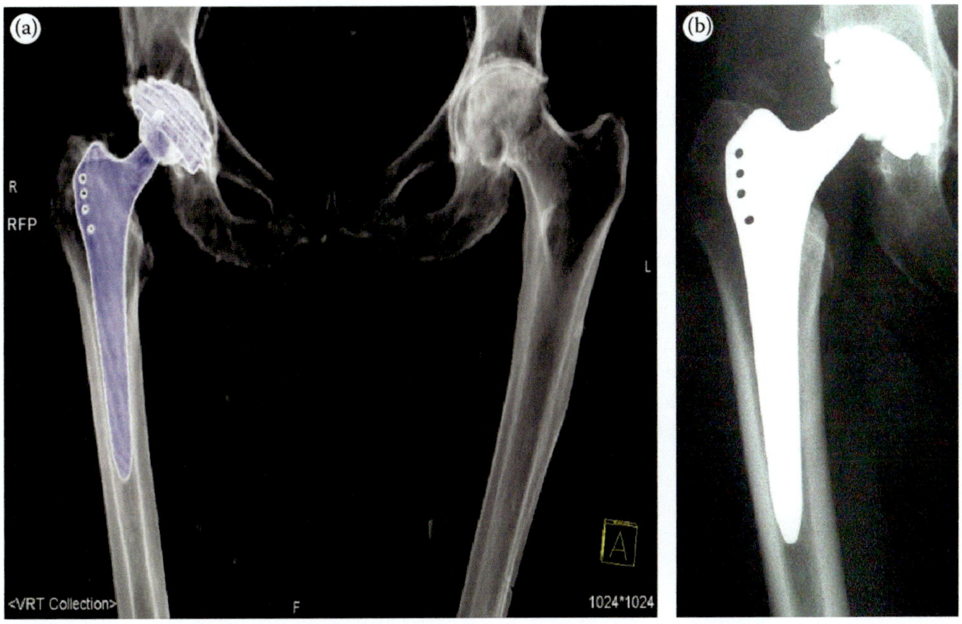

FIGURE 11.6 Incidental postmortem computed tomography (CT) finding supporting the identification. Total endoprosthesis of the right hip postmortem (a) and antemortem (b). Because the 3D CT data can be turned around and a superposition analysis is possible, it is easier to do the identification analysis process independent of the direction of the x-ray beam and the antemortem 2D x-ray.

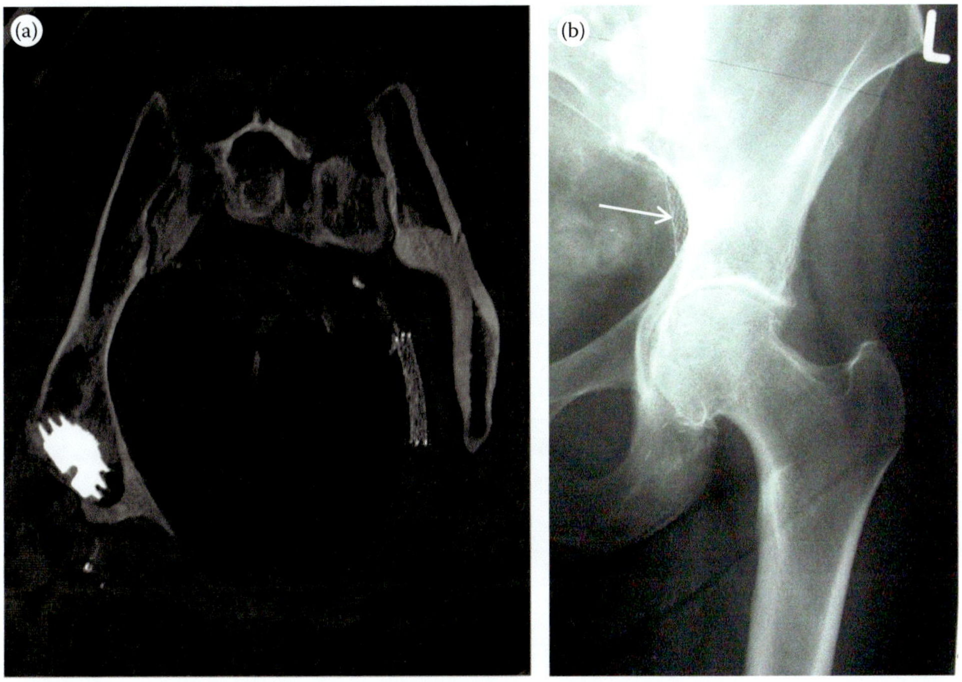

FIGURE 11.7 Incidental postmortem CT finding supporting the identification. Stent in the iliac artery postmortem (a) and antemortem (b).

corpses. There is no need for special positioning techniques (rubber bands or density substitutes for missing tissue if the samples are macerated because of odor effects) as was used to produce postmortem periapical, bitewing films and classic panoramic radiographs, because the Dentalscan is an *in situ* documentation method. A further advantage of the *in situ* documentation process is that there is no secondary damage, which is often a problem, for example, in charred bodies. The data can be stored directly on the hard disc or a CD, and transferred through the Internet or exported to a modern graphic dental identification program (e.g., WinID). In addition to the unlimited data storage space, every possible two- or three-dimensional (2- or 3D) reconstruction of the teeth or other body areas is possible (see Figures 11.1 through 11.7).

BIBLIOGRAPHY

Jackowski, C., Aghayev, E., Sonnenschein, M., Dirnhofer, R., and Thali M. J., Maximum intensity projection of cranial computed tomography data for dental identification. *Int J Legal Med*, 120(3), 165–7. Epub 2005 Oct 20, 2006.

Jackowski, C., Lussi, A., Classens, M., Kilchoer, T., Bolliger, S., Aghayev, E., Criste, A., Dirnhofer R., and Thali M. J., Extended CT scale overcomes restoration caused streak artifacts for dental identification in CT–3D color encoded automatic discrimination of dental restorations. *J Comput Assist Tomogr*, 30(3), 510–3, 2006.

Jackowski, C., Wyss, M., Persson, A., Classens, M., Thali, M. J., and Lussi A., Ultra-high-resolution dual-source CT for forensic dental visualization-discrimination of ceramic and composite fillings, *Int J Legal Med*, 122(4), 301–7, 2008.

Pfaeffli, M., Vock, P., Dirnhofer, R., Braun, M., Bolliger, S. A., and Thali, M. J., Post-mortem radiological CT identification based on classical ante-mortem X-ray examinations, *Forensic Sci Int.*, 13, 171(2–3), 111–7, 2007.

Sidler, M., Jackowski, C., Dirnhofer, R., Vock, P., and Thali, M., Use of multislice computed tomography in disaster victim identification—advantages and limitations, *Forensic Sci Int.*, 4, 169(2–3), 118–28, 2007.

Thali, M. J., Markwalder, T., Jackowski, C., Sonnenschein, M., and Dirnhofer, R., Dental CT imaging as a screening tool for dental profiling: Advantages and limitations, *J Forensic Sci.*, 51(1), 113–9, 2006.

Thali, M. J., Yen, K., Plattner, T., Schweitzer, W., Vock, P., Ozdoba, C., and Dirnhofer, R., Charred body: Virtual autopsy with multi-slice computed tomography and magnetic resonance imaging. *J Forensic Sci.*, 47(6), 1326–31, 2002.

12 Radiological Identification of Individual Remains

B.G. Brogdon

CONTENTS

In the ideal medicolegal facility the use of medical imaging should be as routine as the autopsy and, in point of practical fact, may be used as a substitute for the autopsy under certain situations.

J. F. Edland (1980)[1]

INTRODUCTION

The radiological identification of individual human remains depends entirely on matching specific and unique visual findings or features on both antemortem and postmortem radiological images of that person (Figure 12.1). Postmortem findings confirming sex, age, stature, or race may be either confirmatory or exclusionary. Postmortem radiographic evidence of a specific injury, disease, or congenital anomaly known to have been present in a particular missing person can lead to presumptive identification. However, positive radiological identification requires comparative matches of anatomic features on pre- and postmortem radiological examinations. Sometimes a cluster of relatively common or nonspecific anatomic changes can establish identification beyond reasonable doubt; on other occasions a single unique finding is sufficient.

Example Case 11-1. A 49-year-old male was abducted from his home at gunpoint, in front of a witness, and driven away in his own automobile. On the following morning, forest rangers discovered the smoldering hulk of a car in a remote site but left without examining the vehicle after determining that it represented no fire hazard. Later that day, children at play reported bones in the trunk of the burned-out auto, and the sheriff's department undertook a more thorough investigation. The vehicle was identified as that of the abductee. The trunk contained a fragmented, partially carbonized, and calcined human skeleton that was removed to the laboratory of the Alabama State Medical Investigator and

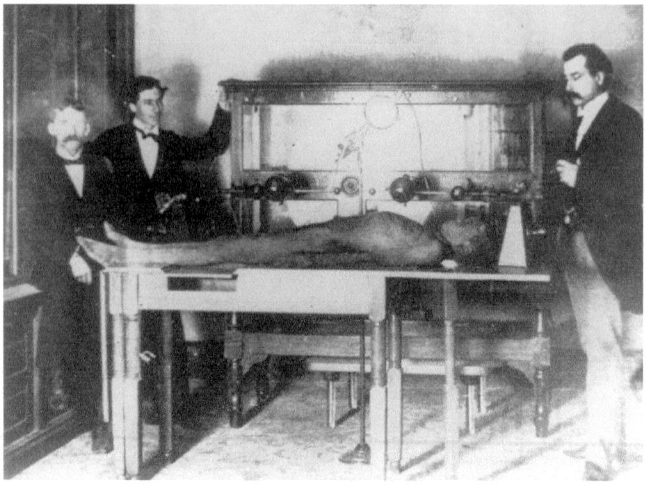

FIGURE 12.1 The earliest photograph of roentgenography of a cadaver found by the author. It was made in 1898 at the American School of Osteopathy, Kirksville, MO. (Courtesy of the Center for the American History of Radiology, Reston, VA.)

radiographed. Dissection of some remaining soft tissue in the bony pelvis revealed a prostate, thus confirming that the victim was a male. Fragments of the skull and mandible were edentulous, as had been the owner of the car. A bullet wound in the posterior skull suggested the cause of death (Figure 12.2). The gunshot and/or the intensity of the fire had destroyed most of the frontal bone and the frontal sinuses. (Figure 12.3). A roentgenogram of a charred fragment of the right foot and ankle revealed a talotibial fusion (Figure 12.4); the presumed decedent had had such an operation 15 years earlier, but the antemortem films had been destroyed. Consequently, only presumptive identification was possible. Finally, another search of the nooks and crannies of the automobile trunk turned up a shrunken, calcined, but intact left patella with a punched-out defect on its dorsal surface, which was recognized by the radiologist as a classic, albeit uncommon, patellar lesion (Figure 12.5). At about that time a radiological examination of the car owner's left knee was discovered at a nearby community hospital where he had

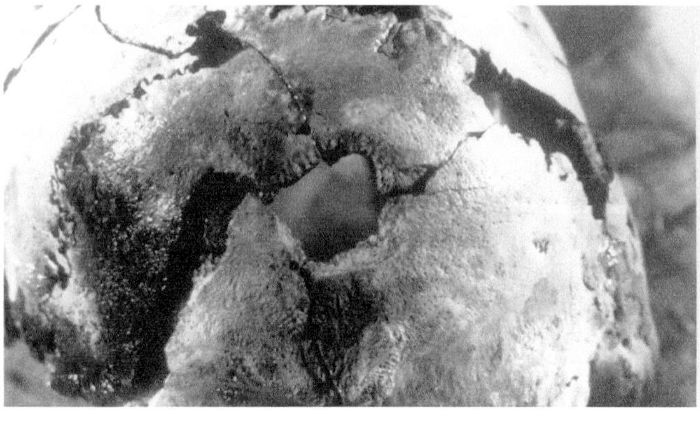

FIGURE 12.2 Case 11-1: gunshot exit wound in posterior skull.

FIGURE 12.3 Case 11-1: Frontal sinuses destroyed by gunshot and/or interior fire. The frontal part of the skull (above) is resting on a bed of sand (below) during the process of reconstruction of fragments.

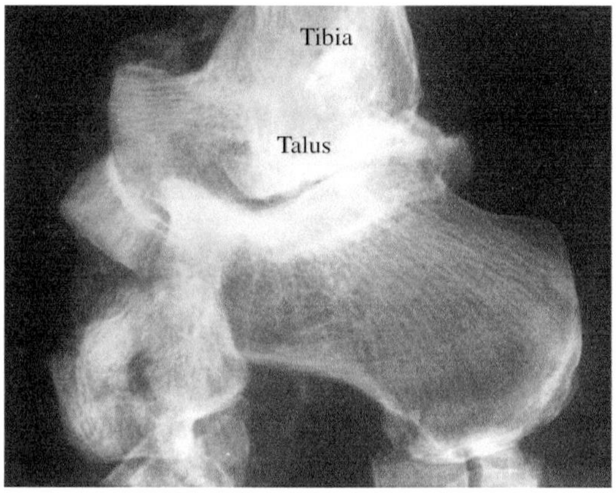

FIGURE 12.4 Case 11-1: this second-largest (after a portion of the pelvis) of all fragments recovered from the trunk of the car shows fusion of the joint between the talus and the distal tibia as a result of previous surgery. (Reprinted from Riddick, L., et al., *J. Forensic Sci.*, 28, 263, 1983. © ASTM. With permission.)

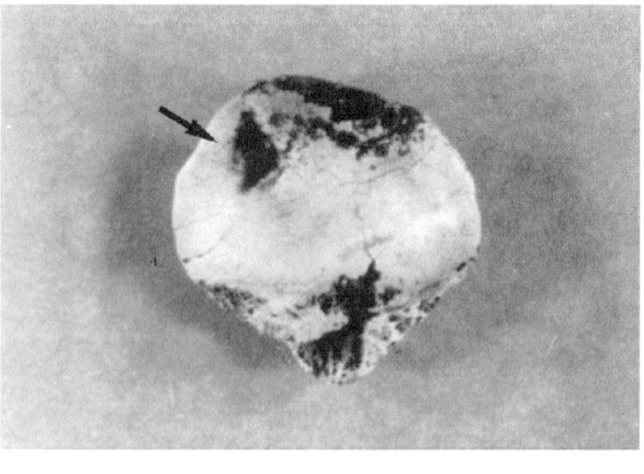

FIGURE 12.5 Photograph of the posterior surface of the left patella found in a second search of the trunk. Arrow marks the "dorsal defect." (Reprinted from Riddick, L., et al., *J. Forensic Sci.*, 28, 263, 1983. © ASTM. With permission.)

been treated following a motor vehicle accident two years earlier (Figure 12.6). Radiographs of the isolated patella (Figure 12.7), corrected for size, could be absolutely superimposed on the patella (which also contained a classic dorsal defect) as seen on the frontal view of the knee (Figure 12.8). Thus, positive identification could be based on the precisely matching comparison of a single patella containing a lesion seen in 1% or fewer of the general adult population.[2,3]

POSTMORTEM RADIOGRAPHY

In the ideal situation, soon after unidentified remains are brought to the morgue, good police work or other information will come up with a presumed identification. Shortly thereafter, antemortem roentgenograms of the presumed decedent are discovered at a nearby institution and brought to the morgue for evaluation and are found to reveal several unique anatomic features. These antemortem roentgenographs can be duplicated with postmortem studies of appropriate body parts using a radiographic unit permanently housed within or convenient to the autopsy suite. Antemortem views can be replicated by careful positioning of the unknown remains, and positive identification can be quickly substantiated or disproved.

Unfortunately, those ideal conditions are rarely obtained.

Sometimes it is possible to retain the unidentified remains almost indefinitely in the holding vaults or drawers while waiting for antemortem materials to surface. In any case, the postmortem radiographs ideally should precede the autopsy, which,

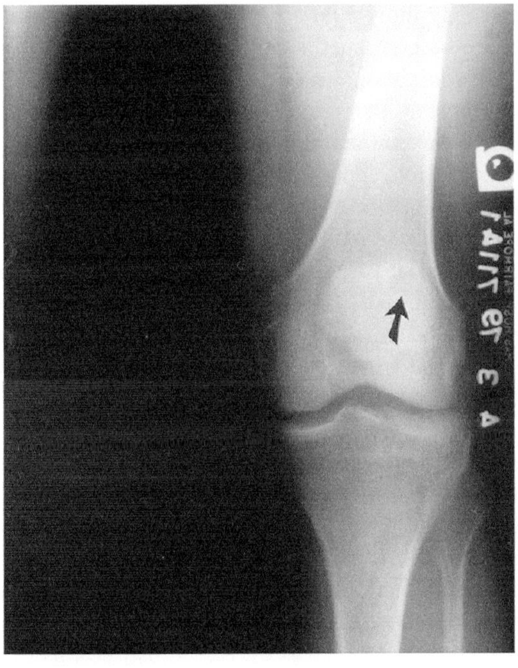

FIGURE 12.6 Case 11-1: frontal view of left knee of the presumed decedent obtained after an automobile accident 2 years prior to his abduction. A dorsal defect of the patella is present (arrow). (Reprinted from Riddick, L., et al., *J. Forensic Sci.*, 28, 263, 1983. © ASTM. With permission.)

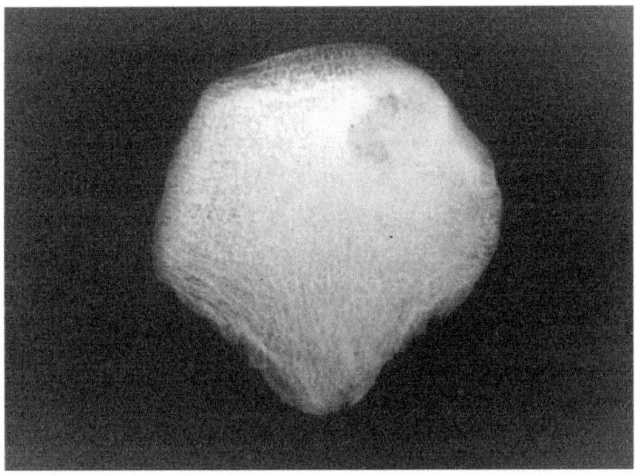

FIGURE 12.7 Case 11-1: roentgenogram of left patella retrieved from trunk. The dorsal defect is obvious in the upper outer, quadrant. (Reprinted from Riddick, L., et al., *J. Forensic Sci.*, 28, 263, 1983. © ASTM. With permission.)

at the very least, will disturb the continuity, relationship, and configuration of the skull and anterior thoracic cage.

More often, and especially in mass casualty situations, there will be pressure to process and release the body so that the opportunity for postmortem radiography is fleeting. Since the availability of antemortem studies for eventual comparison is unpredictable, and since those examinations may be quite limited in scope, whole-body radiography of the remains should be undertaken. Furthermore, it is essential that standard radiographic positions or projections be employed in the postmortem radiography in order that the resultant images can be compared with whatever antemortem radiographs may subsequently be discovered (see Chapter 39 for technical details).

Whole-body radiography takes time and requires a large number of films. In the Air India crash off the coast of Ireland, 12–14 large-sized films were required to examine each adult body. The numbers are fairly constant; in the first mass casualty situation in which radiography was used extensively, 13 films per body were required.[4] Thus there may be a temptation to take shortcuts. Furthermore, most morgues do

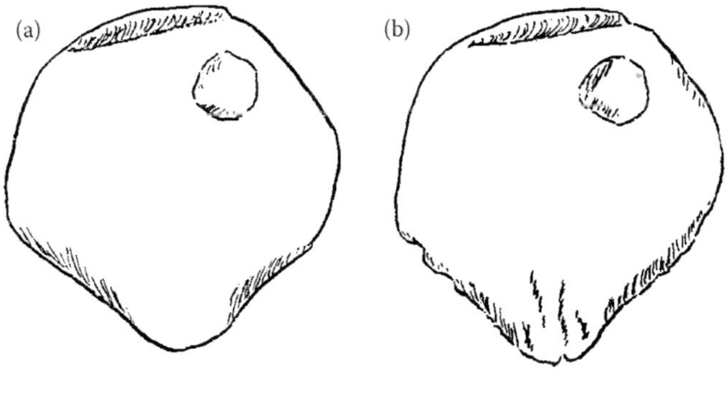

FIGURE 12.8 Case 11-1: Tracings of patellar roentgenograms (corrected to size) from (a) the knee x-ray, and (b) the carbonized and calcined patella from the trunk are exactly superimposable, with identical features and configuration. (Reprinted from Riddick, L., et al., *J. Forensic Sci.*, 28, 263, 1983. © ASTM. With permission.)

not have the convenience of a fixed radiographic installation and radiological capabilities are limited to mobile equipment brought into the autopsy suite or to the often erratic availability of adjacent hospital facilities.

Hospital personnel will not be eager to share mobile equipment that is mostly used for patients in critical care units. There will be even less enthusiasm for bringing bodies or parts, often unsightly and malodorous, into patient care areas and examining rooms, especially during normal duty hours. Such arrangements may demand enormous tact, diplomacy, and negotiating skills on the part of the medical examiner and/or the forensic radiologist. Nevertheless, it is important to obtain the most complete postmortem radiography possible, as soon as possible. An opportunity missed or postponed is unlikely to return.

At the time of the first edition of this book, more than 80% of antemortem radiological examinations obtained for clinical purposes comprised routine x-ray examinations or roentgenograms. The remainder was made up of other techniques and modalities (Table 12.1). Catheter procedures, nuclear medicine studies, and magnetic resonance (MR) examinations collectively added up to about 9% of total examinations and, for the most part, were useless for identification by comparison. Computed tomography (CT) accounted for about 9% of all radiological studies and some of those could be used for comparison with other CTs or even some conventional roentgenograms.

Of the conventional roentgenograms, almost 8% were studies of the breast (mammograms) and are useless for purposes of identification.

The breakdown of conventional roentgenographic statistics showed that more than 40% of all such examinations were chest films. Chest films show remarkable consistency of bony structure appearances over time, but faulty positioning of postmortem films can seriously jeopardize the chance for comparative identification.

In conventional x-ray studies, the extremities account for about one-fifth of the total and are almost equally divided between upper and lower limbs. Because of their propensity for injury, congenital malformation, or degenerative change, the extremities may contain extremely useful roentgenographic features. Unfortunately, the extremities of unidentified remains are often useless for comparative purposes because of separation and scattering, incineration, decay and decomposition, and by the activities of carnivores.

Almost 5% of clinical x-rays are of the head and neck, and the skull contains many features appropriate for comparison identification.

In our experience, the spine and pelvis are most likely to survive as useful postmortem material and are included (one or the other or both) in about 20% of antemortem roentgenographic examinations. Hence, if post mortem films *must* be limited in number, the odds would favor using them for the carefully positioned chest (including the thoracic spine and lower cervical spine) and the abdomen (to include the lumbosacral spine and pelvis). That strategy will backfire, however, in many individual cases and the foregoing statements should not be construed as an endorsement of limited postmortem examinations.

Unfortunately, the frequency with which various body areas and parts are examined radiographically for medical purposes is not directly related to the availability of those areas or parts for radiological comparison for purposes of identification.

In Bass and Driscoll's experience with incomplete skeletons in Tennessee,[5] the skull or skull bones, femora, mandibles, and innominate bones were most commonly retrieved (Table 12.2). This may reflect durability, size, or the likelihood of chance discoverers of remains to recognize certain bones and overlook others.

In a series of 30 identifications by antemortem and postmortem radiological comparisons,[6] Murphy and coworkers found the chest most useful and the pelvis least useful (Table 12.3).

The distribution of radiological procedures and the frequency of sectional imaging have changed remarkably in the past decade. Statistics for the same medical teaching center from which Table 12.1 was derived illustrate this evolution. Sectional Imaging (CT and MRI) has more than doubled to 22% of the total examinations. Routine roentgenography has diminished to 69%.

Budgets for coroners and medical examiners will rarely include the luxury of a position for a trained x-ray technologist

TABLE 12.1
Distribution of Radiologic Exams by Body Part and Modality

		Distribution	
Type Exam	**Body Part**	**%**	**Number of Cases**
Roentgenograms (X-rays)	Chest	43	36
	Lower extremity	11	9
	Upper extremity	10	8
	Spine	8	7
	Breast (mammogram)	8	7
		7	6
		5	4
		4	3
		4	2
		100	82
Other modalities			
Nuclear medicine			4
MRI			4
Catheter procedures			1
CT	Head/neck	44	
	Abdomen	22	
	Pelvis	10	
	Thorax	6	
	Spine	9	
	Other	6	
	Total	100	9

TABLE 12.2
Skeletal Elements Present in 58 Fragmented Skeletons

Bone	Percent
Skull or skull bones	66
Femora	48
Mandibles	41
Innominates	40
Tibias	38
Ulnas	33
Humeri	29
Fibulas	28
Scapulae	28
Clavicles	22
Radius	22
Sacrum	17
Patella	13
Sternum	12

Source: Reprinted from Bass, W. M., and Driscoll, P. A., *J. Forensic Sci.*, 28, 159, 1983. © ASTM. With permission.

or radiographer although this is most desirable. In many facilities an interested assistant or diener can learn on the job to make usable images in frontal and lateral projections (see Chapters 35, 38, and 39). Sometimes a trained professional hospital or clinic radiographer will become sufficiently interested in forensic work to do it voluntarily in slack or free time, or on a part-time basis, or for periodic rewards such as travel expenses to forensic meetings. However obtained, however trained, and however compensated, the person who actually positions and exposes the roentgenograms is absolutely critical to the success of the entire endeavor. The educated, experienced and sophisticated eye of the radiologist or other professional observer may be required to detect and interpret the subtle nuances recorded on the film,[7] but without adequate technical support that eye will be blinded. Technical support becomes more critical as medical Examiner systems gradually acquire access to sectional imaging modalities.

TABLE 12.3
Positive Radiologic Identification by Anatomic Region

Region	Number[a]	Percent
Chest	16 of 30	53
Skull	6 of 30	20
Extremities	6 of 30	20
Lumbar spine	5 of 30	17
Cervical spine	3 of 30	10
Pelvis	1 of 30	3

Source: Reprinted from Murphy, W. A., Spruill, F. G., and Gantner, G. E., *J. Forensic Sci.*, 25, 727, 1980. © ASTM. With permission.

[a] Some bodies could be identified by comparison of more than one region.

FILM PROBLEMS OR PROBLEM FILMS

We previously have alluded to the problem that both the antemortem and postmortem radiographs available for comparison may be of sub optimal quality, and that there often is no opportunity to repeat them or obtain better examples. Archival films that are simply dirty often can be cleaned with Kodak film cleaner or by running them through the terminal wash and dry sections of an automatic film processor.

Special attention is called to the paper by Fitzpatrick et al. who were able to overcome the problems of seemingly inadequate comparison films by utilization of selected adjunctive techniques.[8] These include optical (slide projection) and photographic enlargement, photographic contrast enhancement, digitization, and digitization with computer enhancement (Figure 12.9).

Sometimes, injudicious marking of processed radiographs with pens or markers will obscure significant details or, more often, seriously impair the radiographs for reproduction or use as an exhibit or illustration. Sometimes the offending marks can be removed by gentle wiping with a lintless cloth or soft paper after spraying the area with hairspray. (We have found Aqua Net one of the better and cheaper products for this purpose.)

IDENTIFICATION BY COMPARISON OF SOFT TISSUES

Soft tissues (nonskeletal tissues not ordinarily radiopaque) may play a role in comparative identification with radiological techniques. Murphy[6] found calcifications in the chest helpful or specific in several cases (Table 12.4). Vascular calcifications have been matched on occasion. Calcified scars or post traumatic calcifications/ossifications (e.g., post-traumatic myositis ossifications) can be distinctive. Certain physiological calcifications may serve for identification; Messmer found a calcified falx cerebri useful in one case (personal communication).

There is increasing use of opaque clips, sutures, stents, filters, and connectors in surgical procedures throughout the body, and these may be useful for future identification.

Inclusions of foreign material in soft tissues (bullets, shrapnel, glass, gravel, etc.) may have unique appearances and locations.

Enteric accretions (gallstones,[9] kidney stones, bladder stones, phleboliths, parasitic encrustations, etc.) can be used for identification.

IDENTIFICATION BY COMPARISON OF SKELETAL TISSUES

Except for teeth, bones are the most durable of body tissues and are the basis for the overwhelming majority of nondental radiological identification. Fortunately for us, bone is also a rather dependable tissue. It is more consistent than most other organs and tissues in its response to all of the insults that befall it—growth and development, disease, trauma, nutritional and

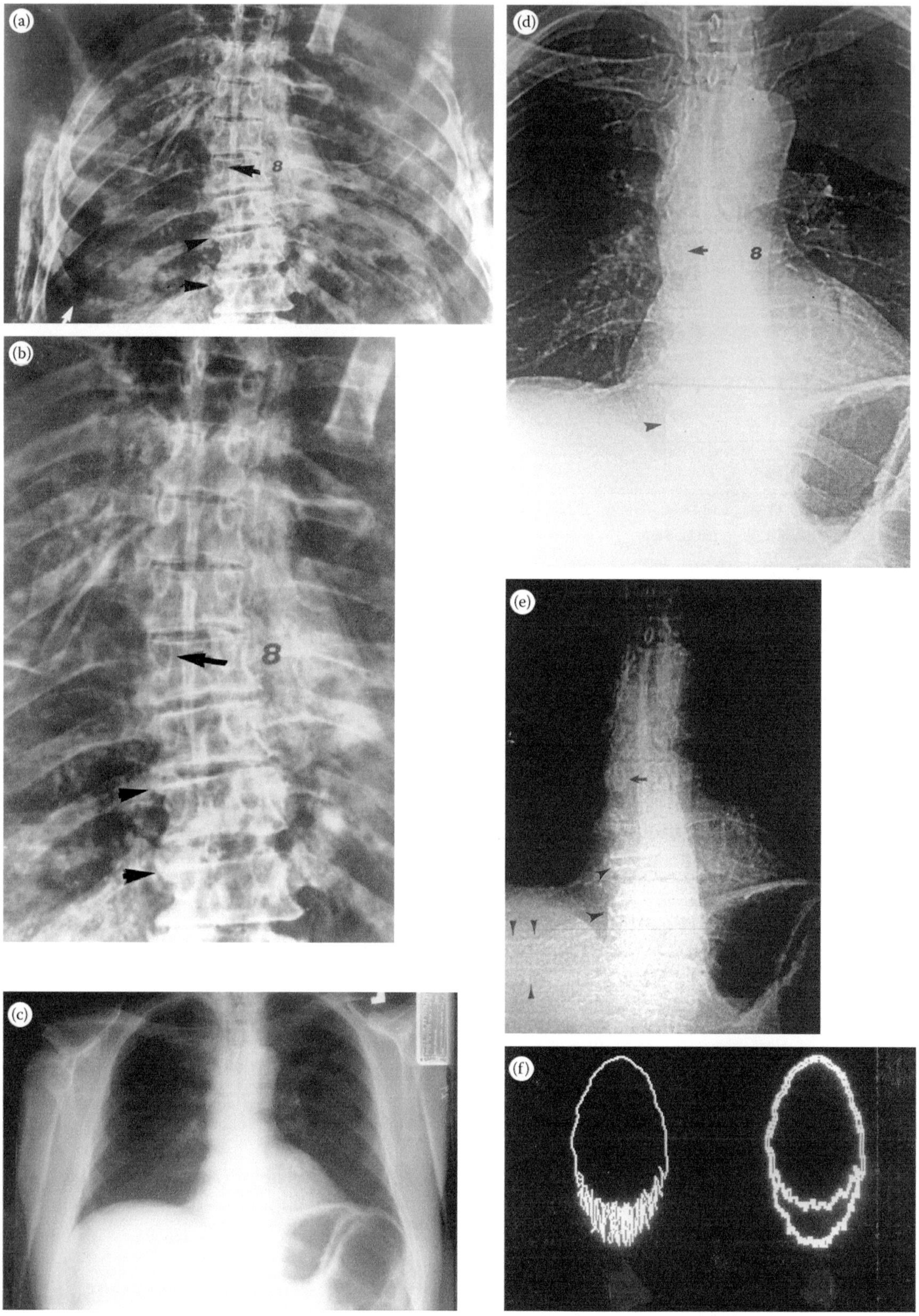

FIGURE 12.9 (a) A postmortem chest roentgenogram shows several features for a potential match including an unusual pedicle on T-8 vertebra (arrow), spurring of vertebral body margins (arrowheads), and costal cartilage ossification (white arrow). (b) Close-up shows some of the above features with better definition. The body was buried. Two years later, (c), an antemortem chest radiograph was found. The bony features could not be seen on this radiograph. (d and e) Show digitized and computer-enhanced versions of the antemortem chest radiograph, which bring out the matching features. (f) diagrammatic analysis of enhanced T-8 pedicle shows the crescentic density of the inferior margin on the left-hand image; on the right the crescent is highlighted by edge enhancement. (Reprinted from Fitzpatrick, J. J., et al., *J. Forensic Sci.*, 41, 947, 1996. © ASTM. With permission.)

metabolic conditions, aging (degeneration), and thermal injury. Consequently, the skeleton also is a good historian.

> "... bones make good witnesses—although they speak softly, they never lie and they never forget. Each bone has its own tale to tell about the past life and death of the person whose living flesh once clothed it. Like people, some bones impart their secrets more readily than others; some are laconic; others are positively garrulous."
>
> **—Clyde Collins Snow and John Fitzpatrick**[10]

Upon cessation of skeletal growth, the general configuration of a bone, the shape and direction of its various processes and protuberances, and the pattern of its major

TABLE 12.4
Radiological Observations Useful for Identification

11	Chest calcifications
9	Normal anatomic variations
7	Evidence of surgery
5	Fractures
2	Congenital anomaly
2	Abdominal calcifications
2	Arthritis

Source: Reprinted from Murphy, W. A., Spruill, F. G., and Gantner, G. E., J. Forensic Sci., 25, 727, 1980. © ASTM. With permission.

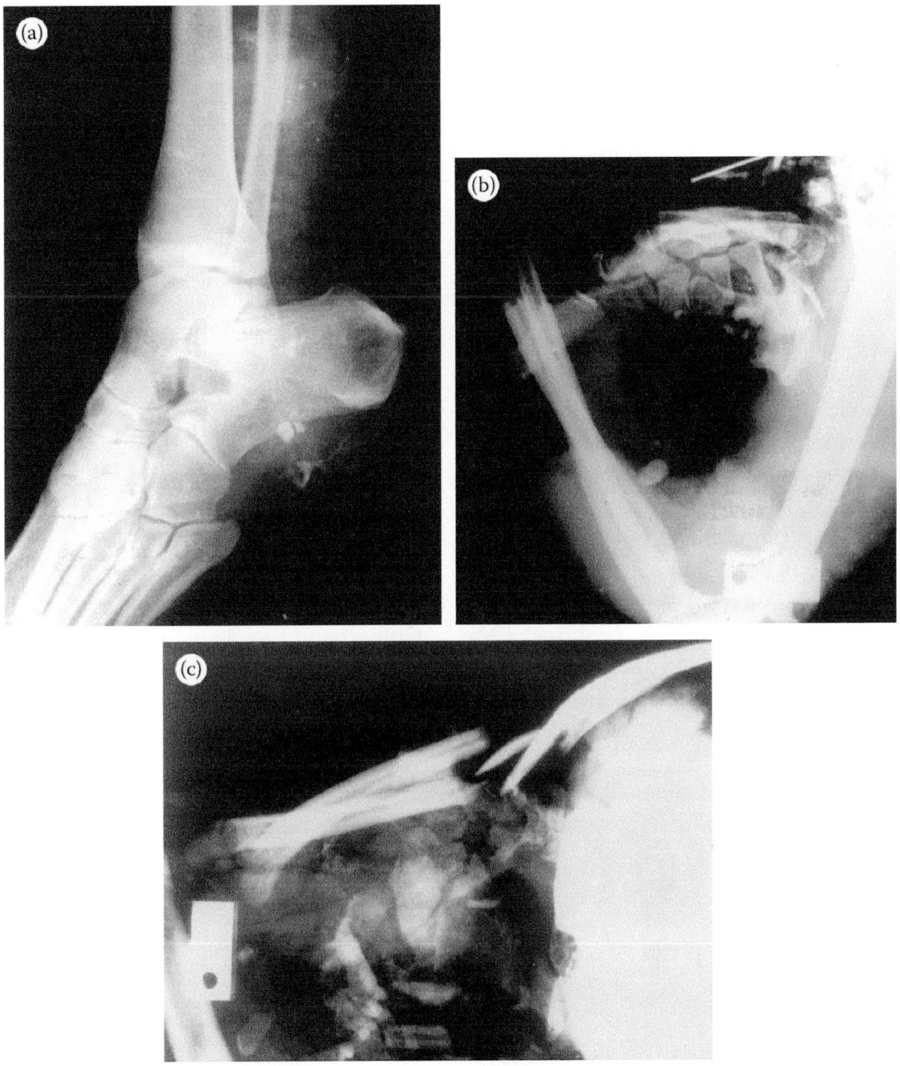

FIGURE 12.10 Bodies that have been burned severely show a degree of tissue destruction that is a function of temperature and time. Distortions of body position by shrinkage of flexor muscle groups produce the pugilistic attitude, which may complicate radiological evaluation. The bones and teeth are most likely to survive thermal destruction. The bones burned through the flesh tend to shrink as temperatures increase to 1100°C. Bones at high temperatures may develop multiple perpendicular fractures to the long axis of the bone. The long bones may show warping or bowing as they cool. Defleshed dry bones tend to develop longitudinal fractures or striae. Although it may be difficult, one must try to differentiate thermal fractures from impact fractures in crash and burn situations or where the fire may be intended to hid other evidence. (a) Mild flexion deformity as a result of flexor group shrinkage during the fire. (b) Transverse fracturing and marked pugilistic position of the upper extremity after a high intensity fire in an aircraft accident. (c) Transverse fracturing and bowing of upper extremity bones from a high temperature fire.

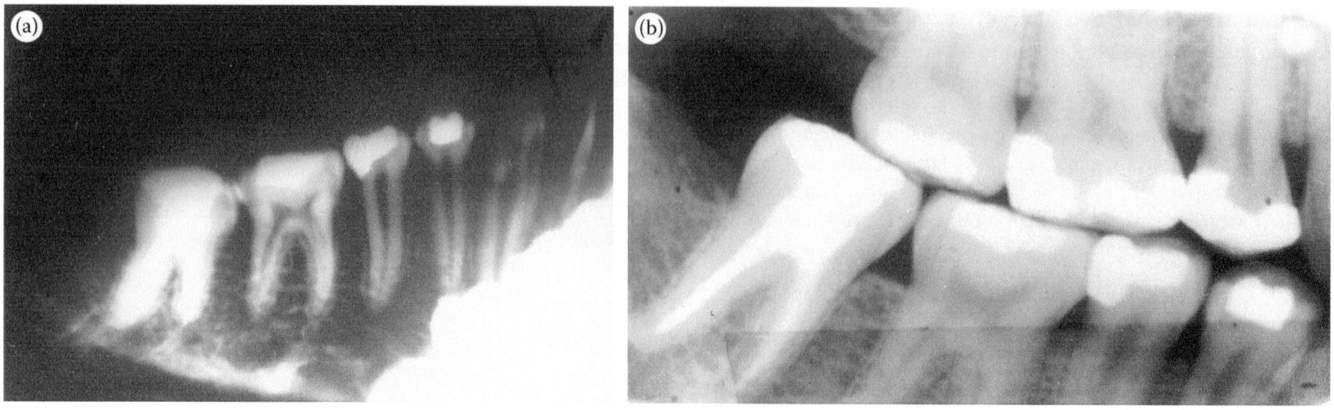

FIGURE 12.11 (a) Postmortem radiograph of mandibular fragment compared with (b) antemortem bitewing radiograph. The root canal work and restorations are identical.

trabeculae and vascular structures remain relatively unchanged for a "normal" life. Trauma, destructive disease, and surgery may modify the bone substantially. The slow and insidious changes of aging and wear and tear will gradually alter the configuration of a given bone, but not to the extent that the passage of several years will produce confounding change.[11]

Bones that are burned in the flesh at temperatures of 800°C or less (ordinary house fires) will show shrinkage of about 1% or less and may sustain cracks perpendicular to the long axis of the bone. Burned dry bones may exhibit longitudinal striae. Warping is more common in fat- and muscle-encased incinerated bones (Figure 12.10). Shrinkage increases between 800°C and 1100°C (cremation temperatures) then levels off. Even at those temperatures, the bone, which will be grayish white in color and extremely fragile, will preserve configuration and internal structures sufficient to be used for comparison identification (Figures 12.5 through 12.8). The amount of shrinkage is negligible when compared to magnification factors affecting the antemortem bone encased in soft tissues[12,13].

REGIONAL CONSIDERATIONS

Skull

Dental Arches

When dealing with a body not clearly recognizable by surface inspection, the contents of the maxillary and mandibular arches, the teeth, are the most productive anatomic areas for individual identification, whether by direct comparison with dental records or by radiological comparison. This is largely the province of the forensic odontologist and has been extensively covered in Chapter 10. Occasionally, radiography of the unidentified head or skull will reveal dental findings of such characteristic individuality that the expertise of the dentist is not required (Figures 12.11 and 12.12).

Paranasal Sinuses

Culbert and Law[14] are credited with the first identification of human remains by radiological comparison. Culbert had operated for mastoiditis upon a man who, years later, disappeared in the Indus River in India. A skeleton was retrieved

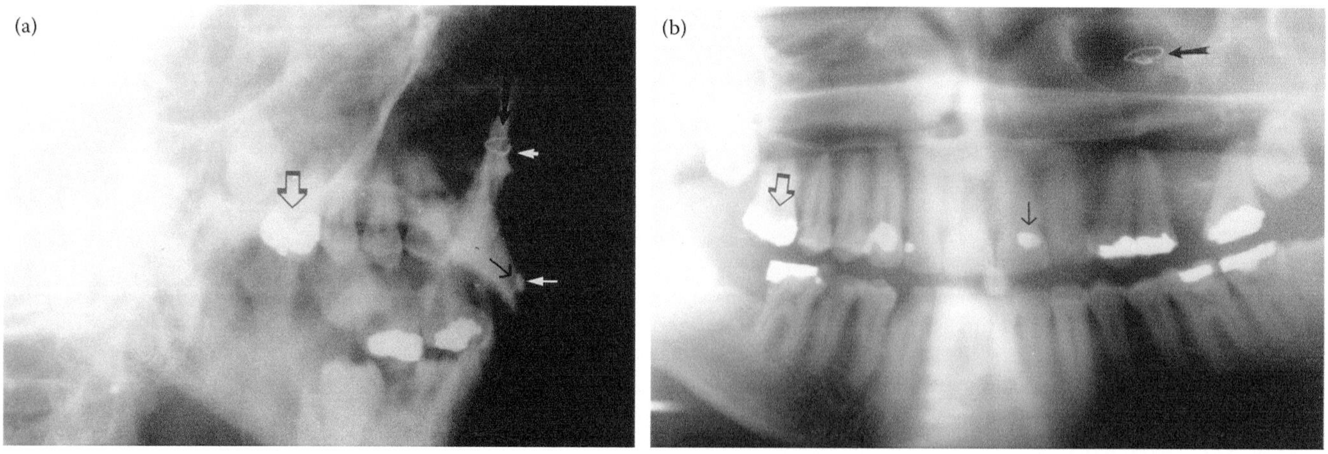

FIGURE 12.12 (a) Postmortem facial roentgenogram shows unique restorations and a wire suture in the orbital floor. (b) antemortem panoramic dental examination shows identical findings.

from the river two years after the disappearance, and Drs. Culbert and Law were able to confirm the identity by comparison of antemortem and postmortem roentgenograms of the man's frontal sinuses and his postoperative mastoid processes. Since their publication in 1927, the value of frontal sinus patterns, especially, has become widely known.

Pneumatization of the maxillary and ethmoid sinuses progresses from birth through the end of the second decade of life. The frontal sinuses develop as extensions of the superior ethmoid group and their appearance may vary from six months to, more commonly, two years of age. Frontal sinus development may be unilateral and is totally absent in about 5% of the population. The sphenoid sinus develops last by extension from the posterior ethmoid cells, and there is considerable variation in time of appearance and ultimate size and configuration in the sphenoid bone.

The frontal sinuses, especially, develop unique scalloped margins with internal septae and pseudoseptae. The frontal sinuses are as unique to the individual as his fingerprints.[15] Even identical twins will have different frontal sinus patterns.[16] The other paranasal sinuses also have individual variations but they are less striking and more difficult to compare.

Comparison of frontal sinus configuration is easy on frontal view radiographs of the skull, even with considerable variations in angulation or projection between two radiographs (Figure 12.13). Elaborate systems of mensuration and classification of the sinuses have been proposed,[17,18] particularly by the anthropologists, but such elaboration are not really required. Simple "eyeball" comparisons will suffice in virtually every case, as reconfirmed by a recent prospective study.[19]

Once paranasal sinus growth has stopped, the appearance will remain relatively stable throughout life. The generalized osteoporosis or absorption of bone that occurs with aging may enlarge the sinuses slightly but will not substantially change the overall pattern. Trauma, tumor, destructive disease, and acromegaly can change the size and configuration of the sinuses substantially. Fortunately, except for trauma, those instances are quite uncommon.

Unfortunately, the frontal sinuses are susceptible to severe fracturing and distortion in automobile and aircraft crashes, and in conditions of extreme heat.

Even fragments of frontal sinuses may contain such distinctive patterns that comparative identification is possible[20] (Figure 12.14).

Mastoids

The air cell development in the mastoid processes, the squamous portion of the temporal bone, and the petrous ridges may show characteristic individuality that can lead to successful radiological identification.[21,22] These areas may be seen on lateral views of the skull or cervical spine and on the Townes view of the skull (Figure 12.15). The mastoids usually are not shown in as good detail as the frontal sinuses and evaluation for identification consequently is more difficult and less often successful.

Sella Turcica

Voluter[23] suggested comparison of size and configuration of the sella tursica for purposes of identification, pointing out

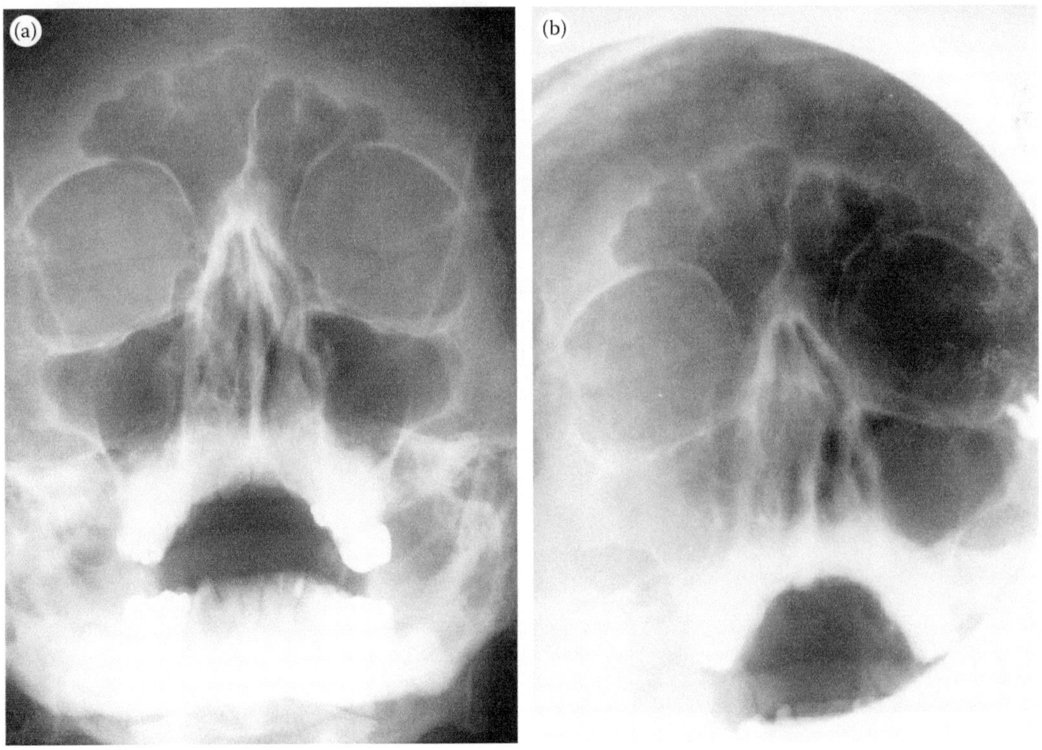

FIGURE 12.13 (a and b) Water's view of the paranasal sinuses taken 15 years apart and with slight variations in positioning and tube angulation. Still, there is no doubt that the septation and lobulation of the frontal sinuses are identical.

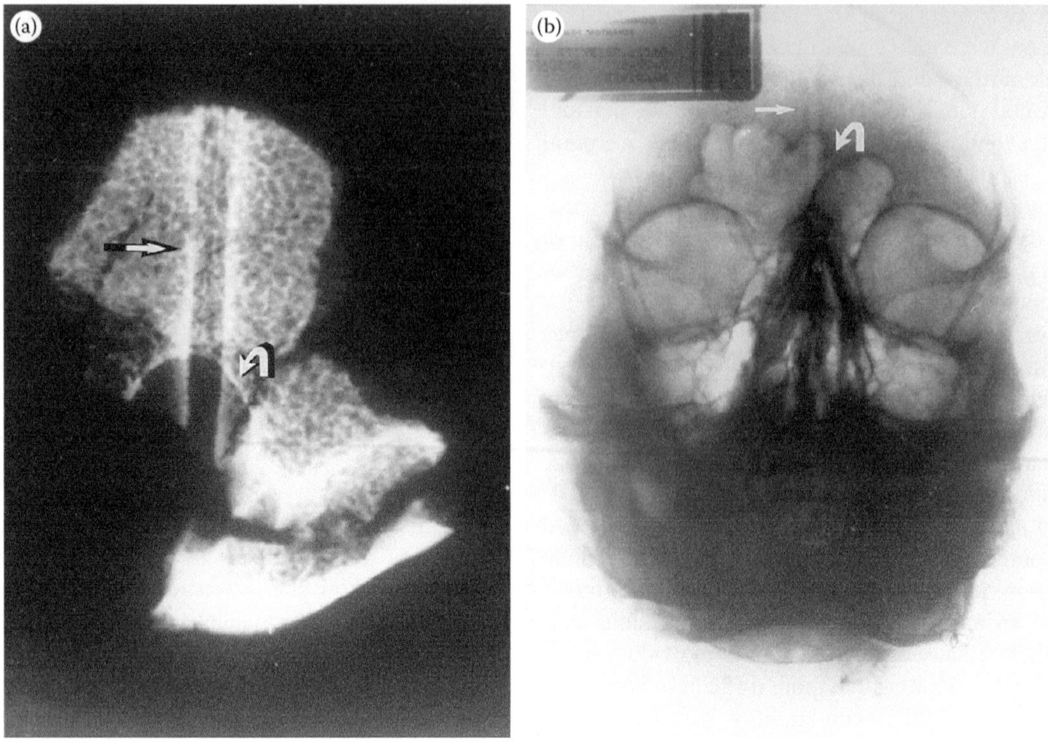

FIGURE 12.14 (a) Fragments of a frontal bone found at a cremation site in Guatemala. (b) Antemortem Water's view of the skull belonging to a journalist who disappeared in Guatemala four years earlier. A deep sulcus in the frontal crest and the lobulation of the frontal sinuses establish a match. (Reprinted from Owsley, D. W., *J. Forensic Sci.*, 38, 1372, 1993. © ASTM. With permission.)

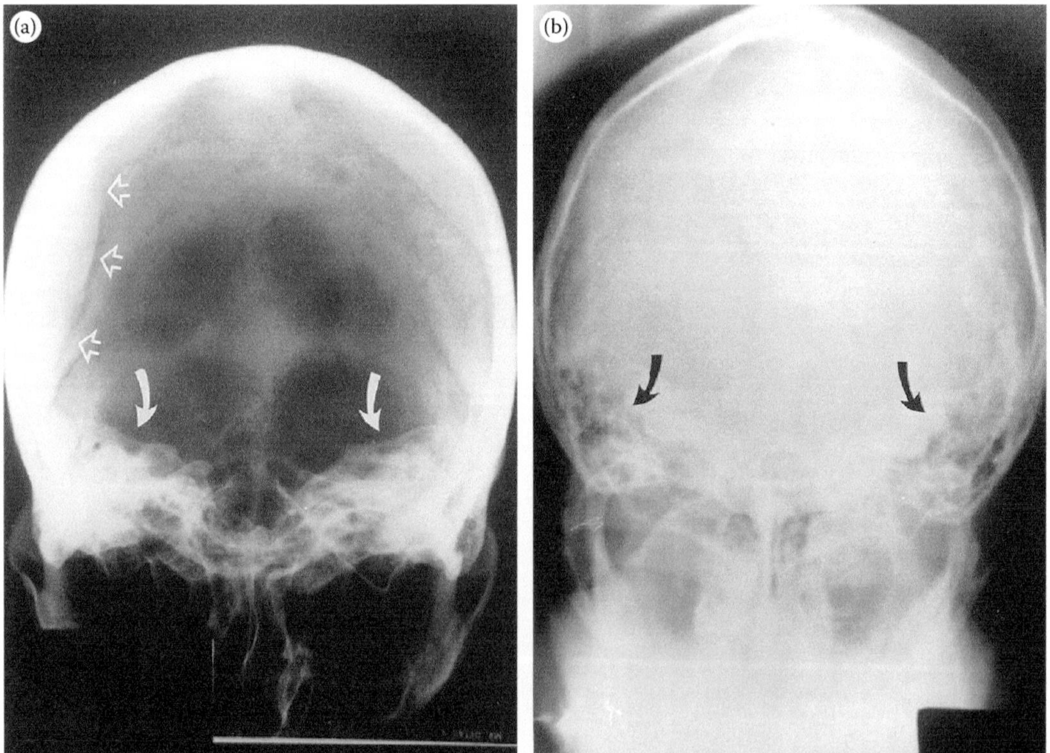

FIGURE 12.15 (a) Postmortem and (b) antemortem Townes view of the skull were submitted for possible identification. This clearly is a nonmatch. Note the difference in mastoid air cell development and distribution (arrows). Also, the postmortem skull has an unusual feature, an old calcified subdural hematoma (open arrows) on the right.

that its protected central position in the skull base often preserved it even in the face of extreme trauma, incineration, or decomposition. Actually, Singleton[24] had anticipated Voluter by using radiological comparison of the sella turcica in victims of the *Noronic* disaster.

These major features of the skull, the *frontal sinuses*, *mastoids*, and *sella tursica* are equally useful in disproving a presumed identification, and this is often possible without the fine radiographic detail required for a positive identification (Figure 12.16).

Other Identifying Features in the Skull

The general configuration of the skull, protuberances such as the brow ridge and the inion, vascular grooves,[25] surgical lesions,[26] cranial sutures,[27] trauma, hyperostosis interna frontalis,[6] and disease states may prove useful in any given case of identification through comparison of skull images (Figures 12.17 through 12.20). The mandible may contribute features. Also, the upper segments of the cervical spine usually are included in lateral views of the skull. Note the nonmatch of the upper cervical segment in Figure 8.16a and b.

Chest

Since the chest is the area of the body most often examined radiologically, it frequently affords positive comparative radiological identification.

The pattern of costal cartilage ossification may be unique. Murphy found it most useful.[6] Martel and coworkers[28] in a prospective study, believed they found unmistakable costal cartilage markers. However, they divided their test cases by sex and may have fallen into the same trap as Vastine et al.,[29] who believed identical twins had identical costal cartilage ossification patterns. The similarities of pattern may have been sexually determined rather than individually unique. Ossification of costal cartilage is a dynamic progressive process and attempts to match patterns over time lapses are fraught with difficulty.

Calcification within the lungs or pleura may be shape, size, and/or location specific but tend to be displaced on postmortem radiographs as the lung collapses.

Anomalies, diseases, tumors, and traumatic lesions of the ribs may contribute to identifying matches.

Sternal configuration on lateral views has contributed to identification of unknown remains in at least two cases.[30,31] A cross-table lateral view of the body is required to compare with an antemortem lateral view of the chest (see Chapter 39).

Scapular configuration in postmortem and antemortem radiographic studies allowed Ubelaker[32] to make a positive identification (Figure 12.21).

The vertebral segments of the thoracic spine and the costovertebral joints usually are obscured on chest radiographs by the heart and mediastinum, although special techniques may bring them into view.[8] However, the cervicothoracic junction of the vertebrae and upper ribs is seen fairly well on most chest radiographs and may serve for comparison identification.[33,34] The positioning of patients for the ubiquitous posteroanterior view of the chest is highly standardized nationwide, and anatomic structures at the thoracic inlet and in the pectoral girdles are quite reproducible (Figures 12.22 through 12.24). Sanders et al.[35] used a variant configuration of the medial end of a single clavicle to effect radiological identification.

However, there is a serious recurring problem in comparing features in postmortem and antemortem chest radiographs. The most common antemortem chest radiograph, by far, is obtained in the posteroanterior position with the patient erect or standing with the shoulders thrust forward against the film holder and the arms akimbo (Figure 12.25a). The usual postmortem chest radiograph is obtained with the body supine, the shoulders back, and the arms alongside the

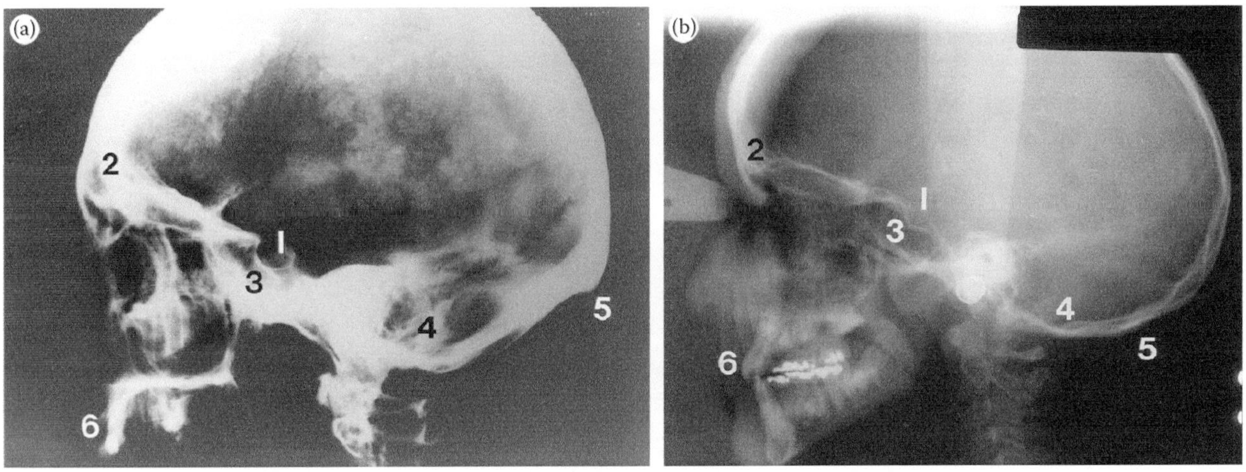

FIGURE 12.16 A 16-year-old girl disappeared from her home in Mississippi. She had had an orthometric skull examination a year earlier. Two years after her disappearance, skull roentgenograms from an unidentified young female body were sent for comparison. (a) Postmortem study. (b) Antemortem x-ray. There is no match. Note differences in the sella turcica (1), frontal (2) and sphenoid (3) sinuses, and general configuration of the frontal slope, depth of posterior fossa (4), inion (5), and angle of maxillary incisors (6).

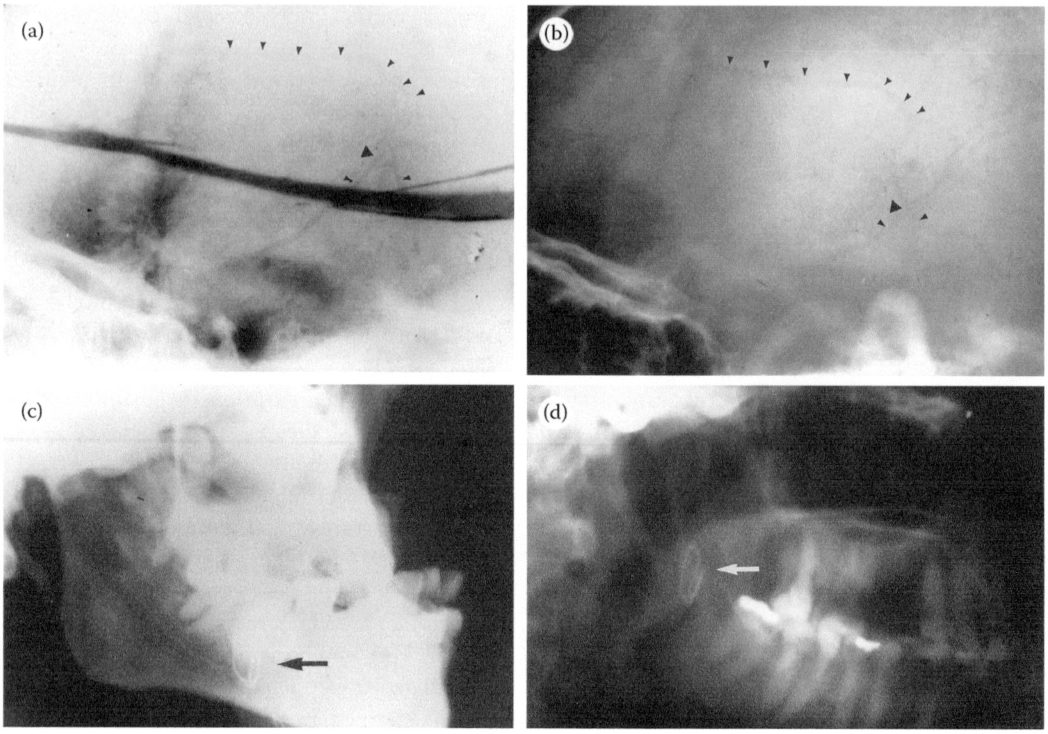

FIGURE 12.17 This body lay in the morgue for two years before antemortem films from a likely "missing person" became available. (a) Close-up lateral view of the postmortem, post-autopsy skull could be matched with (b) the antemortem study, by comparing vascular grooves. This person also had an unusual "pig-tail" wire suture in his jaw as an additional matching feature. (c) Postmortem, (d) antemortem.

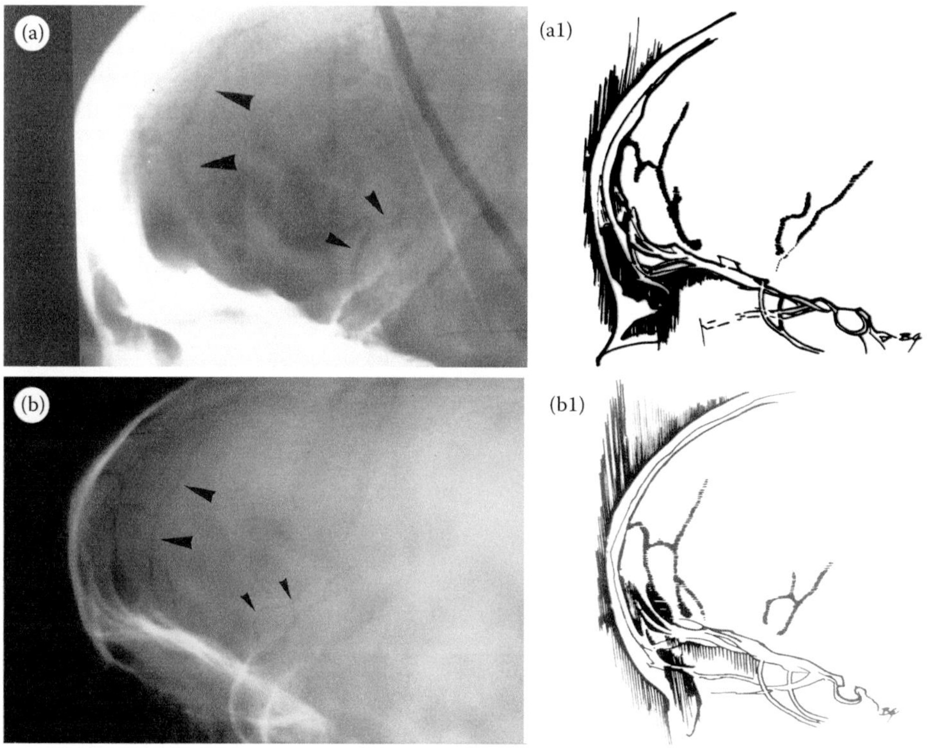

FIGURE 12.18 Messmer and Fierro's case[24] shows that vascular grooves, once established in the skull, undergo little change with growth and development. (a) Skull radiograph of decomposed remains of a 13-year-old female found 3 months after her disappearance. (b) Skull radiographs from an antemortem examination at age 6. The pattern of vascular grooves (arrowheads) is identical. (a1 and b1) tracings to assist the viewer. (From Messmer, J. M. and Fierro, A. F., *Am. J. Forensic Med. Pathol.*, 7, 159, 1986. With permission.)

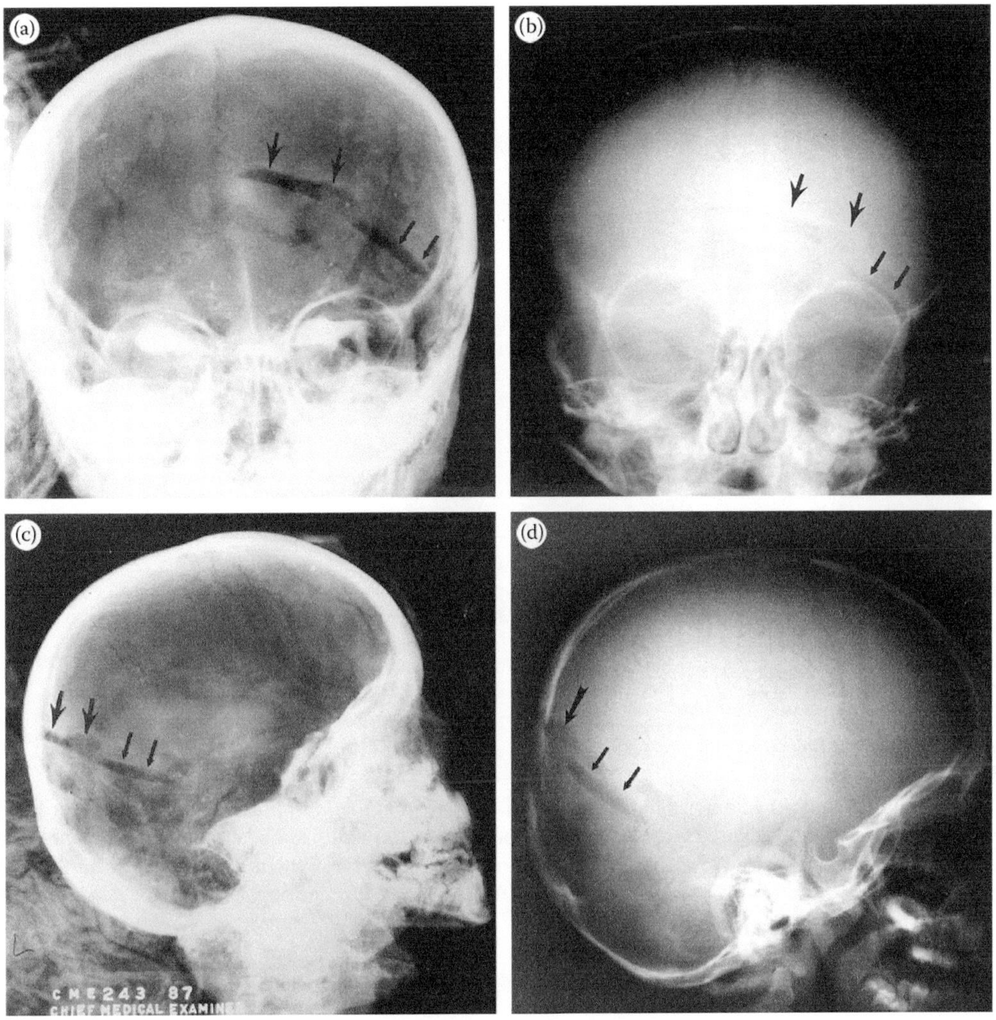

FIGURE 12.19 Identification of (a, b), decomposed postmortem remains by comparison with (b, d), antemortem skull examination showing surgical defects from a lambdoid synostectomy. (Reprinted From Hogge, J. P., Messmer, J. M., and Fierro, M. F., *J. Forensic Sci.*, 40, 688, 1995. © ASTM. With permission.)

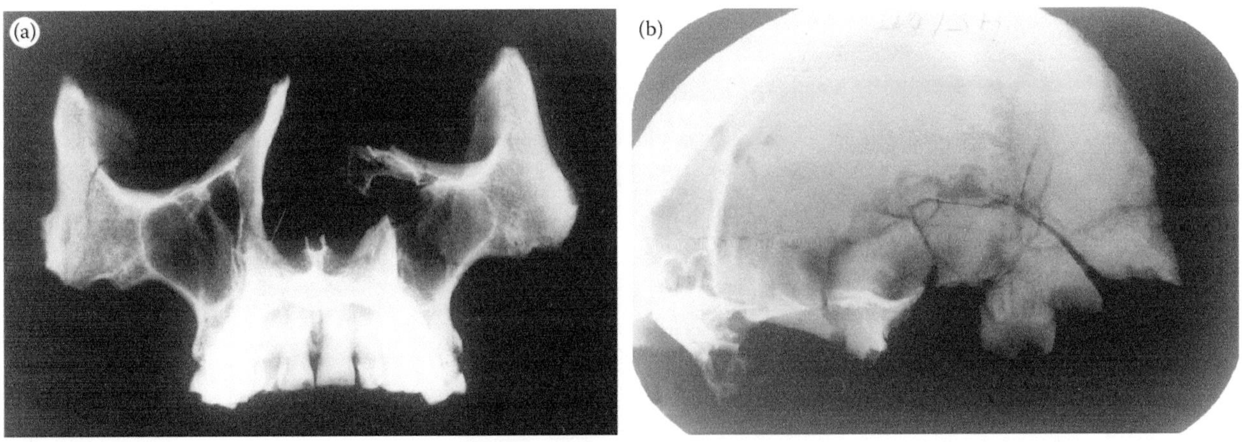

FIGURE 12.20 Professor Walter Bessler of Winterthur recently identified the exhumed remains of the Swiss hero, Jürg Jenatsch, from skull fragments shown (a) in frontal view, and (b) in slightly oblique lateral view with the frontal bone to the left. Jenatsch had freed the canton of Graubünder from foreign occupation. He was killed on January 24, 1639, by the stroke of an axe (perhaps wielded by a former lover). The left side of the cranial vault and left orbit are destroyed and fissures are visible in both maxillary regions and the right zygoma. This led Professor Bessler to assume that Janatsch either received a blow to the left temporal region, then fell on his face, or was struck twice, the second blow to the left orbital region. (Original images courtesy Professor Dr. Walter Bessler, Winterthur, Switzerland, 1997.)

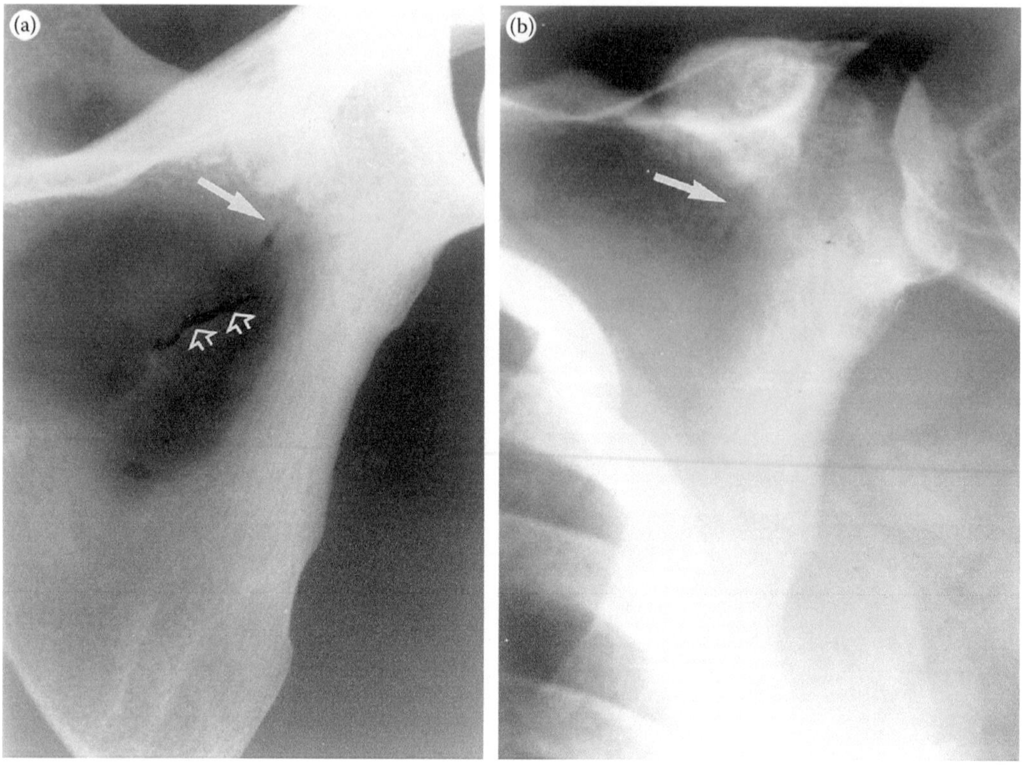

FIGURE 12.21 (a) Postmortem and (b) antemortem radiographs of the scapula. The morphology of the lateral border with its undulations is identical and was not matched when compared with 100 other scapulae from the Smithsonian collection. Not mentioned, but also contributing to the match, is a vascular channel (arrows) present in both images. There is a postmortem fracture in A as well (open arrows). (Reprinted from Ubelaker, D. H., *J. Forensic Sci.*, 35, 466, 1990. © ASTM. With permission.)

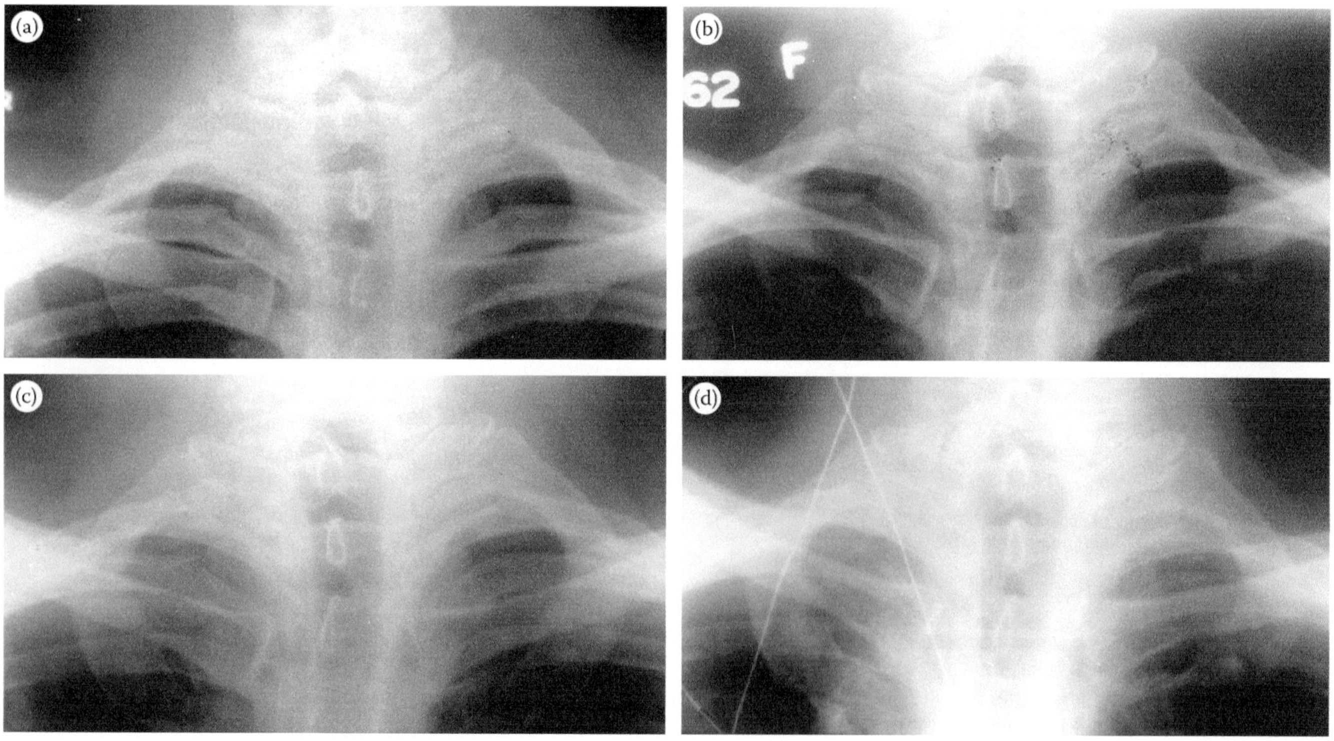

FIGURE 12.22 (a–d) Detail of the cervicothoracic junction of posteroanterior chest radiographs of the same person over a 50-year interval from 1944 to 1994. Note the identical configuration of the bony landmarks. (The first costochondral cartilage became ossified during this period of observation.)

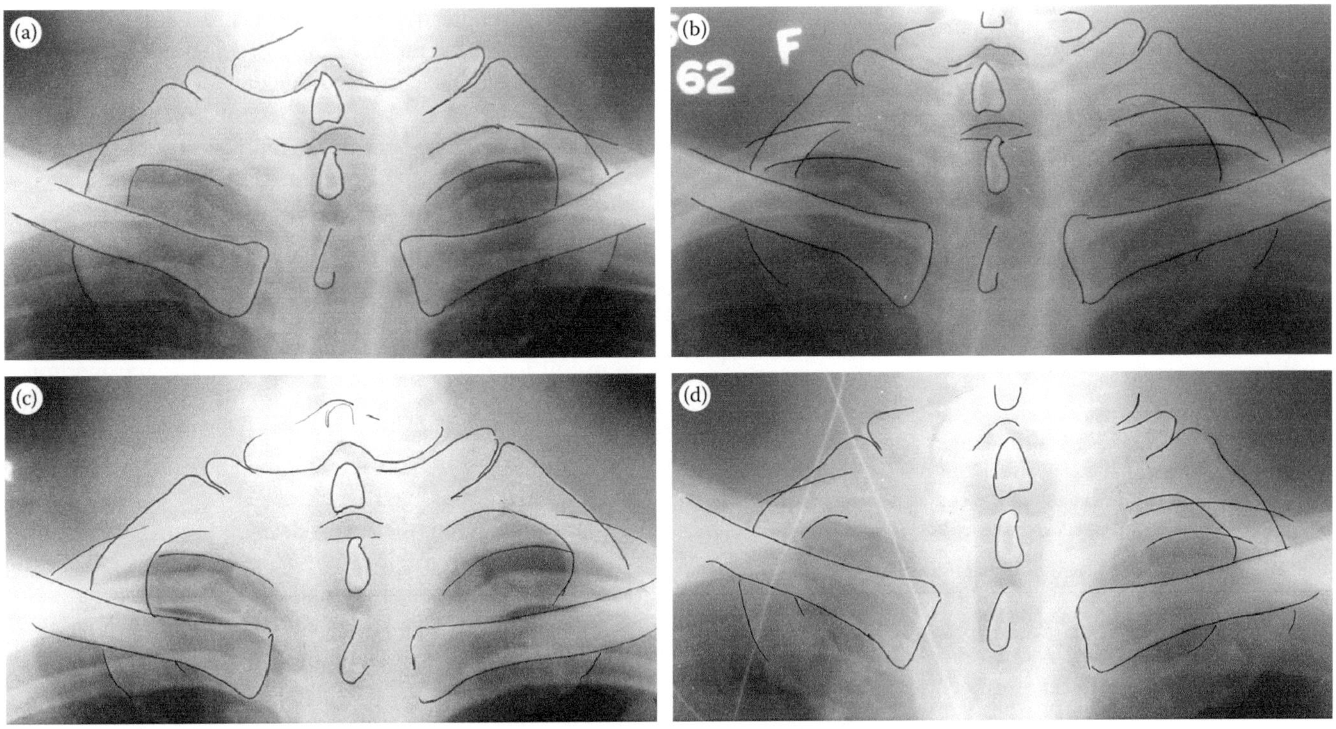

FIGURE 12.23 (a–d) Identical images as in Figure 12.20. Outlines of bony landmarks are traced in ink to assist inexperienced viewers in seeing similarities.

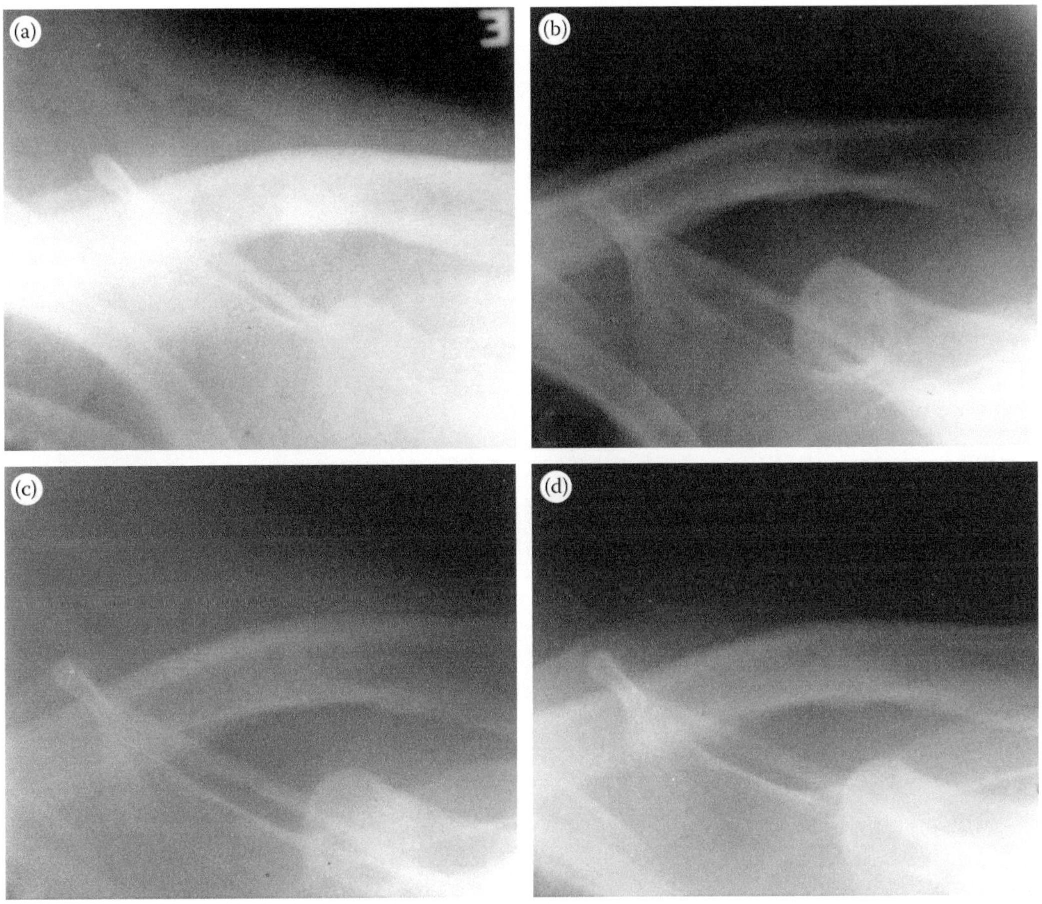

FIGURE 12.24 (a–d) Detail of the superior portion of the scapula and adjacent clavicle in same individual and same time interval as Figure 12.21. The matching features are obvious.

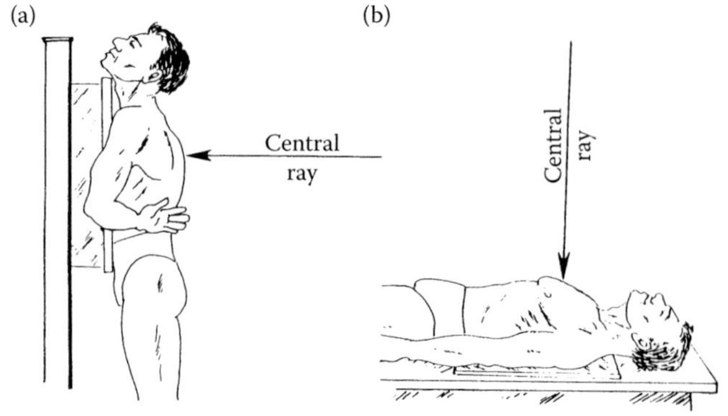

FIGURE 12.25 (a) Positioning for a posteroanterior erect view of the chest. (b) Positioning for an anteroposterior supine view of the chest. It is obvious that the position and projection of bony structures of the pectoral girdle and at the cervicothoracic junction will be quite different from one view to the other.

body (Figure 12.25b). The result is a quite different, often incomparable, distortion and projection of the cervicothoracic bony structures and the bony components of the pectoral girdle. If the antemortem examination was performed with the patient supine, the problem is obviated, of course. When antemortem radiographs are not available before disposal or disposition of the body, it is wise to get postmortem radiographs comparable to both erect posteroantero and supine anteroposterior views.

Finally, the proliferation of open heart surgery presents another feature for radiological identification. Wire sutures used to close median sternotomies and vascular clips offer opportunity for antemortem and postmortem matches, and in this case the antemortem radiographs will most often have been made with the patient supine (Figure 12.26).

Abdomen and Pelvis

The abdomen and pelvis are the components of the human body most likely to survive catastrophe sufficiently intact to be useful for radiological identification. The retroperitoneal tissues, the lumbosacral spine, protected by surrounding heavy musculature and ligaments often remains articulated and easy to position and radiograph for comparison with antemortem studies.[6,36,37]

Since each lumbar vertebra develops from three primary and five secondary ossification centers, there is great variation in the eventual size and configuration of the individual components of each vertebra and between the different vertebra making up the spinal column (Figure 12.27). The spine also is subject to disease, trauma, tumor, and degeneration. A study of the lumbar spine in 936 healthy asymptomatic young men who were candidates for the Air Force Academy or Air Cadet training showed at least one roentgenologically identifiable abnormality in 60% of them[38] (Table 12.5). Consequently, identification by comparison of individual vertebral characteristics often is relatively simple.

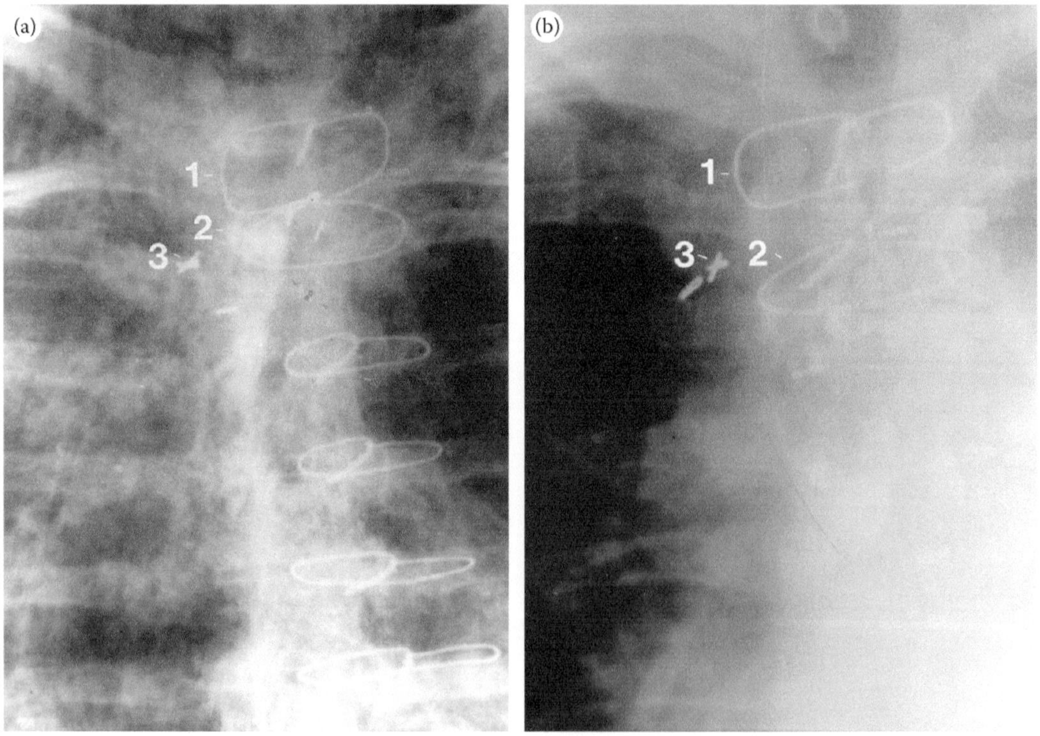

FIGURE 12.26 (a) Postmortem and (b) antemortem supine chest radiographs showing matching patterns of wire sutures in the sternum. (There is enough variation in angulation to shift the deeper clips so they do not match precisely. Others may move out of position with death and collapse of the thoracic viscera.)

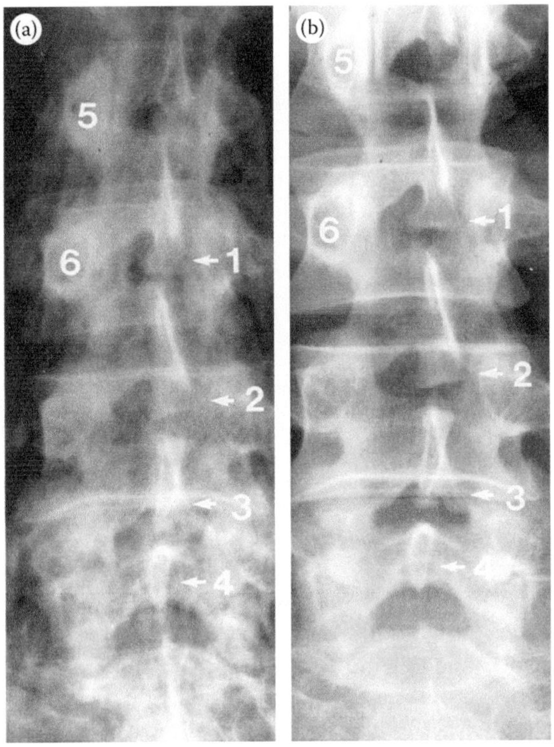

FIGURE 12.27 (a) Postmortem and (b) antemortem frontal views of the lumbar spine. Highly individualistic configurations of the spinous processes enable an easy positive match. Other spinal elements can also be matched (see corresponding numbers).

Example Case 11-2. A severely decomposed body (Figure 12.28a) was found in southern New Mexico. A chest radiograph (Figure 12.28b) revealed the probable cause of death. Law enforcement officers suggested a possible victim who had undergone an earlier gastrointestinal (GI) series. Enough

TABLE 12.5
The "Normal" Lumbosacral Spine (936 Cases)

Entity	Number	Percent
Spina bifida occulta	334 total	35.7
Multiple	40	4.3
Previous Scheuermann's disease	194 total	20.7
With Schmorl's nodes	96	10.3
Scoliosis greater than 1 in.	5 total	0.5
Rudimentary ribs or ununited transverse process	60 total	6.4
Transitional vertebrae	108 total	11.5
Sacralization	31	3.3
Bilateral	20	2.1
Unilateral	11	1.2
Lumbarization	77	8.2
Bilateral	56	6.0
Unilateral	21	2.2
Spondylolysis	71 total	7.6
Bilateral	57	6.1
Unilateral	14	1.5
Spondylolisthesis	42 total	4.5
Limbus vertebrae	9	1.0
Hemangioma	4	0.4
None of above variations	376	40.2

Source: From Crow, N. E. and Brogdon, B. G., *Radiology*, 72, 97, 1959. With permission.

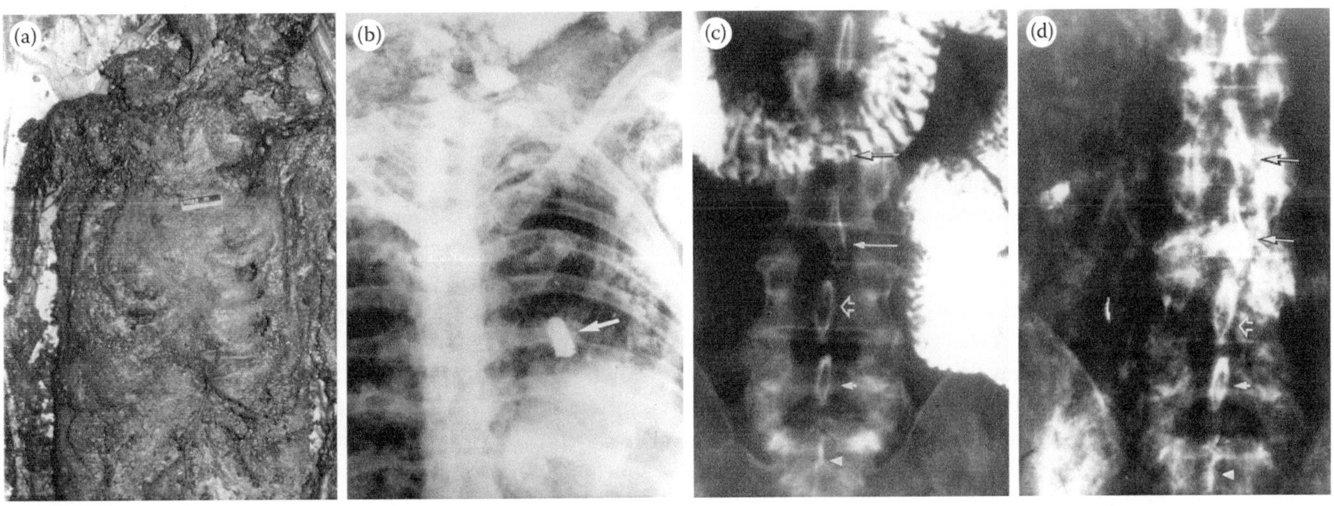

FIGURE 12.28 (a) Decomposed body. (b) Large caliber bullet in chest (arrow). (c) Abdominal roentgenogram from antemortem GI series reveals vertebral elements that can be matched with (d) the postmortem film. Note the two spinous processes slanted to viewer's right (arrows), the spinous process shaped like an exclamation point (open arrow), the elliptical process below (short arrow), and the vertebral bar on the first sacral segment (arrowhead).

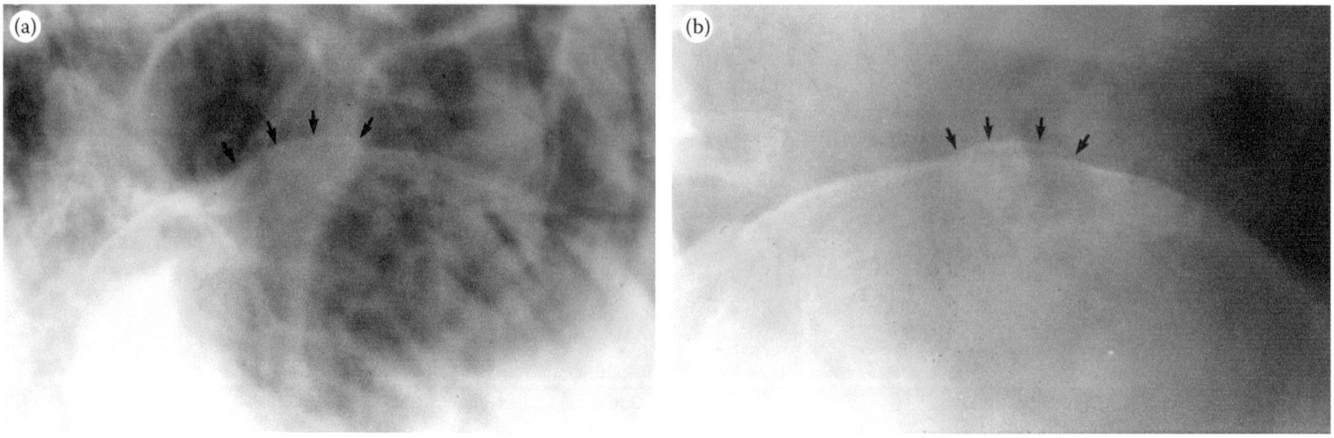

FIGURE 12.29 (a) Postmortem and (b) antemortem pelvic radiographs matched by an unusual bony excrescence on the left iliac crest (arrows).

vertebral elements were unobscured by barium on a film from the GI series (Figure 12.28c) to permit a positive match with the postmortem radiograph of the lumbar spine (Figure 12.28d).

The pelvis also is an osseous survivor with many distinctive features for matching.[6,33] Figure 12.29 shows a match on a "floater" from Elephant Butte Lake using a peculiar bony excrescence on the iliac crest. Congenital acetabular dysplasia with hip dislocation on a postmortem roentgenogram led to identification of one case.[39]

OTHER BONES

Other bones have been used for radiological identification: the patella,[2] the elbow and assorted long bones in extremities,[34] and the hand and wrist. Greulich[40] studied radiographs of the hands and wrists of 70 pairs of same-sex twins, of whom 40 pairs were believed identical, and found that all had individual distinguishing features.

SINGLE BONE IDENTIFICATION

When dealing with unidentified human remains, any single bone is a potential matching identifier by radiological examination if the antemortem counterpart can be located, hence the recommendation that the entire body be radiographed before disposal or release.

Features leading to identification by matching radiographs or other images of individual bones can be classified as follows: (1) anomalous or unusual development; (2) disease or degeneration; (3) tumor; (4) trauma; (5) iatrogenic interference; and (6) vascular grooves and trabecular patterns.

In the happy instance when antemortem radiological studies of the presumed decedent are available while the body or bones are still at hand, then concentrated efforts to create exactly matching pairs of images is both practicable and rewarding. This may be done with trial and error positioning (and repositioning) of the remains or, if skeletonized, by using the shadow positioning technique of Fitzpatrick and Macaluso.[41] This involves making an underexposed duplicate of an antemortem film showing a promising bony feature. The outline of the skeletal part on the duplicate is used as a "mask" and placed on top of the film cassette. The x-ray tube is positioned at the same distance from the film as when the antemortem exposure was made. (This is usually 40 in. except for erect PA chest films at 72 in. and supine AP chest films at 36–40 in.) The skeletal part to be duplicated is held under the tube so that the positioning light casts the bone's shadow onto the cassette. Manual manipulation can then exactly superimpose the shadow onto the "mask" and an identical projection obtained—if there is, indeed, a match.

Anomalous or Unusual Development

The dorsal defect of the patella (Example Case 11-1), the variations in vertebral configuration (Figure 12.27), and the rhomboid fossa of the clavicle in Sanders et al.[35] are examples of this category.

Disease or Degeneration

Judging from the literature, matching skeletal remains by lesions secondary to a disease process is rare. However, degenerative changes are frequently helpful or definitive.

Example Case 11-3. A body was found in early springtime beneath the loading dock of the feed store in a small Western town. Having frozen and thawed several times, and having suffered the depredations of a variety of varmints, the body was unrecognizable. Radiographs were obtained before burial at county expense. Finally, someone remembered that the town drunk hadn't been around much since the first blizzard of the previous winter. After some search, an antemortem view of the abdomen was found at the regional Veterans Administration Hospital, and matched perfectly with the postmortem film (Figure 12.30).

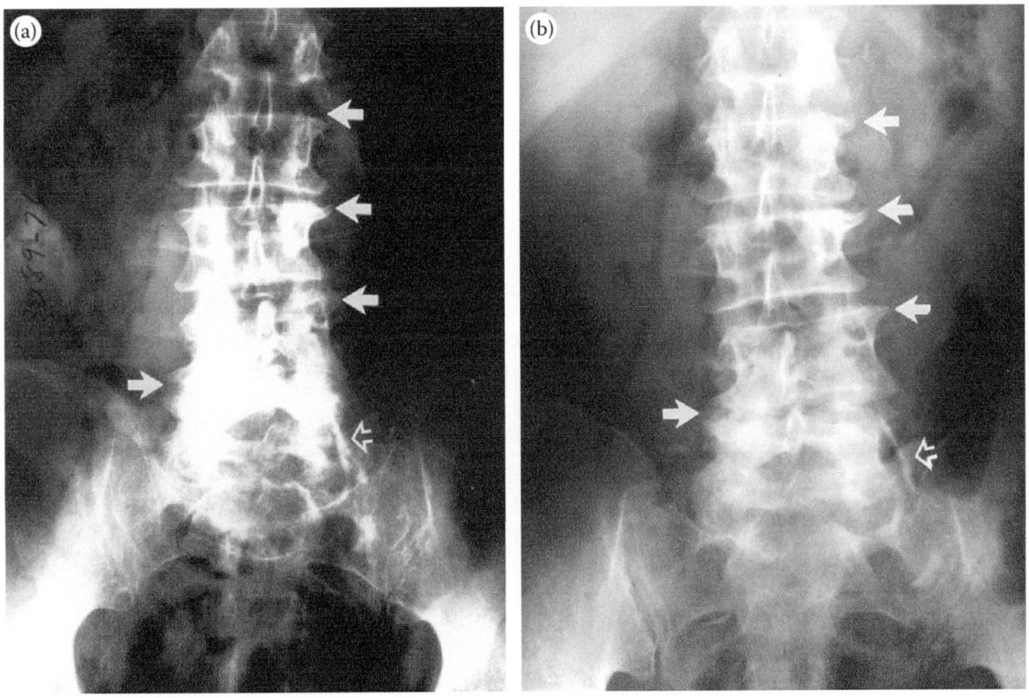

FIGURE 12.30 (a) Postmortem and (b) antemortem roentgenogram of the abdomen. Note the matching degenerative spurring and lipping of vertebral body margins (arrows). Additionally, the spinous processes match, and there is identical atherosclerotic calcification in the left common iliac artery (open arrows).

Tumors

Bone tumors are relatively uncommon. In young persons, bone islands, nonossifying fibromata, or osteochondromata might sometime be found in a decedent. The tumefaction in Figure 12.29 would fit that category.

Trauma

As predicted by d'Courmelles in 1898, traumatic lesions are helpful and fairly common features for identity matches.

Example Case 11-4. A bag of bones was sent to Dr. Weston (see Preface to first Edition) in the hope that he could assist in the identification of this case (Figure 12.31a). The bones were those of a young adult male with no unusual features except for a healed fracture of the left clavicle with some residual deformity (Figure 12.31b). Law officers had a presumptive identification for the skeleton and found a radiograph of the left shoulder of that individual at age 14 with a fresh fracture of the left clavicle, but this only added to the *presumptive* identification. Finally, almost two years later, good investigative work turned up a chest film of the presumed decedent as an adult. High up in the corner of that film was a deformed left clavicle (Figure 12.31c). We were able to position the dried clavicle precisely enough (Figure 12.31d) to exactly reproduce the image of the clavicle on the chest film (Figure 12.31e), thus making a positive identification.

Iatrogenic Interference

Doctors and dentists leave their indisputable markers in some bodies and they are a godsend for identification purposes. Figures 7.2, 12.17c and d, and 12.26 are examples already shown. Figures 12.32 and 12.33 furnish additional examples of this category.

Vascular Grooves and Trabecular Pattern

Vascular foramina and grooves, and the pattern of bony trabeculae, are critical to the radiographic matching of some antemortem and postmortem images, particularly if the residual skeletal material is fragmentary. Matching requires precise positioning of postmortem specimens and excellent exposure technique. The detail required for identification purposes by direct inspection of radiographs often is difficult to reproduce in publications. We have demonstrated matching vascular patterns earlier (Figures 12.17 and 12.18).

Kahana and Hiss[42] have reported a system of matching bony trabecular patterns using computerized densitometric line maps or densitographs. We believe going to these lengths are, rarely if ever, necessary, given good radiographic technique and position coupled with educated "eyeball" comparison. We have had good experience with direct trabecular comparisons, as has Murphy[6,34] and others.[22,28,35]

Example Case 11-5. Skeletonized human remains were found in coastal wetlands. The skull was missing. The ends

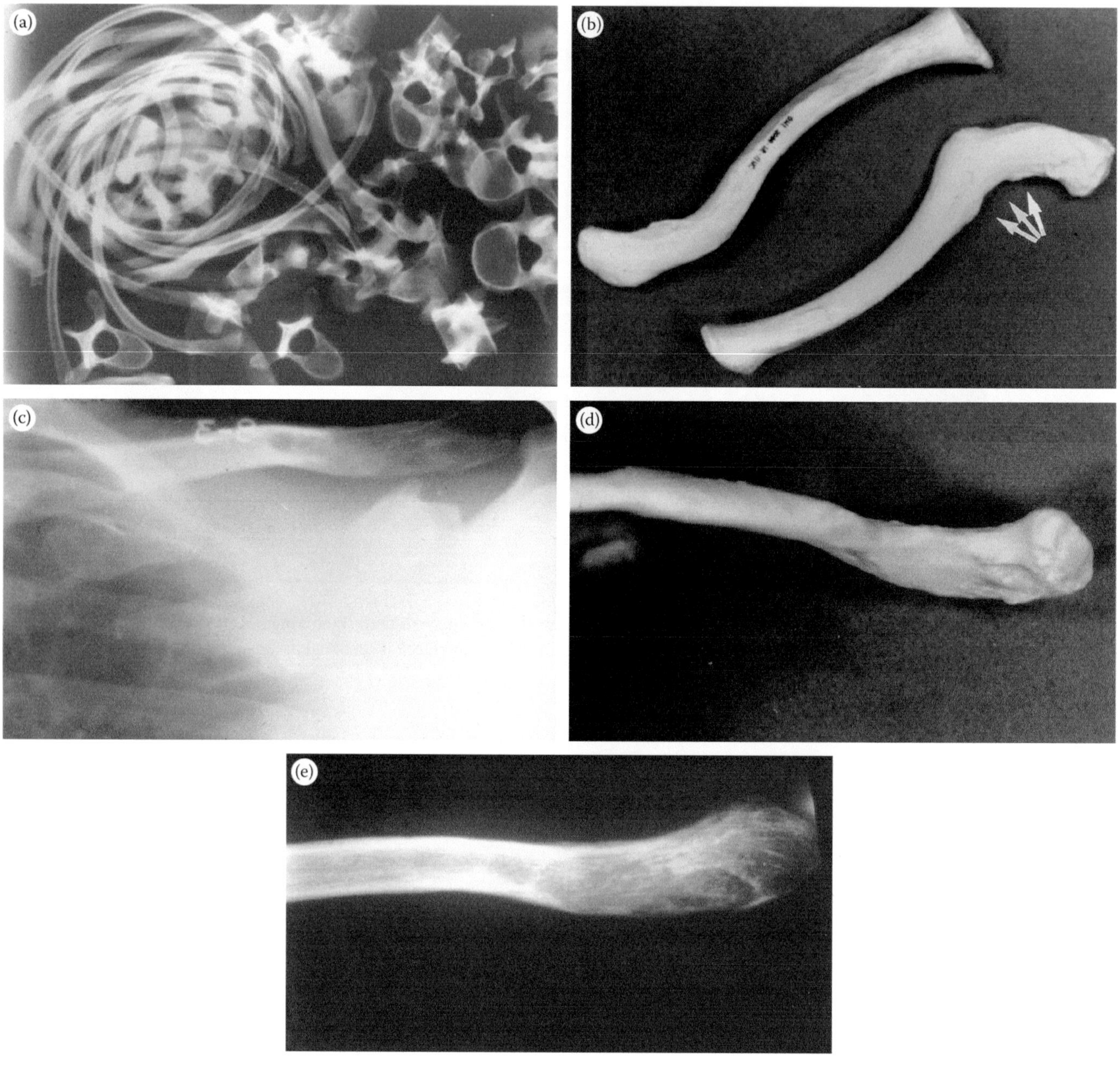

FIGURE 12.31 (a) Dried skeletal remains of a young adult male. (c) Slight deformity of the left clavicle from an old healed fracture of the lateral third. (c) Close-up of left clavicle in corner of antemortem chest radiograph. Note the irregular configuration and altered trabecular pattern related to the old healed fracture. (d) With careful positioning of the dried clavicle a radiograph, (e) reproduces the findings seen in the clavicle on the chest radiograph.

of the long bones were mostly destroyed. An innominate bone (hemipelvis) was the largest intact bone. The remains were identified as those of a young adult female. The Air Force had out a Missing Person report on a similar person. The usual pre-induction chest film had somehow been omitted, but a radiograph of the pelvis had been obtained during a bout of pelvic inflammatory disease. The innominate bone was meticulously positioned to match the antemortem image and a perfect match of contour, vascular (nutrient) groove, and trabecular pattern was obtained (Figure 12.34).

Because of its mass, the innominate bone frequently escapes destruction and has many identifiable features. Moser and Wagner[42] have emphasized the importance of the nutrient canal as a forensic marker and describe three patterns: parallel (as in Case 8-5 above), V-shaped, and Y-shaped. The supraacetabular, suprapubic portion of the innominate shows configurational variability and is rich with coarse trabecular patterns.

Example Case 11-6. An almost totally skeletonized female body was retrieved from a pond near a home whose occupant had been missing for some months. The only antemortem radiograph of the missing woman was a pelvimetry study during an earlier pregnancy. A carefully positioned image of the right innominate bone exactly matched the counterpart

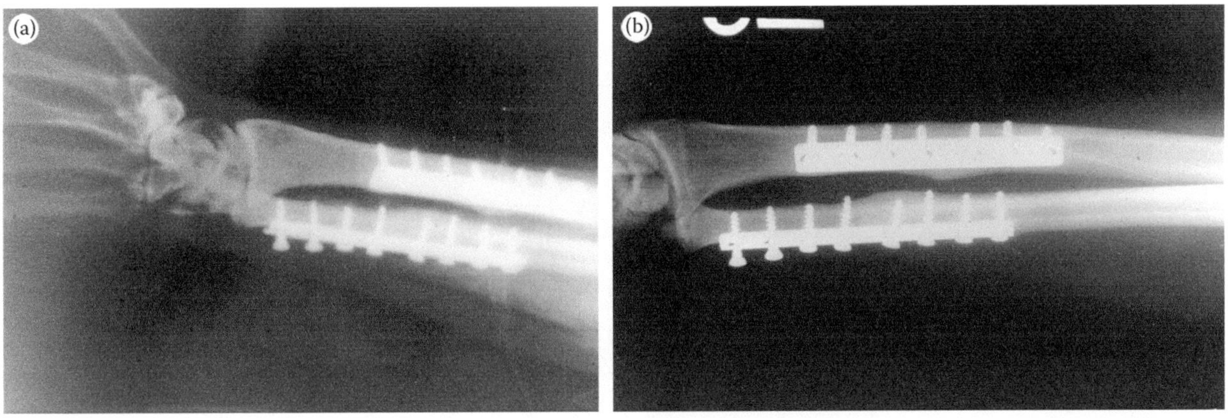

FIGURE 12.32 (a) Postmortem and (b) antemortem roentgenograms of the forearm of a air crash victim with "plate and screws" fixation devices in place.

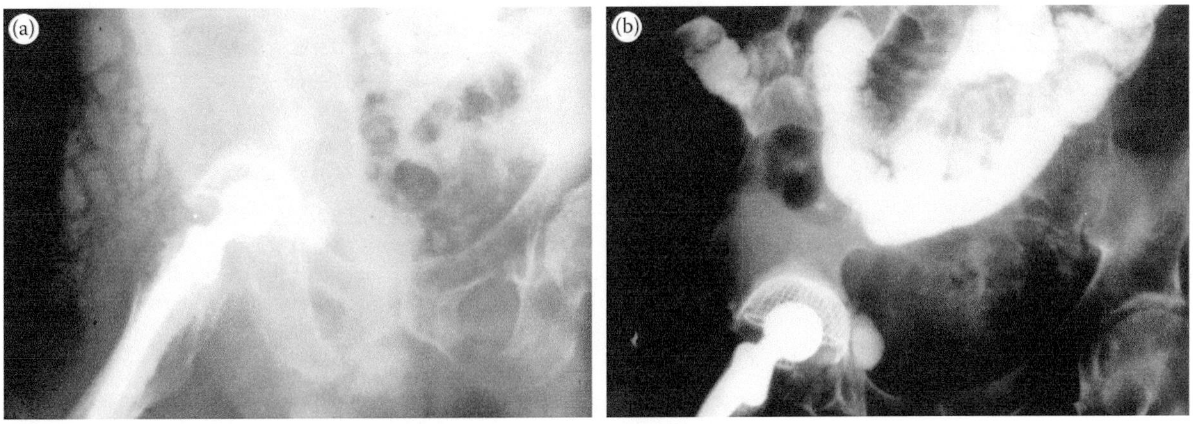

FIGURE 12.33 (a) Postmortem and (b) antemortem roentgenograms of an air crash victim who had undergone hip replacement surgery.

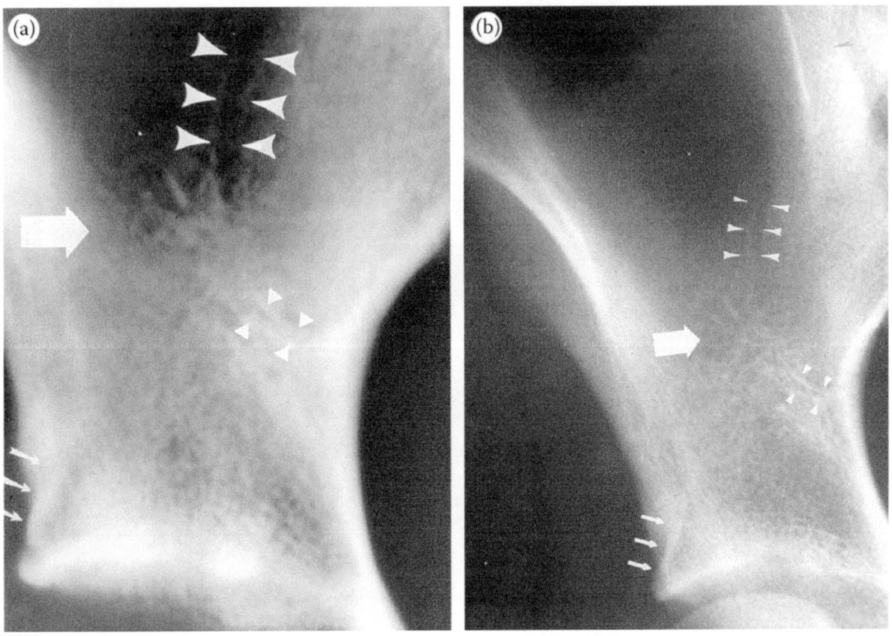

FIGURE 12.34 (a) Slightly enlarged detail of postmortem x-ray study of the innominate bone. (b) Detail from antemortem pelvic radiograph. There are many matching features: general configuration, large vascular or nutrient groove (arrowheads), linear trabecular pattern (triangles), coarse trabecular pattern (large arrows), and focal contour feature (small arrows).

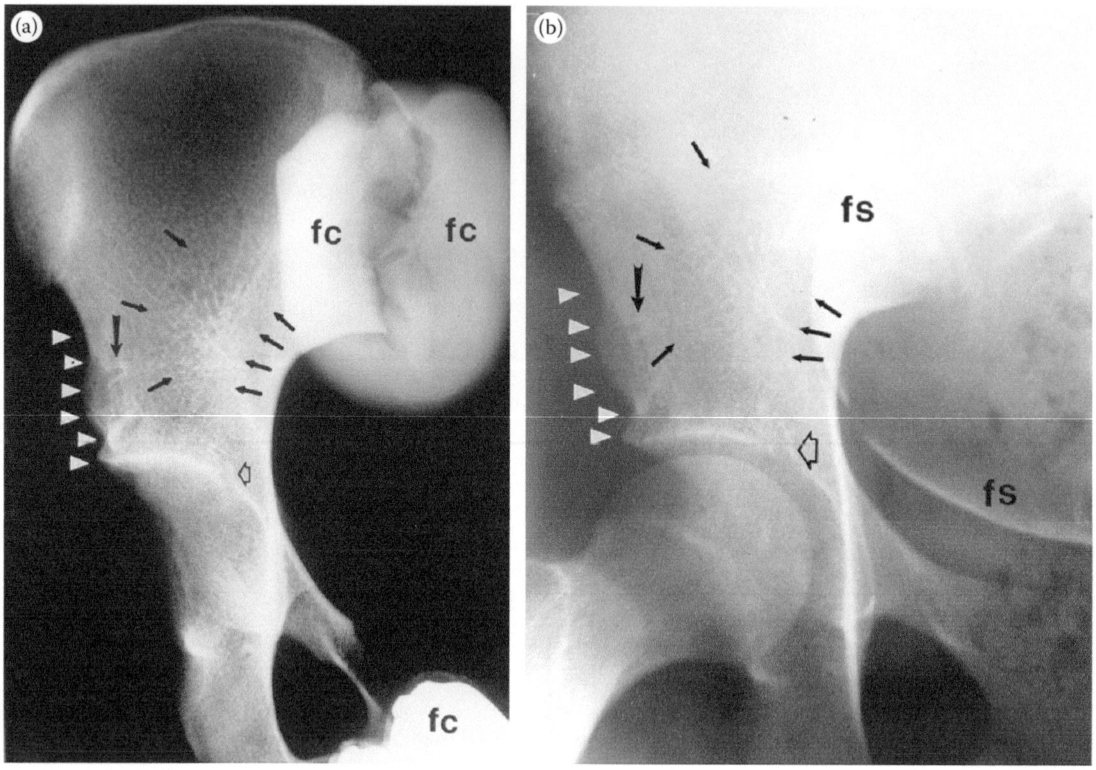

FIGURE 12.35 (a) Roentgenogram of innominate bone. (b) Detail of innominate bone included in earlier pelvimetry of the presumed decedent. There are matches in configuration (triangles), an acetabular notch (open arrow), and trabecular pattern (arrow); "fc" = florist's clay supports; "fs" = fetal skull.

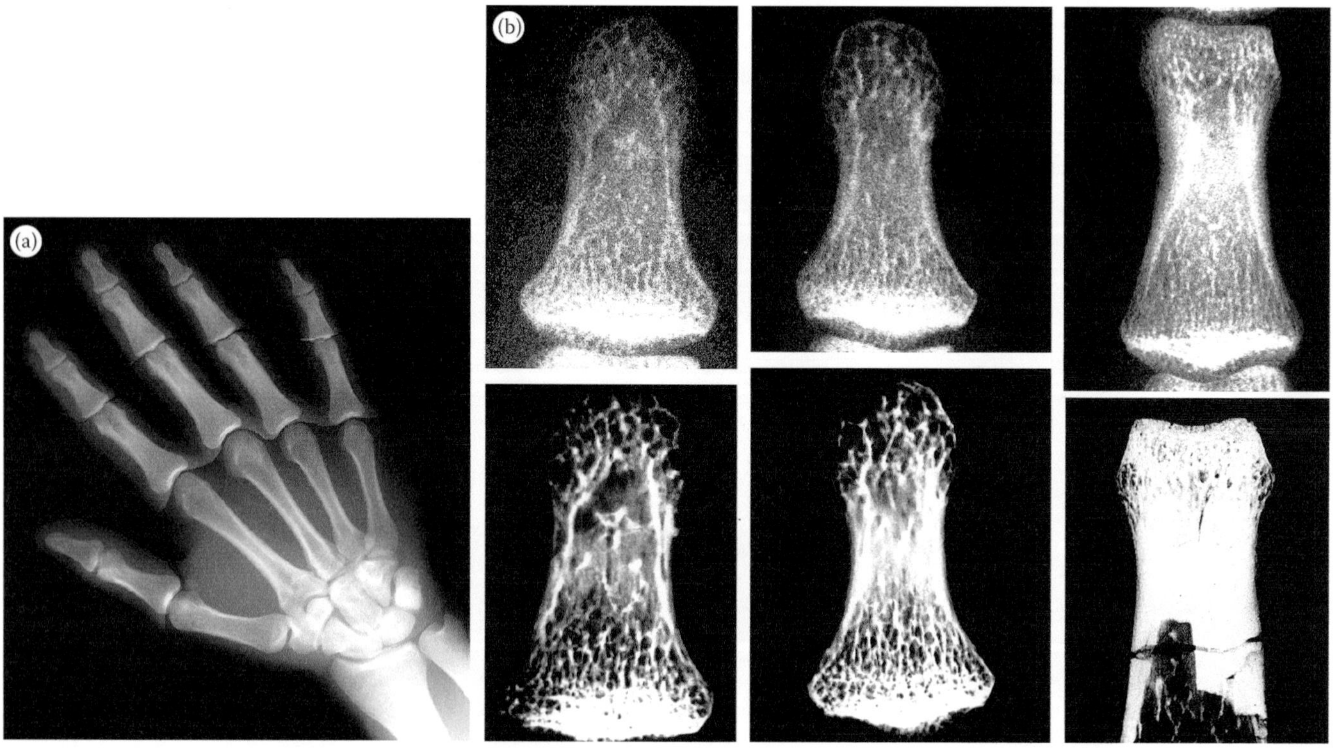

FIGURE 12.36 (a) Antemortem radiograph of left hand and wrist of a young male reported missing. (b) Top row: close up of terminal phalanges of index and ring fingers and fragment of middle phalanx of index finger enlarged from antemortem hand wrist film. Bottom row: postmortem radiographs of burned bones retrieved from fire-pit. (c) Top row of antemortem phalanges and bottom row of postmortem phalanges with arrows indicating identical unique trabecular patterns allowing positive identification.

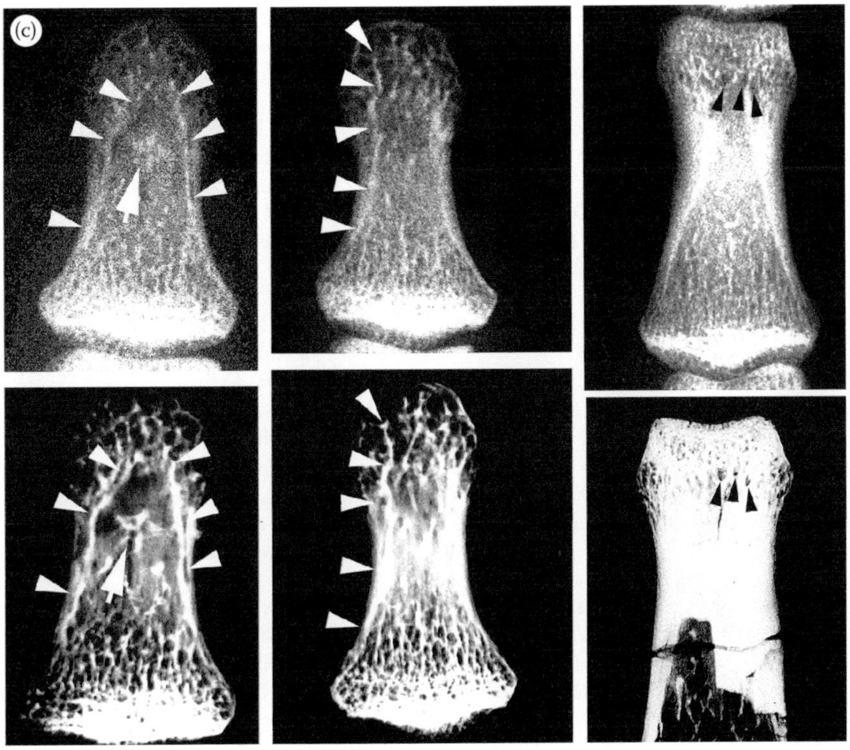

FIGURE 12.36 *(Continued)*

structures on the pelvimetry study in both configuration and trabecular pattern (Figure 12.35).

Example Case 11-7. A young man reported missing was last seen in the area of a rural property owned by a reclusive woman with a history of turbulent relations. A warranted search party discovered burned bones in several fire pits and a burn-barrel from which 30 five-gallon buckets of incinerated refuse was collected. Two physical anthropologists meticulously collected from this daunting refuse, the commingled remains of a young adult human male and eight other animal species. The burned human remains were inadequate for dental or DNA identification. Thirty-three small jewelers' boxes, each containing a bone or tooth fragment, were sent for possible identification by comparison with a few antemortem radiographs of the presumed victim. None of the burned fragments were large enough for this purpose except for a few intact and fragmentary finger bones. Three of these could be successfully matched with an antemortem radiograph of the left hand and wrist of the presumed victim. (Figure 12.36) This positive identification was validated when the perpetrator stipulated the murder and the identification of the victim when pleading guilty by reason of insanity.[44]

REFERENCES

1. Edlund, J. F., Some general considerations in the use of diagnostic imaging in forensic medicine, in *Legal Medicine with Special Reference to Diagnostic Imaging*, James, A. E., Ed., Urban & Schwarzenberg, Baltimore, 1980, p. 244.
2. Riddick, L., Brogdon, B. G., Laswell-Hoff, J., and Delmar, B., Radiographic identification of charred human remains through use of the dorsal defect of the patella, *J. Forensic Sci.*, 28, 263, 1983.
3. Johnson, J. T. and Brogdon, B. G., Dorsal defect of the patella: Incidence and distribution, *Am. J. Roentgenol.*, 139, 339, 1982.
4. Brown, T. C., Delaney, R. J., and Robinson, W. L., Medical identification in the "Noronic" disaster, *J. Am. Med. Assoc.*, 148, 621, 1952.
5. Bass, W. M. and Driscoll, P. A., Summary of skeletal identification in Tennessee, 1971–1981, *J. Forensic Sci.*, 28, 159, 1983.
6. Murphy, W. A., Spruill, F. G., and Gantner, G. E., Radiologic identification of unknown human remains, *J. Forensic Sci.*, 25, 727, 1980.
7. Hogge, J. P., Messmer, J. M., and Quynh, N. D., Radiographic identification of unknown human remains and interpret experience level, *J. Forensic Sci.*, 39, 373, 1994.
8. Fitzpatrick, J. J., Shook, D. R., Kaufman, B. L., Wu, S. J., Kirschner, R. J., MacMahon, H., Levine, K. J., Maples, W., and Charletta, D., Optical and digital techniques for enhancing radiographic anatomy for identification of human remains, *J. Forensic Sci.*, 41, 947, 1996.
9. Nye, P. J., Tytle, T. L., Jarman, R. N., and Eaton, B. G., The role of radiology in the Oklahoma City bombing, *Radiology*, 200, 541, 1996.
10. Snow, C. C. and Fitzpatrick, J., Human osteological remains from the Battle of the Little Big Horn, in *Archaeological Perspectives on the Battle of the Little Big Horn*, Scott, D., Ed., University of Oklahoma Press, Norman, OK, 243, 1989.
11. Sauer, N. J., Brantley, R. E., and Barondess, D. A., The effects of aging on the comparability of ante-mortem and post-mortem radiograph, *J. Forensic Sci.*, 33, 1223, 1988.

12. Holland, T. D., Use of the cranial base in the identification of fire victims, *J. Forensic Sci.*, 34, 458, 1989.
13. Kennedy, K. A. R., The wrong urn: Commingling of remains in mortuary practice, *J. Forensic Sci.*, 41, 689, 1996.
14. Culbert, W. L. and Law, F. M., Identification by comparison of roentgenogram of nasal accessory sinuses and mastoid processes, *J. Am. Med. Assoc.*, 98, 1634, 1927.
15. Ubelaker, D. H., Positive identification from radiograph comparison of frontal sinus patterns, in *Human Identification*, Rathbun, T. A. and Buikstra, J., Eds., Charles C. Thomas, Springfield, IL, 1984, chap. 29.
16. Asherson, N., *Identification by Frontal Sinus Prints*, H. K. Lewis, London, 1965.
17. Yoshino, M., Miyaraka, S., Sato, H., and Seta, B., Classification system of frontal sinus patterns by radiography. Its application to identification of unknown skeletal remains, *Forensic Sci. Int.*, 34, 289, 1987.
18 Krogman, W.M., Iscan, M.Y. *The Human Skeleton in Forensic Medicine*, 2nd ed. Charles C. Thomas, Springfield, IL, 1986, Chapter12.
19. Smith, V. A., Chistensen, A. M., and Myers, S. W., The reliability of visually comparing small frontal sinuses, *Proceedings Am. Acad. Forensic Sci*, Colorado Springs, CO, 2009, p. 135.
20. Owsley, D. W., Identification of the fragmentary, burned remains of two U.S. journalists seven years after their disappearance in Guatemala, *J. Forensic Sci.*, 38, 1372, 1993.
21. Rhine, S., Radiographic identification by mastoid sinus and arterial pattern, *J. Forensic Sci.*, 36, 272, 1991.
22. Adkins, L. and Potsaid, M. S., Roentgenographic identification of human remains, *J. Am. Med. Assoc.*, 240, 2307, 1978.
23. Voluter, G., The "V" test, *Radiol. Clin.*, 28, 1, 1959.
24. Singleton, A. C., Roentgenological identification of victims of "Noronic" disaster, *Am. J. Roentgenol.*, 66, 375, 1951.
25. Messmer, J. M., and Fierro, A. F., Personal identification by radiographic comparison of vascular groove patterns of the calvarium, *Am. J. Forensic Med. Pathol.*, 7, 159, 1986.
26. Hogge, J. P., Messmer, J. M., and Fierro, M. F., Positive identification by postsurgical defects from unilateral lambdoid synostectomy: A case report, *J. Forensic Sci.*, 40, 688, 1995.
27. Sekharan, F. C., Identification of skull from its suture pattern, *Forensic Sci. Int.*, 27, 205, 1985.
28. Martel, W., Wicks, J. D., and Hendrix, R. C., The accuracy of radiological identification of humans using skeletal landmarks: A contribution to forensic pathology, *Radiology*, 124, 681, 1977.
29. Vastine, J. H., Vastine, M. E., and Orango, O., Genetic influence on osseous development with particular reference to the disposition of calcium in the costal cartilages, *Am. J. Roentgenol.*, 59, 213, 1948.
30. Tsunenari, S., Uchimura, Y., Yonemitsu, K., and Oshiro, S., Unusual personal identification with characteristic features in chest roentgenograms, *Am. J. Forensic Sci. Pathol.*, 3, 357, 1982.
31. Rougé, D., Telmon, N., Arrue, P., Larrouy, G., and Arbus, L., Radiographic identification of human remains through deformities and anomalies of postcranial bones: A report of two cases, *J. Forensic Sci.*, 38, 997, 1993.
32. Ubelaker, D. H., Positive identification of American Indian skeletal remains from radiographic comparison, *J. Forensic Sci.*, 35, 466, 1990.
33. Hyma, B. A. and Rao, V. J., Evaluation and identification of dismembered human remains, *Am. J. Forensic Med. Pathol.*, 12, 1991.
34. Murphy, W. A. and Gantner, G. E., Radiologic examination of anatomic parts and skeletonized remains, *J. Forensic Sci.*, 27, 9, 1982.
35. Sanders, I., Woesner, M. E., Ferguson, R. A., and Noguchi, T. T., A new application of forensic radiology: Identification of deceased from a single clavicle, *Am. J. Roentgenol.*, 115, 619, 1972.
36. Owsley, D. W. and Mann, R. W., Positive personal identity of skeletonized remains using abdominal and pelvic radiographs, *J. Forensic Sci.*, 37, 332, 1992.
37. Ikeda, N., Umetsu, K., Harada, A., and Tsuneo, T., Radiological identification of skeletal remains: A case report, *Jpn. J. Legal Med.*, 41, 270, 1987.
38. Crow, N. E. and Brogdon, B. G., The "normal" lumbosacral spine, *Radiology*, 72, 97, 1959.
39. Varga, M. and Takács, P., Radiographic personal identification with characteristic feature in the hip joint, *Am. J. Forensic Med. Pathol.*, 12, 328, 1991.
40. Greulich, W. W., Skeletal features visible on the roentgenogram of the hand and wrist which can be used for establishing individual identification, *Am. J. Roentgenol.*, 83, 756, 1960.
41. Fitzpatrick, J. J. and Macaluso, J., Shadow positioning technique: A method for post-mortem identification, *J. Forensic Sci.*, 30, 1226, 1985.
42. Kahana, T. and Hiss, J., Positive identification by means of trabecular bone pattern comparison, *J. Forensic Sci.*, 39, 1325, 1994.
43. Moser, R. P., Jr. and Wagner, G. N., Nutrient groove of the ilium, a subtle but important forensic marker in the identification of victims of severe trauma, *Skeletal Radiol.*, 19, 15, 1990.
44. Brogdon, B. G., Sorg, M. H., and Marden, K., Fingering a murderer: A successful anthropological and radiological collaboration, *J. Forensic Sci.,* 55(1), 248–250, 2009.

13 Radiology in Mass Casualty Situations

Mark D. Viner and Joel E. Lichtenstein

CONTENTS

INTRODUCTION

Disasters of many kinds require processing and identification of multiple victims. Railroad and aircraft accidents, unfortunately, are common and come to mind immediately. Natural disasters such as earthquakes, floods, and hurricanes are recurrent problems. Hurricane Katrina and the South East Asia tsunami of 2004 are recent examples where radiology was a key component of the response.[1,2] Collapse or fire in high-occupancy buildings such as hotels and factories provide other examples, such as the 1948 dockside explosion of fertilizer ships in Texas City, TX, with 561 fatalities; the nuclear power plant accident at Chernobyl in the Ukraine in 1986; and the 1987 toxic chemical release at Bhopal in India, causing approximately 3000 deaths.

Mass casualty incidents are unfortunately not restricted to naturally occurring or accidental causes. Terrorist incidents including the April 1995 bombing of the Federal Building in Oklahoma City, the events of September 11, 2001, and the London bomb attacks of July 2005 are all recent examples in which radiology played a significant role in both the identification of victims and investigation of the incidents.[3–7] Deliberate single acts of mass murder, such as these, together with sustained campaigns of atrocities against populations, such as the genocides of Rwanda and Yugoslavia, and the human rights abuses associated with repressive regimes often result in multiple mass casualty events. While the causative factors and the state of preservation of remains vary between them, the scientific principles employed in their investigation remain the same, and radiological methods are invaluable.[8]

Mass casualties, by their nature, tend to involve emergencies that are unexpected by those who are called upon to respond. They are stressful situations in which even those with little prior interest or experience may be called upon to help. Radiology can be enormously helpful in the task of identifying victims. In an emergency, radiographers and general radiologists may be asked to aid in a multidisciplinary team, usually headed by a forensic specialist. The basic principles of radiographic identification are the same as those applied to individual victims, but the circumstances and conditions are generally much different. Despite the presence of international guidelines for the management of such incidents and the identification of the deceased,[9] improvisation by inexperienced workers has usually led to the somewhat painful evolution of remarkably similar procedures and principles, even in seemingly different circumstances. Hopefully, a recounting of the lessons learned from past events will provide a useful starting point for those called upon in the future and will minimize the need to keep "reinventing the wheel."

The most extensive experience to date is that gleaned by government agencies and is based largely on investigations of aviation accidents. The lessons learned, however, are applied readily to other mass disasters, as well as to smaller-scale events. The former Aviation Pathology Department (now the Armed Forces Medical Examiner) at the Armed Forces Institute of Pathology (AFIP) in Washington, DC, aids in the investigation of all US military aviation fatalities and is a consultant to the National Transportation Safety Board. As part of a joint committee that coordinates all aviation pathology in the United States, Canada, and the United Kingdom, its representatives are participants or observers in most mass casualty aircraft-accident investigations. Full-body radiography of victims is a routine part of the investigation of US military aviation fatalities.

Identification of victims of mass casualties is important for a variety of reasons, some of which may not be immediately obvious. Usually, it will be known that certain parties are among the victims of an event, but the degree of confusion as to precisely who is, or is not, included is often surprising. Passenger manifests and occupancy lists are notoriously inaccurate. Even in a closed population, where the names of all the victims are thought to be known, it is important to provide positive identification of individual remains if possible. Humanitarian and psychological motives drive efforts to release remains to grieving relatives rapidly, but with the assurance of accurate identification.[10,11] Legal and insurance requirements for accurate identification are clear. The scientific importance of accurate individual identification may be less obvious, but is vital as a source of data for safety engineering and accident prevention.

HISTORY

The first use of x-rays for the identification of multiple casualties was in the 1949 fire, aboard the Great Lakes liner *Noronic* in Toronto, Canada.[12,13] Survey radiographs were obtained on 79 of 119 fatalities, and antemortem radiographs were obtained for 35 of these. Many additional radiographs were obtained as needed for comparison, and eventually 24 cases were positively identified by radiology alone (Figures 13.1 and 13.2). In many other cases, x-rays were helpful in supporting or excluding identifications suggested by other techniques.

The largest systematic application of radiology in a mass casualty involved the March 1977 crash of two 747 jumbo jets in the Canary Islands, resulting in 576 fatalities. An American plane on the ground was struck by a Dutch aircraft taking off. The majority of the 326 American victims were partially cremated, but there was relatively little fragmentation or commingling of remains. A team from the AFIP investigated the bodies after their return to a mortuary at Dover Air Force Base, Delaware. This government mortuary serves the eastern part of the United States as well as all of Europe and the Middle East. All remains were radiographed during initial processing and selected bodies later were studied further as required by the identification process.[14–16] Radiology provided the source of positive identification in 25 cases.

The 1979 crash of a wide-body jet at O'Hare Airport in Chicago resulted in 274 fatalities with more severe fragmentation. Similar techniques were employed, but a single radiologist, John J. Fitzpatrick was called in relatively late in the investigation when 50 victims remained to be identified. Working mostly after hours and on weekends in addition to his normal duties as teaching faculty at Cook County Medical Center, he was able to make 20 additional identifications.[16,17] This identification effort saw one of the first applications of digital computers to facilitate the tedious sorting and processing of massive amounts of ante- and postmortem data.

A team from the AFIP also investigated the more than 900 victims of the November 1978 Jonestown, Guyana tragedy, although radiology played a relatively minor role.

A similar AFIP team identified all 256 victims of a military plane crash in Gander, Newfoundland in 1985, at which time the radiology facilities at Dover were expanded and upgraded.[18]

On June 23, 1985, Air India flight 162 (a Boeing 747) disintegrated at 31,000 ft as a result of a bomb explosion. There were 329 passengers aboard. Many were ejected from the aircraft, and 131 bodies were recovered from the sea off the coast of Cork, Ireland. The bodies were taken to the Cork Regional Hospital, a modern university teaching facility with 600 beds and an up-to-date full-service Department of Radiology. Total body radiography was carried out. Nearly

FIGURE 13.1 Dr. Singleton was a prominent Canadian radiologist who was Professor and Head of the Department of Radiology at the University of Toronto, and Past President of both the Canadian Association of Radiologists and the American College of Radiology. The Attorney General of Ontario asked him to conduct roentgenological examinations of the victims of the Noronic disaster in an attempt to assist in identification. Thus, Dr. Singleton became the father of mass casualty radiography. (From Hall, M.R. and Arthur C. Singleton, *Am J. Roentgenol.*, 105, 210–211, 1969. With permission.)

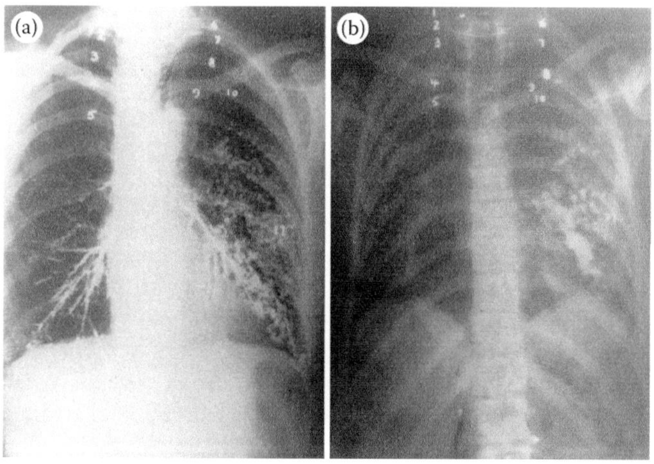

FIGURE 13.2 Perhaps the easiest of Dr Singleton's identifications was the patient who had undergone a bronchogram, a procedure in which a contrast medium is instilled into the bronchial tree. [2] The opaque shadows in the lung are seen clearly on both (a) antemortem and (b) postmortem radiographs. There were additional matching bony points. (From Singleton, A.C., *Am J. Roentgenol.*, 66, 375, 1951. With permission.)

all of the victims were Indians, most were young adults, 30 were children. Despite the massive radiological effort, all bodies were identified by non-radiological methods.[19,20]

On April 19, 1995, a 4800 lb car-bomb exploded in front of the Alfred P. Murrah Federal Building in Oklahoma City, killing 168 people and injuring many more. Radiology played a key role in the identification of the deceased. Within 10 days, 165 of the 168 deceased had been examined by full-body x-ray series, an operation that involved 60 radiographers taking an average of 15 films per case. Equipment was supplied by vendors and the national guard to supplement the basic facilities at the Medical Examiners Offices. A team of 10 radiologists evaluated the images, on a rotating basis, in conjunction with other forensic personnel. Radiology examinations accounted for six positive identifications after findings from dental examination, fingerprinting, and other methods proved inconclusive.[5]

Following the terrorist attacks on September 11, 2001, the victims of the attack on the Pentagon were taken for identification to the Port Mortuary, Dover Air Force Base, Dover, Delaware. Forensic radiology was a key component of the system of casualty victim identification. A full-body x-ray examination was undertaken in the two radiographic suites using film developed using a daylight processing system and radiology reports were recorded on skeletal diagrams that accompanied the cadavers through the identification process. Radiological methods proved useful for differentiation of commingled remains, identification and location of personal effects, and victim identification as well as location and retrieval of items of forensic significance.[6]

On July 7, 2005, three terrorist bombs exploded on underground trains and a further bomb on a bus in London. Fifty-six people were killed in these explosions and several hundred people were injured. Fifty-six cadavers were recovered from the scenes of the explosion together with a large number of fragmented remains. In accordance with the London Resilience Mass Fatality Plan, the Association of Forensic Radiographers (AFR) mobilized its forensic radiography response team and equipment, some of which were provided on loan by the medical supply industry. A total of 27 radiographers worked in teams of between 6 and 8 for 12 h per day during the subsequent 11 days. Two mobile C- arm fluoroscopy units, a digital direct radiographic (DR) unit, a computed radiography (CR) unit, together with plain film radiography and dental radiography were employed. The majority of identifications were made by dental methods or fingerprints, but radiological methods once again proved very effective for differentiation of commingled remains, identification and location of personal effects, and victim identification, as well as location and retrieval of items of forensic significance.[7]

FACILITIES

Mass disasters often require temporary morgues and improvised field x-ray operations. The details will depend upon the nature of the casualty and upon the number and condition of the victims. Although it will be much more convenient for the radiographer and radiologist to use pre-existing, permanent radiology facilities in hospitals or in the medical examiner's offices, the scale of the incident and/or the nature of the event may render this impossible for logistical or security reasons. For the radiology team, the advantages of familiar surroundings, facilities, and personnel are difficult to overestimate, and in some cases involving smaller incidents of limited complexity, it may be possible for clinical radiology departments to be used while avoiding normal patient traffic areas or, when practical, by working at off-peak hours. In larger-scale events, whole facilities may be commandeered.

In the 1990s, the dogma regarding the use of pre-existing clinical facilities shifted somewhat because of public health and environmental issues. Concern over blood-borne pathogens has led to stringent regulations restricting access to disaster sites and to any material potentially containing human remains. Requirements include using isolation garments or decontamination suits and make it logistically difficult to bring remains into working clinical facilities. The forensic investigation of accidents that may involve acts of negligence or malicious criminal acts introduces further restrictions relating to the potential contamination of forensic evidence. Such constraints have prompted a return to the use of field facilities and temporary morgues. In recent years, and particularly following the events of September 11, 2001, government agencies in many countries have made considerable investment in emergency planning infrastructure. In some cases this has included the procurement of imaging equipment, specifically for use in mass casualty incidents. Such equipment may be expected to be deployed as part of a local response, or as part of a wider coordinated response by dedicated teams such as DMORT or the national disaster victim identification (DVI) teams that exist in other countries.[21] Wherever and however such planning takes place, it is important for the radiologist and radiographer to be consulted in the planning for such eventualities in order to ensure that optimum facilities are put in place for the identification of the deceased.

In any event, security and privacy should be the major concerns. In that regard, some degree of isolation or the ability to cordon off facilities is a great advantage. Police or military personnel will be required to control the press and curious onlookers as well as grieving survivors.

The 1977 Canary Island crash provided a prototype for establishing an emergency investigative facility. A field x-ray department was set up in a warehouse area of the mortuary at Dover Air Force Base, Delaware. It was maintained for three weeks, during which time all remains were radiographed. The location was chosen for its security capabilities as well as for its facilities and nearness to transportation.

Radiologists were part of a multidisciplinary team of over 130 military and federal civilian workers, which included experts in fingerprinting, dental analysis, anthropology, blood chemistry, toxicology, medical photography, and personal effects investigators as well as forensic pathologists. An "assembly-line" system was established so that each group

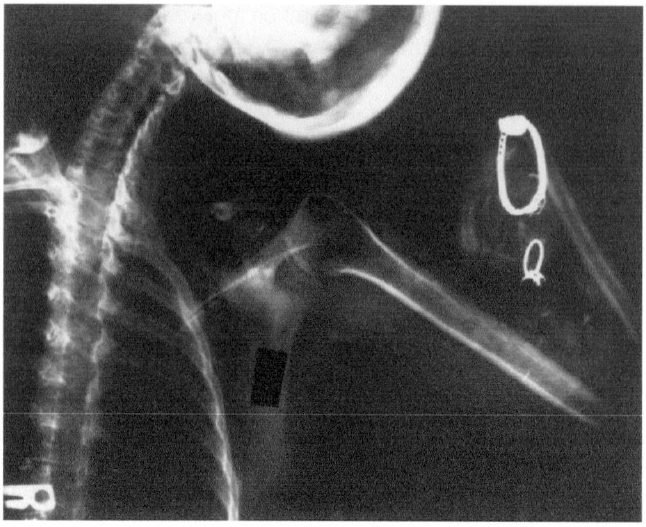

FIGURE 13.3 Radiograph of severely burned remains on which no personal effects were evident on external examination. A wristwatch and ring, clearly seen on the radiograph, were not found on initial autopsy. When recovered, both items were instrumental in identifying the victim, thus illustrating the value of obtaining radiographic data prior to pathologic examination. (From Lichtenstein, J.E. and Maxdewell, J.E. *Der Radiol.*, 22, 352, 1982. With permission.)

gathered whatever data were available in their discipline from each victim's body. Whenever possible, identifications were confirmed by multiple modalities. If one group, such as the FBI fingerprint team, made an apparent identification, all other modalities were cross-checked for consistency of their data. Radiological examination of all remains was an early step, so that films could be processed and reported prior to the pathologic examination, which was performed near the end of the process. Thus, details that might be hidden on gross external examination, but that were revealed radiologically could be sought and confirmed at autopsy (Figure 13.3).

Field radiography is discussed in detail in Chapter 36. Many of the challenges and solutions are applicable to mass fatality incidents and these are discussed in the following section.

EQUIPMENT

When faced with an unexpected emergency, there is a tendency to adopt a "camp-out" mentality in which portable equipment and resources are assembled for use in field facilities. When morgues do not have fixed radiographic facilities, portable hospital x-ray units with grid cassettes are commonly used. In an emergency, there is a temptation to take such units into the field (Figure 13.4). However, many of these machines are battery operated and have a limited working time on battery power and then must be recharged. They lack the power and the motorized scatter-reduction grids of fixed units, making it difficult to achieve consistent film quality. The use of grid cassettes requires careful alignment

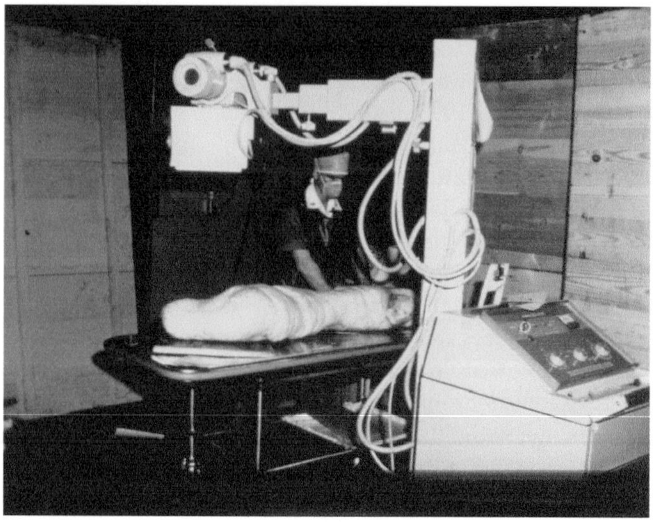

FIGURE 13.4 A rechargeable battery-powered x-ray unit being used to radiograph a wrapped casualty victim. A scatter-reduction grid cassette is positioned beneath the body and carefully aligned with the beam center to avoid artifacts. The body must be lifted and the cassette repositioned for each exposure. (From Lichtenstein, J.E. et al., *Aviat. Space Environ. Med.*, 1004, 1980. With permission.)

to prevent artifacts and the subject must be lifted and moved before and after each exposure.

During World War II, rugged, transportable field x-ray units were developed, which could be powered from field generators or commercial power lines and which included built-in, under-table scatter-reduction grids (Figure 13.5).

In recent years, rugged, transportable portable units have been developed for both military and veterinary use. Such

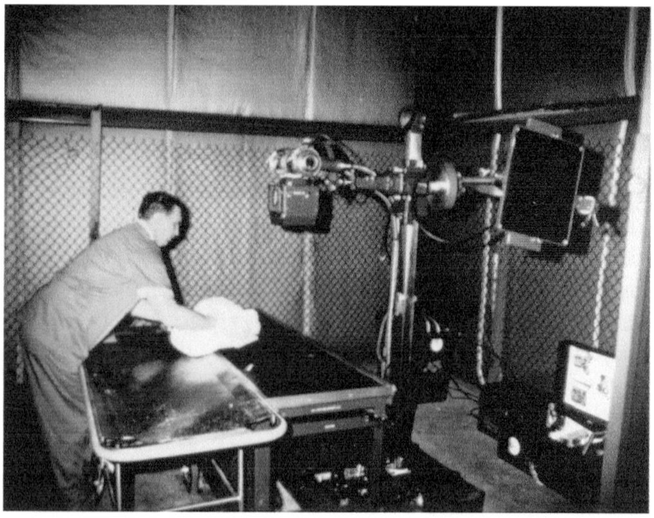

FIGURE 13.5 An old military field x-ray unit being used to radiograph a wrapped body part during a mass casualty identification effort. Note that the machine is situated in an outside-corner of a metal structure. Portable lead shields have been assembled to separate the machine from adjacent work areas. (From Lichtenstein, J.E. and Madewell, J.E., *Der Radiol.*, 22, 352, 1982. With permission.)

(a) (b)

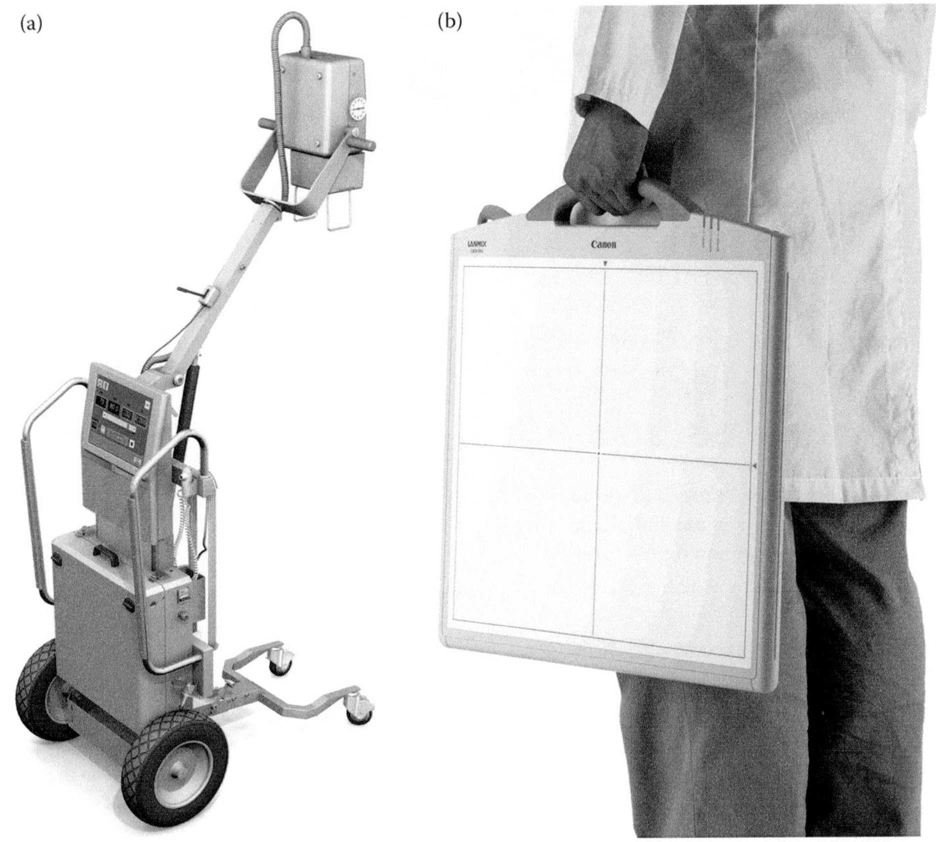

FIGURE 13.6 (a) Mobile rugged direct digital x-ray units are now widely used for military, forensic, and humanitarian applications. The high resolution, portability, and robustness of these units make them ideal for mass casualty morgue deployment, and a number of national DVI teams have invested in such equipment for this purpose. (b) Direct digital detectors provide an "instant" high-resolution imaging system, and can be obtained in a range of sizes. The latest units employ wireless technology offering greater flexibility. (From Xograph Healthcare Ltd., With permission.)

units may either be used with a field deployable CR systems or be combined with a DR x-ray system (Figure 13.6). These units are much better suited to deployment in the emergency situation, but they may be unfamiliar to hospital-based radiographers and technologists or morgue technicians. When contemplating use of unfamiliar equipment, sources of experienced technical assistance must be considered. Senior technologists or manufacturer's representatives may be most helpful.

The investigation of the Gander, Newfoundland, crash used the same morgue at Dover AFB, and a similar team approach as that used for the Canary Island investigation. The radiologists employed more modern, completely self-contained, air-transportable field x-ray facilities, which included film-processing capability, eliminating the need to transport cassettes (Figure 13.7).

The facilities at Dover were further expanded and modernized during the Persian Gulf conflict in 1992. Digital C-arm fluoroscopes were obtained and roller-bearing conveyors were installed to ease handling of heavy remains (Figure 13.8). An innovation was the installation of baggage-scanning radiographic units for screening of arriving remains. Similar units were also installed in the Persian Gulf to permit screening remains for hidden live ammunition before shipment to Dover (Figure 13.9). They are also useful to rapidly screen large quantities of debris for any potentially human material or any other specific items that might have distinctive radiographic characteristics. Both C-Arm fluoroscopes and slit-beam scanning equipment such as the "Lodox

FIGURE 13.7 External view of air-transportable self-contained field x-ray facilities expanded and in use with generator cables and air-conditioning hoses.

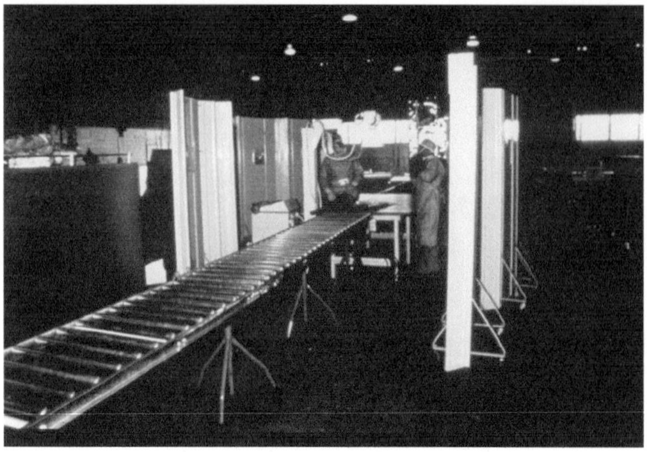

FIGURE 13.8 Roller-bearing conveyors arranged to aid in transporting heavy remains. Note the use of several individual x-ray shields arranged to provide radiation protection for other teams.

Statscan" (see Chapter 34) can also be used to undertake such initial screening process.

C-Arm fluoroscopes were routinely deployed by the United Nations Forensic Teams in the investigation and identification process of several thousand victims of the genocide in the former Yugoslavia. They proved extremely useful in the rapid identification and location of hazardous objects, ballistic material, and personal effects prior to pathology examination and for screening large quantities of exhumed debris for any potentially human material (Figure 13.10).[22]

C-Arm fluoroscopes, CR, digital DR, and film radiography were employed to examine and identify the remains of the 56 people who died in the terrorist attacks in London on July 7, 2005,[7] (Figure 13.11). Fluoroscopy was employed for the triage of the cadavers while both digital radiography (DR) and CR were employed for triage of the body parts. The use of CR was abandoned after the first day as DR proved to be far more rapid than CR in this particular application. Two radiographers examined each case using correct anatomical positioning where possible, which was facilitated by the use of transparent body-part bags in many cases. Images were stored on a workstation accessible to pathologist and anthropologist for evaluation, and written to CD Rom as a permanent record. The use of digital technology allowed images to be displayed to evaluate both soft tissue and bony elements and facilitated rapid triage of commingled body parts for anthropological analysis and DNA testing. In many cases, subsequent radiography for analysis of bony elements was not required due to the high definition nature of the images and the use of correct anatomical positioning.[7]

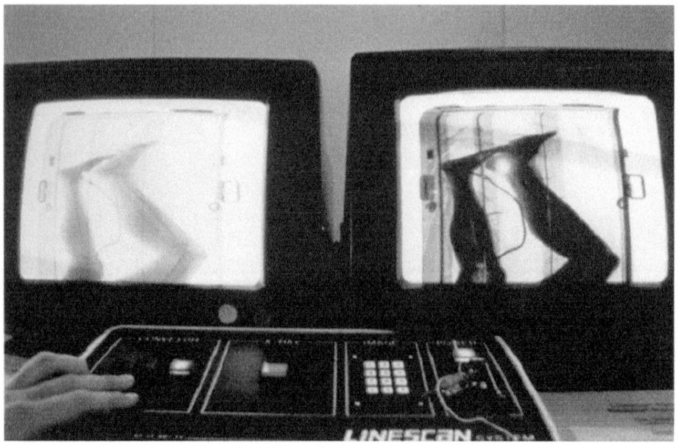

FIGURE 13.9 Linescan low dose baggage-scanning fluoroscope in use to screen remains still in body shipping caskets.

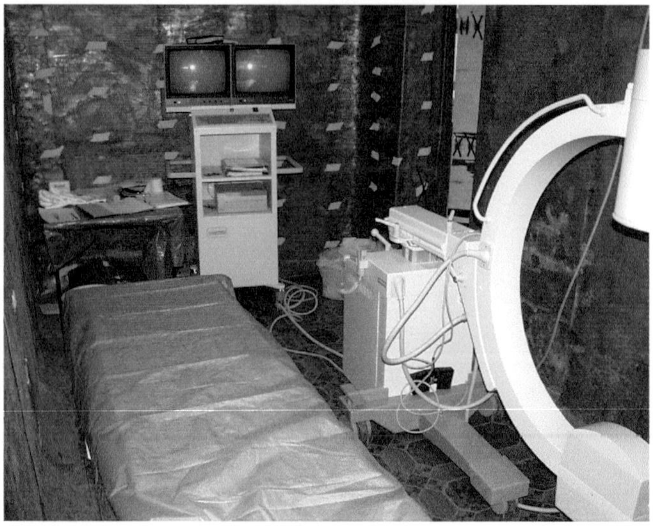

FIGURE 13.10 Digital C-Arm fluoroscope in use by the Forensic Teams of the United Nations International Criminal Tribunal for the former Yugoslavia during 2001 to examine the remains of several thousand victims of genocide. Its television- monitor display is useful for initial screening and the images can be stored on hard disc for later analysis, exported to a picture archive and communication system (PACS), or printed out in a variety of formats. (Note the lead-lining that has been affixed to the walls of the emergency mortuary in order to provide appropriate radiation protection.)

The use of CT scanning in mass fatality incidents is discussed in more detail in Chapter 14 Trailer-mounted CT scanners provide a fully transportable imaging facility, complete with control area, reporting room, and on-board generator and air cooling systems. Images can be transmitted for remote evaluation by means of a digital or satellite link. Applications of this technology in the mass casualty situation are in their infancy but early experiences are proving very positive.[23,24] Incident response deploying this technology together with fluoroscopy and DR is likely to become the "gold standard" in the future.

The advent of DR has largely eliminated the requirement for film processing in most medical facilities, however, many medical examiners offices still rely on film-based imaging systems for their routine investigations. The deployment of film-based imaging into the field situation, while by no means impossible, presents considerable logistical challenges. Darkrooms for handling and processing film, and the associated chemical and water supply issues, are major problems in

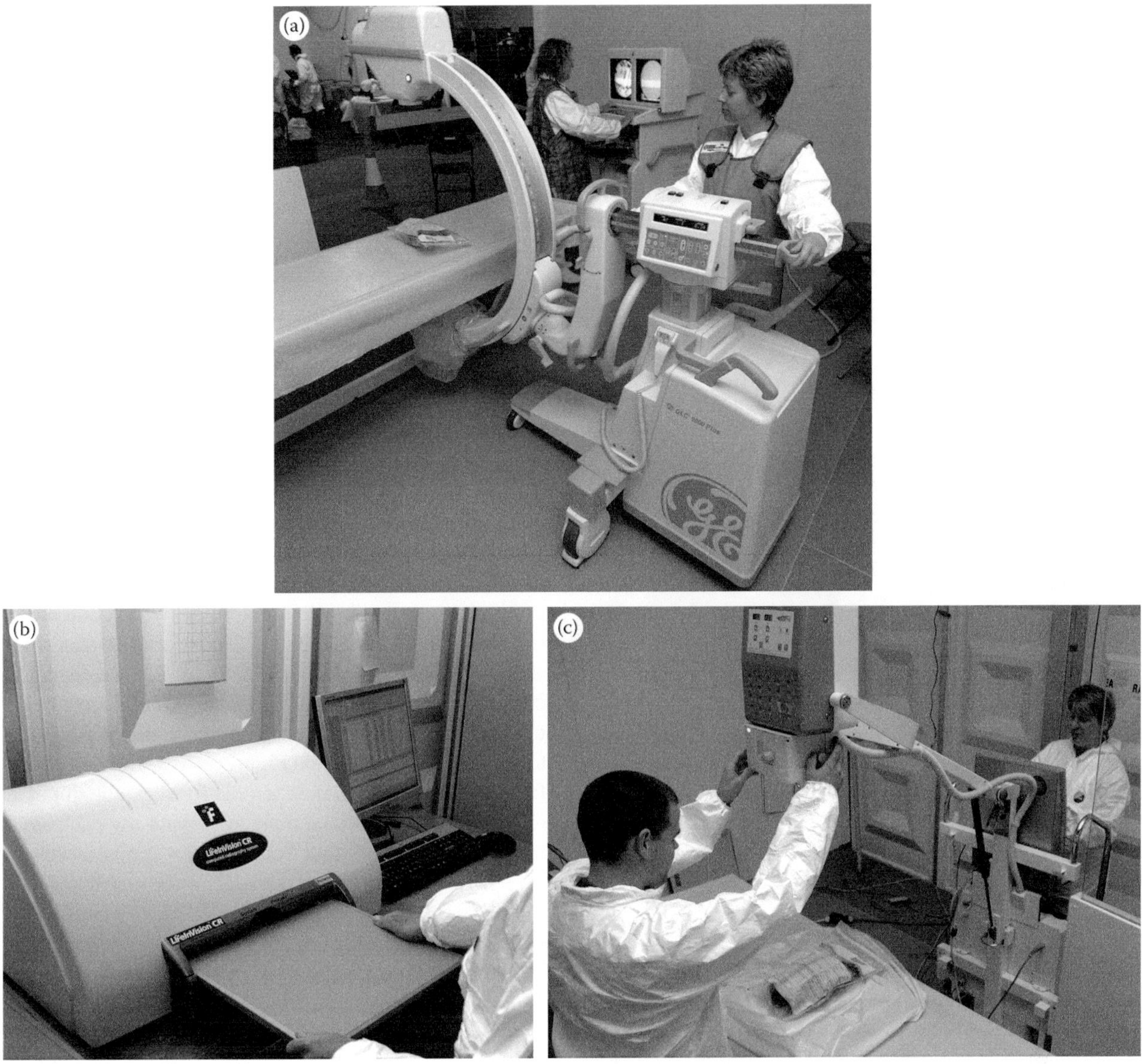

FIGURE 13.11 (a) Fluoroscopy being used as a triage method for rapid examination of cadavers and body parts following a mass fatality incident. (b) Portable CR readers offer the same flexibility and mobility as film-screen systems, but without the chemical-processing requirements. (c) Direct DR units offer a rapid high-resolution option for examining cadavers, body parts, and artifacts. (From Metropolitan Police, Scotland Yard. With permission.)

field operations. In some cases, pre-existing fixed facilities in nearby hospitals have been used to load and unload cassettes, which are then transported to and from the investigation site. In such cases, great care must be taken to properly label individual films and collate film packages. A compulsive, reliable member of the team should be stationed at the processing facility to assure quality control. Many of these problems are avoided if the modern military field units can be made available. If not, smaller tabletop processors may be acquired for small-scale operations (Figure 13.12). Some of these can be used without permanent plumbing for water supply by means of water recirculation. Sometimes, darkrooms have been "jury-rigged" and temporary plumbing installed. In the case of the DC-10 crash at O'Hare Field, a darkroom was improvised inside a cargo compartment of a wide-bodied jet.

The radiologists should set up a temporary office at the morgue where they can supervise the radiography and interpret the films while maintaining immediate contact with other investigators. The viewing equipment employed will need to be consistent with the imaging modality in use, but a key requirement is to be able to permit comparison of large numbers of pre- and postmortem images while minimizing repetitive image handling (Figure 13.13). If film is to be used, a bright light is especially useful because technologists may have difficulty optimizing exposures and extremes of density range are likely.

It may be desirable to use standard full-size 14 × 17 in. (35.6 × 43.2 cm) detector/film exclusively for optimum coverage and ease of handling. Permanent metallic markers should always be used to label images at the time of exposure to

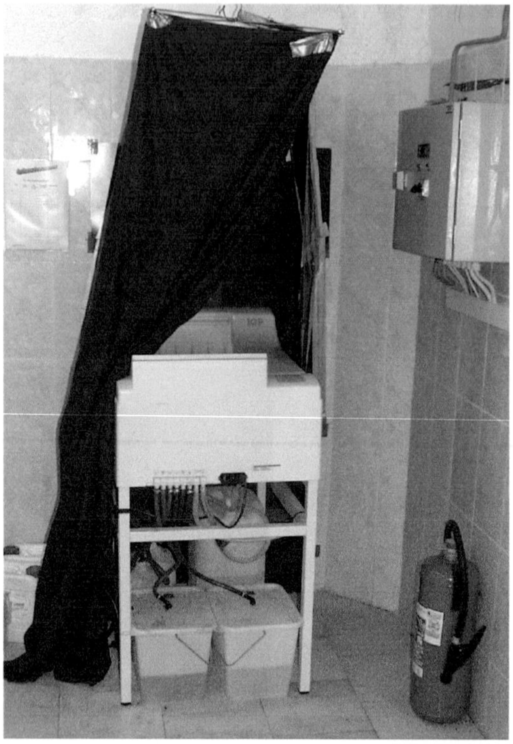

FIGURE 13.12 A small, self-contained tabletop film processor used by the Forensic Teams of the United Nations ICTY in Bosnia in 2000. These units require minimal installation, using a single 13 amp power supply and water recirculation from a tank beneath the unit.

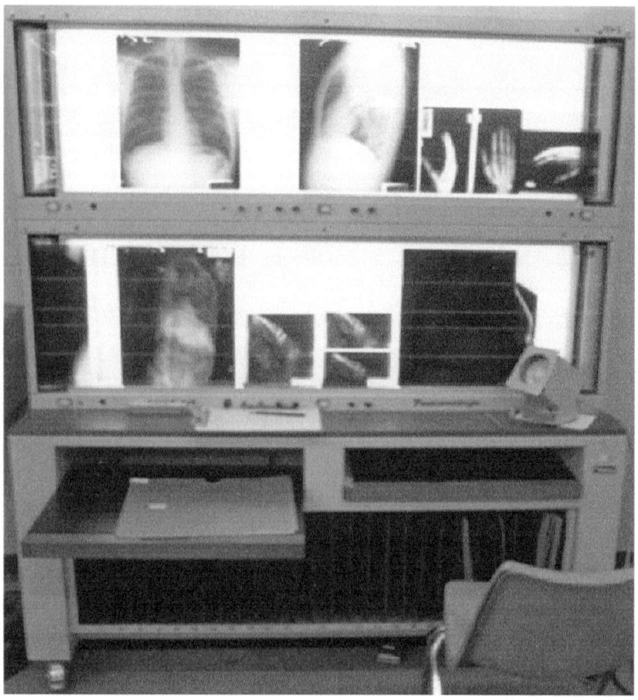

FIGURE 13.13 "Rotoscope" film viewer. The scroll system of mounting the films is especially useful for directly comparing large numbers of pre- and postmortem images for potential matching. Note the bright light (arrow) for viewing overexposed radiographs.

eliminate any questions engendered by digital annotations or adhesive labels/ hand-scrawled markings applied after processing. The technologists' names, date, and time of the image should also be recorded on the image either at the time of exposure within the DICOM header or photographically via actinic marker prior to development in the case of film.

Radiation exposure to team members should be minimized by placing the x-ray exposure facility within a controlled area that is either adequately shielded or physically separated from other team activities by means of an exclusion zone. An outside corner of a building may be used, with portable prefabricated lead shields erected around the radiographic units (Figure 13.14). For all operations, the advice of a radiation protection specialist should be sought and in the ideal situation, a radiation physicist should be consulted in the compilation of the emergency plan. Radiation monitoring badges should be issued to all workers in the immediate area and radiation levels monitored.

Routine methods of film reporting are usually impractical in a field environment, and although some government emergency response agencies may deploy a dedicated digital records system, arrangements for dictation, transcription, and secretarial support are likely to be either significantly different from normal clinical practice, cumbersome, or entirely unavailable. In a true field situation, a single-page, handwritten x-ray report is useful (Figure 13.15), and a photocopier, in fact, is almost essential. A simple form can be duplicated and one copy attached to the outside of the x-ray package for easy reference, while another accompanies the body to permit its review by pathologists at the time of the autopsy.

Yet another copy may be included in a master folder. A simple diagram (Figure 13.16) to indicate portions of the body present or missing is helpful.

METHODOLOGY

While the main goal of the initial medical investigation of mass casualties is the rapid identification of the victims, rather than determining the mechanism of death, the precise forensic protocol adopted will be dependent upon the nature of the incident. Mass casualty incidents arising from natural disasters are likely to present relatively straightforward issues of victim identification, whereas other criminal incidents (e.g., terrorist incidents) will present conflicting priorities of incident investigation and identification of the deceased.

Whatever forensic protocol is adopted, the use of imaging techniques for mass fatalities is best deployed along the same lines as for the management of major trauma in the emergency room[25]:

- *Primary Survey*—Initial triage and assessment
- *Secondary Survey*—Standard examination of specific body parts (e.g., dentition)
- *Tertiary Examinations*—Specific examinations performed in response to findings during primary or secondary surveys or during pathology, odontology, or anthropology assessment. (Figure 13.17)

FIGURE 13.14 (a) Portable lead-lined x-ray shield being assembled. Designed to protect an individual technologist, several of these may be combined to form a wall to provide radiation protection for nearby workers. (From Lichenstein, J. E., et al., *Aviat. Space Environ. Med.*, 51, 1004, 1980. With permission.) (b) Physical barriers being used to create an exclusion zone around the perimeter of the radiation area. (From Metropolitan Police, Scotland Yard. With permission.)

X-RAY REPORT

I.D. # ______
NAME _________

GROSS FEATURES:

APPROX. AGE: Child
Teen
Adult
Without Degen. Changes
With Degen. Changes

BONE STRUCTURES: Small Large

POSSIBLE SEX: Male __ Female __, ? __

POTENTIALLY USEFUL IDENTIFYING FEATURES:

(1) Foreign bodies
(2) Personal effects
(3) Appliances
(4) Surgical changes
(5) Anomalies or unique skeletal features
(6) Dental structures: Not seen __ Seen __
(7) Sinuses: Not seen __ Seen __

ADDITIONAL VIEWS SUGGESTED:

COMMENTS:

FIGURE 13.15 A simple one-page x-ray report form for field use when screening casualty victims. It can be duplicated with one copy accompanying the body and the other being kept with the image files. (From Lichtenstein, J.E. et al., *Aviat. Space Environ. Med.*, 51, 1004, 1980. With permission.)

Primary Survey (Triage)

Ideally, all forensic autopsies should include a preliminary radiographic screening or "Primary Survey." The help these data provide the pathologist may be an end in itself, but they also provide the first step in primary radiographic identification. Important clues may be demonstrated if the victims are radiographed initially with clothing and personal effects in place. Such preliminary screening may be undertaken using radiography, fluoroscopy, or CT scanning, providing a screening study for the pathologist by emphasizing dental and surgical artifacts, injury patterns, as well as descriptions of observed personal effects and foreign bodies.

The purpose of the primary survey is to undertake an initial assessment of the remains. This should be considered essential in all cases but is particularly important for circumstances involving extensive fragmentation, decomposition, or intermixing with debris. Imaging examination of the sealed body bag will yield all or some of the following information about the contents of the body bag:

- Recognition of body parts with discernable anatomical landmarks that can be used for body part identification, which is especially useful with cases of fragmentation and decomposition.
- An indication of whether the remains of more than one individual are present.

(a)

Site Code ____________________

Body Number ________________

RADIOGRAPHY SURVEYS

Whenever possible the body, body part or individual bones should be examined prior to the post-mortem examination. Please record the following information:

Primary Survey: *Initial Fluoroscopy of the Bag or Box*

Date of examination:	**Time of examination:**	
Radiographer 1	Print:	Signed:
Radiographer 2	Print:	Signed:
Bag/box (Please tick one)	☐ Opened prior to examination ☐ Opened at examination	☐ Unopened

Record of images submitted for evidence:

Image No.	Exhibit No.	Features seen

Image format:	Handed to: (Please print) Signed: Position:

Recorded by ____________________

Signature ____________________

Date ____________________

FIGURE 13.16 Examples of radiology survey recording forms developed by the Inforce Foundation and the Association of Forensic Radiographers. (a and b) Primary Survey Recording Form, (c and d) Secondary Survey Recording Forms, and (e) Tertiary Examination Recording Form. (From Inforce Foundation & The International Association of Forensic Radiographers. With permission.)

(b)

Site Code ____________________

Body Number ____________________

INJURY/PATHOLOGY LOCATION DIAGRAM

Top of Bag

Right side of bag

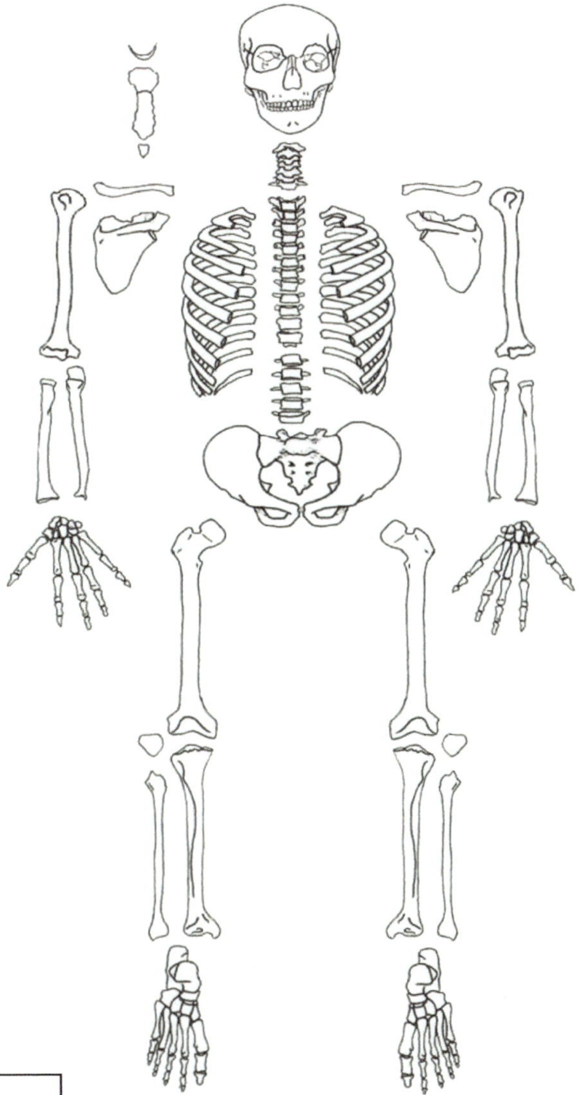

Left side of bag

KEY

#: Fracture
—: Amputation
P: Pathology
M: Metal work

Recorded by ____________________

Signature ____________________

Date ____________________

FIGURE 13.16 (*Continued*)

(c)

Site Code ____________________

Body Number ____________________

Secondary Survey: *a) Skeletal Survey*

Date of examination:		
Radiographer 1	Print:	Signed:
Radiographer 2	Print:	Signed:

Skeletal Survey

Examination Performed	View	Side	Exhibit No.
Skull			
Face			
Mandible			
Chest / Ribs			
Abdomen			
Pelvis / Hip			
C Spine			
T Spine			
L/S Spine			
Humerus			
Radius & Ulna			
Hand			
Femur			
Tibia & Fibula			
Foot			

Image format	Handed to: (Please print) Signed: Position:

Recorded by ____________________

Signature ____________________

Date ____________________

FIGURE 13.16 *(Continued)*

(d)

Site Code ____________________

Body Number ____________________

Date of examination:		Time of examination:
Radiographer 1	Print:	Signed:
Radiographer 2	Print:	Signed:

Secondary Survey: *b) Dental Survey*

Peri-apical/Bitewings

Circle each tooth or area radiographed

18	17	16	15	14	13	12	11	21	22	23	24	25	26	27	28
48	47	46	45	44	43	42	41	31	32	33	34	35	36	37	38

R L

Occlusal Views:	Exhibit No.

Mandible Views	Exhibit No.

Comments:

Image format:

Handed to: (Please print)

Signed: Position:

Recorded by ____________________

Signature ____________________

Date ____________________

FIGURE 13.16 *(Continued)*

(e)

Site Code ____________________

Body Number ____________________

Tertiary Survey:

Body Part/Pathology

Ante-Mortem/Peri-Mortem Trauma and/or Pathology

Date of examination:		**Time of examination:**
Radiographer 1	Print:	Signed:
Radiographer 2	Print:	Signed:

Findings of Examination

Description of Body Part/Part Examined	**Image No.**	**Exhibit No.**

Image format:	Handed to: (Please print) Signed: Position:

Recorded by ____________________

Signature ____________________

Date ____________________

FIGURE 13.16 *(Continued)*

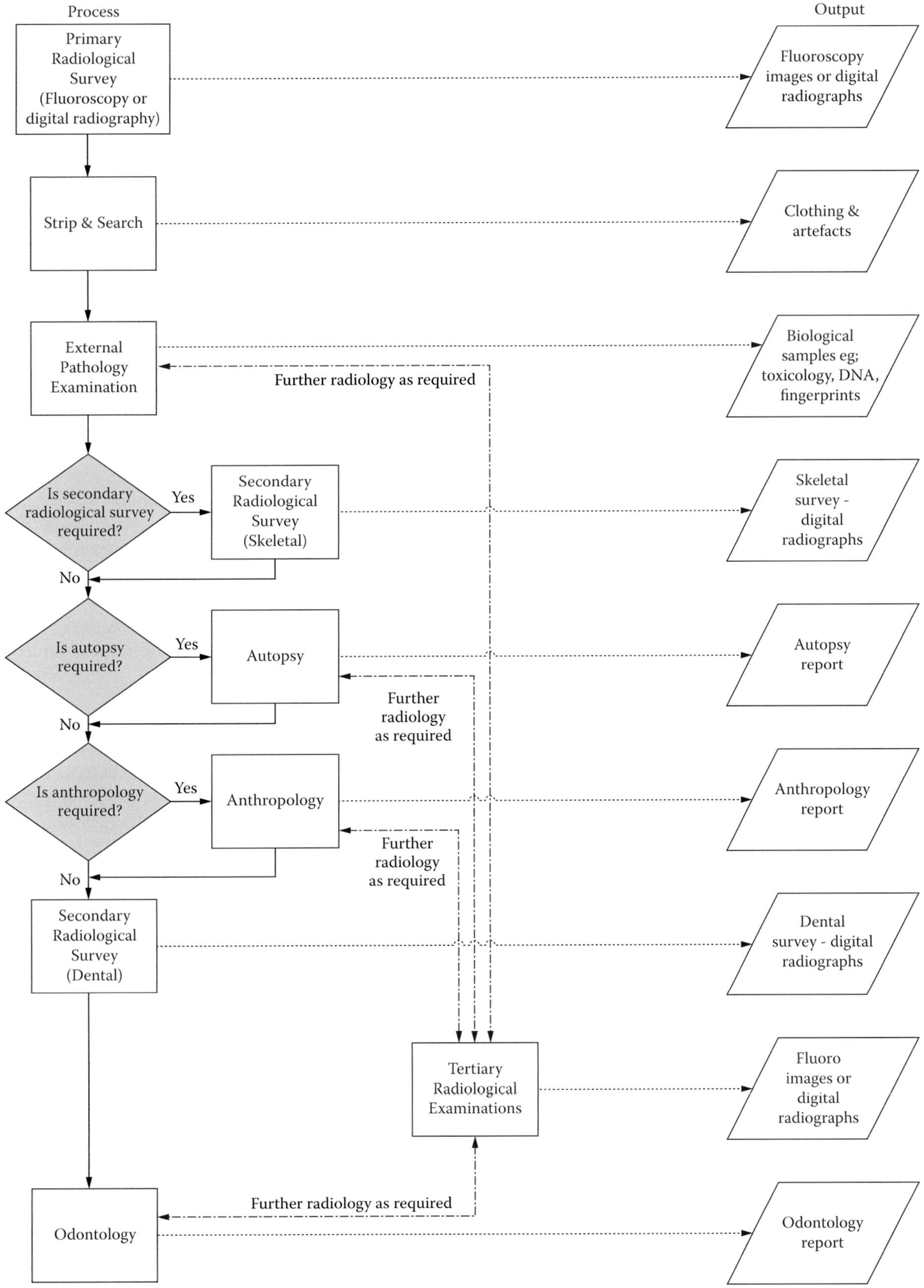

FIGURE 13.17 Flowchart showing the workflow within the emergency morgue and the integration of radiology examinations at key stages within the process. (From Viner, M. D., In Adams, B. J., Byrd, J. E., Ed., *Recovery, Analysis and Identification of Commingled Human Remains*, Humana Press, Totowa, NJ, 2008. pp. 169–174. With permission.)

- The location and nature (if possible) of any hazardous material—unexploded ordnance, metallic sharps, glass, and so on.
- The location and degree of skeletal trauma. This may include the location of any projectile fragments with possible associated bony injury.
- The location of unassociated teeth and other small body parts useful for identification.
- The location of personal effects, for example, jewelry, cigarette lighters, keys, wallets, and so on. (This may be particularly useful in cases of burned or exhumed remains, where these artifacts may be difficult to locate.)
- The presence of any unique identifying features (e.g., prosthetic hip replacement), which may require further radiographic investigation following autopsy and/or odontology.
- The presence of previous healed fractures and other preexisting pathological or anatomical features that may be useful for identification purposes.
- The presence of dental work (bridges, crowns, root canal treatments, etc.).

The information obtained from the primary survey should be recorded on a report form to aid subsequent pathology, anthropology, and odontology examinations.

If fluoroscopy is used, it is recommended that the entire examination is recorded using video or similar technology. Although hard or soft-copy images should be taken of significant findings, fluoroscopy is a dynamic real-time examination and it is advisable that the process is undertaken by a radiographer or radiologist along with a forensic pathologist or anthropologist and appropriate law enforcement personnel to ensure continuity. The use of either film or DR as a primary survey tool will result in a series of hard or soft-copy images as a permanent record.

In cases with large amount of fragmented commingled remains, a primary survey of body parts using DR can be used to construct an image database to aid the identification process. Images can be categorized and filed according to observed anatomical parts, facilitating rapid retrieval for later comparison with antemortem radiographs, should this prove necessary. It should be remembered that in most cases radiographs produced at the primary survey stage will not be of sufficient quality to permit accurate evaluation by a radiologist, due to the random nature of the anatomical positioning of the body parts within the body bag. In some cases, where there is minimal disruption of the cadavers, it may be possible to ensure that radiographs taken during the primary survey are in the true anteroposterior position, which will limit any radiographs required at the secondary stage.

Further radiography examinations may be required as part of the investigative process as detailed below. These examinations, performed subsequent to the primary survey should in all cases be correctly anatomically positioned so that the resultant radiographs replicate as far as possible the standard views that would be undertaken on a live subject.

Secondary Survey (Standard Radiographic Examination)

The value of a routine radiographic skeletal survey (a "Secondary Survey" in which films are obtained in as near anatomic position as possible, following removal of artifacts and clothing) will depend upon the nature of the incident and will be decided by weighing the usefulness of the information gained versus the time taken by the procedure. In some cases, this procedure may be employed routinely for all victims, while in others it may be used only for specific cases—for example, aircraft pilots but not passengers. Such a survey can provide additional useful information for the pathologist emphasizing antemortem anatomical variations and enabling estimates of age, sex, and stature.

The secondary survey should be undertaken after the initial strip and search and external examination so that standard positions can be replicated without overlying clothing and other artifacts. In most cases, examinations can be undertaken following autopsy, but this will be dependent upon the precise nature of the examinations to be performed. Examinations of the skull, for example, are in most cases best undertaken prior to the cranial vault being opened.

The use of imaging to obtain standard projection radiographs as a routine part of the examination protocol should be restricted to those cases that are likely to yield the greatest benefit from the deployment of resources required. In the case of a mass fatality investigation, advances in other identification methods (e.g., DNA) have largely negated the need for the full skeletal-survey examination employed in previous incidents.[5,18] In such cases, secondary surveys are now usually restricted to routine dental surveys as a means of collecting postmortem data for later comparison with antemortem films. As the incidence of dental x-ray examinations and availability of dental records in Western populations is high, the likelihood of such antemortem data being available makes the routine dental survey worthwhile. In other populations, where dental treatment is either rare or poorly documented, it may be decided that a routine dental radiography survey is not indicated. However, there may be value in routinely taking radiographs of the mandible in the region of the third molar in order to provide data for age estimation.

The precise requirements for dental radiography surveys will be determined by the working practices of the forensic odontologist. In the absence of a presumptive identification with antemortem data for comparison, a full sequence of intraoral periapical films showing the entire dentition, together with bilateral bitewing examinations represents a thorough survey from which comparison can be made with antemortem data. In the case of large-scale examinations involving multiple fatalities, the secondary dental survey represents a significant and time-consuming part of the postmortem data collection process. A team approach, involving odontologists and experienced dental radiographers and nurses working together in the incident mortuary, can greatly increase the speed of the identification process and negate the requirement for body parts to be examined elsewhere.[7]

Similar to the dental images, there may be other indications for including a series of cranial and postcranial projections in the routine postmortem examination protocol. For example, in the case of examining the remains from suspected atrocity crimes, the possibility of systematic antemortem torture and/or beatings may indicate routine examination of body parts, such as skull, limbs, and ribs for evidence of healed or healing fractures at the time of death.[26,27]

In all events, the precise protocol employed will be dependent upon a number of factors unique to the circumstances surrounding the death of the individuals. The protocol will, thus, need to be agreed in consultation with the coroner, medical examiner, odontologist, anthropologist, radiologist, and radiographer in order to achieve the maximum benefit while limiting examination time and resources.

Victims successfully identified by other means (e.g., dental anatomy, fingerprints, etc.) may receive no such further radiographic investigation except to check for inconsistencies of data between the various techniques. Those bodies not identified by other means may then undergo a secondary survey and/or be reexamined in additional projections or by using more elaborate (Tertiary Examination) techniques for comparison and matching with available antemortem films.

Tertiary Examinations (Special Circumstances)

A range of medical imaging techniques and examinations may be useful in determining the identity of an individual, their anthropological profile, or in determining the cause and manner of their death. The techniques employed will vary from case to case and will be determined by the nature, or suspected nature, of the incident under investigation. In most instances, the requirement for further specific imaging will be determined as a result of data obtained at primary radiological survey or from the pathology, odontological, anthropological, or crime scene examinations. It should again be remembered that radiographs produced at the primary survey stage will not be of sufficient quality to permit accurate evaluation by a radiologist due to the random nature of the anatomical positioning of the body parts within the body bag. Examples of possible indications are:

- Any unique skeletal features or pathological conditions seen during the primary survey or identified during examination by the pathologist, anthropologist, or odontologist that may be useful for identification.
- To replicate poor quality antemortem dental radiographs by undertaking subsequent examinations using substandard angulations to facilitate accurate comparison.
- In cases where evidence of trauma identified by the pathologist, anthropologist, or odontologist may indicate further imaging investigations to determine the nature of the injury or weapon used.
- To detect, locate, and retrieve items of forensic evidence seen during the primary survey but not located during examination by the pathologist.
- For those cases, which are proving difficult to identify via other means, a full skeletal survey may be useful in determining age, sex, stature, and so on, or for detecting unique skeletal features, which have been previously documented in antemortem records.
- To document injuries and injury patterns for the purposes of the criminal investigation or as a means of negating the requirement for full autopsy where the cause of death is known.

In the last two cases, imaging examinations may be particularly useful in cases of fleshed remains where anthropological examination is difficult.

IDENTIFICATION

Individual identification may be established by a variety of means including visual recognition, analysis of personal effects, fingerprints, dental charting, and so on. The identification process in mass fatality incidents is far more complex, due in part to the numbers of individuals involved and in part to the nature of the incident. In most mass fatality incidents, for example, visual identification is not advisable, due to the increased possibility of misidentification, personal effects belonging to one individual may become mingled with those of another, and bodies may suffer trauma, disfigurement, and dismemberment, rendering some identification methods unreliable.

Identification in multiple fatality incidents is thus accomplished by painstaking scientific analysis of all available evidence and confirmed by means of establishing a number of primary and secondary indentifying features. The primary identifying methods usually are: dental radiology, fingerprinting, and DNA analysis, and these are supported by secondary methods such as examination of personal effects and medical records, anthropological profiling, and so on.

Radiology is usually referred to as a secondary method of identification, although in certain circumstances it may be acceptable to some authorities as the fourth primary identifier. In cases of severe tissue damage and mutilation, radiology may become the primary (and often the only) means of positive identification. Even when teeth remain, forensic odontologists use x-rays as an extension of dental charting because radiographs provide almost unlimited points of identity. Matches may be based upon very tiny details of dental filling shapes and trabecular patterns in surrounding alveolar bone.

Sometimes, surgical changes, effects of old injury, or anomalies demonstrated by the radiographic data may be specific enough for identification based upon history alone. However, when dealing with many victims, the sine qua non is the availability of antemortem radiographs for direct comparison. In addition to physicians and dentists, local medical societies, employers, and practitioners such as chiropractors and podiatrists should be contacted as potential sources of antemortem x-rays.

The number of exactly matching features required for a positive identification depends upon the number of victims involved. Often, complex anatomic patterns found in ante- and postmortem material can be superimposed radiographically to provide indisputable evidence of identity even from very small fragments. Potentially matching remains can be reexamined in various projections after clearing away soft tissue and debris to permit such superimposition (Figures 13.18 and 13.19). The spine and individual bones such as the clavicles are particularly amenable to this technique.[28]

Even when complete superimposition is impossible, x-ray data often will support or, by disclosing obvious mismatches, refute a tentative identification suggested by another technique. A basic principle is the requirement of confirmation by as many different techniques as possible, and the careful cross-checking of each technique for exclusionary data, before making a final certification of identity.

In the case of burning and mutilation, there are often some remains for which too few data are available for positive identification by the ordinary means of directly matching ante- and postmortem features. Then reliance must sometimes be placed on a process of exclusion. When a small number of such bodies remain to be identified, they can be separated by gross exclusionary features such as sex, stature, and blood type, which would not by themselves be definitive. The computer can be a useful tool in handling multiple potentially identifying features among multiple bodies or parts (Figure 13.20). The inferences drawn from this process can lead to definitive identification, but only so long as it is known absolutely that each body must correspond to one of a known group of victims.[11]

The technique was employed in the last stages of identifying 256 bodies from a 1985 military transport plane crash in Gander, Newfoundland in which it was thought that all the passengers and crew were known and that no other victims were involved.[18] It was also successfully employed in the positive identification of fragmentary human remains recovered from the scene of a terrorist bombing in the Jewish

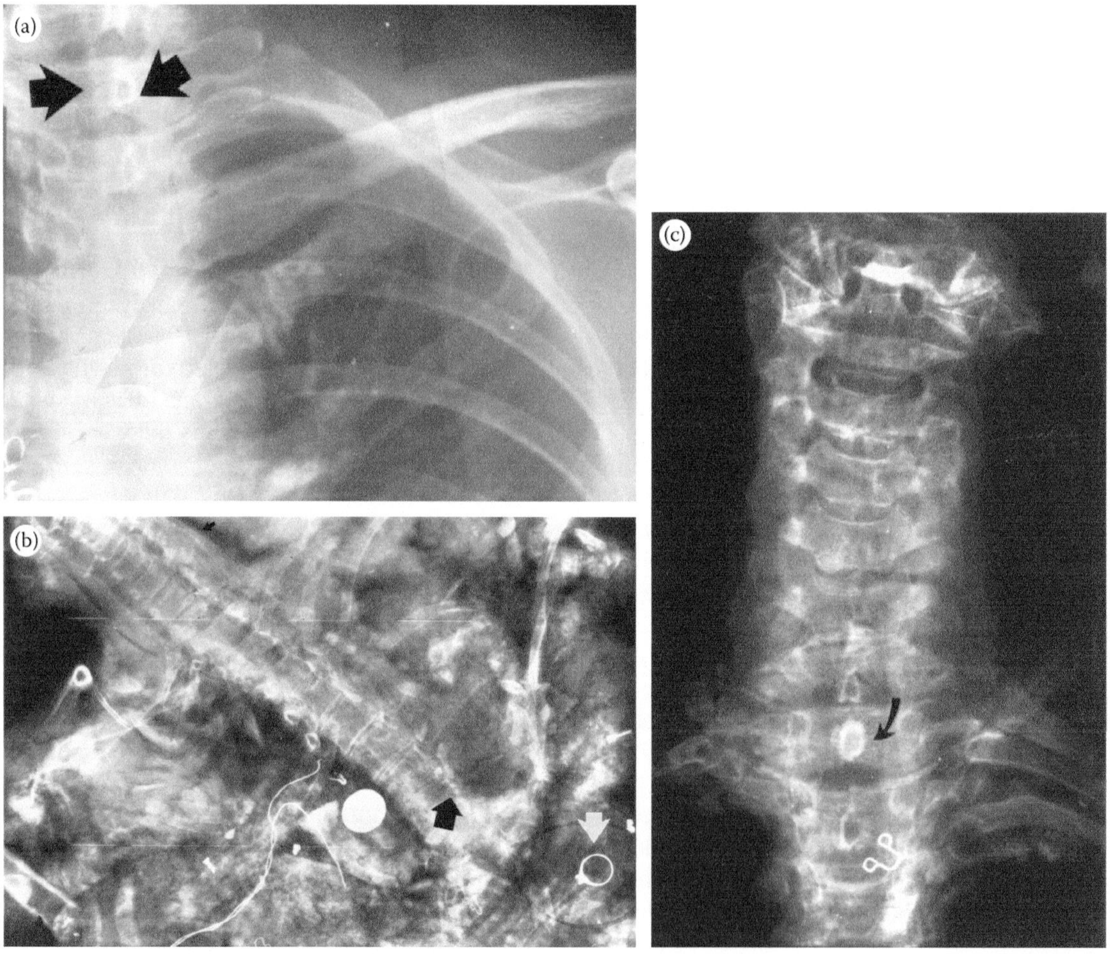

FIGURE 13.18 (a) An antemortem film of a known victim of the Chicago DC10 crash. Arrows indicate tracheal calcifications and a T-1 spinous process with dense cortex relative to that of T-2. (b) Postmortem film taken with remains in a body bag shows severe disruption of the skeleton and multiple foreign bodies including aircraft parts. A ring (white arrow) was noted and recovered. It was later identified as belonging to the victim seen in A. A black arrow points to calcified tracheal cartilages similar to those seen in A. (c) the upper cervicothoracic portion of the skeleton was cleaned and radiographed in standard anatomical positions. Note the density of the T-1 spinous process (curved arrow) and the shape of the C-7 spine—findings that match those in the antemortem film A. (From Lichtenstein, J. E., Fitzpatrick, J. J., and Madewell, J. E., *Am. J. Roentgenol.*, 150, 751, 1988. With permission.)

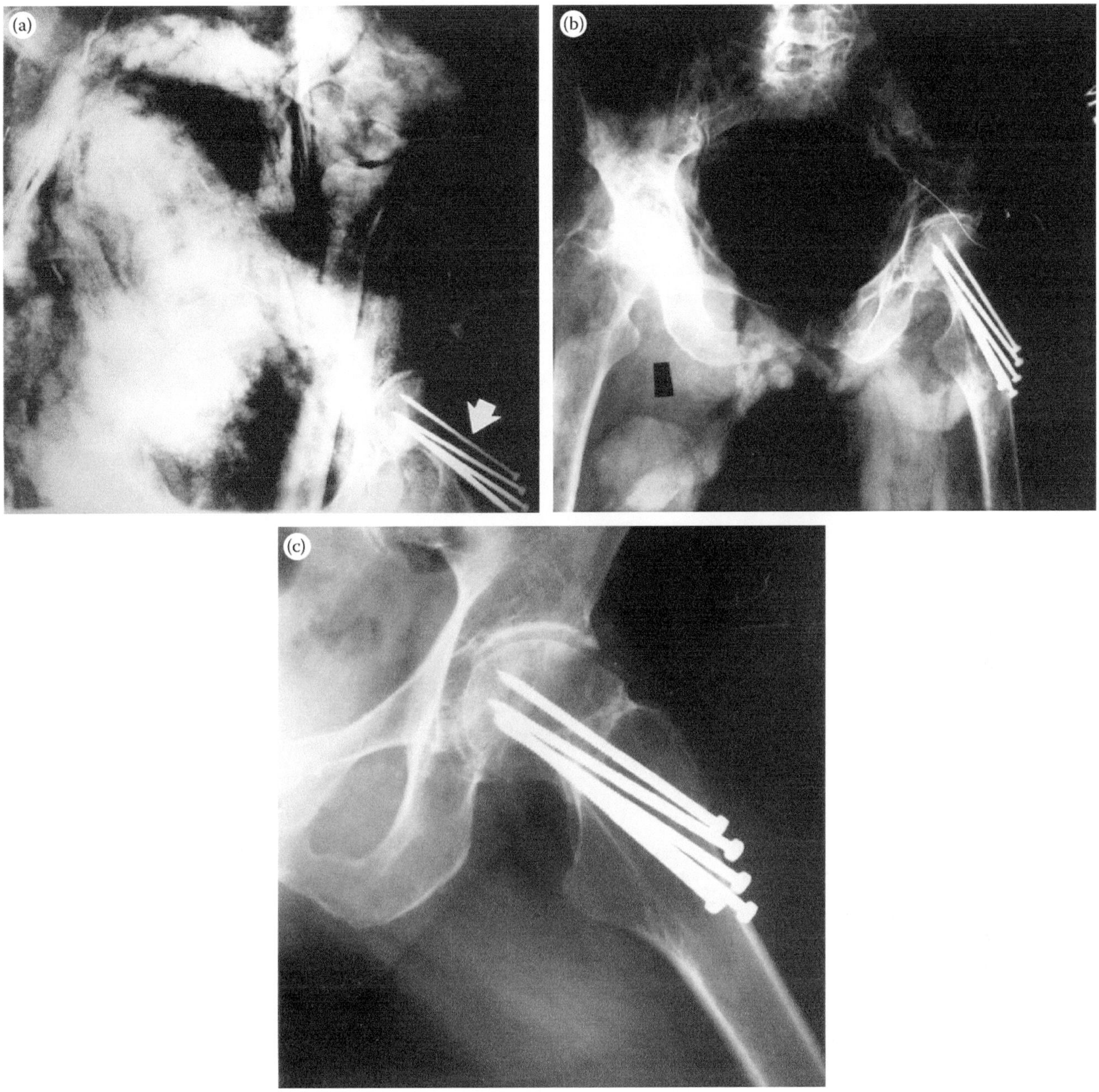

FIGURE 13.19 (a) Initial screening study of wrapped remains showing metallic screw fixation of a proximal femur fracture (arrow). (b) Note improvement achieved by reexamination in anatomic position after unwrapping. Superimposition with antemortem film. (c) Permitted positive radiographic identification. (From Lichtenstein, J. E., Fitzpatrick, J. J., and Madewell, J. E., *Am. J. Roentgenol.*, 150, 751, 1988. With permission.)

Argentine Mutual Association Center in Buenos Aires, Argentina, in July 1994.[29]

INJURY MECHANISM AND PATTERN ANALYSIS

While identification itself is important for humanitarian and legal reasons, scientific interest lies in the need to identify victims to correlate their injury mechanisms with their relationship within the fatal environment.[30–40]

If the nature and cause of an event is to be analyzed, the detailed mechanisms of injury and death of the victims must be determined. That requires that the location and roles of individuals within the fatal environment be known. This, in turn, requires documentation of the site of discovery of the remains and their accurate identification as specific individuals. There is a special need to identify and analyze the remains of those in control of the environment, such as pilots or plant operators. Patterns of injury to hands and feet of aircrew members, for example, may establish the status of controls and suggest the position or actions of the victims. Such analysis is useful not only for investigation of specific events, but more importantly to correlate patterns found in similar incidents in an attempt to prevent them in the future.

Name	Sex	Race	Age		Height	Weight	Hair Color	Features		
*	M	White	47		65	155	Brown&Red	Knee injuries		
	*	Sex	Race	Age	Stature	Weight	Hair Color			
25	955	M	White	45 ± 10	66 ± 2	160 ± 10	Brown&Red	Post/Teeth		
25	1013	M	White	45 ± 10	67 ± 2	150 ± 15	Brown&Red	Muscular		
22	941	M	White	28 ± 4	68 ± 3	155 ± 15	Brown&Red	Post/Teeth		
22	988	M	White	55 ± 10	66 ± 2	170 ± 15	Brown&Gryng	Circ		
22	994	M	White	55 ± 10	67 ± 2	190 ± 10	Brown&Red			
22	1031	M	White	45 ± 10	68 ± 2	150 ± 20	Brown&Gryng	Fx/Up/Extr		
22	1036	M	White	40 ± 10	68 ± 4	170 ± 15	Brown&Red	Post/Teeth		
22	1064	M	White	33 ± 6	68 ± 3	165 ± 15	Brown&Red	Post/Teeth Circ	Fx/Back	
21	792	M	White	50 ± 10	71 ± 4	230 ± 30	Brown&Red	Hirsute		
21	1025	M	White	50 ± 15	72 ± 4	200 ± 20	Brown&Red	Med. Bld		

Name	Sex	Race	Age		Height	Weight	Hair Color	Features		
*	M	White	37		65	155	Brown&Red	Knee injuries		
	*	Sex	Race	Age	Stature	Weight	Hair Color			
25	799	M	White	40 ± 10	69 ± 3	160 ± 10	Brown&Red			
25	3	M	White	35 ± 10	70 ± 3	150 ± 15	Brown&Red	Metalplate Uncirc	Hirsute	Large/Head
25	1036	M	White	40 ± 10	68 ± 4	155 ± 15	Brown&Red	Post/Teeth	Fx/Back	
25	1064	M	White	33 ± 6	68 ± 3	170 ± 15	Brown&Red	Post/Teeth CiRc		
23	910	M	White	50 ± 10	67 ± 3	190 ± 10	Brown&Red			
23	941	M	White	28 ± 4	68 ± 3	150 ± 20	Brown&Red	Post/Teeth		
23	944	M	White	35 ± 10	69 ± 3	170 ± 15	Brown&Red	Hirsute		
23	1031	M	White	45 ± 10	68 ± 2	150 ± 20	Brown&Red	Fx/Up/Extr		
22	988	M	White	55 ± 10	66 ± 2	170 ± 15	Brown&Red	Circ		
22	990	M	White	45 ± 10	70 ± 4	150 ± 20				

FIGURE 13.20 Example of computerized sorting for identifying features of multiple bodies in a mass casualty situation.

Events, where forces are on the borderline of survivability, are of particular interest. Important questions concern the factors determining why some survive, or do not, in potentially survivable situations. Head or lower-extremity injuries can explain failure to escape from otherwise survivable situations such as postcrash fires. Separating mechanical from thermal injury patterns and determining whether burning was the cause of death or occurred postmortem after other debilitating injuries can provide important data. These data are vital in studying patterns of mass casualties in order to improve engineering and safety procedures.

REFERENCES

1. Adams, N. S., My life as a forensic radiographer, *Journal of Radiology Nursing*, 22 (2), 56–59, 2007.
2. Dawidson, I., Identification of the Swedish tsunami victims in Thailand, *Forensic Science International*, 169 (1), S47–S48, 2007.
3. Anon., *ACR Bull.*, 6, 25, 1995.
4. Allen, E. W., Scenes from the Oklahoma City bombing, *J. Nucl. Med.*, 36, 30, 1995.
5. Nye, P. J., Tytle, T. L., Jarman, R. N., and Eaton, B. G., The role of radiology in the Oklahoma City bombing, *Radiology*, 200, 541, 1996.
6. Harcke, H. T., Bifano, J. A., and Koeller, K. K., Forensic radiology, response to the Pentagon attack on September 11, 2001. *Radiology* 223(1), 7–8, 2002.
7. Viner, M. D., Rock, C., Hunt, N., Martin, A. W., and MacKinnon, G., Forensic radiography, response to the London suicide bombings on 7th July 2005, *Proceedings of The American Academy of Forensic Sciences 58th Scientific Meeting*, Seattle, WA, p. 176, 2006. (Paper presented at the American Academy of Forensic Science 58th Scientific Meeting, Seattle, WA, 2006).
8. Anderson, A., Cox., M., Flavel, A., Hanson, I., Hedley, M., Laver, J., Perman, P., Viner, M., and Wright, R., Protocols for the investigation of mass graves, in *The Scientific Investigation of Mass Graves: Towards Protocols and Standard Operating Procedures*, Cox, M., Hanson, I., Flavel, A., Laver, J., and Wessling, R., Eds., Cambridge University Press, New York, NY, 2008.
9. Interpol, *Disaster Victim Identification Guide*, http://www.interpol.int/Public/DisasterVictim/Guide.asp.
10. Gross, E. M. and Blumberg, J. M., Identification and injuries of air-crash victims, *Arch. Environ. Health*, 13, 289, 1966.
11 Tarlton, S. W., Identification in aircraft accidents, in *Aerospace Pathology*, Mason, J. K., and Reals, W. J., Eds., College of American Pathologists Foundation, Chicago, 53, 1973.
12. Singleton, A. C., The roentgenological identification of victims of the "Noronic" disaster, *Am. J. Roentgenol.*, 66, 375, 1951.
13. Brown, T. C., Delaney, R. J., and Robinson, W. L., Medical identification in the "Noronic" disaster, *J. Am. Med. Assoc.*, 148, 621, 1952.
14. Lichtenstein, J. E., Madewell, J. E., McMeekin, R. R., Feigin, D. S., and Wolcott, J. H., The role of radiology in aviation accident investigation, *Aviat. Space Environ. Med.*, 51, 1004, 1980.
15. Lichtenstein, J. E. and Madewell, J. E., Role of radiology in the study and identification of casualty victims, *Der Radiologie*, 22, 352, 1982.
16. Lichtenstein, J. E., Fitzpatrick, J. J., and Madewell, J. E., Role of Radiology in fatality investigations, *Am. J. Roentgenol.*, 150, 751, 1988.
17. Joyce, C., and Stover, E., *Witnesses from the Grave*, Little, Brown, Boston, 94, 1991.
18. Mulligan, M. E., McCarthy, M. J., Wippold, F. J., Lichtenstein, J. E., and Wagner, G. N., Radiologic evaluation of mass casualty victims: Lessons from the Gander, *Newfoundland Accident, Radiology*, 168, 229, 1988.
19. Hill, I. R., The Air India jet disaster Kalishna-injury analysis, in *Uses of the Forensic Science*, Caddy, B., Ed., Scottish Academic Press, Edinburgh, 121, 1986.
20. Harbinson, J. F. A., Pathology organization and victim identification after the losses of Air India flight 182, 23/6/1985, *Can. Soc. Forensic Sci. J.*, 20, 16, 1987.
21. Interpol, Resolution on Disaster Victim Identification, 1996.
22 Viner, M. D., Forensic investigation: The role of radiography, *European Congress of Radiology, European Radiology, Supplement to Volume 11, Number 2, February 2001*, 2001, p. 95. (Paper presented at the European Congress of Radiology, Vienna 2001.)
23. Rutty, G. N., Robinson, C. E., BouHaidar, R., Jeffrey, A. J., and Morgan, B., The role of mobile computed tomography in mass fatality incidents, *Journal of Forensic Sciences* 52 (6), 1343–1349, 2007.
24. Rutty, G. N., Robinson, C., Jeffrey, A., and Morgan, B., Mobile computed tomography for mass fatality investigations, *Forensic Science, Medicine, and Pathology* 3 (2), 138–145, 2007.
25. Viner, M. D., *The Use of Radiology in Mass Fatality Events*, in Adams, B. J., Byrd, J. E., Ed., *Recovery, Analysis and Identification of Commingled Human Remains*, Humana Press, Totowa, NJ, 2008, pp. 169–174.
26. Tonello, B., Mass grave investigations, *Proceedings of 1998 Imaging Science and Oncology;* British Institute of Radiology, 1998, p. 69. Paper presented Imaging Science & Oncology, Birmingham, UK, 1998.
27. Viner, M. D., The radiographers role in forensic investigation, *Hold Putsen: Journal of the Norwegian Society of Radiographers*, Nr 9/2001 October, Oslo, 2001.
28. Sanders, I., Woesner, M. E., Ferguson, R. A., and Noguchi, T. T., A new application of forensic radiology: Identification of deceased from a single clavicle, *Am. J. Roentgenol. Radium Ther. Nucl. Med.*, 115, 619, 1972.
29. Kahana, T., Ravioli, J. A., Urroz, C. L., and Hiss, J., Radiographic identification of fragmentary human remains from a mass disaster *American Journal of Forensic Medicine and Pathology* 18 (1), 40–44, 1997.
30. Barrie, H. J. and Hodson-Walker, N., Incidence and pathogenesis of fractures of the lumbar transverse processes in air crashes, *Aerosp. Med.*, 41, 805, 1970.
31. Besant-Matthews, P. E., Photography and radiography in aircraft accident investigation, in *Aerospace Pathology*, Mason, J. K. and Reals, W. J., Eds., College of American Pathologists Foundation, Chicago, 1973, 177.
32. Dunne, M. J., Jr. and McMeekin, R. R., Medical investigation of fatalities from aircraft-accident burns, *Aviat. Space Environ. Med.*, 48, 964, 1977.
33. Fatteh, A. V. and Mann, G. T., The role of radiology in forensic pathology, *Med. Sci. Law*, 9, 27, 1969.
34. Gable, W. D., Pathology patterns in aircraft accident investigation, *Aerosp. Med.*, 39, 638, 1968.
35. Krefft, S., Estimation of pilot control at the time of crash, in *Aerospace Pathology*, Mason, J. K., and Reals, W. J., Eds., College of American Pathologists Foundation, Chicago, 96, 1973.
36. Krefft, S., Who was at the aircraft's controls when the fatal accident occurred? *Aerosp. Med.*, 41, 785, 1970.

37. Mason, J. K., Passenger tie-down failure. Injuries and accident reconstruction, *Aerosp. Med.*, 41, 781, 1970.
38. McMeekin, R. R., An organizational concept for pathologic identification in mass disasters, *Aviat. Space Environ. Med.*, 51, 999, 1980.
39. Rhodes, R. S., Misleading injury patterns, *Aerosp. Med.*, 41, 794, 1970.
40. Simson, L. R., Jr., Roentgenography in the investigation of fatal aviation accidents, *Aerosp. Med.*, 43, 81, 1972.

14 New Approaches to Radiology in Mass Casualty Situations*

Angela D. Levy and Howard T. Harcke

CONTENTS

INTRODUCTION

The application of cross-sectional imaging techniques to forensic medicine represents a new approach to radiological imaging in disasters and mass-casualty events and affords an opportunity to increase the contributions of imaging to the forensic evaluation of mass casualty situations. The role of radiology in disasters outlined in the previous chapter does not change when cross-sectional imaging is added to the radiological evaluation of the victims of mass casualties. In our opinion, radiology becomes more effective and brings the potential to increase both the speed and accuracy of radiological support to the forensic pathologists and anthropologists supporting a disaster response.

The disaster imaging protocols used by the Office of the Armed Forces Medical Examiner System are outlined in this chapter, and the rationale for the adoption of these protocols and the communication of imaging results in a disaster operation are discussed. These protocols reflect the incorporation of Multidetector Computed Tomography (MDCT) into a fixed mortuary that is designed to expand and contract in response to the number of casualties and nature of the incident. Radiographs and MDCT images illustrating findings that are likely to be found in disaster cases are reviewed.

INTEGRATED IMAGING IN MASS-CASUALTY SITUATIONS

Our current strategy for disaster radiology has evolved from experience at the Charles C. Carson Center for Mortuary Affairs, Dover Air Force Base, Dover, Delaware. Pivotal elements of this experience were (1) the mass-casualty operation that followed the September 11, 2001 terrorist attack, (2) the design and opening of a new mortuary facility in 2003, and (3) the installation of MDCT in 2004. The response to the terrorist attack on the Pentagon in 2001 represented a unique mass-casualty disaster from several aspects. It was terrorist-initiated rather than accidental or natural and both an aircraft and a building were involved with fire. The event produced civilian, military, and terrorist decedents that required identification (Harcke et al., 2002). The lessons learned were carried forward in the design of a new mortuary facility. The addition of cross-sectional imaging necessitated revisions for the processing protocols. In the event of a future mass-casualty disaster, the facility can operate with the advantages afforded by an existing facility and established procedures.

The first consideration in disaster imaging is the need for "safety screening" prior to forensic assessment of the remains. Human remains from a mass-casualty disaster may contain hazardous material such as unexploded ordnance or sharp

* The opinions and assertions contained herein are the private views of the authors and are not to be construed as official or as reflecting the views of the Departments of the Army or Defense.

objects that pose a danger to the workers who handle the remains. While this risk is low in natural disasters, it may be sufficient in some situations to warrant scanning of the human remains with a low-resolution device similar to the type that is used for airport security screening. Because of the military focus at Dover, all incoming material is scanned before forensic processing begins (Figure 14.1).

Standard anterior–posterior (AP) radiographs serve an important role in the triage of recovered material in a mass casualty investigation. Recovered material includes human remains, which may be intact bodies or dissociated parts, and dissociated personal effects. Ideally, imaging takes place after recognizable human remains have been labeled and forensic pathologists or anthropologists have studied the mixed material and separated the human remains. This eliminates unnecessary radiographs and allows arrangement of material in an optimal display for imaging. However, it is not always possible to separate the mixed material. In these cases it is beneficial to x-ray all material that is received to identify human remains, recoverable evidence, and personal effects from the debris associated with the disaster.

We routinely perform radiographs (including MDCT imaging) with all clothing and personal effects in place. While this preserves evidence in an undisturbed state, it may create artifacts and possible confusion on the radiographs. If necessary, the body is returned to radiology for repeat imaging after external examination and cleaning. The spectrum of human remains sent for radiography ranges from intact bodies to small body fragments that are unrecognizable from their gross appearance and mixed with debris from the disaster site. All material received is uniquely labeled, which is a critical practice for radiology and the tracking of specimens in the morgue facility. Radiographs are clearly labeled such that the image matches a specific specimen. It is the responsibility of radiology personnel to ensure correct labeling of each image and to annotate reference points. Right and left markers are also used on intact human remains. If there is doubt on sidedness, the images are labeled as indeterminate. When obtaining the serial AP radiographs from head to toe, there must be sufficient anatomic overlap of radiographs so that no area is missed and serial images are easily matched together.

The same radiological equipment may not be available for each event. The type of equipment available will depend on the facilities and resources that are available at the time of the event. While conventional film-screen based radiographic equipment is still in use in many forensic facilities, state-of-the-art digital equipment is more optimal. Dark room or daylight chemical processing is necessary for film-screen radiography, which limits the ease of use and speed of obtaining radiographs. Digital radiography is strongly recommended for disaster radiology. Computed radiographic (CR) or direct radiographic (DR) digital systems are available in configurations suited for either fixed or temporary mortuary facilities. A DR system is faster and more efficient. The DR unit used at our facility is the Swissray ddrRMulti that occupies a dedicated radiographic room (Figure 14.2). Radiographs are displayed in a picture archiving and communication system (PACS) network.

Based upon radiographic findings the specimen is sent to MDCT (GE Lightspeed16Xtra, General Electric Medical Systems, Milwaukee, WI) (Figure 14.3) or directly to the autopsy room. Intact bodies are routinely imaged with MDCT. The decision to perform MDCT on dissociated human remains depends upon the specific body part and potential to recover evidence, specifically metallic fragments. All intact bodies are scanned isotropically from head to toe with 0.625 mm collimation. Images are reconstructed with a 0.625-reconstruction interval and 0.625-reconstruction thickness. Most bodies can be scanned from head to toe on our scanner because it has extended table travel. Scanners

FIGURE 14.1 A screening scanner is used to screen all remains for unexploded ordnance, sharp objects, or other hazardous materials prior to forensic assessment.

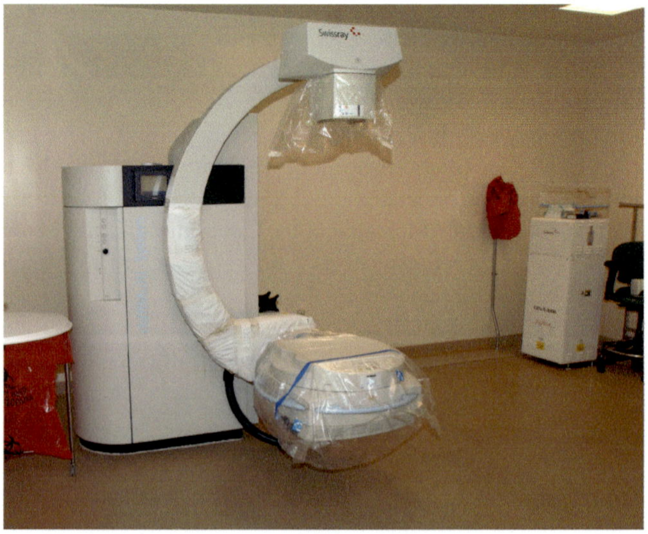

FIGURE 14.2 Digital radiographic room.

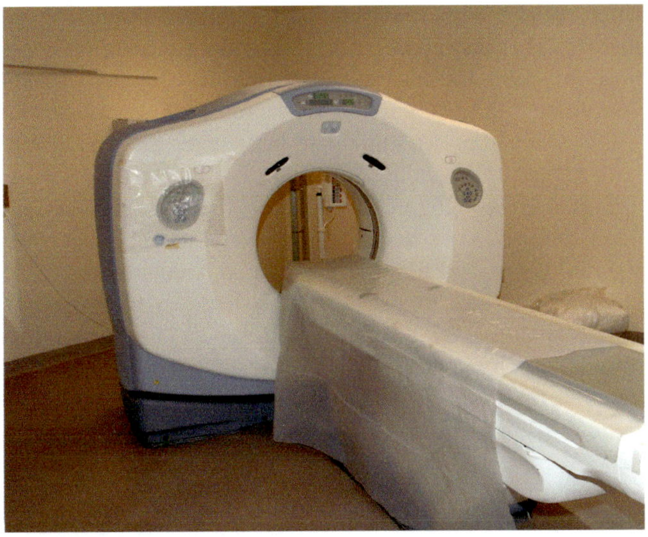

FIGURE 14.3 MDCT scanner.

with standard table travel require two series in order to obtain images of the entire body.

We consider radiography to be essential even when MDCT is available. Because of its excellent resolution and lack of metallic artifact, radiographs complement MDCT. Furthermore, imaging of small, dissociated parts and parts mixed with debris is more efficient with radiography. Both radiography and MDCT images are stored and viewed on a PACS network in the radiologist's work and autopsy room (Figure 14.4). Multiplanar and 3D reformations are performed by the radiologist on a 3D workstation. We use a GE Advantage Workstation 4.2 (General Electric Medical Systems, Milwaukee, WI) (Figure 14.4a) and Vitrea Workstation (software version 4.0.0.0, Vital Images, Inc, Minnetonka, MN). Although our experience with MDCT is in a fixed facility, mobile CT units are available for temporary facilities or to augment a fixed facility (Rutty et al., 2007).

In the autopsy suite, a mobile C-arm fluoroscopy unit is used to locate small metal fragments when the MDCT localization was not sufficient for retrieval (Figure 14.5). The C-arm is also useful to check debris-filled bags for the presence of human tissue when the radiography room is busy and for limited angiographic studies and when digital radiography and MDCT are nonoperational. All personnel heed radiation safety precautions when using the C-arm. Protective equipment (e.g., lead aprons, thyroid shields, and lead gloves) must be available and worn when using the equipment.

Forensic operations during a mass-casualty disaster often require extended work hours (Bluth et al., 2007; Heffernan et al., 2007). The facility may operate with one or two 12-h shifts per day for 7 days per week. We recommend two radiology technologists and one CT technologist per shift with an additional supervisor who oversees the area and troubleshoots. Nonradiology personnel who assist by moving gurneys and transferring of bodies to the radiography and CT tables can augment the radiology technical staff. When all personnel are trained and practiced, the workflow through radiology is approximately seven to eight cases per hour. The number of radiologists required depends upon the number of forensic pathologists. Our experience suggests a ratio of one radiologist for every three forensic pathologists.

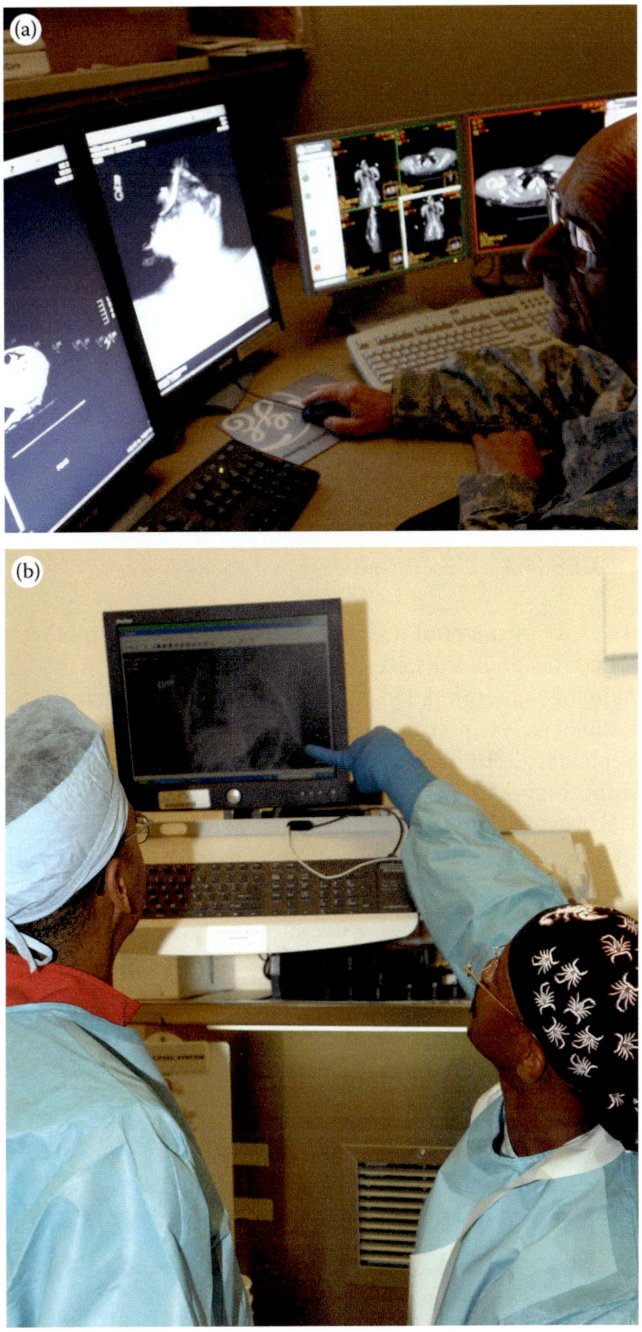

FIGURE 14.4 Imaging viewing. (a) Radiologist reading room with a PACS and 3D workstation. (b) Wall-mounted computer adjacent to the autopsy table allows images to be compared directly to the autopsy images.

REPORTING IMAGING RESULTS

The value of radiographic and MDCT imaging depends upon timely and effective communication of the results to those making decisions on the direction for further processing of

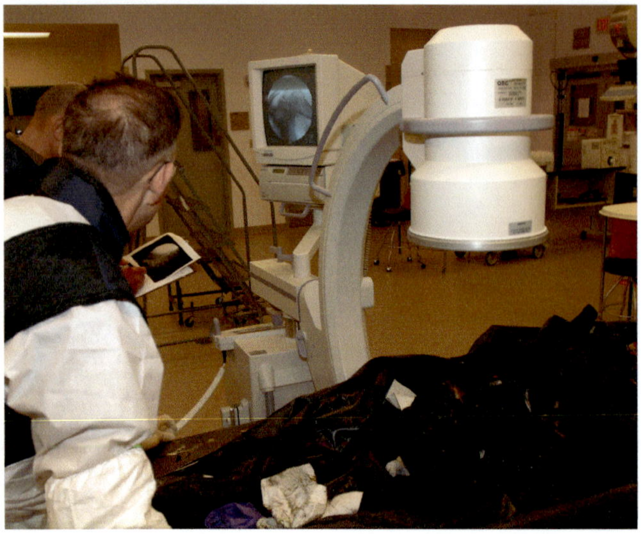

FIGURE 14.5 C-arm fluoroscopy in the autopsy room.

the body or material imaged. Because there is the potential for confusion in a disaster operation, it is our opinion that the radiology interpretation should be a report written by the radiologist or person interpreting the study. During the September 2001 mass casualty, a multipart radiograph request form composed of sheets of paper separated by carbon paper was used for handwritten reporting. The original report was filed with other documents attached to the remains and the chief technologist and radiologist retained copies of the reports. This proved helpful for accounting and consulting purposes after the case moved to the next forensic station. Our current radiology report is written on a worksheet that was developed to accommodate brief notes associated with both radiographic and MDCT findings that are annotated on a human figure (Figure 14.6). This is similar in format to other autopsy worksheets. Verbal communication between the radiologist and forensic pathologist is equally as important as the written report. This is particularly true when the radiologist observes a finding that may dictate the need for special handling of the body. Fortunately, in our facility the location of the autopsy room immediately adjacent to the radiology work area makes this communications possible. Also, there is a PACS workstation in the autopsy room and computer monitors to view images adjacent to each autopsy table (Figure 14.4b). In high volume situations, we typically have one radiologist working in the autopsy area to provide immediate consultation.

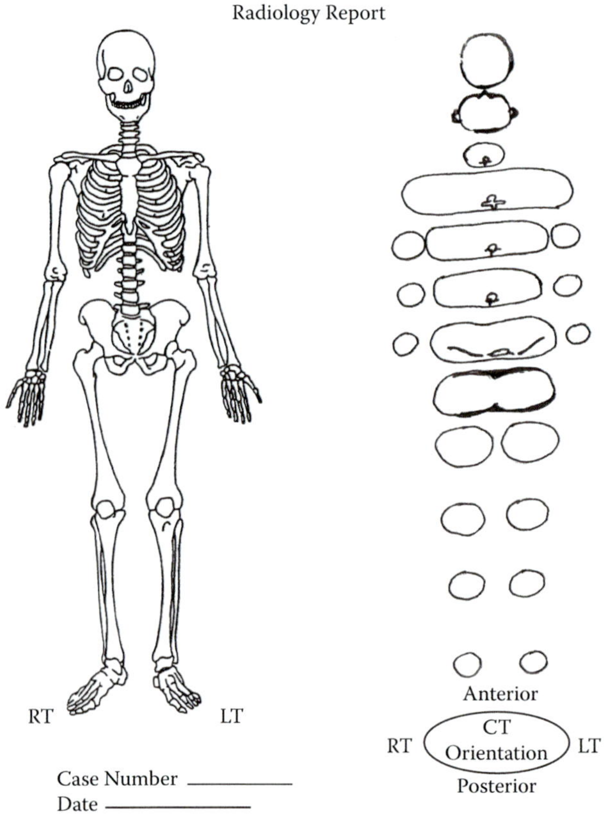

FIGURE 14.6 Radiology worksheet for charting the radiographic and MDCT findings.

INTERPRETING RADIOGRAPHY AND IMAGING FINDINGS IN MASS CASUALTIES

The radiographic and MDCT findings in mass-casualty deaths reflect a range of injury mechanisms and causes of death that include blast injury, blunt trauma, ballistic injury, thermal injury, and drowning. Because human remains may be recovered intact or fragmented, imaging should be tailored to the specific event and remains recovered. Radiography is the method of choice for imaging fragmented human remains with the purpose of establishing identity by recognition of specific skeletal markers, teeth, and/or tissue that may be suitable for DNA recovery (Harcke et al., 2009). Both radiography and MDCT are useful when intact bodies, torsos, or large body parts are recovered. MDCT is uniquely suited to identify unsuspected injury and metallic fragments that may be recoverable and important for incident investigation. MDCT may be useful to help determine the cause of death.

IMAGING FINDINGS IN MASS-CASUALTY VICTIMS

Postmortem Change and Decomposition

Accurate interpretation of postmortem radiography and MDCT requires thorough knowledge of the radiographical and imaging features of postmortem change and decomposition. Postmortem change, namely lividity, and decomposition begin to occur immediately after death and are considered normal findings on postmortem imaging and should not be mistaken for injury or disease (Levy et al., 2010).

Lividity is not apparent on radiography. On MDCT, it causes increased attenuation of the affected organs, vasculature, and tissues (Shiotani et al., 2002). This is most prominently observed in the cerebral dural sinuses, cardiac chambers, great vessels, and lungs. The degree of decomposition on postmortem imaging is highly variable because the rate of decomposition is unpredictable and dependent upon the cause of death and external environment. If there is a delay in recovering bodies after a mass-casualty event or the environmental conditions are hot and humid, there may be a significant amount of decomposition present in the recovered human remains. Putrefactive gas is the hallmark of decomposition on radiographs and MDCT. It is first observed in the intestinal wall and mesenteric and portal venous system. As decomposition progresses, gas enters all vascular structures and potential anatomic spaces (Figure 14.7). MDCT may also show other features of decomposition such as autolysis and liquefaction in advanced stages of decomposition.

Blast Injury

Blast injuries may be the consequence of suicide bombs, car bombs, and other types of explosions. Blast injuries are classified as primary, secondary, tertiary, and quaternary blast injuries. Primary blast injury is caused by the blast wave and often results in barotrauma or fragmentation of the body. In most cases it is not possible or practical to differentiate

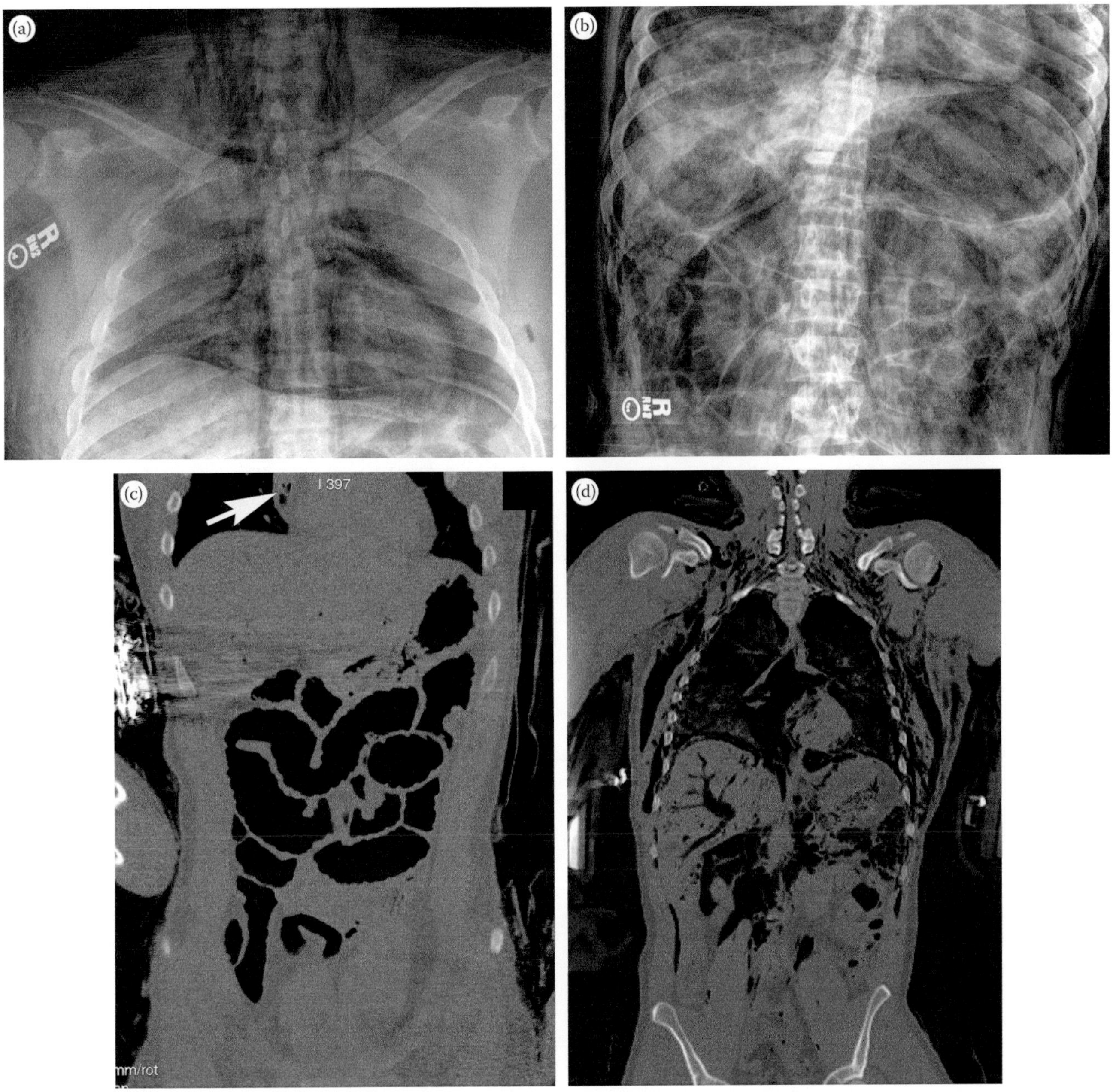

FIGURE 14.7 Imaging features of decomposition in four different subjects. (a, b) Radiographs of the chest and abdomen in two different subjects with advanced decomposition show gas-filled vasculature and anatomic spaces. (c) Coronal MDCT of the abdomen and lower chest in a subject with mild decomposition shows distended intestine and gas in hepatic vasculature and inferior vena cava (arrow). (d) Coronal MDCT of the chest and abdomen in a subject with moderate decomposition shows gas in the subcutaneous tissues, vasculature, heart, abdominal organs, and pleural and peritoneal spaces.

primary blast injury on imaging studies. For example, pulmonary laceration and hemorrhage occurs in both primary and tertiary blast injury.

Secondary blast injury is ballistic injury. Both penetrating and perforating wounds may occur and all sizes, shapes, and types of materials can be found as projectiles. The composition of some materials is more easily recognized on radiographs compared to MDCT because radiographs have superior edge detail. On MDCT, the attenuation of projectile fragments is also variable because both natural and man-made materials may be projectiles.

Tertiary blast injury is blunt force injury that occurs when the body is propelled through the air and collides with a secondary object or a moving object collides with the body. MDCT is the most useful imaging modality for evaluating blunt trauma because occult fractures such as those in the spine and pelvis may be difficult to thoroughly evaluate at autopsy and may have caused or contributed to death (Figures 14.8 and 14.9). Multiplanar and 3D MDCT analysis of fracture patterns may allow the direction of blast to be determined in many cases.

Quaternary blast injury includes all other blast effects such as thermal and inhalation injury, as well as contamination from chemical, biological, or radiological hazardous materials. These injuries are often found in conjunction with injuries from the other mechanisms. Burns are one of the most common quaternary injuries. The severity of primary, secondary, and tertiary blast injury should always be carefully considered when determining the cause of death in a blast victim that has significant burns because fire-related injury may have occurred after death.

Projectile Injury

Projectile injury in mass casualties may be from gunshot wounds, blast fragments, ballistic projectiles, or the result of impalement. Full-body radiography is used in the forensic assessment of projectile injury to document and locate all metallic fragments. Orthogonal radiographic projections (frontal and lateral views) are the most optimal method of precise localization if radiography alone is being used. C-arm fluoroscopy may augment localization if lateral views cannot be obtained. MDCT provides precise 3D localization of projectile fragments. It has been shown that this technique is an effective method not only for localization of bullet fragments but also for documenting the gunshot-wound track and evaluating internal organ injury prior to autopsy (Harcke et al., 2007; Levy et al., 2006; Thali et al., 2003).

Evaluation of projectile injury requires determination of the entry wound, exit wound (if present), and wound track. Using MDCT alone to locate and classify entry and exit wounds on the skin surface can be problematic. Direct visualization of the skin surfaces at autopsy or even by photograph is more accurate. In high velocity gunshot wounds, the direction of the bullet path can only be ascertained radiographically, by analysis of the surrounding bone and soft tissue structures or if metallic bullet fragments or bone fragments have been deposited along the path of the bullet. The wound track is the visible remnant of the laceration, shock wave, and cavitation created by passage of the projectile. Gas and hemorrhage are the principal imaging findings of gunshot-wound tracks in soft tissue. Fractures are the manifestations of gunshot-wound tracks in bone. There is a high degree of variability in the appearance of gunshot-wound tracks depending upon the type of tissue the bullet passes through and the velocity of the bullet. Bone findings are often the most helpful because fragments of bone are often left in the direction the projectile traveled through the body.

Blunt-Force Injury

Blunt-force wounds are classified into four major categories: abrasions, contusions, lacerations, and skeletal fractures. Postmortem MDCT is useful to visualize and reconstruct blunt -injury patterns prior to autopsy (Donchin et al. 1994). In some cases, multiplanar and volumetric reformatted MDCT images may provide better visualization of blunt-traumatic fractures than autopsy. A 3D display of head, spine, and pelvic injuries may facilitate the understanding of the mechanism of injury. The head and chest are the most common sites for lethal blunt-force injury. In the head, the spectrum of injury ranges from scalp lacerations and skull fractures to intracranial hemorrhage and cerebral contusion. Radiography and MDCT are useful to show clinically significant pneumothoracies and hemorrhage within the chest. However, vascular injury and visceral contusion is not readily detected on MDCT. MDCT is very useful in the diagnosis of spine, pelvis, and extremity fractures; 3D MDCT is very helpful to visualize the entire injury pattern if analysis of the injury mechanism is necessary.

Thermal Injury

In fire-related deaths, radiological findings do not contribute to the diagnosis of smoke inhalation and COHb intoxication. Imaging is best used to evaluate burned or charred human remains for occult injury and the remnants of unsuspected projectiles. The radiographic and MDCT features of thermal injury range from minimal irregularities of the skin surface in partial thickness burns to thermal amputation and fracture. The imaging hallmark of severe burns is loss of the dermal layer with exposure of the underlying subcutaneous fat and/or muscle. In severely charred victims, skeletal muscle is exposed and retracted. Shortening and thermal destruction of muscle also cause the muscle to pull away from the distal ends of the bone such that soft tissue does not cover the distal bone. These findings may be seen in the torso and the extremities.

Traumatic fractures and amputations in severely burned remains may cause interpretive difficulties because heat-related cortical fractures and amputations are common in severely charred remains. Thermal fractures are linear cortical fractures in bone uncovered by soft tissue or bone. They are usually found in areas of severe charring and exposure to heat. In contrast, traumatic fractures are found in unexposed bone

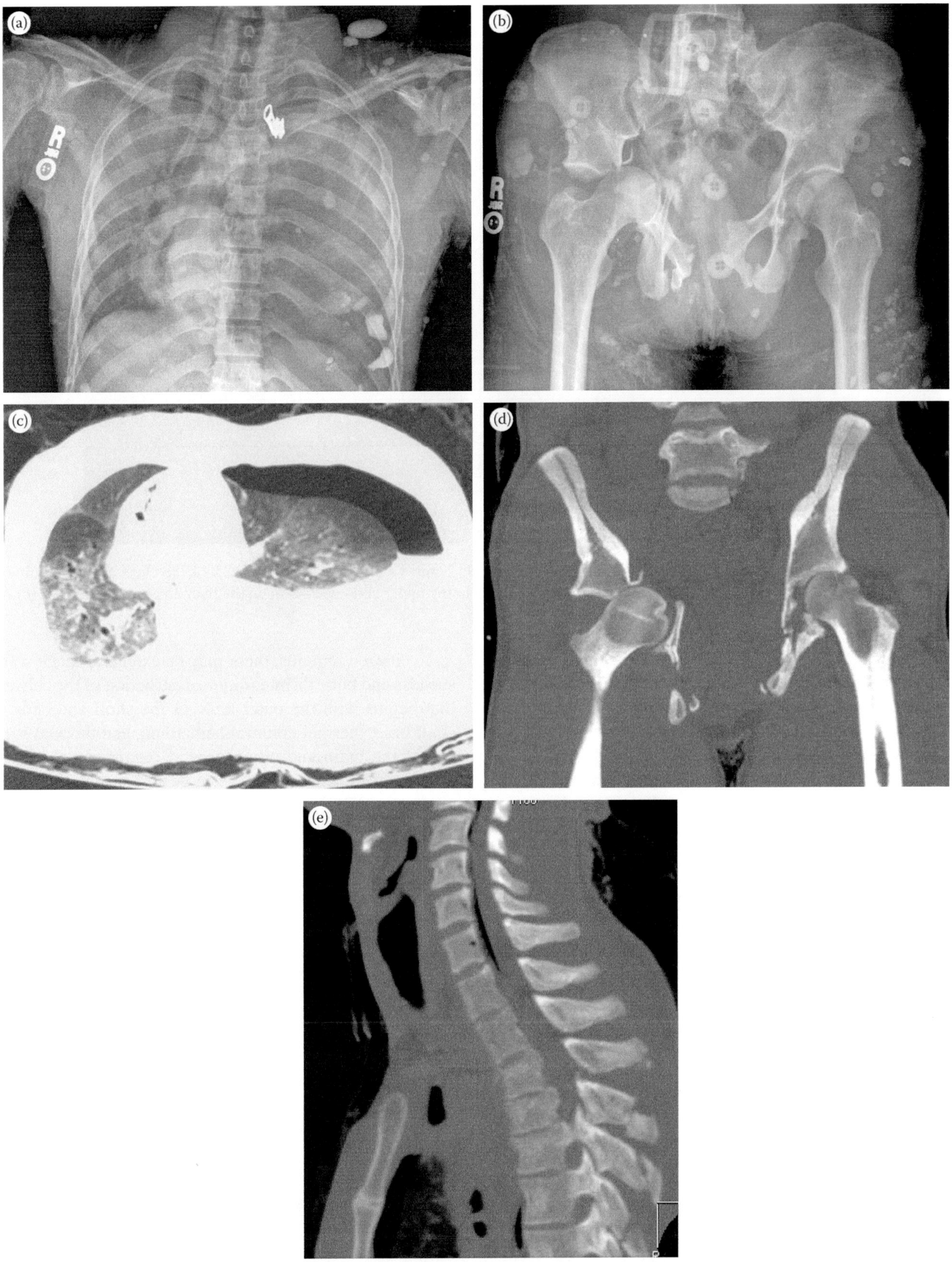

FIGURE 14.8 Blunt-force injury in a suicide bomb victim. (a, b) Radiographs of the chest and abdomen show a left tension hemopneumothorax and complex pelvic fractures. External debris projects over the chest and abdomen. (c) Axial MDCT of the chest shown in lung windows confirms a left tension hemopneumothorax. (d) Coronal maximum intensity projection image shown in bone window displays the complex pelvic fracture. (e) Sagittal MDCT of the cervical and upper thoracic spine shows a multiple vertebral body and spinous process fractures.

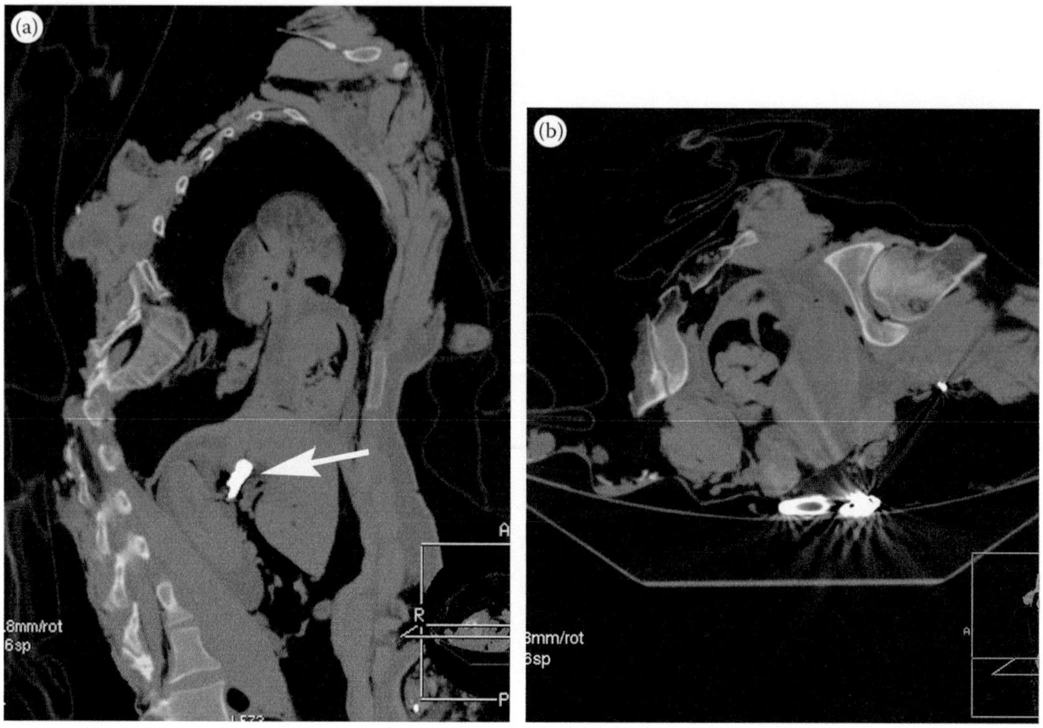

FIGURE 14.9 Blunt-force, projectile, and thermal injury in a suicide bomb victim. (a) Sagittal MDCT of the chest and upper abdomen show extensive thermal and blunt force injury. A metallic projectile is in the upper abdomen between the liver and kidney (arrow). (b) Axial MDCT of the pelvis shows thermal tissue loss and amputation.

and are typical of mechanical injury such as a spinal compression fracture, complex pelvic fracture with involvement of the sacroiliac joints and symphysis pubis, and oblique and/or comminuted fractures of the distal extremities and ribs. Thermal amputations have smooth transverse or angulated margins that are not covered by skeletal muscle because of thermal related skeletal muscle shrinkage and retraction. In contrast, traumatic amputations have sharp, angulated margins or evidence of comminution. Another finding that is indicative of heat-related bone injury is the finding of mottled lucency in the marrow space on multiplanar 2D images (Levy et al., 2009).

In charred remains, there may be extensive facial soft tissue loss and bone disintegration, destruction of the calvarium that begins with the outer table of the skull and ends with skull base, thermal epidural hematoma, and thermal shrinkage of the brain and other visceral organs. As the internal organs shrink and contract from heat exposure, their attenuation on MDCT increases. Specifically, in the lung, the attenuation change should not be mistaken for underlying pathological processes such as pulmonary edema or pneumonia. The thermal changes that occur in visceral organs limit the detection of injury and underlying disease. However, the

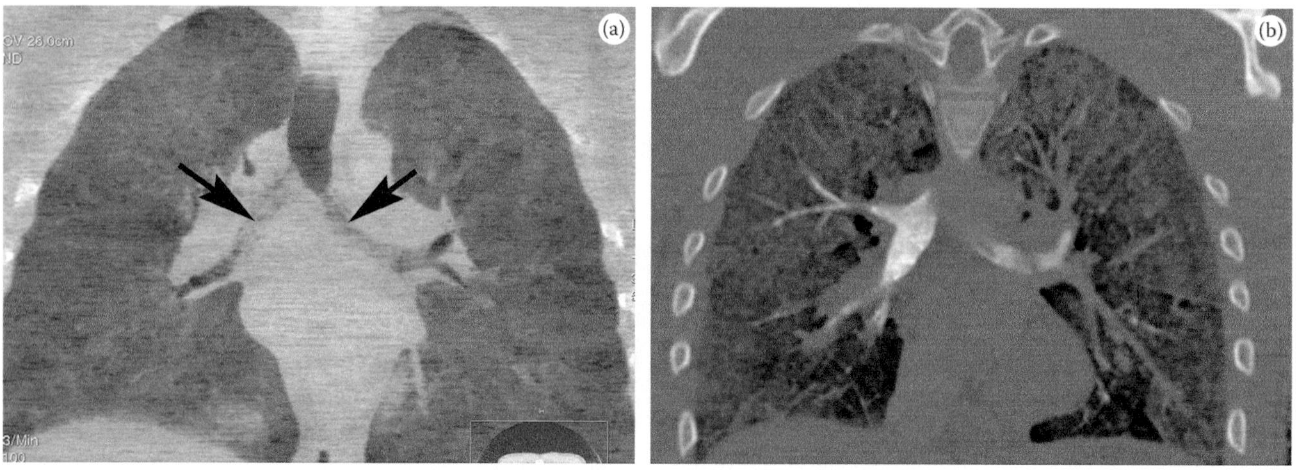

FIGURE 14.10 Pulmonary MDCT findings in two different drowning victims. (a) Coronal MinIP of the chest in a victim of freshwater drowning shows frothy fluid in the main stem bronchi (arrows) and bilateral pulmonary edema. (b) Coronal MDCT of the chest in a saltwater drowning victim shows high attenuation sand throughout the bronchi bilaterally.

finding of unsuspected traumatic injury or metallic fragments from a ballistic injury can facilitate the determination of the cause of death (Figure 14.9).

Drowning

Pulmonary edema is the most prominent finding of drowning on plain film radiography. The distribution of pulmonary edema may be perihilar, apical, or dependent. The severity of edema may range from the appearance of interstitial and septal lines on radiography to confluent alveolar opacification. Fluid within the tracheobronchial tree may be difficult to observe on radiography because of the supine positioning of the body. It is almost always present on MDCT. Radiography and MDCT may also show high-density material in the airways and stomach that may represent inhaled or swallowed sand or silt.

MDCT closely parallels autopsy for the depiction of the anatomic findings that are supportive for the diagnosis of drowning. Sinus fluid, mastoid fluid, subglottic tracheal and bronchial fluid, and pulmonary ground glass opacity are consistently present on MDCT. Minimum intensity projection images (MinIP) are useful to evaluate the extent of fluid-filled bronchi because the attenuation difference between two low attenuation structures (air in the lung and air in the bronchi) is maximized. Since many of these findings are nonspecific and found in other causes of death, the presence of airway froth and sand may be helpful indicators of drowning in the appropriate setting (Figure 14.10) (Levy et al., 2007). Other findings such as pleural effusions; dilated and engorged right-side cardiac chambers and great vessels; and fluid, sand, or debris in the stomach may be present (Christe et al., 2008).

CONCLUSIONS

The Charles C. Carson Center for Mortuary Affairs, Dover Air Force Base, Dover, Delaware was specifically designed to expand and contract its operational capacity. In this respect, comprehensive radiological imaging for mass-casualty disasters of varying size and complexity can be performed. While most incidents are supported by forensic facilities with fewer resources, we feel the inclusion of cross-sectional imaging in the radiological portion of the forensic autopsy proves worthwhile. Timely access to MDCT data gives the forensic pathologist an opportunity for faster recovery of evidence, the ability to tailor the autopsy, and in cases where MDCT and other information is definitive as to the cause and manner of death, to eliminate invasive autopsy.

REFERENCES

Bluth, E. I., Kay, D., Smetherman, D., et al., Managing in a catastrophe: Radiology during Hurricane Katrina, *AJR Am J Roentgenol*, 188, 630–2, 2007.

Christe, A., Aghayev, E., Jackowski, C., Thali, M. J., and Vock, P., Drowning—post-mortem imaging findings by computed tomography, *Eur Radiol*, 18, 283–90, 2008.

Donchin, Y., Rivkind, A. I., Bar-Ziv, J., et al., Utility of postmortem computed tomography in trauma victims, *J Trauma* 37, 552–5; discussion 555–6, 1994.

Harcke, H. T., Bifano, J. A., and Koeller, K. K., Forensic radiology: Response to the Pentagon attack on September 11, 2001, *Radiology*, 223, 7–8, 2002.

Harcke, H. T., Levy, A. D., Abbott, R. M., et al., Autopsy radiography: Digital radiographs (DR) vs multidetector computed tomography (MDCT) in high-velocity gunshot-wound victims, *Am J Forensic Med Pathol*, 28, 13–9, 2007.

Harcke, H. T., Monaghan, T., Yee, N., and Finelli, L., In press, Forensic imaging guided recovery of nuclear DNA from the spinal cord, *J Forensic Sci*, 54, 1123–6, 2009.

Heffernan, T. E. T., Alle, S., and Matthews, C. C., Weathering the storm: Maintaining an operational radiology department at Ochsner Medical Center throughout Hurricane Katrina, *Radiology*, 242, 334–7, 2007.

Levy, A. D., Abbott, R. M., Mallak, C. T., et al., Virtual autopsy: Preliminary experience in high-velocity gunshot wound victims, *Radiology*, 240, 522–8, 2006.

Levy, A. D., Harcke, H. T., Getz, J. M., et al., Virtual autopsy: Two- and three-dimensional multidetector CT findings in drowning with autopsy comparison, *Radiology*, 243, 862–8, 2007.

Levy, A. D., Harcke, H. T., Getz, J. M., and Mallak, C. T., Multidetector computed tomography findings in deaths with severe burns, *Am J Forensic Med Pathol*, 30, 137–41, 2009.

Levy, A. D., Harcke, H. T., and Mallak, C. T., in press, Postmortem imaging: MDCT features of postmortem change and decomposition, *Am J Forensic Med Pathol,* 31, 12–7, 2010.

Rutty, G. N., Robinson, C. E., Bouhaidar, R., Jeffery, A. J., and Morgan, B., The role of mobile computed tomography in mass fatality incidents, *J Forensic Sci*, 52, 1343–9, 2007.

Shiotani, S., Kohno, M., Ohashi, N., Yamazaki, K., and Itai, Y., Postmortem intravascular high-density fluid level (hypostasis): CT findings, *J Comput Assist Tomogr*, 26, 892–3, 2002.

Thali, M. J., Yen, K., Vock, P., et al., Image-guided virtual autopsy findings of gunshot victims performed with multi-slice computed tomography and magnetic resonance imaging and subsequent correlation between radiology and autopsy findings, *Forensic Sci Int*, 138, 8–16, 2003.

Section IV

Gunshot Wounds

Apparently the first gunshot wound to be detected by Professor Rontgen's "new kind of ray" was Professor Wright's unlucky rabbit (see chapter 2). As for the first x-ray examination of an acute gunshot wound leading to extraction, primacy seems to belong to M.L. Pupin of New York City. Prescott Hall Butler, Esq., was wounded by the accidental discharge of a shotgun at short range into his hand. The roentgen plate exposed on February 14, 1896, clearly showed multiple pellets there, which were removed by the surgeon, William Tillinghart Bull. Six days earlier, experiments at Chicago's Western Electric plant had eventuated a roentgenogram of the Chief Engineer's hand. Word spread, so on February 10, 1896, a Mr. Louis Burkhart presented himself at the plant requesting an x-ray on *his* hand. A bullet embedded there several years earlier in Strassburg, Germany, was revealed and successfully removed by Dr. James Burry the following day.[1]

But the earliest gunshot wound to be radiographed was sustained by King Charles XII of Sweden on November 30, 1718. He was felled by a fatal shot to the head during the siege of fortress Frederiksten in Norway. At the time of the fourth examination of his body in the ensuing two centuries, the multiple fractures of the king's skull were reassembled, photographed and x-rayed in 1917. A through-and-through bi-temporal wound path from left to right was thus revealed. (Figure IV.1). A detailed forensic and historical analysis[2] proved with reasonable certainty that the monarch was shot at short range by one of his own people using a specially made musket ball.

Nowadays, the forensic radiological experience is replete with evaluation of both fatal and non-fatal gunshot wounds, their effects and sequellae. Some of the analytical parameters are not much changed since the early days, but the availability of modern modalities and techniques has brought new developments to the process.

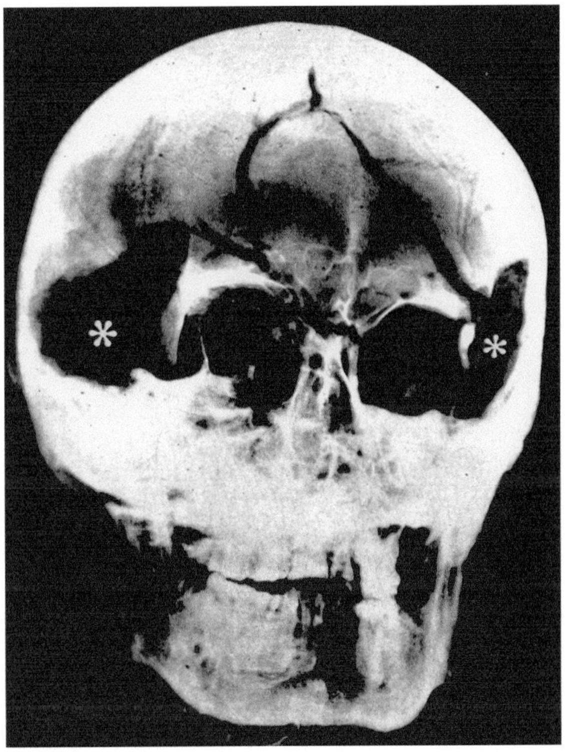

FIGURE IV.1 Frontal view of the skull of King Charles XII of Sweden, died 1718, radiographed 1917. The smaller entrance wound is in his left temporal area (small asterisk) and the larger, externally beveled exit wound is on his right (larger asterisk).

REFERENCES

1. Grigg, E.R.N., *The Trail of Invisible Light*, C.C. Thomas, Springfield, IL, 1965, pp. 29,30.
2. Hougen, H.P. and Munck, O., The death of King Charles XII—a forensic evaluation, *Proc. Am. Acad. Forensic Sci.*, Colorado Springs, CO, 1999, p. 161.

15 Forensic Radiology of Gunshot Wounds

B.G. Brogdon and James M. Messmer

CONTENTS

INTRODUCTION

The most recent Uniform Crime Reports (2007)[1] lists 16,919 murders, or 5.6/100,000 or 46/day in the United States, the highest homicide rate in any Western industrialized country. Firearms account for approximately 60% of those. Long-term longitudinal studies, reports, and internet postings by the Center for Disease Control, Center for Injury Prevention and Control, and the Department of Justice show that gunshot homicides are predominantly in young adults, while suicides by firearm slightly exceed homicides and are most likely in older age groups. Suicide and homicide greatly outnumber accidental deaths by gunshot. Gunshots are second only to automobile accidents as a cause of juvenile injury-related deaths. There are more than 700 nonfatal injuries by firearms for every fatality.[2]

Consequently, the informed forensic radiologist has much to offer the forensic pathologist and/or public official in the investigation and analysis of gunshot wounds whether fatal or nonfatal, intentional or accidental.[3] This chapter will provide an overview of the spectrum of radiological information and considerations pertinent to the conventional evaluation and analysis. Chapter 16 will offer new insights, modalities, and methodology recently developed for forensic gunshot investigation.

BASIC RADIOLOGICAL INFORMATION

The forensic pathologist uses x-rays in evaluating gunshot wounds in several ways.[4,5] First and foremost, the pathologist is interested in the *location of the bullet.* While this may seem straightforward from external inspection of the body, bullets will frequently end up in a site far distant from their entrance points, particularly if they have struck a bone. The natural curvature of the ribs and the skull can cause bullets to change trajectory, significantly. The availability of fluoroscopy in the autopsy suite is especially valuable for allowing the pathologist to scan the body for unsuspected bullets. Knowing the specific location of a bullet saves the pathologist much time and may avoid needless effort in searching for bullets that are inaccessible. Figure 15.1 demonstrates a case where only one bullet was easily accessible.

X-rays may also reveal whether there are bullets of a different *caliber* present. This can be valuable in cases where multiple weapons are involved. The *number of bullets* is also important and must be correlated with the entrance and exit wounds. A discrepancy may lead to a search for bullets at the scene. More than one bullet may enter through a single entrance wound, particularly when automatic weapons are used.

X-rays may also reveal information about the *angle* and *direction of fire* (Figure 15.2). Small metallic fragments

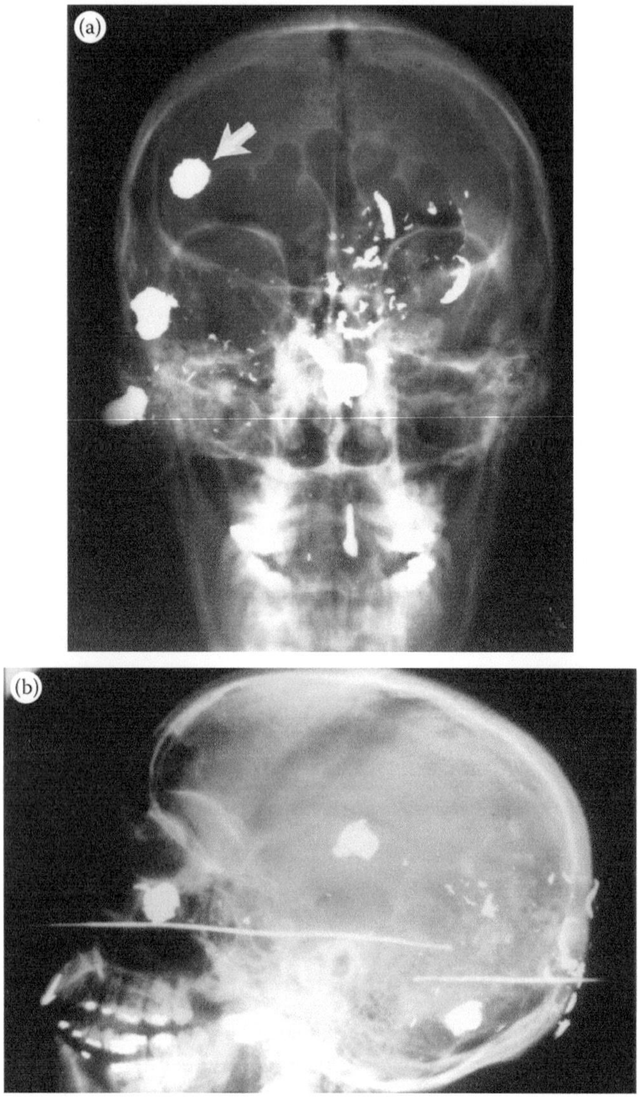

FIGURE 15.1 (a) An execution-style murder resulted in five bullets, but only the bullet overlying the right frontal sinus (arrow) was easily accessible. The remainders were lodged within the facial bones. (b) Shows the futility of probing for bullets. (From Messmer, J. M. and Fierro, M. F., *RadioGraphics*, 6, 457, 1986. With permission.)

produced when a bullet strikes bone, and subsequent scattered bone fragments, may lead directly to the bullet and clearly indicate the bullet's path. Correlating this information with the scene of the crime helps recreate the relative positions of victim and assailant. While the type of weapon can frequently be determined by eyewitness reports or recovery of the weapon from the scene, the radiographs may reveal clues as to the *type of weapon*. For example, high-velocity hunting ammunition wounds can leave a characteristic "lead snowstorm" radiographic picture because of extensive fragmentation of the unjacketed bullet (Figure 15.3).

X-rays may be the first *indication that a crime has been committed* when decomposed bodies are discovered. The normal putrefaction that occurs, with its attendant bloating of the tissues and deformity of the body, can easily mask an entrance wound. (Figure 12.28). Bodies partially destroyed by fire or skeletonized remains should always be radiographed to determine unsuspected foul play (Figure 3.32). Retrieval of the metallic fragments can even help in the identification of remains if there is a gunshot wound in the past history of the presumed decedent.

BALLISTICS

A general understanding of bullets, weapons, and the tissue damage bullets inflict is valuable in film interpretation.[6–13] The three general types of guns are handguns, rifles, and shotguns. The muzzle velocity of rifles is higher than that of handguns or shotguns. Table 15.1 gives a list of common weapons and their muzzle velocities. The bullets fired by handguns and rifles are similar, but are distinctly different from those fired from shotguns. Each type of weapon has characteristic wounding patterns. In general, higher-velocity bullets cause more damage than slower ones, but speed is not the only factor in tissue destruction. The weight of the bullet, its internal construction, and the amount of pitch and yaw during the flight of the bullet also play an important role in the amount of damage.

TYPES OF BULLETS

The type of ammunition used for handguns and rifles is a cartridge, which consists of a cartridge case (usually made of brass), primer, powder charge, and the bullet.[4] The caliber of a bullet is expressed as a decimal corresponding to the diameter in inches, or by the actual diameter in millimeters. Bullets of identical calibers may have different weights, and the forensic pathologist measures both weight and caliber to define the weapon more accurately.

Lead is the most common base metal used for making bullets. Antimony or tin is usually added to the lead to increase the hardness. The lead bullet can be fully or partially covered by other metals. Copper or copper alloys are frequently used in the production of such "full metal-jacketed" or "partially metal-jacketed" bullets. This coating, called gilding, both hardens and lubricates the bullet as well as prevents leading of the action and barrel, which can cause the weapon to jam. A partially metal-jacketed bullet exposes the lead tip, which can then be hollowed out to increase the mushrooming effect of the bullet as it enters tissue.

The barrels of rifles and handguns have grooves along their length to impart a rotational spin along the long axis of the bullet, which stabilizes the flight of the bullet. These grooves cause unique markings on either the lead or metal jacket of the bullet (Figure 15.4). These "class characteristics" may indicate the make and model of the weapon. More subtle imperfections in the barrel and grooves cause unique markings and are the basis for determining whether a particular weapon fired a particular bullet.

In the case of the partially metal-jacketed bullet, the metal jacket may separate from the lead portion of the bullet and can be readily identified radiographically by its lower

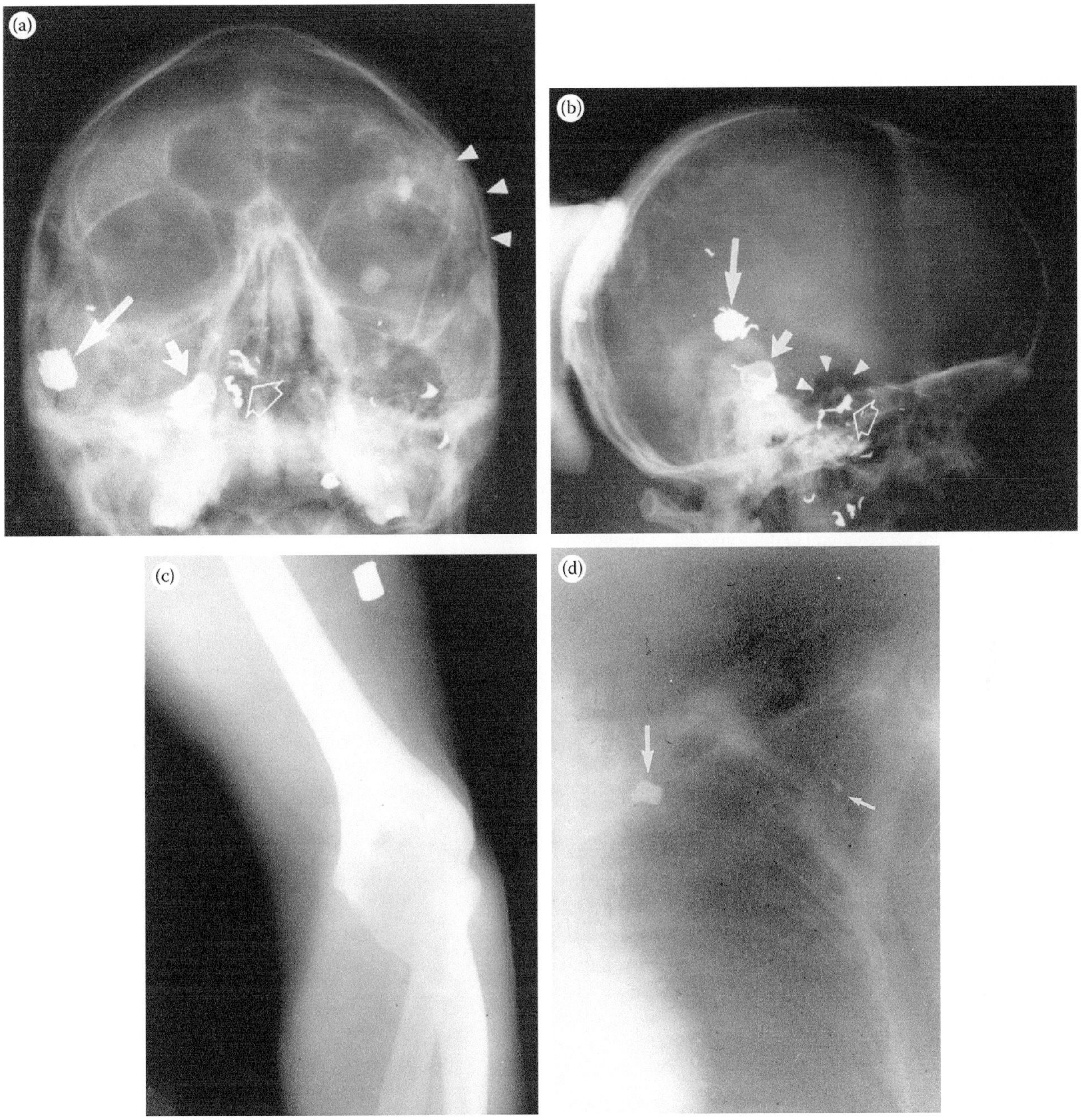

FIGURE 15.2 Case of the fleeing felon: a body with multiple short-range gunshot wounds was found beside the highway. The location of wounds suggested that the victim might have been in the right-hand seat of an automobile, his assailant in the driver's seat. Later, the driver was stopped for speeding in the blood-stained car, and confessed. The victim was a hitch hiker, fleeing escape from prison, but who unfortunately, got picked up by an even meaner felon who robbed and killed him. (a) Frontal and (b) lateral view of the skull show a left temporal wound of entry (arrowheads). There are scattered bone and bullet fragments throughout. The bullet bounced off the sella (open arrow). The jacket (short arrow) separated, and the bullet (long arrow) came to rest against the right parietal bone posteriorly. (c) Nonfatal gunshot to the left upper arm. (d) Nonfatal bullet wound, trajectory from left axilla (small arrow) to left mediastinal border (large arrow). A smaller-caliber bullet found radiographically in the left leg proved to be from a foiled robbery years earlier.

radiodensity than the lead component (Figure 15.5). It is important to remember that in partially metal-jacketed bullets the ballistic markings are on the metal jacket and not the lead. Hence, it is the jacket that must be retrieved for testing.

Radiologists should always note the location of fragments of a sufficient size that might contain ballistic information. Accurate location of bullets can save the forensic pathologist significant time.

Information about the type of bullet can be determined by its radiographic appearance. "Mushrooming," which is the characteristic flattening of one end of the bullet when it strikes flesh, indicates a solid lead or partially metal-jacketed bullet. A bullet with a unique appearance that was pulled from the market in 1993, but can occasionally be seen, is the Black Talon. The design of this bullet is a hollow-point tip with sharp metal points that unfold upon impact (Figure 15.6).

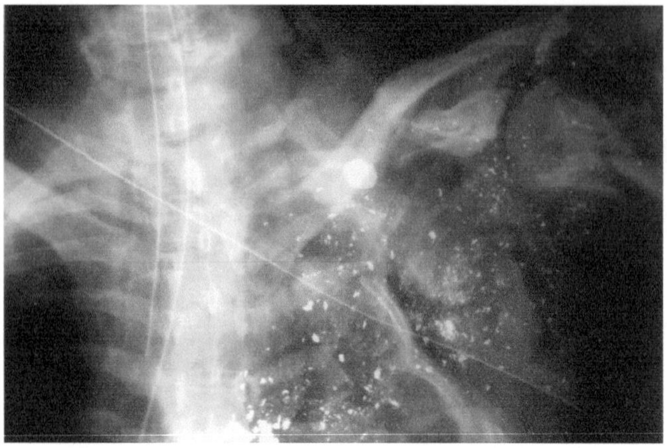

FIGURE 15.3 A high-velocity rifle wound to the chest left a typical "lead snowstorm" of fragments. The victim was accompanying her daughter who had just won a divorce. The enraged exhusband shot and killed his exwife as she exited the courthouse, then killed his exmother-in-law as she cowered behind a car. (Note the spread of the "snowstorm" from superolateral to inferomedial.) The shooter was judged not guilty by reason of insanity in the first event, and guilty of murder in the second event. (This judicial result may be the first of its kind for two killings occurring within a matter of seconds.)

The concept was that the sharp metal points would cause more damage than the simple mushroom effect of the plain lead bullet. In fact, despite the formidable appearance, the differences in wounding effect are negligible.[14] It is important, however, for the radiologist to be aware of this type of bullet and alert the forensic pathologist or surgeon to its presence, since the sharp metallic tips can easily penetrate a rubber glove and injure the investigating physician.

The Glaser safety slug is a bullet designed to impart all of its kinetic energy into the tissue. Consisting of multiple small lead pellets encased in a copper cup with a Teflon plug at the tip, these bullets are designed to incompletely penetrate the victim, therefore, eliminating the possibility of ricochet or injury to bystanders; hence the "safety" name. The radiographical appearance mimics that of a shotgun wound and tissue damage can be extensive, particularly at close range[15] (Figure 15.7a). The copper cup may be visible radiographically, but the Teflon plug is not.

TABLE 15.1
Some Common Weapons and Muzzle Velocities

Cartridge	Muzzle Velocity (ft/s)
0.22 short HV	1125
0.22 long HV	1240
0.30–0.30 Winchester	2410
0.357 Magnum	1550
0.38 S and W	685
0.44 special	755
M-16	3250

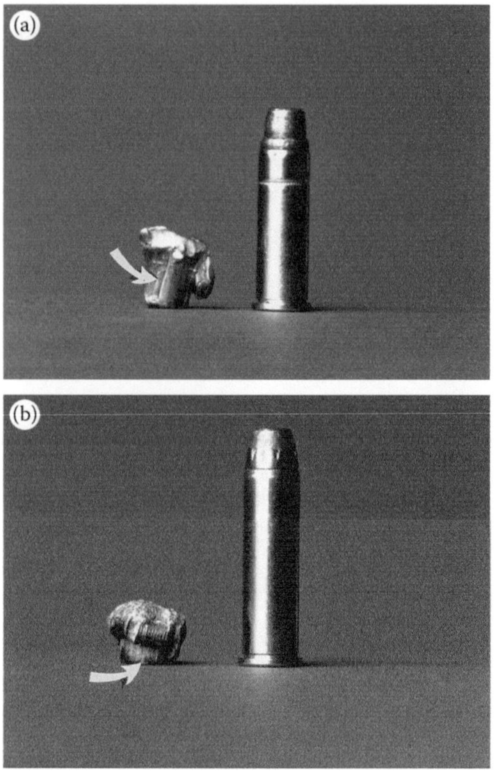

FIGURE 15.4 (a) A solid lead bullet demonstrates the characteristic ballistic markings (arrowhead). (b) In the partially jacketed bullet the crucial ballistic information is on the copper jacket rather than the lead component (arrowhead). (From Messmer, J. M. and Fierro, M. F., *RadioGraphics*, 6, 457, 1986. With permission.)

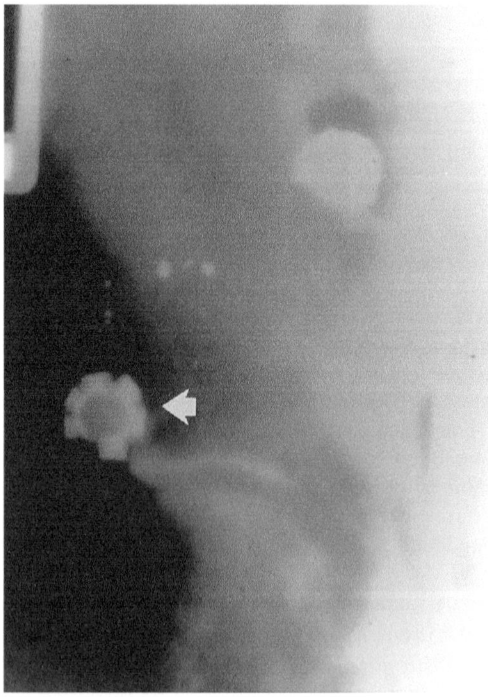

FIGURE 15.5 In this gunshot wound to the left hip, the lead and jacket components have become separated. The important ballistic information is on the jacket (arrow). (From Messmer, J. M. and Fierro, M. F., *RadioGraphics*, 6, 457, 1986. With permission.)

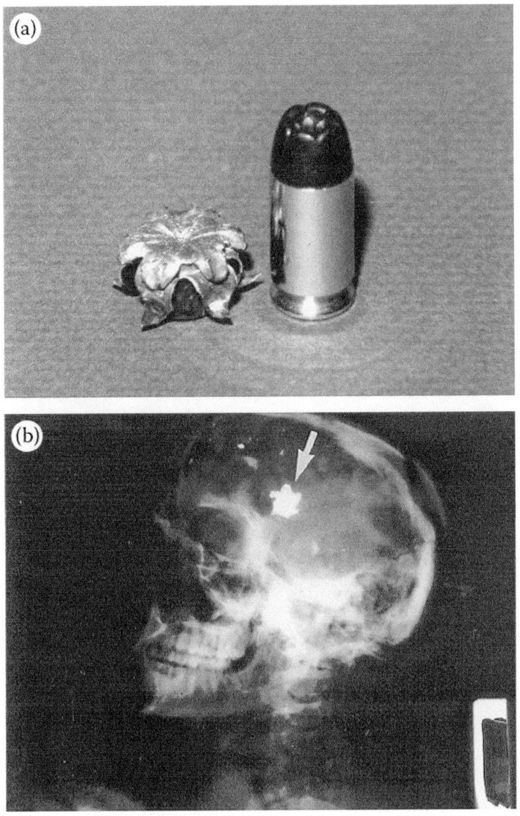

FIGURE 15.6 (a) Fired Black Talon demonstrates the characteristic sharp projections that are exposed as the bullet mushrooms. (b) The radiograph of this homicide victim demonstrates the characteristic appearance of the Black Talon (arrow) and allows for a radiographical determination of the type of bullet. (It also warns the autopsy surgeon of a potential puncture wound.)

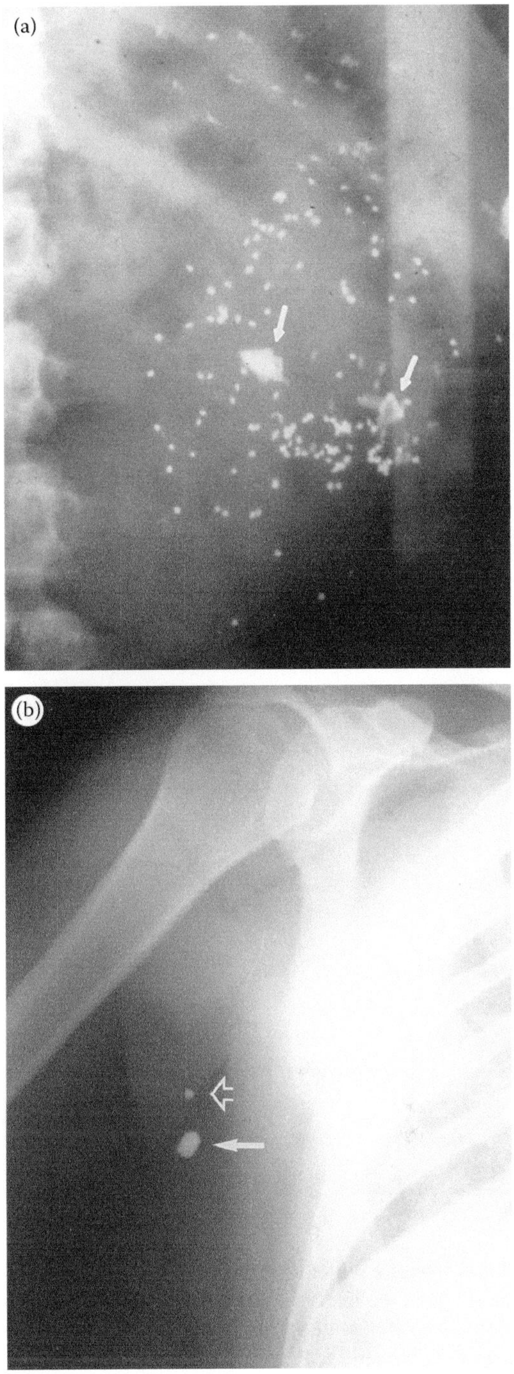

FIGURE 15.7 (a) A fatal wound to the flank from a Glaser safety slug, which mimics a shotgun wound radiographically. Extensive damage to the retroperitoneum and kidney resulted in a rapid death from exsanguination. Arrows indicate fragments of the cup. (b) Winchester-Western 0.25 caliber centerfire handgun load consisting of a copper-coated lead hollow-point bullet (arrow) containing a single No. 4 steel birdshot pellet (open arrow). (Courtesy of Dr. James C. Downs.)

Another handgun load with a radiographic "signature" is the Winchester Western 0.25 caliber centerfire cartridge introduced in 1981. It contains a copper-coated lead hollow-point bullet filled with a single No. 4 steel pellet.[16] The radiographic finding of a single small bullet accompanied by a single small shot is unique (Figure 15.7b).

Bullets, projectiles, made of rubber, plastic, or ceramic are rare in the United States, but, when encountered can have quite distinctive radiological appearances and produce unusual wound patterns (Figure 15.8).

TISSUE DAMAGE

The amount of tissue damage that occurs is proportional to the kinetic energy of the bullet expended in the tissue. While the literature has emphasized the importance of bullet velocity in tissue damage, the weight of the bullet, internal composition and configuration, and yaw in the flight path are also contributing factors.[6,7,10] Yaw is the angle of the long axis of the bullet with its path of flight. A bullet entering tissue at 90° of yaw would present a significantly larger surface area and, therefore, cause more tissue damage. The rifling on the inside of the barrel imparts a spin to the bullet, which stabilizes its flight. As the bullet exits the barrel the emerging gas may produce minimal yaw during the flight path, but once the bullet enters the tissue it can rotate and cause greater damage.

A bullet that expends all of its kinetic energy in a body imparts its maximum potential of stopping power. A bullet that passes completely through a body intact, retaining its

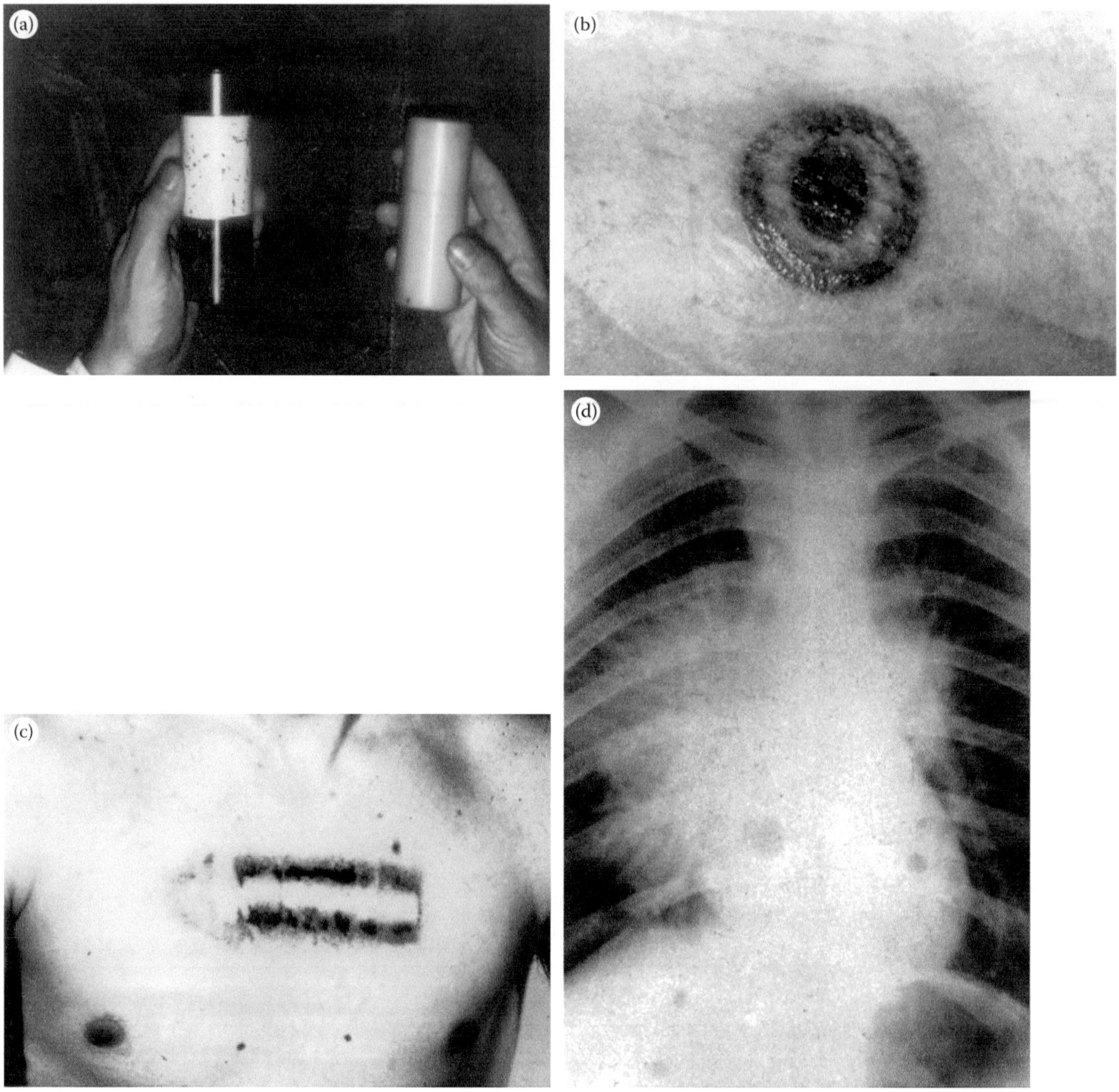

FIGURE 15.8 (a) Plastic bullets (Northern Ireland). (b) Bull's-eye or target lesion from plastic bullet hitting end-on. (c) Typical bruise from lateral hit of a plastic bullet. (d) Focal pulmonary contusion from end-on hit from plastic bullet. (e) Frontal and (f) lateral views of a plastic bullet penetrating the left orbit and entering the sphenoid bone. (g) Nonpenetrating rubber bullet producing massive fractures of the skull (Israel). (h) Two rubber bullets in the thigh. One can see the halo of the rubber coating around the metallic core. (i) Bullet In the thigh from a disintegrating ceramic bullet. (j) Intact ceramic bullet in the hand having traveled more than 100 m. The projectile had sufficient energy to injure the bone, but did not fragment.

kinetic energy, will cause less damage than a bullet that enters and fragments.

As noted above, the composition of the bullet is also important in determining the amount of tissue damage. The Hague Peace Conference of 1899 stipulated that bullets used in war should be protected against deformity by a copper jacket. Ironically, the bullets now frequently used in peacetime have more destructive potential than those used in war.[10]

Other factors that contribute to tissue damage are the elasticity of the tissue struck, the production of bone fragments, and temporary cavitation. Less-elastic tissue such as the brain, liver, or spleen will incur more damage than the more elastic muscle of the extremities. When a bullet strikes bone, its path can be deflected and fragmentation can occur, producing bone fragments that act as additional small projectiles, thus increasing tissue damage.[17] As the bullet passes through the body it pushes tissue away from its path. In the first several microseconds a temporary cavity forms, which can damage relatively unyielding tissue but has less effect on elastic tissue such as a muscle.

Extensive skull fracturing occurs with higher-caliber gunshot wounds to the head (Figure 15.9a). If the fracture lines extend across the intracranial vascular sinuses and the wound is open to the outside, the negative hydrostatic pressure draws

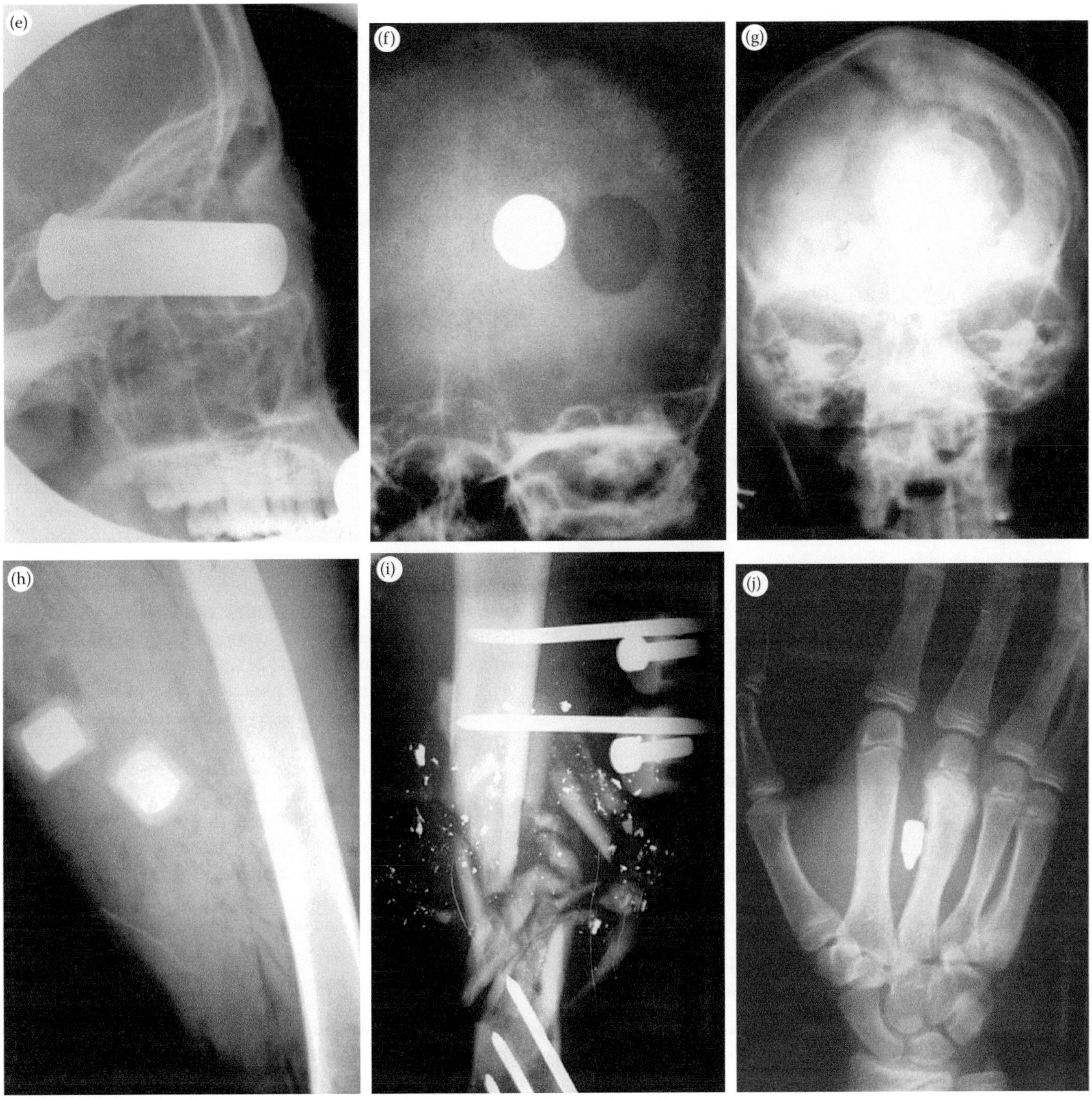

FIGURE 15.8 *(Continued)*

air into the vascular system as the heart continues to beat. This air can occasionally be visualized radiographically in the vascular sinuses and the occipital veins (Figure 15.9b). If the amount of air is sufficiently extensive, postmortem views of the chest will show air in the great vessels, heart, and pulmonary outflow tract[18] (Figure 15.10). The presence of the air may be a causative factor in the eventual cessation of the heart beat.

Air embolism can be venous or arterial. Up to 100 cc of venous air can be handled by the body without fatality if the air is introduced slowly.[19] Arterial air emboli can occur when air enters the left side of the heart, usually through a patent foramen ovale or penetrating lung trauma.[20–25] Arterial air can be more deadly, with as little as 2–3 cc of air causing death experimentally in animals.[26,27] The explanation for the disparity in the clinical pictures is that arterial air can enter the coronary arteries, causing fatal arrhythmias, or enter the cerebral arteries producing cerebral insufficiency. The presence of air exclusively in the circulatory system differentiates if from the type of air pattern seen in the normal putrefactive process, which is more extensive and involves both the vascular system and soft tissues.

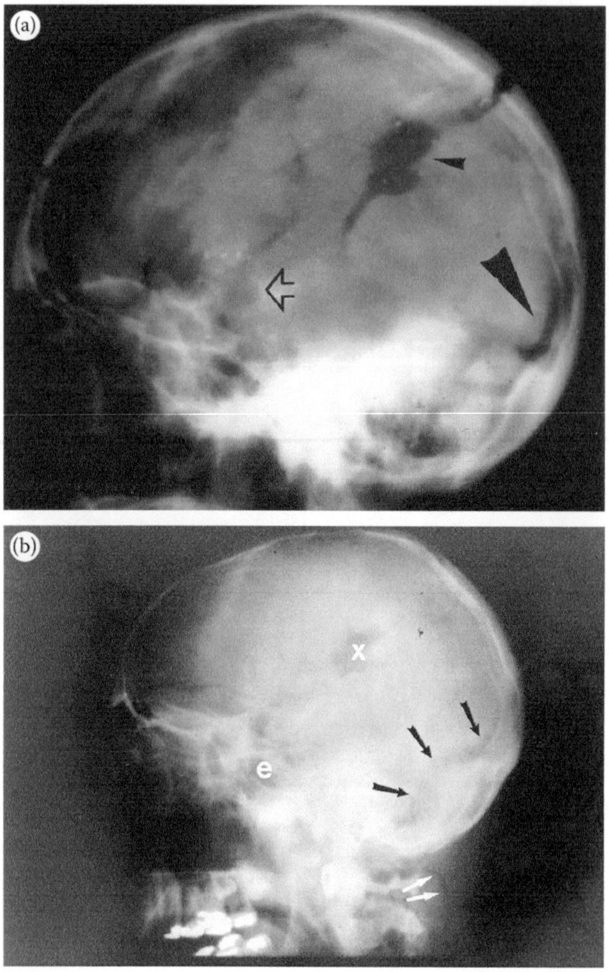

FIGURE 15.9 (a) A lateral skull film with a large-caliber suicide gunshot wound demonstrates the entrance wound (arrow) and the larger exit wound (arrowhead). Note the extensive fracturing. There is also air within the sigmoid sinus (large arrowhead). (b) A self-inflicted 0.45 gunshot wound to the head with a through and through injury (e = entrance, x = exit). Note the air in the sigmoid sinus (arrows) and occipital vein (white arrows). (From Messmer, J. M. and Fierro, M. F., *RadioGraphics*, 6, 457, 1986. With permission.)

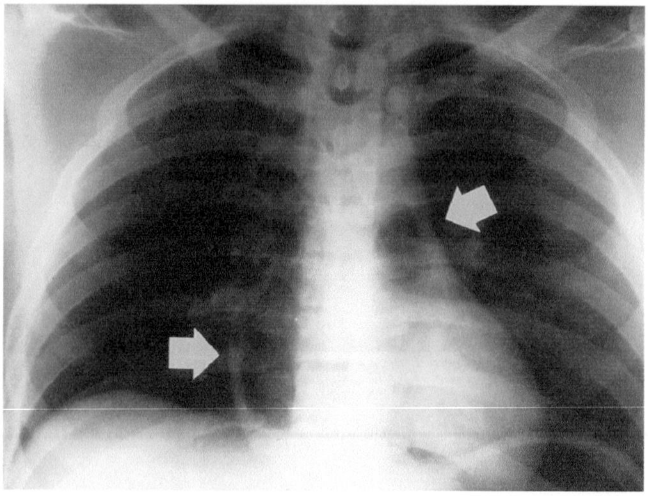

FIGURE 15.10 A postmortem view of the chest in a patient with a 0.38 caliber suicide gunshot wound to the head. There is air in the right atrium and pulmonary outflow tract (arrows), which entered the vascular system through the intracranial sinuses. (From Messmer, J. M. and Fierro, M. F., *RadioGraphics*, 6, 457, 1986. With permission.)

SHOTGUNS

Shotguns differ from handguns and rifles in that the inside of the barrel is smooth and the missile consists of a few to hundreds of metal spheres packed in a paper or plastic tube. The pellets, or shot, emerge from the barrel in a mass that disperses into a pattern determined by the range of fire, barrel length, and degree of "choke," which is partial constriction of the bore of the shotgun at its muzzle end to control the distribution of the shot. The spherical shape of shot adds to instability in flight. As a result of all these factors, the effective range for shotguns is usually measured in tens of meters, as opposed to rifles, which have ranges of hundreds to thousands of meters.[10,28]

The caliber of shotguns is measured in gauges rather than hundredths or thousandths of inches as rifles and handguns are measured. The range of gauge is from 8 (largest) to 0.410 (smallest). (The latter designation is actually the caliber of the barrel measured in thousandths of an inch: there is also a 9-mm shotgun, named for its barrel diameter[6]). For the other gauges, the number represents the number of lead balls of a diameter equal to the diameter of the barrel that would weigh one pound.

There are two basic sizes of shot, birdshot and buckshot. The size of the individual shot is expressed by a number ranging from 12 (smallest) to 000 (largest). The total number of shot in any one shell depends on the size of the shot and the gauge of the shell. For example, more size 9 birdshot pellets will be contained in a 12-gauge shell than in a 20-gauge shell. The largest 00 and 000 buckshot have diameters of 0.33 and 0.36 in., respectively, making them equivalent in size to handgun bullets.

The shotgun slug is a single lead projectile sized to fit the various gauges of shotgun barrels. These were three designs, which vary somewhat in configuration and arrangement, in type of wadding and, in one type no longer manufactured, a plastic sabot. Obviously these slugs are heavy with a rapid loss of velocity, which reduces the likelihood of injury to nontargets, as they remain in the target body. Massive internal injuries are produced before the slug breaks into a few large pieces or stops as a flattened disk of lead. At short ranges, the wad may make an external mark or enter the body along with the slug.

Typically, wounds from shotguns are among the worst seen in civilian wounds.[11,29] At ranges of 1–2 yards the entrance wound is usually a single large, jagged-edged wound. As the range increases to 3–4 yards, and depending on the gauge of the weapon and the size of the pellets, there may be single pellet wounds around the periphery of the larger wound. At distances beyond 20 yards the full pattern develops and there will be multiple small pellet wounds (Figure 15.11).

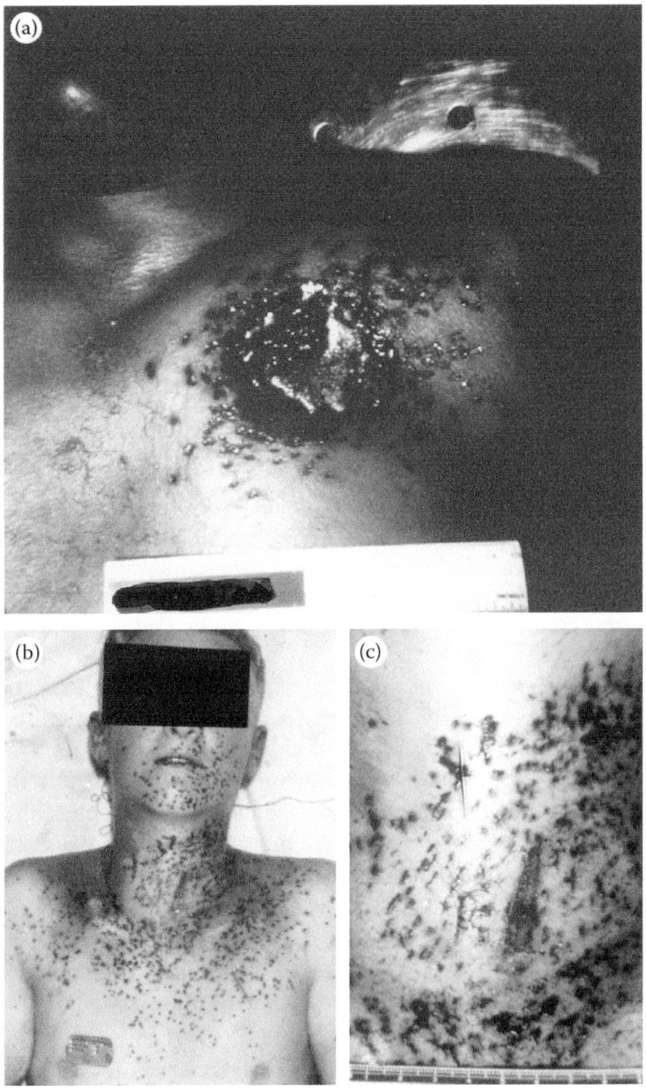

FIGURE 15.11 (a) A fatal shotgun wound from several yards away shows that some of the pellets have begun to form the pattern, but the bulk have entered as a mass causing the larger wound centrally. (b) A fatal shotgun wound to the upper chest and face shows multiple entry points indicating the weapon was fired from a distance of approximately 40 ft allowing the pattern to develop. (c) Close-up of neck with inch ruler at bottom showing shot spread distance. (From Messmer, J. M. and Fierro, M. F., *RadioGraphics*, 6, 457, 1986. With permission.)

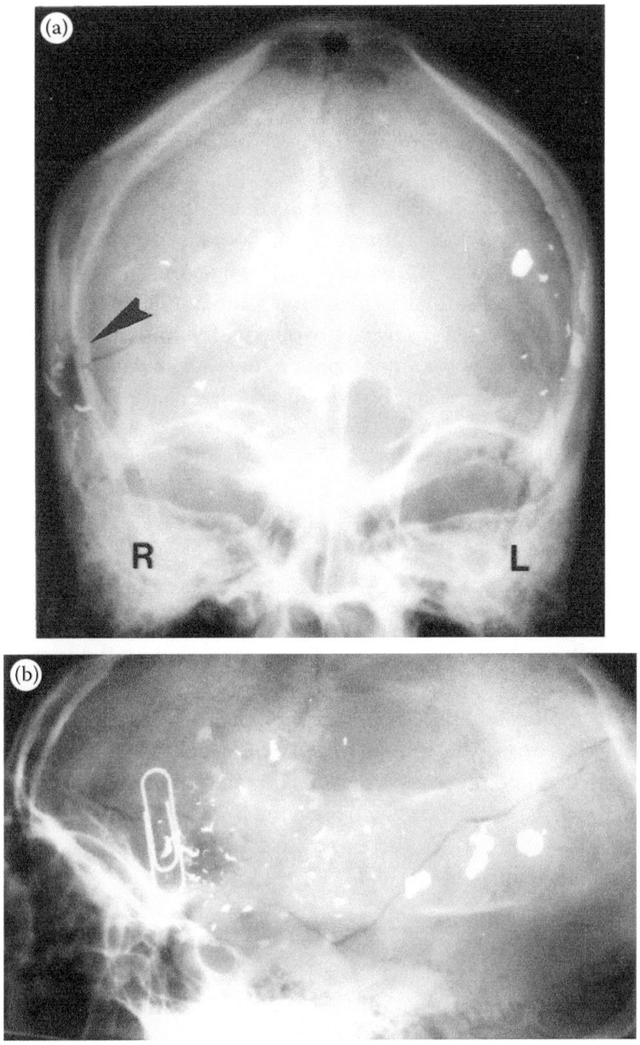

FIGURE 15.12 (a) The frontal radiograph of the skull demonstrates the entrance wound (arrowhead) in the right temporal region with metallic fragments scattered throughout the brain tissue. The major fragment comes to rest inside the skull at a higher level than the entrance wound. (b) The lateral radiograph shows the anterior to posterior distribution to the metallic fragments. A paper clip marks the entrance wound. Note also the minimal calvarial fracturing.

SUICIDE BY GUNSHOT

Annually, the number of suicides exceeds the number of murders in the United States, and firearms are involved in up to two-thirds of suicides. A general knowledge of the characteristics of suicide gunshot wounds can help avoid unwarranted conclusions.

The majority of suicide gunshot wounds are contact wounds, usually to the right side of the head since most people are right-handed. Therefore, wounds to the right temple are common and usually have a posterior and superior trajectory (Figure 15.12). However, the nondominant hand is used often enough that it cannot be considered exceptional or beyond the realm of the possible.[30] Suicide wounds to the top or back of the head have been reported. Suicide gunshot wounds in or near the eye are uncommon and raise the possibility of a homicide. Likewise, suicide wounds to the mouth generally do not involve the tongue or the teeth. When the tongue is involved, the possibility of homicide is raised. The mechanism of action is proposed to be the victim's attempt to push the weapon from the mouth with the tongue.[5]

In suicide by gunshot—whether handgun, rifle, or shotgun—the favored targets are the head, chest, and abdomen, in that order.[30] In shotgun wounds to the head, the handedness of the suicidal person does not correlate as well as with the temple wound, since one hand must hold the muzzle to the head while the other pulls the trigger. The unconscious selection of priorities for these tasks by the shooter is unpredictable. However, because of the contortion required to shoot oneself in the chest or abdomen with a long gun, "trigger-handedness" usually correlates with trajectory; downward and to the left for a right trigger hand, and downward and to the right for lefties.

The presence of multiple gunshot wounds, or the radiological finding of several bullets, does not preclude suicide. The victim may have had previous nonfatal gunshot wounds, for instance. There are reports of nonfatal gunshot suicide attempts followed by success with poisoning or hanging.[28] About 2% of suicides involve multiple gunshot wounds. These remarkable cases can involve faulty ammunition, poor aim resulting in a nonfatal wound, fleeting survival or reflex action producing more firings, or the use of automatic weapons.[30–32] There are reports of up to four suicidal gunshot wounds to the head[33] and nine suicidal gunshot wounds to the chest.[32]

PITFALLS

In the radiological evaluation of gunshot wounds a number of pitfalls await the unwary, the overenthusiastic, or the inexperienced viewer. One must be careful not to exceed the limitation of the method. Some of the difficulties in evaluating suicide, or in discriminating between suicide and homicide, have been pointed out already. The radiologist should detect and demonstrate the evidence on the film; further speculation and assumption are the province of other members of the forensic team.

SIZE OF THE MISSILE

There is great temptation to estimate the caliber of a bullet or the size of shot by "eyeballing" a radiograph. It is a temptation to be resisted. Any missile radiographed within the body will be magnified to some degree, and only a small degree of magnification destroys any hope of "eyeball" accuracy (Figure 15.13).

A review of the characteristics of firearms involved in fatalities in Milwaukee during a 5-year period[34] showed that 26% involved either 0.22 or 0.25 caliber weapons in almost equal distributions, 67% involved weapons varying from 0.312 to 0.357 in. in diameter, and 7% were larger. It is obvious that the differences within the first two subgroups are minimal.

If there are two views taken at right angles of the bullet within the body, and accurate knowledge of the focal spot-to-bullet distance and the bullet-to-film distance, then the approximate caliber of the bullet can be calculated using the formula presented in Figure 8.6. With this degree of accuracy one may distinguish an undeformed 0.22 caliber bullet from a 0.25 and a 0.38 from a 0.45.[35] A 0.38 caliber bullet (actually 0.357 in. in diameter) cannot be distinguished from a 0.357 or a 9-mm caliber (0.355 in.). The designations of caliber are misleading since a 0.38 is not really 0.38 in. in diameter and a 0.44 is really 0.4295 in. in diameter.[35,36]

A rather elaborate radiographical method of estimating caliber and weight of both deformed and undeformed bullets

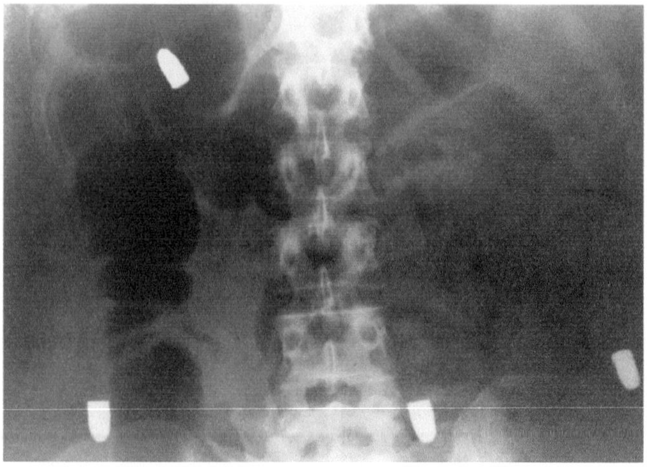

FIGURE 15.13 Effect of magnification. Bullets of 0.32 and 0.38 caliber were placed at different positions on a body and radiographed. Because of the variation in magnification due to different object-film distances, all of the bullets appear to be about the same size. (From Messmer, J. M. and Fierro, M. F., *RadioGraphics*, 6, 457, 1986. With permission.)

has been reported by Bixler et al.[37] Three orthogonal radiographs, fluoroscopically aligned, are used to determine the cross-sectional area of the bullet, and the weight determination is dependent on a data base derived from 48 bullets previously removed from humans. This technique was accepted for purposes of an internal police investigation but, when reported, had not yet been accepted in a US court.

Apparently, one federal court has admitted a quantitative method of bullet identification by means of the ratio of bullet diameter to length.[38]

Attempts have been made by Berryman et al.[39] to determine the caliber of a missing bullet. The entrance wounds in cranial bone were examined to investigate relationships between wound diameter and bullet caliber. They found so many variables affected the equation (large variety of bullets involved, variation in shape and surface, gyroscopic instability, intermediate targets, and tangential entries) that accurate caliber determination was undependable. They believe that this method could be improved sufficiently to allow elimination of suspect weapons, but not to identify a specific caliber.

An entrance wound in skin cannot be used to determine caliber. It may be even smaller than the diameter of the bullet or, if entering through a skin fold or crease, may even be slit-like in configuration.[40]

A bullet entering the body ordinarily will travel in a straight line until it comes to rest, strikes bone, or exits the body. Radiological evaluation can be helpful in any of these instances, and in the exceptions to the rule.

Determination of entrance and exit are usually the province of the attending clinician or the autopsy surgeon. Collins and Lantz[41] found that even trauma specialists had difficulty interpreting entrance and exit wounds and number of projectiles: 74% of multiple gunshot wounds were interpreted incorrectly, and 37% of exiting single gunshot wounds were misclassified. We have personal experience of one nonfatal shooting where six experienced observers (sworn officers, emergency technicians, nurses, and physicians) who saw the patient's naked body disagreed in every instance on the number and location of bullet wounds.

Determination of entrance and exit wounds in the skull by the rule of intersecting fractures has been described already in Chapter 3 (see Figure 3.7).[42] If a bullet strikes a bone, its subsequent path may be indicated by metal and bone splinters (see Figure 3.5, where a fleeing burglar was shot in the back). Small lead particles show how the bullet was deflected around the curve of the rib before coming to rest in superficial soft tissue. Often, however, the scattered pattern of bone and metal fragments is so disorganized that the pathway is indeterminate.

Canadian investigators[43] hypothesized that a centerfire rifle projectile might produce a cone-shaped fragment pattern with the apex at the entry site, thus allowing interpretation of the trajectory from routine radiographs. This could be useful when entrance and exit wounds are altered or destroyed by decomposition or other causes. They found it difficult to accurately describe a three-dimensional cone from two-dimensional radiographs. An incorrect opinion of bullet direction was rendered 57% of the time when the entry site was unknown, and in 19% of the cases even when the location of wounds was known.

Gas or air may be introduced into the body of a gunshot victim from the outside or from the gastrointestinal tract or respiratory system, but will not mark the bullet pathway dependably. Rather, the air in solid tissues will tend to distribute along fascial planes or in body cavities by gravitational forces.

In the living, computed tomography (CT) is invaluable in the analysis of gunshot wounds of the head and neck or body.[36] Foreign materials carried into the wound may simulate tissue densities; wood splinters may be confused with air shadows. CT of the abdomen ideally is performed before peritoneal lavage. Angiography may be required to evaluate possible damage to the cardiovascular system, and contrast studies of the esophagus are recommended in midline wounds of the neck or thorax. Remember that the digital scout film or topogram, preliminary to a CT scan, is useful in localization of missiles.

Large metallic fragments in the body may produce large "star patterns," which obscure detail on CT. Magnetic resonance imaging (MRI) then may be particularly helpful in assessing injury to neural and vascular structures and to solid organs. However, if the metallic fragments are ferrous or paramagnetic (i.e., some forms of nickel, cobalt, or iron), MR artifacts also may obscure detail. Ferrous metals in the brain, spinal cord, or eye are contraindications to MR examination, as is the presence of a cardiac pacemaker.[35]

Steel shot, required by law when hunting waterfowl, are affected by large magnetic fields. Usually, they can be differentiated radiographically from lead shot by their resistance to deformation (see below in the section "Pellet Problems").

NUMBER OF BULLETS

Determination of the number of bullets expected or found within the body may be surprisingly difficult. There may be confusion between entrance and exit wounds. It is possible that a bullet will enter the same wound made by a round fired earlier in a sequence. Rarely, a bullet will lodge in the barrel of the weapon after firing; a second firing may propel both bullets out of the barrel. These "tandem bullets" can enter the body through the same entrance hole.[44]

Other metallic densities may be confused radiographically with a bullet (Figures 15.14 and 15.15). Further, not all projectiles shot from guns are bullets (Figures 15.16 and 15.17).

Example Case 14-1—This young man was shot in the left chest and developed pneumothorax, which necessitated

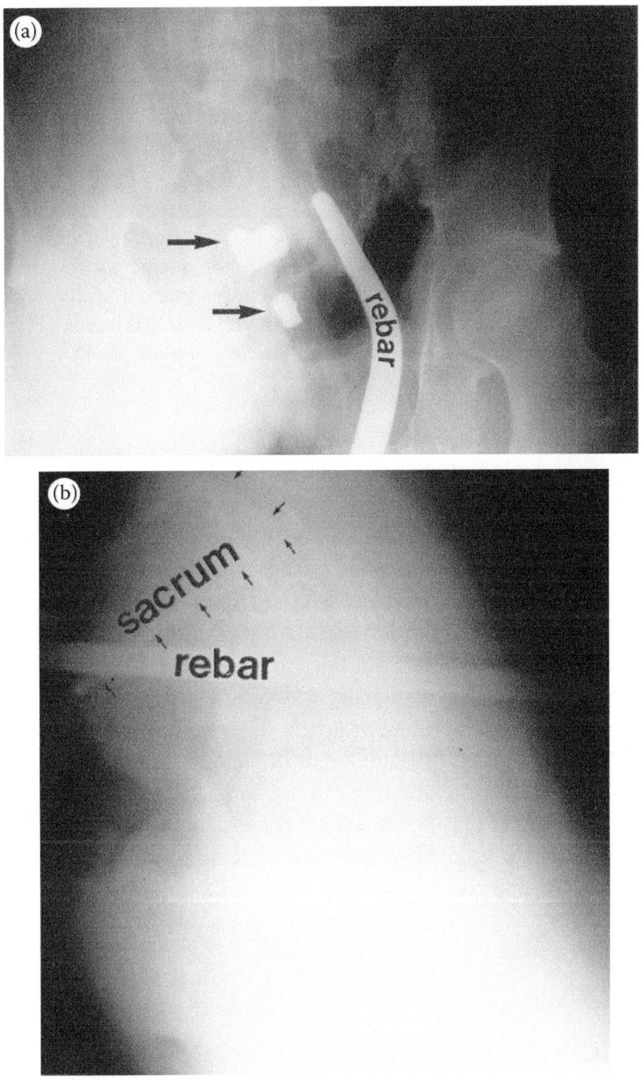

FIGURE 15.14 Articles of clothing can be confused with bullets. (a) Arrows point out a zipper pull and a waist fastener that could be confused with bullets. Actually, this biker crashed into a bridge under construction and impaled himself, (b) on a length of rebar that was cut off and brought in with him. He survived. Deceased gunshot victims should be radiographed initially with their clothes on in order to detect any spent bullets lying loose in the clothing.

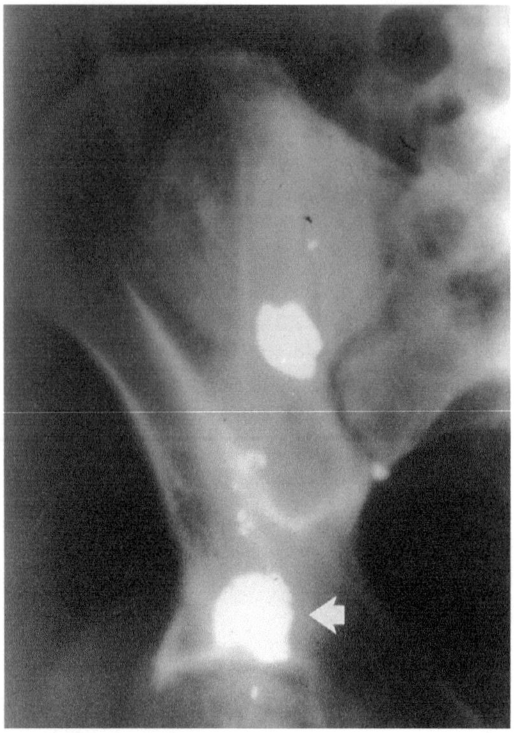

FIGURE 15.15 This is the same gunshot wound as shown in Figure 15.5. On this frontal view (which was taken first) the separated jacket (arrow) might be confused with a second bullet. (From Messmerr J. M. and Fierro, M. F., *RadioGraphics*, 6, 457, 1986. With permission.)

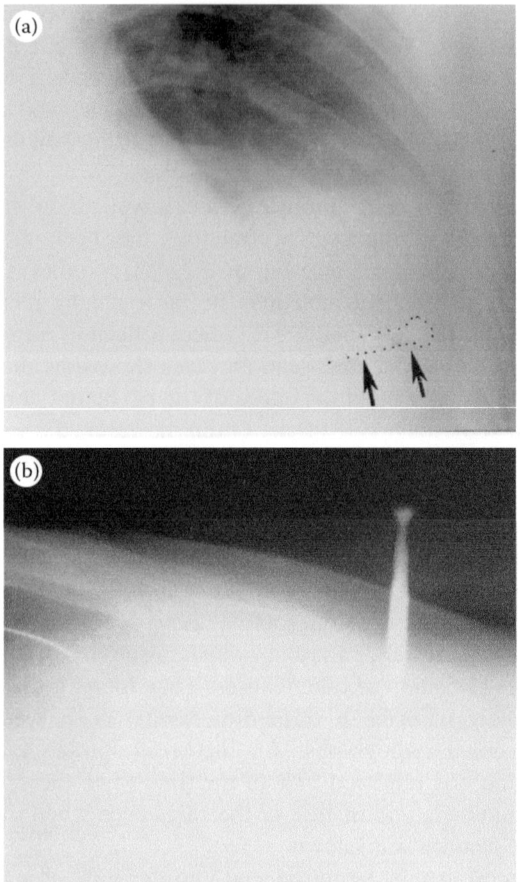

FIGURE 15.16 Wound from a stud gun. (a) The missile (arrows) is barely seen within the density of the liver and blood or fluid in the left pleural space and is outlined for the reader's benefit. (b) Cross-table lateral view of chest. (head toward viewer's left). The stud has entered the chest at the level of the tip of the xiphisternum.

placing a pleural drainage tube. The tube appeared to be malfunctioning, and a frontal view of the chest showed a bullet-shaped density at the end of the chest tube (Figure 15.18a). The question was raised whether the bullet had become lodged on the chest tube. A frontal view of the abdomen revealed an apparent second bullet, although the victim claimed to have been shot only once. Figure 15.18b showed that the bullet-shaped density was an integral part of the tube tip, designed to facilitate insertion. The real bullet had slipped into the abdomen through a rent in the diaphragm.

More common, and more important, is the "missing" bullet. The bullet may be "missing" because some other injury is mistaken for a gunshot wound, or because an exit wound is overlooked or disguised. More often, the bullet is "missing" because it is not in its expected location due to migration or embolization.

MIGRATION OF MISSILES

Movement of foreign objects within the body can occur in tubular structures such as the vascular system, the bronchial tree, the alimentary canal, the urinary tract, and the neural canal. They can also travel within less confined spaces such as the pleural space or the peritoneal cavity. Because they are of metallic density, bullets and pellets are readily visualized radiographically. Any missile not quickly located near its wound of entry or along an obvious tract requires radiological localization. In the dead, this can save time for the forensic pathologist in recovering essential evidence and information. In the living, rapid and accurate location of migrating missiles can be life saving as well as time saving.

Factors determining or modifying missile migration include force and direction of blood flow, gravity and position, pressure changes associated with the Valsalva maneuver, variation in vascular anatomy, the site of entry, and the weight of the projectile.[45,46]

Patients with missile migration may demonstrate absence of a projectile from the general area of an entrance wound, change in position or number of bullets or pellets (or fragments thereof), lack of an exit wound, and in some cases suggestive clinical findings such as signs of a distal vascular occlusion.[45] It must be remembered that a bullet found in a location inconsistent with the entry wound bullet path may not reflect either migration or embolization; one must investigate the possibility that it resides there as a result of an old gunshot wound in the remote part.[35]

Not infrequently, serial radiographic studies will be necessary to reveal movement of the bullet (Figure 15.19).

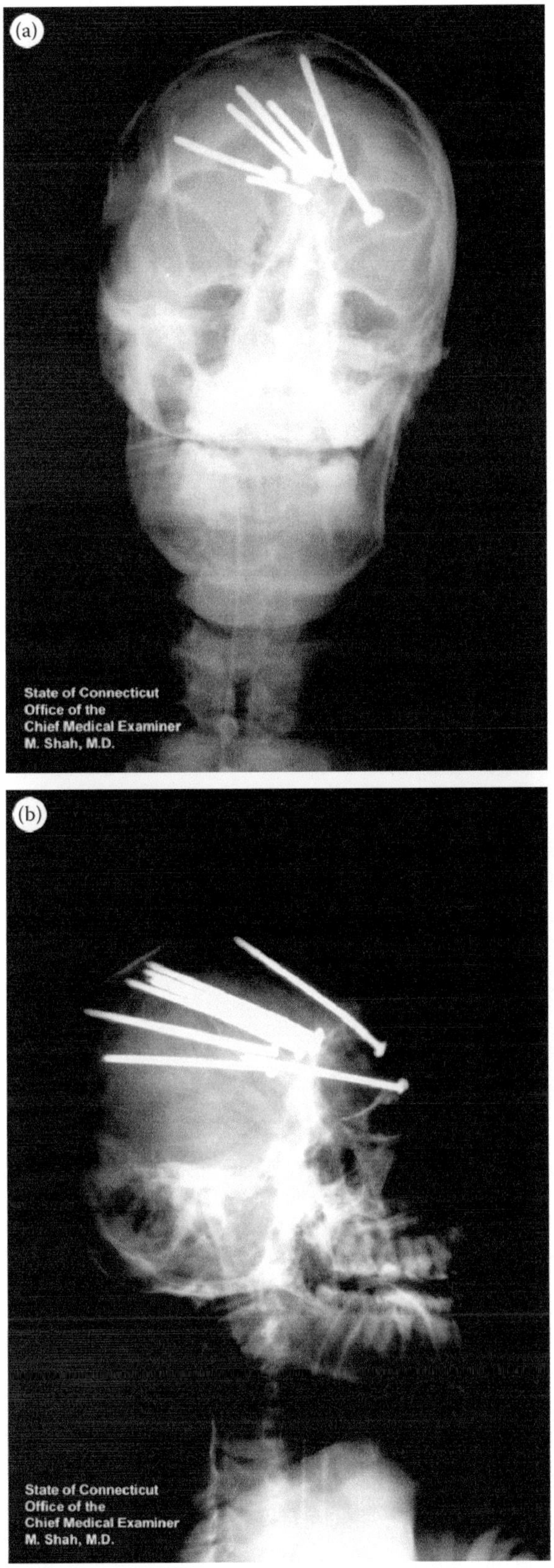

FIGURE 15.17 Murder with a nail gun. (a) Frontal view. (b) Lateral view. The victim was shot eight times through the left orbit with this tool ordinarily used in framing houses. (Courtesy of Dr. Malka B. Shah.) (From the office of the Chief Medical Examiner of the State of Connecticut. With permission.) (M.C. Sanntag, Forensic Photographer.)

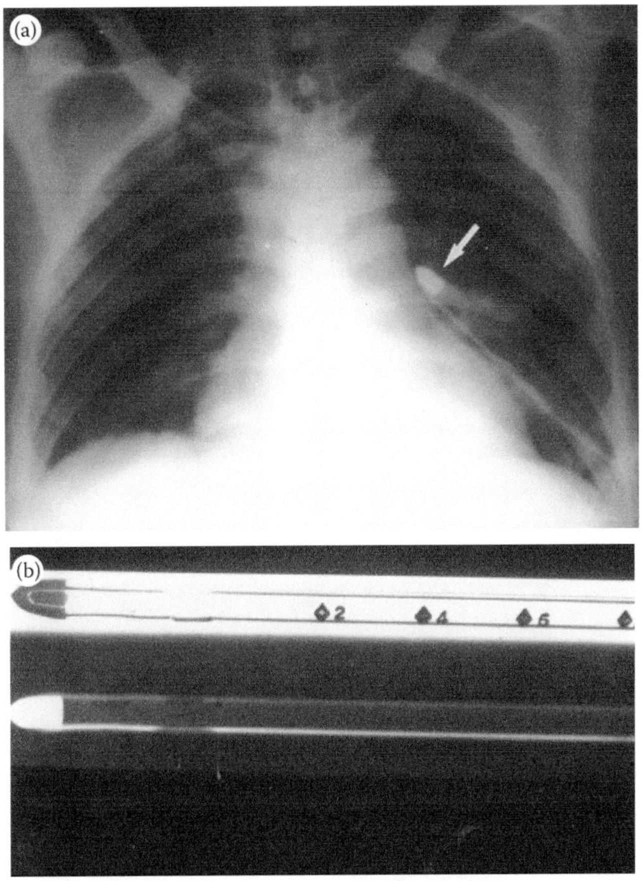

FIGURE 15.18 (a) Chest radiograph shows bullet-shaped density at tip of chest tube. (b) Photograph (top) and radiograph (bottom) show that the bullet-shaped density is an integral part of the tube. (From Messmer, J. M. and Fierro, M. F., *RadioGraphics*, 6, 457, 1986; Messmer, J. M. and Wadsworth, J. D., *J. Forensic Sci.*, 29, 340, 1984. With permission.)

Bullets fired into the spine may lose velocity abruptly against the strong bone structure there and sink for surprising distances into the spinal subarachnoid space. Radiological examination offers the only rapid and easy way to locate such missiles, and will prevent laminectomy at the wrong level (Figures 15.20 and 15.21).

The urinary tract is a rare location for missile migration. Bullets or pellets can gain access through wounds in the kidney or bladder (Figure 15.22).

Wounds to the face can result in bullets being swallowed[46] (Figure 15.23). Projectiles fired into the chest or abdomen can be introduced directly into the alimentary canal.[47] The bullets usually will pass uneventfully in the stool, but the thoracic or abdominal cavities may be contaminated by a leakage from the perforated gut.

Example Case 14-2—A young male was shot in the right chest just beneath the nipple. There was no exit wound. A chest radiograph (Figure 15.24a) showed pneumothorax but no intrathoracic bullet; however, a bullet (arrow) was seen in the right upper abdominal quadrant. Upon the assumption that the diaphragm had been penetrated, a laparotomy was performed, but the diaphragm and all abdominal organs

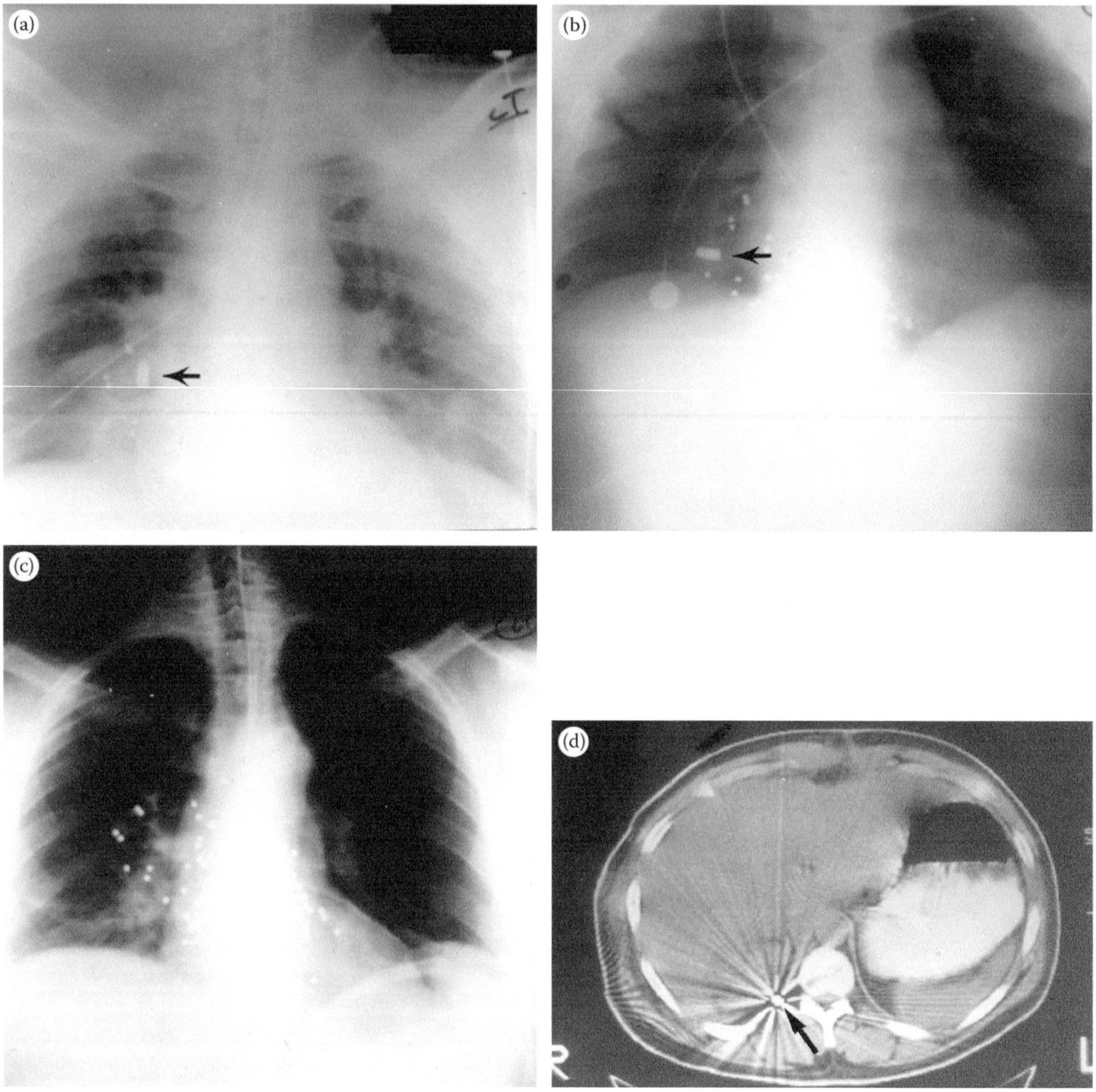

FIGURE 15.19 (a) Supine chest radiograph showing bullet (arrow) from acute gunshot wound. Small shot are from an old shotgun injury. It was not possible to do a lateral or erect view because of the patient's clinical condition. (b) Supine chest two days later. Bullet has rotated and moved—obviously in pleural space. (c) Two days later bullet disappears from chest radiograph. (d) CT reveals bullet deep in posterior costophrenic sulcus (star pattern). Bullet obscured on routine chest film by density of full-thickness liver.

were found to be intact and normal. Five days later a CT was done to localize the bullet, now thought to be within the liver. Rather, the topogram (Figure 15.24b) showed the bullet in the right lower quadrant, and the CT scan (Figure 15.24c) localized it within the cecum. This prompted a contrast study of the esophagus (Figure 15.24d), which demonstrated a leak (arrow) at the bullet entry site. The bullet passed in the fecal stream.

We have also seen a case where shotgun pellets introduced directly into the stomach through an abdominal wound migrated into the esophagus by the mechanism of gastroesophageal reflux.

Missiles can enter the tracheobronchial tree by direct penetration or aspiration (Figure 15.25).

Once its energy is spent, a bullet can come to rest in one of the components of the cardiovascular system, or it may migrate to a distant site. The eventual resting place of a bullet that enters either the arterial or venous system depends closely upon its site of entry, the position of the victim, and the weight of the projectile.[48–56] Approximately 80% of vascular emboli are arterial and 20% are venous.[57] Vascular emboli are most often seen in adults, although there are reports in children.[58]

In general, bullet emboli are associated with small-caliber, low-velocity bullets. Heavier projectiles tend to "sink" and are more likely to be found in the dependent portions of the body (Figures 15.26 and 15.27). In a collective review of 36 cases of cardiac wounds associated with peripheral

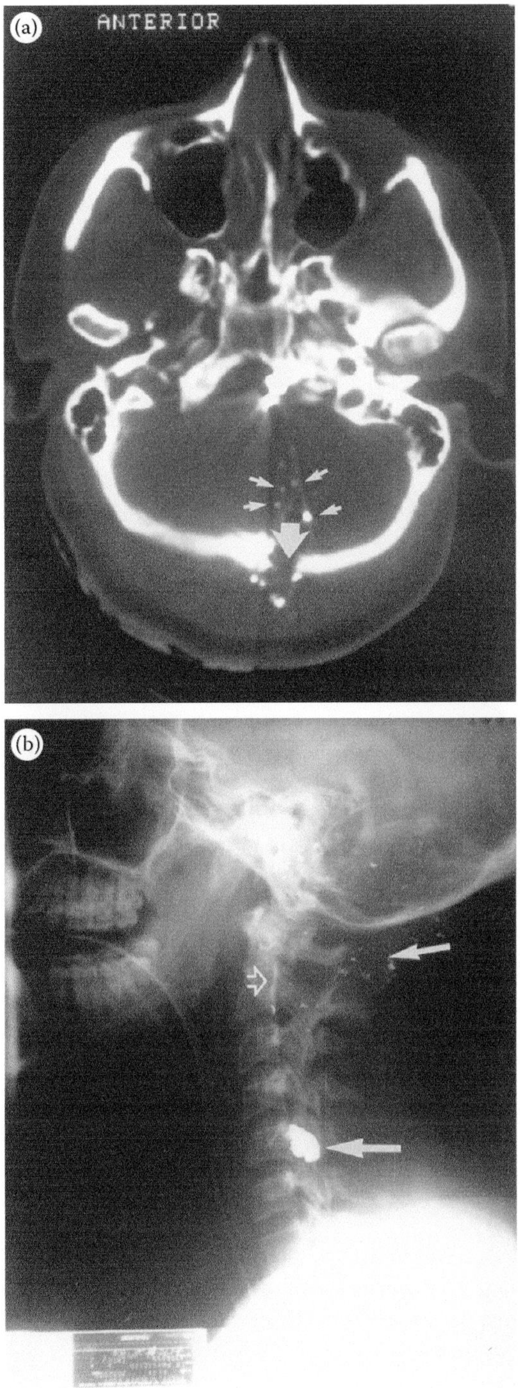

FIGURE 15.20 (a) Bullet fired into the base of the skull cut a groove in the occipital bone (large arrow) scattering fragments (small arrows) into the posterior fossa as shown on CT examination. (b) Lateral view of cervical spine shows that the bullet traversed the posterior elements of the C-1 vertebra (small arrows), impacted on the posterior body of C-2 (open arrows), then dropped in the spinal canal before coming to rest at the C-5 level (large arrow).

embolization, 75% involved embolization to the lower extremities and the remainder to the neck and visceral vessels.[59,60] Lighter projectiles such as shotgun pellets can be swept superiorly, against gravity, by arterial flow pressure. There are several case reports of intracerebral vascular accidents caused by migrating shotgun pellets, where the site of entry is usually either the neck or the chest.[61–65] We have personally seen only one such case.[66]

Example Case 14.3—A young male was shot in the chest and abdomen with a shotgun. At surgery he was found to have four puncture wounds in the heart, hemopericardium, and multiple puncture wounds in the spleen, pancreas, colon, and small bowel. [(Figure 15.28a) shows the shot pattern over the heart postoperatively.)] His postoperative progress was satisfactory for three days, then suddenly he developed dyspnea, chest pain, and cardiac collapse. Although resuscitated, he remained comatose and developed a dense left hemiplegia. A repeat chest radiograph (Figure 15.28b) showed that one of the pellets formerly overlying the heart shadow was gone. A carotid arteriogram (Figure 15.28c and d) showed that it had migrated to the right middle cerebral artery, which was totally occluded. The patient died, and the charge was changed from attempted murder to murder.

Pellets and small bullets may be transported by venous flow, even against gravity, and cause great mischief (Figures 15.29 through 15.31). Even large missiles occasionally follow venous flow (Figure 15.32).

Missile penetration of a major vessel, even the aorta, may not lead to fatal or even serious exsanguination, but distal occlusion is virtually inevitable.

Example Case 14.4—A young male was shot in the left flank with a handgun, but no bullet was present on the admission radiograph of the abdomen. Physical examination revealed no pulse in the left leg. A radiograph (Figure 15.33a and b) showed the bullet in Hunter's canal along the course of the superficial femoral artery. An aortogram (Figure 15.33c and d) demonstrated a pseudoaneurysm originating on the posterior wall of the aorta, indicating the site of bullet entry.

In general, symptomatic missiles should be removed. Left atrial or ventricular missiles floating free should be removed prophylactically. Missiles floating in the right heart may be removed or watched. Embedded missiles in the heart are relatively safe. Bullets that have traversed the bowel before embedding may be removed to prevent potential infection. Missiles next to major vessels may be removed to avoid risk of erosion into the vessel.[35,36] Cardiac neurosis from knowledge of a bullet within the heart may be a patient-management problem.[67]

To summarize the problems of missile migration, we would reiterate that radiographic detection of missile migration depends upon careful examination of serial studies and knowledge of the manifestations of missile migration. These include a change in the number or positions of missile fragments, absence of a missile from the general area of the wound, or the development of signs of distal vascular occlusion. We recommend radiographic examination immediately prior to operation to confirm the anatomic location of missiles that need to be removed, since failure to find a missile at operation may require multiple intraoperative radiographs to survey the body for the final resting place of a bullet embolus.

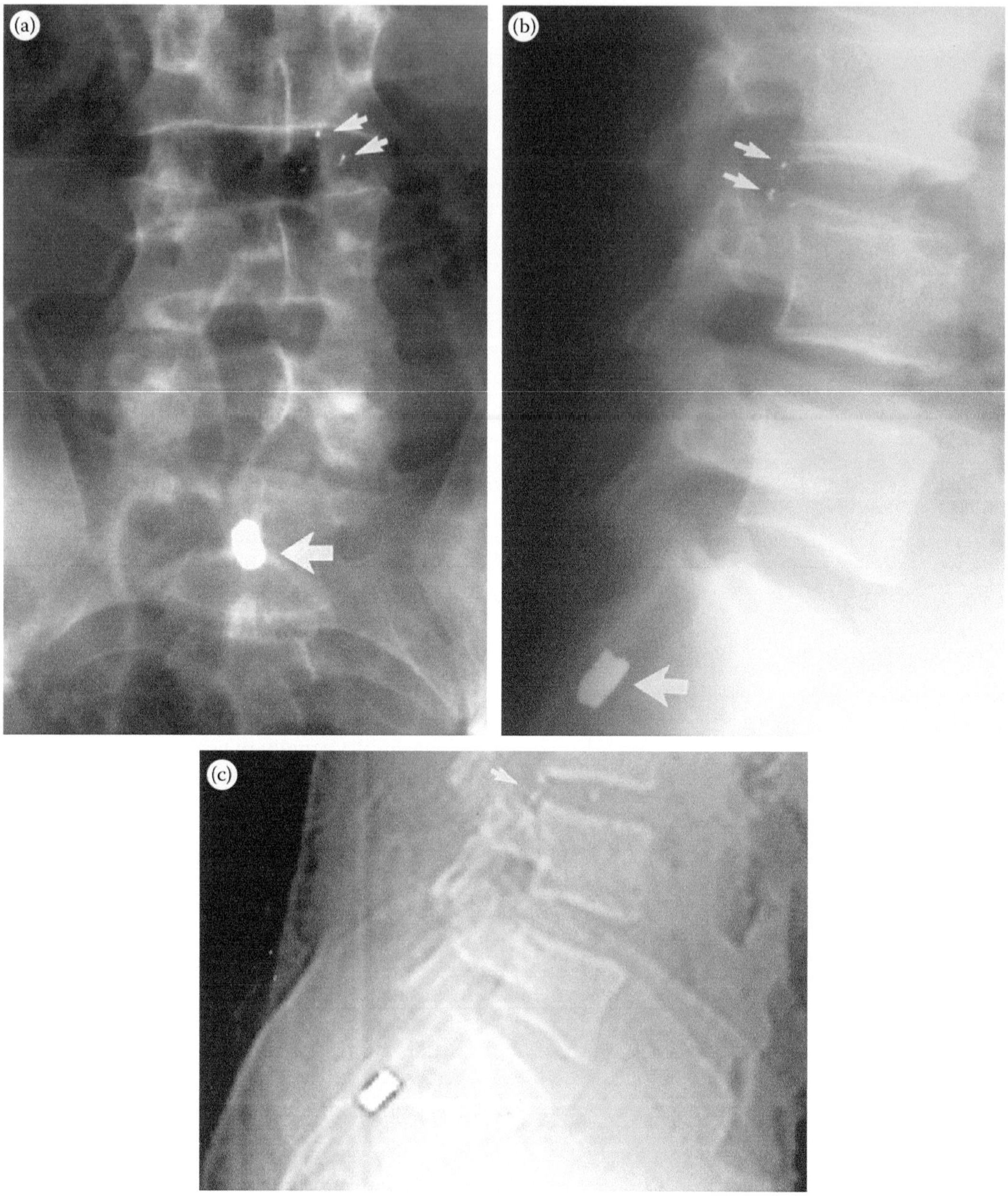

FIGURE 15.21 Gunshot wound to the back (a) frontal and (b) lateral views show the bullet struck the left inferior articular process of the L-3 vertebra, leaving two small metal fragments (small arrows), then dropped into the subarachnoid space, coming to rest in the caudal sac (arrow). (c) Note that the topogram or digital scout film for a subsequent CT is just as useful as the lateral radiograph.

PELLET PROBLEMS

Deformation

Except for the steel loads mandated for waterfowl hunting, shotgun pellets of all sizes are subject to deformation. Froede et al. have pointed out that deformation of round pellets, particularly flattening, should be recognized as "a gross radiographic pitfall."[29] Large shot, in particular 0 or 00, upon flattening and/or fragmentation may bear no resemblance to shot and, rather, suggest large-caliber bullets or jackets. Lead shot will deform upon impacting firm body tissues, especially bone. Shot ricocheting off hard surfaces back into the body may be especially confusing. Radiography in more than one projection will help sort out these problems. In one case, lack of deformation of round shot was helpful in evaluating a death scene.

Example Case 14.5—The partially clad, mostly decomposed body of a young male was found in southern New Mexico in the springtime. The remnants of clothing and, more importantly, antemortem and postmortem roentgenographic matches of a previously fractured arm, enabled positive identification of the body. He was a young Navajo subteenager who had run away from a Bureau of Indian Affairs boarding school during a blizzard the preceding

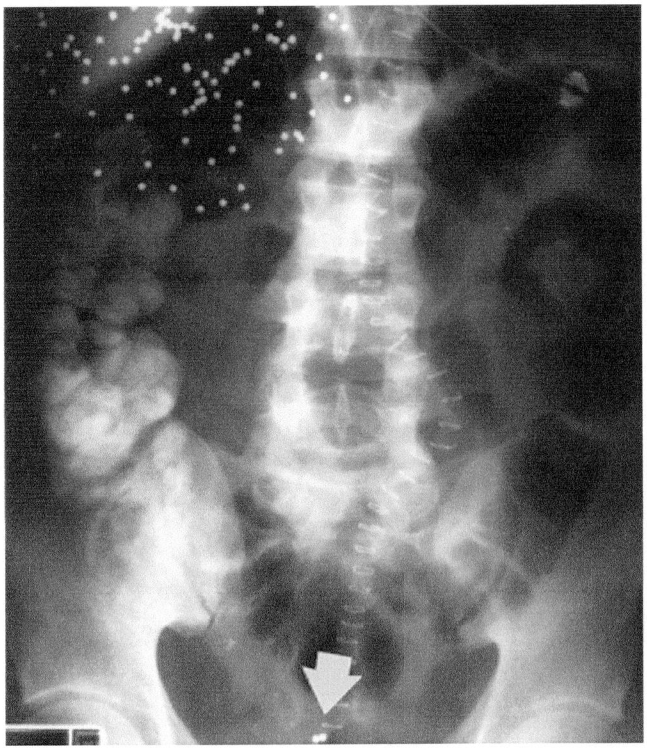

FIGURE 15.22 A shotgun wound to the back injured the right kidney. The two pellets overlying the lower pelvis entered the urinary bladder (arrow) after passing through the right ureter.

winter—presumably to return to his home on an Arizona reservation. The problem was that the roentgenological studies of the body, as found, showed numerous very small shotgun pellets. Was he a homicide? It was noticed (Figure 15.34a) that *none* of the pellets were deformed, and that the shot pattern was unusual as compared to the usual shotgun victim's roentgenogram (Figure 15.34b). The body had been found in the periphery of the Sheriff's Posse Skeet and Trap Range, which had been heavily used all winter. It is believed that the young lad died of exposure on his first night out, and that the spent shot from skeet loads had drifted down onto, and into, the decomposing body and its clothing throughout the long winter of his disappearance.

PELLET PATTERN: RANGE OF FIRE

The spherical pellet fired from a nonrifled shotgun is aerodynamically disadvantaged. It slows down rapidly in air or tissue. It is not a long-range missile. Pellets exiting a shotgun muzzle gradually spread out. There are many systems for determining the range of fire of shotguns using information on gauge, shot size, choke, distance, and so on. Most are relatively inaccurate. The radiologist needs to know that the shot pattern is compacted at contact or close range where other materials (wad, shot cup, plastic, cardboard) may appear in the wound and on the radiograph, then spreads with increasing range or distance (see Figure 15.11). There is a large pitfall in this generalization. Whereas making conclusions about the range of shotgun fire from external inspection of the wound is usually accurate, drawing conclusions from radiographs is another potential pitfall. Radiographically, the dispersal of pellets may suggest that the weapon was fired from a distance, when in actuality the weapon was fired from close range (Figure 15.35). The phenomenon responsible for this is known as the "billiard ball" effect (Figure 15.36).

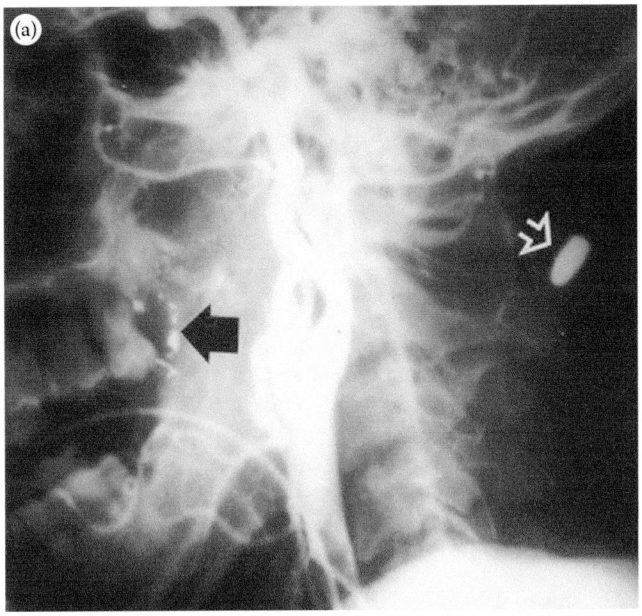

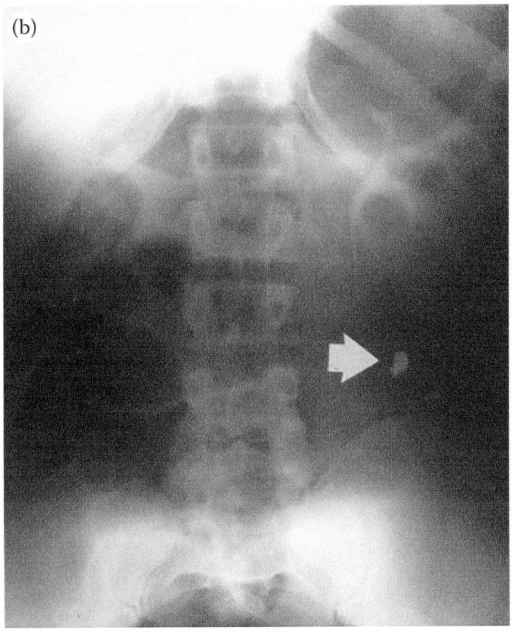

FIGURE 15.23 (a) This victim was shot twice in the face with a 0.25 caliber handgun. A lateral view of the neck following injection of contrast into the common carotid artery shows one bullet in the soft tissues of the posterior neck (open arrow) and metallic fragments (arrow) over the coronoid process of the mandible. (b) A supine view of the abdomen shows a bullet (arrow) in the left abdomen. This bullet had been swallowed and was eventually recovered in the stool. (From Messmer, J. M. and Fierro, M. F., *RadioGraphics*, 6, 457, 1986. With permission.)

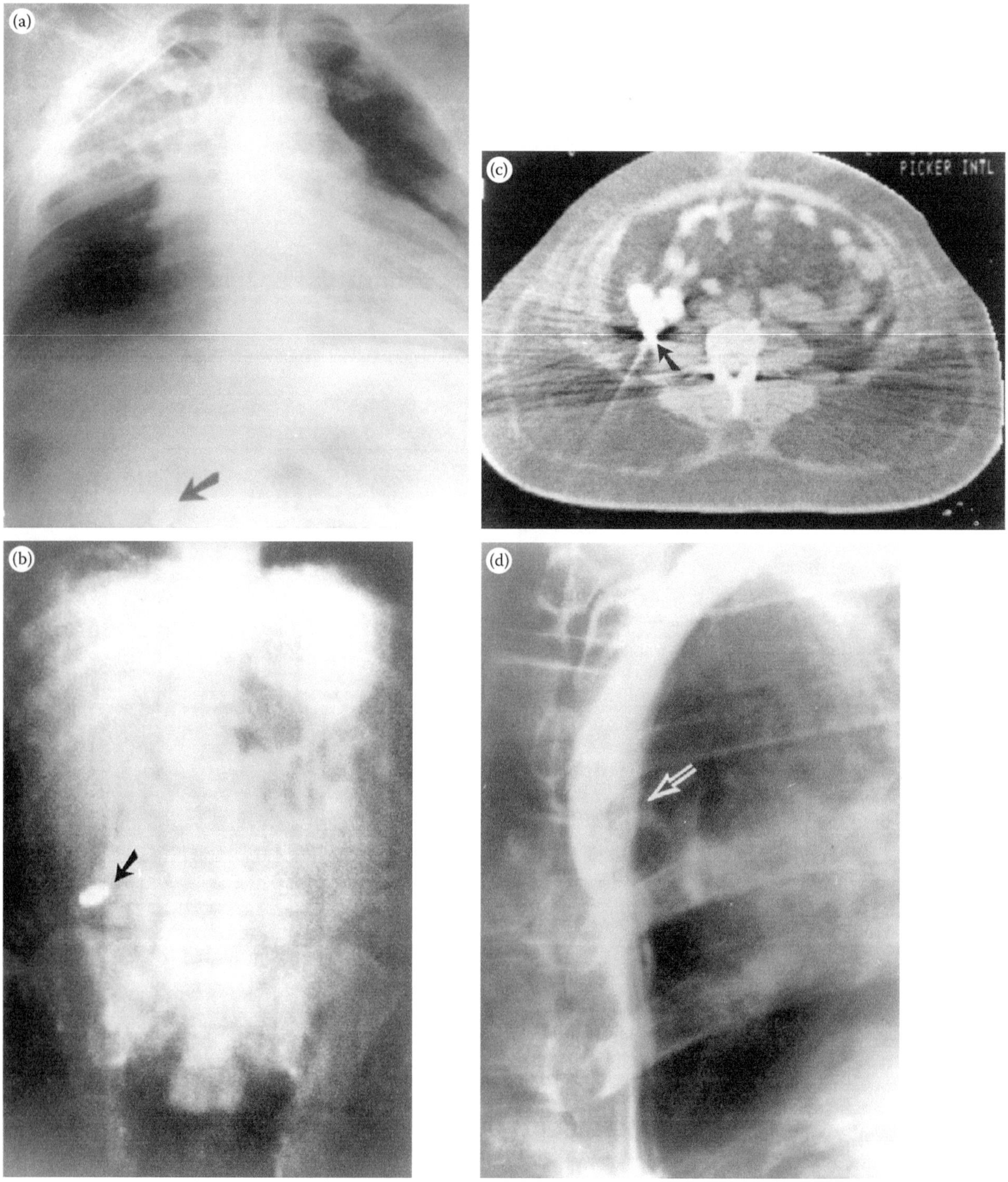

FIGURE 15.24 (a) Chest radiograph: no intrathoracic bullet, but one seen faintly beneath the diaphragm (arrow). (b) Digital scout film prior to CT shows bullet in RLQ (arrow). (c) Bullet creates star pattern in cecum on CT scan. (d) Wisp of water-soluble contrast medium escapes esophagus at site of bullet injury. (From Hughes, J. J., *J. Trauma*, 27, 1362, 1987. With permission.)

When a mass of shotgun pellets enters tissue, the first pellets which penetrate are slowed and struck from behind by the following pellets, which can cause them to scatter through the tissue much like a cluster of billiard balls struck by the cue ball. The radiographic image must be correlated with the physical examination of the skin surface wound pattern in order to avoid incorrect range estimates.[35]

BULLET WOUNDS: RANGE OF FIRE

The range of fire assessments of bullet wounds may depend on the composition and relative amounts of residue found on the skin. Primer residue usually contains some combination of lead, antimony, or barium. Bullet residue from nonjacketed bullets is 70–90% lead. Coated bullets leave an admixture of

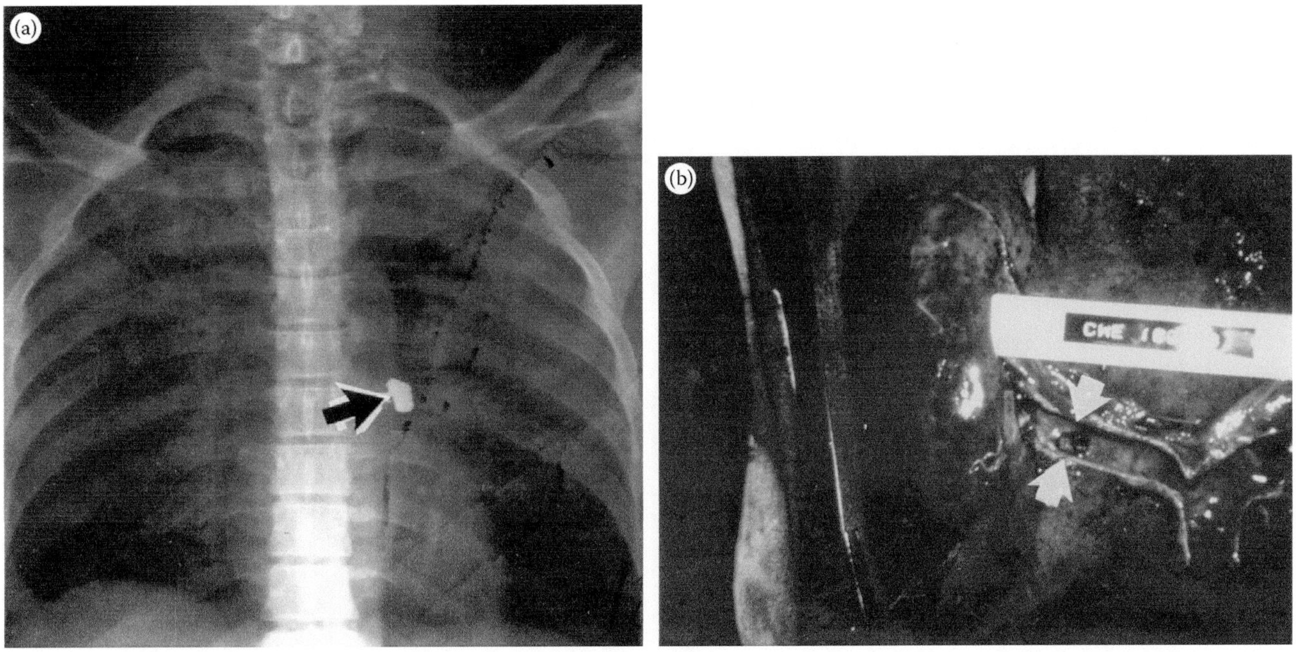

FIGURE 15.25 (a) This victim suffered two gunshot wounds to the face. A radiograph of the head (not shown) demonstrated only one bullet. A frontal view of the chest demonstrates a bullet (arrow) overlying the left hilum. (b) The autopsy findings showed the blackened bullet that had been aspirated within the left main stem bronchus (arrows).

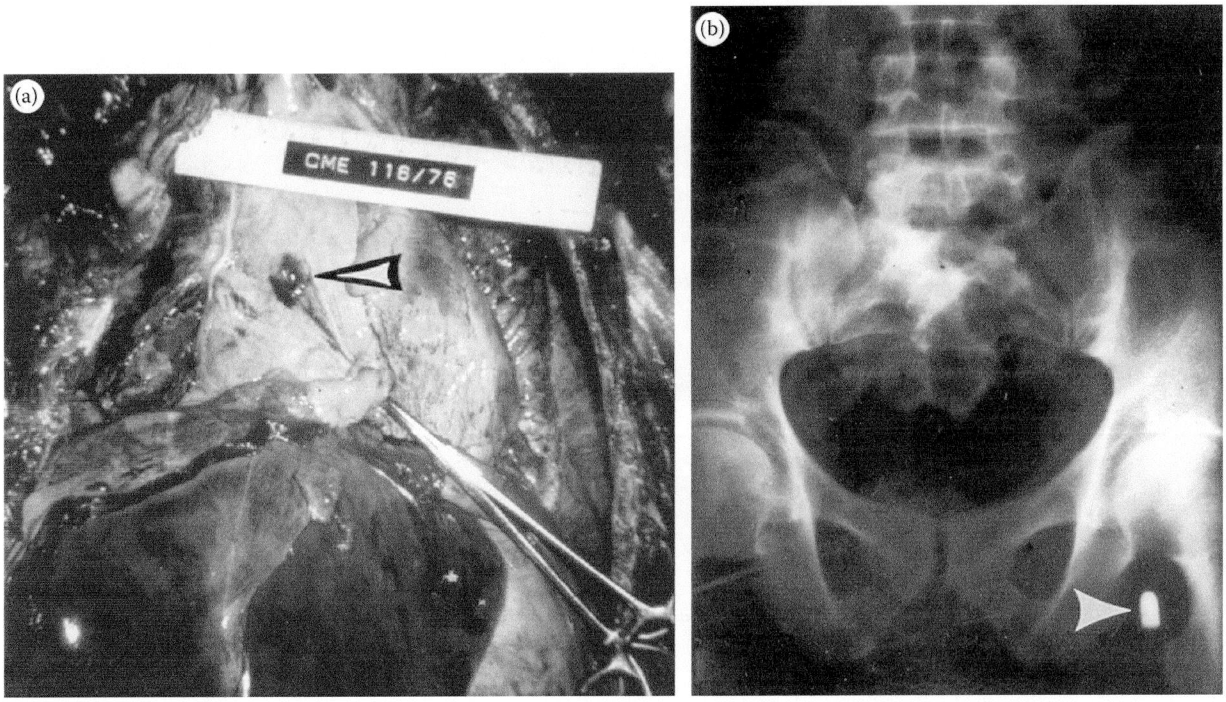

FIGURE 15.26 (a) Although the autopsy revealed a gunshot wound to the anterior surface of the heart (arrowhead), a frontal view of the chest had revealed no bullet. (b) A radiograph of the pelvis showed a bullet overlying the left groin. Postmortem dissection showed it to be in the left femoral artery. Forces of both gravity and flow probably effected this migration. (From Messmer, J. M. and Fierro, M. F., *RadioGraphics*, 6, 457, 1986. With permission.)

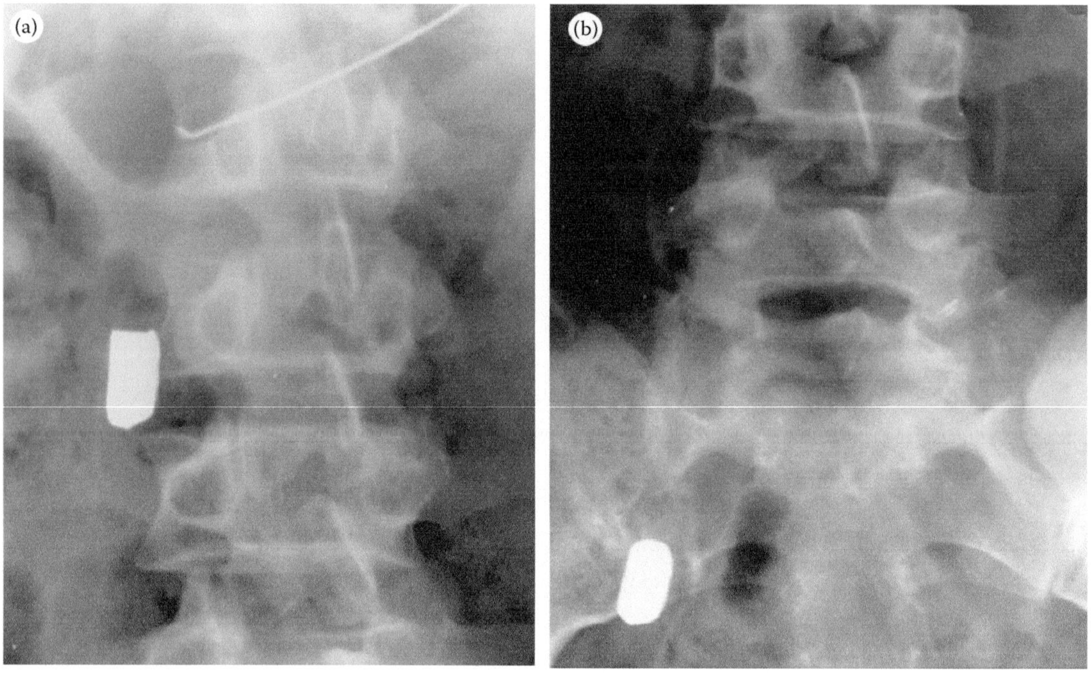

FIGURE 15.27 (a) Abdominal roentgenogram of a young male immediately after admission for a gunshot wound to the mid-abdomen shows the bullet just to the right of L-3. (b) Subsequent scout film prior to an excretory urogram shows the bullet has migrated from its initial position (consistent with the inferior vena cava) *against* venous flow to a location consistent with the right iliac vein.[12]

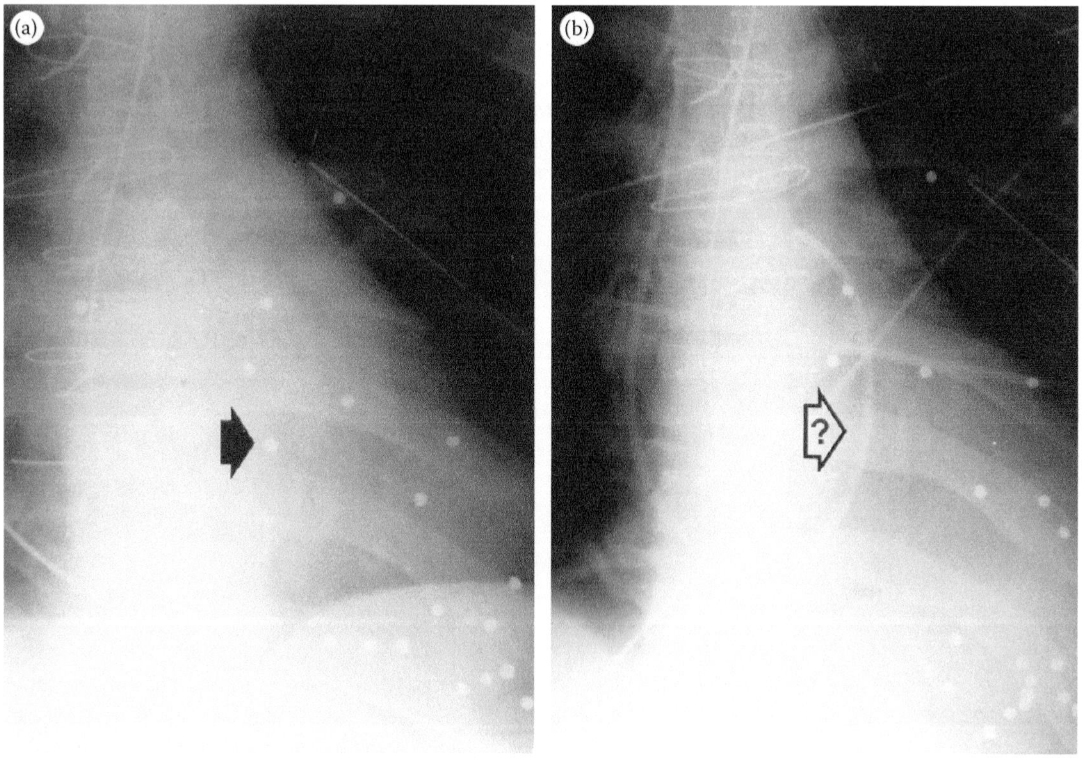

FIGURE 15.28 (a) Postoperative chest with shot pattern over the heart. Arrow marks critical shot, which is missing three days later. (b) When the patient undergoes a sudden clinical crisis. (c and d) Right carotid arteriogram shows occlusion of the right middle cerebral artery by the pellet embolus (arrow). (From Kase, C. S., et al., *Neurology*, 31, 458. With permission.)

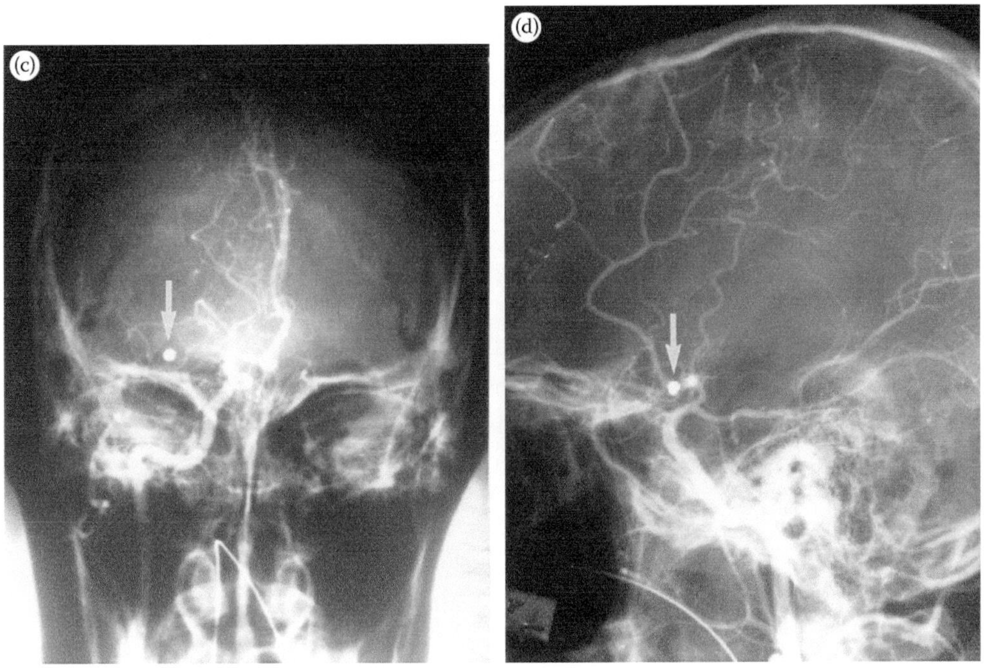

FIGURE 15.28 *(Continued)*

lead and copper. Jacketed or semijacketed bullets produce the least amount of lead and only small amounts of copper-containing residue.[68]

Residue deposits on excised skin have been examined with soft x-rays generated by a Faxitron® unit (Faxitron x-ray LLC, 575 Bond St. Lincolnshire, IL) (see Chapter 29) for localization, then further studied by scanning electron microscopy and energy-dispersive analysis of x-rays.[69]

Clothing can redistribute gunshot residue and even prevent it from reaching the skin. Thus, examination of the clothing may be as important as examination of the body.[69] Clothing may be so blood-soaked as to render direct examination impossible. Faxitron examination of the clothing can be extremely helpful in defining the location and distribution of gunshot residue, thus, allowing further collection and analysis (Figure 15.37).

TEAM EFFORT

The radiographic method has been shown to be useful in determining the location, size, type, number, and migration of bullets lodged within the body. Often x-rays will suggest

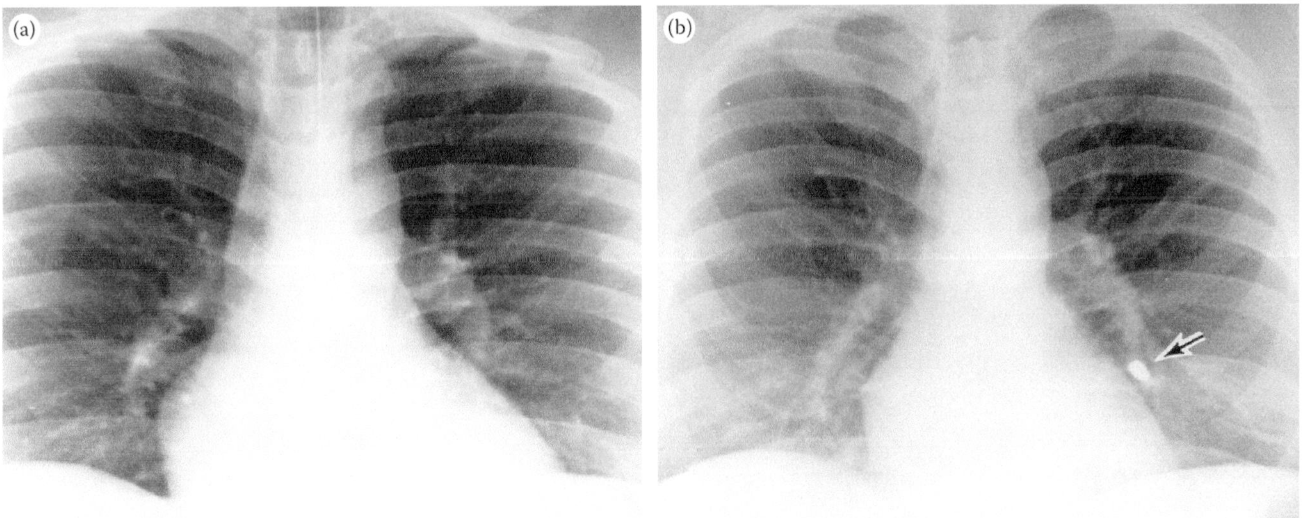

FIGURE 15.29 (a) Normal supine postoperative chest radiograph of young male following gunshot wound to abdomen. A tear in the inferior vena cava had been found at laparotomy. Several days later signs of pulmonary embolism developed. A repeat study (b) showed the bullet (arrow) embolized to a branch of the left pulmonary artery. (Adapted from Sellier, K. G. and Kneubuehl, B. P., *Wound Ballistics and the Scientific Background*, Elsevier, Amsterdam, 1994.)

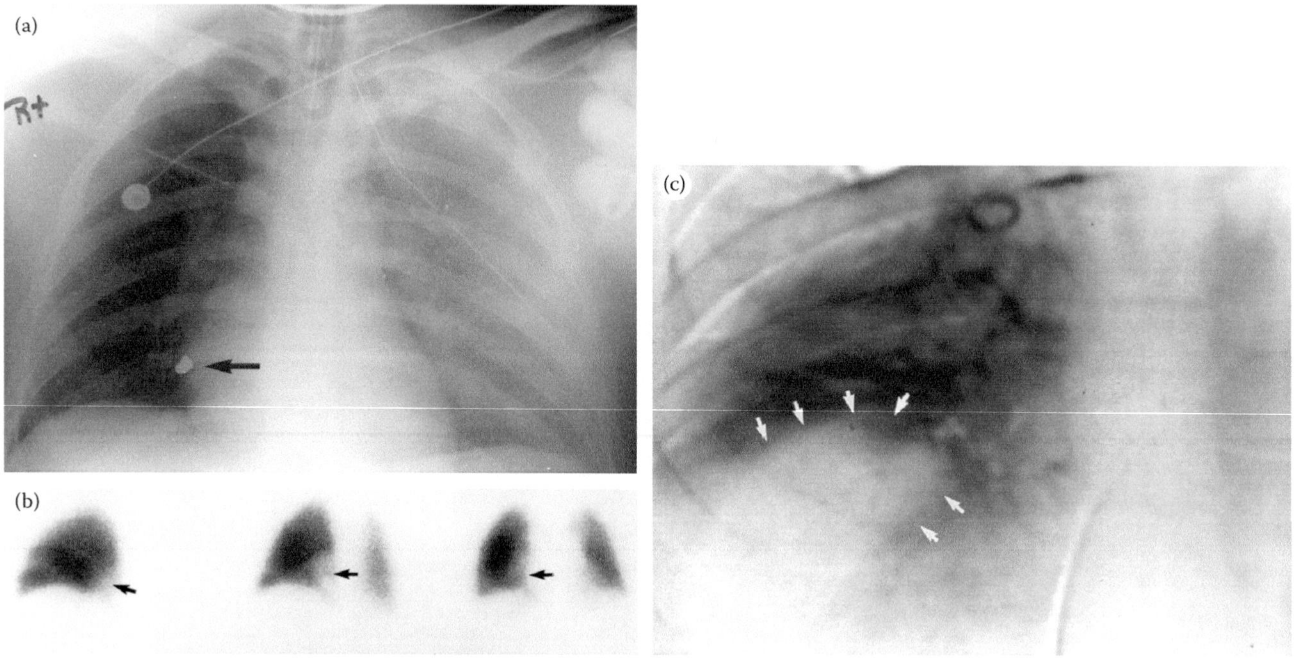

FIGURE 15.30 Another inferior vena cava to pulmonary artery embolus. (a) Chest radiograph. (b) Positive radioisotope lung scan showing perfusion defect in the right lower lobe posteriorly (arrows). (c) Pulmonary arteriogram. Contrast is reversed in this subtraction image so that perfused lung is black, nonperfused lung is clear (arrows).

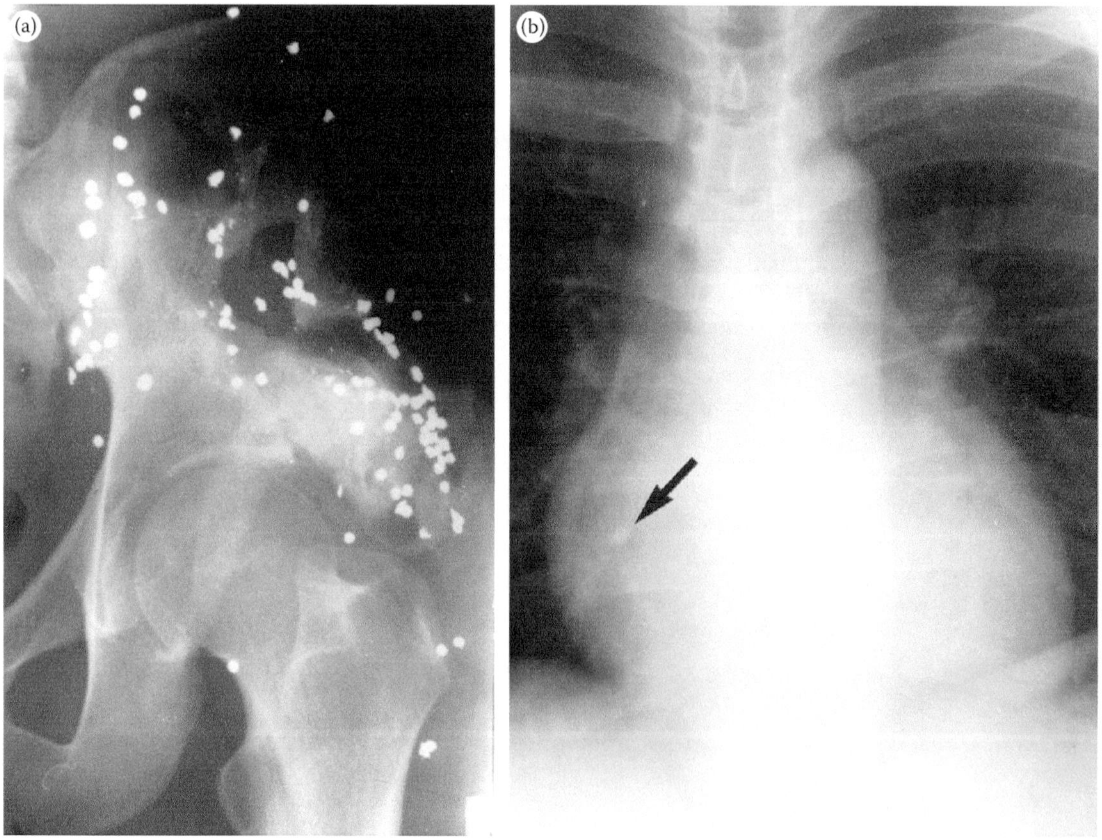

FIGURE 15.31 (a) Shotgun wound to the left hip. (b and c) Routine PA and lat. chest radiographs reveal a pellet lying within the heart. (d) Spot-film taken during fluoroscopy shows to-and-fro blurring of the shot as it moves with the heart beat. This means it is either in the wall of the right ventricle or trapped in the chordae tendineae inside the right ventricle. A shot moving freely in the ventricular cavity would describe a more circular motion.

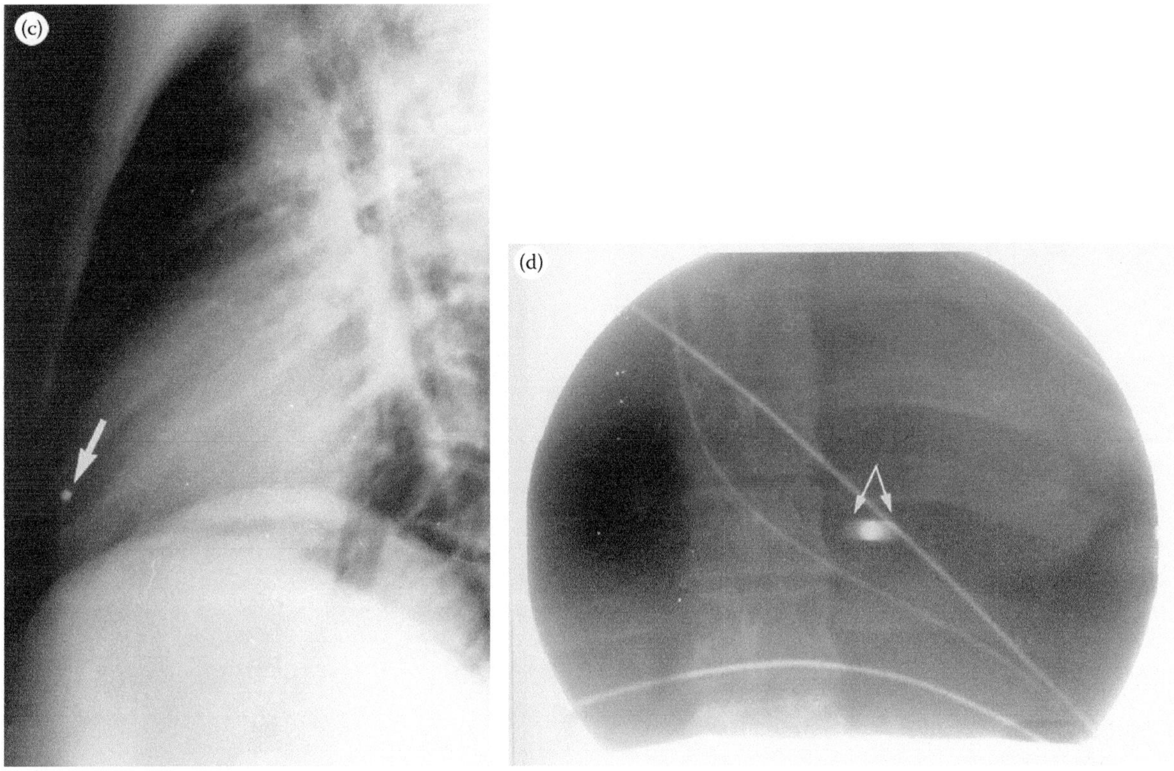

FIGURE 15.31 *(Continued)*

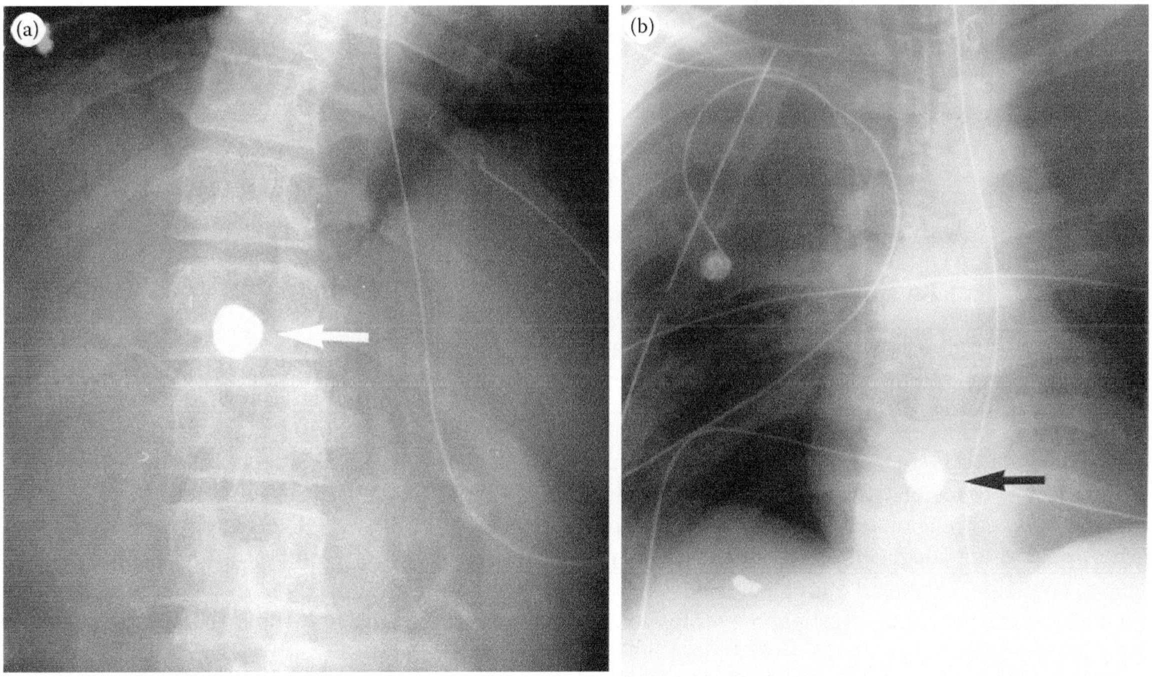

FIGURE 15.32 (a) A 0.45 caliber handgun wound to the back introduced a large bullet into the inferior vena cava. (b) Serial film studies showed it moved slowly into the right ventricle (arrow) where it stayed until removed with a stone-basket introduced into the right ventricle via the jugular vein and superior vena cava.

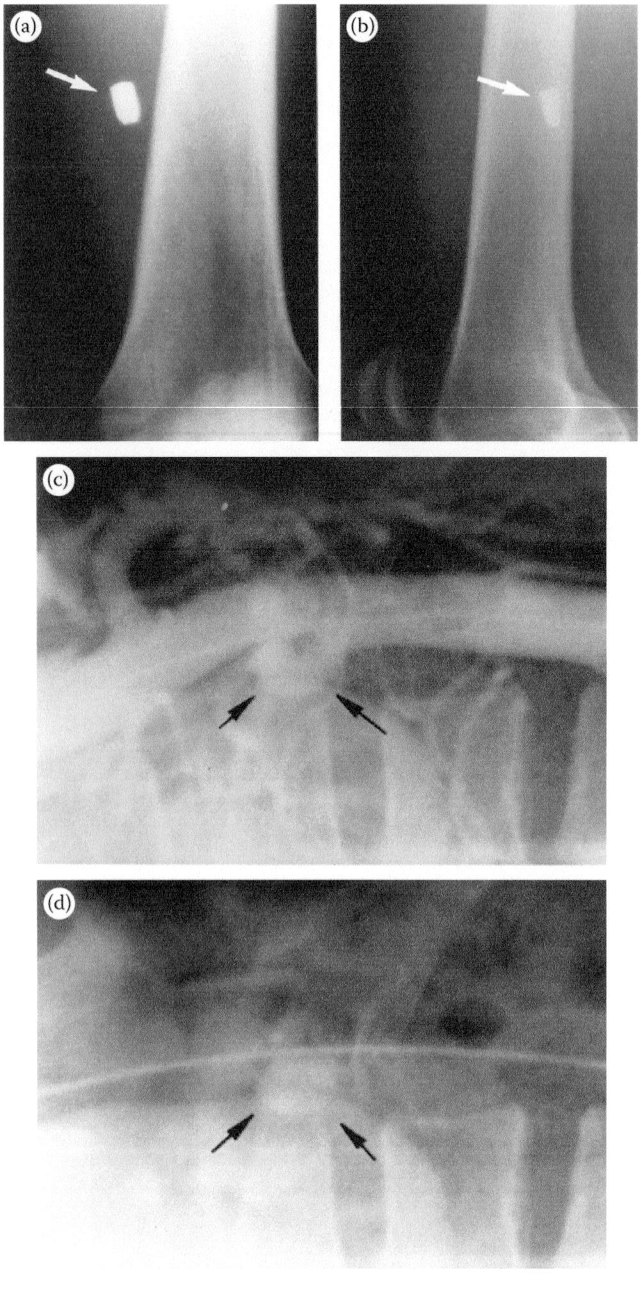

FIGURE 15.33 (a and b) Bullet (arrow) from left flank wound now lodged in the superficial femoral artery in the thigh. (c) and (d) Early and late films from a lateral aortogram show a pseudoaneurysm (arrows) arising from the posterior wall of the aorta where the bullet entered the systemic circulation.

the angle and direction of fire and, sometimes, the range of fire. It may be possible to recreate the relative position of the shooter and victim by correlation of these factors. This type of evidence is most often sought in cases of suicide or, in criminal cases, homicide, attempted murder, and assault.

The value of good crime scene investigation, professional photography, accurate laboratory work, and detailed police and medical records to the radiological evaluations of gunshot wounds cannot be overemphasized. Interdisciplinary coordination and cooperation are essential to the successful understanding of many, perhaps most, fire-arms-related cases.

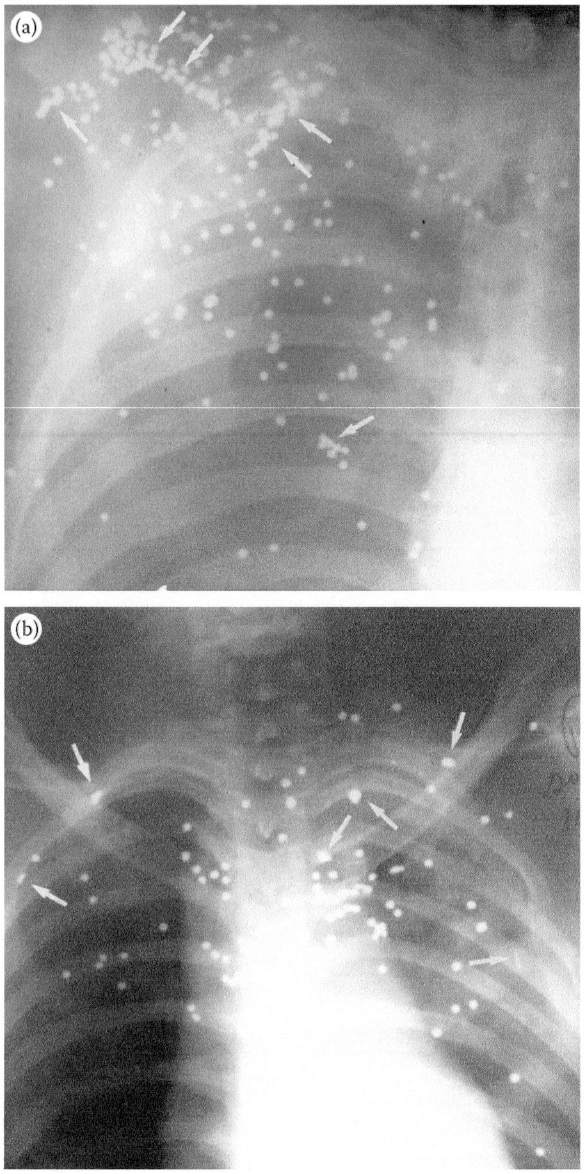

FIGURE 15.34 (a) Shot pattern over the body of a young Native American boy is unusual (see arrowed clusters) and NO pellets are deformed. (b) Typical shot pattern of a relatively long-range shotgun wound. Note the many deformed pellets (arrows).

Example Case 14.6—This case has to do with a young man shot during an arrest and later brought to criminal court on charges of resisting arrest and menacing. The defendant was in the bad graces of his boss because of some missing checks. The boss, who knew his employee was a scofflaw, tipped the police that he might be found at his girlfriend's house. Officers were dispatched there to arrest him on the basis of many outstanding traffic warrants. Frightened by a bevy of officers at the door, the young man fled downstairs to a basement laundry room, then into a dead-end furnace room. Hiding behind the furnace he superficially slashed his wrists with a small knife grabbed on his way through the kitchen (Figure 15.38a).

Four police officers occupied the basement laundry room (Figure 15.38b): Patrolman Able on the stairs; Sergeant Baker

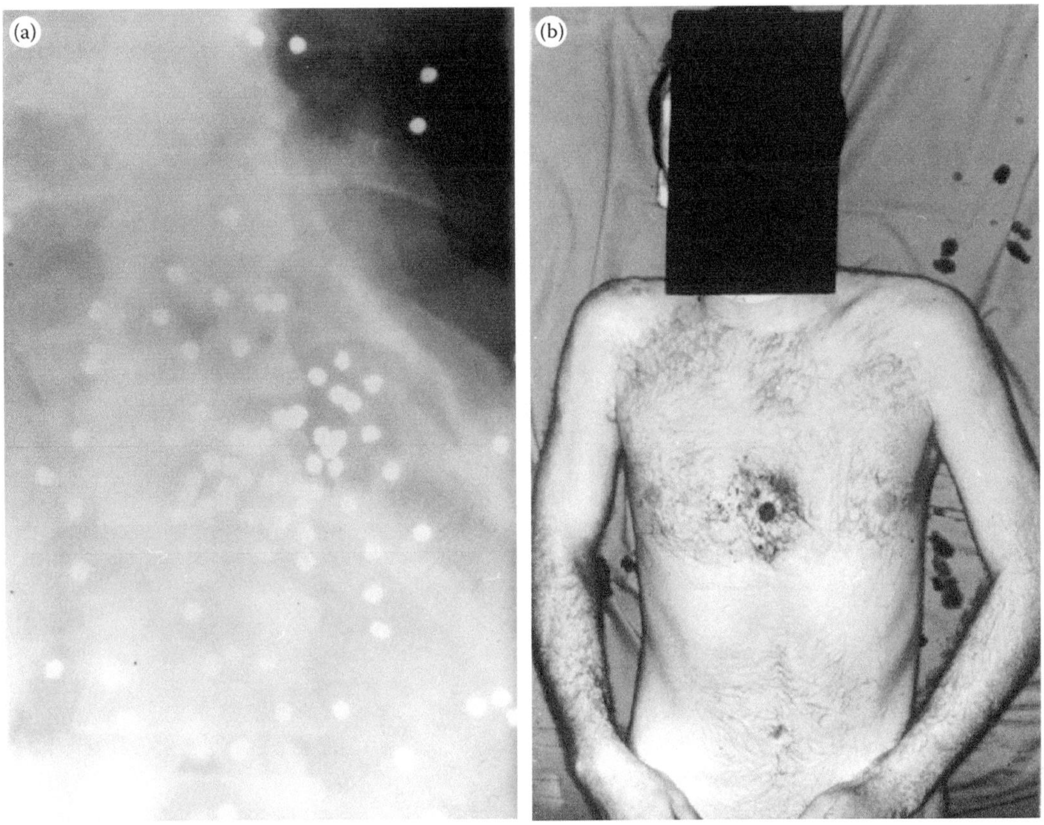

FIGURE 15.35 (a) A postmortem radiograph of the midbody area shows a spread out pattern of shotgun pellets suggesting that the weapon had been fired from a distance. (b) The wound was, in fact, a contact suicide wound of the epigastrium. (From Messmer, J. M. and Fierro, M. F., *RadioGraphics*, 6, 457, 1986. With permission.)

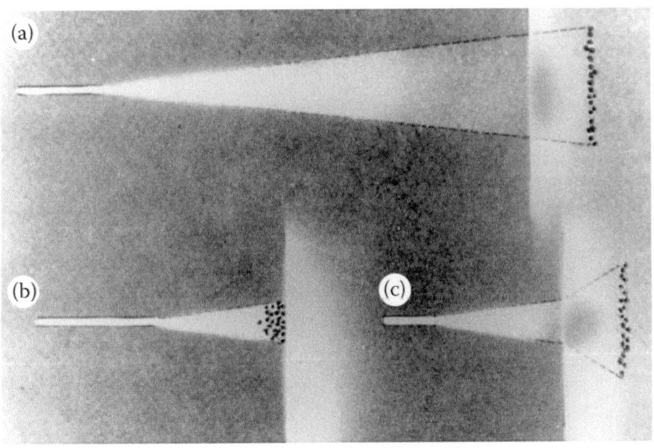

FIGURE 15.36 Schematic drawing illustrating the "billiard ball" effect. (a) If a shotgun is fired at intermediate to long range, the pattern has a chance to form. The pellets enter the body through individual openings. A spread pattern of pellets will be seen on a radiograph. (b) If the shotgun is fired at close range, the leading pellets are slowed as they enter the body and are struck by the trailing pellets. (c) The colliding pellets ricochet to spread the shot pattern so that it may simulate the pattern of longer-range fire. This is known as the "billiard ball" effect. (Redrawn from Messmer, J. M. and Fierro, M. F., *RadioGraphics*, 6, 457, 1986. With permission.)

at the foot of the stairs with his weapon braced on the handrail; Patrolman Charlie on the back wall facing the furnace room door; and Patrolman Delta to Charlie's right. Excellent crime scene photography allowed detailed recreation of the scene. Each officer had a different weapon and each weapon had a different load; thus every recovered casing and bullet could be traced to its shooter.

When the young man appeared at the door, there was a taped record of confusing and contradictory orders for the young man to "Come out!," "Get back!," "Don't move!," and so on, before there was a sudden flurry of four gunshots in less than two seconds (Figure 15.38c). The policemen stated they had been attacked by the man with a knife. He is down on the laundry room floor just beyond the door to his furnace room haven. He is severely wounded with five bullets wounds in his body. Two entrance wounds are located close together side by side in his left flank. One is in his lower back to the left of the spine. One is high of the left shoulder posteriorly. There is one exit wound in his right upper abdominal quadrant. Four casings are on the floor along with one bullet later identified as coming from Sgt. Baker's weapon. Three bullets remained in the body (Figure 15.38d). One of them, removed at surgery from the right lower quadrant of the abdomen, is from Ptl. Delta's weapon (Figure 15.38e). Since Baker is the only one who fired twice, the bullet entering from the left side and tracking across and slightly posteriorly to lodge in the lumbar vertebra (Figure 15.38f) has to be his

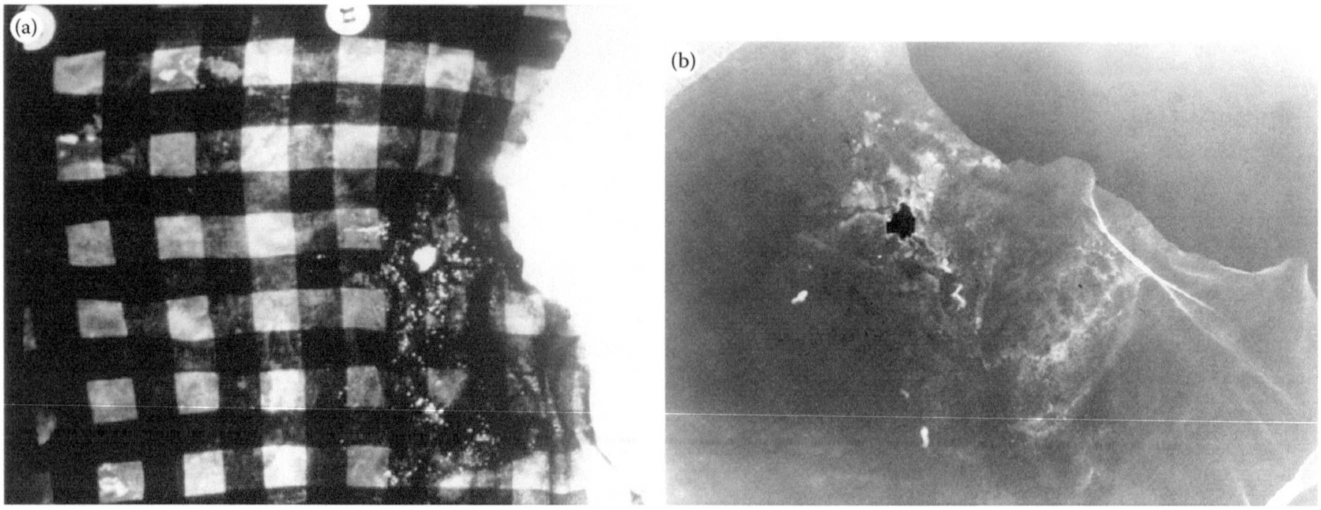

FIGURE 15.37 (a) A blood-soaked red and black plaid flannel shirt is a difficult field to examine for gunshot residue. (b) A low-energy radiograph with the Faxitron unit discloses the location and distribution of residue.

FIGURE 15.38 (a) Path of fugitive down stairs, through basement laundry room to location behind furnace where he superficially slices wrists with small knife grabbed as he passed through upstairs kitchen. (b) Positions taken by four armed Police Officers. (c) Fugitive returns to furnace room door (holding knife) to a barrage of conflicting order. (d) Four shots are fired in 2 sec. One bullet exits body to floor. Three bullets remain In body. (e) Sgt. Baker fires first as victim turns to reenter the furnace room. Ptl. Delton fires almost simultaneously. His bullet #1 is removed from lower abdomen at surgery. (f) Baker's first shot #2 strikes lumbar vertebra. (g) Ptl. Able fires last from stairs. His round #3 traverses left shoulder, anterior chest wall and comes to rest (h) in abdominal wall. (i) Baker's second shot entered the left back, traversed vital organs (j) and exited low in the right chest to be retrieved from the floor.

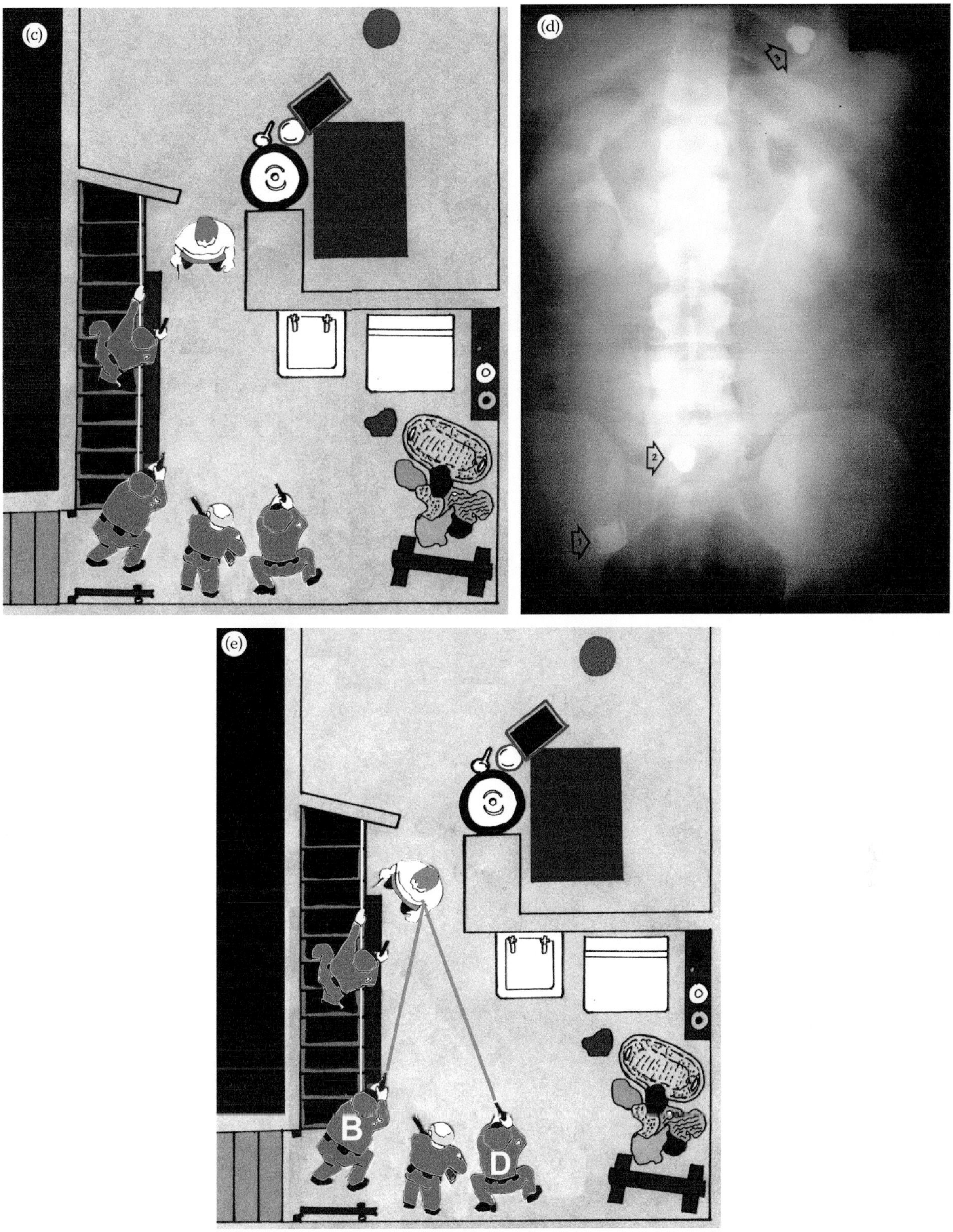

FIGURE 15.38 (*Continued*)

as well, because only Ptl. Able (Figure 15.38g) could have fired the bullet entering high on the left shoulder posteriorly, coursing along the anteromedical ribcage to come to rest in the left anterior soft tissues at the thoracoabdominal junction (Figure 15.39h).

Thus, the officers were not attacked by the man with a small knife (which ended up beneath the top of the stairs), but fired when he abruptly turned to his right to reenter the furnace room. As he was spinning around, the sequence of shooting had to be: first shot from Baker to the lumbar spine,

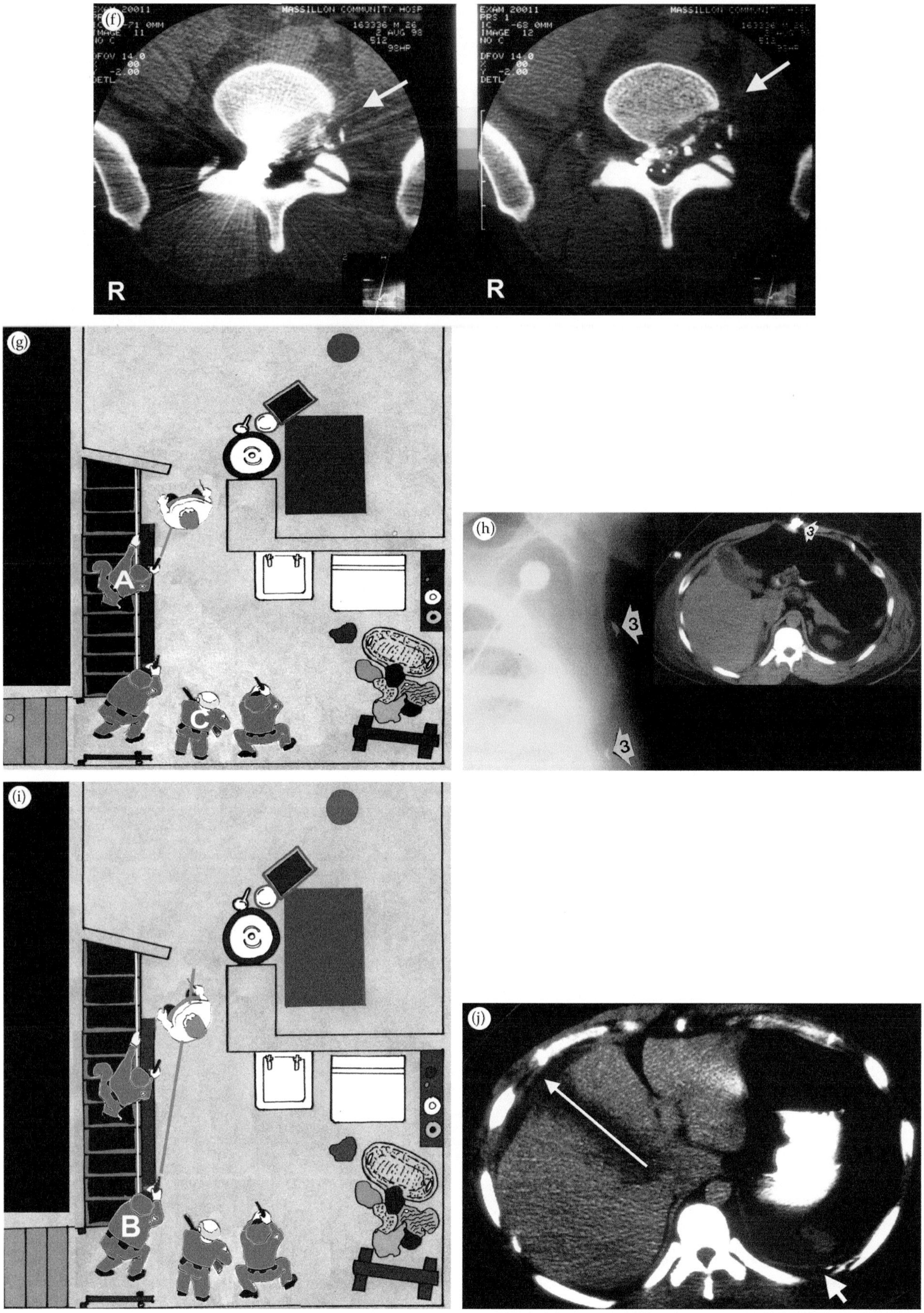

FIGURE 15.38 *(Continued)*

followed almost immediately by Delta's shot, which goes almost straight across to the right lower quadrant. (Figure 15.38e) Baker's second shot (Figure 15.38i) enters the back, exits the right anterior abdomen (Figure 15.38j) to be recovered from the floor. Able fires down from the stairs as the victim has virtually completed his turn to the furnace room door (Figure 15.38g). Fortunately, Officer Charlie never fired his shotgun.

The defendant was acquitted of the "menacing" charge, served a short sentence for resisting arrest, and received a settlement from the city after filing a complaint entered in District Court for "unreasonable force and assault and battery under the Fourth, Fifth, and Fourteenth Amendments and Section 42 of the U.S. code."

REFERENCES

1. Federal Bureau of Investigation, Crime in the United States 1992, Uniform Crime Reports, U.S. Department of Justice, Washington, DC, 2007.
2. Teret, S. P., Wintemute, G. J., and Beilenson, P. L., The firearm fatality reporting system, *J. Am. Med. Assoc.*, 267, 3073, 1992.
3. Messmer, J. M., and Fierro, M. F., Radiologic forensic investigation of fatal gunshot wounds, *RadioGraphics*, 6, 457, 1986.
4. Di Maio, V. J. M., *Gunshot Wounds, Practical Aspects of Firearms, Ballistics, and Forensic Techniques*, Elsevier, New York, 1985, chap. 1.
5. Fatteh, A., *Medicolegal Investigation of Gunshot Wounds*, Lippincott, Philadelphia, 1976.
6. Hollerman, J. J., and Fackler, M. L., Gunshot wounds: Radiology and wound ballistics, *Emergency Radiol.*, 2, 171, 1995.
7. Fackler, M. L., Wound ballistics. A review of common misconceptions, *J. Am. Med. Assoc.*, 259, 27–30, 1988.
8. Hollerman, J. J., Fackler, M. L., Coldwell, D. M., and Ben-Menachem, Y., Gunshot wounds. Bullets, ballistics, and mechanisms of injury, *Am. J. Roentgenol.*, 155, 685, 1990.
9. Hollerman, J. J., Fackler, M. L., Coldwell, D. M., and Ben-Menachem, Y., Gunshot wounds. Radiology, *Am. J. Roentgenol.*, 155, 691, 1990.
10. Swan, K. G., and Swan, R. C., Principles of ballistics applicable to the treatment of gunshot wounds, *Surg. Clin. North Am.*, 71, 221, 1991.
11. Wilson, J. M., Shotgun ballistics and shotgun injuries, *West. J. Med.*, 129, 149, 1978.
12. Sellier, K. G., and Kneubuehl, B. P., *Wound Ballistics and the Scientific Background*, Elsevier, Amsterdam, 1994.
13. International Committee of the Red Cross, *Wound Ballistics, an Introduction for Health, Legal, Forensic, Military and Law Enforcement Professionals (film with additional Information)*, ICRC, Geneva, 2008.
14. Wilbur, C. G., Letter to Editor, *J. Forensic Sci.*, 40, 722, 1995.
15. Jones, A. M., Reyna, M., Sperry, K., and Hock, D., Suicidal contact gunshot wounds to the head with .38 special Glaser safety slug ammunition, *J. Forensic Sci.*, 32, 1604, 1987.
16. Leffers, B., and Jeanty, D., Handgun pellet ammunition ("snake shot") wounds: Report of three cases, *J. Forensic Sci.*, 27, 433, 1982.
17. Fackler, M. L., Surinchak, J. S., Malinowski, J. A., and Bowen, R. E., Bullet fragmentation: A major cause of tissue disruption, *J. Trauma*, 24, 35, 1984.
18. Messmer, J. M., Massive head trauma as a cause of intravascular air, *J. Forensic Sci.*, 29, 418, 1984.
19. Erben, J., and Nadvornik, R., The quantitative determination of air embolus in certain cases of fatal trauma, *J. Forensic Med. Pathol.*, 10, 45, 1963.
20. Graham, J. M., Beall, A. C., Matlox, K. L., and Vaughan, G. D., Systemic air embolism following penetrating trauma to the lung, *Chest*, 27, 449, 1977.
21. Meier, G. H., Wood, W. J., and Symbas, P. N., Systemic air embolization from penetrating lung injury, *Ann. Thorac. Surg.*, 27, 161, 1979.
22. Smith, J. M., Richardson, J. D., Grover, F. L., Arom, K. V., Webb, G. E., and Trinkle, J. K., Fatal air embolism following gunshot wound to the lung, *J. Thorac. Cardiovasc. Surg.*, 72, 296, 1976.
23. Thomas, A. N. and Roe, B. B., Air embolism following penetrating lung injuries, *J. Thorac. Cardiovasc. Surg.*, 66, 533, 1973.
24. Thomas, A. N. and Stephens, B. G., Air embolism: A cause of morbidity and death after penetrating lung trauma, *J. Trauma*, 14, 633, 1974.
25. Westcott, J. L., Air embolism complicating percutaneous needle biopsy of the lung, *Chest*, 63, 108, 1973.
26. Durant, T. M., Oppenheimer, M. J., Webster, M. R., and Lang, J., Arterial air embolism, *Am. Heart J.*, 38, 481, 1949.
27. Rukstinat, G., Experimental air embolism of coronary arteries, *J. Am. Med. Assoc.*, 96, 26, 1931.
28. Di Maio, V. J. M., *Gunshot Wounds. Practical Aspects of Firearms, Ballistics, and Forensic Techniques*, Elsevier, New York, 1985, chap. 15.
29. Froede, R. C., Pitt, M. J., and Bridgemon, R. R., Shotgun diagnosis: "it ought to be something else", *J. Forensic Sci.*, 27, 428, 1982.
30. Di Maio, V. J. M., *Gunshot Wounds. Practical Aspects of Firearms, Ballistics, and Forensic Techniques*, Elsevier, New York, 1985, chap. 14.
31. Introna, F. I., Jr. and Smialek, J. E., Suicide from multiple gunshot wounds, *Am. J. Forensic Med. Pathol.*, 10, 275, 1989.
32. Habbe, D., Thomas, G. E., and Gould, J., Nine-gunshot suicide, *Am. J. Forensic Med. Pathol.*, 10, 335, 1989.
33. Jacob, B., Barg, J., Haarhof, K., Sprick, C., Wšrz, D., and Bonte, W., Multiple suicidal gunshot wounds to the head, *Am. J. Forensic Med. Pathol.*, 10, 289, 1989.
34. Hargarten S. W., Karlson, T. A., O'Brien, M., Hancock, J., and Quebbeman, E., Characteristics of firearms involved in fatalities, *J. Am. Med. Assoc.*, 275, 42, 1996.
35. Hollerman, S. W. and Fackler, M. L., Gunshot wounds: Radiology and wound ballistics, *Emergency Radiol.*, 2, 171, 1995.
36. Hollerman, S. W. and Fackler, M. L., Bullets, pellets, and wound ballistics, in *Radiologic Guide to Medical Devices and Foreign Bodies*, Hunter, T. B., and Bragg, D. B., Eds., C.V. Mosby, St. Louis, 1994, chap. 19.
37. Bixler, R. P., Ahrens, C. R., Rossi, R. P., and Thickman, D., Bullet identification with radiography, *Radiology*, 178, 563, 1991.
38. Molnar, S., Identification of bullet caliber from x-ray film, *Assoc. Firearm Tool Mark Examiners News.*, 12, 45, 1971.
39. Berryman, H. E., Smith, O. C., and Symes, S. A., Diameter of cranial gunshot wounds as a function of bullet caliber, *J. Forensic Sci.*, 40, 751, 1995.
40. Di Maio, V. J. M, *Gunshot Wounds*, Elsevier, New York, 1985, chap. 4.
41. Collins, K. A. and Lantz, P. E., Interpretation of fatal, multiple and existing gunshot wounds by trauma specialists, *J. Forensic Sci.*, 39, 94, 1994.

42. Smith, O. C., Berryman, H. E., and Lahren, C. H., Cranial fracture patterns and estimate of direction from low velocity gunshot wounds, *J. Forensic Sci.*, 32, 1416, 1987.
43. Straathof, D., Bannach, B. G., and Dowling, G. P., Radiography of centrefire gunshot wounds, in *Proc. Am. Acad. Forensic Sci. Annu. Mtg.*, New York, 1997, p. 138.
44. Di Maio, V. J. M., *Gunshot Wounds*, Elsevier, New York, 1985, chap. 10.
45. Hughes, J. J., Brogdon, B. G., and Eichelberger, R. P., Migrating missiles, *Ala. J. Med. Sci.*, 21, 416, 1984.
46. Messmer, J. M., and Fierro, M. F., Radiologic forensic investigation of fatal gunshot wounds, *RadioGraphics*, 6, 457, 1986.
47. Hughes, J. J., Bullet injury to the esophagus detected by intestinal migration, *J. Trauma*, 27, 1362, 1987.
48. Ledgerwood, A. M., The wandering bullet, *Surg. Clin. North Am.*, 57, 97, 1977.
49. Di Maio, V. J. and Di Maio, D. J., Bullet embolism: Six cases and a review of the literature, *J. Forensic Sci.*, 17, 394, 1972.
50. Fatteh, A. and Shah, Z. A., Bullet embolus of the right profunda femoris artery, *J. Forensic Sci.*, 15, 139, 1968.
51. Kelley, J. L., A bullet embolus to the left femoral artery following a thoracic gunshot wound, *J. Thorac. Surg.*, 21, 608, 1951.
52. Morton, J. R., Reul, G. L., Arbegast N. R., Okies, J. E., and Beall, A. C., Bullet embolus to the right ventricle, *Am. J. Surg.*, 122, 584, 1971.
53. Padula, R. T., Sandlet, S. C., and Camishion, R. C., Delayed bullet embolization to the heart following abdominal gunshot wound, *Ann. Surg.*, 169, 599, 1969.
54. Saltzstein, E. C. and Freeark, R. J., Bullet embolism to the right axillary artery following gunshot wound of the heart, *Ann. Surg.*, 158, 65, 1963.
55. Sclafani, S. J. and Mitchell, W. G., Retrograde venous bullet embolism, *J. Trauma*, 21, 656, 1981.
56. Symbas, P. N., Hatcher, C. R., and Mansour, K. A., Projectile embolus of the lung, *J. Thorac. Cardiovasc. Surg.*, 5, 97, 1968.
57. Rich, N. M., Collins, G. J., Andersen, C. A., McDonald, P. T., Kozloff, L., and Ricotta, J. J., Missile emboli, *J. Trauma*, 18, 236, 1978.
58. Massad, M. and Slim, M. S., Intravascular missile embolization in childhood: Report of a case, literature review, and recommendations for management, *J. Pediatr. Surg.*, 25, 1292, 1990.
59. Shannon, J. J., Vo, N. M., Stanton, P. E., and Dimler, M., Peripheral arterial embolization: A case report and 22 year literature review, *J. Vasc. Surg.*, 5, 773, 1987.
60. Ward, P. A. and Suzuki, A., Gunshot wound of the heart with peripheral embolization: A case report and review of literature, *J. Thorac. Cardiovasc. Surg.*, 68, 440, 1974.
61. VanGilder, J. C. and Coxe, W. S., Shotgun pellet embolus of the middle cerebral artery, *J. Neurosurg.*, 32, 711, 1970.
62. Dadsetan, M. R. and Jinkins, J. R., Peripheral vascular gunshot bullet embolus migration to the cerebral circulation: Report and literature review, *Neuroradiology*, 32, 516, 1990.
63. Dada, M. A., Loftus, I. A., and Rutherford, G. S., Shotgun pellet embolism to the brain, *Am J. Forensic Med. Pathol.*, 14, 58, 1993.
64. Oser, A. B., Moran, C. J., Cross, D. T., and Thompson, R. W., Shotgun pellet embolization to the intracranial internal carotid artery: Report of a case and review of the literature, *J. Trauma*, 1, 200, 1994.
65. Jones, B. L. and Tomsick, T. A., Shotgun pellet embolism to the basilar artery, *Am. J. Roentgenol.*, 165, 744, 1995.
66. Kase, C. S., White, R. L., Vinson, T. L., and Eichelberger, R. P., Shotgun pellet embolus to the middle cerebral artery, *Neurology*, 31, 458, 1981.
67. Bland, E. F. and Beebe, G. W., Missiles in the heart, *N. Engl. J. Med.*, 274, 1039, 1966.
68. Di Maio, V. J. M., *Gunshot Wounds*, Elsevier, New York, 1985, chap. 12.
69. Lang, P. E., Jerome, W. G., and Jaworski, J. A., Radiopaque deposits surrounding a contact small-caliber gunshot wound, *Am. J. Forensic Med. Pathol.*, 15, 10, 1994.
70. Messmer, J. M. and Wadsworth, J. D. *J. Forensic Sci*, 29, 340, 1984.

CREDITS

From Vogel, H., Gewalt im Rontgenbilt, acomed verlagsgesellshaft mb H, Landsbert/Lech, 1997, with permission Figures 14.8A–G, I, J; From Brogdon, B. G., Vogel, H., McDowell, J. D., A *Radiologic Atlas of Abuse, Torture, Terrorism, and Inflicted Trauma*, CRC Press, Boca Raton, FL, 2003, with permission Figures 14.8h, 14.17, 14.38b,c (Redrawn) 14.38d, f, h–j.

16 New Developments in Gunshot Analysis

Stephan A. Bolliger, Beat P. Kneubuehl, and Michael J. Thali

CONTENTS

INTRODUCTION

Apart from identification procedures, gunshot injuries are one of the foremost fields of postmortem forensic radiology. Postmortem radiology serves to locate the projectile, depict the bullet track, and may help in identifying the ammunition and the weapon type used. This facilitates the retrieval of the bullet and of potentially important fragments (Di Maio, 1999). The retrieval of such foreign bodies is essential as these objects may display unique rifling characteristics and therefore help identify the individual weapon used. The knowledge of the bullet course through the body is also of utmost importance, as this may be of great value when reconstructing the crime scene and the position of the perpetrator and the victim.

Postmortem radiology on gunshot victims is generally performed using conventional x-ray machines (see Chapter 15).

With the advance of medical technologies such as computed tomography (CT), introduced by Hounsfield and Cormack in the early 1970s, new possibilities opened up for forensic pathologists. Indeed, the first CT scan was performed on a victim of a gunshot injury to the head as early as 1977 (Wullenweber et al., 1977).

With the invention of spiral CT, three-dimensional (3D) reconstructions of radiological images became possible. Such multislice CTs (MSCT), which have become the everyday clinical standard, have been implemented in forensic pathology by different groups with promising results, especially with regard to the evaluation of gunshot injuries (Thali et al., 2003a, 2003b; Levy et al., 2006; Harcke et al., 2007; Andenmatten et al., 2008).

Over recent years, x-ray examinations have been gradually replaced by CT scans (Schumacher et al., 1985; Stein et al., 2000; Thali et al., 2001, 2003a, b, 2007).

CT scans have several advantages over conventional x-ray imaging. The latter lacks the three-dimensionality needed to determine the exact location of the projectile in the body. In a conventional x-ray, at least two radiographs have to be performed from different angles, and this requires that either the x-ray film plate or the corpse has to be moved. X-ray imaging is far more time consuming than performing one rapid MSCT. Furthermore, the size of an object cannot be assessed accurately with the conventional x-ray.

The advantage of the MSCT is that the forensic pathologist can rotate the 3D image prior to, or even during, autopsy and thus gain an exact knowledge of the projectiles position.

Furthermore, MSCT can differentiate different structures with regard to their radiopacity. An x-ray will depict structures of different radiological density in various shades of black, gray, or white.

Although this suffices to determine whether foreign objects are within the body, x-rays cannot tell the examiner just how radiopaque the foreign body happens to be. Obviously, not all foreign objects in gunshot incidents are projectiles. Most objects found in gunshot incidents are bone fragments, which are, strictly speaking, not "foreign," but certainly do not belong to soft tissues or organs. Especially if the bullet has passed through an obstacle prior to hitting the victim, different objects may be lodged in the victim. For example, when shooting through a window pane, glass particles of the shattered pane may penetrate the body and be readily confused with bullet fragments in conventional x-rays. With this knowledge prior to autopsy, tedious searches for projectiles in corpses, only to discover that the x-ray image of the assumed bullet fragment was, in fact, only a piece of bone or some other fragment of higher radiopacity than the surrounding tissue, can be avoided.

MSCT can differentiate these foreign objects by virtue of their radiopacity or Hounsfield unit (HU) characteristics Bolliger et al., 2009.

GENERAL CLASSIFICATION OF GUNSHOT WOUNDS

Gunshot wounds can graze, penetrate, or perforate. Grazing occurs in cases where the bullet strikes the victim extremely tangentially. The resulting abrasion may be easily mistaken for a classic abrasion due to blunt trauma (i.e., nonlethal ammunition). Penetration is defined as the bullet striking and entering, but not exiting the body. Perforation is reserved for all cases in which the bullet exits.

Depending on the distance, gunshot injuries can be divided into three categories: contact, intermediate, and distant. Some authors include a fourth category, the near-contact wound, which fits neither in the definition of contact or intermediate gunshot wounds. The morphology of gunshot wounds is described exhaustingly in a multitude of forensic textbooks. We therefore refrain from describing these in further detail.

FORENSIC ASPECTS OF GUNSHOT INJURIES

In cases of gunshot injuries, several aspects are of utmost importance.

These are, besides the obvious question of a possible third-party involvement, the following:

- Weapon handling
- Firing distance
- Penetrating or grazing injury
- Entrance and exit wound
- Striking angle and bullet course through the body.
- Gunshot priority in cases of multiple gunshot wounds
- Lodging of the bullet or other foreign bodies (i.e., in bullets passing through intermediate targets) in the corpse
- Cause of death and vitality of the injuries
- Bullet type and size, identification of the individual weapon

These aspects are dealt with in detail below.

Weapon Handling

In contact wounds, the muzzle imprint may give clues as to how the weapon was held against the body upon firing. This telltale superficial abrasion around the entrance wound can be digitalized with photogrammetry-based 3D surface scanning and matched to the likewise scanned weapon in question. Not only can—under ideal circumstances—a type of firearm be excluded or confirmed, but the positioning of the weapon onto the skin at the moment of firing can also be reconstructed (see also Chapters 24 and 29). This thus-obtained firing position is an important clue in the assessment of contact gunshots. In a suicidal shooting, the victims generally assume a comfortable firing position. Weapon handling that is uncomfortable for the victim must always give rise to suspicion of third-party involvement.

Firing Distance

Apart from contact wounds, in which the typical wound morphology proves the distance, the firing distance determination usually belongs to the field of crime scene investigators. Gunpowder tattooing, namely superficial abrasions due to unburnt or burning gunpowder particles striking the skin, and, rarely, the presence of such particles on the skin may give rise to a rough estimate of the firing distance . Comparison with such a "tattooing" in a firing reconstruction using the original weapon and a reference surface can render the distance estimation more precise. Other methods involve the chemical detection of gunpowder residues on the skin. However, such an examination requires physical contact with the body, which, in the case of surviving victims, cannot always be performed due to the risk of wound infection.

Here again, forensic imaging may be of assistance (Stein et al., 2000) showed that gunshot residues can be detected with CT. The metallic particles, especially lead, which are produced by firing, are radiopaque and therefore easily detectable with high-resolution scanners (Figure 16.1). In an experimental setting on porcine skin, shots of more than 10 cm could be distinguished from contact gunshots regardless of the type of bullet (i.e., jacketed or solid lead) used.

Penetrating or Grazing Injury

Of great importance is the discrimination between penetrating and grazing wounds. The former may be lethal, whereas the latter are usually harmless. Furthermore, penetrating wounds may result in a bullet lodging in the body, which can be examined and provide clues as to the weapon involved. Additionally, a grazing wound (Figures 16.2 and 16.3) consists basically of a combination of an entry and exit wound, a characteristic a true penetrating wound does not. This difference may lead to confusion regarding the total number of gunshots sustained, as well as the firing position. However, the differentiation between actually penetrating or grazing gunshot wounds is not always easy. The morphology of the wound may be similar in both cases, especially

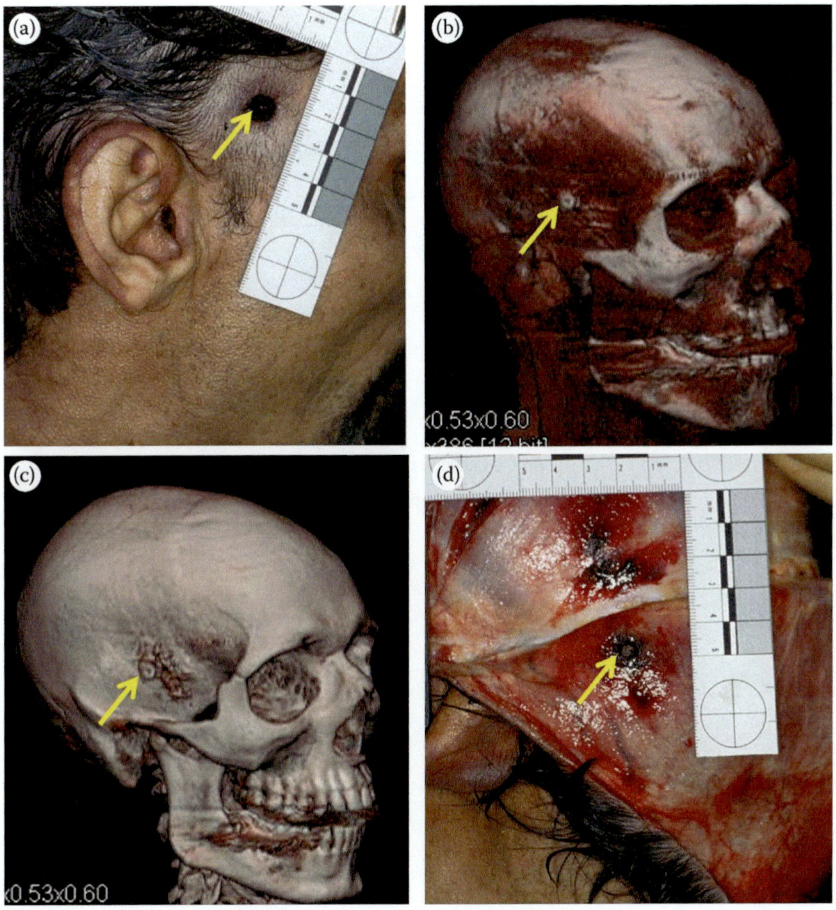

FIGURE 16.1 Suicidal gunshot to the right temple. (a) Photograph of the entrance wound (arrow) with gunpowder residues. (b and c) 3D CT reconstruction of the skull with and without muscles. The radiopaque gunpowder residues are clearly visible (arrow). (d) Autopsy photograph of the inside of the scalp showing gunpowder residues.

if the penetrating gunshot strikes the victim tangentially. Putrefaction and maggot infestation can further obscure the wound morphology, making this differentiation more difficult.

Postmortem MSCT can, however, readily distinguish a penetrating wound from a grazing wound. By virtue of metallic and bone fragments lining the bullet channel, a true penetrating wound can readily be seen in MSCT. MRI, the imaging method of choice can visualize soft-tissue lesions, and therefore prove a penetrating injury. Obviously, if a lodged bullet is detected by either MSCT or MRI within the body, and only one wound is present, the penetration is clear.

Entrance and Exit Wounds

The differentiation of entrance and exit wounds is extremely important, as it is absolutely essential for all further case reconstructions. For example, a gunshot to the back with an exit wound at the front of the chest creates a completely different situation than the opposite. Usually, the wound morphology (and in cases of short distance or contact gunshots also gunpowder residues) suffices to determine whether a wound is due to the entering or exiting of a projectile. However, in badly damaged corpses such as charred bodies or putrefaction, neither the wound morphology nor gunshot residue analysis helps in distinguishing an entrance wound.

With MSCT, different additional criteria can be used to address this question: apart from the trail of metal and bone fragments (Figure 16.4) along the bullet course, osseous and cartilaginous lesions can provide proof as to the traveling direction, and therefore the entering direction, of the bullet (Figure 16.5). In the skull, cone-shaped beveling helps distinguish an entrance from an exit wound. This phenomenon is described in detail below.

Bullet Course through the Body

Another very important aspect is the determination of the course of a bullet through the body (Figures 16.6 through 16.8). First, this can give clues as to how the victim was shot. A cranial to occipital sinking bullet course may, for example, be due to a gunshot originating from above the victim, that is, from a window or balcony, or against a kneeling or otherwise bent downward victim to mention just a few possibilities. Obviously, the finding of a gunshot wound to the back, as opposed to the front of a victim, has immense judicial consequences. A self-defensive action is hardly plausible in a gunshot to a victim's back.

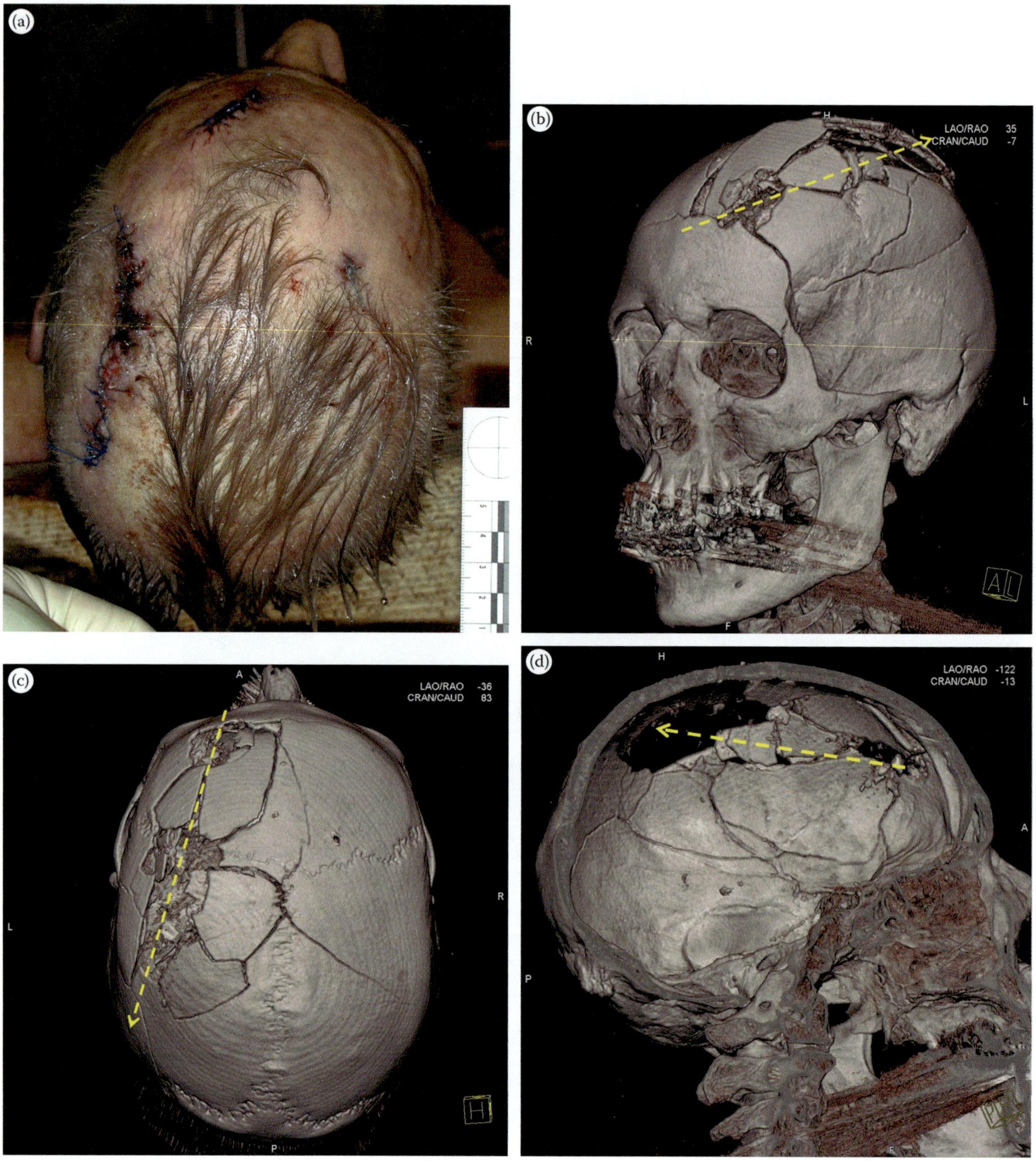

FIGURE 16.2 (a) Photograph of a gunshot injury to the forehead of a victim who died in hospital shortly after the incident. Note the sutured wounds to the forehead and the vertex. (b–d) 3D CT reconstruction of the skull showing extensive fractures along the path of the bullet (dashed arrow).

Forensic laymen often unjustly assume the bullet course to be along the direct pathway between entrance and exit wound. This is, as every experienced forensic pathologist can confirm, not always, or even rarely the case. A change in tissue texture, especially from a soft tissue to hard bone may sometimes deflect the bullet. The fragmenting of a bullet may give rise to two or more pieces traveling at different angles through the victim. Such unexpected bullet courses are indeed a challenge for the forensic pathologist and may change the previously assumed cause of death completely. Other, albeit rarer, "wild" courses are bullet embolisms. In one case, a gunshot to the head traveled through the base of the skull and neck,

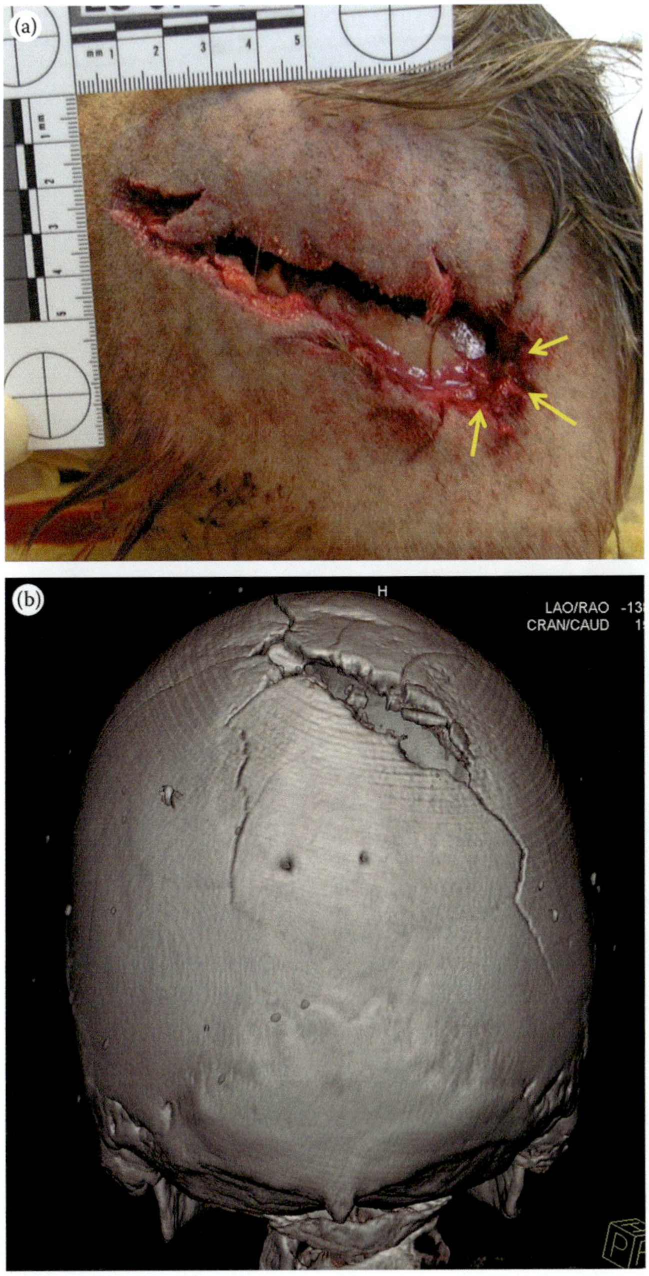

FIGURE 16.3 (a) Photograph of the victim of a shooter running amok showing a deep laceration of the scalp at the vertex. The partly torn, partly abraded wound margins, with an underlying osseous, of this grazing shot can be confused with a laceration due to blunt trauma. In this case, the abraded region (arrows) indicated the area where the bullet struck the head. (b) CT reconstruction of the skull with a furrow-like fracture of the vertex.

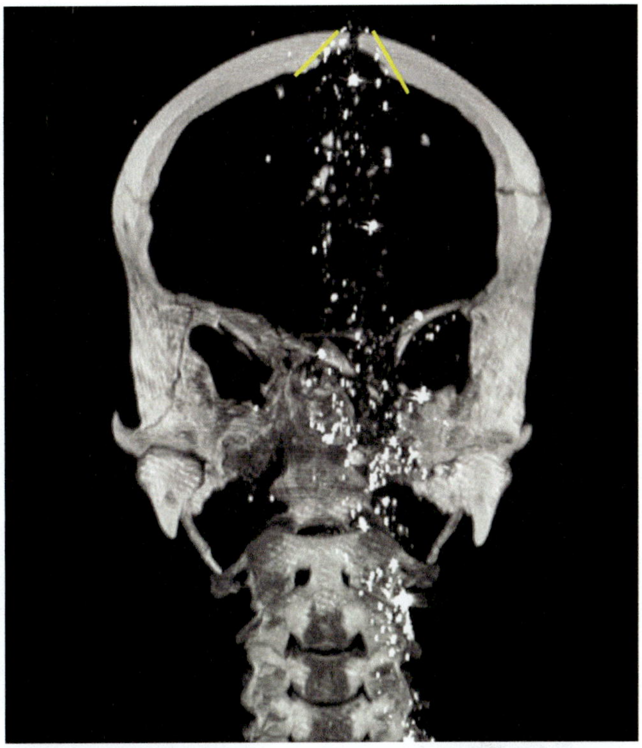

FIGURE 16.4 CT, coronar reformation of a gunshot injury to the vertex. Note the inward beveled entrance wound (highlighted with lines) and the comet-tail-like bullet track through the head and neck downwards.

and entered the thoracic aorta (Figure 16.9). There, it rested upon cardiopulmonary collapse and gradually sunk with the blood to the level of the abdominal aorta.

Conventional autopsy procedure in determining the bullet course consists of plane radiographs and a time-consuming layer-by-layer autopsy. Sometimes, a probe is inserted into the wound to determine the general course of the bullet. This blind probing of wounds (i.e., prior to autopsy and trace collection) is not only obsolete but also downright careless. The possible additional gain of information, which would have been evident at a later stage of the autopsy anyway, does not compensate for the danger of displacing traces (i.e., gunpowder residues, textile fibers, fragments of intermediate targets) and of harming previously uninjured structures.

MSCT is capable of depicting the general direction of the bullet by showing which bones were perforated. Often, bone fragments of damaged or shattered bones and bullet fragments "pave" the bullets pathway much like a comet's tail (Figures 16.4 and 16.9a). This bone-fragment trail gives a general impression of the wound direction.

However, the state-of-the art postmortem imaging of bullet pathways in soft tissues is clearly the MRI (Figure 16.10). As stated before, MRI is capable of visualizing soft-tissue lesions with a high degree of accuracy and is therefore better suited for such examinations.

However, caution should be exercised in the interpretation of bullet courses through highly mobile organs such as the lungs. The postmortem position of the lungs in an open thoracic trauma, as gunshot wounds to the chest often are, does not correlate to the antemortem position, as the lungs may have collapsed after injury.

Gunshot Priority

The term "gunshot priority" refers to the succession of gunshot injuries to the body. In cases of several gunshot injuries

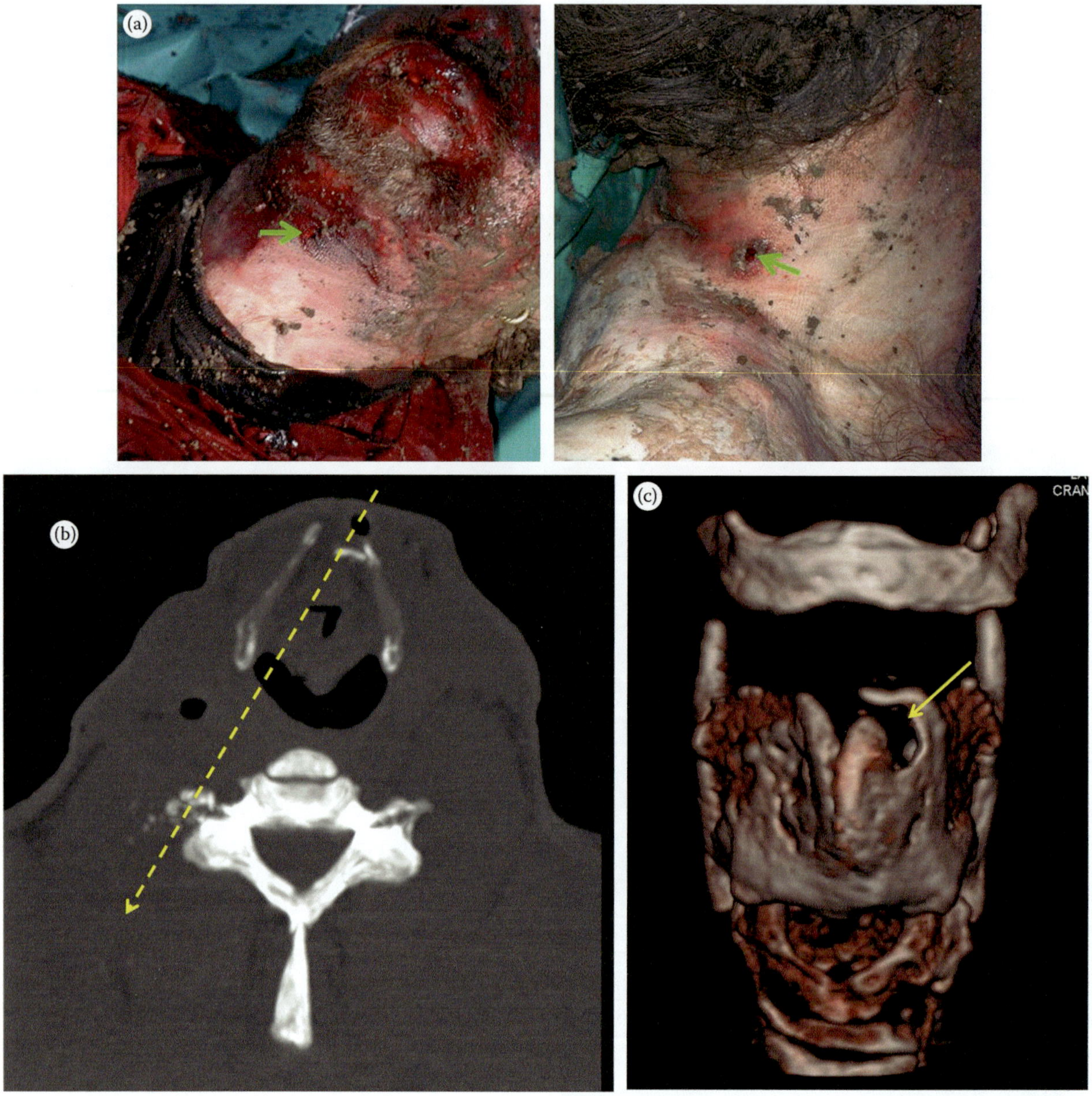

FIGURE 16.5 (a) Photographs of a decomposing homicide victim found buried in the woods with gunshot injuries to the throat and the right side of the neck (arrows). A differentiation between entrance and exit wound is not possible due to putrefactive changes. (b) CT, axial image of the neck. Note the inbent thyroid cartilage fragment and the fractured transverse process along the bullet track through the neck. The bullet path through the neck can be reconstructed (dashed arrow), therefore permitting a differentiation between entrance and exit wounds.

sustained from different angles, it is of great reconstructive relevance to determine in which succession these were fired. If a victim displays injuries to the back and the front of the chest for instance, then the question will arise as to whether the victim was first shot in the back and then, after turning around, in the chest. Differentiating between this scenario and the opposite, namely a first gunshot to the chest and then perhaps into the fleeing victim's back will have obvious judicial consequences. Furthermore, if different persons shot the victim, it may be of relevance to determine which person shot the victim first.

This gunshot priority evaluation is not always easy or possible and requires a great amount of experience from the forensic examiner. Nevertheless, certain scenarios allow for a clear evaluation. In cases of lung perforation, a gunshot priority assessment is possible. In the first phase of multiple gunshot injuries to the chest, the lungs collapse. The subsequent bullet injuries to the lung are therefore more or less in accordance with the thoracic lesions. Therefore, the penetrations that are not in accordance with the postmortem collapsed condition are probably the first injuries, whilst the opposite is often true for later injuries. Here again, great

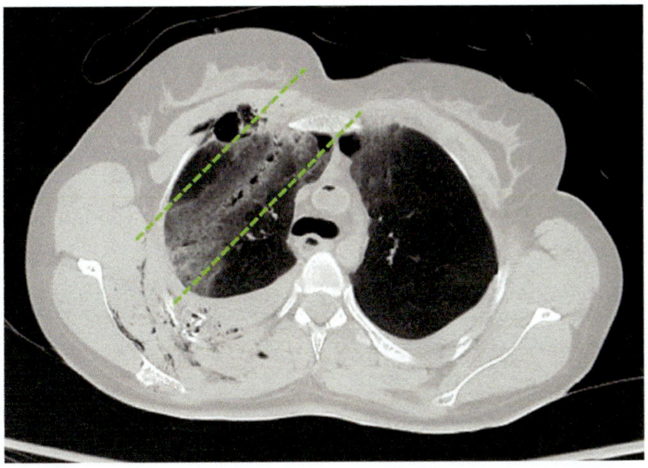

FIGURE 16.6 Axial CT image of the thorax showing a bullet track through the lung (highlighted by dashed lines).

caution should be exercised in interpreting these possible clues; a hemothorax or tension pneumothorax will obviously deem this general rule incorrect.

Another important clue regarding gunshot priority concerns Puppe's Rule. This rule, described in further detail below, relies upon the fact that a fracture line will not cross a pre-existing fracture.

Lodging of the Bullet or Other Foreign Bodies

Radiological imaging has proven invaluable in the detection of foreign bodies, foremost projectiles, in the body.

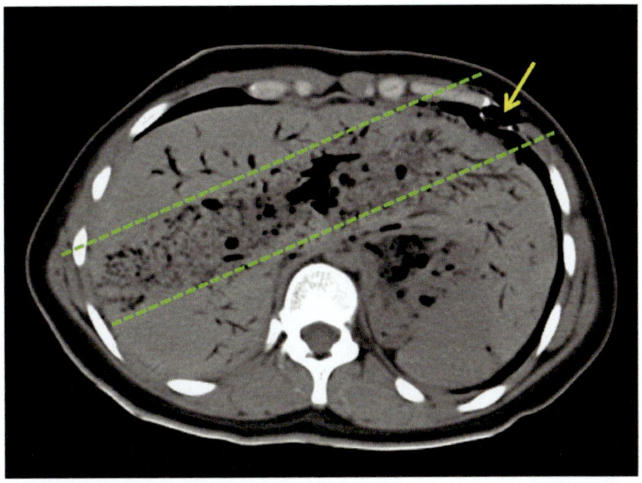

FIGURE 16.7 Axial CT image of the abdomen. A bullet track (highlighted by dashed arrows) running through the liver is clearly visible. Note also the abdominal wall defect, in this case the entrance wound (arrow).

Occasionally, the presence of bullet particles can even prove or disprove a certain type of ammunition. In one case (Kneubuehl and Thali, 2008), a police officer was caught in crossfire between his colleague who used standard police issue 9 mm full-metal jacket ammunition and a perpetrator who fired 45 auto full-metal-jacketed bullets. X-ray showed a shattering of both femora with tiny radiopaque structures in the immediate vicinity of the fractures. Conventional examination techniques could not determine from which weapon the officer was injured. By firing at synthetic bones embedded in gelatine, only the 45 auto left the above-mentioned

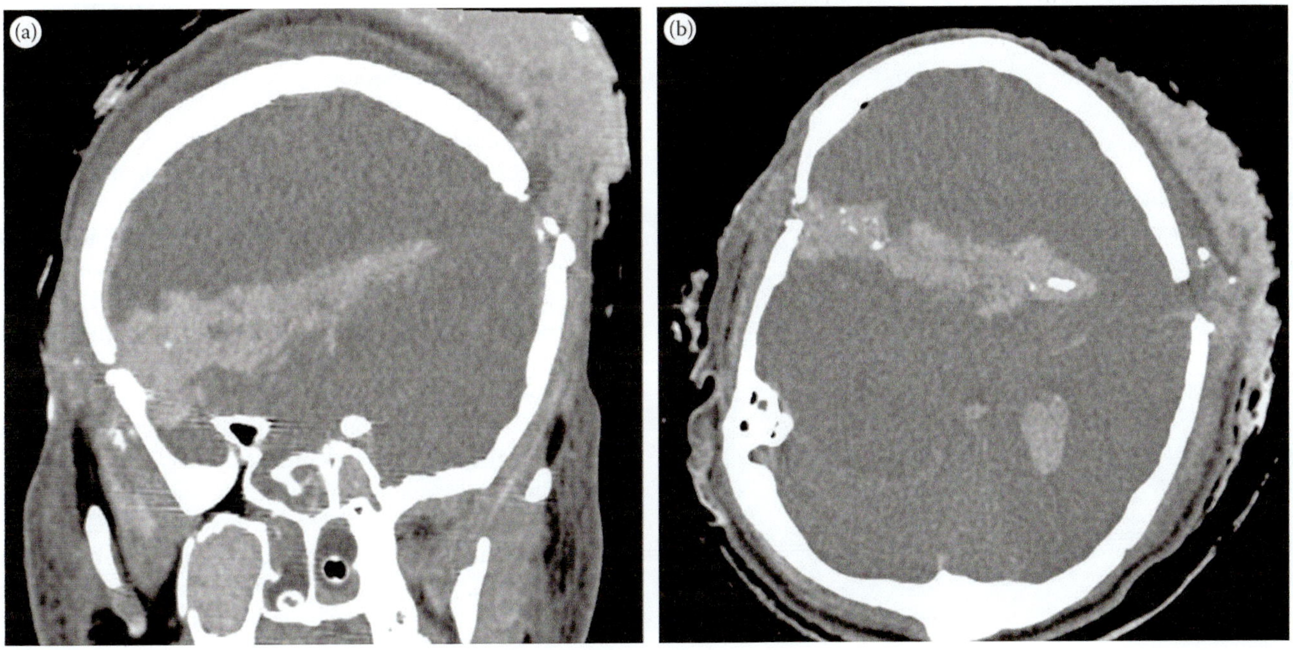

FIGURE 16.8 CT, coronal view (a) and axial view (b) of the head of a victim of a suicidal gunshot to the right temple with a clearly depicted bullet track through the brain. The light gray color of this bullet track is due to the influx of blood into the crushed brain tissue of the temporary wound cavity.

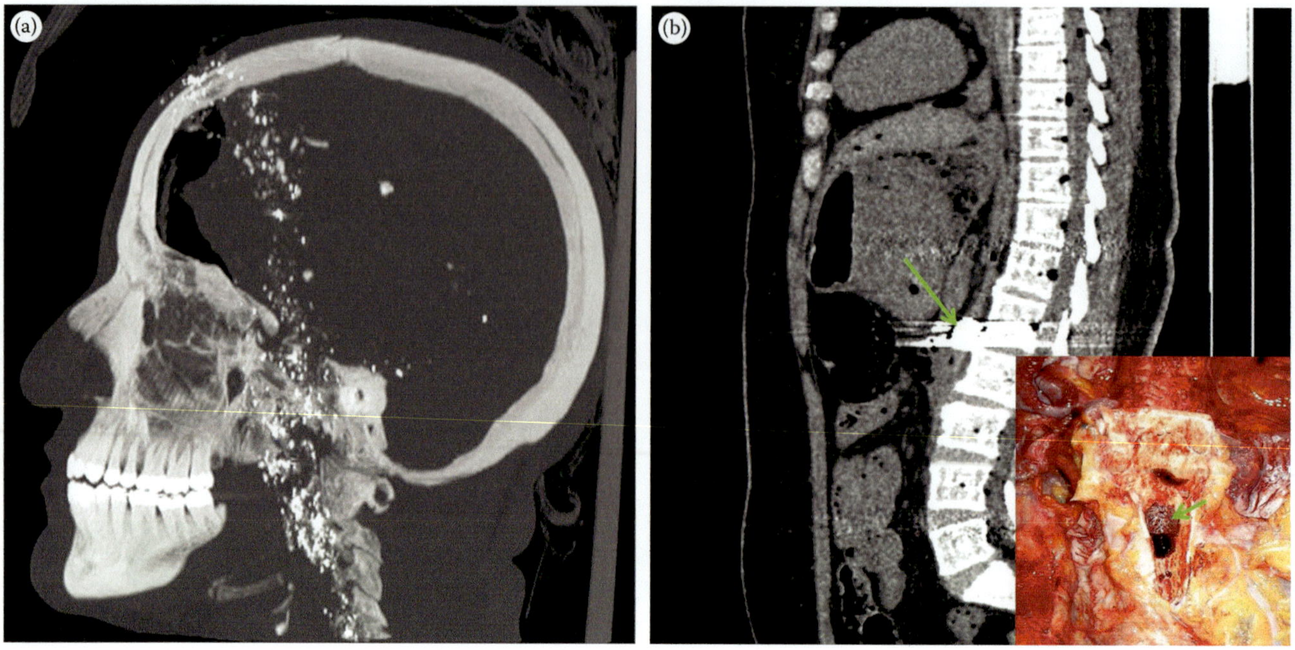

FIGURE 16.9 (a) Sagittal CT image of a victim of a homicidal gunshot to the vertex. Note the fragment trail along the bullet track from the vertex through the skull base into the neck. (b) Sagittal CT image through the trunk of the victim seen in (a). A bullet (arrow) is seen resting in the abdominal aorta. Insert: Autopsy photograph of the opened aorta with the bullet (arrow) within.

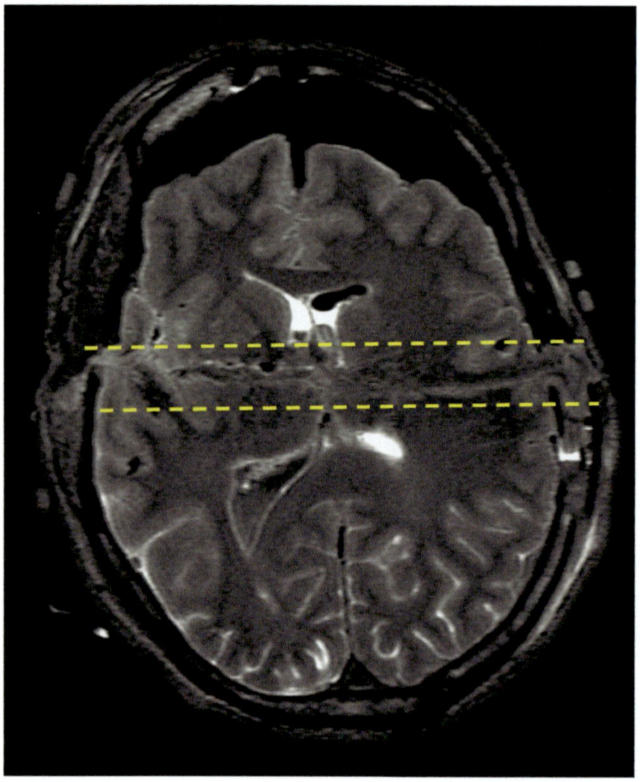

FIGURE 16.10 MRI, coronar, T1 weighted image of the brain showing a bullet track (highlighted by dashed lines) traversing the brain.

fragments, whereas the police ammunition repeatedly failed to so (Figure 16.11). It was therefore concluded that the officer was injured by the perpetrator and not by the other police officer.

As stated above, the main advantage of MSCT compared to conventional x-ray images lies in its capacity to depict findings three- dimensionally and in different radiopacity gradients (Figure 16.12).

The 3D depiction of a bullet facilitates the localization of such objects of interest enormously; one flick of a button can locate the projectile in the body accurately. In conventional x-ray examinations, this undertaking required at least two radiographs from different angles. This meant that either the x-ray film plate or the corpse had to be moved. Taking conventional x-rays is far more time consuming than performing one rapid MSCT and less precise regarding size assessment of foreign objects.

Apart from single bullets within the body, the three-dimensionality of MSCT examinations has also proven to be invaluable in assessing shotgun casualties. It is an almost Herculean undertaking to retrieve or show the pellets within the body. With x-rays, the general location can be assessed. However, a 3D MSCT reconstruction can depict their distribution within the body accurately, thus making an autoptic assessment superfluous (Figure 16.13).

The capability of MSCT to discriminate between different radiopacities is also of great help in preautopsy examinations. Conventional x-rays, although undoubtedly sufficient

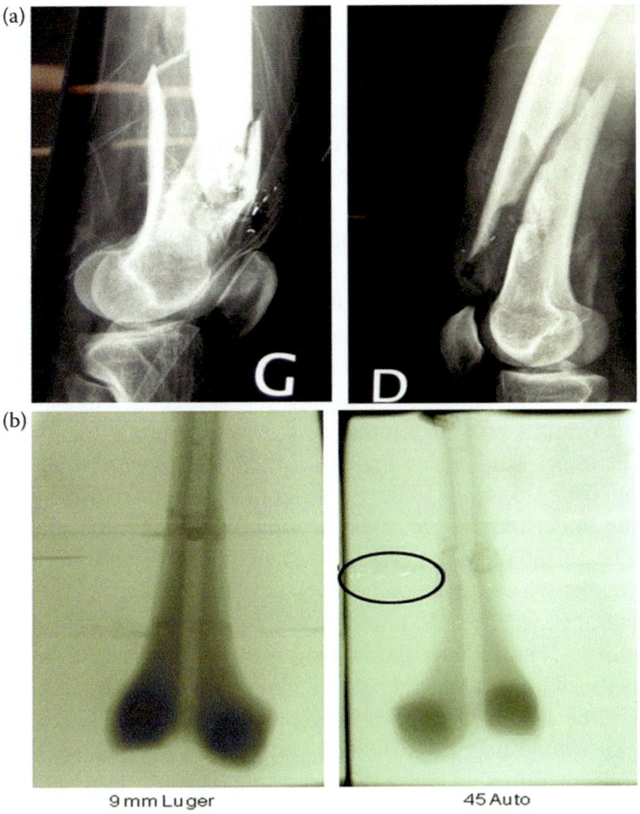

FIGURE 16.11 (a) X-rays of the femurs of a police officer caught in crossfire between his colleague and a suspect. Note the tiny radiopaque fragments near the fractures. (b) X-ray images of synthetic bone models in gelatine. The tiny radiopaque structures are only seen in the bottom right image, in which the victim was shot at with 45 autoammunition used by the suspect. The bone model seen in the bottom left, which was shot at by police ammunition (9 mm Luger) does not show such fragments.

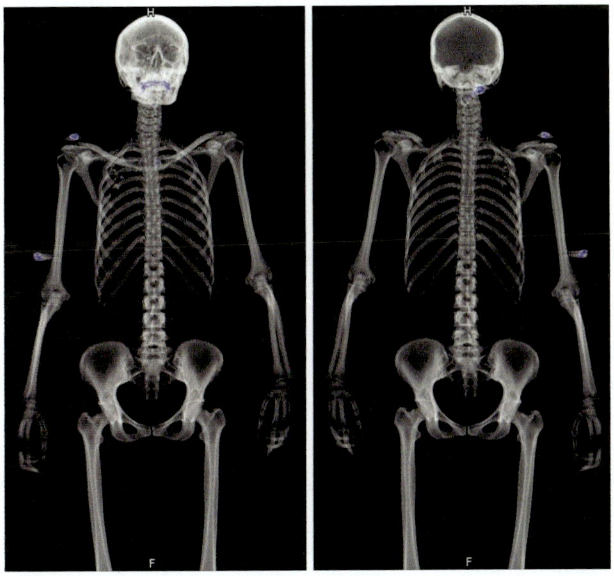

FIGURE 16.12 Semitranslucent 3D CT bone reconstruction of a homicide victim. Objects with high radiopacity are colored blue due to ramp settings. Note the multiple blue objects (projectiles).

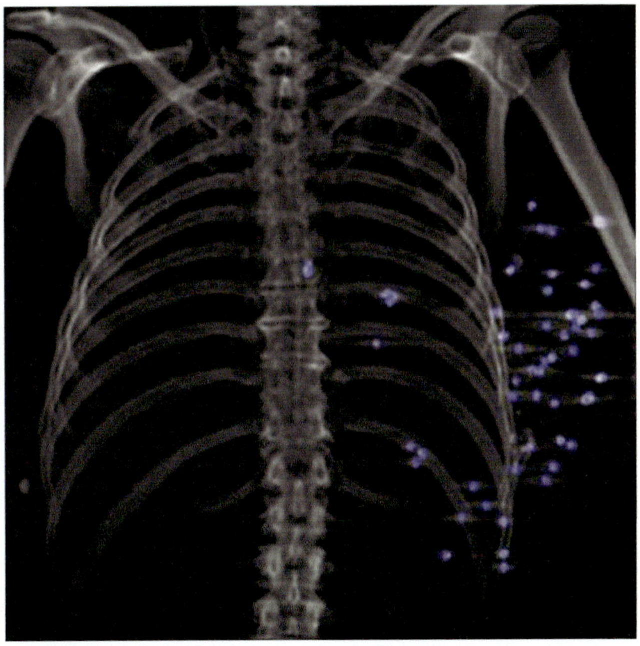

FIGURE 16.13 Semitranslucent 3D CT reconstruction of the ribcage of a man who committed suicide by shooting into his chest with a shotgun. The lead shotgun pellets are colored blue due to the ramp settings.

in detecting foreign bodies by virtue of their greater radiopacity than physiologically normal osseous structures, can visualize these only in a black and white manner. A further determination of the kind of foreign body, that is, bullet fragment, glass, and so on, is not possible.

Unfortunately, it is these other foreign objects that are under certain circumstances invaluable for case reconstructions. If the projectile is fired through an intermediate target such as a window, tiny fragments may lodge in the body. The identification and retrieval of such fragments with subsequent analysis may give clues to the crime scene, which obviously is not always identical with the location the corpse was found in.

Cause of Death and Vitality of the Injuries

Depending on the organs injured in the course of the bullet through the body, a multitude of different causes of death are possible. Of these, death due to craniocerebral trauma, cardiovascular injuries with exsanguination, and gas embolism are the most frequently encountered causes of death.

Another aspect is the so-called vitality of the injuries. It is of obvious judicial relevance whether the sustained gunshot injuries occurred when the victim was alive or already dead.

Bullet Type and Size, Identification of the Weapon Type

The crime scene investigators often need information as to the type of bullet involved as soon as possible in order to include or exclude different weapon types. To date, this meant waiting until the projectile or fragments of it were retrieved at autopsy. Great caution should be exerted when trying to determine the size of a projectile in conventional x-ray examinations; depending on the location it may appear larger or smaller.

With MSCT, the size and form of the more or less intact projectile can, however, be assessed accurately prior to retrieval. Furthermore, under ideal conditions, the type of the bullet with respect to it being jacketed or not and even the metal used for the jacket can be identified with MSCT (Jackowski et al., 2006). Therefore, the crime scene investigators have additional information regarding the weapon-type involved even before autopsy. Due to the limited resolution of MSCT today, the rifling of the bullet cannot be assessed accurately enough in the body, making bullet or fragment retrieval necessary.

The retrieval of the whole bullet or bullet fragments is of utmost importance, as the rifling on the projectile may lead to the identification of the individual weapon and therefore lead to the perpetrator. This undertaking is enormously facilitated by postmortem radiology, especially MSCT. MSCT suffices in locating the projectile within the body accurately and therefore allows for a minimally invasive extraction before or even instead of an autopsy.

HEAD INJURIES

Although the head compromises only a small percentage of the human body, head injuries due to gunshot wounds are a frequent finding in forensic practice and therefore warrant a separate section. Because head injuries often lead to rapid death, this small body region is frequently targeted in suicidal as well as homicidal gunshots.

Gunshots to the head leave a multitude of traces and clues that are also generally easily detectable with postmortem imaging procedures as described below.

The wound morphology gives clues as to the distance and—in cases of contact wounds—also of the weapon type. By virtue of the rigid skull, the presence of other clues is a rule.

If a bullet strikes a skull tangentially, so-called "gutter wounds" arise (La Garde, 1916). These may be confined to the outer table or even the inner table. In such cases, very low-velocity bullets can slip between the skull and the tough tissue of the scalp to a completely distant part of the head. Without prior radiography, these gunshots give rise to certain moments of surprise at autopsy. Indeed, the localization of such ultimately flattened bullets between the scalp and the skull is difficult, as they frequently drop out of their position when the scalp is drawn off the skull at autopsy. Postmortem MSCT is able to locate such projectiles rapidly and therefore helps to avoid time-consuming searches for possibly paper-thin bullets.

If the bullet strikes the head at a shallow angle, it may split into two fragments whilst penetrating the bone. One part can then glance off the skull leaving only a superficial bone lesion, whilst the other penetrates the skull. This gives rise to key-hole-shaped wounds (Dixon, 1982).

A more classical form of gunshot injury to the head is the penetrating and often perforating trauma. Here, the bullet typically enters the skull, transverses through the brain, and may or may not—depending on the energy of the bullet—exit the skull. Obviously, the differentiation of exit and entrance wound is immensely important for reconstructive purposes.

The skull, by virtue of it consisting to a large extent of more or less flat bones, makes the differentiation between entrance and exit wounds fairly easy. When a bullet passes through these flat bones at a more or less perpendicular angle, the result is a round or round-oval defect (primary fracture) in the bone. This is true for exit and entrance wounds. On the opposite side of the bone, the inner table is beveled away from the bullet impact site, thus creating a cone-like appearance. The tip of the cone points towards the gun, and therefore enables a differentiation between entrance and exit wound. In entrance wounds, the inner table is beveled out ("inward beveling") (Figure 16.4). Exit wounds display the opposite: here it is the outer table that is beveled out ("outward beveling").

The skin-wound morphology with regard to entrance and exit wound may be difficult to interpret with the naked eye for a number of reasons (extensive lesions, putrefaction, insect and other animal involvement). However, postmortem imaging is capable of demonstrating the beveling features of the bone with sufficient accuracy as to determine exit and entrance wounds.

Another feature that gunshot wounds to the skull often possess is the presence of secondary fractures. As the skull is a rigid structure, a sudden increase of the intracranial pressure as in penetrating gunshot injuries to the head may give rise to a secondary fracturing of its most fragile areas. Such secondary fractures depend on the firing distance and the bullet's kinetic energy. Contact wounds give rise to extensive secondary fractures because of the gas-pressure from the muzzle produced by the discharge entering the head (Figure 16.14).

The most extreme form of secondary fracturing is the so-called "Kroenlein" gunshot. Here, a gunshot with high energy transfer is applied directly to the head. The result of this explosion-like force to the head is a complete destruction of the skull with cerebral exenteration.

The forensic relevance of such secondary fractures is the so-called gunshot priority in multiple wounds to the head. Fracture lines will not pass pre-existing fractures, as the strain on the bone—the essential prerequisite for fracture formation—ceases at the pre-existing fracture. This rule, also known as "Puppe's rule" after its first describer, allows for the determination of the gunshot priority in multiple wounds to the head. Obviously, such telltale fractures are not detected at external examination.

These secondary fracture lines are clearly visible in MSCT. Therefore, the gunshot priority can be assessed before autopsy.

As the brain is almost completely encapsulated by the skull, a bullet will invariably have to penetrate the skull in

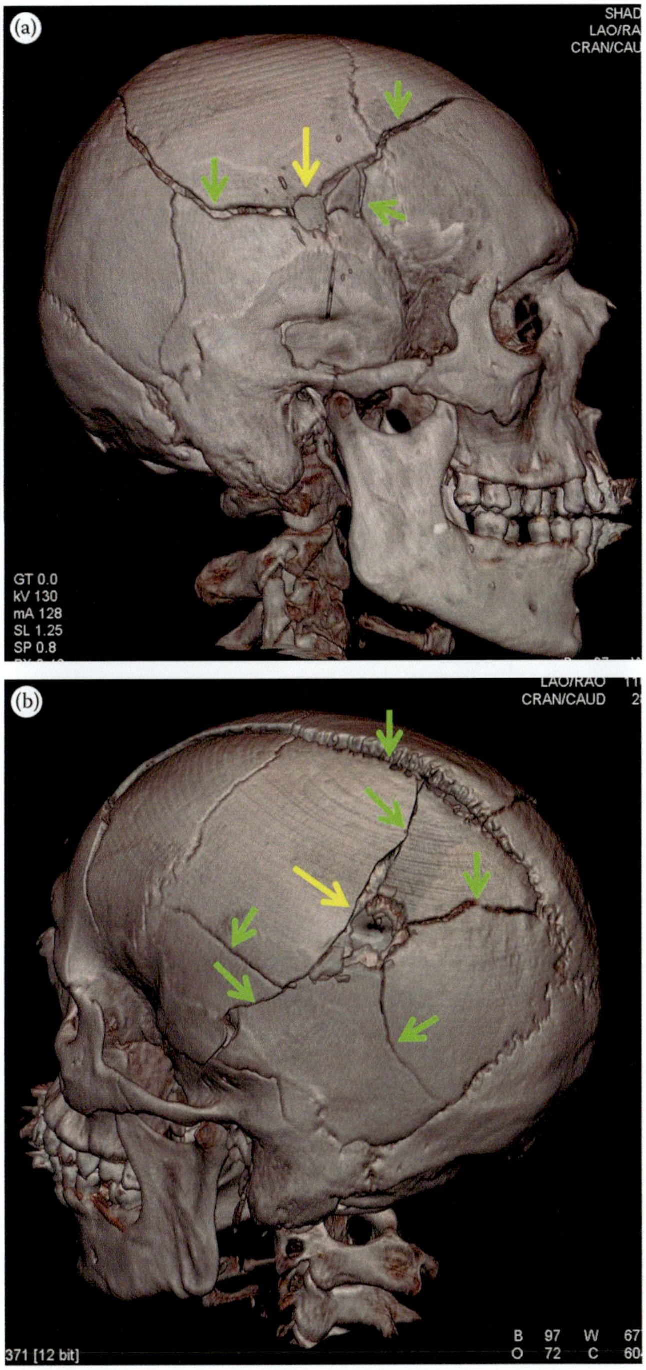

FIGURE 16.14 (a) 3D CT reconstruction of the skull of a suicidal gunshot injury to the right temple. Note the secondary fractures (green arrows) radiating from the entrance defect (yellow arrow). (b) Here, the exit defect (yellow arrow) is seen, again with multiple secondary fractures (green arrows). Note the burst sagittal suture and the clearly visible outward beveling of the exit defect.

order to create cerebral injuries. The chances for the bullet to leave a trail of lead, steel or bone fragments are therefore fairly great. Due to this fragment trail, MSCT can depict the bullet course through the brain easily. However, caution should be exerted in overestimating the accuracy of such a bullet course through the brain; it does not necessarily have to correspond exactly to the bullet course through the head. The injured brain may, if the victim survives for a certain amount of time, swell. This cerebral edema can be one-sided, giving rise to a shift of the cerebral structures and therefore of the bullet course. Oppositely, a shrinking of the brain, as encountered in a pneumocephalon will alter the course of the bullet through the brain compared with the direct path between entrance and exit defect.

Postmortem MRI is not effective in displaying hard foreign particles, as these generally do not contain water. Therefore, bullets, glass and, to some extent, bone will remain as signal-silent structures. The cerebral tissues and their respective damage, that is, contusion, hemorrhages, and so on. are clearly distinguishable. The bullet course can, even with the above-mentioned caveats, also be depicted.

CONCLUSION

We believe that postmortem imaging, ideally a combination of MSCT and MRI, is a helpful adjuvant to the state-of-the-art classic forensic examination of ballistic trauma.

MSCT can differentiate between entrance and exit wounds, detect bullets and other foreign bodies in the corpse, provide information on the injured organs and therefore the cause of death, and give a general overview of the bullet course. Furthermore, MSCT can address the question of gunshot priority in multiple wounds.

MRI gives, besides a general impression of the bullet course, detailed information on soft-tissue injuries. The cause of death is therefore more accurately detailed in MRI than in MSCT examinations. However, MRI does not depict hard objects such as the bullet or other foreign particles sufficiently.

REFERENCES

Andenmatten, M. A., Thali, M. J., Kneubuehl, B. P., et al., Gunshot injuries detected by post-mortem multislice computed tomography (MSCT): A feasibility study, *Leg Med (Tokyo)*, 10, 287–92, 2008.

Bolliger, S. A., Oesterhelweg, L., Spendlove, D., et al., Is differentiation of frequently encountered foreign bodies in corpses possible by Hounsfield density measurement? *J Forensic Sci*, 54, 1119–22, 2009.

Di Maio, V. J. M., *Gunshot Wounds: Practical Aspects of Firearms, Ballistics, and Forensic Techniques*, CRC, Boca Raton, FL, 1999.

Dixon, D. S., Keyhole lesions in gunshot wounds of the skull and direction of fire, *J. Forensic Sci.*, 27, 555–66, 1982.

Harcke, H. T., Levy, A. D., Abbott, R. M., et al., Autopsy radiography. Digital radiographs (DR) vs multidetector computed tomography (MDCT) in high-velocity gunshot-wound victims, *Am J Forensic Med Path*, 28, 13–9, 2007.

Jackowski, C., Lussi, A., Classens, M., et al., Extended CT scale overcomes restoration caused streak artefacts—3D color encoded automatic discrimination of dental restorations for identification, *JCAT*, 30, 510–3, 2006.

Kneubuehl, B. P., and Thali, M. J., Experimentelle rekonstruktion in *Wundballistik-Grundlagen und Anwendungen*, Kneubuehl, B. P., Coupland, R. M., Rothschild, M. A., and Thali, M. J., Eds., Springer, Berlin, 2008.

La Garde, L. A., *Gunshot Injuries*, William Wood and Co., New York, 1916.

Levy, A. D., Abbott, R. M., Mallak, C. T., et al., Virtual autopsy: Preliminary experience in high-velocity gunshot wound victims. *Radiology*, 240, 522–8, 2006.

Schumacher, M., Oemichen, M., Konig, H. G., et al., Computer tomographic studies on wound ballistics of cranial gunshot injuries, *Beitr. Gerichtl. Med*, 43, 95–101, 1985.

Stein, K. M., Bahner, M. L., Merkel, J., Detection of gunshot residues in routine CTs, *Int J Legal Med*, 114, 15–8, 2000.

Thali, M. J., Kneubuehl, B. P., Bolliger, S. A., et al., Forensic veterinary radiology: Ballistic-radiological 3D computer tomographic reconstruction of an illegal lynx shooting in Switzerland, *Forensic Sci Int*, 171, 63–6, 2007.

Thali, M. J., Schweitzer, W., Yen, K., et al., New horizons in forensic radiology: The 60-second digital autopsy-full body examination of a gunshot victim by computed tomography, *Am J Forensic Med Path*, 24, 22–7, 2003.

Thali, M. J., Watzke, O., and Kachelriess, M., Bullets and metal artefacts-state of the art, *Rechtsmedizin*, 11, 193, 2001.

Thali, M. J., Yen, K., Vock, P., et al., Image-guided virtual autopsy findings of gunshot victims performed with multi-slice computed tomography (MSCT) and magnetic resonance imaging (MRI), and subsequent correlation between radiology and autopsy findings, *Forensic Sci Int*, 138, 8–16, 2003.

Wullenweber, R., Schneider, V., and Grumme, T., [A computer-tomographical examination of cranial bullet wounds], *Z. Rechtsmed.*, 80, 227–46, 1977.

Section V

Radiology of Abuse

> Forensic Science is used to predict not the future but the past
>
> **Henry C. Lee**

Nowhere in forensic radiology is Dr. Lee's provocative aphorism less applicable than in the field of abuse. Here lies the opportunity to go beyond the limits of the necropsy "where death delights to help the living." The early identification and proper management of the victims of abuse while they are still living cannot only predict, but also can modify, the future. Life can be preserved, even enhanced, and necropsy can be averted.

Radiologically detectable abuse simplistically can be defined as an improper, usually intentional, action leading to physical injury. But abuse does not rest as much on definition as on conceptualization. The concept of abuse is inextricably entangled in history, religion and culture. Hence, its definition may vary widely within geographical boundaries, the passage of time, and the evolution of religious and societal mores.

Even now, there is no universal agreement upon what contributes physical or emotional abuse of human beings from infancy through old age. What we might consider extreme examples of abuse are tolerated in some parts of our world and segments of its inhabitants. Genital mutilation, purdah, deformation or mutilation of body parts, physically or emotionally painful interrogation, enslavement, even mass murder, may be considered acceptable, desirable or even obligatory in some cultures and religions, but abhorrent, detestable and odious in other societies. Abuse of spouses or intimate partners is a modern concept only recently recognized and still quite limited in range. In many parts of this 21st Century world, such partners still can be battered, burned or stoned to death without fear of societal or legal rebuke. The almost world-wide historical tradition of veneration of elders, even in primitive cultures, has been eroded by a spate of "granny bashing" which in some "enlightened" societies is almost equal to child abuse in frequency. This section will illustrate the radiological manifestation of pan-generational inflicted trauma generally accepted in Western countries as abusive. It must be recognized that not all forms of abuse are revealed by radiological investigation.

B. G. B.

17 Child Abuse

B.G. Brogdon

CONTENTS

HISTORICAL PERSPECTIVE

Virtually every application of radiology within the forensic sciences was initiated or predicted within one year of Röntgen's discovery. The conspicuous exception is perhaps radiology's greatest contribution to the forensic sciences. That it took 50 years and the reluctant conclusions of an observant pediatric radiologist to awaken public conscience and consciousness about one of its greatest evils is a sad commentary on humanity through the ages.

The very idea that there is such a thing as child abuse is a relatively modern concept arising from an ignoble history.[1] From biblical times through the age of industrialization, paternal power was absolute. A father could abandon a child, abuse it, sell it into slavery, put it to death, or cut it in half. The father's right to correct or discipline was limited only by his conscience, and this right extended *in loco parentis* to all adults involved in rearing or supervising the child, including teachers, trainers, masters of apprentices, workhouse bosses, factory foremen, and superintendents of children's "asylums."

The industrial revolution with its unceasing demand for cheap labor exacerbated the problem as poor couples learned that their best money crop was the fruits of their loins. The excesses attendant to child labor and the workhouses did stir the social consciousness of some influential writers, notably Dickens in England and America. Some societal response began to emerge. It is interesting to note that the first Society for the Prevention of Cruelty to Children had to be established under the aegis of the Society of the Prevention of Cruelty to Animals, but only after arguing successfully that children were members of the animal kingdom.[1] The annual meetings of the Mobile (Alabama) Society for the Prevention of Cruelty to Animals and Children were regularly reported in the local newspaper *The Mobile Register*. A 1902 article lauds the passing of the horse-drawn streetcar, "one of the most prolific sources of cruel treatment of horse and mules." The plight of children was not mentioned. A few child labor laws were enacted. Thus children gradually came under the protection of the law, in public at least, but not necessarily at home.

In 1860, Ambriose Tardieu (1818–1879) was a French physician specializing in pathology, public health, and legal medicine (Figure 17.1). A year later he would become Professor of Legal Medicine at the University of Paris, a post he held until his death. But in 1860, Tardieu published an article (Figure 17.2) on the abuse and maltreatment of children; it was reprinted in a book on wounds (Figure 17.3) published a year after his death. Unfortunately, this treatise seems to have had a much greater impact on modern historians[1–3] of child abuse than on Tardieu's contemporaries: unfortunate because in his 32 cases Tardieu set forth all of the salient features of child abuse—sociologic, demographic, and medical—except for the radiological. He recognized care givers as the perpetrators, described the typical injuries, and observed the emotional responses of the victims. Still, the abused child as a clinical entity was largely unrecognized until the essential elements of this syndrome, both clinical and radiographic, started to surface in the 1930s, notably in a few institutions such as Babies Hospital in New York.[4,5]

FIGURE 17.1 Ambroise Tardieu, 1818–1879. (From Silverman, F. N., *Radiology*, 104, 337, 1972. With permission.)

MÉDECINE LÉGALE.

ÉTUDE MÉDICO-LÉGALE

SUR LES

SÉVICES ET MAUVAIS TRAITEMENTS

EXERCÉS SUR DES ENFANTS,

Par le Dr Ambroise TARDIEU,

Professeur agrégé de médecine légale à la Faculté de médecine.

Parmi les faits si nombreux et de nature si diverse dont se compose l'histoire médico-légale des coups et blessures, il en est qui forment un groupe tout à fait à part, et qui, laissés jusqu'ici dans l'ombre la plus complète, méritent à plus d'un titre d'être mis en lumière. Je veux parler de ces faits qualifiés sévices et mauvais traitements, et dont les enfants sont plus particulièrement victimes de la part de leurs parents, de leurs maîtres, de ceux en un mot qui exercent sur eux une autorité plus ou moins directe.

FIGURE 17.2 First page of Tardieu's 1860 paper on child abuse. (From Silverman, F. N., *Radiology*, 104, 337, 1972. With permission.)

ÉTUDE MÉDICO-LÉGALE

SUR

LES BLESSURES

COMPRENANT

LES BLESSURES EN GÉNÉRAL

ET

LES BLESSURES PAR IMPRUDENCE

LES COUPS ET L'HOMICIDE INVOLONTAIRES

PAR

AMBROISE TARDIEU

Professeur de médecine légale á la Faculté de médecine de Paris.

PARIS

LIBRAIRIE J.-B. BAILLIÈRE ET FILS

19, RUE HAUTEFEUILLE, 19

1879

FIGURE 17.3 Title page of Tardieu's 1879 book on wounds, which included a reprint of his 1860 article on child abuse. (From Silverman, F. N., *Radiology*, 104, 337, 1972. With permission.)

John Caffey (1895–1978) received his MD from the University of Michigan in 1919. He spent almost 3 years in postwar Eastern Europe, then returned for a career in pediatrics, eventually being appointed to the full-time staff of Babies Hospital in New York City.[3] When that hospital finally installed a "modern" radiographic and fluoroscopic unit in 1929, Dr. Caffey was placed in charge of it. With no formal training in radiology, but with the encouragement and support of Dr. Ross Golden (Chairman of Radiology at the College of Physicians and Surgeons of Columbia University), Dr. Caffey became the father of Pediatric Radiology, the first recognized subspecialty in diagnostic radiology (Figure 17.4). Dr. Caffey had a very long and active career at Babies Hospital and, after his retirement there, at the Children's Hospital of the University of Pittsburgh. He made many contributions, but perhaps his most celebrated one had its genesis in an article published in 1946.[6] This landmark paper described a peculiar association of multiple fractures of the long bones of children suffering from chronic subdural hematoma.

To digress for a moment, the history of subdural hematoma is one of interesting paradox. Both Paré and Vesalius recognized the traumatic basis of subdural hematoma in the case of Henry II of France who died after being injured in a tournament celebrating the marriage of his daughter to Phillip II of Spain.[1] However, in 1856, Virchow suggested an infectious cause for the condition. Since Virchow was an unarguable authority, his opinion prevailed for the next 70 years, and the condition was most commonly spoken of as "pachymeningitis interna hemorrhagica." However, Sherwood in his classic 40-page paper in 1930[7] reemphasized the likelihood of traumatic origin, but even so, in his conclusion he stated, "... the etiology is obscure. It was unusual to find that in 5 of the 9 cases described the patients were cared for in institutions or by foster mothers. Trauma due to injury at birth or other means is a possible factor, although not proved in the series of cases reported." In 1939, Ingraham and Heyl[8] reported that subdural hematoma appears more frequently in undernourished children and, in the majority of instances, there is a history of trauma.

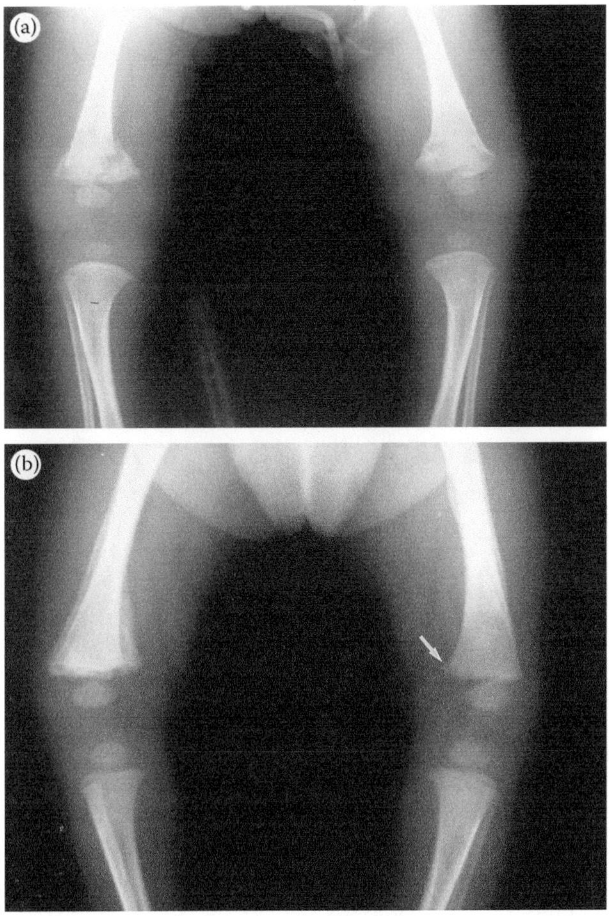

FIGURE 17.5 (a) Severe bilateral metaphyseal fragmentation in the distal femora. The "corner fractures fragments" are larger than usual. (b) Small "corner fracture" on left; "involucrum" on right.

FIGURE 17.4 John Caffey, the father of pediatric radiology. (Reprinted from the Center for American History of Radiology, Reston, VA. With permission.)

To return to Caffey, in 1946 he described patients with subdural hematomas who also had multiple fractures in long bones. The bony lesions were somewhat unusual with metaphyseal fragmentation (Figure 17.5) and formation of what he called large involucrums (because they resembled that manifestation of chronic osteomyelitis, but which really is calcification in subperiosteal hemorrhages) (Figure 17.6). He also noticed a pattern of fractures in stages of healing (Figure 17.7).

Some of Dr. Caffey's puzzlement shines through as he says,

> For many years we have been puzzled by the roentgen disclosure of fresh, healing and healed multiple fractures of long bones of infants whose principal disease was chronic subdural hematoma. In not a single case was there a history of injury to which a skeletal lesion could reasonably be attributed. No predisposing generalized or localized disease for pathologic fracture was present. ... the fractures appear to be of traumatic origin but the traumatic episodes and the causal mechanism remain obscure.

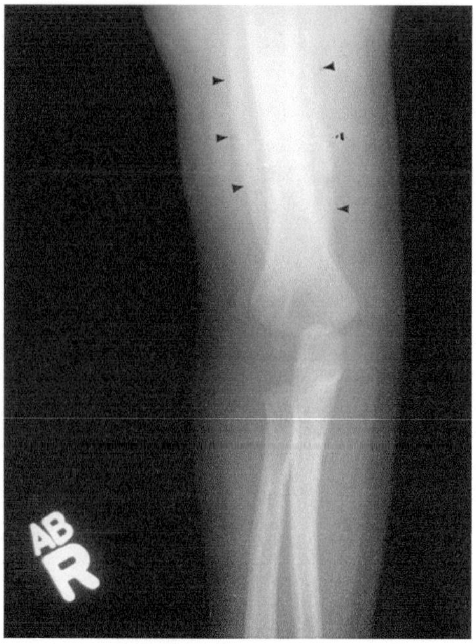

FIGURE 17.6 Large "involucrum" (actually a calcified subperiosteal hematoma) around the shaft of the humerus.

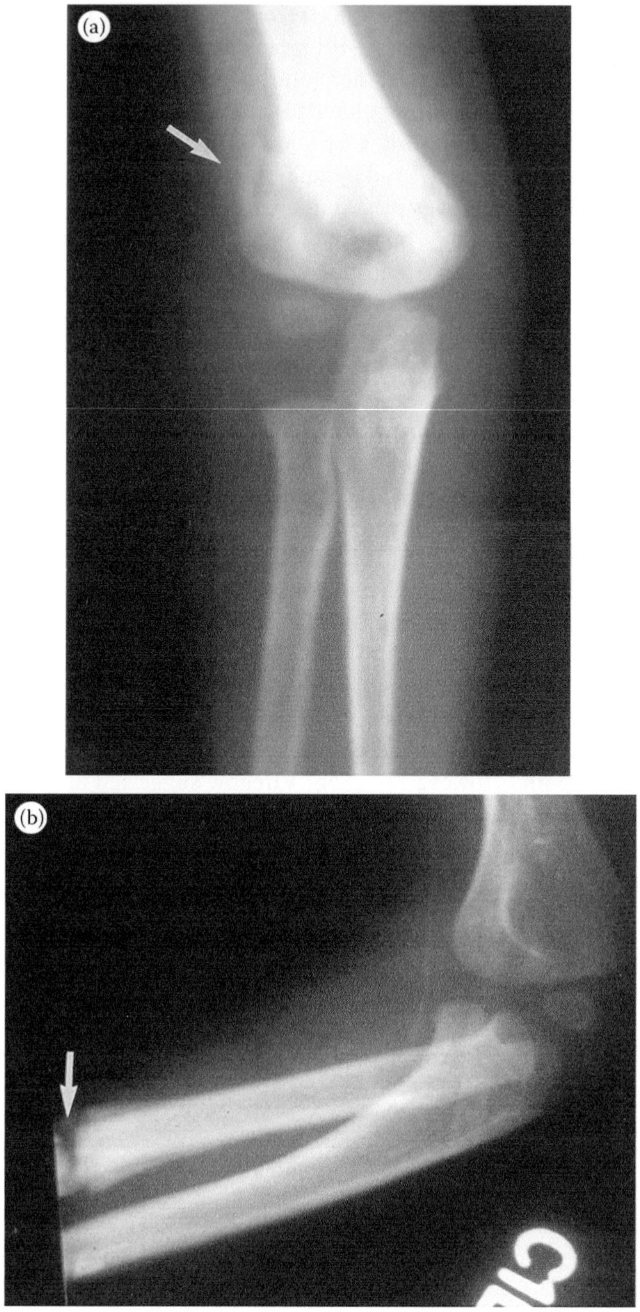

FIGURE 17.7 This is the initial presentation of the child whose terminal radiograph is shown in Figure 3.3. She first presented at age one with (a) an acute fracture of the distal metaphysis of the humerus (arrow). (b) On the edge of the lateral view a second fracture, a nonunion of the radius was seen (arrow). The child was removed from the home. The mother and father divorced. The father remarried and got the child back. The stepmother killed her with a fist to the epigastrium at age two.

Nevertheless, the essential roentgen essence of child abuse had been documented. Dr. Caffey confirmed his earlier findings in a distinguished lectureship in 1957[9] and illustrated some new radiographic features: traumatic bowing of the ends of the diaphyses due to metaphyseal infraction (Figure 17.8), metaphyseal cupping, and ectopic ossification centers (Figure 17.9)—all associated with the previous findings of involucrum formation, metaphyseal fragmentation, and fractures of differing duration. The description was now more complete.

In a third communication in 1965,[10] Dr. Caffey spoke of the relative diagnostic values of the history, physical examination, laboratory tests, biopsy findings, and radiographic observations. Caffey noted that the history of trauma was frequently withheld. Giving the benefit of doubt, he opined that the history of trauma was sometimes unknown to the family. He concluded that, "the radiographic changes are pathognomonic of trauma, but they never identify the perpetrator of the trauma or his motive." He still believed that "the great majority of simple, even serious traumatic episodes to children . . . are accidents for which no one is responsible."

There still were careful disclaimers in excluding underlying processes such as infection, malnutrition, avitaminosis, metabolic bone disease, and the like as contributory factors. Dr. E. B. D. Neuhauser, Radiologist to the Children's Hospital in Boston, noted, "every case ... has been an example of needless expense to the hospital or to the patient in our hospital, with endless blood cultures, many studies, search of old films to see if it didn't really resemble scurvy, and the roentgen department has forgotten what scurvy looks like, not having seen a case for some months . . . ."[11]

Disciples of Caffey in Pittsburgh and Cincinnati further studied the problem, and Silverman's paper of 1953[12] can be credited with finally focusing the attention of the radiological community on the entity.

Still, a variety of somewhat euphemistic names were applied to the syndrome—mostly suggesting undetected trauma of some kind such as hidden, concealed, denied, unsuspected, unrecognized, or clandestine. Caffey suggested

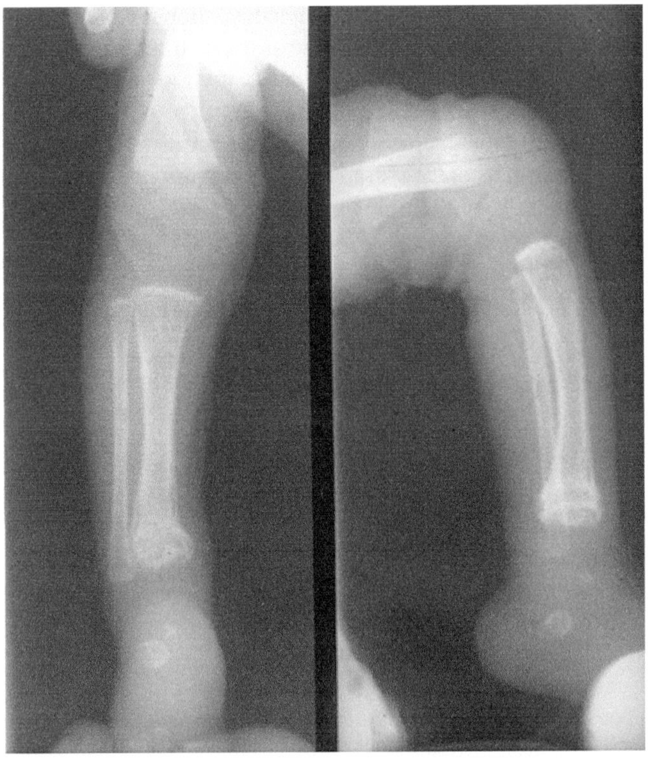

FIGURE 17.8 Example of what Caffey called "traumatic bowing."[6]

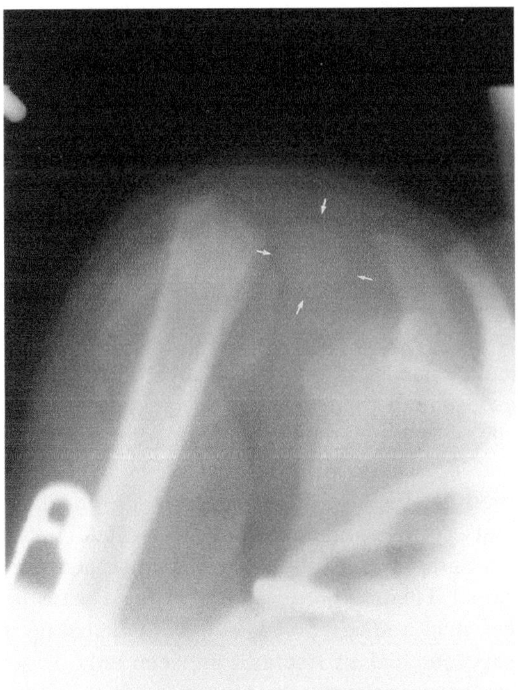

FIGURE 17.9 Example of what Caffey called "ectopic ossification centers."[6] The proximal humeral epiphysis (arrows) is widely separated from the metaphysis, which shows a large calcifying subperiosteal hemorrhage. The glenohumeral joint is wide but not dislocated.

parent infant trauma stress syndrome (PITS) long before the expression "it's the pits" existed.[2] Galdston in 1965 offered the phrase *Parental Dysfunction* which causes no antiparent bias and suggests the psychoemotional storms that paralyze parental self control. Several eponymic designations were proposed without general adoption: "Le Syndrome de Silverman" in France, "Le Syndrome de Caffey," "Caffey's Third Syndrome" and "The Syndrome of Ambroise Tardieu".[1] Finally, in 1962, Kempe[13,14] intentionally coined the inflammatory epithet, *Battered Child Syndrome*, to attract attention to this neglected clinical and social problem. This provocative and anger-producing term was successful in gaining widespread public attention. This was enhanced by a science writers' forum on the subject sponsored by the American College of Radiology (ACR) in 1962, featuring a radiologist, a pediatrician, and a distinguished jurist on the panel.

Bert Girdany, who went to Pittsburgh following his residency with Dr. Caffey at Babies Hospital, recognized the inadequacy of a single physician in dealing with the problem and organized a trauma team made up of a pediatrician, pediatric psychiatrist, pediatric radiologist, and social worker. The value and results of this team approach were reported in Elizabeth Elmer's book, *Children in Jeopardy*.[15]

All of the pertinent findings finally had been established with the addition of another factor, that is, injury inappropriate in view of the history, circumstances, or stage of development of the involved child.[16]

OVERVIEW

It is recognized now that child abuse is an umbrella term covering a broad spectrum of intentional harmful interference with the health, happiness, and development of an infant or child. There are at least six forms of child abuse.

SPECTRUM OF CHILD ABUSE

For those who would want a much more thorough, in-depth account of child abuse than is possible within the limitation of a single chapter, the books by Kleinman[17] and Reece[18] are recommended.

Physical Abuse

Abuse resulting in physical injury was the first form recognized, and the most deadly. Radiological findings are central to the diagnosis of the condition and to the prosecution of the perpetrators in the majority of the cases. Physical abuse is the principal concern of this chapter and will be discussed in some detail further on.

Nutritional Deprivation

Radiographically, one may see osteoporosis of bone and diminished body fat in children who are starved. These are subtle findings and are likely to be unremarked unless there is stimulus from the history or other findings to search for them.

Occasionally, infants are placed on faddist diets or are overmedicated at home. One can see evidence of hypervitaminosis or hypovitaminosis (i.e., vitamin A intoxication or rickets).

Emotional Abuse

There is a clinical entity variously known as maternal deprivation syndrome, psychosocial dwarfism, deprivation dwarfism, abuse dwarfism, or Kaspar–Hauser syndrome.[19] Those afflicted may show growth retardation, retarded bone age, and osteopenia. Treatment will result in reversal of these findings, and there may even be spreading of cranial sutures from rapid brain growth.

Neglect of Medical Care or Safety

This can result in all forms of trauma that are literally unintentional, but are actually due to neglect (i.e., massive injuries resulting from a vehicular accident in which the child is not secured in an approved seat).

Intentional Drugging or Poisoning

There are no specific radiographic findings. Arsenic has been seen in the stomachs of poisoned adults, but no examples in children are known. Children may be given drugs by parents or others. Prenatal drugging of infants is an increasing problem. Somewhat akin to poisoning is the abusive introduction of foreign objects into the esophagus of infants or children. Nolte[20] reviews the subject and reports instances of coins, beans, a cooking spatula, a teething ring, and broken glass being fed or forced-fed to the victims by adults or other children.

Sexual Abuse

Sexual abuse does not come to the radiologist's attention unless there is massive injury such as colorectal or vaginal laceration with or without pneumoperitoneum.

INCIDENCE OF CHILD ABUSE[21]

It is estimated that five children die every day in the United States from abuse or neglect by parents or care givers. This amounts to some 2000 deaths per year, of whom the vast majority are younger than 4 years old. Most are less than 2 years of age; 40% of the children who die from abuse and neglect are under 1 year of age. About 10 times as many children survive abuse as die from it. An estimated 18,000 children per year are permanently disabled by abuse or neglect. Actually, it is not known how many disabled persons have been made so by abuse. For instance, there are 90,000 Americans with brain damage from head injuries; no one knows how many were the victims of intentional childhood trauma.

Most physical abuse comes from the father or male care giver (stepfathers, boyfriends, close friends, brothers, cousins), not from the mother as would be the conventional wisdom. It is the enraged, stressed male who is most likely to beat the child, shake it, or suffocate it—most often to stop the infant from crying. Mothers are more likely to be responsible for child neglect deaths from drowning, fires started by unsupervised children, or dehydration, and starvation. Most babies cry about 30% of the time and physical abuse to stop the crying becomes almost self-perpetuating since the battered, dazed, or brain-damaged child is indeed quite likely to stop crying for a while after the assault.

RADIOLOGY OF PHYSICAL ABUSE

Abusive head trauma is the leading cause of death of all trauma-related deaths of children.[22] Skeletal injuries, particularly of the appendicular, are most likely to bring abuse to our attention and to document the problem. However, musculoskeletal injuries are rarely fatal. Injuries to the rib cage are quite common (perhaps up to 25% of all skeletal fractures[23]) but frequently are missed on routine radiography if acute or fresh. Spinal fractures are rare but may have serious consequences. Fractures in the shoulder girdle are highly suspicious of abuse. Skull fractures are not necessarily an indication of abuse unless associated with intracranial damage and/or neurologic findings. Intentional trauma to the abdomen or thorax of the infant or child is less common than skeletal trauma but carries a 40–50% mortality rate.[21]

PROTOCOLS FOR EXAMINATION

There is no universal agreement on the proper system of examination of an infant or child suspected of being physically abused. Table 17.1 reproduces the ACR Appropriateness Criteria® for examination of infants and children who are suspected of being subjected to physical abuse. These are categorized under four variant clinical conditions. The appropriateness criteria are intended as guides and are not to be construed as having any other weight.

Levitt et al. suggest in Reece's book[22] that in demonstrating lesions of abusive head injury CT is preferable for subarachnoid hemorrhage; MR is preferable for subdural hematoma, concussive injury, and shear injury; and CT and MR are equal in their efficacy for demonstrating epidural hematoma. CT is preferable for fracture detection.

Sty has been a long-time advocate of bone scintigraphy in evaluation of the suspected abused child, and has cogent arguments.[24,25] Kleinman has compared skeletal surveys with bone scintigraphy and Table 17.2 shows that there are certain trade-offs.[26] There is no question that bone scans are more sensitive than radiographic surveys in detecting skeletal lesions. About 10% of fractures are seen only on scintigraphy. The exception is the skull fracture, which is more readily seen by radiography. Radiography has more specificity in that certain fracture patterns are virtually diagnostic of abuse. Further, the normally high uptake in the growing ends of the bones of infants and children, where many abuse lesions are found, is a distinct problem, especially if there are

TABLE 17.1
American College of Radiology ACR Appropriateness Criteria

Radiologic Procedure	Rating	Comments	RRL*
Clinical Condition: Suspected Physical Abuse—Child			
Variant 1: Child 2 years or less, No Focal Signs or Symptoms			
X-ray skeletal survey	9	Includes at least two views of the skull	Med
MRI head	5	For evidentiary purposes only	None
NUC Tc-99m bone scan whole body	4	May be useful in selected cases. For evidentiary purposes only	Med
CT head	2		High
US abdomen	2		None
Rating scale: 1 = Least appropriate, 9 = Most appropriate			*Relative radiation level
Variant 2: Child 2 Years or Less, Head Trauma with History, No Focal Findings, No Neurologic Abnormality			
X-ray skeletal survey	9	Includes at least two views of the skull.	Med
MRI head	7		None
CT head	6		High
NUC Tc-99m bone scan whole body	4	May be useful in selected cases. For evidentiary purposes only.	Med
US abdomen	2		None
Rating scale: 1 = Least appropriate, 9 = Most appropriate			*Relative radiation level
Variant 3: Child up to Age 5, Seizures or Neurologic Signs and Symptoms, with or without Physical Findings			
CT head	9		High
X-ray skeletal survey	9	Includes at least two views of the skull	Med
MRI head	8	May be appropriate as alternative to CT or following CT	None
NUC Tc-99m bone scan whole body	4	May be useful in selected cases. For evidentiary purposes only	Med
US head	2		None
Rating scale: 1 = Least appropriate, 9 = Most appropriate			*Relative radiation level
Clinical Condition: Suspected Physical Abuse—Child			
Variant 4: Child of any Age, Visceral Injuries, Discrepancy with History, Physical and Laboratory Examinations Inconclusive			
X-ray skeletal survey	9	Includes at least two views of the skull.	Med
CT abdomen and pelvis with contrast	9		High10
MRI head	2		None
MRI abdomen and pelvis	2		None
CT head	2		High
US abdomen and pelvis	2		None
CT abdomen and pelvis without contrast	2		High
Rating scale: 1 = Least appropriate, 9 = Most appropriate			*Relative radiation level

Source: Reprinted from the American College of Radiology. No other representation of this material is authorized without expressed, written permission from the American College of Radiology. Refer to the ACR website at http://www.acr.org/ac for the most current and complete version of the ACR Appropriateness Criteria. With permission.

Note: An ACR Committee on Appropriateness Criteria and its expert panels have developed criteria for determining appropriate imaging examinations for diagnosis and treatment of specified medical condition(s). These criteria are intended to guide radiologists, radiation oncologists and referring physicians in making decisions regarding radiologic imaging and treatment. Generally, the complexity and severity of a patient's clinical condition should dictate the selection of appropriate imaging procedures or treatments. Only those exams generally used for evaluation of the patient's condition are ranked. Other imaging studies necessary to evaluate other coexistent diseases or other medical consequences of this condition are not considered in this document. The availability of equipment or personnel may influence the selection of appropriate imaging procedures or treatments. Imaging techniques classified as investigational by the FDA have not been considered in developing these criteria; however, study of new equipment and applications should be encouraged. The ultimate decision regarding the appropriateness of any specific radiological examination or treatment must be made by the referring physician and radiologist in light of all the circumstances presented in an individual examination.

bilateral injuries. Scintigraphy demands good positioning and, consequently, sedation. The whole-body radiation dose is 2.5 times higher with scintiscanning, the examination is more expensive than routine radiography, and it takes much longer to get the results. Further, throughout the land it is relatively easy to obtain radiographs at any hour of the day or night with reasonably competent personnel to perform them. While this is true of scintigraphy in certain institutions, it is by no means universally available. Consequently, in most areas of the United States, radiography is the examination of choice for injuries of the musculoskeletal system with supplemental CT or MRI as indicated for head injuries. In institutions where radiography and scintigraphy are equally available, then the choice of modality rests entirely with the radiologist in charge, as it should with all other similar diagnostic decisions.

TABLE 17.2
Comparison of Skeletal Surveys and Bone Scintigraphy

	Skeletal Survey	Bone Scintigraphy
Sensitivity	Moderate	High
Specificity	High	Low
Sedation	Rare	Common
Dose		
Gonodal	Very low	Low
Metaphyseal	Very low	Moderate
Availability	High	Varies
Need for additional studies	Occasionally	Always
Cost	Low	70–300% higher
Technical factor dependency	Moderate	High
Interpreter dependency	Moderate	High

Source: From Kleinman, P. K., *Diagnostic Imaging of Child Abuse*, Williams & Wilkins, Baltimore, 1987, p. 25. With permission.

Follow-up or repeat skeletal surveys two weeks after suspicious findings on an initial skeletal survey, other imaging studies, history or physical examination, are strongly advocated by Kleinman and associates.[27] They found that follow-up studies increase the number of definite fractures by 27%.

CT is of inestimable value in evaluating trauma to the thoracic and abdominal viscera. CT is also excellent in disclosing rib fractures hard to see on routine radiography.

The advantages of CT and MR in the evaluation of brain injuries have already been mentioned. In the rare case of spinal injury in a child, MR is an ideal supplementary or complementary examination to evaluate spinal cord damage.

Ultrasonography is of value in studying abdominal and retroperitoneal visceral injuries and has had some limited applications in the musculoskeletal system (Figure 17.10).[28]

MR is useful in evaluating injuries in and around joints. It is not a convenient screening procedure, however, and will be mostly used in follow-up and in complementary studies.

Optimally, the radiologist should tailor each examination protocol to fit the demands of the clinical problem as it is presented, allowing for the parts of the body or organ systems affected or suspected of injury. If a "routine" skeletal survey by x-ray is needed, it should include AP supine and lateral chest, AP views of the upper extremities with PA hands, AP lumbar spine and pelvis, AP views of the lower extremities, AP feet, and frontal and lateral skull.[26,29] High-detail radiographs with good collimation are required. The all-inclusive "babygram" is to be avoided. If there are positive or suspicious findings for child abuse, a brain CT or MRI is advisable.

SKELETAL INJURIES

The frequency of skeletal injury in cases of child abuse varies widely in the literature, ranging from 11% to 55%.[30] The vast majority of fractures occur in patients under 3 years of age, and half of them are in infants. The extremities are convenient "handles" by which the child can be grabbed, swung, shaken, or pulled. Hence, extremity fractures are most common. Certain fractures and other skeletal injuries are particularly suggestive of intentional trauma.

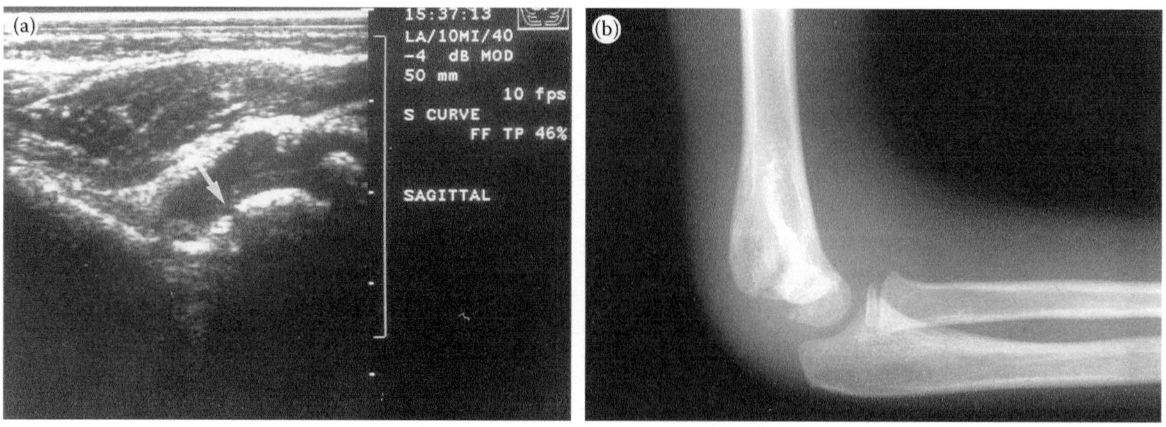

FIGURE 17.10 (a) A supracondylar fracture of the distal humerus demonstrated by ultrasonography (arrow). (b) Confirmatory radiograph.

Metaphyseal Injuries

The metaphyseal lesion of child abuse, first described by Caffey, is virtually pathognomonic. This fracture extends transversely across the extreme end of the metaphysis separating a disk of bone from the primary spongiosa of the metaphyses and the zone of provisional calcification of the physis. This disk is usually thicker in its periphery than in its centrum and, according to the projection, may appear as a transverse fracture line, as metaphyseal chip fractures, as a so-called bucket-handle fracture,[30] or as a disk (Figure 17.11). This metaphyseal injury is rarely accompanied by periosteal reaction; vascular injury and interference with growth may cause bowing of the extremity as described in Caffey's second paper.[9] Metaphyseal injuries are ordinarily seen in children who don't yet walk and are not associated with normal handling, rough play, or accidental falls. The most common locations for these metaphyseal injuries are the knee, ankle, and distal humerus.

Periosteal New Bone

The periosteum is loosely attached to the underlying bone in infants and children and is easily separated by twisting and pulling. This results in subperiosteal hemorrhage which will calcify. These lesions are usually silent lesions but if the bleeding is massive there may be palpable swelling and pain. This finding was referred to as involucrum in Caffey's original paper.[6] The subperiosteal calcification can be only a subtle thin line or may be of massive proportions (Figure 17.12). Shopfner[31] was first to point out that a single thin line of periosteal calcification can exist normally in infants 1–4 months of age, is invariably bilateral, and by itself is not diagnostic of abuse.

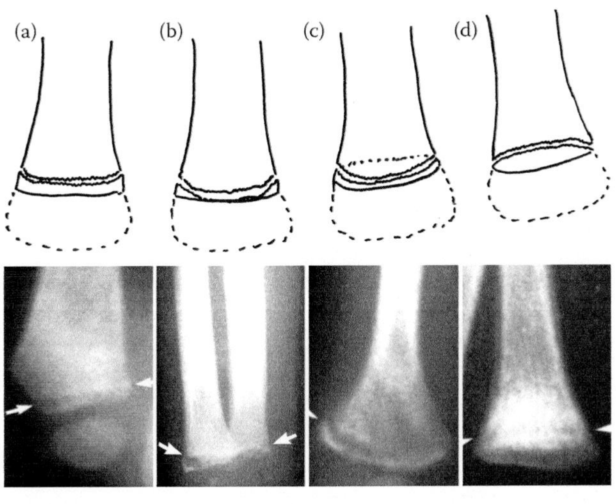

FIGURE 17.11 Top: schematic, bottom: actual illustration of the four possible appearances of a typical metaphyseal fracture according to the projection: (a) transverse, (b) corner fractures, (c) "bucket-handle", (d) discoid. (Top row schematic from Brogdon, B.G., chap. XXI: Forensic aspects of radiology, in Spitz, W.U., Ed., *Medicolegal Investigation of Death*, 4th ed., C.C. Thomas, Springfield, IL, 2006, 1135, With permission.)

Diaphyseal Spiral Fractures

Oblique long bone fractures were found in 15% of children radiographed for suspected abuse in Hilton's experience,[32] but only 5% in the series of Klineman et al.[33] They are highly suggestive of abuse, particularly in the nonambulatory child (see Figure 3.6b). Like the "toddler's fracture" (see Figure 3.6a) they may be difficult to see when fresh and only become apparent when there is associated periosteal reaction. There may be extensive associated periostitis if treatment and immobilization is delayed[32] (Figure 17.13). These fractures apparently result from twisting or torsion forces.

Transverse Long Bone Fractures

We have found these have a high specificity for child abuse in the nonambulatory child especially, and are a fairly common finding. They seem to be related to abusive grabbing and swinging forces which cleanly snap the bone (Figure 17.14).

Dislocations

True joint dislocations are rarely seen in child abuse cases although dislocation of a secondary ossification center is not unusual. When dislocations are seen, they are usually associated with massive trauma and there should be a good explanatory history.

Rib Fractures

Klineman et al. found rib fractures to be even more common than fractures of the long bones[33] (Figure 17.15). When fresh they may be very difficult to see, but usually heal with abundant calluses and become quite obvious on delayed studies. Posterior rib fractures are particularly suggestive of child abuse and are thought to result from grasping the child with anterior compression of the chest. Applying force from front to back may cause more lateral fractures. Rib fractures are practically never seen after resuscitative efforts in children. Fractures of the first rib are highly suggestive of abuse. Scintiscans are ideal for detecting rib fractures (Figure 17.16).

Hand Fractures

Except for fractures of the distal phalanx of the fingers from closing doors (but hardly ever automobile doors), hand and feet fractures are quite rare in infants and children and highly suspicious for abuse[32] (Figure 17.17). They were found in 4% of the series of Klineman et al.[33]

Clavicle Fractures

Clavicular fractures are the most common perinatal fracture, occurring typically in the midshaft. Such fractures are practically never seen in child abuse although injuries to the lateral end of the clavicle may be seen as a component of shaking.[34]

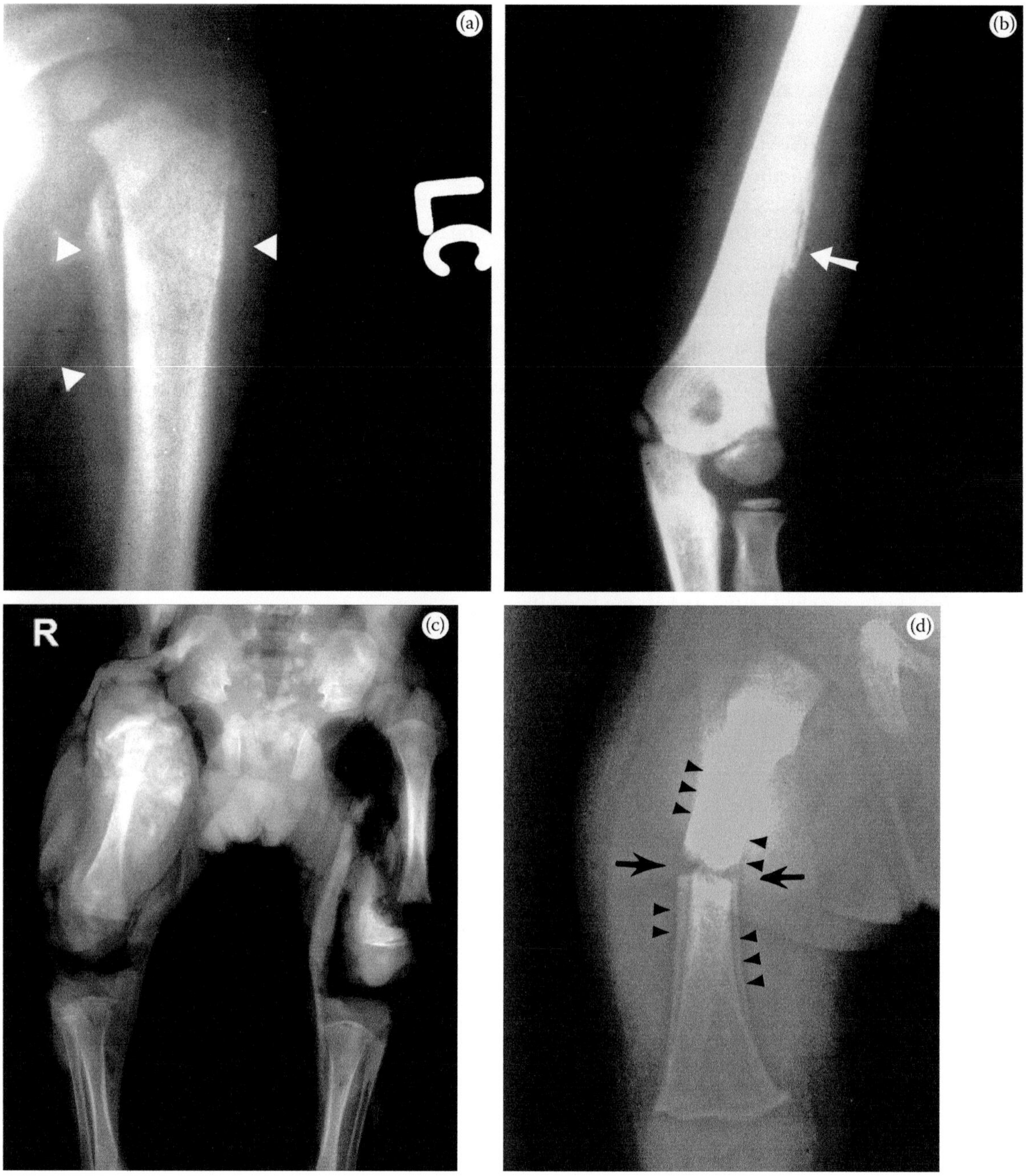

FIGURE 17.12 (a) Calcifying subperiosteal hematoma without fracture due to separation of periosteum from bone by twisting or pulling. (b) Localized calcified hematoma from direct blow. (c) Autopsy radiograph of fatally abused infant. Massive calcification of hematoma/callus on the right, around and below an epiphyseal dislocation, with periosteal stripping. Fresh spiral fracture on the left. (d) Evidence of injuries of different age on a single radiograph. The fresh transverse fracture is through laminar periosteal reaction from an earlier torsion injury. Periostitis of the newborn is excluded because the opposite femur has none.

Scapular Fractures

Fractures of the scapula are highly suspicious for child abuse and usually involve either the blade of the scapula or, more commonly, the acromion. Care must be taken to differentiate a true fracture from an ununited apophysis (Figure 17.18).

Rare Fractures

A high specificity for child abuse is found in fractures of the sternum and the spinous processes of the spine.[30] Vertebral body fractures or subluxations are rare. This author has seen only one instance of lateral spine subluxation in 57 years of practice and that resulted from the child being swung against

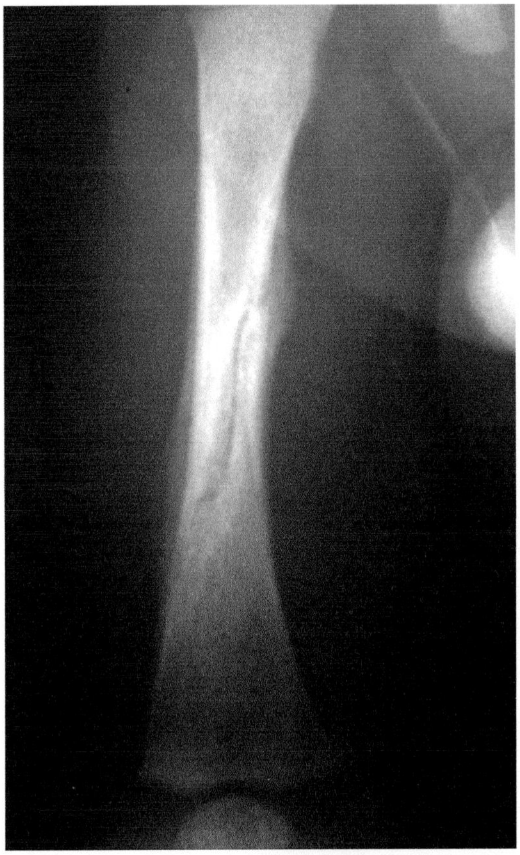

FIGURE 17.13 Spiral fracture of the femur from torsion with thickened "swedged" periosteal thickening at the proximal end of the fracture; thick unilaminar periosteal healing elsewhere. This fracture is several weeks old.

a wall (Figure 17.19). Cervical injuries are slightly more common.

Multiple Fractures and Fractures of Different Ages

Again, first mentioned by Caffey,[9] multiple fractures and fractures in various stages of healing have a high specificity for child abuse if there is an absent or inconsistent history of trauma (Figures 17.20 and 17.21). In a study of 165 fractures in 31 infants who died of child abuse,[33] all but 2 had at least 1 healing fracture present; 36 fractures were acute and 13 were of indeterminate age.

Confusing Bone Lesions of Nontraumatic Origin

A number of congenital disease or disease processes may simulate to some extent the bony injuries due to intentional trauma.[32,35,36] Most of these "resemblences" are pretty far-fetched but may be confusing to the inexperienced observer. They will be familiar to most pediatric radiologists. Any radiologist serving as an expert in a court case involving child abuse should study them because opposing attorneys are likely to throw them up as substitutes for abuse. The most popular entity for this purpose is osteogenesis imperfecta. Others include congenital syphilis, ricketic conditions, Caffey's disease, leukemia, prostaglandin E therapy to keep the ductus open in congenital heart disease, Menkes syndrome (kinky-hair disease), neuroblastoma, metastases, vitamin A intoxication, scurvy, osteomyelitis, methotrexate therapy (uncommon now), Larsen's Syndrome, arthrogryposis, myelodysplasia, congenital indifference to pain (very rare), Schmid-like metaphyseal chondrodysplasia, Dilantin® therapy, and normal variants. We have already spoken of physiologic periosteal new bone. One may also see some spurring and cupping of metaphyses in healthy infants during the early months of life. Fractures of the extremities during childbirth, especially breech deliveries or versions of transverse lies may simulate the fractures of child abuse but are rarely seen in these days of high Caesarian rates. Equally rare is the bucking or "ping-pong ball" fracture, usually of the parietal bone, formerly most often seen associated with forceps deliveries (Figure 17.22).

SKULL AND FACIAL FRACTURES

Linear skull fractures (Figure 17.23) are not highly suggestive of child abuse. Accidental head injuries in children are not often serious. Falls from baby chairs or tables, sofas, and beds rarely cause a linear fracture and this is usually not associated with intracranial damage. There is an extensive literature on the results and significance of falls from various heights onto different surfaces.[22,37–42] Again, experts going to court with child abuse cases should familiarize themselves with this literature, as it is almost always brought up if there is any head injury involved. If one can draw a generalization from the various reports it would be that head injuries resulting from falls from beds or sofas or down stairs yield a rare skull fracture and relatively minor trauma. Falls from extreme heights, falls onto extremely hard surfaces, or high-velocity impact injury provide the opportunity for serious injury and usually have a corroborative history. Children who fall from heights from 10 ft to up to 5–7 storeys rarely die from those falls and, in fact, rarely sustain life-threatening injuries. Consequently, serious intracranial injuries and complex fractures of the skull resulting from "short falls" or accidental "bumping" strongly suggest intentional trauma (Figure 17.24). Accidental fractures of facial bones are extremely unlikely in infants and children without a history of massive trauma (Figure 17.25).

INTRACRANIAL INJURIES

Intracranial injury is frequently associated with abusive head trauma and includes both subarachnoid and subdural hemorrhage, intracerebral and intracerebellar hemorrhage, and massive edema (Figure 17.26). Combinations may exist,

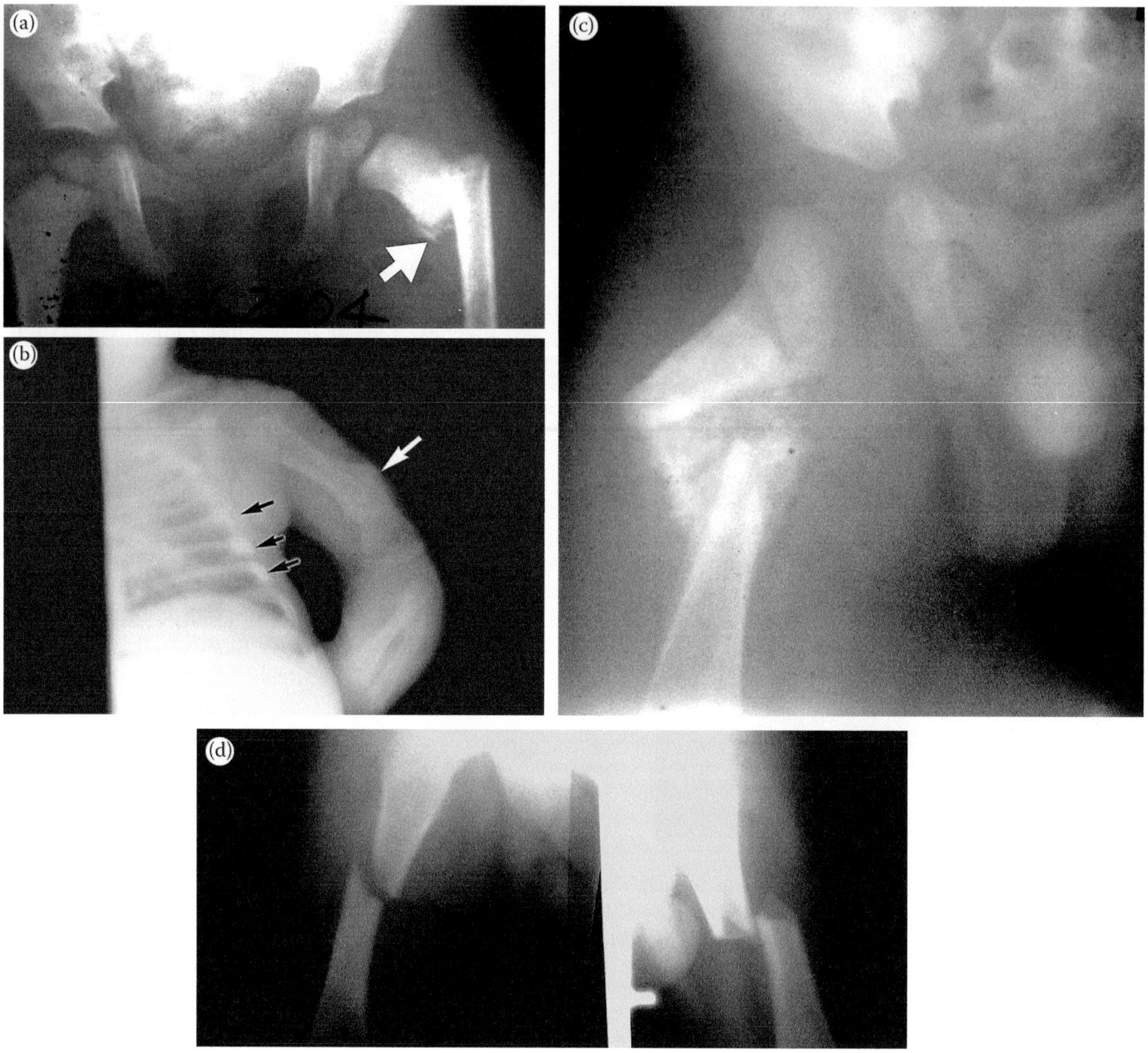

FIGURE 17.14 (a) Displaced transverse fracture of the proximal femur in a nonwalking child. (b) Healing displaced transverse fracture of the femur in a 4-month-old abused infant. (c) An almost completely healed transverse fracture of the humeral diaphysis with residual angulation (large arrow). There are lateral rib fractures as well (small arrows). (d) Bilateral transverse femoral fractures in a nonwalking abused child.

of course. Klineman[37] suggests that any case involving a complex skull fracture or a neurologic finding occurring in an infant or child after a fall from a height reported to be 90 cm (2.9 ft) or less should be regarded as possible abuse.

Radiographically demonstrated severe or complex skull fracture should be followed up with CT or MRI immediately (Figure 17.27). Even without skull fracture, CT or MRI is recommended when child abuse is strongly suspected from other findings.

SHAKING INJURIES

Guthkelch is credited with first linking subdural hematoma to shaking forces—a situation we now recognize as the "Shaken Baby Syndrome."[32,43] However, that claim of primacy may belong to Weston, who described three instances of subdural hematoma in infants who had been violently shaken.[30,44]

The head of the infant or child is large and heavy relative to the rest of its body, and sits at the end of a relatively long, narrow, and weak neck. When the baby is grasped by the chest, shoulders, or arms, and shaken, the head can reach high levels of translational and rotational velocity. The brain inside the calvaria will move at a different speed and become asynchronous with the bony envelope. Tearing of bridging vessels can ensue, producing subdural hemorrhage. It has been shown that the rotational velocity or acceleration is causative. Translational motion alone (as in a woodpecker) will not produce the injury. Thus one can see massive intracranial hemorrhage and brain swelling with spreading of the intracranial suture, unassociated with skull fracture or soft tissue bruising (Figure 17.28). However, shaking is also frequently associated with direct injury to the head along with

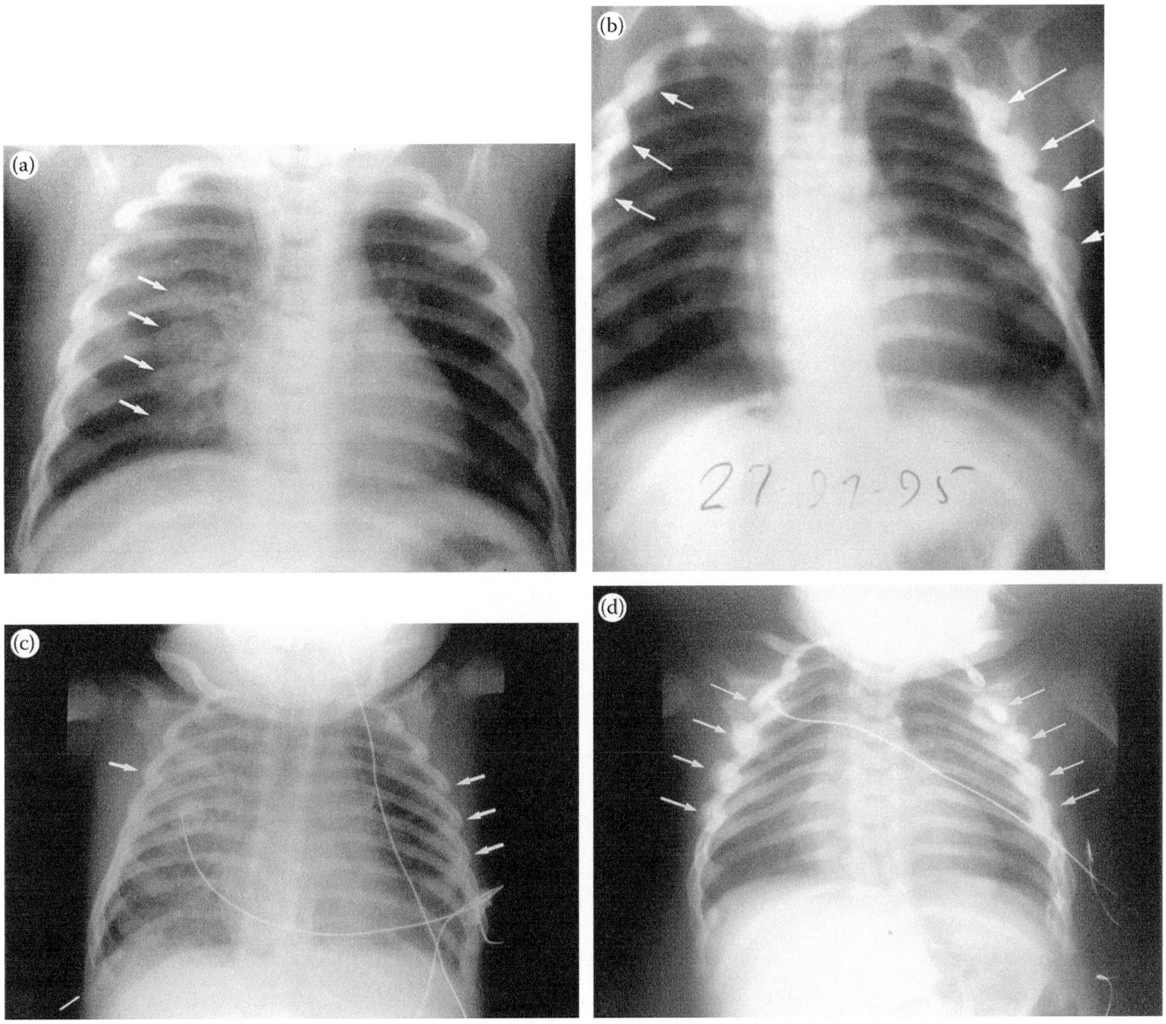

FIGURE 17.15 Rib fractures. (a) Typical healed posterior fracture from AP compression. (b) Healed lateral rib fractures. (c) Acute rib fractures (arrow) were missed and the baby sent home; the baby then returned, (d) with multiple bilateral healing fractures (note hazy callus surrounding ribs). At this time the infant also had a skull fracture. (c and d Courtesy of Dr. Damien Grattan-Smith.)

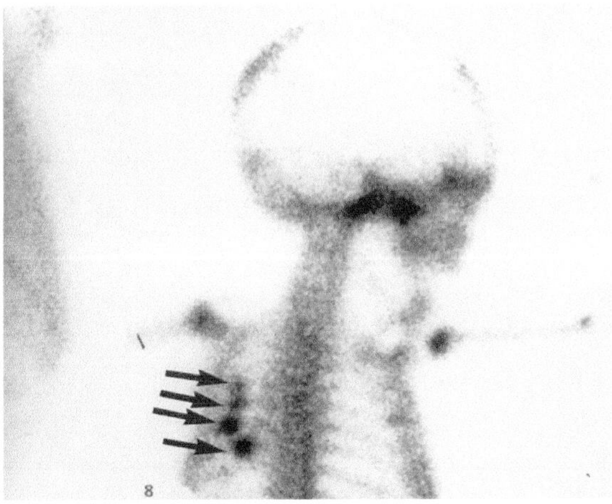

FIGURE 17.16 Nuclear bone scan is ideal for early detection of rib fractures (arrows). Note normal high uptake at growth plate in the shoulder. Symmetrical, bilateral metaphyseal injuries would be difficult to appreciate here. The infant is slightly rotated so the proximal humeral uptake is slightly asymmetrical. (Courtesy of Dr. Damien Grattan-Smith.)

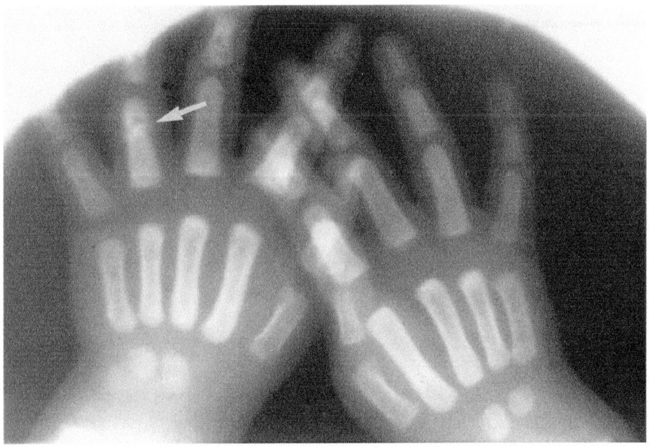

FIGURE 17.17 This 22-month-old Navajo is the same patient as seen in Figures 17.9 and 17.11a. The broken proximal phalanx (arrow) is further evidence of abuse. When social workers went to the home, they found a brain-damaged abused sibling.

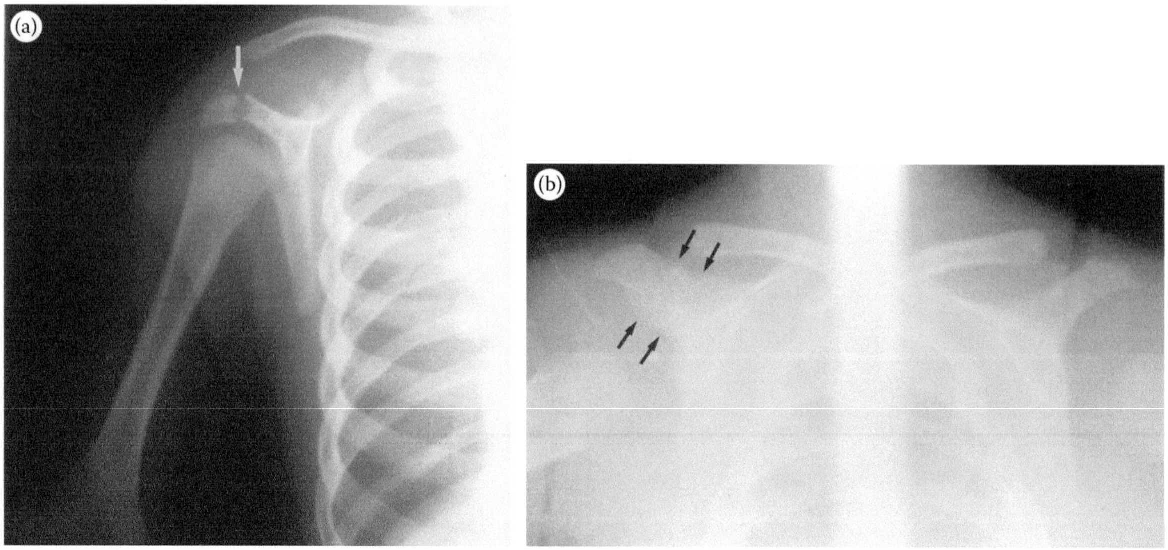

FIGURE 17.18 (a) Acute fracture of the acromial process of the scapula. (b) Healed fracture of the right acromion and normal opposite side.

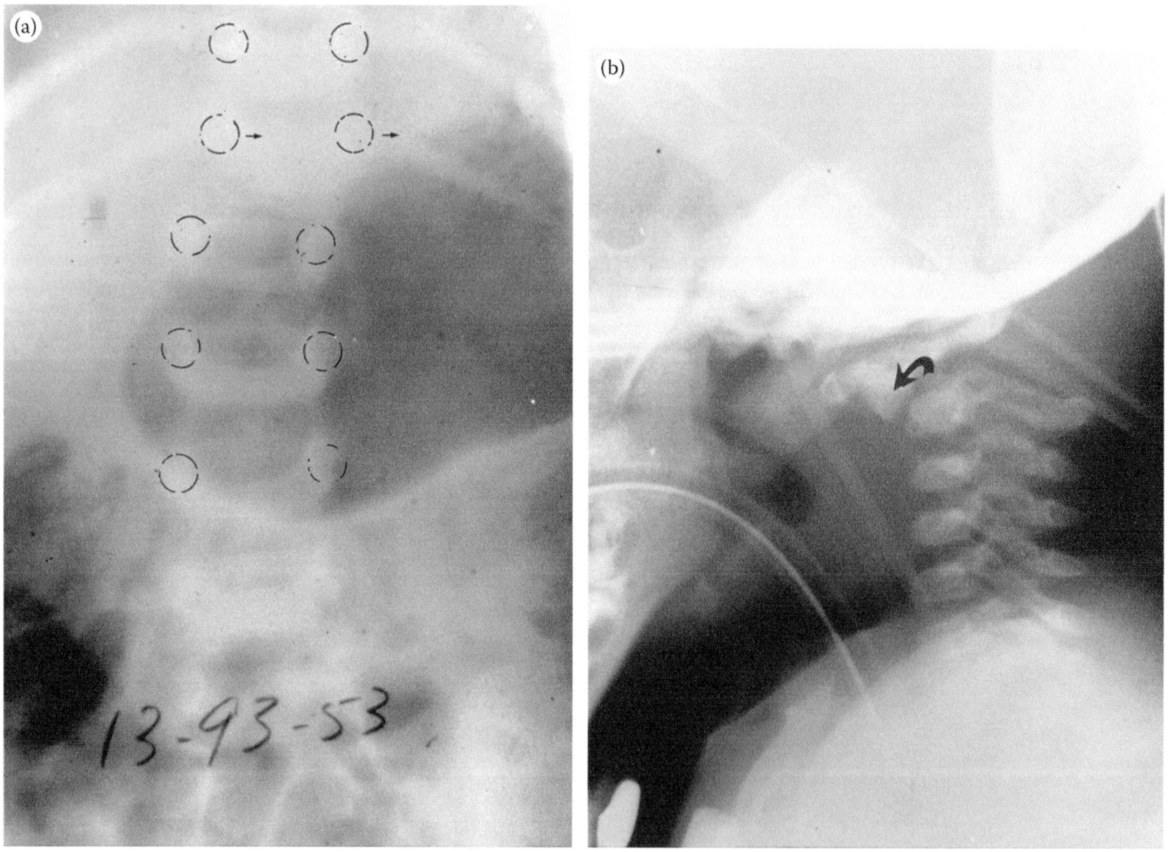

FIGURE 17.19 (a) There is a lateral dislocation of T-12 on L-1. The pedicles are marked so the malalignment can be better appreciated on this low-contrast radiograph. This 2-year-old was brought in because she wouldn't "pass her water." She also had a skull fracture. She had been slammed against a wall. (b) Another infant, who was hit in the back of the head with a blunt weapon, sustained a separation of the odontoid process from the body of C-2 with anterior subluxation of C-1. Note separation of the spinous processes at the level of injury.

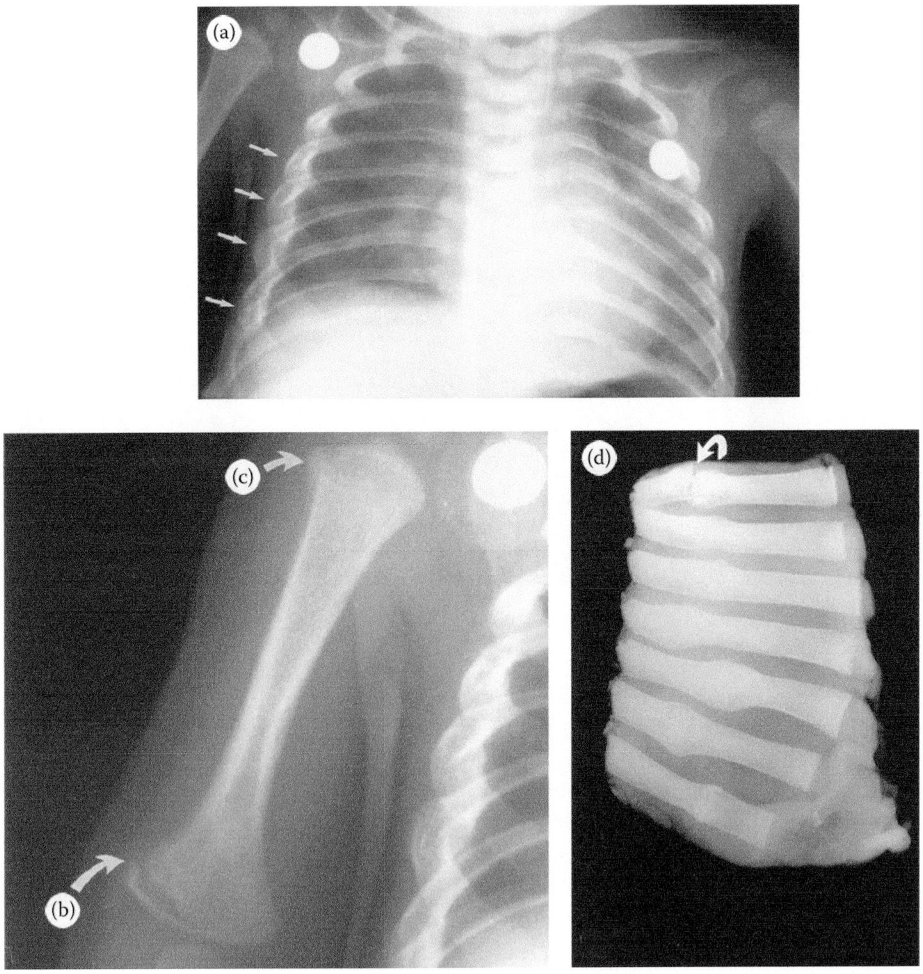

FIGURE 17.20 Multiple injuries in various stages of healing. (a) Old rib fractures (arrows), (b) new "bucket-handle fracture of distal humerus (long arrow), (c) "corner fracture" proximal humerus, and (d) postmortem faxitron study of chest wall preparation shows a new rib fracture (arrow) through one of the old healed fractures. (Courtesy of Dr. J.C.U. Downs.)

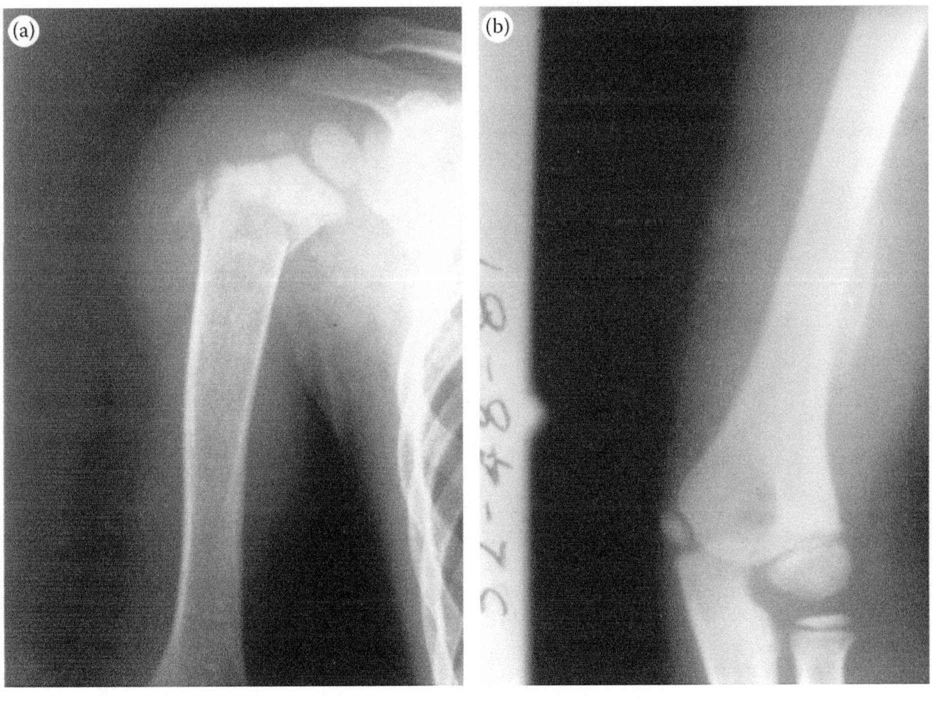

FIGURE 17.21 Multiple injuries in a 7-year-old girl who said her father beats her when he gets drunk. (a) Old Salter 2 fracture of the proximal humerus. (b) Myositis ossificans in the left arm. (c) and (d) Bilateral epiphyseal injuries of the radius and ulna at the wrist with resorption of bone.

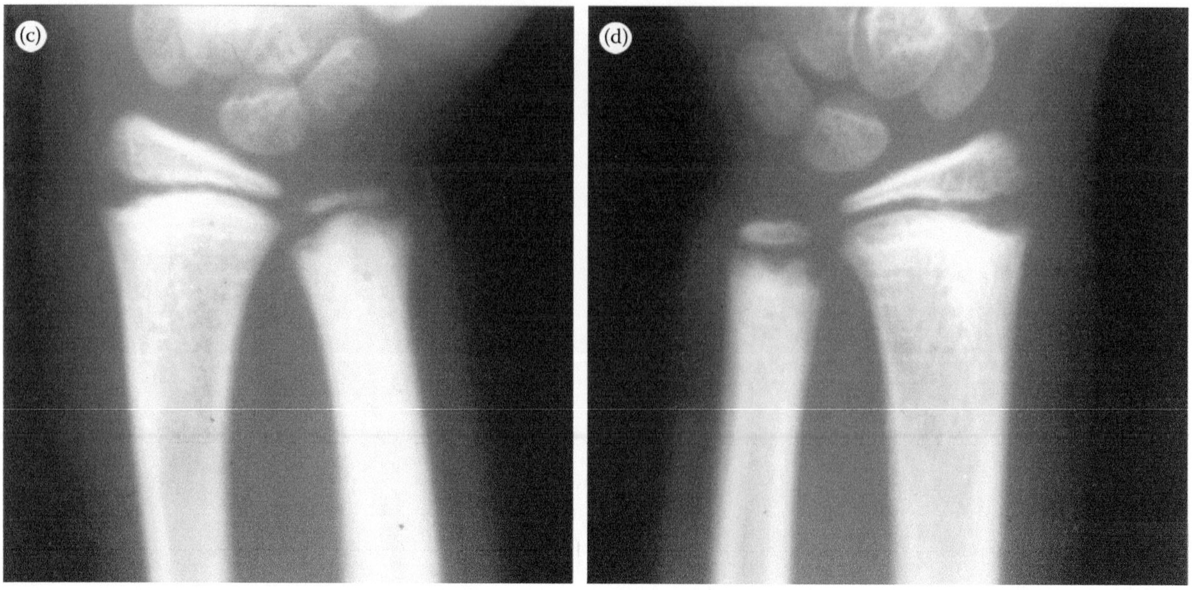

FIGURE 17.21 (*Continued*)

the shaking. There has been controversy as to whether intracranial injuries can be accounted for by shaking alone, without a direct impact blow to the head, but this possibility is accepted by most authorities at the present time.[22,45,46] Of interest are reports that for sometime the Israeli Security Service included the techniques of shaking the upper torso of suspects during interrogations. This occasionally produced intracranial lesions similar to those found in the Shaken Baby Syndrome. (The practice has since been outlawed by the Israeli Supreme Court.[47])

VISCERAL TRAUMA

Thorax

Although the thoracic bony cage is one of the most frequent sites of abusive injuries, the thoracic viscera are rarely reported as being involved. Occasionally, one sees some pleural fluid or blood associated with rib fractures. Pulmonary contusions are rarely reported. Injuries to the heart and mediastinum from a blunt force are extremely rare as a manifestation of child abuse.[46]

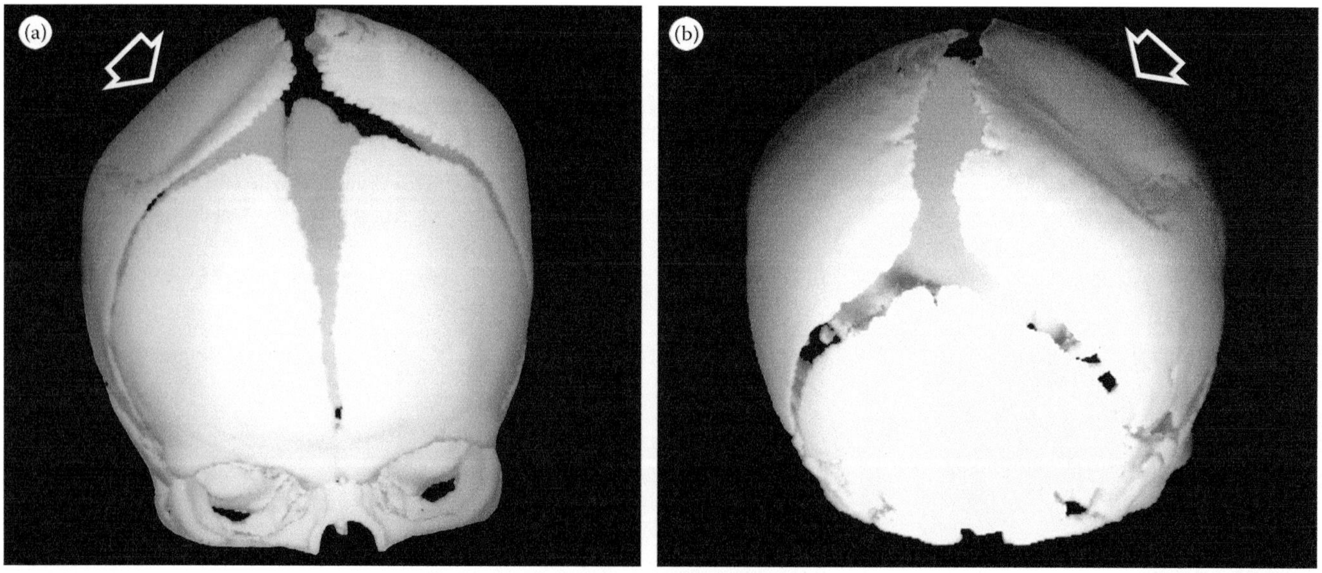

FIGURE 17.22 Intrapartum injury. The ping-pong ball fracture is a smooth, rounded depression in the thin elastic newborn skull without a fracture line. Frequently seen after forceps extraction, when they were common. This newborn's head was already in the birth canal when a cesarean section was begun. The obstetrician pushed it back from below into the uterus. (a) Right parietal ping-pong fracture in frontal view. (b) Occipital view, 3-D CT reconstruction.

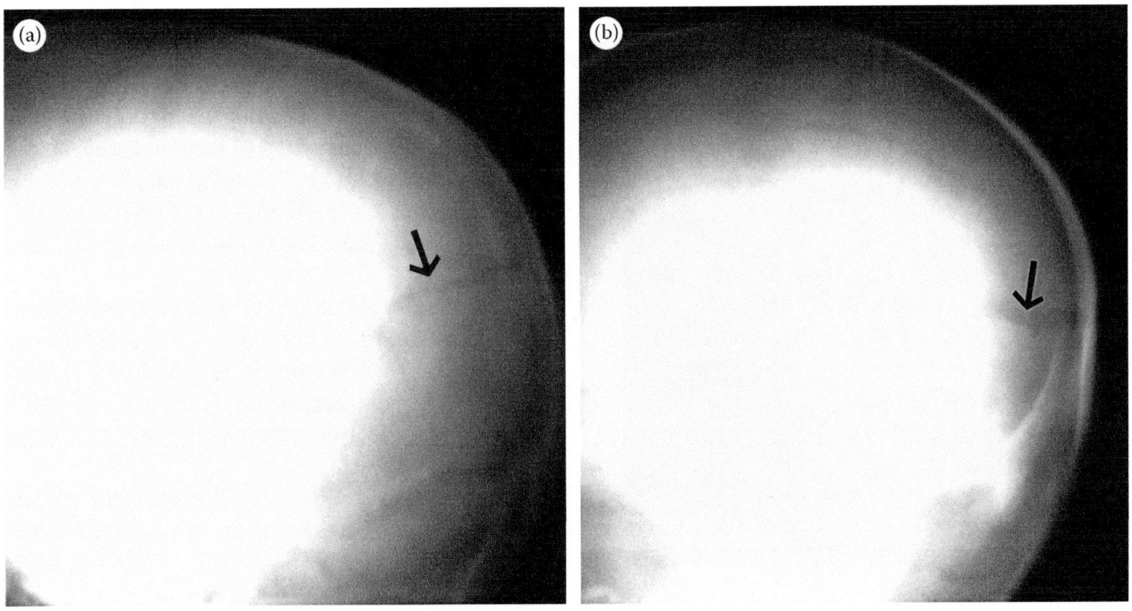

FIGURE 17.23 (a and b) Simple parietal skull fracture.

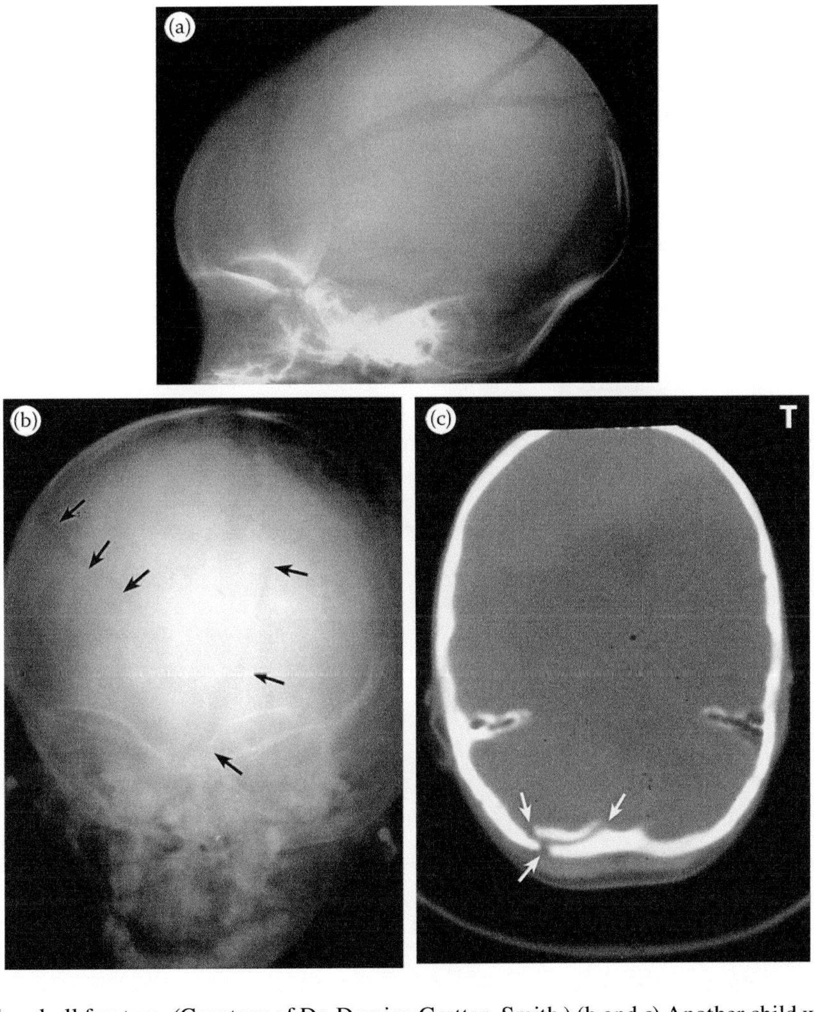

FIGURE 17.24 (a) Complex skull fracture. (Courtesy of Dr. Damien Grattan-Smith.) (b and c) Another child with complex (black arrows) and depressed (white arrows) skull fractures. (The boyfriend claimed he accidentally hit the child's head against the door-frame while carrying her from room to room.)

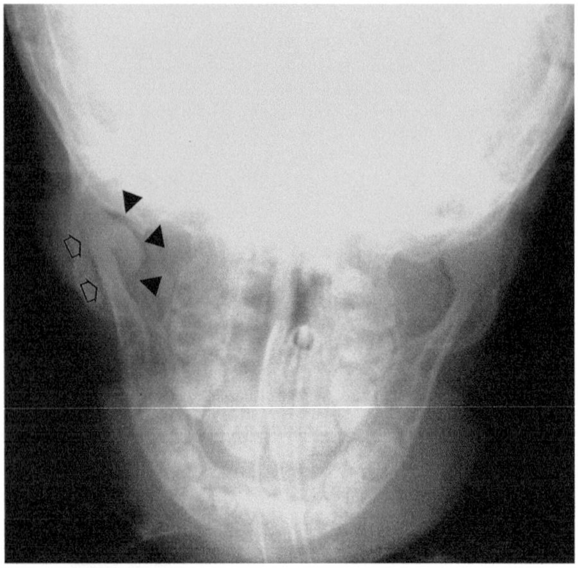

FIGURE 17.25 Fractured right mandibular neck (arrows) with dislocated mandibular head (arrowheads). (Courtesy of A. M. Kroman and Dr. S. A. Symes.)

ABDOMEN

On the other hand, the abdomen is a fairly frequent site of abusive trauma, particularly after children become ambulatory. Blunt force in the form of a fist or knee can cause severe damage to intra-abdominal viscera and is associated with a high mortality rate. External physical evidence of trauma is usually lacking.[49] In the absence of history of a major traumatic episode, such as a vehicular accident, abuse should be suspected when these injuries are seen in young children.

Intramural hematoma of the duodenum is the most common of these injuries from child abuse.[48] Frequently, the child will be brought in with vomiting as the duodenum becomes obstructed. The blow to the epigastrium compresses the retroperitoneal portion of the duodenum against the spine to produce the injury. The radiographic findings are typical on gastrointestinal (GI) series or CT (Figures 17.29 and 17.30). There may be associated injuries in other areas or organ systems.

Lacerations of the liver and pancreas may result from similar trauma. Lacerations of these organs have already been shown in Chapter 3, Figure 3.3. A pseudocyst may develop in the traumatized pancreas and can be demonstrated by GI series, ultrasound, or CT[46,47] (Figures 17.31 and 17.32).

Perforation of the small bowel, stomach, or colon can result from abusive blunt force of the abdomen although this is not quite as common as the preceding entities[49,50] (Figure 17.33).

Despite its vulnerability in other settings, the spleen is not often injured in child abuse. The kidneys, ureters, and bladder also are usually spared.[48]

SOFT TISSUES

Often, children with radiographically demonstrable injuries of child abuse will have associated soft tissue findings, particularly bruises and burns. These usually are not demonstrable radiographically, of course, but may be apparent to the radiographer who should alert the radiologist to the finding (Figure 17.34).

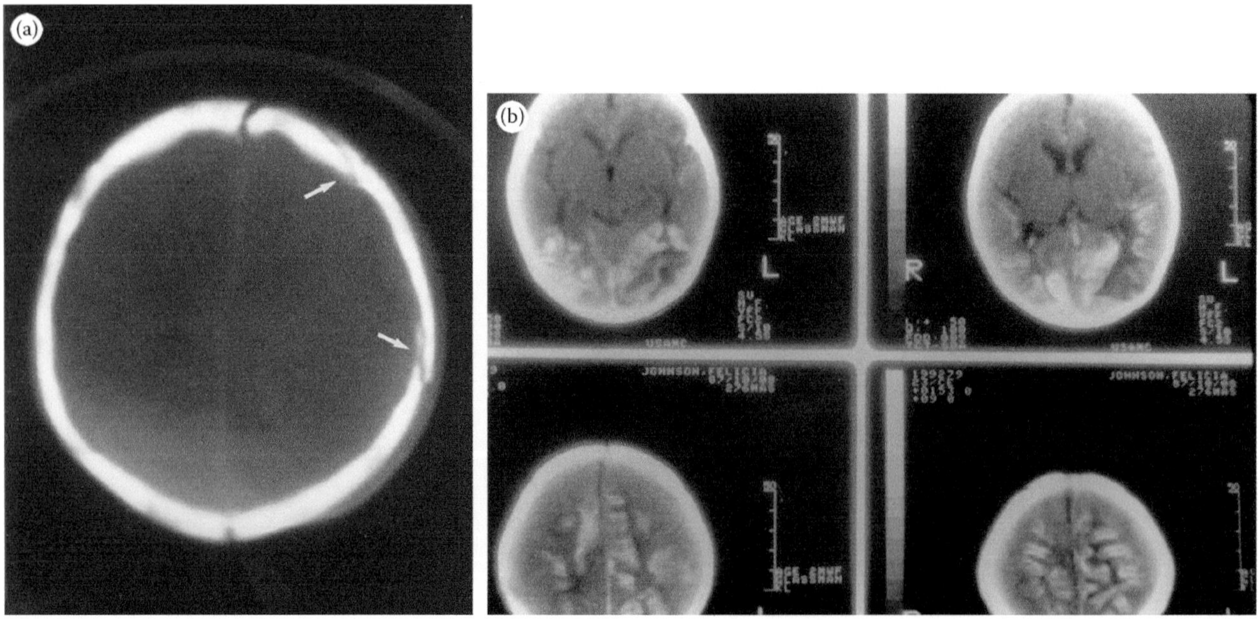

FIGURE 17.26 Same baby as in Figure 17.24 with fatal brain hemorrhage. (a) CT with bone windowing shows culvarial fractures (arrows). (b) Brain windowing shows extensive internal and extra-axial hemorrhage.

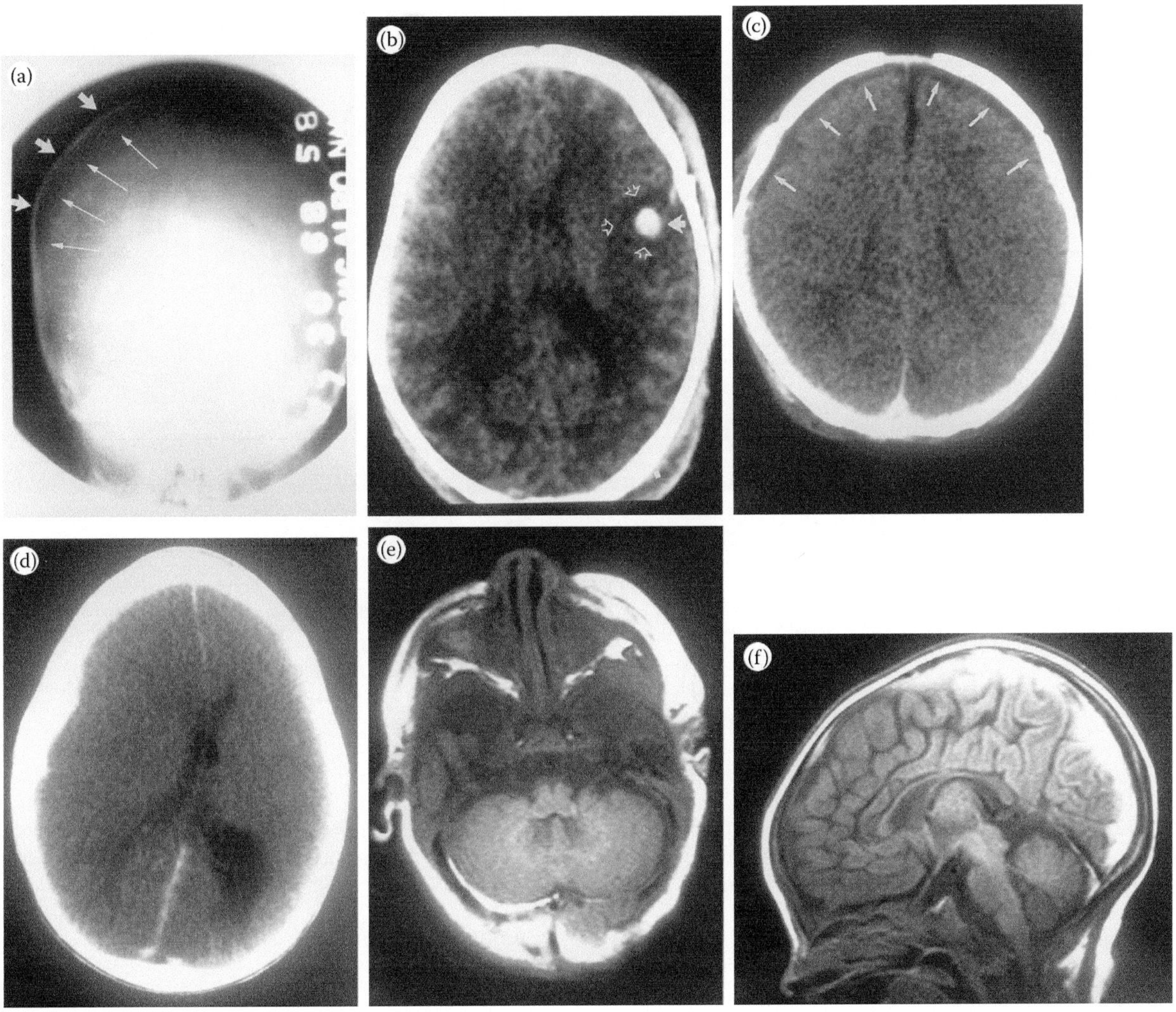

FIGURE 17.27 Intracranial injuries. (a) Acute subdural hematoma shown by angiography as the dark space between the bony calvaria (short arrows) and the vascularized cortex (long arrows). (b) Hemorrhagic contusion (arrow) with surrounding edema (open arrows) on unenhanced CT. (c) Subdural hematoma around frontal lobe (arrows) on unenhanced CT. (d) Unenhanced CT shows blood in interhemispheric sulcus, extra-apical blood on left, massive edema of right hemisphere with shift to the left, obstruction of left lateral ventricle at the foramen of Munro, and cisterns. (e and f) MRI shows subdural blood around the cerebellum and temporal, frontal, and occipital lobes.

VIRTUAL AUTOPSY IN INFANTS AND CHILDREN

It is difficult to assign primacy in postmortem, preautopsy investigations with CT. The Japanese, at the Tsukuba Medical Center Hospital, performed more than 500 postmortem CT examinations between February 1985 and July 2003.[51] The age range of those sub jects is unknown to us.

Ros and colleagues in 1990[52] appear to be the first to perform postmortem, preautopsy MRI in the pediatric age group. They studied six cadavers including three stillborn infants ranging from 29 to 42 weeks gestational age and one 14-day-old live birth. They found autopsy superior in detecting small abnormalities, but MRI superior in detecting gross cranial, pulmonary, and vascular abnormalities. Additionally, MRI was superior to detecting air and fluid in potential body spaces. Thus, this study, although very small, was predictive of future comparative studies in this area.

Between 1990 and 1993, Hart et al.[53] studied 11 cases of unexplained death or suspected child abuse in children 2 years of age or younger. All received preautopsy MRI of the head. They found MRI to have advantages over autopsy and CT in some areas: mastoid fluid, focal cortical and shearing injuries, small tears that might be confused with an artifact of removal or brain cutting (Figure 17.35), and intraparenchymal hemorrhage. However, preautopsy imaging was found an advantageous supplement to the autopsy by directing the attention of the pathologist to areas of special interest or abnormality.

Subsequent experience has confirmed the initial conclusion that postmortem, preautopsy CT and MRI each has advantages in detecting certain abnormalities and that combining the two sectional imaging modalities with conventional autopsy can increase total yield.

Postmortem, preautopsy MRI is much less common worldwide than postmortem CT, mostly because of problems

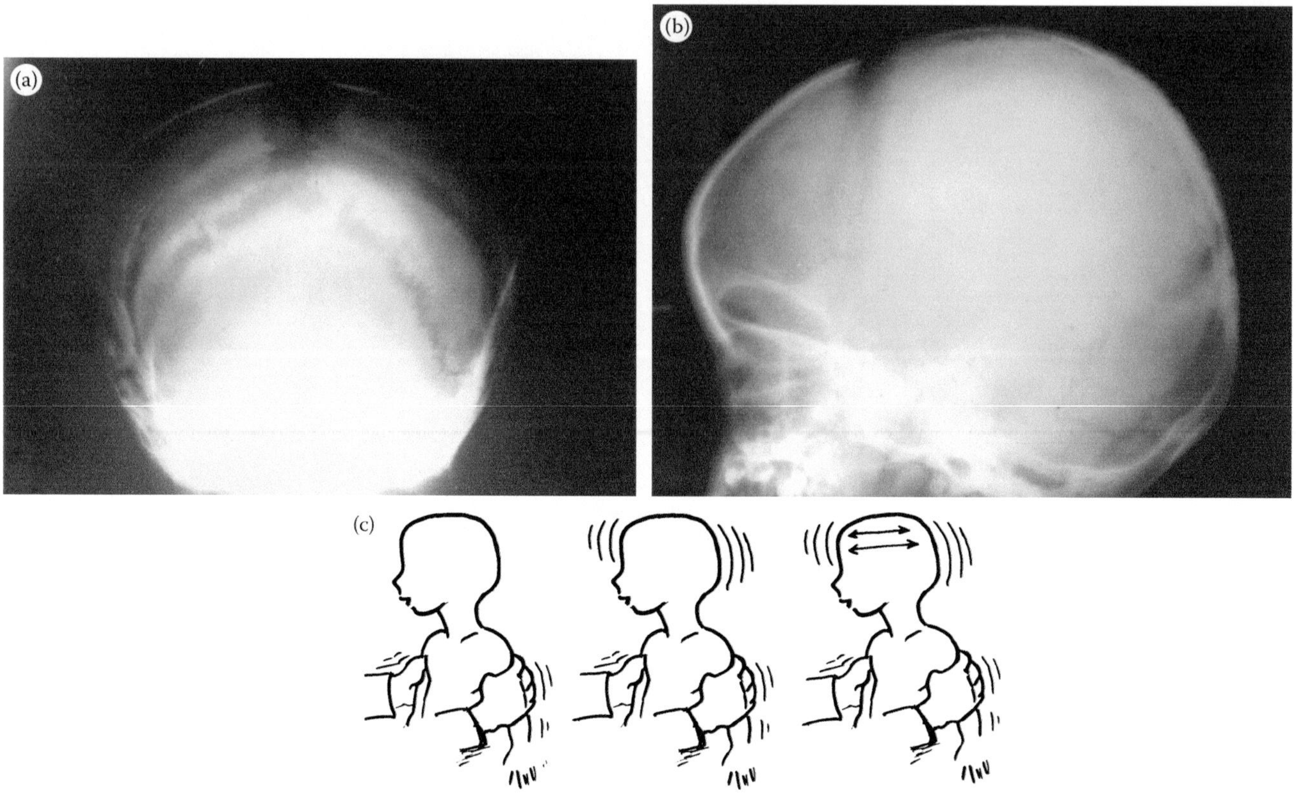

FIGURE 17.28 (a and b) Shaken baby syndrome. No fractures, but marked sutural spreading from subdural hematoma and cerebral edema. This baby died within minutes of admission to hospital. (c) Head and brain eventually oscillate asynchronously with both translational and rotational motion, shearing vessels and causing intracranial bleeding.

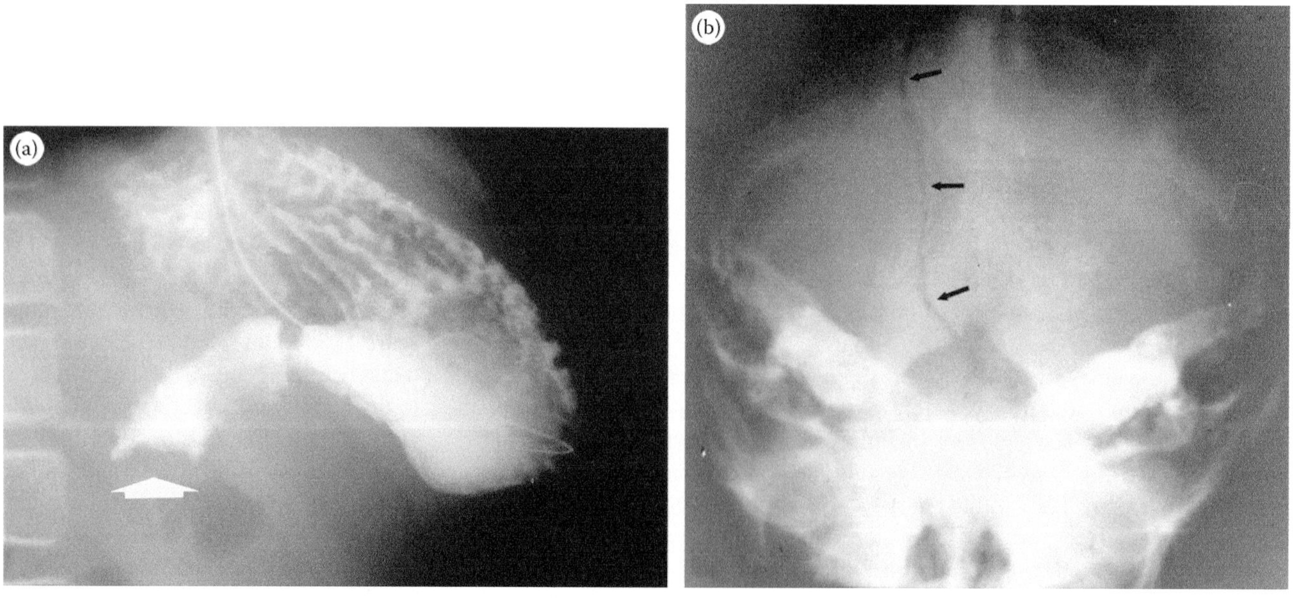

FIGURE 17.29 (a) GI series shows obstruction of barium (broad arrow) at the second portion of the duodenum by intramural duodenal hematoma. "Pad sign" on stomach suggests pancreatic edema or hemorrhage. (b) Child also had an occipital skull fracture.

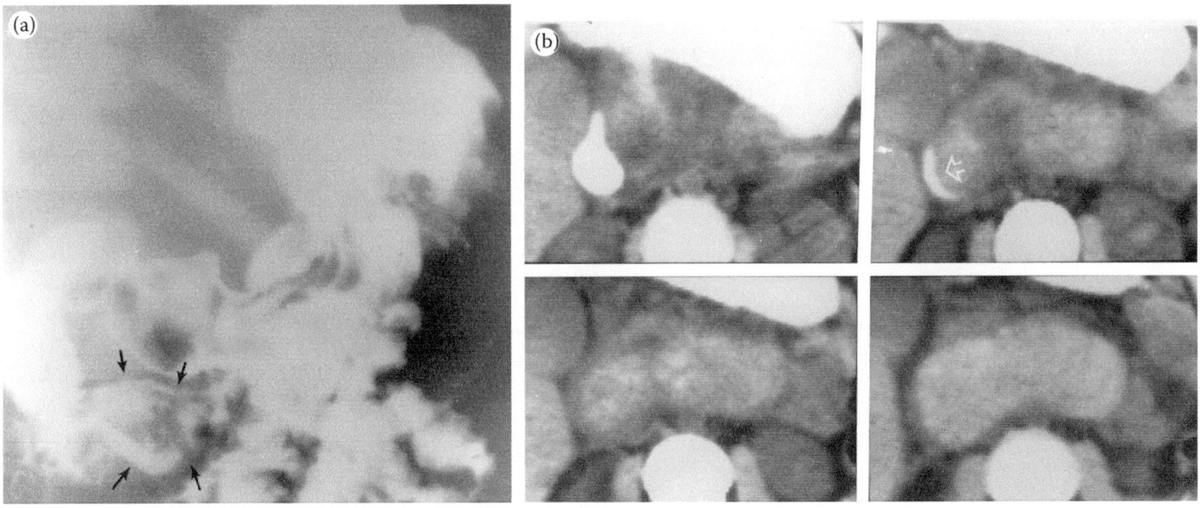

FIGURE 17.30 (a) GI series shows widening and incomplete filling of the third portion of the duodenum (arrows) due to intramural hematoma. (b) CT scans showed mottled densities of blood and contrast material in duodenum. Open arrow shows narrow channel remaining open for flow of contrast medium. (From Hughes, J. J. and Brogdon, B. G., *J. Comput. Tomogr.*, 10, 231, 1986. With permission.)

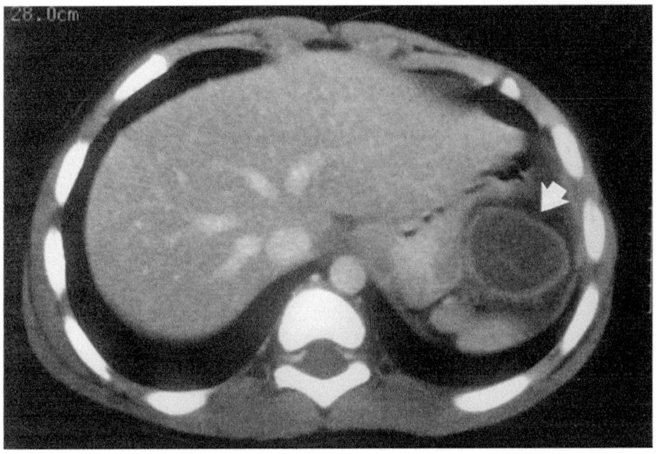

FIGURE 17.31 Enhanced abdominal CT shows pseudocyst in tail of pancreas (arrow) from abusive trauma to abdomen. (Courtesy of Dr. Damien Grattan-Smith.)

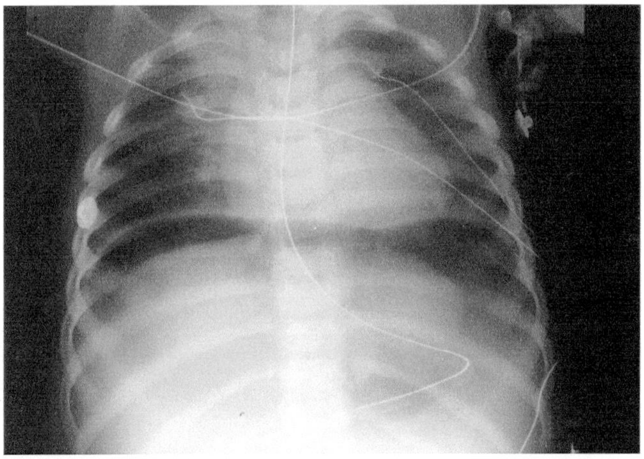

FIGURE 17.33 Free intraperitoneal air under diaphragm from traumatic gastric rupture. (Courtesy of Dr. Damien Grattan-Smith.)

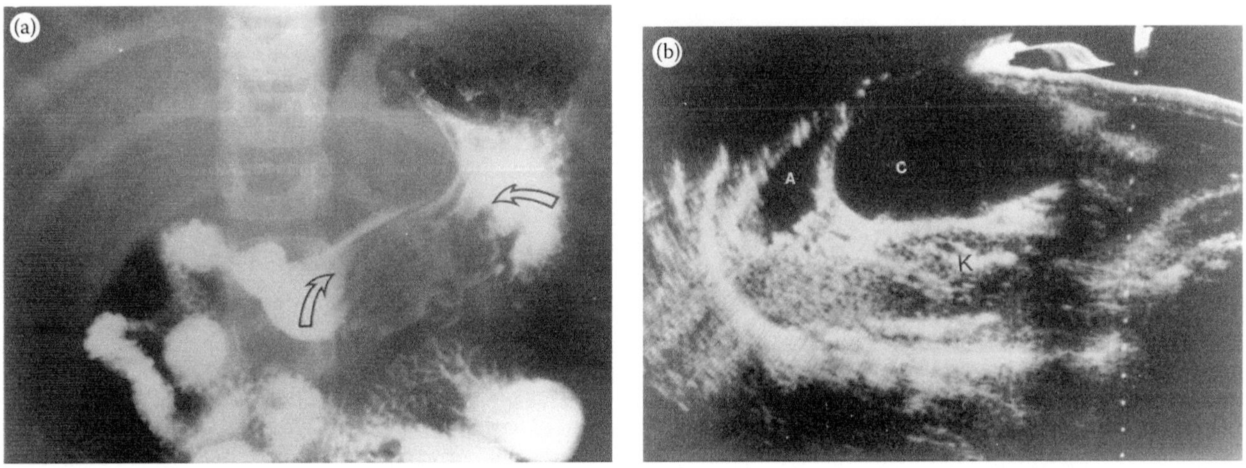

FIGURE 17.32 (a) GI series shows impression on stomach (arrows) from pancreatic pseudocyst in a vomiting 4½-year-old child. (b) Ultrasonography shows anechoic mass (C) anterior to left kidney (K) and ascites (A). (From Kleinman, P. K., Raptopoulos, V. D., and Brill, P. W., *Radiology*, 141, 393, 1981. With permission.)

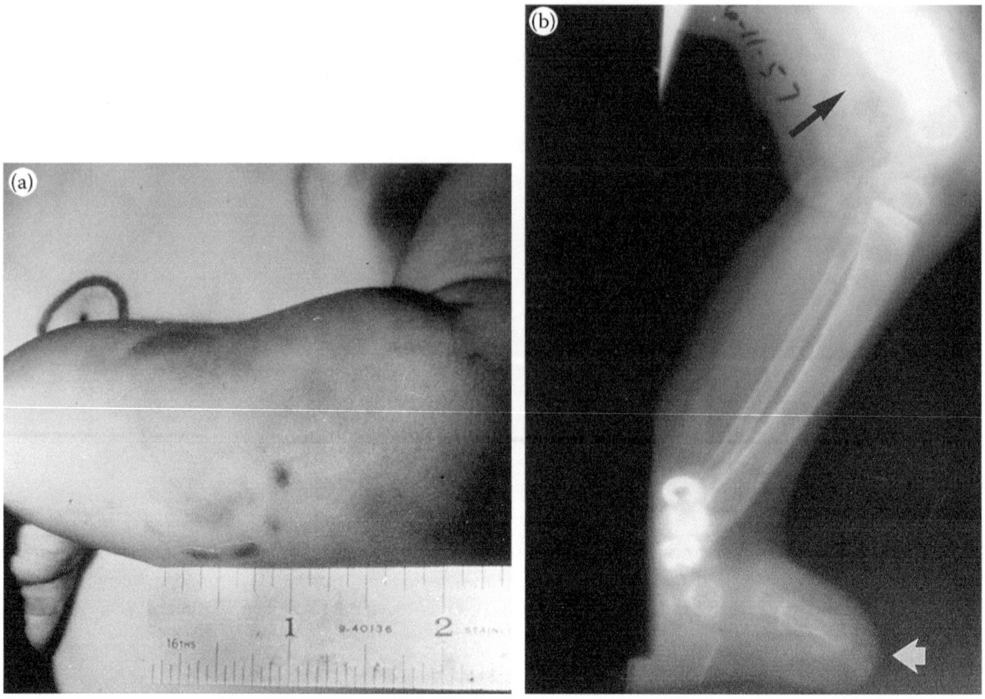

FIGURE 17.34 (a) Bruises and burns on the upper extremity of child shown in Figures 16.22B–E. (b): 11-month-old girl from hippie commune. Fracture of distal femur is evident (black arrow). She literally had her feet held to the fire, and many of her toes were burned off (white arrow).

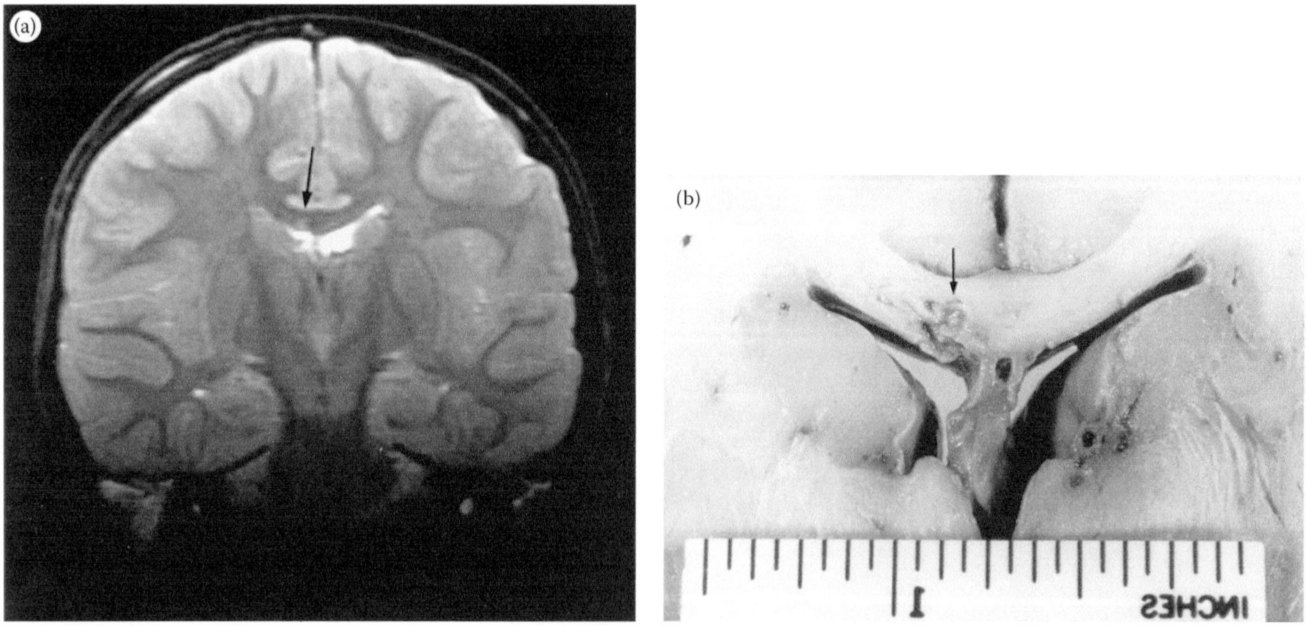

FIGURE 17.35 This 19-month-old boy had suffered cardiopulmonary arrest during transport to the hospital after reportedly being beaten. (a) a coronal T2-weighted image from post-mortem MRI shows a small focus of increased signal intensity in the body of the corpus callosum (arrow). (b) careful sectioning of the brain at the corresponding location disclosed the tear (arrow). (The photograph has been reversed to correspond with radiographic convention used in MRI.) (From Hart, B. L., Dudley, M. H., and Zumwalt, R. E., Post-mortem cranial MRI and autopsy correlation in suspected child abuse, poster presentation at 1994 RSNA annual meeting, Chicago. With permission of the Radiological Society of North America.)

of cost and availability, although in Japan MRI is disfavored primarily because it is "too slow" (S. Shiotani, personal communication).

Most investigations and utilization of virtual autopsy and anthropology have been done in adults (see Chapter 8 and Section VII), but have been useful in evaluating infants and children for abuse, sudden death, accidental trauma, skeletal age, air embolism, and intracranial and soft-tissue change before and after death.[51–59]

REFERENCES

1. Silverman, F. N., Unrecognized trauma in infants, the battered child syndrome, and the syndrome of Ambriose Tardieu: Rigler Lecture, *Radiology*, 104, 337, 1972.
2. Caffey, J., The parent–infant traumatic stress syndrome: (Caffey–Kempe syndrome), (battered baby syndrome), *Am. J. Roentgenol.*, 114, 217, 1972.
3. Merten, D. F., Cooperman, D. R., and Thompson, G. H., Skeletal manifestations of child abuse, in *Child Abuse: Medical Diagnosis and Management*, Reece, R. M., Ed., Lea & Febiger, Philadelphia, 1994, chap. 2.
4. Evans, K. T., and Knight, B., Forensic radiology, *Br. J. Hosp. Med.*, 17, 17, 1986.
5. Silverman, F. N., Presentation of the John Howland Medal and Award of the American Pediatric Society to Dr. John Caffey, *J. Pediatr.*, 67, 1000, 1965.
6. Caffey, J., Multiple fractures in long bones of children suffering from chronic subdural hematoma, *Am. J. Roentgenol.*, 56, 163, 1946.
7. Sherwood, P., Chronic subdural hematomas in infants, *Am. J. Dis. Child.*, 39, 980, 1930.
8. Ingraham, F. D., and Heyl, H. L., Subdural hematoma in infants and childhood, *J. Am. Med. Assoc.*, 112, 198, 1939.
9. Caffey, J., Some traumatic lesions in growing bones other than fractures and dislocations, clinical and radiological features. The Mackenzie Davidson Memorial Lecture, *Br. J. Radiol.*, 30, 225, 1957.
10. Caffey, J., Significance of the history in the diagnosis of traumatic injury to children. Howland Award Address, *J. Pediatr.*, 68, 1008, 1965.
11. Neuhauser, E. B. D., Discussion of paper by F. N. Silverman, *Am. J. Roentgenol.*, 69, 413, 1953.
12. Silverman, F. N., The roentgen manifestations of skeletal trauma, *Am. J. Roentgenol.*, 69, 413, 1953.
13. Kempe, C. H., Silverman, F. N., Steele, B. F., Droegemueller, W., and Silver, H. K., The battered child syndrome, *J. Am. Med. Assoc.*, 181, 105, 1962.
14. Kempe, C. H., Pediatric implications of the battered baby syndrome. Windermere Lecture, *Arch. Dis. Child.*, 46, 28, 1971.
15. Elmer, E., *Children in Jeopardy*, University of Pittsburgh Press, Pittsburgh, 1967.
16. Brogdon, B. G., Child abuse: The radiologists' role, *Pathologist*, 31, 134, 1977.
17. Kleinman, P. K., *Diagnostic Imaging of Child Abuse*, Williams & Wilkins, Baltimore, 1987.
18. Reece, R. M., *Child Abuse: Medical Diagnosis and Management*, Lea & Febiger, Philadelphia, 1994.
19. Kleinman, P. K., *Diagnostic Imaging of Child Abuse*, Williams & Wilkins, Baltimore, 1987, chap. 9.
20. Nolte, K. B., Esophageal foreign bodies as child abuse, *Am. J. Forensic Med. Pathol.*, 14, 323, 1993.
21. Report of the U.S. Advisory Board on Child Abuse and Neglect, A Nation's Shame: Fatal Child Abuse and Neglect in the United States, U.S. Department of Health and Human Services, Washington, DC, 1995.
22. Reece, R. M., *Child Abuse: Medical Diagnosis and Management*, Lea & Febiger, Philadelphia, 1994, chap. 1.
23. Kleinman, P. K., *Diagnostic Imaging of Child Abuse*, Williams & Wilkins, Baltimore, 1987, chap. 4.
24. Sty, J. R., and Starsbuk, R. J., The role of bone scintigraphy in the evaluation of the suspected abused child, *Radiology*, 146, 369, 1983.
25. Sty, J. R., Radiological imaging applications in forensic medicine, workshop at the *Annu. Meet. Am. Acad. Forensic Sci.*, New Orleans, February 14, 1994.
26. Kleinman, P. K., *Diagnostic Imaging of Child Abuse*, Williams & Wilkins, Baltimore, 1987, chap. 1.
27. Kleinman, P. K., Nimkin, K., Sprivak, M. R., Rayder, S. M., Madansky, D. L., Shelton, Y. A., and Patterson, M. M., Follow-up skeletal surveys in suspected child abuse, *Am. J. Roentgenol.*, 167, 893, 1996.
28. Nimken, K., Kleinman, P. K., Teeger, S., and Spevak, M. R., Distal humeral physeal injuries in child abuse: MR imaging and ultrasonography findings, *Pediatr. Radiol.*, 25, 562, 1995.
29. Kleinman, P. K., Refresher course on child abuse, *Annu. Meet. Int. Skeletal Soc., New Orleans*, October 21, 1995.
30. Kleinman, P. K., *Diagnostic Imaging of Child Abuse*, William & Wilkins, Baltimore, 1987, chap. 2.
31. Shopfner, C. F., Periosteal bone growth in normal infants: A preliminary report, *Am. J. Roentgenol.*, 97, 154, 1966.
32. Hilton, S. V. W., Differentiating the accidentally injured from the physically abused child, in *Practical Pediatric Radiology*, Hilton, S. V. W., and Edwards, D. K., III, Eds., W. B. Saunders, Philadelphia, 1994, chap. 14.
33. Kleinman, P. K., Marler, S. C., Jr., Richmond, J. M., and Blackbourne, B. D., Inflicted skeletal injury: A postmortem radiologic–histopathologic study in 31 infants, *Am. J. Roentgenol.*, 165, 647, 1995.
34. Kogutt, M. S., Swischuk, L. E., and Fagan, C. J., Patterns of injury and significance of uncommon fractures in the battered child syndrome, *Am. J. Roentgenol.*, 121, 143, 1974.
35. Brill, P. W., and Winchester, P., Differential diagnosis of child abuse, in *Diagnostic Imaging in Child Abuse*, Kleinman, P. K., Ed., William & Wilkins, Baltimore, 1987, chap. 11.
36. Brogdon, B. G., Vogel, H., McDowell, J. D., *A Radiologic Atlas of Abuse, Torture, Terrorism, and Inflicted Trauma*, CRC Press LLC, Boca Raton, FL, 2003, chap. 1, pp. 37–44.
37. Kleinman, P. K., in *Diagnostic Imaging in Child Abuse*, Williams & Wilkins, Baltimore, 1987, chap. 8.
38. Helfer, R. E., Slovis, T. L., and Black, M., Injuries resulting when small children fall out of bed, *Pediatrics*, 60, 533, 1977.
39. Reiber, G. D., Fatal falls in childhood: how far must children fall to sustain fatal head injury? Report of cases and review of the literature, *Am. J. Forensic Med. Pathol.*, 14, 201, 1993.
40. Root, I., Head injuries from short distance falls, *Am. J. Forensic Med. Pathol.*, 13, 85, 1992.
41. Weber, W., Experimentelle untersuchungen zu schädelbruchverletzungen des säuglings, *Reichsmedizin*, 92, 87, 1984.
42. Nimityongskul, P., and Anderson, L. D., The likelihood of injuries when children fall out of bed, *J. Pediatr. Orthoped.*, 7, 184, 1987.

43. Guthkelch, A. N., Infantile subdural hematoma and its relationship to whiplash injuries, *Br. Med. J.*, 2, 430, 1971.
44. Weston, J. T., The pathology of child abuse, in *The Battered Child*, 3rd ed., Kempe, C. H., and Helfer, R. E., Eds., University of Chicago Press, Chicago, 1968, 77.
45. Gilliland, M. G. F., and Folberg, R., Shaken babies—some have no impact injuries, *J. Forensic Sci.*, 41, 114, 1996.
46. Collins, K. A., and Nichols, C. A., Pediatric homicide by shaking: A ten-year prospective study, *Proc. Annu. Meet. Am. Acad. Forensic Sci.*, Washington, DC, February 21, 1997, p. 139.
47. Sontag, D. A., A strike against brutality, *Mobile Register*, September 9, 1991.
48. Kleinman, P. K., Visceral trauma, in *Diagnostic Imaging of Child Abuse*, William & Wilkins, Baltimore, 1978, chap. 7.
49. Kleinman, P. K., Raptopoulos, V. D., and Brill, P. W., Occult non-skeletal trauma in the battered child syndrome, *Radiology*, 14, 393, 1981.
50. Fossum, R. M., and Descheneaux, K. A., Blunt trauma of the abdomen in children, *J. Forensic Sci.*, 36, 47, 1991.
51. Kaga, K., Aizawa, T., Yoshiyuki, T., et al., Autopsy imaging—the present situation of postmortem computed tomography (PMCT) of our institution (in Japanese), *J. Japan Assoc. Radiol. Technol.*, 51, 118, 2004.
52. Ros, P. R., Li, K. C., Vo, P., Baer, H., Staab, E., Preautopsy magnetic resonance imaging: Initial experience, *Magnetic Reson. Imaging*, 8, 303, 1990.
53. Hart, B. L., Dudley, M. H., Zumwalt, R. E., Post-Mortem Cranial MRI and Autopsy Correlation in Suspected Child Abuse, poster presentation at 1994 RSNA annual meeting, Chicago.
54. Dedouit, F., Sévely, A., Costagliola, R., et al., Reversal sign on ante and postmortem brain imaging in a newborn: Report of our cases, *Forensic Sci Int.*.182, ell, 2008.
55. Clarot, F., Proust, B., Eurin, D., Vaz, E., LeDosseur, P., Sudden infant death syndrome and virtual autopsy: Scalpel or mouse? *Arch. de pédiatr.*, 14, 636, 2007.
56. Wright, C., Lee, R. E., Investigating parietal death: A review of the options when autopsy consent is refused, *Arch. Dis., Child Fetal Neonatal Ed.*, 89, 285, 2004.
57. Dedouit, F., Guilbeau-Frugier, C., Caputani, C., et al., Child abuse: Practical application of autopsy, radiological and autopsy studies, *J. Forensic Sci.*, 53, 1, 2007.
58. Sowell, M. W., Lovelady, C. L., Brogdon, B. G., Wecht, C. H., Infant death due to air embolism from peripheral venous infusion, *J. Forensic Sci.*, 52, 183, 2007.
59. Deduoit, F., Auriul, J., Braga, J., et al., Age determination by magnetic resonance imaging: A preliminary study, Presented at the *Am. Assoc. Physical Anthropologists*, Chicago, April, 2009.

CREDITS

From Brogdon, B.G., Vogel, H. and McDowell, J.D., Eds., A Radiological Atlas of Abuse, Torture, Terrorism and Inflicted Trauma. CRC Press, Boca Raton, FL, 2002. Figures 17.12a & b, 17.13a, 17.14a & d, 17.22–23, 17.25.

18 Abuse of Intimate Partners and of the Elderly

An Overview

B.G. Brogdon and John D. McDowell

CONTENTS

INTRODUCTION

Domestic violence in all of its forms—child abuse, partner abuse, and elder abuse—is pervasive in most "civilized" societies. Many persons have learned from within the family that violence is a means by which long- or short-term goals may be accomplished. In Western society, the family is the single most common locus of violence. It is impossible to separate one form of domestic violence from another. Abused children are likely to have abused mothers. Both men and women in a dysfunctional relationship may abuse each other. Spousal abuse does not magically begin or end at age 65; it is merely reclassified as elder abuse.

Violence, while not uniquely American, is a learned behavior that is intergenerationally transmitted. Recent high-profile cases involving American celebrities have sensitized most of us to the potential morbidity and mortality associated with dysfunctional family relationships. Americans are more likely to suffer injuries from domestic violence than they are likely to be injured by a person outside the home.[1,2] Prevalence estimates of domestic violence in the United States indicate that as many as 50% of all families have experienced some form of intrafamily violence, and 25–50% of couples interviewed report at least one incident of physical abuse involving a family member. Despite this large number of suspected domestic violence cases, adequate surveillance of domestic violence does not exist. Estimates of domestic violence based on reported cases might probably underestimate the true prevalence rate. A Lewis Harris Association poll of 1973 randomly surveyed women in Kentucky showed that 21% reported being physically abused by her partner, and nearly two-thirds of divorced or separated women reported being the recipient of nonaccidental trauma. Szihovacz found that one in four surveyed women in Pennsylvania reported being assaulted by her intimate male partner.[1–9]

Until relatively recently, social attention in abuse beyond childhood was directed almost exclusively at the plight of women physically abused by men. Abuse was identified as the most common etiology of trauma in women, exceeding motor vehicle accidents, muggings, and rapes combined.[10,11]

BATTERED WOMEN

The term "battered women" was first used in 1974 by Pizzey to describe the female victim of violence within marriage or cohabitation.[12] A commonly accepted definition of the term battered women is a woman who has received "deliberate, severe, repeated, demonstrable, physical injury from her partner."[13] In general terms an abused or battered woman is one who is subjected to serious and/or repeated physical injury as a result of intentional, deliberate assaults by her male companion.[2]

Physically dominating one's wife (including "wife beating") has had, until relatively recent times, both legal and social sanction. The husband's authority to chastise his wife was explicitly written into the laws of church and state and later incorporated into English common law, which American law generally follows.[3,8] The husband's right to hit his wife was written into the law of the United States in 1824 with one limitation—that he was restricted from using a switch larger in diameter than his thumb. This law was overturned in 1874, but the ruling was qualified with the statement that if no permanent injury had been inflicted, nor malice, cruelty, or dangerous violence had been shown, that it is better to leave the parties to forgive and forget.

It is only within the last few decades that assault in the home has been recognized and prosecuted as criminal behavior. Contemporary American society began examining the issue of violence within the family in the mid-1960s when a national survey showed that 20% of all Americans approved of slapping one's wife.[11] Surprisingly, approval for physically striking one's mate increased with income level and education; 25% of college-educated men believed striking one's mate was acceptable behavior.

There is a strong relationship between spousal abuse and child abuse.[8,14,15] Studies have shown that between one-third and one-half of the families in which women have been battered also have children present who have been physically and/or sexually abused.[8,12,16] Physical abuse during pregnancy is not uncommon. Both the pregnant woman and her unborn child are potential victims of intentional trauma.

Whereas battering in the nonpregnant female is more commonly directed against the head and neck,[17,18] abuse during pregnancy is frequently directed toward the breasts and abdomen.[3,19–21] The aggressive behavior directed against the abdomen of pregnant women with its potentially harmful effect on the fetus has engendered the term "prenatal child abuse." There is reportedly an increased rate of abortions and premature births found in women who give a positive history of physical abuse.

Cate et al. first used the term "premarital abuse" to describe the physical violence associated with dating and courtship.[22,23] Their study reported that 22% of the respondents had been the victims of premarital violence or had been violent toward a premarital partner.

The permutations of intimacy have become so broad and nongender specific for the problem—as it involves "adults" in the somewhat indeterminate time span between childhood and elderhood—that it is now gathered under the umbrella term "abuse of intimate partners."

PROBLEMS IDENTIFYING AND TREATING VICTIMS OF DOMESTIC VIOLENCE

The recognition and reporting rate of adult domestic violence victims is quite low when compared to the actual prevalence. The reporting rate for battered women may be as low as 2% in large metropolitan emergency departments.[24] Also, some health care providers blame the victims for provoking the violence and hold them responsible for their own injuries.[25]

Recognizing and treating the domestic violence victim is the important first step toward intervention on behalf of the victim.

ASSOCIATING FRACTURES WITH DOMESTIC VIOLENCE

Radiographic examinations can provide evidence of intentional trauma in battered women. The majority of assaults in nonpregnant battered women are directed at the head, neck, and face. Ellis et al.[26] reviewed mandibular fractures occurring in Glasgow, Scotland, from 1974 to 1983 and found that alleged assaults were the most common cause of fractures in males and females and that greater than 60% of the assaults of females occurred at home. A second 10-year review of fractures of the zygomatico-orbital complex showed that, in females, most (54%) assaults occurred at home, although women may give conflicting histories.[27]

INJURY PATTERNS

The two most common causes of facial trauma in adults are motor vehicle accidents and domestic violence. McDowell analyzed hospital records of adult women to determine if any patterns specific to domestic violence could be determined and used to identify the battered women presenting for care of maxillofacial trauma (facial fractures).[28]

Certain variables were solely observed in one of the two study groups. LeFort 1 and LeFort 2 fractures wvere seen *only* in victims of motor vehicle accidents (Figure 18.1). Fractures of the mandibular ramus and fractures of the mandibular body/ipsilateral angle were seen only in battered women (Figure 18.2). Previous emergency department visitation and prior facial fracture (by patient history or radiographic report) were seen only in battered women[29] (Figures 18.3 through 18.8).

FRACTURES

A total of 175 facial fractures were seen in the 114 records reviewed (the total number of fractures exceeds the number of records because some patients had more than one fracture associated with their presenting injuries): A total of 89 fractures were found in the 58 victims of motor vehicle accidents and 86 fractures were found in the 56 battered women.

There were 117 cases of mandibular fractures (single and multiple sites), representing 66.86% of the total of 175 facial fractures. Of the mandibular fractures (single and multiple sites), 69 (58.97%) were seen in battered women.

In this study, fractures involving both the mandibular body and angle were the most common facial fractures, representing 33.71% (59/175) of all facial fractures. Fractures of the mandibular body and angles were seen much more frequently in battered women than in motor vehicle accident

(a) (b) (c) (d) (e) (f)

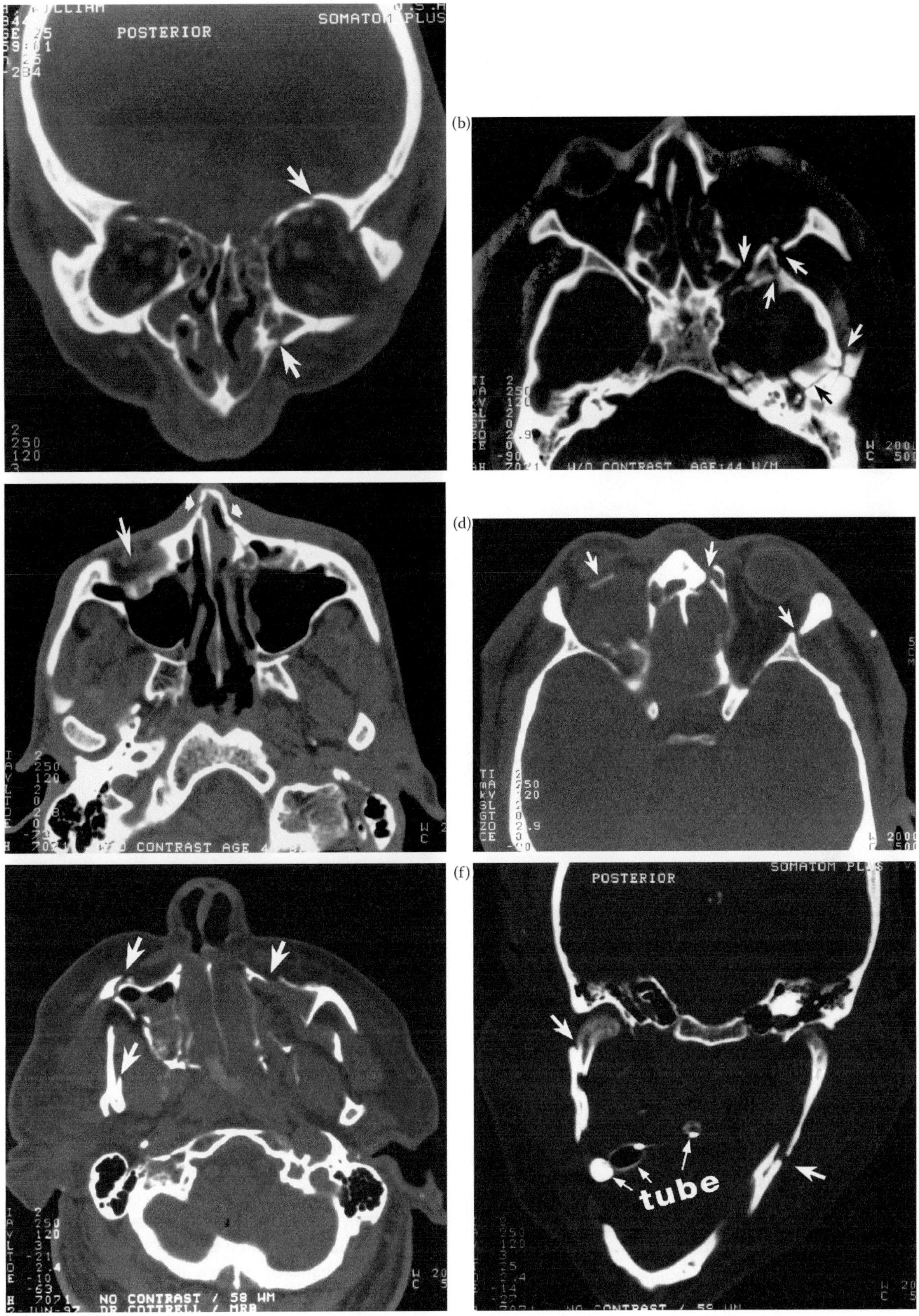

FIGURE 18.1 (a–f) Examples of massive injuries to the nasal bones, maxilla, mandible, orbits, and base of skull from automobile accidents as shown by CT examination. Note how in several cases the paranasal sinuses and nasal passages are filled with blood.

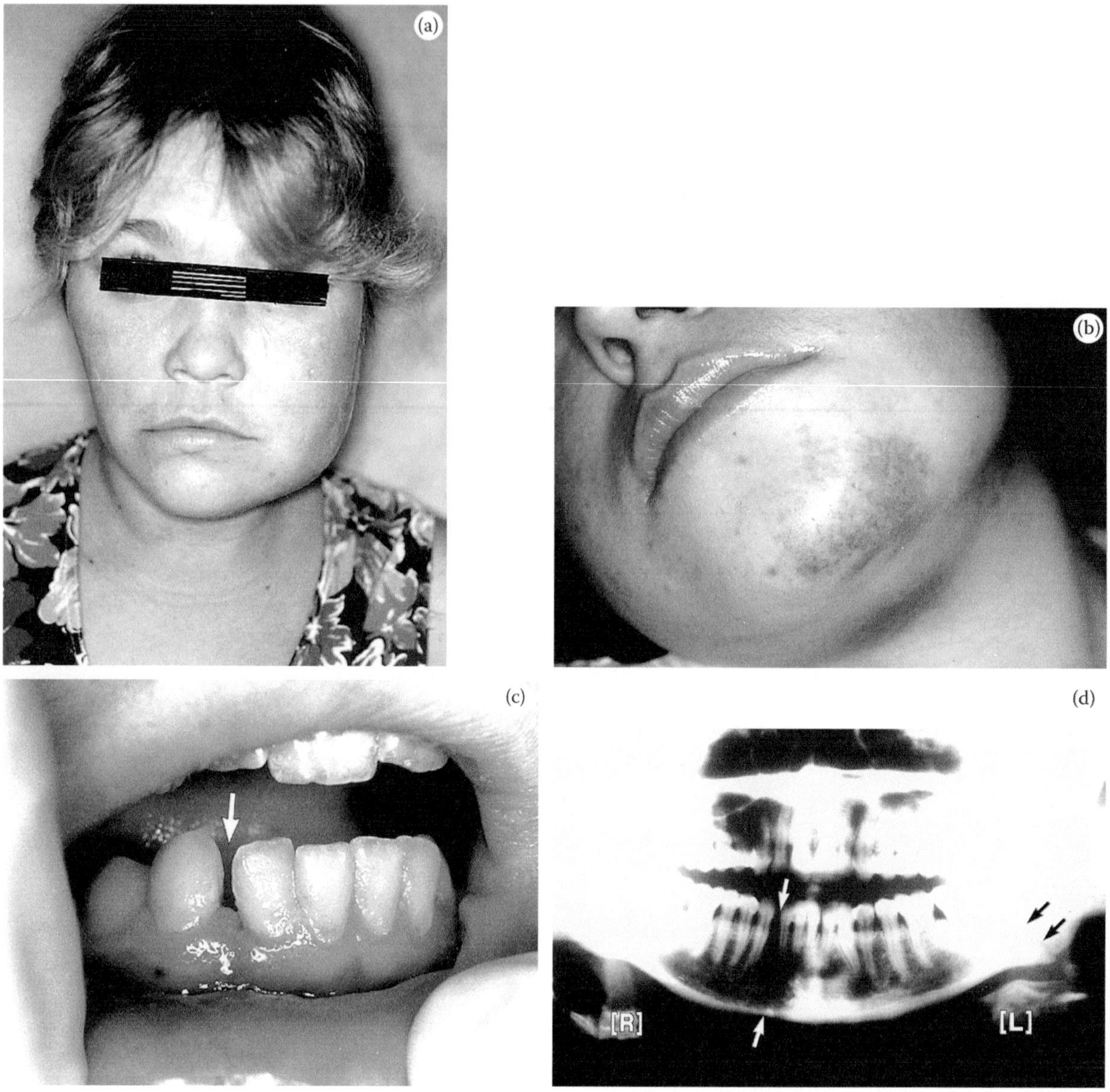

FIGURE 18.2 (a) Woman, 43, hit with husband's fists, has massive swelling over left jaw. (b) Bruise and abrasions to chin. (c) Separation of teeth at fracture site. (d) Panorex study shows fractures through the left mandibular angle and right mentalis.

victims. Of the mandibular fractures involving the body and angle, 71.18% (42/59) were seen in battered women. Battered women presented with 73.53% (25/34) of the fractures involving the mandibular angle.

In battered women, besides zygomatico-facial fractures, nasal bone fractures, orbital fractures, and so on. (Figures 18.9 through 18.11), injuries at other sites may also be seen, of course. Defensive injuries (fending fractures of the hands and forearms) are not uncommon (Figures 18.12 and 18.13). Soft tissue injuries may be striking (Figure 18.14). Blows to the chest (breast) are frequent but, in our experience, rib fractures are not common. Abdominal injuries require more sophisticated studies (i.e., CT); for instance, we have seen a ruptured renal cyst from a kick.

Delayed Presentation for Care

As previously described in the literature, victims of intentional trauma frequently report for care on a delayed basis—57% of battered women presented for care more than 24 hours after their injuries were reportedly inflicted. Only 14% of women victims of motor vehicle accidents presented for care more than 24 hours after their injury.

Zachariades and co-authors[30] examined the medical records of battered women to determine the weapon and location of facial trauma associated with domestic violence. In over 70% of the facial injuries, the weapon was the husband's or boyfriend's hand. Mandibular fractures were the most common presenting facial injury, representing 39% of

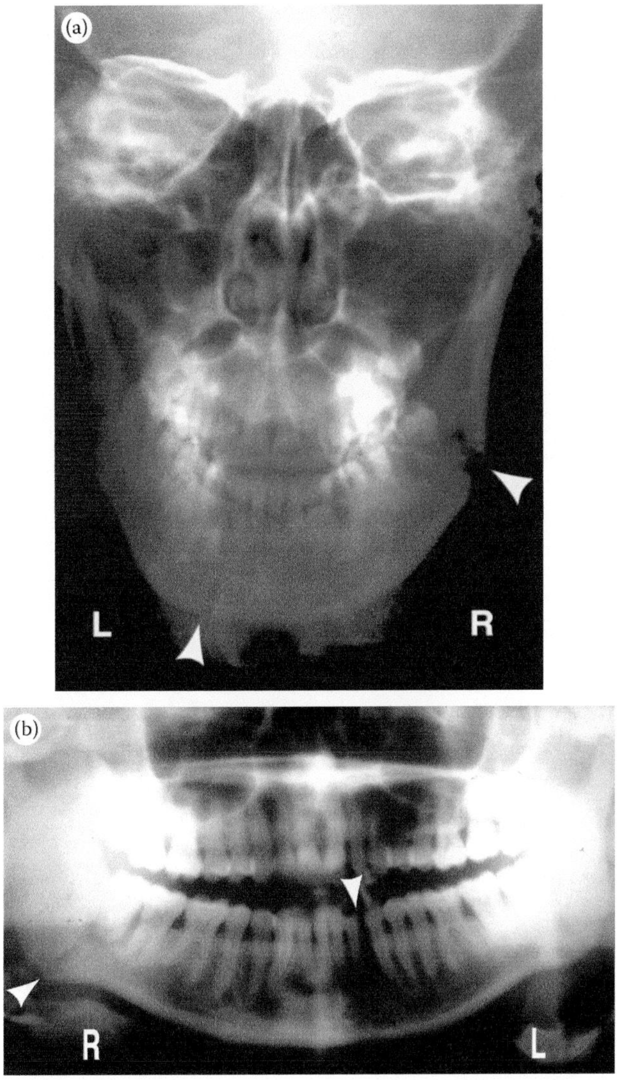

FIGURE 18.3 Typical battering fractures in a 53-year-old female. (a) Frontal (PA) radiographs of the mandible show a right angular fracture (arrowheads) and, faintly, a left mental fracture (arrow). (b) Panoramic view shows the angle fracture and separation of the left mandibular lateral incisor and canine (arrows).

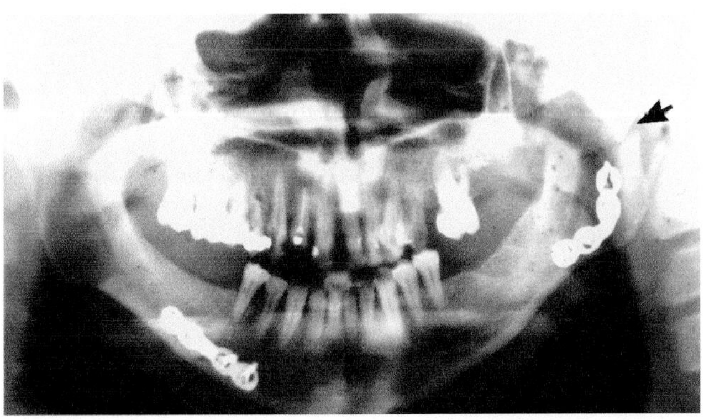

FIGURE 18.4 This 51-year-old female presented after a recent battering. Examination revealed old metallic fixation devices stabilizing typical battering fractures from a previous episode. There is a new fracture of the left mandibular neck.

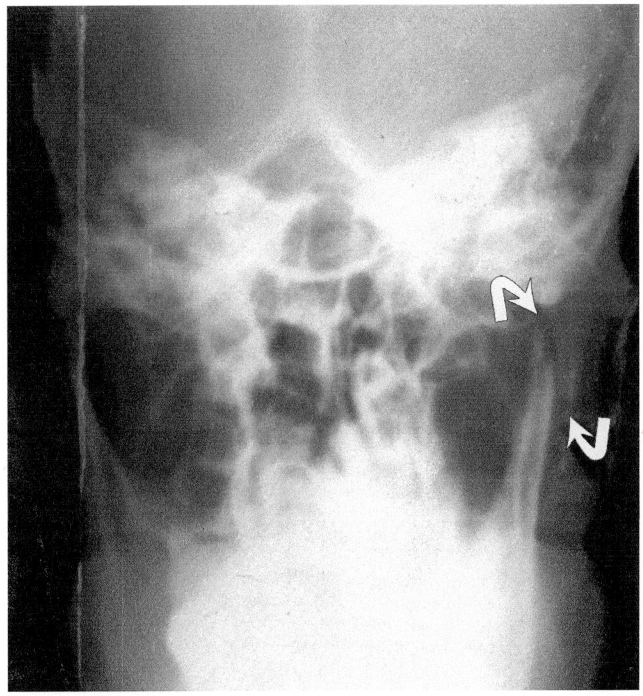

FIGURE 18.5 Oblique fracture of the left mandibular neck in a battered woman.

all facial injuries in women assaulted by their husbands or boyfriends.

Most states presently have enacted and are enforcing legislation protecting persons abused by family or household members.[24,25] While there continues to be a positive trend toward actions taken on behalf of the abused/neglected child, legislators and health care providers have not responded as effectively on behalf of adult victims of domestic violence. While child abuse has been described as a major health problem that requires the highest priority, population-based estimates indicate that battered partners are far more common than abused children.[31–33]

Until recently, the primary focus of intimate partner abuse has focused on the defenseless woman victimized by her aggressive male.[33] Contemporary social studies have necessitated a revision of this concept.[34–38] Among high school and college couples, surveys show incidence of violence received to range from gender equality to a majority of males. Similarly, and surprisingly, females were more likely to initiate violence against their male partners, including pushing, shoving, holding down, slapping, hitting and biting. In one survey, hitting with the fist, kicking, and biting was seven times more likely from women than from men.[39,40]

Consideration of publication dates of articles dealing with dating abuse suggests that female generated violence against heterosexual partners is increasing. One report of sampling over a 10-year interval showed that the overall rate of partner abuse had doubled and violence perpetrated by women was twice the rate of male perpetrators.[41,42]

A survey of lesbian, gay, and bisexual persons reports more childhood psychological, physical, and sexual abuse,

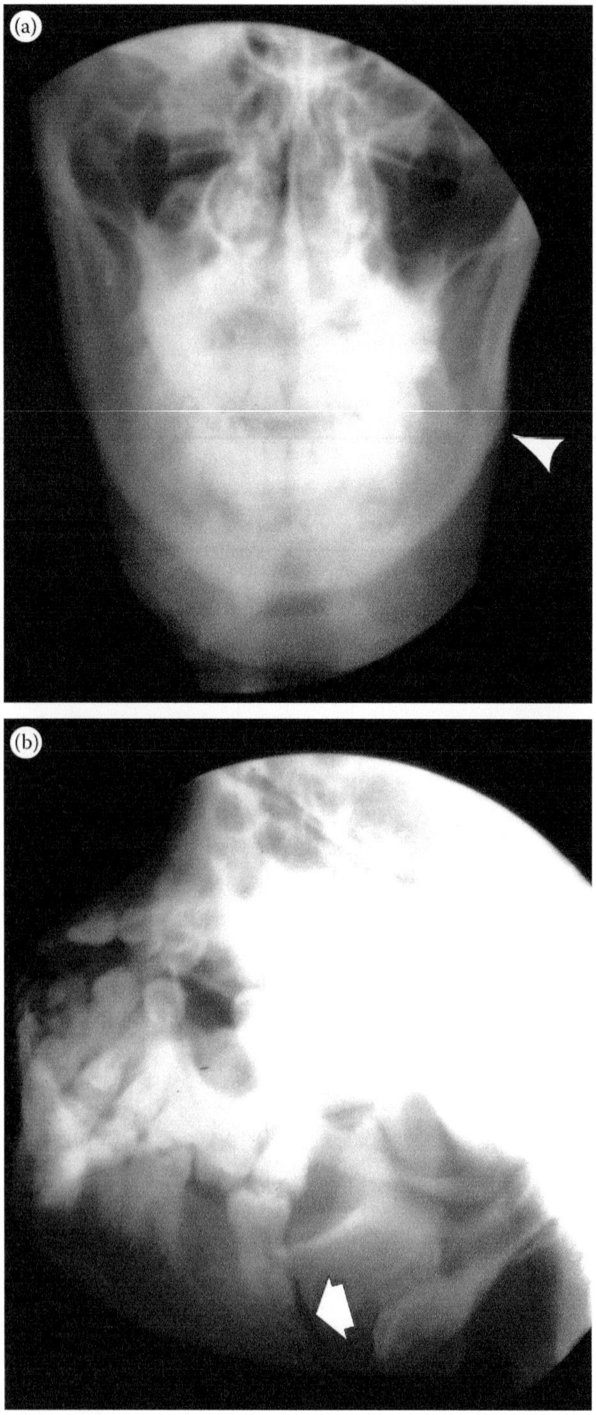

FIGURE 18.6 Typical mandibular angle fracture through the socket of the third molar. (a) Frontal view. (b) Oblique view.

more partners' psychological and physical victimization as adults, and more sexual assaults in adulthood—and the victimization was greater in men than among women.[43] Among men who have sex with men, partner abuse ranges from 12% to 36%, and this is lower than estimation of violence among lesbian partners.[44]

Among nonstudent heterosexual partners, indications are that there is either remarkable increase in violence by women

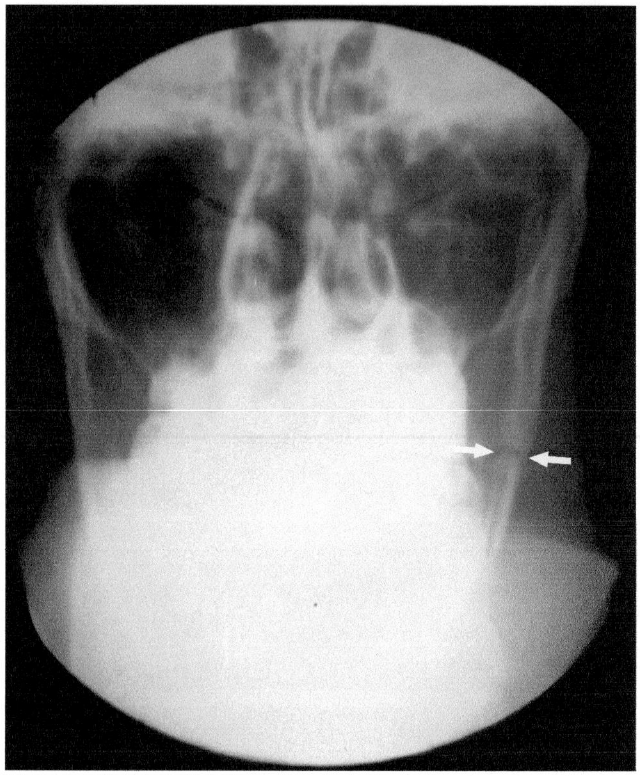

FIGURE 18.7 Fracture of the left mandibular body.

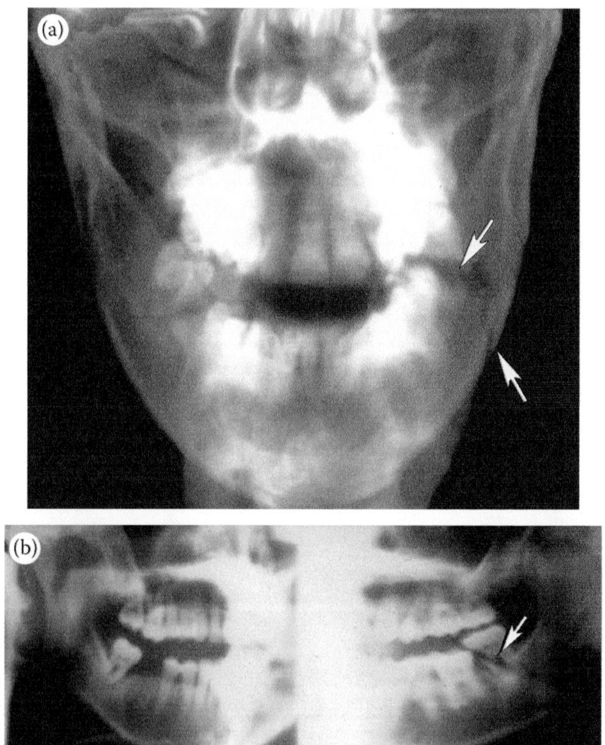

FIGURE 18.8 Left mandibular angle fracture with fractured third molar. (a) Frontal view. (b) Oblique views.

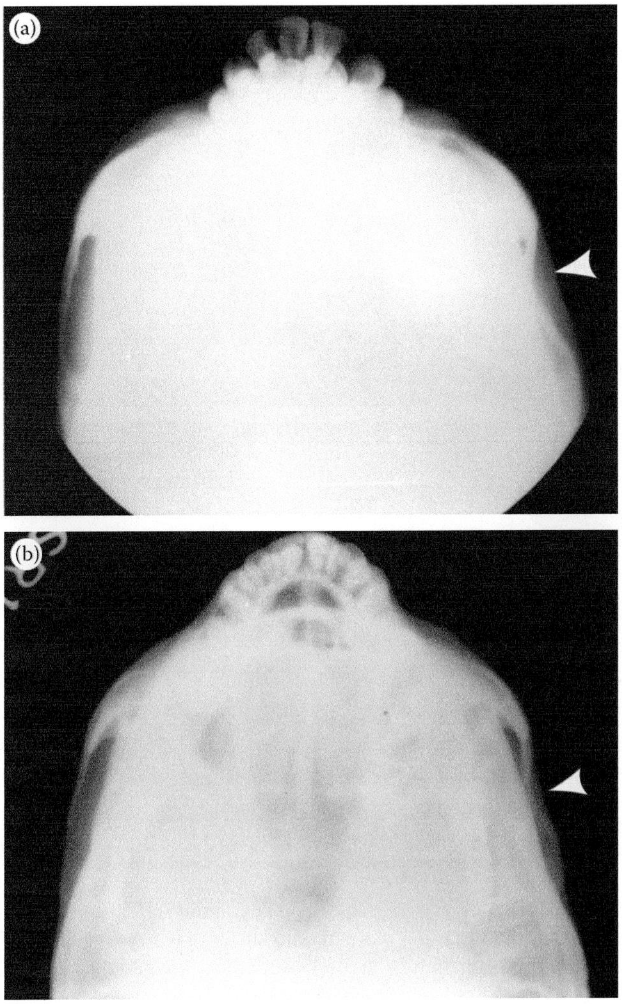

FIGURE 18.9 (a and b) Almost identical depressed fractures of the zygomatic arch in two battered women.

against their men or else a change in the awareness and/or reportage of it in recent times. Reports of cross-gender violence at a trauma center admission were evenly split, but the severity score of injury for men was almost twice that of women. The majority of men were admitted for stab wounds, the women for assault.[45] Another study showed that men used weapons 25% of the time, but female assailants used weapons 86% of the time; of 74% of men sustaining injury, 84% required medical care. The conclusion was that male victims are injured more often and more seriously than female victims of domestic abuse.[46]

Partner abuse can rise to deadly levels (Figures 18.15 and 18.16). Data from several cities over a 10-year period indicated that for every 100 men who killed their wives, 75 women killed their husbands, but in some venues the ratio of wives as perpetrators exceeds that of husbands.[47]

In a similar study over the same time period, Black husbands were at greater risk; spousal homicide among Blacks was more than 8 times that of White couples. Homicide rates were almost 7 times higher for spouses in inter-racial marriages. For both Black and White couples, the risk increased

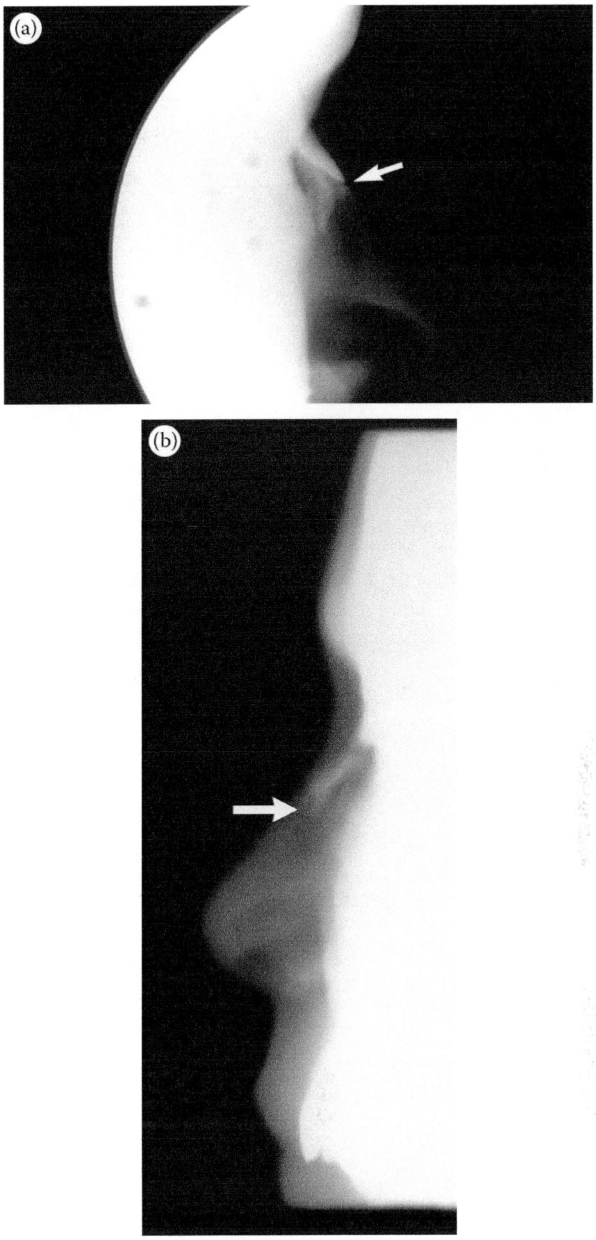

FIGURE 18.10 (a and b) Almost identical fractures of the nose of two women from battering.

as age differences increased. Wives and husbands were equally likely to be killed by firearms (about 72% of the time), but husbands were more likely to be stabbed and wives more likely to be beaten to death. Escalation of arguments accounted for 67% of spousal homicides.[48]

The "brutality" of killing is higher for: (a) gays and lesbians relative to heterosexuals; (b) men relative to women; (c) gay relative to heterosexual men; and (d) lesbians relative to heterosexual women. Homicide rates are highest in gays relative to heterosexual and lesbian couples. Lesbian couples have the lowest homicide rate, but women kill their partners more brutally than men.[49]

Although we have found no studies of "patterned" violence in domestic abuse couplings beyond that of men on women, it

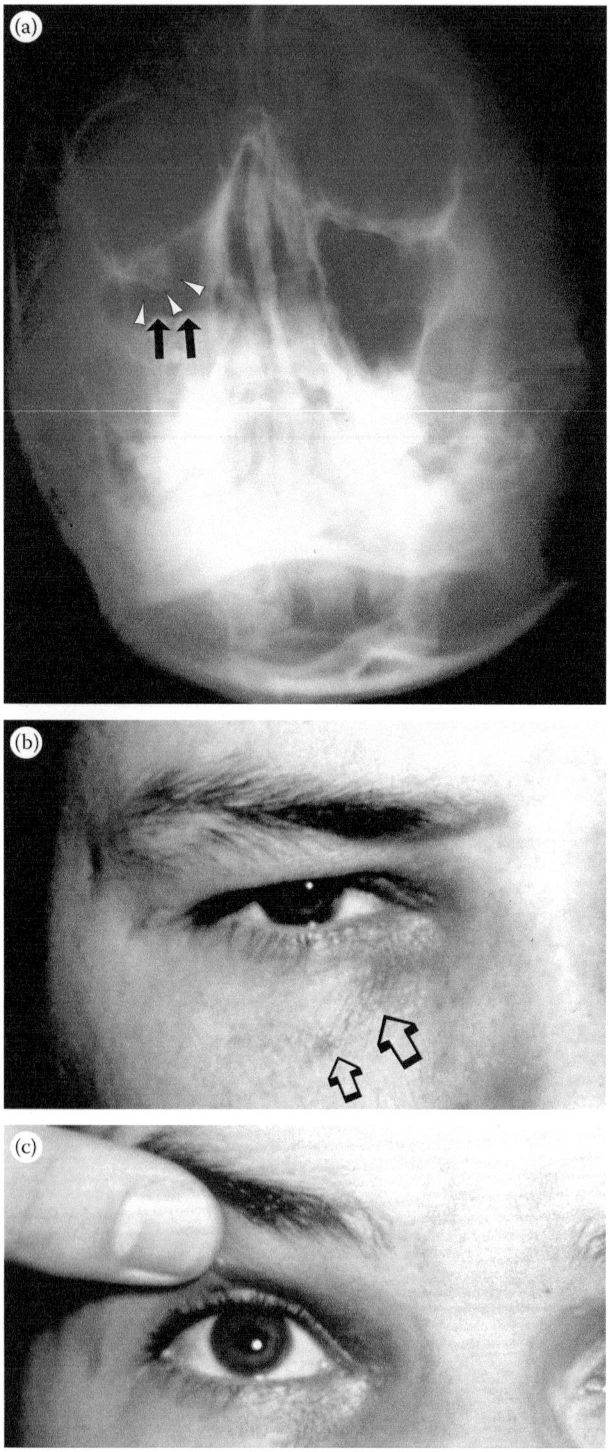

FIGURE 18.11 Blowout fracture of the orbital floor from being struck with a fist. (a) Waters view shows herniation of soft tissues from the orbit into the right maxillary sinus (white arrowheads) and an air-fluid level indicating fresh blood in the maxillary sinus (black open arrows). (b) Photograph of victim showing narrowing of the palpebral fissure from swelling. Note ecchymosis (subcutaneous blood) beneath the eye (arrows) due to bleeding from the orbital floor fracture. (c) Loss of upward-outward gaze on the right due to herniation of extraocular muscle into the floor fracture. The victim has diplopia (double vision) as a result.

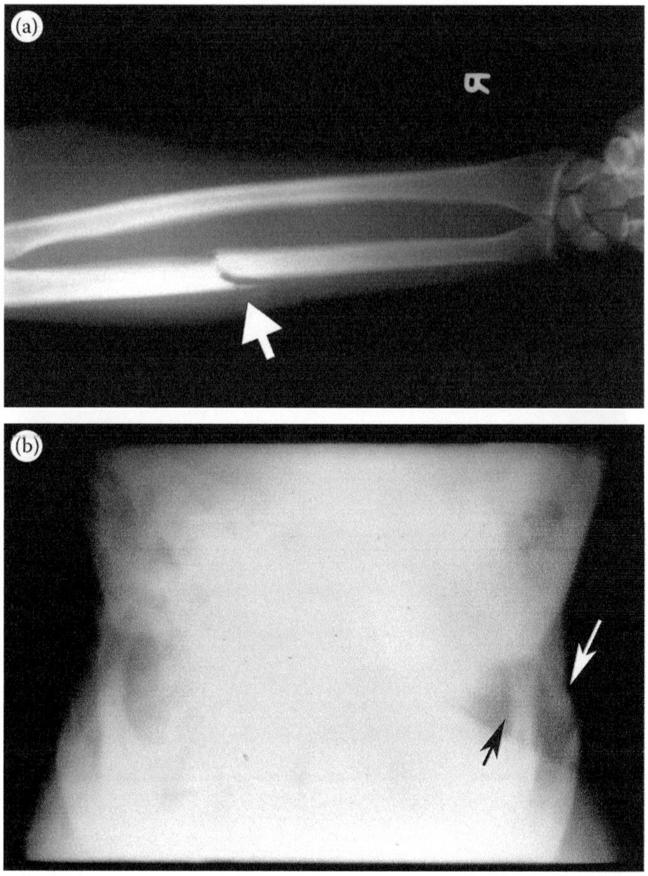

FIGURE 18.12 (a) Typical defensive injury, fending fracture (arrow) of the ulna in a woman who suffered (b) A fracture of the mandibular neck (white arrow) overriding the ascending ramus (black arrow).

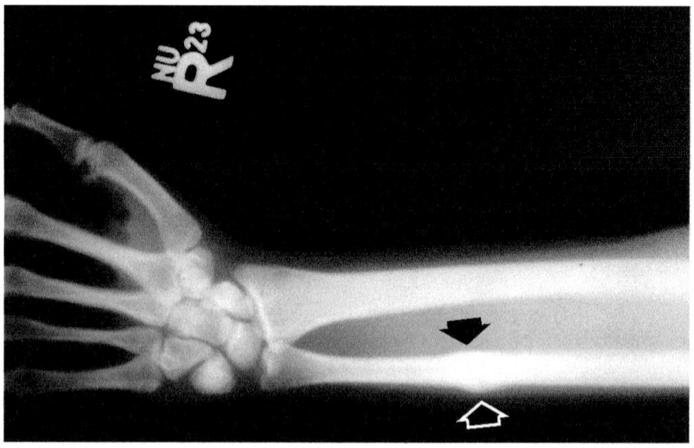

FIGURE 18.13 This woman who came in with fresh contusions from a recent beating was found to have a healed fending fracture (arrows), no longer symptomatic, from a previous episode of battering.

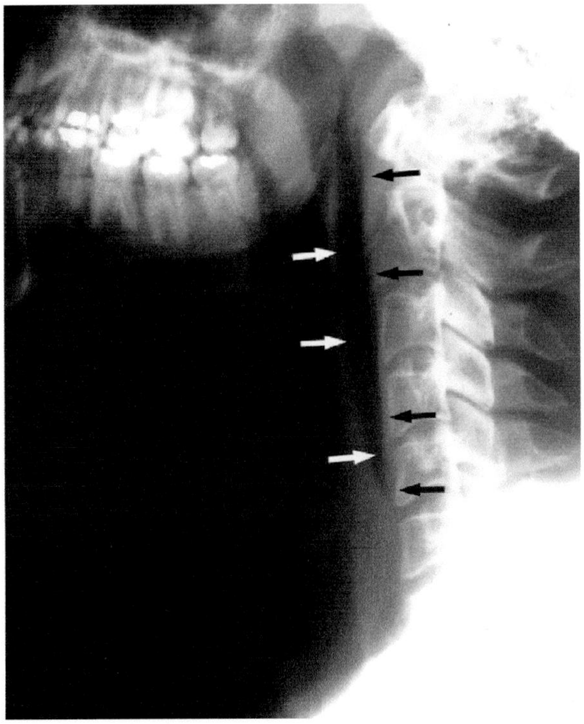

FIGURE 18.14 This young woman was hit in the throat by her boyfriend, sustaining a rupture of the larynx with subsequent dissection of air into the paravertebral space (arrows).

is expected that the head and neck would be major target areas; hence, dentists and oral surgeons along with emergency physicians will be on the front lines of diagnosis and treatment.[50] Specific investigations of pattern injuries are needed to better understand and detect this parvenus problem.

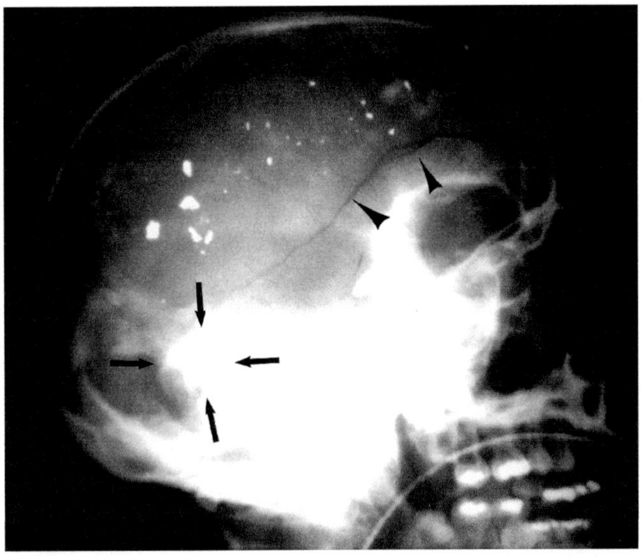

FIGURE 18.16 Fatal partner dispute. A male abuser shot his girlfriend twice, once through the mouth and then through the posterior parietal area. The first bullet, little deformed (arrows), is almost hidden by the overlying petrous bone. The second bullet fragmented upon striking the skull, revealing its track. Either gunshot wound would have been fatal. Note the skull fracture (arrowheads) associated with the second wound of entry and massive sudden increase in intracranial pressure.

Abuse of the Elderly

Worldwide Issue

The elderly are the fastest growing age group in the United States and, indeed, in most of the top tier countries in the

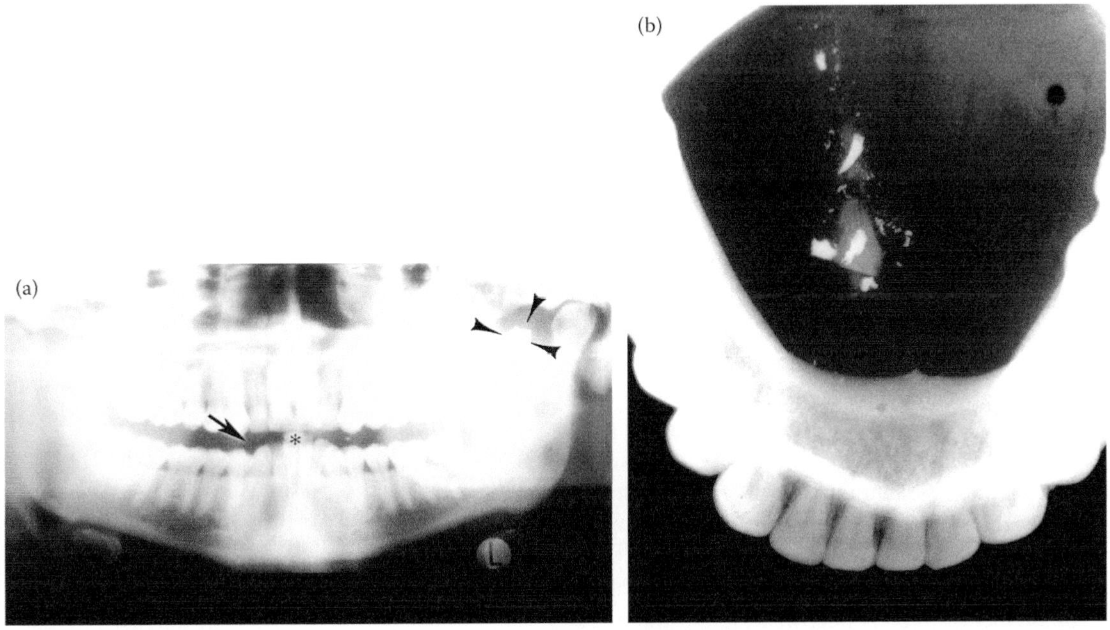

FIGURE 18.15 This young male abuser was shot from a distance of 10 to 12 ft with a Saturday Night Special in the hands of his regular victim. (a) The nonlethal bullet entered the mouth, sheering the right lateral mandibular incisor (arrow) and coming to rest in the left oropharynx of the abuser (arrowheads). (b) A subsequent dental examination (occlusal film) shows tooth flakes and bullet fragments embedded in the tongue. One of the largest fragments can be seen between the teeth in A and is marked with an asterisk.

rest of our world. There are more than 35 million people in the United States (about 13% of the population) 65 years of age or older and that figure is projected to double by 2030. The post-WWII "baby boomer" generation of approximately 77 million Americans will start turning 65 in 2011. The fastest-growing segment of the U.S. population is the 85 and older age group. There are about 76,000 centenarians in the United States, predicted to reach 2.5 million over 100 years of age by 2060.[51–53] And there are several countries whose citizens enjoy greater longevity than Americans!

The aging of the population will have a tremendous impact on medical and social services. In the United States, approximately 80% of persons of age 65 have at least one chronic disease, 50% have at least two.[52] These, of course, may lead to severe disability. Further, each year one-third of Americans beyond the age of 65 will suffer a fall; 20–30% of them will sustain injuries serious enough to affect their independence and mobility, increase the risk of premature death, and elevate the requirement and cost of their healthcare and social support.[53]

However, it is the incidence and effect of nonaccidental trauma to the elder population that engages our concerns in this chapter.

Almost 30 years after Caffey's first paper on abused children, Burston introduced the concept of elder abuse in his letter to the *British Medical Journal* on granny battering. It was treated as a new phenomenon, but it was not. Elder abuse simply had been largely unremarked and ignored as a societal problem. Actually, it is almost as common as child abuse and, as aging of the population continues, may soon exceed it. Statistical studies on elder abuse are suspect because it is estimated that only one in five cases is reported to the authorities.[54–58]

Earlier literature would seem to indicate that domestic abuse has been prevalent in Western society with a strong American predilection. As many as 50% of all American families have experienced some sort of inter-family violence.[7–9] Now we are learning that the problem extends beyond the Western world, even being reported in Eastern cultures famous for filial piety and veneration of elders.[59] Abuse of the aged is of worldwide interest. As this is written, a search in PubMed for "elder abuse" reveals in the first 40 titles, 13 papers from outside the United States, including Japan, Mexico, Norway, Taiwan, and the United Kingdom. By 2030, the worldwide population aged 65 or older will increase to 937 million, or 12% worldwide. The largest percentage of aged people will be found in Europe and North America, the lowest in Sub-Saharan Africa.[51] As the number and percentage of elders increases, more cases of maltreatment can be expected. The typical victim is older than 75 years, often older than 80 years.[60] Most studies of elder abuse show the incidence to be gender neutral, but there are studies showing two-thirds of victims to be female, others showing a higher incidence in males because they are less capable of independent living and thus more at risk from family and caregivers.

DETECTION OF ABUSE AND ABUSERS

Elderly abuse is approximately as common as child abuse, affecting about 6% with physical abuse.[52] Those numbers are "soft" because large epidemiological studies are lacking, and both the victim and the perpetrators tend to deny its existence or belittle its seriousness.[52,61]

The elderly, like children, are at risk for several types of abuse. These have been variously categorized but would include: (1) physical abuse; (2) mental, emotional, or psychological abuse; (3) neglect; and (4) economic abuse such as theft or misuse of the elder's assets.[62] Some would include self-abuse or self-neglect, mostly involving the reclusive, sometimes incompetent, older person usually living alone. Neglect is the most common form of abuse, even though physical abuse is most often apparent.[63]

Physical abuse of the elderly is usually received from the victim's spouse (50%), less often from the victim's child or children (23%) and, contrary to popular opinion, only in 17% of cases is the physical violence at the hand of nonfamily care givers.[64]

The actual act of physical violence can range from a push or a shake to assault with a deadly weapon. In a Boston survey,[62] the most common act of physical violence was pushing, grabbing, or shoving (63%), followed by having something thrown at the victim (46%); slapping was involved in 42% of incidents; 10% of the physically abused elderly in another series had been hit with a fist, kicked, or bitten.[64]

The injuries sustained by elder victims are nonspecific and have some commonality with those seen in both child abuse and abused women. Maxillofacial injuries rank high[65] (Figures 18.17 through 18.20), as do defensive injuries and those related to grasping, squeezing, or forcible restraint (Figure 18.21).

One must be cautious. The mere presence of injury in an elderly person receiving home care or domiciled in a nursing facility is not proof of abuse. The elderly person with chronic disease, malnutrition, senile osteoporosis, and disuse atrophy can be extremely fragile. Handling injuries (fractures) can occur here just as they do in the premature nursery (Figures 18.22 and 18.23). One must be watchful for other signs of abuse or neglect, be alert to injuries inappropriate to the patient's level of activity, and one must be inquisitive. If possible, the suspected victim should be gently interrogated alone or where he is free of fear of retaliation or vengeance. The victim often is embarrassed by his situation and also afraid that a complaint may result in being torn from the abusive home situation to be thrown into a worse situation in the care of strangers.

Estimates of incidence and detection rates of elder abuse are low, and abuse may be difficult to prove. The victims may be unable to relate the circumstances of the injury. If able the victim may not reveal the problem because of fear of reprisal, embarrassment, lack of privacy when interviewed by a physician, dentist, or social worker, fear of being sent to a nursing home, or fear of being kicked out of one. The much feared, and often maligned, nursing home is responsible for only a

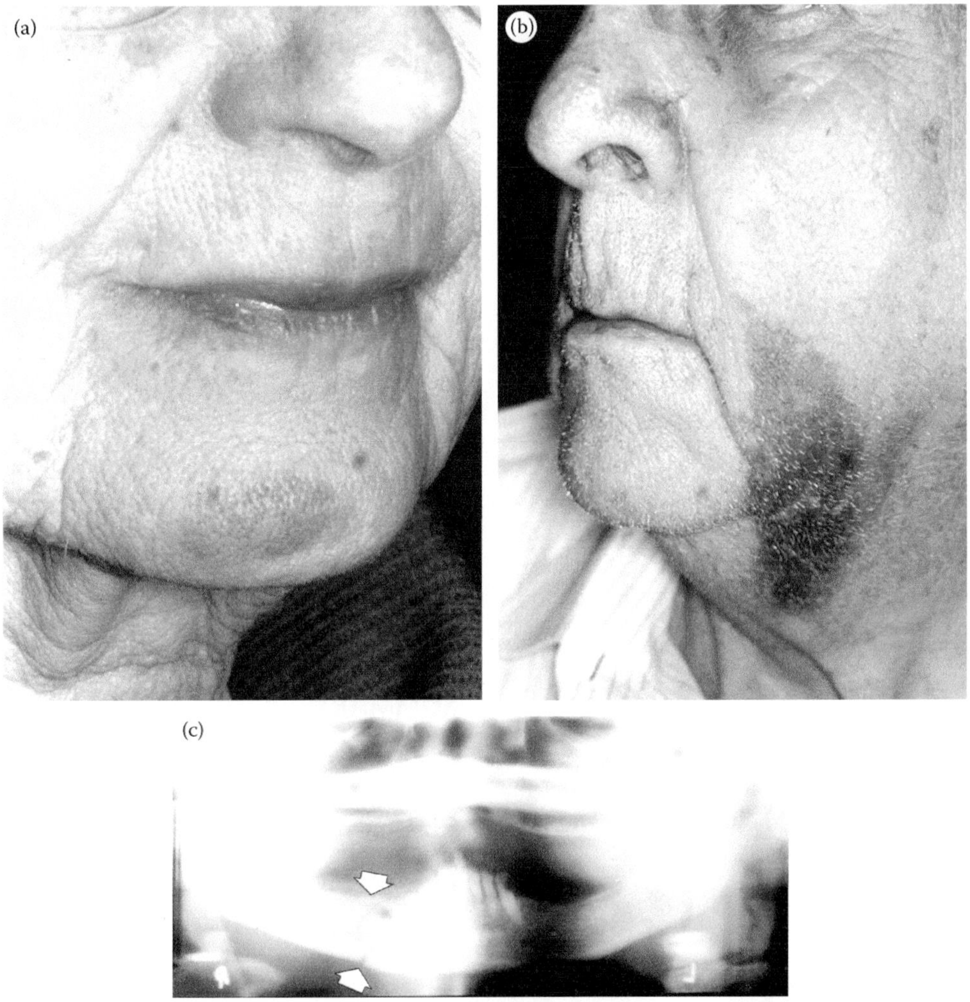

FIGURE 18.17 Battered elder woman. (a) Shows a bruise on the right jaw at the site of a mental fracture. (b) Shows a large hematoma draining down from a left mandibular angle fracture. The flattened left cheek prominence suggests a possible left zygomatic or malar fracture, but this was not investigated. (c) Panoramic view defines the right mental fracture well (arrow). The left mandibular angle fracture was demonstrated on another film (not shown).

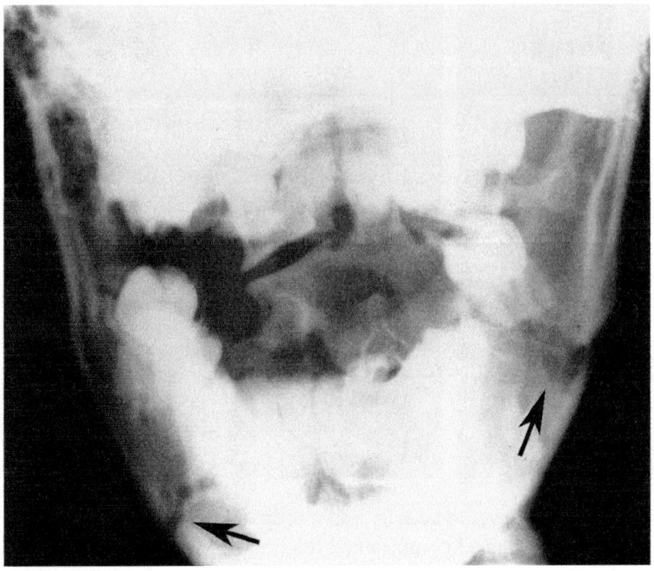

FIGURE 18.18 Typical mandibular fractures (arrows) in a battered elderly man.

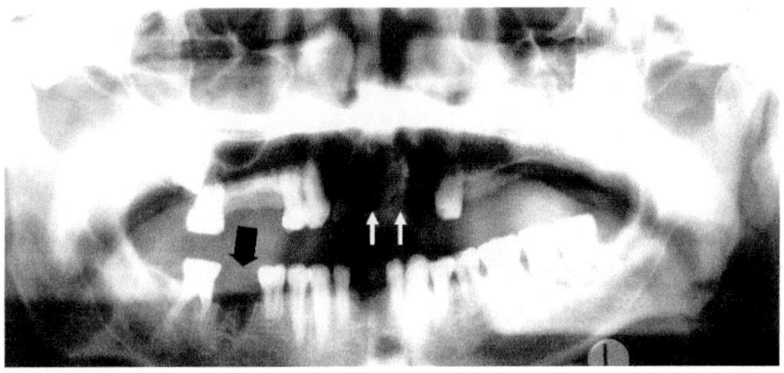

FIGURE 18.19 This 67-year-old woman was assaulted by her adult son. Her maxillary incisors were avulsed (small arrows) and her right mandibular first molar also was avulsed (arrow) in an unusual location for avulsion from a blow.

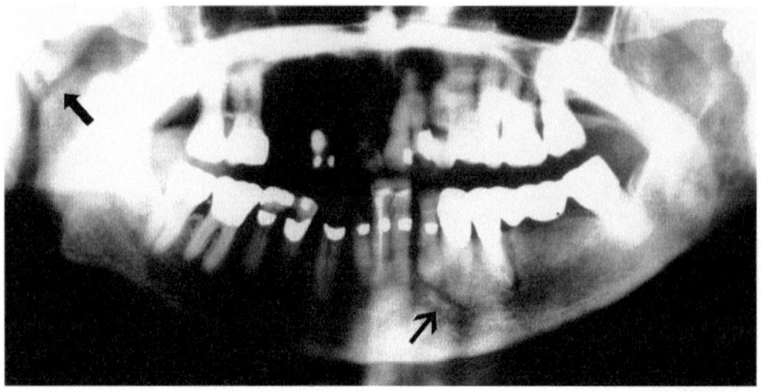

FIGURE 18.20 The woman, 64, was beaten by her husband. Examination shows fresh fractures of the right mandibular neck (arrow) and of the left mentalis with extension through the root of the lateral incisor (broad arrow).

small percentage of elder abuse. In the United States, only about 8–10% of the elderly population reside in institutions.[51] The risk of abuse increases with caretaker stress and depression and victim dependency, all more likely to occur with in-home care.[63,66]

Why do health professionals fail to diagnose and report abuse of the aged? The victim's failure or inability to communicate, the victim's lack of privacy with the health professional, the professional's failure to recognize the problem, equivocal history and physical findings, misdirection by caregivers—the "fox-in-the henhouse" effect[67] (see caption for Figure 18.24).

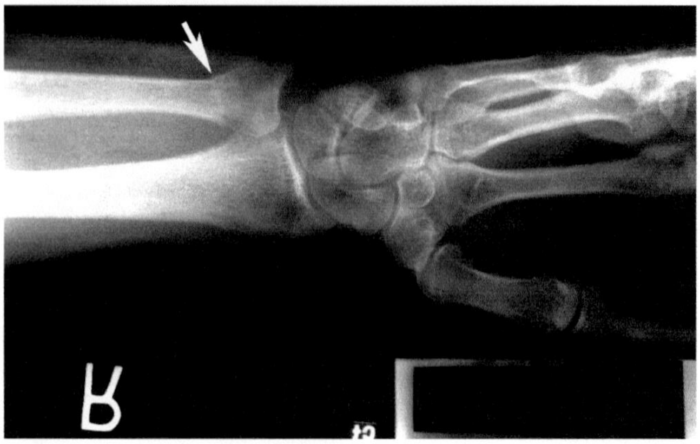

FIGURE 18.22 Minimal displaced fractures of the distal end of the ulna in an elderly nursing home patient. Is it a fending fracture, a twisting or restraint injury, or simply a result of extreme osteoporosis and normal handling? (From Brogdon, B.G., *Medicolegal Investigation of Death*, 4th ed., Spitz, W.U., Ed., C.C. Thomas, Springfield, IL, 2006, chap. XXI. With permission.)

SUMMARY

Domestic violence is a serious public health problem with a potentially fatal outcome. Battered women, children, and elderly persons occur in all segments of society. Without intervention, the frequency and severity of assaults tends to increase. Physicians, dentists, and other health care providers

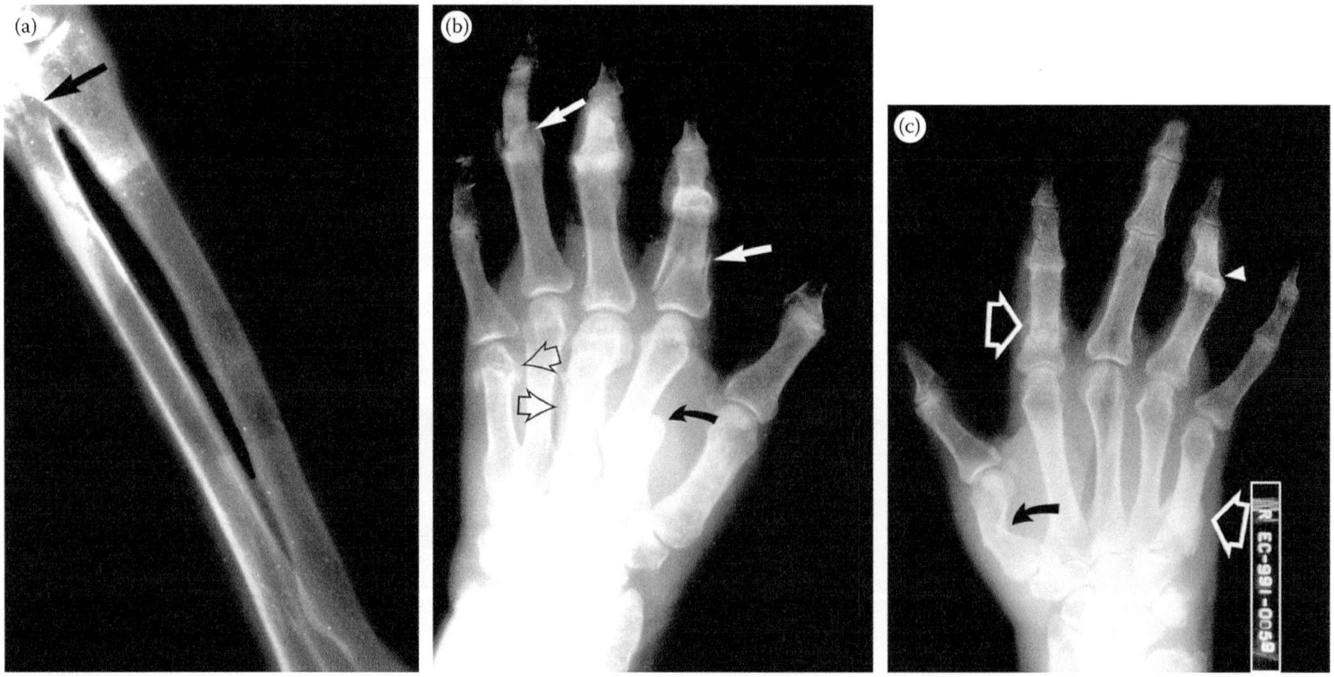

FIGURE 18.21 This octogenarian female was bedfast. Injuries inappropriate to her level of activity and multiple injuring in various stages of healing were signs of abuse. (a) Forearm with marked osteoporosis and with a displaced comminuted fending fracture of the distal ulna at the edge of the picture (arrow). (b and c) Both hands show new fractures (arrows), healing fractures (open arrows), and healed fractures (curved arrows) with residual deformity and dislocation (triangle). The fending fractures of the ulna are again seen (open curved arrow) in (a) (Courtesy of Dr. M.G.F. Gilliland.)

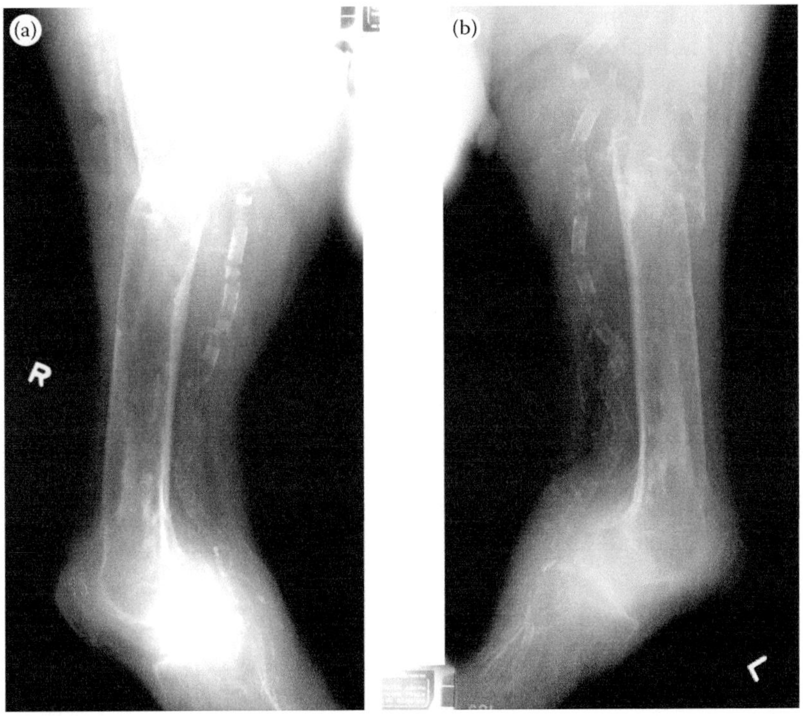

FIGURE 18.23 This obese elderly male diabetic had kidney failure, profound osteoporosis, and bilateral below-the-knee amputations. He was quite difficult to lift or move. His lower extremity stumps necessarily were used as handles, but were unable to take the strain. (a) The right femur was fractured while being cared for at home. (c) The left femur was fractured while being lifted for transport to the hospital by trained emergency medical service personnel. Impression: no abuse. (Courtesy of Dr. J.C.U. Downs.)

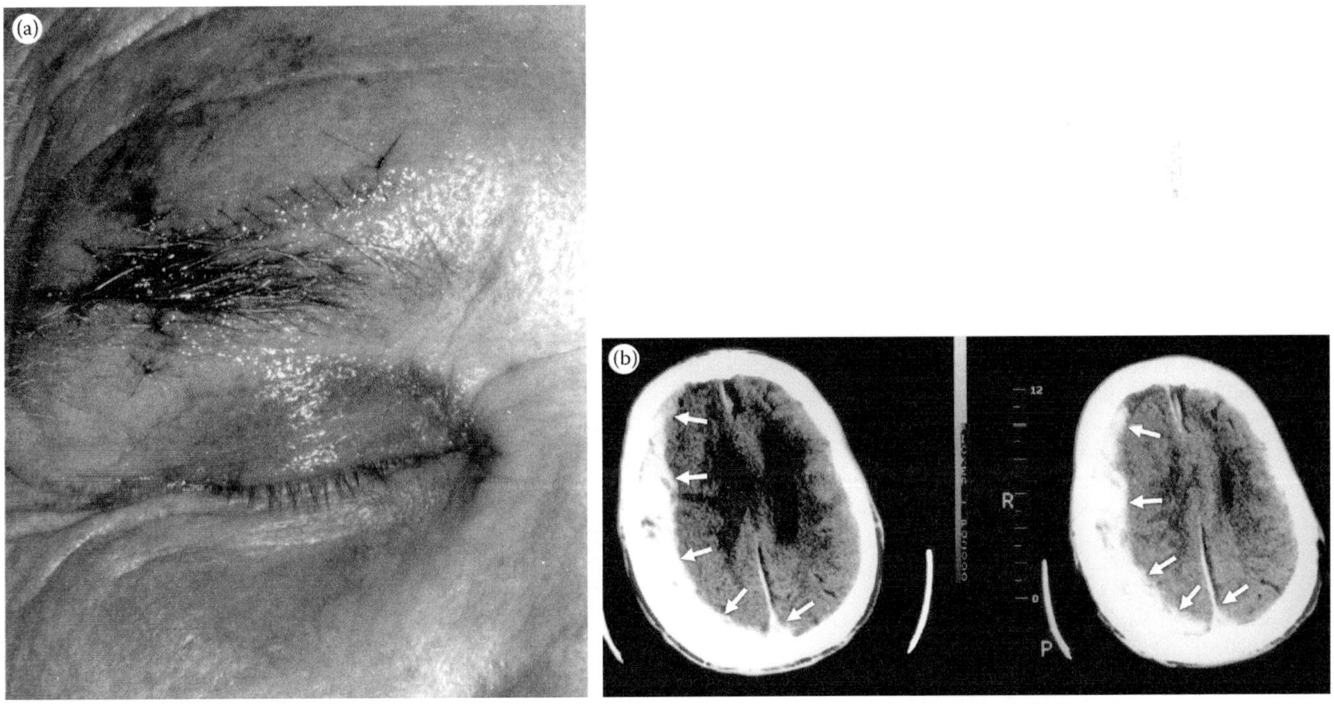

FIGURE 18.24 This elderly man with dementia was taken from the nursing home where he was domiciled to the hospital. (a) He had scrapes and bruises about the forehead and one eye and a split lip. His attendant suggested that the patient either fell down or against something while walking or had fallen out of bed. The patient was treated and released back to the nursing home. (b) Two days later, he was returned to the hospital with massive subdural hematoma (arrows) that proved fatal. The attendant abruptly left town. Investigation revealed that the victim was not ambulatory and that his mattress was already on the floor to prevent his falling out of bed! (From Brogdon, B.G., in *Medicolegal Investigation of Death*, 4th ed., Spitz, W.U., Ed., C.C. Thomas, Springfield, IL, 2006, chap. XXI. With permission.)

can learn to recognize the signs and symptoms of nonaccidental trauma. Once recognized, the victims of nonaccidental trauma can be made aware of their rights under the law.

Every health care provider should be sensitized to the possibility that the patient may be a victim of intentional trauma. Competent clinicians would not consider discharging a patient presenting with a life-threatening condition, yet many battered adolescent adults and elderly persons are discharged from the emergency department without any arrangement being made for their safety.[3,24] Treating physical injuries without offering essential support to a domestic violence victim has been described as being "simply bad medicine."[31]

REFERENCES

1. Staus, M. A., Gelles, R. J. and Steinmetz, S. K., *Behind Closed Doors: Violence in the American Family*, Sage Publications, Newbury Park, CA, 1988.
2. Stark, E. and Flitcraft, A., *Spouse Abuse in Surgeon General's Workshop on Violence and Public Health: Source Book*, Centers for Disease Control, Leesburg, VA, 1985, p. SA1.
3. Hilberman, E., Overview: "The wife-beater's wife" reconsidered, *Am. J. Psychiatr.*, 137, 1336, 1980.
4. Gelles, R., *The Violent Home: A Study of Physical Aggression Between Husbands and Wives*, Sage Publications, Beverly Hill, CA, 1974.
5. Dobash, R. E. and Dobash, R. P., The case of wife beating, *J. Fam. Issues*, 2, 439, 1981.
6. Krugman, R. D., Advances and retreats in the protection of children, *N. Engl. J. Med.*, 320, 531, 1989.
7. Centers for Disease and Control, Education about adult domestic violence in U.S. and Canadian Medical Schools 1987–1988, *Morbid. Mort. Wkly Rep.*, 32, 17, 1989.
8. Walker, L. E., *The Battered Woman Syndrome*, Springer-Verlag, New York, 1984.
9. Szihovacz, M. E., Using couple data as a methodologic tool: The cases of marital violence, *J. Marriage Fam.*, 45, 633, 1983.
10. Drossman, D. A., Leserman, J., Nachman, G., Li, Z. M., Gluck, H., Toomey, T. C., and Mitchell, C., Sexual and physical abuse in women with functional or organic gastrointestinal disorders, *Ann. Int. Med.*, 113, 828, 1990.
11. Stark, R., and McEvoy, J., Middle-class violence, in *Change, Readings in Societal and Human Behavior*, CRM Books, Del Mar, CA, 1972, p. 272.
12. Pizzey, E., *Scream Quietly or the Neighbors Will Hear*, R. Enslow Publishers, Short Hills, NJ, 1977.
13. Gayford, J. J., Battered wives, *Med. Sci. Law*, 15, 237, 1975.
14. Swanson, R., Battered wife syndrome, *Can. Med. Assoc. J.*, 130, 709, 1984.
15. ACOG, The battered woman, *Am. Coll. Obstet. Gynecol. Tech. Bull.*, 124, 1, 1989.
16. Chez, R. A., Woman battering, *Am. J. Obstet. Gynecol.*, 158, 1, 1988.
17. Berrios, D. C. and Grady, D., Domestic violence: Risk factors and outcomes, *West. J. Med.*, 155, 133, 1991.
18. Shepherd, J., Gayford, J., Leslie, I., and Scully, C., Female victims of assault, a story of hospital attenders, *J. Cranio-Maxillofac. Surg.*, 16, 233, 1988.
19. Schei, B. and Bakketeig, L., Gynaecological impact of sexual and physical abuse by spouse. A study of a random sample of Norwegian women, *Br. J. Obstet. Gynaecol.*, 96, 1379, 1989.
20. Sampselle, C. M., The role of nursing in preventing violence against women, *J. Obstet. Gynecol. Neonatal Nursing*, 20, 481, 1991.
21. Bewley, C. and Gibbs, A., Violence in pregnancy, *Midwifery*, 7, 107, 1991.
22. Makepeace, J. M., Courtship violence among college students, *Fam. Relat.*, 30, 97, 1981.
23. Cate, R. M., Henton, J. M., Koval, J., Chistopher, F. S., and Lloyd, S., Premarital abuse, a social psychological perspective, *J. Fam. Issues*, 3, 79, 1982.
24. Greany, G. D., Is she a battered woman? A guide for emergency response, *Am. J. Nursing*, 84, 724, 1984.
25. Ghent, W., DaSylva, N., and Farren, M., Family violence: Guidelines for recognition and management, *Can. Med. Assoc. J.*, 132, 541, 1985.
26. Ellis, E., Moos, K., and El-Attar, A., Ten years of mandibular fractures: An analysis of 2,137 cases, *Oral Surg. Oral Med. Oral Pathol.*, 43, 120, 1985.
27. Ellis, E., El-Attar, A., and Moos, K., An analysis of 2,067 cases of zygomatico-orbital fractures, *J. Oral Maxillofac. Surg.*, 43, 417, 1985.
28. McDowell, J. D., *A Comparison of Facial Fractures in Women Victims of Motor Vehicle Accidents and Battered Women*, Master's thesis, University of Texas Graduate School of Biomedical Sciences, San Antonio, Texas, 1993.
29. McDowell, J. D., Kassebaum, D. K. and Stromboe, S. E., Recognizing and reporting victims of domestic violence, *J. Am. Dent. Assoc.*, 123, 44, 1992.
30. Zachariades, N., Koumoura, F. and Konsolaki-Agouridaki, E., Facial trauma in women resulting from violence by men, *J. Oral Maxillofac. Surg.*, 48, 1250, 1990.
31. McLeer, S. V. and Anwar, R. A., Education is not enough, a systems failure in protecting battered women, *Ann. Emerg. Med.*, 18, 651, 1989.
32. Dewsbury, A. R., Battered wives. Family violence seen in general practice, *R. R. Soc. Health*, 95, 290, 1975.
33. Collier, J., When you suspect your patient is a battered wife, *RN*, 50, 22, 1987.
33. Hamel, J., Toward a gender-inclusive treatment of intimate partner violence research and theory: Part 1—Traditional perspectives, *Internal J. Men's Health*, 6, 36, 2007.
34. Gryl, F. E., Stith, S. M., and Bird, G. W., Close dating among college students. Differences by use of violence and by gender, *J. Social Personal Relation.*, 8, 243, 1991.
35. Hines, D. A. and Sandino, K. J., Gender differences in psychological, physical and sexual abuse among college students using the revised Conflict Tactus Scale, *Violence and Victims*, 18, 197, 2003.
36. Burston, G. R., Granny-battering, *Br. Med. J.*, 3, 592, 1975.
36. Hird, M. J., An empirical study of adolescent dating aggression in the U.K., *J. Adolescence*, 23, 69, 2000.
37. Butler, R. N., *Why Survive?: Growing Old in America*, Harper & Row, New York, 1975.
37. Pedersen, P., and Thomas, C. D., Prevalence and correlation of dating, violence in Canada, *Canad. J. Behav. Sci.*, 24, 490, 1992.
38. Henton, J., Cate, R., Koval, J., Lloyd, S., and Christopher, S., Romance and violence in dating relationships, *J. Fam. Issues*, 4, 467, 1983.
39. Ridley, S. A. and Feldman, C. M., Female domestic violence toward male partners: Exploring conflict responses and outcome, *J. Fam. Violence*, 18, 157, 170.
40. Plasr, M. S. and Glessner, J. C., Violence in courtship relations: A southern sample, *Creative Sociol.*, 11, 198, 1983.

41. Stacy, C. L., Schandel, I. M., Flannery, W. S., Conlon, M., and Milardo, R. M., It's not all Moonlight and roses: Dating violence at the University of Maine, *College Student J.*, 28, 2, 1994.
42. Straus, M. A., Prevalence of violence against dating partners by male and female university students worldwide, *Violence Against Women*, 10, 790, 2001.
43. Balsm, K. F., Rothblum, E. D., and Beauchine, T. P., Victimization over the life span: A comparison of lesbian, gay, bisexual, and heterosexual siblings, *J. Consult., Clin. Psychol.*, 73, 477, 2005.
44. Creenwood, G. L., Relf, M. V., Huang, B., Pollack, L. M., Canchoia, J. A., and Catunia, J. A., Battering victimization among a probability-based sample of men who have sex with men, *Am. J. Public Health*, 92, 1964, 2002.
45. Vasquez, D. and Falcone, R., Cross-gender violence, *Am. Emerg. Med.*, 29, 427, 1997.
46. McLeod, M., Women against men: An examination of domestic violence based on an analysis of official data and national victimization data, *Justice Quarterly*, 1, 171, 1984.
47. Wilson, M. I. and Daley, M., Who kills whom in spouse killings? On the exceptional sex ratio of spousal homicides in the United States, *Criminology*, 30, 189, 1992.
48. Mercy, J. A. and Saltzman, L. E., Fatal violence among spouse in the United States, 1975–85, *Am. J. Public Health*, 79, 595, 1989.
49. Mize, K. D. and Shackleford, T. K., Intimate partner homicide methods in heterosexual, gay and lesbian relationships, *Violence Vict*. 23, 98, 2008.
50. McDowell, J. D., Diagnosing and treating victims of domestic violence, *N.Y. State Dent. J.*, 62, 36, 1996.
51. Center for Disease Control, Public health and aging: Trends in aging—United States and worldwide, *MMVR Weekly*, 52, 101, 2003.
52. Cooper, C., Selwood, A. and Livington G., The prevalence of elder abuse and neglect, *Age Ageing*, 37, 151, 2008.
53. Brock, J. N., The evolution of the aging population, *State of Business Magazine*, XI, 3, 1998.
54. Burston, G. R., Granny-beating, *Br. Med. J.*, 3, 592, 1975.
55. Wolf, R. S., Elder abuse: Ten years later, *J. Am. Geriatr. Soc.*, 36, 758, 1988.
56. American Medical Association, Council on scientific affairs report: Elder abuse and neglect, *JAMA*, 257, 966, 1987.
57. Pillemer, K. and Finkelhor, D., The prevalence of elder abuse: A random sample survey, *Gerontologist*, 28, 51, 1988.
58. Lett, J. E., Abuse of the elderly, *J. Fla. Med. Assn.*, 82, 675, 1995.
59. Yan, E., Tang, C, S.-K., and Yeung, D., No safe haven: A review of elder abuse in Chinese families, *Trauma, Violence and Abuse*, 3, 167, 2002.
60. Collins, K. A., Elder maltreatment: A review, *Arch. Path. Lab. Med.*, 130, 1290, 2006.
61. A. M. A., Council on Scientific Affairs Report: Elder abuse and neglect, *J. Am. Med. Assoc.*, 257, 966, 1987.
62. Wolf, R. S., Elder abuse: Ten years later, *J. Am. Geriatr. Soc.*, 36, 758, 1988.
63. Jayawardena, K. M. and Liao, S., Elder abuse at the end of life, *J. Palliat., Med.*, 1, 127, 2006.
64. Pillemer, K. and Finkelhor, D., The prevalence of elder abuse. A random sample survey, *Gerontologist*, 28, 51, 1988.
65. McDowell, J. D., Elder abuse, the presenting signs and symptoms in the dental practice, *Texas Dent. J.*, 107(2), 29, 1990.
66. Cooper, C., Selwood, A., Blanchard, M., Walker, Z., Blizzard, R., and Livingston, G., The determinants of family carer's abusive behavior to people with dementia, *J. Affect. Disord.*, Epub: May, 2009.
67. Brogdon, B. G., Forensic aspects of radiology, in *Mediolegal Investigation of Death*, 4th ed., Spitz, W. U., Ed., C. C. Thomas, Springfield, IL, 2006, chap. XXI.

CREDITS

From Brogdon, B. G., Vogel, H., and McDowell, J. D., Eds., *A Radiologic Atlas of Abuse, Torture, Terrorism, and Inflicted Trauma*, CRC Press, Boca Raton, FL, 1990. Figures 18.1, 18.3—18.9b, 18.10b—18.16, 18.18–18.20.

Section VI

Radiology in Nonviolent Crimes

Excluding tort actions for personal injury and/or malpractice, the major thrust of forensic radiology has been the evaluation of violence and violent crime. It is used to evaluate missile injuries and other traumatic lesions. It has established the age, and thus sealed the fate, of convicted murderers. It discovers abuse. It seeks to explain accidental or unattended death. It is employed to identify remains that have been mutilated, incinerated, separated, commingled, decomposed, or dehydrated and are beyond recognition by more conventional means.

But as we have already seen in Chapter 2, the radiological pioneers also applied Röntgen's rays to such relatively peaceable criminal pursuits as mail fraud, adulteration of foodstuff, and forgery of legal documents and other items.

In this section, three modern examples of the use of imaging techniques in detecting nonviolent crimes are presented.

B. G. B.

19 Smuggling/Border Control

B.G. Brogdon, H. Vogel, and P.R. Algra

CONTENTS

INTRODUCTION

As mentioned in Chapter 2, as early as 1897 the French customs service (*la Douane*) was using fluoroscopic x-ray equipment to apprehend smugglers of contraband seemingly widely disparate in importance by today's standards, ranging from jewels to cigarettes and matches (which were monopolies of the government).[1]

THE "BODY PACKER"

Beginning in the 1970s, a new breed of smuggler began to be recognized and apprehended by the radiological method.[2] This was the "body packer," a specialized type of "mule," who smuggled contraband drugs (mostly cocaine) across borders in specially devised packages secreted in the carrier's alimentary canal by ingestion or, in the case of larger packages, by insertion into the vagina or rectum.

Many of these smugglers were users or addicts earning the wherewithal to feed their own habit. A few were simply bringing in a supply for their own use. Others were commercial carriers in it for the money or the excitement.[3] The monetary rewards are not insubstantial. The value of purchases from developing countries can increase almost tenfold with successful (undetected) importation.[4] Whatever the motivation the purpose is the same—to transport concealed narcotics across borders and through customs without detection.

Packages stored for the "short haul" in the vagina or rectum are easily accessed and discovered by manual examination in a body search. This probably prompted the more common practice of swallowing the packages of contraband since no authority has immediate access to the gastrointestinal tract.[5] Body packing quickly became a worldwide problem.

Early cases of body packing were brought to official attention when the perpetrators became ill or died from an overdose of the transported drug (Figure 19.1). Other cases came to light because of obstruction of the alimentary canal. One inept smuggler succeeded in lodging the packet in the cervical esophagus, with total esophageal obstruction and was caught because he was unable to swallow his own saliva.[3] Thus, apprehension of the smuggler is important not only from the standpoint of law enforcement, but it may also save the life of the miscreant. A package that does not pass through the gastrointestinal tract may break open and cause death immediately, or as long as ten days after ingestion.

PACKAGING

Cocaine, heroin, amphetamines, hashish, and marijuana have all been transported in the alimentary tract.[3,6–9] Initially the drug is wrapped in several layers of latex by using condoms, the fingers of surgical gloves, or even toy balloons (Figure 19.2). Sometimes inner layers of other materials such as carbon paper, cellophane, and plastic wrap were included (Figure 19.3). The ends of the packages were securely tied like a sausage and the resultant package is swallowed or secreted in the rectum. Up to 214 such packages have been found in a single "mule."[10] These packages usually contained 3–7 g of narcotics.[11] They were round or oval in shape and most were 1–2 cm in diameter.[5]

Some condoms may act as semipermeable membranes. Once fluid from the gastrointestinal tract gains access to the package, the law of Gibbs–Donnan equilibrium takes effect; cocaine hydrochloride, a salt, may diffuse out into the gut or may cause additional fluid to be drawn into the package to the

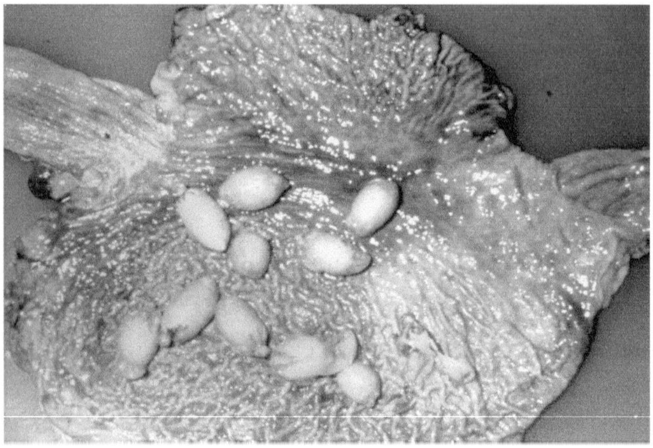

FIGURE 19.1 Latex-covered narcotic packages are shown in the opened stomach of a body packer who died of an overdose when one of the packages ruptured. (Courtesy of James M. Messmer, MD.)

point of rupture. In either event, cocaine toxicity can ensue rapidly.[12] Cases of clinical obstruction or drug overdose led to abdominal radiography, and this disclosed the classic radiological finding in the body packer. During wrapping, air gets captured between the layers of packaging in the majority of cases. Gastrointestinal gas may pass into a deteriorating package. Gas may be generated inside the package by fermentation of plant material at body temperature as in the case of marijuana.[8,13]

DIAGNOSIS/DETECTION

Radiographically one looks for regularly shaped round or oval foreign bodies outlined by arcuate or encircling thin air

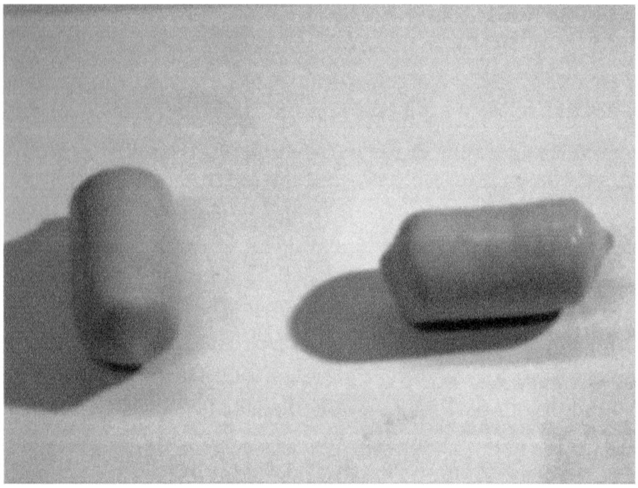

FIGURE 19.2 Modern hand-wrapped latex drug packages (called "bolitas" in the Caribbean) are multilayered and may incorporate foil or transparent cellotape to lessen air-trapping between layers or failure with spillage. (From Algra, P. R., Brogdon, B. G., and Marugg, R. C., *Am. J. Roentgenol.*, 189, 331–336, 2007. With permission.)

FIGURE 19.3 Machine-wrapped bolitas use combinations of paraffin, electrician's tape (which blackens the packages) or candlewax over hardened cocaine pellets to eliminate air-trapping, failure or obstruction. These capsules are hardened to resist breakage. (From Algra, P. R., Brogdon, B. G., and Marugg, R. C., *Am. J. Roentgenol.*, 189, 331–336, 2007. With permission.)

shadows, sometimes in multiple layers (Figure 19.4). The entire package may be outlined by intramural gas in the gut.

The bag form may taper slightly from its fundus to the neck, which may have a rosette form where it has been tied off.[14]

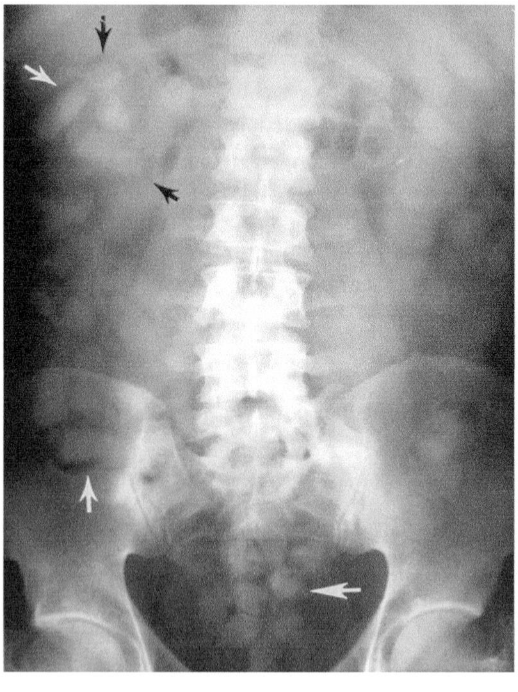

FIGURE 19.4 Abdominal radiograph of a body packer showing rounded and ovoid, slightly hyperdense, packages, some of which are surrounded by a halo of entrapped gas or air (some marked with arrows). (Courtesy of Richard N. Azpuru, MD.)

The increased sensitivity of computed tomography (CT) is useful in evaluating equivocal cases (Figure 19.5). Water-soluble iodinated contrast material has been given orally to confirm or exclude body packages and also is used to follow the elimination of drug packets in smugglers treated conservatively rather than with surgery.[15] Some authors suggest routine urine testing of suspects to diminish false negative or positive findings, and as an indication for the necessity of further radiological studies.[12,16,17]

ADVANCED PACKAGING AND DETECTION TECHNIQUES

As radiological detection of intracorporeal contraband became more prevalent and accurate, packaging techniques improved. Many drug packages ("bolitas" in the Caribbean trade) are machine wrapped, smooth, ovoid, and of slightly larger capacity (10–12 g of pure narcotic). (Figure 19.6) Wax, paraffin, electrical tape, aluminum foil, cellophane tape, and rubber can be found in various combinations[12,18] (Figures 19.2 and 19.5). The bags may be of variable density: hyperdense, hypodense, or isodense (Figure 19.7).

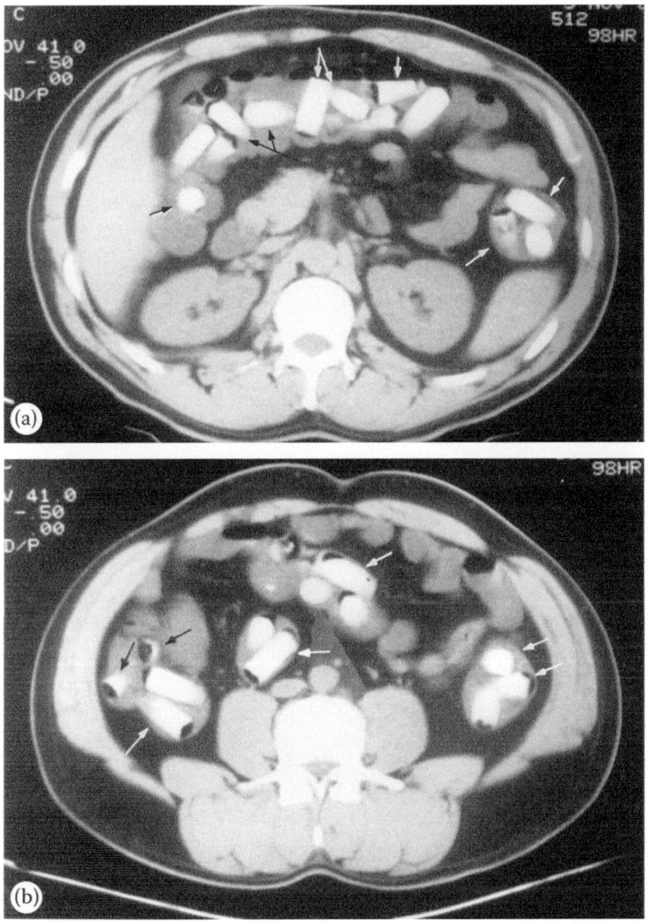

FIGURE 19.5 (a) and (b) CT images of a body packer showing multiple drug packages somewhat denser than bowel content. Most contain entrapped air at the ends of the packages. (Courtesy of Richard N. Azpuru, MD.)

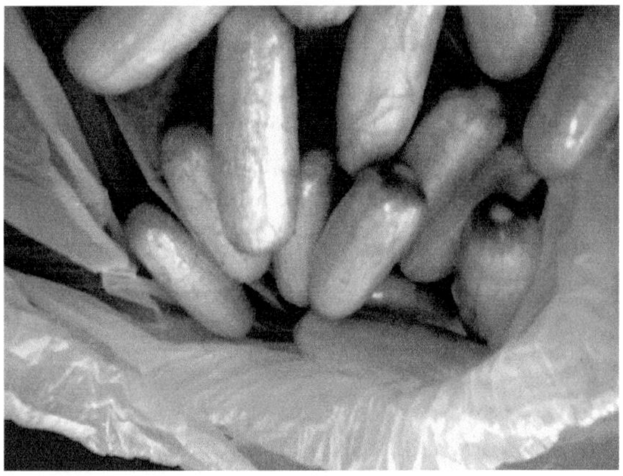

FIGURE 19.6 Machine-wrapped bolitas of colorless paraffin "streamlined" for easy passage through the gut. (Courtesy R.C. Marugg, MD.)

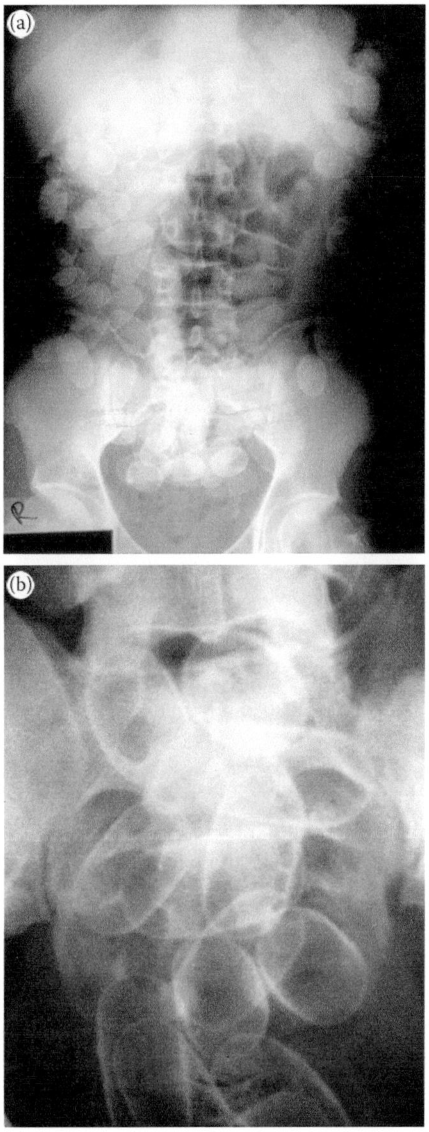

FIGURE 19.7 (a) Latex-wrapped packages slightly more dense than mixed bowel contents. (b) This machine-wrapped hard cover is unusually dense.

Of the three major drugs transported internally, cocaine's density ordinarily is closest to water or feces. Usually heroin in less dense and cannabis more dense than feces, but this density may be altered by the degree of purity or the composition of additives.[18] Cocaine is the principal target of radiological investigation. Diagnostic characteristics of positive findings vary according to the type of packaging; the size, number, and distribution of packages; and the habitus, habits, and cooperation of the smugglers.[4,19–21]

Modern packaging has greatly reduced the telltale evidence of obstruction, and leakage is rare.[4,12]

Swallowing of bolitas is sometimes facilitated by first drinking a "Colombian cocktail" of cocaine in a soft drink.[20] It has been reported that newly recruited mules undergo training by learning to swallow large grapes whole.[22]

By swallowing these separate batches at four-hour intervals, the entire gastrointestinal track can be loaded. Parasympathomimetic drugs may be used to inhibit peristalsis

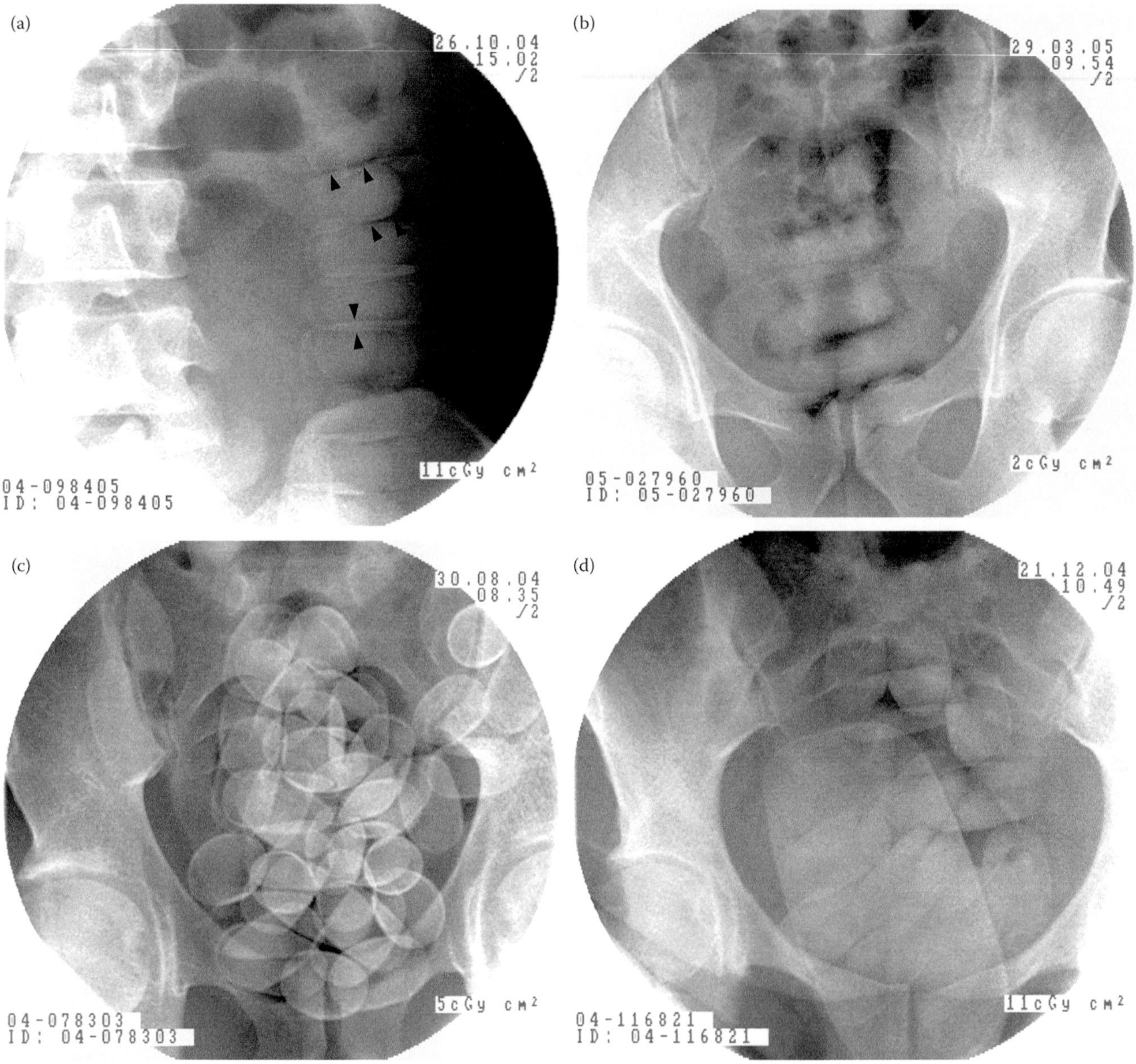

FIGURE 19.8 (a) Diagnostic features. A 31-year-old man with hand-wrapped latex bolitas in descending colon arranged in "stack of coins" configuration. Air trapped between layers causes the dark halo of the "double condom" sign (arrowheads). (b) Diagnostic features. A 23-year-old woman with cellotaped latex bolitas inserted rectally (a "pusher"). Note: Bolitas for rectal insertion are larger than swallowed bolitas. (c) Diagnostic features. A 21-year-old man contains ideally imaged 2 × 3 cm machine-wrapped ovoid bolitas with hard capsule slightly more dense than tissues, feces, or latex wrap, but with no halo of trapped air. This "pusher"/"packer" had bolitas throughout the bowel. (d) Diagnostic features. A 31-year-old female carrying multiple machine-wrapped rectally inserted bolitas plus a large vaginal bolita which contained 37 grams of pure cocaine. (From Algra, P. R., Brogdon, B. G., and Marugg, R. C., *Am. J. Roentgenol.*, 189, 331–336, 2007. With permission.)

because premature defecation in flight requires the carrier to clean and reswallow the package en route. A passenger found dead on arrival at his destination was found at autopsy to have a failed bolita spilling a fatal dose into the esophagus. Presumably the reswallowed package had been weakened by its previous exposure to peristalsis and digestive juices[20].

Images of the left upper abdomen, the right mid-abdomen, and the pelvis obtained quickly with an upright digital x-ray machine have proved effective in scanning for drug packages from the stomach to the rectum and vagina (Figure 19.8). Since many body-baggers are also addicts and consequently chronically constipated, their rounded compact stools (scybala)[12] may resemble drug packages (Figure 19.9). A false-positive interpretation of an abdominal plain film can result from this.[12]

Carriers and their employers use many ploys in an attempt to degrade the image or distract the interpreter.[24] Most of these are unsuccessful. Movement during exposure or standing away from the film or receptor reduces image sharpness, but these maneuvers are usually detected by the radiographer (Figure 19.10). A widely held but mistaken belief embraced by traffickers is that metallic embellishment of the clothing on body, ingested metallic objects, or positive contrast media will distort or confound detection by obscuration or distortion of the contraband images (Figures 19.11). On the contrary, these ploys increase detection rates; at Schiphol Airport, finding an intra-abdominal coin correlated 100% with intra-alimentary or intracavitary drugs in 36 consecutive cases (Figure 19.12). The most-nearly effective evasion was to purposely pack air and sand between layers of the package to simulate feces (Figure 19.13).

The increased sensitivity of sectional imaging, CT, or MRI, will resolve equivocal finding[24] but, so far, is impractical for routine drug screening at airports and other points of entry (Figure 19.14).

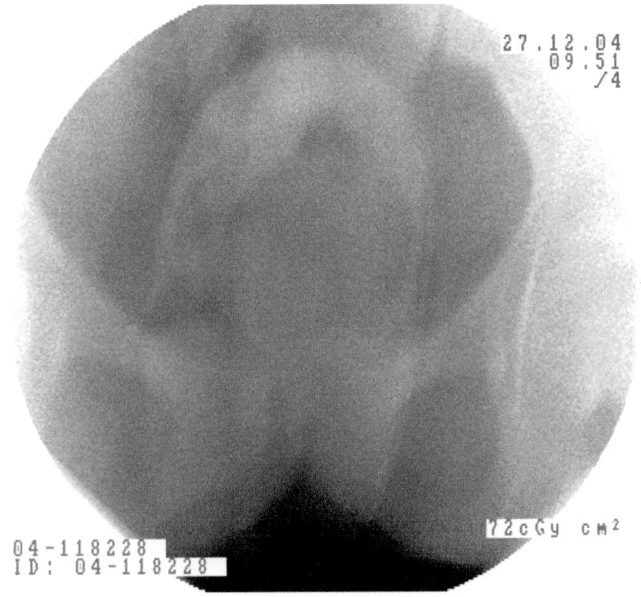

FIGURE 19.10 Evasive tactics. A 23-year-old man. Note general degradation of the image. This "pusher" with a large rectal package deliberately disobeyed instructions to stand motionless against the image receptor. (From Algra, P. R., Brogdon, B. G., and Marugg, R. C., *Am. J. Roentgenol.*, 189, 331–336, 2007. With permission.)

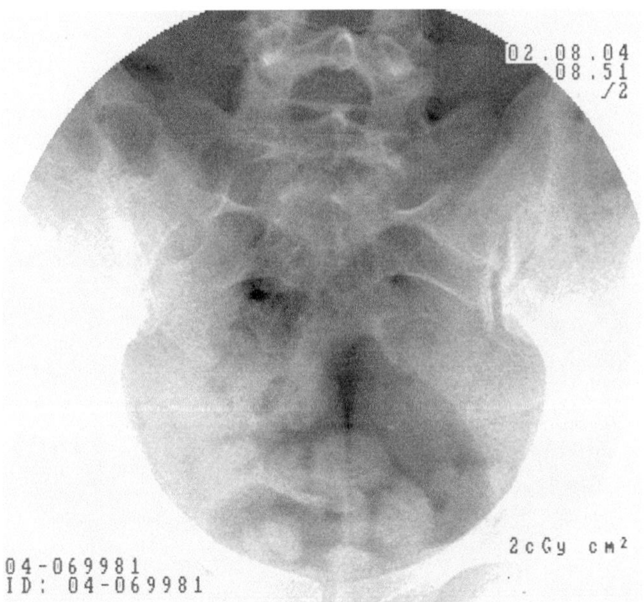

FIGURE 19.9 False positive. A 31-year-old man with slightly hyperdense stool in rectum confused with bolitas. "Mules" frequently are constipated, being themselves users, or may use medication to prolong transit through the bowel. (From Algra, P. R., Brogdon, B. G., and Marugg, R. C., *Am. J. Roentgenol.*, 189, 331–336, 2007. With permission.)

BORDER CONTROL

In recent years, illegal immigration and threats to security by terrorists have increased incentives to better control traditional smuggling of drugs and objects of value at national ports of entry. An array of sophisticated equipment and methods have been developed to meet this challenge, thus stimulating a new industry producing high-tech radiation-based security devices to search persons, luggage, cargo, and transport. Both x-rays and gamma-rays are employed in a variety of ways to search for drugs, explosives, weapons, cigarettes, alcohol, stolen vehicles, jewels, books, tapes, videos, protected animals, proscribed plants, and undocumented humans.[23–27]

SEARCH OF THE PERSON[4,26,27]

Earlier in this chapter we have dealt with methods of detecting contraband carried inside the body. Heretofore, discovering of concealed materials carried on or outside the body has required a magnetic search (only good for materials), a pat-down search, a strip search or a cavity search.

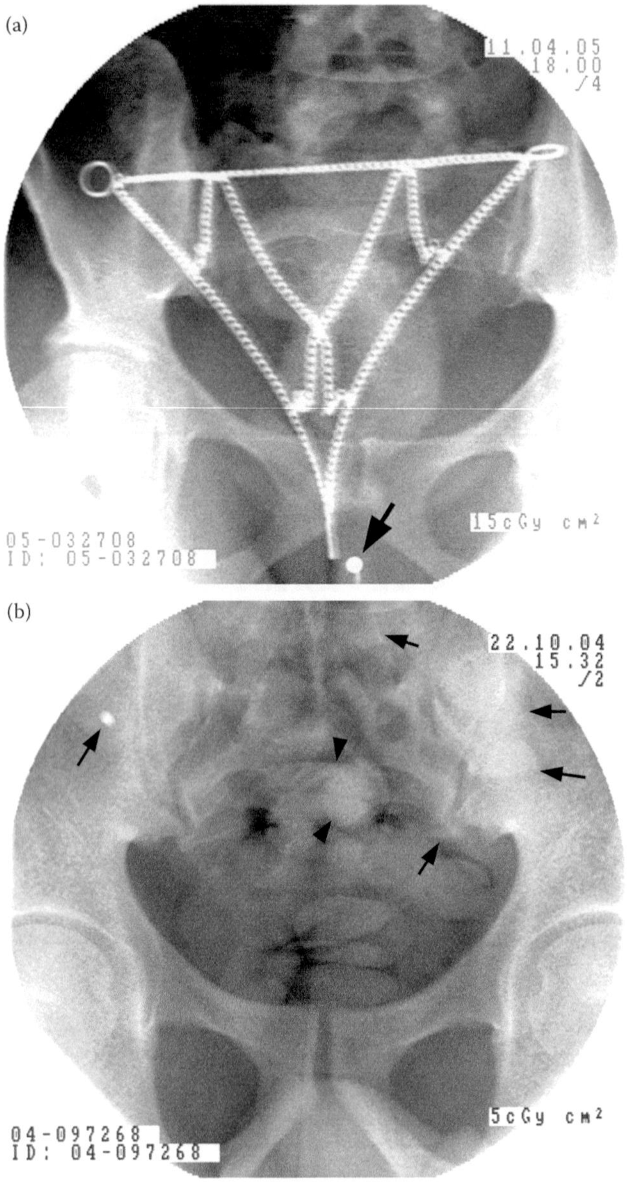

FIGURE 19.11 (a) Evasive tactics. A 20-year-old woman "pusher" with metallic decorations in undergarment for distractions. Note labial stud. (arrow). (b) Evasive tactics of a 19-year-old male with metallic pellet in right colon (arrow), contrast material (arrowheads) in left colon, and bolitas in the rectum. (From Algra, P. R., Brogdon, B. G., and Marugg, R. C., *Am. J. Roentgenol.*, 189, 331–336, 2007. With permission.)

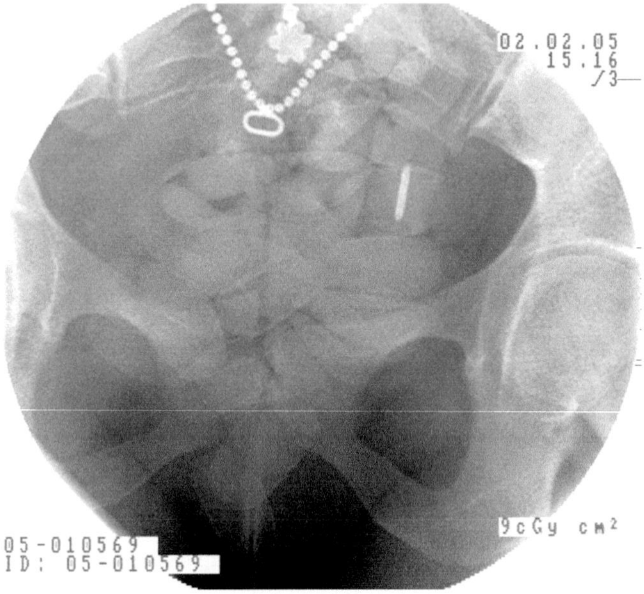

FIGURE 19.12 Evasive tactics. A 25-year-old woman with a coin in the colon which "tips off" the sigmoid bolitas. Note the pendant from a navel piercing and chain dangling from belt. (From Algra, P. R., Brogdon, B. G., and Marugg, R. C., *Am. J. Roentgenol.*, 189, 331–336, 2007. With permission.)

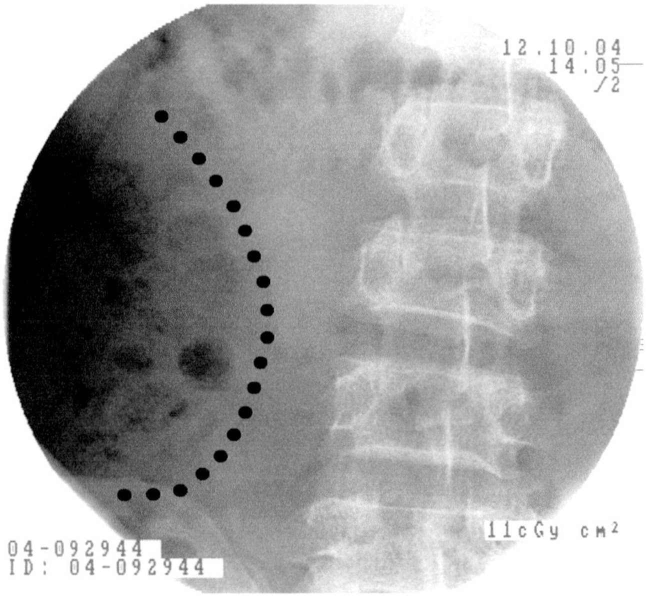

FIGURE 19.13 Evasive tactics. A 44-year-old woman. In the right colon are multiple condom-wrapped bolitas intentionally layered with air and sand (circled) to closely simulate feces (almost successfully). (From Algra, P. R., Brogdon, B. G., and Marugg, R. C., *Am. J. Roentgenol.*, 189, 331–336, 2007. With permission.)

Radiation-based modalities employed in searching the person include backscatter imaging, transmission/fluoroscopic imaging with routine medical equipment[20] (already illustrated), transmission summation images produced by scanners dedicated for specific search targets, dual-energy scanners, and CT scans (requiring transport to a medical facility).

MODALITIES

Backscatter imaging uses the scattered ("Compton") radiation from low-energy (50 k Vp) x-rays that hardly penetrate

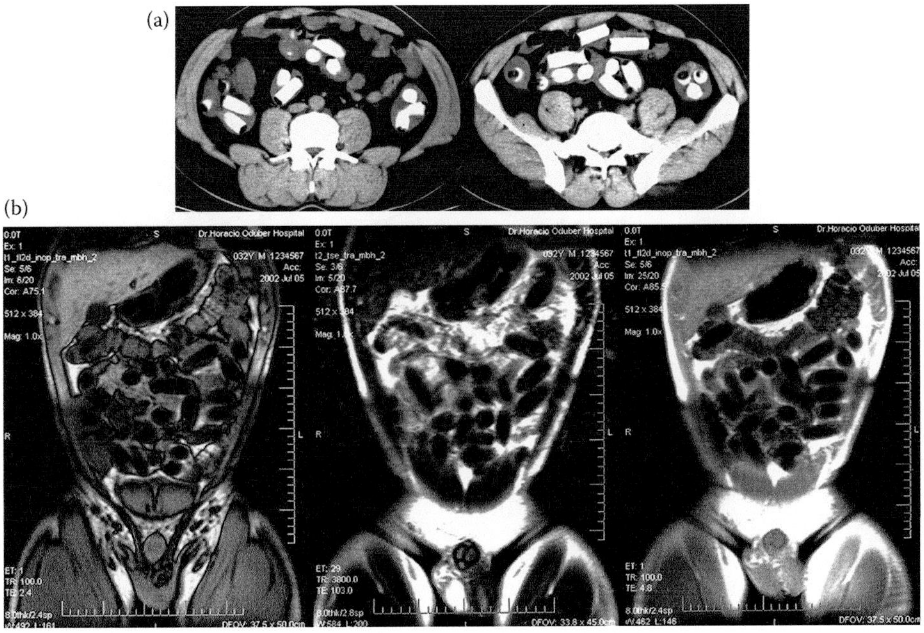

FIGURE 19.14 (a) Sectional imaging. CT images show the cocaine packages to be slightly higher than tissue density, and the trapped air in the condoms becomes more obvious. (Courtesy of Richard N. Azpuru, MD.) (b) Sectional imaging. MR images using different protocols clearly depict the machine-wrapped bolitas. [Left image: Flash 2D, out of phase (TR 100, TE 2.4): Middle image: T2 TSE turbospin echo (TR 3800, TE 103): Right image: Flash 2D, in phase (TR 100, TE 4.8] (Courtesy R.C. Marugg, MD, radiologist at the Dr. Horacio E. Oduber Hospital in Aruba. He reports that he finds more drug packets with MR than with CT examination of the same subject.)

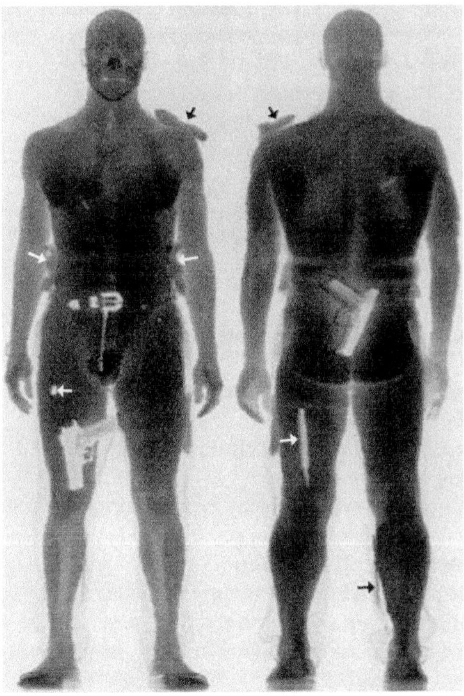

FIGURE 19.15 Frontal and rear backscatter images show only the body surface and items hidden upon it, or in/beneath clothing. This model's image displays on the frontal surface: (1) simulated cocaine package on left shoulder and others on both sides above belt; (2) coin in trouser pocket; (3) handgun. Posterior image shows: (1) Handgun with plastic grip at waist; (2) file taped to rear thigh; (3) plastic knife medial side right calf; and again, simulated drug package on left shoulder. (From A.S. & E., Inc., With permission.)

the body. The Compton scatter back x-ray produces an image of the body's surface and metallic, plastic, or vegetable materials concealed upon it (Figures 19.15 and 19.16). Internal structures are not revealed, but a rather accurate reflection of surface features, including genitalia, is displayed; consequently, there are privacy issues with backscatter searches.

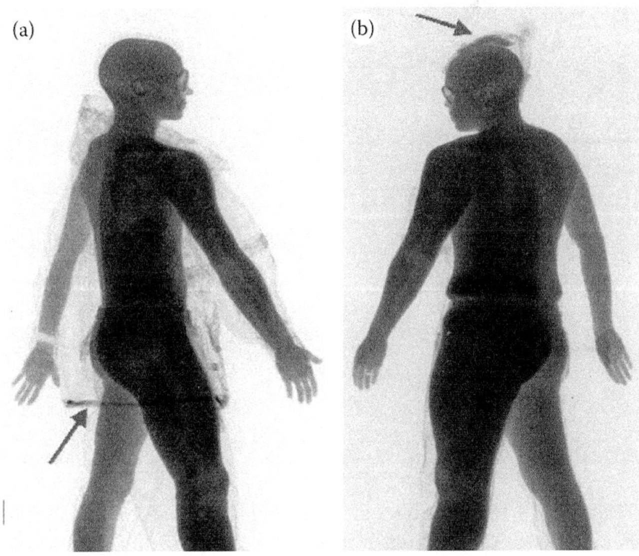

FIGURE 19.16 (a) Oblique backscatter image shows drugs in the hem of the parka that had been dipped in coca pulp. (From A.S. & E., Inc., With permission.) (b) Drug package hidden in large coiffure or hairpiece. (From Compass Security., With permission.)

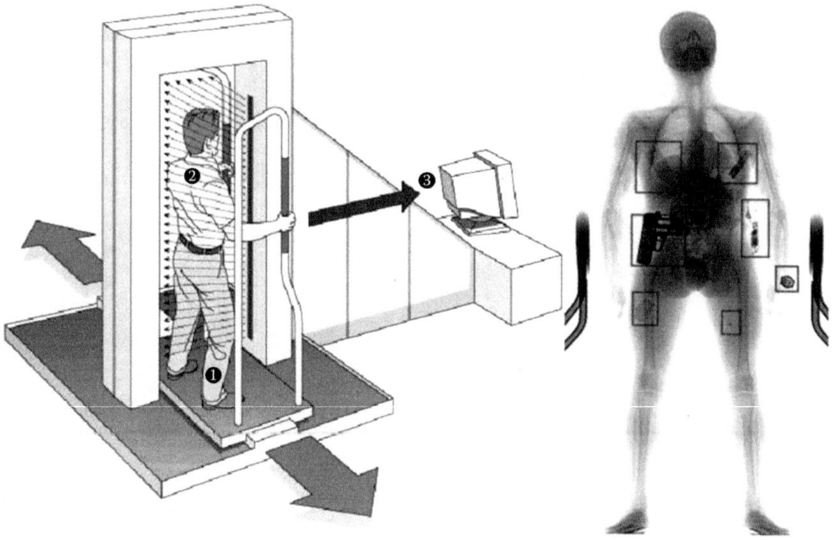

FIGURE 19.17 Schematic of transmission scanning equipment: The person in his usual upright position is moved laterally, while standing on a moving frame or belt (1). A conventional x-ray tube produces a pencil beam, which moves rapidly from top to bottom, and which is adjusted on a line of detectors (2). A transmission/summation image is produced and displayed on a screen (3).

Transmission imaging can be done with standard medical equipment (Figure 19.17 and 19.18). This gives a high-definition summation image of internal structures and contents, and is particularly effective in detecting intracorporeal drugs as already demonstrated. Radiation doses range around 3.2–6 μSv.

Transmission imaging equipment is also designed for nonmedical use but with specific targets in mind, that is, metal, plastic, raw diamonds, etc. These units necessarily are simplified for on-site field work (one-button, no adjustment, low dose, and vandalism proof). The x-ray energy is fixed to a maximum performance of the detector at a scanning dose of less than 3 μGy whenever and for whatever purpose. Obviously there are compromises involved. While the radiation dose is lower than that from medical imaging equipment, the detail of near-isodense materials within the body is not as good (Figures 19.19 through 19.23). On the other hand, there

FIGURE 19.18 Photograph of a Soter RS bodyscanner, a fan-beam transmission scanner installed worldwide by ODSecurity. The traveler is standing on the platform moving from side-to-side facing the fan-beam which is directed to a vertical row of detectors behind him. The operator stands to left and sees the resultant image on the screen in front of him. The attendant, right front, positions the traveler. (Courtesy of Soter, ODSecurity. With permission.)

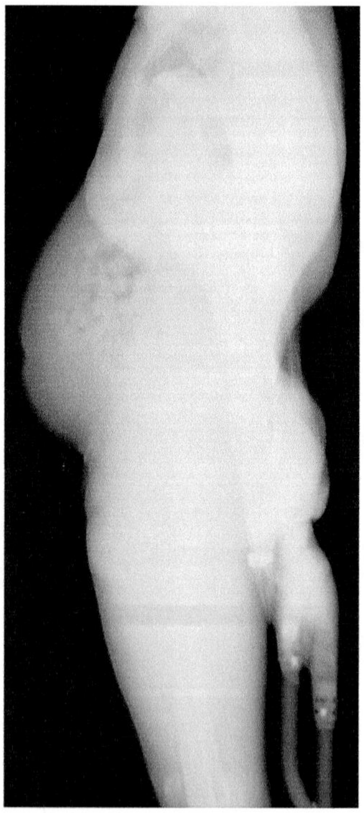

FIGURE 19.19 This protuberant abdomen is better imaged with low-dose transmission in the lateral projection. Multiple cocaine packages are seen with surrounding air. (Courtesy of Soter, ODSecurity. With permission.)

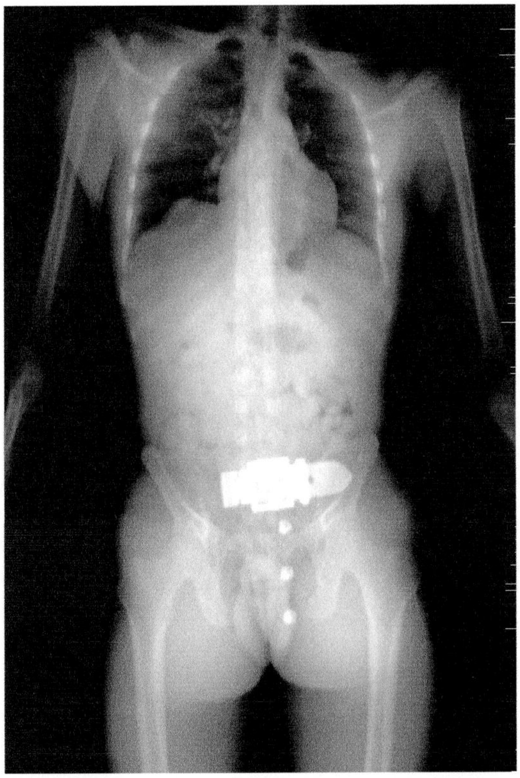

FIGURE 19.20 Low-dose transmission image shows large belt-buckle, three metallic trouser buttons and cocaine bolitas faintly in right mid-abdomen, better on the left, with "double condom" sign. (Courtesy of Soter, ODSecurity. With permission.)

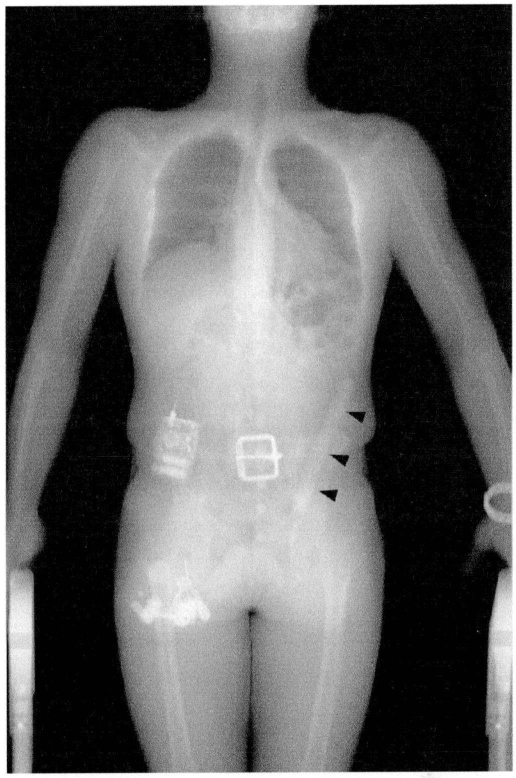

FIGURE 19.22 Low-dose transmission scan showing coins or contraband in pocket, an electronic device on the belt, and a stick of dynamite hidden beneath the clothing, left flank. (Courtesy of Soter, ODSecurity. With permission.)

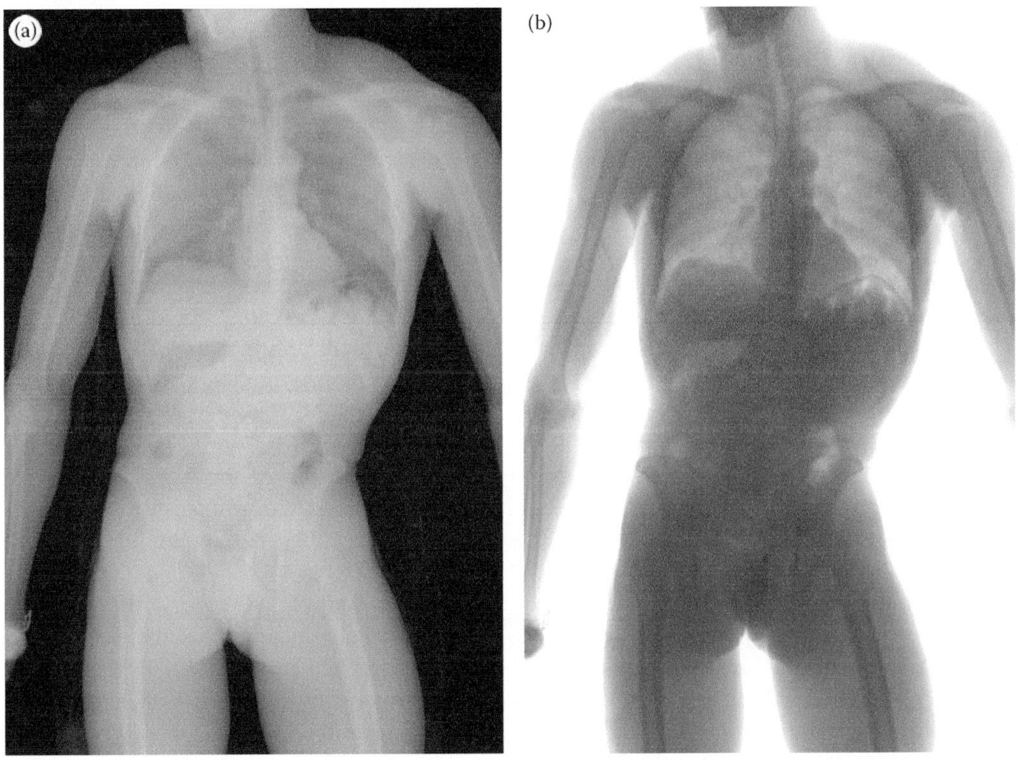

FIGURE 19.21 (a) "White bone" and (b) "Black bone" displays of low-dose transmission beam showing cocaine packages in the air bubble at the top of the stomach. Switching between the two types of displays may be helpful. (Courtesy of Soter, ODSecurity. With permission.)

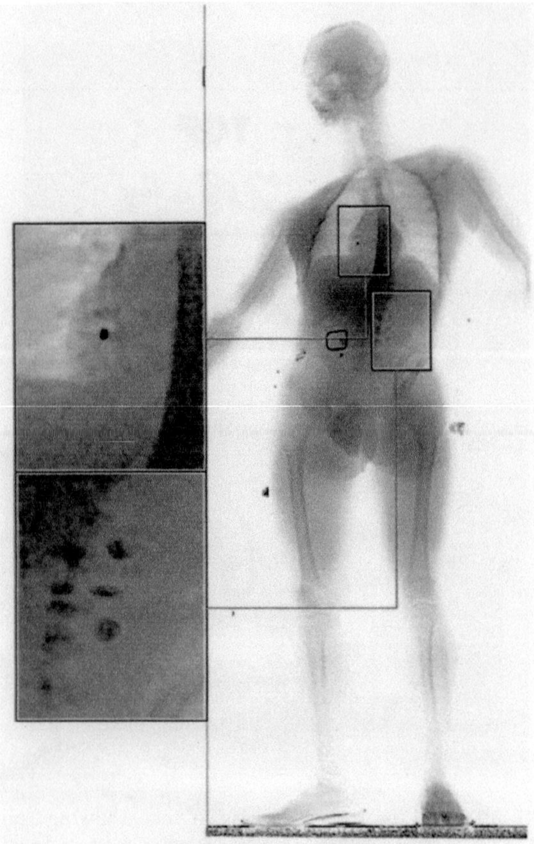

FIGURE 19.23 Full-body fluoroscopic scan showing raw diamonds and/or semiprecious stones in the esophagus and gastrointestinal tract.

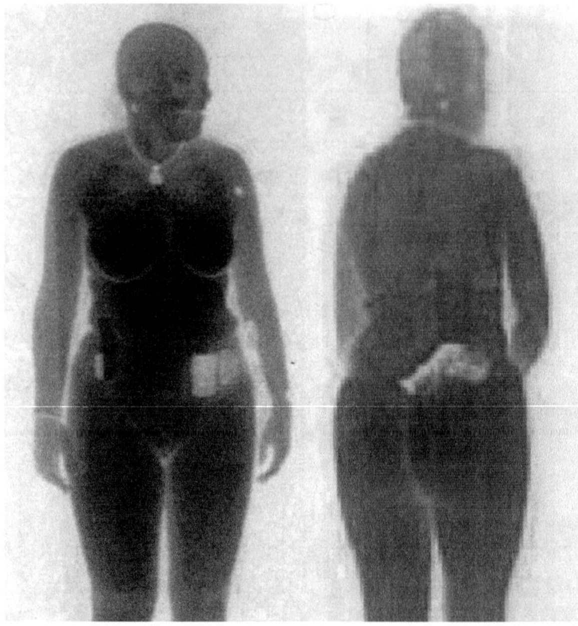

FIGURE 19.24 At the inauguration of backscatter imaging at an airport, an elegantly dressed official of the manufacturer tested the device while wearing a handgun. The test was perfect, but she was embarrassed by the display of her silhouette beneath the finery.

is essentially the same effectiveness as backscatter scanning for objects attached to the skin or hidden in clothing but is effective to the point that, if there are many packages in the alimentary canal, some will be visible.

OTHER CONSIDERATIONS

The use of ionizing radiation for purposes of search and control raises the issue of exposure for nonmedical purposes. This is illegal in some jurisdictions (Germany, for instance) and creates ethical concerns among some physicians and scientists. In some locales, radiation exposure is accomplished without the knowledge or permission of the person examined. In other places, the person is given the choice between radiation search, manual search, or detention until drug-free stools are passed or other tests are performed.

The backscatter system, in particular, raises issues of privacy, since breast and genital areas may be displayed. This simply offends the modesty of some by revealing general body contours, even if the genitals and breasts are obscured (Figure 19.24). In others, backscatter imaging may offend strong religious or cultural prohibitions against display of the uncovered body. Each jurisdiction must resolve these issues with some compromise between the requirements of security, the law, and pubic opinion or acceptance.

SEARCH OF LUGGAGE, TRANSPORT, AND CARGO

X-rays and gamma rays are utilized at international borders and internal checkpoints to inspect and control not only people, but also their hand-baggage and checked luggage, Transshipped goods and the transport and the carrier transporting them—automobiles, trucks, railway carriages, shipping pallets, and containers also are inspected with radiation-based systems (Figures 19.25 and 19.26).

Systems employed range from medical and industrial type x-rays through acceleration with 5–10 μeV energy and Cobalt-60 or Cesium-137 sealed sources. This variety can produce fluoroscope images, or used for computer-analysis of forward- and backscatter, for CT scans, or for dual-energy separation of different materials in the same container by analysis of atomic numbers. Computerized colorization of the response of different materials can aid in their rapid detection and identification (Figures 19.27 and 19.28).

Obviously some of the examined vehicles will contain people, either known or unsuspected. Those people may or may not be forewarned of impending exposure (Figures 19.29 and 19.30).

Combinations of dual-energy, transmission, and backscatter imaging can eliminate most of the visual "clutter" of superimposed objects in the same image field of view (Figure 19.31).

Systems using fan-beam and detector combinations can image large vehicles: trucks or railway cars even while they are in motion at speeds of 10/miles per hour or less. Conversely, mobile units can inspect stationary targets in a "drive-by" mode (Figure 19.32).

FIGURE 19.25 Transmission image of trailer-truck loaded with citrus boxes.

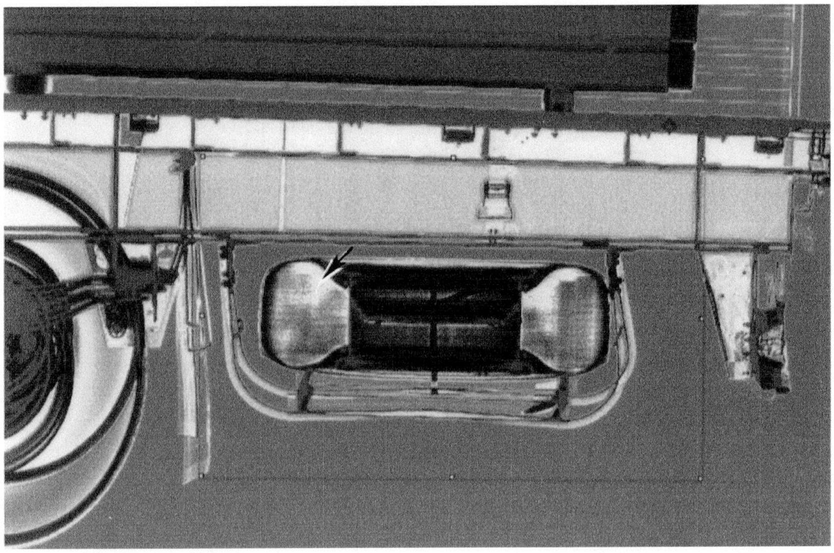

FIGURE 19.26 Rectangular cigarette cartons smuggled inside an air-filled spare-tire carried horizontally beneath a truck. Compare with air-filled tire on the ground in front (left) of the spare. Transmission image with 10 μeV acceleration beam.

FIGURE 19.27 Salamanders (a protected species) smuggled in a suitcase.

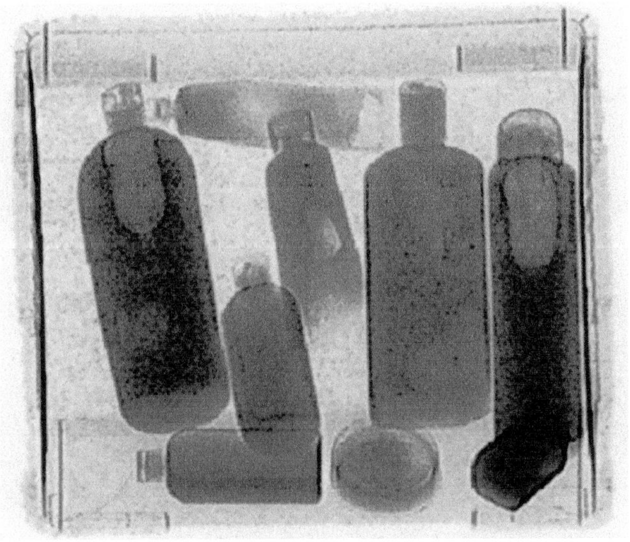

FIGURE 19.28 Perfume bottles in a box. Some contain drug-filled condoms fitted in the bottle's neck. This reuse is supposed to thwart drug-sniffing dogs.

FIGURE 19.29 Backscatter image of people hidden behind and under cargo in a 19-wheeler truck.

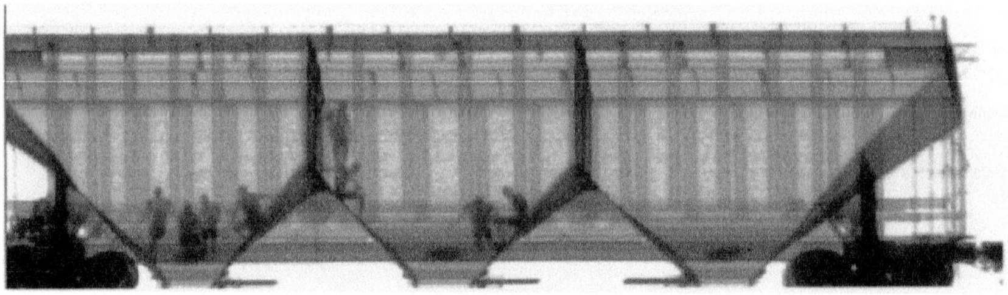

FIGURE 19.30 Stowaway illegal immigrants imaged in a railway hopper car at the Honduran/Mexican border. Fan-beam from a Cobalt-60 source permits image capture while the train is moving at a constant speed of 10 mph or less.

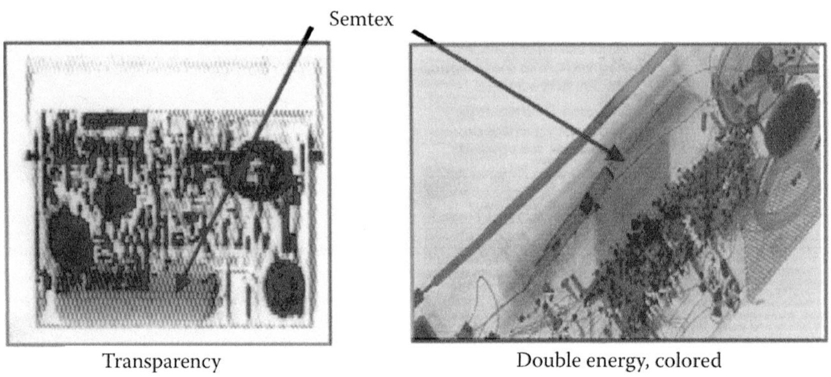

FIGURE 19.31 Left: Semtex plastic explosive can only be suspected in this radio studied by transmission or fluoroscopic technique. Right: Dual energy analysis makes separation from "clutter" and detection easier. Computer-added coloration will further highlight the semtex.

FIGURE 19.32 Backscatter imaging, from a moving vehicle, of a row of parked automobiles. Note the person hiding in the trunk of the automobile on the left. Only vehicle parts closest to the backscatter device can be analyzed. (Courtesy of A.S. & E., Inc., With permission.)

FIGURE 19.33 Liquid in containers inside a truck makes a wavy pattern as it sloshes around. It is illegal alcohol being hauled in a truck supposedly hauling solid freight.

FIGURE 19.34 (a) Closer scanning image of the same trailer-truck shown in Figure 19.25, which is supposed to contain boxed wrapped lemons. The middle three tiers look different. (b) Close-up showing two tiers of "funny" lemons on left, "normal" lemons on right. They actually were packages of cocaine shaped and wrapped like real lemons.

In general, these various search modes are designed to reveal densities where there should be voids, motion where these should be stillness, asymmetry where these should be order, and unexpected, unexplained, or ominous silhouettes (Figures 19.26, 19.33 and 19.34).

CONCLUSION

Security and border control systems based on radiation are being developed and deployed worldwide with ever-increasing sophistication and dedication of purpose. Unfortunately, it is uncertain at this time whether the bottlenecks and slowdowns experienced with current security measures will be alleviated or aggravated as a result.

REFERENCES

1. Collins, V. P., Origin of medico-legal and forensic roentgenology, in *Classic Descriptions in Diagnostic Radiology*, Vol. 2, Bruwer, A. J., Ed., Charles C. Thomas, Springfield, IL, 1984, p. 1593.
2. Freed, T. A., Sweet, L. N., and Gauder, P. S., Balloon obturation bowel obstruction: A hazard of drug smuggling, *Am. J. Roentgenol.*, 127, 1033, 1976.
3. Dunne, J. W., Drug smuggling by internal body concealment, *Med. J. Aust.*, 2, 436, 1983.
4. Pinsky, M. F., Ducas, J., and Ruggere, M. D., Narcotic smuggling: The double condom sign, *J. Can. Assoc. Radiol.*, 29, 79, 1978.
5. Pamilo, M., Suoranta, H., and Suramo, I., Narcotic smuggling and radiography of the gastrointestinal tract, *Acta Radiol. Diagn.*, 27, 213, 1986.
6. Beerman, R., Nunez, D., Jr., and Weth, C.V., Radiographic evaluation off the cocaine smuggler, *Gastrointest. Radiol.*, 11, 351, 1986.
7. Suarez, C. A., Arango, A., and Lester, J. L., III, Cocaine-condom ingestion, *J. Am. Med. Assoc.*, 238, 1391, 1977.
8. Dassel, P. M., and Punjabi, E., Ingested marijuana-filled balloons, *Gastroenterology*, 76, 166, 1979.
9. Karhunen, P. J., Penttila, A., and Panula, A., Detection of heroin "body-packers" at Helsinki airport, *Lancet*, 1, 1265, 1987.
10. Gherardi, R. K., Baud, F. J., Leporc, P., Marc, B., Dupeyron, J.-P., and Diamant-Berger, O., Detection of drugs in the urine of body-packers, *Lancet*, 1, 1076, 1988.
11. McCarron, M. M., and Wood, J. D., The cocaine "body-packer" syndrome: Diagnosis and treatment, *J. Am. Med. Assoc.*, 250, 1417, 1983.
12. Wetli, C. V., and Mittleman, R. E., The "body-packer" syndrome—Toxicity following ingestion of illicit drugs packaged for transportation, *J. Forensic Sci.*, 26, 492, 1981.
13. Karhunen, P. J., Suoranta, H., Panttila, A., and Pitkäranta, P., Pitfalls in the diagnosis of drug smuggler's abdomen, *J. Forensic Sci.*, 36, 397, 1991.
14. Sinner, W. N., The gastrointestinal tract as a vehicle for drug smuggling, *Gastrointest. Radiol.*, 6, 319, 1981.
15. Caruana, D. S., Weibach, B., Goerg, D., and Gardner, L. B., Cocaine-packet ingestion: Diagnosis, management, and natural history, *Ann. Intern. Med.*, 100, 73, 1984.
16. Marc, B., Baud, F. J., Aelion, M. J., Gherardi, R., Diamant-Berger, O., Blery, M., and Bismuth, C., The cocaine body-packer syndrome: Evaluation of a method of contrast study of the bowel, *J. Forensic Sci.*, 35, 345, 1990.
17. Gherardi, R., Marc, B., Alberti, X., Baud, F., and Diamant-Berger, O., A cocaine body packer with normal abdominal plain radiograms: Value of drug detection in urine and contrast study of the bowel, *Am. J. Forensic Med. Pathol.*, 11, 154, 1990.

18. Hergan, K., Kofler, K., and Oser, W., Drug smuggling by body packing: What radiologists should know about it. *Eur. Radiol.*, 14, 736–742, 2004.
19. Boguz, J. J., Althoff, H., Erkens, M., Maier, R. D., and Hoffman, R. Internally concealed cocaine, analytical and diagnostic aspects, *J. Forensic Sci.*, 40, 811–815, 1995.
20. Algra, P. R., Brogdon, B. G., and Marugg, R. C., Role of radiology in a National Initiative to Interdict Drug Smuggling: The Dutch experience, *Am. J. Roentgenol.*, 189, 331, 2007.
21. Brogdon, B. G., and Vogel, H. Search of the person, in *A Radiologic Atlas of Abuse, Torture, Terrorism, and Inflicted Trauma*, Brogdon, B. G., Vogel, H., and McDowell, J. D., Eds., CRC Press, Boca Raton, FL, 2002, 301–307.
22. Marston, J., *Maria Full of Grace* (award winning film), Journeyman Picture Production, 2004.
23. Brogdon, B. G., and Vogel, H. Border control and internal security, in *A Radiologic Atlas of Abuse, Torture, Terrorism, and Inflicted Trauma*, CRC Press, Boca Raton, FL, 2002, p. 299.
24. Brogdon, B. G., and Vogel, H. Search of the person, in *A Radiologic Atlas of Abuse, Torture, Terrorism, and Inflicted Trauma*, CRC Press, Boca Raton, FL, 2002, p. 301.
25. Vogel, H., and Brogdon, B. G., Search of luggage, cargo, and transport, in *A Radiologic Atlas of Abuse, Torture, Terrorism, and Inflicted Trauma*, CRC Press, Boca Raton, FL, 2002, p. 301.
26. Vogel, H., and Vogel, B., The invisible sentry: Radiological imaging in terror and terror prevention, *Forensic Sci. Med. Pathol.*, 2, 3, 2006.
27. Vogel, H., Ed., *Violence, War, Borders, X-rays: Evidence and Threat*, Shaker Verlag, smbH, Aachen, 2008, chaps. 1–8.

CREDITS

From Brogdon, B.G., Vogel, H. and McDowell, J.D., Eds., A Radiological Atlas of Abuse, Torture, Terrorism and Inflicted Trauma. CRC Press, Boca Raton, FL, 2002. Figures 19.7, 19.23, 19.25–30.

20 Forensic and Clinical Usage of X-rays in Body Packing

Patricia M. Flach, Steffen G. Ross, and Michael J. Thali

CONTENTS

HISTORICAL BACKGROUND OF BODY PACKING

Roentgen-ray examination was brought into use in a customs house in France that used the fluoroscopy x-ray equipment to detect the so-called "infernal machines" in packages.[1] These infernal machines could possibly be an explosive book filled with mercury fulminate and scraps of iron such as nails, screws, or even a revolver cartridge. In 1896, such an infernal machine was forwarded to two prominent deputies of the French chamber. As a result, the French called attention to the importance of Roentgen examinations at customs. Thus, the use of x-rays as forensic proof of hidden substances embedded in sealing wax, books, newspapers or parchment began.

Of course, economic considerations found their way into x-ray use as well. Octroi was a local tax collected on luxury goods (including jewels, cigarettes, and matches, which were monopolies of the government[2]) and substances used for food and drink that were brought into the district of Paris for consumption. In the nineteenth century, officers of octroi were posted at the entrances of Paris to inspect goods. The obtained revenue was used on city expenditures by the municipality. According to a newspaper article, "Custom Houses and the X rays" of the *New York Times* (July 31, 1897), a man who tried to smuggle cigars into Paris and a woman who was hiding a bottle on her body, with assumed absinthe as content, were both detected by x-rays. The *New York Times* predicted in late 1897 that swallowing contraband substances would be detected by x-ray.

The *New York Times* article turned out to be correct, as a radiologist experienced in detection now uses the appropriate modality to diagnose suspected contraband. In many cases, however, there are pitfalls to this process.

In 1973, balloon obturation bowel obstruction was first reported in the literature as a hazard of drug smuggling; meanwhile, those hiding goods within their body began to be recognized and apprehended by radiological methods.[3] Since 1972, fluoroscopy and plain films have been used to image suspected packages, but the first reported computed tomography (CT) scan of a drug courier was not performed until January 1990. That year, 15 miles short of New York's Kennedy International airport, Avianca flight 52 crashed on the north shore of Long Island while traveling from Bogota to New York. This airline was known as "Air Cocaine" among federal agents due to the number of smugglers who were detained. One of the survivors egested multiple packs of drug-filled condoms into a bedpan. In order to rule out any internal abdominal injuries after the trauma, an abdominal radiograph and the first CT of a body packer were completed. This 46-year-old drug carrier concealed a total of 29 packages; he was convicted and sentenced to six years in prison.[4]

DEFINITION OF DRUG CARRIERS AS "BODY PACKERS"

Drug carriers, the so-called "mules," smuggle contraband drugs (mostly cocaine) across borders in specially devised packages

TABLE 20.1
Definition of Drug Carriers

Body packer	Slang terms: "mule," "swallower," "courier," "internal carriers" A drug carrier who swallows specially prepared drug packets in order to evade detection of illicit narcotic drugs while smuggling them across borders in the gastrointestinal tract
Body pusher	A drug carrier who conceals drug containers within body cavities such as the rectum or vagina.
Body stuffer	Slang terms: "quick swallower," "mini packer" A person (typically a trafficker or user) who ingests small amounts of loosely wrapped drug, plastic pouches or small pellets upon an unexpected encounter with law enforcement for fear of arrest

within the body itself. Currently, the generic term "body packer" is commonly used to describe a drug courier who swallows the drug and hides it in the alimentary tract. To increase specificity, however, two additional terms have become widely accepted: "body pusher" and "body stuffer." The "body pusher" inserts the drug packages rectally or vaginally; these packages are usually bigger than those that are swallowed. The "body stuffer" ingests small drug packets (in general, small round or oval packs) to try to avoid being tried for a crime (Table 20.1). In many cases, an individual may serve simultaneously as a body pusher and body packer. The literature reports cases of more than 200 packages at once in an individual.[5] Rare reports indicate additional concealment sites for drugs, such as the mouth, the esophagus, ear, and even foreskin.[6,7]

In recent years, body stuffing has become more popular because of the difficulty of detecting the small packs radiologically. Body stuffers typically deal with cut cocaine that is not as accurately wrapped as the bigger packages of a body packer or body pusher. Leakage from a ruptured package may not necessarily lead to intoxication because of the smaller amounts of drug and lower purity, but is more probable because of poor fabrication of the pack. After a quick discharge from custody, the packs are regurgitated or collected from feces after gastrointestinal passage.

APPEARANCE OF "BODY PACKERS" AND THE TRANSPORT OF COCAINE ACROSS BORDERS

Incorporated drug smuggling deals primarily with cocaine; therefore, our focus here will be on this substance class.

The majority of drug carriers are young males, predominantly of colored skin. Their native country is often the same as that where the plant of the narcotic drug is found or cultivated. Nevertheless, there is a new trend to traffic cocaine from South America via West Africa to Europe.

In Europe, primarily Spain and Great Britain, increasing demand for cocaine (13% of global cocaine seizures in 2003 were from Europe) is a result of changes in international commerce, likely an effect of increased availability and lower price. On the other hand, demand for cocaine has moderately dropped in the United States, which still remains the world's largest cocaine market (28% of global cocaine seizures in 2003).[8]

Traditionally, drug traffickers shipped contraband goods by boat or by drug-carrying "body packers" from Latin America to Europe or the United States via the Caribbean. Usually, the smugglers tried to put many body packers on a commercial flight so that as many as possible would go undiscovered according to the "sprinkler principle." Because of augmented border patrol by customs—especially since 9/11—that has resulted in 100% control of selected flights, the smugglers altered their routes around 2005. According to the European Schengen treaty, customs officials do additional patrol within borders, for example, on trains.

As of 2008, about 20% of cocaine is brought from Africa, according to the United Nations Office on Drugs and Crime (UNODC). Smuggling routes involve a detour in the transportation of the narcotic drug, for example, from Colombia or Bolivia to the western borders to the Atlantic coast of South America. The cocaine is then taken to West Africa by fishing boats or small private planes, which are eviscerated to transport as much gas and cocaine on board as possible. Most confiscations of cocaine in Africa occur in Nigeria. Seizures of cocaine base and salts, including crack cocaine, showed an alarming increase from 2005 to 2006 (e.g., in Nigeria in 2005, 395.910 kg were seized; in 2006 that rose to almost 14,435.990 kg; compare this to the highest seizures in Peru and Colombia in 2006, which ranged from 181,310.215 kg to 19,452.717 kg).[8] After trading and fixing vendees, the drug is transferred from interim storage in Africa to the target country in Europe. The contraband is typically brought to Europe unaltered, either by boat or by drug mules on a commercial flight. Aware of this problem, customs authorities attempt to enforce control at international airports by dragnet investigations (Table 20.2). According to a well-known dragnet by customs, drug importers apparently prefer White and European or North American drug mules. For drug mules recruited from Africa or South America, clear motives include money, the fight for basic survival, or threats to the life of next of kin. Drug carriers from developed nations, however, are baited by a luxury vacation, say for example in the Caribbean or with a vacation bonus. At the end of that holiday, the carrier has to transport the illicit drug across the border with a common payment of about €3000 to 4000 (approximately $5500) for whites, while Blacks are paid about €1000. Many of the White drug mules are unemployed and live on social welfare; unsuspecting women are sometimes lured into this by Black dealers with alleged romantic intentions.

Once assigned as a drug courier, the person must undergo a brief training period in which he or she practices swallowing as many grapes, plums or other similarly shaped harmless objects as possible, perhaps including condoms filled with powdered sugar. A body packer usually tries to swallow as many packages as possible; a common amount would be about

TABLE 20.2
Identification of a Drug Courier

Departure	West African, Dom. Republic or northern/eastern countries of South America
Bookings	Bookings are made shortly before departure, with return ticket after 3 weeks at the latest. The drug courier often does not know where and when the booking was made. Ticket payment in cash
Aviation route	Multiple stopovers at transit airports
	Destination ambiguous or contradictory
Nationality	Africa, northern parts of South America, Eastern Europe, nationalized Netherlands and increasingly inconspicuous white drug carriers
Gender	About 80% male, about 20% female
Age	Typically ranging from 17 to 35 years of age
Passport	Non or few entry/exit visas
Appearance	Lower social status
	Insecure, nervous attitude; improper sweating; intoxication symptoms
	Poorly fitting and/or new clothes
	Long sleeves to hide puncture sites if intravenous drug addict
	No or little food and liquid intake during flight (flight crew should be trained to inform customs about conspicuous behavior of passengers) and halitosis due to reduced food and liquid intake
	Light baggage
	Round and high sum of cash, no credit or bank cards
	Cooperative with customs or border patrol
Medication	In order to reduce bowel movements: Loperamide (opioid receptor agonist).
	To ease ingestion: lubricants or coconut oil
	Latex gloves, condoms, cellophane sheaths or other paraphernalia

70 to 100 packs or the equivalent of about 0.7–1 kg of cocaine at a high purity level.[9] According to the World Drug Report 2008, this amount is equivalent to a wholesale price of US $41,000/kg in Switzerland or almost $47,000/kg in the United States, and a street price of US $74/g in Switzerland or US $86/g in the United States. The packs that are transported within a drug courier usually contain high- purity substance. During flight, the drug carrier follows a strict diet or eats nothing. To reduce bowel peristalsis, the body packer uses constipating medication such as loperamide or diphenoxylate to keep the rectal tenesmus as low as possible for the subsequent 36 hours. Transit times may range from 2 days to 3 weeks. Often the drug smuggler carries sprays and deodorants to cover bad smells after vomiting or flatulence and coal tablets as first aid medication in case of rupture. During training, they are advised to keep the abdomen stretched while sitting and counseled about how to act when crossing borders. Clearly, the body packer is well aware of the total number of packs he has swallowed, since he must collect them from his own feces and deliver the correct quantity of cocaine. Once the drug carrier has passed the borders unmolested, mild laxatives or enemas are used to accelerate container retrieval from feces.

Drug mules, when caught, seem educated in strategies to avoid good diagnostic images on CT. Usually, the drug carrier initially acts cooperative and consents to the radiological procedure, but during the CT scan, the suspect starts to get agitated and begins rapid respiration and abrupt movements (Figure 20.1). The repeated evidence of this behavior substantiates the official suspicion of professional instruction in how to avoid detection. Plain abdomen radiographs are becoming less important, probably due to radiolucent wrappings of newer drug containers, especially when the drug mule is very constipated or carries most of the load in the upper alimentary tract. Drug carriers even swallow sand or other radiopaque substances in order to blur radiographs. Moreover, it is widely known that summation and well-manufactured or small packages can easily be overlooked. Aluminum foil has been recommended for better radiolucency, which is more of a myth as it does not change the appearance of a pellet on plain x-ray or CT scan.

Currently, customs authorities try to use new screening methods, such as ion mobile spectrometer (IMS) analysis when they have a suspect in custody. For IMS, a sweat

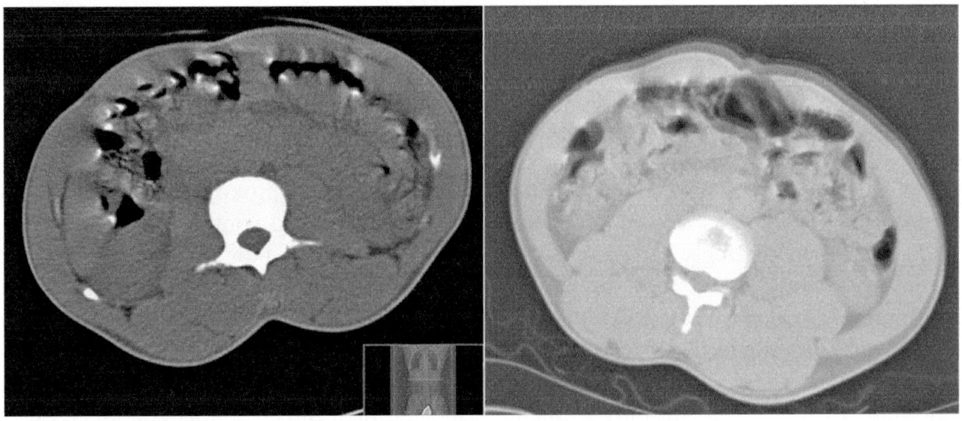

FIGURE 20.1 Axial plane (left 64 MDCT, right 16 MDCT). Exaggerated breathing resulting in motion and beam hardening artifacts in well-trained or agitated patients or, rarely, due to a language barrier and lack of understanding of the breathing instructions.

sample is obtained and analyzed. Current studies propose a specificity of 75% to detect whether the suspect has swallowed any cocaine packs. The analysis is still limited by an inability to distinguish between a simple user, a user who is also a body packer, or a contaminated person. Of course, older methods, although well-established and still used, are prone to being more expensive or unreliable, time consuming, and invasive: drug testing in urine, blood, and hair; radiological imaging; and collecting feces and its foreign objects as evidence.

MANUFACTURING PACKAGES

For a radiologist, it is crucial to know how drug containers appear, which is best understood in the context of existing package types and manufacturing methods. Drug containers are known by a variety of names, such as bolitas, bolos, pellets, finger stalls, and condoms, but all indicate any kind of incorporable drug pack.

Possible types of packing vary widely. Form and size depend on whether they are handmade or produced mechanically. Drug containers may be of enormous length, up to 15 cm; these are usually pushed rectally or vaginally. Current pellets for body packing are usually mechanically manufactured, with a uniform shape, whereas handmade ones have variable shape.[5]

Unfortunately, these packages cannot be feasibly classified, although some classifications are attempted in the literature.[10] These categorizations do not describe the variety of pack wrappings that appear on radiological examination. Drug traffickers try to make the packages more secure for the body packer and continuously change the appearance of the packs, at least to increase their own financial benefit by avoiding easy detection by the authorities.

The packets are known to be multilayered. The inner layer is a tightly filled pouch made of balloons, condoms, finger stalls, or common plastic bags for food, either tied with bindings or more professionally with a twisted and heat-sealed cone-shaped ending. This first layer is covered by multiple additional layers of latex or polyethylene, again with a heat-sealed ending, rubber (caoutchouc) or hard wax/paraffin coating. Intermediate layers may include aluminum foil, stannoil, carbon paper, more plastic and cellophane, toilet paper, filter paper, glassine paper, and lots of tape. Other materials may be incorporated to alter the radiodensity in an attempt to limit the risk of detection.[11]

There are two versions that are still commonly used. The "older" type of a drug packet is the well-known "condom package" (Figure 20.2), but the increasingly common taped pellets have multiple layers of polyethylene food wrapping, tape and plastic bags (Figure 20.3). Some of these body packs are additionally wrapped with aluminum in order to change the radiodensity (Figure 20.4).

These orally incorporated packs differ not only in their manufacturing but also in their size; the average length ranges from 3 cm to 6.5 cm, with a mean diameter of 1 to 2 cm. The mean length is approximately 4.5 cm. The packs are round or oval in shape and contain between 10 and 12 g of cocaine. The condoms tend to be a little smaller, with an average filling of 8–9 g of compressed cocaine. Certainly, even larger packages have been described, but those are mainly found in body pushers.

FIGURE 20.2 (a–c) A tightly filled condom tied with bindings on the right end, creating the so-called rosette sign on plain radiographs (see key findings pp. 318–334). Note the typical reservoir on the left side, which can usually be identified on radiological imaging due to residual air in the point. The condom is sealed with caoutchouc in this case. (c) Shows the coated reservoir of the condom on the right side. Note the neat manufacturing of the heat-sealed end (arrow) of the plastic-wrapped drug container on the left.

Some of the covers used for body packing are semipermeable materials, such as condoms, or are not resistant to gastric acid, such as some plastic foils. Therefore, persons smuggling drugs in this manner might be contaminated and

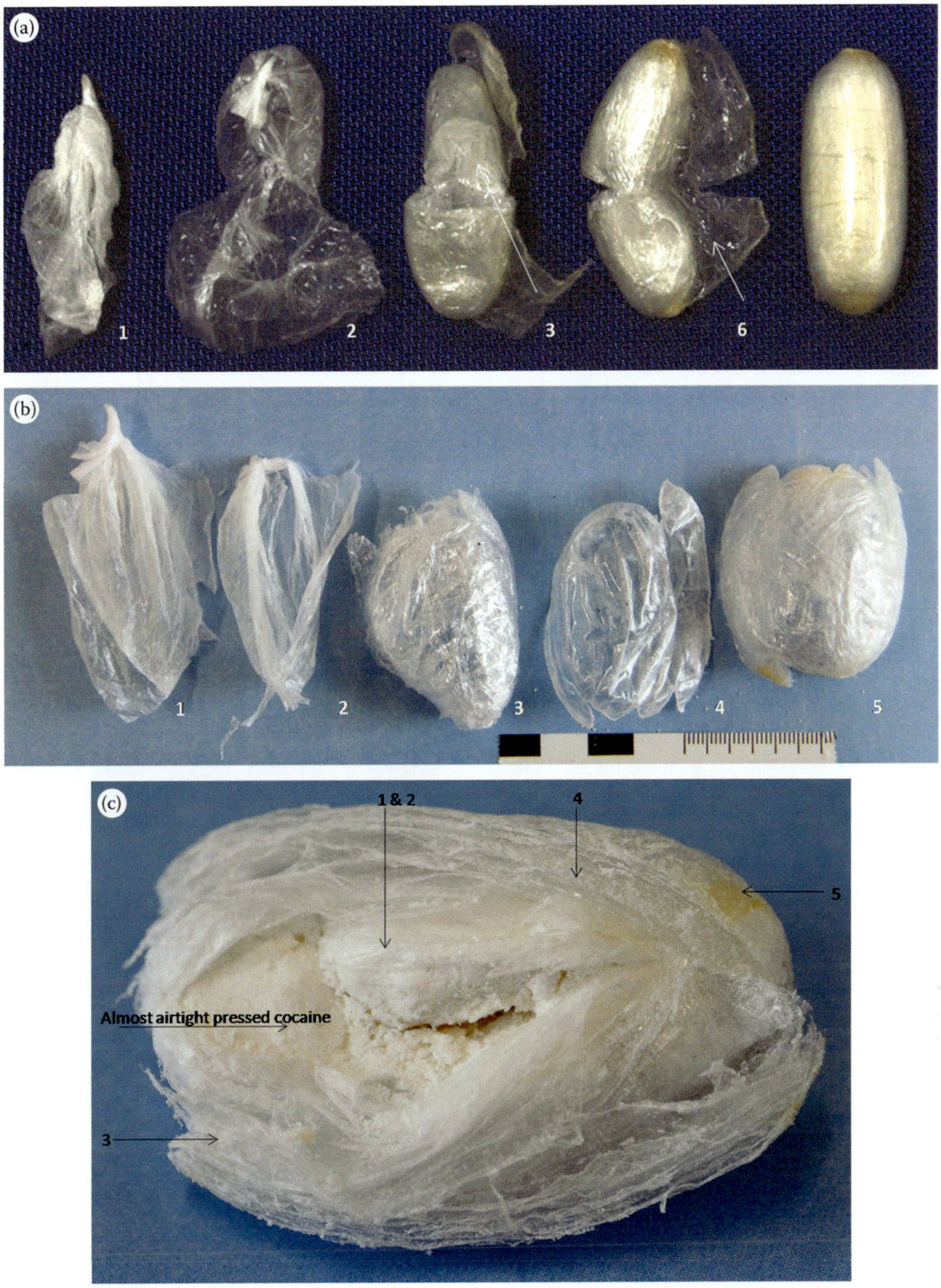

FIGURE 20.3 (a–b) Two different opened body packs, probably mechanically manufactured. The inner sheath is shown on the left (1), which is a common thin plastic bag for food knotted at the point as a single end and heat-sealed. This is again wrapped with another plastic bag (2) and closed in the same manner. Cellophane is taped as an intermediate layer (3), sometimes with small tape or one broad adhesive foil, and encased in a thicker plastic sheath (4). Additionally, there is some taped plastic foil (5) with an outer layer of heated and sealed polyethylene food wrapping on one single end. (c) Shows a magnified view of a longitudinally opened package.

test positive in drug screening, even if they are not users, or might show signs of intoxication. Rupture or leakage can occur in two ways: by mechanical movement or by chemical digestion. Acid in the stomach may invade the packs based on relative osmolarity, after which the fluid dissolves the cocaine powder and disintegrates the pellets from the inside, and the drug may diffuse as leakage (Figure 20.5). Real rupture of packages seems to occur when the packages are still in the stomach, with the pylorus being a presumable obstacle for large foreign bodies. As reported previously, the ileocecal valve seems to be no obstruction.[12] Of course, there are predisposing factors for obstruction, such

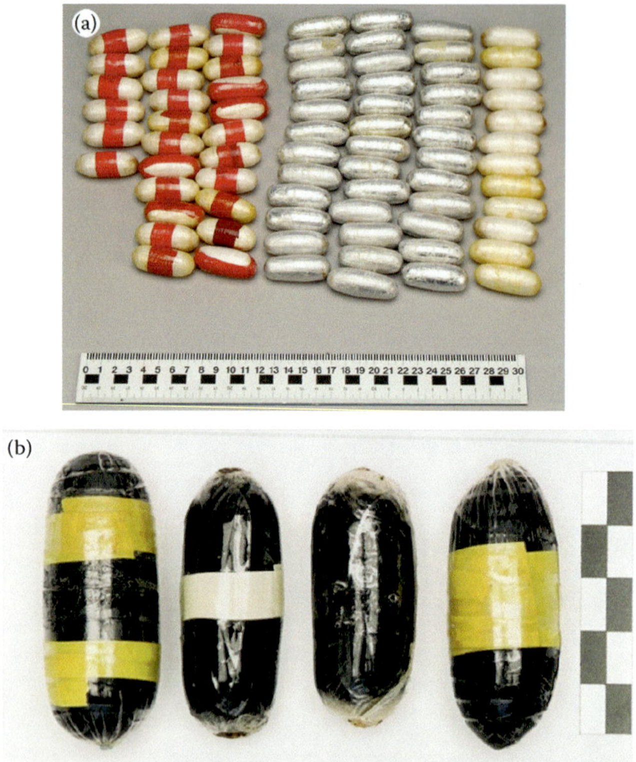

FIGURE 20.4 (a) Note the aluminum foil-wrapped packs in the middle of the picture. The red tape probably indicates the end of the adhesive bonding, presumably to ease access after excretion. (From The State Police, Basel-Landschaft, Switzerland. With permission). (b) shows only a different colored wrapping with the same technique as mentioned above. All of the packs correlate to about 10 grams of cocaine with a high degree of purity. (From The State Police, Bern, Switzerland. With permission).

as intestinal paralysis due to medication or opioid misuse. Certainly, the size and accumulation of a large number of packages and the manner of wrapping will all lead to obstruction. Taped packages tend to open by chemical digestion, whereas the fingers of rubber gloves or packages using condoms tend to rupture by mechanical means due to force against limited elasticity.

FIGURE 20.5 Autopsy specimens with a display of the gastric content. Note the intraluminal staining of the packages due to fluid intrusion with massive disintegration of the taping and dissolved content followed by the death of the female body packer.

Body stuffers transport other sizes of drug containers, ones which usually have a maximum length of 1.2 cm and a width of 1 cm at most if oval-shaped (Figure 20.6a). Whereas the smaller round beads are a maximum of 1 cm in diameter; these are kept in the cheek pouch and swallowed when required by risk of a potential roundup (Figure 20.6b). Body stuffers are more likely to show symptoms of intoxication than body packers because of the poor and unsafe fabrication of pellets filled with cocaine. Radiological detection of small globules is particularly difficult owing to their size and low radiodensity (see key findings, pp. 318–334).

In order to prevent drug smuggling, ongoing studies are examining better radiological or toxicological analyses and detection possibilities. As improvements are being made, however, drug traffickers have already begun using new

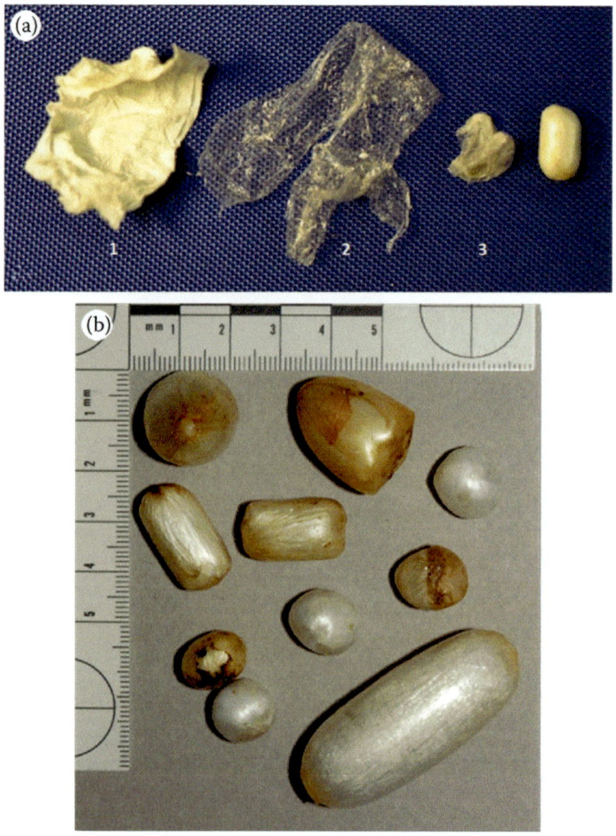

FIGURE 20.6 (a) Toilet paper is shown as an inner layer (1). The cocaine is compressed and covered by a plastic sheath as the core element of a small oval packet. The next layer consists of polyethylene food wrapping (2) with a plastic bag around it (3). (b) The stained, browner pellets have already been excreted and kept for a court exhibit or analysis. Note the variability of the package shapes. The cone-shaped packages could be ingested by body packers and body stuffers due to their intermediate size, with a length of 2.5 cm and a maximum diameter of 2 cm. The smaller spherules and oblong, ovoid pellets are of typical body stuffer size. To compare the differences in size, note the typical body-packing drug container with a length of 5 cm and a width of 2 cm in the lower right corner.

approaches to conceal packages. According to a recent Interpol cocaine alert in May 2008, three incidents involved the swallowing of cocaine in liquid form (presumably with an oily substance or ethanol as a solvent and carrier substance). If this becomes a new trend radiological imaging must be adapted. This smuggling technique poses a new significant threat to the authorities.

DRUGS AND THEIR EFFECTS

Smugglers mainly work with cocaine and heroin, and to a minor degree with opium; therefore, the emphasis of this section will be on the substance classes of cocaine and heroin.

In order to give a short overview (further information, e.g., on http://www.drugabuse.gov/Infofacts/), all common drug types in the Western drug scene are mentioned. The following substances play a minor role in drug smuggling by body packers, due to little demand, synthetic production possibilities and/or a small margin of profit:

- Cannabis products like hashish or marijuana
- Stimulant drugs like amphetamines or their derivatives (e.g., Ecstasy or MDMA = 3,4-methylenedioxy methamphetamine)
- Club drugs like GHB (gamma-Hydroxybutyrate = Xyrem)
- Rohypnol (Flunitrazepam) and Ketamine (dissociative anesthetic)
- Hallucinogens like LSD (D-lysergic acid diethylamide = indolalkaloid of ergot "secale cornutum")
- Peyote (mescaline = ingredient of cactus)
- Psilocybin (4-phosphoryloxy-*N*,*N*-dimethyltryptamine = "magic" mushrooms)
- PCP (phencyclidine = primarily developed as an i.v. anesthetic)
- Various inhalants
- Khat (a stimulant drug derived from a shrub "catha edulis")
- Anabolic-androgenic steroids (AAS = substances related to male sex hormones)

OPIATES

Opium is the dried chyle of still-verdant fruit capsules of the opium poppy, *Papaver somniferum*, which is rough opium. Opium contains morphine, codeine, noscapine, papaverine, and thebaine. All but thebaine, which lacks an analgesic effect, are used clinically as analgesics to reduce pain without loss of consciousness. Heroin is a synthetic morphine (appearing as white to brownish powder), which can be manufactured either directly from opium or from semi-purified morphine.[13] Heroin was initially used to suppress the urge to cough, especially due to the raging tuberculosis at the turn of the century. Tragically, heroin was also used to fight morphine addiction, just as morphine was used to overcome opium addiction. It was quickly clear that heroin exceeded the narcotic and addictive potential of morphine or opium, because of its quick metabolism and faster penetration of the blood–brain barrier.

Heroin overdose produces depression of the central nervous and respiratory system, and treatment for symptomatic body packers is naloxone as an antidote to reverse the effects of the drug. As a common side effect of naloxone, noncardiogenic pulmonary edema has been described.[14] About 50–250 mg of intravenously administered street heroin is a common dose, with a potency of 5–10%.[15] The minimum lethal dose is 200 mg, but users can tolerate a much higher dose and still survive.

COCAINE

The coca plant *Erythroxylon coca* has small blossoms that mature into red berries. For cocaine production, only the green leaves that contain the alkaloid cocaine are needed. Freshly harvested leaves smell like tea; if chewed, they numb the mouth. Traditionally, these leaves were chewed together with citron to achieve better potency and to suppress famine, thirst, pain, or drowsiness. The original version of Coca Cola had coca leaves as an ingredient along with the cola nut, but due to its addictive effect, this was prohibited in the early twentieth century. After harvesting, the leaves are dried and pulverized and, thereafter, extraction of the cocaine hydrochloride begins. Crack appears to be the base of cocaine hydrochloride and is mixed with an inorganic base into a paste. This extracts the hydrochloric acid and crystallizes the crack (with high purity). If baking soda is used instead of the inorganic base, the result is cut crack. Cocaine has an enormous potential for psychic addiction but not physical addiction.

Cocaine overdose produces strong stimulation of the central nervous system and heart, which leads to seizures, hypertension, cardiac ischemia and arrhythmias, and life-threatening hyperthermia. There is no antidote like naloxone for cocaine intoxication. Immediate surgical intervention is indicated (endoscopy is definitely not the method of choice), followed by symptomatic therapy such as cooling the body, monitoring and, if necessary for cardiac ischemia, administering nitroglycerin or nitroprusside.[14] The use of beta blockers is not called for, as vasoconstriction and cardiac ischemia may be increased. Cocaine packs swallowed by body packers or eventually inserted by body pushers usually contain drug with a high level of purity (60–80%), whereas body stuffers swallow the regular cut product with very different potencies ranging from 20% to 70%. The mean intravenous dose is about 10 mg (1–16 mg), while the mean oral dose is 20–50 mg. The maximum single dose is about 100 mg, which is actually about one-tenth of the small pellets containing about 700–1000 mg (Figures 20.6B and 20.21). Users may tolerate up to 30 g/day.[15] In this context, then, just one small pellet swallowed and ruptured by a body stuffer may be lethal. For suggested algorithms for the management and treatment of body packers, refer to further literature.[11,16,17]

KEY FINDINGS IN RADIOLOGICAL IMAGING

Of course, detection of the packs varies widely depending on the experience of the radiologist and his knowledge of the appearance of incorporated drugs. Many false diagnoses in reading the images of drug mules could be reduced by considering some simple rules. First, the radiologist should know the patient's history, for example, if the police observed him swallowing drug packages and, if so, does the patient complain of pain or prolonged constipation due to the swallowed packages. Knowing the time of ingestion could help the examiner to know which part of the gastrointestinal tract (e.g., the stomach for recent packing or the colon for earlier packing) should be the focus. Diagnostic findings of plain radiographs and CT scans are summarized in Tables 20.3 and 20.4.

Plain Radiograph

According to the traditional diagnostic multi-modality approach, the first examination should be a conventional (computed) radiograph (CR) or Lodox (LSDR: low-dose linear slit digital radiography), if available. LSDR is a whole-body scanning technique developed as a security detector. This device was initially designed to spot diamonds being smuggled out of mines and made a crossover from a diamond detector to a low-dose trauma scanner (Figure 20.7). Its reduced radiation dose and high quality make it a good alternative to CR, but it has some drawbacks, such as a slightly distorted image, giving no reliable length measurements (Figure 20.8). Since most institutions only have plain radiographs, the radiologist must decide what kind of abdominal x-ray position is needed to diagnose the packs. If no indication (such as free abdominal air or ileus) besides body packing is present, the supine position, including the lower rib cage and small pelvis (lower image border should be the inferior pubic ramus), should be favored. In the upright position, image information from the lower pelvis will be lost and with it the depiction of presumed packages in the rectum and sigmoid, which are common locations for packages. Besides, the intestines descend and overlap in upright radiographs (Figures 20.9 and 20.25). There is no additional benefit in imaging the left lateral position, and even strict lateral imaging does not show any greater likelihood of diagnosing packs (Figure 20.10).

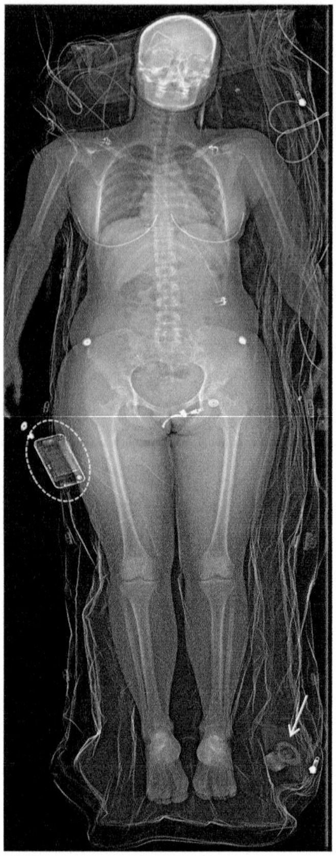

FIGURE 20.7 Trauma scanning after a parachute accident. Performing an LSDR in a trauma room, the radiologist and clinicians are able to get an easy overview of the skeletal status, lung, and mediastinum and are even able to detect foreign bodies. There is summation of the external metal parts of the bra, ECG electrodes, and zipper and even an iPhone is depicted (oval circle). No fractures or parenchymal pathologies are seen. Note the vacuum mattress with the plug valve at the lower end on the right (arrow).

TABLE 20.3
Reading Plain Radiographs

- Plain radiograph—Lodox (LSDR)
- Supine > upright > left lateral position > strict lateral
- Magnify
- Inversion: black bone converted to white bone
- Rosette sign
- Double condom sign
- Suspect gas formation
- Pitfalls: e.g., scybala, nephrolithiasis, other foreign bodies or summation

TABLE 20.4
Reading Computed Tomography

- Unenhanced CT Abdomen; low-dose CT
- No contrast medium needed (i.v., oral or rectal)
- Scout view as first overview
- Suspicious gas formation, double condom and rosette sign, air filled caps
- Hard and soft kernel
- Abdomen and lung window
- Reconstructions (coronal/sagittal)
- Curved or 3D reconstruction
- Assess stomach to rectum, even distal esophagus
- Do not forget vaginally inserted packs

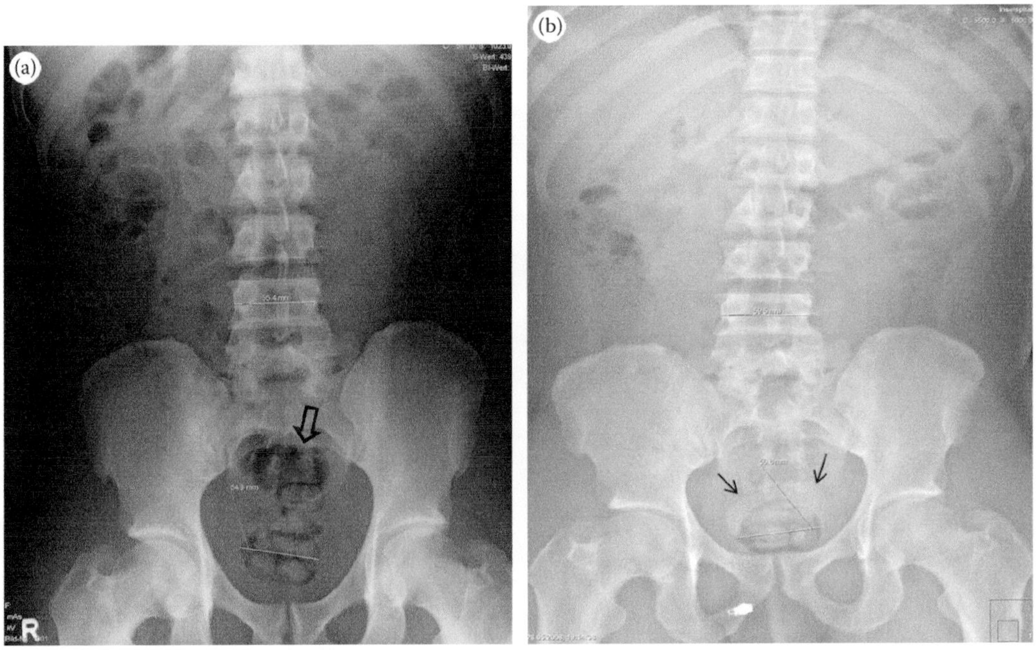

FIGURE 20.8 (a) Computed radiograph, (b) LSDR. Note the different length measurements and the distortion of the pelvis in the same patient. There were 2.5 days between the prior CR and the subsequent LSDR. (a) The condom pack projecting on the rectum is filled with multiple small pellets neighboring other packs (also note the suspicious gas bubble corresponding to another pack, arrowhead). Due to the air in the prominent, measured pack, the observer is clearly able to identify the shape of a condom with the reservoir pointing to the patient's right side, showing a clear double-condom sign. Investigating the follow-up (b) a little closer, the observer is able to detect a longitudinal pack that was not visible before (black arrows). Keep in mind that packs might have different densities and may be blurred.

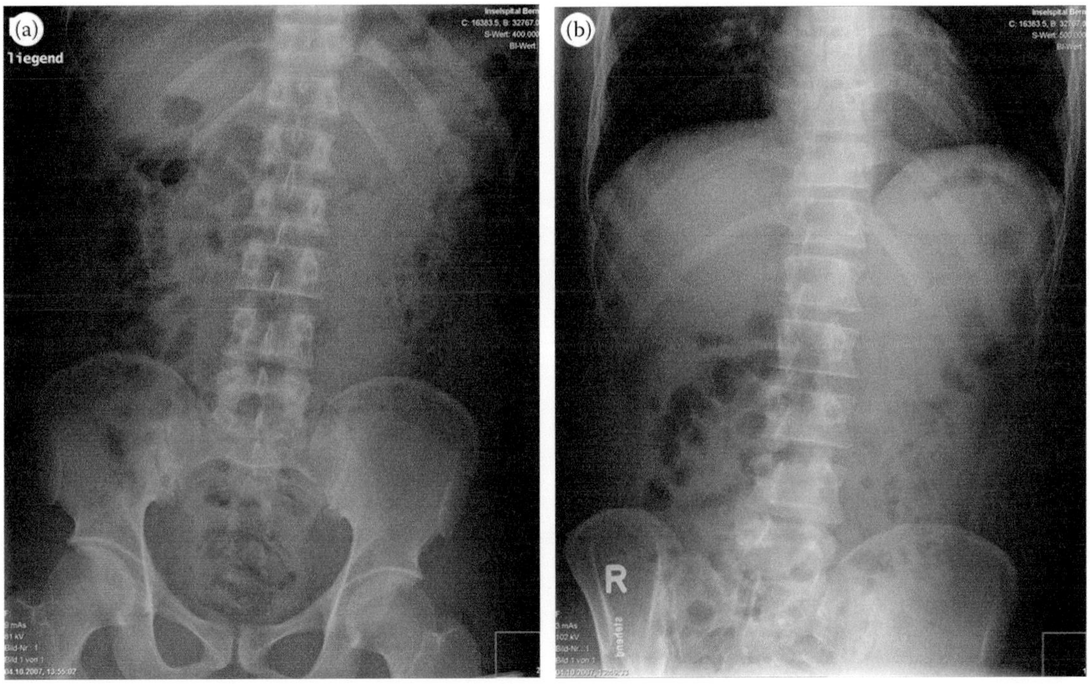

FIGURE 20.9 CT scan (a) supine; (b) upright. Clearly depicted body packs projecting on the rectum in the supine radiograph. Note the upright radiograph taken prior to the supine image; at first, the radiologist would have missed the obvious packs.

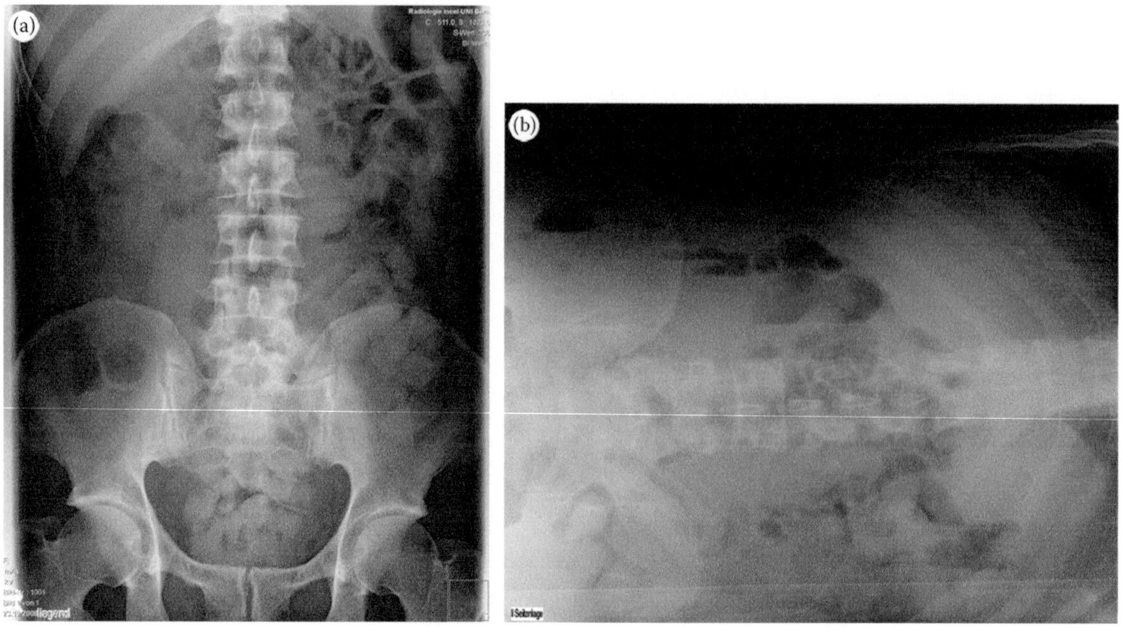

FIGURE 20.10 CT scan: (a) supine; (b) left lateral position. (a) Classic example of a body packer with about 100 packs, which are easy to diagnose by plain radiograph in the supine position. Due to suspected free abdominal gas, an additional image in the left lateral position was obtained. (b) The radiologist is still able to diagnose packs, but it is much harder to detect the vast extent of incorporated packs.

In detecting body packs, magnification and white bone conversion to black bone might help (Figure 20.11). There are well-known key findings such as a slim gas halo "double condom sign" around a pack due to inevitable air inclusion during manufacturing or a specific appearance such as a "rosette sign" from the twisted end of a package. Newer packs are heat-sealed; therefore, the rosette sign is not seen very often anymore. The double condom sign, however, is still frequently seen (Figure 20.12). Atypical or shaped gas formations should be considered likely indicators of packing. These are quite often mistaken and overlooked as usual intestinal gas, but could appropriately lead to a diagnosis of body packing. Common pitfalls in reading body packing images are scybala, calcifications, normal intestinal air, and other foreign bodies (Figure 20.13), as described and depicted in Algra 2007.[18] Of course, the patient's history should point the way forward in these cases. Drug mules often try to achieve artifacts with poor outcomes by superimposing metal coins or other radiopaque structures. A well-informed body packer will try to reduce bowel movements with constipating medication, as mentioned before. This aggravates detection on plain radiographs. Swallowed pellets in body stuffers, or even in body pushers, are not always reliably detectable on plain radiograph (Figure 20.14) since they are too small, too radiolucent and often superimposed in the stomach. The gold standard of radiological imaging for body stuffers is CT. In very rare cases, x-ray fluoroscopy might be considered when stuck packs persist with dysphagia in the pharynx.

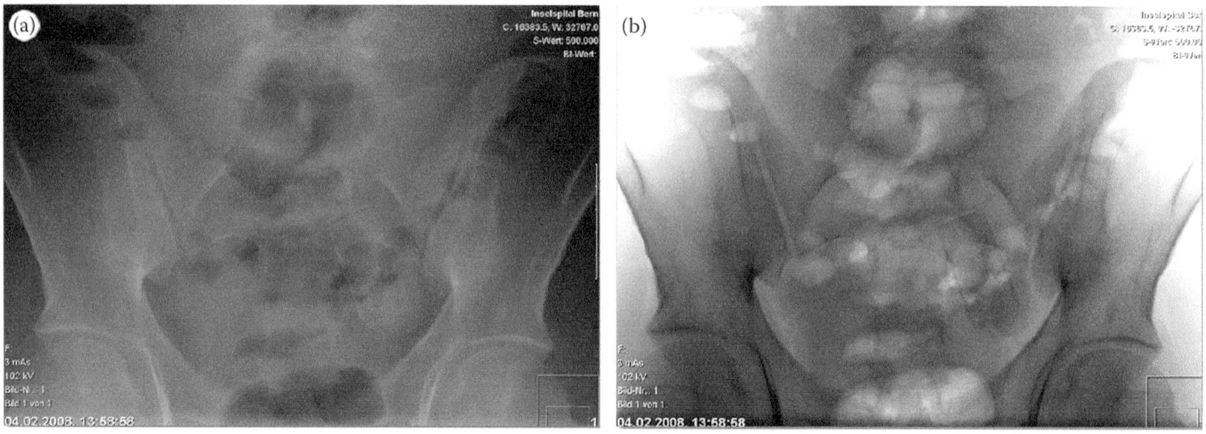

FIGURE 20.11 CT scan: (a) "white bone;" (b) "black bone." In this magnified view, there is a 5.7-cm-long pack detectable pack on the left lateral mass of the sacrum. It is more easily identified if the investigator switches from white bone to black bone and back.

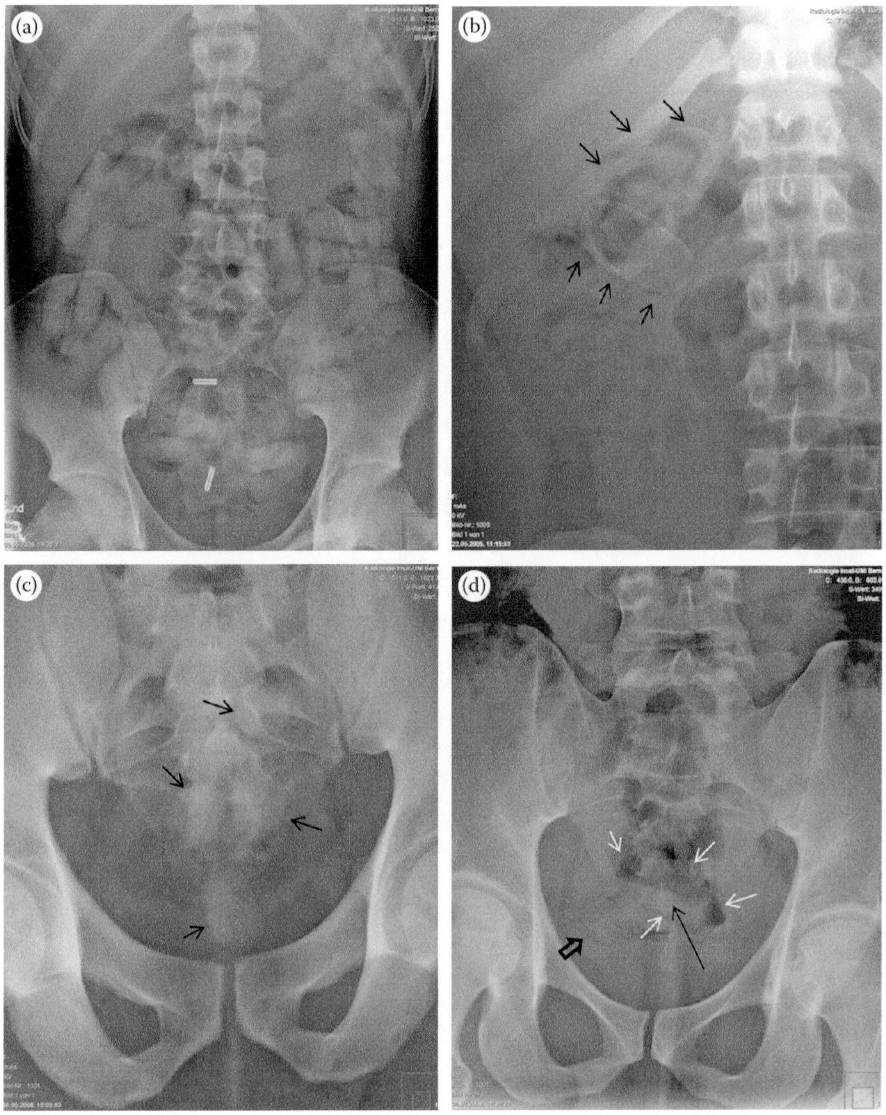

FIGURE 20.12 CT scan (a) A classic body packer with typical air trapped between layers in numerous "newer" (see Manufacturing Packages, p. 314) packs projecting on the colon and rectum. (b) Two packages superimposed on each other with a double condom sign. (c) Multiple machine-manufactured packs in the recto-sigmoid, only discriminable from the sacrum and feces due to the tiny gas halos of the "double condom sign." (d) In this magnified view of a body pusher, there two packs are detectable: one is small and cone-shaped (short black arrow), while the other is a larger pack with a rosette-shaped end (white arrows). The coccyx joint (long black arrow) should not be mistaken as a double condom sign.

Computed Tomography

Currently, almost every primary health care center has the ability to perform a CT scan. Therefore, the management of drug mules relies more and more on the specificity and sensitivity of CT. The traditional therapy for drug mules is guarded excretion under mild laxatives and pain medication until three drug-free stools are delivered. Since drug mules are very often constipated, this may take awhile, especially in cases with over one hundred packs. At best, there is a frequently occupied holding cell with a special drug toilet to collect the feces in a professional and guarded way; therefore, fast management complemented by radiological imaging is appreciated. The guarded suspect still has the possibility of manipulating the evidence by means of manual extraction of rectal packages or regurgitation of swallowed ones, since European countries are not permitted to put detainees under 24-h surveillance by camera or man. A half-blinded glass, as limited by basic rights, enables a guard to see the upper parts of the patient who is using the drug toilet. The suspect is alone in his cell, but could still dissolve the extracted cocaine powder into his coffee or yogurt or even rubbing it into small gaps in the wall of the usually well-prepared cell. Some patients try to scrub their feces with drug into their blanket. Based on Swiss law, the drug mule can only be charged for the amount of listed evidence, which means that the penalty depends on the collected cocaine. This explains the urge to get rid of the cocaine in any way possible or even to re-ingest the extracted packs. The authorities running the guarded ward tend to rely more and more on CT diagnoses either to discharge a suspect

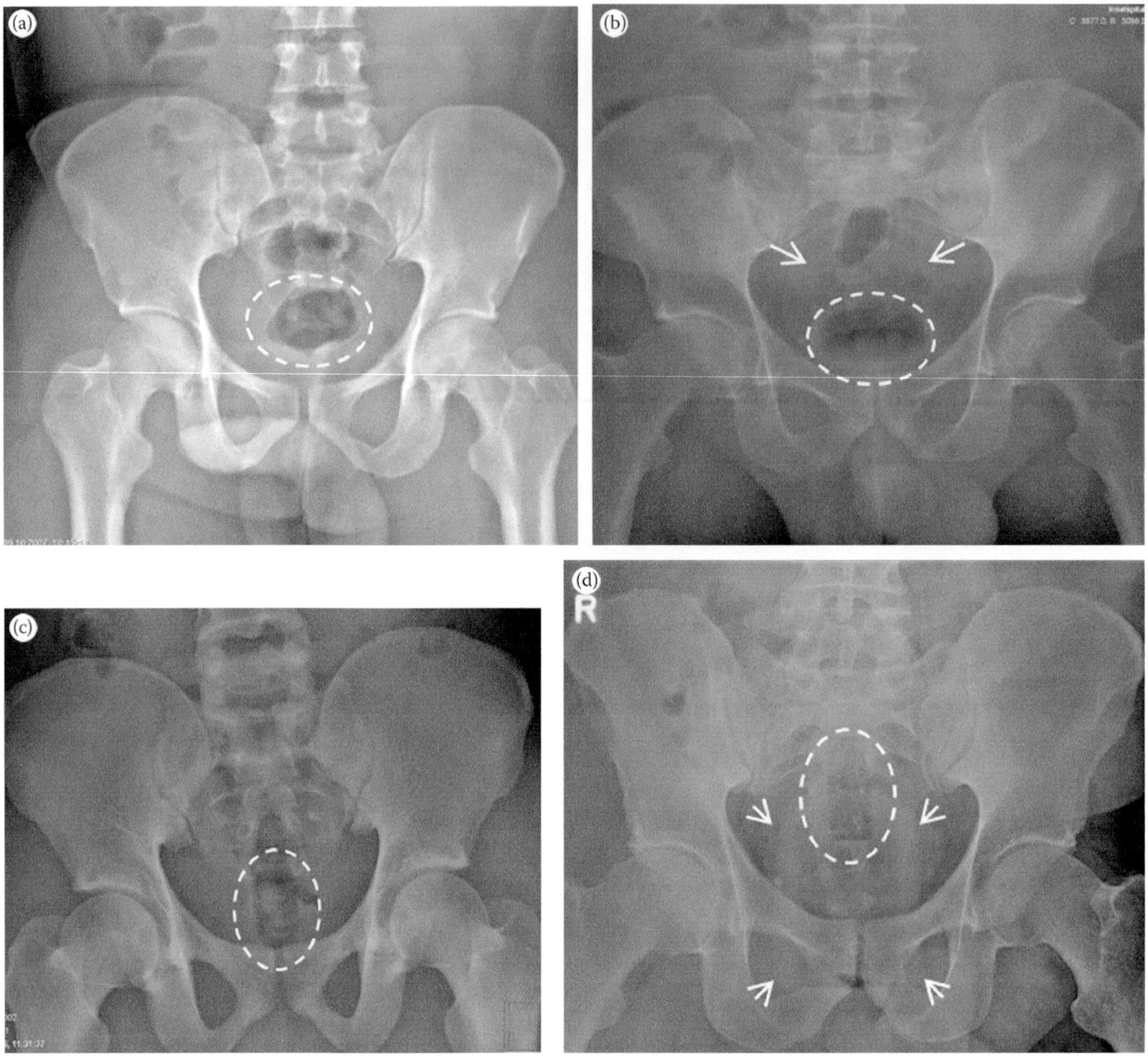

FIGURE 20.13 (a and c) An atypical gas formation with a radiolucent cocaine package mimicking gas in the rectum. (b) Similar-looking gas formation (circle) of a rectally inserted vaseline dose (arrows). (d) Same patient as in (b), only one year later, now with a soap dispenser inserted rectally. (e) Classic body packer with multiple packages (see placed arrow). (f) Note the similar appearance of scybala in the ascending colon (arrow) and the rectum (circle). (g) Radiolucent small body packs (circle). (h) Body pusher with multiple rectally inserted cocaine packs. (i) Almost the same appearance of lost love beads in the recto-sigmoid colon. (j) Shaped gas formation with slightly opaque pellets in the air-filled condom. (k) Air-filled shaped structure of a bottle cap inserted into the rectum.

if negative or to redeploy the convicted felon to a state prison even before delivering three drug-free stools in order to prepare space for a new suspect. Consequently, there is a medical and legal need for proper and fast diagnoses on CT,[19] making critical the knowledge of the incorporated drug container shapes and the competency to read a CT. Body stuffers, in particular, need extra care, as they are often quite hard to detect. It is important to assess the complete gastrointestinal tract carefully, from the esophagus to the anus, while focusing on the stomach in body stuffers. In females, the vaginal cavity must be checked for body pushing, although this is usually digitally examined if suspected. A close look should also be taken at the airways (if within the scan range) if the patient complains of dyspnea, as at least one drug trafficker ran away from a round-up and tried to swallow a pack but instead aspirated it (Figure 20.15).

Apart from organizational and forensic motives, there are medical indications of the utmost importance for radiological imaging. Early radiological detection and confirmation of the diagnosis of the body packer is crucial in cases of suspected intoxication either by leakage or rupture of a drug container (Figure 20.16). A CT scan should be performed, and the patient should undergo immediate surgery for removal after imaging to be sure that no pack is overlooked. Asymptomatic patients usually undergo conservative management. Common additional indications for CT include a localization of the

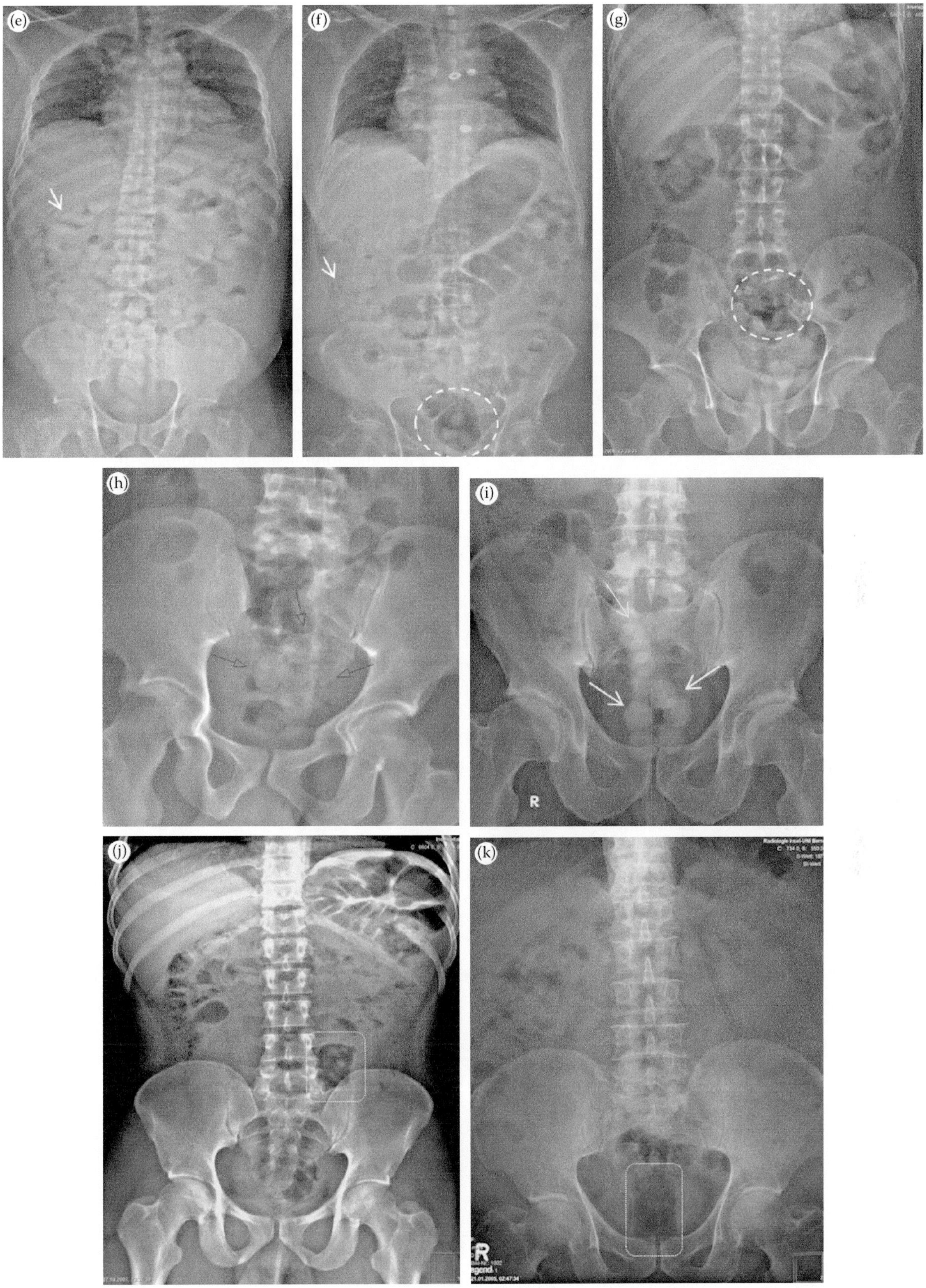

FIGURE 20.13 (*Continued*)

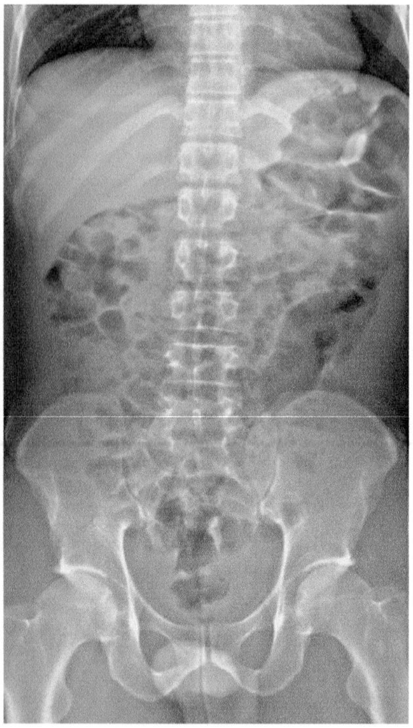

FIGURE 20.14 An LSDR scan of a body stuffer with four small ovoid pellets in the stomach and duodenum. Obviously, no detection of these packs is feasible, even though the patient is not constipated.

packs inducing an obstruction resulting in ileus, with even ulceration, perforation, and subsequent peritonitis described in the literature.[4] Any of these may lead to death. Nevertheless, the radiologist must verify the indications using radiation exposure. Low-dose protocols need to be implemented in further imaging of drug mules. An abdomen CT-protocol with 1.2 mm collimation; 120 kV ref. 120 mAs could be lowered to 80 kV ref. 250 mAs in order to reduce radiation significantly in the future management of drug mules. Male gonad capsules may be used if not extending into the scan range, as they produce large streak artifacts (Figure 20.17).

Before reading the CT images, the scout CT scan should be used for a first overview, especially in cases where there is no prior plain radiograph (Figure 20.18). Here again, one should look for suspicious gas formations or odd densities as described above (see plain radiograph, pp. 318–320) and consider the time of ingestion. Eventually, there will only be detectable air-filled caps at the end of each pack.

First and foremost, the radiologist must know about the density appearance of drug packages on CT, which quite often correspond with opacities on plain radiographs. There are reports describing measured Hounsfield Units (HU) to determine what kind of drug is being smuggled.[11,20–22] Using CT density measurements, however, is the only reliable detection method for cocaine-contaminated bottles in cargo.[23] Otherwise, relying on density measurements is critical, as one can see in the scan of these two drug packages (Figure 20.19), both containing cocaine but wrapped differently. Therefore, it is clear that the density of the cocaine packages strongly

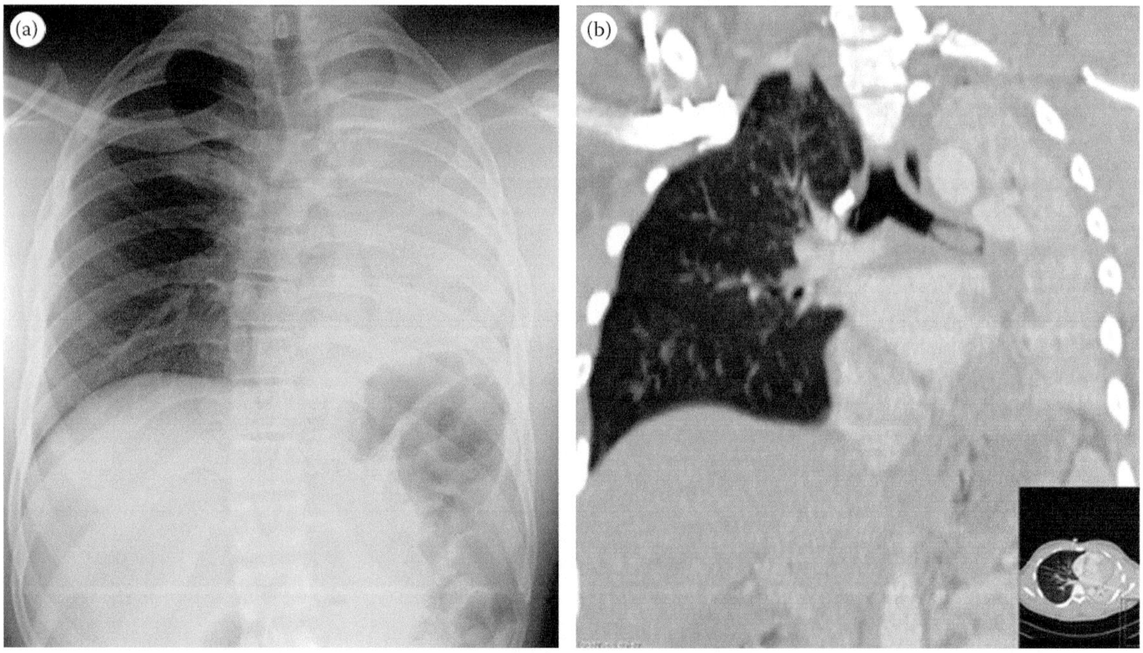

FIGURE 20.15 (a) CR of the thorax with opacification of the left lung. Clinical suspicion of pneumothorax, which was not confirmed by imaging. (b) One hour later, a CT scan was performed. In the coronal reconstructions, a body pack clearly obliterates the left main bronchus, resulting in atelectasis of the left lung with mediastinal shift to the left side. This pack was removed by bronchoscopy, and the patient was discharged to state prison. (c) Axial view, hard kernel and lung window. (d) Axial view, soft kernel and abdominal window, with good depiction of the air crescent around the cocaine in the pack, making the double condom sign. (e) Volume rendering of the lung alone, showing air-filled structures in purple.

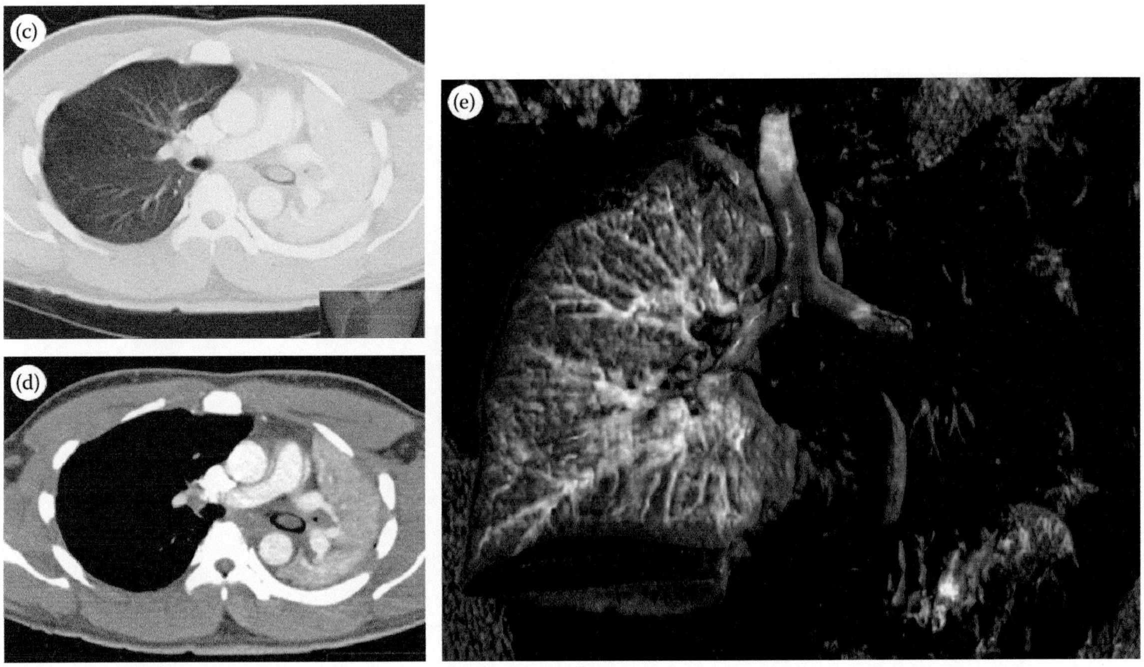

FIGURE 20.15 *(Continued)*

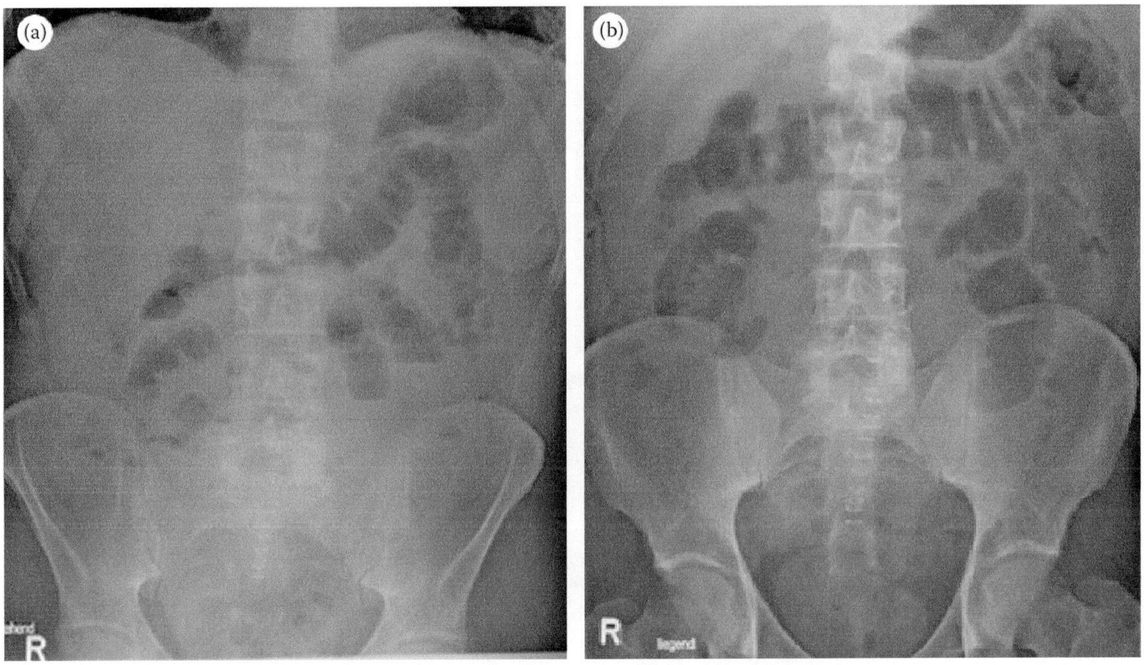

FIGURE 20.16 (a) Upright (b) supine CR; this patient underwent immediate laparotomy after showing signs of cocaine intoxication. No imaging was performed prior to the operation. (a and b) were taken 6 days after surgery with suspicion of an ileus, which could not be confirmed by plain radiograph. The patient's history of body packing was clear, but no foreign object could be detected after surgery. (c) upright (d) supine CR; the patient complained of increasing abdominal pain, meteorism, and flatulence. Over the course of about 2 weeks post-op, a pack overlooked in surgery could be detected and identified as the cause of ileus. This shows how easily packs can be missed by summation in the previously acquired images. (e) Volume reconstruction in the coronal plane depicting the remaining hyperdense (HU 195) cocaine pack in the terminal ileum. (f) 3-D volume reconstruction (Figure 20.18a is the corresponding scout view). (g) Axial plane displaying a transverse view of a body pack (circle). Note the scar of the median laparotomy (white arrow) and the contrast-filled descending colon (arrowhead) with density similar to the pack.

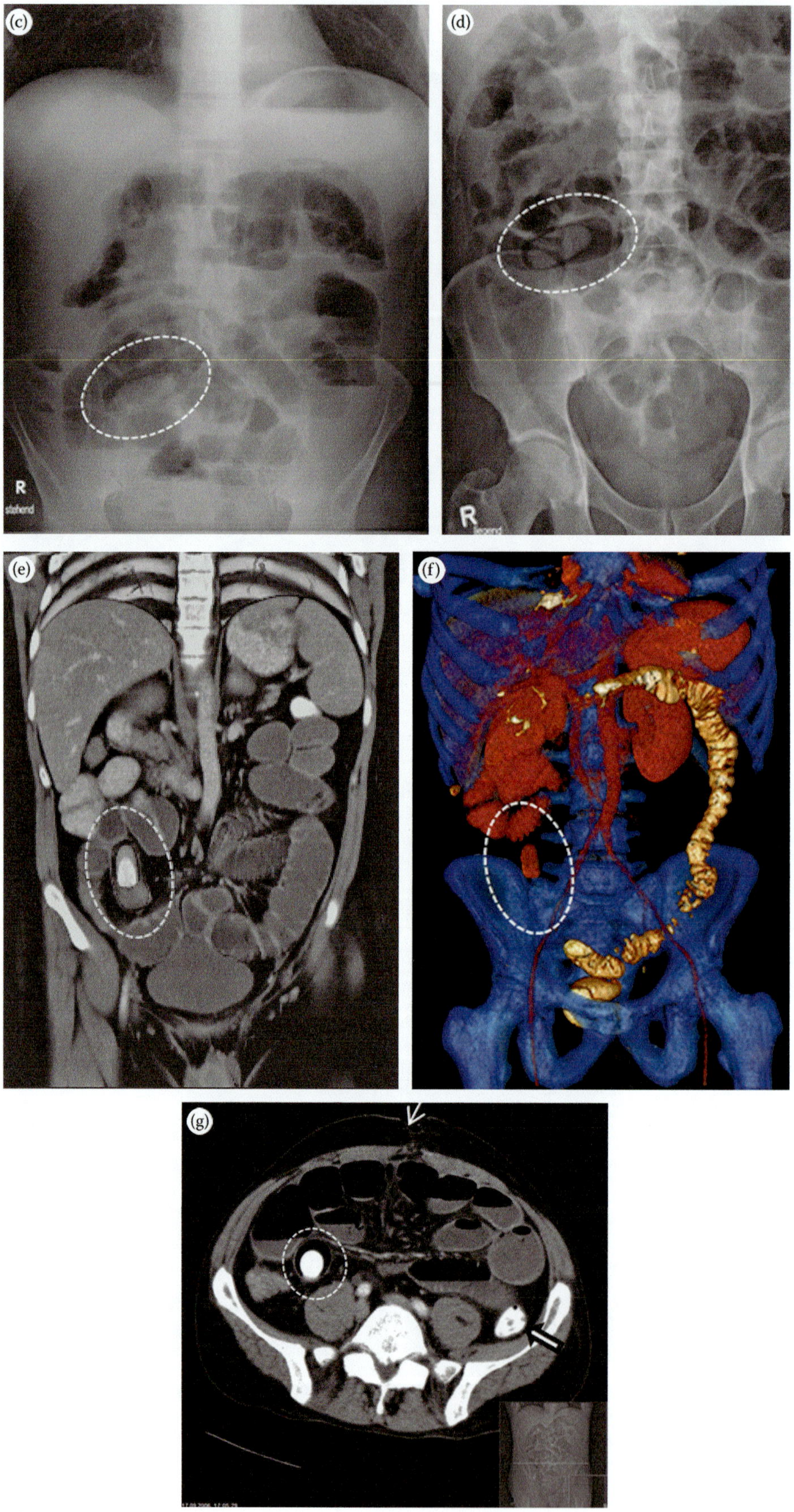

FIGURE 20.16 (*Continued*)

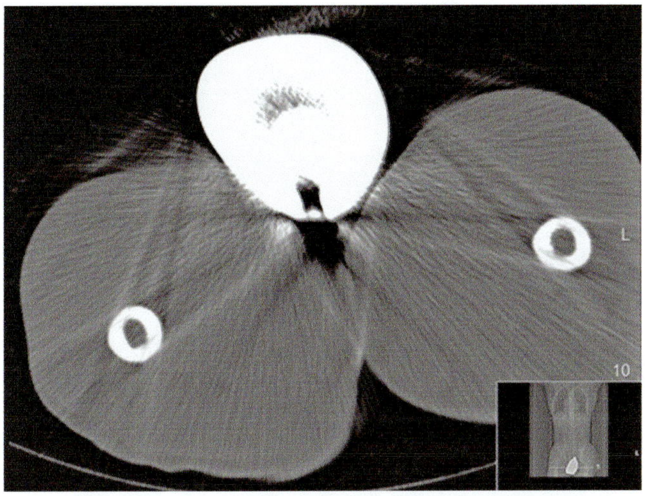

FIGURE 20.17 Axial plane with depiction of an applied gonad shield, in order to reduce radiation, with typical streak artifacts.

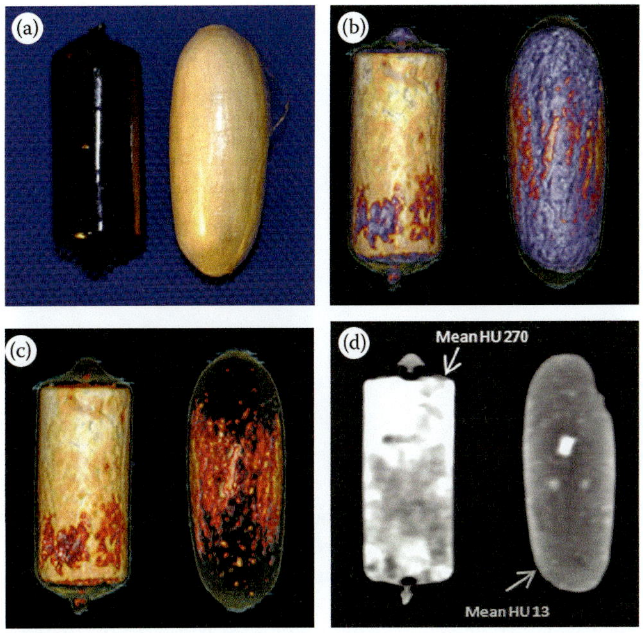

FIGURE 20.19 (a) On the left side is a black "old-fashioned" caoutchouc-coated condom pack next to the light-colored "current" pack wrapped in plastic foil and heat sealed. (b–c) 3-D volume rendering of those packs with a good depiction of the clearly different densities on the CT scan, both containing cocaine, with an almost radiolucent appearance of the plastic foil-wrapped pack. (d) Coronal multi planar reconstruction (MPR) of the packs with density measurements. Note the small air-filled caps of the condom in the dense pack. Clearly, the crystallized particle in the middle of that pack would measure much denser.

depends on the grade of its compression, the substance form (e.g., meaning compressed powder or crystallized substance, Figure 20.20), and the manufacturing and coating (Figures 20.19 and 20.21). "Older" packs were usually wax or rubber-coated (see Manufacturing Packages, 314) and, therefore, were very dense, whereas the "newer" packs appear isodense or hypodense compared to adjacent scybala. Small pellets (e.g., 1 g) obviously appear very hypodense on account of the small compressed center, with the adjacent wrapping being more radiolucent. A hypodense pack might not contain a

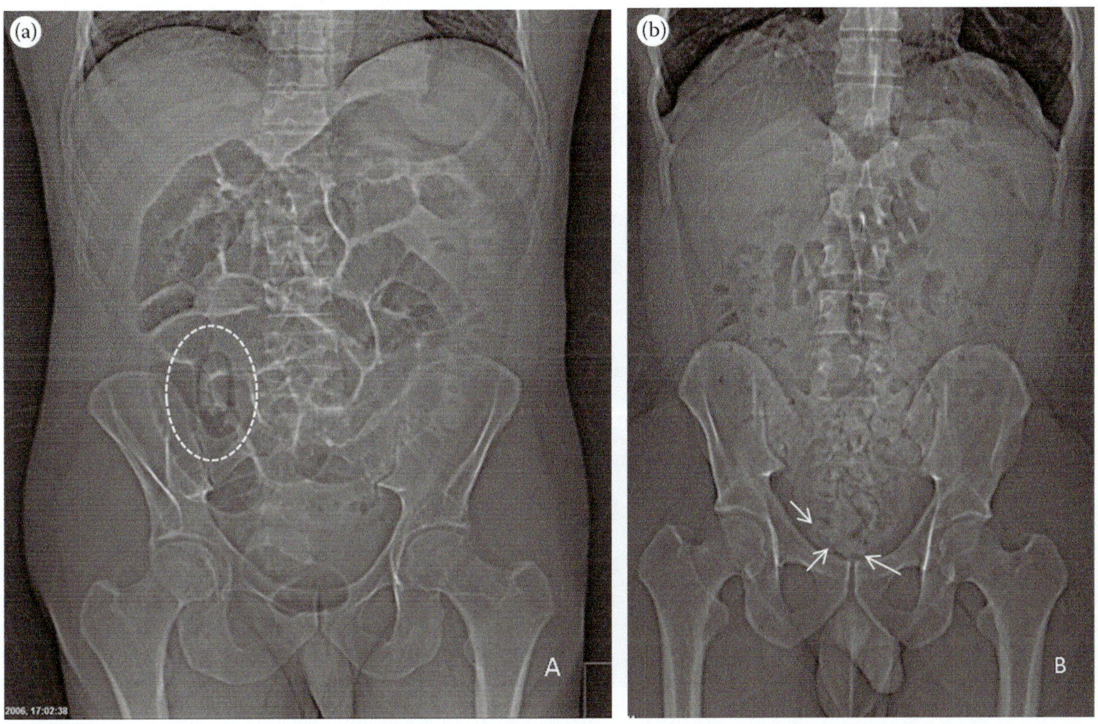

FIGURE 20.18 (a) In the scout CT scan, the investigator is already able to detect a radiolucent condom filled with an opaque pack (circle) projecting on the small bowel. (b) Classic body packer with up to 30 radiopaque packs in the recto-sigmoid and descending colon (one pack is marked by arrows).

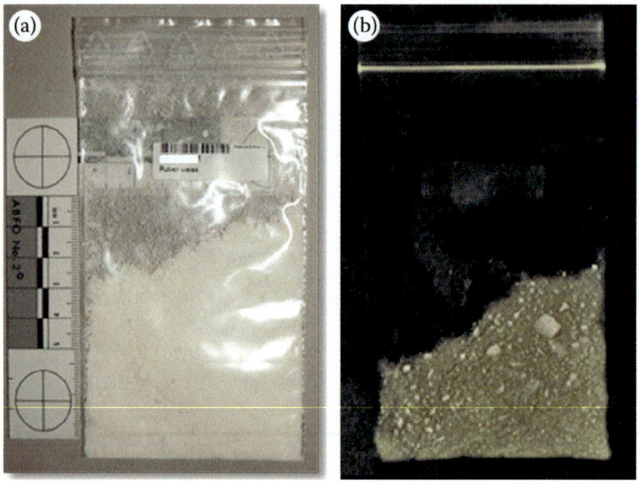

FIGURE 20.20 (a) A bag of about 10 g of loose cocaine powder. (b) In the 3-D reconstruction, there is a good depiction of solid and powdery parts, depending on compression and crystallization.

wax-like substance such as heroin or opium, but might instead be loosely wrapped cocaine. Studies are needed to create a definite and reliable differentiation between incorporated substance classes. Aluminum foil-wrapped packs do not obscure radiological diagnoses and only result in slightly more beam-hardening artifacts (Figure 20.22).

Packs with a higher HU than soft tissue (HU 30–70) are easily identified (Figure 20.23). Nevertheless, the primary clue in detecting even hypodense packets is window-leveling.[24] Scans that appear initially negative reveal the presence of a concealed drug container under window-manipulation to a common lung window (Figure 20.24). The typically used abdominal window has a width of 450 HU with the center at 50 HU; the examiner needs to alter this window to a width of 1500 HU with the center at −500 HU. By applying a lung window, the investigator will be able to detect even small pellets or larger hypodense packs (Figure 20.25). For

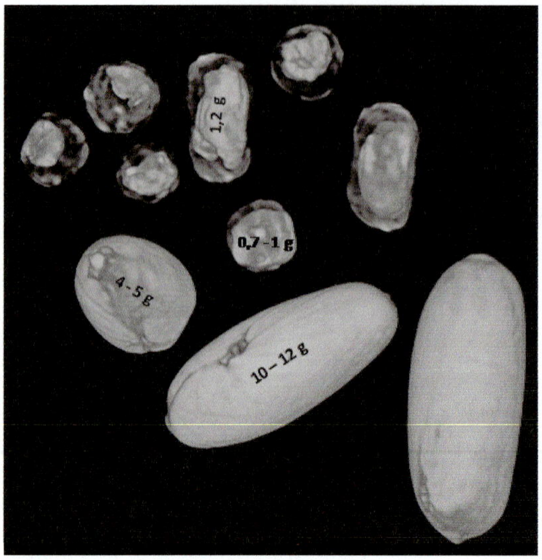

FIGURE 20.21 3-D reconstruction depicting the size varieties of current packs labeled with mean weight. (This image matches the photograph in Figure 20.6b). *Note:* The radiolucent wrapping of the small pellets due to toilet or filter paper. It is important to know that the small packs appear very radiolucent at the rim and only opaque in the very center.

a further improved diagnostic approach, there is the possibility of not only reconstructing the soft kernel (e.g., B30) with an abdominal window, but also calculating an additional data set reconstructed in a hard kernel (e.g., B70) with sharp edges to gain a better resolution, with the acceptance of a higher signal-to-noise ratio (Figure 20.26). As in every diagnostic approach for radiological data, reconstructions in the sagittal and coronal planes add information, while curved reconstructions help to visualize the number and localization of the packs (Figure 20.27a and b). Three-dimensional visualization is very convenient for laymen and courts (Figure 20.27c–e). The slice thickness should be

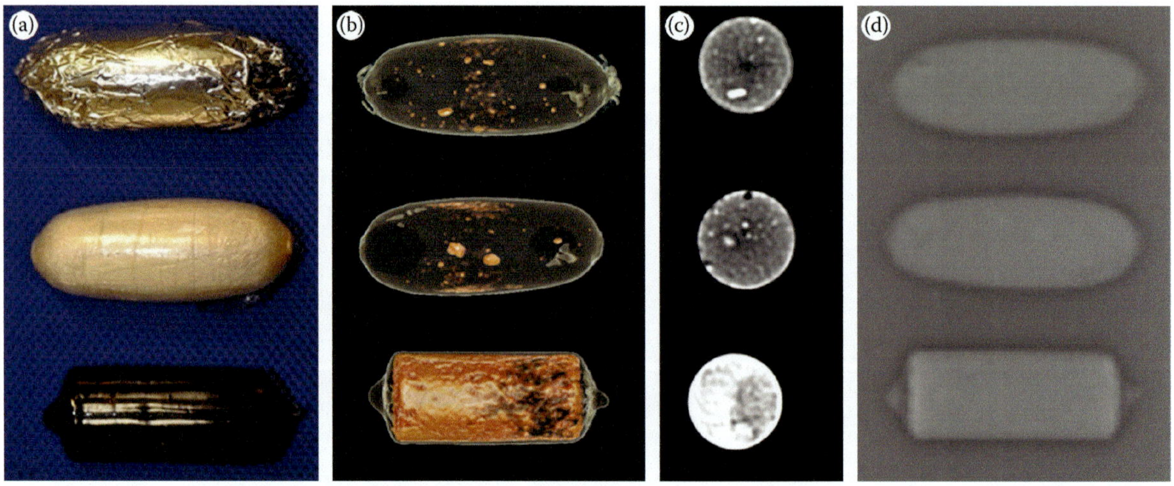

FIGURE 20.22 (a) At the top, an aluminum foil-wrapped plastic-taped pack; in the middle, a common plastic-taped pack; and at the bottom, a rubber-coated pack. (b) 3-D reconstruction showing no significant difference between the aluminum foil-wrapped pack and the one below. (c) Axial view of the packs in the same order. The aluminum foil-wrapped pack only shows slightly sharper margins. The aluminum foil does not obscure the pack. (d) Showing the corresponding pack by aluminum foil unaltered CT scout view.

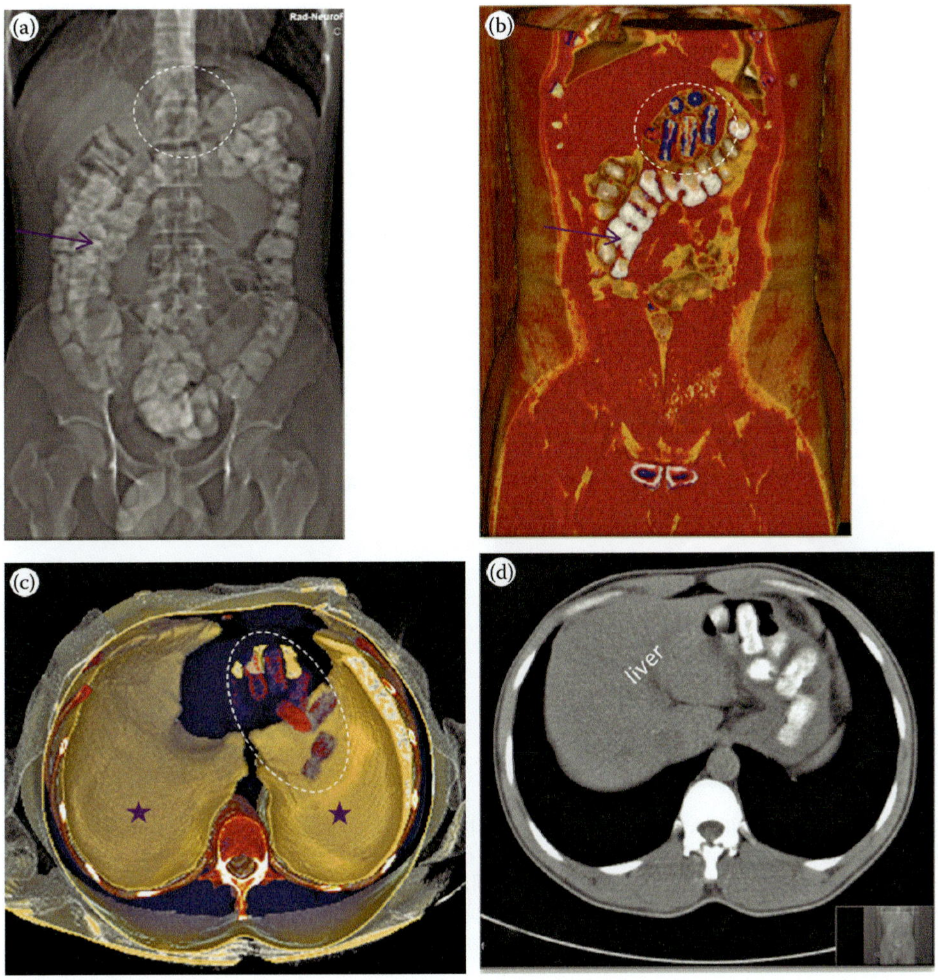

FIGURE 20.23 (a) CT scout view wherein the white circles mark the hyperdense packs projecting on the stomach. A contrast agent was administered rectally (the mauve arrow marks the contrasted colon). (b) 3-D multiplanar reconstruction in the coronal plane. Three longitudinal and three axial cut cocaine packs in the stomach. Compact structures like bone or the coating of the drug containers appear purple according to the dense HU. (c) Axial plane, perspective from bottom up with view of both lungs (asterisk). The packs are encircled with some adjacent truncated contrast media of the colon (mauve arrow). (d) Axial plane, view from the feet to the top. Note the hyperdense packs in the stomach and the duodenum. This patient swallowed 13 drug containers.

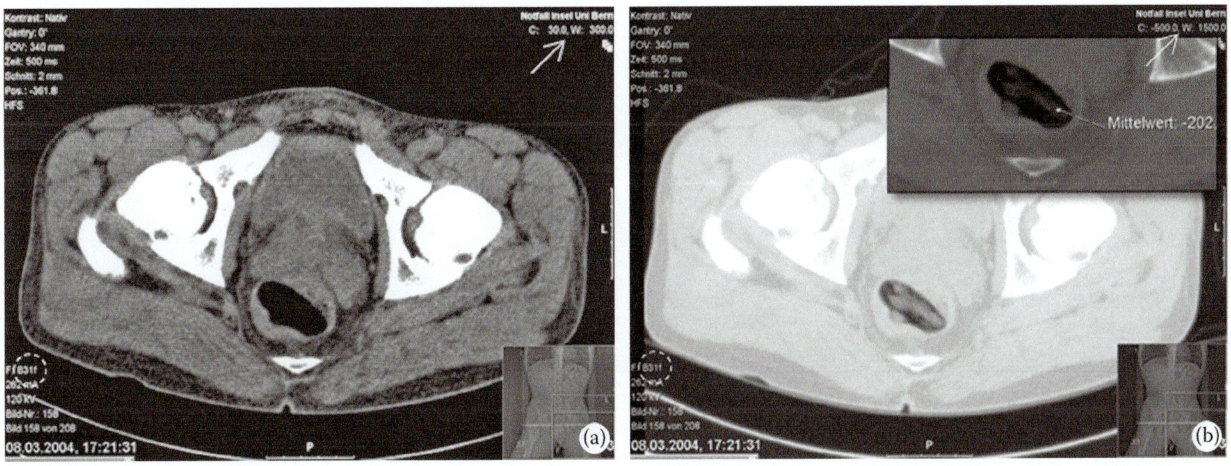

FIGURE 20.24 Both axial images in soft kernel (white circle). (a) Suspiciously shaped gas formation in the rectum that is easily underdiagnosed using only the abdominal window (arrow). (b) After changing to the lung window, the rectally located pack catches the examiner's eye. The content seems not to be very compact; therefore, there is a negative measurement of about –200 HU. This pack was confirmed to contain cocaine.

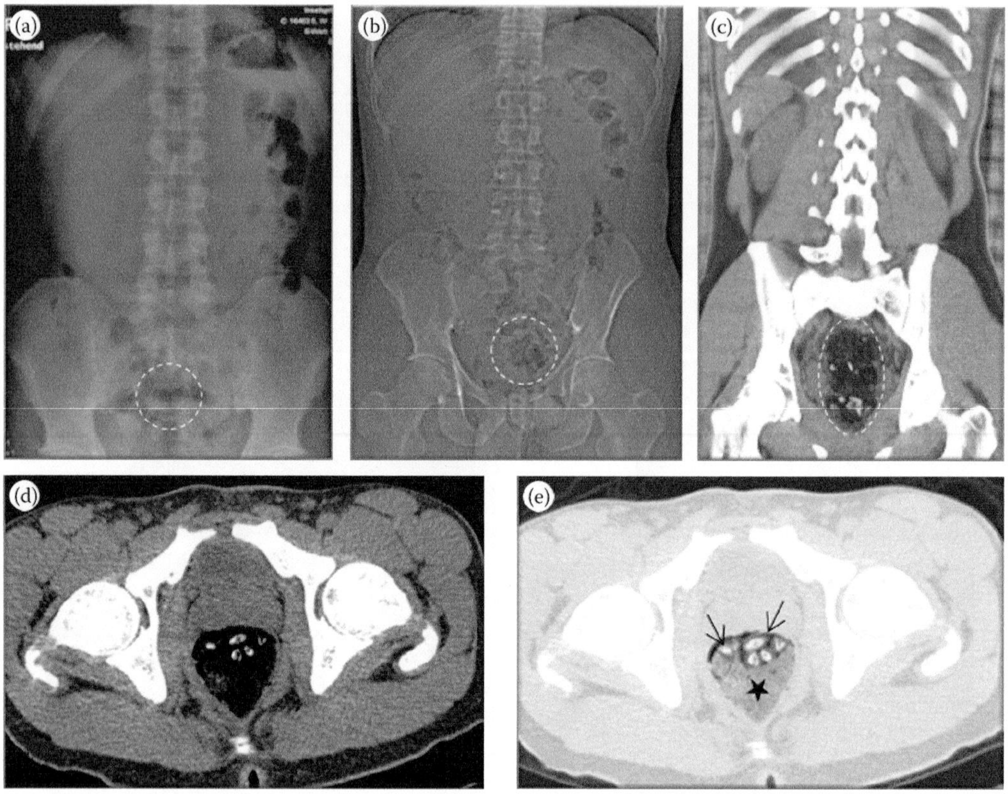

FIGURE 20.25 (a) CR in the upright position. Retrospectively detectable packs projecting on the rectum (white circle). (b) CT scout view. More easily detectable packs due to the supine position. There were only 9 min between the CR and the CT. (c) Coronal MPR depicting potential body packs. (d) Axial plane: abdominal window, soft kernel, slice thickness of 1.5 mm, still suspicious hyperdense structures in the rectum. (e) Axial plane, same setting except for a change to the lung window, with a good depiction of one small ovoid pellet and a condom filled with five small pellets (arrow). There were 11 small pellets in this feces-filled rectal ampulla (asterisk) of a body pusher.

1.5 mm in order to detect small pellets and 5 mm in routine diagnosis (Figures 20.28 through 20.32).

If no indication other than body packing is given, there is no need for oral or rectal contrast agent administration; in some cases, it even appears to be unhelpful due to similar density (Figure 20.33). In addition, contrast agents accelerate peristalsis and produce osmotic inflow in the intestines, which would be contraindicated in bowel obstruction by body packs. Intravenous contrast media should be given if there are other indications such as ileus. Still, there are common pitfalls such as dense scybala with beam-hardening artifacts that simulate dense cocaine, in addition to shaped food, such as big noodles,

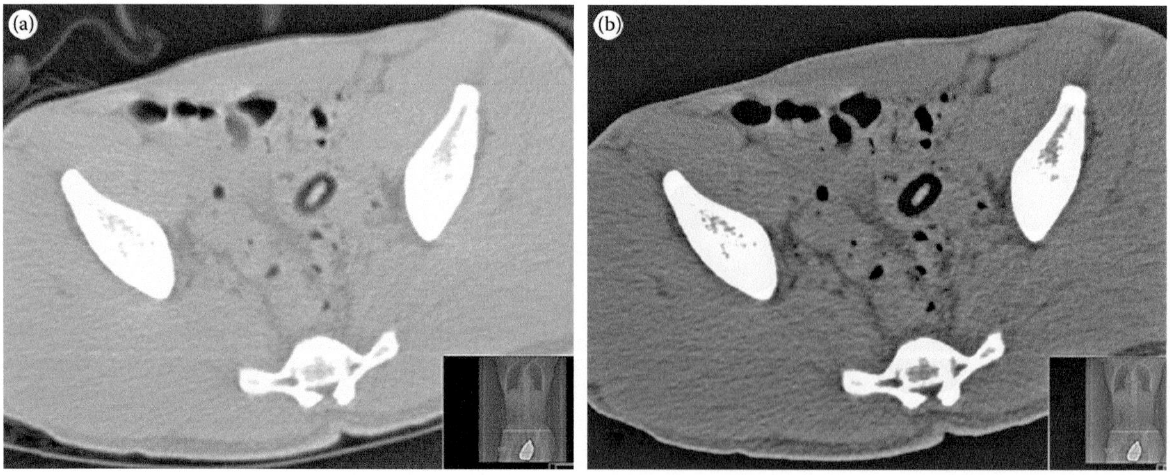

FIGURE 20.26 (a–c) Axial plane with a view from bottom up depicting one small pack in the sigmoid colon. (a) Same stack as in (b). Both reconstructed in soft kernel, with (a) showing the lung window with blurred margins. (b) Abdominal window; only the opaque center is observable. (c) Same data reconstructed in a hard kernel with a sharper and better defined delineation of the pack. Even the fold of the inner plastic wrapping is visible. Regrettably, there is increased noise. (d) Color-coded MPR of the pack for easier assessment.

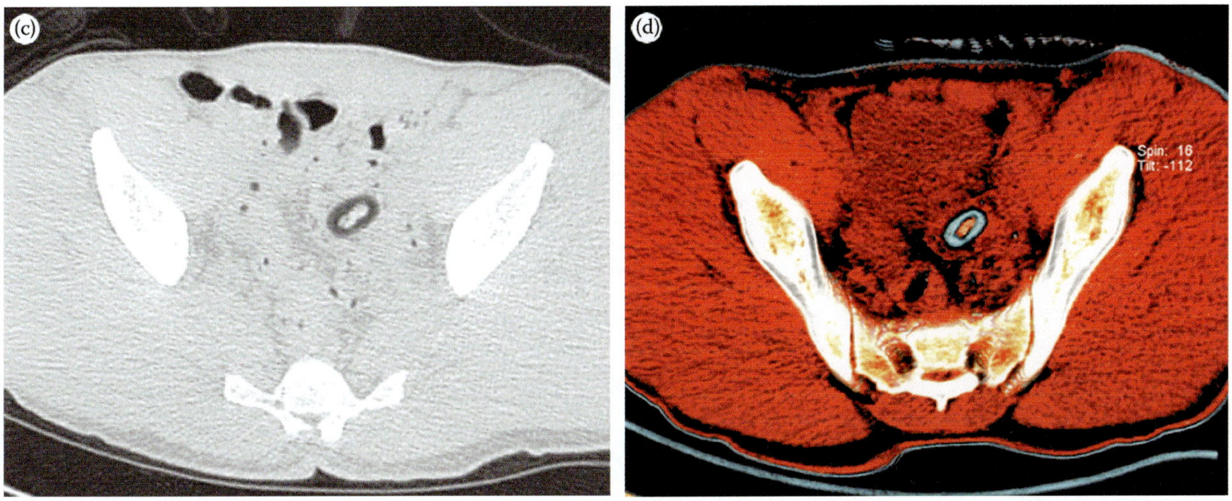

FIGURE 20.26 (*Continued*)

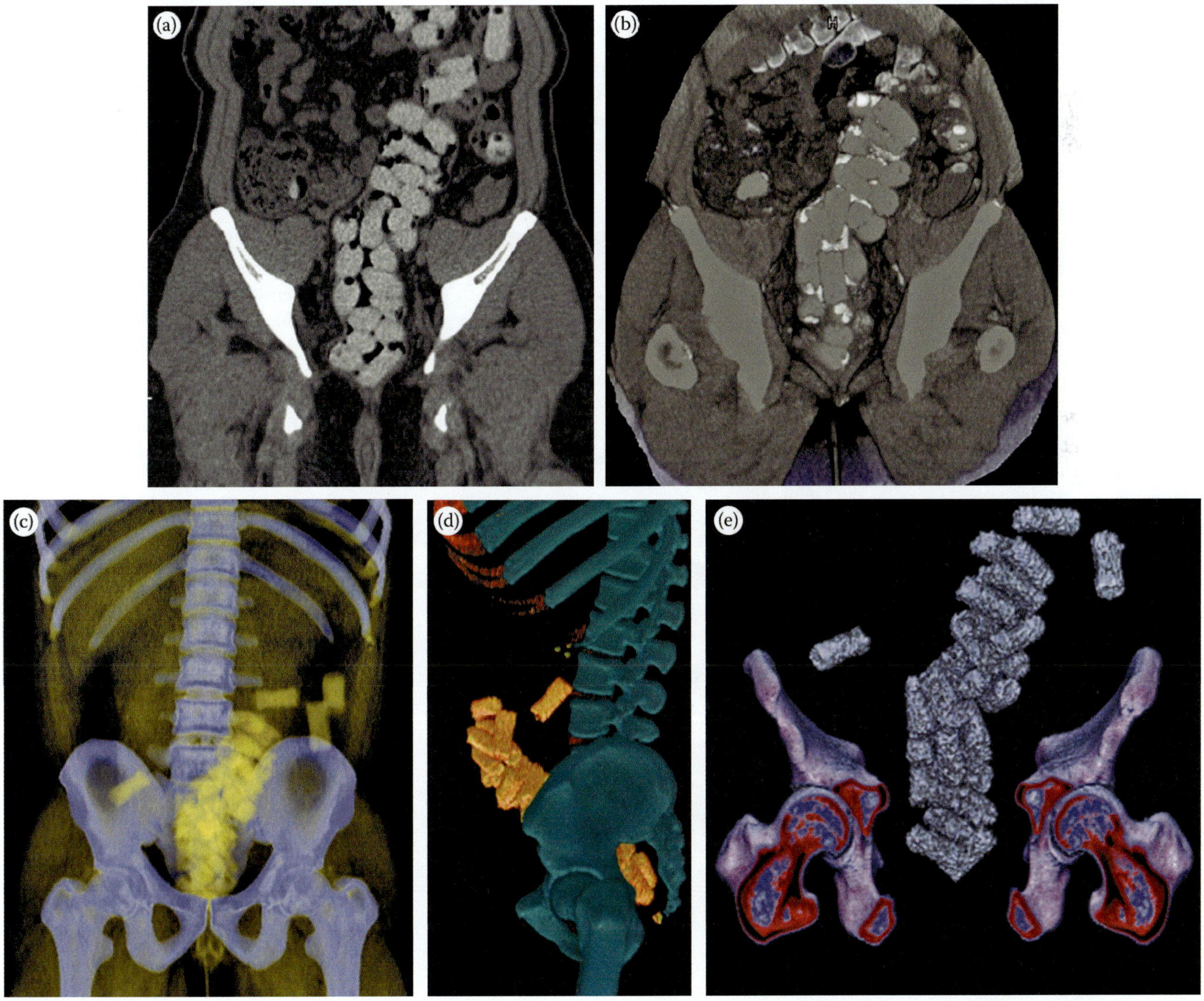

FIGURE 20.27 (a) Curved coronal reconstruction along the descending colon to the anus, showing the enormous extent of packs. (b) Oblique multiplanar reconstruction using a volume rendering technique, demonstrating the truncated packs in the recto-sigmoid colon. (c–e) 3-D volume reconstruction for a clear picture of the findings.

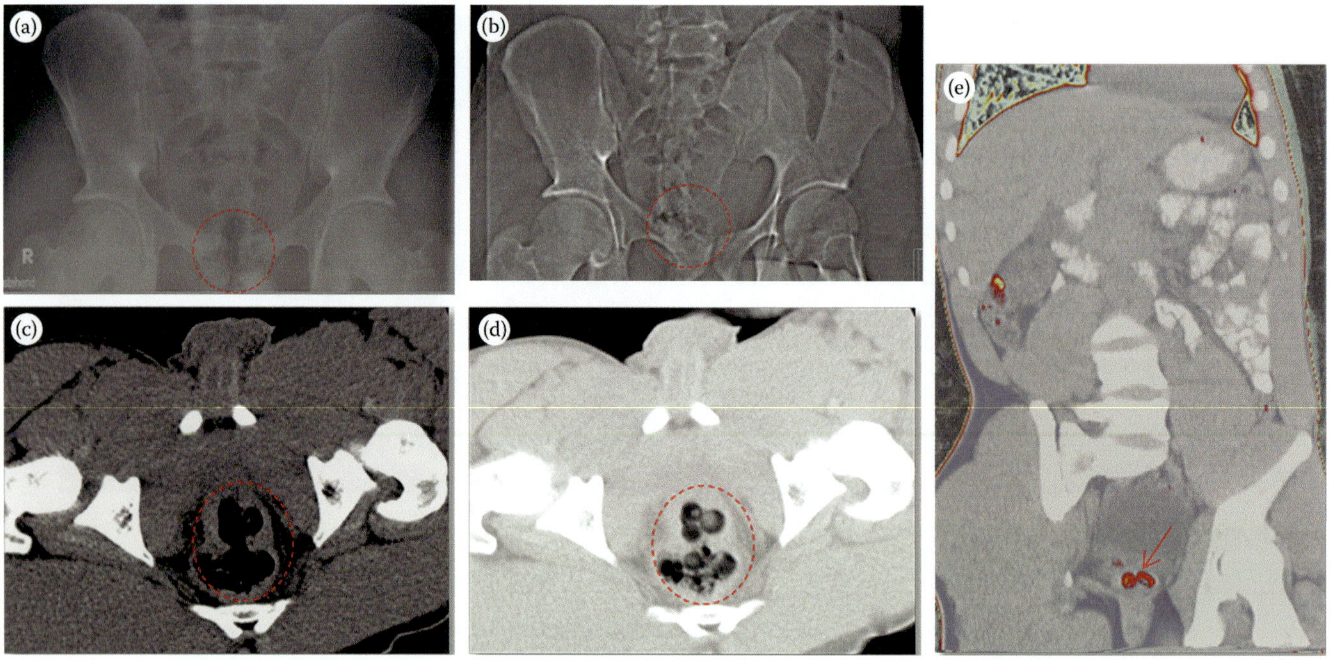

FIGURE 20.28 (a) CR in the upright position shows superimposed packs due to descended intestines. The packs were not detectable on plain radiograph with the patient in upright position. (b) Scout view in the supine position showing atypical gas formation with slightly denser regions. (c) Axial plane using an abdominal window; this finding will definitely be underdiagnosed if not seen by manipulating the window. (d) Axial plane with a wider window applied showing multiple round pellets in a body pusher. (e) Oblique multiplanar color-coded reconstruction depicting two packs in the rectum.

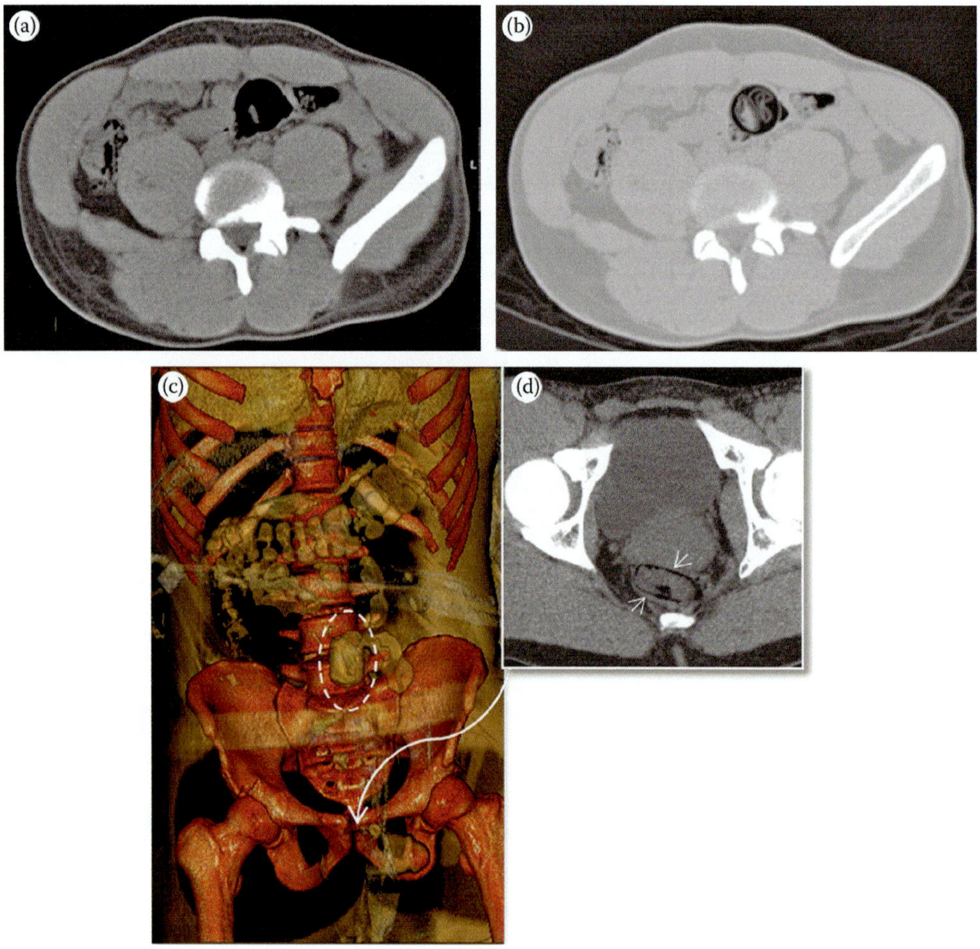

FIGURE 20.29 (a) Axial plane depicting a suspicious round distension of the jejunum. (b) Axial plane revealing an air-filled condom with two pouches of cocaine powder inside it. (c) Volume rendering, especially showing air-filled structures such as the condom (white circle). (d) Drug mules may have swallowed different kinds of packs. The one in the rectum was initially overlooked (arrows).

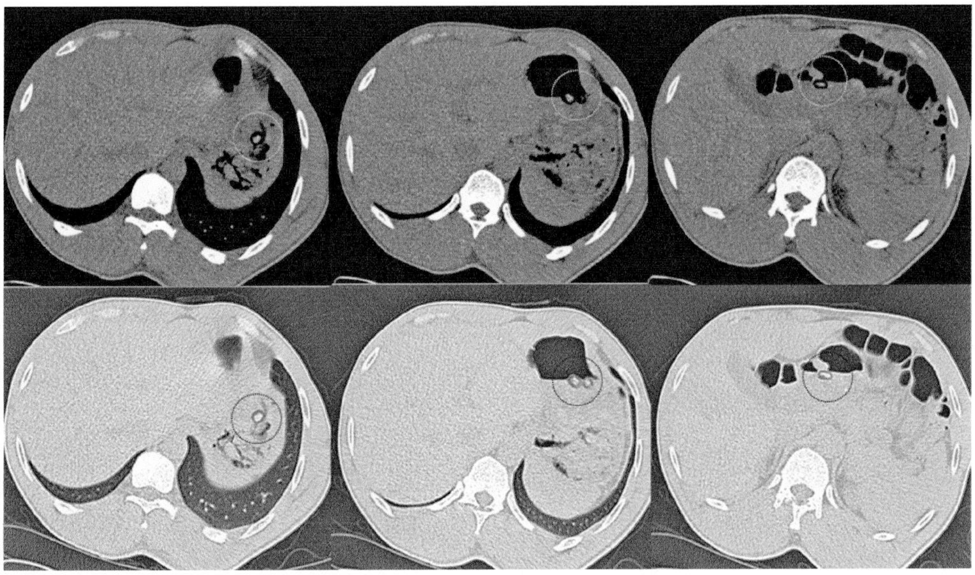

FIGURE 20.30 Axial plane (1.2 mm) with a view from the bottom up; cross-sectional images in descending order from stomach to duodenum. The CT corresponds to Figure 20.14 (LSDR prior to CT). The upper row demonstrates unidentified, small hyperdense "dots" in the stomach and duodenum. The lower row in hard kernel with lung leveling allows diagnosis of body stuffing.

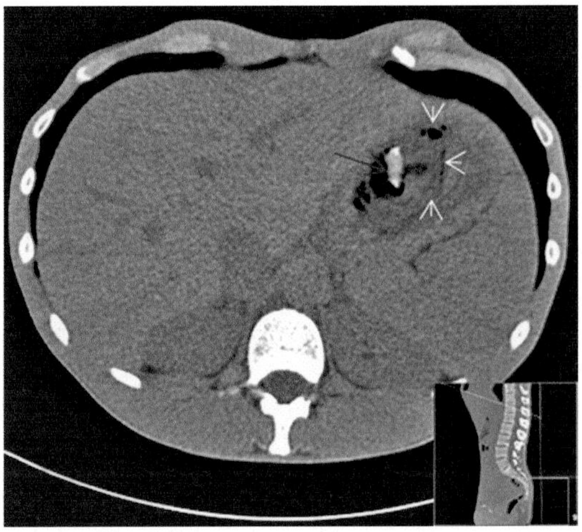

FIGURE 20.31 Axial plane, view from the bottom up. Clearly visible hyperdense trapezoid-shaped cocaine pack in the stomach (red arrow). Neighboring this pack there is one more that appears isodense to chyme. Only the air-filled cap with a narrow air crescent makes that container detectable (white arrow).

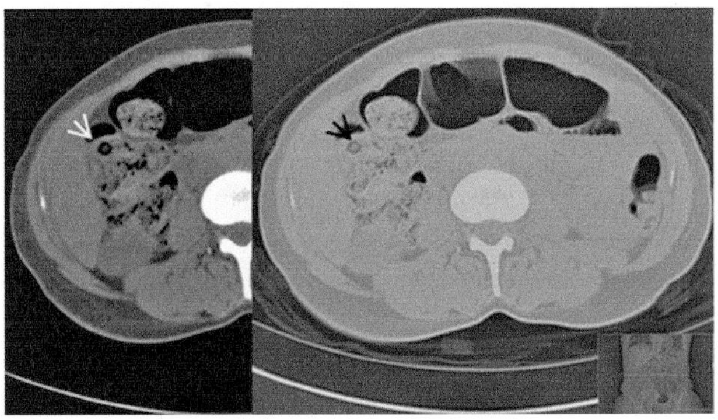

FIGURE 20.32 Axial plane, view from the bottom up. Single spherules (<1 cm, arrow) detectable in the ascending colon after identifying a trafficker.

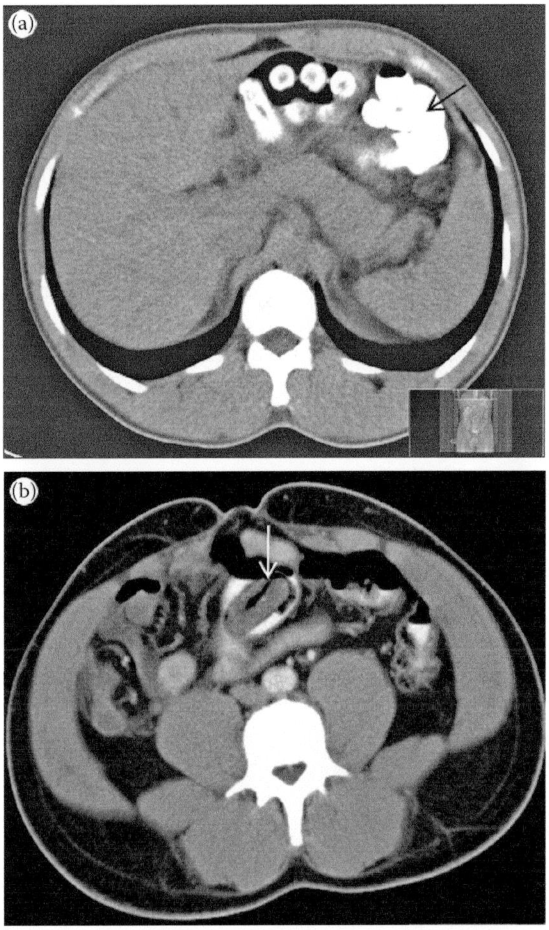

FIGURE 20.33 Axial plane with a view from the bottom up. (a) Hyperdense packs in the stomach with no adjacent contrast agent. In this case, contrast was only applied rectally; therefore, there is some contrast located in the transverse colon (black arrow). (b) Hypodense pack in the transverse colon; in this case, contrast eased the delineation of the packs but only due to its low density.

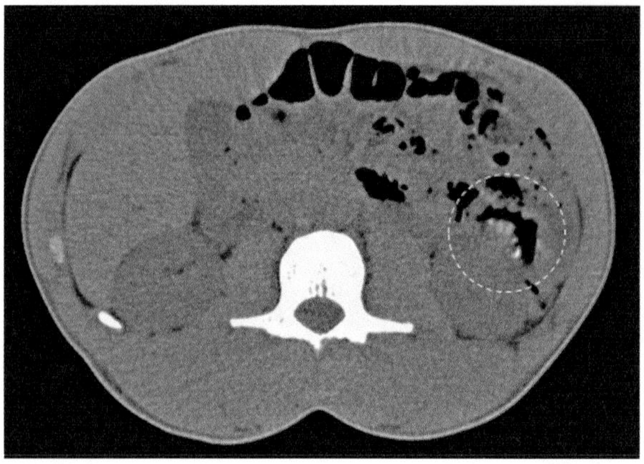

FIGURE 20.34 Axial plane with a view from the bottom up. The encircled area shows beam hardening of feces, which can easily be mistaken for a drug pack.

pulse, grains, stones of fruits, or just other swallowed foreign bodies, especially those located in the stomach (Figure 20.34). Formed scybala may also be problematic.

OTHER IMAGING MODALITIES

Of course, other imaging modalities are also used, such as the aforementioned x-ray fluoroscopy or sonography. Ultrasound is a fast and radiation-free alternative,[25,26] especially in cases of a noncompliant patient, as there is no point in performing a CT scan in a fidgety or nonconsenting suspect (Figure 20.1). Sonography can confirm swallowed packs, but a negative examination is inconclusive and does not mean that the patient did not conceal incorporated drug containers. The packs appear on sonography as oval or round, smooth-surfaced, hyperechogenic structures at the margin, with dorsal echo extinction due to absorption. A pitfall in ultrasound is obviously compact stool, but this can be differentiated by close inspection of the surface of the suspected pack vs. stool. Feces show an irregular surface unlike drug containers.[26] This examination is highly dependent on the experience of the investigator and is not reproducible.

MRI has several drawbacks such as its long examination time, its need for a cooperative patient, its high cost (in these cases mainly at the expense of the state, as most suspects are asylum-seekers), and its unavailability in emergency situations.[1,18] Nevertheless, MRI is an excellent tool to visualize newer incorporable liquidized cocaine packs such that, this time, the radiologists might be one step ahead!

REFERENCES

1. Brogdon, B. G., Vogel, H., and McDowell, J. D., *A Radiologic Atlas of Abuse, Torture, Terrorism, and Inflicted Trauma*, CRC Press, Boca Raton, FL, 2002.
2. Collins, V. P., Origin of medico-legal and forensic roentgenology, in *Classic Descriptions in Diagnostic Radiology*, Vol. 2, Bruwer, A. J., Ed., Charles C. Thomas, Springfield, IL, 1984, p. 1593.
3. Deitel, M., and Syed, A. K., Intestinal obstruction by an unusual foreign body, *Can Med Assoc J*, 18, 221–4, 1973.
4. Miller, J., Hendren, S., and Liscum, K., Giant gastric ulcer in a body packer, *Journal of Trauma* 45(3), 617–619, 1998.
5. Bulstrode, N., Banks, F., and Shrotria, S., The outcome of drug smuggling by "body packers"—the British experience, *Ann R Coll Surg Engl*, 84, 35–38, 2002.
6. Cordero, D. R., Median, C., and Helfgott, A., Cocaine body packing in pregnancy, *Ann Emerg Med*, 48(3), 323–5, 2006.
7. Beck, N. E., and Hale, J. E., Cocaine "body packers", *Br J Surg*, 80(12), 1513–6, 1993.
8. United Nations Office on Drugs and Crime (UNODC); http://www.interpol.int/public/drugs/cocaine/default.asp (02.08.2007 last modified).
9. De Probst, N., Lefebvre, A., Questel, F., et al., Prognosis of cocaine body—packers, *Intensive Care Med*, 31(7), 995–8, 2005.
10. Kelly, J., Corrigan, M., Cahill, R. A., and Redmond, H. P., Contemporary management of drug-packers, *WJES*, 2, 9, 2007.
11. Traub, S. J., Hoffman, R. S., and Nelson, L. S., Body packing—The internal concealment of illicit drugs, *N Engl J Med*, 349, 26, 2003.
12. Heinemann, A., Miyaishi, S., Iwersen, S., Schmoldt, A., and Püschel, K., Body-packing as cause of unexpected sudden death, *Forensic Sci. Int.*, 92, 1–10, 1998.
13. Karch, S. B., *The Pathology of Drug Abuse*, CRC Press, Boca Raton, FL, 2000.
14. Goertemoeller, S., Behrman, A., and Muller, A. A., The risky business of body packers and body stuffers, *J Emerg Nurs*, 32, 541–4, 2006.
15. Madea, Praxis Rechtsmedizin, Springer Verlag 2003, ISBN3-540-43885-8.
16. Traub, S. J., Kohn, G. L., Hoffman, R. S., and Nelson, L.S., Pediatric "body packing", *Arch Pediatr Adolesc Med*, 157, 174–177, 2003.
17. Greenberg, M. I., and Shrethra, M., Management of the pregnant body packer, *J Toxicol Clin Toxicol*, 38, 176–7, 2000, Abstract.
18. Algra, P. R., Brogdon, B. G., and Marugg, R. C., Role of radiology in a National Initiative to Interdict Drug Smuggling: The Dutch experience, *AJR* 189, 331–336, 2007.
19. Schmidt, S., Hugli, O., Rizzo E., et al., Detection of cocaine filled packets—Diagnostic value of unenhanced CT, *EJ Rad* 67, 133–138, 2007.
20. Hergan, K., Kofler, K., and Oser, W., Drug smuggling by body packing: What radiologists should know about it, *Eur Radiol* 14, 736–742, 2004.
21. Taheri, M., Hassanian-Moghaddam, H., Birang, S., et al., Swallowed opium packets: CT diagnosis, *Abdom Imaging* 33,262–266, 2008.
22. Meyers, M. A., The inside dope: Cocaine, condoms, and computed tomography, *Abdom Imaging* 20, 339–340, 1995.
23. Grabherr, S., Ross, S., Regenscheit, P., et al., Detection of smuggled cocaine in cargo using MDCT, *AJR*, 190, 1390–1395, 2008.
24. Sengupta, A., and Page P., Window manipulation in diagnosis of body packing using computed tomography, *Emerg Radiol* 15, 203–205, 2008.
25. Hierholzer, J., Cordes, M., Tantow, H., Keske, U., Mäurer, J., and Felix, R., Drug smuggling by ingested cocaine-filled packages: Conventional x-ray and ultrasound, *Abdom Imaging* 20, 333–338, 1995.
26. Meijer, R., and Bots, M. L., Detection of intestinal drug containers by ultrasound scanning: An airport screening tool?, *Eur Radiol* 13, 1312–1315, 2003.

21 Larceny

B.G. Brogdon

CONTENTS

LARCENY BY INGESTION

In 1896, C. Thurston Holland (1863–1941) was a London physician who often assisted his friend, Sir Robert Jones, the famous pioneering orthopaedist, in his free Saturday morning clinic. His introduction to radiology was quite accidental, having to do with a young patient who had a retained pellet from a shotgun wound in the hand. Nevertheless, Holland's interest was piqued by the novel experience, and he went on to become one of England's premier pioneers of radiology and well known throughout the scientific world. He became President of the First International Congress of Radiology in 1925 in London. He published a short memoir of the first year of radiology entitled, "X-rays in 1896" in *The Liverpool Medico-Chirurgical Journal*, 45, 61, 1937.[1]

In telling of the hysteria associated with the first news of Röntgen's discovery, Dr. Holland recounts several amusing examples, including the well-known advertisement for x-ray-proof underclothes for young ladies. He recalls that the *Strand Magazine* carried a series of stories (obviously fictional) illustrating "the Adventures of a Man of Science." In the July 1896 issue a man suspected of having swallowed a diamond is lured by the "Scientist" into his laboratory where the culprit is undressed and carefully positioned so that the rays would pass through his body. "… My electrical battery worked well. The rays played admirably in the vacuum tube. I removed the cap from the CAMERA…". Get the idea? The climax, of course, was that the photographic plate clearly showed the diamond "just below the region of the ileo-caecal valve"! This undoubtedly is the first of many stories of *larceny by ingestion* thwarted by the x-ray.

Dr. Holland had noticed in a roentgenogram of a lady's hand that the diamonds in a large marquise ring did not "cast a shadow; the mounting appeared empty on the plate." So, on August 7, 1896, he exposed a diamond star and a paste brooch (Figure 21.1). The diamonds were radiolucent; the paste was densely opaque!

Despite Dr. Holland's finding, for a century now we have continued to be regaled periodically with sensational stories of "The Swallowed Diamond." Attempts to follow up on these press releases have been unsuccessful and puzzling. Either the radiographs seem never to have existed, or else they fail to show the ingested gem stones.

A case in point is illustrated in Figure 21.2. This newspaper clipping[2] has all the usual features of the genre: the distracted salesperson; the clumsy thief; the obligatory x-ray; the triumphant discovery, "He was x-rayed and there it was…"; the twist at the end of the tale—the poor rookie detective given the sordid task of retrieving the evidence; and, especially, the missing denouement. The radiology group covering the institution where the examination was performed were tracked down. They were most cooperative and sent the original radiograph and a copy of their report. There was no foreign body of any kind visible within the abdomen. They could offer no explanation for the wire service story.

Another, more spectacular, case of larceny by ingestion came on the wire from Milan[3] (Figure 21.3). Think of it—11,216 stones worth a million dollars! Perhaps the gems were outlined by air in the plastic bag? Dr. Leonardo Lovisatti, a friend and well-known radiologist in Northern Italy, was contacted. Some weeks later he replied, "… the police will not give any information…. I have been in touch with the heads of the most important Radiology Departments in Milan, but none of them can help. The Press Agencies have no lead on where the x-rays had been carried out."

Now if the thief swallows the mounting as well as the diamonds, the story is entirely different and credible (Figure 21.4).[4] Any metal used to mount precious stones will be apparent radiographically.

Only one case is known to the author in which a positive radiograph was obtained after ingestion of an unmounted jewel. In that case, Röntgen's rays betrayed not one crook, but two!

A woman who had been viewing a group of loose stones in a jewelry store was seen to put her hand to her mouth and cough moments before abruptly arising from her seat to leave the store. She was detained by the proprietor who had her arrested. Subsequently, a radiograph of the abdomen was obtained and clearly showed an opaque foreign body in the typical shape of a brilliant-cut diamond (Figure 21.5). The charges against the woman were dropped, however, immediately after the radiologist pointed out to the arresting officers

FIGURE 21.1 Holland's 1896 roentgenograph of a diamond star and a paste brooch. The diamond mountings appears empty, the paste "gems" are densely opaque. In 1896 this required a two-minute exposure. (From Holland, C. T., In *Classic Descriptions in Diagnostic Radiology*, Vol. 1, Bruwer, A. J., Ed., Charles C. Thomas, Springfield, IL, 69, 1984. With permission.)

that real diamonds are not radiopaque and that the jeweler might be guilty of fraud!

Our experience and/or frustration with "swallowed diamond" stories led to the undertaking of a group of experimental radiographs. A couple of tissue-isodense plastic containers assembled like a "double-boiler" were set up so that specimens of real and *faux* gems could be suspended in 20 cm of water (Figure 21.6). This simple phantom reproduced the situation of a gem inside a fluid-filled viscus inside a reasonably trim body. It could now be shown that diamonds

4A The Houston Post/Sun., January 18, 1987 *

X-rays of woman's stomach provide police sparkling results

MILAN, Italy (UPI)—A woman swallowed 10,999 diamonds and 217 emeralds in an attempt to smuggle them into Italy, but she was arrested after X-rays revealed the jewels in her stomach, police said Saturday.

Magdalena De Vree of Belgium flew into Milan's Linate Airport from Zurich, Switzerland "a few days ago," police said.

Customs police said the woman appeared nervous, but a search of her luggage turned up no contraband.

Smugglers, however, occasionally swallow drugs and other contraband, so police obtained a court order permitting them to take the suspect to a hospital for an X-ray examination.

The tests revealed the presence of two small plastic bags containing 10,999 diamonds and 217 emeralds in the woman's stomach. Police estimated the value of the jewels at about $1 million and arrested the woman.

FIGURE 21.3 Newspaper clipping of a spectacular story of grand larceny by ingestion.

Man swallows stolen diamond

EULESS, Texas (AP)—Officers in this Fort Worth suburb were using a potent laxative today in their attempt to recover a stolen $10,000 diamond swallowed by a Dallas man.

The 3-carat diamond had not been recovered by early today.

OFFICERS SAID the man entered the Euless Gold & Silver Exchange about 2 p.m. Tuesday and asked to look at diamonds. When clerks dealt with other customers, he grabbed a stone and walked out of the store.

An employee stopped the man just outside the store, but police said he had swallowed the diamond by then.

The man was first questioned by police and then taken to Harris Hospital in nearby Bedford where staff confirmed the theft by X-raying his stomach.

"HE WAS X-RAYED, and there it was," said Euless Police Sgt. Bill McClendon. "They gave him something to make him throw up, but by then the diamond had moved to the intestines and they said there was only one way to get it then."

Rookie detective Bob Freeman was given the chore of administering the laxative and checking for the diamond. A second investigator took over the chore later.

Charges will be filed when the diamond is recovered, McClendon said.

FIGURE 21.2 Newspaper clipping of typical "swallowed diamond" story.

Alabama's oldest newspaper

MOBILE REGISTER

May 12, 1994
Since 1313

THURSDAY

PORTLAND, Ore.—After waiting four days, deputies finally got the evidence in the disappearance of a diamond ring.

Steven Wade Hamilton produced the wedding band Tuesday, the sheriff's department said. He is accused of stealing it from a store Friday and then swallowing it, along with a less expensive ring.

Originally, the store estimated the value of the diamond ring at $2,300, but the $3,605 pricetag was still attached after deputies recovered the evidence at the Multnomah county jail.

On Sunday, Wade produced a ring valued at $160, plus two dimes and a nickel.

Hamilton, 32, faces robbery and criminal mischief charges.

—From wire reports

FIGURE 21.4 Newspaper clipping of a thief whose taste ranged from diamond rings to small change.

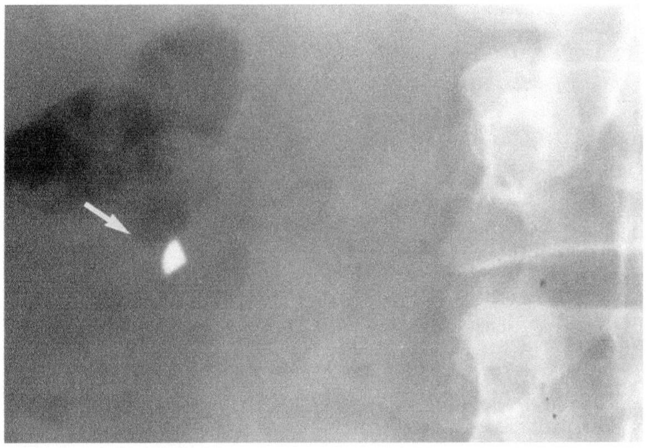

FIGURE 21.5 Radiograph of the abdomen of a suspected jewel thief. Note the dense object (arrow) in the shape of a brilliant cut diamond. (Courtesy of Robert R. Roper, MD)

FIGURE 21.6 Water-phantom for roentgenology of real and *faux* jewels. A ring (arrow) is positioned at mid-depth.

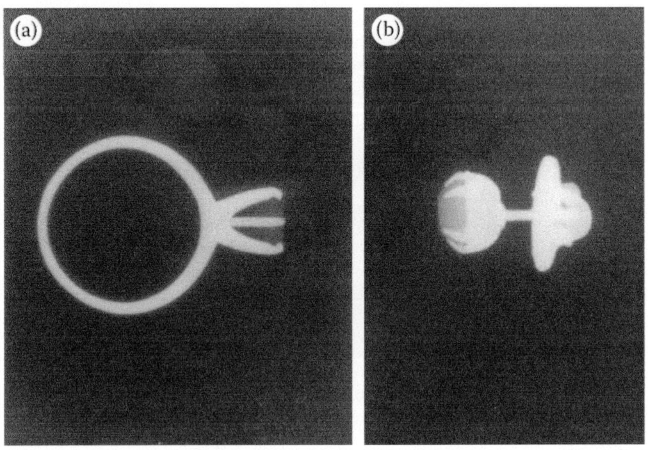

FIGURE 21.7 (a) Diamond solitaire ring, and (c) fake solitaire tie-tack in 20 cm water-phantom.

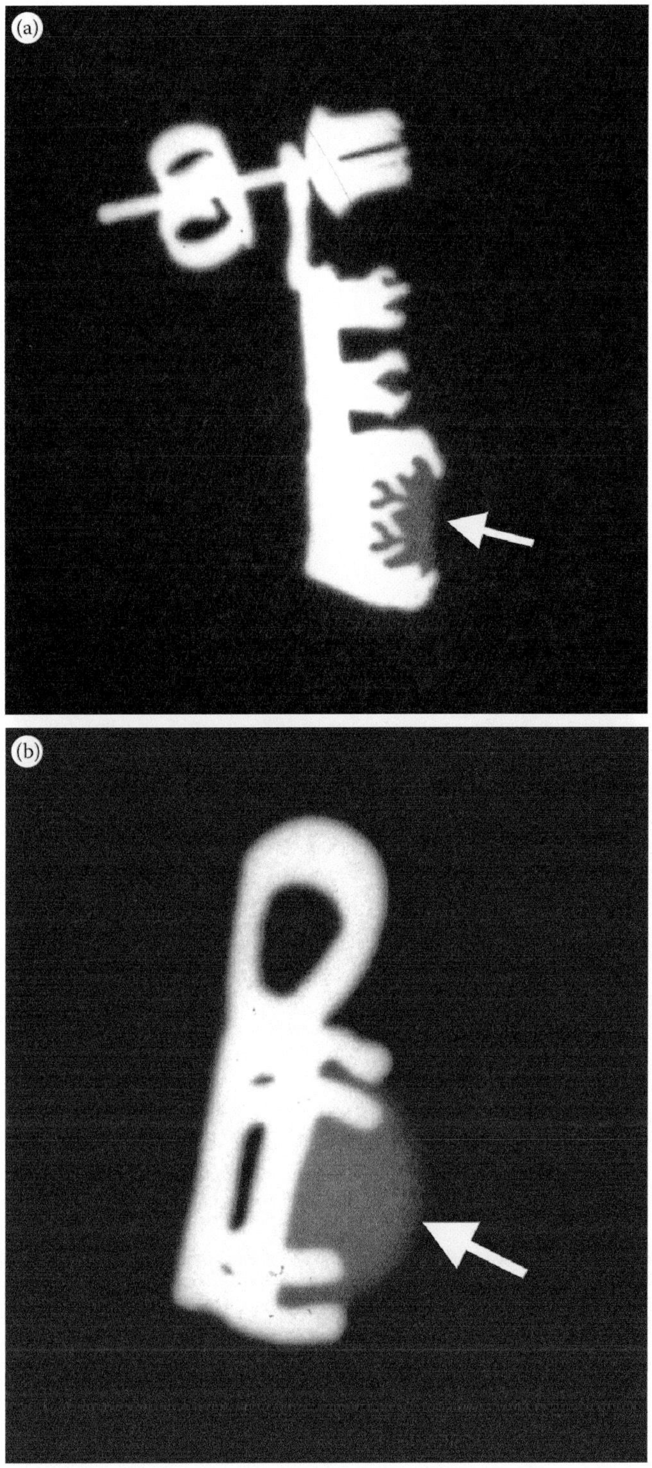

FIGURE 21.8 Other stones in a 20 cm water phantom (a) Emerald; (b) Star Ruby; (c) Purple Amethyst; (d) Rhinestone.

were not sufficiently opaque to be visible inside the water-density structures of the human body by conventional roentgenology (Figure 21.7), although the Vogels indicate that fluoroscopy may reveal diamonds when covered by 14 cm or less of water.[5] While it is conceivable that their silhouettes might be apparent if suspended in an air-filled viscus, with

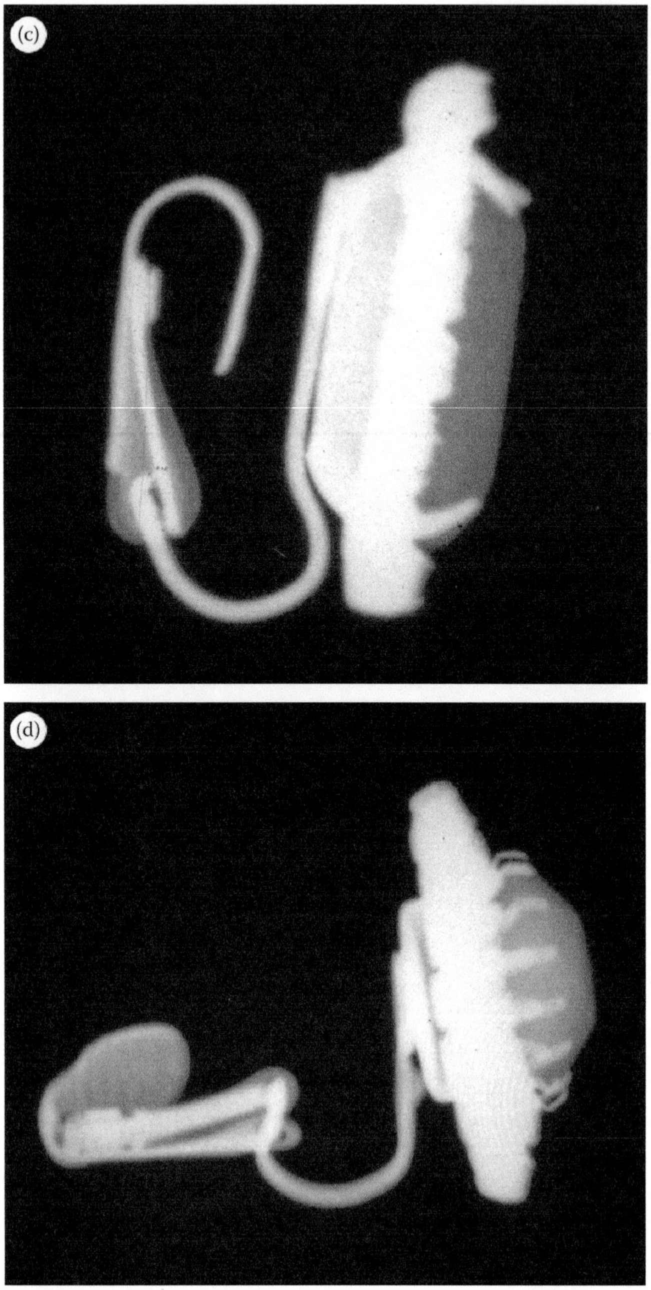

FIGURE 21.8 (*Continued*)

routine positioning the gem would sink beneath any fluid present in even a partially aerated bowel. The radiographic density of gemstones is a function of their atomic number. Thus, rubies and emeralds may or not be visible in tissue-equivalent material depending on tissue thickness and the quality of the beam. Other gemstones are faintly to moderately radiodense (Figure 21.8). Many of the false gems were radiopaque. High-quality glass is quite dense. False pearls could be distinguished from real ones.

Our conclusion is that conventional roentgenography or fluoroscopy cannot be depended upon to reveal swallowed precious stones. Negative findings in such cases do not preclude the presence of ingested gems. Demonstrated opaque "jewels" are likely semiprecious, at best.[6]

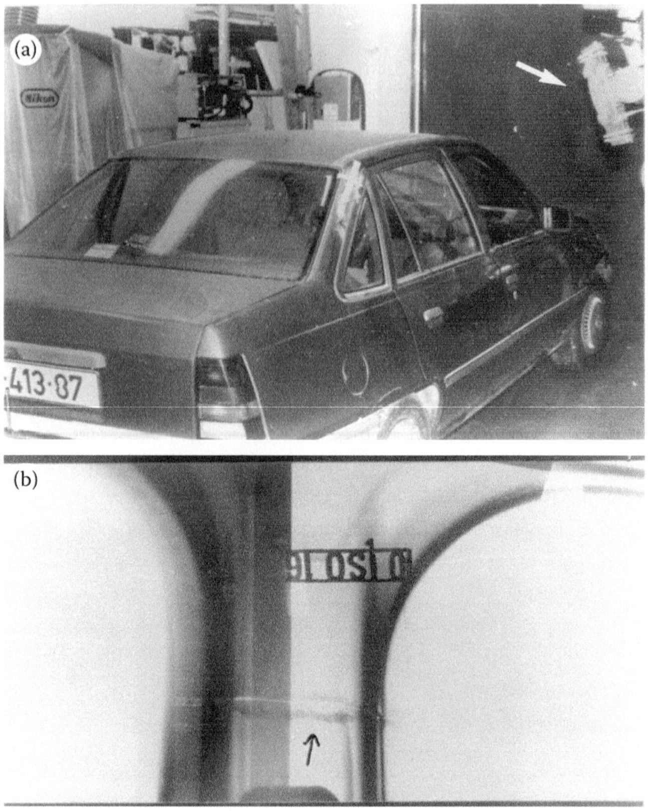

FIGURE 21.9 (a) Industrial x-ray tube in upper right corner of photo (arrow) is positioned to radiograph the door post of a suspected stolen and re-identified auto. (b) Positive print of radiograph shows a nonoriginal welding seam (arrow) on the door post below the number. (Reprinted from Springer, E. and Bergman, P., *J. Forensic Sci.*, 39, 751, 1994. © ASTM. With permission.)

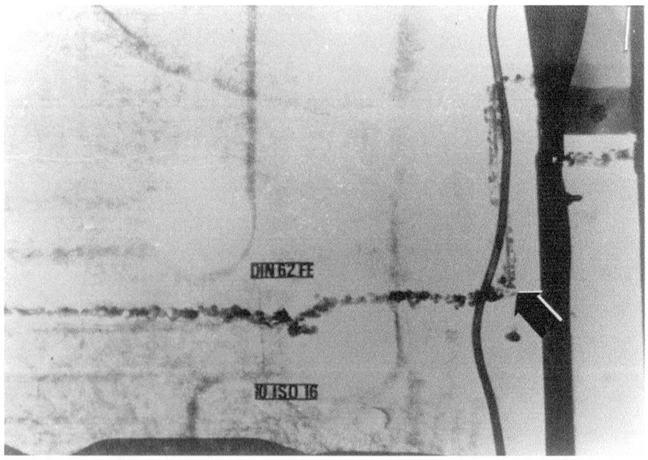

FIGURE 21.10 Positive print of a radiograph of flooring in the front passenger compartment showing nonoriginal welding. The arrow points to a common nail used as filler in the weld seam. (Reprinted from Springer, E. and Bergman, P., *J. Forensic Sci.*, 39, 751, 1994. © ASTM. With permission.)

It is possible that computed tomography (CT), combined scanning methods, or dual energy techniques will improve detection of ingested gems, especially diamonds. Search of the diamond miners in Africa by radiation depends largely on the contamination or accretion of other minerals on the "raw" diamonds.[5]

RADIOLOGY AND AUTO THEFT

When Professor Röntgen opened to view the inner works, recesses, and loads in his prized shotgun with his new-found rays, he opened up a new field: *nondestructive testing*. The German and Austrian military were the first to employ this opportunity to examine armaments and other materials for quality control or defects without destroying the test object. Now nondestructive testing methods emphasizing a variety of radiant energy forms and sources are used widely in many industries. But industrial nondestructive testing is beyond the scope of this book.

However, the Israel Police adapted the nondestructive testing method to the detection of the common nonviolent crimes of auto theft and auto forgery.[7]

Often a new identity is created for a stolen vehicle by one of three main methods: (1) simple alteration of documents; (2) obliteration and restamping of the engine and chassis vehicle identification number (VIN); or (3) transfer of key parts from a wrecked vehicle to a stolen one. In the latter instance, the part containing the VIN of the wrecked vehicle is cut out and welded onto the stolen vehicle. Ordinarily, the finish must be removed from the suspect vehicle in order to reveal the forgery. This is quite destructive, of course, and is quite undesirable, especially if no forgery is discovered.

The Israel Police have access to an industrial x-ray machine and have found that properly positioned radiographs readily display the tell-tale welding seams of the forged VIN on the stolen auto (Figure 21.9). The patterns of the original spot-welding and the secondary seam-welding with a filler rod are different (Figure 21.10).

REFERENCES

1. Holland, C. T., X-rays in 1896, in *Classic Descriptions in Diagnostic Radiology*, Vol. 1, Bruwer, A. J., Ed., Charles C. Thomas, Springfield, IL, 1984, 69.
2. Anon., Associated Press release, *Mobile Press*, October 16, 1985, 34A.
3. Anon., United Press International release, *Houston Post*, January 18, 1987, 4A.
4. Anon., Wire report, *Mobile Register*, May 12, 1944, 2A.
5. Vogel, H. and Vogel, B., The invisible sentry: Radiological imaging: Radiological imaging in terror and terror prevention, *J Forensic Med Path*, 2, 1, 3, 2006.
6. Brogdon, B. G., Larceny by ingestion, Presented at the *Annu. Meet. American Academy of Forensic Sciences*, Cincinnati, February 23, 1990.
7. Springer, E. and Bergman, P., Applications of non-destructive testing (NDT) in vehicle forgery examinations, *J. Forensic Sci.*, 39, 751, 1994.

22 Radiology of Fakes and Forgery in Art

A. Everette James

CONTENTS

INTRODUCTION

The use of radiographic techniques to detect fraud and forgeries in the art world has great historical precedence. Within months of Röntgen's discovery, W. König of Frankfurt was using roentgen rays to examine oil paintings in order to detect alterations in them. Almost 70 years ago Elliott published a 12-page article in the *American Journal of Roentgenology* on the use of roentgen rays and the scientific examinations of paintings (Figure 22.1). With the advance in digital imaging and other technologies, the opportunities to document both authenticity and fakery are enhanced. In this chapter we will use our personal experience with valuable paintings and antique waterfowl decoys as paradigms to illustrate basic principles and to provide some insight into approaches that have proven useful by illustrative vignettes.

AUTHENTICATION OF ANTIQUES

The evidentiary nature of a visual image such as a radiograph, and the ability to produce an immediate permanent archive, are very compelling features as evidence. The same rules of evidence and documentation apply whether one is authenticating a $300,000 Elmer Crowell preening pintail decoy done at the turn of the twentieth century (Figure 22.2), a genre scene produced in the middle of the nineteenth century by John George Brown (Figure 22.3), the dental radiographs of Dr. Mengele, or the skull of *homo erectus*. This image process represents a unique method of documentation in which the final form provides any trier of fact an opportunity to visually examine the inner composition of the object at issue and is truly compelling legal evidence if properly introduced into the court and interpreted by an expert witness.

Antique waterfowl decoys represent a uniquely native North American art form. Since most of our cultural and artistic heritage is largely derivative, these objects are different as they are products of a true American heritage and tradition.

The first people to use decoys (a Dutch word for trap) were our Native Americans, who taught our colonial forebears to use them in luring waterfowl into close gunning range. Although the first decoys were fashioned from straw and earth, by the mid and late nineteenth century they were carved from wood. Later other materials were employed in their construction which can be used to "date" a decoy in questions of authenticity. After all, if the decoy was made of cork and Joe Lincoln of Accord, MA died before the use of cork—he did not make the decoy.

Decoys were carved according to certain species which varied in different geographic locations, and the materials and type of wood used reflected what was available in that area. The heads were carved separately from the bodies and were affixed with nails inserted in an often characteristic pattern, or fitted with dowels. For example, Ned Burgess of Waterlilly, NC drove three nails in the base of the neck to affix the head. They were driven obliquely, two posterior and one anterior. If one is evaluating the authenticity of a "Ned Burgess" decoy and does not see this pattern, it is very likely not by the famous carver. Weights and ties were applied to the decoy to provide balance so that it would float properly in the water, and to secure the decoy when set out as part of a rig (large grouping of decoys placed on the water in a particular pattern to appear attractive to the waterfowl). Weights are often unique to a particular foundry and can be dated. Tie nails and patterns for fixation also are identifiable.

The wooden construction of a decoy is, indeed, a fortunate circumstance since with use certain characteristic changes due to aging and damage from weather conditions will be reflected by lines and fissures (checks) in the wood, and these can be seen on the radiograph. The hardware employed in the construction process also provides excellent data of an evidentiary nature as the same kind of nails were used during the same time period in cabinetry and building construction and can be quite accurately dated (Figure 22.4). As noted, weights were often produced in foundries near the waterfowl area where the decoy carvers lived. Thus, a weight made for the Ward brothers in Crisfield, MD will not be like that made for Joe Lincoln of Accord, MA, or Ned Burgess of Waterlilly, NC.

Just as one evaluates structural alterations due to aging and disease processes in human medicine, we can evaluate

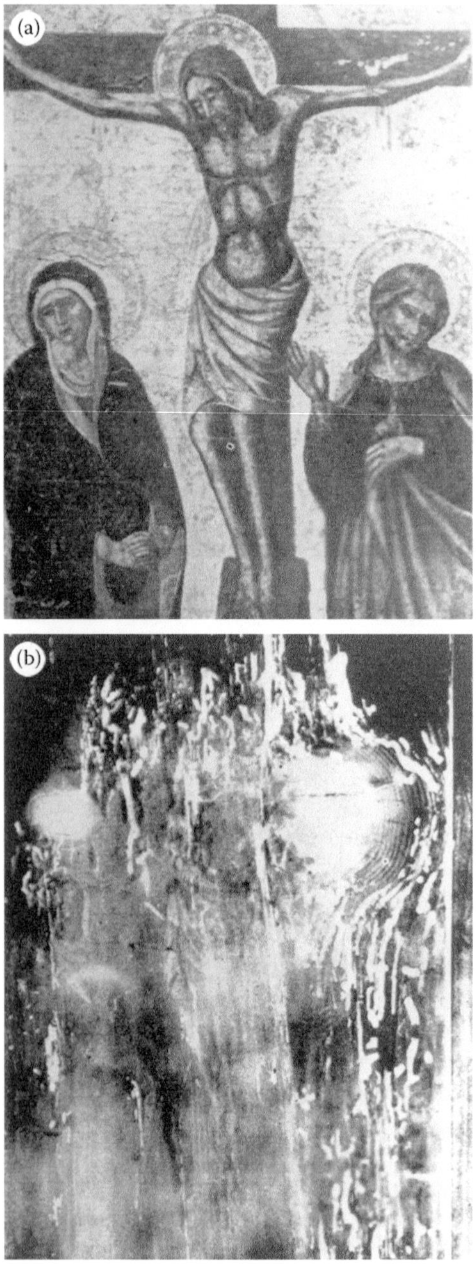

FIGURE 22.1 (a) Elliott's example of a forgery after the style of Luca di Tommé. (b) A roentgenogram shows that the wooden panel was old and full of worm holes before the paint was applied. No artist would have painted a picture on a wormy piece of wood. Paint added after worm holes have appeared is a certain sign of forgery. (From Elliott, W. J., *Am. J. Roentgenol.*, 50, 779, 1943. With permission.)

the *equivalent* changes in a decoy. These may be quite determinative in assessing its authenticity and value. Since it is no longer uncommon for individual decoys to sell for $25,000 (and a significant number have fetched over $100,000), the practice of "doctoring" and even faking decoys has become much more common. At these prices, it becomes "worth the effort" for the dishonest carver or fabricator.

While one can certainly carve a decoy body and apply a head that will present a form that upon visual inspection

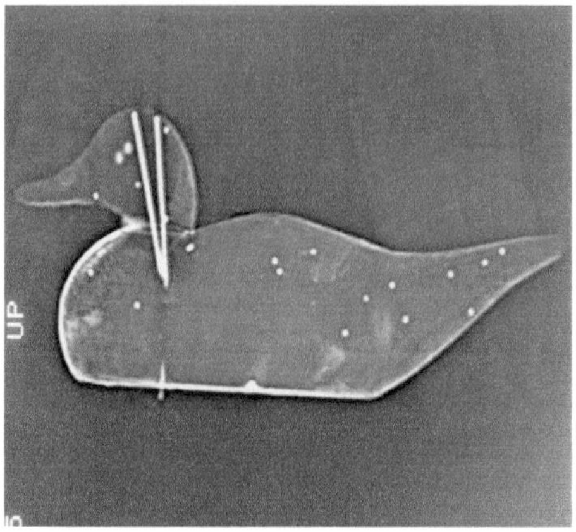

FIGURE 22.2 CT scout view or topogram of a pintail decoy by Elmer Crowell (ca. 1910). There is shot in the body and head of the decoy. The neck has been fractured and shortened with pins inserted to strengthen this vulnerable area. The decoy is genuine but not pristine. An Elmer Crowell decoy brought $314,000 at auction. (Collection Ward Museum, gift of A. Everett James and Son.)

alone appears authentic (Figure 22.5), the internal structure of a 50-plus- year-old object composed of various materials is much more difficult. Nails from businesses that supply materials for restoration (usually furniture) will appear as "vintage" nails. However, introducing these in such a manner that produces the "rust tracking" on the radiograph is another matter (Figure 22.6). Since decoys were used over long periods of time one would expect to see radiographic evidence of repositioning of the hardware, and if this is not present one should look for other signs of "newness" such as the lack of many layers of paint applied at different times. Decoys were

FIGURE 22.3 Painting attributed to John George Brown (ca. 1850).

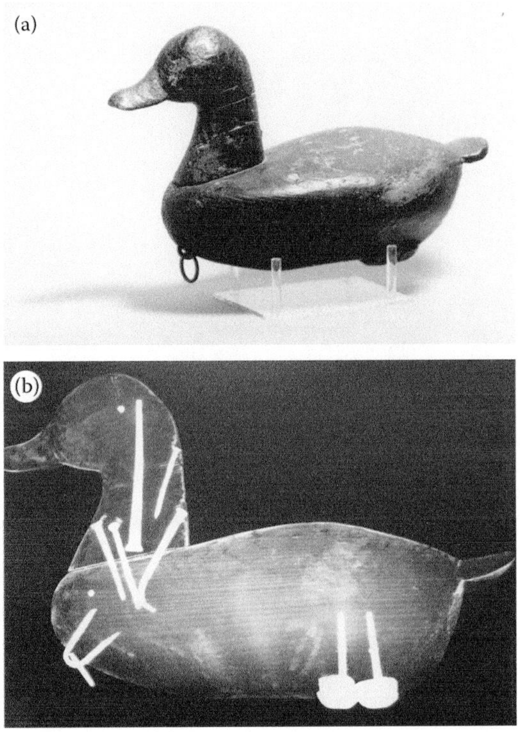

FIGURE 22.4 (a) A Ruddy Duck decoy by Alvirah Wright (*ca.* 1890). (b) Radiograph. The erect head, chubby body, and flat tail are characteristic of this rare and desirable carver. Note the extensive nail pattern to reinforce the neck due to in-use cracks. The density in the middle of the body is due to over-paint. The unique weight placement and tie line attest to the genuine nature of this decoy. Another member of this rig sold for $32,000 at auction, and many reheads or fakes have subsequently been fabricated. (James Collection.)

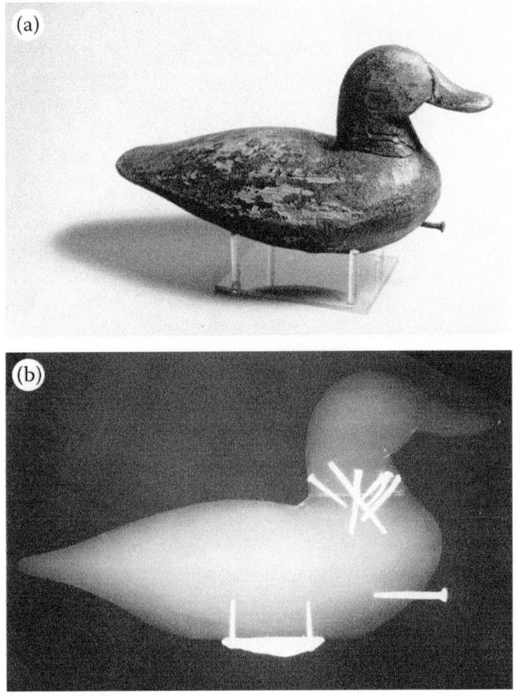

FIGURE 22.5 (a) A Ruddy Duck fashioned by an expert contemporary carver with aged wood, antique hardware, and style typical of a Knotts Island, Back Bay carver. The surface paint was applied in several coats and also aged. To visual inspection this has the appearance of a very desirable species which at auction might bring $15,000 to $25,000. (b) Roentgenogram reveals no reinsertion of hardware, age rust internally, true cracks and checks through the wood, or layering of the paint that would be expected if the decoy was truly from the 1900 to 1920 era.

used by guides, market gunners, and avid hunters for years, and they would be repainted to make them more visible.

In determining authenticity, one must appreciate that "authenticity" has several connotations. Many collectors will not accept a decoy that has had another head inserted (rehead), presumably due to damage of the original. The "rehead" can usually be detected by the difference in the grain pattern between the wood of the body and the inserted head, and the pattern of the nails to affix the head to the body (Figure 22.7). If the decoy was allegedly fashioned before the *Waterfowl and Migratory Game Acts* of 1914–1917, there often will be shot in the decoy; the legendary "market hunters", in an effort to be more efficient in their harvesting practices, shot the waterfowl while they were on the water in large groups. If one can detect shot in the body, then shot should also be present in the head (Figure 22.8).

"Pristine birds" are much more valuable than those that have been restored no matter how well the repair is done. Checks and cracks will often be filled in and paint applied over the area to conceal the repair. Fractures to the wood are most common in the midportion of the neck due to the hunter grasping this part of the decoy to set it out or to retrieve it after the hunt. Bills are often broken in use and storage because of their location where the wood is thin, making this a vulnerable area (Figure 22.9). No matter how skilled the deception, these repairs can be detected by radiographic studies.

Just as in almost all collectibles, authenticity and condition are the primary determinants of value. When the value

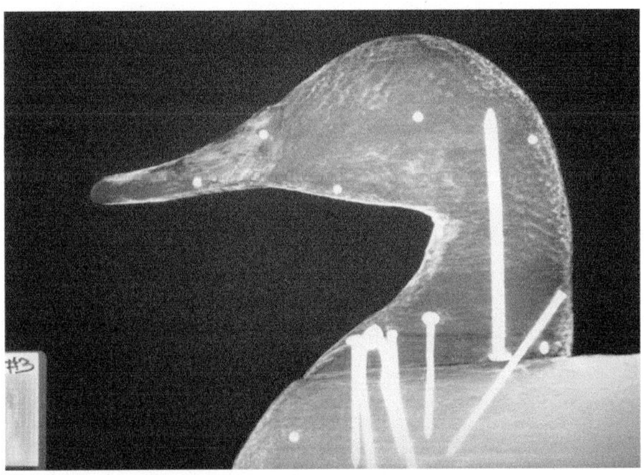

FIGURE 22.6 Decoy by Wilton Walker, Barca, NC (ca. 1930). Note the faint halo of rust around the central nail used to strengthen the neck. The linear opacities on the neck represent paint in rasp marks, an old technique to create a nonreflective surface.

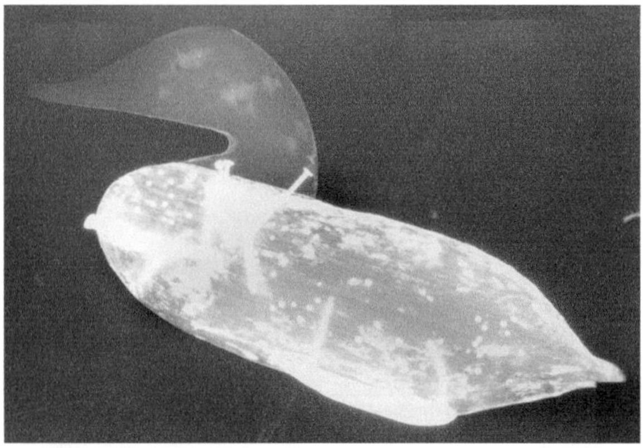

FIGURE 22.7 Radiograph of a decoy showing an antique body and a replaced head. There is an obvious difference in the paint layering and in the quality of the wood.

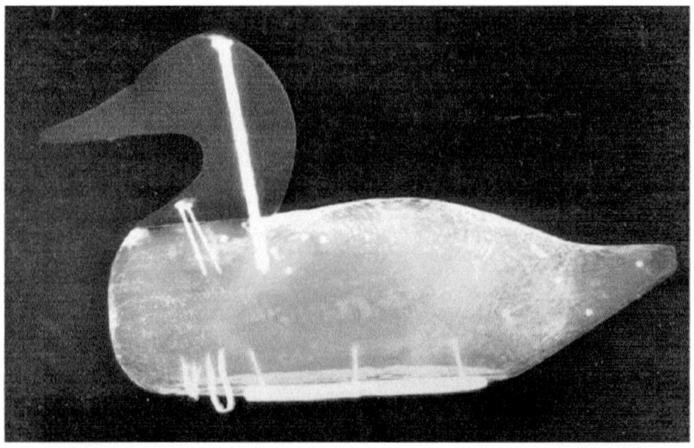

FIGURE 22.8 Another reheaded decoy. The new wood of the head and neck is almost entirely radiolucent. The aged paint and wood of the body is in sharp contrast. Note also the lead shot in the body but none in the head.

reaches a certain level, the temptation of forgery and misrepresentation become compelling to some. While these attempts at perfidy may not initially be very sophisticated, and can readily be discovered, radiographic techniques may become the standard form of analysis to uncover and document the evidence that a crime of substantial money value has occurred. Astute collectors have come to rely upon radiographic studies as part of their standard evaluation practice prior to acquisition of a valuable decoy. After all, would it not be embarrassing to pay $30,000 for an Alvirah Wright that was not "right"? While the prices of collectable decoys have increased substantially, these pale by comparison with those of fine paintings. If one sits with a bidding paddle at a Sotheby's or Christie's auction and sees many paintings selling for more than $1 million, one can understand that the temptation to produce a fake can be very compelling.

ANALYSIS OF PAINTINGS

In discussing the radiographic analysis of paintings these remarks will deal primarily with oil paintings that are applied on canvas, as these are the most commonly encountered

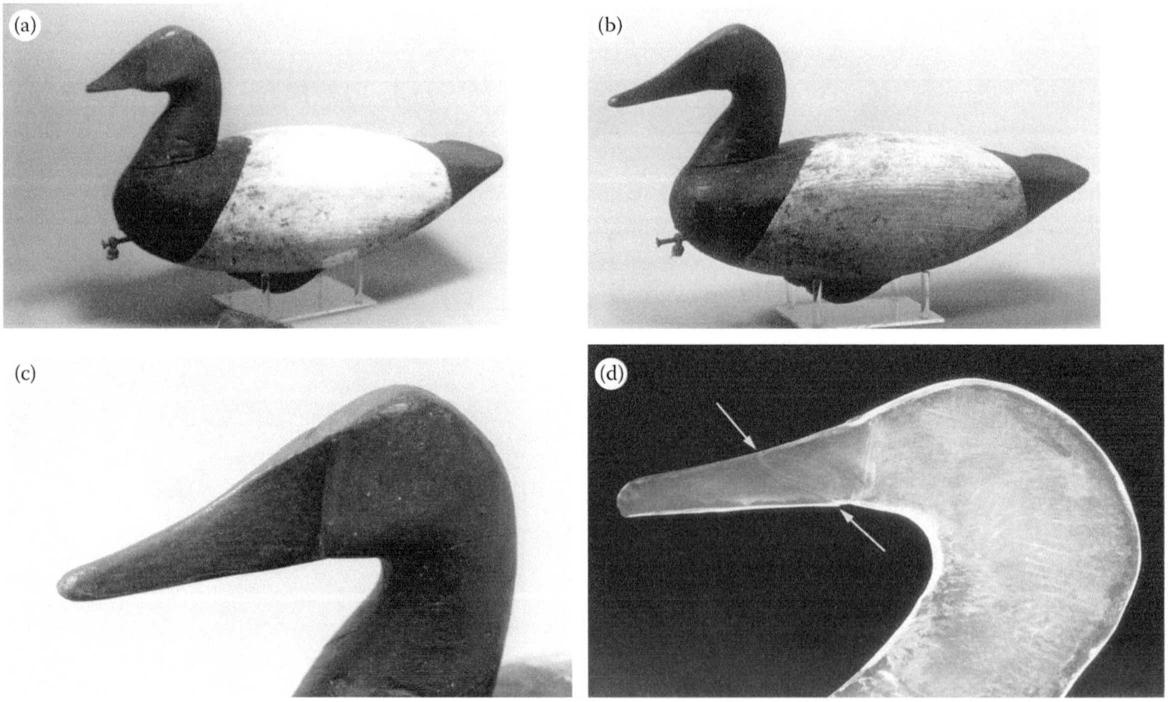

FIGURE 22.9 (a) A broken Canvas Back decoy by Ben Ethridge of the North Carolina Outer Banks (ca. 1920). (b) An expert bill replacement was done which is undetectable, (c) even by close visual inspection. (d) Radiography discloses the hairline fracture between the arrows. According to the tenets and strict criteria of decoy evaluation, this fracture reduces the importance of this decoy several-fold. Radiography was the determinant in this evaluation process.

combination of medium and support. We will mention paintings applied to paper, board, panel, and masonite. While radiographic studies are most often used to assess the condition of a painting prior to acquisition or restoration, these diagnostic techniques apply to determining fakes and forgeries as well.

Oil paintings may contain as many as seven layers that have their own characteristic appearance during the aging process. The response to the external elements as well as internal changes over time give a painting its look or patina. Thus, the successful faker or forger must not only duplicate the compositional materials but the aging process as well. While it may be readily feasible to meet the challenge of a visual or raking and UV light inspection, the differential absorption of the x-rays by the compositional materials can provide powerful documentary evidence that a painting is not what it is purported to be.

As noted earlier, there is an aging process that should be expected with paintings, and particularly with certain types of compositional techniques and materials. As an example, the "visionary artists" such as Albert Pinkham Ryder and Ralph Blakelock are among the most "faked" artists. They painted at the end of the last century and the first decade of the present, producing dark, moody paintings composed of many layers of varnish and pigment. Their works will show abundant crackleur (crackle or simply cracks) that should penetrate to the support, whether linen fibers of canvas, board, or wooden panel. The application of many pigment layers characteristically makes these paintings unstable, and the radiograph will show the cracks and fissures (Figure 22.10). If this is not seen, then the chance that the work is a painting by one of the visionary group is almost nonexistent.

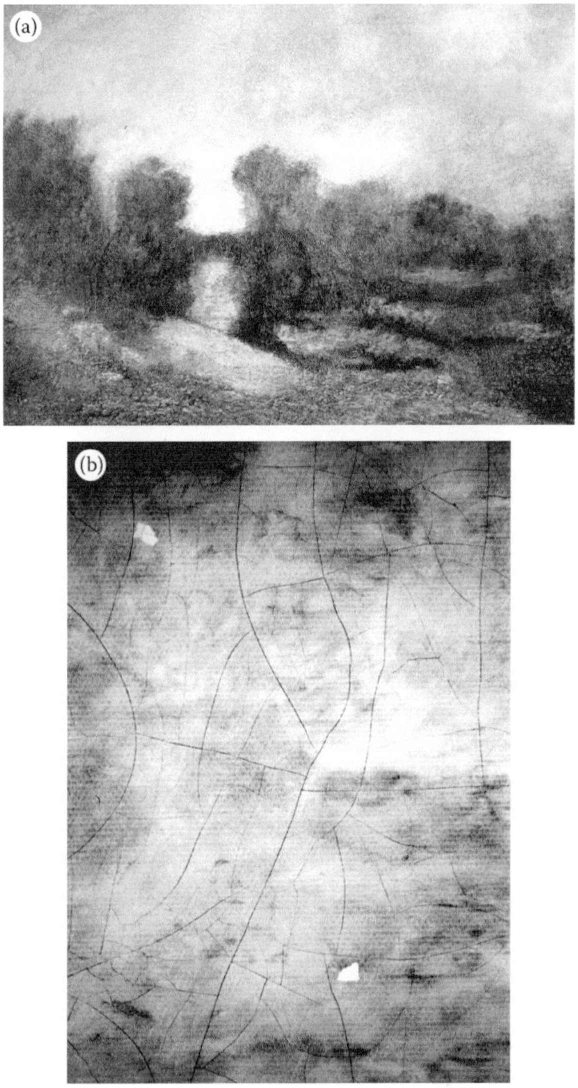

FIGURE 22.10 (a) *Creek at New Bedford* by Ralph Blakelock. (b) Magnification radiograph discloses the typically abundant crackleur of this artist.

Pigment materials can be assessed very well by radiographic techniques. One should recall that with older works repairs may have been necessary, and these areas will have different composition materials than the original. Since the radiographic image most often will encompass the entire painting or at least a large area of the work, one can determine by attenuation of x-rays where to sample in order to evaluate the compositioned material (Figure 22.11). Blind multiple biopsies and sample selection for chemical or dating analysis can produce inaccurate results if the samples are unwittingly taken from an area of repair.

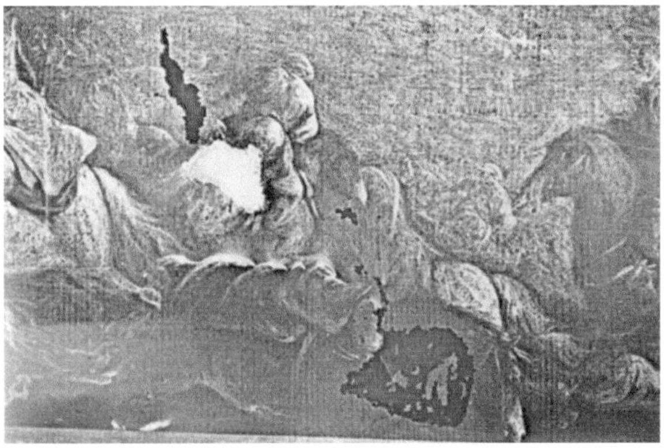

FIGURE 22.11 Detail radiograph of a Saracen painting, artist unknown, shows damage to the linen canvas repaired with a lead-based filler, consistent with an old repair.

Once, we were called by a major American museum to examine a quite valuable painting because the restorer realized the surface appearance was changing. The work had never been properly cleaned and varnished since its acquisition in the 1920s. On raking light examination, the surface was very uneven and there was specific crackleur at the apex of ridges that were not present in previous descriptions by curators during the 60 years the museum had owned the painting. Although we performed a CT to look at the individual pigment layers, plain radiography at different kilovoltage levels revealed the etiology of the changes—there was one, and possibly two, paintings underneath the one producing the external visual image (Figure 22.12). This led to a lively and enthusiastic discussion by the museum staff as how

FIGURE 22.12 Radiograph of a valuable painting revealing one, possibly two, paintings beneath the external visual image.

to "rescue" the images under the one they had (which was valued at above $1 million). Recognizing that neither our malpractice insurance nor emotional composition made it appropriate to be a party to this type of activity, we made a discreet exit and never attempted to find out what transpired as a result of the radiographic findings.

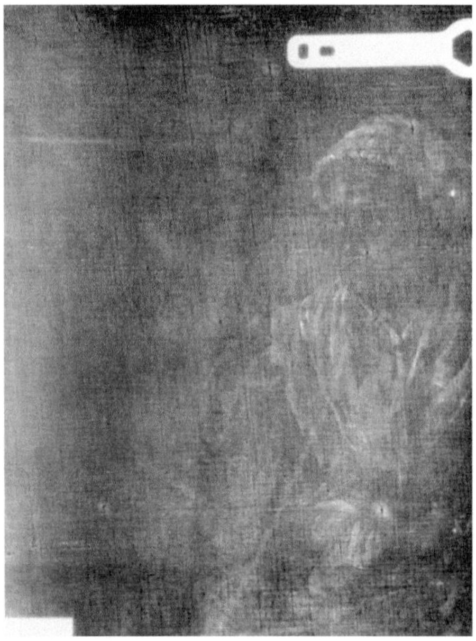

FIGURE 22.13 Radiograph of the John George Brown painting shown in Figure 22.3. The right-hand side of the picture shows a composition pigment layer containing lead, whereas the left-side pigment is acrylic. The early painting had been unfinished and was "completed" recently by another hand.

We were once requested by a museum and an auction house to evaluate a genre painting allegedly by a famous artist of the mid-nineteenth century, John George Brown (Figure 22.3). The painting had been consigned by the grandson of the artist. Upon inspection by the curator of the museum that might acquire it and the staff of the painting department of the auction house, there was concern about the authenticity of the painting as to whether it had been significantly restored or in fact almost entirely inpainted, in which case it would not be a desirable example of the artist's work for public display.

As one obtains a relevant history in any type of examination and gleans whatever information is obtainable by visual inspection, the investigation of a painting should proceed in a like manner. On viewing the work, the right side of the painting containing a seated female figure was very competently rendered in the manner of John George Brown. This part of the canvas displayed the expected crackleur and patina of a mid-nineteenth century work using pigment containing some amount of lead. However, the left side of the painting contained a window through which one viewed a bright, almost garish landscape composed with little sense of perspective and completely out of concert with the excellent draftsmanship for which John George Brown was known. The museum and the auction house wanted to declare the entire work a fake, but we were intrigued by the provenance (pedigree) and the fact that only the left side of the painting had elements that were not the work of John George Brown.

We took the work to the New York Medical Center (Cornell) and persuaded our colleagues to allow us to do a scanning CT and plain radiography using different kilovoltage settings. We hoped to detect lead in the pigment composition, certainly in the right side of the work if not the left. The CT was obtained to see if the so-called Hounsfield numbers would identify lead.

The radiograph (Figure 22.13) revealed quite strikingly that the right side of the composition pigment layer contained lead and that the left-side pigment was probably of acrylic, which was later confirmed by chemical analysis. This material was only available many decades after the death of John George Brown. Thus, the right side of the painting was stylistically characteristic of John George Brown and the pigment layer was composed of materials consistent with the middle of the nineteenth century, whereas the left side of the painting was obviously of fairly recent vintage and composed by a rather unskilled hand. The radiograph demonstrated that the left side was not painted over the original John George Brown as the pigment layer was entirely of the acrylic application. Further inquiry revealed the true cause of this diagnostic dilemma. We had thought the scenario probably was that the left side of the work had been severely damaged, with removal of the pigment and repainting of the scene afterward. In fact, the painting had been inherited unfinished and was "completed" by another hand shortly before it was consigned to auction.

In the nineteenth and early twentieth centuries, smuggling of masterpieces from Europe became an all too common practice. One of the methods to escape detection was to paint an image over the Old Master and then remove it later. Radiographs, of course, would easily detect this deceitful practice today.

The ability to transform the images of art objects from the analog to the digital format opens up great possibilities in detection of frauds and forgeries. Digital radiography in this context is meant to describe any method of radiographic image production in which the silver halide-based film is replaced by an electronic system to either produce or display an image. We have experimented with film-based, digital fluoroscopic, scanned projection, and scan point source systems.

Digital radiography opens up possibilities such as subtraction of various pigment layers or separation of the pigment layer image from the support upon which this layer rests. Digitalization may well allow more accurate characterization of the compositional materials to establish the authenticity, or lack thereof, of a particular work. In theory, the brush- stroke pattern of a particular artist might be characterized by the scanning digital technique. By studying a body of the artist's work in this manner one might develop a signature or profile of this technique. The technology utilized in crime labs and in determination of chromosome abnormalities, such as deletions, might be applied using digital imaging methodology.

In the detection of fakes and frauds as well as in the determination of composition of many art objects, a technique providing visible images that can be observed by the lay public and appropriately described and interpreted by experts provides unique and compelling data. The role of newer technologies, such as MRI , is yet to be explored. This discussion has centered upon the applications with which we have had personal experience, and biomedical imaging technology that is widely available. Hopefully , this has provided some insight into the possibilities of radiographic imaging taken in its largest context.

ACKNOWLEDGMENTS

I am indebted to Felix Fleischner, who taught me to be inquisitive about what one does not presently know, and to Russ Morgan, who provided me the environment as a graduate student to learn the basics of imaging technology. To my colleagues in Radiological Sciences at Johns Hopkins and the Center for Medical Imaging Research at Vanderbilt, as well as coinvestigators such as Mitch Bush at the Smithsonian, I owe a great debt of gratitude. The Eastman Corporation has kindly allowed republication of several illustrations used in monograph Vol. 63, No. 1, *Medical Radiography and Photography*. Finally, I would like to note the inspiration of my mother, Pattie Royster James, and Aunt Geneva James Weaver, who taught me to appreciate beauty in many forms.

BIBLIOGRAPHY

1. Bridgman, C. and Keck, S., The radiography of paintings, *Med. Radiogr. Photogr.*, 37, 3, 1961.
2. Elliott, W. J., The use of the roentgen ray in the scientific examination of paintings, *Am. J. Roentgenol.*, 50, 779, 1943.
3. Falke, T. H. M., Zweypfenning-Snijders, M. C., Zweypfenning, C. V. J., and James, A. E., Jr., Computed tomography of an ancient Egyptian cat, *J. Comput. Assist. Tomogr.*, 11, 745, 1987.
4. Glasser, O., *Wilhelm Conrad Röntgen and the Early History of the Roentgen Rays*, Charles C. Thomas, Springfield, IL, 1934, chap. 18.
5. Held, J., Alteration and mutilation of works of art, *So. Alt. Q.*, 62, 1, 1983.
6. James, A. E., Jr. and Gibbs, S. J., Radiographic analysis of paintings, *Med. Radiogr. Photogr.*, 63, 1, 1987.
7. James, A. E., Jr., Gibbs, S. J., Sloan, M., Price, R. R., and Erickson, J. J., Digital radiography in the analysis of paintings: A new and promising technique, *J. Am. Inst. Conservation*, 22, 41, 1982.
8. James, A. E., Jr., An inside look at decoys, *Wildfowl Art*, 24, Summer, 1987.
9. James, A. E., Jr. and Heller, R. M., Western Egyptian desert expedition, *Explorers*, September, 1981.
10. James, A. E., Gibbs, S. J., Sloan, M., Erickson, J., and Diggs, J., Radiographic techniques to evaluate paintings, *Am. J. Roentgenol.*, 140, 215, 1983.
11. James, A. E., Jr., Waddill, W. B., and Feazell, G. L., The medical imaging technologies as evidence, *J. Contemporary Law*, University of Utah College of Law, 11(1), 1984.
12. James, A. E., Not every old canvas is down for the count, *Med. Econ.*, September 3, 1984.
13. James, A. E., Imaging the images, *Am. Med. News*, March 28, 1986.
14. James, A. E., Decoys yield up secrets to radiographic imaging, *Diagn. Imaging*, February, 1985.
15. James, A. E., X-ray analysis of paintings, *Maine Antique Digest*, April 1987.
16. James, A. E., The use of x-ray imaging techniques in appraisal of paintings, *Valuation*, June 1985.
17. James, A. E., The healing of paintings, *South. Med. J.*, 84, 1231, 1991.
18. James, A. E., The radiography of art and artifacts, *Electromedica*, 44, 144, 1988.
19. James, A. E., The radiographic analysis of decoys. *Ward Foundation*, Fall 1996.
20. James, A. E., Is your decoy real? *Decoy Hunter*, 1989.
21. Keck, C. K., *The Care of Paintings*, Watson-Guptill, New York, 1967.
22. Osterman, F. A., James, A. E., and Heshiki, A., Xeroradiography in veterinary radiography: A preliminary study, *J. Am. Vet. Radiol. Soc.*, 16, 143, 1975.

Section VII

Virtual Imaging

23 Reporting and Finding Expert Opinion

Virtopsy and the Logic of Scientific Discovery of K.R. Popper

Richard Dirnhofer

CONTENTS

I do not make any hypotheses, I do not sell any opinions, I have only written, what I have seen.

E.v. Störck—Maria Theresa's private doctor

For nothing is more difficult than to distinguish plain things from doubtful things which the mind of itself adds at once.

Lucretius

INTRODUCTION

In the administration of justice, the general mandate to the expert, namely to report "finding and expert opinion," and these according to their "best knowledge and conscience," contains a clear separation between objective, correctly reported grounds for the findings and the resulting conclusions derived from them; that is, between the so-called protocol that is and the opinion of the expert. Thus each expert, in his or her capacity as the "extended brain of the court," should feel obligated to know the epistemological precepts of K. R. Popper's critical rationalism, namely a clear division between the objective (findings) and the subjective (conclusions arrived at in the expert opinion). This is because experts can only comply with their duty to the truth under this methodological premise.

The three basic elements of Popper's research logic [1], namely objectivism (there is an absolute truth that one can approach), realism (a reality exists independent of us), and critical discussion (statements are to be verifiable) are also the foundations of court procedure-oriented thinking when preparing the objective elements of the facts of the case and the normative decision(s) derived from it.

The differences between legal and medically scientific ways of thinking that are emphasized again and again in the relationship between experts and judges first emerge at the level of the normative decision process when, for example, the so-called "limit value thinking" or the various legal causality theories (e.g., equivalence and adequacy theories) are brought to bear.

But also at this level, the core element of critical rationalism, namely the respective verification of the legal results done independently from one another by the various official channels of jurisdiction, is institutionalized.

However, during the mutual, empirical verification of the relevant facts—that is, the disclosure of the real structure of the case with the aid of an expert—the way of thinking remains the same. For this reason it is also justified that the administration of an appearance of the crime scene lies with the organs of the judicature.

The scientific results learned from the Virtopsy Project (http://www.virtopsy.com), which involves the objective determining of findings from autopsies via data from modern imaging diagnostics and other objective procedures, were so encouraging that from now on the judicature is compelled to concern itself with the questions and methods of the project for the purpose of addressing the reliability of the procedures that are used and submitting the results to a critical verification.

VIRTOPSY: NEW WAYS FOR REPORTING FINDINGS

The Change in Method

Over the last century, scientific-methodical innovations and thus associated progress in increasing the value of forensic evidence have constantly appeared in numerous subareas of medical forensic sciences.

Exemplary for this are the new methods in toxicology with their now highly specific and sensitive analysis results and the revolutionary methodological transformation in forensic genetics from blood group to DNA analysis.

This is not so in the core area of forensic medicine, namely in corpse openings (autopsies). To date, in this field the reporting of findings is still done with the same instrumentation, the "autopsy instruments" developed in the eighteenth and nineteenth centuries, and the results are documented in written autopsy protocols.

This traditional procedure is augmented by hand sketches, schematic drawings, and two-dimensional (2D) photographic documentation. The representational shortcomings with regard to their two- dimensionality, their inadequate detection of morphologic details, and their dissecting method in the sense of—often repulsive—"bloody documentation" of the forensic photograph collections, are sufficiently well known. Once in a while (i.e., also far too infrequently), supportive aids for the unaided eye, such as, for example, operation microscopes and stereo magnifying lenses are also utilized.

The exponential further development of medical imaging already begun in the 1970s, together with computer-assisted photogrammetry and surface scanning developed in the 1990s, verified these procedures with regard to their usefulness, including in the autopsy area, which was necessary with a view towards implementing them in forensic practice as well.

The accelerated deployment of radiological methods in forensic pathology was already being urged by G. Brogdon in his standard work [2]. Following the beginning of the computer tomography (CT) era in the mid-1970s and the development of magnet resonance tomography (MRT, MRI), it was mainly the introduction of multislice spiral technology in CT at the end of the 1990s that suggested that it could be used in forensic practice, since sectional images of the whole cadaver could now be generated in any desired orientation and in 3D in a timely manner.

At around the turn of the preceding century, it was primarily Japanese, German, and Swiss research groups that once again set out on the path of postmortem imaging, autopsy imaging, and visual autopsy. "Once again" because, already in 1898, 3 years after being discovered by Röntgen, a cadaver was examined with x-rays [2].

Where the Japanese and German research groups aimed at establishing imaging methods as a supplement to autopsies, the research approach of the Swiss group under the direction of R. Dirnhofer [3] at the Institute for Forensic Medicine (IFM) of the University in Berne went further, namely towards the developing of a minimally invasive autopsy, the so-called Virtopsy. The initial results by the Berne Virtopsy group already had an impact and became the foundation for the activities of further research groups in the United States, Australia, Denmark, Sweden, France, China, and in cities like Lausanne and Zurich.

It appears that now is the right time for jurisprudence to also discuss the Virtopsy procedure in comparison with the conventional autopsy techniques and become informed about the new possibilities of imaging methods for forensic pathology. This is against the background that today several of the imaging methods have already found their way into forensic expert opinions [2].

Which Imaging Methods have Supplanted the Classic "Autopsy Instruments"?

The basic tools for performing a virtual autopsy are already installed at the IFM in Berne and are shown in Figure 23.1.

They consist of [4]:

- Robot-guided surface scanning for 3D, undistorted, colored documenting of the body surface. In classic autopsy procedure this corresponds to the so-called external examination of the deceased body.
- Multi-slice spiral CT and MRT for a layer-by-layer display of the body at all spatial levels, optionally also in 3D, corresponding to the so-called internal examination of the corpse in an autopsy.
- A virt-angio (postmortem angiography) machine with which the heart and vessel system of the cadaver can be displayed. Based on this methodology, it is also possible to carry out minimally invasive interventions within the framework of the autopsy, whereby the integrity of the cadaver, unlike in an autopsy, essentially remains intact.

Image- and robot-guided, contamination-free specimen removal for a wide variety of supplementary forensic analyses, such as histological, bacteriological, virological, toxicological, and diatomic examinations, replacing the usual preservation of specimen material from the corpse.

The objective of the Virtopsy project is to convert "corpse evidence" into a permanent system of findings that can be archived and called up at any time, the data of which are available as layered images in any desired spatial orientation and, when needed, can also be transformed into a 3D model. On these corpse models that correspond to real conditions, findings can be displayed, documented, and eventually archived.

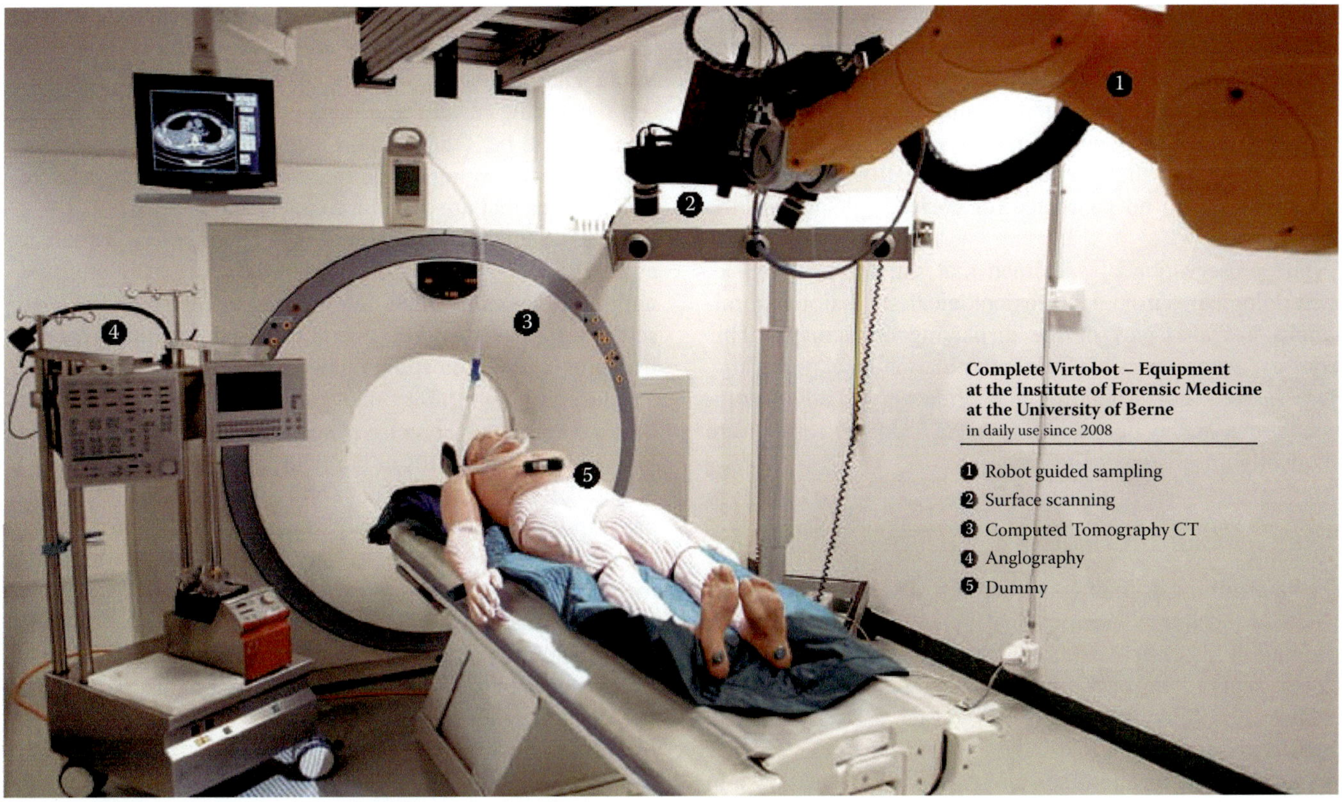

FIGURE 23.1 Virtopsy equipment at the Institute of Forensic Medicine in Bern.

WHAT DOES IT MEAN TO "REPORT THE FINDINGS"?

With this mandate for securing evidence, the judge expects a clear, vivid, and complete description of the "corpse object," which can be obtained via the immediate and competent perception of the expert.

This is achieved in four steps:

- A correct displaying of the findings; that is, bringing the findings into view with the greatest detail possible.
- An interpretation-free determining of the findings; that is, the precise registering of the facts of the findings.
- An objective documenting of the findings; that is, the written record of what was perceived.
- Reliable storing and archiving of the findings; that is, ensure a reliable opportunity for recourse to the data.

The Correct Displaying of the Findings

In order to bring the corpse findings into view, a dissecting method is required.

Depending on the criminological question being investigated and the objective, special dissecting techniques are used for this [5].

That is why corresponding, strategic decisions must already have been made before the autopsy for a purposeful, dissecting procedure. Prior to opening the corpse, information about "the circumstances of the case," especially the so-called evidence subject, must be available.

Exemplary for this include the techniques of sectioning for different populations: newborn sectioning to answer questions concerning age, birth injuries, and whether the newborn was alive; traffic accident case sectioning concerning questions such as the direction from which a pedestrian was hit; and poisoning sectioning concerning determination of the so-called "three poison paths."

In addition to this, there are special procedures for displaying thrombosis sources, shot and stab channels, the findings for strangulation victims, or when there is suspicion of air embolism, for example, during medical interventions or traumatic influences.

Moreover, in the display of findings following the external examination of the corpse, special methodological procedural approaches are also to be considered: for example, this might include waiting for the superficial skin lesions to dry in order to better determine their form, extent, and severity.

An autopsy is thus a unique, irreproducible act, such that important findings that are not at the beginning well and carefully prepared are lost. As is well known, there are enough examples of examinations that have gone wrong.

The result also still depends on the prevailing technical, dissecting skill of the pathologist. Not least, the corpse alterations with their various stages of decomposition, as well as piety considerations, and with them the preparative sparing of certain bodily areas, set limits to clearly displaying the findings.

For these reasons, the often varying legal regulations for autopsy procedures, such as the predefined guidelines for autopsies within the framework of quality management systems, are also no guarantee for correctly displaying the findings.

The preservation of specimen material for histological, bacteriological, virological, toxicological, or diatom examinations are also part of the displaying of findings. This requires, where possible, a contamination-free procedure. Here as well, corresponding and often very costly recommendations are made, which, as seen from the results, are frequently unsatisfactory.

Determining of the Findings

The determining of findings is that step of securing evidence, which in the immediate personal perception of the expert, using all of his or her senses, should lead to the unerring and precise registration of the facts, that is, the findings on the cadaver. For this, vision is primarily used, as well as touch, smell, and hearing.

In order to record findings unerringly, it is necessary—in Lucretius' sense [6]—to distance oneself from interpretational steps of a diagnostic procedure and simply register only the facts of the findings such as the shape, size, weight, surfaces, cut area, interfaces, colors, and other subtle details. This also includes evaluating the consistency and coherence of the organs, as determined via touch, the interfaces of the organs with regard to color, markings, and tissue fluids and, with the hollow organs, observing their capacity, their walls, and their contents.

According to Charles Lewis Peirce, every sensorial perception on the retina is also always immediately interpreted with eidetic personal expectations. Thus, from that which is perceived, that is, from the electromagnetic waves with wavelengths between 400 and 800 nm that impinge on the retina, a subjective perceptual judgment results.

Gertrude Stein has trenchantly formulated this individually subjective process: "In the things of the mind, you will want the things you know."

This natural and inherent tendency of the human observer's perception to convert the morphological details quickly and interpretatively into a diagnosis can, through an approach like that described by Schwarzacher [7] in the middle of the previous century, be impeded:

"When we look at an object, this can happen in totally different ways. Either we let our attention scan over it quickly and are striving to obtain an overall impression or we proceed in such a way that we systematically submit it to a precise observation, point by point. The latter, namely the systematic, step-by-step observation, appears to be more certain, but runs the risk of the larger context being lost such that one cannot see the forest for the trees. The most advantageous, therefore, is a suitable combination of both ways. Above all, one must learn to see as naively as possible and yet, like in an atlas with the right page open, to have all the old memory images in mind. The more successful the separation and this comparison of looking inward and outward is, the more correct the findings will be. One should make this naive seeing a habit, even in uninteresting cases, in order to find the correct solution in cases that are difficult to interpret."

When Schwarzacher thus speaks about seeing as naively as possible, he means—probably in the sense of Lucretius—nothing other than when determining findings, to free oneself from quick (because one is often under time pressure), interpretative-diagnostic prejudgments.

The most important thing about determining findings is thus a meticulously precise observation of all particularities: that is, the master of his or her field is distinguished by a love of detail. This is because observing the findings and, that is, orienting oneself to the morphological features and details means, in the truest sense, to "pay attention" in the sense of a mentally concentrated and careful observing.

In summation, the determining of findings is admittedly an immanently subjectively influenced process, which, with an appropriately disciplined procedure on the part of the observer, as Schwarzacher described, should ensure that the findings are determined in as much detail as possible.

The Documentation of the Findings: The Autopsy Protocol

The autopsy protocol is the core document within the framework of the legal autopsy. It serves not only as the foundation for the conclusions of the experts' opinions and mainly their reasons, but also for their critical verification by the organs of the judicature. Other experts also require the protocol for a possible second or senior expert opinion.

The protocol thus becomes objective knowledge in the sense of K. R. Popper: "The documentation is the step over the abyss of unexpressed thoughts to the same expressed protocol sentences that have been formulated in writing, which are now made available to a general, intersubjective critical discussion," that is, also independent of pathologists.

In order to fulfill this central purpose essential for judicial truth-finding, for example, the Austrian code of criminal procedure requires that "the protocol to be made about the autopsy should be so definitive and is to be laboriously composed such that it affords a complete and true conception of the viewed object."

In order to achieve this, the following principles must be observed:

- Describing the corpse findings that result from external and internal examination.
- Describing them in a form that allows the later reader to "make an image" of that which the expert perceived; allowing others to virtually mentally participate in the observation using one's mind.
- Describing the autopsy technique and more specifically, the dissecting procedures.

- Having a command of the formalized, descriptive protocol language.
- Limiting oneself to facts; no diagnoses belong in the protocol.
- Giving a complete description, that is, also determining negative findings and those that are not pathological.
- Not using preprinted protocol forms.

A few explanatory remarks about these principles:

The goal of making a protocol of the external and internal examination is the objective written record of those impressions that our senses give us when a corpse is opened.

In the description we should thus only express that which we actually see and avoid all interpretations. The very best is when we "photograph with words." Because this is only possible to a limited extent and exact local descriptions of the findings are often difficult, one is happy to avail oneself of graphic–schematic and photographic representations of the findings.

Such sketches should make the description easier but not replace it. Photography is also a part of documenting, but always stands within the context of the written protocol.

For this reason, a description of the autopsy techniques should be mentioned in the protocol because they clearly reflect the strategic, dissecting decisions of the pathologists and their methodological expertise with regard to displaying the findings. A second or senior expert can then also read up on the methods that were followed.

Short and concise protocols can have a larger amount of information than a protocol with many pages. A prerequisite for this is a command of the descriptive, formalized protocol language. In doing so, a special verbal manner of expression has been agreed upon in order to obtain a transparent conformity between the description and the facts, that is, between the protocol sentences and the actual findings. This can be illustrated by means of color perception.

Description is never done with pure spectral colors, but with color mixtures. For this reason, two basic colors have been agreed upon and then a third hue is added. In this way, using the formalized, descriptive protocol language, the normal color of the liver and kidneys is gray-red because gray, as the inherent color of the parenchyma, and red, as that of blood, have been agreed upon. Liver anemia is thus expressed by "palc gray-rcd."

Moreover, concepts such as pin-head sized or pea-sized referring to formalized descriptive surface areas such as coins or palms have been included in the protocol language.

The principle of "no diagnoses belong in the protocol" is frequently not adhered to.

When a diagnosis occurs in a protocol, every experienced person knows that what was recorded in the protocol was not seen with one's own eyes but, based on numerous observations, has already been distilled, abstracted, and generalized.

This is because a diagnosis names and interprets the perceptions according to the current state of knowledge. In addition, due to their descriptions, which are mostly vague because they are generalized, a wide range of later interpretive possibilities remains open.

There are numerous examples of this error in protocols:

- The liver is acutely congested, instead of the liver is discolored dark blue red.
- The liver is fatty instead of yellowish-gray-red and doughy; written in an anticipation of a microscopic diagnosis.
- Cauterization of the mucous membrane, instead of reddening, swelling, and a breaking up of the mucous membrane.
- Traces of abrasions on the left side of the back instead of widespread, brown-red, dried skin in bands.

It thus belongs to the principles of protocol writing that in the documentation of findings in protocols no diagnoses are permitted and in the diagnoses, no descriptions. The use of already set diagnoses principally sheds a bad light on the person who wrote the autopsy protocols.

The description of negative as well as nonpathological findings shows that the pathologist has paid special attention to them. Conditions that were not found are also often important and valuable grounds for the conclusions reached in expert opinions.

Preprinted autopsy forms principally have no place in forensic–pathologic protocols because they promote thoughtless execution and allow the feeling of responsibility for transparent, information-rich protocol-writing to atrophy.

In summary, H. Hamperl [8] should be quoted concerning the relevance of the autopsy protocol: "It is clear that at the moment the doctor is mainly interested in the diagnosis. However, this should not hide the fact that the more perishable of the two documents is concerned. Just as we today possibly look down with compassion, based on today's knowledge, on the interpretations and diagnoses of our scientific ancestors of 100 or 50 years ago, the same will most likely happen with us—as we hope in our belief in constant progress. What remains is what was correctly seen and recorded in the findings report, whether it was put in writing 50 years ago or today. Thus one might say that coming generations look over the shoulder of whoever writes a findings report and expect precision and conscientiousness in his work so that with a clear conscience it can be used as the foundation for new scientific interpretations."

THE STORING OF THE FINDING

Storing of the finding involves the written and digital archiving of all records, primarily:

- Circumstances of the case: criminal, judicial information.
- Protocol: written recording of the findings, supplemented with drawings, diagrams, photos, and videos.

- Expert opinion: preliminary expert opinions, definitive expert opinions.
- Courtroom protocol: following oral discussion of the expert opinions.

THE MEANING OF OBJECTIVE KNOWLEDGE FOR THE EXPERT AND THE COURT

The concept of objectivity is connected as closely as possible with the professional understanding of everyone, including medical experts.

Compared to the organs of the judicature, the expert with his objectivity does not have to guarantee truth, but rather the careful search for truth. This is analogous to those doctors who work clinically and who are not only concerned with healing but also with the careful treatment of their patients.

For this task, various facets of objectivity (Table 23.1), as discussed by K. R. Popper, are needed.

Basic prerequisites are the personal, temperamental mindset of the expert in the sense of incorruptibility, impartiality, and resistance to being influenced. In order to fulfill this Ciceronian "sine ira et studio," all personal references and subjective mindsets must be neutralized through the utilization of intellectuality and empiricism. Popper subsumed these prerequisites under "personal, subjective objectivity."

Due to the temperamental and personal weaknesses of human beings, which also include poor training, lacking knowledge, insufficient experience, and inadequate skills, the objectivity of the results that the expert develops still cannot be guaranteed.

This objectivity of the results—which the organs of the judicature indeed expect—is only attained through an objective procedure in which the results are offered for discussion as simply and substantially formulated assertions in clear, verbal and written language and are independently verifiable by various persons as to whether they are in agreement with reality. K. R. Popper understood this under the intersubjectivity of critical rationalism.

As objective knowledge, this part of objectivity is independent of human temperament, personal characteristics, and abilities.

He provides an example for this: "A verdict is not objective because the judge is objective, but because every able jurist would have come to the same verdict."

TABLE 23.1
Objectivity and the Expert

The Facets of Objectivity

- Impartial, unswayable, unerring
- Committed to intellectualism and empiricism

→ Sine ira et studio: personal objectivity

Objectified knowledge—clear, simple, substantial

→ Verifiable through critical, intersubjective discussion → revisable quality

This objectivistic approach is also a prerequisite for the search for the truth at the legal level. This is because the organs of the judicature endeavor, through the critical consideration of the expert opinion's evidence as objective knowledge, to solve the respective case. The process of eliciting evidence in a practical, forensic medicine case shows the care that must be taken before it comes to a verdict of "guilty" or "not guilty."

For comparison, the intersubjective verification procedure of scientific results is shown in Table 23.2. Via the various sections of critical discussion, it passes through the corresponding steps of the hypothesis levels, improving them each time, as K. R. Popper presented scientific progress with his well-known formula $P_1 \rightarrow VT_n \rightarrow FB_n \rightarrow P_{2n}$ in his book *Objective Knowledge* [9].

The intersubjective verification procedure in a concrete case for an expert opinion in forensic medicine is accompanied by a similar, albeit not more intensive, critical–rational discussion. As an example, the sequence is shown in Tables 23.3 and 23.4.

Following the discovery of a body and the problem of suspicion of a criminal act that is connected with it, the first criminalistic hypotheses and the questions posed by the investigative authorities are formulated based on the first visible alterations of the corpse and the information of the local appearance of the crime scene. The well-known "criminalistic Ws" of Who, What, Where, When, Why, and How, serve nicely as a framework for this.

The subsequent autopsy leads to an improved hypothesis II by the first pathologist. This will then be critically discussed with the second pathologist and other experienced forensic doctors. For hypothesis III or an investigative opinion, a conference of those dealing with the case, those involved in the manhunt, and the technical-criminalistic study of clues are brought together for discussion.

As things further develop, hypothesis IV is generated as a preliminary but already written expert opinion.

The results of possible further supplementary examinations, for example, from the fields of toxicology, histology, bacteriology, as well as experimental investigations still remain reserved. At this level, negative examination results also play an important role in the sense of the diagnosis by exclusion procedure.

Once these results are available, hypothesis V, namely as a "definitive, written expert opinion," together with the protocol records (examination report, sketches, drawings, photos), is delivered to the court.

At this level, which is decisive for the normative verdict, a critical examination within the framework of the so-called appraising of the evidence by the judge, public prosecutor, and defense attorney takes place. This process often stands in direct connection with a supplementary "oral discussion of the expert opinion" together with the experts.

From this, hypothesis VI finally results as a corrected, amended expert opinion. As part of the overall so-called evidential landscape, it serves the court's verdict.

With this, the critical verification procedure by forensic medical expertise has not yet been completed. It remains

TABLE 23.2
Intersubjective Forensic Verification Procedure

A. RESEARCH

Scientific problem
↓
Hypothesis I
↓
Discussion with co-authors
↓
Hypothesis II—abstract for conference → Rejection
↓
Presentation
↓
Discussion with the scientific community
↓
Hypothesis III—publication → Rejection
↓
Peer review → Rejection
↓
Revision → Rejection
↓
Hypothesis IV—publication
↓
Subsequent falsification

TABLE 23.3
Intersubjective Forensic Verification Procedure (Part 1)

B. CASE WORK

Cadaver and discovery location (informative local appearance)
↓
Problem (suspicion of a criminal act)
↓
Hypothesis I—autopsy according to Peirce; conjecture according to Popper
↓
Illustrating the problem—argument (7 Ws)—judge's mandate (question to the experts)
↓
Autopsy
↓
Hypothesis II—(1st pathologist)
↓
Critical discussion at the dissecting table with two pathologists, … etc.
↓
Hypothesis III—"investigative opinion"
↓
Discussion in conference of those dealing with the case (manhunt, criminologists)
↓
Hypothesis IV—preliminary, written expert opinion
↓
Supplementary examinations (toxicology, histology, experiments, etc.) (by exclusion procedure)
↓
Hypothesis V—definitive, written expert opinion with protocol

TABLE 23.4
Intersubjective Forensic Verification Procedure (Part 2)

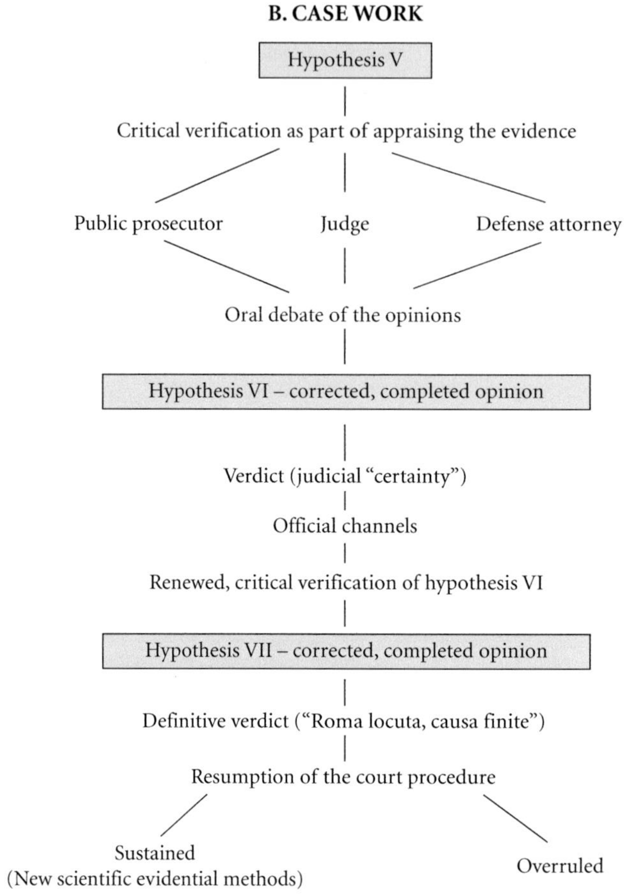

possible for it to still be subjected to examination by the next higher court, until a definitive verdict in the sense of "Roma locuta, causa finita" is handed down.

In summation, this concerns the judicial reaching of a verdict with the goal of judicial certainty and the search for the impossibility of it being otherwise, a classic "conjecture and refutation" process, in the sense of K. R. Popper [10].

"VIRTOPSY": THE BETTER SOLUTION FOR OBJECTIVE KNOWLEDGE

Images in Science

According to Dürrenmatt [11], there are the two expressions in the German language: "sich ein Bild machen" (to visualize) and "im Bilde sein" (to be informed). "We are never informed about this world if we have not made some inner picture of it. This 'making' is a 'creative act'."

With this, Dürrenmatt has expressed a human need that is also reflected in the century-long technical development phases of imaging procedures. Already in the seventeenth century, there was the desire to make the invisible visible, enlarge the small, and pull in that which lay far away. For this purpose, microscopes and telescopes were developed, with which Galileo Galilei discovered Jupiter's moons and made astronomy a science 400 years ago.

Photography, developed in the middle of the nineteenth century, became a milestone of pictorial culture. At the end of the nineteenth century, Konrad Wilhelm Röntgen then discovered a radiation that could make findings under human skin visible, the beginning of proper medical imaging. Through further developments with radiology, ultrasound, computer tomography, magnetic resonance tomography, as well as the refinement of pathological microscope technology with electron microscopy, scanning electron microscopy, and right up to the Hubble telescope, an increasingly "iconic penetration of the reality" [12, 13] of the human body has taken place. Even the atomic level has been imaged with scanning tunneling microscopy. In medicine, with the development of molecular imaging, with the aid of modern procedures, and nuclear medical methods, another new era has just begun.

The image is no longer a supplement, but rather has become part of the research process. It has established itself in the scientific knowledge process. It no longer merely serves the purpose of information and communication for transporting knowledge, but it has in fact become a constitutive element in science itself; images have become a medium of knowledge [13]. This phenomenon is today seen in large interdisciplinary research projects. Today the significance of perception for the human mind connects various disciplines: the natural sciences, the human disciplines, and the liberal arts [14].

In the forensic area, this development of scientific imaging has led the image to have a documentary, registrational function. Through the mechanical, direct, and automatic registering of physical data (nuclear spin, radiation intensities, etc.), an image is no longer seen as iconic, that is, as something similar, but rather corresponds—in the sense of mechanical objectivity [15]—to an undistorted recording of reality, that is a technical copy of reality.

As a result, the forensic–pathological findings are depicted in a way that is independent of human interventions and personal experiences and mindsets.

Moreover, circumstance of which we previously could make no, or only inadequate images, can now be graphically displayed. Exemplary for this are, for instance, the various gas embolisms in the human body, the display of complex fracture systems in bones, as well as shot and stab channels. This is also true for forensically important findings in somatic regions, such as the facial region, which for reasons of piety must be protected from a structure-destroying preparation.

Moreover, regions that up until now have always been technically difficult to prepare, such as the Atlanto-occipital joint or the pelvic region, have become accessible for recording findings through graphic representations.

This "delegation of vision to the machine" [14] has, however, not only extended our visual capacity, but also has the decisive advantage of the machine being more reliable than our tired, inexperienced eyes.

From a historical point of view, these modern imaging technology possibilities fulfill a constant desire by the forensic, pathologic–anatomic, and anatomic–technical fields to have a graphic documentation of the findings.

Eduard v. Hofmann [16] has already held that "the protocol to be made about the autopsy is to be so definite and laboriously recorded such that it affords a complete and detailed conception of the observed object. The protocol is to be written such that one, in the absence of his own perceptions, can make an accurate image of the findings, that is, can also mentally participate in the observation using one's mind." For this reason, he in no way felt himself obligated to go along with a purist culture of exclusively written and verbal protocol-writing and enriched his protocols with very vivid sketches that he did himself [17].

Schwarzacher [7] also expressed the wish that "when writing a protocol one must photograph with words."

For photographic documentation in forensic pathology Anton Werkgartner [18] held that "the description of the skin wounds should always be supplemented with images, above all when from the form, length and distribution of the injuries conclusions can be made regarding their creation, the type, form and size of the weapon, and how it was employed."

From this claim, it is already apparent that the image not only serves as illustrative material to be looked at, but rather can serve the central purpose of forensic reconstruction. That is, it can provide an additional gain in knowledge.

As a pathologist, H. Hamperl [8] formulated desire for images as : "A protocol should be written such that one can graphically reconstruct the thoughts of the pathologist."

Finally, in anatomy the desire for depictions to always be truer to reality was also realized by the matchless "Photographic Picture Atlas of Practical Anatomy" by Walter Thiel [19]. In this work, the anatomic structures depicted are shown by photographic documentation as they are really seen and "thus to a large extent correspond to the visual impressions obtained when a cadaver is being prepared and thus also during operations." These pictures can thus be immediately applied to practical–clinical realities, which is why W. Thiel chose the expression "practical anatomy" for his book.

A COMPARISON OF THE CONVENTIONAL AND THE IMAGING AUTOPSY METHOD

In the following, the advantages of the virtoptic versus the classic conventional autopsy method will now be discussed and compared (Tables 23.5 through 23.7 and, summarizing, Tables 23.8 and 23.9).

In Table 23.5 the following relevant differences in the displaying of the findings become apparent:

Using surface scanning, Virtopsy makes a complete recording of the bodily surface findings in their exact measurements possible. This, in turn, makes exact 3D real data reconstructions possible with regard to the various forms of the weapon used in the crime as well as the course of the criminal events.

The internal viewing of the cadaver not only refers to the so-called three somatic cavities but to a complete displaying of the findings from "head to toe." Through this, somatic regions are also included in the displaying of the findings that normally are either spared for reasons of piety or, due to their difficult preparatory accessibility, like for instance the Atlanto-occipital joint and the pelvic region, make the recording of findings difficult.

The recording of findings using equipment that delivers images replaces the individualistic, very diverse manual dexterity of the pathologists and additionally permits, independent of the various autopsy tenets in the conventional method, a standardized procedure according to Standard Operation Procedure (SOP).

The "virtual knife" of imaging also permits sectioning through decomposition-altered tissue and thus the ability to display forensically relevant findings, even with advanced corpse alterations.

The fundamentally destructive procedures that occur during dissection—for example, through the use of hammers and chisels or special preparation procedures, such as maceration—are replaced by a nondestructive, digital slicing procedure that allows for the selection of the sectional thickness.

TABLE 23.5
Improved Findings Reporting (Part 1)

Autopsy	Virtopsy
Display of Findings	
• Incomplete, external cadaver viewing	• Complete recording of the cadaver surface
• Not "head to toe" ("three body cavities", piety)	• Complete display of the body, 1500–2000 sections
• Dependent on manual skill	• Reliable machine—SOP
• Differing "autopsy schools and methods"	
• Display of the findings depends on the condition of the cadaver (corpse alterations)	• Virtual scalpel at all levels ~ 0.2 cm–1.2 cm sectional thickness
• Instrumentarium—destructive procedure	• nondestructive procedure
• Contaminated acquisition of specimens	• Contamination-free specimens (tox., histo., infect., diatoms, etc.) – reliable analysis results

Robot- and image-steered specimen extraction not only serves diagnostic safety but also, above all, a contamination-free procedure, whereby the expressiveness of the subsequent analyses with regard to toxicology, infectious diseases, and diatom examinations are substantially increased.

In Table 23.6, the advantages concerning the determining of the findings are listed. Indeed, the decisive step with regard to improving quality is the broadening of the number of people who can be involved in the determining of the findings and/or the process of determining the findings. The determining of the findings is no longer only dependent on the respective subjective, visual memory of the first and second pathologist, but is also broadened by the possibility of including several experts in the sense of an intersubjective visual memory. Through this, just as Schwarzacher postulated, extensive experiences contributed by paging through the individual, visual atlases, together with naive seeing, complement each other superbly.

TABLE 23.6
Improved Findings Reporting (Part 2)

Autopsy	Virtopsy
Determining the Findings	
• Dependent on subjective, visual memory ("4 eyes")	• Extended through "intersubjective" visual memory ("X eyes")
• Light-dependent recordings of the findings	• Not light reflections but rather registering of the atoms, absorption of the radiation
• Fatigue—reduced alertness and attention	• Also possible the next day
• Second opinion no longer on the real object but via protocol	• Second opinion on virtual, one-to-one 3D object

TABLE 23.7
Improved Findings Reporting (Part 3)

Autopsy	Virtopsy
Documentation of the Findings	
• Oral / written recording of the findings	• Uniform, pictorial evidence—increased informational content
• Loss of the descriptive formalized protocol language	• No "immunization" of the protocol
• Lack of time for making the protocol	• Time saving; only 15′ acquisition and 15′ data calculation
• Extent of the protocol making: dependent on legal principles	• Independent, worldwide
• No 3D description	• 3D display for reconstruction
• "Clean" documentation	• bloody documentation (photographs)
Storing the Findings	
• Incomplete archiving	• Complete archiving (decomposition, cremation—no exhuming)
• Protocols, pictures, sketches	• The corpse as evidence can be called up as a virtual one-to-one object and in 3D
Confirmable – evidential	

In addition, the determining of findings is no longer dependent on the respective lighting conditions or the state of visual attentiveness, for example, fatigue.

In summation, one thus arrives at a sure, detailed and, in view of the informational contents, richer determining of findings.

In Table 23.7 a comparison with regard to the documenting and storing of findings is detailed.

The significant step taken by Virtopsy in the documenting of findings is moving from making verbal-written protocols of the findings to the uniformly image-based documenting of the entire body. The evidence of the dead body is available as a virtual object in an image and no longer in—what is in many fields—obsolete descriptive formalized protocol language. At the same time, the picture has a registrative function because it corresponds to reality and this in an effective way. This is in contrast to imprecise, vague, and difficult-to-discuss verbal formulations in protocols that make a relevant criticism impossible. With this the "immunization of the protocol" that often set limits on a critical discussion is eliminated.

This is because the protocol, in the form of a graphic statement, generally increases its informational content with regard to the scope, import, and significance of the findings.

Moreover, with regard to the effort involved and the scope of writing the protocol—depending on the respective legal principles of the codes of criminal procedure—significant improvements come about. This is because, following a short acquisition phase and calculation of the data, documenting can result that is independent of the legal principles.

TABLE 23.8
Virtopsy—The Better Objective Knowledge (Part 1)

Evidence—Corpse (Contains All of the Information)	
Classic Autopsy	**Virtopsy**
Instruments: destructive	*Surface scanning, radiation absorption, nuclear spin:* non-destructive
↓	↓
Technical skill	*Machine as a virtual scalpel – SOP*
↓	↓
Partial displaying (putrefaction, piety, body regions)	*Complete displaying* (from "head to toe")
↓	↓
Light reflections	Physical data + computer (image processing)
↓	↓
Retinas of 1st and 2nd pathologist	
↓	
Subjective image of a real object subjective perceptual potential: • training, knowledge, experience, own cases • dependent on: light, alertness (fatigue)	*Objective image* generated from the physical set of data—formatted in 3D"mechanical objectivity" independent of the pers. references and subj. attitudes
↓	↓
Personal subjective description of the subjective image Object and description coalesce into the protocol "trust" in the expert	*Personal subjective description of the objective image* • Subjective description and object remain separate • Trust in the expert and the virtual object
↓	↓
"Bloody" documentation (photographs)	*Clean documentation*

Finally, a 3D, graphic display facilitates better replicability for lay persons and, in addition, provides impressive, real databased, forensic reconstructions.

It is also not insignificant that the documenting follows certain implicit esthetics. This is because all that is organic is suppressed, such as, for example, bodily fluids and blood. Thus a "clean," extreme objectifying of the documenting occurs. A psychological pre-incrimination of the accused by a "bloody act" is eliminated.

In summation, through the imaging procedure, a form of documenting in the sense of Lessing is available with this by means of which one can form an objective judgment. This is because it is possible to "orient oneself on that which is there," namely on the findings in their appropriate reality, even when through the digital transformation of the real physical data (such as radiation absorption or nuclear spin) the findings have been converted from a real into a virtual existence.

Ultimately, with regard to storing the findings, a huge advantage arises in that incomplete archiving by means of protocols, pictures, and sketches is replaced by complete archiving of the "corpse as the means of evidence" as a virtual, undistorted object in a 3D dataset, which can be freshly accessed at any time and is thus available for new opinion-related deliberations. As a result, cremations no longer prevent court-ordered examinations and, in the future, exhumations will no longer be required.

From what has been presented, there is no problem in seeing that the image in medicine must naturally also be included in evidence collecting in order to be able to satisfy society's demand that the judicature, in its methodological processes, must avail itself of the established scientific insights if an increase in legal certainty for society can arise from this.

CONCLUSIONS

According to K. R. Popper, science is not defined by its results, but rather by its objective methodology. The scientific development of the Virtopsy research project confirms this claim in that it creates "objective knowledge" at the various levels of reporting findings (see Tables 23.8 and 23.9). In the area of forensics, this leads to a hollowing out of the subjectivity that up until now was inherent mainly in the subjective documenting of an autopsy. This is because Virtopsy, as an imaging procedure, creates pictures that correspond to the displayed circumstances in the sense of an "adequatio ad rem." This is because the generated images are "true" in the form of a detail-rich, visual representation of reality with high informational content. From this, an independent reality can be created by the expert with the possibility of critically questioning it.

Along with this, however, language as a medium of discursive thought exchange among experts and the judge still remains completely preserved.

The display using imaging suppresses the descriptive, formalized protocol language, the awkward coding of which is

TABLE 23.9
Virtopsy—The Better Objective Knowledge (Part 2)

Classic Autopsy	Virtopsy
Loss of the real object	*Preservation of the real object in its virtual existence*
↓	↓
Expert opinion – forensic reconstruction based on the protocol, sketches and photos	*Expert opinion – forensic reconstruction* based on more exact, real 3D data – experimentally reproducible
↓	↓
Reduced replicability	*Improved replicability*
↓	↓
Storing	*Storage*
Archiving: "personal – subjective" documentation	Archiving: objective 3D dataset of the body (PACS)
↓	↓
Preservation of contaminated specimens (Toxicological, bacteriological, histological, ...)	*Preservation of contamination-free specimens*
↓	↓
A second opinion is difficult • "Immunized" protocols	***Second opinion*** based on *"objective" knowledge* • critical review possible (teleforensics)
↓	↓
Psychological stress for the relatives	*Reduced psychological stress for the relatives*
↓	↓
Rejection by religious groups	*Acceptance by religious groups*

then eliminated. The insights from the image data are traceable to their sources at any time in their deduction by experts, so that procedures of arriving at conclusions are also potentially comprehensible years later. At the same time, the importance of the relevant results, the skills of the pathologist, and the importance of his or her language remain completely preserved since there can be neither automated guidelines nor a dictionary for the interpretation of the images. Only the spoken conclusions of the experts can give the images their comprehensible meaning.

The dialogue between the judge and the experts is facilitated with this methodology, as it is more easily understood by the jurist. On the other hand, the pathologist is increasingly compelled to avoid conclusions that are not based on the elicited findings.

In this way, Virtopsy accommodates the human being as a "homo pictor," who, according to Dürrenmatt, is "able to visualize." However, one is not spared from reflecting on the images, that is, from image criticism and competence, analogous to protocol criticism and competence in classic procedures.

From the perspective of forensic science, Virtopsy also means progress in the field of forensic pathology morphology. This is because the results of all international research groups show [20] that technically generated images not only prove their worth for scientific information and communication, but also as a promising instrument for further, more significant forensic insights in the future. For this reason, nothing stands in the way of a future, fascinating expansion of the prop room of evidential possibilities [21].

REFERENCES

1. Popper, K. R., *Logik der Forschung*, J.C.B. Mohr (Paul Siebeck), Tübingen, 1994.
2. Brogdon, B. G., *Forensic Radiology*, CRC Press, Boca Raton, FL, 1998.
3. Dirnhofer, R., Jackowski, C., Vock, P, Potter, K., and Thali, M., Virtopsy: Minimally invasive, imaging- guided virtual autopsy, *RadioGraphics*, 26, 1305–1333, 2006.
4. Thali, M. J., Ross, S., Oesterhelweg, L., Grabherr, S., Buck, U., Naether, S., Jackowski, C., Bolliger, S. A., Vock, P., Christe, A., and Dirnhofer, R., Virtopsy—Working on the future of forensic medicine, *Rechts- medizin* 2007, 17, 7–12, 2006.
5. Madea K-St Saternus, U. B., Gerichtliche Obduktion – Umgang mit dem toten Menschen und Obduktionstechnik, *Research in Legal Medicine*, Vol. 36, Schmidt-Römhild Verlag, Lübeck, 2007.
6. Titus Lucretius Carus, *De rerum natura*, Ph. Reclam Jun. Stuttgart, 1973.
7. Schwarzacher, W., Befund und Gutachten. *Wien Klin Wschr*, 59, 1–6, 1953.
8. Hamperl, H., *Leichenoeffnung, Befund und Diagnose*, Springer-Verlag, Berlin, 1964.
9. Popper, K. R., *Objektive Erkenntnis*, Hoffmann und Campe Verlag, Hamburg, 1973.

10. Popper, K. R., *Vermutungen und Widerlegungen*, Mohr Siebeck Verlag, Tübingen, 2000.
11. Keel, D., *Denken mit Dürrenmatt*, Diogenes Verlag AG, Zürich, 1982.
12. Doelker, Ch., Gschwendtner-Wölfle, R., and Lürzer, K., *Sehen ist lernbar. Beiträge zur visuellen Alphabetisierung*, Sauerländer Aarau Verlag AG, 2003.
13. Burri, R. V., *Doing Images—Zur Praxis medizinischer Bilder*, Transskript Verlag, Bielefeld, 2008.
14. Böhm, G., *Wie Bilder Sinn erzeugen. Die Macht des Zeigens*, Berlin University Press, 2007.
15. Daston, L., Galison, P., *Objektivität*, Suhrkamp Verlag, Frankfurt am Main, 2007.16. v. Hofmann, E., *Lehrbuch der Gerichtlichen Medizin*, Urban und Schwarzenberg Verlag, Wien-Leipzig, Berlin, p. 28, 1898.
17. Hausner, E., *Die historische Sammlung des Institutes für gerichtliche Medizin in Wien*. Hausner, Vienna, 2008.
18. Thiel, W., *Photographischer Atlas der praktischen Anatomie*, Springer Medizin Verlag, Heidel berg, 2005. 20. Thali, M. J., Dirnhofer, R., and Vock, P., *The Virtopsy Approach*, CRC Press, Boca Raton, FL, 2009.
21. Weber, M., Virtopsy and the Swiss legal System: New evidence—Law in forensic medicine? in *The Virtopsy Approach*, Thali, M. J., Dirnhofer, R., and Vock, P., Eds., CRC Press, Boca Raton, FL, 2009.

24 Photogrammetry-Based Optical Surface Scanning

Silvio Näther, Ursula Buck, and Michael J. Thali

CONTENTS

INTRODUCTION

The photogrammetric three-dimensional (3D) documentation of patterned injuries of the skin and bones in forensic medicine can help researchers to determine the shape and size of the injury and enables the true-to-scale correlation of the wound and inflicting tool. Advantages of the method include its temporal independence and the preservation of facts (patterned injuries and traces), which allows for analyses at any time.

Digital close-range photogrammetry is a technique for measuring objects in 3D from photographs taken from different perspectives [1]. Three-dimensional coordinates of discrete points of an object can be determined by taking measurements from at least three photogrammetric images (Figure 24.1). Before photographing, reference targets need to be stuck or painted on and around the object, and one precise measurement of distance (i.e., the reference line) within the area of the object has to be established as a known distance. Additional "natural marks" (e.g., dot-sized blood stains or skin impurities) can be used as reference targets (points). The more points of the object surface are acquired, the more detailed is the resulting 3D model, and the more detailed it can be subsequently represented. Marcel Braun introduced this procedure at the Institute of Forensic Medicine (IFM) at the University of Bern in 1996 [2,3]. Despite many advantages of photogrammetry, the limited number of acquirable surface points, the long analysis time, and the limitation to a few relevant areas of the object are some of the disadvantages.

In our view, the optical 3D surface scanning method is efficient and useful for whole-body documentation. Since 2003, the IFM in Bern has been using the GOM ATOS XL-3D (GOM mbH, Braunschweig, Germany) digitizer [4] to image the surfaces of bodies, tools, and vehicles[5,6]. External injuries of accident or crime victims, weapons, injury-inflicting tools, and damage to accident vehicles are digitized and electronically stored for further investigations and analyses. Thus, it is possible to compare patterned injuries with assumed injury-causing tools and to reconstruct courses of events.

With the GOM TRITOP/ATOS (GOM mbH, Braunschweig, Germany) digitizing system, the process is subdivided into two steps. Photogrammetry is performed to predefine reference targets that support 3D surface scanning in the second step of digitalization.

PHOTOGRAMMETRICAL 3D DIGITALIZATION WITH GOM TRITOP

To acquire the images, we use a digital single-lens reflex (SLR) camera with a fixed focal length of 24 mm. A ring flash ensures constant exposure without shade effects. The camera and the lens are not calibrated. A simultaneous calibration is processed automatically every time a series of

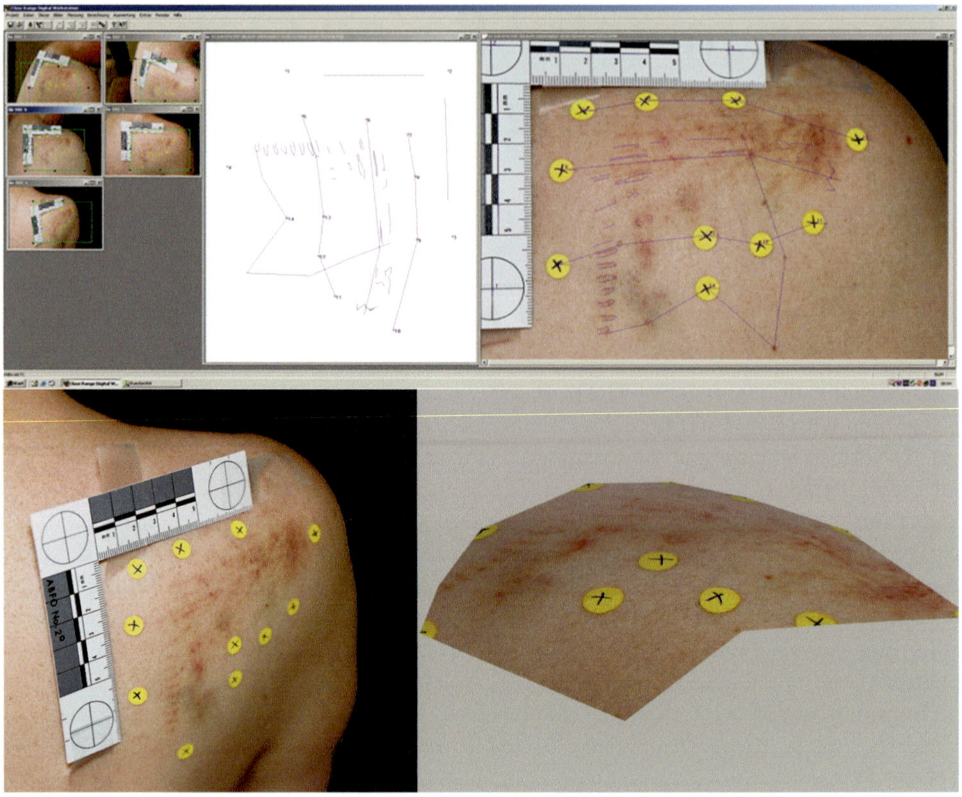

FIGURE 24.1 By measuring discrete points and drawing injury lines in 3D-CAD, a surface model of the area of interest can be generated.

images is transferred to the computer and computed with the TRITOP software.

Before taking photographs from different views, the body has to be prepared with different types of reference targets:

- *Coded markers* are applied for automatic computation of the image assembly. These markers feature a circular black-and-white pattern around the white center point. Each pattern encodes a point number ranging from 1 to 428. For scaling, two coded scale bars that feature a precisely determined distance between two coded markers are applied to the object area.
- *Noncoded circular markers* support the orientation of single measurements of the surface scanner, so that a full 3D model can be generated step-by-step.
- To match body surface scan data with radiological data (multislice computed tomography [MSCT] and magnetic resonance imaging [MRI] data), various so-called multimodality *radiographic markers* are applied to the body. These markers are captured by the surface scanning process, and they are visible in the radiological images, where they serve as reference targets.

The fundamental photogrammetry principle is triangulation. The photographic projection can be mathematically described as a central projection of the object onto the film plane. Every single object point is mapped to an image point, which is the intersection of the projection beam with the image plane. The spatial location of a point can be determined at the intersection of the projection beams of this point from several images. With the reversion of the mapping, the size and shape of the object can be reconstructed accordingly.

For the TRITOP analysis, every image series has to consist of at least six pictures. The images should be taken from several positions with convergent views (Figure 24.2). Because the markers are coded, the orientation and calculation of the images and the computation of the coordinates runs automatically.

The results are 3D coordinates of the coded and noncoded reference targets as well as the camera positions (Figure 24.3). In the next surface- scanning step, these reference data can be imported by the ATOS scanning software.

OPTICAL 3D SURFACE SCANNING WITH GOM ATOS

The GOM ATOS 3D digitizer relies on the principle of triangulation. A projection unit in the middle of the sensor projects a white light fringe pattern onto the object. Two charge-coupled device cameras capture the fringe pattern that moves over the object's surface in different phases (Figure 24.4). The

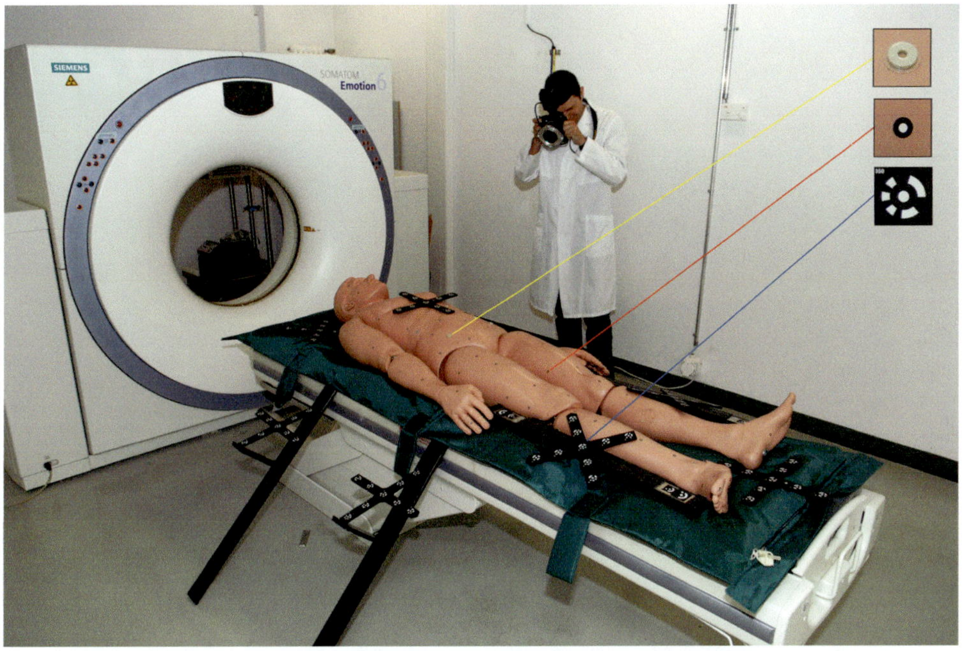

FIGURE 24.2 Equipment for photogrammetric documentation: A digital single-lens reflex camera with ring flash, plastic crosses with coded reference markers (blue), coded scale bars, noncoded reference targets (red), and radiographic markers (yellow).

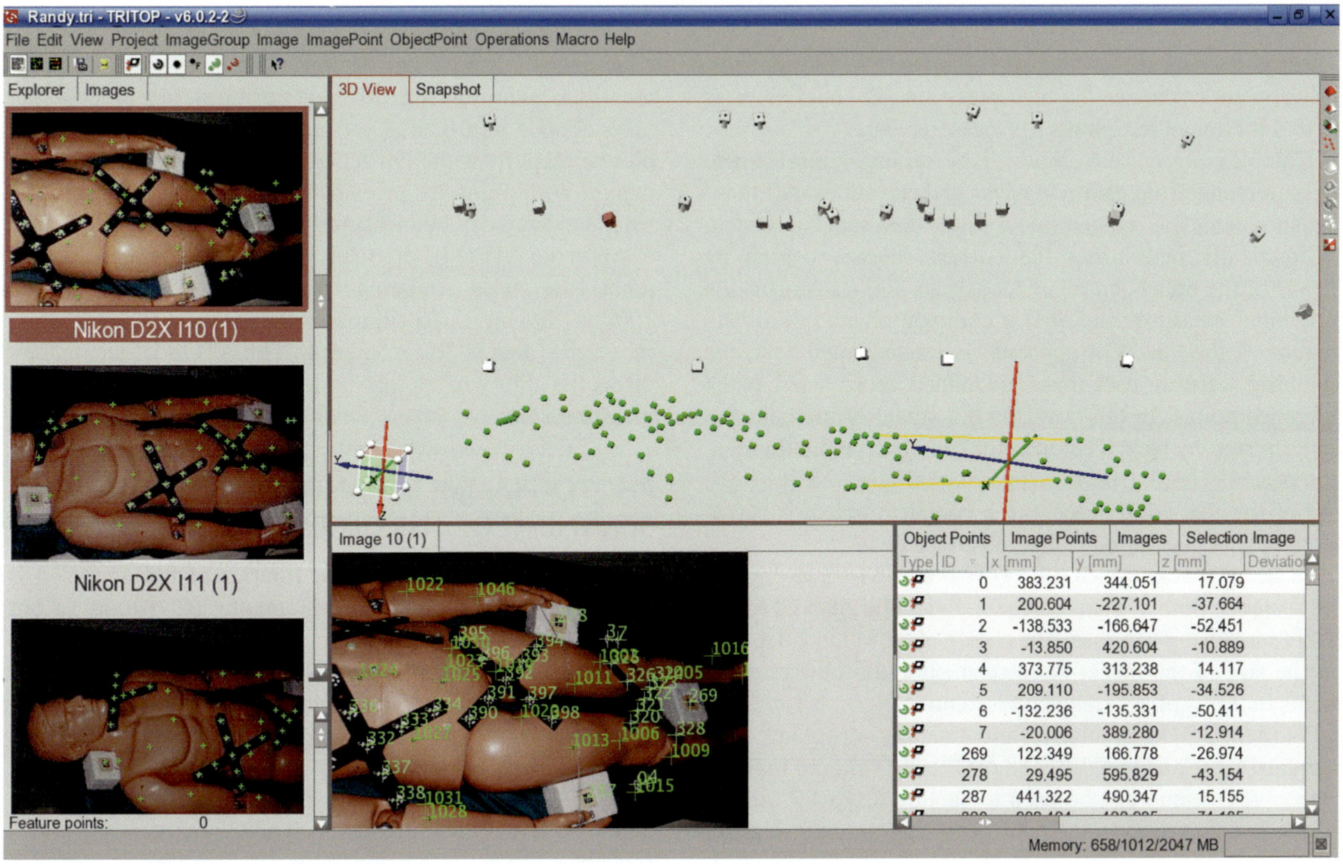

FIGURE 24.3 Results of the photogrammetric image series analysis. On the left side are the original images, on the bottom right are the measured 3D coordinates of the coded and noncoded reference points, and on the top right is the point cloud, rendered in 3D. The green points are the reference markers, the yellow lines are the scale bars, and the gray boxes are the camera positions.

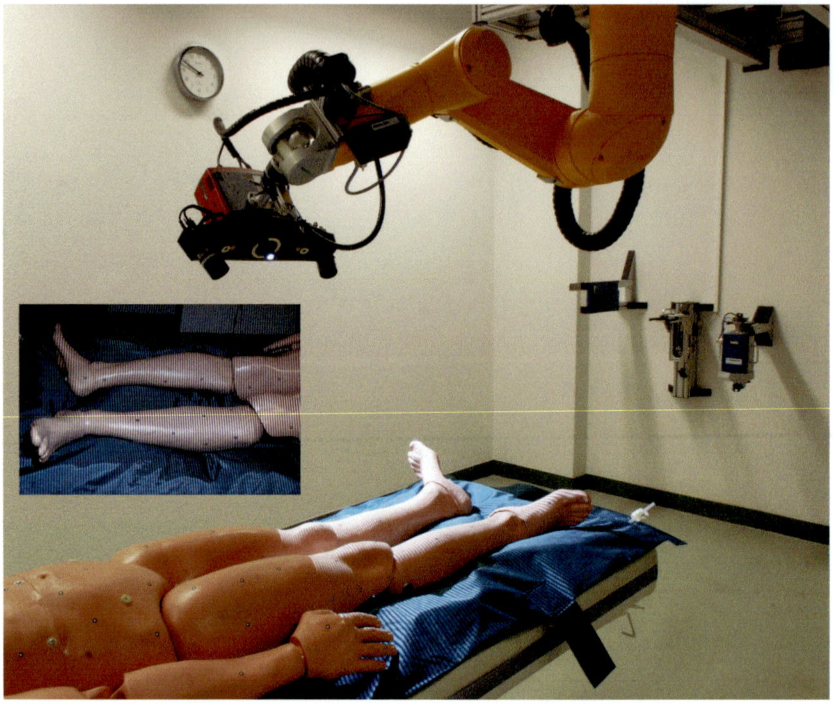

FIGURE 24.4 3D documentation of the external injuries of the deceased with the high-resolution surface scanner, performed on the CT table. The ATOS III sensor is mounted on a 6-axis robot for automated surface documentation. Fringe patterns are projected onto the body surface under white light projection; the two cameras record the fringes. Within seconds, the software calculates the high-precision 3D coordinates of up to 4 million object points per measurement.

sensor is connected to a high-end computer. Within seconds, the software calculates the high-precision 3D coordinates of up to 4 million object points per measurement.

Complex objects are digitized by taking several single measurements from different views and elevations. All these measurements are merged into a single data set. The system automatically determines the current sensor position by means of the noncoded reference targets and transforms the individual measurements into a common object coordinate system. If the TRITOP approach is implemented first, the scanning software uses these predefined coordinates as the reference points. In this way, the 3D model of the object is built up step-by-step. The operator can observe the digitization process continuously on the screen (Figure 24.5). At the end of the digitizing process, the complete object is represented by precise high-resolution surface data.

In Bern, we use the GOM ATOS II and ATOS III digitizing systems. Both systems can be individually adapted to different measuring volumes. The choice of the volume depends on the size of the object to be scanned and on the requirements in terms of accuracy and resolution.

With the ATOS II, three different measuring volumes in a standard setup and two different measuring volumes in small objects (SO) setup can be used to capture up to 1.3 million points per measurement. Each measuring volume requires special lenses, and the angles of the cameras and the measuring distance have to be adjusted accordingly. Before starting the measuring process, the system has to be calibrated with the help of calibration objects to ensure dimensional consistency. Using a calibration object, different measurements are acquired from different positions, in a process that takes about 5 min. The SO systems are mainly used for complex small parts with high accuracy and data quality requirements. With point distances of less than 0.02 mm, even the smallest detail can be measured (Figure 24.6)[7]

With the ATOS III digitizer, four different measuring volumes are used (ranging from 150×150 mm up to 1500×1500 mm). To digitize human bodies, the normal measuring area is 500×500 mm. The ATOS III technology offers the ability to capture up to 4 million object points per measurement.

PHOTOGRAMMETRY AND 3D SURFACE SCANNING OF THE DECEASED

The documentation of external injuries of the deceased is performed in the computed tomography (CT) room at our Center of Forensic Imaging and Virtopsy. The bodies are placed on the CT table. At first, the reference targets are attached to the corpse. The noncoded markers are uniformly distributed, and radiographic markers are affixed to the body. These markers are detected by CT and MRI, and they are also captured by the surface scanner. For the photogrammetry process, plastic crosses with coded markers and coded scale bars are placed on and around the corpse.

After successful computation of the photogrammetric images, the calculated coordinates of the reference targets are transferred into the ATOS scanning software. After

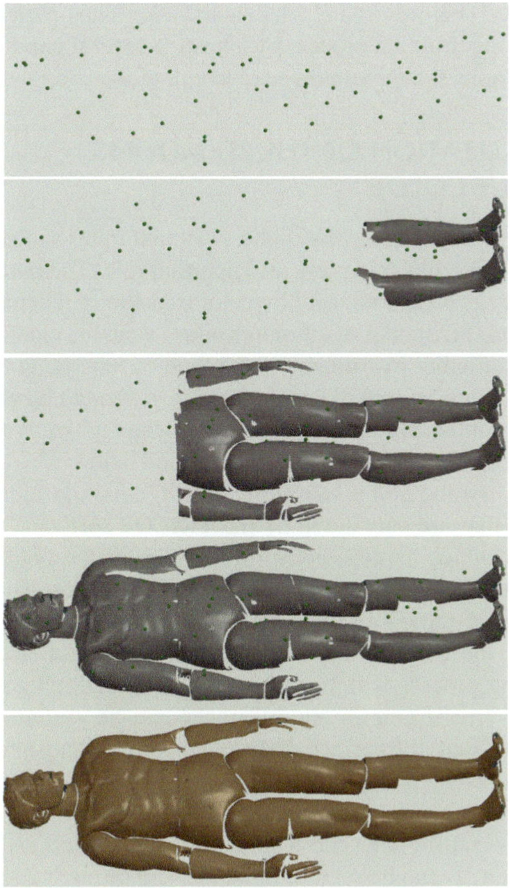

FIGURE 24.5 Creating the 3D surface model with the optical surface scanner: Each additional measurement is transformed fully automatically into a common object coordinate system. The green dots are the imported noncoded reference targets calculated from the photogrammetry project.

removing the coded markers and scale bars, the scanning can commence.

When the scanning of one body side is complete, a CT scan is performed. The position of the body is not changed between these procedures so that the subsequent fusion of both data sets is easier. After the surface and CT scan of one side, the body has to be rotated so that the whole procedure can be repeated with the rear of the body.

DIGITIZING INJURIES OF LIVING PERSONS

When patterned injuries of a living person are documented, the scanning can be processed without performing photogrammetry. Noncoded reference markers are applied to the skin where the patterned injury is located. The areas of interest are captured with several measurements. In the first scan, the coordinates of the reference markers are determined. During the scanning process, the patient has to avoid any motion. Due to this requirement the patient is laid down on an examination couch in our institute or in the hospital. To fully document the injuries, one to four measurements are performed, a process that requires between 5 and 10 min to complete. After the scanning, close-up photographs of the injury are taken.

GENERATING 3D MODELS FROM THE SURFACE SCANNING DATA

The result of the optical surface scanning of a measured object is a point cloud of millions of points—each scan generating up to 4 million—that represents the scanned object. To work with the data, perhaps measure certain dimensions,

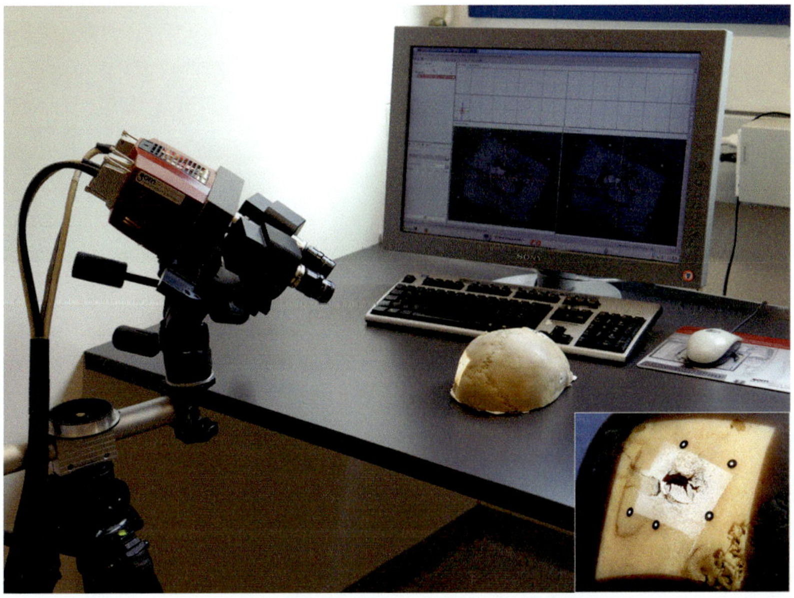

FIGURE 24.6 For the detailed measurement of fine injuries with high requirements in terms of the point density and accuracy of the resulting 3D model, the ATOS SO ("Small Objects") sensor head configuration is used. In this example, a bone lesion on a skullcap is scanned to enable an accurate comparison with the presumed injury-inflicting instrument.

a polygon mesh has to be computed. The steps for generating these 3D models include alignment, polygonization, mesh smoothing, and thinning [8].

In the alignment process, the individual measurements are finely aligned based on the surface. Measuring deviations are reduced. A color view and a scale represent the deviations (Figure 24.7). When the deviations are acceptable, the point clouds can be polygonized.

Polygonization means the conversion of the measuring point cloud into a mesh of nonoverlapping triangles. The mesh exhibits different densities depending on the curvature of the object. The operator can influence the density of the mesh by adjusting the so-called polygonization raster.

During the polygonization process, the individual measurements are recalculated using the highest point resolution. Any overlapping areas are deleted, and the resulting single measurements are merged into one polygon mesh (Figures 24.8a, 24.8b, and 24.9a).

Smoothing eliminates measuring noise. During smoothing, the points are shifted so that they better integrate into the environment. The operator can control the smoothing by adjusting several parameters. We recommend a smoothing intensity that eliminates the point noise but preserves small object details that are important for analysis (Figures 24.8c and 24.8d).

For certain applications (i.e., visualization and animation), the amount of polygon mesh data (in the 3D model) should be reduced by thinning. Thinning takes the curvature of the object into consideration. Strong curvature areas are thinned less than flat areas. This ensures that important details are preserved (Figures 24.8e, 24.9b and 24.9c). The operator can control the thinning process by adjusting several parameters, for example, the maximum edge length of the triangles.

VISUALIZATION OF THE 3D MODELS IN TRUE COLOR

For the forensic documentation of external injuries, the color of the injury and skin plays an important role. The pattern of an injury is often indicated by color transitions. Therefore, a true-color 3D model of an injury offers valuable clues to the injury-inflicting instrument or the type of impact. For these reasons, high-quality 3D color data are generated by projecting color information from images onto the digitized surface. Two methods for doing this are described below.

The first method is to use the captured photogrammetric images that are calculated by the TRITOP software. These images, which capture color information of the object, are taken in addition to the normal TRITOP images. They are taken after removing the coded crosses that cover the surface of the corpse. The resulting polygon mesh from the scanning process is imported into the TRITOP photogrammetry system. One or more images can be automatically projected onto the surface of the 3D model. By means of this process, every triangulation point is assigned a color value from one or more images.

The resulting data file features excellent shape and color definition, as can be seen in Figures 24.10 through 24.12.

With the second method, the 3D model is imported into the 3D Studio Max (Autodesk, Inc., CA, USA) animation

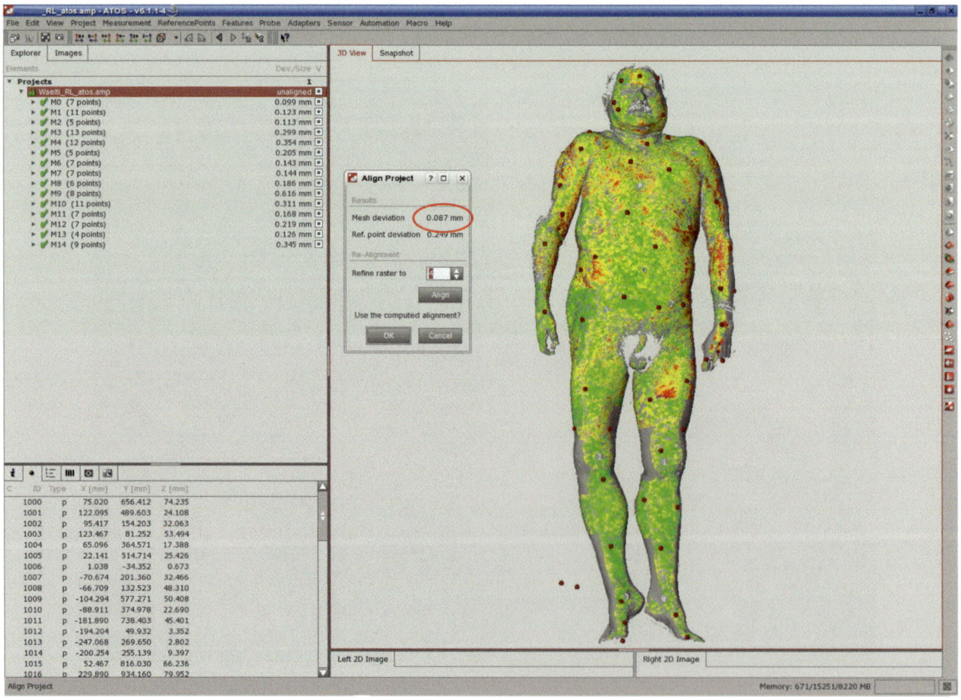

FIGURE 24.7 With the fine-alignment of all single measurements, any remaining small gaps can be closed. The mesh deviation is displayed in the 3D color view with values in the dialog box (red ellipse). The resulting data represent the remaining noise across all meshes.

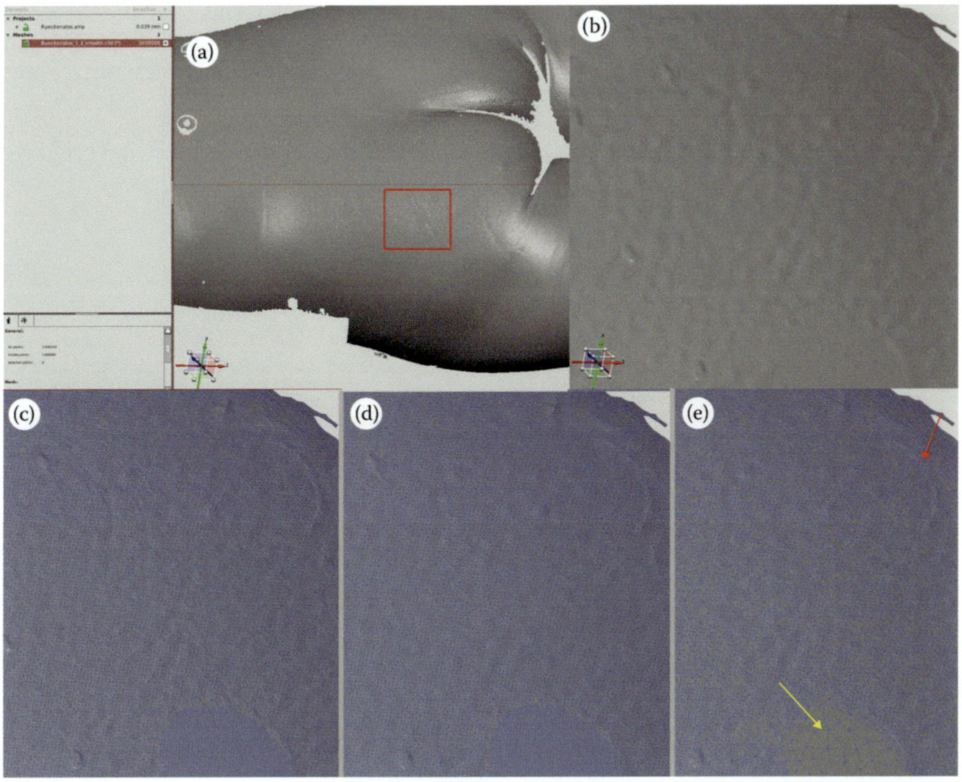

FIGURE 24.8 Representation of the polygon mesh structure of the 3D model scanned with a 500 × 500 mm measuring volume. (a) Polygonized mesh of the backside of a body with a raster setting of 1:1 (b) Polygon mesh in detail (red rectangle in 3a) (c) Triangles within the detailed mesh (d) After smoothing the mesh to eliminate the measurement noise (e) After thinning the mesh without loss of important details (red arrow signals a hair on the body, the yellow arrow points to a strongly thinned area of the cut and a filled reference marker)

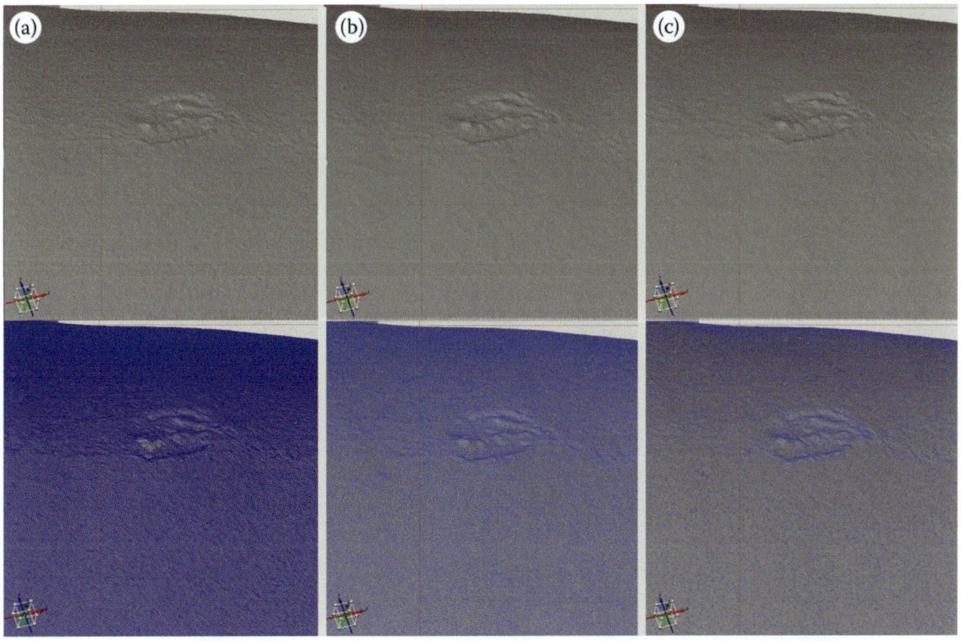

FIGURE 24.9 (a) Polygonized mesh of a small skin injury with a raster setting of 1:1 (length of 5 mm). (b) After smoothing the mesh to eliminate the measurement noise (c) After thinning the mesh without loss of important details

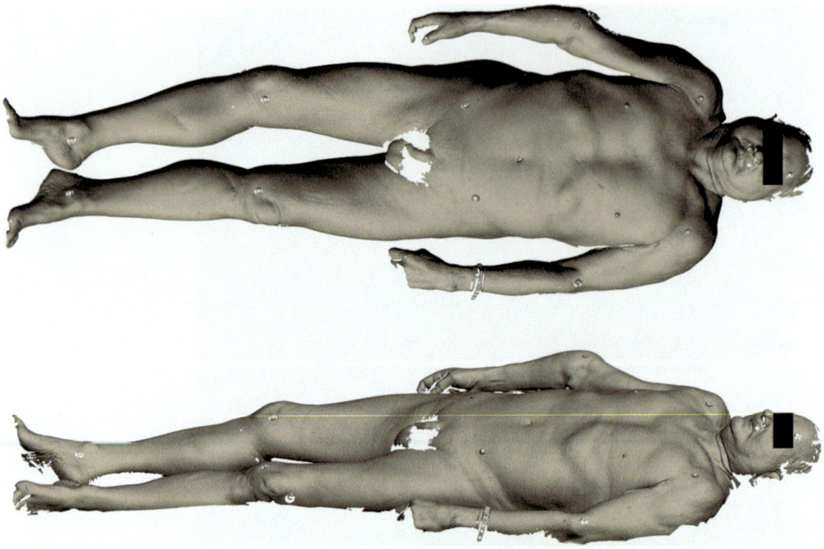

FIGURE 24.10 Gray-shaded polygon mesh model of the front side of the body, calculated from the optical surface scan data.

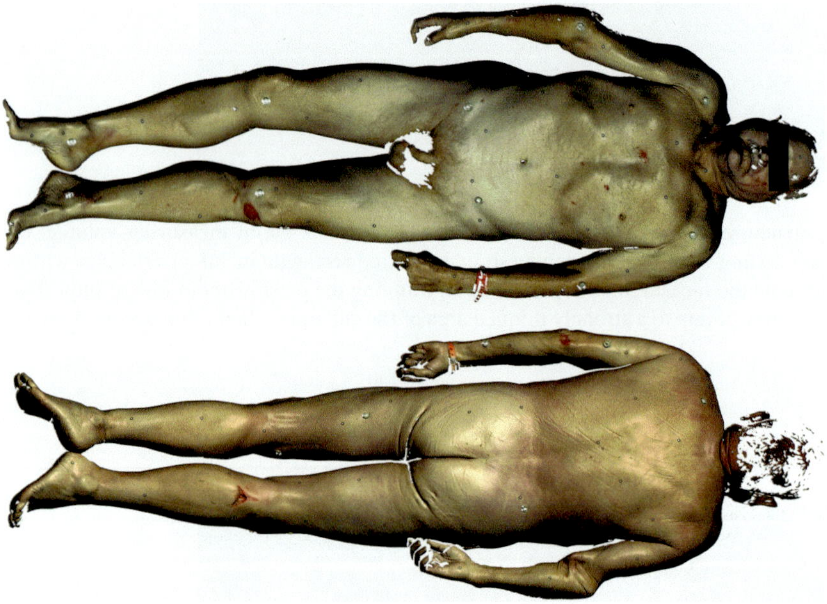

FIGURE 24.11 3D surface models of the whole body of the accident victim with all relevant patterned injuries in true color.

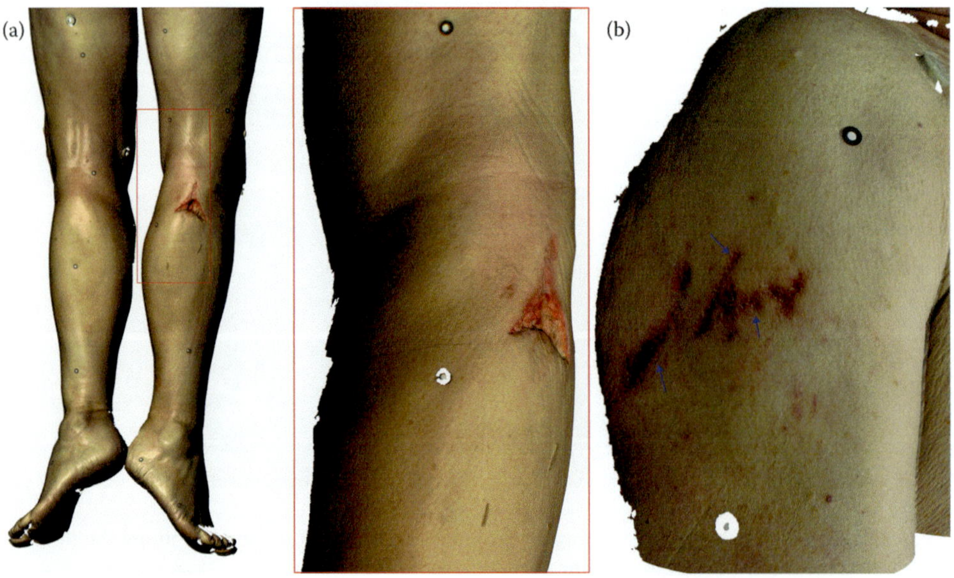

FIGURE 24.12 The high-resolution true-color surface model displays the structure of the skin including abrasions and lesions. (a) Injury on the posterolateral side of the right lower leg. (b) Abrasion on the lateral side of the right shoulder. Even the flaps of the epidermis are visible.

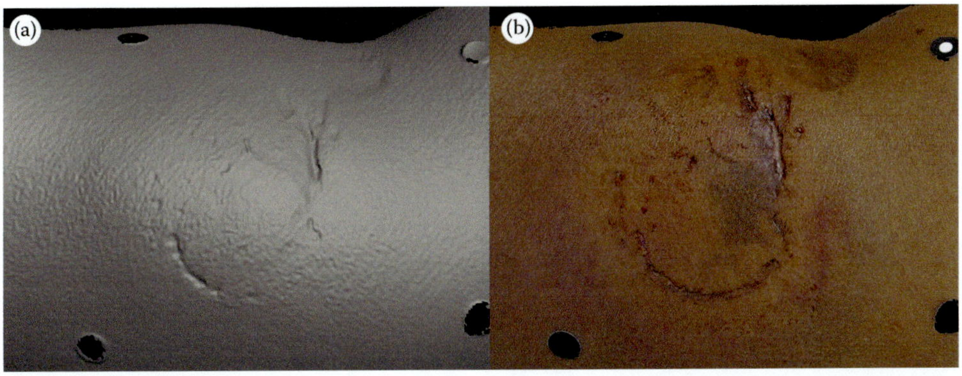

FIGURE 24.13 3D model of a bite mark on the upper arm of a living person: (a) High-resolution surface data from a digitizing process using the ATOS III system. (b) The true-color 3D surface data of the bite mark, generated by projecting a digital photograph onto the gray-shaded 3D model.

software[9]. Well-illuminated photographs of the object are projected onto the 3D model. By means of the reference targets, measured in the 3D model as well as in the photographs, the camera position and the focal length are calculated. The photographs are defined as a mapping material and projected onto the 3D surface model using the required camera information (Figure 24.13). The result is high-resolution color information that is not dependent on the degree of model optimization. This allows us to optimize the surface data to achieve faster workflow and to secure high-resolution data from the texture material.

This second method is best used for real data-based animation with complex data sets and to analyze the bodies of living persons in the absence of photogrammetry (Figure 24.14).

MORPHOLOGIC–GEOMETRIC COMPARISON OF PATTERNED INJURIES WITH THE ASSUMED INJURY-CAUSING TOOL

Ursula Buck, Silvio Näther, Marcel Braun, and Michael J. Thali

The morphology of a patterned injury offers hints about the injury-inflicting instrument. It is rarely possible to identify

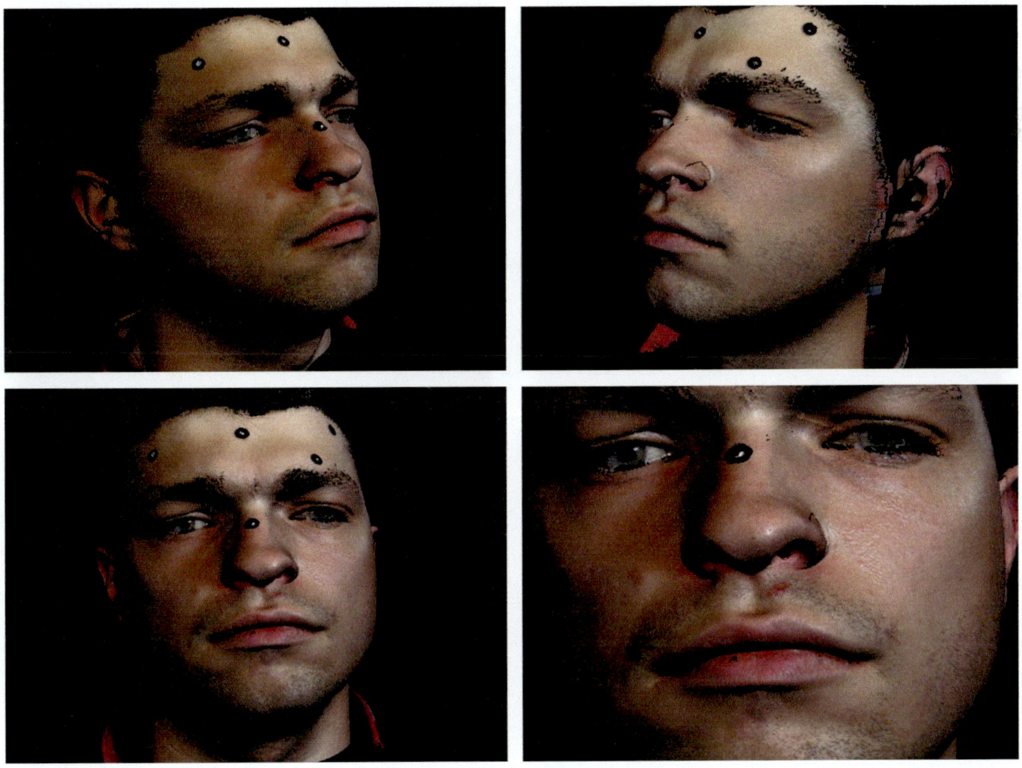

FIGURE 24.14 Different views of a true-color 3D model of the face of a living person.

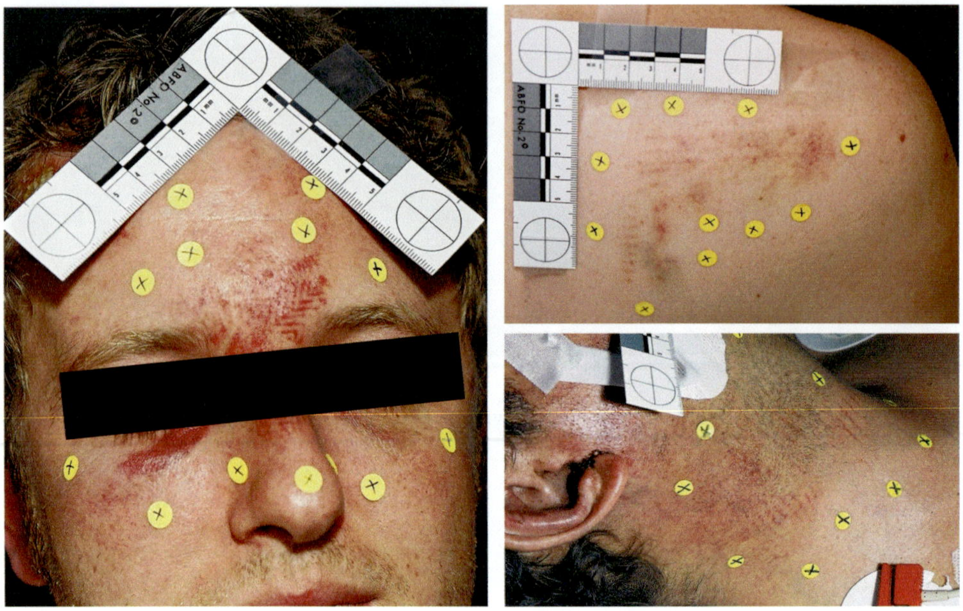

FIGURE 24.15 Patterned injury on the forehead, on the back, and on the neck of another victim (case 1).

the injury-inflicting instrument solely with the naked eye by examining the morphology of the injury. In the majority of cases, it is only possible to obtain information about the injury-causing tool after detailed examination and analysis. A direct comparison of the presumed injury-inflicting instrument with the injury at the time of autopsy is almost impossible. The presumed injury-inflicting instrument is often unknown, and direct contact between the two could mix important traces of evidence (e.g., DNA). The new process of 3D digitizing enables a one-to-one computerized comparison of the patterned injury with the injury-causing instrument. This method is noncontact and time-independent.

The virtual true-to-scale colorized 3D models of the injury and the presumed instruments can be compared and contrasted using modern computer software. Consequently, we can assess evidence about the injury-inflicting instrument and how it was used. The analyses and results of real injuries on the skin and bone of living and deceased persons caused by different injury-inflicting instruments are illustrated herein.

Blunt Trauma (Case 1)

In everyday forensic practice, the question often arises as to which of several injury-inflicting instruments caused certain blunt trauma injuries to the victim. From this reconstructive point of view, both forensic photogrammetry and surface scanning play an important role. With the computer-based correlation of appropriate true-to-scale 3D models, which are produced by the Virtopsy process, it becomes possible to link an injury-inflicting instrument to an injury [10–14]. This process and the results of the correlation can be demonstrated in the context of real case studies.

Several individuals had physically assaulted and kicked two men. The men had shoe-sole-patterned injuries on the face, back, and neck (Figure 24.15). The injury-causing shoes had to be linked to the injuries, using a pool of seven shoes belonging to seven suspected offenders. All patterned skin injuries were documented using 3D photogrammetry at the hospital. The 3D model of the injury on the face is shown in Figure 24.16. The suspected shoes were digitized using the GOM surface-scanning system (Figure 24.17). The person, who had been kicked in the face, was examined with a clinical CT scan. Using anatomical landmarks, the resulting CT model from the clinical scan was fused with the 3D model of the part of the face with the external injury (Figure 24.18). Based on the true-to-scale 3D models, the patterned injuries of both men were compared with all suspected shoes and could be linked to the injury-causing shoes. Figures 24.19,

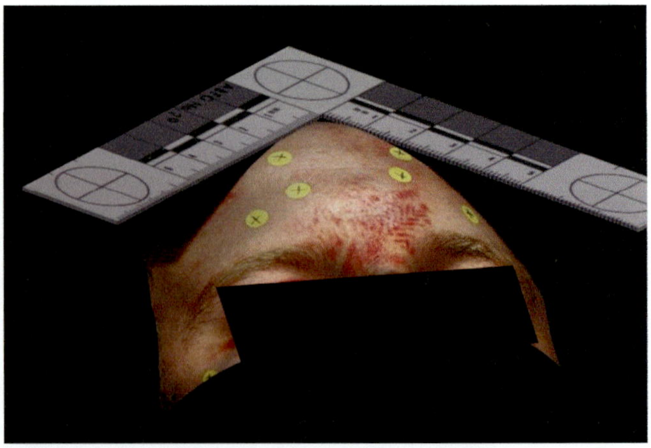

FIGURE 24.16 3D model of the face. A ruler was placed on the head for scale definition in the context of photogrammetric analysis (case 1).

FIGURE 24.17 The resulting 3D models of the shoe soles of the suspected perpetrators (case 1).

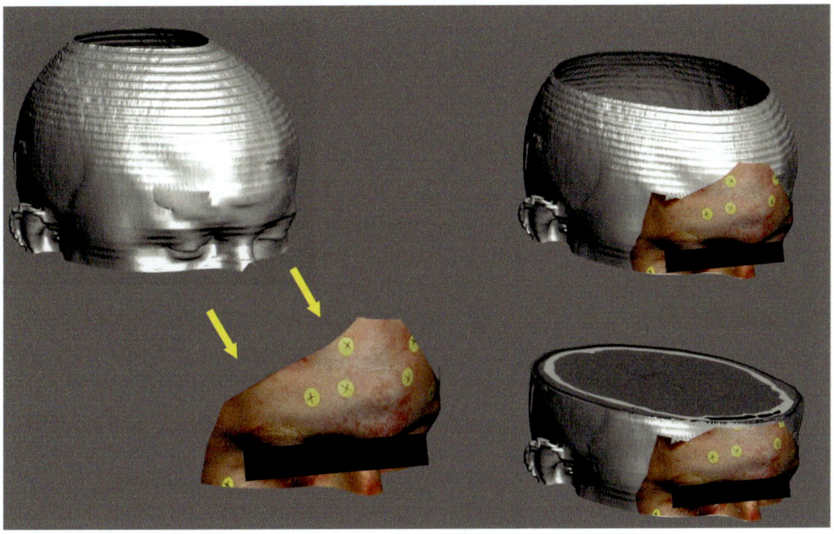

FIGURE 24.18 Fusion of the surface data of the exterior facial injury with data from the clinical CT (case 1).

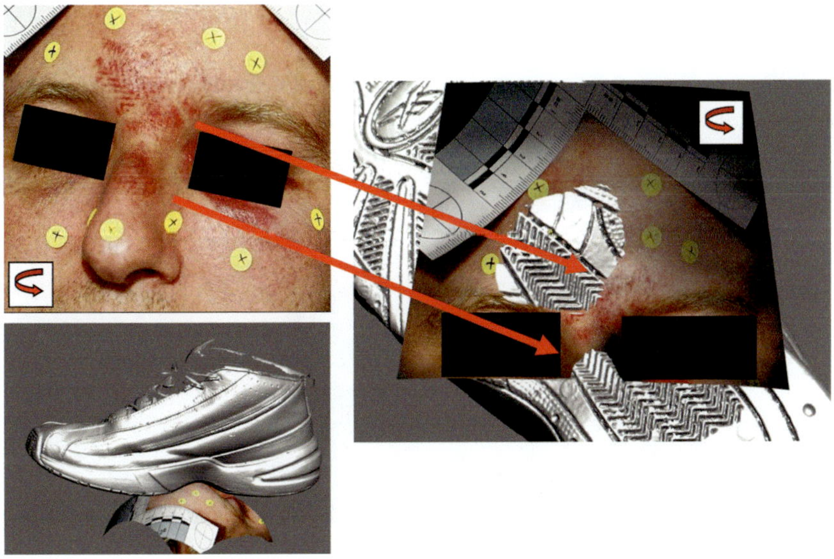

FIGURE 24.19 3D match analysis of the patterned injury in the face with the suspected injury-causing sole of a shoe (case 1). The zig-zag-pattern of the sole matched the pattern of the injury in shape and size.

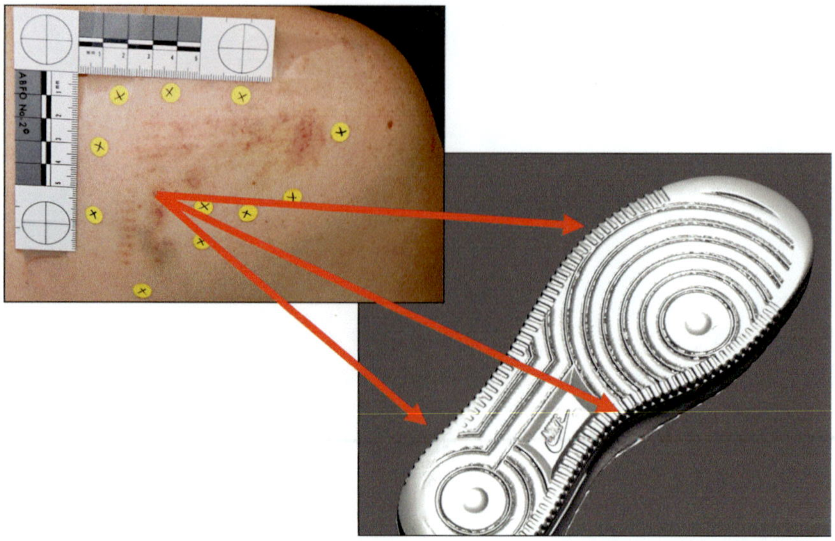

FIGURE 24.20 With the 3D match analysis the pattern of the suspected injury-causing shoe sole (case 1) could be linked to the injury on the shoulder (red arrows).

Figure 24.20, and Figure 24.21 show the results of our match analysis of the patterned injuries with the suspected injury-causing shoes.

Sharp Trauma

Knife Crime (Case 2)

In a homicide case, a man was stabbed with a knife. He presented with sharp force injuries to the skin, soft tissue, and skeletal tissue (pelvis). There was no longer any DNA on the possible injury-causing knives. Solely examining the skin injuries and the surface of the bone injury provided no clear clue as to the type of knife used (Figure 24.22). When a knife penetrates bone, the sharp tool normally leaves an impression in the bone. The characteristics of shape and size indicate the type of injury-inflicting knife. To analyze patterned injuries of tool marks made in bone, we used noninvasive micro-CT. Using computer software, a comparison between the bone injury and the presumed injury-causing instrument was possible. In this case, the bone injuries were examined with a micro-CT system at the Institute of Medical Physics in Erlangen, Germany (Figure 24.23). The plane 2D images of the micro-CT volume data sets that optimally showed the wound profile of the knife were visualized using a computer-assisted design (CAD) program. To compare the bone injuries with the knife profiles, the thicknesses and forms of the knives were measured and visualized using the same software. The match analysis of the presumed knives with the

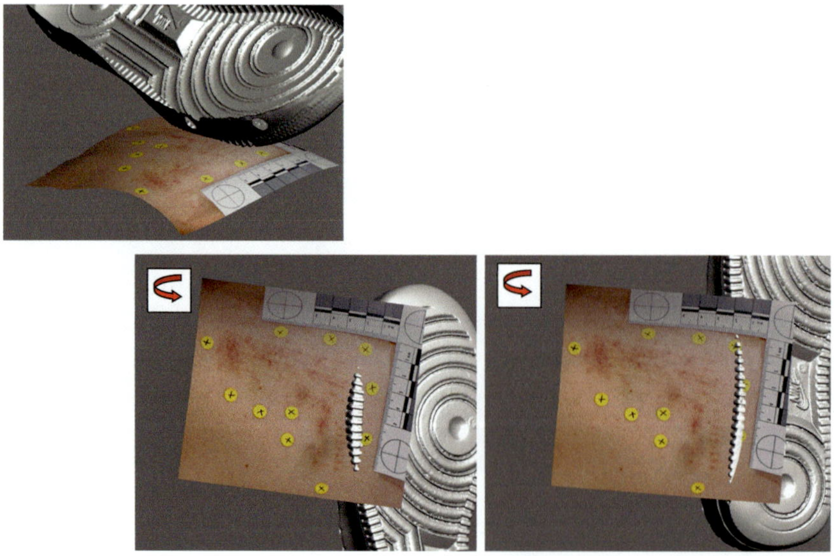

FIGURE 24.21 3D match analysis of the patterned injury on the shoulder with the suspected injury-causing sole of a shoe (case 1). The sole of the second shoe in Figure 24.17 matched the patterned injury on the shoulder. The first shoe is of the same type but has a different size, so it does not match.

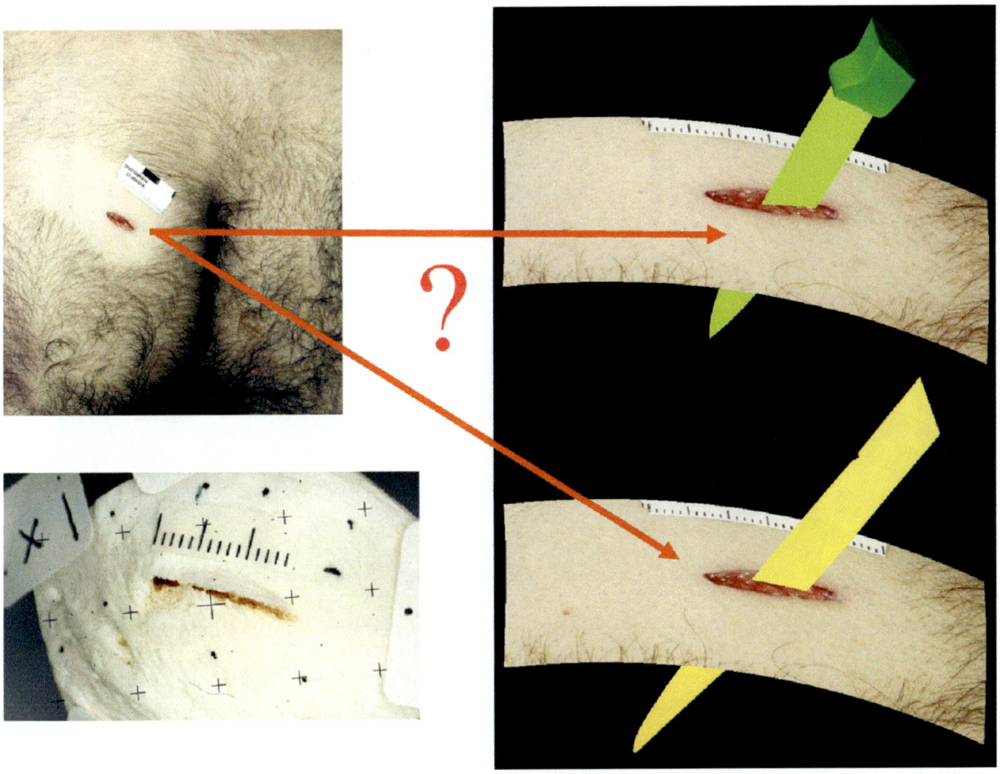

FIGURE 24.22 Sharp force injuries to the skin, soft tissue, and skeletal tissue (pelvis) (case 2). Solely examining the skin injuries and the surface of the bone injury provided no clear clue as to the type of knife used.

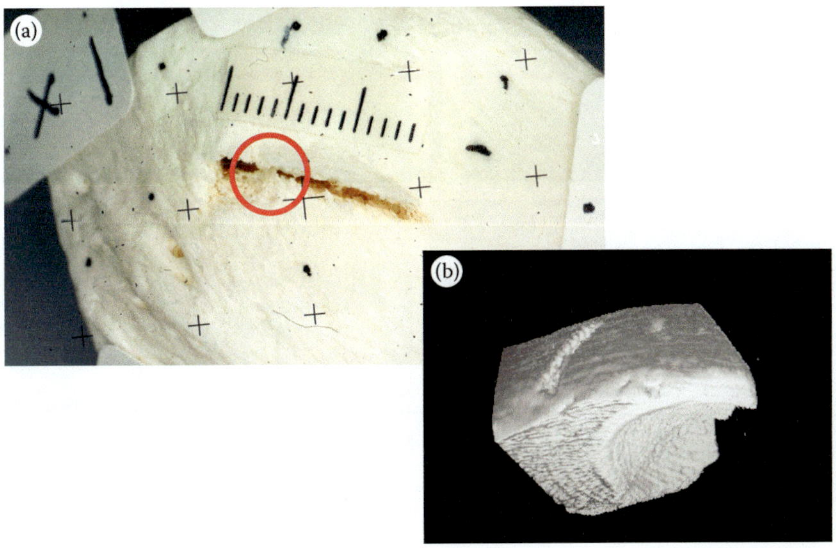

FIGURE 24.23 Digitalization of a bone part with an impression caused by a knife: (a) With a Micro-CT system at the Institute of Medical Physics in Erlangen, Germany. (b) 3D surface model of the scanned bone part, reconstructed using the CT data (case 2).

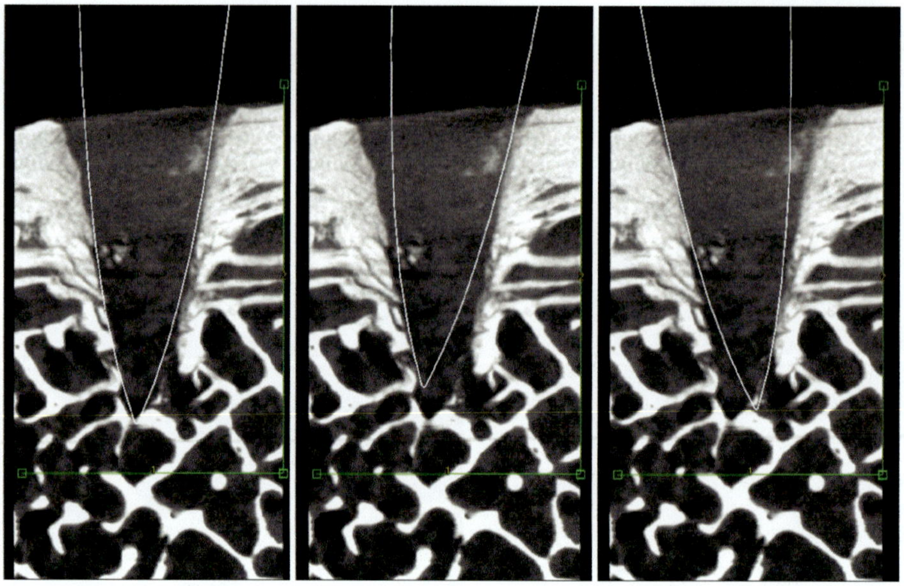

FIGURE 24.24 Superimposition of a plane Micro-CT image, displaying the penetration injury of the bone part, with the CAD drawing of the presumed injury-causing knife (case 2).

wounds is shown in Figures 24.24, 24.25, and 24.26. The shape of the injury-inflicting knife was determined, and the match between bone injury and knife was confirmed.

Muzzle Imprint (Case 3)

In gunshot cases with muzzle imprints, it is important to know the position of the gun while firing the shot and to identify the injury-causing gun. To answer these questions, an analysis of the wound is necessary.

This case involved a muzzle imprint on the right side of the head sited above the ear (Figure 24.27) and an exit wound on the left side of the head. At the scene of the crime, two weapons were found, as well as 20 bullets that had been fired into a park bench next to the victim. The internal autopsy

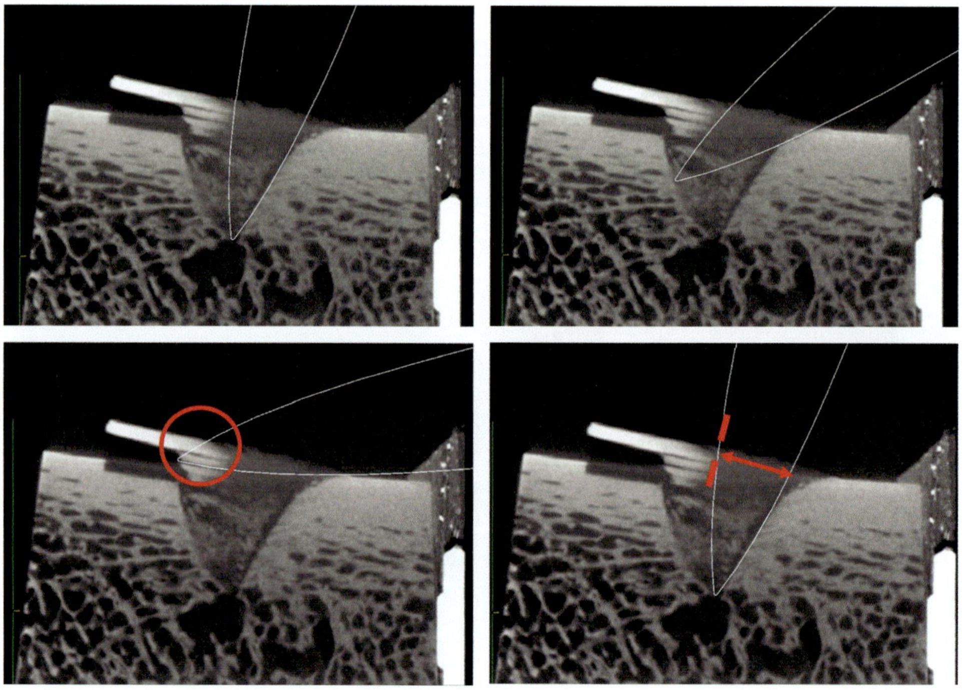

FIGURE 24.25 A knife-induced injury to the bone with a rotational component (case 2). Superimposition of the CT image with the CAD drawing of the knife.

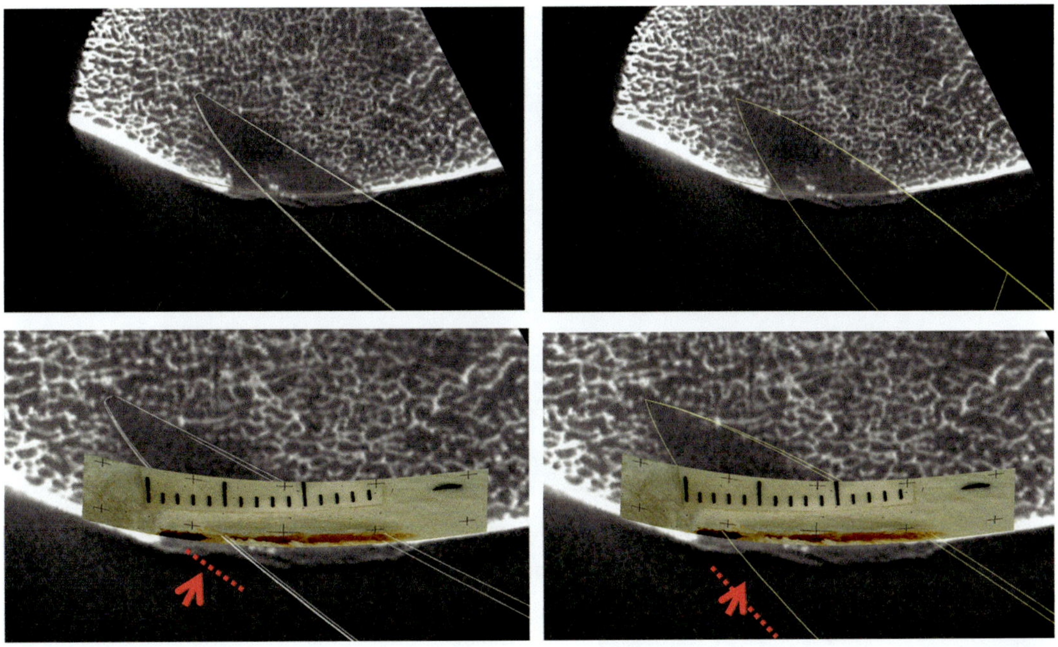

FIGURE 24.26 A knife-induced injury to the bone with a rotational component (case 2). Superimposition of the CT image with the CAD drawing of the knife.

findings of the deceased were documented using MSCT and MRI. The external injuries to the head of the deceased, the presumed injury-inflicting gun, and the projectile were digitized using the GOM ATOS II system. The gun was digitized when the slide was in the front and back positions, to capture the complete geometric form of the gun (Figure 24.28). The required 3D models of both positions were merged using reference targets attached to the gun. Finally, the gun slide was animated in the appropriate manner using the software. This was necessary to capture all the possible positions of the gun slide while a shot was fired for the purpose of comparison with the muzzle imprint. Using a computer-assisted comparison of the 3D model of the muzzle imprint with the muzzle of the gun, we determined the exact positioning in terms of how the gun was held while firing the shot. The shot direction was reconstructed with the entrance and exit holes of the bullet on

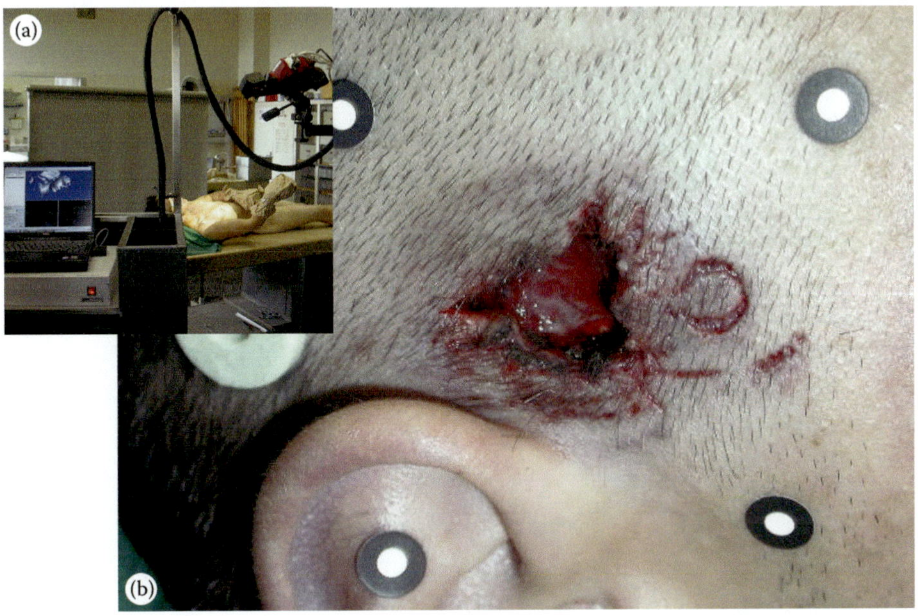

FIGURE 24.27 3D surface digitization of a muzzle imprint (case 3): The scanner setup and the body, placed on the examination table in the autopsy room (inset, a). Digital photograph of the muzzle imprint (b).

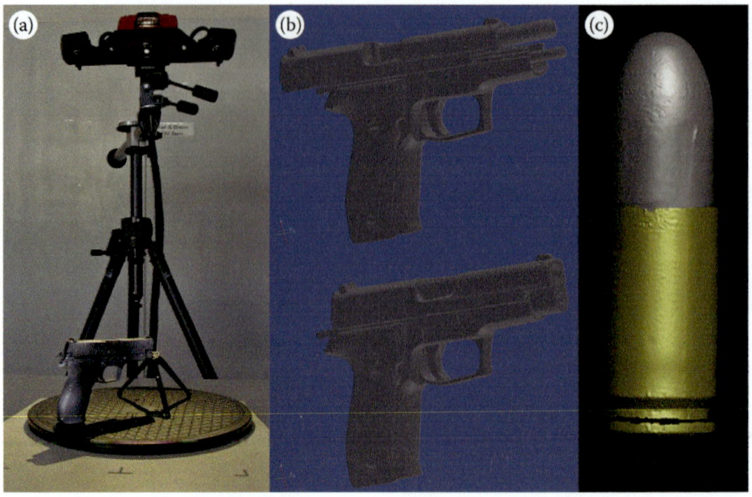

FIGURE 24.28 3D surface scan of the pistol (case 3). The ATOS II 3D Digitizer on a rotation table (a), the 3D model of the pistol with the slide pulled back and in front position (b), and a 3D model of the round (c).

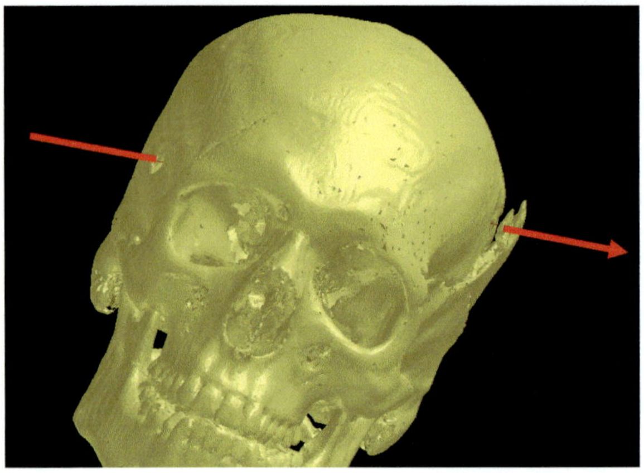

FIGURE 24.29 Based on the entrance and exit holes in the 3D model of the skull, the shot direction was reconstructed (case 3).

the 3D model of the skull generated from the CT data (Figure 24.29). The match of the form of the muzzle of the gun with the injury lines was clearly visible (Figure 24.30). In addition to 3D models of the external injuries and the injury-inflicting gun, the 3D model of the head was sliced virtually, in the axial and sagittal directions, to show the bullet channel through the brain on the internal CT and MR images (Figures 24.31 and 24.32).

Bite Marks (Case 4)

Bite marks can be found on food, on malleable objects, or on skin. In this section, the focus is on bite marks on the skin. Geometric bite-mark analysis for identification purposes is based on the individuality of a dentition profile, which is used to match a bite mark to a suspected perpetrator. This matching

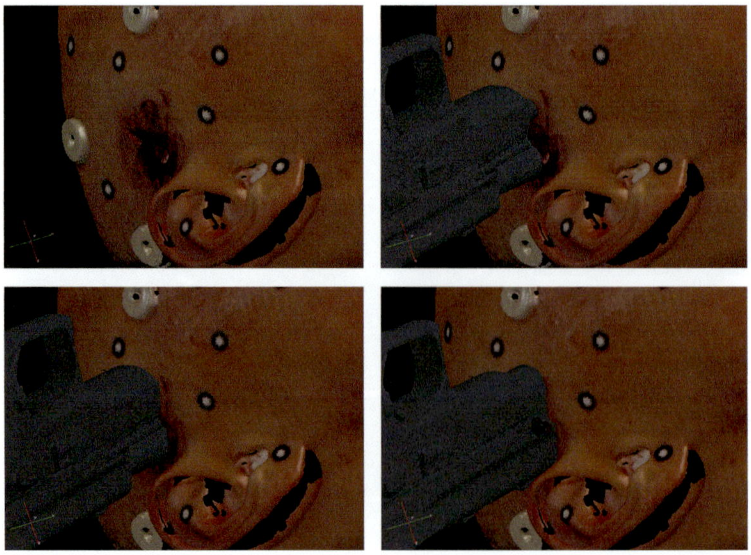

FIGURE 24.30 By comparing the 3D models, the muzzle imprint could be matched with the pistol (case 3). With knowledge of the shot direction, the accurate position of the pistol on the head could be determined.

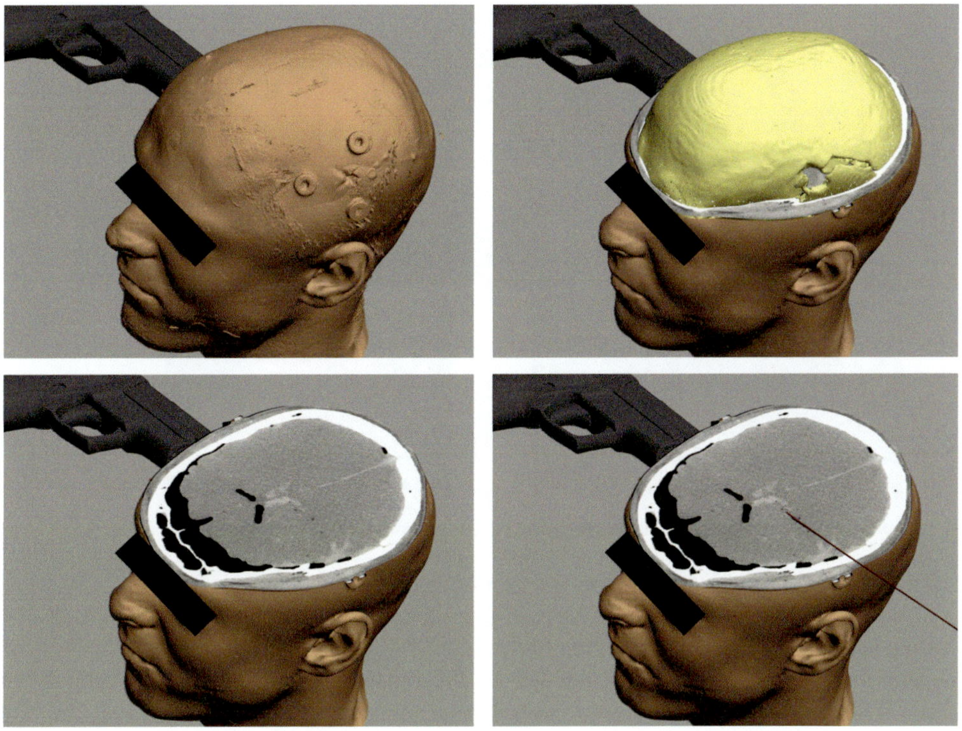

FIGURE 24.31 The position of the pistol during the shot was reconstructed using the data of 3D surface scanning and CT scanning (case 3). The generated 3D models display the bone and skin defects, the plane CT images display the defects in the brain tissue.

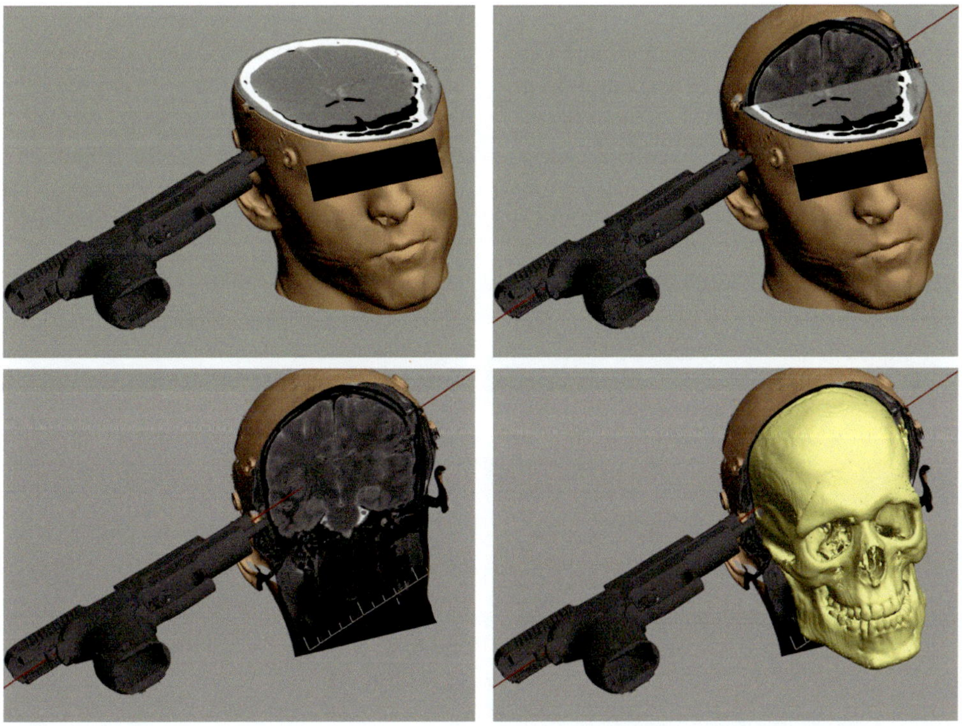

FIGURE 24.32 The position of the pistol during the shot was reconstructed using 3D surface scanning, CT and MRI data (case 3). The resulting 3D models display the bone and skin defects, the plane CT images display defects in the brain tissue. MR images exhibit better tissue resolution. In the lower left-hand corner an MRI image combined with the 3D model of the head displays the bullet pathway.

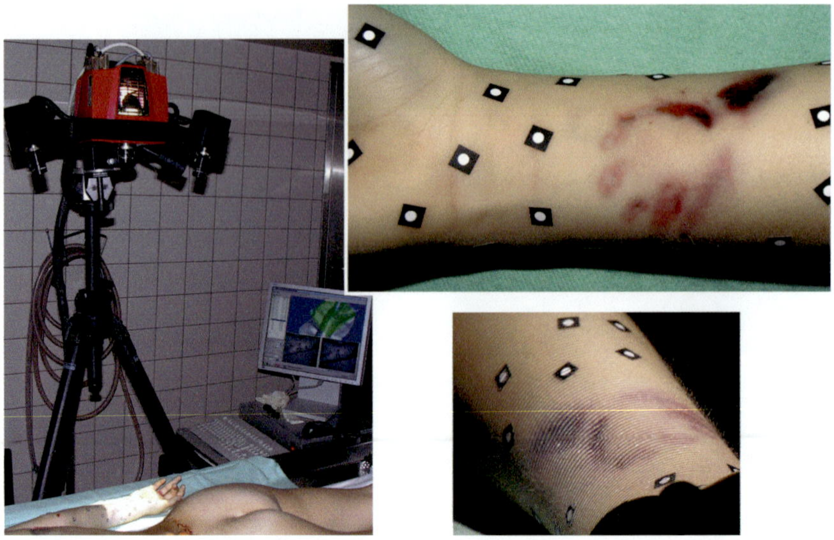

FIGURE 24.33 3D documentation of a bite mark injury on the right forearm of the victim (case 4).

has to consider that biting is a dynamic process involving two moving systems. The mandible and the maxilla meet together during the biting action. If the bite is applied to skin, the skin will move too, depending upon the location of the part of the body and the tissue constitution at this site. Additional components of the bite pattern could be caused by victim–biter interaction or movement.

In this homicide case, a woman was stabbed with a knife. She died due to the fatal stab injuries. Furthermore, she was bitten on the right arm. The bite-mark injury was documented using TRITOP photogrammetry and the ATOS II high-resolution surface scanner (Figure 24.33).

The 3D model of the bite mark was compared and contrasted with 3D models of the plaster casts of the dentitions of two suspected perpetrators (Figure 24.34). The teeth of one suspected perpetrator viewed in gray were not matched to the bite mark (Figure 24.35). The comparison of the wound with the dentition of the other suspected perpetrator showed a match between the teeth of the mandible and maxilla to the bite mark on the arm of the deceased (Figures 24.36, 24.37, and 24.38). The match between the front teeth of the mandible and the bite mark was apparent on the true-color model of the skin, marked with orange, red, and blue lines, as well as on the gray-shaded 3D model of the wound. The 3D model showed the depth of the impression on the skin, which correlated with the form of the teeth (Figure 24.39). Figure 24.40 shows the configuration of the perpetrator and the victim when her right arm was bitten.

Strangulation (Case 5)

Investigations of strangulation incidents, involving either living or deceased persons, are commonly performed in forensic medicine. In addition to examining the internal autopsy findings and collateral signs, examining the patterned skin injury can give clues as to the strangulation tool. In this case, a woman claimed that she was manually strangulated by an unknown person. She had superficial patterned injuries on the neck. The presumed injury-causing instrument was a necklace (Figure 24.41).

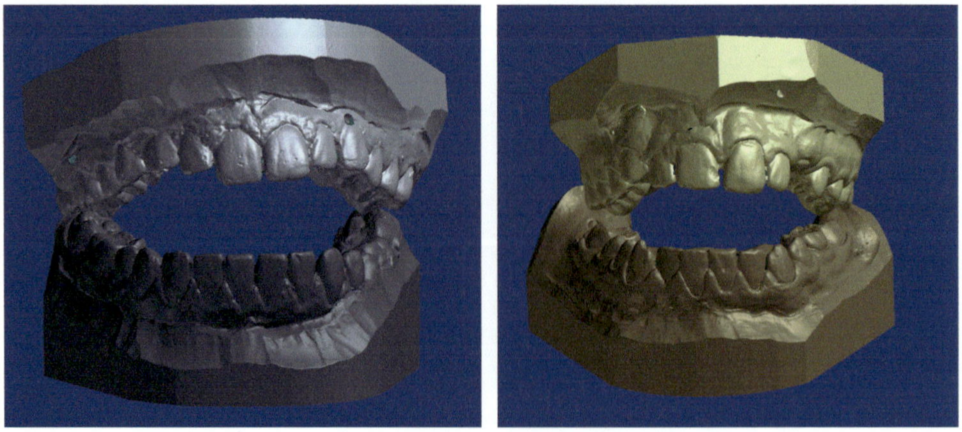

FIGURE 24.34 The dentition casts of the suspected offenders were digitized and 3D models were generated (case 4).

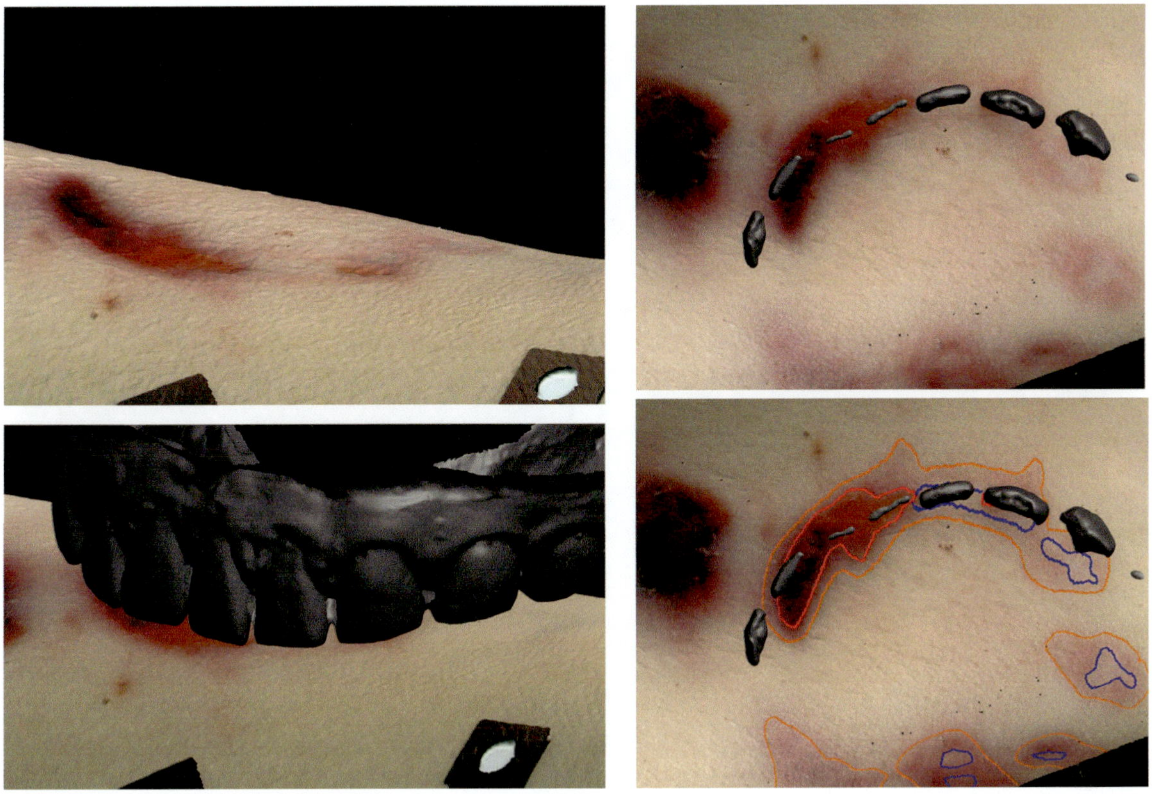

FIGURE 24.35 When we compared the surface model of the bite mark with the dentition models of the first set of suspected offenders, a correlation could not be determined (case 4).

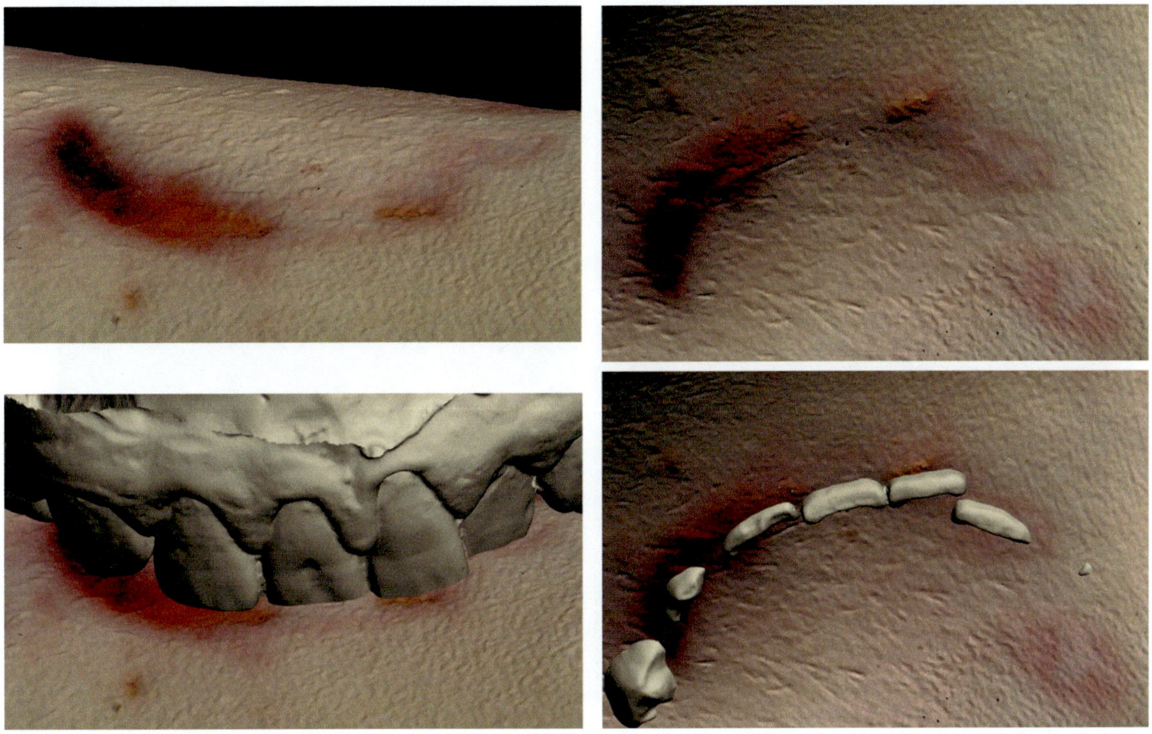

FIGURE 24.36 In a comparison of the surface model of the bite mark with the dentition models of the second suspected offenders a correlation could be determined (case 4).

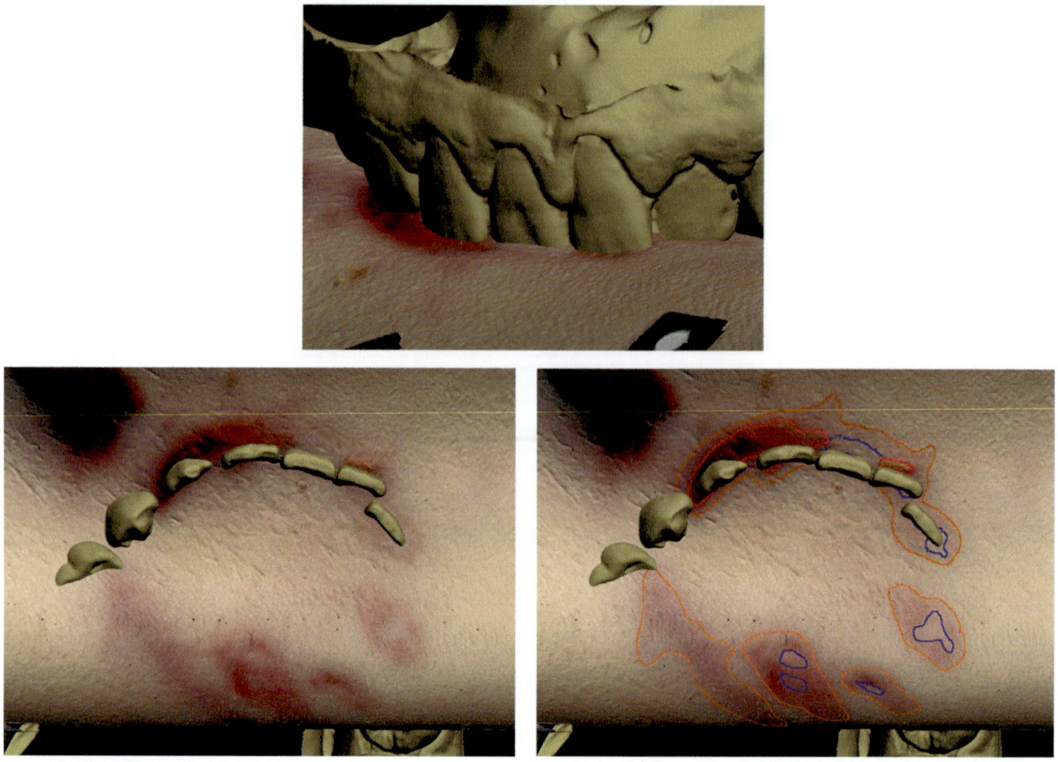

FIGURE 24.37 When we compared the surface model of the bite mark with the mandible model of the second set of suspected offenders, a correlation could be determined (case 4).

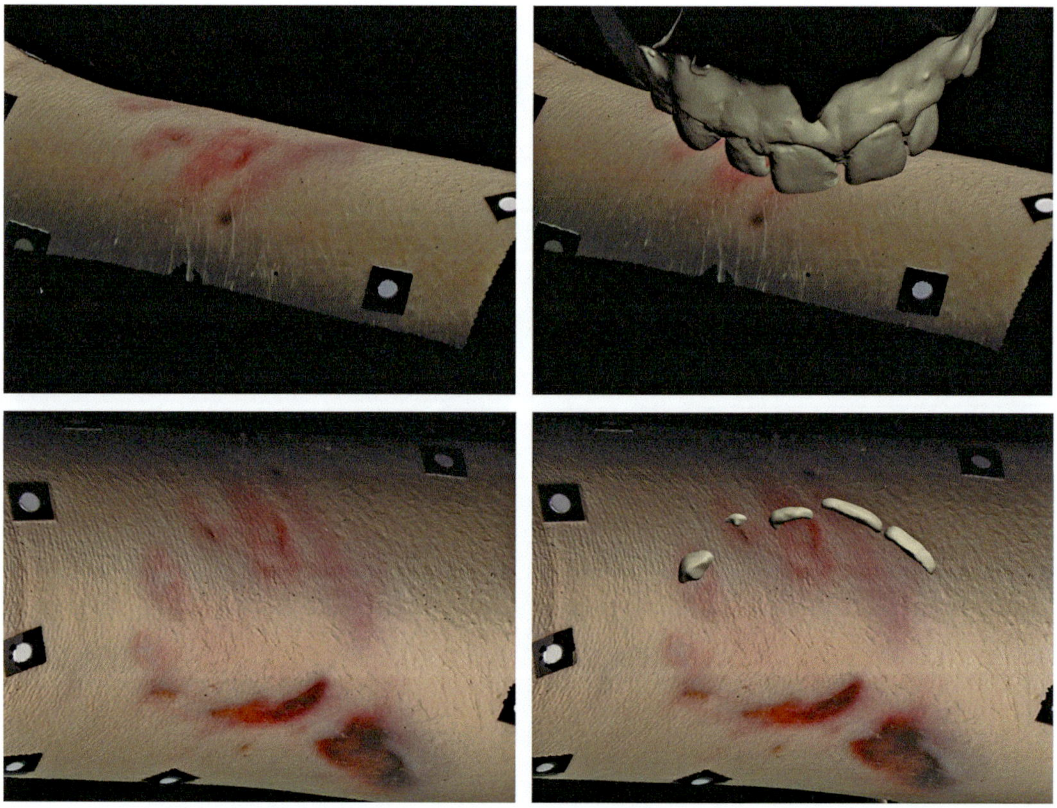

FIGURE 24.38 Also the maxilla of the second set of suspected offenders matched the bite mark (case 4).

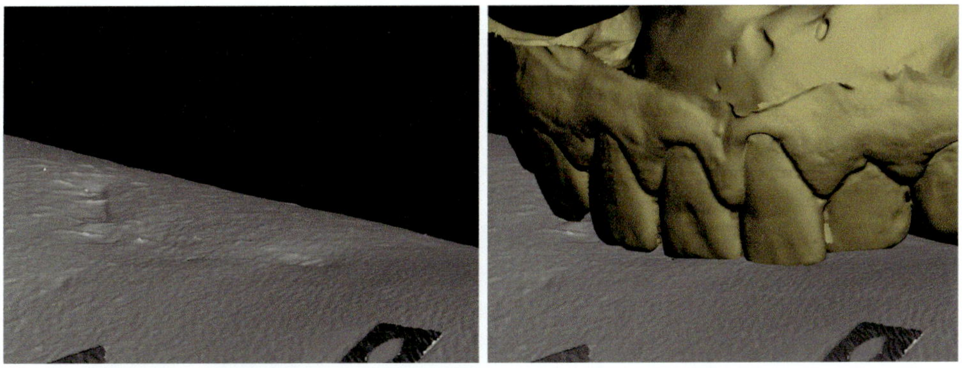

FIGURE 24.39 The uncolored 3D model of the bite mark displays the impressions of the teeth (case 4).

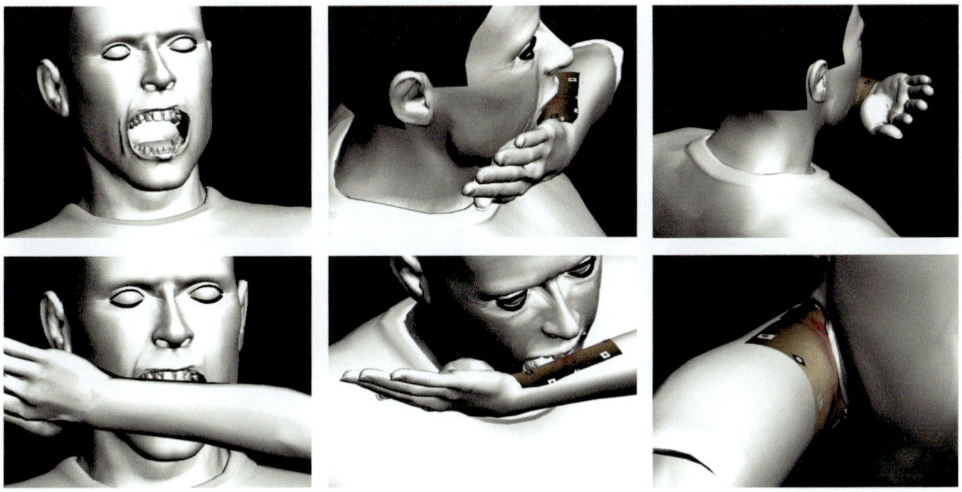

FIGURE 24.40 This graphic shows a possible position of the offender while biting the victim (case 4).

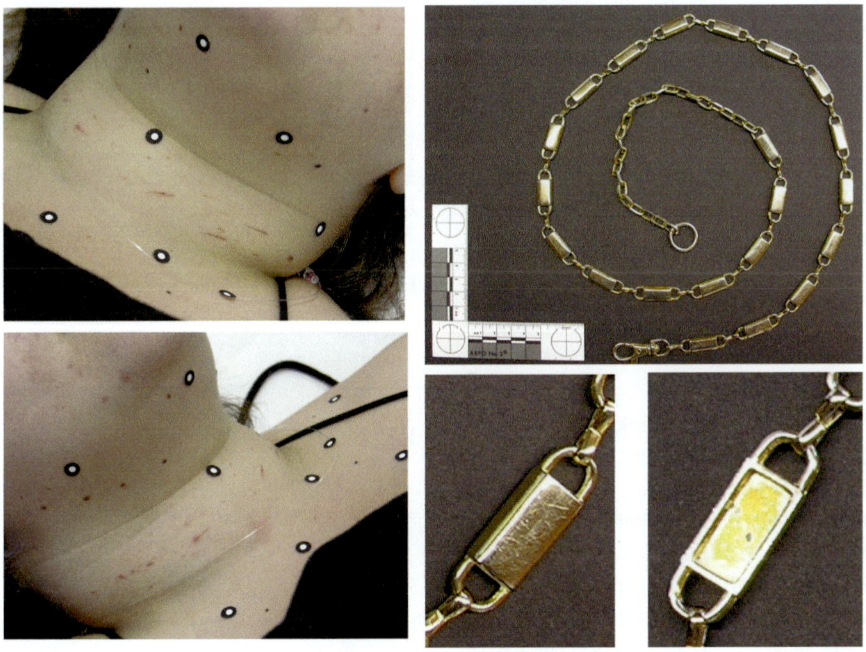

FIGURE 24.41 A woman with superficial patterned injuries on the neck (case 5). The presumed injury-causing instrument was a necklace. The skin injury and the necklace were documented with the 3D surface scanner.

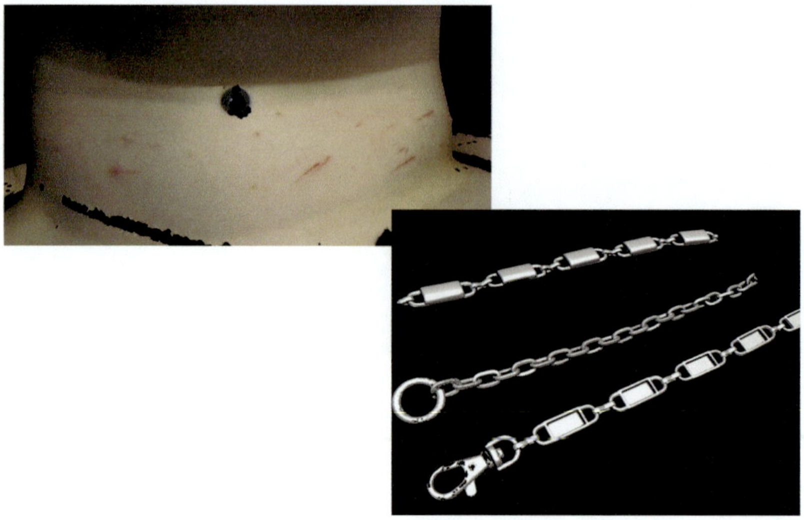

FIGURE 24.42 3D models of the skin injuries and the necklace (case 5).

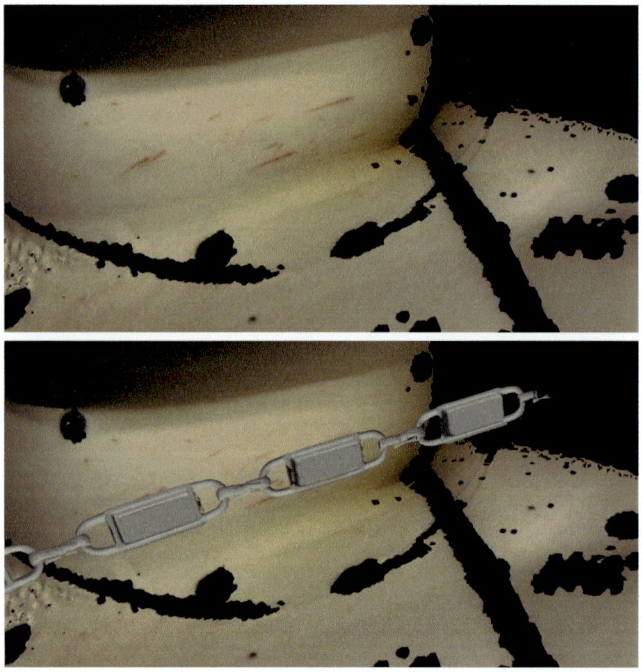

FIGURE 24.43 Match analysis of the necklace and the superficial skin injuries (case 5).

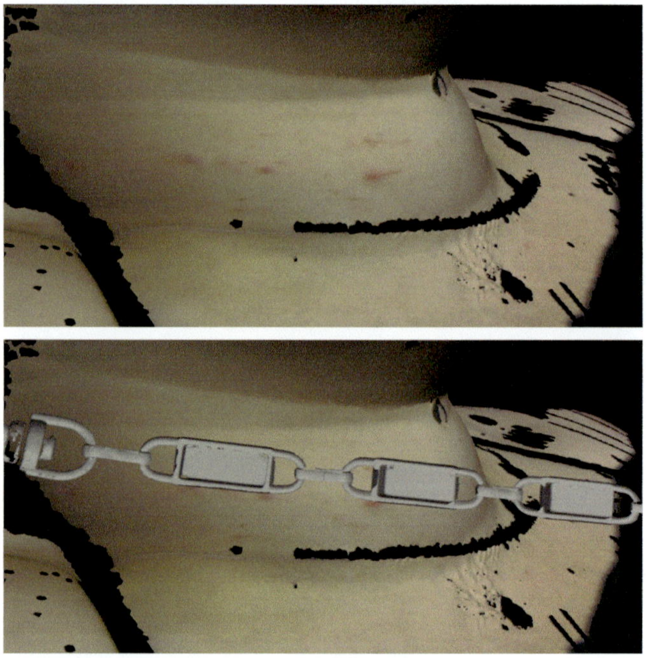

FIGURE 24.44 Match analysis of the necklace and the superficial skin injuries (case 5).

The necklace and the injuries on the neck of the victim were digitized with the 3D surface scanner. A photograph was taken of the injuries on the neck and was projected onto the surface data. The result was a 3D model of the neck injuries and the necklace as shown in Figure 24.42. The match analysis concluded that the width of the links of the necklace could possibly fit the patterned injuries (Figures 24.43 and 24.44), but the evidence was not sufficiently conclusive because the necklace injuries were only superficial.

REFERENCES

1. Luhmann, T., Robson, S., Kyle, S., and Harley, I., *Close Range Photogrammetry: Principles, Techniques and Applications*. Whittles Publishing, UK, 2006.
2. Brüschweiler, W., Braun, M., Fuchser, H. J., and Dirnhofer, R. Photogrammetrische Auswertung von Haut- und Weichteilwunden sowie Knochenverletzungen zur Bestimmung des Tatwerkzeuges - grundlegende Aspekte, *Rechtsmedizin*, 7, 76–83, 1997.

3. Brueschweiler, W., Braun, M., Dirnhofer, R., and Thali, M. J., Analysis of patterned injuries and injury-causing instruments with forensic 3D/CAD supported photogrammetry (FPHG): An instrument manual for the documentation process, *Forensic Sci Int*, 132 (2), 130–38, 2003.
4. GOM, Germany, Braunschweig, http://www.gom.com, 2009.
5. Thali, M. J., Braun, M., and Dirnhofer, R., Optical 3D surface digitizing in forensic medicine: 3D documentation of skin and bone injuries, *Forensic Sci Int*, 137 (2–3), 203–08, 2003.
6. Thali, M. J., Braun, M., Buck, U., Aghayev, E., Jackowski, C., Vock, P., et al., VIRTOPSY—scientific documentation, reconstruction and animation in forensic: Individual and real 3D data based geo-metric approach including optical body/object surface and radiological CT/MRI scanning, *J Forensic Sci*, 50 (2), 428–424, 2005.
7. Naether, S., Buck, U., Werner, B., Zingg, C., and Thali, M. J., Non-contact documentation of physical characteristics of ecstasy tablets, hemp coins and imprint punches by using 3D optical surface scanning, *Can Soc Forensic Sci J*, 41 (4), 191–198, 2008.
8. *GOM ATOS User Manual,* http://www.gom.com, 2009.
9. Autodesk, 3ds max, http://autodesk.com/3dsmax, 2009.
10. Thali, M. J., Braun, M., Buck, U., Aghayev, E., Jackowski, C., Vock, P., Sonnenschein, M., and Dirnhofer, D., VIRTOPSY—Scientific documentation, reconstruction and animation in forensic: Individual and real 3D data based geo-metric approach including optical body/object surface and radiological CT/MRI scanning, *J Forensic Sci* 50(2), 428–24, 2005.
11. Thali, M. J., Braun, M., Markwalder, T. H., Brüschweiler, W., Zollinger, U., Malik, N. J., Yen, K., and Dirnhofer, R., Bite mark documentation and analysis: The forensic 3D/CAD supported photogrammetry approach. *Forensic Sci Int*, 135(2), 115–21, 2003.
12. Thali, M. J., Braun, M., and Dirnhofer, R., Optical 3D surface digitizing in forensic medicine. 3D documentation of skin and bone injuries, *Forensic Sci Int*, 137(2–3), 203–08, 2003.
13. Thali, M. J., Braun, M., Wirth, J., Vock, P., and Dirnhofer, R., 3D Surface and body documentation in forensic medicine: 3D/CAD photogrammetry merged with 3D radiological scanning, *J Forensic Sci*, 48(6), 1356–65, 2003.
14. Thali, M. J., Taubenreuther, U., Karolczak, M., Braun, M., Brueschweiler, W., Kalender, W. A., and Dirnhofer, R., Forensic microradiology: Micro-computed tomography (micro-CT) and analysis of patterned injuries inside of bone, *J Forensic Sci*, 48(6), 1336–42, 2003.

25 Cross-Sectional Imaging and Swiss Virtobot Documentation and Analysis

Work Flow and Procedure

Michael J. Thali, Steffen G. Ross, Stephan A. Bolliger, Tanja Germerott, Patricia M. Flach, Lars C. Ebert, Thomas Ruder, Michael Bolliger, Ulrich Preiss, Sandra Mathier, Laura Filograna, Gary Hatch, and Garyfalia Ampanozi

CONTENTS

Albert Einstein once said, "One recognizes a really good idea by the fact that, from the outset, making it a reality seems impossible." In addition to the interest in research, this statement has again and again been the *feu sacre* of the Bern Virtopsy Research Group (see http://www.virtopsy.com as well as Thali, et al., 2009.).

The Virtopsy project arose from an application-oriented research endeavor as one had to match an implement to the form of the injury. Over the years, besides three-dimensional (3D) surface documentation, somatic interior documentation using computed tomography (CT) and magnetic resonance (MR) were added. The fusion of 3D surface data and 3D radiological interior data was the next step. Further projects followed, with the introduction of postmortem biopsies. Postmortem biopsies are necessary because at the present time, even with elaborate radiological systems such as micro-CT and micro-MR, not all somatic findings can be visualized at the cellular level. It is necessary, therefore, that a postmortem biopsy be performed and specimens are examined under the microscope.

As a supplement to postmortem biopsies, postmortem angiography was introduced into the Virtopsy project, which makes it possible to display the smallest vessel injuries that cause leakages. On the other hand, it is also possible to display perfusion and vessel occlusion, such as is necessary, for instance, in the area of cardiac infarctions or occlusions of cerebral vessels. In the following pages these procedures are shown in detail.

It was always important in the Virtopsy project that the postmortem findings, which are also shown in the following images and displayed in their totality in *Virtopsy Approach*, could be visualized. The findings of the postmortem insights could also be transferred to the living. The initial step was taken in the analyses of strangulation cases, in which the Bern Virtopsy Research Group accomplished pioneering work. Today MR examination findings on strangulation victims in Switzerland are already included in judgments about mortal danger and qualifying the effects of violence to the neck. There have already been sentences handed down that were based only on MR examinations and the injury findings made with them.

Within the Virtopsy project, process optimization was important (Figures 25.1 through 25.4). The VIRTOBOT concept that we presented in 2003 to the National Institute of Health in Washington DC/Bethesda has been realized through the years. Since November 2009, the Bern Institute for Forensic Medicine (IFM) has had a forensic examination process line at its disposal that makes it possible to subject a corpse to a robot-guided 3D photogrammetry-supported surface area scan and afterwards to a CT examination as well as a postmortem biopsy and angiography. As a supplement, the Bern Institute, at the time of this writing, has a whole body MR system at its disposal that is also able to display the minutest injuries in soft tissue and organs. Besides the institutional vision and advanced technologies by means of which the Bern Institute has distinguished itself over the past few years, it has also put together an advanced research team, which is just as important. At the moment, the fourth Virtopsy generation—consisting of graduate engineers, radiologists, imaging specialists, and forensic doctors—is at work at the Bern Institute under the direction of Michael J. Thali. Important for the functioning of a team is an "open minded" attitude as well as inter- and transdisciplinary cooperation.

With regard to the use of these technologies, the following procedures are recommended:

1. CT scanning
2. Postmortem biopsy
3. Postmortem angiography

If these have been introduced, then about 60–80% of the causes of death and the reconstruction problems from the

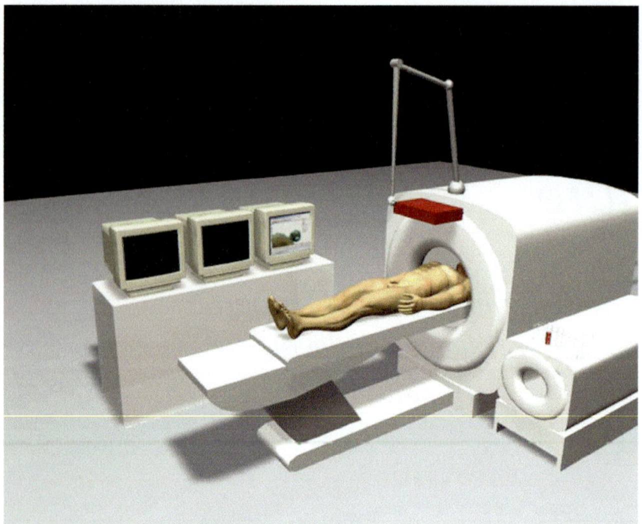

FIGURE 25.1 Virtobot, concept presented at NIH, Washington DC/Bethesda in 2003.

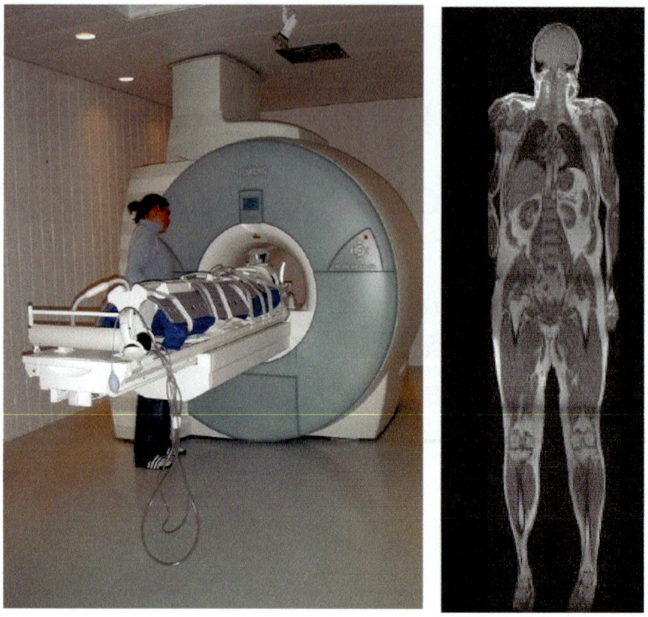

FIGURE 25.3 The Virtopsy Team has since 2009 owned a TIM-MRI scanner at the Forensic Institute, which allows the scanning of body soft tissue and organs with high resolution in half-an-hour.

forensic case material that as a rule arrives at the IFM can be answered.

For a further buildup of a similar Virtopsy postmortem research group, the following modalities are recommended:

4. Surface area scanning
5. MR scanning

Over the last 10 years, the Virtopsy Project, with its systematic integration of various technologies and modalities, such as photogrammetric 3D surface scanning and CT and MR scanning, has been perceived in professional circles as being revolutionary. Great strides have been made also in the areas of clinical and postmortem forensic medicine, postmortem biopsy and angiography, and synthetic somatic modeling. After a decade, the technologies used at the Forensic Institute in Bern have been integrated as an (r)evolutionary process development in daily forensic practice.

Over the next years, we can expect a further increase in the possibilities of answering forensic problems and causes of death through continuing technological improvements. Research areas include further development dual-energy instruments that already exist, better resolution in CT scanning, further development of MRI instruments to higher Tesla levels, and also—especially—software developments such as synthetic MRI and parallel imaging, etc. As our experience in Switzerland shows, noninvasive methods are valued highly by the courts and organs of jurisprudence, especially because of the exceptional 3D visualization and reproducibility. Those belonging to special cultural and

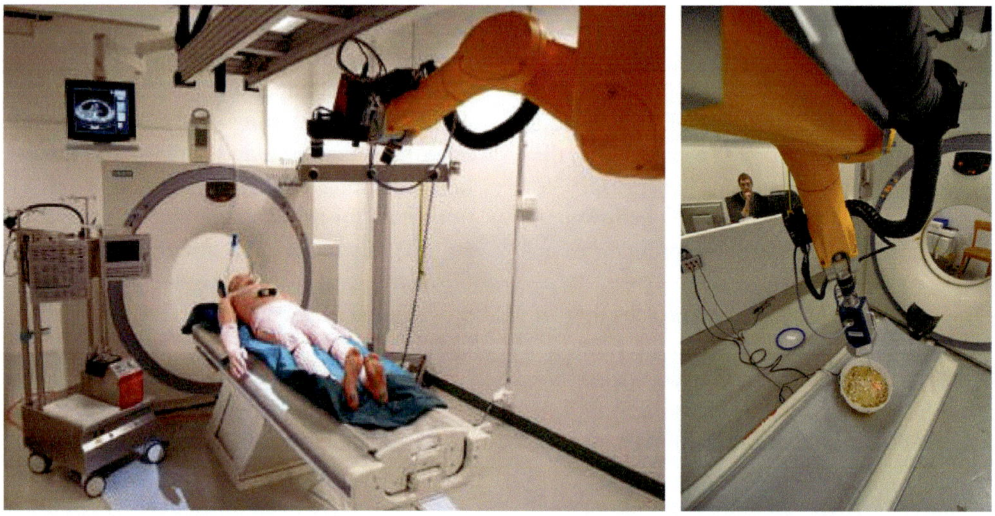

FIGURE 25.2 Virtobot of the Virtopsy Team in 2009: system integration of navigation controlled, robotic, and photogrammetry-based 3D surface scanning and CT with postmortem angiography and robotic biopsy system.

FIGURE 25.4 Some of the Virtopsy team members since 2009: Team and Technology are both important.

religious groups that oppose autopsies also value the development of noninvasive autopsies.

Significant future developments are those made with critical minds, always with a view to increasingly eliminate mistakes, so that no erroneous judgments occur. Even after about 10 years of successful development at the Bern Institute in Switzerland, the field is still relatively new. Newer groups that enter this field are warned not to begin this without the necessary caution, because mistaken conclusions in forensics can have enormous negative effect on those affected by them, not only in the postmortem area but also in the growing market for clinical forensic medicine, that is, the examination of living individuals who require forensic evaluation. The imaging field offers an enormous potential, not only with regard to research but also in forensic services and teaching.

In the following, a number of illustrations from CT and MR imaging are shown (Figures 25.5 through 25.57). For the sake of completeness, it must be mentioned that detailed data can also be found in the *Virtopsy Approach* book.

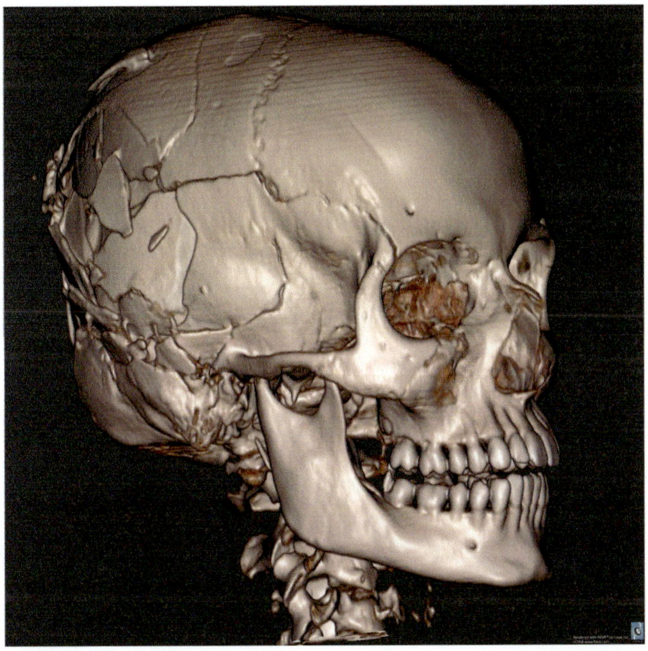

FIGURE 25.5 MSCT 3D bone reconstruction of the head of a homicide victim. A distinctive fracture system after a blunt force trauma to the head is shown, as viewed from anterior and right lateral.

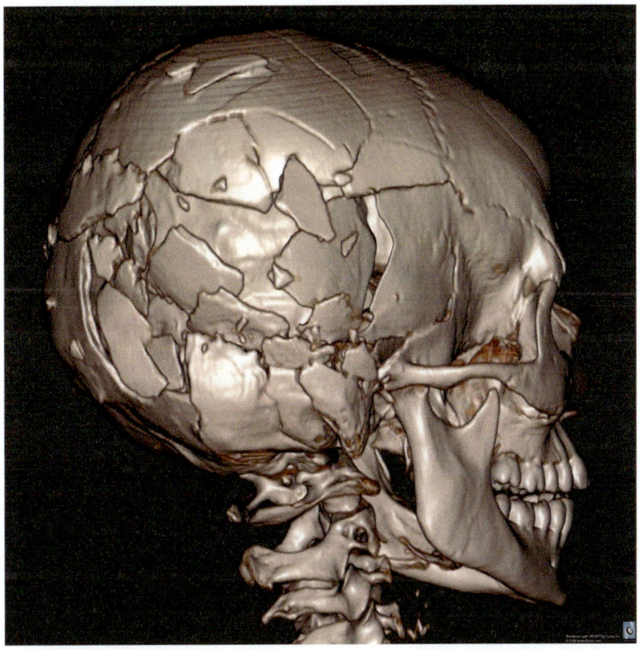

FIGURE 25.6 MSCT 3D bone reconstruction of the victim in Figure 25.5, showing the distinctive fracture system to better advantage, as viewed from posterior and right lateral.

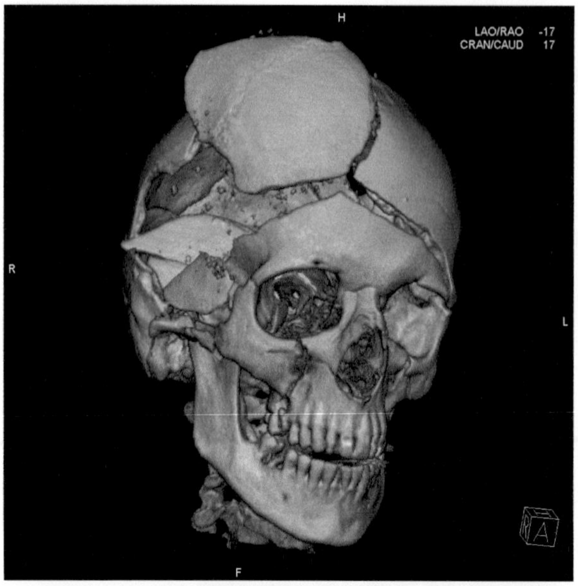

FIGURE 25.7 MSCT 3D skull reconstruction of a man who shot himself while playing Russian roulette. Note the extensive damage of skull arround the entrance defect. The skull literally burst open due to the hydronynamic effect of the 9 mm bullet.

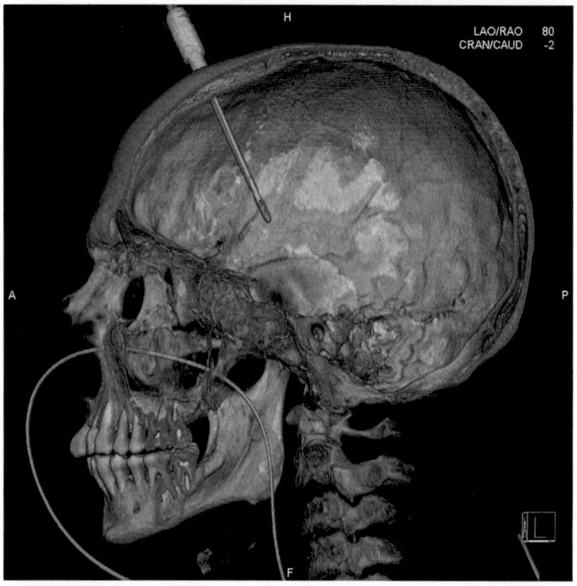

FIGURE 25.8 MSCT 3D reconstruction of a skull with a measuring probe for intracranial pressure and a nasogastric tube.

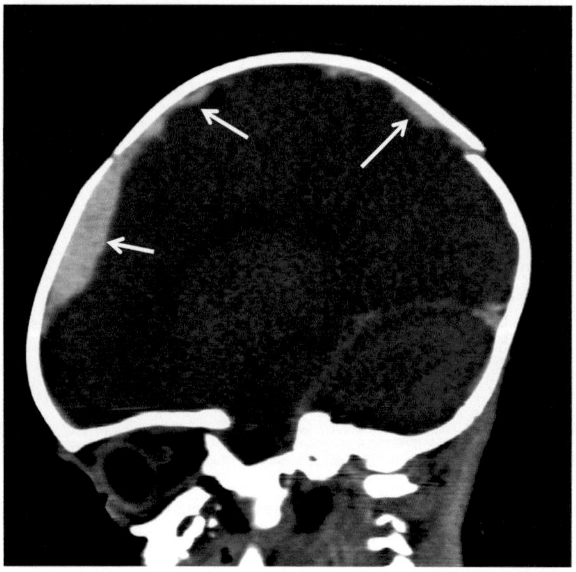

FIGURE 25.9 MSCT, sagittal image of the head of a 6-week-old baby. A subdural hemorrhage of the brain (yellow arrows) is shown.

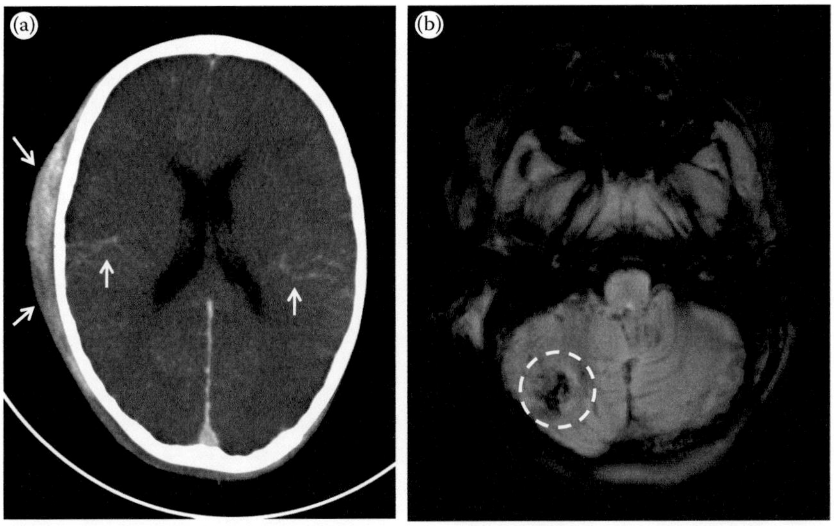

FIGURE 25.10 (a) Axial MSCT illustrating a subarachnoid hemorrhage with fresh blood in the gyri of the brain (red arrows) and signs of brain edema. The yellow arrows indicate a hemorrhage of the epicranium on the right side. (b) MRI image of the head showing an injury of the cerebellum (perforated circle).

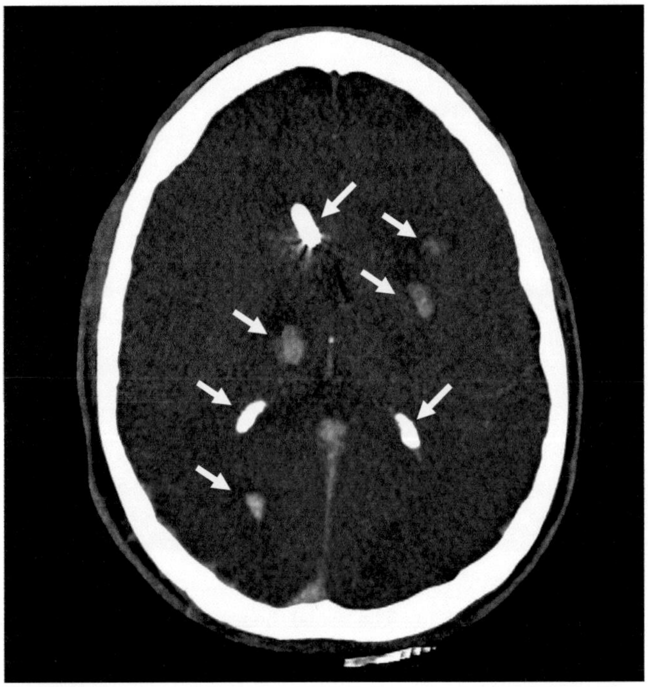

FIGURE 25.11 In this axial reconstructed MSCT image of the head of a man after a skyglide accident, multiple hemorrhages in the brain parenchyma are clearly visible (red arrows). Note also the blood in the ventricle system (green arrows) and the measuring probe for intracranial pressure (yellow arrow).

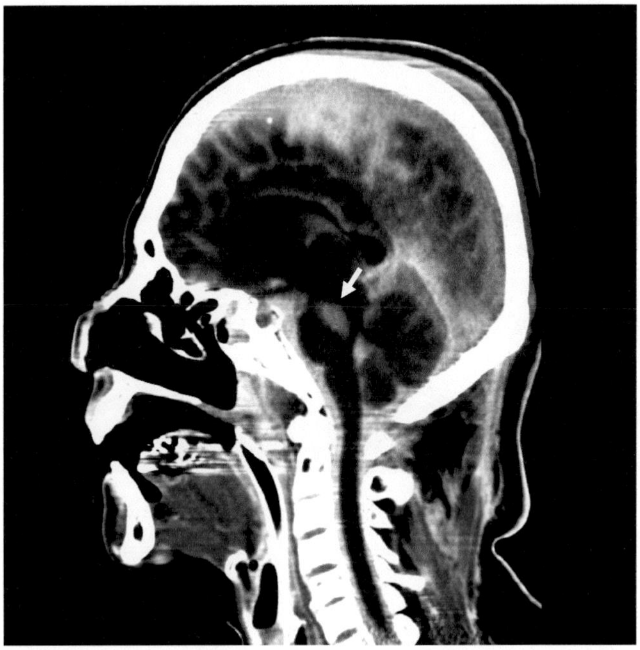

FIGURE 25.12 Sagittal reconstructed MSCT image of the head of a 72-year-old woman who died after a collapse at home. Subdural and subarachnoidal hemorrhages are clearly visible. Note the hemorrhage in the brainstem (red arrow).

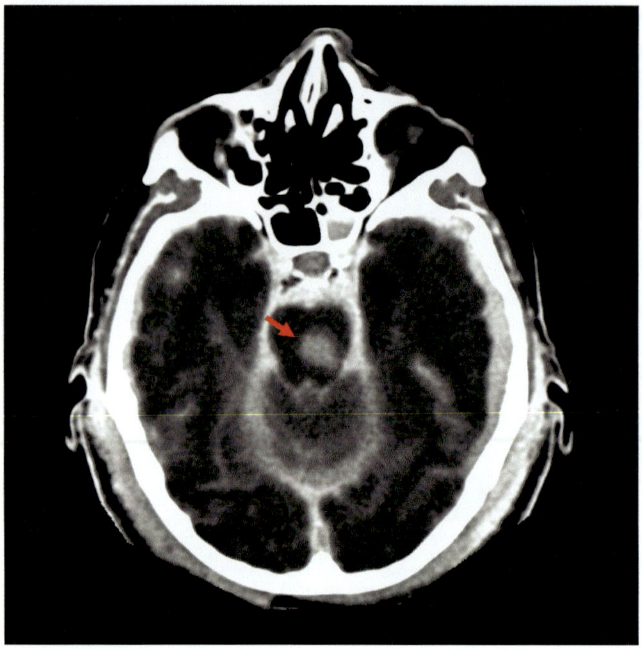

FIGURE 25.13 Axial reconstructed MSCT image of the head of a 72-year-old woman who collapsed at home. Subdural and subarachnoidal hemorrhages are clearly visible. Note the hemorrhage in the brainstem (red arrow).

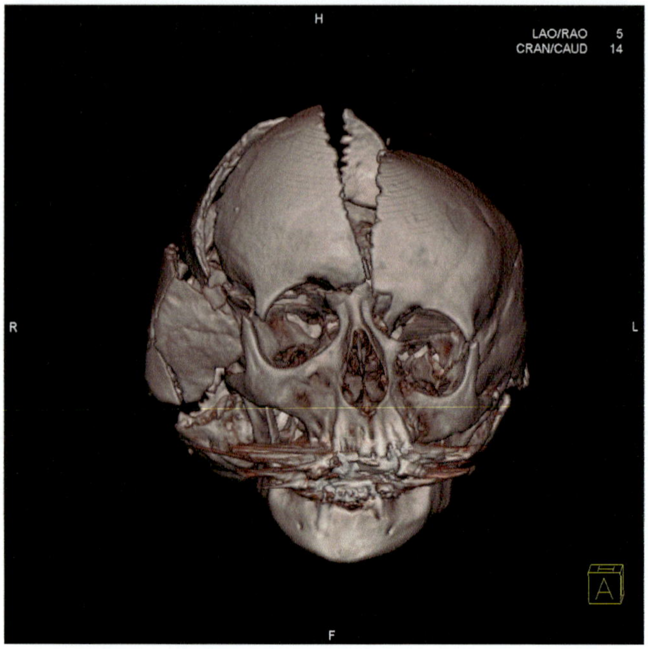

FIGURE 25.14 Anterior view of a MSCT 3D skull reconstruction of a woman who jumped off a high bridge and landed on her back. Extensive fracturing of the skull.

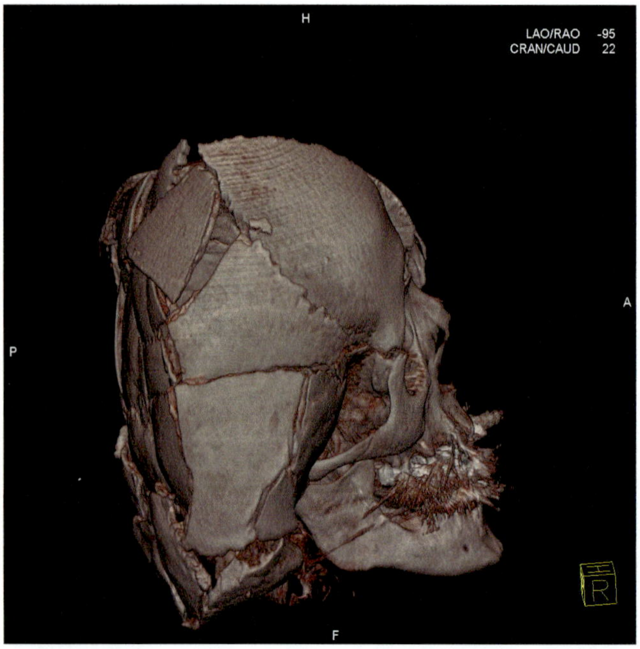

FIGURE 25.15 Lateral view of the woman shown in Figure 25.14.

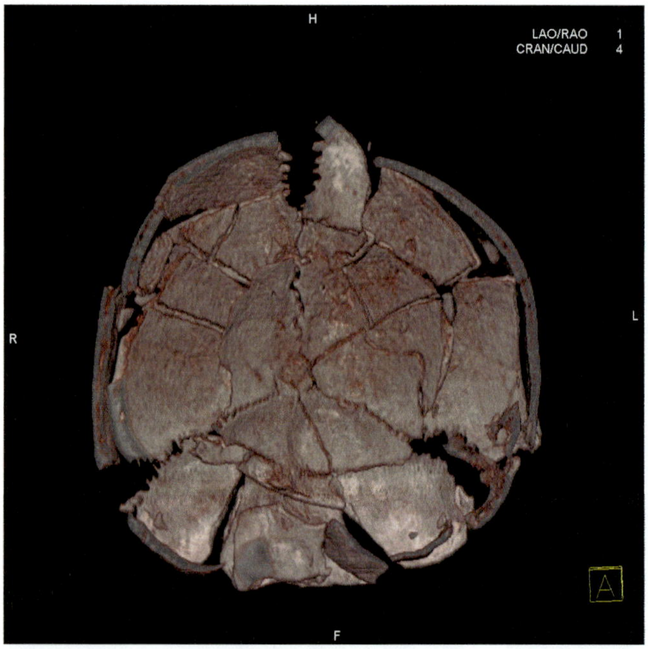

FIGURE 25.16 MSCT, 3D skull reconstruction of the woman shown in Figure 25.14. The facial bones have been virtually removed; the cranium is seen from an anterior view.

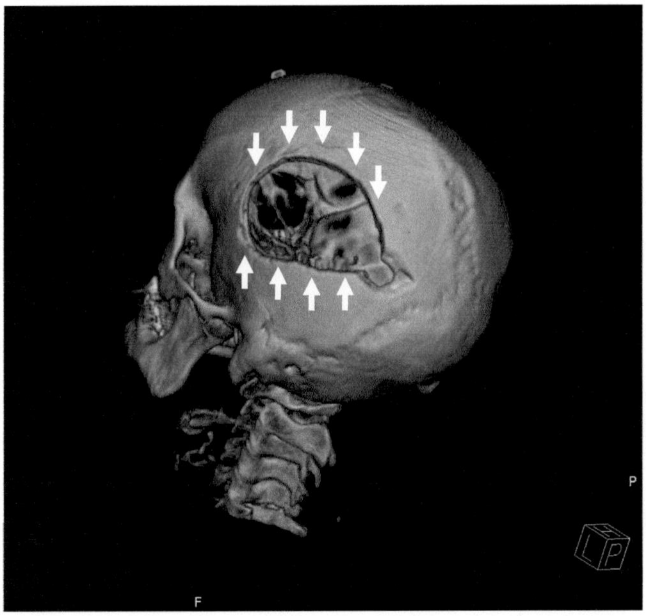

FIGURE 25.17 MSCT, 3D reconstruction of the skull of the victim seen in Figure 25.14. Note the skull defect due to the craniotomy (red arrows).

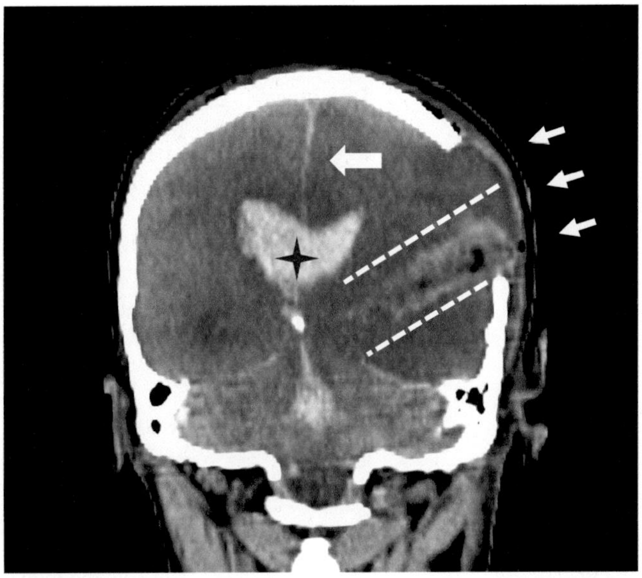

FIGURE 25.18 MSCT, coronal image of the head of a homicide victim who was stabbed to death. The wound channel (yellow lines) and blood within the ventricular system are clearly visible (black star). Notice the craniotomy (small green arrows) on the left side due to an increasing and ultimately lethal cerebral edema and a midline shift toward the right (green arrow).

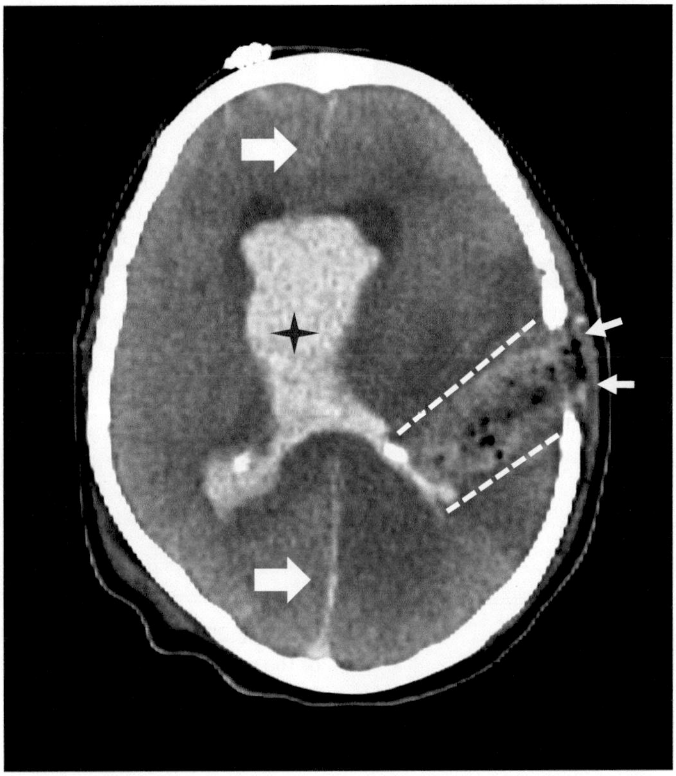

FIGURE 25.19 MSCT, axial image of the head of the homicide victim shown in Figure 25.18.

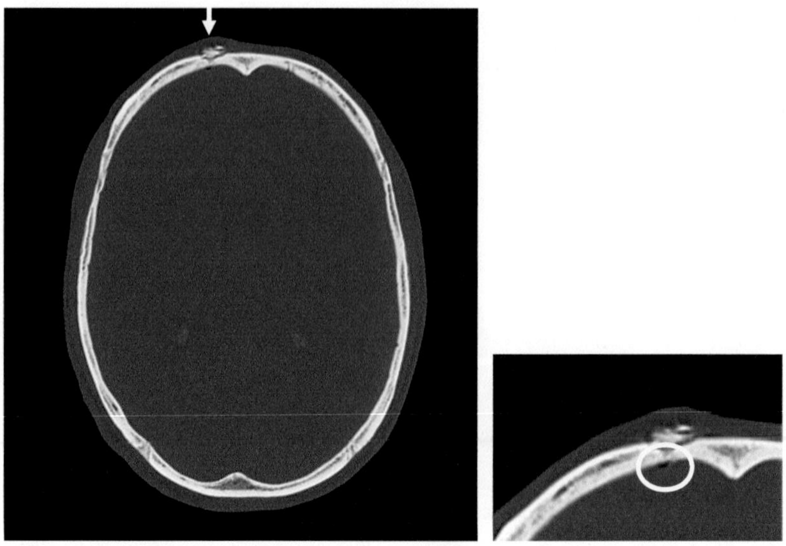

FIGURE 25.20 MSCT, axial image of the head of a young man who suffered a stab wound to the forehead. Note the defect of the cranial bone (red arrow) and the small air bubble (red circle) .

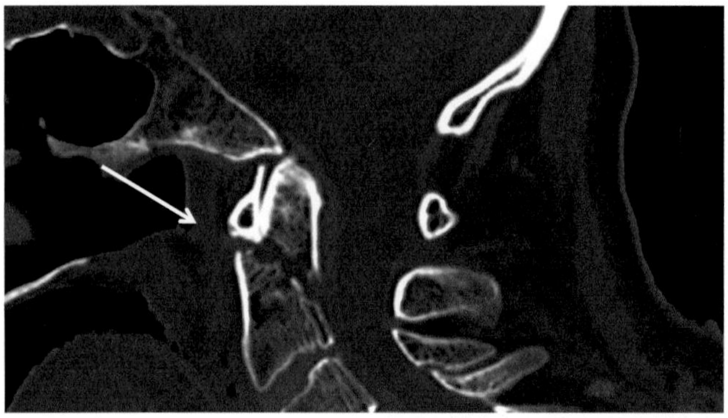

FIGURE 25.21 MSCT, sagittal image of the upper cervical spine of a 91-year-old man. The man collapsed at home and suffered an odontoid fracture. The image shows the slightly dislocated dens without relevant compression of the cervical spine.

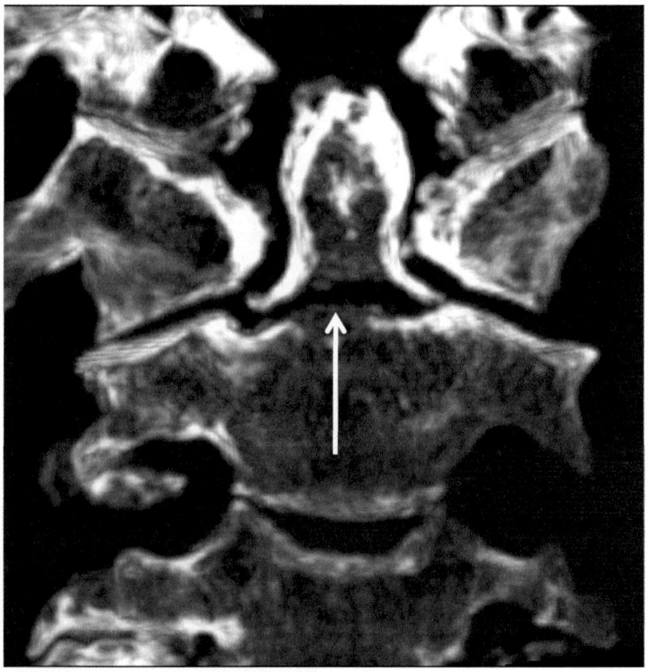

FIGURE 25.22 MSCT, coronal image of the upper cervical spine of the man shown in Figure 25.21.

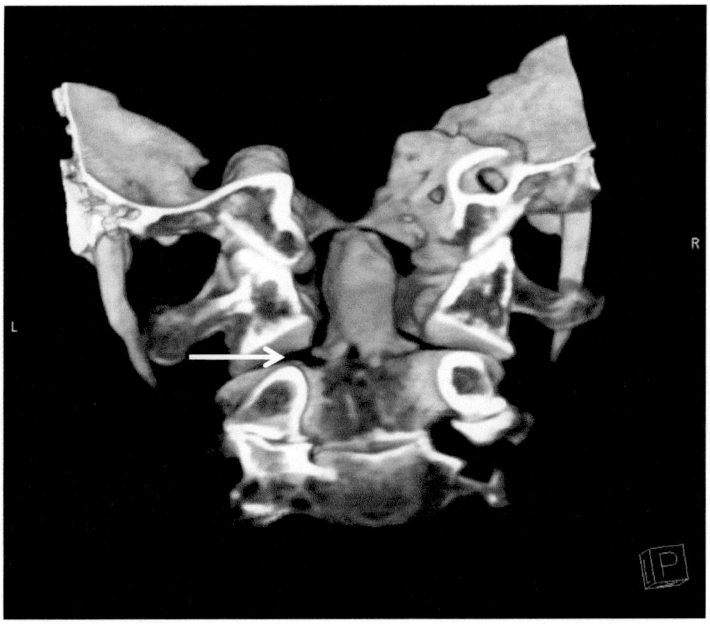

FIGURE 25.23 MSCT, 3D reconstruction and virtual dissection demonstrating the fracture seen in Figure 25.22 but with preservation of the 3D relationships of the bones (arrow).

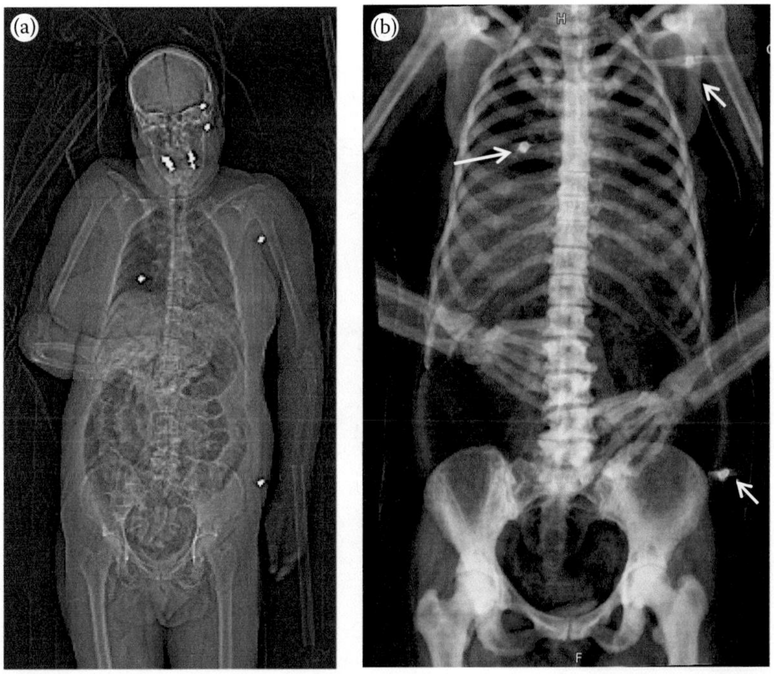

FIGURE 25.24 Topogram and MSCT reconstruction of a female homicide victim. (a) The topogram shows five bullets in the trunk (arrows) and the head. (b) In this MSCT reconstruction the position of three bullets in the trunk is displayed.

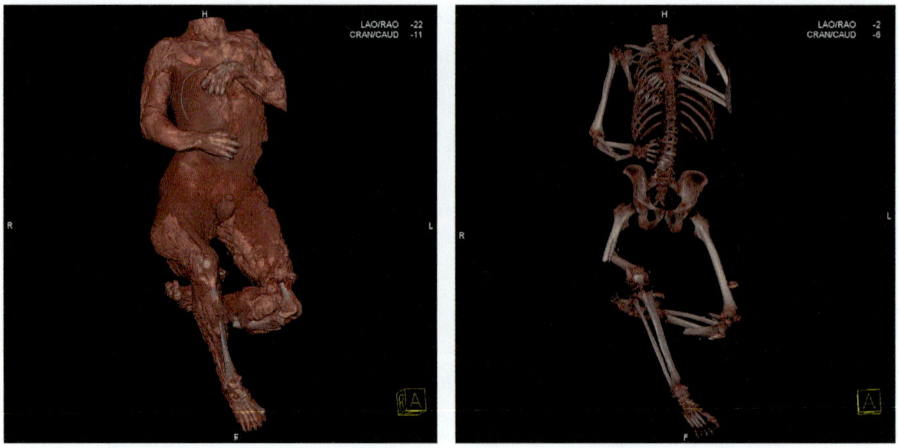

FIGURE 25.25 MSCT 3D reconstruction of a victim of an airplane crash. Volume renderings of a whole-body scan with different voxel opacites for soft-tissue and bone.

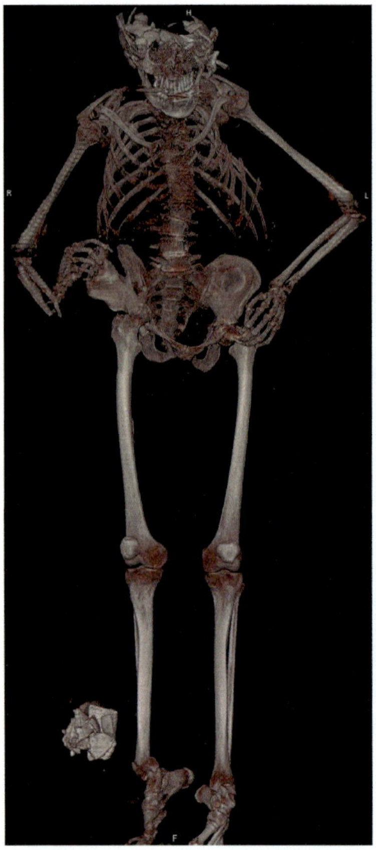

FIGURE 25.26 Whole-body MSCT reconstruction of the skeleton after massive blunt trauma. Parts of the destroyed skull are visible beside the right leg.

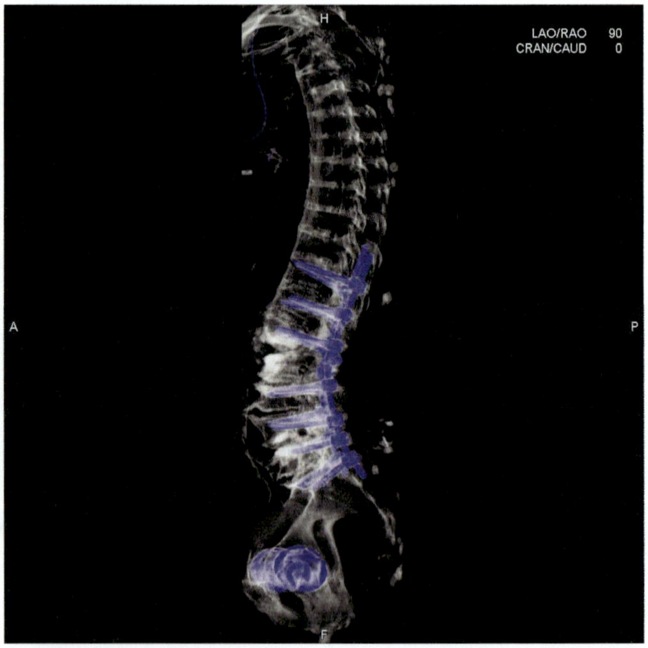

FIGURE 25.27 Incidental postmortem CT findings. Note the surgical plates and screws in the spine. Furthermore, there are two acetabular cups clearly visible.

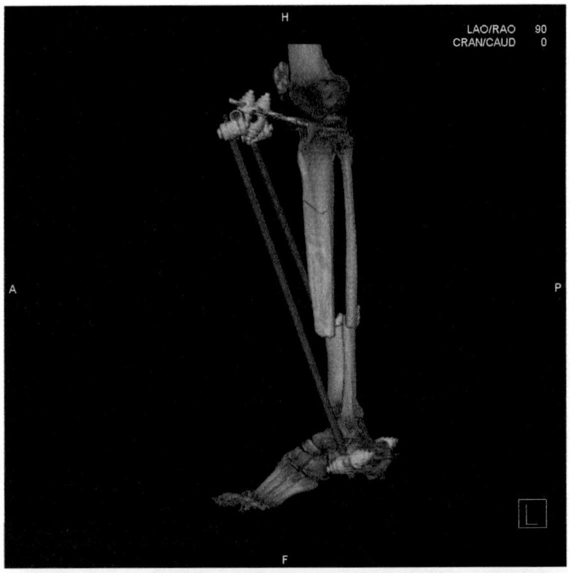

FIGURE 25.28 View from the left side: On this 3D volume rendering model of the left leg, an external fixation is displayed. Fractures of the tibia and the fibula were the result of being overrun by a tractor.

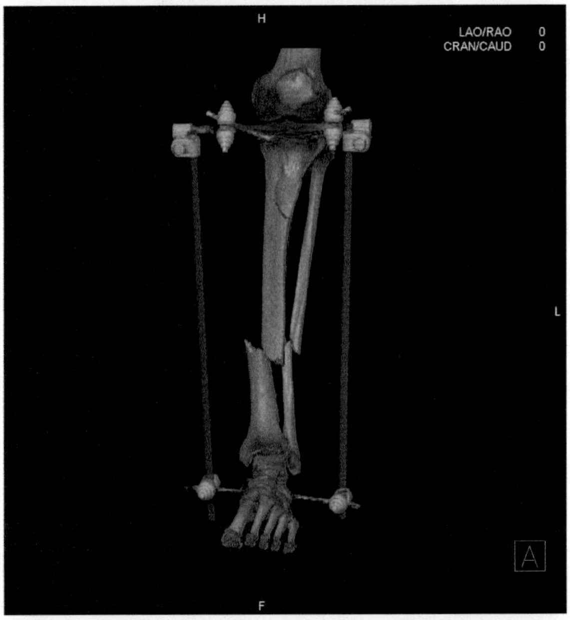

FIGURE 25.29 Anterior view of victim shown in Figure 25.28.

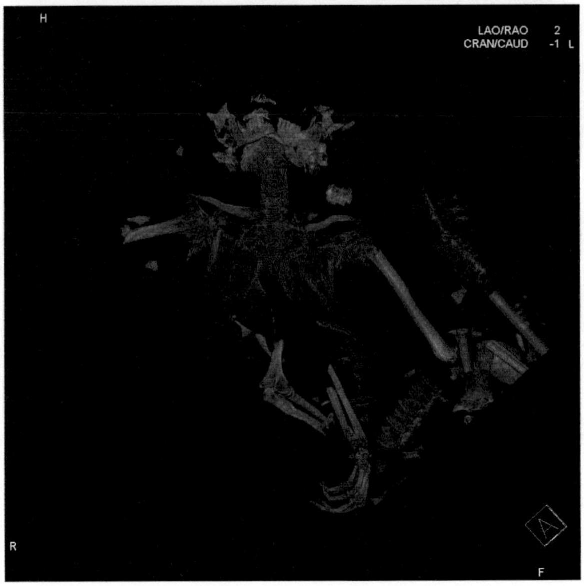

FIGURE 25.30 MSCT whole-body 3D reconstruction. Massive destruction of the skeleton of a person who was run over by a train in a suicide.

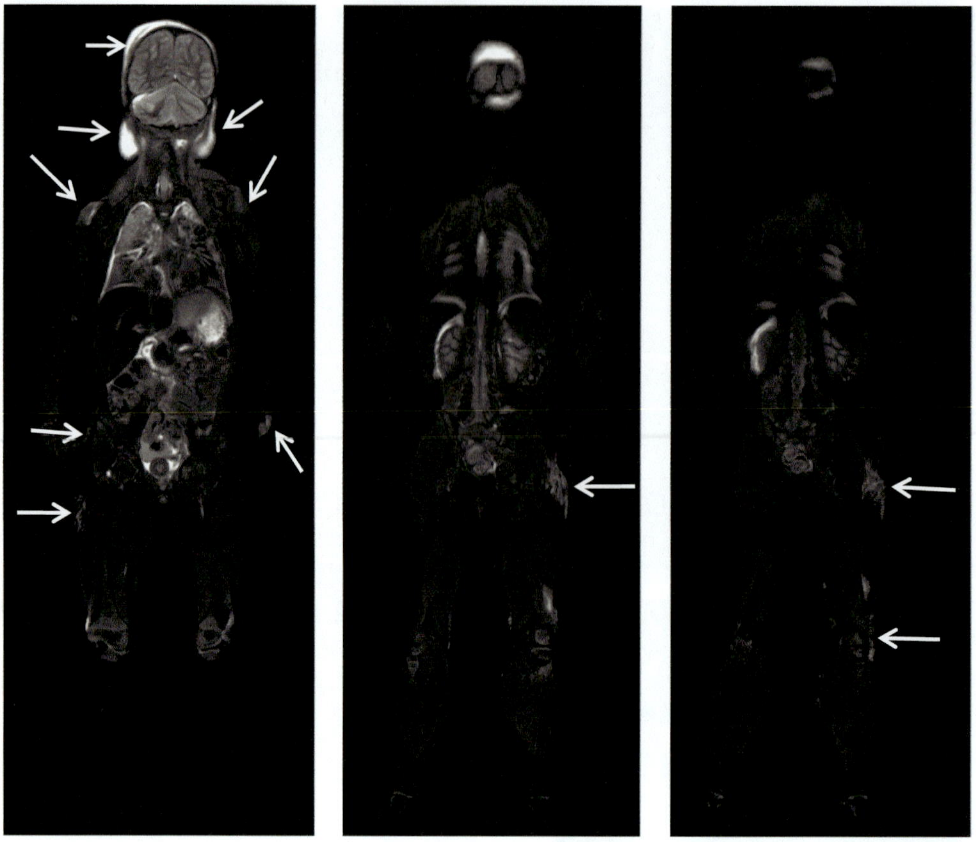

FIGURE 25.31 Total imaging matrix (TIM), T2-weighted MR images. Note the multiple hematomas in the soft tissue (red arrows).

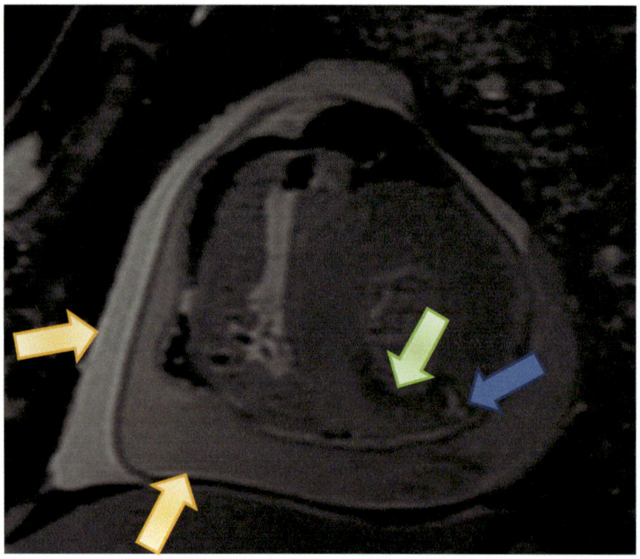

FIGURE 25.32 71-year-old male with myocardial infarction. Note hypointense tissue at the posterior wall of the left ventricle (green arrow). There is rupture of the ventrical wall (blue arrow) and extensive pericardial effusion (orange arrows).

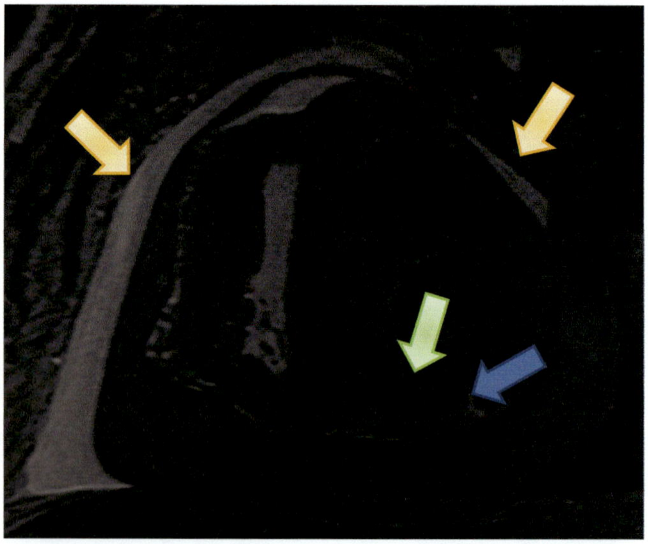

FIGURE 25.33 A different sequence in the same MR study as Figure 25.32 redemonstrates the effusion and infarct (orange and green arrows), but displays the rupture to better advantage (blue arrow).

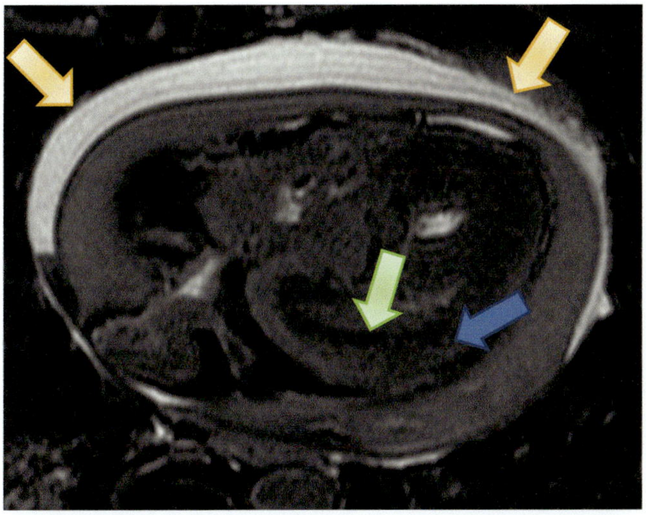

FIGURE 25.34 Axial MR image of the victim from Figures 25.32 and 25.33 displays the effusion (orange arrows) and infarct (green arrow) to better advantage, but the ventricular rupture is less well seen (blue arrow).

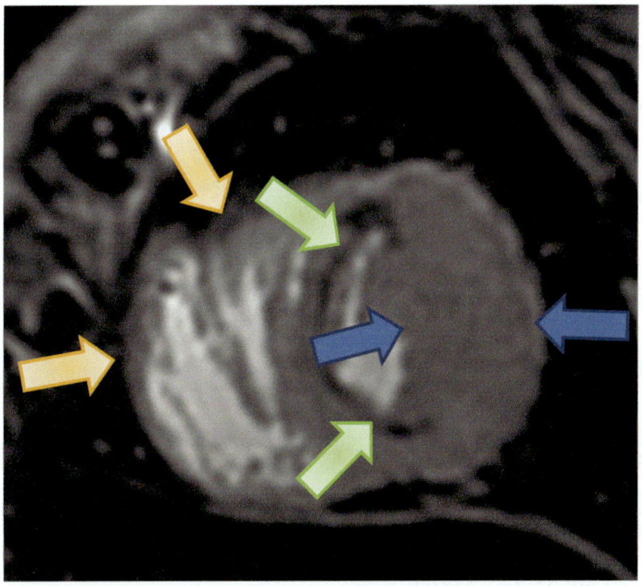

FIGURE 25.36 Short axis MR image of victim in Figure 25.35 with severe hypertrophy of the left ventricular wall (blue arrows) and thinning of right ventricular wall (orange arrows) with scarring tissue (black line, marked with green arrows) after old myocardial infarction.

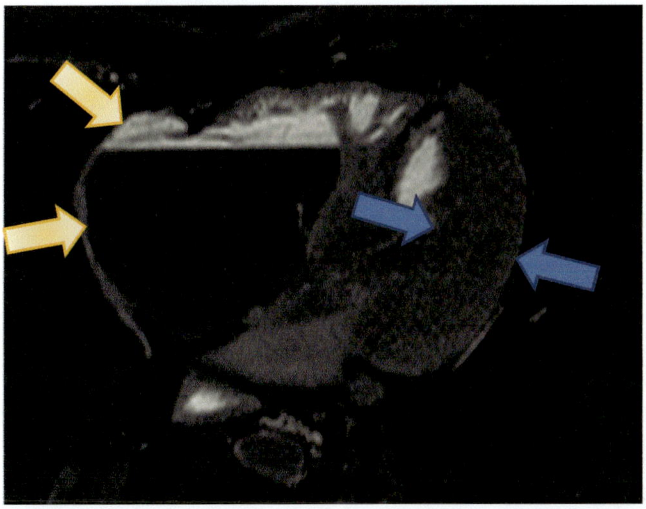

FIGURE 25.35 Axial MR image of a 47-year-old male with severe hypertrophy of the left ventricular wall (blue arrows) and thinning of right ventricular wall (orange arrows).

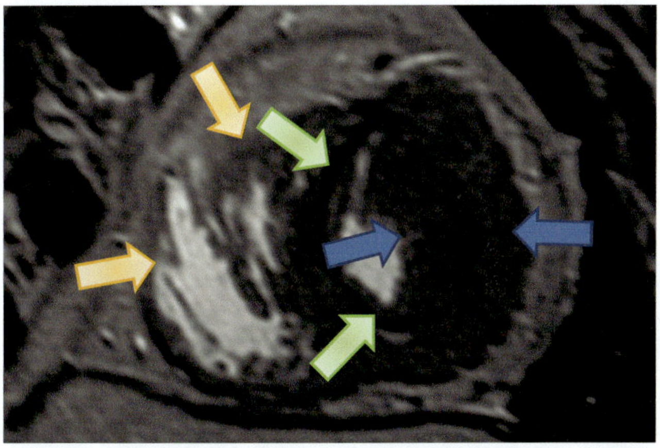

FIGURE 25.37 Same case continued: 47-year-old male with severe hypertrophy of the left ventricular wall (blue arrows) and additional short axis sequences from the case in Figures 25.35 and 25.36 display the ventricular thinning and thickening (orange and blue arrows respectively) equally well, but contrast of the scar is poor—illustrating the importance of sequence selection.

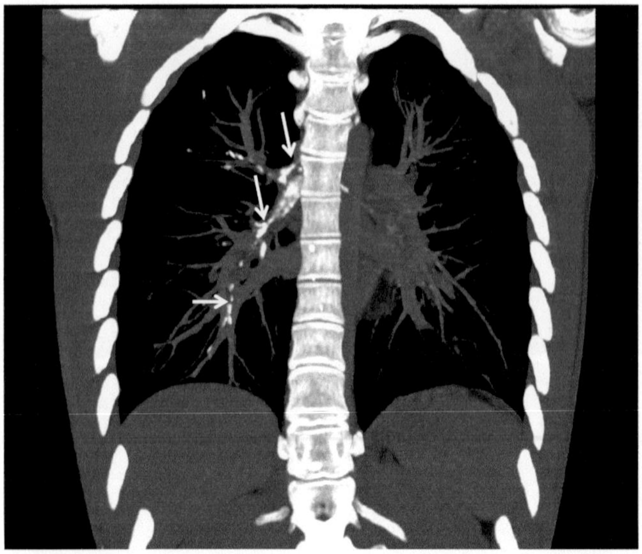

FIGURE 25.38 MSCT, coronal maximum intensity projection of a drowned man. The bronchial system is filled with water. On the right side, radiodense zones (red arrows) representing sand aspiration are clearly visible.

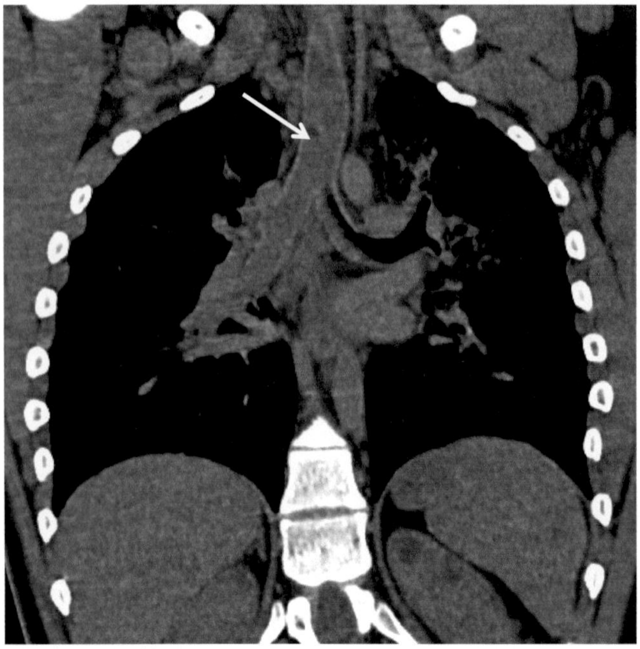

FIGURE 25.39 In this coronal MSCT image, fluid in the trachea of a drowned man is displayed (red arrow).

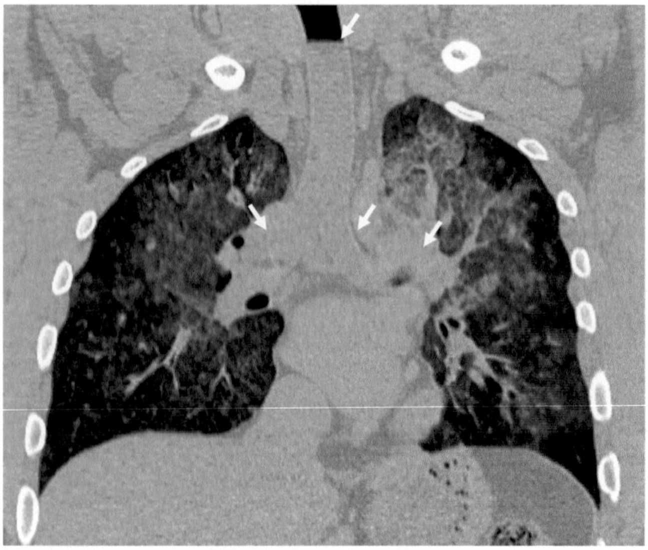

FIGURE 25.40 MSCT, coronal image of a drowned man. Note the fluid in the main and middle bronchi (yellow arrows) and the fluid level (red arrow) in the trachea. The lung lobes show a mosaic pattern with hyperdensities, compatible with aspiration.

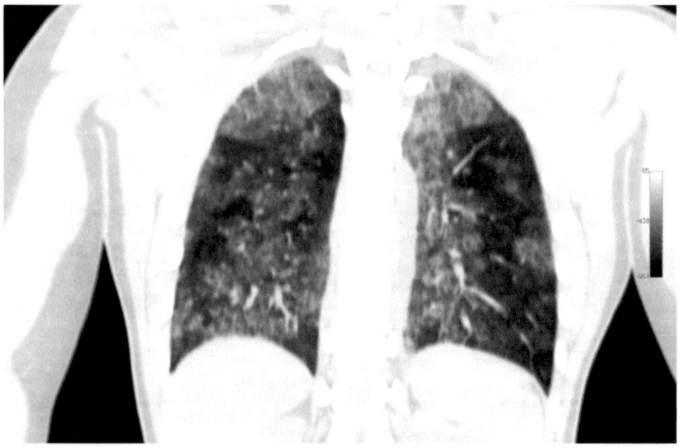

FIGURE 25.41 MSCT, coronal image of a drowned man. The lung lobes show a mosaic pattern with hyperdensities, compatible with water aspiration.

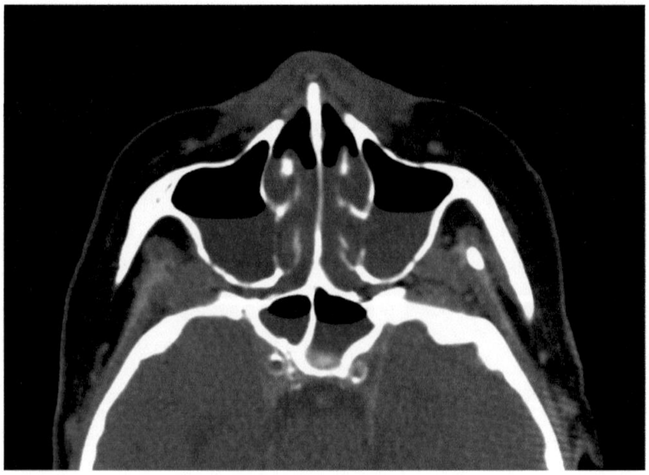

FIGURE 25.42 This axial MSCT image of the head of a drowned man shows fluid levels in the paranasal sinuses.

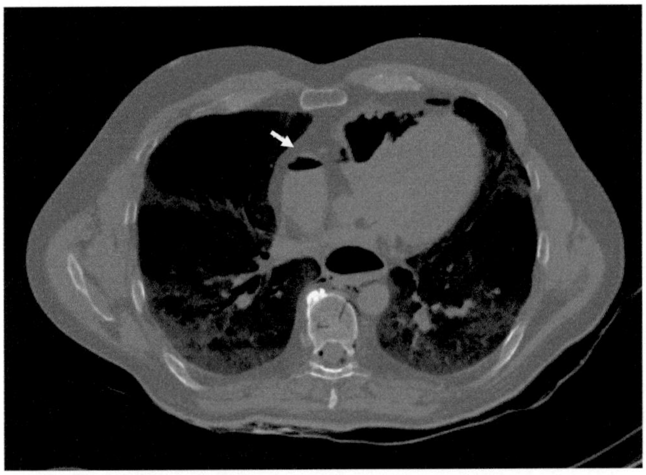

FIGURE 25.43 Axial reconstructed MSCT image showing a gas embolism with air in the right atrium and ventricle (red arrow).

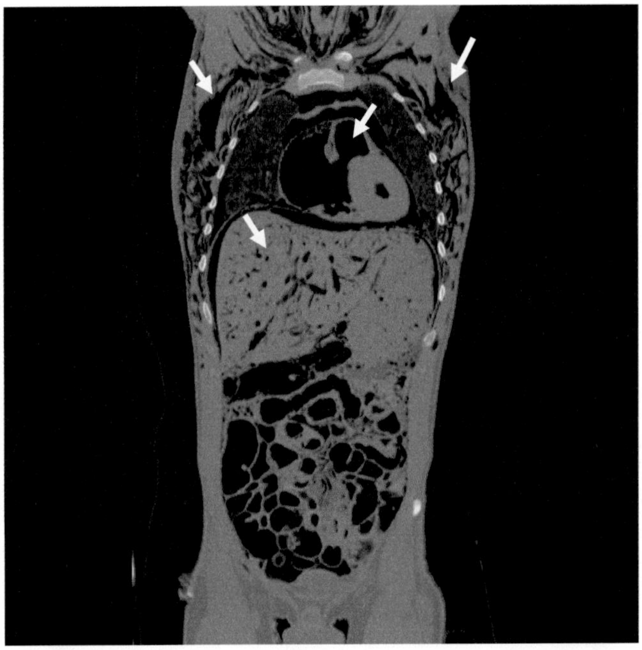

FIGURE 25.45 This coronal MSCT image shows the abdominal organs in a decomposing body. The organs can be readily located and assessed with regard to injuries. Note the accumulation of gas in the liver, in the heart, and the subcutaneous tissues (all indicated by red arrows).

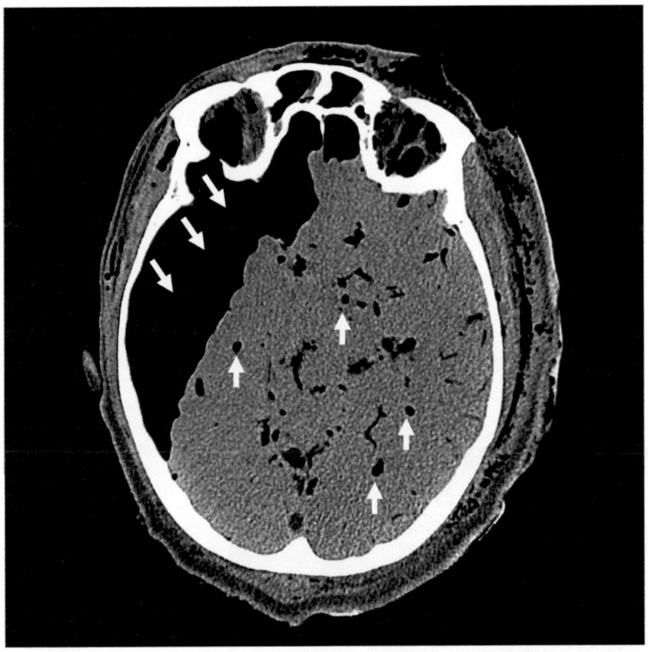

FIGURE 25.44 MSCT, axial image of a decomposed head. Note the small bubbles in the cerebral parenchyma and the free extracerebral gas.

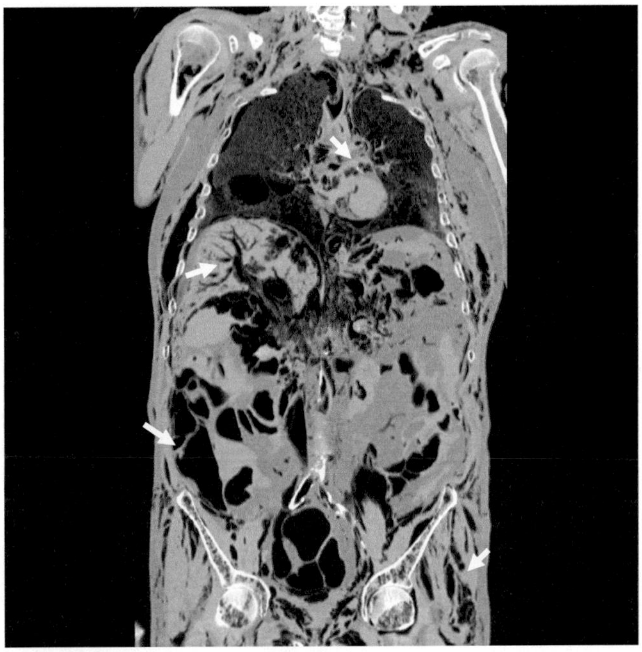

FIGURE 25.46 This coronal MSCT image shows the abdominal organs in a decomposing body. Note the accumulation of gas in the liver, in the heart, and in the gastrointestinal tract. Also visible is the feathery apperance of the musculature (red arrows). This corpse is in a moderate state of decay.

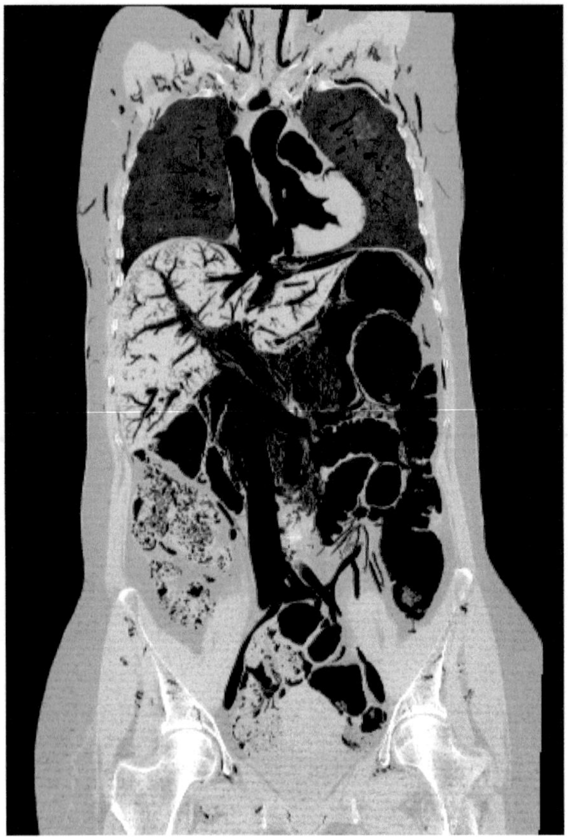

FIGURE 25.47 Coronal MSCT image demonstrates mild–moderate decomposition change.

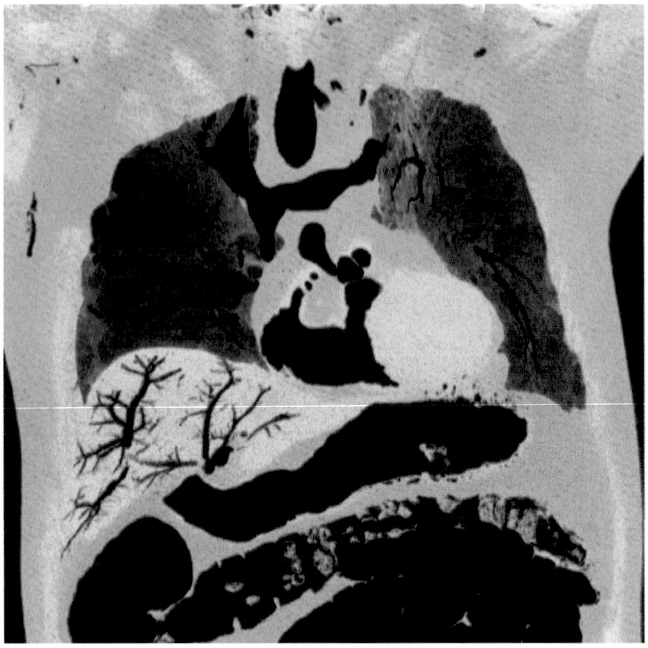

FIGURE 25.48 Another case of mild–moderate decomposition clues that the air collection are not related to embolism include the extensive degree of intrahepatic gas and the intramural gas in the stomach wall.

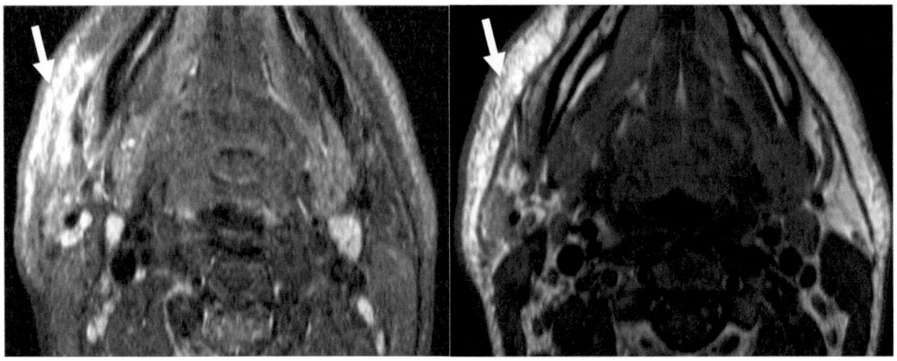

FIGURE 25.49 MRI axial images of the neck, TIRM, and T1-weighted, showing subcutaneous hemorrhage and edema of the soft tissue on the right side in a strangulation victim.

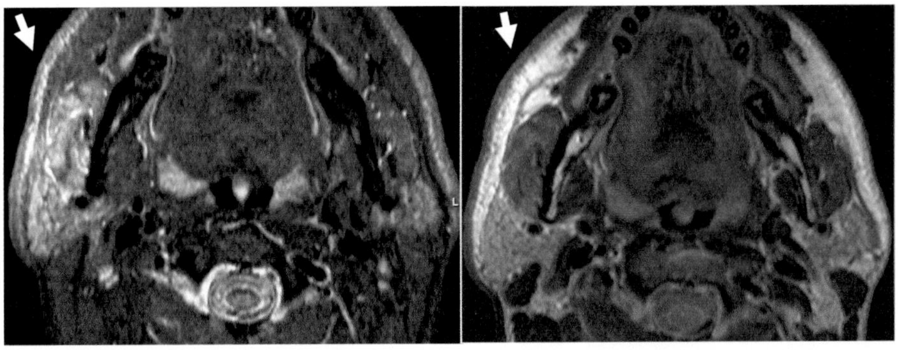

TIRM T2

FIGURE 25.50 MRI axial images of the neck, TIRM, and T2-weighted showing an intracutaneous hemorrhage and edema (red arrows) in a strangulation victim.

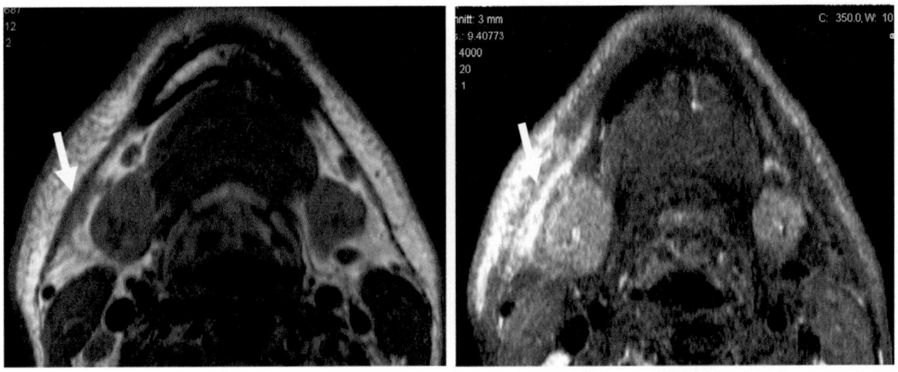

T1 TIRM

FIGURE 25.51 MRI axial images of the neck, T1-weighted, and TIRM in a strangulation victim. The platysma on the right side is thickened with blurred margins (red arrows).

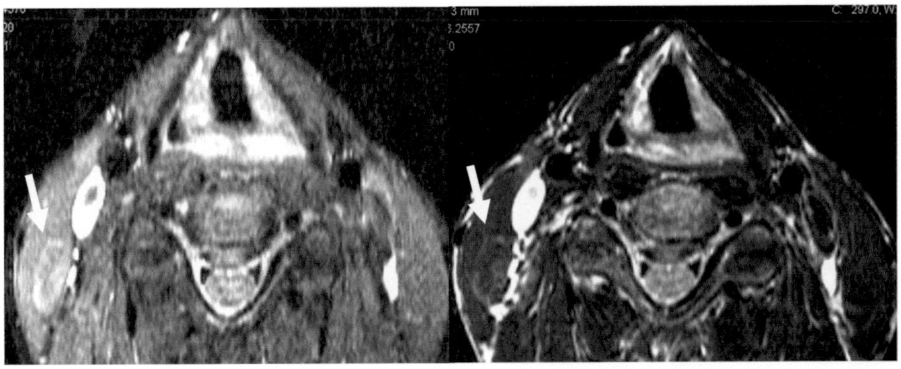

TIRM T2

FIGURE 25.52 TIRM and T2-weighted axial MRI images of the neck of a young man who had suffocated the previous day. Note the hyperintense region of the intramuscular hematoma on the right side (yellow arrows).

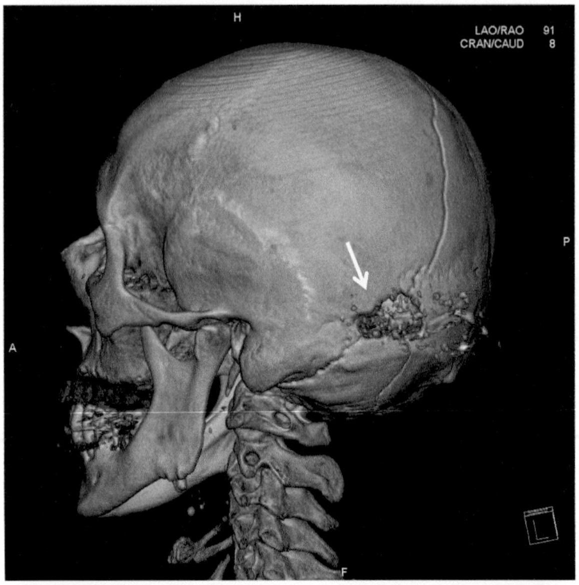

FIGURE 25.53 MSCT 3D reconstruction of the skull of a homicide victim. Note the bone defect at the occiput due to a graze shot.

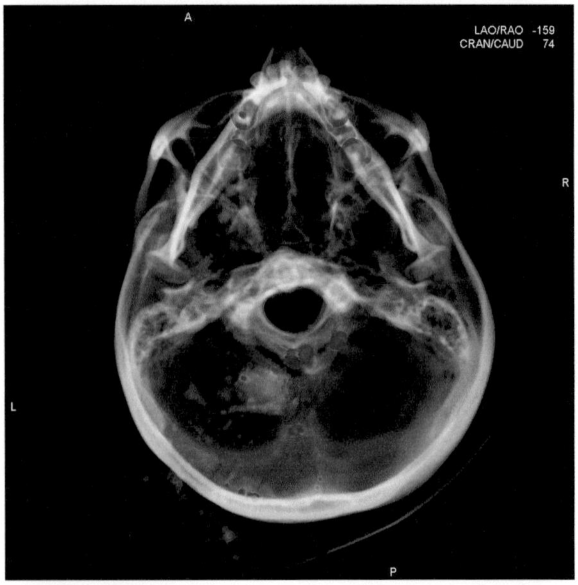

FIGURE 25.54 MSCT 3D reconstruction of the head of a homicide victim who was shot. The entrance wound of the bullet is located in the occiput. The metallic fragments of the bullet with high HU density are coloured blue and can be easily detected in the area of the entrance wound and inside the skull. Note also the blue staining of the dental work in the upper jaw.

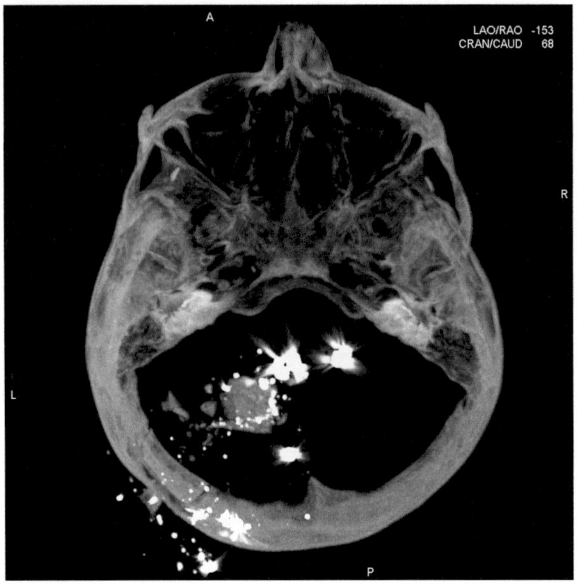

FIGURE 25.55 MSCT 3D reconstruction of the skull of a homicide victim. Note the bone fragments inside and the skull. The fragments of the projectile cause streak artifacts due to their metallic composition.

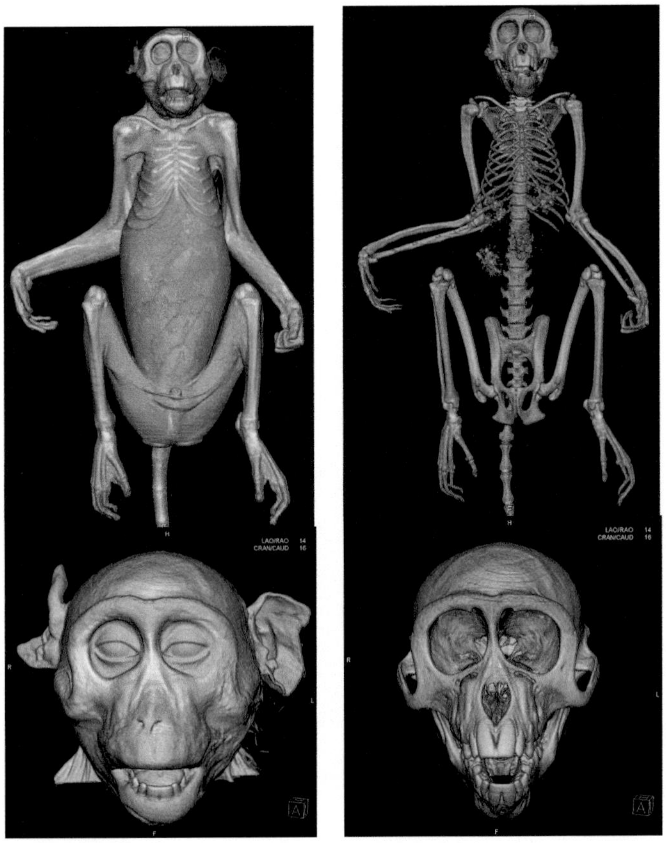

FIGURE 25.56 3D reconstruction of the soft tissue and the skeleton of a Javanese ape.

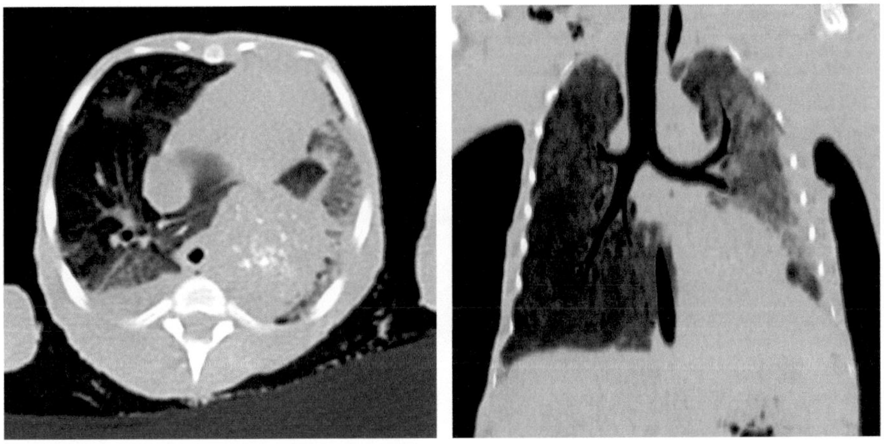

FIGURE 25.57 MSCT, axial and coronal reconstrucion of a Javanese ape. Note the tumor in the lower lobe of the lung and the occlusion of the bronchus caused by the tumor bulk.

REFERENCE

Thali, M. J., Dirnhofer, R., and Vock, P., 2009. *Virtopsy Approach: 3D Optical and Radiological Scanning and Reconstruction in Forensic Medicine*, CRC Press, Boca Raton, FL.

26 Clinical and Forensic Radiology Are Not the Same

Patricia M. Flach, Steffen G. Ross, Andreas Christe, and Michael J. Thali

CONTENTS

INTRODUCTION

Medicine today relies heavily upon radiological imaging in daily clinical practice. Accurate diagnostic imaging is fundamental for the effective treatment of several pathologies. Cross-sectional imaging has gained greatly in importance in recent decades, with the improvements in modalities such as computed tomography (CT) and magnetic resonance imaging (MRI). Moreover, cross-sectional imaging has revolutionized forensic medicine, not only for diagnostic purposes but also for documentation, as it is admissible in court and is well-suited to show complex anatomical relationships to laymen in a three-dimensional (3D) and bloodless manner. In contrast to autopsy, radiological imaging is nondestructive; it allows archiving of complete information from an examination and from additional objective evaluations at a later date.

With the ongoing adoption of cross-sectional imaging in forensic pathology, a formerly almost neglected subspecialty in radiology has become a hot topic: forensic radiology (Figure 26.1). Clinical radiologists tend to misconstrue findings in postmortem imaging. A clinical radiologist with no experience in postmortem imaging will swiftly stumble upon multiple pitfalls and learn that postmortem imaging and clinical radiology are not the same. The examiner must take into account vast postmortem changes due to gravity, autolysis with breakdown of barriers, edema and blood extrusion, and coagulation or gas formation.[1,2]

At the inception of postmortem cross-sectional imaging,[2] the appearance of the organs of a corpse, with the sedimentation effect of the blood, gas formation caused by decomposition, altered body temperature, and especially the different forensic approaches to diagnosis, was not well understood. Furthermore, clinical radiologists were much more accustomed to contrast-enhanced imaging methods than to unenhanced examination like that mainly used in postmortem imaging. Although postmortem computed tomography angiography (pmCTA) offers very good results in depicting vascular pathologies or areas that are hard to dissect manually (see Chapter 28, *Postmortem Angiography*), this technique has not yet been widely adopted by the forensic community.[3–5] Thus, the following text will focus on the differences between clinical radiology and noncontrast postmortem imaging. Basic postmortem changes that occur in cadavers cause image findings that mimic well-known pathologies in clinical imaging but are perfectly normal in postmortem imaging. Over the past few years, as forensic radiologists have gained more experience, several postmortem radiological signs have been established.

Diverse Settings of Forensic Imaging Compared to Clinical Radiology

Compared to imaging in living patients, there are, of course, no movements or respiration artifacts or adverse reactions to contrast media in forensic imaging (Table 26.1). In most cases, much smaller scanning volumes are used in living patients due to radiation hygiene (Figure 26.2). Living patients require indication-adapted dynamic imaging, such as additional nephrographic or renal excretion phases. In forensic imaging, the appearance of pmCTA resembles that of antemortem CTA, except that there is no dynamic contrast enhancement. Oral or rectal filling with contrast is not feasible postmortem due to absent peristalsis, resulting in the accumulation of contrast medium in the stomach with only passive fluid transfer in the duodenum (position-dependent) and rectal outflow of contrast because of the absent anal sphincter tone. Rectal contrast medium

TABLE 26.1

Clinical	Forensic
Motion artifacts (heart, lung, intestines)	No motion-related artifacts
Small scan volumes due to radiation protection	Whole-body imaging with no need for low radiation dosage
Enteral contrast agent	Not feasible
Dynamic contrast phases	Unenhanced imaging (arterial and venous injection in pmCTA)

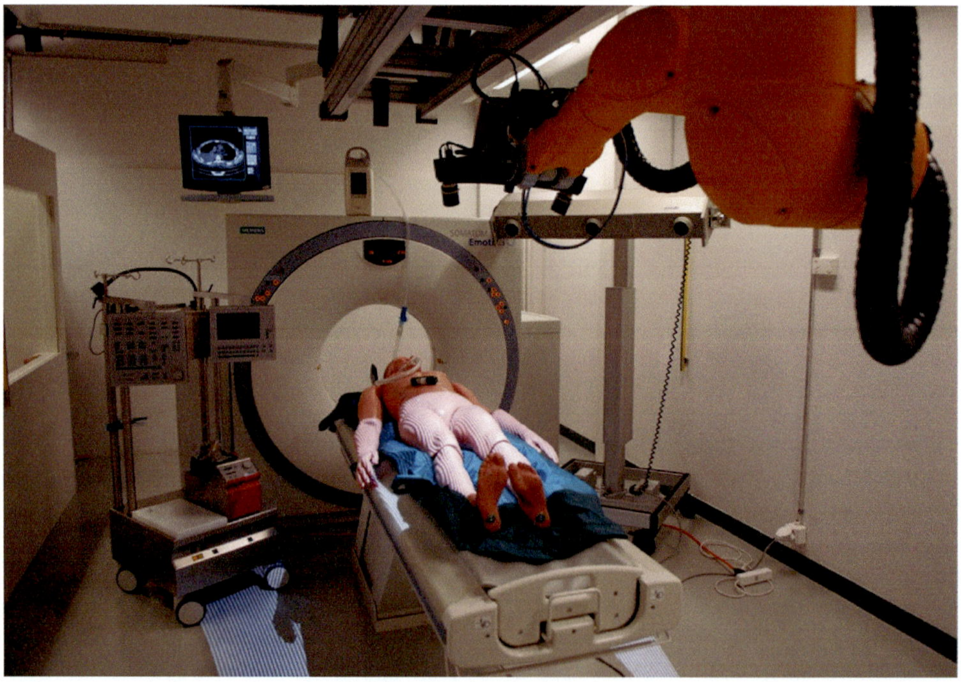

FIGURE 26.1 Forensic imaging unit: 6-slice CT with supplementary equipment, such as the modified heart–lung machine on the left (for CTA), the orange robot-guided biopsy device, surface scanning on the right, and a fluoroscopy monitor for image-guided biopsy.

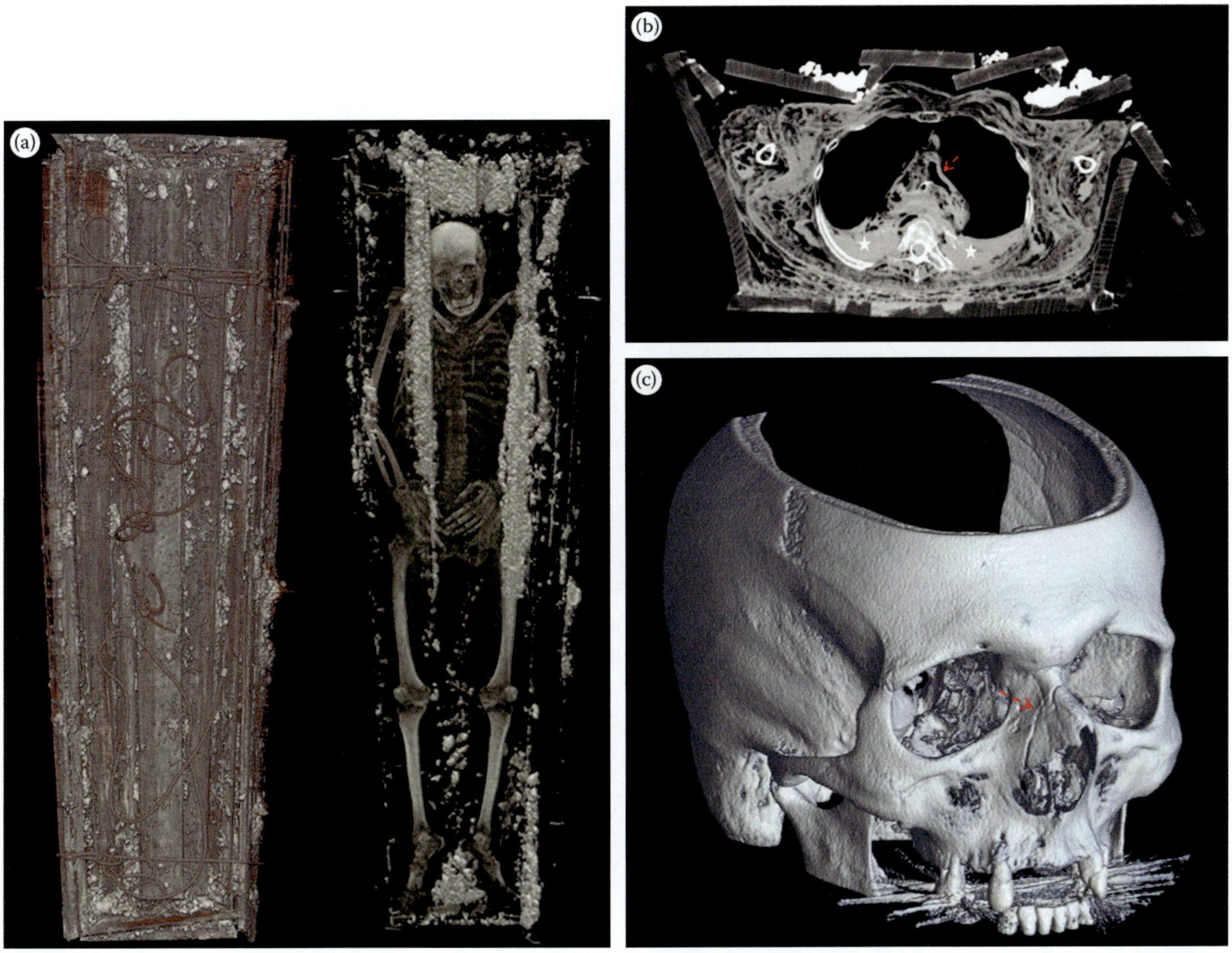

FIGURE 26.2 (a) Extensive scanning volume: a wooden coffin after being exhumed (five months after burial). The corpse was virtually preserved *in situ* in the casket. Additional subsequent scans of the corpse were performed after exhumation. (b) Axial plane through the thorax with vast symmetric putrefaction of the soft tissue and decomposition of the inner organs, such as the heart (red arrow) and both lungs (white asterisks). Note the dense soil on top of the coffin. (c) A clinical scan of the paranasal sinus depicting a fracture of the nasal bone (red arrow). Obviously, a much smaller scanning volume with a much lower dose of radiation was used.

can be applied through a large bore lockable catheter such as an endotracheal tube. The examiner should focus on positioning the deceased in a manner comparable to clinical scans as this facilitates later reading and reconstruction. Limiting factors are rigor mortis; frozen, charred, or vastly destructed bodies due to trauma (Figure 26.3); or decomposition. Rigor mortis of the arms should be broken in order to elevate the upper extremities to avoid streak artifacts, especially in the upper abdomen caused by the elbows (Figure 26.4). The investigator must keep in mind that raising the arms will result in a position change of the shoulder girdle, possibly causing pathologies to slip out of the scan range (Figure 26.5). Therefore, in cases of suspected trauma of the upper extremities, an additional scan with lowered arms should be performed, supplemented by an extended field of view. The thorax and abdomen should be scanned as a whole, as it is cumbersome to read images if the thorax and abdomen are "cut" into two different parts. A typical whole body scan is shown in Figure 26.6. A quick look at the scout view is an easy way to rule out removable, potentially artifact-causing objects, such as stones or other foreign objects in the body bag. According to forensic guidelines, medical foreign material, such as tracheal tubes or vascular catheters, should not be removed prior to the scan in order to document potential malpractice issues (Figures 26.7 and 26.8). In addition, especially in cases that will undergo pmCTA, samples of peripheral blood and urine have to be collected. Image-guided biopsy of lymph nodes, small nodules in the lung or other regions of interest can be performed after diagnostic imaging (see Chapter 27, *Postmortem Biopsy*) to obtain minimally invasive histology. In order to ease the work of the prosector or pathologist and to avoid false-positive findings, puncture sites must be marked.

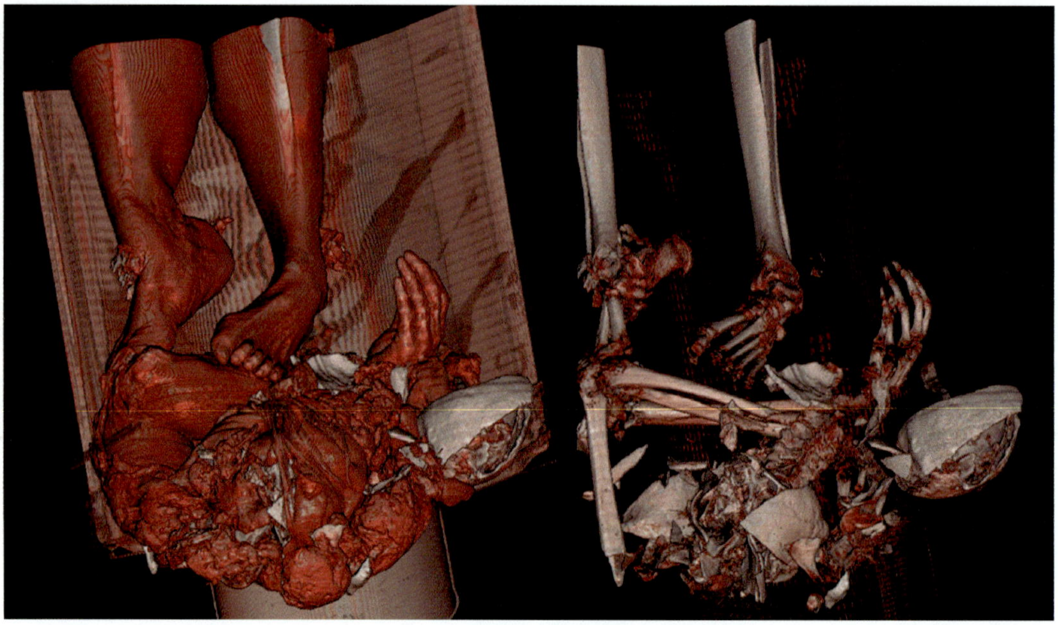

FIGURE 26.3 Included in the scanning volume of the lower extremities are the collected body parts of a suicide victim who was run over by a train. The corpse was decapitated while lying between the wheel–rail system, resulting in a destroyed shoulder girdle, amputation of the right arm, and destruction of the cervical spine and the skull.

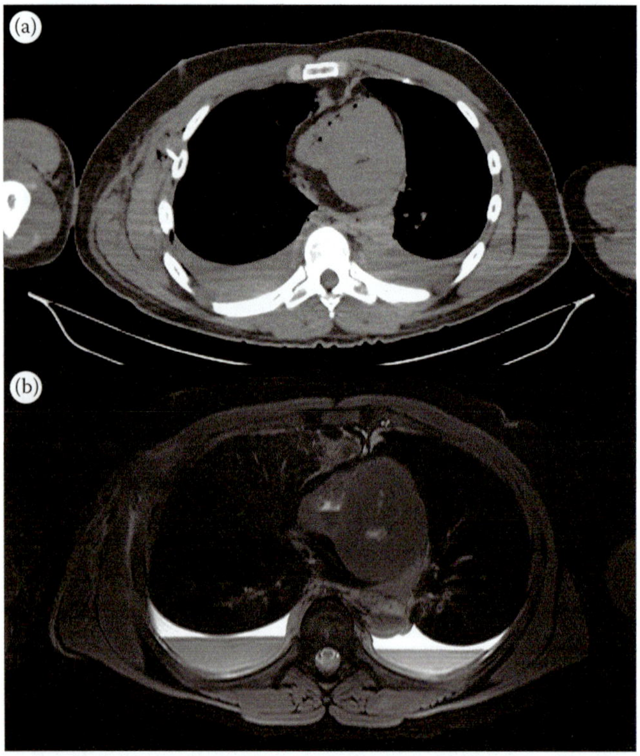

FIGURE 26.4 (a) Aortic tear with bilateral hemothorax. Note the artifacts caused by lowered arms. Missing fat plane between aorta and mediastinal hemorrhage in unenhanced pmCT is suggestive of aortic rupture. (b) Better soft-tissue contrast in this T2w image.

Technical Settings in Forensic Imaging

In routine postmortem imaging (within the Virtopsy project in Bern), the deceased body will undergo a classic forensic autopsy combined with external inspection, postmortem multislice computed tomography (pmCT) and eventually magnetic resonance imaging (MRI) of the whole body without contrast media. If indicated, the corpse will undergo pmCTA as well (see Chapter 28, *Postmortem Angiography*). Currently, a sufficient tool for pmCT is a 16-row multidetector CT, especially with a large bore, but even a 4- or 6-row multidetector CT will work, with major concessions. Whole-body data should be acquired with a slice thickness of less than 3 mm in axial sections, and sagittal and coronal reformations should be calculated. Selected regions (e.g., the larynx, small implants, or coronaries) should be additionally scanned at a sub-millimeter slice thickness. Image reconstruction in soft tissue and bone-weighted kernel is necessary. Kilovolts and mAs (milliampere-seconds) are only limited by the technical parameters of the utilized CT scanner. Data analysis (preferably by a radiologist experienced in forensic imaging) and postprocessing, such as 3D reconstructions (e.g., for court) of CT and MR imaging data, can be carried out on any CT workstation that can process large image stacks (e.g., Leonardo, Siemens Medical Solution or OsiriX Imaging software). For intra- and interindividual comparison, a picture archiving and communication system (PACS) workstation should be used.

If additionally needed, MRI is currently best performed using a whole-body imaging MRI unit (TIM, Total Imaging Matrix) with a wide inner bore diameter. Coronal, sagittal, and axial images with different signal weightings are acquired

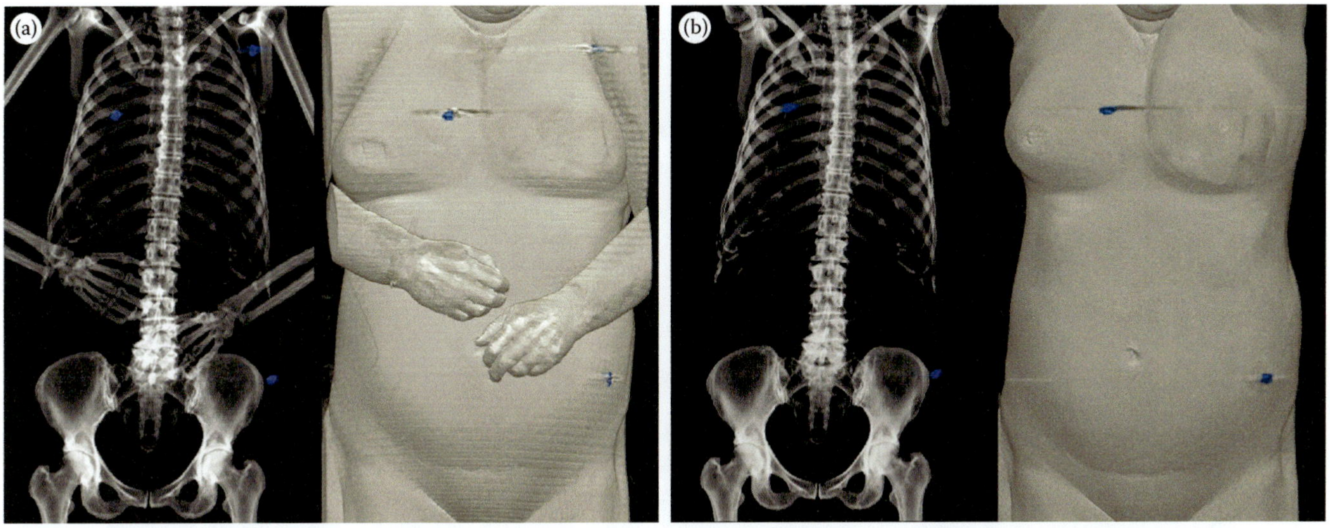

FIGURE 26.5 (a) A homicide victim with three projectiles (indicated in blue) in the torso. A first scan with lowered arms and an extended field of view (resulting in poorer image quality) is shown. (b) A scan of the same victim with raised arms does not show the projectile in the left axillary region due to the position change of the upper extremities. The body was not moved between the two scans.

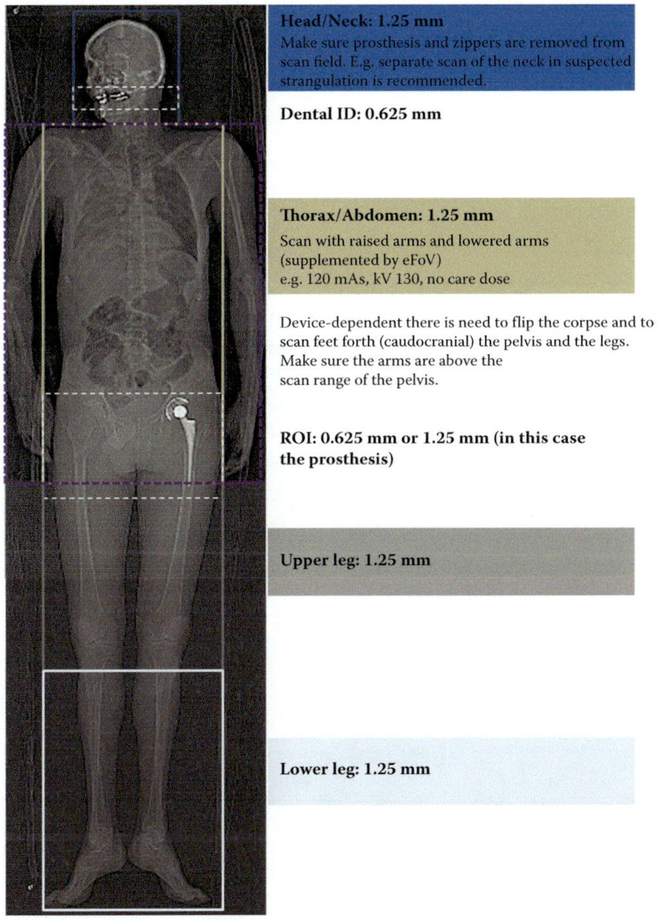

FIGURE 26.6 An example of how to plan and perform a postmortem whole-body scan. Both a soft and a hard kernel must always be reconstructed (e.g., 30 s and 70 s).

(T1-weighted spin–echo and T2-weighted fast spin–echo sequences, with and without fat saturation; turbo inversion recovery sequences; and gradient-echo sequences adapted to indication).

pmCT can be performed at any stage of decomposition, with the tradeoff that findings of the parenchymal organs will be increasingly superimposed in cases with advanced putrefaction. Identification of a person by means of dental ID or other characteristics, such as metallic implants or old fractures, is still possible, even in a strongly decomposed cadaver (Figures 26.9 through 26.11). The mean duration of a pmCT depends strongly on the cooling capacity of the x-ray tube of the CT scanner. Scan times for a whole body pmCT will range between less than a minute with a state-of-the-art 64-slice CT to almost half-an-hour with an outdated 4-slice CT. MRI-acquisition times also depend on the type of tomograph and the desired anatomical coverage and sequences. An experienced radiographer or health physicist is needed to adjust the default sequences to the requirements of postmortem imaging.

If feasible, the radiologist responsible should attend the complete autopsy; first, to add to the knowledge gained by imaging, and second, to obtain a better correlation between the radiological findings and the reality seen at autopsy. Only a direct correlation between radiological imaging and autopsy, including histology, allows for a proper learning curve for the forensic radiologist.

General Postmortem Changes

After death, the body's tissue starts to change its structure: a breakdown of cellular barriers and evolving tissue edema change the appearance of different tissues in postmortem imaging. This process is followed by early putrefaction, usually starting with gas accumulation in the portal venous and

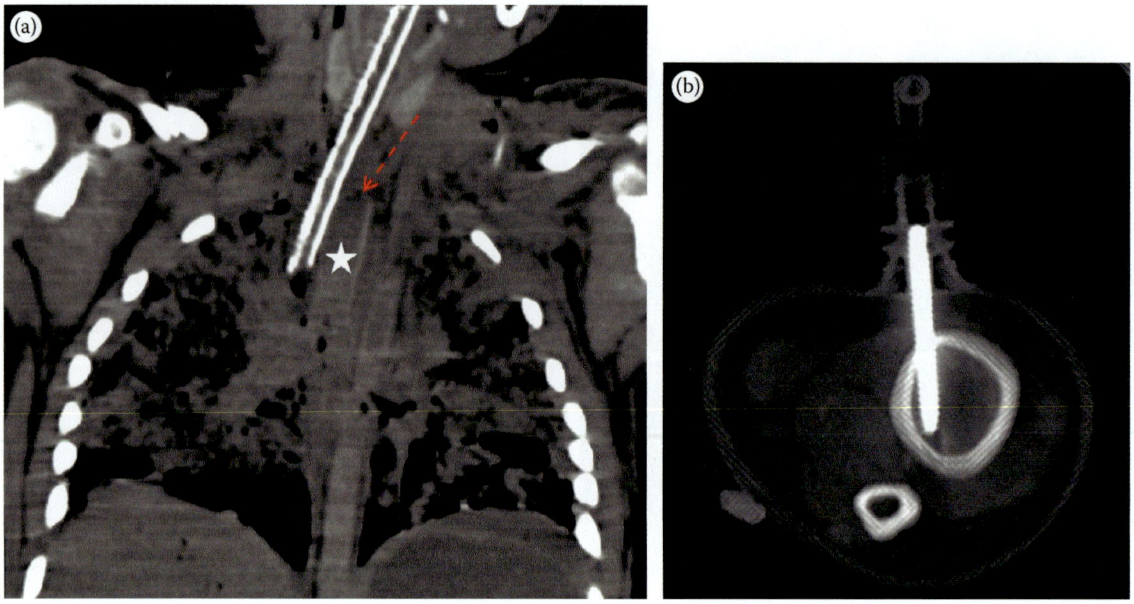

FIGURE 26.7 (a) Complete rupture of the trachea with massive aspiration of blood. The distal part of the endotracheal tube (inserted during resuscitation efforts) is ineffective due to tracheal rupture (red arrow) and is placed paramediastinally. The distal trachea is blood-filled (white asterisk), and the lung tissue shows typical signs of extensive blood aspiration. (b) Intraosseous access device into the bone marrow of the tibia showing correct placement.

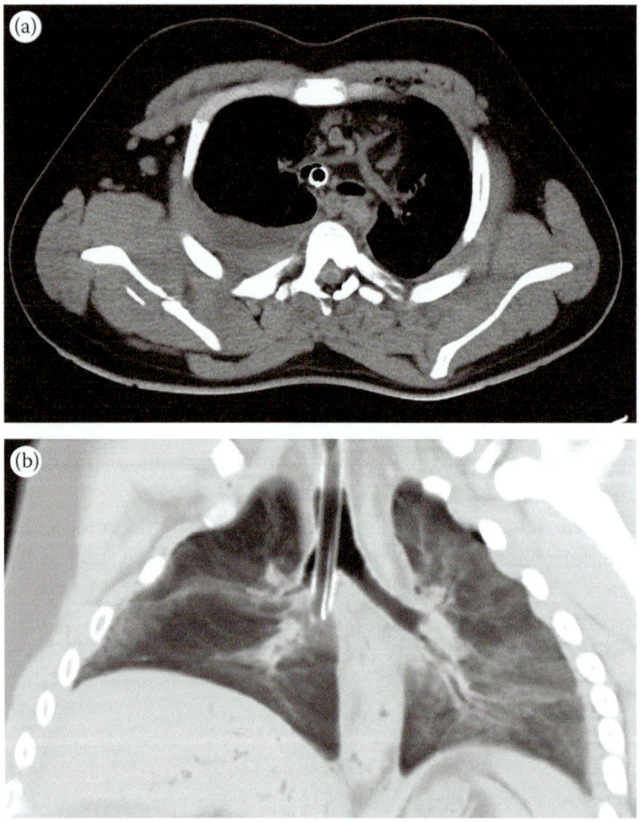

FIGURE 26.8 (a) "Vanishing aorta" and collapsed pulmonary arteries with right-sided hemothorax and massive osseous trauma. Note the endotracheal tube in the right main bronchus in a case of fatal exsanguination. (b) Coronal reconstruction with depiction of the tip of the tube in the bronchus of the right lower lobe.

hepatobiliary systems caused by the very fast reproduction of gas-producing bacteria in the gastrointestinal tract of the corpse. Due to the high concentration of digestive enzymes, the pancreas is also affected by early decomposition. Proper differentiation between air embolism (Figure 26.12) as a vital reaction and gas accumulation due to decomposition is not easy for a radiologist who is not aware of the postmortem changes caused by decomposition (Figure 26.13).[6,7] The

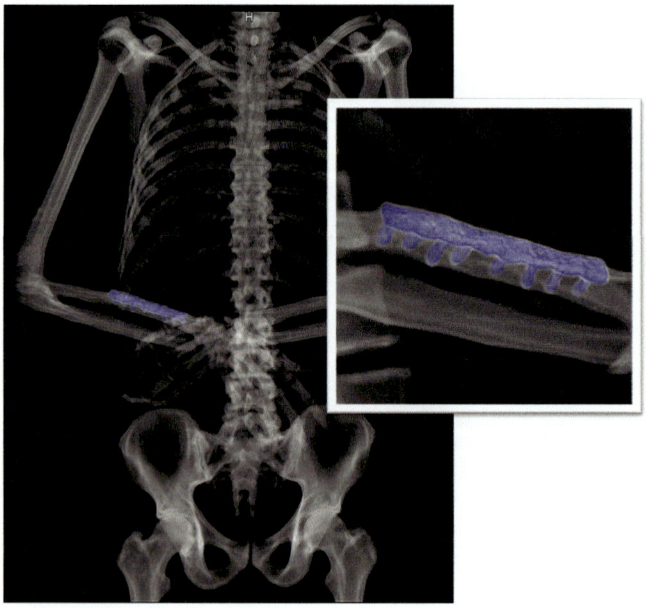

FIGURE 26.9 Metal implants such as this fixation device might help to identify the deceased.

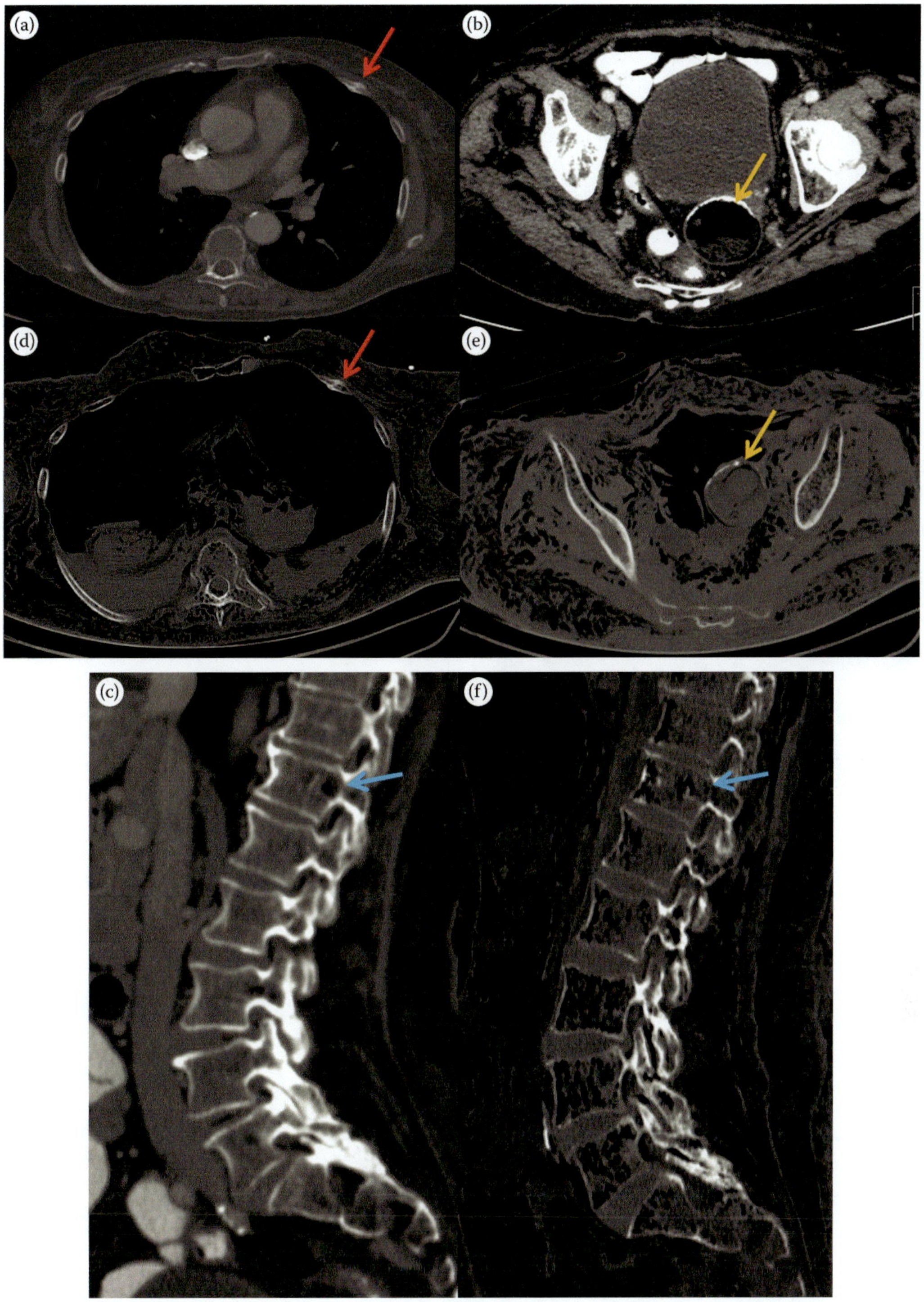

FIGURE 26.10 (a–c), Antemortem images; (d–f), pmCT. (a) Axial plane of an antemortem scan clearly depicting an older rib fracture (red arrow) with callus, corresponding to the identical fracture on pmCT. (b) Axial plane of the pelvis showing a tumorous marginal calcified mass in the antemortem scan. This patient had a known teratoma (yellow arrow) that is still well preserved postmortem due to calcification. (c) Increased trabeculation of a vertebral hemangioma (blue arrow) in the 12th thoracic vertebra.

distribution of gas caused by decomposition involves the whole body and is dependent on corpse position and gravity, in contrast to air embolism, which appears only intravascularly (Figure 26.14).[8]

In order to evaluate postmortem images properly, the investigator must keep in mind that observation of certain postmortem findings is position-dependent; for example, relapse of the brain in putrefaction, air-fluid level in drowning/hematosinus, or internal livor mortis of the lung. Hence, one might be able to speculate on the predominant postmortem position of the corpse. This is only possible with detailed knowledge of the circumstances of the case. The investigator

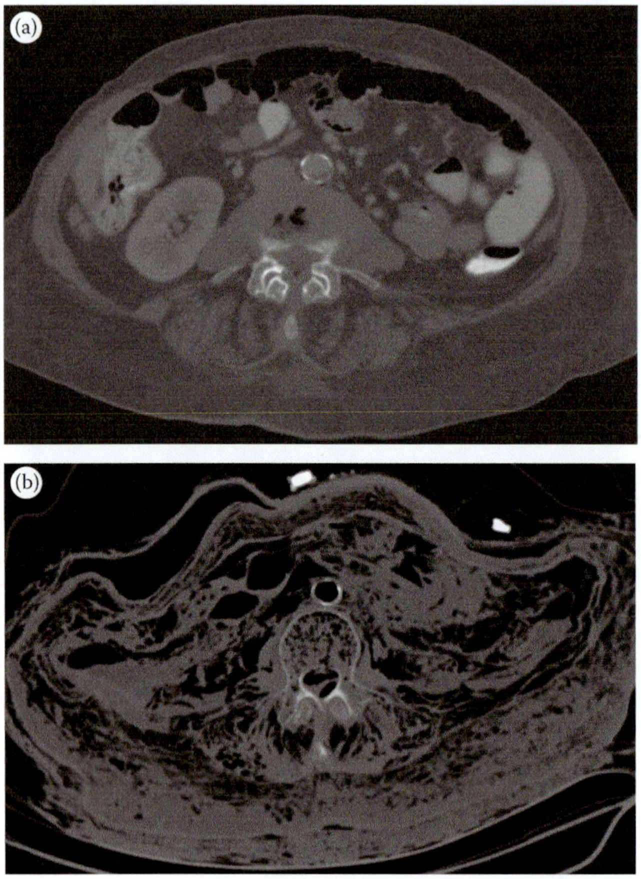

FIGURE 26.11 (a) Antemortem CT. (b) pmCT. Even at late stages of decomposition, arteriosclerosis prevents vessel collapse and might be helpful for identification.

also must keep in mind that injuries inflicted to the lung with hemo- or pneumothorax will result in consecutive dislocation or even collapse of the parenchyma. The position of the body during the CT or MRI scan in most cases is not the same as that at the moment when the injury was inflicted. The examiner must keep this in mind during virtual reconstruction of wound channels.

IMAGING OF THE HEAD

Loss of Cortico-Medullary Differentiation

The loss of cortico-medullary differentiation appears immediately after death and increases in severity with time, but typical signs of brain edema do not necessarily show up. For the brain, the act of dying is pathophysiologically like suffering a stroke: postmortem hypoxia leads to edema formation, which obscures the border between white and grey matter. The difference in attenuation or intensity between the cortex and medulla of the brain is less obvious on CT than it is on MRI due to the better soft tissue contrast of the latter. Therefore, the loss of differentiation is visible earlier on CT (Figure 26.15). A loss of cortico-medullary differentiation is also seen in living patients, but it is not compatible with regular brain function and is usually seen in brain death (Figure 26.16). In the course of decay, putrefaction will increase. As decomposition proceeds, relapse and liquefaction of brain tissue will take place, with limited diagnostic possibilities. With CT or MRI, however, an *in situ* depiction of the brain with its potential pathologies is feasible, even at a late stage of decomposition, unlike autopsy, where the

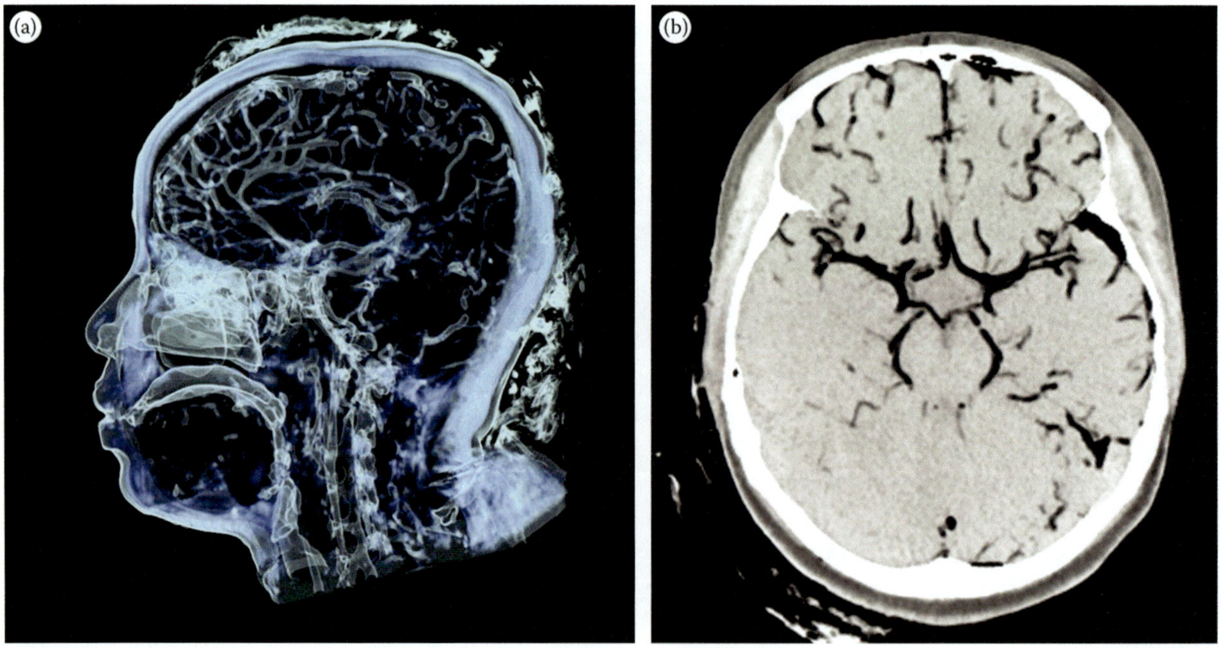

FIGURE 26.12 (a) Volume reconstruction of air-filled structures clearly depicting intravascular air due to a deadly air embolism. (b) Axial minimum intensity projection (MinIP) of the Circle of Willis showing only intravascular air (position-dependent).

brain would flow out (Figure 26.17) after opening the skull. Obtaining findings by autopsy at this stage of decomposition and consecutive dislocation is thus extremely difficult.

Brain Swelling

Postmortem hypoxic edema formation results in brain swelling. The increasing size of the brain narrows the sulci, cisterns, and ventricles (Figure 26.18), as observed in dead brain tissue after a stroke in a living individual. Differentiation between agonal or antemortem brain edema and cytotoxic postmortem loss of cortico-medullary differentiation is crucial but hardly possible given that the two pathologies tend to happen simultaneously.[9] Tonsillar herniation combined with narrowed, sickle-shaped side ventricles and absent basal cisterns is indicative of ante/perimortem brain edema.

Cerebral Infarction

When a person dies immediately after a stroke, the dead brain tissue from the stroke does not have time to generate more edema than the rest of the brain during the act of dying. Acute stroke can only be assumed in pmCT by demonstrating asymmetric swollen hypodense regions of the brain, indicating that the patient lived for hours after the stroke. Infarction in the living is not visible in CT in the first two hours. Older cerebral infarctions are equally depicted on antemortem and postmortem CT (Figure 26.19). MRI, with its much higher soft tissue resolution, is able to depict a stroke after 30 min with diffusion weighted imaging *in vivo*.[10,11,12,13] Unfortunately, the brain always suffers from ischemia during the process of dying, so diffusion weighted imaging is not helpful in pmMRI (Figure 26.20).

Hyperdensity of Venous Sinuses

A hyperdense sagittal sinus is a well-known sign in clinical radiology for thrombosis of the venous sinus. Due to the sedimentation effect of the blood after death, the blood clots and becomes more hyperdense in the vessels of the dependent parts of the body. In pmCT, the posterior superior sagittal sinus in particular becomes hyperdense in appearance since most cadavers are placed in a supine position, simulating antemortem thrombosis of the sagittal sinus (Figures 26.15, 26.18, and 26.21).

Intracranial Gas

Pneumcephalus (Figure 26.22) caused by skull fracture appears the same after death as in the living. Small focally located intracranial gas bubbles indicate an adjacent fracture of the skull. Gas embolism as a potential cause of death shows extensive accumulation of intravascular gas (Figure 26.12). Global putrefaction, on the other hand, shows

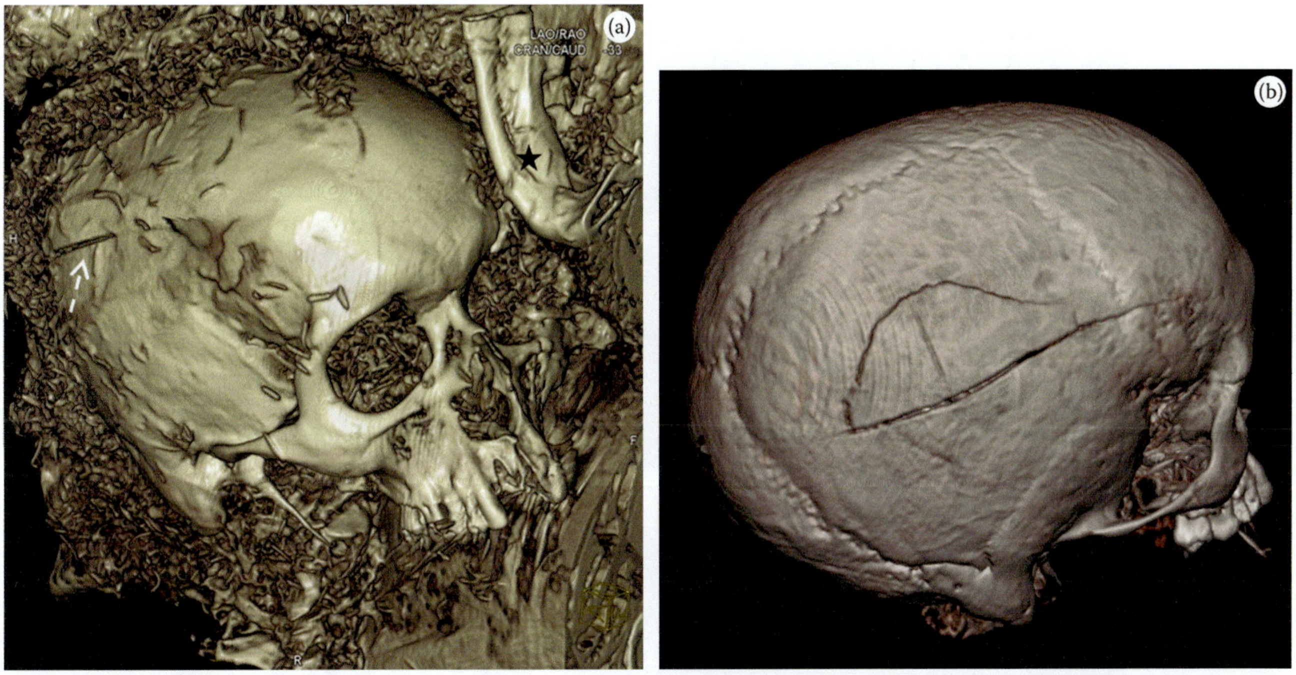

FIGURE 26.13 Panels (a–d) show a single case. (a) 3D reconstruction of the skull and its surface showing a massive infestation of maggots with a dislocated mandibula (black asterisk) sustained in the course of decay. In this case, the parietal skull fracture (white arrow) is much easier to depict pmCT than by inspection. (b) Skull fracture due to chronic alcohol abuse. The corpse was found in an apartment weeks after death. (c) Axial plane of the brain with only little parenchyma left due to decomposition and grubs. (d) After autopsy, a right lateral view of the macerated skull showing the fracture system, similar to pmCT. (e) A case of moderate decomposition with less maggot infestation. (f) Coronal plane, maximum intensity projection (MIP). Extensive gas accumulation is seen in the body cavities, with bloated intestines and gas throughout the soft tissue.

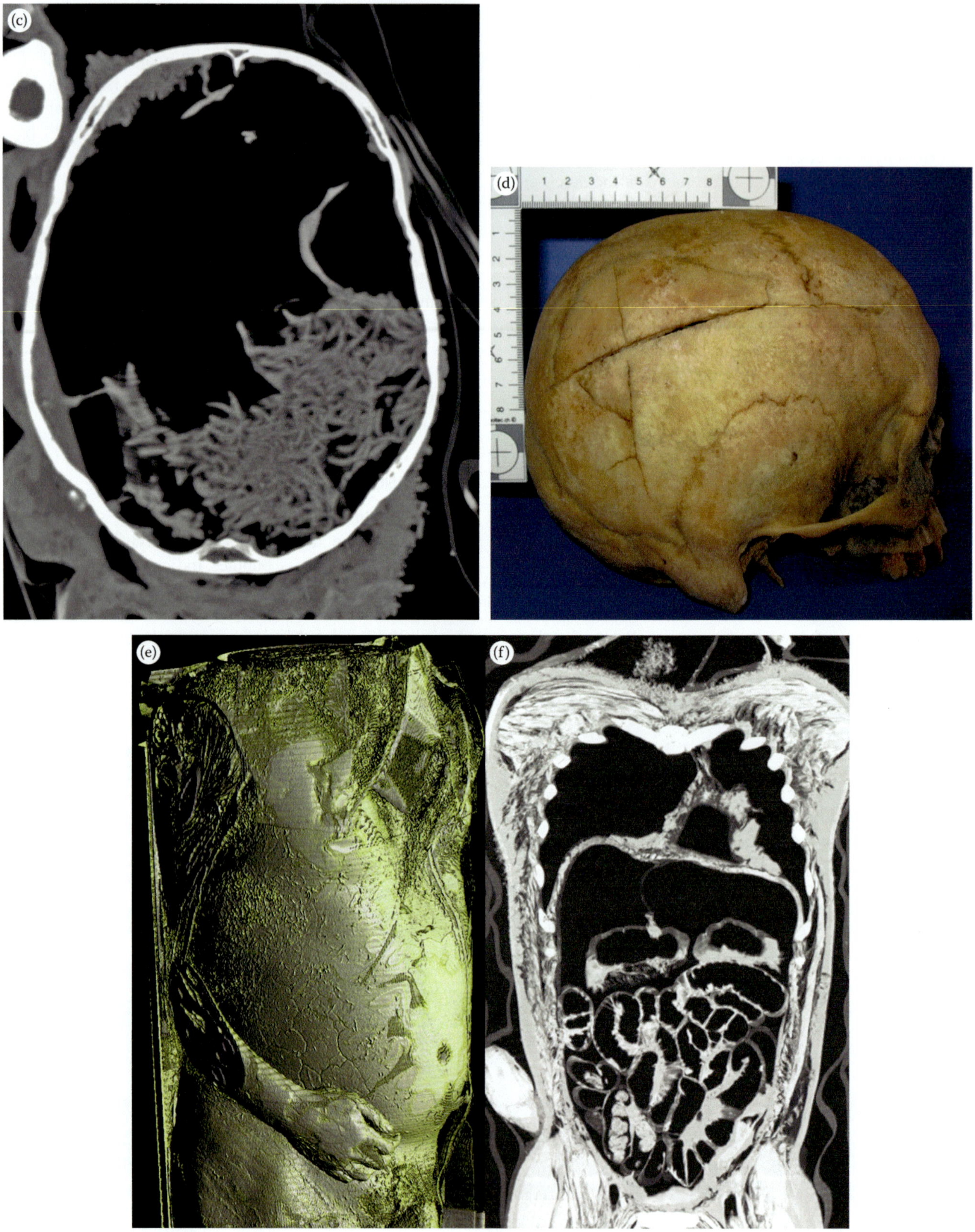

FIGURE 26.13 *(Continued)*

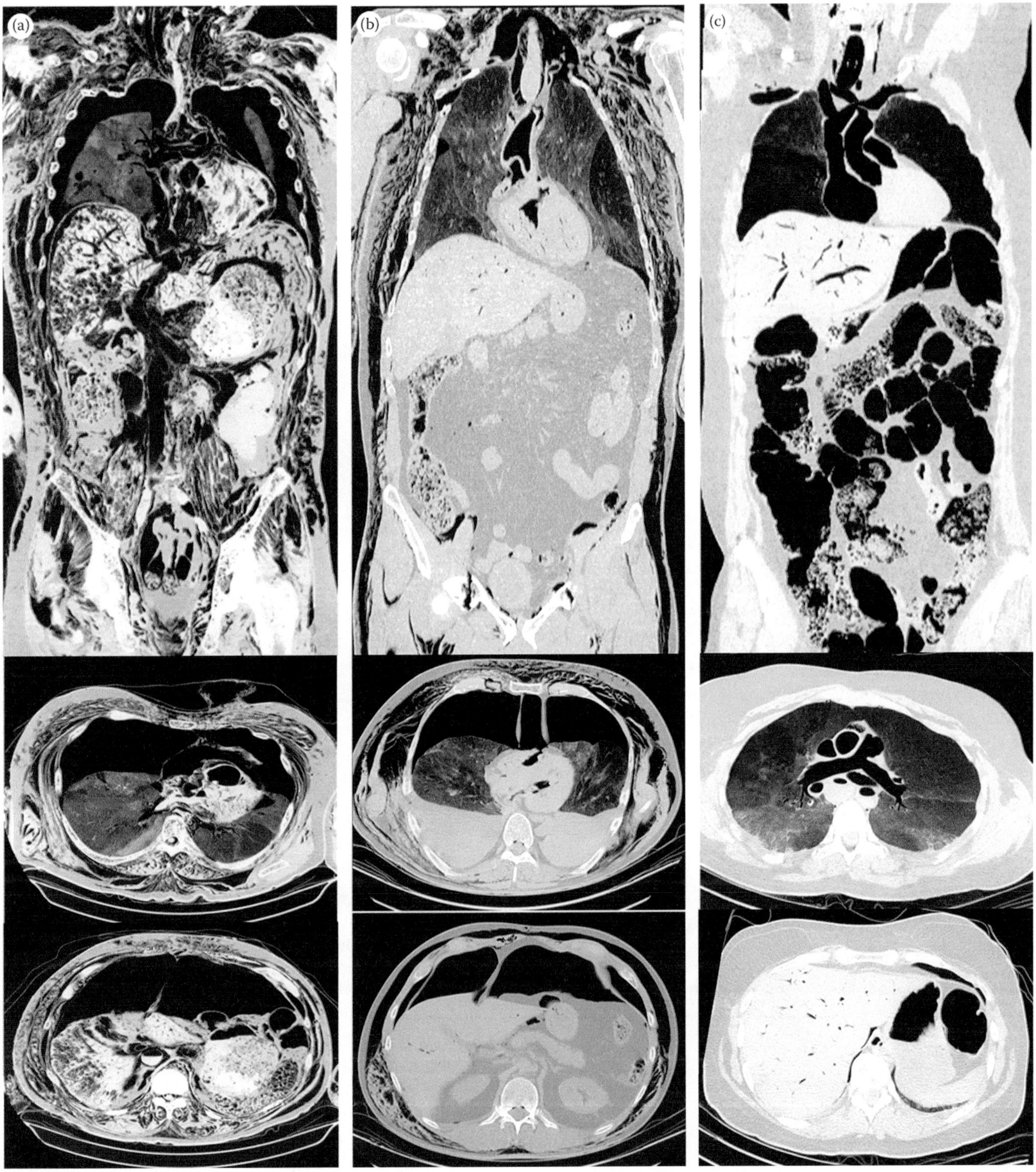

FIGURE 26.14 The upper row shows a coronal reconstruction; the middle row, axial planes at the level of the heart; and the lower row, axial planes of the liver. (a) Vast putrefaction of the body with gas in all body cavities, the vasculature, the hepatobiliary system, and the soft tissue. (b) Extensive soft tissue emphysema and pneumothorax. Note that there is no air in the vasculature and minimal air in the biliary system of the liver. This is a traumatic pattern of air in this recently deceased body. (c) Massive air embolism filling all supra-aortic vessels, the pulmonary arteries, and the abdominal aorta. There are only slight decompositional changes in the hepatobiliary system.

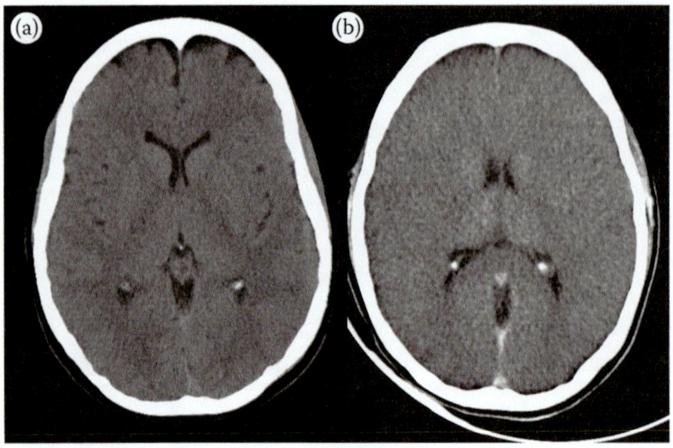

FIGURE 26.15 (a) Antemortem CT of a young man with normal differentiation of the cortex and medulla. (b) Postmortem loss of corticomedullary differentiation in a corpse of the same age as the patient in (a). Note the hyperdense appearance of the superior sagittal sinus as a common finding in pmCT.

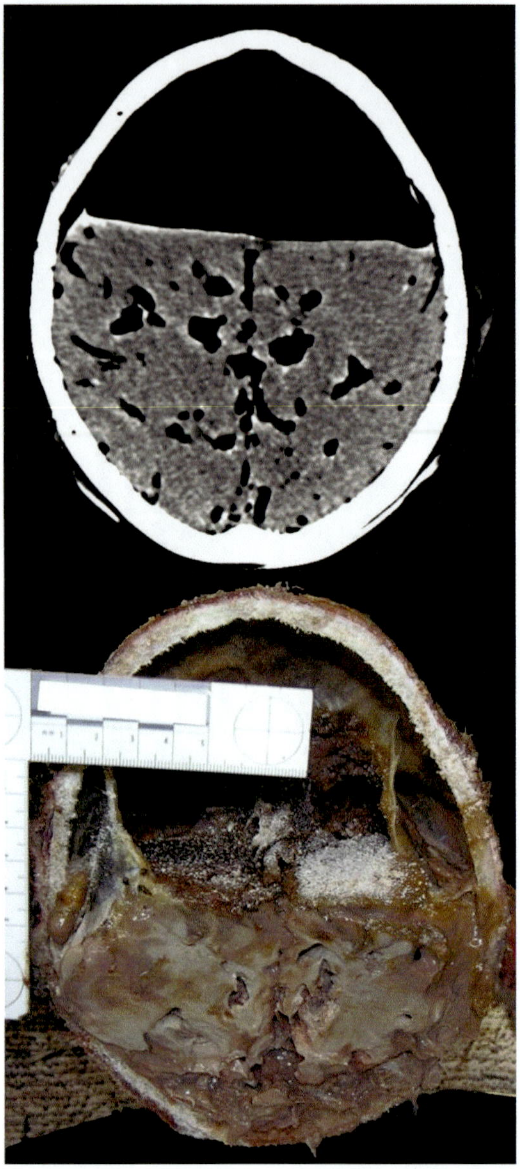

FIGURE 26.17 Advanced stage of decay. Relapse of the brain tissue and initial liquefaction.

position-dependent gas bubbles in the brain parenchyma. At later stages of decomposition, the brain will show position-dependent dislocation due to progressing tissue liquefaction (Figure 26.17).

Intracranial Hemorrhage

There is virtually no difference between imaging intracranial hemorrhage in the living and in the dead. In antemortem and postmortem CT, the blood is still hyperdense to the brain tissue and is therefore easily detectable as epidural, subdural, subarachnoid, or parenchymal hemorrhages (Figure 26.23).[14]

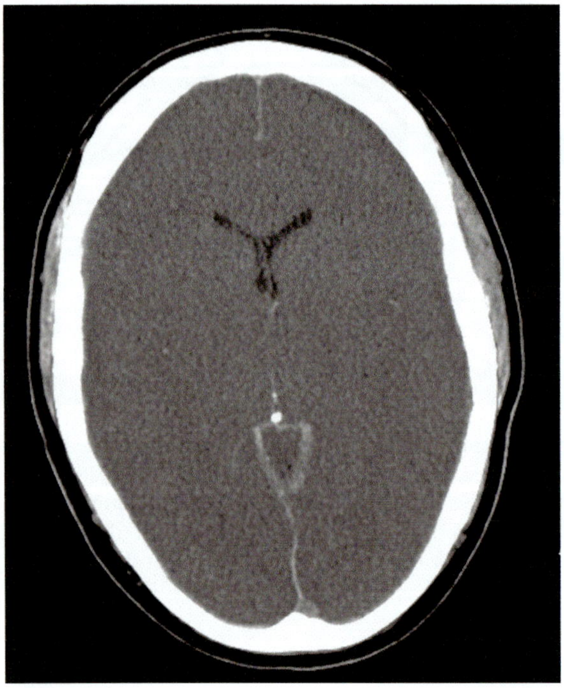

FIGURE 26.16 An antemortem scan of a brain-dead patient.

In MRI, the appearance of blood is dependent on the transformation of hemoglobin to hemosiderin (Figure 26.24).[15,16] Degradation is time-dependent; therefore, MRI is a useful tool to roughly estimate the age of a hemorrhage. The recent literature indicates that the investigator must be very careful when relying on MRI classification of the age of a hemorrhage, as the appearance happens to change very quickly, especially in infants.[17]

Thermal Tissue Changes of the Brain

Charred bodies often show vast tissue loss and changes of the parenchyma due to thermal impact. First, swelling of the brain will occur, followed by shrinkage resulting in the well-known puppet organ. Second, as a consequence of being subjected to high temperatures, the blood will literally boil out

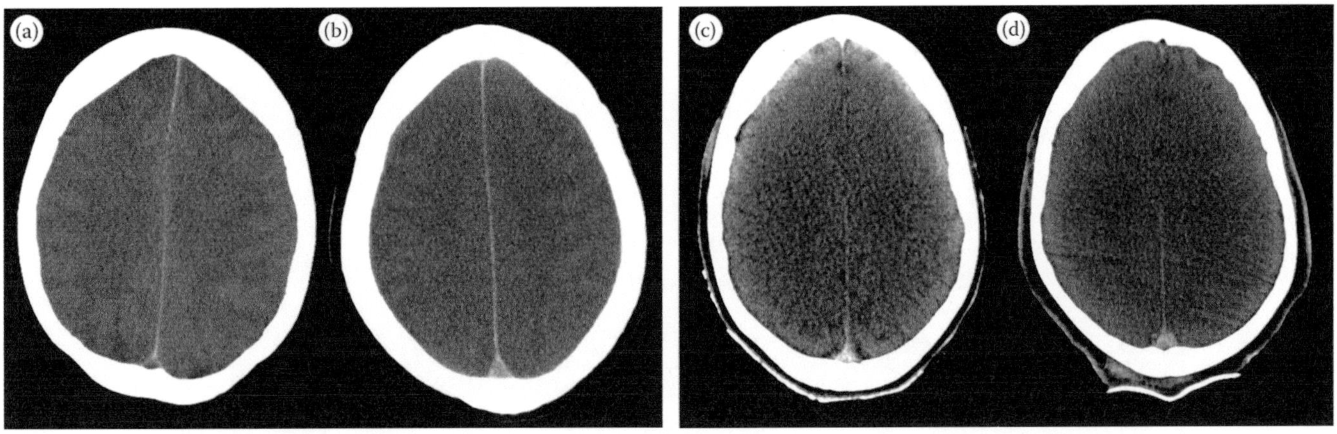

FIGURE 26.18 (a and b), Antemortem CT; (c and d), pmCT. (a) Brain edema with narrowed sulci but still-present corticomedullary differentiation. (b) Extensive brain edema in a brain-dead patient with increasing blurring of the gray and white matter. Note the increasing density of the superior sagittal sinus. (c) A normal pmCT of the brain with still-apparent sulci but a total loss of corticomedullary differentiation. Note the dense superior sinus as a normal postmortem finding. (d) Completely obscured sulci and absent differentiation of the cortex and the medulla in a postmortem scan with brain edema.

of the sinus and appear as a heat-caused epidural hematoma, a so-called "heat epidural," with a brownish, crumbly structure (Figure 26.25).

NECK IMAGING

Strangulation

In strangulation, the medical report can be widely influenced by objective radiological imaging of the deep neck. A combined survey of petechial hemorrhage, witnessed unconsciousness, and clinical MRI is employed to evaluate living victims of strangulation. Specific signs of strangulation on MRI have recently been established.[9,18–20] Some common signs have also been documented from clinical radiology and postmortem imaging: hemorrhage into the skin and subcutaneous and intramuscular hemorrhage/edema. The sternocleidomastoid and platysma muscles are frequently affected. Hemorrhagic lymph nodes can be a sign of strangulation and occur from venous congestion, as do the forensic ophthalmologic petechiae. Life is threatened by bilateral impact to the neck or unilateral impact to the deep nerve-vessel sheath,

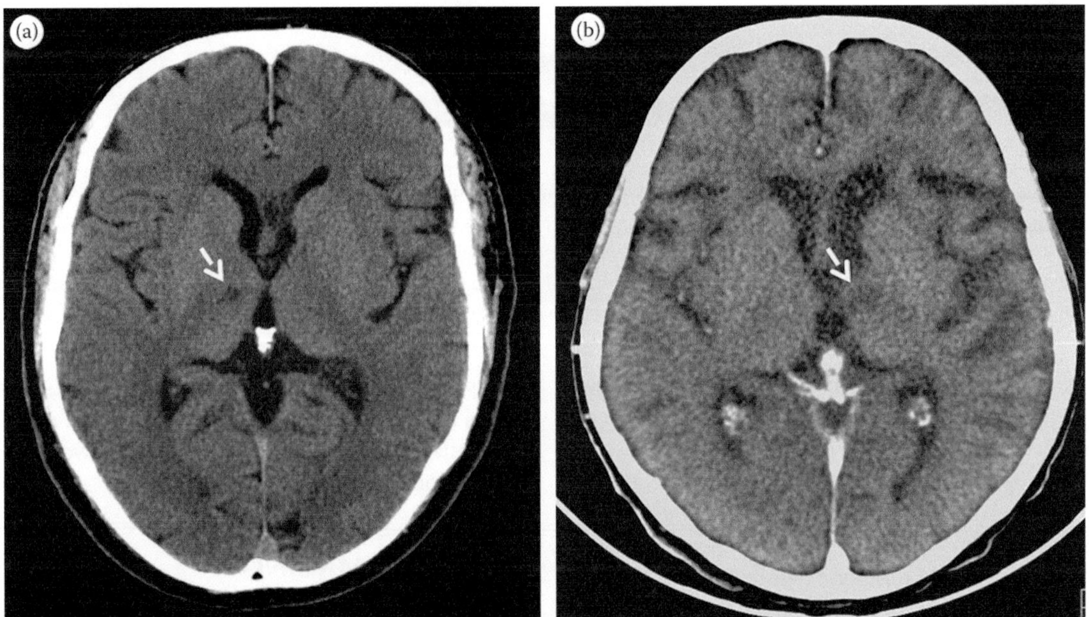

FIGURE 26.19 Both images are cases of almost the same age (approx. 65 years). (a) Antemortem CT with a small infarct area in the thalamus on the right side (white arrow). (b) pmCT showing an infarct of the left thalamus (white arrow). In this case, the sulci are still differentiated postmortem due to brain atrophy.

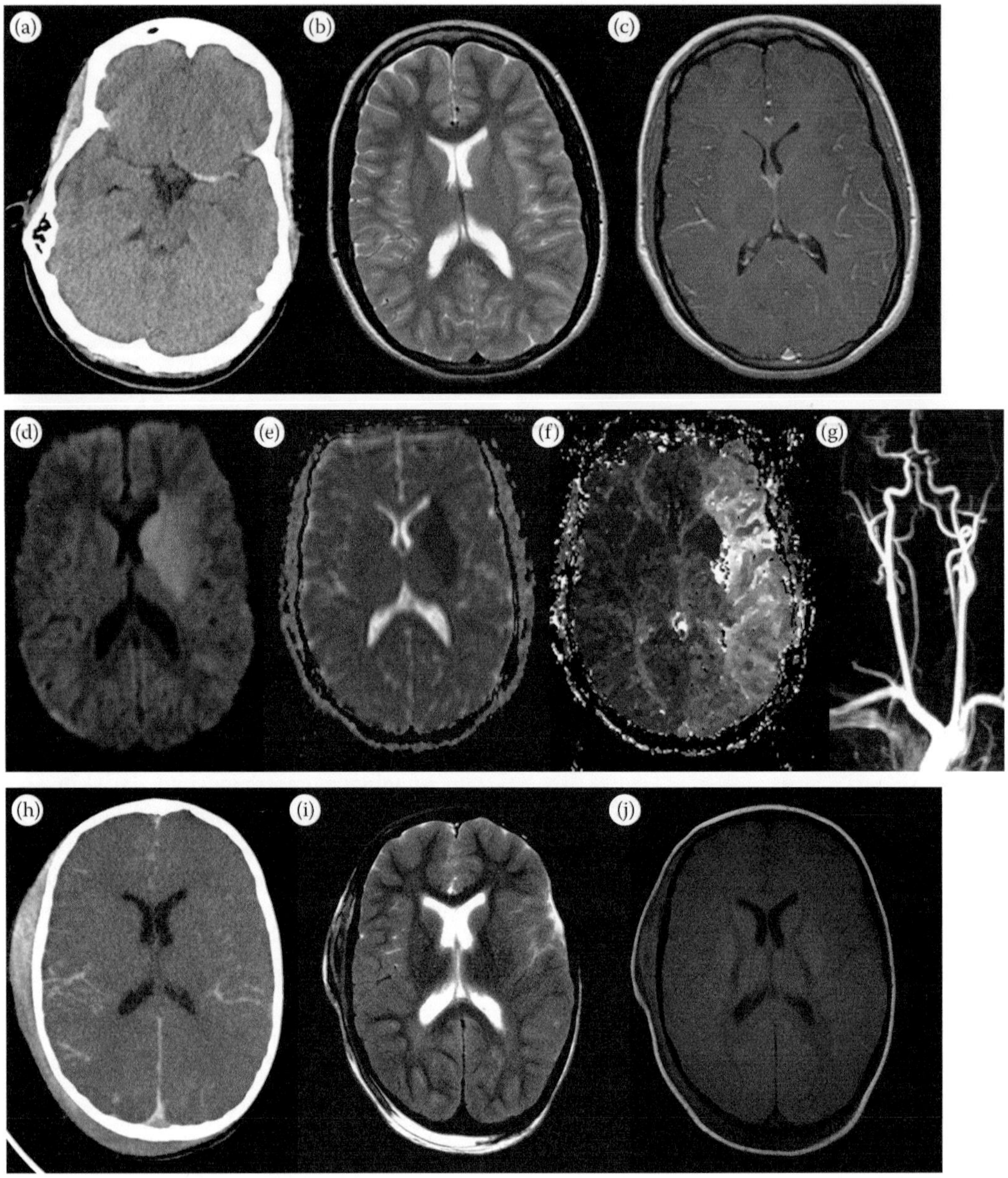

FIGURE 26.20 (a–g), Antemortem, same patient; (h–j), postmortem, same corpse. (a) Dense media sign of the left cerebral artery. (b) T2w tra; no T2-shine-through. (c) T1w tra after gadolinium. (d) Diffusion-weighted imaging showing an acute infarction with restriction of diffusion in the caudate nucleus, internal capsule, and putamen on the left side. (e) Corresponding low apparent diffusion coefficient (f) A deficit of perfusion in the area of the left medial cerebral artery, giving a mismatch of diffusion and perfusion. (g) 3D reconstruction of the supra-aortic arteries showing an occlusion of the left medial cerebral artery. (h) pmCT with diffuse subarachnoid bleeding and cepahlic hematoma on the right side. (i) T2w tra. (j) T1w tra, missing diffusion or perfusion in postmortem imaging.

which can be shown by clinical MRI (Figure 26.26). Victims surviving strangulation rarely show laryngeal cartilage or vessel disruption. These findings result from a more violent act of strangulation (e.g., hanging) that frequently leads to death.

pmCT depicts cartilage and osseous trauma quite well. [21] Dysmorphic cricoid cartilage and displaced upper and lower horns of the hyoid bone are common (Figure 26.27). A close look at a pm CT of the neck is crucial since there might be findings even in cases where strangulation is not suspected (Figure 26.28). A self-inflicted gunshot wound entering the base of the mouth might lead to laryngeal trauma, even if not directly adjacent to the shot canal (Figure 26.29). Of course, CT is superior for depicting the skeletal larynx, but MRI shows excellent results in evaluation of the soft tissue, as in strangulation survivors. Therefore, the two modalities supplement one another.

THORACIC IMAGING

Livor of the Lung versus Pulmonary Edema

Like the external livor mortis of the skin, internal organs like the lung may also develop internal livor mortis in dependent

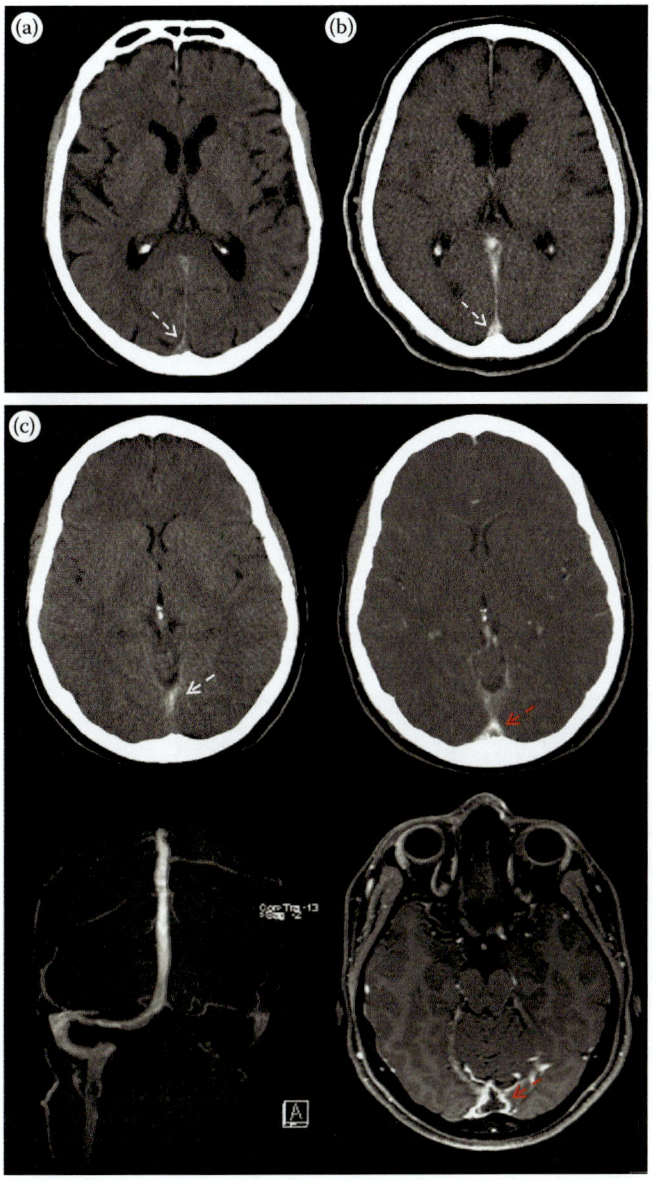

FIGURE 26.21 (a) Antemortem normal CT of the brain with a hypodense superior sagittal sinus (white arrow). (b) pmCT of the brain with a normal postmortem hyperdense superior sinus (white arrow). (c) Antemortem thrombosis of the sagittal sinus: Upper left = unenhanced CT with a slightly hyperdense sinus rectus (red arrow); upper right = CT after contrast medium clearly showing an "empty triangle," indicating thrombosis of the sinus (red arrow); lower left = 3D reconstruction of the veins on MRI, depicting a complete occlusion of the transverse and rectus sinus; lower right = T1 tra MPR > after gadolinium, revealing the thrombosis (red arrow).

lung areas.[16] The typical appearance of internal postmortem lividity in the lung is a dense ground-glass opacity occupying the posterior lobes, with a sharp horizontal demarcation line to the ventral parts of the lung as a result of the corpse lying on its back (Figure 26.30). A mixture of internal livores with pulmonary pathologies tends to be hard to diagnose, particularly since pulmonary edema can also cause ground-glass opacities, although not with such sharp margins as those caused by internal livores of the lung (Figure 26.31). To detect these internal livores mortis in the upper and ventral parts of the lung, a prone position of the deceased has to be assumed; therefore, knowledge of the initial findings at the crime scene is helpful. Internal livores mortis (ground-glass opacity) should not be mistaken for pneumonia (alveolar opacity). Pulmonary edema often appears in postmortem imaging due to fluid effusion from the lung capillaries in the interstitial or alveolar space. Peribronchial cuffing and a signet ring-like pattern of dilated pulmonary arteries appear in these cases, quite similar to that seen in clinical images (Figure 26.32).

Chest Trauma

In postmortem imaging, the investigator typically must deal with much more traumatized cases than in clinical radiology. Obviously, there will quite often be blunt or sharp thoracic trauma [22] that is equally well recognized in antemortem and postmortem CT such as lacerations of pulmonary parenchyma, pneumothorax, hemothorax, or pneumomediastinum (Figure 26.33). Lung contusions are frequent and are usually located in the periphery of the lung, with coarse, patchy, partially confluent consolidations (Figure 26.34). Additional retrograde aspiration may be seen. Tumorous, fibrotic, or emphysematous changes of the lung tissue show similar findings on clinical and pmCT. Blood aspiration shows a characteristic pattern on pmCT (Figure 26.35), and aspiration of blood, stomach contents, or water in drowning is very common, showing peribronchial coarse consolidations (Figure 26.32b). A major pitfall of pmCT is the reconstruction of stab wounds and shot channels through the lung tissue because of collapse of the lungs and displacement of the lung tissue by hemothorax (Figure 26.36). Signs of vascular injury on pmCT include mediastinal hemorrhage or extensive hemothorax; a unilateral left-sided hemothorax is a particularly strong but indirect sign of an aortic injury or rupture of an ipsilateral intercostal artery. Pericardial hemorrhage is also an indirect indicator of myocardial rupture or laceration of the aortic root (Figures 26.37 and 26.38). Bilateral ventral rib fractures or sternal cross fractures are characteristic of resuscitation-related injuries and are not necessarily due to the initial trauma.

Drowning

Drowning shows some specific signs on cross-sectional imaging.[23] Fluid-air levels due to water inflow into the paranasal sinuses are almost mandatory in both the living and the dead (Figure 26.39). Froth and fluid appearing in the pharynx and trachea/bronchi are common. Perihilar opacities of the lung tissue as a result of water aspiration are usually observable. Consecutive bronchospasm causes emphysema aquosum with a low-standing diaphragm in cases of "dry" drowning. The majority of fresh- water drowning cases show a mosaic pattern of the lung parenchyma because of the presence of hypo- and hyperperfused lung areas and of

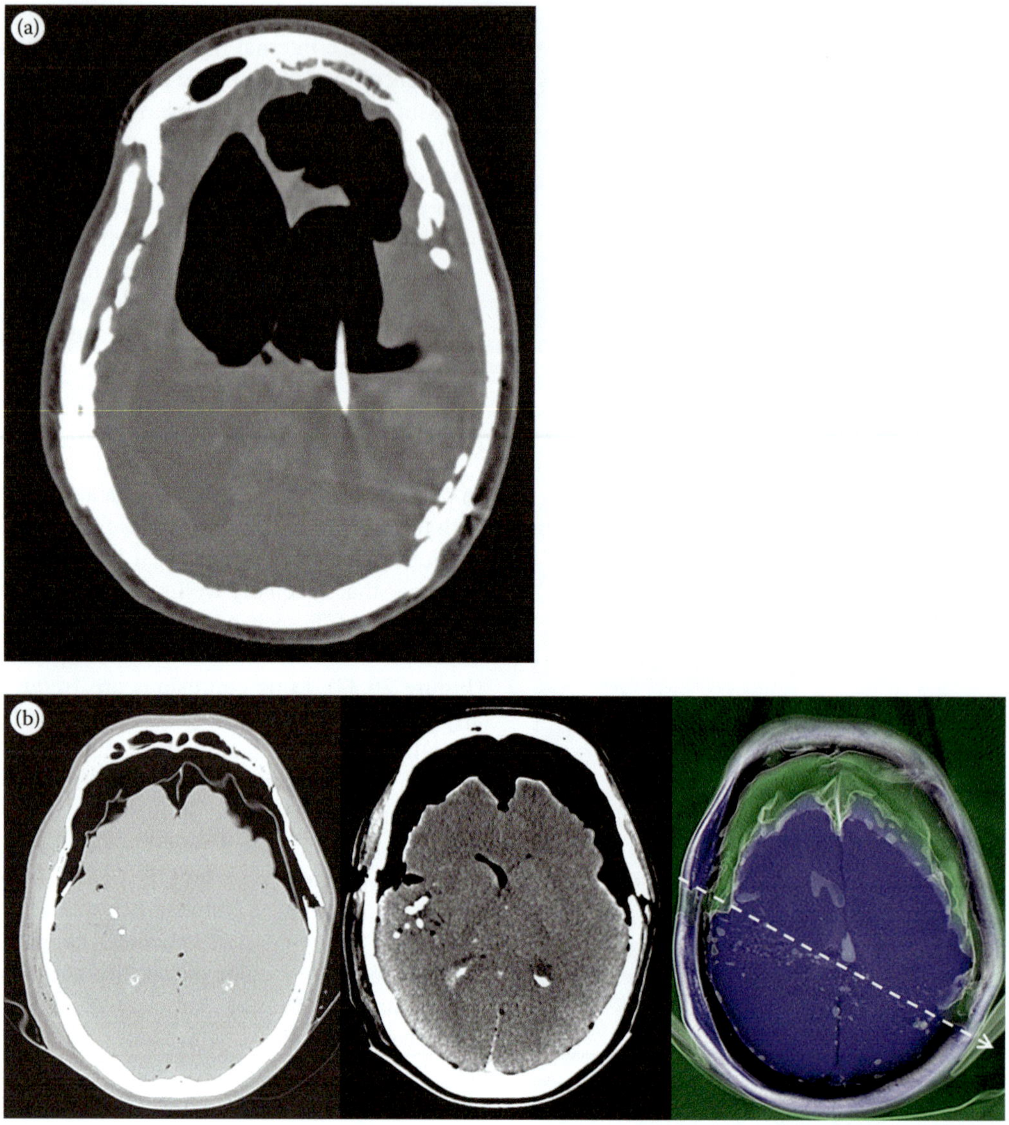

FIGURE 26.22 (a) Antemortem scan after extensive craniocerebral injury with neurosurgical intervention and ventricular drainage posteriorly, presenting with a massive pneumocephalus. (b) Postmortem scan after a gunshot wound to the head (white arrow) with ablation of the meninges, intraventricular blood, and skull fracture.

aspiration (Figure 26.40). The resorption of fresh water by the lung results in hypodensity of the blood, causing hemodilution and possible heart failure. Aspiration of saltwater, meanwhile, leads to signs of pulmonary edema. The first postmortem studies of drowning suggest that a blood density below 55 HU (Hounsfield Units) in the right atrium is indicative of hemodilution caused by resorption of water through the pulmonary circulation.[23] Agonal swallowing of fluid leads to distension of the stomach and duodenum (Figure 26.41). Drowning victims who have undergone intensive medical care usually show the typical radiological signs of adult respiratory distress syndrome (ARDS). The abovementioned froth /fluid in the airways and the fluid distension in the stomach/small intestines may have vanished, depending on the survival interval, on account of medical interventions like suction of the airways by the application of a stomach tube.

Thermal Tissue Changes

In fire victims, subtle changes of the airways are seen. Soot may be detectable on pmCT as hyperdense spots on the tracheal and bronchial mucosa, but a thickening of the pharyngeal mucosa and swelling are more commonly seen (Figure 26.42). How the thermal impact affects the lung tissue probably depends mostly on the degree of burning adjacent to the lung tissue.

Hyperdense Aortic Wall

Many pmCT scans demonstrate a hyperattenuated aortic wall,[24] probably caused by sedimentation of corpuscular and ascending serous components of the blood, leading to a hyperdense impression similar to negative contrast of the aortic wall (Figure 26.43). Moreover, there may be other possible

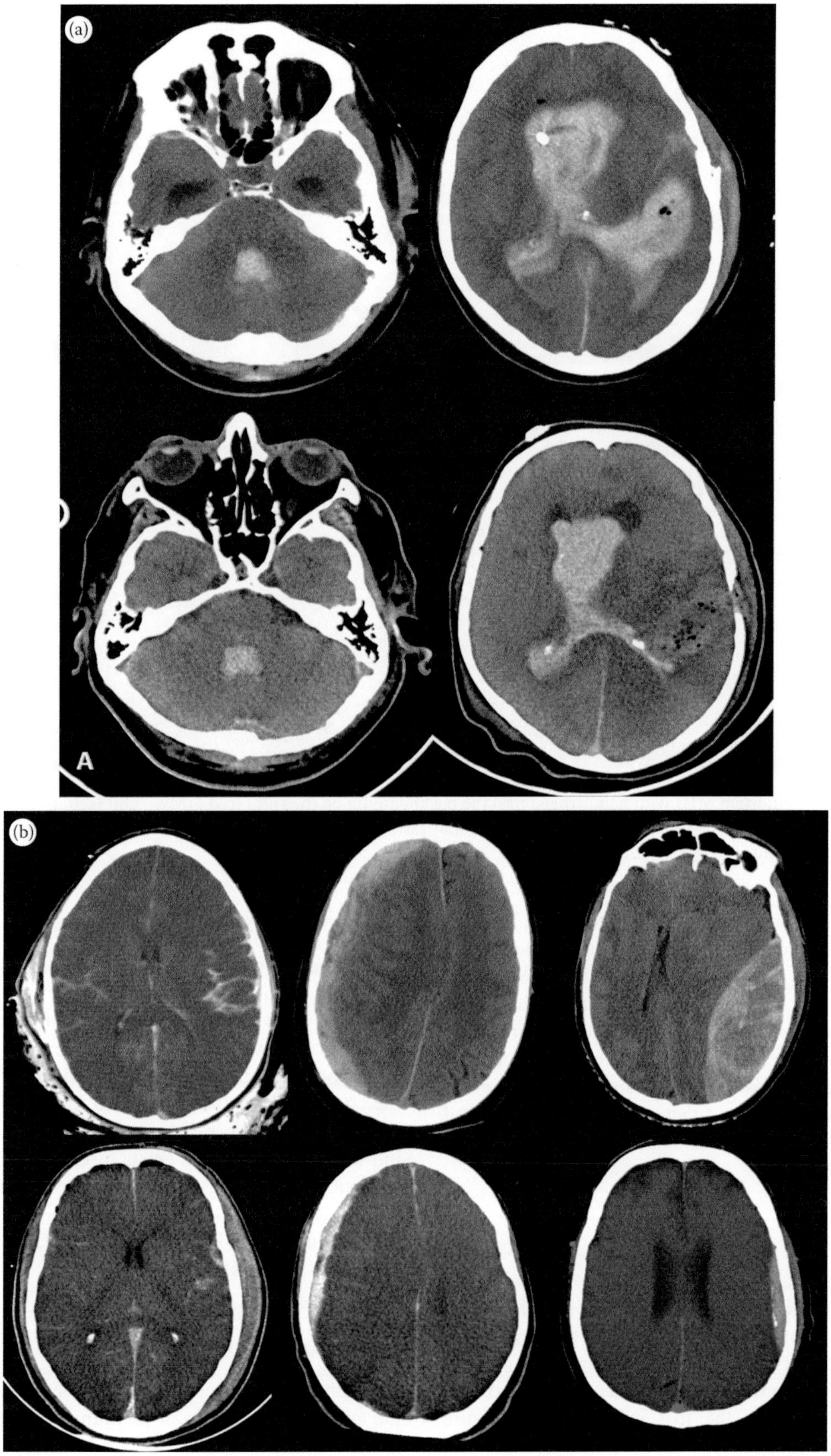

FIGURE 26.23 Upper row = antemortem; lower row = postmortem. (a) Due to a stab wound on the left side, there is intraventricular blood in the lateral ventricles, and the fourth ventricle shows parenchymal hemorrhage along the stab canal. (b) Left, subarachnoid bleeding; middle, subdural hemorrhage; right, epidural bleeding. All types of intracranial hemorrhage show a similar appearance both ante- and postmortem.

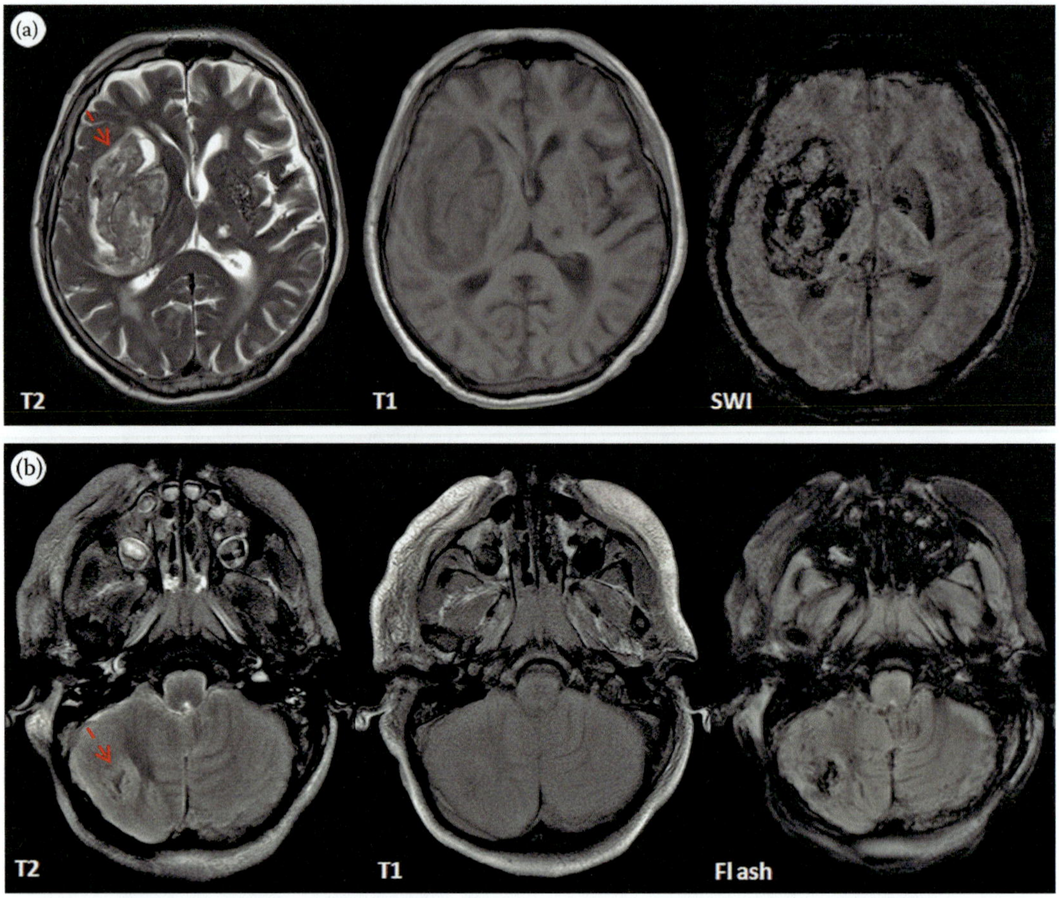

FIGURE 26.24 (a) Antemortem MRI with hemorrhage into the basal ganglia on the right (red arrow). Old lacunar infarct of the left thalamus. (b) Postmortem MRI (TIM) with a hemorrhagic parenchymal lesion of the right cerebellar hemisphere (red arrow) and an adjacent laceration of the sigmoid sinus.

mechanisms, such as contraction of the aortic wall, luminal loss of pressure and decreased attenuation of the lumen due to dilution of the blood after massive infusion during resuscitation. It seems that suddenly deceased victims (e.g., of trauma) tend to develop less postmortem clotting than slowly dying patients (e.g., of chronic illness). Arteriosclerosis presenting as hyperdense and calcified may cause a lack of collapse of the aortic wall by functioning as a bracket and may even be an identification criterion, among other characteristics (Figure 26.11).

Vanishing Aorta

During fatal hemorrhage, the arteries collapse,[25] which is best seen in big vessels like the aorta, especially the aortic arch and the descending aorta, giving rise to the term "vanishing aorta" (Figure 26.8). The pulmonary arteries and even heart chambers will decrease in size after a major loss of blood volume. Of course, the lost intraluminal pressure caused by absent circulation will also affect this; therefore, this sign involves a grey area between normal postmortem changes and major hemorrhage. A collapsed caval vein is an almost universal postmortem finding and is therefore not specific for hemorrhage.

Aortic Rupture

Traumatic rupture of the aorta is best seen on contrast-enhanced CT of survivors, where an active contrast leak from the aorta is evidence for aortic disruption. In postmortem unenhanced imaging, direct identification of an aortic rupture is often not possible. The aorta is usually collapsed, and the course of the disrupted aorta leads directly through the hemorrhage. An aortic rupture can be assumed when a large intrathoracic hematoma exists directly adjacent to the aorta, without demarcating the fat plane to the mediastinal blood (Figure 26.4a). In MRI, aortic tears are slightly better visible than on unenhanced CT because of the much better depiction of soft tissue by MRI (Figure 26.4b).

Pulmonary Embolism

A diagnosis of pulmonary thromboembolism (PE) without the application of contrast medium is practically impossible in both clinical and postmortem radiology. A typical sign in clinical radiology is the development of a peripheral, triangle-shaped pulmonary infarction (Hampton's hump). In most cases of massive fatal PE, there is not enough time to develop this sign, and the examiner of an unenhanced pmCT must

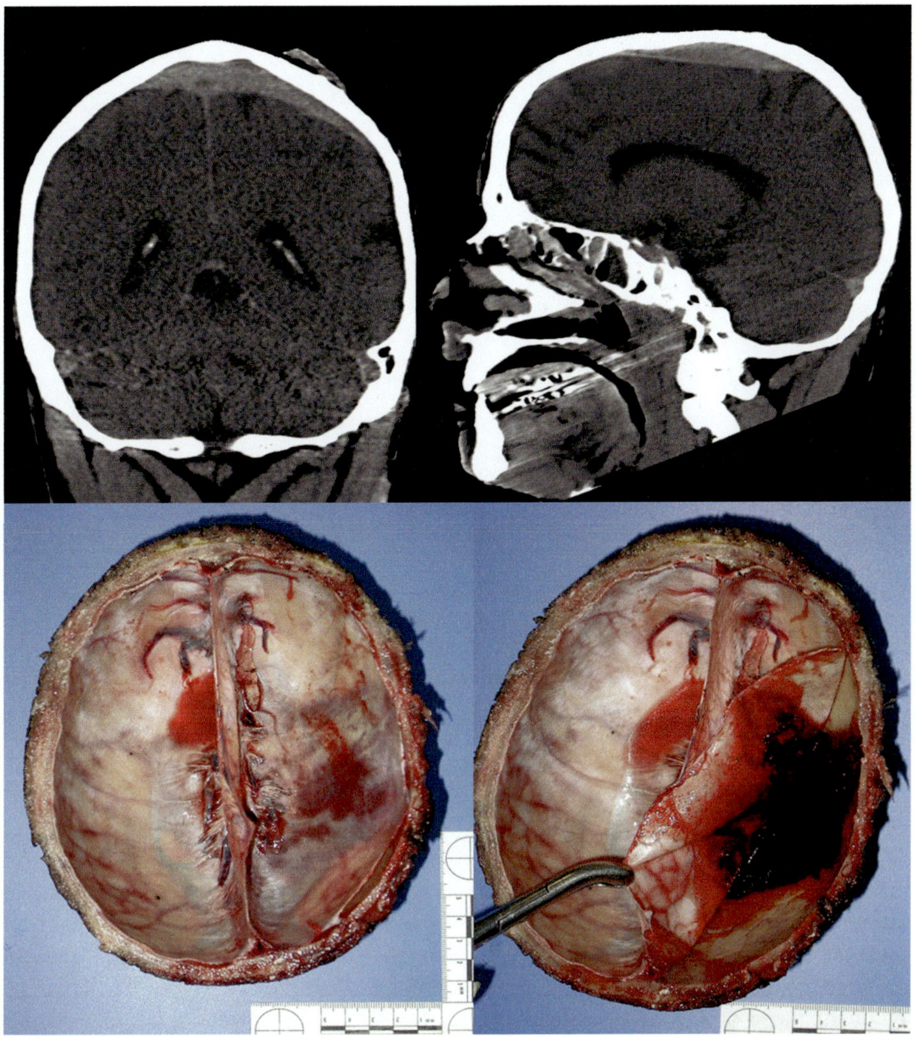

FIGURE 26.25 Heat epidural (note the extension over the cerebral falx).

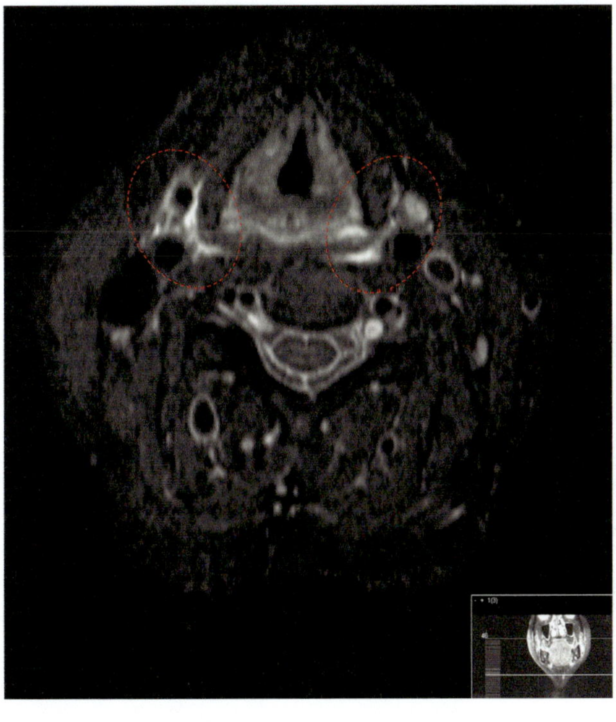

FIGURE 26.26 A clinical MRI of the neck at the level of the vocal cords showing bilateral signal alteration (red circles) in the deep nerve-vessel sheath (carotid space).

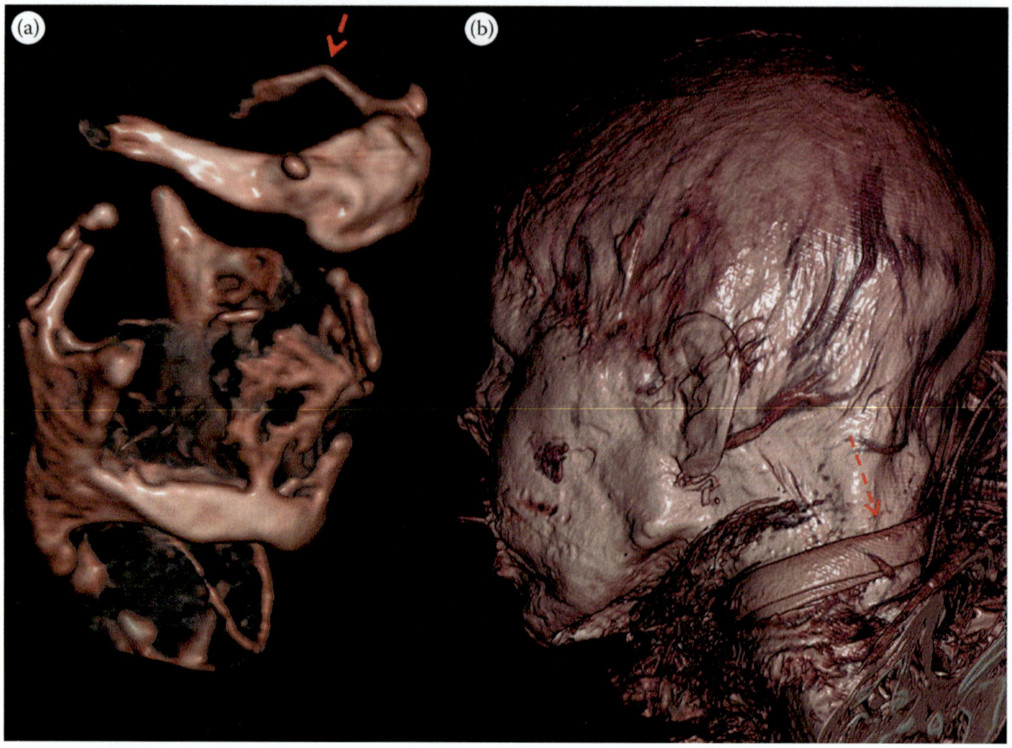

FIGURE 26.27 (a) 3D reconstruction of the larynx showing a fracture of the left greater cornu of the hyoid bone (red arrow). (b) 3D reconstruction of the remaining strangulation tool (red arrow) around the neck.

rely on secondary signs of pulmonary hypertension, such as right-sided heart dilatation, or must perform a pmCTA.

A special case is fatal fat embolism, which appears as a fluid–fluid level with fatty and serous intraluminal parts that are easily detectable by unenhanced pmCT. Due to a lack of circulation, the fat (fatty marrow) will ascend and result in a hypodense (HU about −60 to −100) liquid level caused by characteristic surface tension (Figure 26.44). MRI will probably allow even better visualization of fatal fat embolism due to an even better contrast. Fat embolism manifests more frequently in closed fractures of the pelvis, long bones or after surgical interventions such as endoprosthesis of the hip or marrow nail implantation. However, it can also occur due to amniotic fluid embolism during or after childbirth.

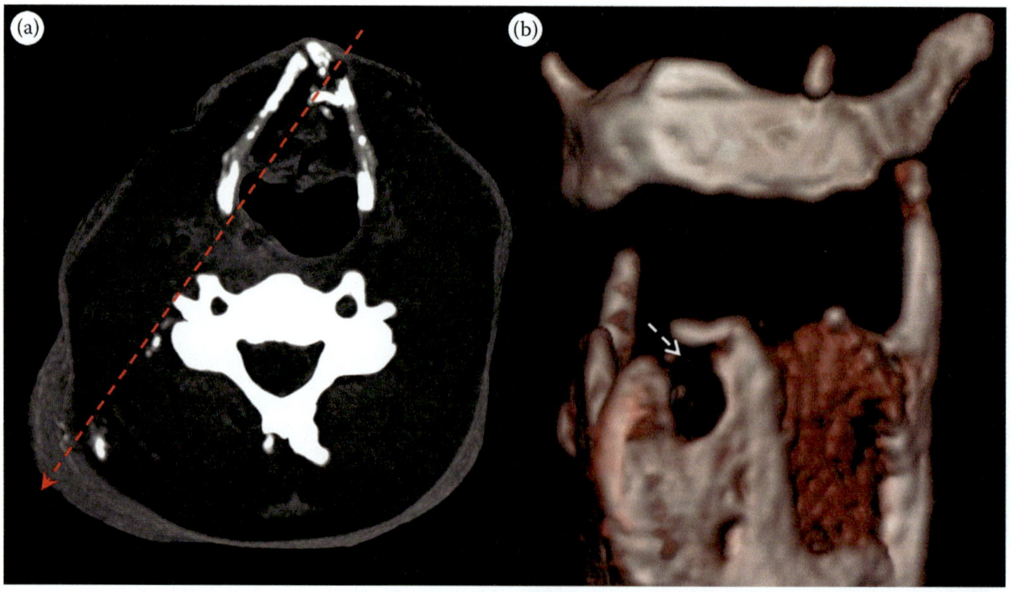

FIGURE 26.28 (a) Axial plane of the thyroid cartilage with entry wound through the left paramedian cartilage due to shooting, with full penetration (red arrow). (b) Left ventral aspect of the larynx with the shot wound (white arrow).

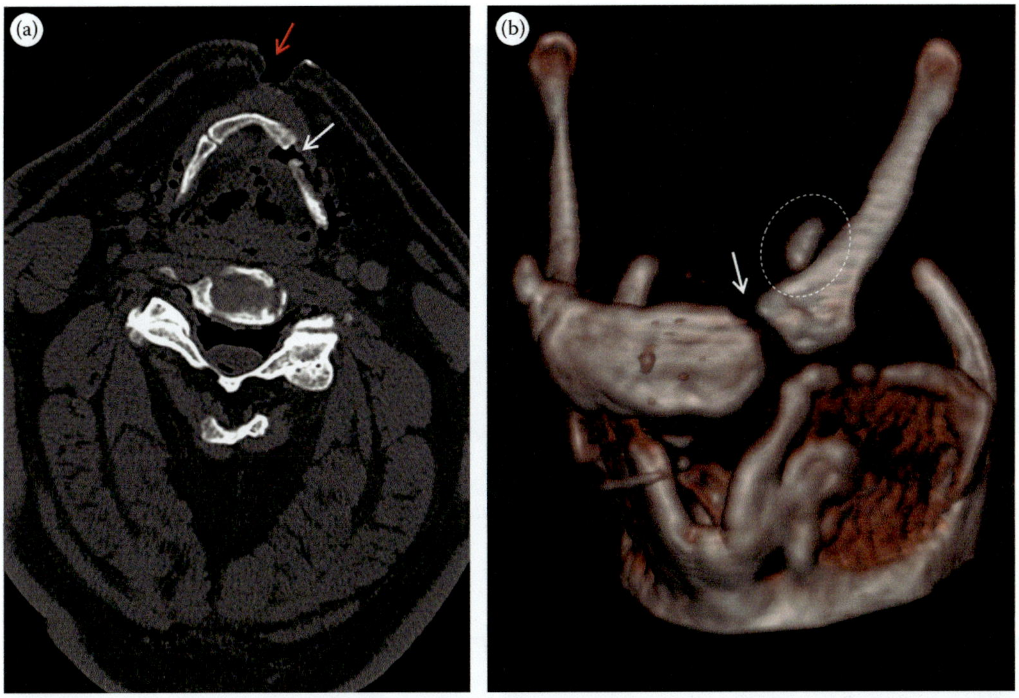

FIGURE 26.29 (a) Axial plane at the level of the hyoid bone. (b) 3D reconstruction of the larynx. Note the displacement of the lateral part of the hyoid bone from the hyoid corpus (white arrow) and the dislocated lesser cornu (white circle). Shot wound defect in the soft tissue (red arrow) with a steep ascending shot canal.

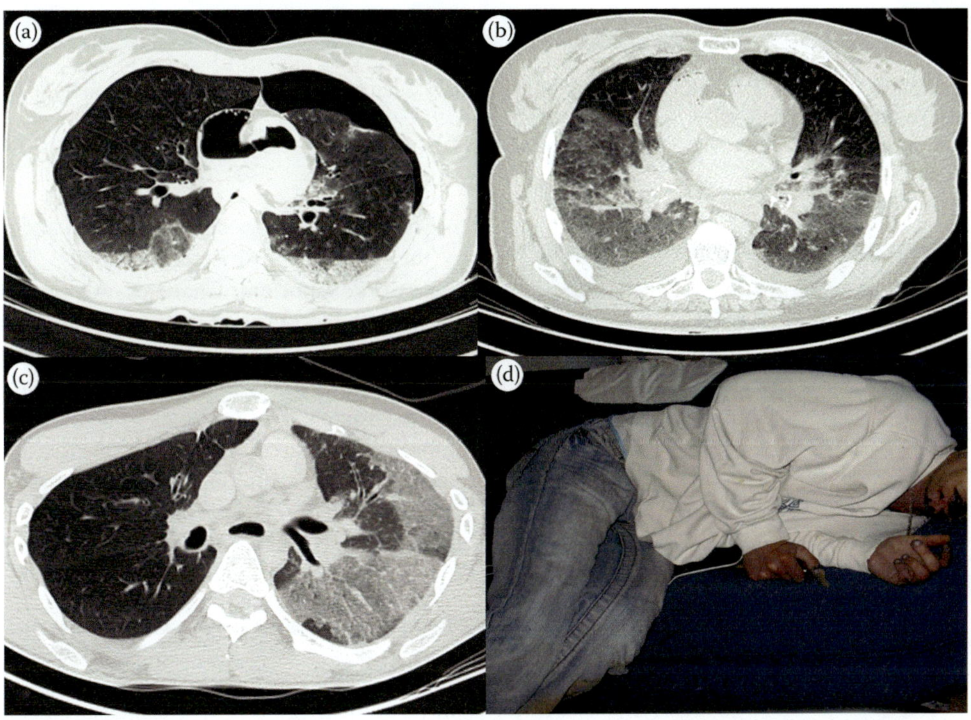

FIGURE 26.30 (a) Sharply demarcated margins of sparse internal livor mortis of a corpse in the supine position. Note the pneumothorax on the left side and the air embolism in the right atrium and ventricle of the heart. (b) Supine position of the corpse with more peculiar internal livores. Note bilateral pleural effusion and fluid in the bronchi. (c) Internal livores, only left-sided. (d) Corresponding crime scene picture with the initial position of the deceased.

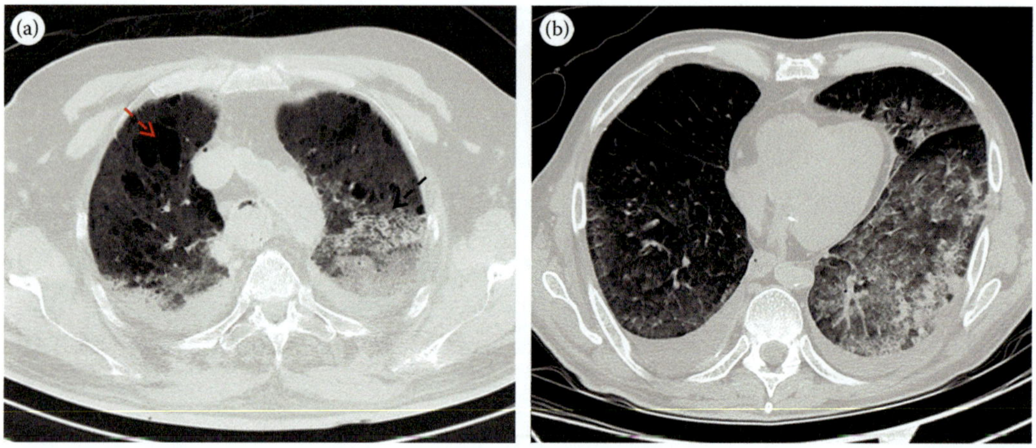

FIGURE 26.31 (a) Emphysematous bullae (red arrow) and fibrotic changes (black arrow) with bilateral pleural effusion and sparse internal livores mortis. (b) Acute pneumonia of the left lung with slight fibrotic changes and bilateral pleural effusion. There are only initial internal livores mortis.

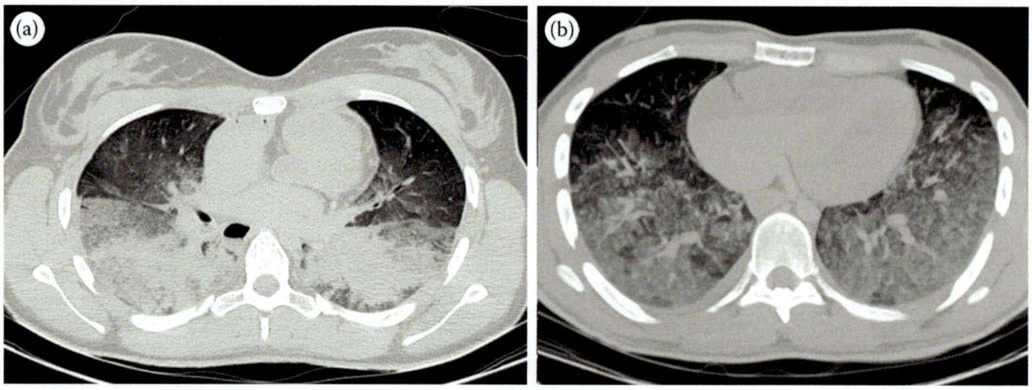

FIGURE 26.32 (a) Combined internal livor mortis and agonal pulmonary edema. (b) A combination of aspiration of stomach contents, pulmonary edema, and postmortem changes.

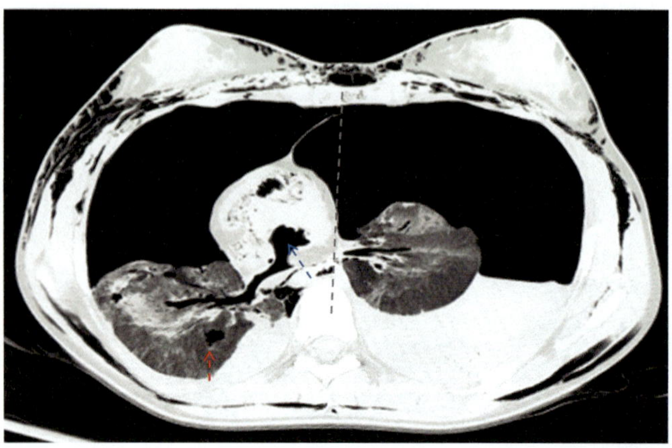

FIGURE 26.33 MinIP depicting broad lacerations (red arrow) of the lung with atelectasis. Bilateral pneumothorax with mediastinal shift and air embolism indicated by the air-filled pulmonary artery (blue arrow) and heart ventricle. Paravertebral moderate pneumomediastinum and predominantly right-sided hemothorax. Note the vast soft tissue emphysema of the young woman.

For pmCT findings of air embolism and putrefaction, refer to the general postmortem changes detailed in Section General Postmortem Changes. Air embolism can occur whenever a blood vessel is open. Usually the pressure gradient in most arteries and veins is greater than atmospheric pressure: this usually impedes air embolism. Of course, open trauma, especially to the head, lung, or heart, can result in air embolism. Pressure changes in scuba diving can also result in gas embolism (pulmonary barotrauma or decompression sickness). Injection of air into a vessel, either accidentally or as a deliberate act, may also lead to a fatal air embolism.

Heart Size, Hypertrophy, or Dilatation of the Heart

As in clinical radiology, the measurement of heart size in relation to the maximal thoracic aperture and the myocardial thickness is possible in postmortem radiology.[26] Unfortunately, in postmortem imaging, the right atrium is sometimes dilated, possibly due to general blood congestion during death

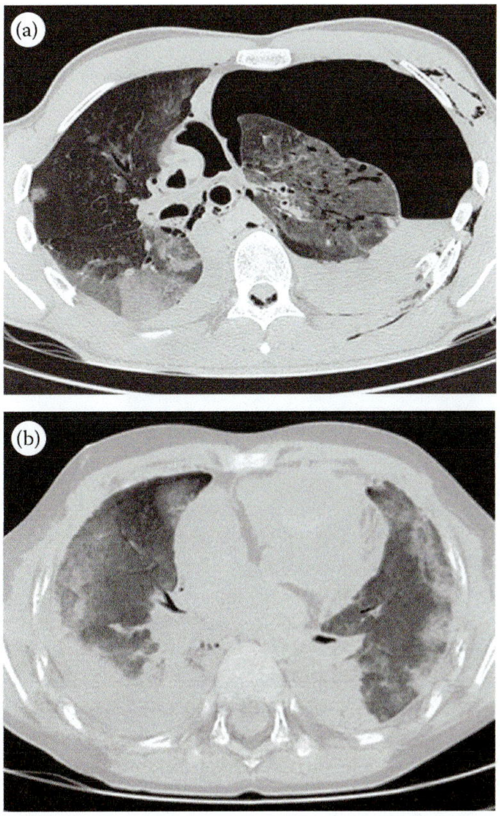

FIGURE 26.34 (a) Small patchy peripheral contusions of the right lung with bilateral hemothorax and internal livores mortis. The left lung shows lacerations and atelectasis with pneumothorax and mediastinal shift. There are also multiple rib fractures on the left. (b) Peripheral band-like lung contusions with bilateral dorsobasal dystelectases.

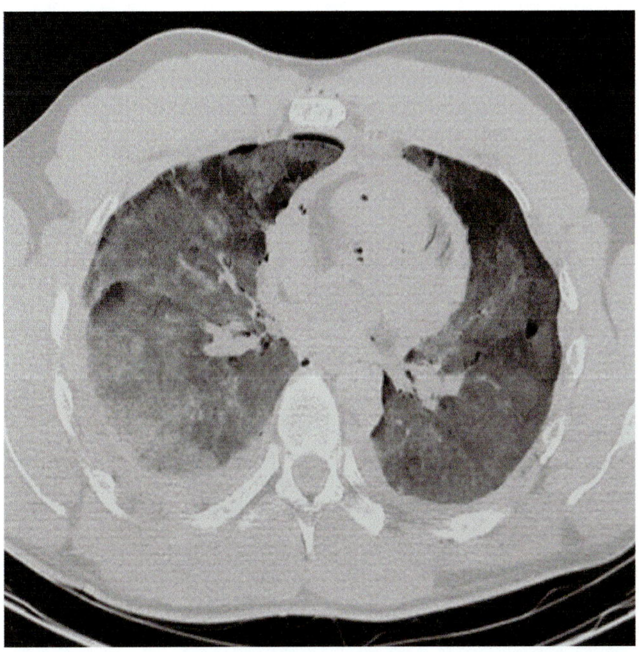

FIGURE 26.35 MinIP depicting aspiration of blood, small parenchymal lacerations, and contusions combined with a slight bilateral hemothorax and a small pneumothorax on the right side. There are sparse internal livores.

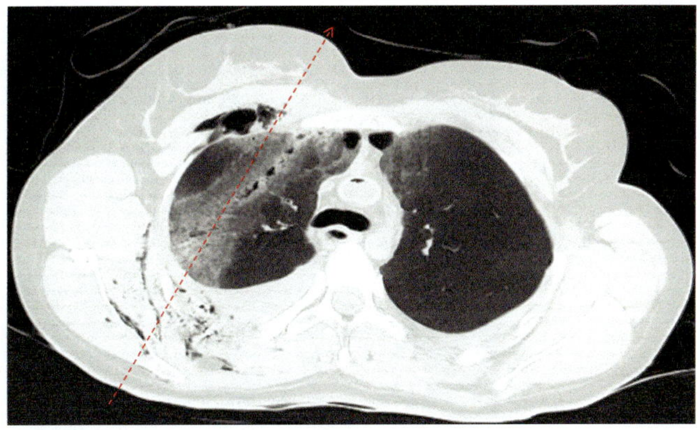

FIGURE 26.36 MinIP depicting a shot wound with complete penetration leading to parenchymal laceration with pneumohemothorax and a slight displacement of the complete right lung. There is soft tissue emphysema along the shot canal.

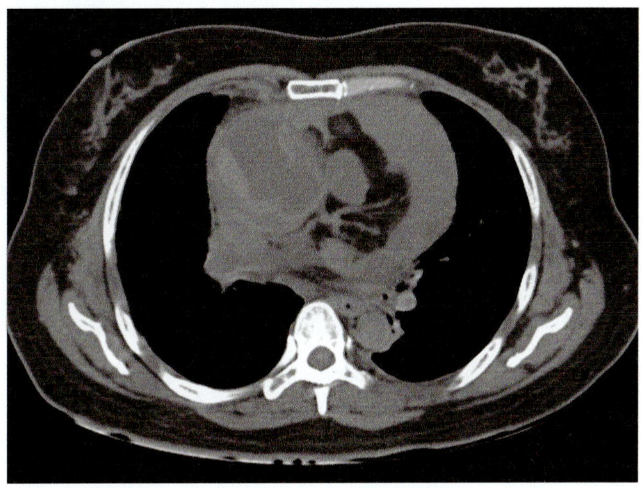

FIGURE 26.37 Pericardial tamponade due to aortic rupture of the ascending portion.

or due to drowning (Figure 26.45).[23,27] This would certainly influence the total heart size; thus, further studies are required. In en- or exsanguination, the opposite effect may appear.[25]

Myocardial Infarction

Age estimation of heart attacks is possible on postmortem MRI: acutely infarcted areas present with decreased signal in necrotic centers and increased signal in marginal edematous myocardial regions on T2-weighted sequences (Figure 26.38). Subacute infarctions show edematous hyperintense regions on T2-weighted sequences and no signal alteration on T1-weighted sequences unless intramyocardial hemorrhage has occurred (T1w hyperintense).[28] Chronic myocardial infarctions show fibrosis with distinctly decreased signals on T1- and T2-weighted sequences. Chronic infarction is depicted similarly on CT, showing a thinning of the wall of the ischemic area. Diagnosing thromboembolic occlusion of the coronaries is not possible without contrast administration.

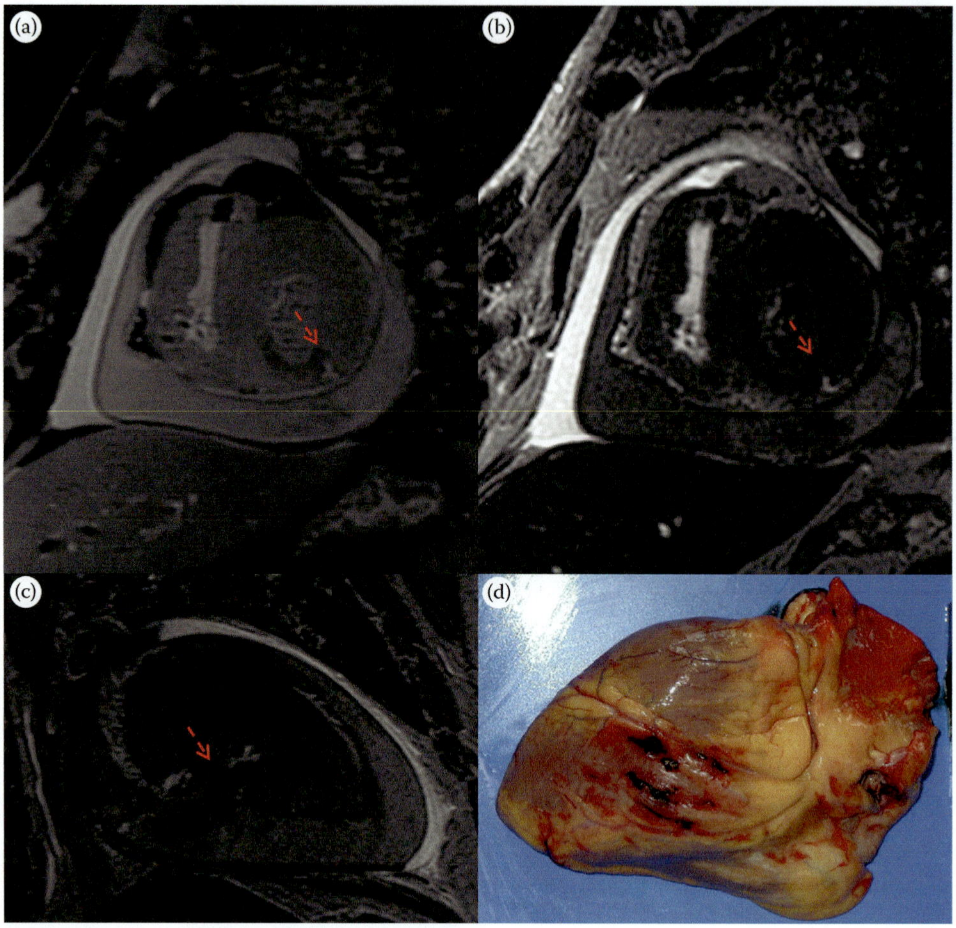

FIGURE 26.38 (a) Short axis of the heart (STIR) depicting signal alterations due to rupture of the posterior wall of the left ventricle (red arrow) in the course of a 2- or 3-day-old myocardial infarction. There is a massive pericardial tamponade caused by the myocardial rupture. (b) Short axis (Flair). (c) Coronal plane, T2w with fat saturation. (d) Autopsy specimen showing the hemorrhagic ruptured scar of the posterior wall.

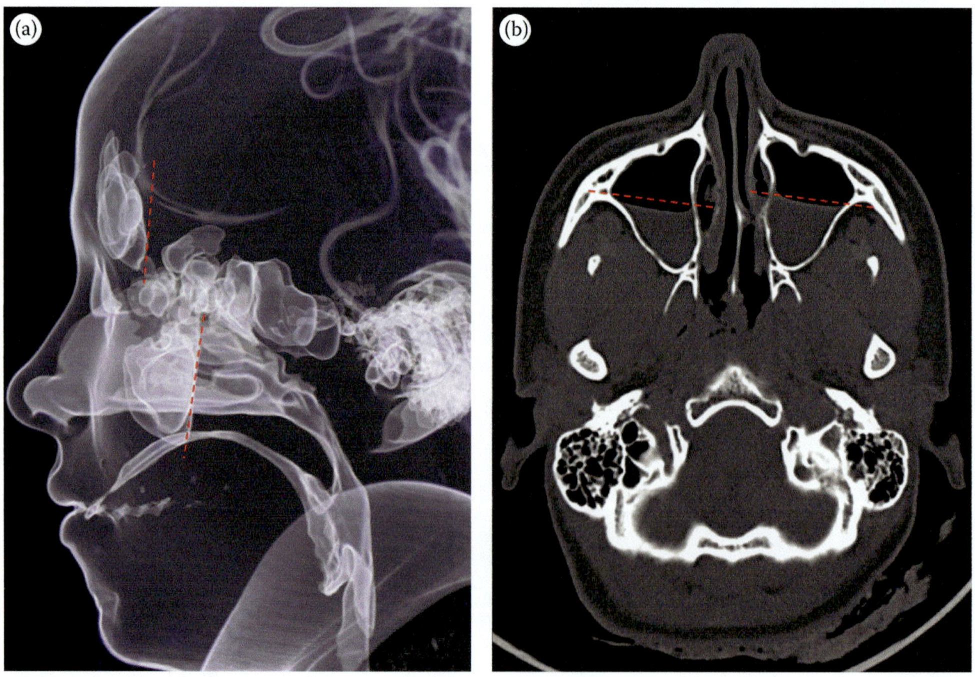

FIGURE 26.39 (a) Volume reconstruction of air-filled structures (b) Axial plane at the level of the paranasal sinus. Fluid-air levels (red lines) in a drowning victim.

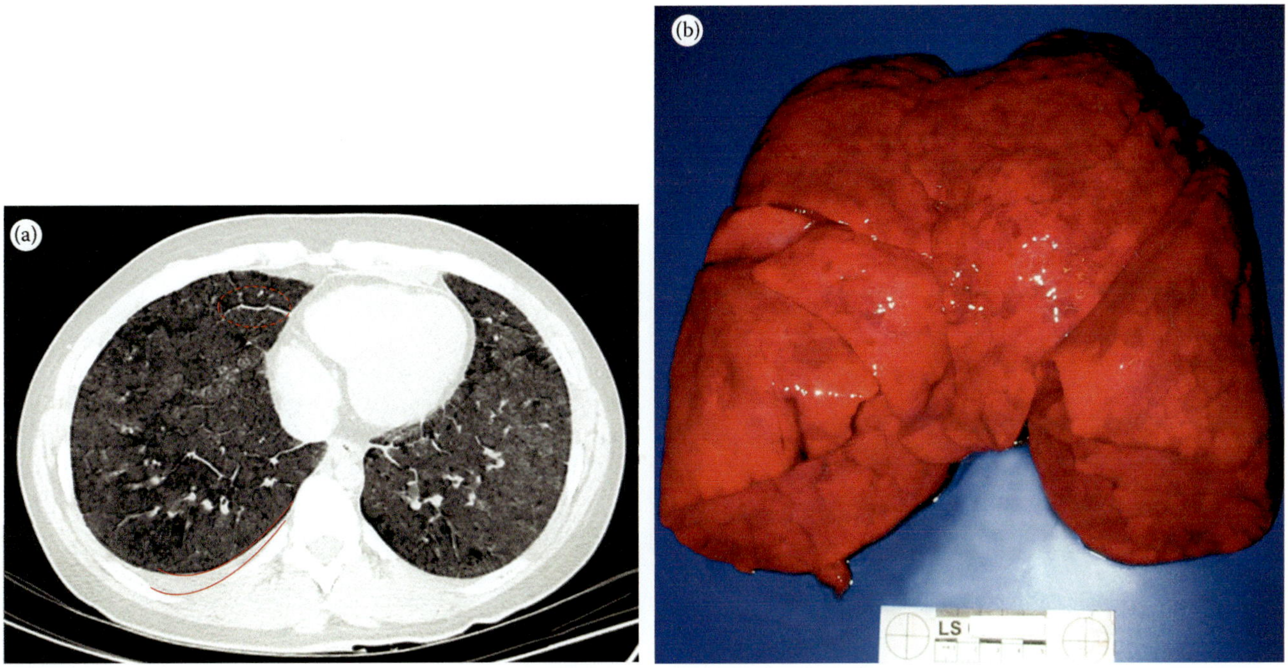

FIGURE 26.40 (a) Typical mosaic pattern in fresh water drowning with thickening of the interstitial space (red oval). Note the small pleural effusion on the right side (marked red). (b) Corresponding autopsy specimen showing lung overdistension, "emphysema aquosum" and subpleural hemorrhages (Paltauf's sign).

ABDOMINAL IMAGING

Abdominal Trauma

On unenhanced pmCT, the sensitivity and specificity of organ injuries are obviously lower for postmortem imaging than for clinical scans due to low soft-tissue contrast (Figures 26.46 and 26.47). In pmCT, the investigator must search for indirect signs of a parenchymal rupture, such as a subcapsular hematoma, heterogeneous parenchyma, or adjacent hyperdense fluid layer. Postmortem MRI shows higher sensitivity when depicting parenchymal pathologies (Figure 26.48).[29,30]

CT in general is highly sensitive to free fluid or gas in the abdominal cavity, although the examiner of a pmCT must take into account that free fluid and gas are not signs of an

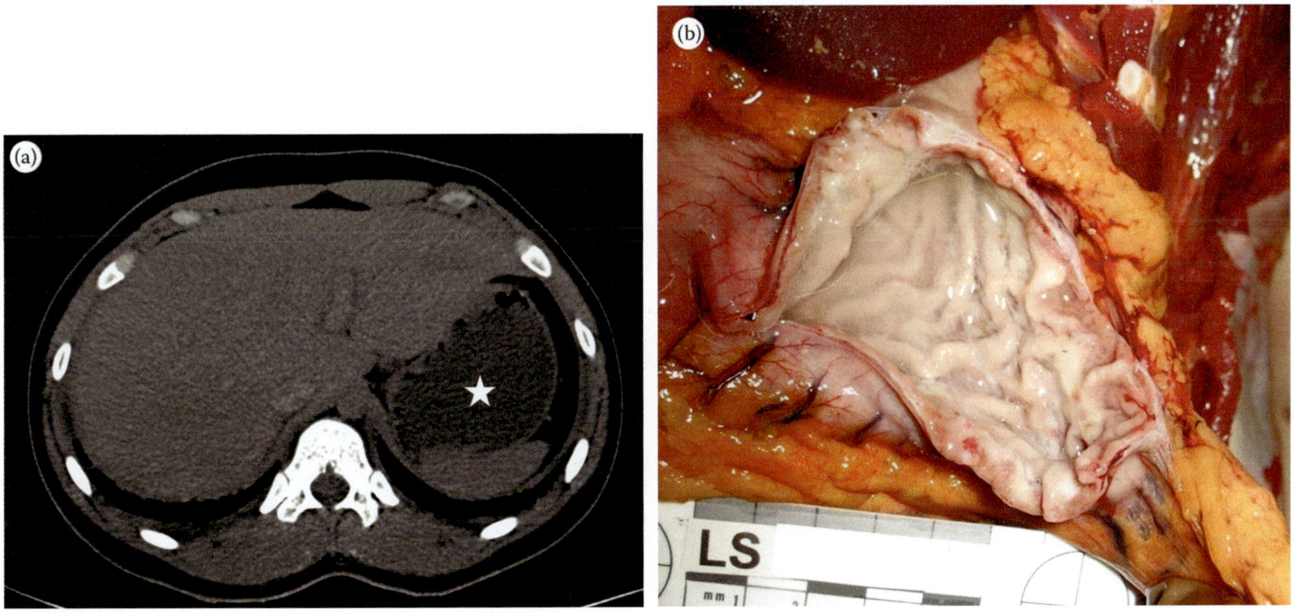

FIGURE 26.41 (a) The white asterisk marks the fluid-filled stomach. (b) Corresponding specimen of the stomach showing clear, sedimented contents.

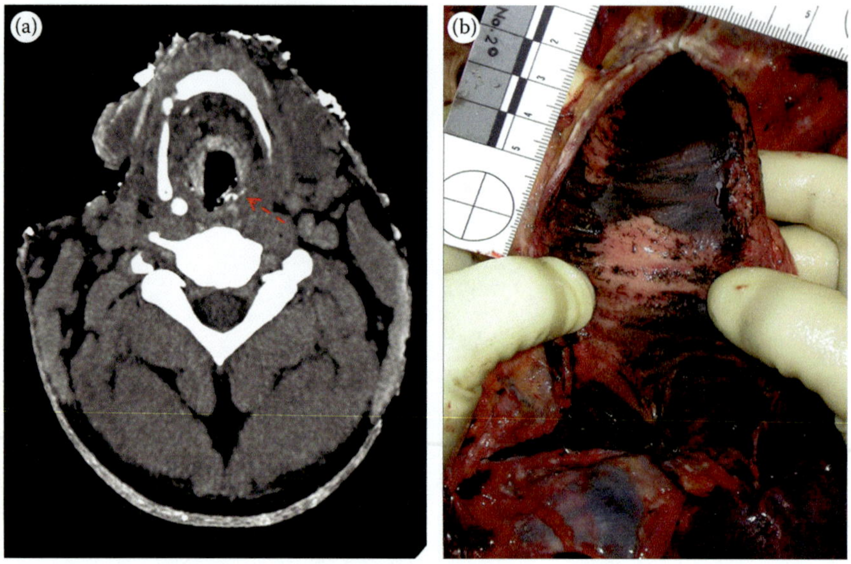

FIGURE 26.42 (a) MIP showing the axial plane at the level of the hyoid bone, depicting hyperdense material in the trachea. (b) Autopsy specimen with soot in the trachea.

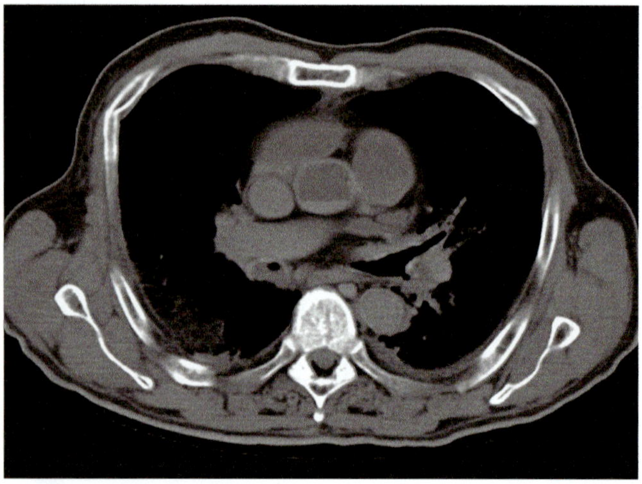

FIGURE 26.43 An inconspicuous pmCT with sedimentation of the corpuscular parts in the thoracic aorta leading to the impression of a hyperattenuated aortic wall.

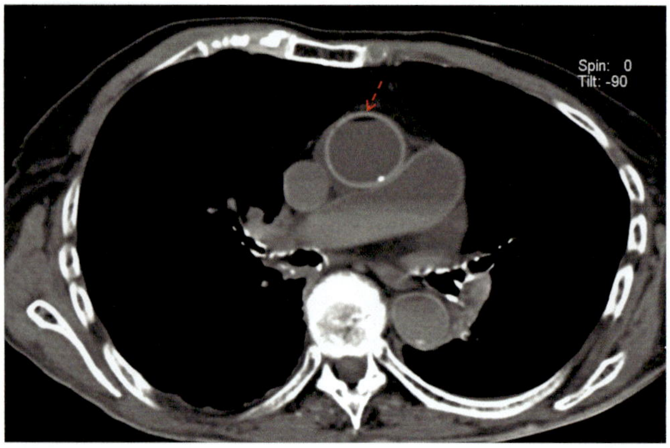

FIGURE 26.44 Fatal fat embolism (red arrow) with a fluid-fluid level in the ascending aorta.

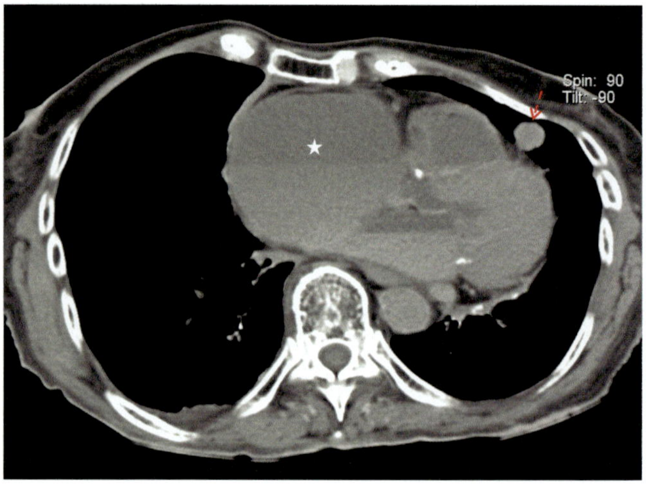

FIGURE 26.45 The right atrium is massively dilated (white asterisk) in this case of cardiac insufficiency caused by right heart congestion. In the inferior lingula, there is incidentally a benign tumor (red arrow).

abdominal trauma in every case as they could also be caused by ongoing decomposition (Figure 26.49). Subcutaneous hematoma or cutaneous lesions are better depicted on pmMRI; nevertheless, CT can still reveal those findings in a limited way and help to reconstruct the inflicted trauma.[29]

Intrahepatic Gas

Antemortem intrahepatic gas is usually found centrally in the liver, mainly in the biliary tree, indicating endoscopic retrograde cholangio-pancreatography, status post surgery, fistula formation in the bowel, or cholangitis. Gas in the portal vein (to the periphery of the liver) occurs in the living either from ischemic colitis or benign intestinal pneumatosis, both of which are rare diseases (Figure 26.50).

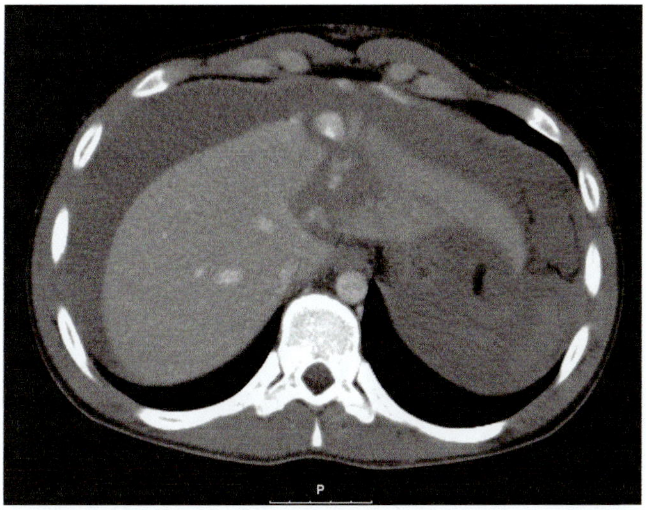

FIGURE 26.46 A clinical scan of a trauma patient with good soft tissue contrast and, therefore, a clear depiction of the central liver rupture with active contrast media extravasation and free fluid in the abdominal cavity. There is also a massive splenic rupture.

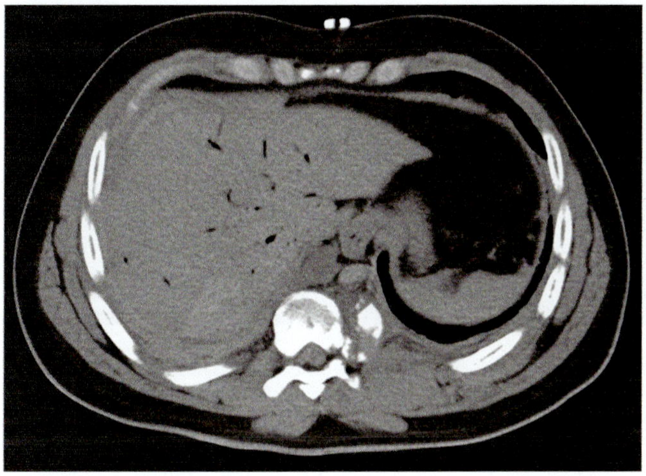

FIGURE 26.47 pmCT of a liver rupture with heterogeneous parenchyma and free fluid adjacent to the liver.

Intrahepatic gas is very common in postmortem imaging as described above (see Section General Postmortem Changes).[31] Arguably, the most frequent cause is putrefaction with initial autolysis of the pancreas eroding the stomach. In the course of decomposition, gas produced by bacteria will probably ascend from the intestines to the liver. Gas in the hepatobiliary system is one of the first indicators of initial autolysis. On further decay, the gas will symmetrically propagate into the whole soft tissue and the vessels (Figure 26.14). The least frequent postmortem causes of gas in the liver are those described above for the living.

Thermal Tissue Changes

Thermal tissue changes cause the skin to burst open and the intra-abdominal pressure to rise, causing the intestines to protrude. Intestinal protrusion is a common finding in charred bodies and is ascribable only to thermal impact (Figures 26.51 and 26.52).

EXTREMITIES

Trauma

In trauma cases, the fracture patterns seen by pmCT are extremely useful in the reconstruction of the location of the

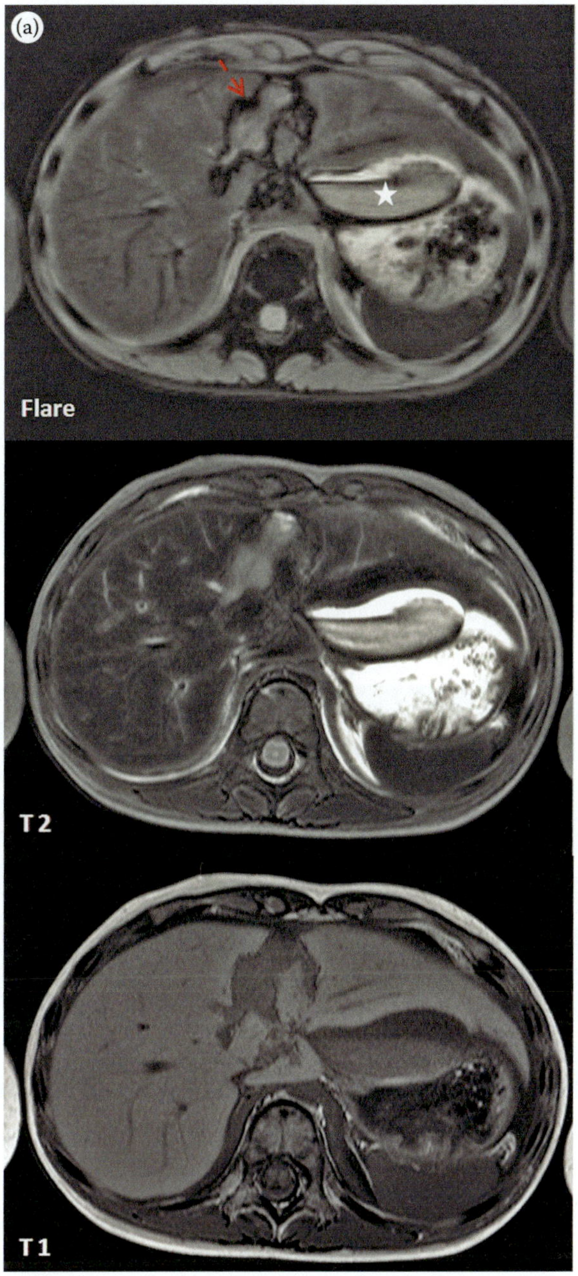

FIGURE 26.48 (a) Four-year-old boy. Postmortem MRI showing a central liver rupture (red arrow) with subcapsular hematoma on the dorsal side of the left lobe (white asterisk). (b) pmCT without contrast and lower sensitivity for parenchymal lesions. (c) Autopsy specimen confirming the radiological findings.

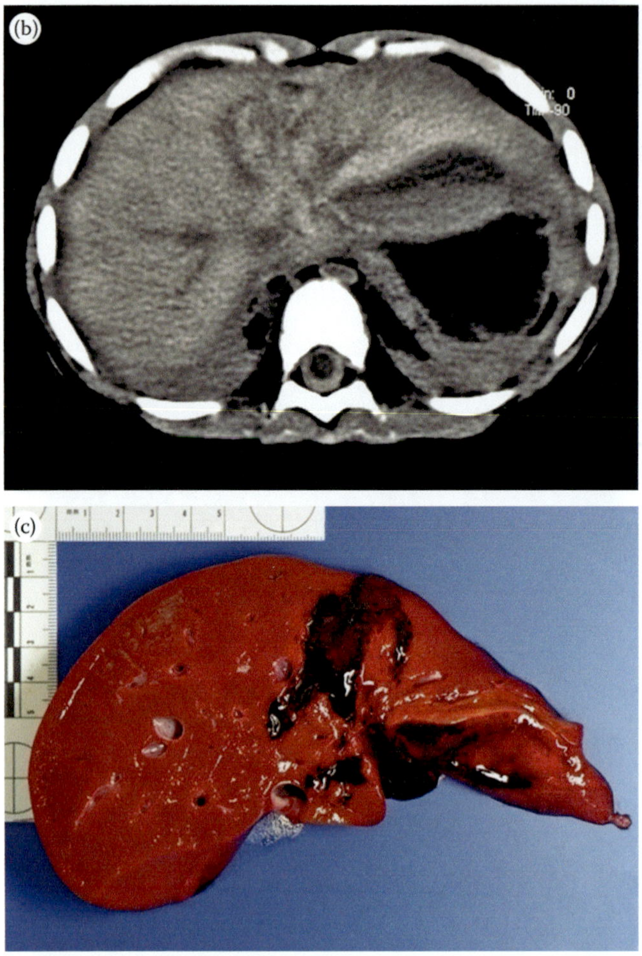

FIGURE 26.48 *(Continued)*

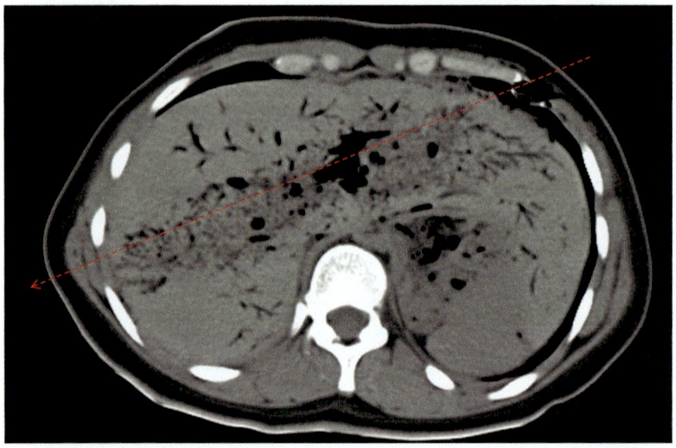

FIGURE 26.49 Laceration of the liver caused by a shot wound with early decompositional changes combined with air in the intrahepatic veins caused by open abdominal trauma and full penetration of the diaphragm.

impact and the body's position during the impact. A classic example is the flexion fracture (Messerer's wedge) of long bones (Figure 26.53). The base of the wedge will indicate the side of direct impact. Other forensic or reconstructive inquiries are clearly depicted with easy-to-assess 3D CT

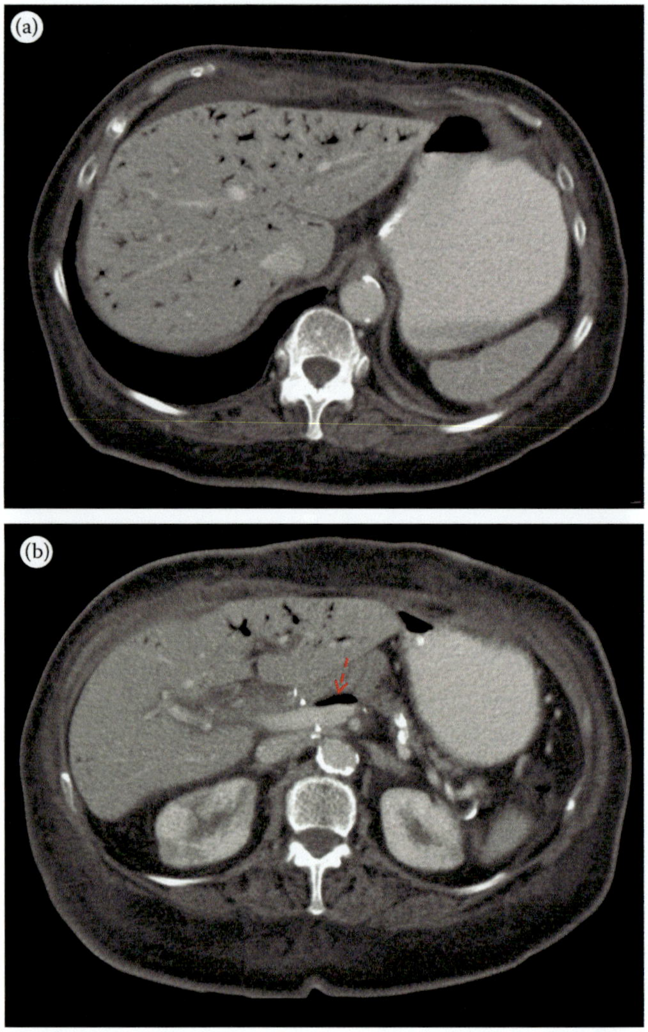

FIGURE 26.50 Antemortem images of ischemic colitis with pneumatosis intestinalis in the same patient. (a) Gas in the intrahepatic portal veins. Note the correctly contrasted hepatic veins. (b) Gas in the portal (red arrow) and mesenteric veins with a slight resemblance to decompositional changes in pmCT.

reconstructions of the extremities (Figure 26.54). Additionally, MRI provides an excellent display of subcutaneous hematoma with possible reconstructive conclusions (Figure 26.55).

Thermal Tissue Changes

Charred victims often show a fencing position of the extremities; therefore, imaging of these corpses with a large field of view (FoV) is strongly recommended (Figure 26.52). The cancellous bone will show rarefication, resulting in heat fractures.

CONCLUSION

The differences between imaging the living and the dead are sometimes not obvious, as in the detection of cranial hemorrhage. At other times, the lack of contrast in postmortem imaging has quite obvious drawbacks, as in the imaging of

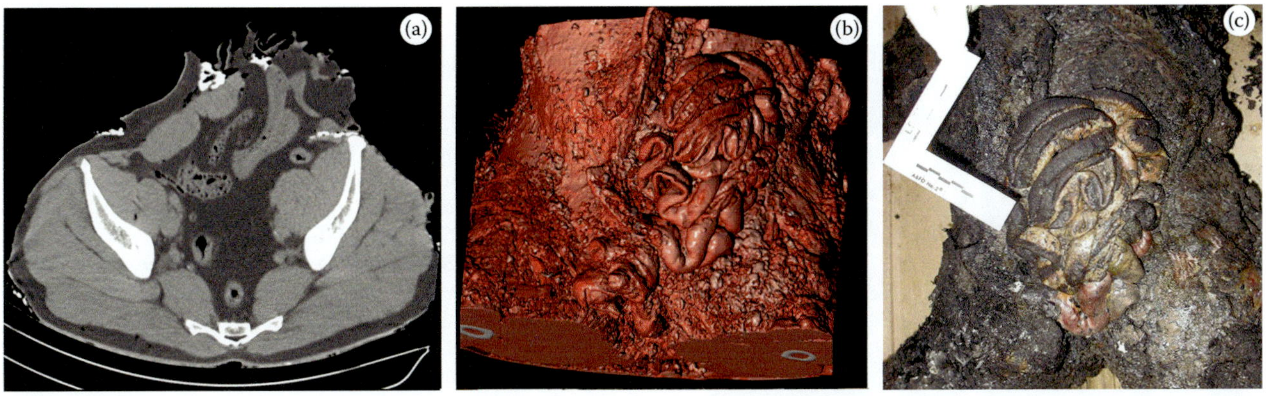

FIGURE 26.51 (a) Axial plane depicting the protrusion of the colon and ileum, and partially the jejunum, with a lateral shift of the abdominal musculature due to thermal impact. (b) 3D reconstruction of the protruded intestines. (c) Photograph of the inspection showing a completely charred body.

vasculature or organs. To overcome the shortcomings of unenhanced pmCT, CTA and MRI can supplement the increasingly routinely utilized pmCT. A broad overview of the main differences between forensic radiology and clinical radiology is intended here, with the aim of facilitating access to postmortem imaging for clinical radiologists and forensic pathologists. Ideally, forensic radiology will eventually be an acknowledged specialization of radiology because clinical and forensic radiology are *not* the same.

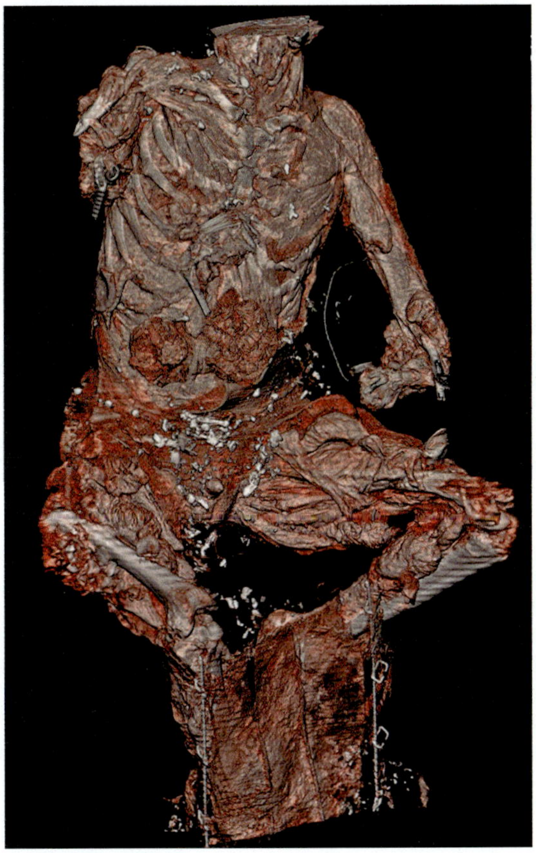

FIGURE 26.52 Fencing position of a charred body, reconstructed with eFoV. Note the protruded intestines.

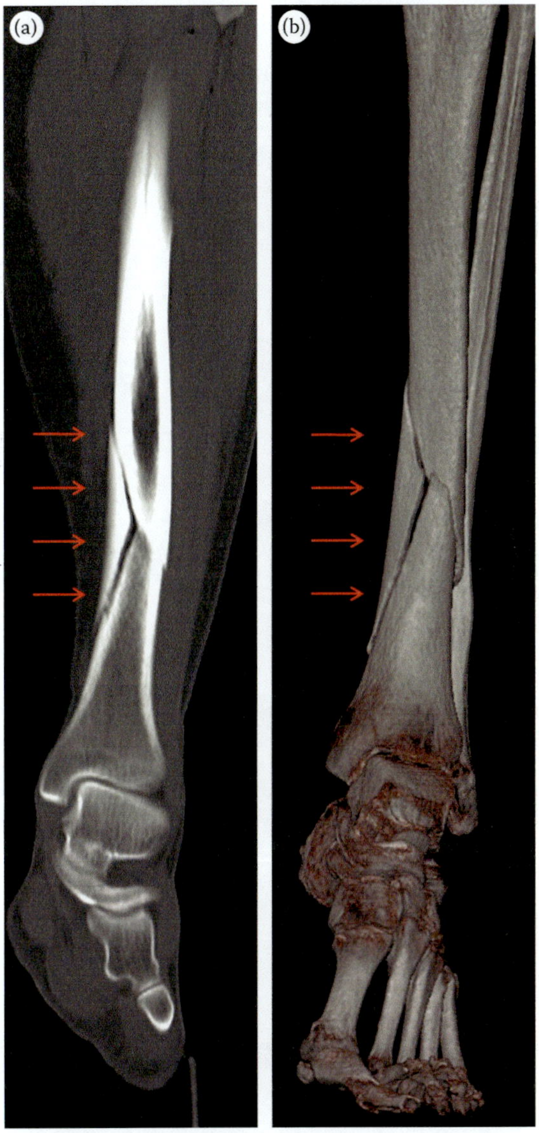

FIGURE 26.53 (a) Coronal MIP of the left lower extremity. (b) 3-D reconstruction. Both images depict direct trauma (red arrows) to the medial tibia indicated by the adjacent base of the "butterfly fragment."

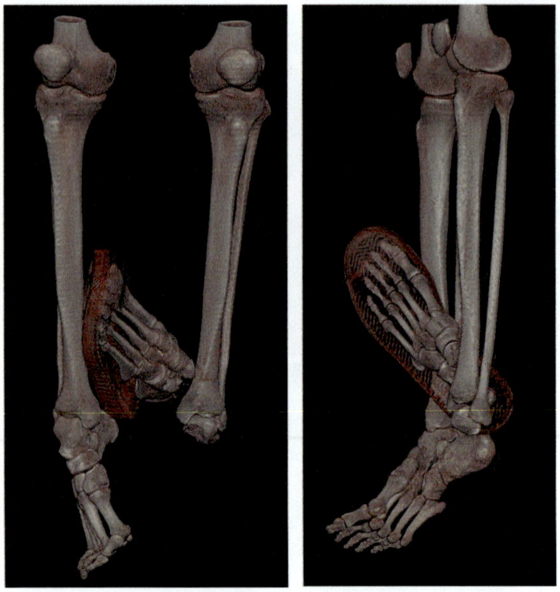

FIGURE 26.54 Subtotal amputation of the left foot at the level of the talus sustained in a parachute accident.

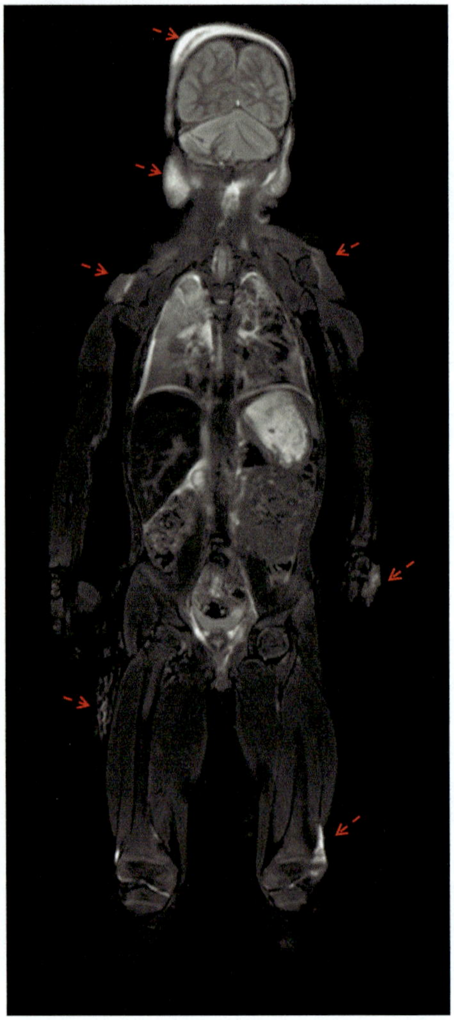

FIGURE 26.55 Coronal turbo-inversion recovery-magnitude (TIRM) sequence (TIM) with multiple subcutaneous and soft tissue hematomas (red arrows) in a child run over by a vehicle.

REFERENCES

1. Thali, M. J., Dirnhofer, R., Vock, P., *The Virtopsy Approach*, CRC Press, Boca Raton, FL, 2009, chap. C1, pp. 145–6.
2. Thali, M. J., Yen, K., Schweitzer, W., et al., Virtopsy, a new imaging horizon in forensic pathology: Virtual autopsy by postmortem multislice computed tomography (MSCT) and magnetic resonance imaging (MRI)—a feasibility study, *J Forensic Sc.*, 48(2), 386–403, 2003.
3. Ross, S., Spendlove, D., Bolliger, S., et al., Postmortem whole-body CT angiography: Evaluation of two contrast media solutions, *Am J Roentgenol*, 190(5), 1380–9, 2008.
4. Grabherr, S., Gygax, E., Sollberger, B., et al., Two-step postmortem angiography with a modified heart–lung machine: Preliminary results, *Am J Roentgenol* 190(2), 345–51, 2008.
5. Jackowski, C., Persson, A., Thali, M. J., Whole body postmortem angiography with a high viscosity contrast agent solution using poly ethylene glycol as contrast agent dissolver, *J Forensic Sci*, 53(2), 465–8, 2008.
6. Patzelt, D., Lignitz, E., Keil, W., et al., [Diagnostic problem of air embolism in a corpse.] *Beitr Gerichtl Med* 37, 401–5, 1979.
7. Pedal, I., Moosmayer, A., Mallach, H. J., et al., [Air embolism or putrefaction? Gas analysis findings and their interpretation.] *Z Rechtsmed*, 99, 151–67, 1987.
8. Keil, W., Bretschneider, K., Patzelt, D., et al., [Air embolism or putrefaction gas? The diagnosis of cardiac air embolism in the cadaver.] *Beitr Gerichtl Med*, 38, 395–408, 1980.
9. Yen, K., Lövblad, K.O., Scheurer, E., et al., Post-mortem forensic neuroimaging: Correlation of MSCT and MRI findings with autopsy results, *Forensic Sci Int*, 15, 173(1), 21–35, 2007.
10. Chien, D., Kwong, K. K., Gress, D. R., et al., MR diffusion imaging of cerebral infarction in humans, *Am J Neuroradiol*, 13, 1097–1102, 1992.
11. Mintorovitch, J., Yang, G. Y., Shimizu, H., et al., Diffusion-weighted magnetic resonance imaging of acute focal cerebral ischemia: Comparison of signal intensity with changes in brain water and Na+, K(+)-ATPase activity, *J Cereb Blood Flow Metab*, 14, 332–6, 1994.
12. Schaefer, P. W., Grant, P. E., Gonzalez, R. G., Diffusion-weighted MR imaging of the brain, *Radiology* 217, 331–45, 2000.
13. Srinivasan, A., Goyal, M., Al Azri, F., et al., State-of-the-art imaging of acute stroke, *Radiographics*, 26, Suppl 1, S75–95, 2006.
14. Añon, J., Remonda, L., Spreng, A., et al., Traumatic extra-axial hemorrhage: Correlation of postmortem MSCT, MRI, and forensic-pathological findings, *J Magn Reson Imaging*, 28(4), 823–36, 2008.
15. Bradley, W. G., Hemorrhage, in *Magnetic Resonance Imaging*, Stark, D. D., and Bradley, W. G., Eds.,. Mosby, St. Louis, pp. 1329–1359, 1999.
16. Jackowski, C., Thali, M., Aghayev, E., et al., . Postmortem imaging of blood and its characteristics using MSCT and MRI, *Int J Legal Med*, 19, 1–8, 2005.
17. Jaspan, T., Current controversies in the interpretation of non-accidental head injury, *Pediatr Radiol*, 38 (Suppl 3), S378–87, 2008.
18. Christe, A., Thoeny, H., Ross, S., et al., Life-threatening versus non life-threatening manual strangulation: Are there appropriate criteria by MR Imaging of the neck? *Eur Radiol*, ahead of print. 2009.

19. Yen, K., Thali, M. J., Aghayev, E., et al., Strangulation signs: Initial correlation of MRI, MSCT, and forensic neck findings, *J Magn Reson Imaging*, 22(4), 501–10, 2005.
20. Yen, K., Vock, P., Christe, A., et al., Clinical forensic radiology in strangulation victims: Forensic expertise based on magnetic resonance imaging (MRI) findings, *Int J Legal Med*, 121(2), 115–23, 2007.
21. Kempter, M., Ross, S., Spendlove, D., et al., Post-mortem imaging of laryngohyoid fractures in strangulation incidents—first results, *Int J Legal Medicine*, ahead of print, 2009.
22. Aghayev, E., Christe, A., Sonnenschein, M., et al., Postmortem imaging of blunt chest trauma using CT and MRI: Comparison with autopsy, *J Thorac Imaging*, 23(1), 20–7, 2008.
23. Christe, A., Aghayev, E., Jackowski, C., et al., Drowning—post-mortem imaging findings by computed tomography, *Eur Radiol*, 18(2), 283–90, 2008.
24. Shiotani, S., Kohno, M., Ohashi, N., et al., Hyperattenuating aortic wall on postmortem computed tomography (PMCT), *Radiat Med*, 20(4), 201–6, 2002.
25. Aghayev, E., Sonnenschein, M., Jackowski, C., et al., Postmortem radiology of fatal hemorrhage: Measurements of cross-sectional areas of major blood vessels and volumes of aorta and spleen on MDCT and volumes of heart chambers on MRI, *Am J Roentgenol*, 187(1), 209–15, 2006.
26. Miller, J. A., Singer, A., Hinrichs, C., et al., Cardiac dimensions derived from helical CT: Correlation with plain film radiography, *Internet J Radiol*, 1(1), 2000.
27. Shiotani, S., Kohno, M., Ohashi, N., et al., Dilatation of the heart on postmortem computed tomography (PMCT): Comparison with live CT, *Radiat Med*, 21(1), 29–35 and 21(2), 2003.
28. Jackowski, C., Christe, A., Sonnenschein, M., et al., Postmortem unenhanced magnetic resonance imaging of myocardial infarction in correlation to histological infarction age characterization, *Eur Heart J*, 27(20), 2459–67, 2006.
29. Christe, A., Ross, S. G., Oesterhelweg, L., Spendlove, D., Bolliger, S., Vock, P., Thali, M. J., Abdominal trauma—sensitivity and specificity of postmortem non contrast imaging findings compared to autopsy findings, *J Trauma*, 66(5),1302–7, 2009.
30. Kelly, J., Raptopoulos, V., Davidoff, A., et al., The value of non-contrast-enhanced CT in blunt abdominal trauma, *Am J Roentgenol*, 152, 41–8, 1989.
31. Jackowski, C., Sonnenschein, M., Thali, M. J., et al., Intrahepatic gas at postmortem computed tomography: Forensic experience as a potential guide for *in vivo* trauma imaging, *J Trauma*, 62(4), 979–88, 2007.

27 Postmortem Biopsy

Steffen G. Ross, Lars C. Ebert, Laura Filograna, and Michael J. Thali

CONTENTS

INTRODUCTION

Postmortem imaging has proven to be a useful tool in the examination of traumatic deaths, and with the application of contrast administration for postmortem angiographies, has dramatically had its diagnostic spectrum extended to identifying natural causes of death. Unfortunately, even contrast-enhanced computed tomography (CT) and magnetic resonance imaging (MRI) fail to completely assess histopathologic changes, wherein the actual image resolution obtained using CT and MRI is not high enough to reliably examine tissue specimens and cellular changes. The first steps toward "virtual histology" has been achieved using micro-CT and micro-MRI (Thali et al., 2004); however, these techniques are still areas of ongoing research and are not yet competitive alternatives to classical histopathologic examination of stained tissue specimens, which is the established gold standard in clinical and forensic pathology. Furthermore, a complete toxicologic screening of body fluids requires a sampling of body fluids, including blood, urine, and cerebrospinal fluid (Figure 27.1). MR-spectroscopy has demonstrated some exciting possibilities for augmenting classic toxicologic analysis by having the ability to identify certain metabolites by their MR behavior (Scheurer et al., 2005), but is also far from being a viable alternative examination method. Image-guided core needle biopsies and fluid sampling are minimally invasive approaches for gathering diagnostically relevant information that is otherwise not provided by CT and MRI, as is the case for myocardial infarctions, malignancy, and injury vitality assessment. Moreover, image-guided needle placement facilitates a sampling that is free of contamination. Image-guided core needle techniques are not time-consuming methods, which allows for fast sampling of the specimen. The speed of these techniques overcomes the problem of rapidly onsetting autolysis and decomposition in situations when the autopsy can only be performed several days after death due to either legal considerations or to excessive workloads in small forensics departments. A minimally invasive approach is of great value in cases where traditional dissection of the body is rejected by relatives for religious or cultural reasons or is refused by the court system (Terry, 1955). Image-guided tissue and fluid sampling is a powerful, minimally invasive tool that can dramatically improve the diagnostic value of virtual autopsies. The first steps toward implementing postmortem image-guided biopsies have been performed and have shown promising results (Aghayev et al., 2007, 2008; Weustink et al., 2009).

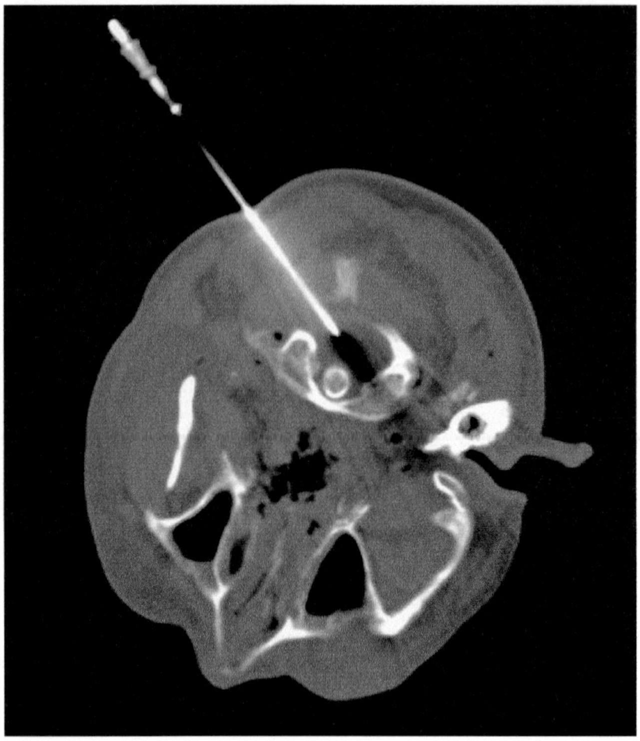

FIGURE 27.1 CT-fluoroscopy-guided intradural placement of a hollow needle at the level of the craniocervical junction for contamination-free sampling of cerebrospinal fluid.

TECHNICAL CONSIDERATIONS

Bioptic Tissue Sampling Techniques and Needles

Fine Needle Aspirations

Fine needle aspiration (FNA) is a percutaneous procedure that uses a fine gauge needle (22G).

Typically, a CHIBA needle (18–22G) with a blind ending inner trocar is used. Since FNA only removes a minimal number of cell clusters, it only allows for cytologic and microbiologic diagnostics. The thin caliber and length of the needle have aggravating dynamic effects in soft tissue, which misguides the inserted needles. Therefore, this method is not suitable for comprehensive histopathologic examination (Hopper et al., 1995).

Tru-Cut or Core Biopsies with Automated Biopsy Guns

The use of automated biopsy devices, or biopsy guns, has become ubiquitous in clinical radiology. There are multiple similar working systems on the market. Automated biopsy devices are easy to use, reliable, and provide generally comparable specimens for histopathologic analysis (Hopper et al., 1993, 1995; de Man et al., 2004). In our group, we use an automatic biopsy gun (Bard Magnum®). Tru-Cut® (Cardinal Health, Dublin, Ohio) biopsy needles are designed to capture high-quality tissue samples by removing a hemicylinder from the targeted tissue volume. Needle calibers typically range from 14 to 18G.

The Coaxial Technique with an Introducer Needle

The coaxial technique uses a hollow needle with a blind-ending inner trocar, which is initially placed under image guidance. The actual biopsy is taken through the lumen of the hollow needle after removing the inner trocar. The advantage of this technique is the capability of taking multiple tissue samples from the same target volume without the need of a new puncture for every biopsy. Since there is often not enough space in the CT gantry to place a long Tru-Cut needle with the connected biopsy gun, the biopsy itself can be performed outside the CT-Gantry by using the easier-to-handle shorter introducer needle as a preformed channel.

Targeting Procedures for Correct Needle Placement

A needle biopsy in a postmortem setup is much easier to perform than in a clinical environment because of the absence of "moving targets" and the possibility of practically unlimited biopsy attempts in the event of a misplaced needle tip. Precision needle placement under CT-fluoroscopy or navigated robotic biopsy allows for the harvesting of tissue samples with an accuracy that is not comparable to classic autopsy techniques. Although needle placement is also possible under sonography (Fariña et al., 2002) or under MR guidance, we will focus in this chapter on CT-guided and navigated biopsies because they have, in our opinion, the greatest potential for a seamless integration into a postmortem imaging workflow.

CT-Guided Biopsies

The main advantages of biopsy planning with CT are high spatial resolution, visualization of the inside of the body without superimposition of structures, and the possibility of nearly real-time needle-control in CT-fluoroscopy. There are several different techniques for performing a CT-guided biopsy.

Classic Step-and-Shoot Technique

Needle approach planning is conducted with cross-sections from a previously acquired CT, while needle position control is maintained with subsequent scans of the region-of-interest. The examining personnel can avoid radiation exposure by leaving the room during the control scans. The needle should be inserted strictly into the scan plane of the CT for proper display of the biopsy needle. Angling the needle in the z-axis, along the CT couch, is made possible by tilting the CT gantry. This technique can be quite time-consuming when multiple biopsies are taken from different organs.

CT-Fluoroscopy

CT scanners have become capable of real-time image-guided procedures with the recent development of rapid CT image reconstruction (Bissoli et al., 2003, Paulson et al., 2001). The main advantage of this technique is the markedly decreased procedure time in comparison with the use of conventional

CT-guidance. Needle approach is also planned with data from a prior CT, but the needle is placed under real-time control with the body in the CT gantry. The examiner is standing next to the CT couch with direct control of needle placement and instant access to the fluoroscopic functions of the scanner. The real-time, simultaneous representation of the target slice and the adjacent superior and inferior slices allows an immediate identification of the needle tip and its direction. Appropriate radiation protection measures must be taken to avoid radiation exposure of the examining personnel.

Navigated Biopsies and Robot-Assisted Biopsies

Most of the authors practicing needle autopsies acknowledge the difficulty of obtaining samples from the desired organs or their parts, which lowers the diagnostic reliability of the method compared with traditional, complete autopsies (West and Chomet, 1957; Underwood et al., 1983; Baumgart et al., 1994; Guerra et al., 2001).

Two developments could potentially overcome this difficulty: surgical navigation and robotic assistance. Surgical navigation uses different imaging modalities and either optical or electromagnetic tracking to generate additional positional information for the surgeon and to enhance the planning process. It has been successfully used in brachytherapy and neurosurgery for precise needle placement (Shimoga and Khosla, 1994). Since then, surgical navigation has become a standard in medical interventions in the fields of neurosurgery and orthopedic surgery (Kettenbach et al., 2008). Several attempts have been made to bring robots into the operating room to increase the precision of interventions. In 1998, a system for robotic biopsies was presented by Bishoff et al. (1998); however, robotic assistance has not yet found its way into forensic autopsies.

Computer-Assisted Navigated Biopsies

CT imaging has proven to be a valuable image modality for positioning biopsy needles, but with drawbacks, such as radiation exposure and a limited needle insertion angle. One way to overcome these limitations without losing the advantages of CT imaging is to use surgical navigation techniques (Figures 27.2 through 27.3). Computer-assisted navigation is being used in daily routines in neurosurgery and orthopedic surgery for precise tool placement.

In computer-assisted navigation, a tracking system tracks the three-dimensional position and orientation of navigation probes that are rigidly attached to all anatomic structures and tools that are involved in the procedure. Different tracking technologies are available. Optical tracking systems that use probes that either reflect infrared light (passive) or emit it (active) are the most common. Electromagnetic tracking systems with sensor coils do not require a line of sight, but are easily disturbed by ferromagnetic materials.

A computer system uses the actual position of the navigation probes as well as the planned needle trajectories to assist the examiner in correctly placing the needle by giving visual feedback about how to move and rotate the needle. Unlike a CT-guided biopsy, a navigated biopsy requires three additional steps. First, a tracking probe has to be rigidly attached to the body. Second, the exact position of the CT slice stack relative to the tracking probe has to be determined. This step is called registration, and involves the identification of the same prominent points in the CT slice stack as well as on the real anatomy with a tracked pointing device. Third, tissue sampling points are defined in the CT slice stack on the navigation system. Finally, a tracked biopsy that is tool-guided by the navigation system is used to place the needle. In a CT-guided biopsy, the greatest accuracy is achieved when the needle is placed entirely in the current CT plane. Needle

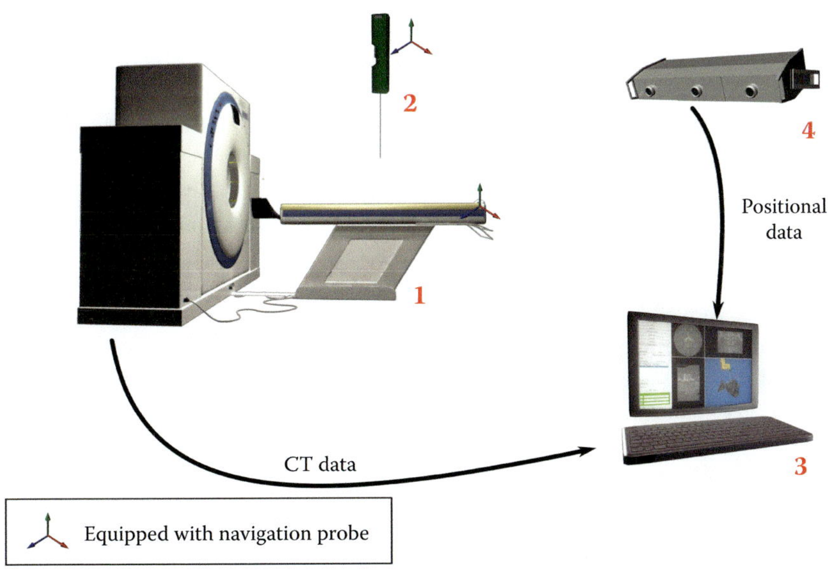

FIGURE 27.2 The system setup for a navigated biopsy. The navigation computer (3) uses the CT volume data generated by a CT scanner (1) and the three-dimensional tracking information gathered by the tracking system (4) to give visual feedback and to allow for intervention planning. The CT table and tools, such as the biopsy pistol (2), are equipped with a trackable navigation probe.

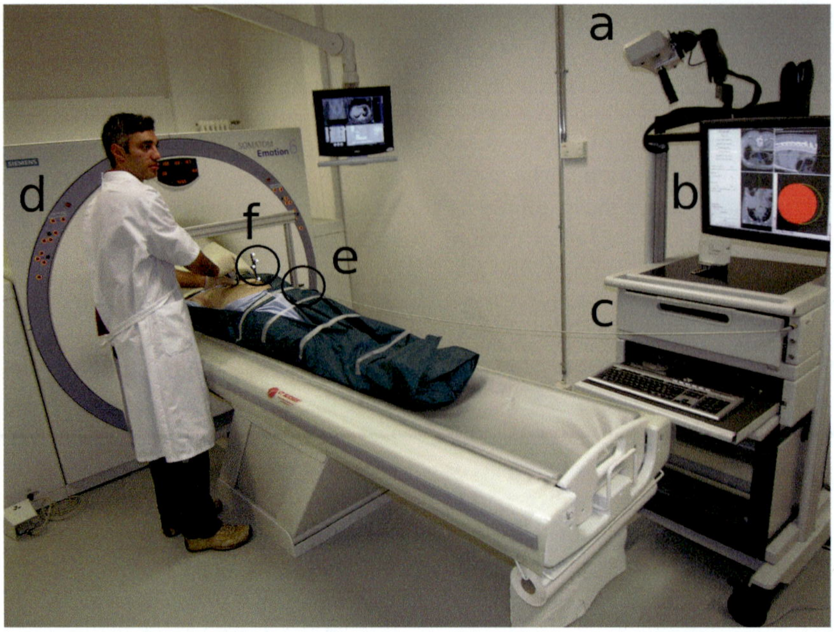

FIGURE 27.3 A navigated biopsy in use: (a) A hybrid (active/passive) camera system Polaris (NDI, Canada), (b) a touch screen for visual feedback, (c) a navigation computer based on the Co-Me medical application framework Marvin (Rudolph et al., 2008), (d) a CT scanner, (e) an active navigation probe attached to the CT table, and (f) a modified biopsy pistol with a mounted navigation probe.

trajectories that are not parallel with the gantry tilt are difficult to place and have a lower accuracy. In contrast, a navigated biopsy uses the entire volume data for visualization, and therefore, the insertion angle is only limited by the visibility of the navigation probes. Radiation exposure is entirely eliminated for the examining personnel. Since the needle is still manually placed, the outcome depends on the training level and skills of the examiner.

Navigated Biopsies with Robotic Assistance

In order to eliminate the human factor, actual needle placement can be automated by combining a robotic system with the navigation system (Ebert et al., 2009). Similar to navigated biopsies, this approach requires navigation probes, a registration procedure of the CT scan to the body probe, as well as trajectory planning (Figure 27.4). This planning includes defining the target of the biopsy, the insertion point on the body surface, and the orientation of the biopsy end-effector. Needle placement is performed by a robotic arm and is monitored by a navigation system that can calculate compensation motion if the body is moved (Figure 27.5). An introducer needle is placed to minimize needle bending and to allow for tissue and fluid sampling; it is used to guide the actual biopsy needle. A second CT scan can be performed to double check the correct needle placement. A robotic system has several advantages. Since the biopsy procedure is independent of the CT plane, it can be performed from any angle that is accessible by the robot. Since there is no radiation exposure to the examiner, more biopsies can be taken without additional risk. The accuracy of the robotic system is independent of the skills and training level of the examiner. Finally, a robotic system allows for tele-biopsies, so the volume can be submitted via the internet and the trajectory planning can be done by an external expert.

APPLICATIONS OF POSTMORTEM BIOPSY IN FORENSIC PATHOLOGY

In this section, we will present the role of core needle biopsy and fluid sampling in diagnosing some causes of death and establishing the vitality of certain lesions.

The correct placement of the biopsy needle is essential for such an undertaking. In many cases, this cannot be guaranteed without image guidance, such as with cross-sectional imaging. Cross- sectional imaging is ideal in identifying the tissue or the anatomic cavity (i.e., heart cavities, bladder, gallbladder, etc.) of interest and in guiding the biopsy needle even if the region of interest doesn't show any imaging abnormality (i.e., due to insufficient resolution of the tomographic image).

Forensic Investigations on the Cause of Death

Defining the cause of death only by postmortem imaging offers evident limitations especially in addressing likely natural deaths that nevertheless have to be frequently investigated by the forensic pathologist. The reason is the insufficient accuracy in visualizing and differentiating some vascular pathologies and microscopic organ alterations.

Among the vascular pathologies, those related to pulmonary thromboembolism and myocardial infarction are the main challenges in postmortem imaging. Postmortem CT-angiography (pmCTA) is able to depict the presence of a

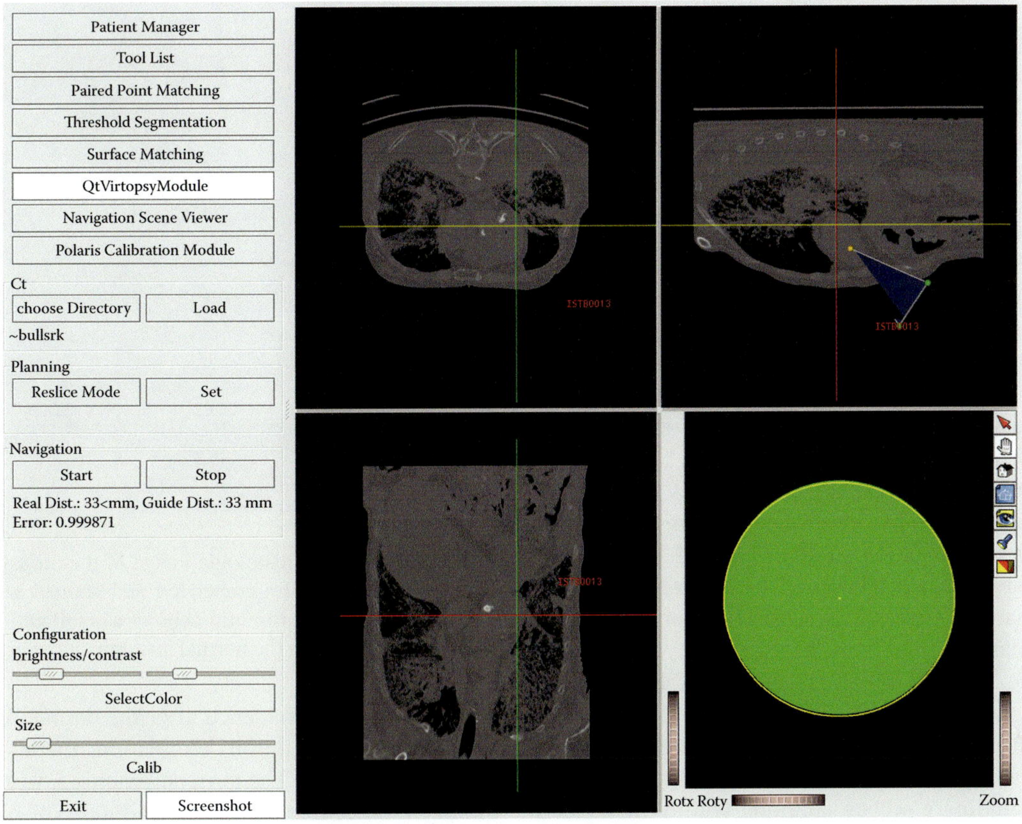

FIGURE 27.4 A screenshot of a navigated biopsy: The MPR re-slices at the defined target. The actual position of the biopsy pistol as well as its orientation is superimposed (top right corner). The targeting aid (bottom right) helps position the biopsy pistol to the planned trajectory. If the inner circle (green) and the outer circle (yellow) entirely overlap, then the biopsy pistol is aligned with the trajectory.

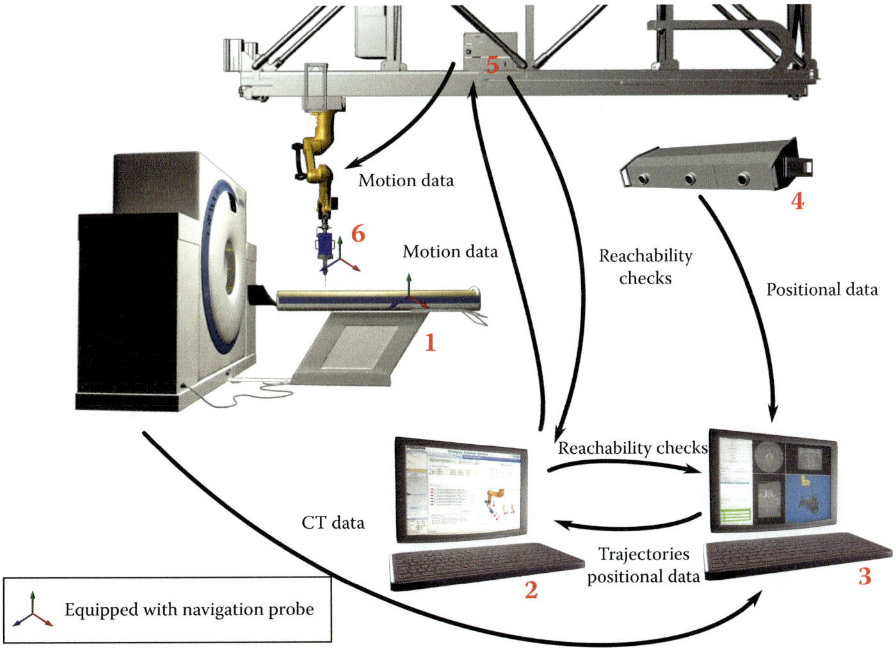

FIGURE 27.5 The Virtobot System setup for robotic biopsy: The entire workflow of a robotic biopsy is coordinated by a control computer that gathers patient data and logs every user interaction and robot motion (2). After initializing a biopsy project on the control computer, a CT scan is performed and the volume data are submitted from the scanner (1) to the planning and navigation station, (3) and are used to define biopsy trajectories. Positional data from an optical tracking system (4) as well as calibration data are used to calculate the final required robot position. A reachability check is performed with the actual robot control (5), and, if possible, the robot with the mounted biopsy end effector (6) is positioned as planned. The position is rechecked and corrected if necessary, and then the needle is placed.

vascular occlusion, but it is not capable of attributing the correct origin. The differentiation between thromboembolic material and postmortem cloths only with imaging is not reliable enough. CT-guided needle biopsies of the suspected thrombotic material provide a sample, sufficient for a histopathologic diagnosis of the materials origin and so also of the cause of death.

According to autopsy examinations, the finding of an atherosclerotic coronary occlusion is not enough to diagnose death due to myocardial infarction. More information is necessary to consider atherosclerotic coronary occlusions that are relevant to the cause of death. Although pmCTA provides a clear visualization of atherosclerotic coronary occlusions, it does not provide histopathologic information about potential myocardial ischemia.

The combination of pmCTA of the coronaries and microscopic analysis of focused myocardial biopsies is fundamental for the correct evaluation of the cause of death in a minimally invasive manner (Figure 27.6). It has to be taken into account that the area of the earliest histopathologic changes in myocardial infarctions is the papillary muscles of the left ventricle, an area which is not accessible by blind biopsies. Other pathologies frequently leading to likely natural deaths examined by a forensic pathologist are tumors. The role of CT imaging in identifying a possibly malignant process and in guiding a proper sampling within the abnormal tissue is clear. Two examples are illustrated in Figures 27.7 and 27.8.

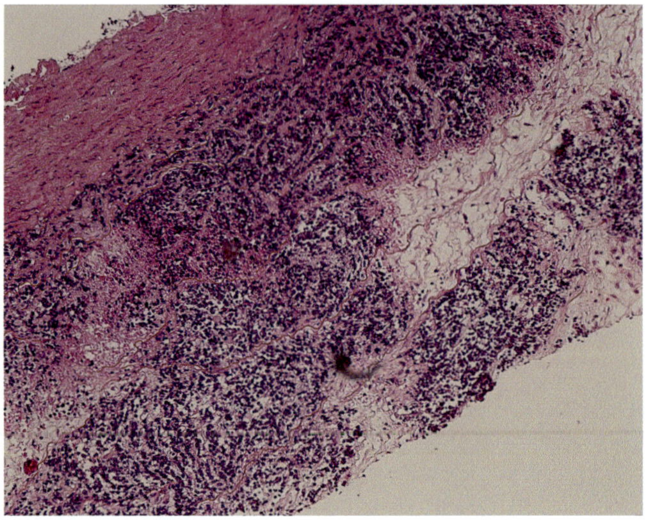

FIGURE 27.7 The histologic image of the biopsy sample of pulmonary mass, formalin-embedded and stained with hematoxylin and eosin, shows the tumoral cells of a small cell lung carcinoma completely subverting the normal anatomy of the lung parenchyma.

Forensic Investigations on Vitality

The main forensic query in which needle biopsies overcome the shortcomings of sectional imaging is the assessment of injury vitality, namely, in addressing whether an injury occurred before or after death. Vital reactions primarily depend on the permanence after the damaging event, of the intactness of respiration, circulation, metabolic processes, and consciousness. The most commonly investigated signs of vitality include hemorrhages, aspiration (of blood or gastric contents), the inhalation of gases (carbon monoxide), embolism (fat, gas, and thrombi), and inflammatory responses at a cellular level.

It has been demonstrated that gas embolism and hemorrhage can be detected by postmortem imaging techniques (Aghayev et al., 2006; Jackowski et al., 2007). Regardless of the other vital reactions, definitive diagnosis remains confined to histopathologic or biochemical analyses, which also occur in traditional techniques of investigation.

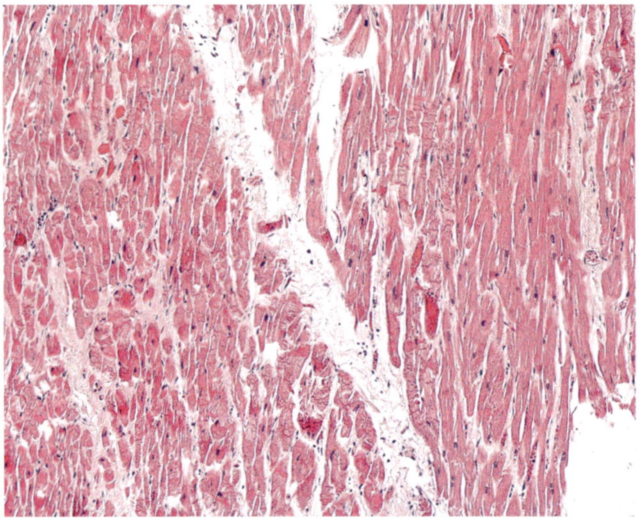

FIGURE 27.6 The histologic image of the biopsy sample of cardiac tissue, formalin-embedded and stained with hematoxylin and eosin, shows the classic appearance of an acute myocardial ischemia, with contraction band necrosis.

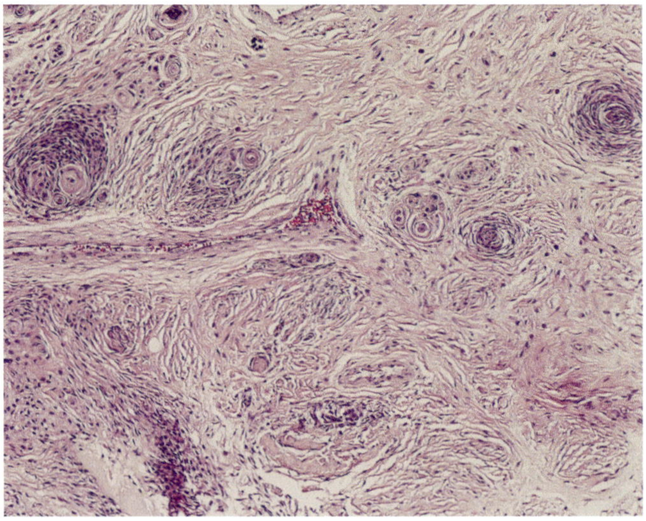

FIGURE 27.8 The histologic image of the biopsy sample of meningeal abnormal tissue, formalin-embedded and stained with hematoxylin and eosin, shows the characteristic presence of psammoma bodies, typically detectable in meningioma.

Aspiration

Postmortem CT imaging of lungs and airways is only able to suggest, with a high degree of probability, the occurrence of aspiration (Thali et al., 2003), especially when inflammatory changes have also occurred. Imaging findings, primarily represented by nodular ground glass opacities or consolidations, are not sufficiently specific to permit a sure diagnosis in the majority of the cases (Webb et al., 2009).

The possibility of correctly attributing a vital reaction to a macroscopic finding is strictly related to the detection of some histologic features of the suspected foci, with some differences depending on the nature and extent of the aspirated material and the survival time after aspiration. Examples include blood cells filling alveolar spaces with intact walls in blood aspiration (Figure 27.9) and lung tissue inflammatory and reactive changes, especially in cases of gastric content aspiration (Janssen, 1984a, Knight, 1975, Saukko and Knight, 2004a).

Inhalation of Gas

Image-guided core needle techniques can significantly improve the diagnostic value of virtual autopsy when gas inhalation may be proof of a vital reaction. For example, in cases of burned bodies, classical analysis focuses on the research of carboxyhemoglobin, preferably in central blood samples, to demonstrate that the victim was still alive when the fire was in progress. The presence of a significant level of carboxyhemoglobin in the blood indicates that the subject was breathing when the fire began (Saukko and Knight, 2004b). Image-guided core needle sampling can provide samples directly from the heart or great vessels for analysis.

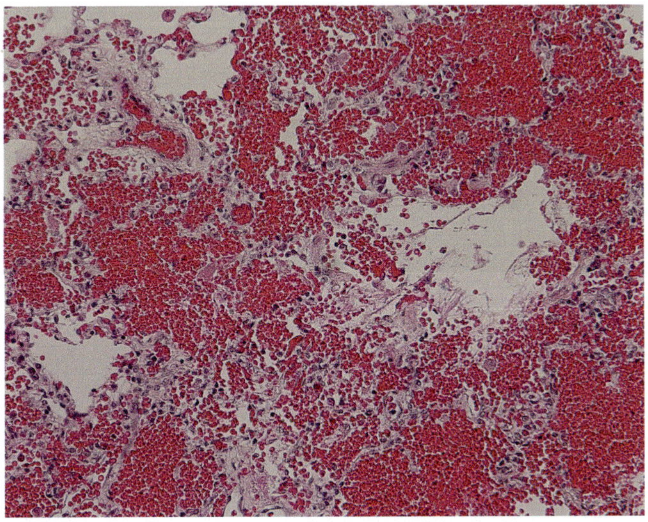

FIGURE 27.9 The image shows the histologic appearance of a biopsy sample (formalin-embedded, hematoxylin and eosin stained) obtained by one of the rounded areas of ground glass opacities evidenced on CT imaging of the lungs in a multitrauma case with multiple cranio-facial fractures. Alveola with intact walls are filled by blood cells, indicating the occurrence of blood aspiration.

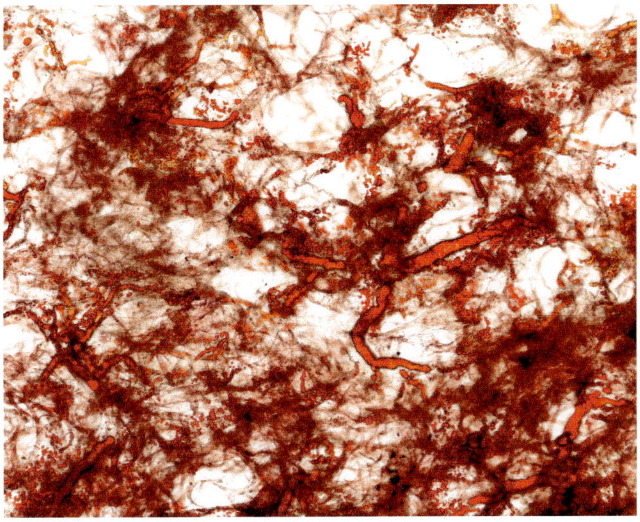

FIGURE 27.10 The image shows the histologic appearance of the biopsy sample of the pulmonary tissue stained with Oil Red, a stain specific for fat, after storage in water, in a multitrauma case. The fat, red-colored, in the pulmonary vessels indicates the presence of fat embolism.

Pulmonary Fat Embolism

Specific findings for severe fat embolisms in postmortem CT (pmCT) cannot be identified, even in cases of massive embolization with associated pulmonary reactive changes (Arakawa et al., 2000; Holbert et al., 2006; Prologo et al., 2004; Ravenel et al., 2002; Woo et al., 2008). A fat embolism itself is an embolic phenomenon of fat droplets within the microcirculation of the lungs. Pulmonary fat embolism (PFE) is a vital phenomenon after trauma, depending on the pumping action of the heart and intact circulation (Janssen, 1984b; Sevitt, 1962) (Figure 27.10).

The documentation and grading of PFE is a crucial medicolegal issue for two reasons: the simple detection of PFE could be interpreted as a sign of antemortem violence, and the documentation of massive PFE, together with the consideration of other data, could suggest a hypothesis to the forensic pathologist on the cause of death. PFE can act as a primary, assisting, or competing cause of death.

REFERENCES

Aghayev, E., Ebert, L. C., Christe, A., et al., CT data-based navigation for post-mortem biopsy—a feasibility study, *J Forensic Leg Med*, 15(6), 382–387, 2008.

Aghayev, E., Sonnenschein, M., Jackowski, C., Thali, M. J., Buck, U., Yen, K., Bolliger, S., Dirnhofer, R., and Vock, P., Postmortem radiology of fatal hemorrhage: Measurements of cross-sectional areas of major blood vessels and volumes of aorta and spleen on MDCT and volumes of heart chambers on MRI, *AJR Am J Roentgenol*, 187(1), 209–215, 2006.

Aghayev, E., Thali, M. J., Sonnenschein, M., Jackowski, C., Dirnhofer, R., and Vock, P., Post-mortem tissue sampling using computed tomography guidance, *Forensic Sci Int*, 166, 199–203, 2007.

Arakawa, H., Kurihara, Y., and Nakajima, Y., Pulmonary fat embolism syndrome: CT findings in six patients, *J Comput Assist. Tomogr.*, 24, 24–29, 2000.

Baumgart, K. W., Cook, M., Quin, J., Painter, D., Gatenby, P. A., and Garsia, R. J., The limited (needle biopsy) autopsy and the acquired immunodeficiency syndrome, *Pathology*, 26(2), 141–143, 1994.

Bishoff, J. T., Stoianovici, D., Lee, B. R., Bauer, J., Taylor, R. H., Whitcomb, L. L., Cadeddu, J. A., Chan, D., and Kavoussi, L. R., RCM-PAKY: Clinical application of a new robotic system for precise needle placement, *J Ensourol*, 12, 82, 1998.

Bissoli, E., Bison, L., Gioulis, E., Chisena, C., and Fabbris, R., Multislice CT fluoroscopy: Technical principles, clinical applications and dosimetry, *Radiol Med (Torino)*, 106, 201–212, 2003.

de Man, R. A., van Buuren, H. R., Hop, W. C. J., A randomised study on the efficacy and safety of an automated Tru-Cut needle for percutaneous liver biopsy, *Neth J Med*, 62, 441–445, 2004.

Ebert, L. C., Ptacek, W., Näther, S., et al., 2009, Virtobot—a multifunctional robotic system for 3D surface scanning and automatic post mortem biopsy, *Int J Med Robot* [Epub ahead of print]. Fariña, J., Millana, C., Fdez-Aceñero, M. J., et al., Ultrasonographic autopsy (echopsy): A new autopsy technique, *Virchows Arch.*, 440(6), 635–639, 2002.

Guerra, I., Ortiz, E., Portu, J., Atares, B., Aldamiz-Etxebarria, M., and De Pablos, M., Value of limited necropsy in HIV-positive patients, *Pathol Res Pract*, 197(3), 165–168, 2001.

Holbert, J. M., Desai, S. R., and Stern, E. J., Section I.1, airspace, vascular, in Diagnostic Imaging: Chest, Amirsys, Gurney, 2006, pp. 40–43.

Hopper, K. D., Abendroth, C. S., Sturtz, K. W., Matthews, Y. L., Hartzel, J. S., and Potok, P. S., CT percutaneous biopsy guns: Comparison of end-cut and side-notch devices in cadaveric specimens, *AJR Am J Roentgenol*, 164(1), 195–9, 1995.

Hopper, K. D., Abendroth, C. S., Sturtz, K. W., Matthews, Y. L., Shirk, S. J., and Stevens, L. A., Blinded comparison of biopsy needles and automated devices *in vitro*: 1. Biopsy of diffuse hepatic disease, *AJR Am J Roentgenol*, 161(6), 1293–7, 1993.

Jackowski, C., Sonnenschein, M., Thali, M. J., Aghayev, E., Yen K., Dirnhofer, R., and Vock, P., Intrahepatic gas at postmortem computed tomography: Forensic experience as a potential guide for *in vivo* trauma imaging, *J. Trauma* 62(4), 979–988, 2007.

Janssen, W., *Forensic Histopathology*, Springer, Berlin, 1984a, pp. 114–123.

Janssen, W., Vital reactions, in *Forensic Histopathology*, Janssen, W., Ed., Springer-Verlag, Berlin, 1984b, pp. 135–141.

Kettenbach, J., Kronreif, G., Melzer, A., Fichtinger, G., Stoianovici, D., and Cleary, K., Ultrasound-, CT- and MR-guided robot-assisted interventions, in *Image Processing in Radiology—Current Applications*, Neri, E., Caramella, D., and Bartolozzi, C., Eds., Springer-Verlag, Springer, 2008, pp. 391–408.

Knight, B. H., The significance of the post-mortem discovery of gastric contents in the air passages, *Forensic Sci.*, 6, 229–234, 1975.

Paulson, E. K., Sheafor, D. H., Enterline, D. S., McAdams, H. P., and Yoshizumi, T. T., CT fluoroscopy–guided interventional procedures: Techniques and radiation dose to radiologists, *Radiology* 220, 161–167, 2001.

Prologo, J. D., Dogra, V., and Farag, R., CT diagnosis of fat embolism, *Am J Emerg Med*, 22, 605–606, 2004.

Ravenel, J. G., Heyneman, L. E., and McAdams, H. P., Computed tomography diagnosis of macroscopic pulmonary fat embolism, *J Thorac Imaging*, 17, 154–156, 2002.

Rudolph, T., Puls, M., Anderegg, C., Ebert, L. C., Broehan, M., Rudin, A., and Kowal, J., MARVIN: A medical research application framework based on open source software, *Com Meth Prog Biomed*, 91(2), 165–74, 2008.

Saukko, P. and Knight, B., Suffocation and 'asphyxia', in *Forensic Pathology*, Ed., Saukko, P., and Knight, B., Arnold, London, 2004a, pp. 361–363.

Saukko, P. and Knight, B., Burns and scalds, in *Forensic Pathology*, 3rd ed., Saukko, P., and Knight, B., Eds., Arnold, London, 2004b, pp. 319–322.

Scheurer, E., Ith, M., Dietrich, D., et al., Statistical evaluation of time-dependent metabolite concentrations: Estimation of post-mortem intervals based on *in situ* 1H-MRS of the brain, *NMR Biomed*, 18(3), 163–172, 2005.

Sevitt, S., In *Fat Embolism*, Butterworth & Co, London, 1962.

Shimoga, K. and Khosla, P., Visual and force feedback to aid neurosurgical probe insertion. *Proceedings of the IEEE International Conference on Engineering in Medicine and Biology*, 1051–2, 1994.

Terry, R., Needle necropsy, *J Clin Pathol*, 8(1), 38–41, 1955.

Thali, M. J., Dirnhofer, R., Becker, R., Oliver, W., and Potter, K., Is 'virtual histology' the next step after the 'virtual autopsy'? Magnetic resonance microscopy in forensic medicine, *Magn Reson Imaging*, 22, 1131–1138, 2004.

Thali, M. J., Yen, K., Schweitzer, W., et al., Virtopsy, a new imaging horizon in forensic pathology: Virtual autopsy by postmortem multislice computed tomography (MSCT) and magnetic resonance imaging (MRI)—a feasibility study, *J Forensic Sci*, 48, 1–18, 2003.

Underwood, J. C., Slater, D. N., and Parsons, M. A., The needle necropsy, *Br Med J (Clin Res Ed)*, 286(6378), 1632–1634, 1983.

Webb, W. R., Müller, N. L., and Naidich, D. P., High-resolution computed tomoraphy findings of lung diseases, in *High-Resolution CT of the Lung*, Webb, W. R., Müller, N. L., and Naidich, D. P., Eds., Lippincott Williams & Wilkins, Philadelphia, 2009, pp. 88–131.

West, M. and Chomet, B., An evaluation of needle necropsies, *Am J Med Sci*, 234(5), 554–560, 1957.

Weustink, A. C., Hunink, M. G. M., van Dijke, C. F., et al., Minimally invasive autopsy: An alternative to conventional autopsy? *Radiology*, 250(3), 897–904, 2009.

Woo, O. H., Yong, H. S., Oh, Y. W., Shin, B. K., Kim, H. K., and Kang, E. Y., Experimental pulmonary fat embolism: Computed tomography and pathologic findings of the sequential changes, *J Korean Med Sci*, 23, 691–699, 2008.

28 Postmortem Angiography

Steffen G. Ross, Patricia M. Flach and Michael J. Thali

CONTENTS

INTRODUCTION

Postmortem investigation of the human vascular system is not a new topic. In the sixteenth century, Leonardo Da Vinci made preparations of the human vascular system by injecting wax into the cardiac chambers of a corpse and then used maggots to free the cast from the surrounding flesh. Only three years after the discovery of x-rays in 1896, the first x-ray-based postmortem angiography was performed by injecting a solution of x-ray opaque material into the coronary arteries of an isolated human heart. The technique of postmortem angiography remained basically the same in the next 100 years, the only variant being the injected contrast medium; the organ of interest had to be removed during autopsy, with successive injection of contrast medium followed by plain x-rays of the specimen. Major drawbacks of this method include the long hours required to prepare multiple organs in the same case for injection and the impossibility of displaying the entire vascular system of the corpse in a three-dimensional (3D) manner. Postmortem whole-body angiography remained an exceptional phenomenon in the twentieth century. The few cases recorded in the literature are exclusively examined animal or human fetuses, embryos, and newborns. An excellent and detailed summary of the history of postmortem angiography was published by Grabherr et al. (2007).

The introduction of computed tomography (CT) in 1974 and its combination with the injection of iodinated contrast medium has revolutionized clinical radiology. CT is much more sensitive to iodine than projection radiography and has the ability to show structures without superimposition. The evolution from single-slice, helical CT to multislice CT (MSCT) in the 1990s had a massive impact on clinical radiology. Diagnostic imaging shifted increasingly away from 2D plain x-ray to 3D MSCT, especially in angiography. A similar trend was observed with a certain time delay in postmortem imaging. Earlier postmortem radiology, which was almost completely plain x-ray-based, adopted more cross-sectional imaging modalities (Dirnhofer et al., 2006). This paradigm shift pioneered forensic imaging during the last decade. In particular, MSCT showed great potential in the imaging of bony pathologies as well as the depiction of air-containing structures, for example, gas embolism or pulmonary pathologies like drowning (Levy et al., 2007) or gunshot injuries (Levy et al., 2006). MRI with its magnificent soft-tissue contrast showed great potential in demonstrating soft-tissue injury, organ trauma, and pathologies of the central nervous system (Añon et al., 2008, Christe et al., 2009). The imaging of the vascular system remained a blind spot in postmortem radiology, however, relying mostly on indirect signs like perilesional hematoma or the collapse of the great vessels (Aghayev et al., 2006).

A clear depiction of traumatic vascular lesions, coronary stenoses, and pulmonary emboli is not possible without the administration of contrast medium. Postmortem CT angiography (pmCTA) not only closes this gap but also provides the unique possibility of examining structures that are not seen in traditional autopsy or conventional plain radiographs. Unlike the plain x-ray postmortem angiography of isolated body parts, minimally invasive pmCTA provides a 3D volume rather than projectional images; it thus provides superior information about overlapping structures. In addition, pmCTA could permit a more detailed diagnosis by displaying structures not readily apparent at autopsy. For instance, regions like the neck, skull base, or pelvis are not reviewable in classic autopsies without causing major damage to the corpse. In 2005, the first minimally invasive whole-body pmCTA of an adult cadaver took place (Jackowski et al., 2005). Since that time, the development of this technology has pushed forward and concentrated primarily on injection techniques and the evaluation of different contrast agents (Grabherr et al., 2008, Jackowski et al., 2008; Ross et al., 2008).

Differences from Clinical (Antemortem) CT Angiography

In contrast to clinical angiography, pmCTA does not involve imaging artifacts caused by heart movement or respiration due to the lack of motion artifacts. There is also no need for precisely timed imaging because there is no circulation with a contrast bolus to chase. Instead, a retrograde (arterial) or anterograde (venous) filling of the vessel lumen takes place. Postmortem clotting is also a problem in postmortem angiography; it must be properly evaluated and should not be generally mistaken as an antemortem thrombembolus. The parenchymal enhancement of different tissues seen in pmCTA is not an exact reflection of the antemortem perfusion.

TECHNICAL CONSIDERATIONS

Contrast Injection Method

The Virtopsy group uses a simple protocol, consisting of a divided injection of a mixture of polyethylene glycol (PEG) and a water-soluble contrast medium into the arterial and venous system without establishing new circulation. This technique was adopted at our institute, and it has proven its potential in more than 100 cases at this time. All of the described findings and images in this chapter are based on the experience gained in cases injected with the above-mentioned contrast mixture of PEG and water-soluble contrast medium.

Access to the Vascular System

The best access to the vascular system is gained by cannulation of the femoral vessels (Figure 28.1). This approach gives fast access (preparation time of less than 10 min) to the relatively large femoral vessels. The insertion of a T-shaped cannula permits the complete perfusion of the otherwise excluded leg at the site where the injection takes place. Other possible target vessels for cannulation are at the neck and the extremities. However, these areas have drawbacks of (1) invasiveness

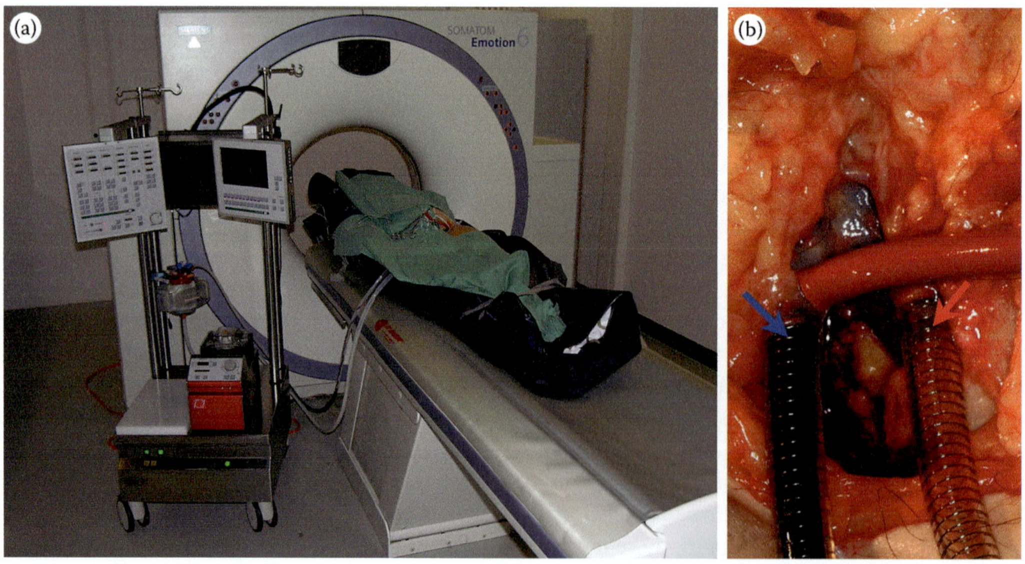

FIGURE 28.1 (a) Setup of the CT angiography. The pressure-controlled heart–lung machine is connected to the cannulated femoral vessels. The corpse lies in an impermeable body bag on the couch of the CT scanner. (b) Minimally invasive access to the vascular system through a femoral incision and cannulation of the femoral artery (red arrow) and vein (blue arrow). Removable ligatures ensure the fixation of the cannulae.

in areas (the neck) that could be seen by the next of kin at the funeral, and (2) potentially problematic small vessel diameters (extremities). The harvesting of body fluids for toxicological examination has to be done before angiography.

Cannulae and Injection Pump

Cannulation of the femoral vessels of an adult corpse can be performed with cannulae with calibers between 16F (arterial) and 20F (venous). A basic flow-calibrated roller pump is sufficient for administration of the necessary volume of contrast in a given time. The use of pressure monitoring in the vascular system provides detailed information regarding the changes of intravascular pressure during the injection. This functionality is not a prerequisite to perform sufficient postmortem angiography, but it opens up new options for the quantification of blood loss in polytrauma cases or for the evaluation of venous backflow in cases of heart failure. This type of information regarding the intravascular pressure status of a corpse is impossible to retrieve when using traditional autopsy.

Contrast Medium

The latest evaluation of two mixtures of contrast media showed clear advantages for an iodinated water-soluble contrast medium mixed with PEG-200 at a ratio of 1:15. This mixture consists of a large polymerized portion (PEG) and an unpolymerized part with small molecular dimensions (contrast medium). After intravascular injection, enhancement is observed in physiologically well-perfused tissues such as of the cerebral cortex, myocardium, pancreatic/splenic tissue, renal cortex, liver, and intestinal wall. This enhancement likely arises from the diffusion of the small molecular hydrophilic contrast medium in the capillary bed into the interstitial and intracellular spaces, whereas the polymerized large PEG molecules have to stay in the intravascular space due to their large size. This mechanism, on the one hand, avoids the unwanted extravasation of contrast in areas with early decomposition (e.g., the gastrointestinal tract) and, on the other, enables the separate visualization of the arterial and venous system by waiting a certain amount of time (15–20 min) between the injections (Figure 28.2). This fact also raises the need for a CT scan immediately after the application of contrast, as there is a constant decrease in intravascular contrast and vessel distention after injection.

CT Scanner and Scanning Parameters

In principle, each helical CT scanner can generate the high-resolution data sets needed for postprocessing by multiplanar reconstructions (MPRs) or the volume rendering technique (VRT). However, the size of the maximally covered volume is highly dependent on the technical abilities of the scanner. With a conventional single-slice helical CT, it is not possible to scan large volumes (e.g., thorax and abdomen or the legs)

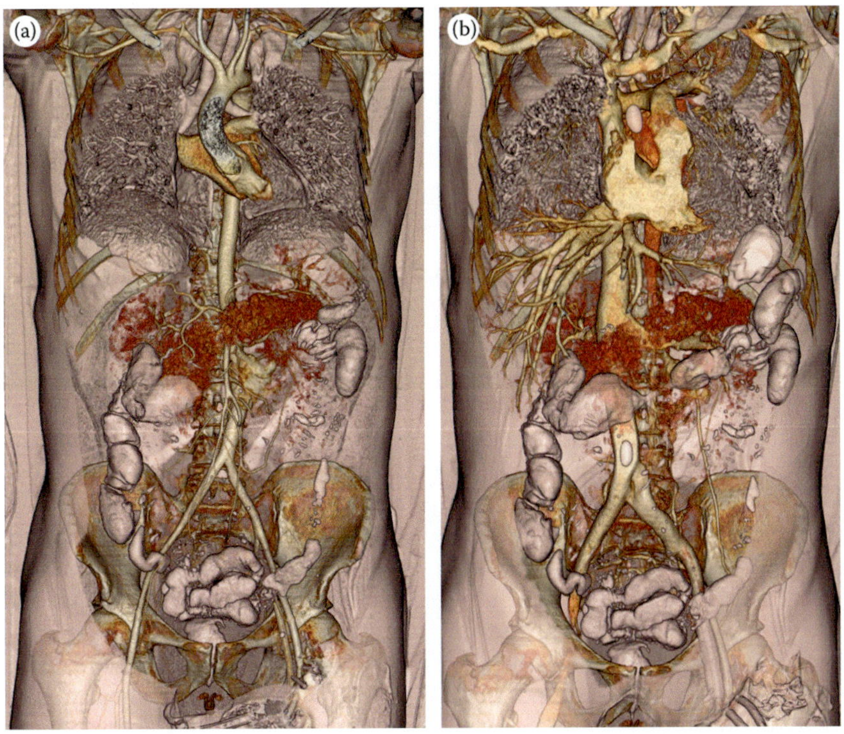

FIGURE 28.2 VRT reconstruction of the trunk after arterial (a) and venous (b) injection. Detailed depiction of the thoracoabdominal vasculature. Note the decreasing arterial enhancement during the interval (15 min) between the injections, allowing an almost separate imaging of arteries and veins.

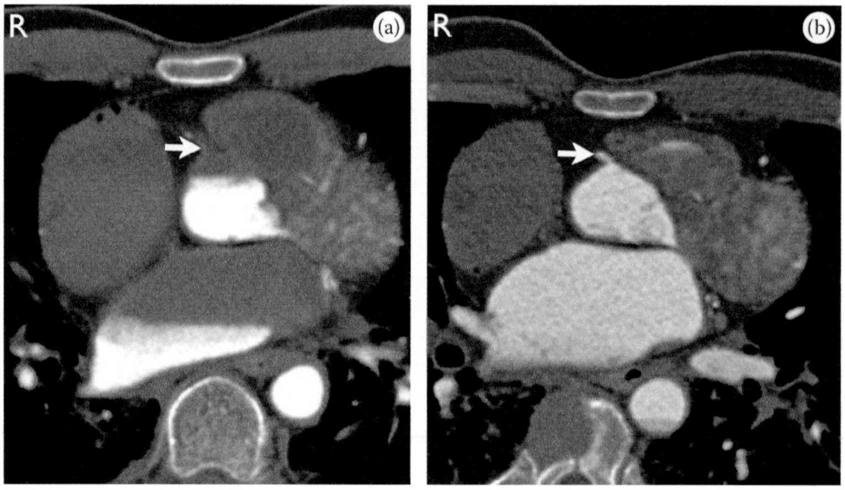

FIGURE 28.3 Incomplete filling of the ventral aortic root after arterial injection in the supine position (a). Note the insufficient perfusion of the right coronary artery (arrow), which shows proper contrast filling after switching to the prone position (b).

in one block and in sufficiently thin slices (<3 mm whole body and sub-millimeter for coronary CTA). The restricted cooling capacity of the x-ray tube is the main factor limiting repeated imaging of large volumes with these scanners, and real whole-body imaging is not feasible. The introduction of MSCT circumvented this limitation by using a new detector design, which enables the simultaneous imaging of multiple slices per rotation of the x-ray tube. Sufficient whole-body pmCTA can be performed with 4- to 8-row MSCT, but which is, at the time of the writing, already outdated. To perform repeated imaging of large volumes at submillimeter thickness without major cooling delays, the deployment of 16-slice MSCT and above is strongly recommended.

Imaging Protocol

The imaging protocol should consist of three basic phases:

1. Unenhanced
2. Arterial
3. Venous

The term "phases" should not be mistaken for the same term in clinical radiology because no new circulation is established in pmCTA. Instead, this term describes the side of the vascular system (i.e., arterial/venous) that is injected with contrast medium.

Administration of contrast should be performed directly on the CT couch before the scan and continued during the scan to maintain proper distension of the vessel lumen. In cases of suspected pathologies of the coronaries, the arterial phase should be divided into a first scan of thorax/abdomen in the supine position and a second scan in prone position. The change in body position from supine to prone in the arterial phase has to be done because of the incomplete filling of the ventral aortic root in the supine position and the consequent insufficient perfusion of the right coronary artery (Figure 28.3). All CT scans of the thorax and abdomen, including those in the prone position, should be performed with the arms raised to avoid streak artifacts at the level of the upper abdomen caused by the elbow joints. The injection volumes have to be adjusted for the physical constitution of the examined body. A rough estimation for an average adult corpse is 1.5l of injection volume each for the arterial and venous systems. In particular, the venous injection should be carefully calibrated to the potential for congestion of the upper venous inflow. An oversized venous injection gives an unwanted boost to the already apparent external signs of congestion and unnecessarily complicates the preparation of the corpse for the funeral. An overview of injection volumes and scan protocols can be found in Table 28.1.

Postmortem Clotting and Intravascular Gas (Putrefaction)

The formation of postmortem blood clots is closely linked to the length of the agonal interval. As a rule of thumb, longer dying processes are related to a greater amount of the anticipated postmortem clotting in the vascular system. In cases with a very short time to death (e.g., in a traffic accident), very few or no clots in the vessels can be found during the postmortem angiography. In contrast, cases with a protracted dying process (e.g., pulmonary embolism, cardiac failure, or cancer) exhibit extensive postclotting. Because the right heart chambers and the great vessels (venous > arterial) are biased toward the formation of postmortem clotting, the diagnosis of pulmonary thromboembolism has to be based on the combination of the radiological findings and the histopathology of the potentially thromboembolic material; this material can easily be obtained with image-guided biopsy.

There is no sharply defined maximum postmortem interval between the time of death and angiography. The speed of decomposition depends to a large extent on external factors, which are different for each case. A postmortem interval of

TABLE 28.1
Imaging Protocol for Arterial/Venous Whole Body pm CT-Angiography

Imaging Phase	Body Position	Imaging Parameters	Injection Volume
Unenhanced	Supine; raised arms for thorax/abdomen	Slice thickness < 3 mm soft tissue/lung kernel	/
Arterial whole body	Supine; raised arms for thorax/abdomen	Slice thickness < 3 mm soft tissue/medium hard (angio) kernel	1.0 l
Arterial cardiac	Prone; raised arms	Slice thickness < 1 mm soft tissue/medium hard (angio) kernel	0.5 l
Venous	Supine; raised arms for thorax/abdomen	Slice thickness < 3 mm soft tissue/medium hard (angio) kernel	1.5 l

Note: The sufficient filling of the right coronary artery demands the switch of the body position to prone. Switch-back to supine before venous injection should be done to promote a proper filling of the pulmonary trunk. The given imaging parameters are by intention relatively imprecise because of the diversity of notation in CT-scanners of different vendors. A 110 kV tube current is suggested for the best iodine x-ray absorption; the mAs setting should be raised to lower the image noise. The given injection volumes evolved from experience and are adjusted on an averaged body model of an adult corpse of approximately 75 kg body weight. The contrast mixture consists of polyethylene glycol and Iohexol (ratio 15:1).

one to two days should not be a problem in terms of diagnostic image quality in most cases. While the increasing volume of intravascular and intraparenchymal gas accumulating during decomposition impairs image quality, the real limiting factor is the presence of a radiologist who is understandably not keen to have a rotten corpse on his or her CT couch.

DIAGNOSTIC SPECTRUM OF POSTMORTEM ANGIOGRAPHY

Cerebral Circulation—pmCCTA

The complete evaluation of the cerebral arteries during an autopsy is a time-consuming procedure that requires extensive neuropathologic knowledge and is accompanied by the risk of destroying findings during the removal of the brain. Angiography permits visualization of the cerebral vasculature of a deceased person *in situ* (Figure 28.4). Even tiny pathologies (e.g., aneurysms in the circle of Willis) are well demonstrated (Figure 28.5). The localization of bleeding sources can be accomplished by tracking the extravasation of the contrast medium (Figure 28.6). Although the pictures are very similar in appearance to clinical angiographies, it is important to bear in mind that the postmortem perfusion status of the brain does not exactly match the *in vivo* conditions. For instance, vascular spasms caused by a subarachnoid hemorrhage can dissolve after death and leave the impression of completely normal brain perfusion. Even under this premise, however, postmortem cerebral angiography provides information that is not retrievable during a classic autopsy. For example, the status of the cerebral circulation after traumatic head injuries can be depicted well with the arterial injection of contrast medium (Figure 28.7). The enhancement of the cerebral cortex after the arterial injection of contrast medium has not yet been thoroughly examined. Further comparative studies between neuropathologic and neuroradiological findings must focus on validation of the angiographic findings.

Neck

Especially in trauma cases, the neck area confronts the pathologist with significant challenges: this narrow field is full of life-eminent structures and only visible with major

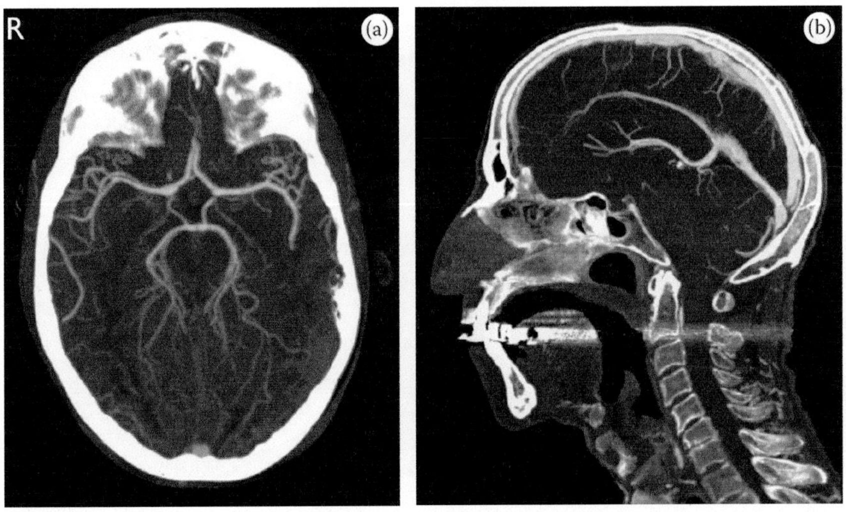

FIGURE 28.4 Maximum intensity projection (MIP) of the cerebral vasculature with no pathology. (a) Axial reconstruction after arterial injection. (b) Sagittal reconstruction after venous injection. Proper depiction of the circle of Willis and the venous sinuses of the brain with no pathology.

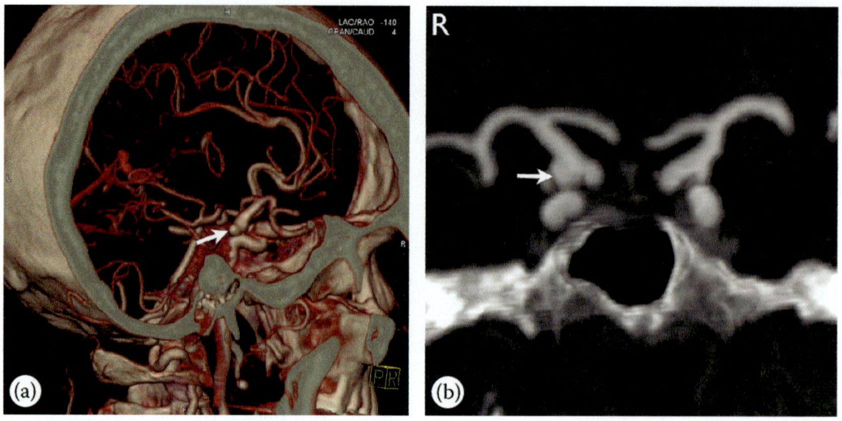

FIGURE 28.5 Arterial vasculature of the brain, (a) VRT (b) Coronal MIP. Depiction of a small, berry-like aneurysm of the proximal right medial cerebral artery.

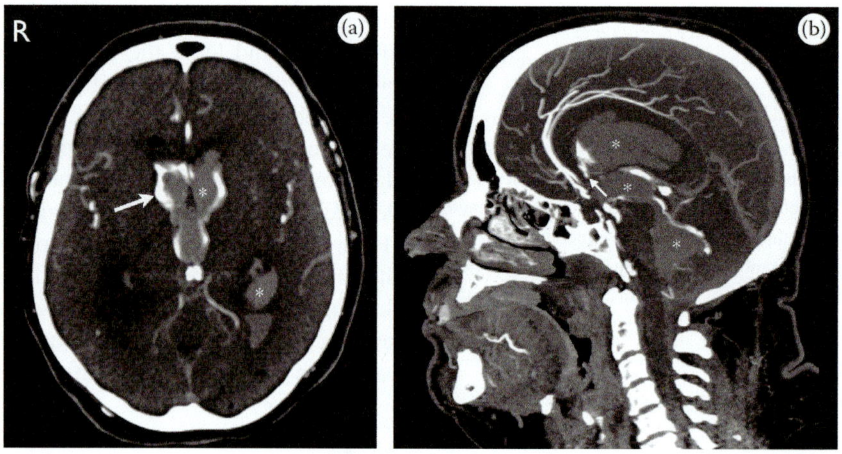

FIGURE 28.6 PmCCTA after arterial injection. (a) Axial cross-section. (b) Sagittal MIP. Massive intraventricular hemorrhage (*). Intraventricular extravasation of the contrast medium (open arrows). The body position during the angiography was supine, so the bleeding source was estimated in the anterior communicating artery. Neuropathologic examination confirmed the diagnosis by depicting a ruptured microaneurysm of the anterior cerebral artery.

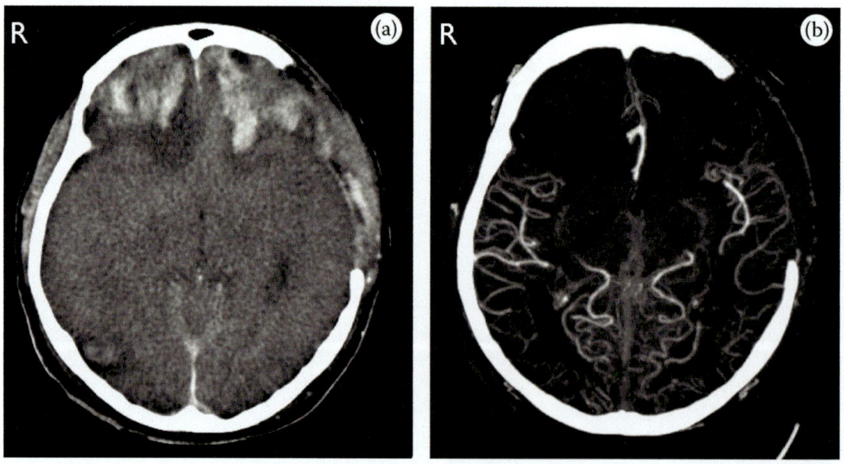

FIGURE 28.7 Evaluation of arterial cerebral perfusion after head trauma. Victim of a motor vehicle accident who died after osteoplastic craniotomy of the left temple. The unenhanced postmortem cranial CT (a) shows massive bifrontal parenchymal hemorrhage and signs of increased intracranial pressure. An axial MIP (b) after arterial injection reveals the complete outline of the anterior arterial circulation cerebral arteries and regular perfusion of the middle and posterior arterial stream beds.

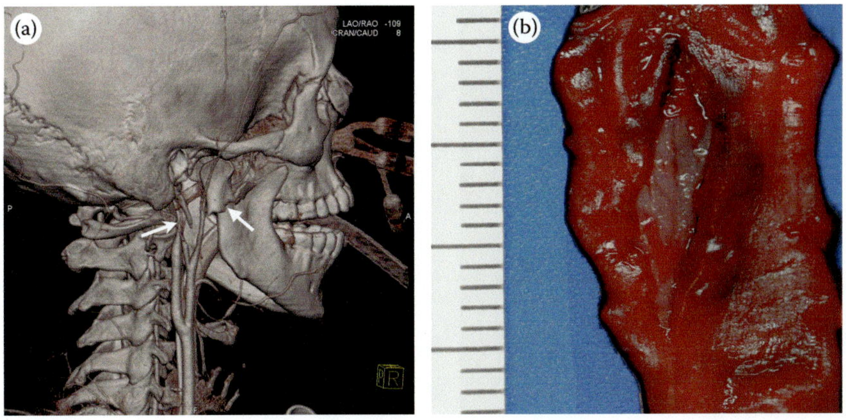

FIGURE 28.8 Victim of a skiing accident with massive blunt trauma to the face and neck. (a) Volume rendering after arterial injection. (b) Autopsy specimen. pmCTA shows a suspected traumatic dissection of the right internal carotid artery with occlusion of the vascular lumen (right arrow). This dissection was confirmed via autopsy, which identified a 3-cm longitudinal laceration of the internal vessel layer. Note the displaced fracture of the right mandibular angle (left arrow).

damage to the corpse itself. Structures like the inferior skull base or the intraosseous course of the internal carotid artery are hardly reachable. Because of this fact, potentially life-threatening pathologies (e.g., short length dissections of the internal carotid and vertebral arteries) are difficult to diagnose in the classical autopsy. pmCTA shows a clear advantage in depicting these pathologies (Figure 28.8).

Thorax

Cardiac CTA

The main key to assess cardiovascular causes of death is the sufficient display of a corpse's coronary system. Autopsy is currently the method of choice to diagnose pathologies of the coronary arteries. The proper preparation of such small vascular structures during autopsy requires meticulous work by the pathologist. Complicated structures such as cardiovascular bypasses or valvular surgery always hold the danger of destroying relevant findings during extensive preparation due to scar tissue caused by the surgery. Minimally invasive postmortem coronary CTA allows the display of the coronary arteries *in situ* with a quality comparable with that of clinical coronary CTA (Figure 28.9). It also enables the 3D depiction of cardiac bypasses without the abovementioned problems inherent in classical autopsy (Figure 28.10). Cardiac pmCTA also permits significant advances in the radiological diagnosis of heart trauma. Myocardial ruptures and lacerations of the coronary arteries are easily revealed by the extravasation of contrast medium (Figure 28.11).

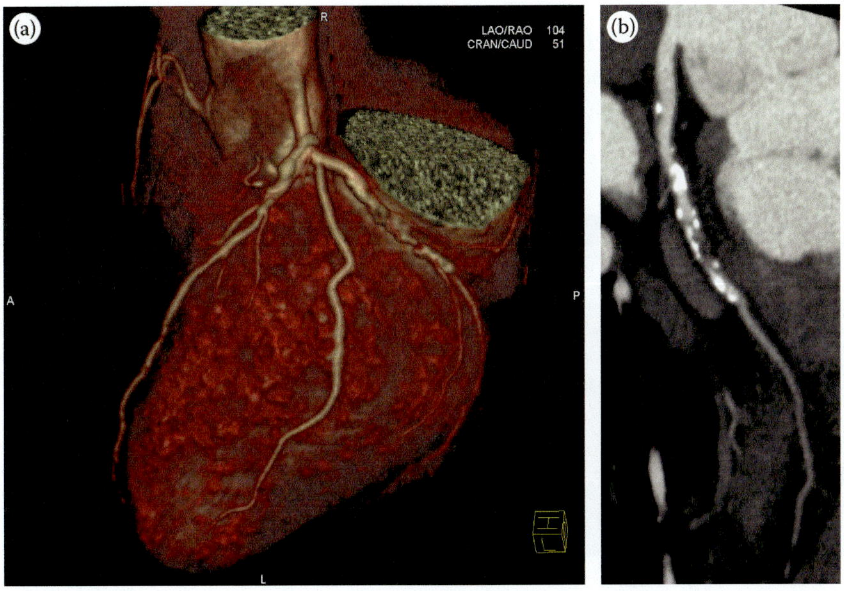

FIGURE 28.9 Postmortem cardiac CTA of coronary artery disease. Volume rendering (a) shows coronary heart disease with massive calcification in the left anterior descending artery (LAD) and subtotal stenosis of the circumflex artery (RCX) that is also depicted in the curved multiplanar reformation of the RCX (b).

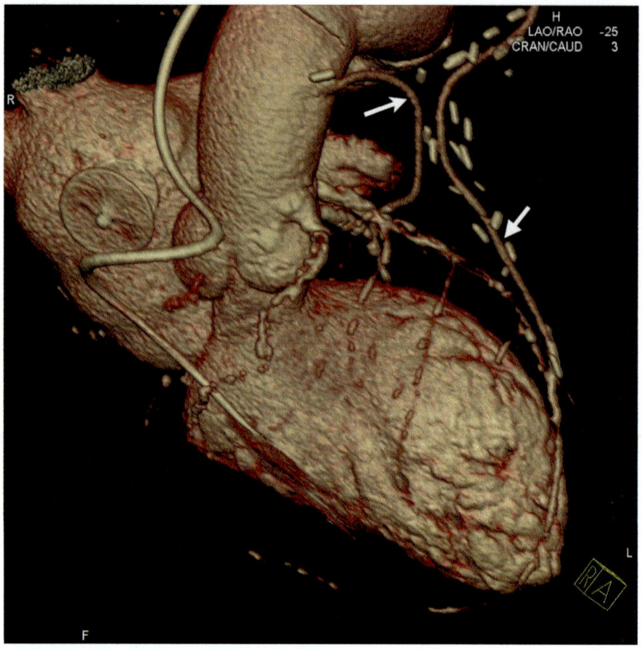

FIGURE 28.10 Postmortem cardiac CTA of cardiovascular bypass in a case with massive coronary artery disease. Volume rendering shows an open double bypass with an aortocoronary bypass (ACB) to the proximal LAD (right arrow) and a left internal thoracic artery (LITA) (left arrow) bypass to the distal LAD. Note the pacemaker electrode in the uncontrasted right ventricle and the amplatzer device in the atrial septum.

Aortic CTA

pmCTA provides an accurate 3D view of the aorta in a minimally invasive manner and is, like its antemortem counterpart, very accurate for the evaluation of aortic aneurysms and dissections. In contrast to unenhanced pmCT, the arterial application of contrast medium allows for the proper evaluation of the whole aorta (including major branches) (Figure 28.12). A narrow location like the pelvic cavity is not easy to prepare during autopsy without causing major damage to vascular structures hidden deep in this place, but pmCTA overcomes this problem.

CT Pulmonary Angiography

Anterograde perfusion of the venous system through the femoral vein permits excellent visualization of the pulmonary arteries via the right atrium and ventricle. It is possible to diagnose pulmonary embolism (Figure 28.13), as long as this diagnosis is confirmed by the histopathologic examination of the suspected central thrombembolus. This can be done with a minimally invasive CT-guided biopsy of the pulmonary trunk. Minor filling defects of the segmental/subsegmental pulmonary arteries are often revealed after venous injection in cases with a prolonged agonal phase. Based on our experience in autopsies of such cases, we note that this effect is caused by the pushing of postmortem clots out of the right ventricle; such a situation should not be mistaken for a real thrombembolus. The pulmonary parenchyma itself shows a significant enhancement in areas of dystelectatic lung tissue and in pneumonic infiltrations. This allows for appropriate differentiation between atelectatic lung tissue and a pleural effusion, a task that is difficult to manage with unenhanced pmCT.

Parenchymal Organs of the Upper Abdomen

The liver tissue shows the maximum enhancement after venous perfusion with contrast medium through the inferior caval vein (Figure 28.14). It is not possible to reach the portal venous system via femoral access to the circulation, so this part of the venous system is outside of the diagnostic reaches of pmCTA. The enhancement of the other upper abdominal organs using pmCTA is generally similar to behavior observed in clinical radiology, with a strong arterial enhancement of the renal cortex, pancreas, spleen (Figure 28.15), and gastric/intestinal walls.

Peripheral Angiography of the Extremities

CT angiography is the examination of choice in the diagnosis of arterial injuries of the extremities. Signs of vascular injury, like ruptures of the vessel wall or extravasations of contrast medium, can be readily detected (Figure 28.16), and additional information regarding osseous and soft-tissue injuries can also be routinely obtained. For the venous

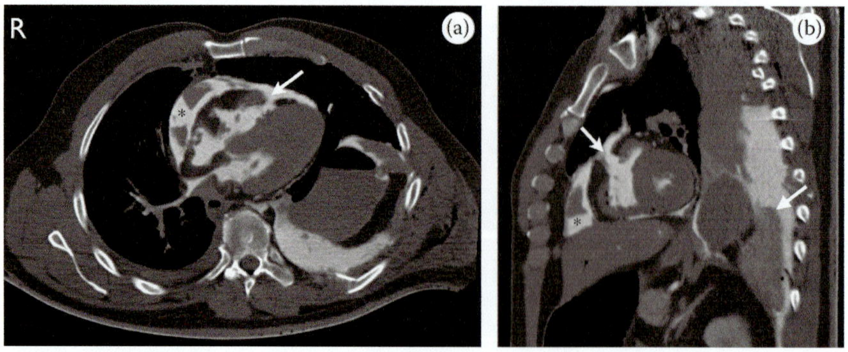

FIGURE 28.11 Victim of a motor vehicle accident with massive blunt trauma to the chest after arterial and venous injection of contrast medium. Axial cross section (a) and short axis MPR (b) show a rupture of the anterior right heart ventricle (open arrow) with extravasation of the contrast medium into the pericardial sack (*) and left thoracic cavity. Note the rupture of the left diaphragm with displacement of spleen and parts of the stomach into the thoracic cavity (closed arrow).

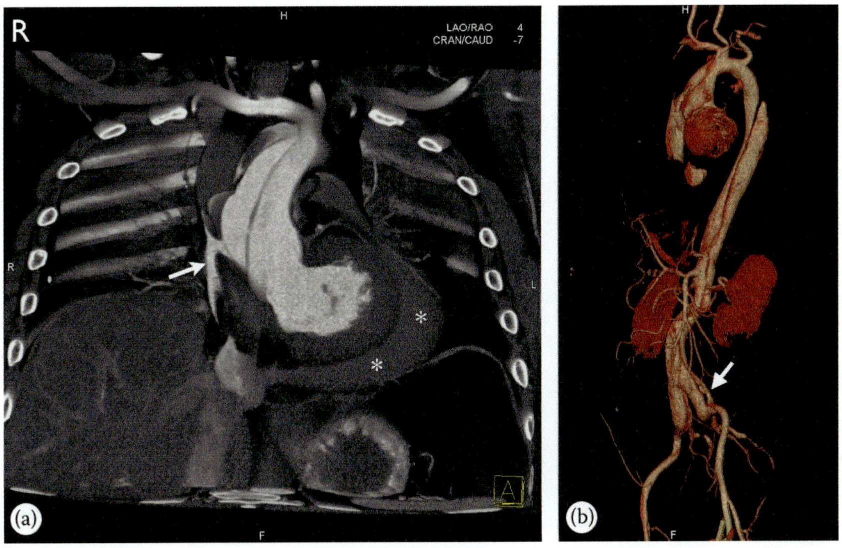

FIGURE 28.12 Thoracoabdominal pmCTA, volume rendering. Stanford I aortic dissection with pericardial tamponade (*) and extravasation of the contrast medium into the pericardium (arrow in a). Note the extension of the dissection down to the iliac arteries (arrow in b).

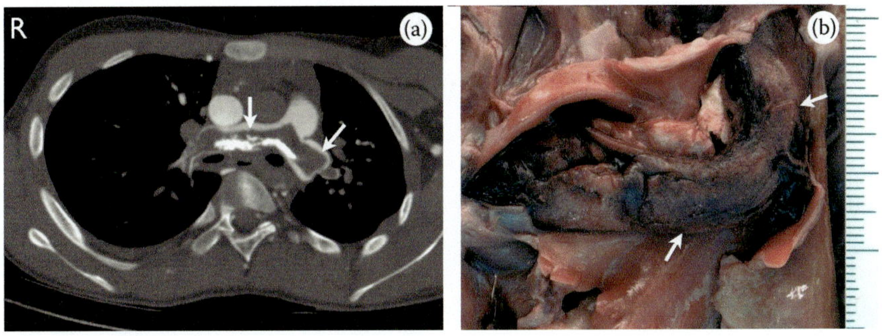

FIGURE 28.13 Pulmonary embolism. (a) Axial pmCTA after venous injection. (b) Autopsy specimen, anterior view of the opened pulmonary trunk. Depiction of a riding pulmonary embolus (arrows) that is partly calcified in the posterior portion.

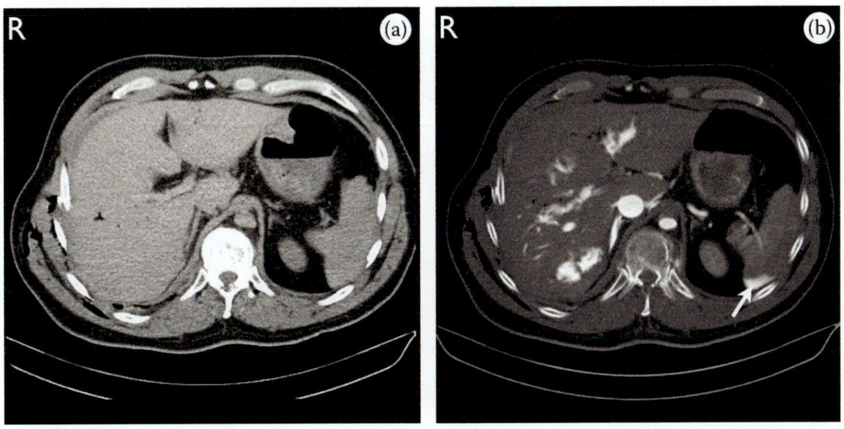

FIGURE 28.14 Blunt trauma of the upper abdomen. (a) Unenhanced pmCT. (b) pmCTA after arteriovenous injection of contrast medium. Note the significantly increased visibility of the liver lacerations after injection of the contrast medium. Perisplenic extravasation of contrast (arrow) is an indicator of splenic laceration.

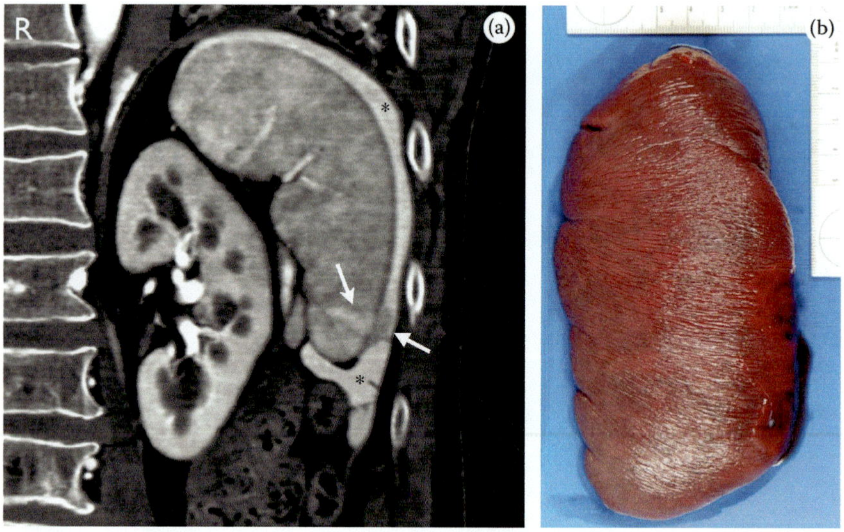

FIGURE 28.15 Abdominal trauma. (a) Coronal MPR of the left upper abdomen after arterial injection. (b) Autopsy specimen. Visible are the parenchymal rupture at the inferior pole of the spleen (open arrow) with adjacent hematoma (closed arrow) and perisplenic extravasations of contrast medium (*). Note the strong enhancement of the renal cortex after arterial contrast application.

system, however, retrograde filling of the venous bed of the extremities with a contrast medium is occasionally unsatisfactory. Depending on the function of the venous valves, sometimes only the proximal portion of the veins of the limbs is filled with the contrast medium. This limits the diagnostic value of CTA in the evaluation of deep venous thromboembolism. A possible method for circumventing this limitation involves antegrade venous injection in the leg using a superficial vein in the foot or ankle. The small diameter of the fragile peripheral veins and the high viscosity of the contrast mixture (compared to blood) are clear limitations of this approach.

CONCLUSION

Minimally invasive whole-body pmCTA shows clear advantages over unenhanced postmortem CT in displaying vascular pathologies. Due to its nondestructive nature, it enables the *in situ* examination of structures that are either not visible or visible only with major destruction of the corpse during traditional autopsy. The combination of this method with both image-guided specimen sampling for histology and toxicology and whole-body MRI may allow it to serve as a viable alternative to autopsy while still guaranteeing the quality of the clinical and forensic assessments.

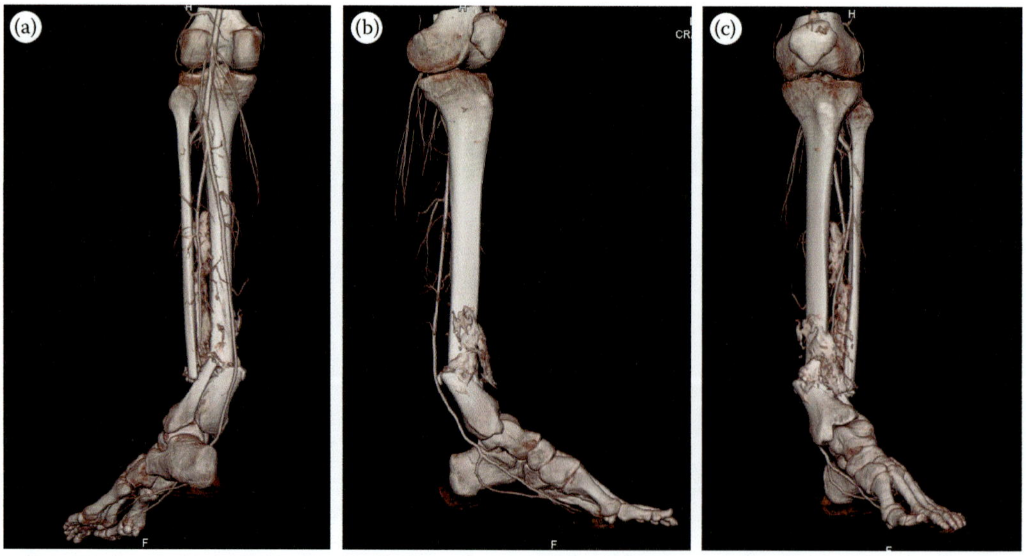

FIGURE 28.16 Extremity trauma. Volume renderings after arterial injection of contrast (a–c). Fracture of the lower leg with rupture of the anterior tibial and fibular arteries. The posterior tibial artery has direct contact with the fractured bone but shows no signs of laceration.

REFERENCES

Aghayev, E., Sonnenschein, M., Jackowski, C., et al., Postmortem radiology of fatal hemorrhage: Measurements of cross-sectional areas of major blood vessels and volumes of aorta and spleen on MDCT and volumes of heart chambers on MRI, *AJR Am J Roentgenol*, 187(1), 209–215, 2006.

Añon, J., Remonda, L., Spreng, A., et al., Traumatic extra-axial hemorrhage: Correlation of postmortem MSCT, MRI, and forensic-pathological findings, *J Magn Reson Imaging*, 28(4), 823–836, 2008.

Christe, A., Ross, S., Oesterhelweg, L., et al., Abdominal trauma—sensitivity and specificity of postmortem noncontrast imaging findings compared with autopsy findings, *J Trauma*, 66(5), 1302–1307, 2009.

Dirnhofer, R., Jackowski, C., Vock, P., Potter, K., and Thali, M. J., VIRTOPSY: Minimally invasive, imaging-guided virtual autopsy, *Radiographics*, 26(5), 1305–1333, 2006.

Grabherr, S., Djonov, V., Yen, K., Thali, M. J., and Dirnhofer, R., Postmortem angiography: Review of former and current methods, *AJR Am J Roentgenol*, 188(3), 832–838, 2007.

Grabherr, S., Gygax, E., Sollberger, B., et al., Two-step postmortem angiography with a modified heart–lung machine: Preliminary results, *AJR Am J Roentgenol*, 190(2), 345–351, 2008.

Jackowski, C., Persson, A., and Thali, M. J., Whole body postmortem angiography with a high viscosity contrast agent solution using poly ethylene glycol as contrast agent dissolver, *J. Forensic Sci.*, 53(2), 465–468, 2008.

Jackowski, C., Sonnenschein, M., Thali, M. J., et al., Virtopsy: Postmortem minimally invasive angiography using cross section techniques—implementation and preliminary results, *J. Forensic Sci.*, 50(5), 1175–1186, 2005.

Levy, A. D., Abbott, R. M., Mallak, C. T., et al., Virtual autopsy: Preliminary experience in high-velocity gunshot wound victims, *Radiology* 240(2), 522–528, 2006.

Levy, A. D., Harcke, H. T., Getz, J. M., et al., Virtual autopsy: Two- and three-dimensional multidetector CT findings in drowning with autopsy comparison, *Radiology*, 243(3), 862–868, 2007.

Ross, S., Spendlove, D., Bolliger, S., et al., Postmortem whole-body CT angiography: Evaluation of two contrast media solutions, *AJR Am J Roentgenol*,190(5), 1380–1389, 2008.

29 Using Real 3D Data for Reconstruction

Ursula Buck, Silvio Näther, and Michael J. Thali

CONTENTS

ANALYSIS, VISUALIZATION, AND RECONSTRUCTION METHODS

Generating 3D Models from Radiological Images

To generate 3D models of the osseous system and skin surface of the deceased from plane multislice computed tomography (MSCT) images, segmentation software is needed. At the Institute of Forensic Medicine in Bern, Switzerland, we used the OsiriX software solution [1]. This program works with the threshold technique for the segmentation of medical two-dimensional (2D) images into a 3D series. The respective object pixels are separated by an adjusted threshold value and are marked in a binary image. After the separation by threshold value, a new binary matrix is created for each image. The slice level allows the 3D model to be computed from these binary images (Figure 29.1).

Fusion of the Surface Data with the Internal Body Data

For the purpose of surface and radiological volume data set fusion, additional multimodality radiographic markers [2] are placed on the body. These markers, visible in the surface-scanning data as well as in the radiological data set, are used to transform the surface model and the magnetic resonance imaging (MRI) data into CT data; this is accomplished by the computer software. After the fusion, it becomes possible to correlate external injuries with the underlying injuries, which are visible in CT and MRI data. In cases where these markers are not used, the surface and radiological data are merged by using "anatomical landmarks," such as prominent skin or bone structures. The merging or fusion process is achieved with specific 3D software programs. Figure 29.2 shows an example of a merged data set. The bones and the surface of the abdomen are generated from the CT images; the 3D model of the back of the body, with the injury in real color, is produced by the surface-digitizing process.

Visualization of Radiological CT and MRI Images in the Context of the Exterior Body (Cut with Virtual Knife, Virtual Cut)

To examine and represent internal injuries, it is necessary to visualize the CT and MR cross-section images (Figure 29.3). In order to map the relevant 2D CT and MR images onto the 3D body model, including external injuries, the digital 3D body model is cut virtually, and the images are placed onto the cut surfaces by computer software (Figures 29.4 and 29.5) [3,4].

Motion of the Virtual Body

To determine how the injury matches with the injury-causing tool (e.g., in a traffic accident), the 3D model of the victim must be placed into various probable positions at the time of the impact.

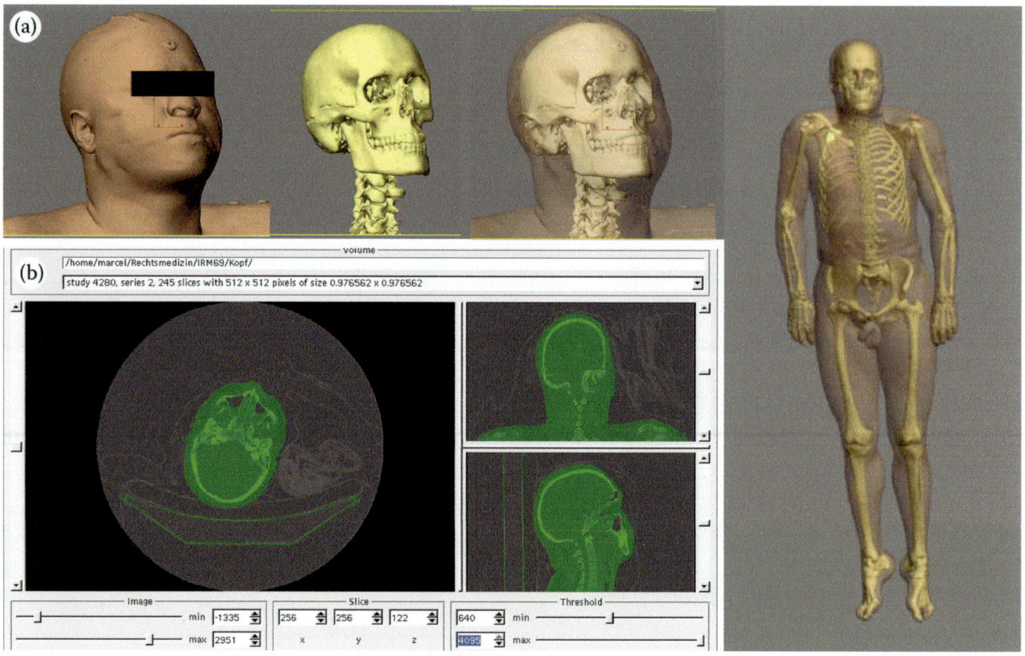

FIGURE 29.1 (a) The 3D models of the skin and the bone structure of the deceased as generated from plane MSCT images by the Sliceviewer software. (b) The CT images are imported into the Sliceviewer software. As the threshold value is adjusted, the data for 3D modeling can easily be extracted.

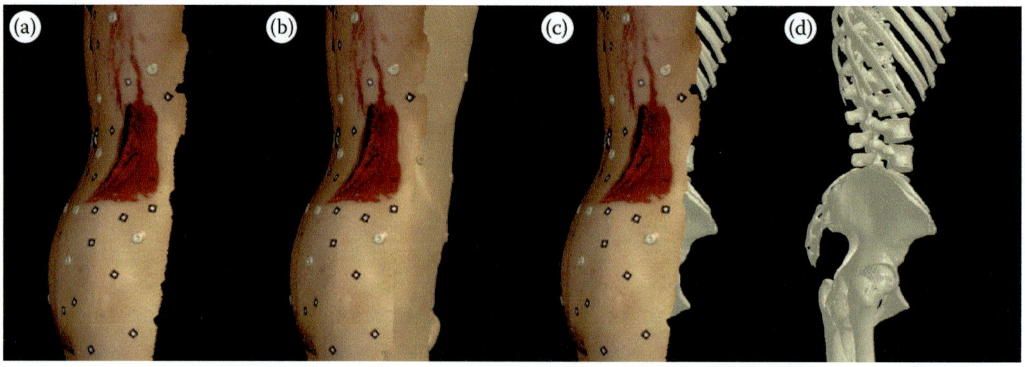

FIGURE 29.2 The 3D model of the external injury on the back and the osseous structure are fused to a real-databased unified 3D model of the abdomen region. (a) Skin model of the injury on the back in real color, generated from digital photogrammetry and 3D optical surface scanning data. (b) Skin model of the back fused with the CT skin model of the front. (c) Skin model of the back fused with the CT bone model. (d) Bone model generated from the CT images.

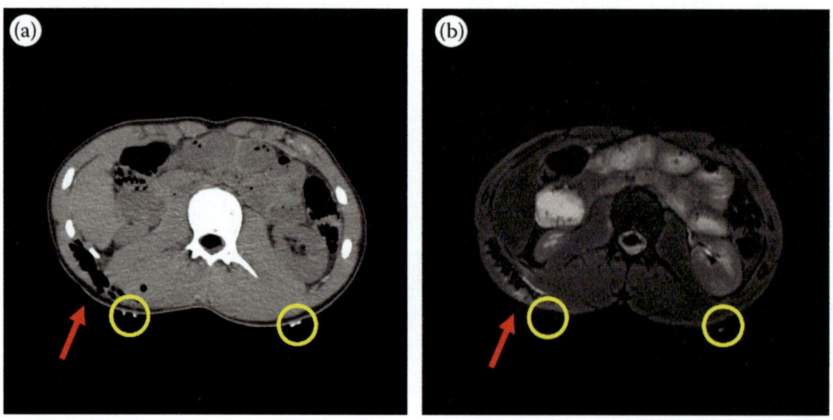

FIGURE 29.3 Axial CT (a) and MR (b) images of the abdomen region. The MR and CT images display the internal soft tissue injury (red arrows). The radiographic markers (yellow circles) are placed on the body to fuse the imaging data with the 3D surface scanning data.

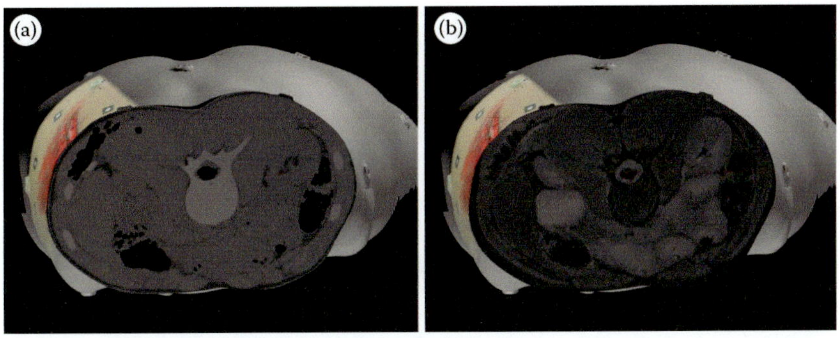

FIGURE 29.4 A virtual cut through the 3D body model, which visualizes the internal autopsy findings combined with the external injuries of the corpse. The CT (a) and MR (b) images, laid onto the cut surface, illustrate the internal findings on the left side.

Using the obtained real-data information about the osseous structure, and especially the joint positions, the surface and skeleton model can be moved virtually and in an anatomically correct manner in order to reconstruct the actual body position during impact. For this purpose, we use the 3Ds max software package [5] together with a modified computer skeleton model (biped). The joints of this computer skeleton model are adapted to the individual bone structure of the CT model by scaling and rotating and then are covered with the skin model created from the CT and surface data (Figure 29.6a). Movements of the skeleton model cause the same movements in the CT models and in the surface model (Figure 29.6b).

3D Reconstruction

3D geometric reconstruction is based on the use of true-color 3D models of the corpse and the presumed injury-causing tools. These true-to-scale 3D models can be moved and rotated in a 3D environment using modern computer software [5,6,7] to compare the actual injuries with the presumed injury-causing tool and to investigate the impact configuration. In cases of virtual incident scene reconstructions, the 3D situation plans are integrated in the analysis and are imported into the virtual scenes.

The aims of the virtual, real-databased, and geometric reconstructions are to compare skin and bone injuries with presumed injury-causing instruments, to visualize shot directions, and to determine various impact positions and directions. The 3D geometric reconstruction process is discussed for several real cases in the following.

Automatic Comparison of Injuries with the Assumed Injury-Inflicting Instrument

In the previously described computer-based manual comparison of accurate true-to-scale 3D models of injuries with likely injury-inflicting instruments, the correlations are

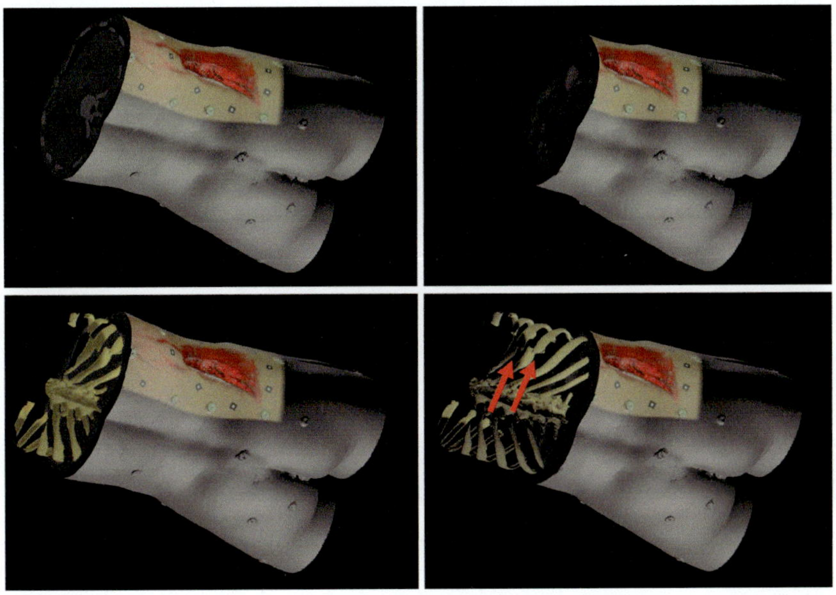

FIGURE 29.5 The unified model of the corpse can be sliced virtually in different positions. This visualization shows the location of the internal and the external injuries, as well as the rib fractures (red arrows).

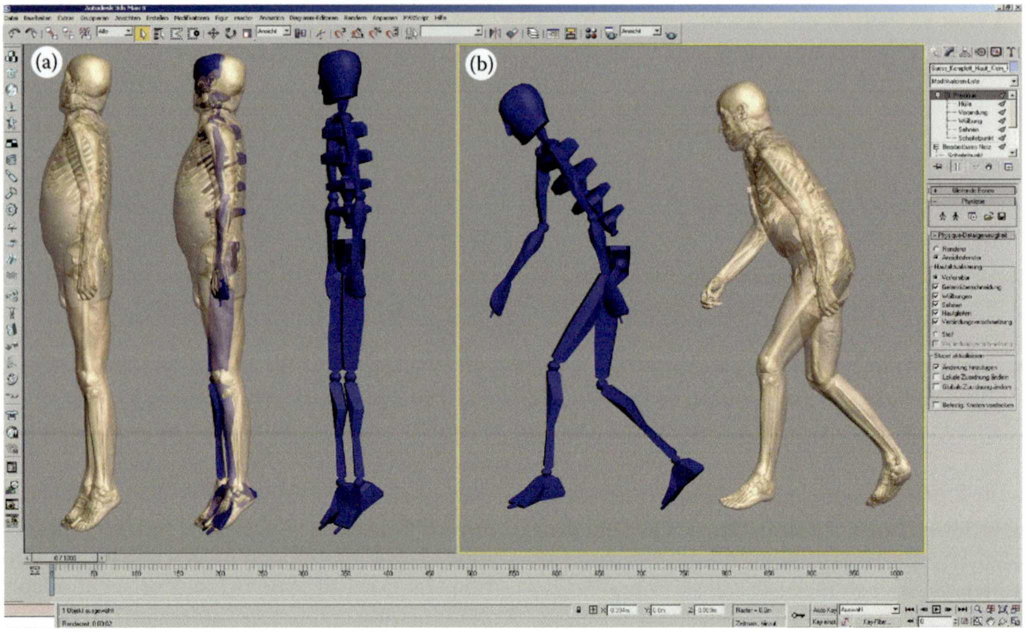

FIGURE 29.6 Animation of the real-data-based 3D-model. (a) The computer skeleton model (blue) is adapted to the individual body structure of the deceased. (b) The movement of the blue computer model causes the same movement of the 3D model of the deceased.

visualized onscreen or in printed documentation. In some cases, it is possible to verify the analysis with an automated mathematical comparison. For this purpose, the presumed correlating meshes are transformed to match the best fit position by the ATOS software [6]. Any deviations between the two meshes can be automatically calculated by the software and may be shown as a color-coded polygon mesh. The color scale represents the amount of deviation.

An automatic comparison is performed between the 3D models of injuries and injury-inflicting instruments, as well as between the 3D models of impressions in different materials and presumed tool marks. Figure 29.7 illustrates an

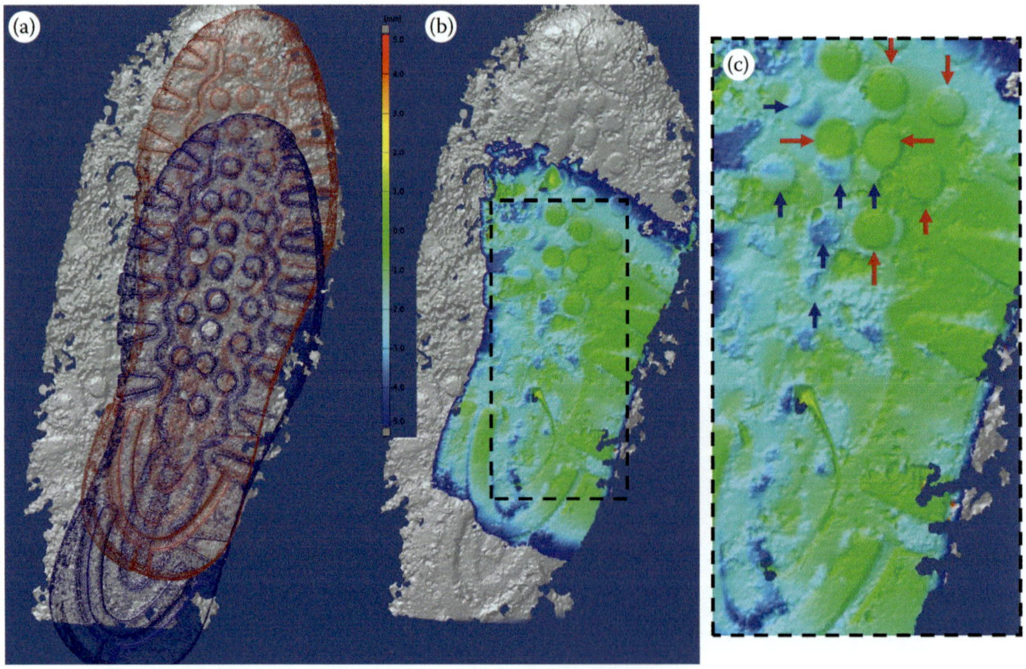

FIGURE 29.7 Computer-aided geometric match analysis of overlapping footwear impressions in snow with the sole of a shoe of the suspected offender. (a) Differentiation of the individual impressions is possible (blue and red soles). (b) The deviation between the sole and the upper impression (from the red sole) can be represented as a color-coded polygon mesh. The color scale represents the extent of any deviation. (c) A detailed view of the mesh shows that the green round structures match the red sole (red arrows); and the blue round structures match the overlapping position of the blue sole (blue arrows).

automatic comparison between a digitized impression in snow and the presumed sole of a shoe [8].

FORENSIC RECONSTRUCTION

Traffic Accidents

When reconstructing traffic accidents between a motor vehicle and a pedestrian or cyclist, we correlate the injuries to the body with the injury-causing object(s) and the accident mechanism. In addition to using photographic documentation and representation to scale of the accident scene and the involved vehicles, we also focus on trace investigations and on the results from forensic medical examinations [9]. The configuration of the vehicles on impact is an important aspect of the analysis. Accident vehicles and associated damage as documented by 3D optical surface scanning (described in Chapter 24) are used, in addition to the exterior and interior body documentation of the deceased. Forensic analysis of traffic accidents includes processing the resulting data to create 3D models, generating 3D representations of the bone fractures (which allows the driving direction of the vehicle to be determined), correlation of injuries with damage seen on the vehicles, and geometric determinations of the impact configuration [10]. In the following case, the methods and results of the forensic 3D accident reconstruction process are illustrated.

Collision between a Car and a Pedestrian

A pedestrian was fatally injured in an accident involving a car. Death occurred at the accident site due to blood aspiration in combination with a profound loss of blood, pneumothorax and fat embolism. The body exhibited collision injuries on the right leg and shoulder. Externally visible injuries were apparent on the left side of the head. Furthermore, the deceased presented extensive chest trauma. It was initially unclear whether the deceased had been knocked over, run over, or rolled over, and his position at the time of the collision was also unclear. The answers to these questions were addressed by accident reconstruction. The internal and external body was documented in 3D using CT, MRI, and optical surface scanning. The front and underbody of the vehicle and the jacket worn by the deceased were also digitized using the optical surface scanner. In such cases, the external injuries can be examined to differentiate between impact and being rolled over or run over, or suffering secondary injuries due to falling onto the road. A right-sided impact was assumed from the collision injuries on the skin in combination with excoriations and contusions of the subcutaneous fat tissue and muscles of the outward side of the lower right leg and the upper right arm (Figure 29.8). On the inner side of the left knee, a large excoriation with underlying pocket-shaped shearing of the skin layers from the periosteum of the knee-cap (decollement) was apparent [11–13]. We noticed a wedge fracture of the right femur and amputation of the right leg as seen in the 3D model of the osseous structure (Figure 29.9). We found no fracture in the bone at the site of the external collision injury of the upper right arm. However, hyperintensity in the T2-weighted fat-saturated MR sequence and hypointensity in the T1-weighted MR sequence in the bone marrow of the right humerus were apparent in the MR images, which indicates the presence of an edema or hemorrhage (contusion) [14]. According to Lysowski and Marek [15], this is characteristic of an impact. These radiological findings in the right humerus, in combination with the wedge fracture of the right femur and the soft-tissue injuries, confirmed the assumption that the collision impacted the right side of the pedestrian. We confirmed that he was knocked over in an upright or bent position before the car ran or rolled him over, as proved by the following findings: a wedge fracture of the femur, a traumatic amputation of the right leg, and an abrasion of the right sole of the shoe [16]. The 3D model of the vehicle digitized by the surface scanner showed that the front part of the car featured only parallel and rectangular structures.

When the 3D model of the injuries of the right leg was examined closely, it became evident that, if the leg had been

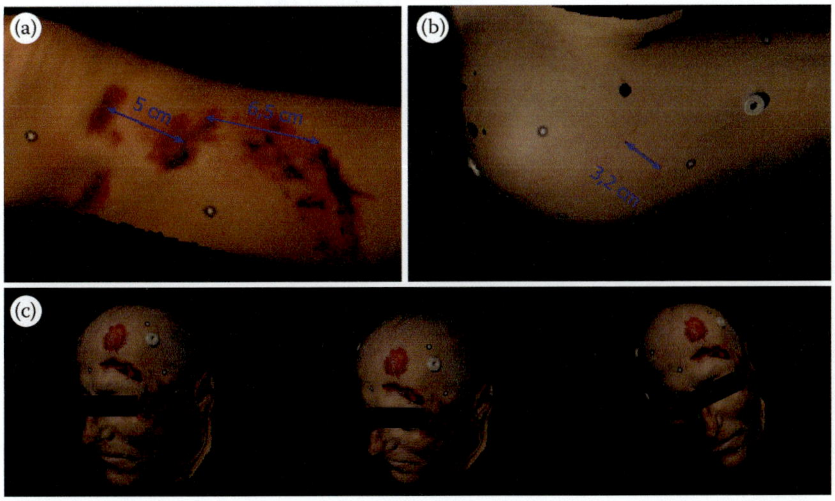

FIGURE 29.8 3D models of the external injuries of the accident victim. (a) Impact injury on the outer right lower leg (b) Impact injury on the right shoulder (c) Injuries to the head caused by secondary ground contact.

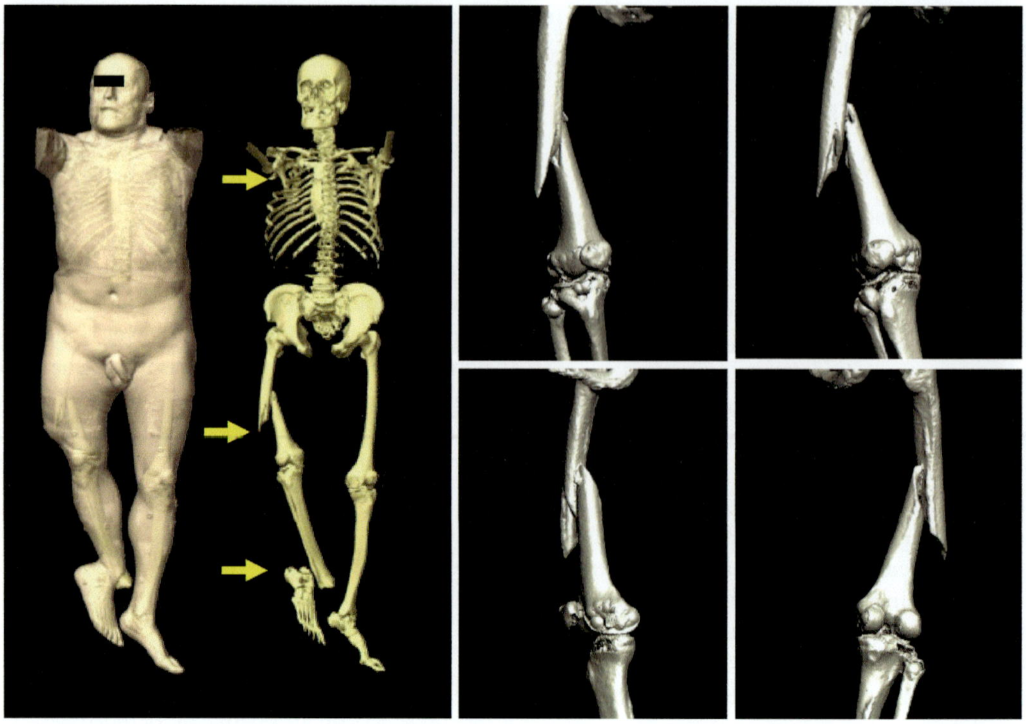

FIGURE 29.9 3D model of the skin and bone structure of the body generated from the CT images. The yellow arrows identify fractures to the right scapula, the right femur, and the right ankle.

in a straight position, the injury lines would not have been parallel (Figure 29.10a). Therefore, the injuries could not have been sustained if the leg had been straight. If the leg had been bent, the injury lines would have run parallel to each other and the vertical abrasion would have been at a right angle to the horizontal lines. Thus, the flexion degree of the leg at the time of the collision could be determined.

This situation was reconstructed on a test subject. The injury lines were drawn onto the leg (Figures 29.10b and 29.11). Figure 29.12 shows the reconstruction of the 3D model of the real-databased right leg. To bring the 3D data set of the body into the assumed impact position, the computer model (biped) was adapted to the skeleton model of the body. The precise impact position was examined with respect to the vehicle front and the injury pattern of the right leg and upper right arm (Figure 29.13). The decollement of the left knee indicated a tangential, shifting force during the right-side impact while the victim was kneeling on his left knee. The fracture of the right femur was then considered. Different positions were programmed into the model until the injury pattern matched the vehicle structures. The decollement, which was seen on the back, and the matching tire imprint on the jacket worn by the man proved that the victim had rolled over. Ultimately, we proved that the man was struck by the car from the right side while kneeling on the tarmac and leaning on his right arm. Because his center of gravity was lower than the height of the vehicle's hood, he was rolled over and was not thrown on top of the car [17].

Crime Scene

Digital Bullet Trajectory Reconstruction in Gunshot Cases

In gunshot cases, the reconstruction of bullet trajectories includes two components: (1) the bullet pathway through the

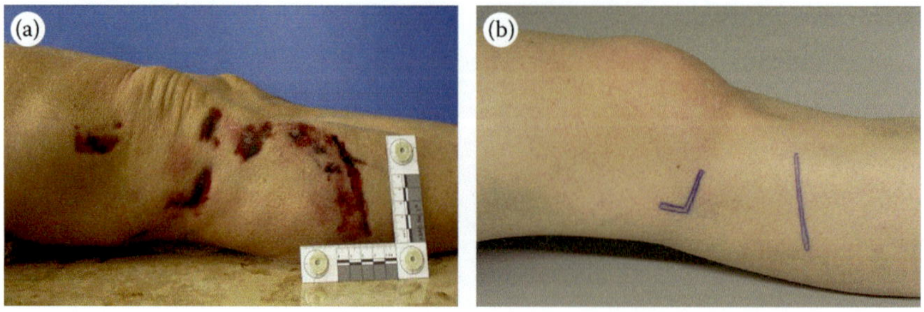

FIGURE 29.10 To reconstruct the impact situation, the injury pattern was drawn onto the leg of a test person. (a) Injury lines on the leg of the deceased (b) Injury lines drawn on the test person

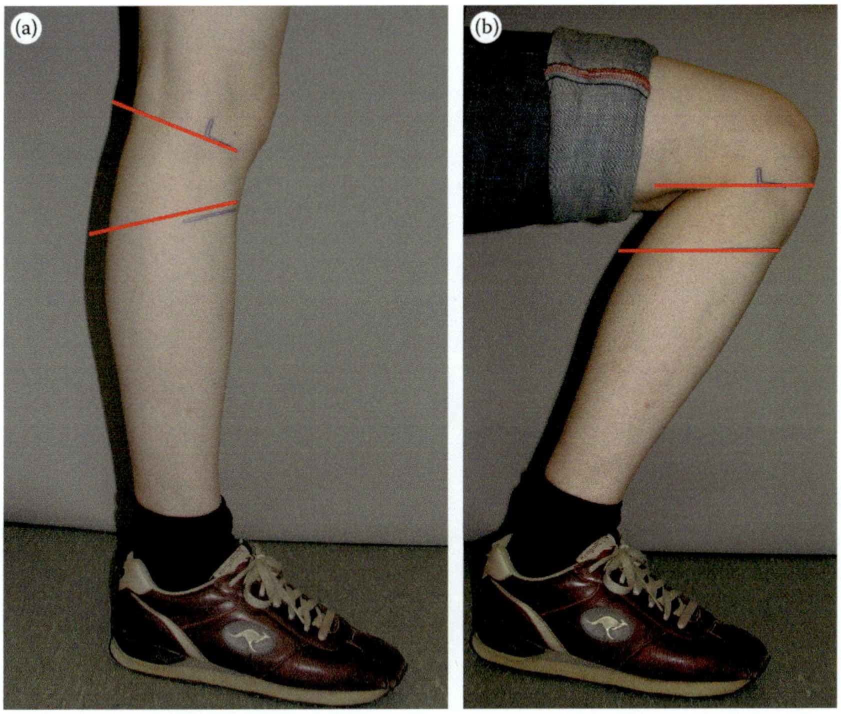

FIGURE 29.11 (a) As shown, the injury lines would not run parallel if the leg had been in a straight position. (b) If the leg is bent, the injury lines run parallel.

body of the deceased (see Section 24) and (2) the direction of the bullet trajectory at the crime scene if the bullet went through an object. The investigation of both components in a unified analysis offers clues as to the configuration and respective locations of the offender and victim. In addition, visualizations from the site and angle of a witness or a bystander's viewpoint are often helpful. Using this approach, cooperation between police and forensic medicine is of great importance. Reconstruction of a homicide by gunshot analyzed by the police and forensic medicine experts was as follows: A father was killed by his son in the parents' bedroom from a gunshot to the head. The shot was heard by the mother from another

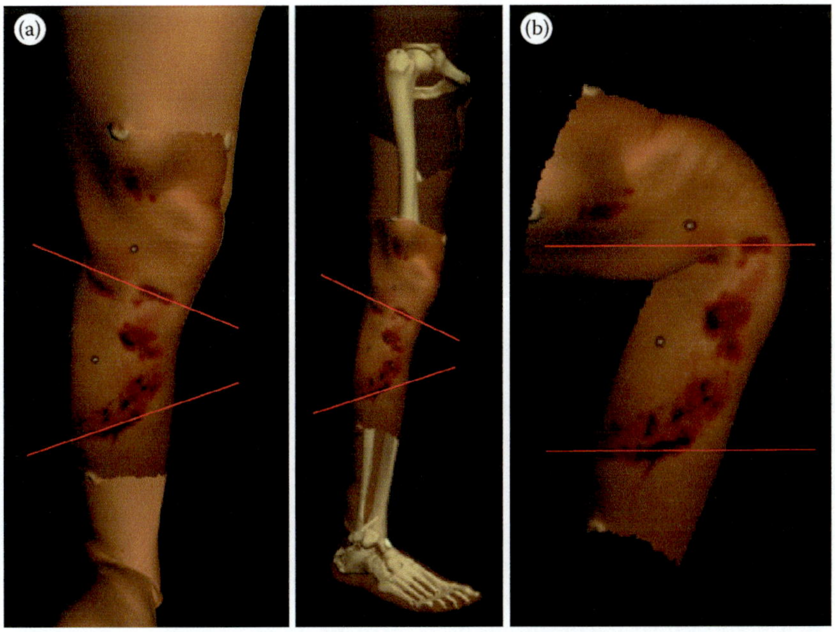

FIGURE 29.12 The real-databased reconstruction of the leg position during the impact: (a) Surface data of the injury while the leg was straight. (b) The injury lines on the virtually bent leg run parallel.

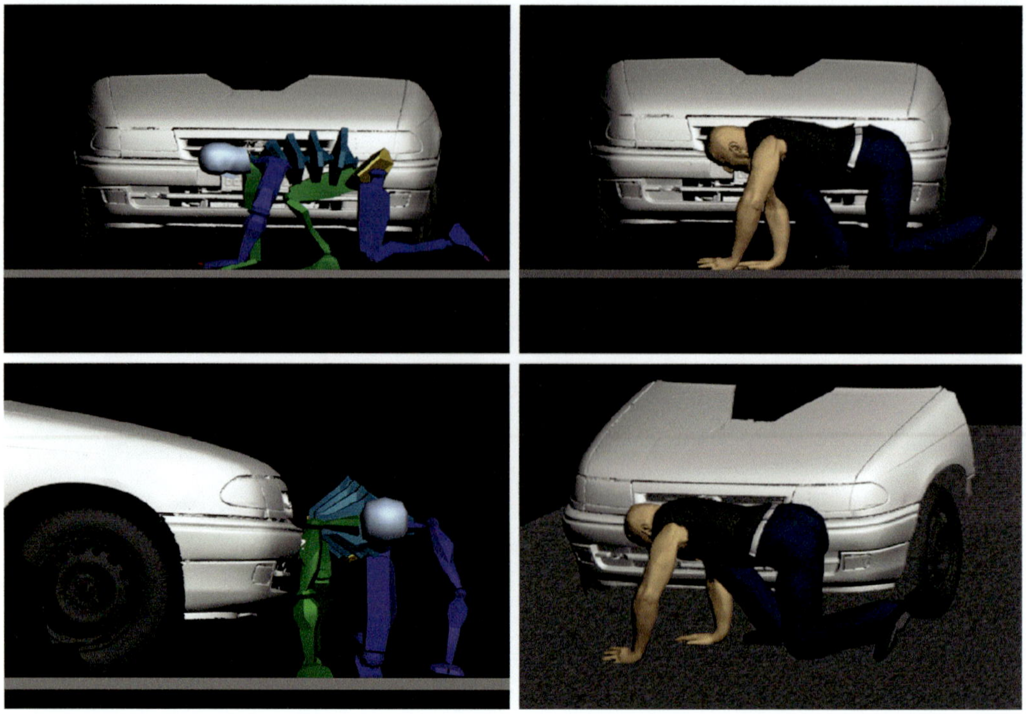

FIGURE 29.13 Considering all injuries and traces, the following impact situation could be reconstructed: the accident victim was in a kneeling position.

room in the house. She found her husband lying on the bed and her son standing in the room with a gun in his right hand. The ambulance team pronounced the victim dead at the scene. The son claimed that, in the course of showing the functions and components of the gun to the father, a projectile was discharged because of faulty gun function. Detailed geometric analysis was carried out to discover whether it was a tragic, accidentally fired shot or a criminal assault.

The autopsy finding was an entrance wound to the left side of the nose and an exit wound on the right side of the nape of the neck. The bullet passed through the soft tissue on the right side of the cervical spine. The man died due to loss of blood caused by the gunshot injury to the head. At the crime scene, the body of the deceased was lying on the bed, and there was a bloodstained quilt and mattress. Moreover, investigators found a mark caused by the bullet in the wooden headboard. The bullet and the gun were sited on top of the cat's climbing frame that was located adjacent to the door. The crime scene was documented in 3D using photogrammetry and tachymetry by police experts. The direction of the trajectory of the bullet through the mattress was measured by means of a rod with a mounted reflector of a known length and of the same diameter as the bullet, namely, 9 mm. This reflector rod was forced through the channel within the mattress to contact the ricochet mark in the headboard. The angles and distances were measured using the tachymeter. A true-to-scale replicated crime scene 3D model was drawn by computer-assisted design (CAD) software using the results of the photogrammetry analysis and the measurements required by the tachymetry. The resulting 3D model of the crime scene included all of the important information for the investigation (Figure 29.14). As is typical in the Virtopsy Project, the corpse was examined using radiological techniques that included MSCT and MRI. The entrance and exit wounds were photographed.

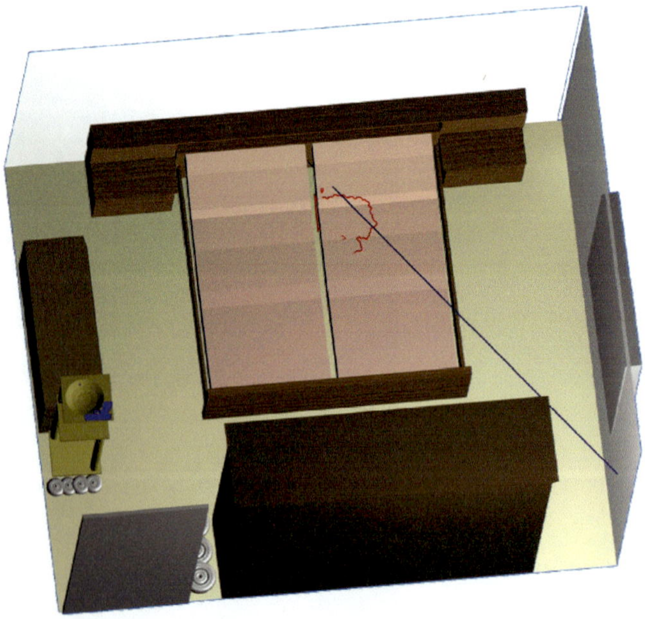

FIGURE 29.14 3D model of the crime scene of a gunshot case. The direction of the trajectory of the bullet (blue line) was determined by tachymetry.

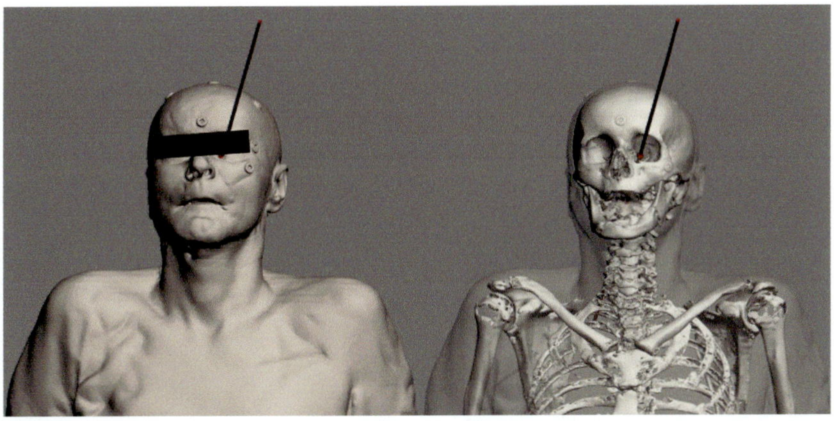

FIGURE 29.15 3D model of the head and skull generated from CT data. Based on the entrance and exit holes in the 3D model of the skull, the shot direction was reconstructed (line).

The 3D model of the deceased generated from the CT images displayed the entrance and exit wounds on the head. The bullet channel through the head of the deceased was drawn as a result of this CT model (Figure 29.15). The 3D model of the corpse was repositioned with computer software to bring the corpse into the position in which it was found and to explore probable positions at the time the shot was fired. Thus, a computer skeleton model was adapted to the bone structure of the CT model of the corpse (Figure 29.16). The computer animated model of the offender was adapted to his body measurements. The next step was the fusion of the 3D scene with the models of the involved persons to reconstruct what happened. All modeled data were combined in one file in the animation software 3ds max. The final scenario at the scene of crime was reconstructed based on photographs that were taken by police officers at the scene. These photographs were integrated into the photogrammetric analysis. After computing the camera positions and the calibration data for the camera used, the photographs were projected into the 3D room by the animation software. The computer model of the corpse was moved into the precise posture and position of the deceased lying on the bed based on the projected photographs (Figures 29.17 and 29.18). The shooting was virtually reconstructed based on the directions of the bullet trajectory in the bedroom and through the head of the victim. The computer model of the corpse was moved to bring it into the probable position when the bullet was fired such that the bullet channel through the deceased's head matched the direction of the

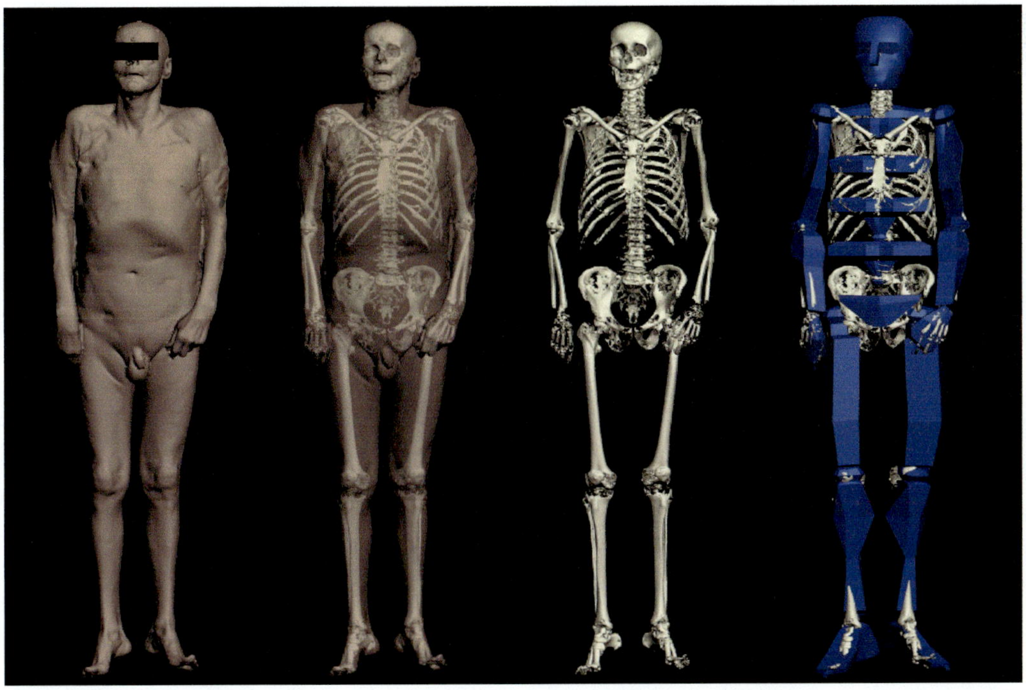

FIGURE 29.16 The 3D model generated from the CT images displays the individual osseous system and skin surface. Based on these models, a computer animation model was adapted to the unique physical parameters of the deceased.

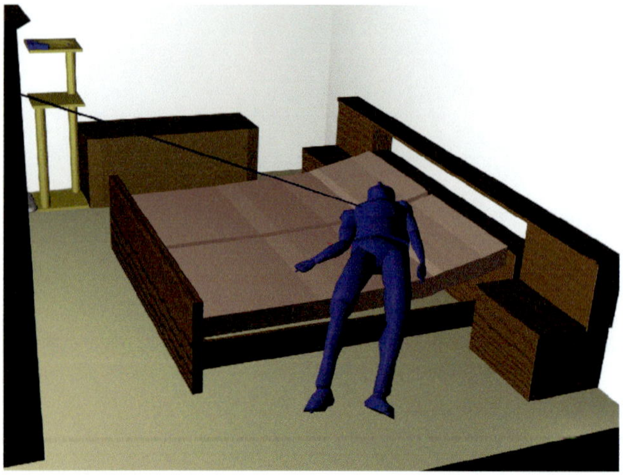

FIGURE 29.17 The computer model of the corpse positioned on the bed in the 3D room. This was the end position in which the deceased was found.

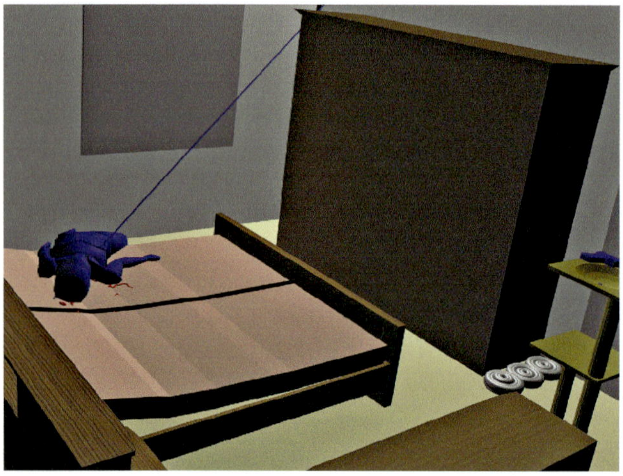

FIGURE 29.18 3D model of the crime scene from a different perspective.

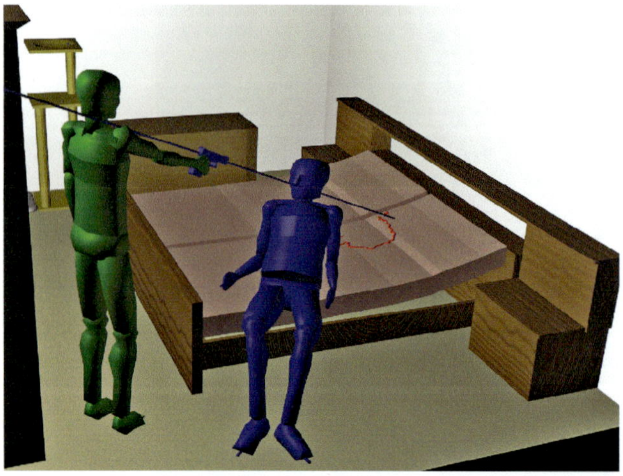

FIGURE 29.19 The presumable offender–victim configuration when the pistol was fired was reconstructed considering the traces, the physical parameters of the victim and offender, the bullet holes in the mattress and head of the bed, as well as the injuries suffered by the victim.

FIGURE 29.20 3D model of the presumable offender–victim configuration from a different perspective.

bullet trajectory in the room. To examine the probable shot positions, the computer model of the offender was moved within the 3D scene (Figures 29.19 and 29.20). In this reconstructed offender–victim configuration, the measured distance between gun muzzle and entrance wound of the victim was equal to the distance determined by the smoke residue analysis results.

REFERENCES

1. OsiriX, Open Source Software, http://www.osirix-viewer.com, 2009.
2. *Multi-Modality Markers for CT and/or MRI*, IZI Medical Products, Baltimore, MD, USA, http://www.izimed.com, 2009.
3. Thali, M. J., Braun, M., Wirth, J., Vock, P., Dirnhofer, R., 3D Surface and body documentation in forensic medicine: 3D/CAD photogrammetry merged with 3D radiological scanning, *J Forensic Sci*, 48(6), 1356–65, 2003.
4. Thali, M. J., Braun, M., Buck, U., Aghayev, E., Jackowski, C., Vock, P., Sonnenschein, M., and Dirnhofer, D., VIRTOPSY—Scientific documentation, reconstruction and animation in forensic: Individual and real 3D data based geo-metric approach including optical body/object surface and radiological CT/MRI scanning, *J Forensic Sci*, 50(2), 428–24, 2005.
5. Autodesk, Inc., Autodesk 3ds max, http://www.autodesk.com/3dsmax. 2009.
6. GOM, ATOS, Germany, http://www.gom.com. 2009.
7. Sensable Technologies, MA, USA, http://www.sensable.com. 2009.
8. Buck, U., Albertini, N., Naether, S., Thali, M. J., 3D documentation of footwear impressions and tyre tracks in snow with high resolution optical surface scanning, *For Science Int*, 171, 157–164, 2007.
9. Metter, D. Rechtsmedizinische Unfallrekonstruktion von tödlichen Fussgänger-PKW-Unfällen, *Z Rechtsmedizin*, 91, 21–23, 1983.
10. Buck, U., Naether, S., Braun, M., Bolliger, S., Friederich, H., Jackowski, C., et al., Application of 3D documentation and geometric reconstruction methods in traffic accident

analysis: With high resolution surface scanning, radiological MSCT/MRI scanning and real data based animation, *Forensic Sci Int*, 170, 20–28, 2007.
11. Metter, D. Das Decollement als Anfahrverletzung, *Z Rechtsmedizin*, 85, 211–19, 1980.
12. Prokop, O. and Radam, G., Rekonstruktion von Verkehrsunfällen aus gerichtsärztlicher Sicht, *Handbuch der Verkehrsmedizin*, p. 952, 1968.
13. Weimann, W., Radspuren an Überfahrenen und ihre kriminalistische Bedeutung, *Arch Kriminol*, 80, 1–6, 1927.
14. Buck, U., Christe, A., Naether, S., Ross, S., and Thali, M.J., Virtopsy–noninvasive detection of occult bone lesions in postmortem MRI: Additional information for traffic accident reconstruction, *Int J Legal Med*, 123(3), 221–26, 2009.
15. Lisowski, Z. and Marek, Z., Hyperpigmentation in the long bones of the lower limbs as a basis for vehicle identification and traffic accident reconstruction, *Forensic Sci Int*, 20, 251–55, 1982.
16. Karger, B., Teige, K., Fuchs, M., and Brinkmann, B., Was the pedestrian hit in an erect position before being run over? *Forensic Sci Int*, 119, 217–20, 2001.
17. Löhle, U., Zur Unfallrekonstruktion bei verschiedenen Unfalltypen, *AJP/PJA*, 4, 524–30, 1995.

30 Applications of Rapid-Prototyping Methods in Forensic Medicine

Lars C. Ebert, Steffen G. Ross, and Michael J. Thali

CONTENTS

INTRODUCTION

The introduction of spiral (helical) computed tomography (CT) in the early 1990s led to exciting new applications in the field of three-dimensional (3D) reconstruction of the former primarily 2D axial CT cross-sections. 3D reconstructed CT data enable the depiction of a variety of pathologies at a glance, even from a radiologically inexperienced viewpoint. Spiral data acquisition allows the display of a whole volume dataset without misperception of anatomic detail. These 3D reconstructions integrate information from a large stack of 2D images, which is often better interpreted than the original cross-sections.

To allow the reconstruction of a 3D model in all three axes (x, y, z) without loss of spatial information, medical volume data sets are usually isotropic. This means that the elements of a volume dataset, the so-called voxels (volumetric picture elements), are cubic.

Different rendering techniques allow for different mapping of a volume data set from 3D to 2D (i.e., Maximum Intensity Projection or different Transfer Functions).

Displaying 3D structures on a 2D screen, however, comes with a loss of information and makes it harder to identify the spatial properties of an object, especially for untrained persons (Bouyssie, 1997). 3D screens provide 3D perception, but this technology is not yet technically mature and is usually not available. Apart from this, it lacks a haptic component.

One way to overcome this problem is rapid prototyping, which allows the production of a physical model based on Computer Aided Design (CAD) data. Standard applications for this technique are found, for instance, in the aerospace and automobile industries, where it is used as a cheap tool to quickly build prototypes (Mankovich, 1990).

RAPID PROTOTYPING IN CLINICAL MEDICINE

Since rapid prototyping and 3D medical imaging have been integrated and made available for medicine (Barker, 1993), these techniques have found many applications. Models generated by CT scanning have applications in maxillofacial (Klimek, 1993; Lambrecht, 1990; Sailer, 1998), spinal (Mizutani, 2008), and orthopedic surgery (Potamianos 1998). CT-based angiographic volume data sets additionally allow for neurovascular and cardiovascular (Kinoshita, 1990) interventions. These models are used to visualize the patient's anatomy and to plan complicated or dangerous procedures (Stoker, 1992), such as tumor resections (Yamazaki, 2007) and reconstructive surgeries. Based on rapid-prototyping models, simulations of interventional procedures can be verified without danger to the patient (Murase, 2008). Increased planning possibilities lead to shorter operation and anesthesia times, decreased blood loss and, therefore, a lower morbidity rate during surgery (Dolz, 2000). Full scale and anatomically correct human phantoms built with rapid-prototyping techniques allow for better training of medical personnel (Holubar, 2009). Stereolithographic models are being used for these purposes, as they are cheaper and easier to handle than cadaveric material, but are still based on real-patient anatomy. Rapid prototyping has also proven to be a valuable tool in implant design (Curodeau, 2000). In biomechanical testing, customized bone systems can also replace work on cadaveric bones (Kakarala, 2006). As the minimal possible structure size with rapid prototyping is 0.1 mm or less, the accuracy of CT data-based modeling is limited by the resolution of the scanners and the segmentation technique rather than the modeling process (Bouyssie, 1997).

On a smaller scale, rapid prototyping is used to create scaffolds with micro- or nano-features for tissue engineering (Peltola, 2008).

POSSIBLE APPLICATIONS OF RAPID PROTOTYPING IN FORENSIC MEDICINE

Digital volume data sets provide an abstract way of depicting pathologies. This fact is of great advantage if a case must be presented in court to medical laymen, who are not used to drawing their conclusions from reports packed with Latin phrases. A few attempts have been made to make use of rapid prototyping in forensic medicine. Abramov et al. were the first to suggest the use of stereolithography for diagnostic and identification purposes (Abramov, 1998). With the increased use of medical imaging techniques such as CT in forensic investigations, the modalities required for rapid prototyping of patient-specific anatomical models are increasingly available (Thali, 2003).

Photographs taken during autopsy may help to demonstrate pathological findings, but their bloody appearance can be quite disturbing and still misunderstood. Full-scale models of pathologies can help to depict the cause of death in a way that is easily understood by medical laymen without bloody autopsy images or abstract volume reconstructions. In conjunction with CT scanning, rapid-prototyping models give a good picture of bone fractures and dislocations (Dolz, 2000). Angiography allows display of the cardiovascular system and internal hemorrhages.

The first time rapid-prototyping methods were applied in conjunction with forensic radiology was the case of the more than 5000-year-old Iceman found in the Tyrolean Alps. In this anthropological case, the body was CT scanned and a stereolithographic model of the skull was printed in order to noninvasively examine the body. Several copies were printed and distributed for scientific studies (zur Nedden, 1994).

PRINCIPLES OF RAPID PROTOTYPING

Traditional Computer Numerical Control (CNC) milling machines provide a means of creating a 3D model out of a material block. This technique, however, fails to create structures that are occluded, such as cavities (Klein, 1992). In 1986, the idea of generating real 3D models based on virtual 3D data evolved into stereolithography (Dolz, 2000). Stereolithography builds up a model layer by layer by hardening a liquid-acrylic resin with an ultraviolet laser beam (Hull, 1986). Unlike CNC milling, this technique is limited to a specific material but allows for the generation of closed cavities and fine structures. Models generated with stereolithography are durable but lack color information. The second most-commonly used method is called 3D printing. It is a relatively new technology and allows the generation of colored 3D models. These systems create layers of powder that are fixed selectively by a binder (Figure 30.1). Comparable to standard ink printers, color can be added to the binder. While stereolithography has an accuracy of 0.1 mm, 3D printing can print structures up to 0.016 mm depending on the particle size of the powder and the binder used. The costs of a 3D printed model are approximately one-third of those for a stereolithographically generated model (Cohen, 2009).

FROM VOLUME DATA SET TO 3D MODEL

Creating a 3D print requires thorough preprocessing of the volume dataset. For rapid prototyping of the anatomies presented, we used a Z510 3D printer (Z Corporation, USA) (Figure 30.2), but the workflow for preparing the dataset was similar to other techniques.

There are some constraints on what can be printed using his technique. Very sophisticated structures such as small vessels can break easily. Closed cavities are possible, but they will be filled with powder without the possibility of removing it. Floating or fragile structures should be equipped with support structures. Large structures such as a full rib

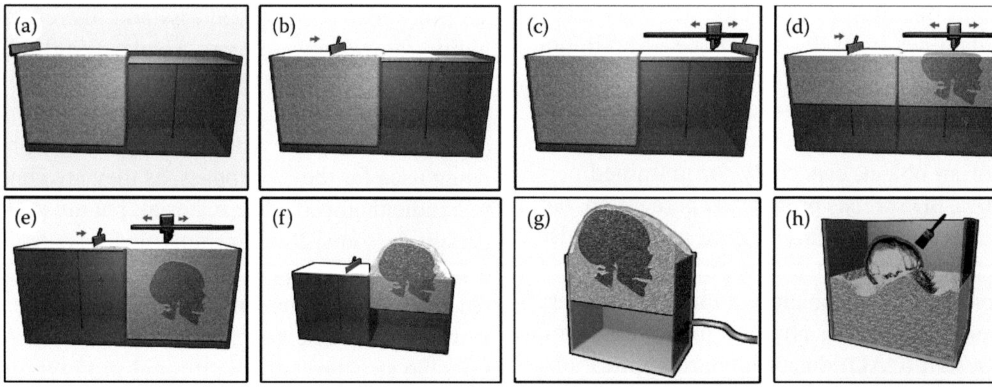

FIGURE 30.1 Process of 3D printing. (a) Left: powder table. Right: printing table. (b) A thin layer of powder is distributed on the printing table. (c) Binder is selectively printed onto the powder layer and cured. (d, e) By gradually lowering the printing table and repeating steps b and c, a 3D model is embedded into the powder. (f) The printing table is removed and (g) put into a cleaning chamber. (h) Excess powder is removed with a brush or air pressure, and the model is finalized with a fixation agent.

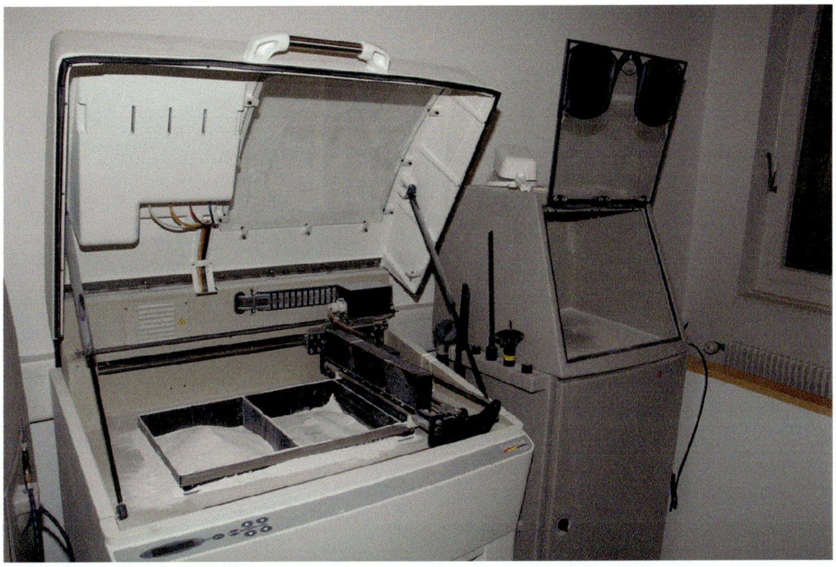

FIGURE 30.2 Z510 3D printer (Z Corporation, USA). Left: printer. Right: cleaning chamber.

cage exceed the possible printing volume and should be split into different prints that can be glued together or scaled down.

Processing a volume data set requires multiple steps (Figure 30.3). The first step after performing the scan is to segment different structures in the voxel-based volume data set. In this step, each voxel is assigned to an anatomical structure. Additionally, some support structures are defined for better model stability (Figure 30.4). Since the software packages of rapid-prototyping machines usually require polygon datasets instead of voxel data, a polygon dataset is generated out of the segmented data. Smoothing filters can be applied to decrease aliasing effects. For large models, algorithms for reducing the number of triangles can be applied. For these steps we used the Amira software package (Visage Imaging, Germany). All models are exported as STereoLithography (STL) files, which is the standard format for rapid prototyping. After importing these files to the printing application ZEdit (Z Corporation, USA), color information for each model is set. If necessary, the model can be scaled. The software allows alignment of the model on the printing table. From this point, the printing layers are calculated automatically and the model is printed. The printing process takes up to six hours depending on the size of the printed model. This includes a drying phase. After freeing the model of unused powder, a hardening agent is applied, which takes an additional half-an-hour to dry.

EXAMPLE CASE 1: CERVICAL SPINE FRACTURE

Following an accidental fall from great height (21 m), a 42-year-old female was documented with minimally invasive postmortem CT angiography 18 h after death. CT scanning was done with a slice thickness of 1.25 mm. Image reconstruction was carried out with a medium soft-tissue kernel. CT of the skull base and cervical spine revealed a bilateral compression fracture of C1 (Jefferson Fracture), a fracture of the spinous process of C2, and a rotational dislocation of C1 against C2 (Figure 30.5).

Cross-sectional CT angiographic data of the skull base and cervical spine were converted into a 3D digital mesh. The vertebral arteries were artificially colored because of the almost similar x-ray density of intravascular compared to bony structures. Additional support structures were added to provide higher stability to the vertebral bodies as well as the skull base. The whole process, including CT angiography, segmentation, and printing, took approximately eight hours (angiography two hours, segmentation two hours,

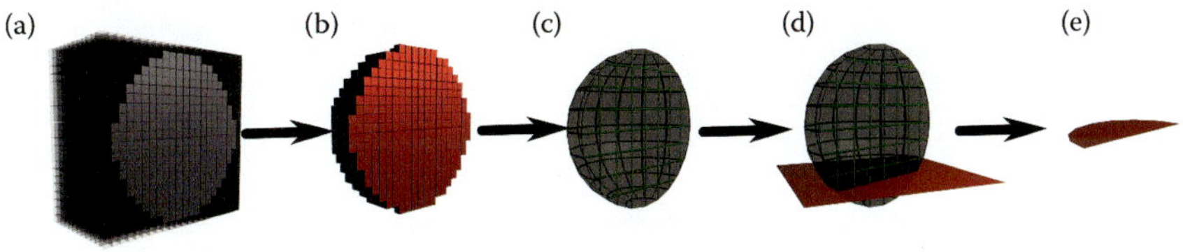

FIGURE 30.3 Processing a volume data set for rapid prototyping. (a) Voxel-based volume data set. (b) Segmentation: each voxel is assigned to a different material. (c) The segmented data is converted into a polygon mesh. (d) Calculating cross-sections for printing. (e) Final printing layer.

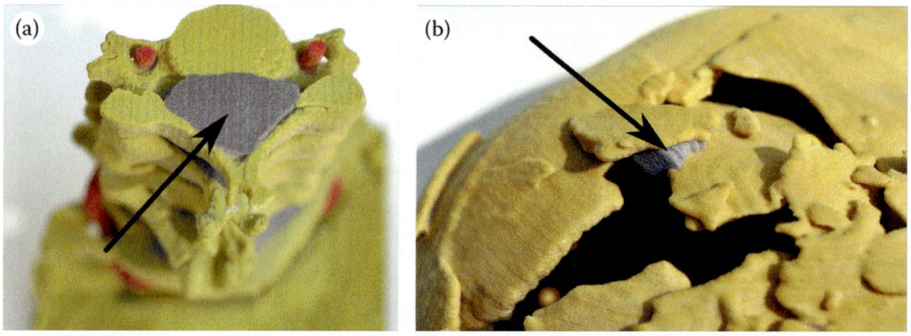

FIGURE 30.4 Support structures in 3D printing (arrows). (a) The spinal canal is filled with gray support structures, fixing the only partly connected cervical spine to the mechanically stable printed model. (b) A support structure has been printed to attach a bone fragment, which would otherwise fall off from the rest of the model after printing.

and printing four hours). In comparison to 2D radiographs or volume rendering, the extent of the damage can easily be understood by laymen, which can be of value if a medical report must be presented in a courtroom. The abstract nature of such a model is a more appropriate way to present medical findings in front of relatives as opposed to autopsy images.

EXAMPLE CASE 2: MULTIPLE SKULL FRACTURES

A 24-year-old female homicide victim suffered massive blunt skull trauma (Figure 30.6). The print was based on a CT scan with a slice thickness of 1.25 mm. The scan was performed 46 h after death. Image reconstruction was carried out in a hard tissue kernel. All bony structures were segmented, and support structures were added to one of the bone fragments as well as the mandible. In order to depict the injury from inside the skull, an opening was created on the opposing side of the fracture. Segmentation of the bone took approximately 20 min, including generation of a 3D mesh and smoothing and reducing the number of triangles. The printing process took eight hours, including drying and fixation of the model.

EXAMPLE CASE 3: ANENCEPHALIC FETUS

An anencephalic fetus from the anatomy department of the University of Bern stored in formalin was documented with an unenhanced CT scan (Figure 30.7). The CT scan had a slice thickness of 0.63 mm, and reconstruction was done with a bone-weighted kernel. The skull, jaw, and cervical spine were segmented. Support structures were added to the spinal canal, mandible, and the temporomandibular joint. Segmentation of the structures took approximately 30 min. The printing process took four hours, including drying and fixation of the model. CT imaging and 3D printing of such a specimen is a valuable contribution to research and medical education without destroying precious and rare preparations.

DISCUSSION AND CONCLUSION

By using virtual autopsy methods, digital high-resolution 3D data sets of anatomical structures are generated. In conjunction with rapid-prototyping methods, this results in the means to quickly generate haptic models based on real anatomy. In contrast to computer-assisted milling methods, cavities and occluded areas are less of a problem. With 3D printing it is also possible to print colored models. In a forensic setting,

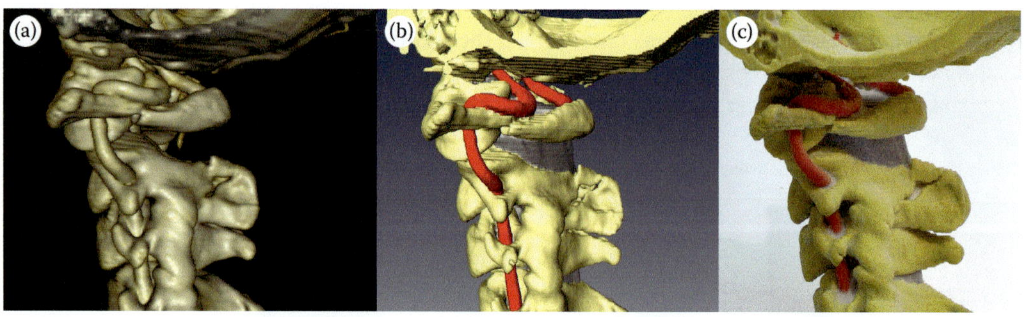

FIGURE 30.5 Postmortem CT angiography of the cervical spine in a case of massive compression and hyperextension of the cervical spine. (a) Volume rendering (Osirix). (b) Segmented polygon model of the cervical spine and skull base (Amira). Yellow: bone, red: vessels, gray: support structures. (c) 3D print of the segmented data. The resulting model nicely depicts the fractures of the left posterior arch of C1 and the spinous process of C2 as well as the course of both uninjured vertebral arteries. Note the rotational dislocation of C1 against C2, giving direct view to the joint surface of the facet joints.

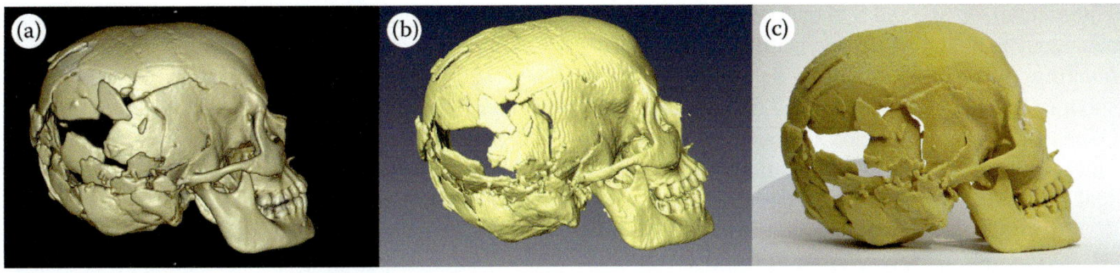

FIGURE 30.6 CT scan of blunt force trauma to the posterior skull. (a) Volume rendering (Osirix). (b) Segmented polygon model of the bone (Amira). (c) 3D print of the segmented data. The stable position of the large partly displaced bone fragments was reached by integrating support structures. This kind of three-dimensional display is not possible in classical autopsy, where alteration during the preparation of the adjacent soft tissue leads to the immediate disintegration of the bony fragments and thus destroys relevant forensic findings.

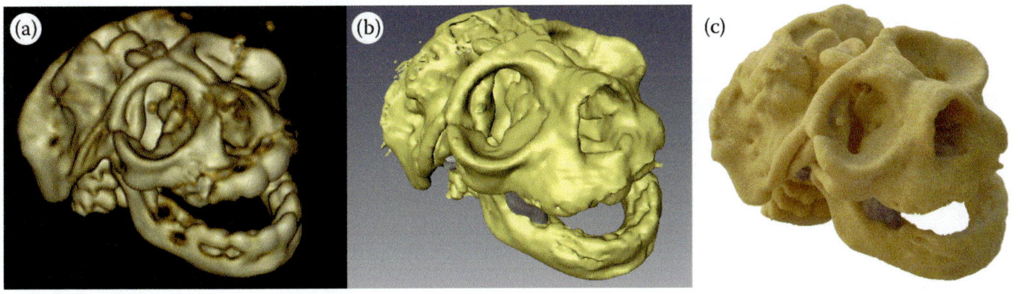

FIGURE 30.7 Cranial CT of an anencephalic fetus. (a) Volume rendering (Osirix). (b) Segmented polygon model of the bone (Amira). (c) 3D print of the segmented data.

this can assist the investigator in presenting evidence to a court in an abstract way without losing the 3D nature of the structures. Therefore, it is more easily understood by laymen than 2D x-ray images, photographs, or 2D volume reconstructions. By utilizing 3D postprocessing software, fractured bones can be reassembled, which can be of use in forensic facial reconstruction. Printed 3D models of anthropological and anatomical specimens can also be used for educational purposes. Depending on the modality used, it is possible to print bone structures, major vessels, or entire organs. Once the image data are gathered, the final model is available after one to two days, depending on its size.

In summary, rapid prototyping is a useful tool to complement medical imaging techniques used in virtual forensic autopsy.

REFERENCES

Abramov, S. S., Boldyrev, N. N., Evseev, A. V., Kotsiuba, E. V., Novikov, M. M., Panchenko, V., Semeshin, N. M., and Iakunin, V. P., The possibility of using a laser stereolithography method in forensic medicine, *Sud Med Ekspert*, 41(3), 13–7, 1998.

Barker, T. M., Earwaker, W. J., Frost, N., and Wakeley, G., Integration of 3-D medical imaging and rapid prototyping to create stereolithographic models, *Australas Phys Eng Sci Med.*, 16(2), 79–85, 1993.

Bouyssie, J. F., Bouyssie, S., Sharrock, P., and Duran, D., Stereolithographic models derived from x-ray computed tomography. Reproduction accuracy, *Surgical & Radiologic Anatomy*, 19(3), 193–9, 1997.

Cohen, A., Laviv, A., Berman, P., Nashef, R., and Abu-Tair, J., Mandibular reconstruction using stereolithographic 3-dimensional printing modeling technology, *Oral Surg Oral Med Oral Pathol Oral Radiol Endod.*, 108(5), 661–6, 2009.

Curodeau, A., Sachs, E., and Caldarise, S., Design and fabrication of cast orthopedic implants with freeform surface textures from 3-D printed ceramic shell, *J Biomed Mater Res*, 53(5), 525–35, 2000.

Dolz, M. S., Cina, S. J., and Smith, R., Stereolithography: A potential new tool in forensic medicine, *Am J Forensic Med Pathol*, 21(2), 119–23, 2000.

Holubar, S. D., Hassinger, J. P., Dozois, E. J., Camp, J. C., Farley, D. R., Fidler, J. L., Pawlina, W., and Robb, R. A., Virtual pelvic anatomy and surgery simulator: An innovati ve tool for teaching pelvic surgical anatomy, *Stud Health Technol Inform*, 142, 122–4, 2009.

Hull, C., Apparatus for production of three-dimensional objects by stereolithography, U.S. Patent 4,575,330, 1986.

Kakarala, G., Toms, A. D., and Kuiper, J. H., Stereolithographic models for biomechanical testing, 13(6):451–4, 2006.

Kinoshita, S., Konishi, G., Takeuchi, S., Ukai, T., and Taniguchi, H., 'Stereovectorcardiogram' made by stereolithography, *Cardiology*, 77(4), 269–71, 1990.

Klein, H. M., Schneider, W., Alzen, G., Voy, E. D., and Günther, R.W., Pediatric craniofacial surgery: Comparison of milling and stereolithography for 3D model manufacturing. *Pediatr Radiol*, 22(6), 458–60, 1992.

Klimek, L., Klein, H. M., Schneider, W., Mosges, R., and Schmelzer, B., and Voy, E. D., Stereolithographic modelling for reconstructive head surgery. *Acta Oto-Rhino-Laryngologica Belgica*, 47(3), 329–34, 1993.

Lambrecht, J. T. and Brix, F., Individual skull model fabrication for craniofacial surgery, *Cleft Palate J.*, 27(4), 382–5, discussion 386–7, 1990.

Mankovich, N. J., Cheeseman, A. M., and Stoker, N. G., The display of three-dimensional anatomy with stereolithographic models, *J Digit Imaging*, 3(3), 200–3, 1990.

Mizutani, J., Matsubara, T., Fukuoka, M., Tanaka, N., Iguchi, H., Furuya, A., Okamoto, H., Wada, I., and Otsuka, T., Application of full-scale three-dimensional models in patients with rheumatoid cervical spine, *Eur Spine J.*, 17(5), 644–9, 2008.

Murase, T., Oka, K., Moritomo, H., Goto, A., Sugamoto, K., and Yoshikawa, H., Correction of severe wrist deformity following physeal arrest of the distal radius with the aid of a three-dimensional computer simulation, *Arch Orthop Trauma Surg.*, 129(11), 1465–71, 2008.

Peltola, S. M., Melchels, F. P., Grijpma, D. W., and Kellomäki, M., A review of rapid prototyping techniques for tissue engineering purposes, *Ann Med.* 40(4), 268–80, 2008.

Potamianos, P., Amis, A. A., Forester, A. J., McGurk, M., and Bircher, M., Rapid prototyping for orthopaedic surgery, *Proc Inst Mech Eng H.*, 212(5), 383–93, 1998.

Sailer, H. F., Haers, P. E., Zollikofer, C. P., Warnke, T., Carls, F. R., and Stucki, P., The value of stereolithographic models for preoperative diagnosis of craniofacial deformities and planning of surgical corrections, *Int J Oral Maxillofac Surg.*, 27(5), 327–33, 1998.

Stoker, N. G., Mankovich, N. J., and Valentino, D., Stereolithographic models for surgical planning: Preliminary report, *J Oral Maxillofac Surg.* 50(5), 466–71, 1992.

Thali, M. J., Yen, K., Schweitzer, W., Vock, P., Boesch, C., Ozdoba, C., Schroth, G., et al., Virtopsy, a new imaging horizon in forensic pathology: Virtual autopsy by post mortem multislice computed tomography (MSCT) and magnetic resonance imaging (MRI)—a feasibility study, *J Forensic Sci*, 48(2), 386–403, 2003.

Yamazaki, M., Akazawa, T., Okawa, A., and Koda, M., Usefulness of three-dimensional full-scale modeling of surgery for a giant cell tumor of the cervical spine, *Spinal Cord.*, 45(3), 250–3, 2007.

Zur Nedden, D., Knapp, R., Wicke, K., Judmaiser, W., Murphy, W. A., Seidler, H., and Platzer, W., Skull of a 5300-year-old mummy. Reproduction and investigation with CT-guided stereolithography, *Radiology*, 193, 269–72, 1994.

31 Facial Reconstruction
New Approaches

Gregory Mahoney and Christopher Rynn

CONTENTS

Facial reconstruction, also known as facial approximation (Clement and Marks, 2006), has been used by law enforcement agencies and archaeologists for decades to provoke a recognition reaction from the public to aid in identification of an unidentified skull, or to visualize the faces of historical figures. Though familiar to many, the accuracy of the process remains a source of debate. The viewer of a facial reconstruction should understand that it is not necessarily an exact portrait, but rather a representation or likeness of the individual as they might have appeared in life based on information derived from the skull. In a forensic context, the intention is to generate leads: thus it is intelligence rather than evidence. Positive identification will ultimately rely upon DNA, fingerprint, or odontological comparison to records obtained by following up these leads. While research continues to lead to better and more accurate approaches for forensic facial reconstruction, it still provides a useful tool today for law enforcement investigators where no other leads are available. For archaeologists, it may provide the only means of visualizing an individual with a reasonable level of accuracy.

The following assumes that the reader will have some knowledge of forensic facial reconstruction and anatomy. There are several books published, and a growing number completely dedicated to the subject. Karen Taylor's *Forensic Art and Illustration* (2001) contains a chapter by Betty Pat Gatliff who, along with Dr. Clyde Snow, co-developed the American method. The method described here is not novel, but rather an adaptation of the American method which is currently in use and accepted within the field, the difference being the tools employed. The method uses computerized three-dimensional (3D) virtual sculpture to allow the practitioner to do more with their sculptures, with more precision and repeatability, and the valuable option of variable transparency, while adhering to any physical method of forensic facial reconstruction.

A major advantage of utilizing any computerized method is that the skull will require far less handling. No matter how the 3D skull model is produced—the best methods to date being an optical laser scan or computed tomography (CT) scan—once scanned, the skull no longer needs to be handled. Of course, for archaeological purposes this is extremely advantageous.

The 3D skull scan data can be reproduced as a solid replica, using a rapid prototype printer, to act as the base for a clay sculpture: however, using the virtual skull model within the computer system provides distinct advantages and possibilities. Many will be evident in the text that follows, but if one thinks of all the possibilities a computer provides today, this will only be magnified in the future. Animation, modeling and realistic texturing, and 3D displays and their impact on recognition are but a few of the areas to be researched.

The method described here utilizes a hardware/software package produced by Sensable Technologies. This system uses a haptic tactile feedback device, in this case the Phantom desktop, in conjunction with their software called Freeform Modeling Plus. This method can be adapted using other systems made by Sensable, or really any other system utilizing a haptic device and 3D sculpting software. The principles remain the same.

The Phantom/Freeform Modeling Plus system was developed primarily for the industrial design and medical fields, and thus has a number of tools within the package that are surplus to requirements regarding facial reconstruction. The following will give a brief description of the tools and functions within the software that are ideal for this purpose. Though awkward at first, with a little practice, a sculptor of clay will find themselves quite at home in the virtual environment and excited by the new possibilities, formerly unavailable to them in the physical world.

Freeform Modeling Plus essentially allows an operator to sculpt virtual "clay" objects in a 3D space. The software is structured in a way that is similar to many photo- editing software programs, where there are multiple layers that can be grouped and turned on or off for viewing. Perspective can be set and turned on or off as needed for viewing. Many of the tools within the workspace would be familiar to a sculptor. There are tools for carving, shaping, and smoothing the clay that look and feel like tools used in the physical world. Other tools are perhaps not so familiar, but extremely valuable. The "tug" tool allows an operator to push or pull areas of clay in any direction and is used more than any other tool. New pieces of clay can be added in specified shapes, custom shapes, or imported from libraries. Flat planes can be aligned in any orientation, can be drawn on, or can display scaled photographs with variable transparency. These planes can also be used to slice or mirror a piece of "clay." The resolution of the "clay" can be changed, and the "clay" can be made harder or softer depending on the desired effect. Tools exist for sketching directly on the surface of an object as well as for measuring. There are many other tools within the software, which may be employed, but those described above are the most useful for facial reconstruction.

Before working in 3D space within Freeform virtual sculpture software, CT data needs to be converted into a "Certified Trust License" (.stl) file type or "Object" (.obj) file type to be imported as a 3D model. Once imported, the skull should be aligned to the Frankfurt Horizontal Plane. Aside from placing the skull in proper orientation for accepted facial reconstruction techniques, it allows an operator to quickly switch between points of view along the X-, Y-, and Z-axis.

It is also advisable to fill a working copy of the virtual skull so that no computing power is required to render interior surfaces of the skull, which are not needed for facial reconstruction. Nevertheless, it is also advisable to keep an unedited skull copy in a separate layer, which can be referenced if necessary. This increases the file size, but if it is turned off (i.e., present but invisible) it is not taxing to the video memory. This can be automated to an extent, but you do have the option to manually fill the skull's interior allowing for more control. A separate piece is created in the virtual space and is blended to fit with the exterior surface. The two pieces are then combined, which effectively fills the voids as required.

With this new combined piece, frontal, posterior, lateral, superior, and inferior planes are attached to the piece, creating an exterior box. These planes can be "sketched" on with drawing tools, or have previously saved "sketches" imported on a selected plane. The model is displayed in orthogonal projection (i.e., perspective turned off so the box sides are at right angles) when sketching on planes. Thus, libraries of tissue depth reference points can be created, saved, and re-used. These "sketched" tissue depth landmarks can be moved on the planes and properly aligned to their corresponding landmarks on the skull behind the plane. Once properly aligned to the individual skull, the reference markers can be projected onto the surface of the skull piece (Figure 31.1). Projected reference points can then be selected and the tissue depth marker extruded from the surface to the appropriate depth of each landmark. This tissue depth marker is then a uniform distance at its most projected surface (Figure 31.2) and precisely perpendicular to the surface of the skull. Using a pre-created box with tissue depth landmarks sketched on the planes to extrude tissue depth markers from a skull in this way can be faster and more accurate than traditional methods.

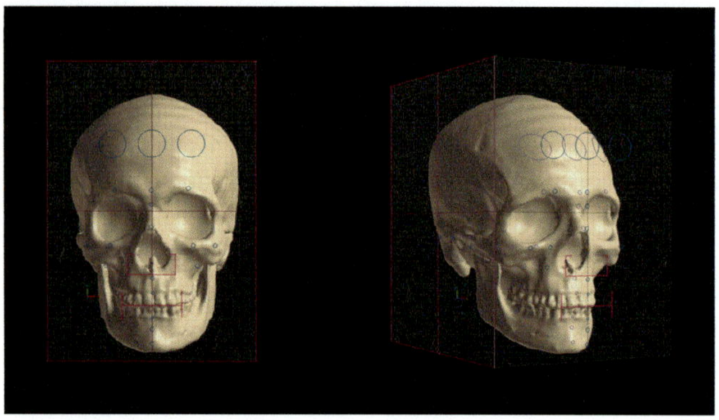

FIGURE 31.1 Sketch planes with tissue depth reference markers positioned on the left image, and projected on the model surface on the right image.

Using a copy of the filled skull, an offset piece can be created at a specific distance from the surface of the skull. Since tissue depth is most consistent and relatively shallow on the calvaria, the new piece is offset to the value associated with tissue depth landmarks around the frontal eminence or vertex. After this piece is created, the surface of the offset piece reflects the shape of the skull exactly, but requires smoothing to more closely resemble skin, while maintaining the set distance from the surface of the skull. This approach also means that any subtleties and asymmetries of the skull are echoed in the soft tissues.

The "tug" tool within the Freeform software essentially consists of a sphere of influence of adjustable size and strength, with the highest degree of influence at the center of the sphere. This tool can be positioned over a tissue depth marker and resized to a point where the radius of the sphere is about two-thirds the distance to an adjoining marker (Figure 31.3). Once positioned properly, the practitioner pulls or pushes the offset "skin" piece until the furthest extent of the tissue depth marker is just visible. This is only done with tissue depth markers that are along the bony eminences of the skull, as the soft tissue around these areas most closely corresponds to the surface of the skull. Areas where there are deep tissue, fat, and more robust muscles will require a different treatment.

The temporalis muscle and its overlying tissue are sculpted based upon the attachment points evident on the 3D skull

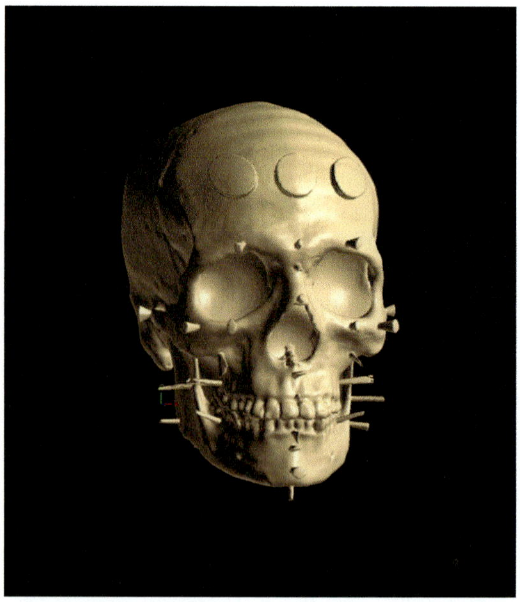

FIGURE 31.2 Tissue depth markers extruded from the model surface.

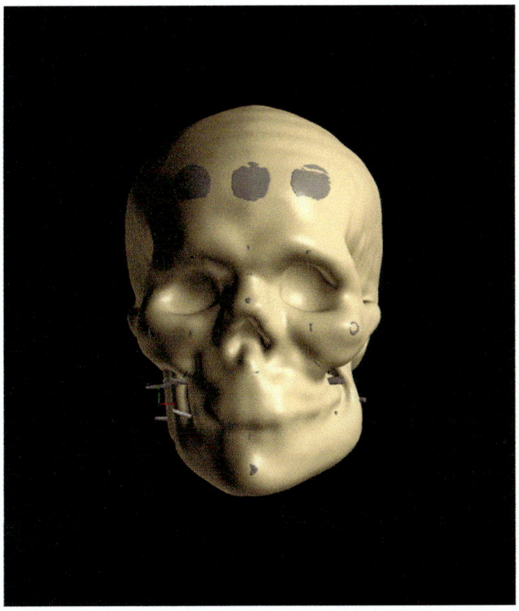

FIGURE 31.3 Offset "skin" layer which has been "tugged" to tissue depth markers. The skin has not been tugged on the subject's right mandible for viewer comparison.

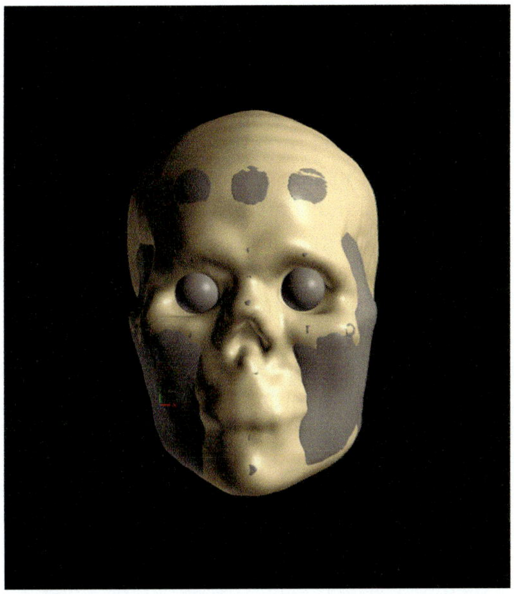

FIGURE 31.4 "Front triangle" and "back triangle" in position, and eyeballs placed in orbits.

model. One end of a column of clay is anchored to a point along the offset tissue over the zygomatic arch and zygomatic process of the temporal bone: the other end of the column is anchored to the point where the muscle would attach along the linea temporalis. This line should still be evident on the surface of the skin created by the offset piece of the skull. If it is not, the operator can either turn the piece invisible or make it transparent, so the skull can always be referenced. The level of transparency can be set, and the piece can still be manipulated as in the opaque view. This is a distinct advantage over applying clay on a skull, where the practitioner loses the ability to directly refer to the skull beneath the clay. When the fan shape of this muscle and tissue is complete, the columns, which are created as one piece, can be smoothed and blended into the surrounding tissue.

In accordance with the American method of facial reconstruction, the "back triangle" is created. This region corresponds with the plane of the lateral cheek and jaw area, superficial to the masseter muscle and parotid gland. The approach is similar to that used for the temporal region: cylinders of clay are attached along the mandibular body, from the gonial angle to the zygomatic arch between the external auditory meatus and the temporozygomatic suture. Again, this cluster of cylinders is subsequently smoothed and blended into the surrounding tissue.

The "front triangle" is created in a similar fashion to the "back triangle." This region corresponds to the soft tissue of the front plane of the face, around the nose and cheeks as well as the anterior half of the body of the mandible. Deep behind the front triangle lies the buccinator, zygomatic, and levator muscles of the mouth, along with subcutaneous fat. Columns are attached along the infraorbital margin, and to the mandible between the gonial angle and approximately halfway along the mandibular body (Figure 31.4). By this point, the face is beginning to take shape.

A neck is imported from a library of presculpted objects, scaled to correspond to the size of the skull and aligned (Figure 31.5). The sternocleidomastoid muscle is attached to the mastoid process, and trapezius muscle is attached to the occipital protuberance, which should be visible. The digastric and mylohyoid muscles and overlying skin can be approximated in the same way as the "front" and "back" triangles, with columns of clay being stretched backwards from their attachments, unless there is sufficient indication do otherwise.

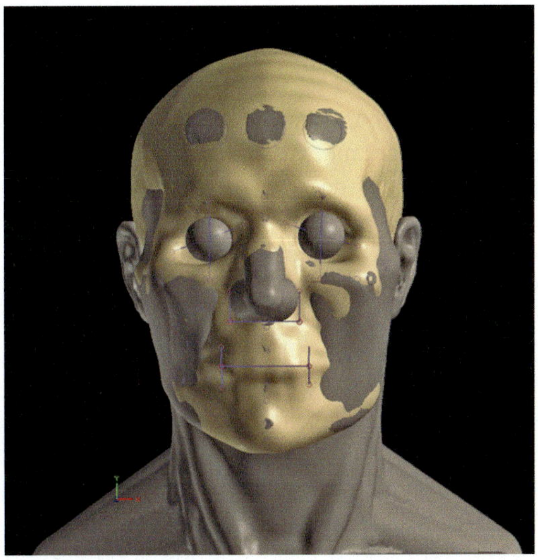

FIGURE 31.5 Neck and ears, scaled and placed after importing. A rough nose has been constructed. Reference lines for feature prediction guidelines are shown.

Libraries of ears can also be utilized, saving an inordinate amount of time when compared to sculpting an ear for each facial reconstruction. The ear chosen can be scaled, distorted, and angled properly according to the bony structure beneath. In order to properly position the ear relative to the external auditory meatus, a thin column is placed in the canal and drawn out and slightly down, to serve as a guide. A pivot point for the ear piece can be positioned at the external auditory meatus, and the ear rotated about this pivot to achieve proper alignment with the mandibular ramus.

The nose can be constructed according to the American method, following Krogman and İşcan's (1986) formulae, but also utilizing Gerasimov's (1955) method. Prior to constructing the nose itself, levator labii superioris is created to help with positioning of the nose, as in the American method. The shape of the muscle can be sketched directly onto the skull, paying attention to the areas of attachment and origin. This area can then be selected and extruded in the same way that the tissue depth markers were created. This approach helps ensure proper orientation of the muscle as it relates to the underlying hard tissue.

Reference drawings can then be sketched on the frontal plane and projected onto the tissue (Figure 31.5). A midline sagittal plane is created which bisects the width of the nasal aperture. Reference lines can be drawn on this plane according to Krogman's and/or Gersimov's prediction method. With all of the reference lines, and an understanding of the skeleton and structures of the nose, a nose is sculpted and positioned.

To place the eyeball, reference lines can be drawn to span the orbit, from the midpoint of the supraorbital margin to the midpoint of the infraorbital margin. The same is done from the medial to the lateral orbital margin, creating a reference for centering the eyeball in the orbit. These reference lines are also used to set the eyeball to the correct depth in the orbit. Once the eyeball has been placed, an offset piece is created from the eyeball, set at the thickness of the eyelid, in the same way that the "skin" piece was created from the skull. The shape of the eye fissure can then be sketched on this offset piece, according to the morphology of the eye socket, aligning the medial and lateral canthii to their respective attachment points. Once the fissure shape has been drawn, the eye area can be selected and extruded back into the piece, at a depth equal to the diameter of the piece, effectively creating the upper and lower eyelids. Either the upper or lower lid can be manually selected and cut and pasted in position as a separate piece. The portion of these newly created eyelid pieces that correspond to the palpebral ligaments can now be tugged to their attachment points. Be careful that the sphere of influence of the tug tool corresponds to the most medial and lateral points on the lid so the majority of the lid adheres to the shape of the eyeball. Be sure that the upper lid overlaps the lower lid at the lateral aspect. The advantage of approaching eyelid construction in this way is that the lid will automatically hug the eyeball, as in reality. To complete the eyes, the soft tissue between the supra and infraorbital margins and between the corresponding upper and lower lids are tugged towards the lids themselves. Be careful to pay attention to the eyelid patterns, which are dictated by the morphology of the eye socket as well as of the fatty tissue and orbicularis oculi (Figure 31.6).

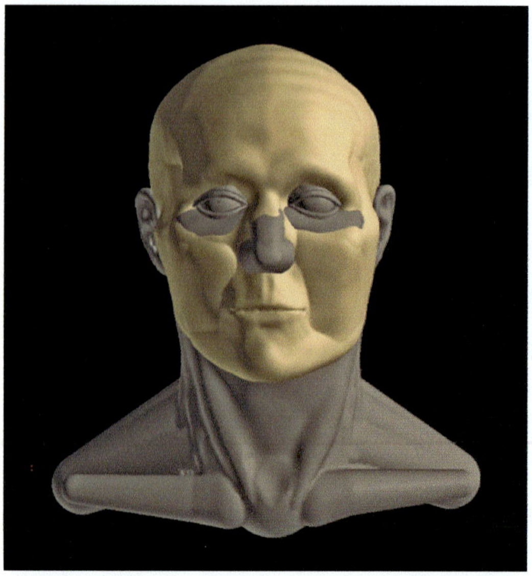

FIGURE 31.6 Eyelids are created using an offset piece from the eyeballs. Eye fissure shape is sketched on this offset piece, and clay deleted from within the fissure outline. Soft tissue above and below the eyelids is also shown.

For the mouth, as with the nose, reference lines can be drawn onto a plane and projected onto the tissues behind the plane. Primarily these are used for placement of the corners of the mouth and thickness of the top and bottom lips. Similar to creating a reference for the placement of the ears, two thin columns are created radiating out from the junctions between the upper canine and first premolar tooth, which helps to position the corners of the mouth. Using these references,

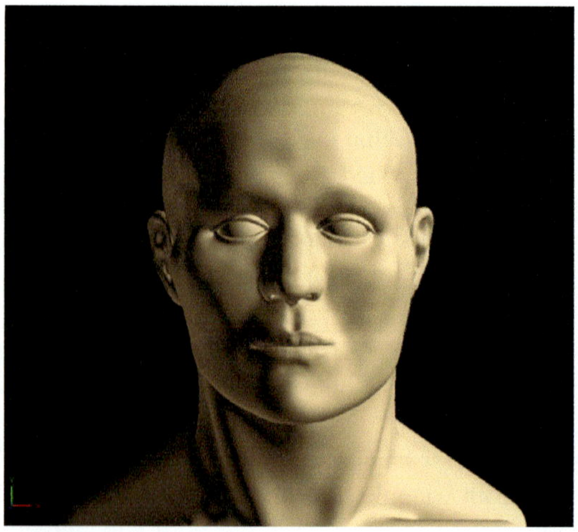

FIGURE 31.7 Final model with all pieces combined together and seams smoothed.

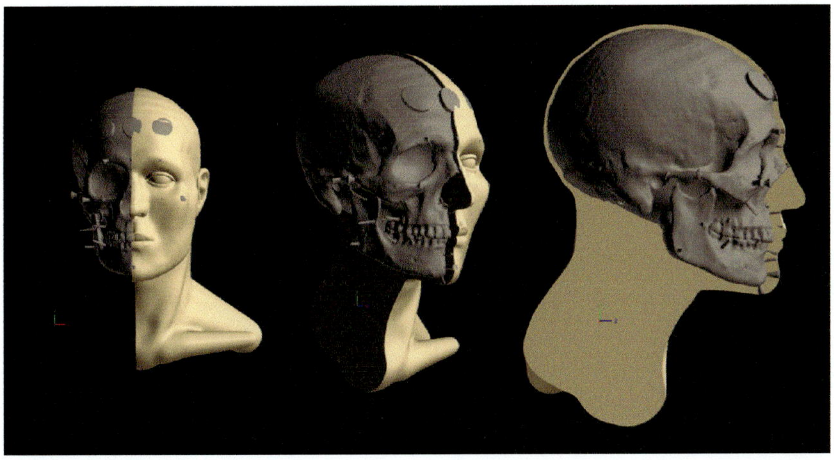

FIGURE 31.8 Final model showing cutaway views to observe soft/hard tissue relationships with tissue depth markers.

areas of the virtual "skin" can be selected, copied, and pasted in position as a new piece. This piece is moved anterior to the existing piece and aligned with the corresponding tissue depth markers. The oral fissure is then carved according to reference lines created earlier, paying attention to the morphology of the dentition and how that translates to the soft tissue. This piece is then blended into the surrounding soft tissue. It is also possible to build a library of mouths, whether sculpted or laser scanned, that can be chosen for similarity, imported, and distorted to fit.

Finally, all pieces are combined into one and smoothed at the seams. After smoothing, any age-related changes such as sagging skin, deep wrinkles, or other fine-tuning can be undertaken (Figure 31.7). At this point, detail can be added by increasing the resolution and hardness of the combined piece. Fine lines, wrinkles, and the oral fissure can all be dealt with in more detail. The virtual clay can be painted within the software; however, this aspect is better dealt with by exporting the model into another 3D modeling/animation program for texturing and rendering. This would depend on budget and time constraints, but allows for one to create potentially photorealistic results. A "lighter" model, smaller in terms of memory storage, can be created by decimating the completed piece with tools within the Freeform software, or creating a NURBS mesh by snapping lines to the surface of the completed piece. This process allows the operator tight control over selective decimation of the polygon mesh, avoiding loss of detail around the facial features. This also means that a polygon mesh can be created of low enough resolution to be suitable for animating the face, based upon typical facial polygon meshes used within animation software.

While this illustrates how the American method was adapted to the use of a haptic device and virtual sculpture software, it should be evident how flexible this approach can be. These same tools have been used quite successfully following the Manchester combination method developed by Richard Neave. Caroline Wilkinson, author of *Forensic Facial Reconstruction* (2004), has adapted the Manchester combination method to the same system. These tools are adaptable to any manual sculptural technique currently practiced, as well as for new data and research that arise over time. It allows practitioners to use the power of a computer to be more consistent with facial reconstructions, while being able to accommodate for the subtleties that the morphology of the skull indicates with regard to its relationship to the overlying soft tissue.

REFERENCES

Clement, J. G., and Marks, M. K., *Computer Graphic Facial Reconstruction*, Elsevier, London, 2005.

Gerasimov, M. M., *The Reconstruction of the Face from the Basic Structure of the Skull*. Nauka, Moscow, 1955.

Krogman, W. M., and İşcan, M. Y., *The Human Skeleton in Forensic Medicine*. C.C. Thomas Publishers, Springfield, IL, 1986.

Taylor, K. T., *Forensic Art and Illustration*, CRC Press, Boca Raton, FL, 2001.

Wilkinson, C., *Forensic Facial Reconstruction*, Cambridge University Press, Cambridge, 2004.

32 Virtopsy and Forensic Tissue Simulation and Synthetic Body Models

Michael J. Thali, Beat P. Kneubuehl, and Stephan A. Bolliger

CONTENTS

In ballistic research, human tissue substitutes are of great importance. Glycerin soap and ordnance gelatine have been used for many years as soft-tissue substitutes. The introduction of polyurethane bone to experimental ballistics has led to the development of skull and other body models (Figures 32.1 and 32.2). On the one hand, anatomic-shaped synthetic body models can be utilized in the reconstructive examination but also—in accordance with the preceding remarks—in simple geometric forms.

Modern imaging procedures can also be employed for the noninvasive examination of soft tissue simulations, such as those made of gelatine and glycerine soap, as well as synthetic body models. In this way the effects of force due to shots, blunt force, and sharp injuries can be visualized radiologically.

Essentially, radiologically documented injuries from real forensic cases can be compared with those produced experimentally on body and tissue models. In the future, forensic hypotheses can be verified or disproved by means of experiments and computed tomography (CT) or magnetic resonance imaging (MRI) documentation on models, comparing them with the corresponding radiological data from real forensic cases.

The following pictures show such visualizations that can be compared with findings in real forensic cases.

Using CT, soap and gelatine blocks and also synthetic body models can be examined (see Figures 32.3 through 32.16).

BIBLIOGRAPHY

Andenmatten, M. A., Thali, M. J., Kneubuehl, B. P., Oesterhelweg, L., Ross, S., Spendlove, D., and Bolliger S. A., Gunshot injuries detected by post-mortem multislice computed tomography (MSCT): A feasibility study, *Leg Med (Tokyo)*, 10(6), 287–92, 2008.

Bolliger, S. A., Ross, S., Oesterhelweg, L., Thali, M. J., and Kneubuehl, B. P., Are full or empty beer bottles sturdier and does their fracture-threshold suffice to break the human skull? *J Forensic Leg Med*, April, 16(3), 138–42, 2009.

Kneubuehl, B. P., Coupland, R. M., Rothschild, M. A., and Thali M.J., *Wundballistik—Grundlagen und Anwendungen*, Springer-Verlag, Berlin, 2008.

Kneubuehl, B. P. and Thali, M. J., The evaluation of a synthetic long bone structure as a substitute for human tissue in gunshot experiments, *Forensic Sci Int*, 17, 138(1–3), 44–9, 2003.

Korac, Z., Kelenc, D., Baskot, A., Mikulic, D., and Hancevic, J., Substitute ellipse of the permanent cavity in gelatine blocks and debridement of gunshot wounds, *Mil Med*, 166, 689–94, 2001.

Korac, Z., Kelenc, D., Hancevic, J., Baskot, A., and Mikulic, D., The application of computed tomography in the analysis of permanent cavity: A new method in terminal ballistics, *Acta Clin Croat*, 41, 205–9, 2002.

Rutty, G. N., Boyce, P., Robinson, C. E., Jeffery, A. J., and Morgan, B., The role of computed tomography in terminal ballistic analysis, *Int J Legal Med*, 122(1), 1–5, 2008.

Thali, M. J., Kneubuehl, B. P., and Dirnhofer, R., A "skin-skull-brain model" for the biomechanical reconstruction of blunt forces to the human head, *Forensic Sci Int.*, 18, 125(2–3), 195–200, 2002.

Thali, M. J., Kneubuehl, B. P., Dirnhofer, R., and Zollinger, U., Body models in forensic ballistics: Reconstruction of a gunshot injury to the chest by bullet fragmentation after shooting through a finger, *Forensic Sci Int.*, 15, 123(1), 54–7, 2001. http://www.virtopsy.com

Thali, M. J., Kneubuehl, B. P., Dirnhofer, R., and Zollinger, U., The dynamic development of the muzzle imprint by contact gunshot: High-speed documentation utilizing the "skin-skull-brain model", *Forensic Sci Int.*, 17, 127(3), 168–73, 2002.

Thali, M. J., Kneubuehl, B. P., Rodriguez, W. R., Smirniotopoulos, J. G., Richardson, A. C., Fowler, D., Godwin, M., Jurrus, A., Fletcher, D., and Mallak, C., Coins as intermediate targets: Reconstructive analysis with synthetic body models, *Am J Forensic Med Pathol*, 30(2), 159–61, 2009.

Thali, M. J., Kneubuehl, B. P., Vock, P., Allmen, G., and Dirnhofer, R., High-speed documented experimental gunshot to a skull-brain model and radiologic virtual autopsy, *Am J Forensic Med Pathol*, 23(3), 223–8, 2002.

Thali, M. J., Kneubuehl, B. P., Zollinger, U., and Dirnhofer, R., A study of the morphology of gunshot entrance wounds, in connection with their dynamic creation, utilizing the "skin-skull-brain model", *Forensic Sci Int.*, 18, 125(2–3), 190–4, 2002.

Thali, M. J., Kneubuehl, B. P., Zollinger, U., and Dirnhofer, R., The "skin–skull–brain model": A new instrument for the study of gunshot effects, *Forensic Sci Int.*, 18, 125(2–3), 178–89, 2002.

Thali, M. J., Kneubuehl, B. P., Zollinger, U., and Dirnhofer, R., A high-speed study of the dynamic bullet-body interactions produced by grazing gunshots with full metal jacketed and lead projectiles. *Forensic Sci Int*, 27, 132(2), 93–8, 2003.

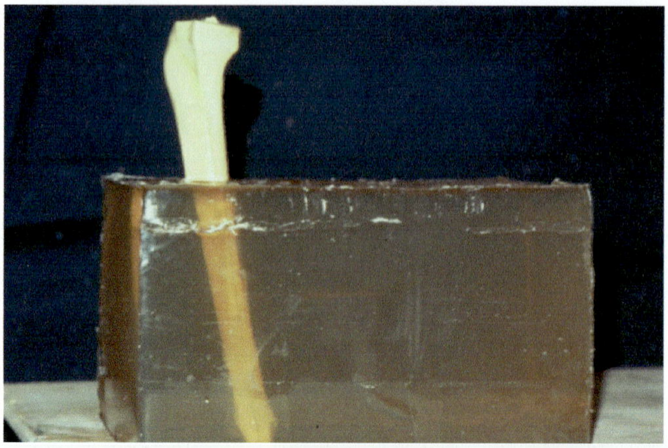

FIGURE 32.1 Soft-tissue model: Synthetic polyurethane bone in a gelatine block.

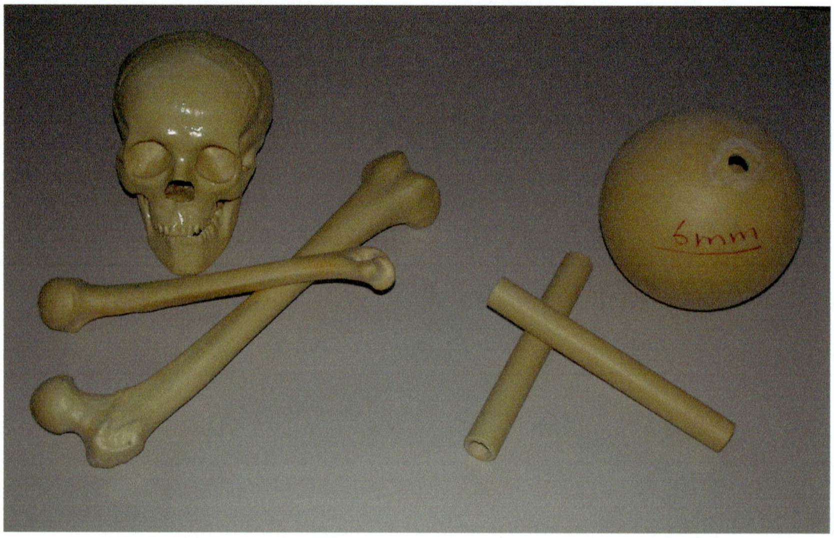

FIGURE 32.2 Synthetic body model in anatomic form and in the corresponding geometric forms.

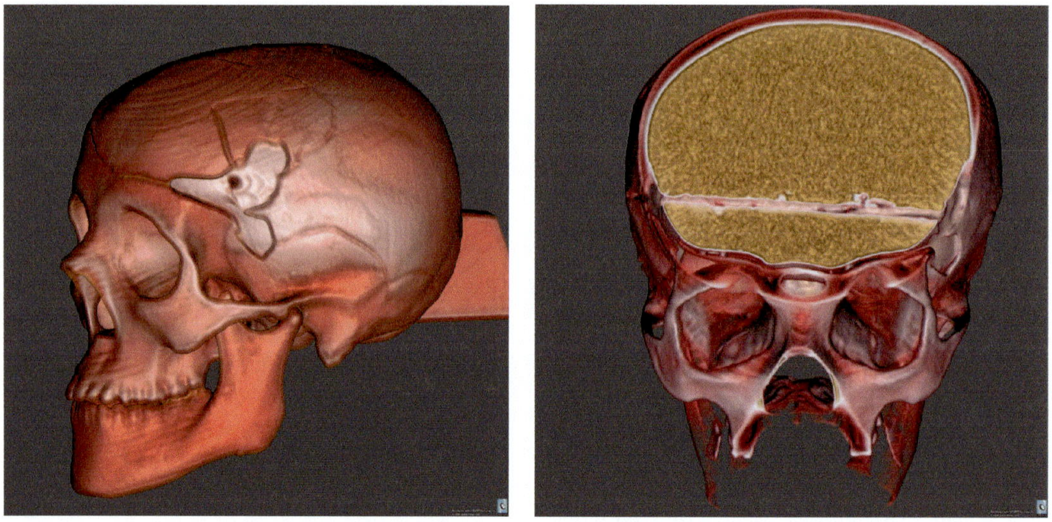

FIGURES 32.3 and 32.4 Gunshot experiment to the synthetic anatomical skull model. Visualization of the shot channel in a synthetic, anatomic head model following virtual cutting open using software.

FIGURE 32.5 Blunt force to the geometric skull model.

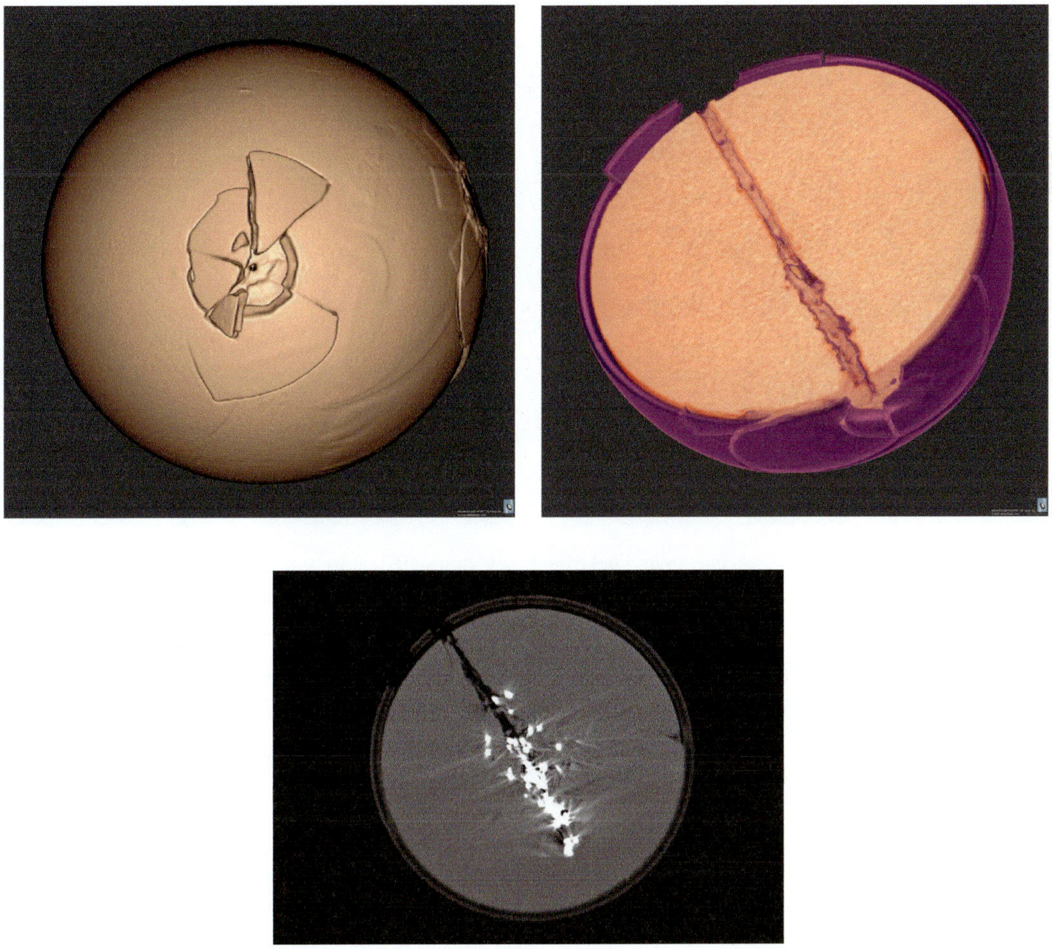

FIGURE 32.6–32.8 Gunshot to the geometric skull model. 3D reconstruction and CT cross sections images.

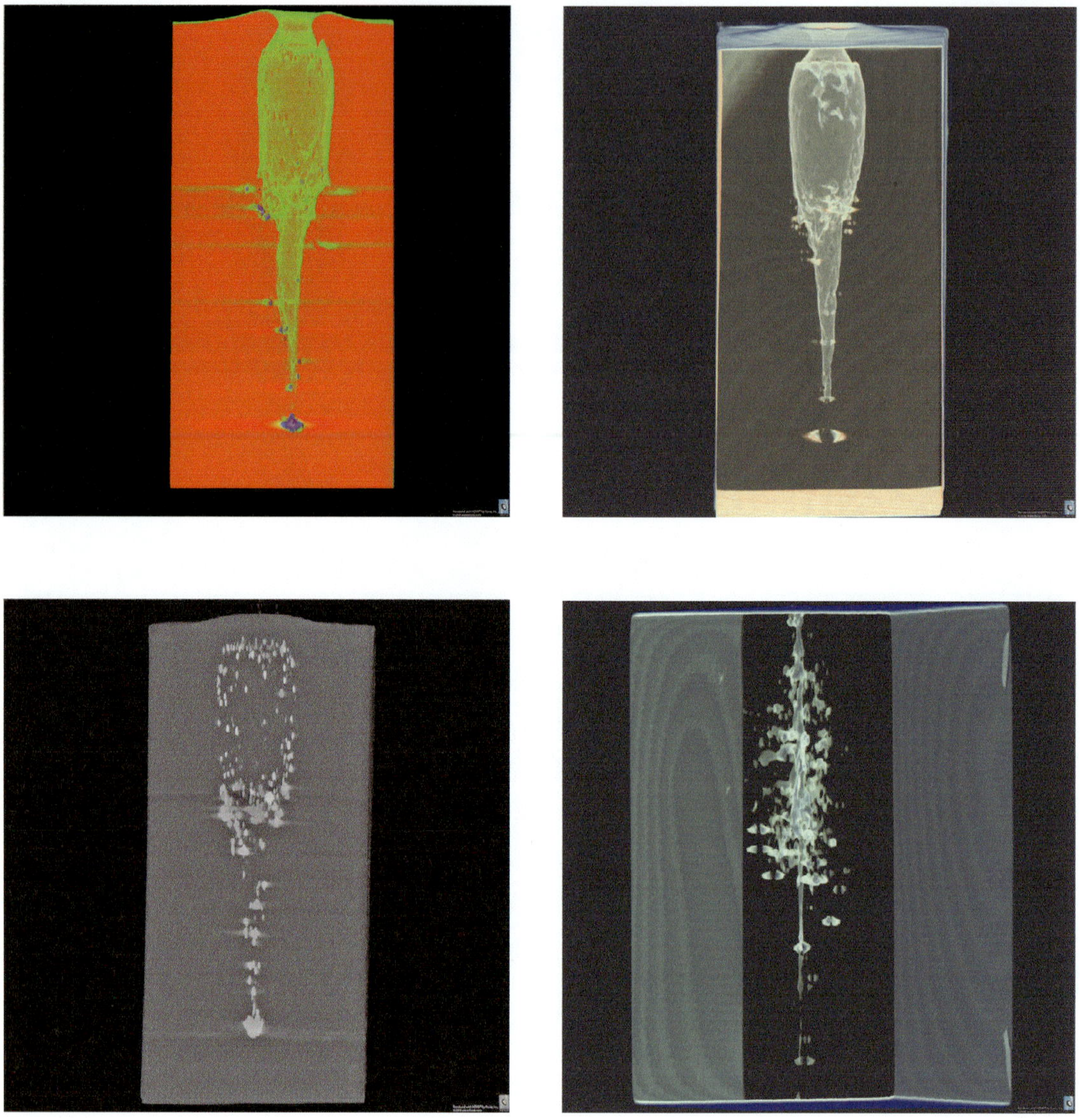

FIGURES 32.9–32.12 Radiological CT of a glycerine and gelatine block shot with a rifle bullet. The position of the shell remainders and the shrapnel released to the shot channel can be easily localized. Clearly visible is also the temporary wound cavity. With the aid of the appropriate software the volume of the temporary wound cavity can be calculated.

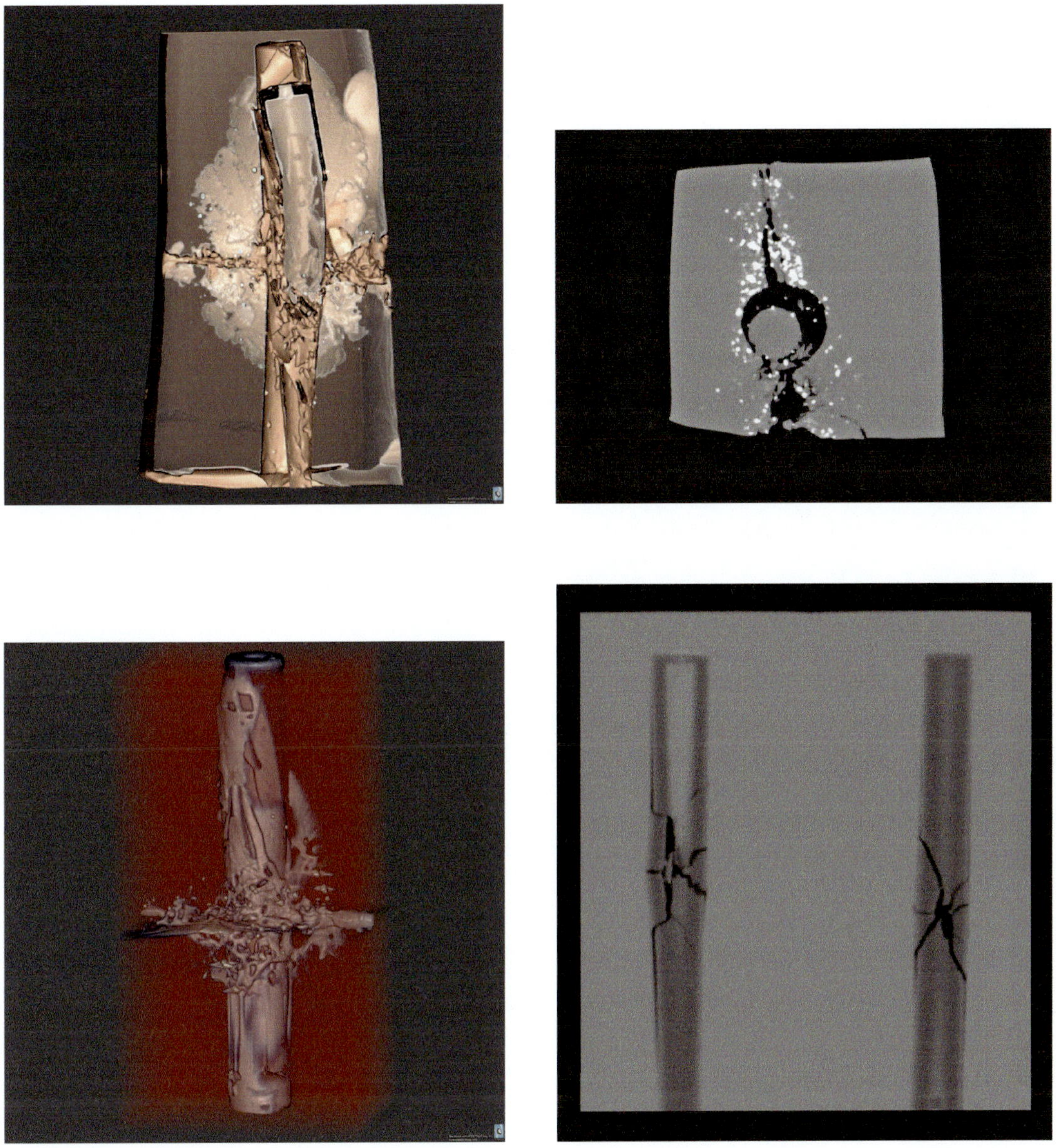

FIGURE 32.13–32.16 CT examination after gun short experiments to gelatine block with synthetic bond simulans.

Section VIII

Facilities, Equipment, and Organization

33 Organization and Management of Forensic Radiology*

Mark D. Viner and Paul F. Laudicina

CONTENTS

INTRODUCTION

As discussed in previous chapters, forensic radiology may be useful to forensic investigations in a variety of different situations, covering the examination of live and deceased human subjects and inanimate objects. The radiographer and radiologist may thus be called upon to provide a forensic imaging service to any one of a number of referring clinicians and investigators and be required to do so both in and out of their normal clinic or hospital surroundings.

Live subjects will normally be referred for examination at the hospital or clinic, for example, in the case of suspected nonaccidental injury in children or other vulnerable groups; following assault or injury due to criminal negligence; whilst in custody on suspicion of drug trafficking. Hospitals and clinics undertaking such cases should thus have appropriate

* Much of the material for this chapter has been drawn from "Guidance for Radiographers Providing Forensic Radiography Services" published jointly by the Society and College of Radiographers (UK) and the International Association of Forensic Radiographers. Their kind permission to use this material in the compilation of this chapter is gratefully acknowledged.

policies and standard operating procedures for conducting such examinations.

In the case of the deceased, the coroner, medical examiner (ME) or forensic pathologist may or may not have a dedicated facility in which x-rays are available, and thus forensic radiology examinations of the deceased may be undertaken in one of many locations. In a region in which there is no dedicated forensic morgue facility, both the x-rays and the autopsy may be completed at the local hospital. In this situation, a radiographer may be required to take a portable/mobile x-ray unit to the morgue to x-ray a body or, in some cases, the body may be transported to the imaging department for examination. It may even become necessary for a radiographer to transport portable x-ray equipment to a funeral home where both the x-rays and autopsy examinations will be completed. Sometimes a county may have a small dedicated facility where the body may be transported, although imaging facilities may only be basic or not available on-site. In larger cities, medical examination facilities will have up to date imaging equipment but they may not have radiographers or technicians on the staff and will rely on hospital radiographers to undertake the examinations.

Irrespective of whether the examination takes place within the hospital or clinic or in the ME facility, the radiographer and/or radiologist working in a forensic environment will be faced with many situations that may be challenging and vastly different to the type of diagnostic imaging that routinely takes place at most imaging centers. In addition to the challenges of working within the legal investigative framework, the radiographer will be required to deal with potentially stressful and traumatic situations. This is particularly true when dealing with deceased subjects where, in addition to the sights and smells with which they are unfamiliar, the radiographer may be required to work with different equipment in unfamiliar surroundings away from their normal base and will be required to adapt their normal technique in order to undertake examinations of the deceased. The conditions of the deceased are quite variable and there are circumstances which necessitate alternate and sometimes unique methods of radiographic positioning and exposure.

In major forensic centers that are equipped with imaging facilities and staffed by radiographers or technicians, forensic imaging examinations of the deceased will be part of the routine work, and staff will thus be more familiar with the challenges presented and the techniques for overcoming them.

In all of the above scenarios, however, it is essential to ensure that appropriate policies and standard operating procedures (SOPs) are in place, both for the conduct of such examinations and for maintaining high standards of image quality, radiation protection, and control of infection. Above all, it is important to remember that the forensic radiology image forms part of a legal investigation. It must therefore be of the highest possible standard, produced in quality controlled and assured conditions and in accordance with established good professional practice. Organization and management of the forensic imaging service should thus be approached with the same rigor that is applied to any other crime lab facility.

ORGANIZATION

Appropriate procedures for forensic radiography undertaken in the hospital or clinic can usually be put in place by adapting existing procedures in place for other examinations. Necessary personnel, equipment, maintenance, and quality assurance programs will already be in place and the addition of a forensic radiology service will mainly be a matter of ensuring an appropriate level of access to equipment, specific training of relevant staff and defining protocols and SOPs.

The degree to which this will be necessary will depend upon the volume and nature of the forensic work undertaken. For example, most centers with an emergency department will normally require procedures for dealing with examinations of assault and nonaccidental injury. Specialist pediatric centers may require supplementary procedures as will departments undertaking examinations of suspected narcotics traffickers, deceased subjects and decomposed material. It should be considered good practice for any imaging department that undertakes forensic radiology to have a named radiologist or radiographer responsible for organizing and managing the service to ensure that appropriate procedures are in place and that legal and imaging processes are subject to regular quality assurance.

The organizational arrangements for radiography in the ME's office will depend upon the number, range and complexity of examinations to be performed, the on-site availability of imaging equipment, the proximity of hospitals and clinics and the funds available for staffing.

If imaging workload is light, it is likely that on-site imaging equipment and staffing will be uneconomical. In such instances, the choices available are:

1. *To transport the subject to a local clinic or hospital for imaging:* In this case, the clinic or hospital department will need to put appropriate procedures and SOPs in place for forensic imaging.
2. *To transport portable imaging equipment to the morgue:* In this case, the ME's office will need to work with a suitable service provider to put in place appropriate procedures for these examinations. This will include the call-out procedure, transportation of suitable equipment AND identification/adaptation of equipment already on-site, identification of a suitable location for imaging examinations to be undertaken, a radiation site-survey, the agreement with local radiation protection rules, and the monitoring of any radiation workers. Procedures for the reporting, viewing and storage of images (either film or digital) should not be overlooked. The importance of qualified radiographers trained and experienced in forensic imaging cannot be underestimated in this scenario, particularly as examinations will be

periodic, the equipment set-up will be temporary in nature, and the radiographer will be required to deal with and adapt to a range of challenging situations in order to provide a quality imaging examination.

If sufficient workload and resources are available, then the most desirable solution is to set up a permanent on-site imaging facility in the morgue. This will entail considerable initial planning which should involve a radiologist, senior radiographer and radiation physicist, together with representatives from the imaging equipment company. Planning for a permanent facility should cover all of the same organizational and health and safety issues detailed above. Although a permanent facility may be established on site, it may not warrant full-time staffing and an agreement may be put in place with a local healthcare facility or with individual radiographers to provide the service on a part-time or call-out basis. If full-time staffing is required, then qualified registered radiographers should be considered as the ideal, with consideration given to cross-training them in other duties within the morgue. An alternative, but a less ideal, situation would be to train existing morgue staff to undertake diagnostic imaging procedures. Whilst it may be a relatively straightforward process to train non-radiographers to take images according to a set protocol, this training will not be underpinned by knowledge of the essential principles of diagnostic radiography and imaging processes. The technician trained in this way should therefore have access to the support of a radiographer who is responsible for the overall quality control and maintenance of the service, and who can be responsible for regular update training and mentoring of staff.

PERSONNEL

Whatever the organizational scenario, any forensic facility that wishes to use medical imaging equipment in support of forensic medicine should involve medical imaging professionals from the outset. In the case of forensic examinations of the living, this will usually happen as a matter of course. For an ME's office wishing to have access to a forensic radiology service, the choices will be either:

- To refer their cases to a hospital or clinic
- To set up a "call out" system where a radiographer, with equipment, attends the morgue on an ad-hoc basis
- To set up a semi-permanent service where on-site equipment is available and radiographers attend when necessary
- To set up a permanent 24-h facility with on-site equipment and radiographers or morgue technicians.

Although it is recognized that budgets and workload will often not warrant the employment of full-time radiographers or radiologists, involving them in the initial stages of setting up a service of any of the types described above to assist with planning, organization, and training is essential. If the service is to be maintained at a high standard commensurate with the requirements of a quality-assured crime lab facility, then the importance of their ongoing involvement for management and quality assurance should not be underestimated.

The Role of the Radiographer

As discussed above, it is recommended that any new service be planned and organized with the assistance of a senior radiographer and radiologist. In the ideal situation, forensic imaging examinations should be performed by experienced registered radiographers who have received additional training in forensic practice. Certification and/or state registration establishes professionalism and quality, which is especially important when x-ray images are to be introduced as evidence in a court of law. In some jurisdictions, radiographers may now be registered as forensic practitioners.[1] It is however recognized that local circumstances may dictate other models of provision involving the use of radiographers from local healthcare facilities or morgue technicians who have received additional training in radiography. Whatever the preferred solution, the requirements for the radiographer or technician will be the same.

The radiographer may be required to be available at all hours. Many radiographers on-call for the coroner usually maintain full-time employment elsewhere. When they are on-call they are subject for availability at any time and are paid on a case-by-case basis.

The radiographer will need to be familiar with all types of imaging including intra-oral dental radiography and be adaptable and resourceful in translating these techniques into the forensic setting. They should have experience in the operation and operator maintenance of all types of equipment that are to be used. The radiographer may be called upon to perform administrative procedures such as ordering and maintaining supplies and filing of diagnostic radiographs and, in centers where case data are computerized, the radiographer should be familiar with computer operations.

The radiographer may be faced with many unpleasant sights and smells. Aside from the many forms of traumatic injury from vehicular accidents, fatal falls, and industrial accidents, bodies may be burned beyond recognition, dismembered and/or may have sustained gunshot, shot gun, or stab wounds from accidental, suicidal, or homicidal situations. Some bodies are not discovered at the time of death but at a later time. Depending upon circumstances of location and temperature, a body may be found in different states of decomposition. The radiographer is likely to be faced with a wide variety of challenging situations and conditions when dealing with cadavers. Bodies may be badly decomposed, dismembered, burned, or skeletalized and may present with terrible odors, insect infestations and generally present an alarming physical appearance. In many situations, the injuries present are quite obvious, and in other situations the body may be unmarked with no evidence of trauma. Certain types of situations may require considerable adaptation of regular diagnostic imaging techniques, and in all cases the radiographer must handle the body carefully when adjusting the anatomy in various positions. The presence of "sharps", such as metal fragments and glass, present a safety hazard. Exposed bone fragments from injuries that may not be apparent under clothing could also produce injury. It is essential that the radiographer is especially careful in body handling. During the examination, some evidence may be revealed, and he/she should be conversant with protocols for documenting and reporting this information.

In some cases the radiographer may need to make compensatory calculations of exposure factors using the inverse square law formula and adjustments for magnification to gather pertinent information, such as establishing the identity of the deceased. The radiographer should therefore be well versed in the technical aspects of diagnostic imaging.

One of the more difficult aspects of forensic radiography is taking x-rays of pediatric deaths. Any death of a child may require a total body x-ray examination regardless of the cause or manner of death. Child abuse cases are especially difficult not only because of the required x-rays but because of the emotional impact the deaths may have upon an individual, and radiographers must be able to complete the examination in an accurate and professional manner so that the necessary information can be obtained.

In addition to the regular assistance provided to the forensic pathologist, the radiographer may be called upon to assist other forensic specialists. In some cases, forensic odontologists will require intra-oral films to be taken that replicate exactly the exposure and angulations of antemortem records in order that they can make a true and accurate comparison for identification purposes. Radiographers will therefore be required to be able to evaluate the antemortem films and replicate the original (sometimes poor!) technique. Forensic anthropologists often work in the identification of skeletalized or partially skeletalized remains and radiographs are often part of that evaluation. Examination of the bones may yield the race, sex, and age of a victim as well as how they may have died, and radiological evaluation may negate the need for further de-fleshing of the remains. Comparison with antemortem films may enable a positive identity to be confirmed but, once again, it will be important for the postmortem images to replicate the antemortem projections exactly.

Forensic imaging therefore presents unique challenges different from routine medical imaging and requires the radiographer to have and develop an additional skill-set.

Of course, morgue technicians will already possess some of the forensic skills that are required, and logistical or financial considerations may necessitate that they take on the role of conducting the x-ray examination. In these circumstances, advice, training, and mentoring by an experienced radiographer will not only enable them to gain competence at routine forensic imaging but will also provide a resource for more complex problems and examinations. Regular, routine quality assurance visits by a radiographer will enable any problems with equipment or technique to be identified and rectified, and will offer the morgue technicians an opportunity to discuss problems and arrange further training or assistance.

The Role of the Radiologist

The role of the radiologist is well covered in other areas of this book and is generally outside the remit of this chapter. However, when setting up a forensic radiology service, it will be essential to involve an experienced forensic radiologist in order to ensure that an appropriate imaging protocol is agreed upon and that the necessary equipment and personnel are available to support it. Medical imaging is a rapidly develop ing field of science and there are an increasingly diverse range of modern imaging techniques that can be applied to the forensic setting. Early discussion with the radiologist will determine which of these is likely to be the most useful in a routine situation and enable the postmortem forensic imaging facility to be designed and managed optimally.

The Role of the Radiation Physicist

Every exposure to ionizing radiation presents a hazard and for this reason the use of x-ray equipment is tightly controlled and regulated in most jurisdictions. Employers and employees using equipment that emit ionizing radiation have a responsibility to use such equipment safely and to ensure that radiation doses to workers and members of the public are kept as low as possible and within certain limits. Radiation protection measures are a combination of equipment design and maintenance, facility design and maintenance, safe operation and appropriate use, together with routine quality assurance and dose monitoring. A full discussion of radiation protection in forensic radiology is outside the scope of this book although some elements related to facility design are discussed later in this chapter and in Chapter 34. It is therefore recommended that any facility seeking to establish or review its forensic radiology provision seeks the professional advice of a qualified medical radiation physicist.

FORENSIC IMAGING PROTOCOL

Every facility that undertakes forensic radiology or provides forensic radiology services to others should have a written forensic imaging protocol covering all types of examinations that are likely to be undertaken. The protocol should be specific to the local service and should adhere to the employing authority's or appropriate organization's policies for the drafting of such documents. The protocol should be developed in consultation with all key stakeholders, which may include senior radiographers, radiologists and hospital managers, coroners, MEs, police forensic coordinators, child protection and elderly care teams, hospital pathologists, occupational health departments, emergency department staff, etc. In their guidance for the provision of forensic radiography services, the College of Radiographers (COR) and the International

Association of Forensic Radiographers (IAFR) suggest the following generic structure of a forensic protocol.[2]

- Introduction
- Statement of intent
- Implementation and review
- Scope of local service
- Referral pathways
- Roles, responsibilities, and interprofessional relationships
- A list of trained radiographers willing to provide the service
- Relevant contact details
- Training/Education
- Welfare of staff
- Quality control and audit

Scope of Practice

In their guidelines, COR and IAFR identify the following four main applications of forensic radiography.[1]

1. *Investigation of nonfatal injuries:* Radiological examination to support the investigation of injury to an individual or individuals. Examples include nonaccidental injury (NAI) of vulnerable individuals (e.g., the young, elderly, and disabled), assault, nonfatal motor vehicle incidents, compensation claims, medical negligence industrial injury or disease, custodial injury, torture or systematic abuse of human rights.
2. *Location of other forensic evidence:* Radiological examination to demonstrate the presence of hidden foreign bodies. Examples include human narcotic packing detection (drug smuggling), nonhuman-narcotic packing detection, other ingested material (e.g., diamond smuggling), ballistic material (shotgun pellets, bullets, shrapnel, arrowheads), nonballistic material (knife blades, needles, etc.).
3. *Cause of death:* The production of evidence to support the investigation of suspicious or unexplained death of an individual or individuals. Examples include road traffic deaths, death following medical intervention, homicide, suicide, custodial death, discovery of decomposed remains, mass fatality incident, genocide or atrocity crime, sudden unexpected death in infants, and sudden infant death syndrome.
4. *Human identification:* The production of evidence to help confirm, determine, or eliminate the identity of both living and deceased persons. Techniques include demonstration of dental structures to facilitate identification through comparative techniques, demonstration of other anatomical structures, trauma and pathological conditions, to facilitate identification through comparative techniques, determination of biological profile (age, stature, sex, socioeconomic status, etc.) through evaluation of skeletal structures, 3D multiplanar reconstruction (e.g., facial reconstruction), demonstration of personal effects (e.g., jewelry), photosuperimposition.

Referral Pathways

As with any radiation examination, it is important that referrals for forensic imaging should be from recognized, approved, referral sources and that such a referral and subsequent examination details should be documented as part of the chain of evidence. The various referral pathways should be identified in advance and incorporated into the SOP in order that appropriate management arrangements can be made and implemented. Referrals may arise from coroners and MEs, forensic pathologists, forensic odontologists, forensic anthropologists, forensic physicians, other hospital physicians, police and security services, customs and excise agents, and so on. In many cases, referrals will form part of an established and predetermined examination procedure and the examination will be conducted according to SOPs agreed in advance between the ME and the radiographer/radiologist. Other referrers should be encouraged to discuss the referral with the radiographer and/or radiologist in order to determine the most appropriate imaging modality and examination. In some jurisdictions, requests for forensic examinations on the deceased may only be submitted when the investigation has been authorized by the coroner or ME, and a procedure will need to be put in place to ensure that this authorization has both been obtained and recorded on the request form. Requests for examinations may also arise as a result of major incidents or mass fatality incidents, and the Major Incident or Mass Fatality Protocol should detail appropriate arrangements for the provision of radiography services on both live and deceased individuals arising from such incidents. It should be noted that, in most if not all cases, mass fatality incidents will be treated as crime scenes and therefore all referrals for examination of the living or the deceased arising from such incidents should be treated as forensic radiography referrals.

Examinations of Live Individuals

Forensic examinations of live human subjects will normally take place in a hospital or clinic although in some cases examinations may be undertaken in a prison or other detention facility having diagnostic radiology capability. Examinations should be undertaken by appropriately trained and licensed staff and conducted in accordance with all existing procedures in place within such facilities. Radiographers should comply with the provisions as set out in the relevant codes of professional conduct and with all relevant regulations for diagnostic imaging and the safe and efficient use of ionizing and nonionizing radiation. The forensic imaging protocol and SOP will need to contain specific procedures for consent, confidentiality, and continuity of evidence, together with detailed imaging protocols for specific investigations, such as suspected nonaccidental injury, and the requirements of any particular care pathways, for example, care of the elderly or child protection.

Consent for Examination

All forensic imaging protocols should address the issues of consent for the examination of both live and deceased subjects. In the case of live subjects, existing hospital, clinic, or other employing authority policies on Consent, together with appropriate organization policies on Mental Capacity and Child Protection should inform the production of the forensic protocol. All forensic requests on living subjects should be treated as nonmedical referrals and full informed written consent obtained. The referring clinician should explain the procedure and obtain consent. It should be noted that, although the law may not actually require consent to be written, radiography for forensic purposes is an area of practice where validity of consent may be questioned and therefore to obtain written consent would be considered best practice. There are no circumstances when implied consent is acceptable.

The consent process should confirm that the consenting individual has been informed about and understands a range of issues that include the purposes of the examination, the nature of the procedure, the duration of the procedure, the risks and benefits of the procedure (including clinical and radiation risk), the persons involved, dignity and privacy issues, and withdrawal of consent.

For individuals unable to give consent, appropriate arrangements should be made to obtain recognized third-party authority. This may, for example, be obtained from parents or legal guardians or individuals appointed by the courts. Third-party consent cannot be given by the referrer. In the case of children and minors or vulnerable adults, such third-party consent may be in the domain of the alleged abuser. Appropriate sensitivity should be considered for obtaining such consent but the parent or guardian must be fully informed. If consent is withheld, the examination cannot be undertaken, and this may lead to circumstances in which these vulnerable individuals may be placed under the protection of the court and third-party consent obtained from the legal guardian. In the case of psychiatric patients, limited powers may be appointed to their clinician and this may extend to consent for such examinations. In cases where consent is withheld by a competent individual and cannot be obtained by a legally appointed third party, the examination should not be undertaken.

Nonaccidental Injury

Examination in cases of suspected nonaccidental injury (NAI) may be requested for children, the elderly or other vulnerable groups, or in other cases of suspected assault. Whilst the examination might also be clinically indicated, all such cases should be treated as forensic examinations. In most cases, they will be performed in the hospital or clinic where protocols specific for the examination of NAI should follow established clinical and professional guidelines.[3–5] It is recommended that any such protocol should both recognize the forensic nature of NAI imaging and cross-refer to the forensic protocol to ensure that the legal requirements of such examinations are met.

In some cases, requests for a NAI skeletal survey examination may arise from an initial clinically justified examination. The skeletal survey in such cases must be regarded as purely forensic in nature and is not clinically justified (i.e., is not expected to affect the clinical management of the patient).

Examinations of the Deceased or of Pathological Specimens

Examination of deceased human subjects or pathological specimens may be undertaken either in the hospital or healthcare clinic or in the ME's morgue. Although there is considerable variation among jurisdictions, it should be considered good practice that imaging procedures should be undertaken by appropriately trained and licensed staff and conducted in accordance with all existing procedures in place within such facilities. Radiographers should comply with the provisions set out in the relevant codes of professional conduct and with all relevant regulations for diagnostic imaging and the safe and efficient use of ionizing and nonionizing radiation. The forensic imaging protocol and SOPs should contain specific procedures for consent, confidentiality, and continuity of evidence, together with detailed imaging protocols for specific investigations such as pediatric deaths, homicide, and so on.

Radiographers undertaking any forensic examination on live or deceased subjects or specimens must be aware of and comply with the local protocol for forensic imaging, which needs to address the following specific issues:

- Authorized referrers
- Continuity of evidence
- Confidentiality
- Health and safety (including infection control and welfare of staff)
- Cultural and religious sensitivities
- Privacy and dignity
- Out-of-hours service provision
- Transfer of cadavers and specimens.

Confidentiality

In all cases, the normal principles of client confidentiality must be maintained in accordance with all relevant standards of professional conduct. In addition, forensic cases will always be regarded as subjudice (i.e., the case is currently under trial or being considered by a court of law or Coroner's court) and must, therefore, never be discussed outside of the court with any person not directly involved in the case until the investigation or inquest has been completed. Furthermore, where the coroner or ME has referred a case for consideration by a court of law, the principles of client confidentiality will continue to apply throughout the proceedings.

Health and Safety

The policies of the hospital, clinic, ME's office or other employing authority relating to Infection Control, Care and Handling of the Deceased, Control of Hazardous Substances and Manual Handling should inform the production of the forensic protocol and provide guidance to radiographers, radiologists, and technicians involved in forensic imaging. It is recommended that, whenever practicable, forensic radiography examinations of cadavers and/or pathological specimens should be conducted in the mortuary or a room specifically set up for that purpose. Should it be necessary to carry out the examination within the Diagnostic Imaging department, appropriate care should be taken to minimize the risk of cross-infection and to ensure that the conduct of the examination causes minimum distress to patients and staff.

All cadavers and human remains must be treated as potentially infected and a risk to the health of the staff. When examining the deceased, imaging staff should adhere to the local mortuary and pathology health and safety policies and take appropriate precautions to minimize any risk of cross-infection during radiography for forensic purposes. Appropriate Personal Protective Equipment (PPE) should be worn at all times when dealing with human remains.[6] It should be remembered that personal protective equipment has two main functions in the forensic setting:

1. To minimize the risk of infection to the operator from pathogens that may be present in or on the subject.
2. To prevent the contamination of forensic evidence on the subject by material from the operator (e.g., hairs, fibers, etc.).

The degree of protection required will depend upon the nature of the remains and the stage of the investigation, that is, whether trace evidence has been collected prior to imaging examination. Imaging staff should therefore seek advice from the ME or other morgue personnel regarding what precautions should be taken. As a minimum, protective clothing should include scrubs, preferably boots or high quality overshoes, apron, and gloves. If there are fluids present, then a gown or water-resistant suit, mask, hat, and glasses/goggles should be worn. In cases where there is a risk of contamination by extremely hazardous material, examinations should only be undertaken by staff who have been trained in the use of specialist protective equipment.

Radiation Protection

The policies of the hospital, clinic ME's office or other employing authority relating to radiation protection should inform the production of the forensic protocol and provide guidance to radiographers, radiologists, and technicians involved in forensic imaging. The advice of a qualified radiation physicist should be sought when setting up the service or when changing any aspect of it, and many jurisdictions will require the appointment of a physicist or other appropriately qualified individual to act as the responsible officer for radiation equipment and procedures.[7]

Forensic radiography examination of cadavers and/or pathological specimens should ideally be conducted in a room specifically set up for that purpose. Should it be done within the Morgue, appropriate care should be taken to minimize the risk of exposure to radiation and to ensure that the conduct of the examination limits the radiation dose as far as is reasonably achievable. Appropriate measures for the safe use of x-ray equipment should be in place. These will include a radiation protection plan and rules/scheme of work for the radiation area, which should detail the extent of the radiation area, the personnel who are permitted to operate the equipment, and procedures for its safe operation. Lead protective screens and other personal protective equipment should be available and utilized where appropriate. Monitoring of those undertaking x-ray examinations will almost certainly be required and arrangements for this should be put in place in consultation with the radiation physicist.

Legal Aspects

It is essential that radiographers, radiologists, and technicians undertaking forensic imaging have an understanding of the medicolegal aspects if the forensic evidence provided is to be of any value to the legal process.

The Nature of Evidence

If a crime is alleged to have been committed, the location in which the act took place, together with all persons and objects present at the time of the incident, is defined as the "Scene of the Crime" or "Crime Scene." Anything or anyone that is subsequently removed from that location still forms a part of that "Scene" and can be examined and presented as evidence in support or defense of a particular legal position. The subject of a forensic radiological examination (live, deceased, or inanimate) is thus part of the "Scene" and may thus in itself constitute evidence.

Any subsequent examination of the physical evidence from the scene can also be presented as evidence and, in the case of forensic imaging, this may include both images and imaging reports. However, before any image, statement, or any other information can be accepted for use in a court of law, it must be judged to be admissible as evidence. To be admissible, the evidence must be properly authenticated and the "continuity" of the evidence must be demonstrated. This means that the evidence presented must be able to be linked back to the "Scene" by means of an unbroken chain of records.

In the case of radiological images, the radiographer or technician supported by an appropriate witness should be able to attest in a court of law that any specific image was produced by them at the date and time indicated, that the image is of the identified evidence, individual or body part, and has not been tampered with during, or as a result of, the image production process. It is therefore essential that robust procedures be followed to ensure authentication and continuity of evidence for diagnostic images.

Continuity of Evidence

In the case of examinations undertaken within the ME's office or other forensic facility, the routine policy and procedures in place to ensure continuity of evidence should be applied. In the case of examinations undertaken in the hospital or clinic setting, or in other facilities where no such policy exists, it is recommended that the following procedures be adopted.

The entire imaging examination must be properly witnessed by an appropriate third party. The witness can be another radiographer or health care worker, healthcare professional, police officer, coroner's officer, social worker, or other member of the child protection team. This person is normally termed the "appropriate witness". The purpose of the appropriate witness is to ensure that images are admissible in law as evidence and to reduce the risk that any challenge to the validity of that evidence will be successful. All actions and communications should be contemporaneously documented by both the radiographer and the appropriate witness, to be presented if or when necessary. All actions should be guided by written protocol. The identities of the radiographer and appropriate witness should be recorded and they should initial any imaging evidence as proof of authenticity. They may also be required to complete a witness statement to support their evidence. The appropriate witness should be present throughout the examination and accompany the radiographer during the development of the radiograph, or production and recording of digital images. Both radiographer and witness should sign or otherwise appropriately authenticate all original analog and digital hard copy images as being an accurate record of the examination.

Continuity of the primary evidence (i.e., the cadavers and/or pathological specimens) must be ensured at all times. Ideally there should be an evidence log recording all movements of the cadaver or specimen and its transfer from the custody of one individual to another. In some cases, the cadaver or specimen will be accompanied by a police officer or coroner's officer who will remain with it at all times, and who will thus retain custody of the cadaver or specimen during the examination. In other cases it will be necessary for the body to be handed over to the radiographer or radiologist. In such cases, the date and time that the cadaver or specimen was received into the imaging department should be recorded in the evidence log, together with the names of the persons handing over and receiving the cadaver or specimen. Radiographers, radiologists, and others undertaking forensic imaging should be aware of their responsibilities for the custody of the evidence during the examination. If they are required to leave the primary evidence to process images, the evidence must be secured, formally returned to custody/care of the mortuary, or left in the presence of a separate appropriate witness, who is sometimes called the "Continuity Officer" or "Responsible Officer". There may also be occasions when a Continuity Officer is required for living subjects undergoing forensic imaging, for example in the investigation of ingestion of narcotics.

The radiographer or radiologist and their employer will be responsible for the security and continuity of all imaging evidence until it is formally handed over to a responsible officer, who must sign to confirm that it has been received. The Responsible Officer may be the coroner, ME, their representative or a member of the police investigation team. In some situations, images are transferred to the responsibility of another investigating discipline. For example, radiographs that are produced may be handed to a radiologist, odontologist, or forensic anthropologist who will use the radiological evidence in preparation of their report for the coroner or ME. In such cases, the rules of continuity will still apply and the radiologist, odontologist, or anthropologist should sign for the images, and the continuity officer should be kept informed all times of their presence for retrieval at a later date.

In many cases, forensic images produced in a hospital or clinic will be reported by a radiologist in the same facility. However, it is still important to maintain continuity of evidence and the reporting radiologist/radiographer/pathologist will need to be satisfied as to the continuity of evidence prior to reporting. For this reason, transfer of original (working copy) images for reporting should be logged and witnessed and the images should not be left unattended or unsecured at any time.

Subject Identity

All subjects must be identified prior to the examination using a unique numbering system and name, if known. Such identification should be used consistently throughout the process and on all documentation and images. Subject identifiers, a locality identifier, and the date and time of examination should be recorded on images using primary identification systems such as lead letters or light markers for film-based systems and DICOM headers or examination data sets for digital imaging systems. Anatomical side indicators must be physically present within the primary radiation beam.

If the appropriate subject and image identification data are not recorded on the image as described above at the time of examination, it is recommended that the examination be repeated. In the event that a repeat examination is not practicable (e.g., for radiation dose reasons in the case of live subjects), any patient or examination identification not included on an analog image should be added using indelible markers and countersigned by a witness. In digital images, this may be added during postprocessing, and a note made on the record to this effect, signed and witnessed. Any patient or examination identification not included on the digital image should be added using post-processing facilities. In either case, a record of this should be made in the radiographer's witness statement. It should be noted that the adding of patient or examination identification after the original exposure should be considered atypical, since good practice requires inclusion at the time of the original examination.

Records

Records should be kept of all forensic radiography examinations. These records should include:

- The identity of radiographer and witness(es)
- The identity and role of any others present and formally involved
- The date, time and location of the examination
- The subject and examination identifiers
- Details of the number and type of projections involved
- The location of any retained evidence
- The handover details for transfer of evidence (including name and signature of recipient)
- The number and location of any copies.

All records should comply with relevant national and local guidelines in regard to maintenance and storage of confidential health records.

It is also recommended that all individuals involved in carrying out the imaging examination should complete their own written documentation which may need to be referred to at a later date. Radiographers and radiologists should keep their own written documentation securely as this also forms a formal record of evidence.

Digital Images

In most jurisdictions, only primary evidence is admissible in court. In film radiographs, this will be the original hard copy image and report. In digital images, a "working copy," rather than the original data set, will be produced to support the report.

It is therefore recommended that a hard copy disk be immediately created of the original raw data image (prior to any algorithm changes or manipulation). This should be authenticated with the initial of the radiographer and witness at the time of production and securely sealed. This will be referred to as the "Master Copy" and will only be referred to if there is a need to demonstrate the integrity of the working copy image information. A second "Working Copy" disk of digital images that have been subject to normal post-processing (including algorithm changes) should be saved separately on another disk at the end of the examination. Any changes made to the original image should be documented.[8] It is recommended that an additional third (backup) working copy should be made in cases where image data is not automatically backed up (e.g., on a PACS system).

Any subsequent algorithm changes and manipulation of images to working copy images (e.g., during reporting) should be documented and the images then saved separately onto another working disk. The final working copy should reflect the reported examination. All previous working copies will be evidence of image manipulation.

Copy Images

- *Film Images:* Copies are secondary evidence and are only admissible where originals can be proven to have been lost or destroyed. Recognizing the risk from the loss of an original film image, it is recommended that a duplicate hard copy image be made at the time of all forensic examination. All copies should be attested by the radiographer who has produced the copies and by a witness. This procedure would need to be agreed to in advance with the coroner, ME, or other relevant legal authority, and the arrangements for dealing with and storing such duplicate images whilst the case is subjudice should be detailed in the written protocol. Any requests for release of copies or for the production of additional copies should be referred to the coroner or ME or the appropriate legal representative with regard to other cases.
- *Digital Images:* Opportunities exist for the later production of hard copy images from digital data and/or the copying of digital data sets. Legal advice should be sought for the production of any such copies and in many jurisdictions special permission from the coroner or ME will be required. Any subsequent copy disks or images should be signed and witnessed by the radiographer who has produced the copies and an appropriate witness.

All image data on any media should be securely stored and all appropriate measures taken to prevent unauthorized access that may compromise the evidence in a court of law.

Education and Training

Those who undertake radiography for forensic purposes should be educated and trained specifically in radiography for forensic practice and receive regular updates as part of continuing professional development (CPD). It is desirable that all forensic radiography examinations be carried out by registered experienced radiographers. As a minimum, they should have relevant and up-to-date knowledge and experience in the following areas in addition to education and training in diagnostic radiography:

- Medicolegal issues relating to the admissibility of evidence
- Imaging techniques for forensic purposes
- Specialist modality imaging; for example pediatric NAI radiography or intra-oral dental radiography
- National and local health and safety regulations, particularly with regard to the handling of deceased subjects
- Statutory legislation, government, and professional and local guidelines related to radiography for forensic purposes
- Different cultural and religious ethics associated with the deceased and their relatives

- Appropriate communication skills for dealing with subjects who have undergone a traumatic experience.

Image Reporting

It is essential that any person carrying out interpretation or reporting of forensic images is trained, competent, and authorized. In most cases, images will be reviewed by the pathologist, anthropologist, or odontologist and used in the compilation of their report. A radiological opinion should be sought where necessary. In some cases, radiographers may be asked to provide a professional opinion on the images that they produce. It is imperative that the radiographer practices within the scope of their competence.

Staff Welfare

Employers should pay particular regard to the potentially distressing nature of some aspects of forensic practice, which could lead to post-traumatic stress disorder (PTSD) in the individual undertaking the examination. Employers have a responsibility to protect the psychological and physical well-being of their employees and should undertake a risk assessment for forensic examinations. In situations where forensic radiography is not part of the routine work of the department (e.g., hospital, healthcare clinic, etc.), it is recommended that employers identify radiographers who are willing to undertake such examinations and ensure that they are appropriately educated and trained for that purpose. Individual radiographers should never be coerced into participating in the provision of forensic radiography service. The forensic protocol must include matters pertaining to the welfare of staff undertaking forensic examinations and this should include information regarding the symptoms and common feelings experienced with PTSD such as basic advice on coping strategies, etiology of PTSD and types of treatment available. It should also include information relating to the availability of support mechanisms (e.g., contact details for support). The emphasis should be on primary prevention and it is recognized that training and education is an integral aspect of minimizing PTSD.[9,10] This includes training specific to radiography for forensic purposes and also raising the awareness of the effects of PTSD.

Equipment Maintenance

As with any other mechanical and electronic apparatus, a regular preventive maintenance schedule should be in place for all parts of the imaging chain: the x-ray equipment, film-processing equipment, digital x-ray systems, viewing apparatus, etc. In many cases, state or national law will require x-ray equipment to be annually inspected and certified by a qualified radiation physicist to ensure that the equipment is functioning safely and that the radiation dose to operators and subjects is kept as low as reasonably achievable. This legislation may not extend to other parts of the imaging chain (e.g., film processing), but it should be remembered that these are just as, if not more, important. Poor film processing can lead to increased radiation doses being given in order to obtain the correct blackening. Hospitals and clinics will have regular maintenance inspections and call-out contracts in place for all of their equipment. Given the nature of the forensic examination and the importance of meticulous attention to detail, it is recommended that similar inspection programs should be in place for equipment in the ME's offices, morgues, etc. This program should usually follow the manufacturer's recommendations for servicing, with maintenance inspections being undertaken annually, biannually, or quarterly. This can be supplemented by regular inspection by the operators or a consulting radiographer. This maintenance program should be backed by a rigorous quality assurance program. These issues are further discussed in Chapter 34.

Quality Assurance

A quality assurance program should be in place to cover all aspects of the forensic imaging service in exactly the same way as for any other investigative procedure within the crime lab. The program will need to cover issues including, but not limited to, the following:

1. Staffing
   - Standards of competence, qualifications, and recruitment
   - Maintenance of training records and continuing professional development.
2. Policies and Procedures
   - Procedures in place for all aspects of the service including radiation protection control of infection, continuity of evidence, imaging protocols, and equipment operation
   - Process for document control, regular review, and records demonstrating that the policies and procedures have been read and understood.
3. X-Ray Equipment
   - Records of regular maintenance inspections and radiation protection certification (annual)
   - Regular checks for radiation output and timer consistency, x-ray tube leakage, beam alignment, collimator alignment, and so on. (quarterly).
4. Image Quality
   - Resolution tests for digital (DR and CR) systems and fluoroscopy units (monthly)
   - Exposure index/exposure value checks for digital imaging systems (daily)
   - Base fog, speed, and contrast index of film-processing systems (daily)
   - Film/screen contact tests (biannually)
   - Regular review of image quality by senior radiographer/radiologist to identify quality issues with equipment, processing, or technique (weekly).
5. Radiation Protection
   - Record of all exposures made, radiation factors set, and dose measurements, if available

- Monitoring of any designated radiation workers (monthly)
- Visual examination and screening of lead rubber aprons and other personal protective equipment (biannually)
- Process for regular review and records demonstrating that policies and procedures have been read and understood by all staff.

ACKNOWLEDGMENT

Much of the material for this chapter has been drawn from "Guidance for Radiographers Providing Forensic Radiography Services" published jointly by the Society and College of Radiographers (UK) and the International Association of Forensic Radiographers. Their kind permission to use this material in the compilation of this chapter is gratefully acknowledged.

REFERENCES

1. www.homeoffice.gov.uk/publications/police/operational-policing/Forensic_Practitiioner_Stan1.pdf?view=Binary
2. The College of Radiographers & Association of Forensic Radiographers, *Guidance for Radiographers Providing Forensic Radiography Services*, SCoR 2008. P2 http://www.sor.org; http://www.afr.org.uk
3. Kleinman, P. K. (ed.) *Diagnostic Imaging of Child Abuse*, 2nd edn., Mosby Inc, Missouri, 1998, p. 168–177.
4. Kleinman, P. K., Nimkin, K., Spevak, M. R., et al., Follow-up skeletal surveys in suspected child abuse. *AJR Am J Roentgenol*,167, 893–896, 1996.
5. British Society of Paediatric Radiologists, Standard for skeletal surveys in suspected nonaccidental injury (NAI) in children, http://www.bspr.org.uk/nai.htm
6. The Royal Institute of Public Health and Hygiene *A Handbook of Mortuary Practice & Safety*, The Royal Institute of Public Health and Hygiene, London, 1994.
7. Ionising Radiation Regulations, Statutory Instrument No. 3232 Crown Copyright, 1999.
8. *Home Office Standards of Digital Evidence: Digital Imaging Procedure. Version 1.0.*, HMSO Home Office, London, 2002.
9. Parkinson, F., *Post-Trauma Stress*, Insight, London, 1993.
10. National Institute for Clinical Excellence Post-traumatic stress disorder (PTSD) NICE Guidelines, March 2005.

BIBLIOGRAPHY

Dimond, B., *Legal Aspects of Radiography and Radiology*, Blackwell Science, Oxford, 2002.

Health Professions Council. Confidentiality: Guidance for registrants—consultation document, 2007.

Ionising Radiation (Medical Exposure) Regulations, 2000 Statutory Instrument No. 1059, Crown Copyright, 2000.

NHS Executive (HSC 1999/053); For the Record: Managing Records in NHS Trusts and Health Authorities, Department of Health, London, UK.

The College of Radiographers, *Professional Standards for those Working as Independent Practitioners*, SCoR, London, 2001.

The College of Radiographers, *The Child and the Law: The Roles and Responsibilities of the Radiographer*, SCoR, London, 2005.

The College of Radiographers, *Health Care Associated Infections (HCAIs): Practical Guidance and Advice*, SCoR, London, 2006.

34 Facility, Equipment, and Radiation Protection

Mark D. Viner, Charles W. Newell, and Chucri M. Jalkh

CONTENTS

INTRODUCTION

The difficulties involved in trying to use hospital or clinical radiographic installations, equipment, and personnel for forensic investigation have already been mentioned in Chapter 33. Attempts to use mobile x-ray equipment temporarily "borrowed" from the clinical facility will meet with similar problems of accessibility and conflicting priorities. If any consistent use or benefit of forensic radiology is expected, then the medical examiner's facility (be it a one-room morgue or a separate building) should be equipped with its own in-house radiological capability. This entails dedicated space (at the very least for repetitive temporary use) for radiographic equipment; x-ray grids; detectors or cassettes and film storage; digital imaging workstations or film-processing and viewing capability; and either an integrated IT system capable of storing, retrieving, and displaying images, or a film identification system (see Chapter 38); and film envelopes and files for storing the archival roentgenograms.

This chapter outlines the equipment that is currently available and discusses its installation, configuration, and requirements for use.

IMAGING TECHNIQUES AND EQUIPMENT

There is now a considerable range of diagnostic imaging equipment available for medical diagnostic use. Not all of the techniques and equipment used are suitable for forensic applications or examination of the deceased, and historically most medical examiners have been restricted to simple x-ray units. However, recent developments in both x-ray imaging and computer technology have opened up new imaging possibilities for the forensic examiner. The main imaging modalities discussed here can be utilized to great benefit in the forensic setting.

PLANAR IMAGING

Traditionally, the radiological image has been a two-dimensional (2D) record of an object, produced by recording the image of the object through which x-rays are transmitted. Structures are thus viewed in one plane only and may be superimposed upon one another. For this reason, it is essential to conduct a full imaging examination by examining the subject in at least two planes: AP and Lateral (see Chapter 35).

Radiography

The production of a single still image—the "radiograph" or roentgenogram—is perhaps the most simple and yet effective application of x-ray technology. A short exposure (usually measured in milliseconds) of an object to a beam of x-rays generated from an x-ray tube results in the x-rays that are transmitted through the object falling upon an image receptor (film or digital detector) and being recorded as an image. The resultant radiograph is a 2D image that records the subject at an instant in time. The size of the image (and therefore of the object that can be recorded on a single image) is limited by the size of detector or film in use and the geometry of the x-ray machine; it is typically a maximum of 17 in. × 14 in. (Figure 34.1).

X-Ray Equipment

Radiographic unit: Regardless of the type of unit employed, all radiographic examinations require the same components to produce radiographic images. Figure 34.2 depicts a typical arrangement of the components necessary to radiograph a patient. Not included in Figure 34.2 is the x-ray control panel, which will be discussed in Chapter 38. The other components are discussed below.

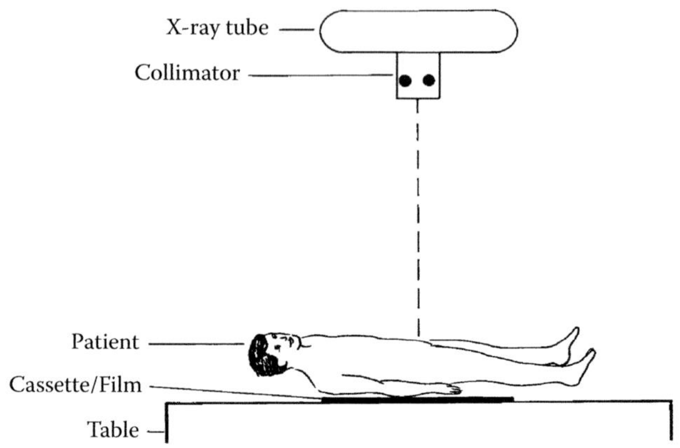

FIGURE 34.2 Schematic drawing of a radiographic unit.

X-ray tube: The function of the x-ray tube is to produce x-rays. The principal components of an x-ray tube are the filament and anode. The anode is the area of the tube where x-rays are produced. The anode is frequently referred to as the focus of the tube, and it plays an important role in producing fine detail in the radiographic image. Most x-ray tubes have two filaments designated as large and small. The small filament causes the anode (focus) to produce fine-detail images. If available, the small filament should be selected whenever fine detail is desirable. High-voltage electricity to produce x-rays is obtained from the standard power supply by means of a step-up transformer. In fixed units, the generator

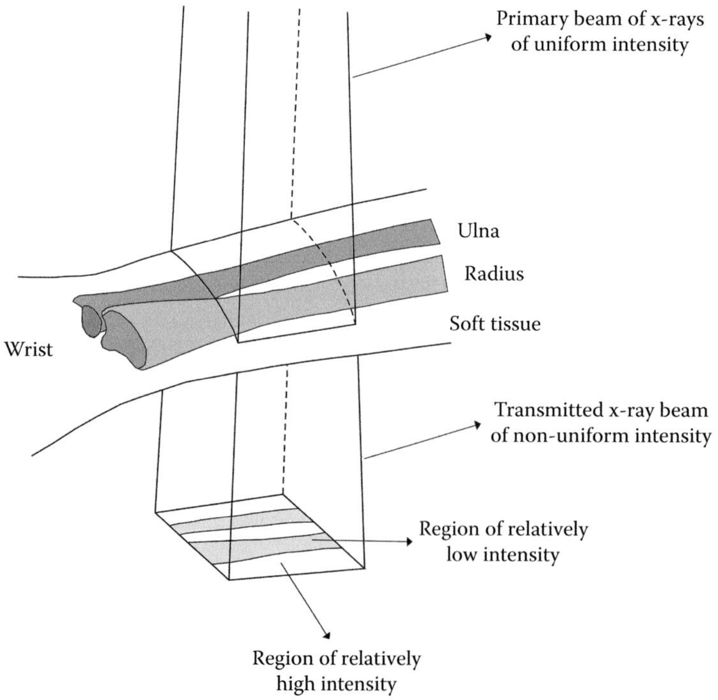

FIGURE 34.1 Principles of x-ray examination. (Viner, M. D., *in* Adams, B., Byrd, J., *Retrieval, Analysis and Identification of Co-Mingled Human Remains*. Humana Press, Totowa, NJ, 2008. With kind permission of Springer Science & Business Media.)

is usually separate and sits on the floor or is mounted on the wall or ceiling. In mobile x-ray units it is self-contained.

Central ray: As depicted in Figure 34.2 the central ray constitutes the center or innermost part of the x-ray beam. It is always desirable to place the central ray in the center of the film, and to the center of the body part being radiographed. Proper centering of the central ray reduces what is referred to as image shape distortion.

Collimator: The function of the collimator is to restrict or limit the size and shape of the x-ray beam. In so doing, the area exposed on the x-ray film is confined to a specific area, for example, 8 in. × 10 in., 11 × 14 in., 14 × 17 in., etc.* The size of the collimator opening can be changed or manipulated by adjusting the horizontal and longitudinal controls located on the front of the collimator. With respect to postmortem radiography, a rule of thumb to follow when making collimator changes is to set the collimator size to correspond with the size of the x-ray film; that is, the collimator setting for 8 × 10 film should be 8 × 10. Following this rule of thumb will enhance image contrast, while reducing the amount of scatter radiation to the medical examiner, the assistant, and other personnel in the autopsy suite.

Detector/Film cassette: At the heart of the planar x-ray system is the image recording system. Traditionally, the x-ray image was recorded on film enclosed in a light-tight envelope or cassette. The film may be sandwiched between two fluorescent "intensifying" screens which emit light when exposed to radiation. In this way, the image on the film is formed mainly by the light emitted from the screens rather than directly by the radiation falling upon the film, a process that significantly reduces the radiation dose and thus exposure times required. Films must be loaded into and removed from the cassette or envelope in darkness and require development (processing) before the image may be reviewed. This process and the equipment requirements will be discussed later in this chapter. In recent years, digital alternatives to film and film cassettes have become available and there are currently two digital imaging solutions suitable for planar radiography Direct Digital Radiography (DR) and Computed Radiography (CR). These will be discussed in the next section.

X-Ray Equipment Options

When considering equipment requirements for planar x-ray, two key components must be considered:

1. The x-ray system (x-ray tube, tube support, generator, etc.).
2. The imaging system (DR detector, CR plate, or film/screen cassette).

* In this chapter, cassette sizes are given in inches. Standard equivalent sizes in centimeters are as follows:
17 in × 14 in—43 cm × 35 cm
14 in × 14 in—35 cm × 35 cm
15 in × 12 in—40 cm × 30 cm
12 in × 10 in—30 cm × 24 cm
10 in × 8 in—24 cm × 18 cm

In most cases, these components can be procured and installed separately from one another and integrated to function as a complete system. However, in the case of DR, most of the units currently available are sold as complete systems; that is, DR detectors are integrated with the x-ray tube and table systems.

1. *X-ray System:* (*Fixed x-ray units*): Ideally, there will be space and money available to have a fixed radiographic installation within, or contiguous with, the autopsy suite (Figure 34.3). Such units need not be very powerful or very expensive. Again, usable second-hand units may be obtained from clinical installations as they purchase newer machines or modalities, merge, or downsize. Sometimes these units can be had for the expense of disassembling, moving, and reinstallation. These fixed units will have (1) an x-ray table for holding the body; (2) a built-in tray for the detector or film/cassette and grid; (3) a transformer; (4) a control panel; and (5) a tube-mount by which the x-ray tube is suspended for positioning over the patient. The latter may be either floor mounted or suspended from the ceiling. Either type should allow the tube to be positioned anywhere over the body for frontal views and anywhere alongside the body for lateral (side) views with the x-ray beam directed horizontally, whilst maintaining a minimum tube focus to detector/cassette distance of 40 in. Ceiling-suspended tube gantries are desirable as they do not take up valuable floor space and allow personnel to work from either side of the radiographic table. They are generally more flexible and permit more rapid examinations, which may be a key consideration in the busy morgue. A cheaper alternative is a wall-mounted unit, which also maximizes floor space but restricts access to one side of the examination table (Figure 34.4).

   If it is not possible to have a fixed radiographic unit in the morgue, it is possible to use either mobile or portable x-ray systems. These are totally self-contained units that simply plug into available power supply convenience outlets. These come in a wide variety of sizes, shapes, and exposure capabilities; the mobile units have wheels and a motor to drive them over short to medium distances. Some portable units are compact and transportable from location to location in the trunk of a car (portable). A typical mobile unit is shown in Figure 34.5. Mobile and portable units may either use film or CR image recording systems, or include an integrated direct digital detector (Figure 34.6).

   Mobile and portable units tend to be reasonably priced, and good usable second-hand units are often available from local hospitals or x-ray supply companies. Safety and dependability are important factors, but high-energy x-ray output is not so important.

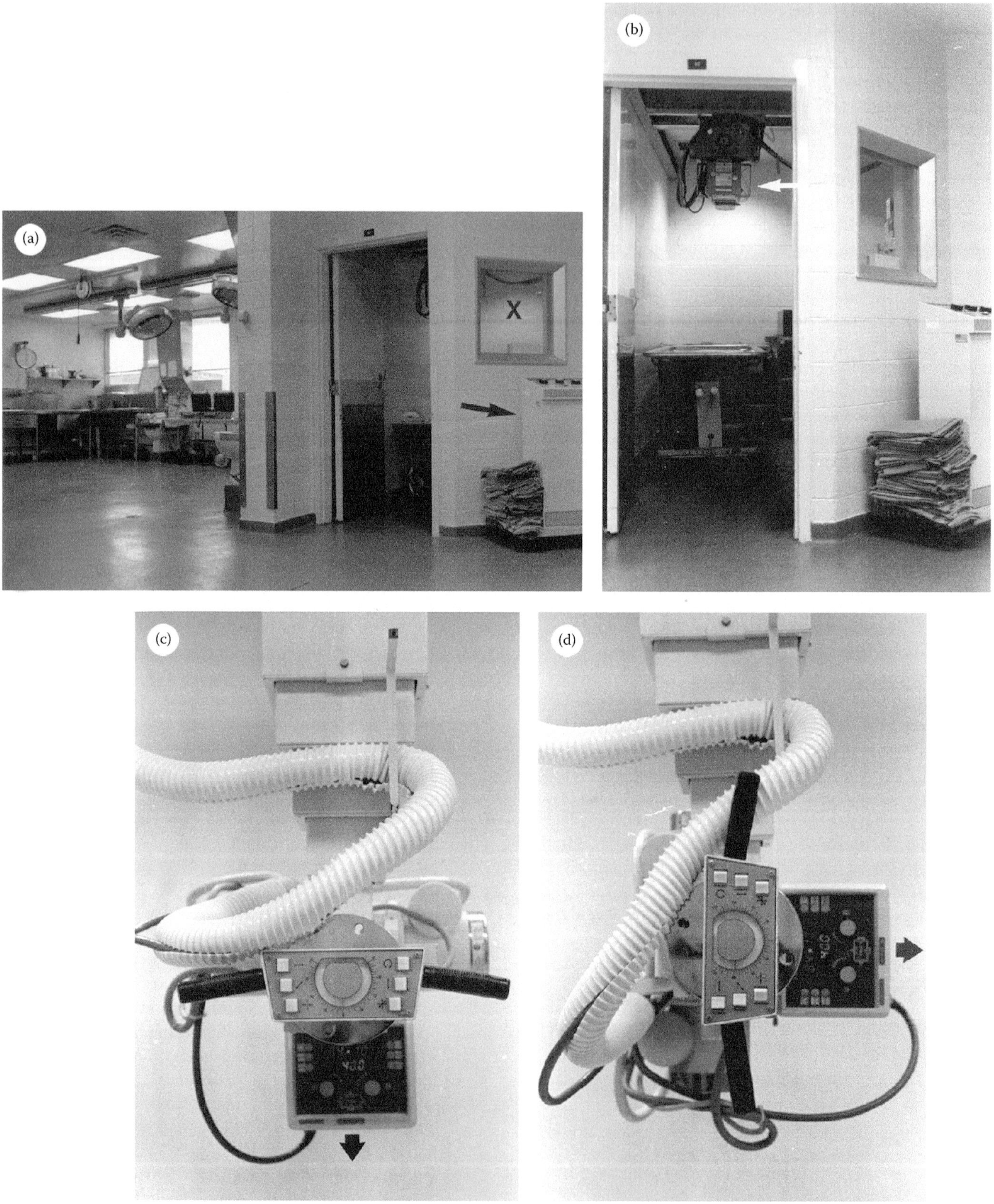

FIGURE 34.3 (a) fixed radiographic installation built into the corner of the autopsy suite of the office of the Medical Examiner of the State of Alabama at the University of South Alabama Medical Center. The control panel (arrow) and viewing window (X) allow the operator to be shielded from the radiation field. (b) Bodies are positioned under a ceiling suspended tube (arrow) that can be raised or lowered. (c) The tube is directed downward (arrow) for frontal or anteroposterior views or (d) to the side (arrow) for cross-table lateral views.

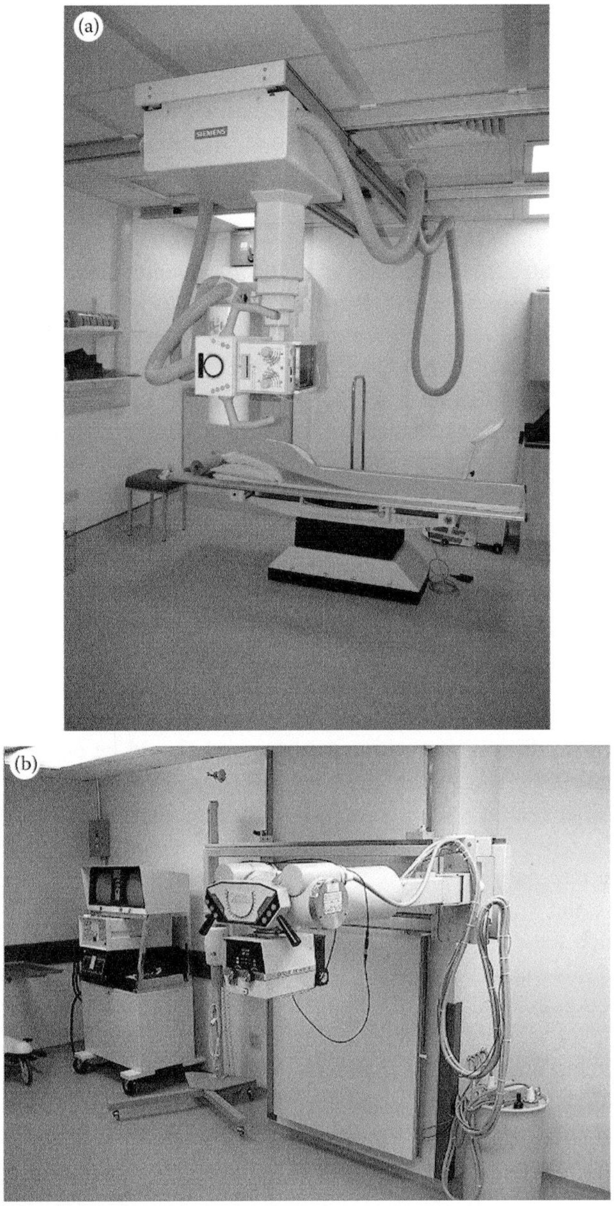

FIGURE 34.4 (a) A ceiling-mounted x-ray gantry offers maximum working space around the table or trolley. (b) A wall-mounted unit offers a flexible space saving alternative. (Image with kind permission of Paul Laudicina & DuPage County Coroners Office, IL).

The main difference in equipment requirements between clinical medical radiology and forensic radiology is that the vast majority of forensic subjects don't move, breathe, pulsate, or have peristalsis. Physical or physiologic motion is simply not a problem. If the exposure factors required for an acceptable single-exposure image exceed the capability of the equipment, then double exposures, or more, are possible.

2. *Imaging Systems:* The obvious function of the digital detector or film is to record and permit the display of the radiographic image. The detector/film is an image receptor and once the image is processed

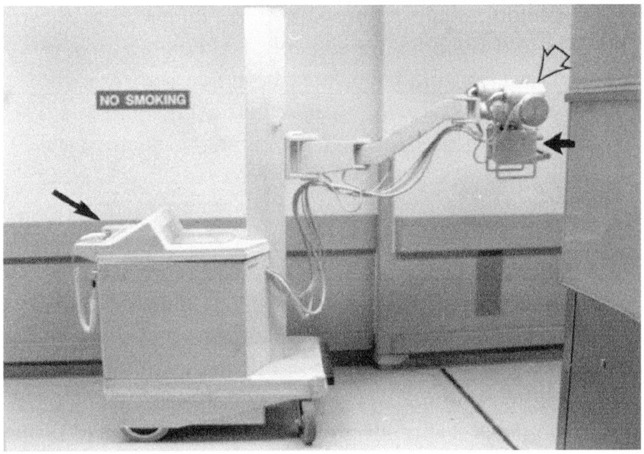

FIGURE 34.5 Typical mobile x-ray unit. The x-ray tube and light-beam collimator (arrowed) are supported by a movable arm. The generator and control panel (arrowed) are contained within a self propelled cart.

or developed and the image is made visible, the image is referred to as a radiograph or a roentgenogram. Throughout this chapter and in following chapters the term "radiograph" will be used when referring to the image as depicted on a viewing screen (digital image) or on an x-ray film (analog image). Images may be recorded using computed radiography plates, direct digital detectors, or film.

3. *Computed Radiography:* A CR cassette takes the place of a traditional film cassette and contains a

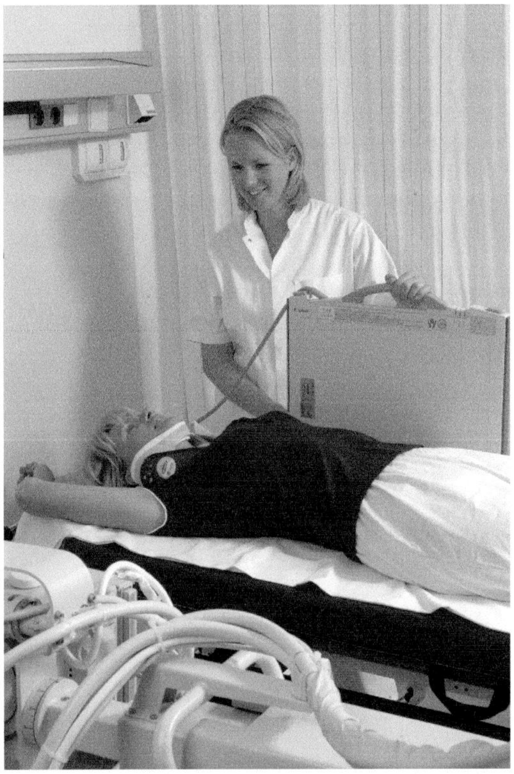

FIGURE 34.6 Mobile DR x-ray unit. (Image with kind permission of Xograph Healthcare Ltd.).

radiation-sensitive phosphor plate in place of the film and intensifying screens. Following exposure to radiation, the plate must be processed in a CR "reader" that electronically scans the phosphor plate for the minute changes that occur in the electrostatic charge as a result of exposure to radiation. These differences in charge are then computed to an x-ray image or radiograph that is viewed on a computer screen (see Chapter 28).

4. *Direct Digital Radiography:* A DR detector takes the place of a traditional film cassette and contains a radiation-sensitive solid state detector in place of the film and intensifying screens. The DR detector is connected directly to a processing computer by means of a wire or wireless network connection. At the instant of exposure, the minute changes in charge on the detector induced by the incident radiation are computed to form an image, which is displayed on a computer screen within milliseconds of the exposure being completed (Figure 34.7).

Film

Whilst film has traditionally been the method for recording the radiographic image, it requires both careful handling and accurately controlled development in order for optimum results to be achieved. For this reason, it requires specific, clean, dark and dry handling areas, well maintained, regularly monitored processing equipment and a supply of appropriately mixed and temperature-controlled chemicals. Whereas these conditions may be cost-effective and relatively easy to maintain within the environment of the busy imaging department or hospital clinic, they are not easy to sustain in many smaller facilities and this may include the majority of medical examiners offices. For this reason, a digital imaging solution should now be considered to be the imaging system of choice. It is recognized that financial resources or local circumstances may not permit such a digital solution, and the following section outlines the requirements for using film-based systems.

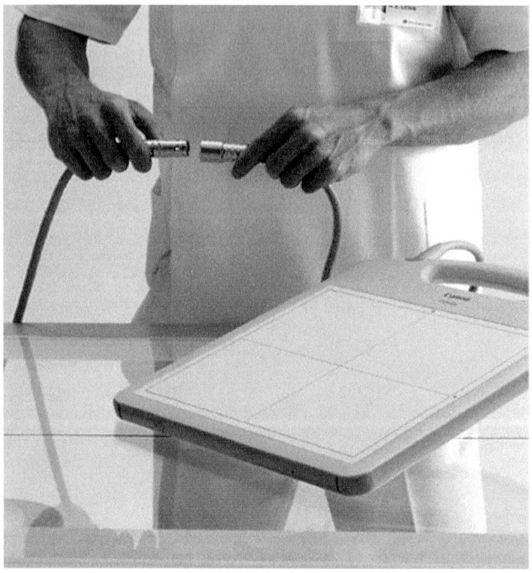

FIGURE 34.7 Direct digital x-ray detector, now available in a range of sizes and with wireless facility. (Image with kind permission of Xograph Healthcare Ltd.).

Storage and handling of unexposed film: Like photographic film, x-ray film is sensitive to, and affected by, a variety of factors. In addition to x-rays, x-ray film is sensitive to almost any color of light and care is therefore required to protect film from light and also to protect exposed or unexposed films in an area where x-ray exposures are being made. X-ray film is also sensitive to heat arising from heating systems, steam pipes, etc. and this may produce excessive film blackening, which often renders the film useless. X-ray film should thus be stored in an area that affords protection from stray radiation, light, and excessive heat and humidity. It should be stored on shelving that allows the film to be kept upright (and not flat as pressure marks may be caused on the film) and off the floor to avoid dampness. X-ray film is usually supplied in boxes of 100 films and can be heavy. It should thus be stored at a height that is consistent with good manual handling practice. Like photographic film, the useful life of an x-ray film is determined by its expiration date. Failure to use film before the expiration date may lessen the quality of the radiograph and an efficient stock control system is therefore necessary.

X-ray film is a sensitive medium and should be carefully handled. Folding, bending, or any form of rough handling may produce artefacts (foreign marks on the radiograph that are not a depiction of the patient's anatomy or pathology) that may interfere with the interpretation of the radiograph. Excessive heat and low humidity may combine to produce conditions in which static electricity is transmitted to the film by the handler and this can cause another artefact referred to as a static mark. Reducing the room temperature and raising the humidity level will alleviate this problem. Automated "daylight" handling systems are available that minimize handling by machine-loading individual x-ray cassettes from a bulk supply of x-ray film that is placed in the machine in normal lighting conditions. If such technology is not available, films will have to be loaded in the cassettes by hand and removed following exposure and loaded into the developing machine. Unexposed films should be handled in a specifically designed and designated darkroom that offers sufficient space for storing smaller amounts of unexposed film, loading and unloading film cassettes, and either feeding these into the automatic film processor or "wet" developing by hand. Dark rooms should have separate wet and dry areas and sufficient bench space to allow cassettes to be loaded and unloaded. They should also be equipped with "safe" lights—lights that emit wave-lengths to which the film is not sensitive—to permit safe working. If film processing is undertaken in the darkroom, a positive pressure air-handling plant that permits a low level fume extraction and several air changes per hour should be installed.

Film Processing: X-ray film and photographic film share another requirement in that they both must be developed to render a visible image. Film development, or film processing as it is most commonly referred to, is perhaps the single most important step in the production of a radiograph. Film quality will be adversely affected when correct film processing is not

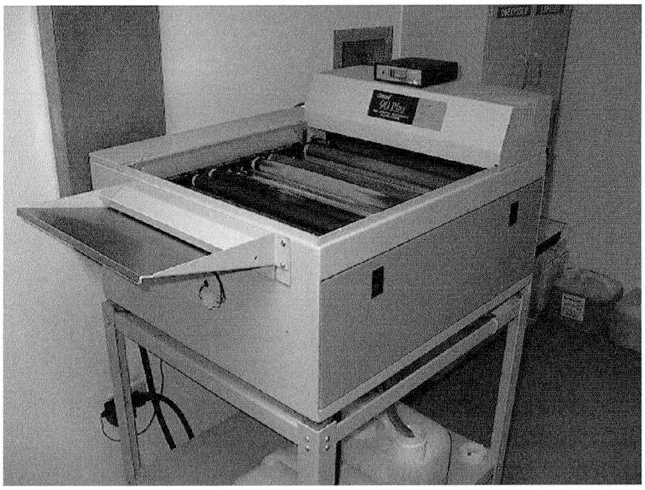

FIGURE 34.8 Automatic film processor suitable for deployment in the morgue. (Image with kind permission of Paul Laudicina & DuPage County Coroners Office, IL.).

attained and this can lead to misdiagnosis or a failure to detect vital evidence. Unless the morgue facility is a particularly busy unit and it has sufficient workload to warrant a dedicated, quality assured, regularly monitored film-processing system, it may be preferable to handle the volume of postmortem radiographs produced in the facility by arranging to have them processed in a hospital radiology department unless precluded by distance, architectural obstacles, or failed negotiation. If such an arrangement is not feasible or unsatisfactory for whatever reasons, small automatic film processors are reasonably priced and easily installed (see Figure 34.8). Should an automatic film processor be located in the autopsy suite, it is essential that the unit is correctly operated, cleaned, maintained, and monitored according to an established quality control procedure. Professional photographic laboratories take scrupulous care to ensure that film-processing conditions remain constant at all times by undertaking quality control checks throughout the day. Radiographic film processing is exactly the same process, and it is essential, particularly in the forensic situation, that the film-processing environment is always maintained in optimum condition in order to ensure consistently high-quality images. The medical examiner is therefore strongly recommended to consult with radiology personnel in order to set up a processor quality assurance program (see Chapter 33).

Cassettes and Screens: A cassette is a light-tight container that holds the radiographic film (Figure 34.9). Cassettes are labeled front and back and the front side of the cassette must always face the x-ray tube. Cassettes typically contain a pair of intensifying screens, the function of which is to emit light when struck by x-rays. The advantage of light emission is that since light produces over 90% of the film blackening, less x-ray exposure is needed to produce an image resulting in less x-ray exposure to the patient and to those who perform the radiographic examination. The color of light emitted by the screens is usually blue or green; it depends upon the phosphor used in the manufacture of the screen and varies

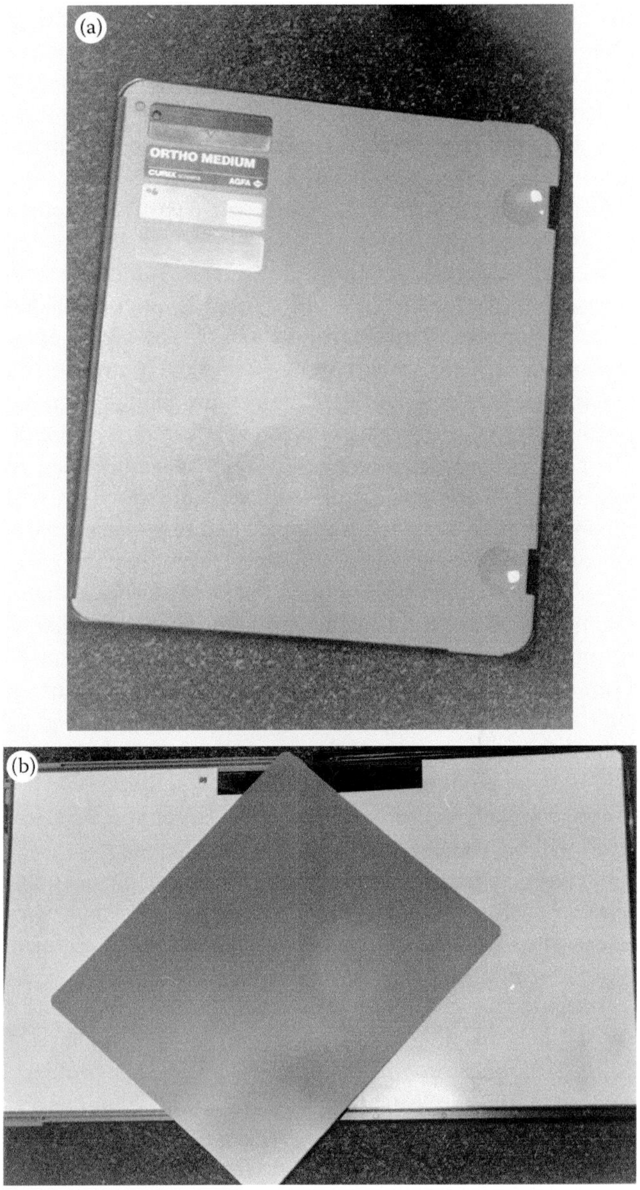

FIGURE 34.9 (a) Closed cassette. (b) Opened cassette showing intensifying screens.

between manufacturers. Film manufacturers also design radiographic film to be sensitive to either blue or green light and it is essential that the correct type of film is matched with the appropriate screens (i.e., blue-sensitive film to blue emitting screens and green to green). With respect to film quality, there is essentially no difference between blue- and green-sensitive film—screen combinations.

In addition to differences in color emission, intensifying screens are also classified according to their speed. The speed of a screen refers to its ability to respond to x-rays and emit light (blue or green). Common screen classifications include slow, medium, regular, and high speed. The significance of speed classifications is that higher-speed screens require less x-ray exposure than slower-speed screens to yield the same degree of film blackening. For example, a medium-speed screen requires twice as much (2×) exposure than a high speed screen. However, high-speed screens reduce the visualization of fine detail in the radiographic image. The slight loss in fine detail, however, is offset by the reduction in x-ray exposure to the patient and personnel who perform radiographic examinations. Converting exposure factors to accommodate differences in screen speed is addressed in Chapter 38.

As previously stated, the film used with a particular screen must be matched to ensure that the film is sensitive to the wavelength of light emitted by the screen. Films are also categorized as being fast or detail—detail refers to the slowest speed—and thus various permutations of film–screen combinations exist. The relative speed of a particular film–screen combination is categorized by manufacturers using a numeric system such that a 100-speed system is referred to as detail, 200 speed as medium, 400 speed as regular, and 800 speed as fast. Within the context of forensic examination it will probably be necessary to have at least two film–screen combinations available. Whilst a medium or fast speed combination will be sufficient for routine "survey" work, a detail system will be required for human identification, pediatric examinations, and other investigations where the resolution of fine detail is required. In certain circumstances, nonscreen film (film contained simply in a light tight envelope) may be utilized for optimum detail. However, the additional radiation exposure required will be significant and this should not be used routinely except when using a cabinet x-ray system, which will be discussed further on in this chapter.

Intensifying screens can become soiled from the accumulation of dust particles or other foreign materials that enter the cassette and settle on the screen surface. Debris deposits on the screen surface prevent light from reaching the film. As a result, an artifact is produced, which always appears as a loss in film blackening corresponding to the size and shape of the dust particle or other debris. To avoid this problem, screens should be cleaned periodically with a commercial screen cleaner.

Grids and Grid Cassettes: The term "scatter radiation" was previously used in the discussion on collimation. The production of scatter radiation occurs when x-rays enter the patient. Upon entering the patient, the x-rays interact with tissue molecules. Many of the x-rays are diverted (scattered) in a different direction. In fact, scatter radiation arises from the patient in virtually any direction or angle. The significance of scatter radiation with respect to radiographic quality is that it produces a fogging effect on the radiograph. The fogging effect increases image blackening, which reduces image quality.

While scattered radiation cannot be totally eliminated, it can be reduced by collimating to the size of the cassette and by employing a grid. The grid absorbs scatter radiation when it is placed between the patient and the cassette/detector. The typical grid consists of extremely thin lead strips standing on their edge. The lead strips are arranged parallel to one another and extend longitudinally from end to end (Figure 34.10). The lead strips are also separated from one another by a spacer (interspace) composed of plastic or aluminum. Since lead is radiopaque (absorbs x-rays), scatter radiation is absorbed by striking the lead grid lines at an angle. In contrast, those x-rays that approach the grid in a perpendicular manner pass through the spaces separating the lead strips. X-rays are able to pass through the interspaces because the interspace material is radiolucent (does not absorb x-rays). The close relationship existing between the lead strips and the interspaces is more fully appreciated by the fact that the number of lead strips can range from 80 to 103 strips per inch. The grid employed in a fixed radiographic installation is permanently mounted beneath the table. In the autopsy suite, one is most likely to utilize a grid cassette or a grid cover that attaches to the front surface of the cassette/detector. In this way, the cassette/detector and grid can be moved about to accommodate all positions and projections. However, grids are not necessary for all examinations and the question of when to use a grid will be resolved through recommendations offered in Chapter 38.

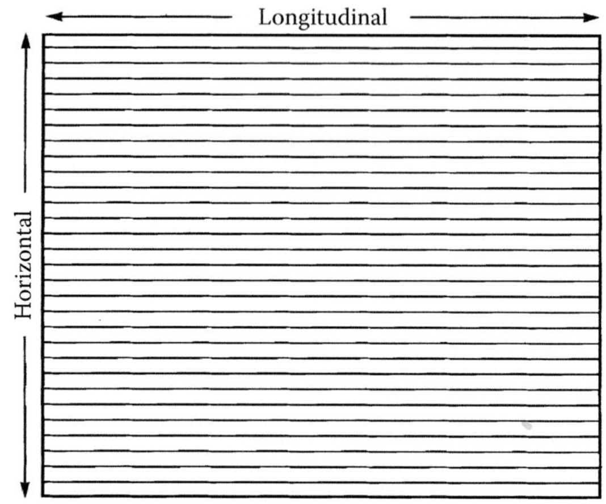

FIGURE 34.10 Schematic drawing of a grid (see text).

Dental X-Ray Systems

Dental radiography equipment is an essential requirement for a medical examiner's facility. While pan-oral radiographs may be routinely used in the living patient, the forensic odontologist or dental radiographer will generally rely on intraoral x-ray equipment to perform a postmortem dental survey. Dental x-ray tubes are a fairly simple piece of apparatus, with a stationary anode tube and a stand- or wall-mounted generator transformer mounted as a single unit device. These units can be operated from a standard 110 V (or 220 V) power supply (Figure 34.11). Handheld battery-powered units are also now available. Image recording is either by means of intraoral dental film (wrapped in a plastic wrapping and available in a range of speeds and sizes) or by means of a direct digital detector system. Dental film does not use intensifying screens, and in order to improve its sensitivity to x-rays, the film emulsion is thicker than normal x-ray film. As a result, the processing chemistry takes time to penetrate the emulsion and the processing times are therefore much longer. . If film is to be used, then a special automatic film processor and dental film chemicals should be used to develop the films (Figure 34.12).

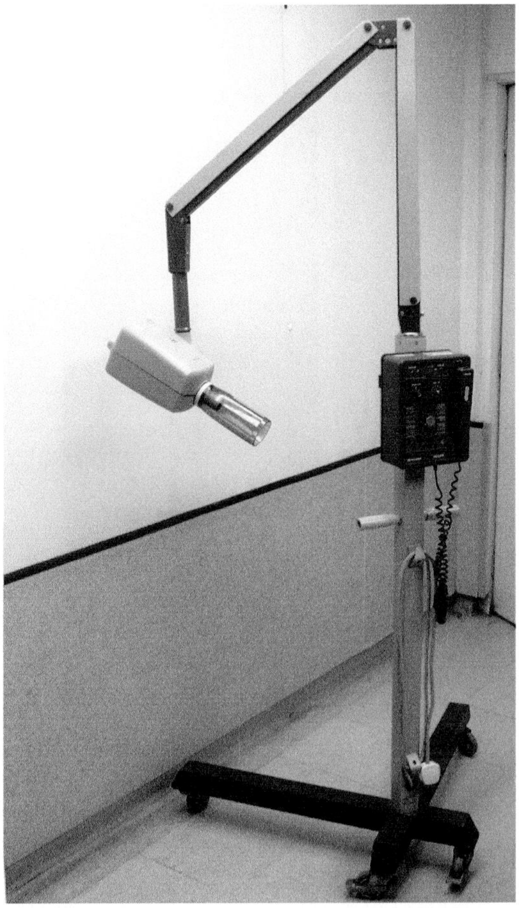

FIGURE 34.11 Stand-mounted dental x-ray tube.

Scanography

It is possible to increase the area of the object recorded on the x-ray image. Traditionally, this has been done using a larger film size and a narrow beam of radiation. The exposure time is much longer and the x-ray machine is moved in a smooth action along the length of the object to be recorded. This technique is known as scanography and has typically been used in orthopedic imaging to assess joint alignment and for accurate measurement of leg length. However, recent advances in digital detector technology allow the x-ray tube and a digital detector to be moved simultaneously along the length of an entire human body in a matter of seconds, producing a single 2D digital image of the subject. This technology has applications in both trauma radiology and in forensic imaging.

Scanography Equipment

The Lodox Statscan® was developed originally for examining suspected diamond smugglers in the mines of South Africa. This machine employs a digital version of the "scanogram" technique described above. A narrow slit beam of radiation passes rapidly over the subject and the resultant image is digitally recorded using 12 solid-state detectors in a linear array detector. An entire body can be examined from head to toe in approximately 13 s and the gantry can be rotated through 90°

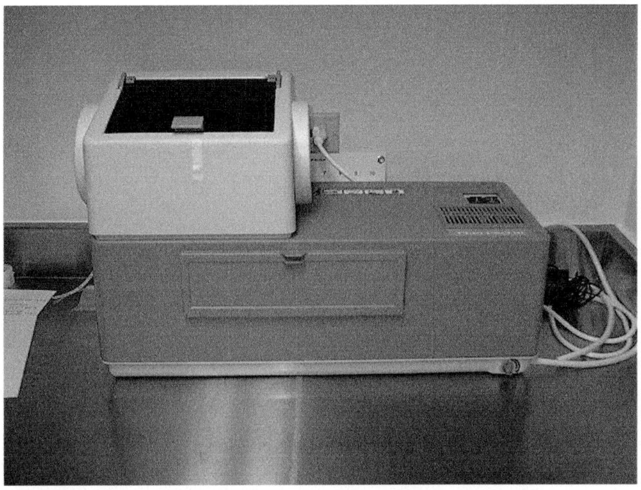

FIGURE 34.12 Dental x-ray film processor. (Image with kind permission of Paul Laudicina & DuPage County Coroners Office, IL.)

(and any angle in between) to give a lateral or oblique image.[1] (Figure 34.13). The Statscan is a relatively new machine that has been employed very successfully for investigation and management of polytrauma trauma centers in the United States, Europe, and Africa. The application of this machine for forensic use is in its infancy but a number of units are now in use in medical examiners' facilities in South Africa.

Cabinet X-Ray Systems

Cabinet x-ray systems are a useful self-contained x-ray device suitable for any radiology, pathology, or forensic science department. The units are available in a range of sizes from those not much larger than a microwave oven to those the size of a large photocopier (Figure 34.14). They have the capability of extremely fine radiographic detail and can produce magnification radiographs and use either film or direct digital technology. They are also capable of very low exposure factors not available on conventional x-ray machines and are thus extremely useful for imaging bone fragments, small parts (e.g., digits, hands, feet), biopsy specimens, and nonhuman materials (e.g., letters, packages, clothing, etc.). The units are self-protected against radiation leakage and operate from standard electrical outlets. Both 110 and 220 V models are available.

Fluoroscopy

It was the light produced by invisible rays falling upon calcium tungstate that led to Röntgen's discovery of the x-ray. X-ray energy causes some compounds to emit light and thus "fluoresce" in proportion to the intensity of the x-ray beam falling upon them. Fluorescent screens are used in the production of "still" x-ray images in both film and digital systems. However, fluorescence is a dynamic process, and as the intensity of the beam falling upon the compound changes, the light emitted changes, and thus the image changes. "Fluoroscopy" is thus the production of a continuous "real-time" image by

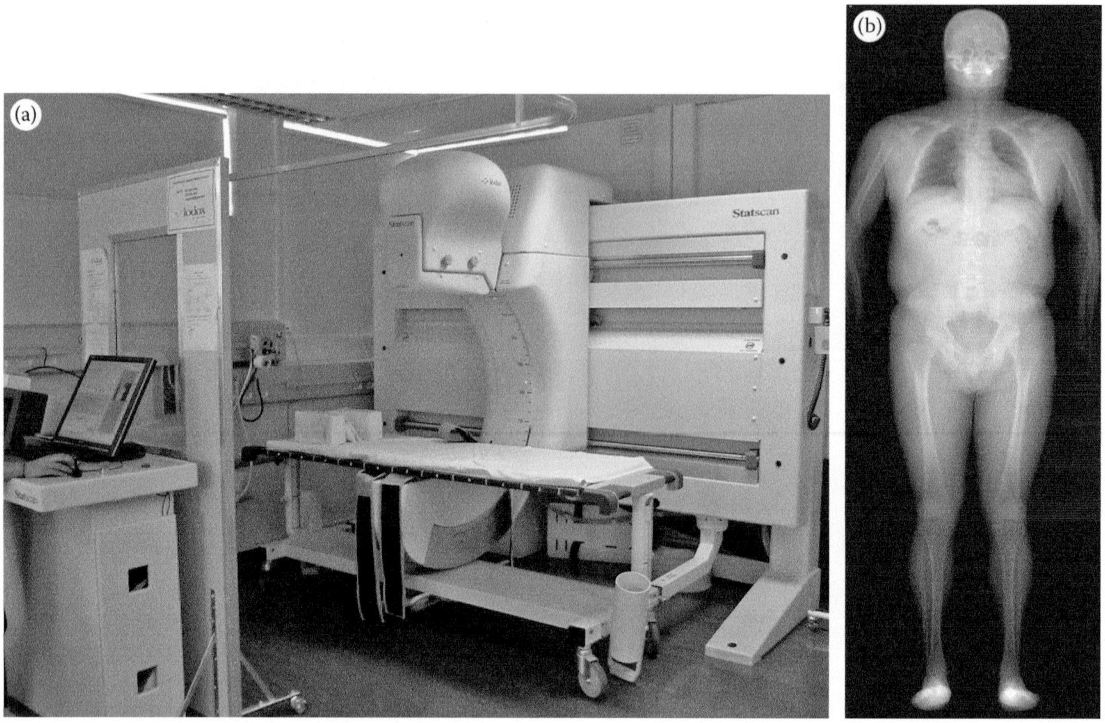

FIGURE 34.13 (a) Lodox Statscan digital scanography machine. (b) Statscan digital scanogram showing full body. (From Viner, M.D., *in* Adams, B. J. and Byrd, J. E., 2008, *Recovery Analysis and Identification of Commingled Human Remains*, p. 162; Lodox Ltd, Johannesburg & the Department of Forensic Medicine, University of Cape Town, South Africa and Springer Science & Business Media. With permission)

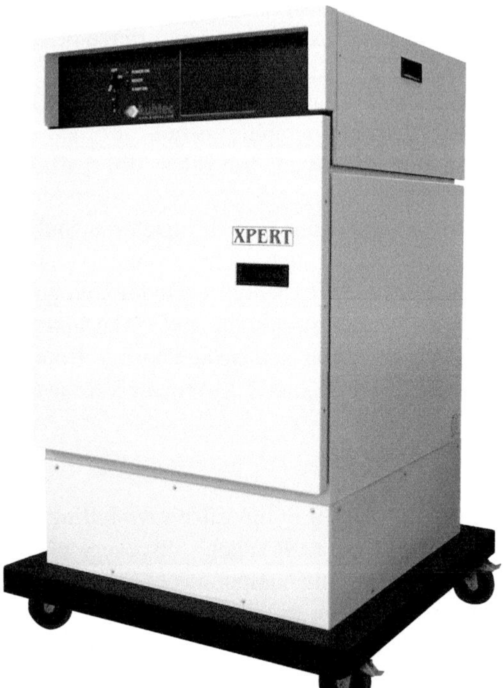

FIGURE 34.14 Cabinet x-ray system. The lead shielded cabinet contains an x-ray source together with both the specimen and a film or detector system. (Adapted from KUB Technologies Inc.).

exposing an object to a continuous (low dose) beam of radiation and recording the changing image produced by the fluorescent screen. The image is limited in size and typically circular (although some digital units now offer a square image format). Maximum image size is typically 14 in. diameter or 14 in. × 14 in. square. Fluoroscopy machines are used in medical imaging in order to examine moving objects or processes such as the transit of barium sulphate through the esophagus and stomach or to assist the surgeon or radiologist to accurately position a prosthesis or catheter. In forensic imaging, fluoroscopy has two main uses: (1) to enable rapid examination and triage of larger objects (e.g., a whole cadaver) by moving the fluoroscope slowly along the length of the object and observing the image, or (2) to permit the rapid location and retrieval of foreign bodies (e.g., ballistic material) under x-ray control, allowing the pathologist to "see" what they are doing by means of the dynamic image (Figure 34.15).

Fluoroscopy Equipment

A fluoroscopic unit in or adjacent to the autopsy suite is a great convenience (Figure 34.16). It allows rapid scanning of bodies for foreign objects, injuries, and unique features useful for comparison identification. The spot-filming or DR capability allows instant radiography to permanently record findings discovered at fluoroscopy. These units also have an overhead tube-mount for conventional radiography, although these are not as flexible as dedicated x-ray systems. Because of the increasing utilization of endoscopy for gastrointestinal

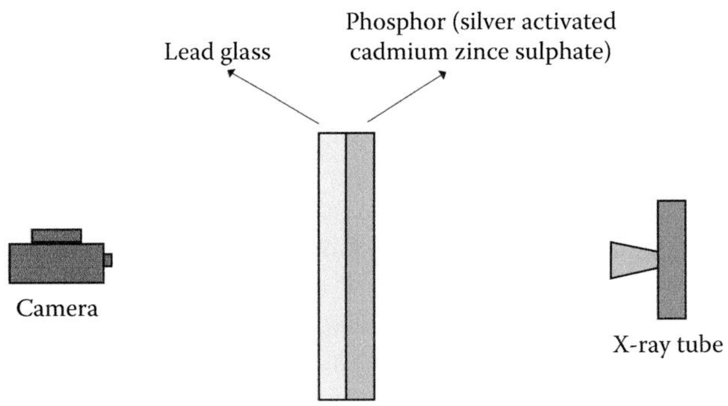

FIGURE 34.15 Principles of fluoroscopy.

diagnosis and the economics of managed care, the use of clinical fluoroscopic units has diminished substantially in recent years. Many hospital fluoroscopy units are underutilized and/or their room space is needed for the newer imaging modalities. Consequently, again, usable second-hand equipment may be obtained at surprisingly low cost. A fluoroscopic installation, though, will be more expensive; require more maintenance, skilful operators, and expert physicians/interpreters; and pose greater radiation risk and protection problems for personnel.

Mobile C-Arm units are well suited for forensic imaging and are employed in a number of forensic examiners offices and morgues (Figure 34.17). They are particularly useful for rapid screening of large numbers of cases and retrieval of ballistic material. For this reason they have been deployed regularly for mass fatality investigations (see Chapter 13).

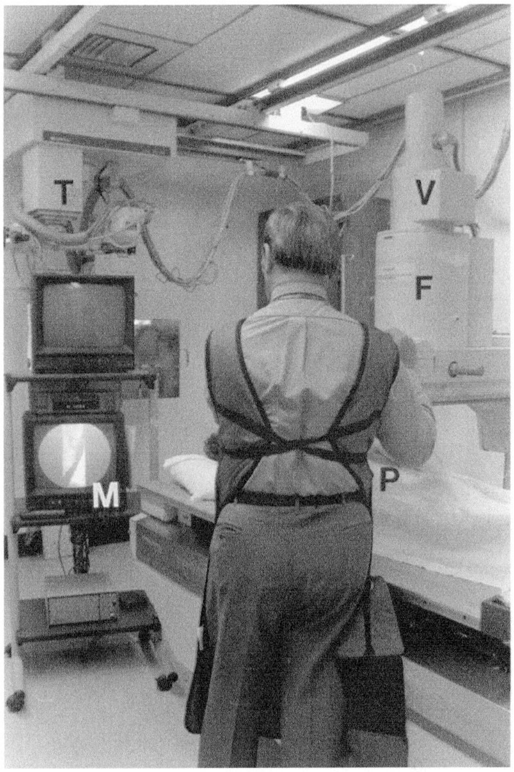

FIGURE 34.16 Radiographic and fluoroscopic unit. The image intensifier in the fluoroscope tower (F) can be moved in three axes to cover the patient (P). A videocamera (V) looks at the output phosphor of the intensifier and transmits the fluoroscopic image to a monitor (M). A ceiling-hung x-ray tube (T) can be moved over the patient or body for routine roentgenograms.

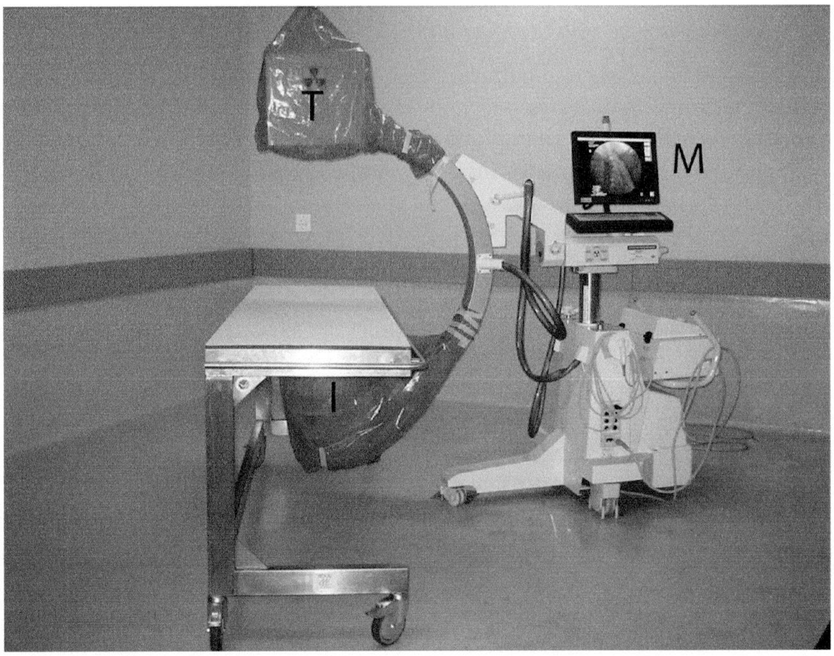

FIGURE 34.17 Mobile C-Arm fluoroscope in the forensic pathology department of the medical examiner's office at Hilbrow Johannesburg, South Africa. This specially designed unit has a rotating anode x-ray tube (T) to reduce heat loading and a 14 in diameter image intensifier (I) for maximum coverage. The viewing monitor (M) and all imaging controls are mounted on the C-Arm, negating the need for an additional monitor cart.

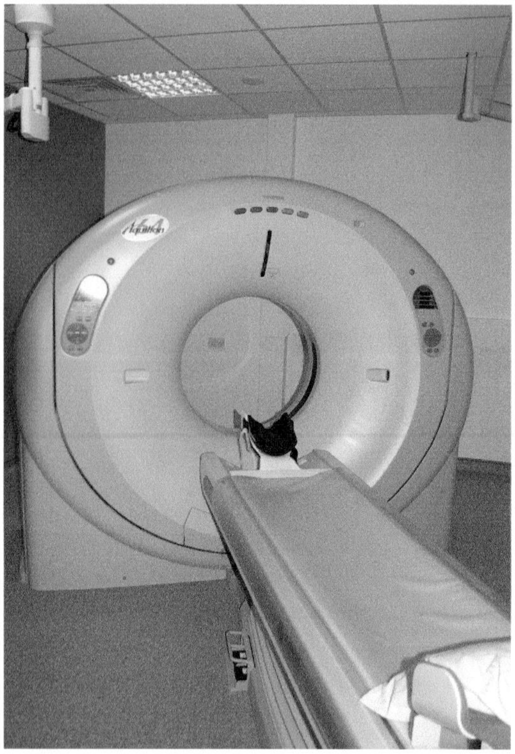

FIGURE 34.18 A modern multidetector CT scanner.

Fluoroscopy relies on a continuous beam of radiation being emitted from the x-ray tube for long periods (up to several minutes) during the examination. Whilst the radiation exposure factors are low, the length of exposure can result in high radiation doses. It is thus essential that the equipment is used in a safe environment, with lead shielding where necessary. All personnel present during the examination should wear lead protective gowns and if close to the tube, thyroid shields and lead protective glasses. When retrieving ballistic material, excellent team work is required between radiographer and pathologist to ensure that the pathologist's hands are not exposed to primary beam radiation. Appropriate training in the use of fluoroscopy is thus an essential requirement.

CROSS SECTIONAL AND MULTIPLANAR IMAGING

Planar imaging, whilst of great benefit in both medical and forensic applications, has limitations. The human body is a 3D structure, and a 2D x-ray process is hindered by the problems of superimposition and the physical and radiation limits of taking an infinite number of radiographic images at different angles in order to fully image 3D structures. Over the years, various methods of minimizing superimposition, for example, linear tomography, or of enhancing certain anatomical structures by the use of contrast media (e.g., air, iodine, barium, etc.) have helped overcome these limitations to some degree.

Over the last 30 years, however, computer technology has revolutionized diagnostic imaging and made it possible not only to image the body in cross-section but increasingly in an almost infinite number of planes simultaneously. By means of highly accurate detection and measurement of both radiation (CT scanning) and magnetic radio-frequency pulses (MRI scanning), rapid computer modeling of the radiographic data permits the reconstruction of 3D images, planar images in any plane, and allows overlying structures to be digitally "removed" from the image, permitting visualization of specific organs and systems. The use of these techniques is covered in depth in Section VII. For the purposes of a discussion on equipment issues, only CT scanning is covered in this chapter.

CAT Scanning: Computed axial tomography (CAT) scanning was developed in the 1970s by a British scientist, Godfrey Hounsfield, working for the EMI Corporation. Hounsfield applied the physics of rotational tomography, going a stage further so that the tube and x-ray detector gantry rotate about a single axis in the body. X-ray detectors replaced film and took a series of measurements continuously while the gantry rotated. A computer analyzed the data recorded and reconstructed them as a digital image of an axial "slice" of the subject displayed on an image matrix on a CRT screen. Each pixel was allocated a level on a gray scale which corresponded to the level of radiation transmitted through the body at that location and thus the relative density of the subject. These values are now known as Hounsfield Units and the scale, in which the density level corresponding to 0 is equivalent to water (mid-gray), spans from dense air, −1024 units (black), to bone, > +400 units or greater, to dense metallic objects of +3071(white) units. Initially deployed as a tool for imaging the head, it enabled radiologists to examine the brain accurately for the first time without the overlying features of the bony skull obscuring vital soft tissue information.

CT scanning, as it is now known, has benefited enormously from advances in technology and computer processing power. The entire body can now be scanned in a matter of seconds and images can be reconstructed in any plane: axial, coronal, sagittal, or oblique. Whilst CT scanning is a routine tool in clinical medicine, it is a relatively expensive technique and has only recently been used for postmortem forensic examination. Studies have shown that it can be a very effective tool at postmortem examination and offers the opportunity to gather data that may otherwise be impossible to collect. The ability of CT scanning to undertake 3D and multiplanar reconstructions allows the pathologist and anthropologist to examine the underlying skeletal structure of fleshed remains, visualizing trauma, degenerative processes, articular surfaces, and taking accurate measurements.

CT Scanning Equipment

Most forensic pathology centers do not yet have the luxury of being able to afford an on-site CT scanner, although there are now a number of centers worldwide that have acquired this facility. The rapid developments in both X-ray detector and computer technology would render any in-depth discussion of currently available equipment out of date by the time that this book comes to print. The precise type and model of scanner will be determined by a number of factors—not least the budget available at the time of procurement, but for maximum benefit, a high-speed multi detector system (MDCT) should be selected. A suitable location will be required for both CT scanner and control room and facilities. A stable high- voltage power supply and cooling plant will also be necessary. A space of around 540 sq. ft (50 m^2) plus space for circulation is recommended for the scanner and control area.[2] It should also be recognized that a sizable capital investment of this nature will also bring with it a sizable on-going expenditure: maintenance contracts, replacement x-ray tubes, trained CT radiographers or technologists, and consumables.

RADIATION PROTECTION

X-rays produced within an x-ray tube represent a form of ionizing radiation. It has long been known that exposure to ionizing radiation may induce various biologic effects, such as malignant disease and birth defects.

There is no absolute evidence of a threshold below which no damage occurs. Even the lowest doses may cause damage to cells which might later lead to malignancy or hereditary effects if the cells irradiated are the germ cells in the gonads.[3] For this reason, radiation exposures should be limited and controlled by appropriate international, national, and state regulations; codes of practice; and schemes of work. X-ray and imaging equipment should be regulated, adequately maintained and inspected, and subject to regular quality-control programs (see Chapter 33). Radiation workers should be appropriately trained in the use of such equipment for the purpose for which it is employed. This is particularly important for fluoroscopy, where the radiation exposure received by operators can be significantly increased by the use of poor techniques and by inexperienced operators using the equipment for longer periods than are necessary to gather the required information.

Provided that appropriately qualified personnel are involved in planning, commissioning, and operation of the equipment, most x-ray techniques can be performed safely in morgue situations. Those who use x-rays in their daily work environment should understand that it is their responsibility to maintain exposure levels As Low As Reasonably Achievable (ALARA) Whilst ALARA typically pertains to antemortem procedures, it is equally important in postmortem procedures from the standpoint of protecting both the operator and others involved.

The purpose of this section is therefore to provide those individuals who participate in postmortem radiography with an understanding of the fundamentals of radiation protection.

Methods of Protection

A fixed radiographic installation is located in a room that typically contains a control booth barrier, which is simply a small wall extending into the room. The control booth barrier, equipped with a lead glass window protects personnel while x-ray exposures are being made and should cover the staff entrance door so that staff entering the area do so behind the protection of the screen (Figure 34.19). Radiation warning lights above or on either side of the entrance doors outside the room should indicate both when the room is in use and when the x-ray exposure is being made (Figure 34.20).

The walls in a radiographic room are usually lined with sheets of lead. Since it is unnecessary for all of the walls to have the same thickness of lead sheeting, the level of protection varies in different areas of the room. The advice of a radiation physicist should therefore be sought during installation and construction of the facility.

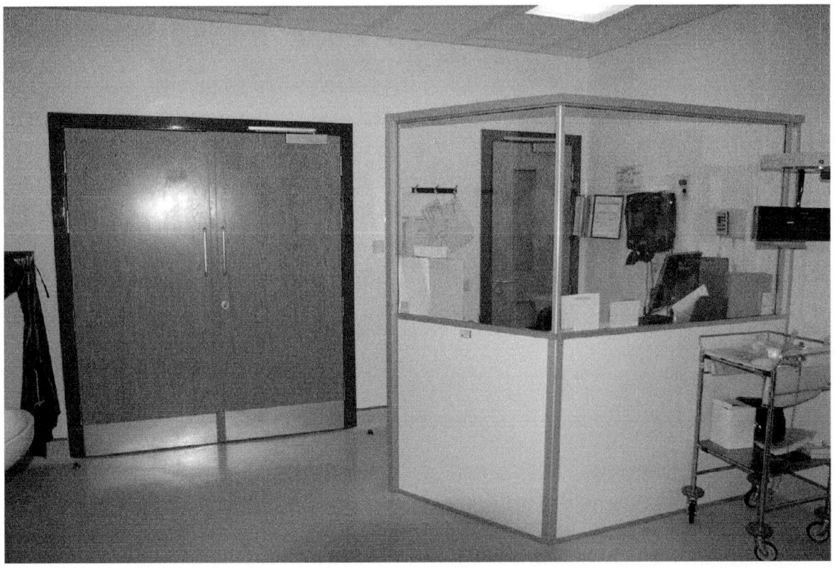

FIGURE 34.19 Control booth with lead protective screen shielding both the control area and the staff entrance door. Note the larger entrance door for cadaver trolleys which should be kept locked during the examination.

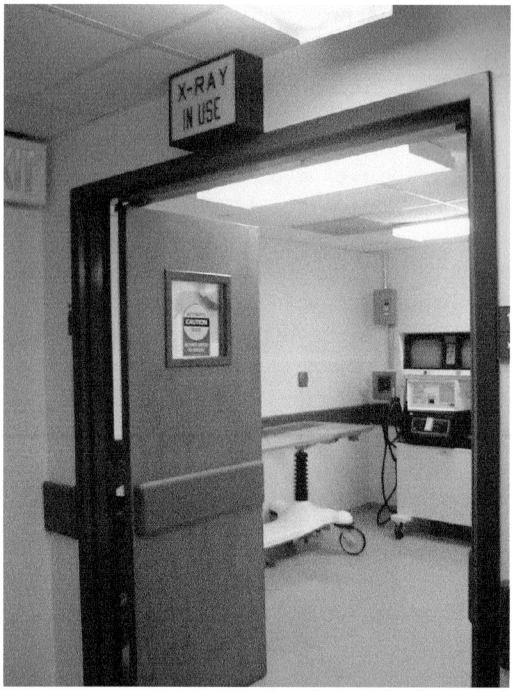

FIGURE 34.20 Entrance to x-ray room showing radiation warning lights above the door. (Image with kind permission of Paul Laudicina & DuPage County Coroners Office, IL.).

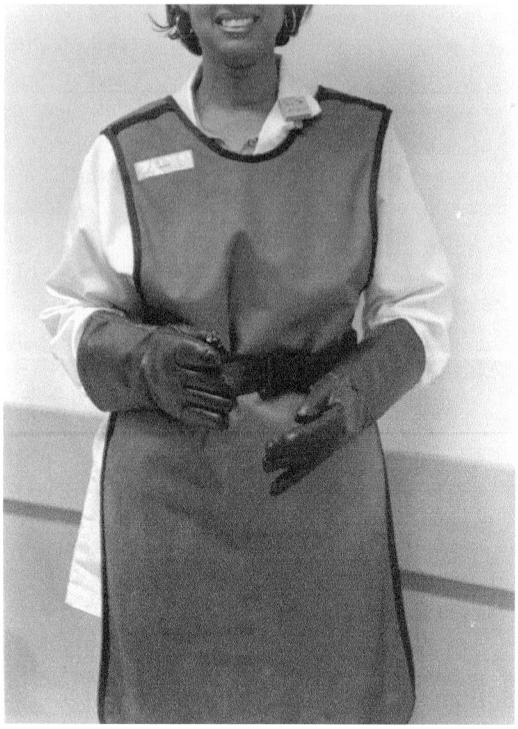

FIGURE 34.21 Lead apron and gloves.

Walls with thicker lead sheeting allow the central ray (primary beam) to be directed toward them, and are referred to as primary barriers. Walls with thinner lead sheeting are referred to as secondary barriers. Secondary barriers are intended to protect against scatter radiation and radiation which may escape from the x-ray tube housing. The primary beam should never be directed to a secondary barrier. The control booth is usually designated as a secondary barrier.

In many autopsy suites, mobile x-ray units are used and a temporary controlled area is established. In this case, the previous discussion regarding primary and secondary barriers may not seem applicable. However, knowledge of primary and secondary barriers will assist in understanding the basic tenets of radiation protection.

There are three fundamental principles employed to reduce radiation exposure to both the patient and personnel. The three principles are time, distance, and shielding.[4] In the absence of a living patient, the following discussion will address reducing radiation exposure to personnel.

Time

Time usually refers to the length or duration of an x-ray exposure. The duration of x-ray exposure usually does not exceed one second. Exposure to radiation can be effectively reduced by maintaining exposure to the shortest time possible. Time also refers to the amount of time spent in an area where exposures are being made. It is therefore important to limit one's presence in the exposure area as much as possible.

Distance

In contrast to source image–receptor distance, distance here refers to the amount of distance between the x-ray tube and the radiation worker. The old adage, "distance is the best protection against radiation," still holds true. In theory, as distance increases, the amount of radiation reaching the patient and personnel decreases. Since mobile x-ray units are equipped with a 6 ft. exposure cord, the operator should extend it to maximum length and stand behind the x-ray unit. This will greatly reduce exposure.

Shielding

As previously indicated, shielding serves as a protective barrier against ionizing radiation. Permanent x-ray installations should be installed in a room equipped with suitably shielded walls. Fixed walls surrounding a radiographic unit may provide an effective barrier depending on the materials from which the wall is built. Additional radiation barriers can be added to existing walls by lead sheathing or barium plastering. Movable barriers can be useful for personnel protection.

In the case of mobile equipment, where structural barriers are missing or impractical, limiting the tube direction can be effective. When otherwise unprotected, personnel should wear protective apparel whenever exposures are being made. Protective apparel consists of aprons and gloves which are impregnated with lead particles (Figure 34.21). Lead aprons and gloves are intended to provide protection from scatter radiation and not to afford protection from the primary beam. Personnel should never position themselves in the direct path of the primary beam. There are no circumstances in which postmortem radiography should require individuals to hold cassettes or body parts in place while exposures are being made, and suitable cassette holders and immobilization devices should be used for this purpose.

While lead aprons and gloves do provide protection against radiation, it should be understood that these devices are subject to cracking and other forms of deterioration. To monitor the effectiveness of these devices, one should make periodic inspections of lead gloves and aprons. Leaded apparel may also be radiographed or screened using fluoroscopy to visualize minute cracks in the leaded material.

In addition to time, distance, and shielding, there are other factors that can be employed to further reduce radiation exposure. These factors are discussed below.

Higher-speed screens require significantly less exposure than slower speed screens to produce the same film density. Any reduction in overall exposure translates into less exposure to personnel.

Maintaining strict collimation settings reduces the amount of scatter radiation. A reduction in scatter radiation means less exposure to personnel.

X-ray equipment requires periodic testing and calibration to ensure it is producing only the required amount of radiation. Failure to maintain appropriate equipment calibration often results in the need to repeat radiographs, and the potential for increased exposure to personnel always exists when radiographs must be repeated. A quality assurance program should therefore be in place for all parts of the imaging process.

Monitoring Radiation Exposure

Personnel should be provided with radiation monitoring devices to determine the amount of radiation received over a given period of time. Such devices do not afford protection. They are however, useful in assessing radiation safety practices, and the possible need to divide the number of radiographic procedures performed amongst other personnel.

Personnel monitoring devices commonly employed today are film badges (Figure 34.22) and thermoluminescent dosimeters (TLD). While each of these devices has certain advantages and disadvantages, both provide a reasonably accurate measurement of the amount of exposure received by personnel. Personnel monitoring devices are routinely employed in radiology departments. Therefore, information concerning the acquisition and proper use of these devices can be obtained by contacting administrative personnel in a radiology department.

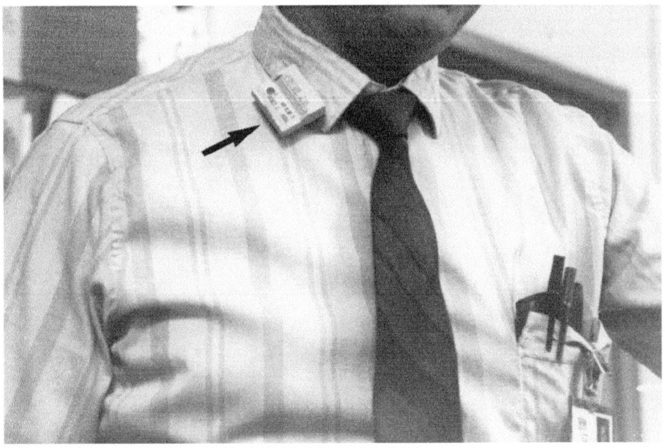

FIGURE 34.22 Film badge (arrow) used to monitor cumulative radiation exposure to personnel.

X-ray Exposure and Pregnancy

The potential to induce congenital abnormalities following exposure to ionizing radiation has been addressed countless times in textbooks and professional journals. However, it should be noted that pregnant radiographers and radiologists routinely perform radiographic examinations and give birth to normal babies. Thus, fetal abnormalities arising from radiation exposure can be avoided. With this in mind, the following considerations are offered to women who perform radiographic examinations.

If an employee knows or suspects she is pregnant, she should immediately inform her supervisor. Once notified, the supervisor will review the appropriate policies and procedures adopted by the institution regarding x-ray exposure and pregnancy.

The employee should conscientiously utilize all protective measures—time, distance, and shielding—throughout her pregnancy.

Female employees should understand that the fetus is most sensitive to ionizing radiation during the first trimester.[5] Therefore, special attention should be given to time, distance, and shielding during this period.

DEPARTMENT DESIGN AND LAYOUT

The physical layout of a radiographic installation in a medical examiner's facility, whether a mobile unit with movable barriers or a sophisticated radiography or fluoroscopy unit in a separate room or even a CT scanner, should be planned with expert advice. It is recommended that an experienced radiology facility manager and a qualified radiation physicist be consulted during the planning and design process to advice on workflow, room size and design, ancillary facilities, radiation protection, equipment selection, and maintenance and quality assurance processes. Detailed advice on the design of diagnostic imaging facilities is available from various sources[6] and this can be adapted to forensic radiology facilities. A qualified radiation physicist should approve all renovation or new construction that will house radiation-producing equipment. This is essential for protection of personnel. Furthermore, a number of states require that all x-ray installations be periodically inspected by a qualified radiation physicist or radiation health physicist. Local requirements can be ascertained by querying the State Department of Health or the State Radiation Control Commission. The administrator or a local medical radiology department is a resource for finding a radiation physicist or radiation health physicist and planning and equipping any facility.

REFERENCES

1. Beningfield, S., H. Potgieter, A. Nicol, S. van As, G. Bowie, E. Hering and E. Latti, Report on a new type of trauma full-body digital X-ray machine, *Emerg Radiol*, 10(1), 23–9, 2003.
2. Department of Health, *Facilities for Diagnostic Imaging and Interventional Radiology*, NHS Estates, London, UK, 2001, p. 176.
3. Engel-Hills, P., Radiation protection in medical imaging, *Radiography*, 12(2), 153–160, 2006.
4. Bushong, S. C., *Radiologic Science for Technologists*, 4th ed., C. V. Mosby, St. Louis, 1988, p. 535.
5. Bushong, S. C., *Radiologic Science for Technologists*, 4th ed., C. V. Mosby, St. Louis, 1988, p. 543.
6. Department of Health, *Facilities for Diagnostic Imaging and Interventional Radiology*, NHS Estates, London, UK, 2001.
7. Viner, M. D., The Role of Radiology, *in* Adams, B., Byrd, J., *Retrieval, Analysis and Identification of Co-Mingled Human Remains*. Humana Press, Totowa, NJ, 2008.

Section IX

Forensic Radiological Technology

35 Imaging in the Medical Examiner's Facility

Tania Blyth, Emily Faircloth, Gerald Conlogue, and Mark D. Viner

CONTENTS

Forensic imaging utilizes many of the basic principles that also apply to medical imaging. The fundamental principles of imaging and a primer of radiological technology for non-radiographers are detailed in Chapters 38 and 39. However, forensic imaging and, in particular, postmortem forensic imaging presents specific challenges to the radiographer which require adaptation of the basic techniques and principles. This chapter discusses the challenges presented in forensic situations and suggests possible strategies and techniques for overcoming problems encountered.

POSTMORTEM IMAGING

Protocols and procedures vary widely among facilities and may not apply to all coroner and medical examiner systems. The principles presented in this chapter are intended as guidance and should be read in conjunction with the protocols of coroner and medical examiner systems.

WHICH CASES ARE EXAMINED?

Diagnostic imaging should form an integral part of the forensic investigation and is performed for numerous reasons. Radiography is often one of the first steps of the autopsy process, providing valuable information to the investigative team. Types of cases requiring radiographic examination are penetrating injuries (e.g., knife and gunshot wounds), sudden death of infants and young children, burn victims and decomposed bodies, hyperbaric deaths, death following surgery.

Gunshot wounds and other penetrating injuries may be radiographed in order to determine the presence or absence of foreign bodies within the victim. With gunshot wounds, information such as the number of bullets within the body, type of weapon or weapons used, location of bullet and bullet jacket fragments, number of weapons used, and angle and direction of fire may be determined radiographically.

Infants and young children are radiographed to rule out signs of abuse. During the autopsy examination the head, neck, thorax, abdomen, and pelvis are examined but the extremities are left intact. As skull, rib, and long bone fractures are commonly a result of nonaccidental trauma, it is important that those anatomical regions are thoroughly examined radiographically as a supplement to autopsy. This should be done according to the recommended pediatric protocol. Leonidas,[1] recommends AP/lat skull, AP/lat thoraco-lumbar spine, AP thoracic cage, and AP extremities to include shoulders, pelvis, hands and feet. Kleinman recommends the use of high-detail, single-screen, single-emulsion systems and low kilovoltages (kVp) with images of the AP appendicular skeleton and AP/lat axial skeleton. In addition, bilateral oblique views of the thorax should be taken when fractures are suspected on the AP view.[2,3]

Coned-down radiographs may be required to clarify suspicious margins of long bone metaphysis to rule out "bucket handle" fractures.

In cases of burn victims and decomposed bodies, identifying features such as fingerprints, tattoos, and facial features may be nonexistent and radiography is often a valuable tool in the identification process. The presence of bullets and/or jackets seen on x-rays suggests a possible crime may have been committed and an attempt made to hide the evidence. Radiographs could also document antemortem dental treatment or reveal prosthetic devices, such as a hip replacement, that can aid in the identification of the individual (Figure 35.1).

PROTECTING YOURSELF ... AND THE EVIDENCE

Before beginning a forensic radiography examination, the situation must be carefully assessed and discussed with the referring medical examiner. In many cases it is necessary to examine the body or remains prior to external physical examination or autopsy. In such cases it is essential to preserve trace evidence by undertaking the examination with the body bag *unopened*. It should be noted that subsequent examinations may thus be necessary following external examination and/or autopsy with the remains removed from the bag so as to obtain true anatomic positioning.

Victims of gunshot wounds may have the hands bagged to preserve evidence. Bags should never be removed and, if

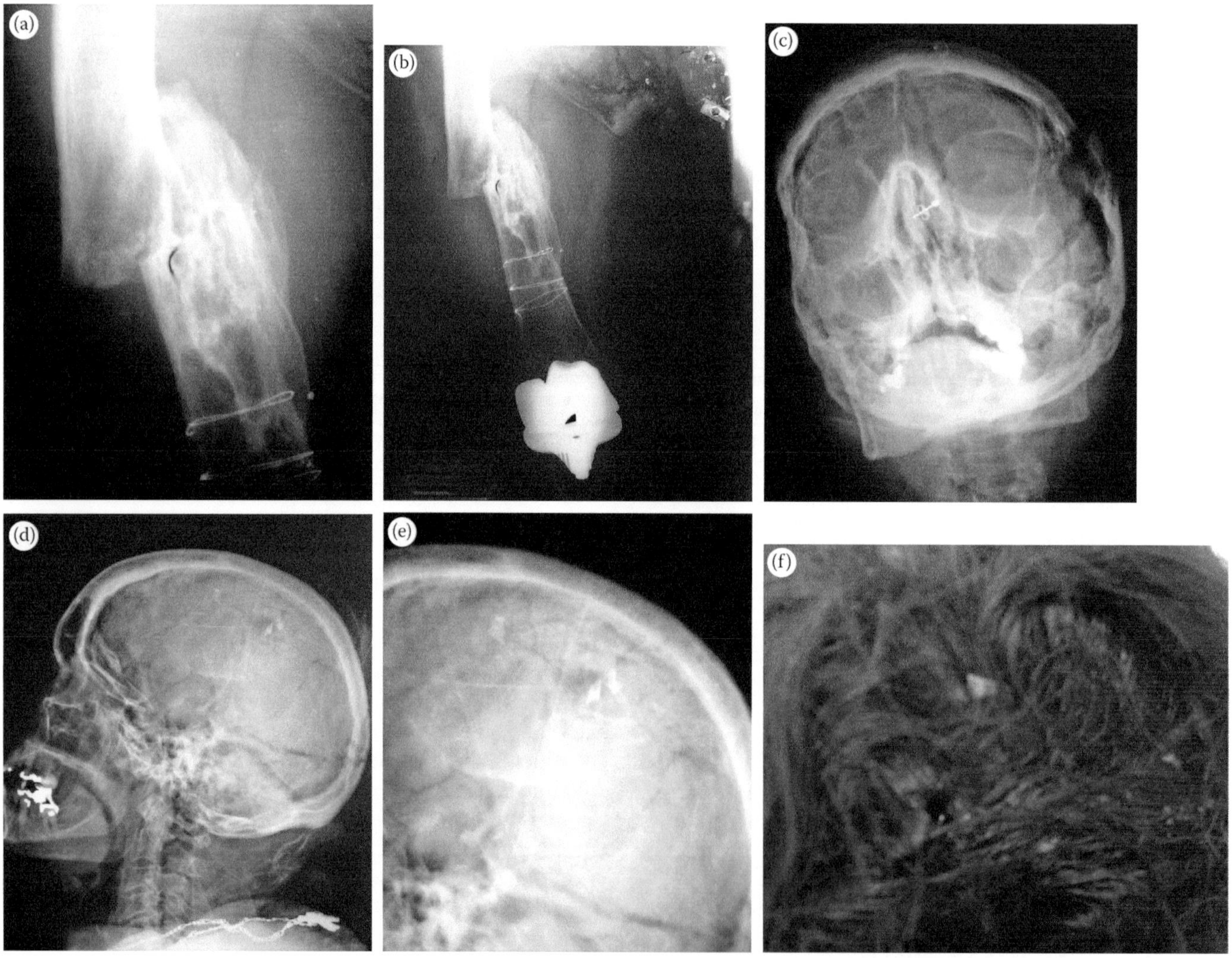

FIGURE 35.1 (a) Old fractures and (b) Prosthetic devices can be used to identify the deceased, as in the case of this burns victim who was easily identified by radiographic means. (c and d) An initial radiological survey is recommended in all cases involving firearms and will frequently enable documentation and retrieval of ballistic evidence, such as these bullet jacket fragments recorded on the skull x-ray (e [coned lateral view]) and found upon examination to lie in the hair (f). (Photos: T. Blyth. With permission.)

x-rays of the hands are necessary, radiographs should be taken through the bag to determine the presence or absence of bullets or bullet fragments. In addition, it is possible for bullets to become dislodged from the body or clothing during transfer or manipulation of the body, and care must be taken to prevent disruption or loss of evidence. In order to preserve evidence and context, clothing should not be removed and initial radiographs should be taken through the clothing.

It is important to be aware of the possible presence of objects that may injure the radiographer, pathologist, and other staff. Bullet fragments, hypodermic needles, and other sharp objects are a danger and their presence should be noted.

Full personal protective equipment (PPE) should be worn. At a minimum this should include scrubs, apron, gloves, boots, and mask. If the body is to be examined during or following autopsy, full head and arm protection and a visor or goggles are recommended (see Chapter 33). Universal precautions should always be practised and each body treated as if it were infected.

Although radiation dose to the decedent is not a concern, it is essential that the radiographer follows appropriate radiation protection practices. Even in postmortem examinations, it is necessary to ensure that the radiation dose is kept *As Low As Reasonably Achievable* (ALARA), not for the decedent, but for the person(s) performing the imaging examination. Radiation monitors must be worn by all imaging personnel and stored away from ionizing radiation when not being worn. With a fixed radiographic unit, the control panel should be positioned behind a protective barrier that will reduce the x-ray dose to zero. The exposure switch on the control panel should not have a cord that will permit the operator to extend the cord into the x-ray room during the exposure. The use of a C-arm for fluoroscopic examinations may be useful in providing a quick screening of the body for foreign objects, but it is essential that the person using the fluoroscope wears protective lead apparel, stands as far away from the radiation source as possible, and keeps the exposure time to a minimum. Personnel working nearby should practise the same safety guidelines.

During postmortem radiographic examination, under no circumstances should a radiographer hold a body. It is never appropriate or acceptable to hold a body while the exposure is taken. Positioning aids such as plastic-coated sponges and sandbags are commercially available and can be utilized in order to obtain the desired projections. The x-ray tube and image receptor can also be moved around the decedent, allowing examination with minimal manipulation of the body. This is particularly helpful during the examination of charred remains. Care must be taken so that the x-ray tube is not pointed directly at the control panel. The control panel is protected by a barrier, but the barrier is not designed to afford protection from the direct primary radiation beam (see Chapter 34).

Caution should be exercised when using tape or other restraints to position a body since the skin may be abraded or damaged. If such restraints are used, the type of restraint used and the anatomy restrained should be noted. In some circumstances and, if permitted by the facility, it may be necessary to break the rigor mortis in order to obtain the desired projections. Permission must be obtained and documentation should be made (Figure 35.2).

IMAGING STRATEGIES

With forensic imaging examinations, it is frequently the radiographer who decides which images are necessary based on the history gathered by the forensic investigators. It is important to take certain factors into consideration. With living patients, standard AP/PA, lateral, and occasionally oblique or axial views are taken. While imaging the deceased, obstacles such as the position of the body, rigor mortis, and twisted or severed anatomy may alter the process. It is desirable to produce a true frontal and lateral projection of the anatomy but this may be impossible given the circumstances and, as a result, it is often necessary to radiograph the body "as it lies."

Pediatric imaging is the exception. Standard AP, lateral, and oblique views are taken when imaging pediatric cases in order to optimize the visualization of fractures. In order to attain the standard AP/PA, oblique, and lateral projections, positioning aids such as plastic-coated positioning sponges should be utilized. A 45° angle sponge is useful for many radiographic examinations (Figure 35.2).

THE IMAGING EXAMINATION

Based on the information provided and the condition of the body, the radiographer should select a course of action and determine the necessary radiographic projections. In cases of gunshot wounds, the area of interest is examined in the frontal (usually AP) and often lateral view. It is important to keep in mind that an entry wound to the chest, for example, does not mean that the exit wound or projectile is in the chest. When a bullet enters the body, a number of things can happen: the bullet can exit one part and enter another; it can ricochet off the bone and end up in a remote site, or it can enter a blood vessel and travel to another location. It may have sufficient velocity on entry to traverse the entire chest or abdominal cavity, but not to exit through the other side. As a result the entire area including both skin surfaces, together with regions directly adjacent to the anatomy of interest, should be included. For example, if the decedent suffered a gunshot wound to the chest, the neck should be visualized on the image and radiographs of the arms and proximal abdomen should also be included in the study to determine the presence or absence of a bullet or bullet fragments. If a foreign body is demonstrated on the AP a lateral view should be taken to localize to the object. In order to determine the exact location of an object within the body, two views, preferably at 90° to each other, should be obtained. If the AP is suboptimal due to exposure, superimposition or body habitus, an additional view taken at 90° to the original should be obtained to prevent the oversight of a foreign body (Figure 35.4).

Since patient dose is not a concern in postmortem imaging, it may be beneficial to avoid collimation and keep the light field to the size of the image receptor. Often a bullet

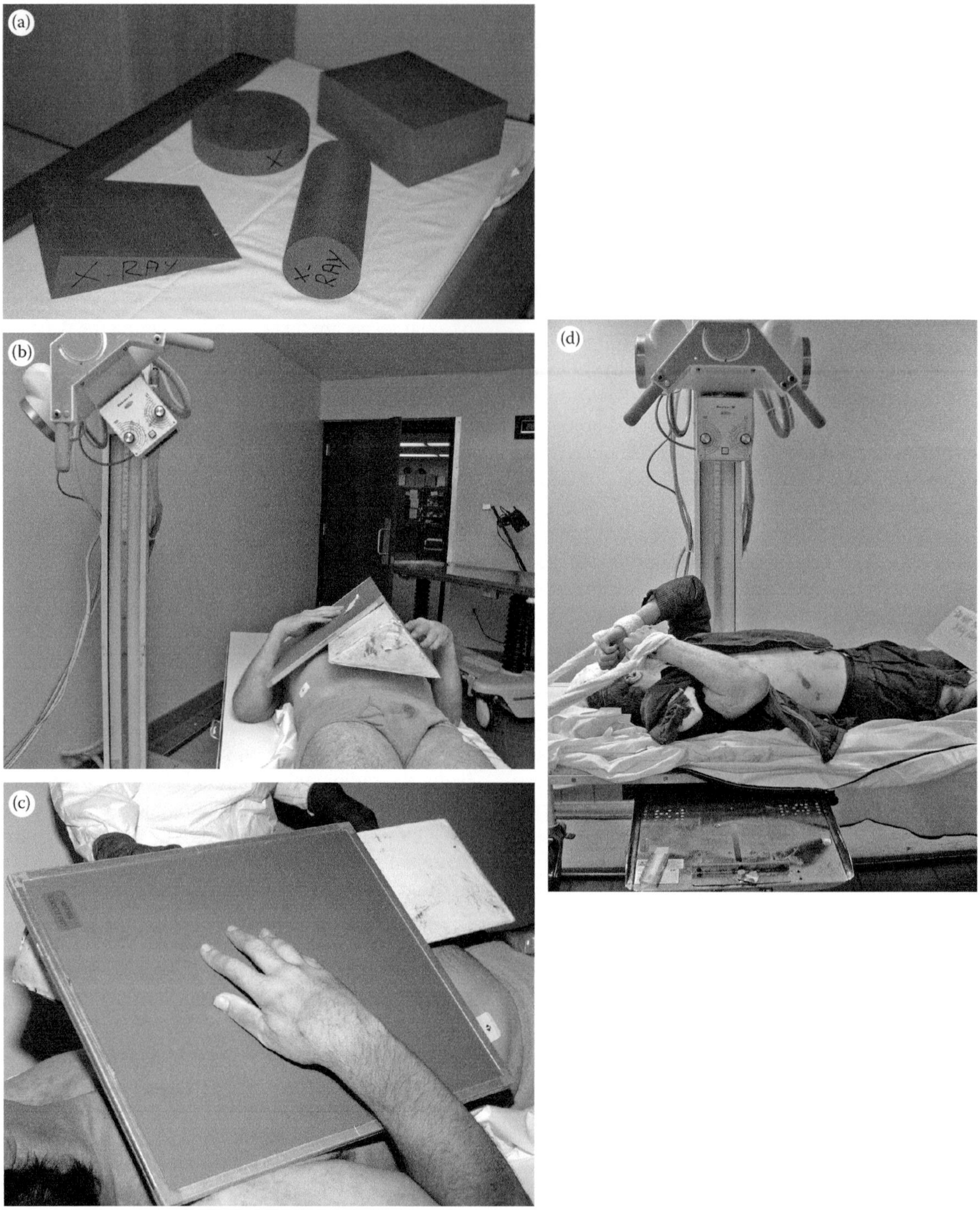

FIGURE 35.2 (a) A range of radiolucent foam pads are useful positioning aids. (From M. Viner. With permission.) (b) In this case, a 45° foam pad has been used to support the cassette against the hand of the subject in order to achieve full pronation (From Mary Sonntag. With permission.) (c) For the Dorsi-Palmar projection note how the x-ray tube has been angled accordingly so that the central ray remains perpendicular to the cassette. (From T. Blyth. With permission.) (d) When necessary, bandage straps can be used to hold the arms clear of the chest and abdomen.

may come to rest in the clothing or fall into the body bag and can easily be missed if the light field is adjusted to the size of the anatomic part. A larger cassette may be used for every body part, including even smaller anatomy such as a hand to include as much information as possible on the image. It is advisable to position the image receptor cross-wise for chest and abdomen radiography in order to include more anatomy and possibly clothing on the image, as a projectile may be located outside the soft tissue in the clothing or body bag. In addition, it is extremely important to overlap the images so no anatomy is missed on the radiographs. If fluoroscopy is used, care should be taken to examine the bag systematically

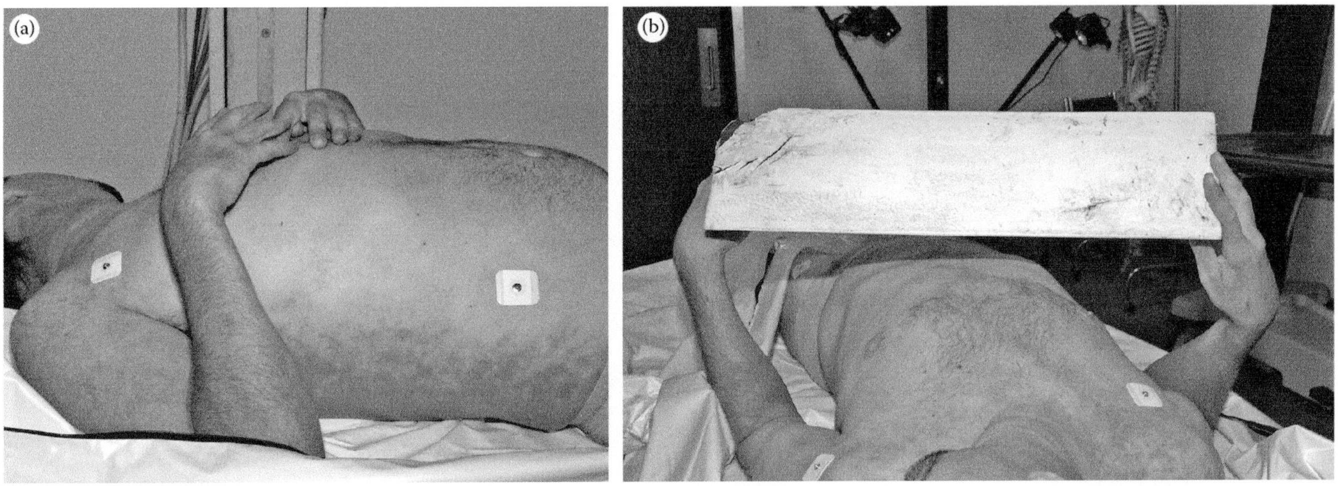

FIGURE 35.3 (a) In this case, the arms of the subject obscure the chest and are difficult to position due to rigor mortis. (b) A long radiolucent foam pad has been used to separate the arms and hold them clear of the chest, enabling a good AP projection of the chest to be obtained. (From Mary Sonntag. With permission.)

to ensure complete coverage of the bag and its contents. This is best achieved by moving the fluoroscope up and down the entire length of the bag in a series of overlapping sweeps along its length.

The relative position of the cassette and central ray may differ from the standard projections on living patients for the reasons previously stated. The central ray is often centered to the cassette rather than to a specific anatomic landmark. In addition, it is necessary to keep in mind the divergent properties of the x-ray beam, which can be used to advantage, or may be a disadvantage, depending on the circumstances. Since the x-ray beam diverges from the source, the center of the beam will consist of x-rays directed perpendicular to the object of interest. The x-rays at the periphery

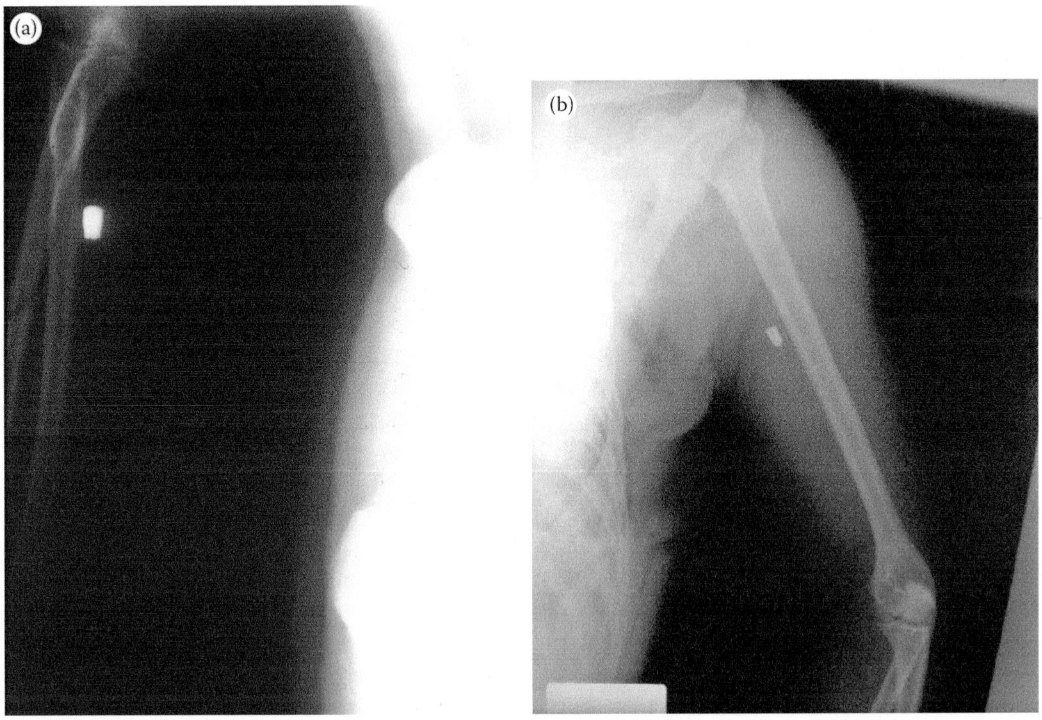

FIGURE 35.4 The importance of a thorough radiographic examination of gunshot victims is illustrated by these cases. (a) A bullet that has entered the left side of the chest cavity has exited through the right side of the abdomen and lodged in the soft tissues of the arm. (From Mary Sonntag. With permission.) (b) A bullet that has entered the chest has punctured the aorta and come to rest in the left brachial artery. (From Mary Sonntag. With permission.) (c) An AP projection of the skull shows one bullet. (From T.Blyth. With permission.) (d) A second is revealed by the lateral projection. (From T.Blyth. With permission.)

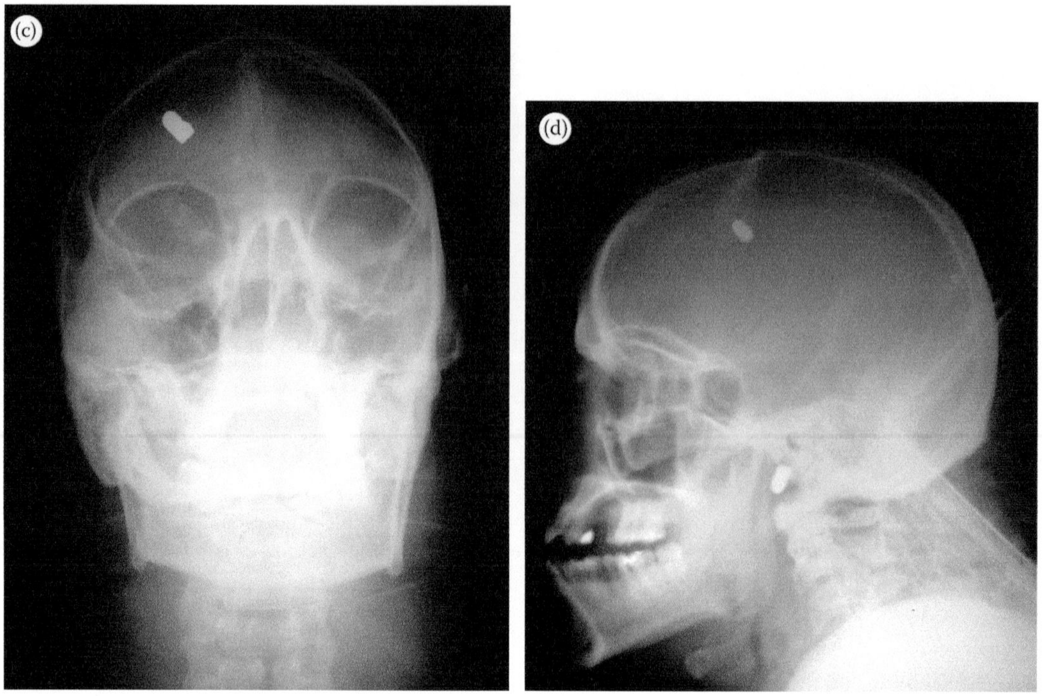

FIGURE 35.4 *(Continued)*

of the beam are directed at an angle. The anatomy located directly in the line of the central ray will be projected in its true position while the anatomy at the edges of the x-ray field will be projected at an angle, causing some distortion. This is a significant disadvantage when x-raying children for evidence of child abuse. The anatomy should be placed directly under the perpendicular portion of the x-ray beam, resulting in a true representation of the anatomy. This is one of the reasons that "babygrams" (a single large-field x-ray exposure to cover the entire body of an infant) should not be performed.[1,4] Kleinman found that, when the thorax and pelvis were imaged together, subtle pelvis fractures were more difficult to visualize.[4]

The divergent beam can be used to the radiographer's advantage while performing a cross-table lateral projection of the skull, chest, abdomen, or pelvis. It is relatively simple to elevate the head from the table surface during a cross-table lateral skull examination; however, it may be difficult to elevate the body off the x-ray table for a cross-table lateral chest, abdomen, or pelvis. A solution would be to lower the central ray so it is not centered to the anatomy, but is just slightly above the table surface. This technique will not remove all of the superimposition from the tabletop but it will minimize it. This technique in combination with the slight (1–2″) elevation of the body from the table surface will effectively remove the superimposition and demonstrate the posterior soft tissues where a projectile may be located (Figure 35.5). If this method is used, it is important to remember to position the stationary grid so the lead strips are perpendicular to the floor to prevent grid cut-off. When using a stationary grid, it is also important to keep the source-to-image distance within the focal range recommended by the manufacturer.

OPTIMIZING RADIOGRAPHIC CONTRAST AND DENSITY

In order to optimize radiographic contrast and density, an appropriate balance between kVp and milliamps (mAs) must be used. Low kVp produces images with high contrast, whereas high kVp settings result in images with low contrast. Low kVp results in increased dose since more radiation is being absorbed by the body and, as a result, mAs would need to be increased in order to provide adequate density on the radiograph. With postmortem imaging, patient dose is not a concern and therefore low kVp settings in combination with higher mAs may be used, resulting in images with increased contrast. High contrast is particularly desirable while imaging postmortem pediatric cases in order to demonstrate subtle metaphyseal fractures. As previously mentioned, it may be advantageous to open the collimation in order to demonstrate the clothing in addition to the anatomy. Decreased collimation results in more scatter and a loss of contrast. Using lower kVp will aid in improving the contrast when the x-ray beam is not collimated.

If the radiographic unit is not equipped with automatic exposure control (AEC), technique charts should be used in order to produce consistent results and minimize repeats. Technique charts list the anatomy to be imaged, appropriate kVp, mAs, source-to-image distance (SID), and specify whether or not a grid should be used. Technical factors required vary depending on the size of the anatomical part

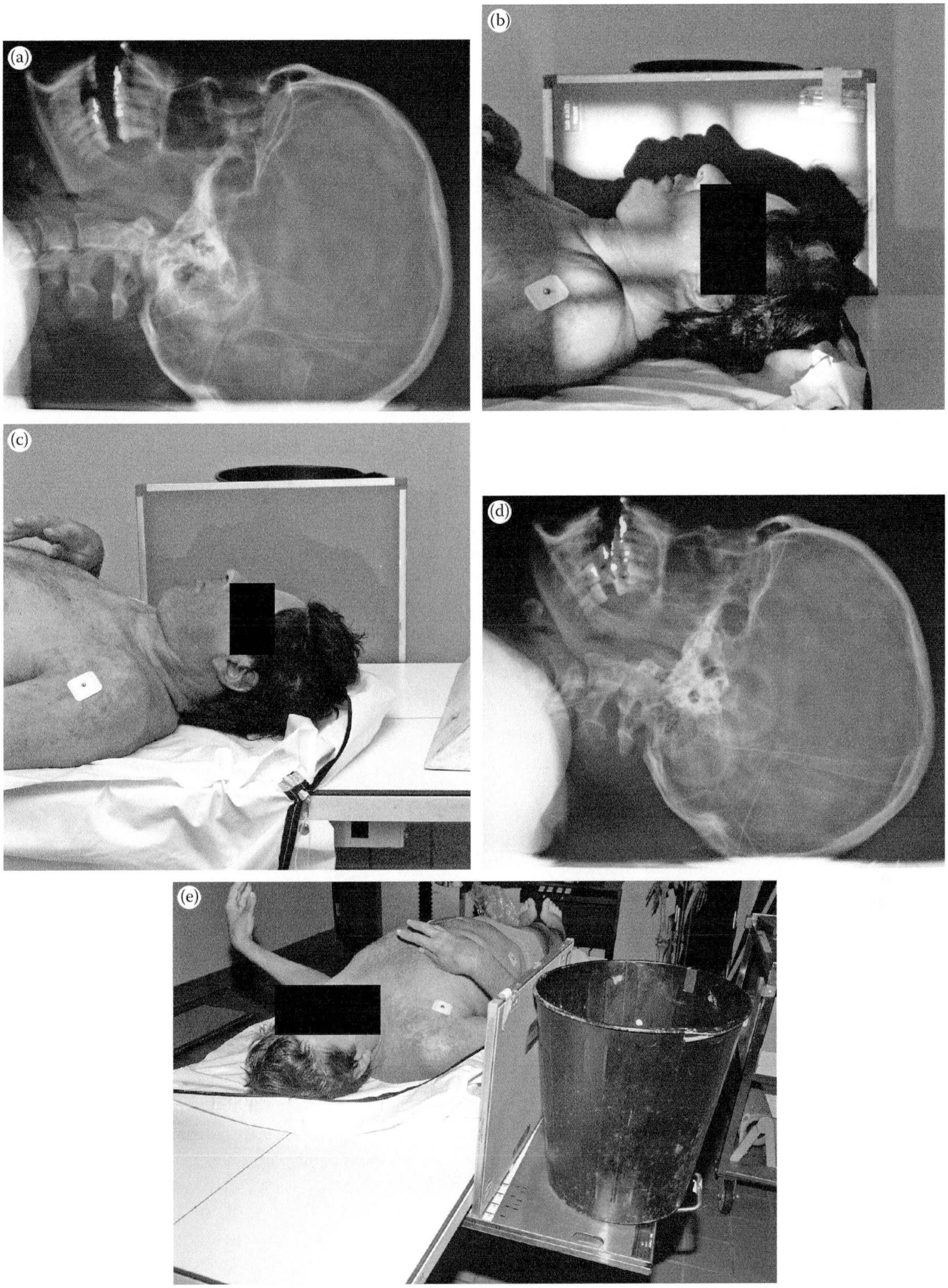

FIGURE 35.5 (a)–(e) demonstrate how the divergent beam can be used to minimize superimposition from the tabletop in cases when it is not possible to elevate the subject from the table using a mattress or foam pad. (For demonstration purposes a horizontal beam lateral skull projection is illustrated, but the principle applies equally to other projections such as lateral chest and abdomen.) (a) Shows the superimposition of the table top along the occipital portion of the skull vault when (b) the central ray is centered to the skull in the normal manner. (c) If the central ray is centered to skim the surface of the table top, the divergent beam projects the skull upwards slightly (note the shadow of the facial features on the cassette when compared to (b). (d) The resultant radiograph shows that the superimposition of the tabletop has been minimized so as to just clear the inner table of the skull vault. (For Figures 35.5a–d; From Mary Sonntag. With permission.) (e) The bucky tray can be used to support the cassette in the lateral position when held in place by a light but rigid object.

being radiographed. Techniques for measuring the relative thickness of body parts may assist in determining the optimum exposure. Calipers may be used for this purpose but in many cases exposure factors can be determined according to the size of the body (small, medium, large, or very large) and this method is sufficient under these circumstances.

PATHOLOGY

While imaging the decedent, the effects of some pathologic changes should be considered in the quest to obtain an optimal image. Often with gunshot wounds, particularly to the chest or abdomen, there is massive blood loss into the thoracic or abdominal cavities, affecting image quality. With increased fluid within the cavity, radiographic density may be insufficient and exposure factors should be increased. In order to maintain adequate radiographic contrast, 80–85 kVp is sufficient for a frontal view of the chest or abdomen, and mAs can be increased to provide adequate density throughout the image. With medical imaging on live patients, a high kVp is often desired to demonstrate lung parenchyma and pulmonary markings. However, with postmortem radiography, rib fractures and projectiles are of concern and excessive kVp may inhibit the visualization of a bullet jacket which is more radiolucent than a lead bullet. In order to optimize the visualization of rib fractures on deceased infants and children, a low kVp of 55–65 should be used.

Postmortem radiography varies drastically from medical imaging when imaging decomposed bodies and charred remains. Depending on the stages of decomposition, exposure factors may be very similar to those used for imaging of live patients or may be drastically different in cases of skeletonized and mummified remains. In the very early stages of decomposition, minimal changes are seen in the body and the exposure factors used would be similar to those required on a live patient. As the body begins to decompose, gases escape into the subcutaneous spaces and body cavities. The body will appear bloated and very large. Caution must be exercised when imaging these bodies because the size of the body is not due to increased fat or muscle mass, but increased gas which is easily penetrated by x-rays. When using the "body size method" to determine exposure in such cases, it is advisable to select one size less than that indicated such that, for example, an individual assessed as "very large" is given the average exposure for a large individual. With extensive decomposition and tissue breakdown, exposure factors will need to be decreased further. With minimal tissue breakdown, a decrease in kVp may be appropriate but with extensive decomposition a decrease in both kVp and mAs will be necessary. With skeletonized and mummified remains, low kVp should be used due to the absence of hydrated tissue. In addition, as a result of the decreased tissue thickness, low mAs will also be necessary.

Similarly, the imaging of decomposed or charred bodies may not require a significant change in exposure factors. Although the outside of the body is severely charred, if the body cavity remains intact, the organs are often well-preserved inside. In the few cases of severely burned bodies, only a slight decrease in kVp may be required in order to achieve the appropriate density.

IMAGE PROCESSING CONSIDERATIONS

If conventional imaging techniques (film and chemistry) are used, it is important that the volume of examinations performed at a facility matches the type of processor used. For example, a low-volume facility would benefit from a small tabletop processor with proportionally small chemical replenisher tanks, compared to a large processor with very large tanks, because the few films being processed do not require as much replenishment of the chemistry, resulting in developer oxidation. Oxidation of the developer chemistry causes a decrease in density and contrast affecting the visualization of recorded detail on the image. Care must be taken to ensure that the proper chemical is added to the appropriate mixing tank. A quality assurance program is not required by law and processors may not be regularly maintained. It is important to be mindful of signs of chemical contamination or exhaustion. When changes are visualized in radiographic contrast or density, the processor is often the culprit and processor malfunction should not be ruled out.

THE RADIOGRAPHER OR TECHNICIAN

The person performing the imaging examination is an important factor in the radiological investigation of the deceased. For reasons of image quality and consistency, safe use of equipment and radiation safety, it is desirable that all forensic radiography examinations of human subjects are carried out by licensed radiographers (see Chapter 33). However, licensed radiographers are not always available and in some jurisdictions they are not required by law for a postmortem imaging investigation. Thus, it is often the autopsy technologist or diener who performs the radiographic procedure. Autopsy technologists have generally learned on the job and have neither the in-depth education nor the experience of a licensed radiographer. They certainly do not possess the skills to achieve optimal image quality in the range of situations that occur in the forensic setting. The autopsy technologist will not have been trained in radiation exposure and processing and usually does not have the ability or experience to select the appropriate image receptor, grid, SID, and focal spot, or to manipulate the exposure factors while changing grids, distance, or image receptor speeds. In addition, it is not unusual for the pathologist to rely on the radiographer for an opinion regarding interpretation. The importance of providing optimal images prior to autopsy cannot be stressed enough and suboptimal images should be repeated. In order to demonstrate the body in its original context, it is necessary to obtain the required images prior to autopsy. Once the autopsy has been performed and the body is buried or cremated, important evidence may be lost or destroyed.

TYPE OF EQUIPMENT AVAILABLE

The type of equipment available for forensic imaging examinations differs drastically from that of many medical imaging facilities. Due to limited funding, medical examiner and coroner offices may rely on antiquated x-ray machines with little or no ancillary equipment such as grids or positioning aids. Regardless of the level of funding, the consultation of a licensed radiographer would prove beneficial in acquisition of the appropriate equipment and may actually minimize the expense. Equipment is discussed in more detail in Chapter 34.

X-ray machines with " bells and whistles" are usually not required for postmortem examinations. A single-phase generator is the least expensive and generally sufficient for postmortem x-ray examinations. Higher mA settings required to produce the extremely short exposure times necessary to eliminate motion in the live patient are not a factor in forensic radiography and, therefore, lower output equipment is more than adequate. If high mA values are necessary for a particular projection, such as a lateral abdomen on a large individual, multiple exposures can be made until the desired density is achieved. The height and length of the table should be considered when choosing radiographic equipment. An adjustable table height aids the radiographer in transferring the body from the gurney cart to the x-ray table and back. X-ray tables are available in various lengths and it is important that the table is long enough to accommodate long bodies. If the x-ray table is too short, the radiographer will need to slide the body along the table so the legs are hanging off the tabletop in order to image the skull. Although very convenient, it is not necessary for the x-ray unit to be equipped with automatic exposure control (AEC). AEC is used to achieve optimal and consistent density on the radiograph. Although it is extremely helpful and takes the guesswork out of exposure factor selection, accurate technique charts can be substituted and, if used properly, can produce similar outcomes. Various types of recording media are available to capture and display the image. Conventional screen/film systems with automatic processing have been successfully utilized for years and provide excellent contrast and spatial resolution, which is particularly desirable for the imaging of pediatrics. Image receptor speeds (or RSI—relative speed index) vary. Screen speeds of 400 and 100 are often used in medical imaging and are readily available for purchase. To maximize spatial resolution, single-screen systems with single-emulsion film may be used and offer a relative speed of 50 RSI. This screen/film combination results in superior recorded detail and may be desired for pediatric and specimen imaging.

Digital imaging systems are now widely available and provide digital images quickly and without the need for a darkroom. Advantages of digital imaging systems include rapid image acquisition and display compared to conventional radiography. Computed radiography (CR) employs imaging plates (IPs) which look similar to cassettes and can be used with the existing radiographic unit. A CR "reader" replaces the darkroom and automatic processor. Direct Digital Radiography (DR) utilizes a detector which is in communication either by a direct (cord) or a wireless connection to the computer and display units. Digital imaging provides faster image display than CR and allows the image to be evaluated without removing the detector from beneath the subject for image processing. It is thus considerably faster and minimizes lifting and handling. It should however be noted that the DR detector is delicate and, if dropped or improperly handled, may be damaged and need to be replaced and the equipment will be temporarily out of service On the other hand, if a CR imaging plate is dropped and broken, it can be replaced relatively inexpensively and the system continues to function while using other IPs. DR units that utilize a cord present additional trip hazards or may become damaged due to inappropriate handling. Image plates, detectors, and cords will also need to be suitably protected from contamination with blood or other bodily fluids.

Troubleshooting

In a perfect world, the equipment works properly every time but this is, of course, not always the case. These situations where equipment fails and radiographs must still be taken are circumstances where forensic radiology differs drastically from medical imaging. With live patients, many of the solutions to the problems described below would be prohibited as a result of increased radiation exposure to the patient. However, postmortem imaging allows for the modification of some of the radiation safety practices designed to limit patient dose, enabling the production of acceptable images during equipment malfunction. As a result of time or logistical constraints and/or the presence of police detectives in the autopsy room awaiting the completion of the x-ray exam and the start of the autopsy, it is often necessary to adopt the mentality of "the show must go on." The imaging examination must be completed quickly and efficiently before the autopsy begins. Some equipment problems result in complete cessation of x-ray exposure but fortunately such situations are rare. More often, equipment problems are minor and exposures can still be taken but with slight modifications.

In circumstances where the field light bulb is burned out, a plumb line can be used to determine where the x-ray beam is centered (Figure 35.6). If the x-ray generator or timer malfunctions and radiation output is decreased, multiple exposures may be taken in order to achieve the desired density. When performing screen–film radiography, the spectral emission of the screen must match the spectral sensitivity of the film or the image density will be reduced. During postmortem radiography, since dose to the body is not a concern, an increase in the mAs can be used to achieve the desired density.

As with some x-ray unit malfunctions, minor processing faults may also be circumvented by adjusting the exposure factors in order to obtain adequate density. If the processor is to blame for increases or decreases in density, the mAs can be adjusted to produce the desired amount of radiographic density. If the developer chemistry is beginning to become exhausted, an increase in the mAs will increase the density

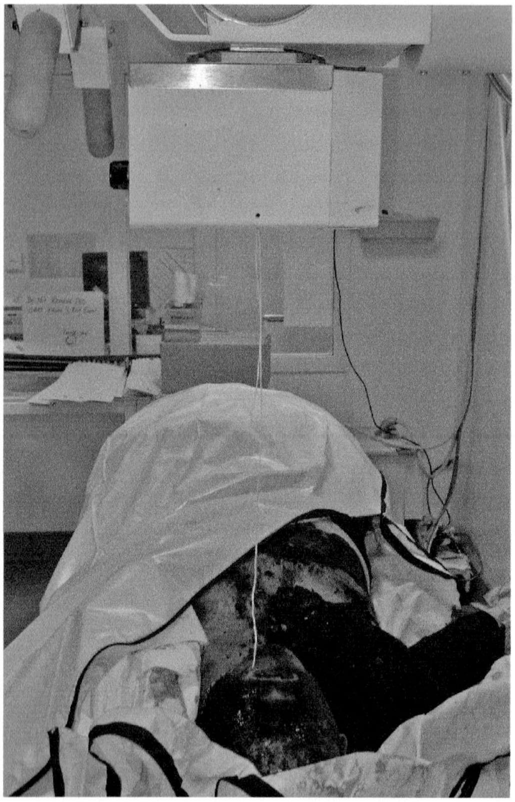

FIGURE 35.6 A plumb line can be used to determine the central ray if the light beam bulb fails.

on the radiograph. If an image must be taken before the processor is adequately warmed up, the mAs may also be increased to produce the desired density; conversely, if the developer temperature is too high, the mAs may be decreased in order to decrease the density on the radiograph. It should be clarified that these adaptations are appropriate for foreign body localization or for determining the presence of hardware or fillings for identification but should not be used in the postmortem examination of infants and young children as radiographic contrast may be compromised. One shortcut used for screen–film systems, which is appropriate for both medical and postmortem radiography, is a method of quickly and inexpensively "seasoning" the developer chemistry. When the developer solution is changed, a starter solution must be added to "season" the chemistry. If starter solution is unavailable, ten consecutive 14 × 17″ films may be run through the processor to provide a comparable effect.

If a grid is not available, the "air-gap" technique may be employed in order to direct scattered radiation away from the image receptor. With the "air-gap" technique, an image receptor is placed 8–12″ from the anatomy of interest, increasing the object-to-image distance (OID) and preventing much of the scattered radiation from reaching the film. The increase in OID also results in an increase in magnification; therefore, an increase in source to image distance (SID) should be utilized to compensate.

In conclusion, many similarities and differences exist between imaging live patients in a medical setting and imaging the deceased in a medical examiner facility. One of the major differences is radiation safety. Patient protection is of enormous concern when imaging live patients, but is not an issue with postmortem radiography; as a result, radiographic positioning, exposure, and protocols can be modified to suit the situation presented to the radiographer during postmortem imaging studies. Postmortem radiography introduces a new array of obstacles that need to be overcome, such as the position of the body, the need to preserve evidence, presence of potentially dangerous objects and often antiquated x-ray machines and limited ancillary equipment. Although not always required by law, it is in the facility's best interest to have a trained, qualified radiographer performing the examinations, ensuring optimal image quality and providing the best possible images under often difficult circumstances.

DENTAL RADIOGRAPHY

Dental Radiography and Postmortem Identification

The dental radiograph is one of the most valuable diagnostic tools available in the practice of dentistry[5] and radiography is frequently used to complement the odontological (or clinical) examination of the teeth, which is considered a primary method of identification of the deceased. Radiography effectively demonstrates dental anatomy, anomaly, pathology, and subsequent dental treatment, if present (Figure 35.7).

The success of dental radiography in postmortem identification relies on the existence of antemortem radiographs or dental records for comparison, but remains a highly accurate way of determining identification of unknown persons. A study by Pretty et al in 2003 reported that experienced dentists accurately identified/excluded individuals through digitized dental radiographs in 91% of cases.[6]

Radiography is a minimally invasive technique providing an objective, permanent record of the dental examination, making it an ideal tool in the forensic or fatality investigation. Dental x-ray imaging equipment is comparatively cost effective, and the rapid acquisition of images is further advantageous for its use within the mortuary.

There are, however, some precautions that need to be taken in the use of dental x-rays, related mainly to the risk of radiation exposure for operating personnel and for those working in the immediate vicinity of the examination area. International regulations relating to the use of x-ray to examine the teeth will vary slightly but all have the common principle that the resultant dose from each x-ray exposure should remain as low as reasonably achievable. The practical application of good radiation protection in dental radiography is described later in this chapter.

Training is required in dental radiography techniques to acquire good quality radiographic images, and is thus operator dependent. Dentists, dental nurses, hygienists, and radiographers or radiological technologists with additional forensic training are the most appropriate personnel to undertake these examinations.

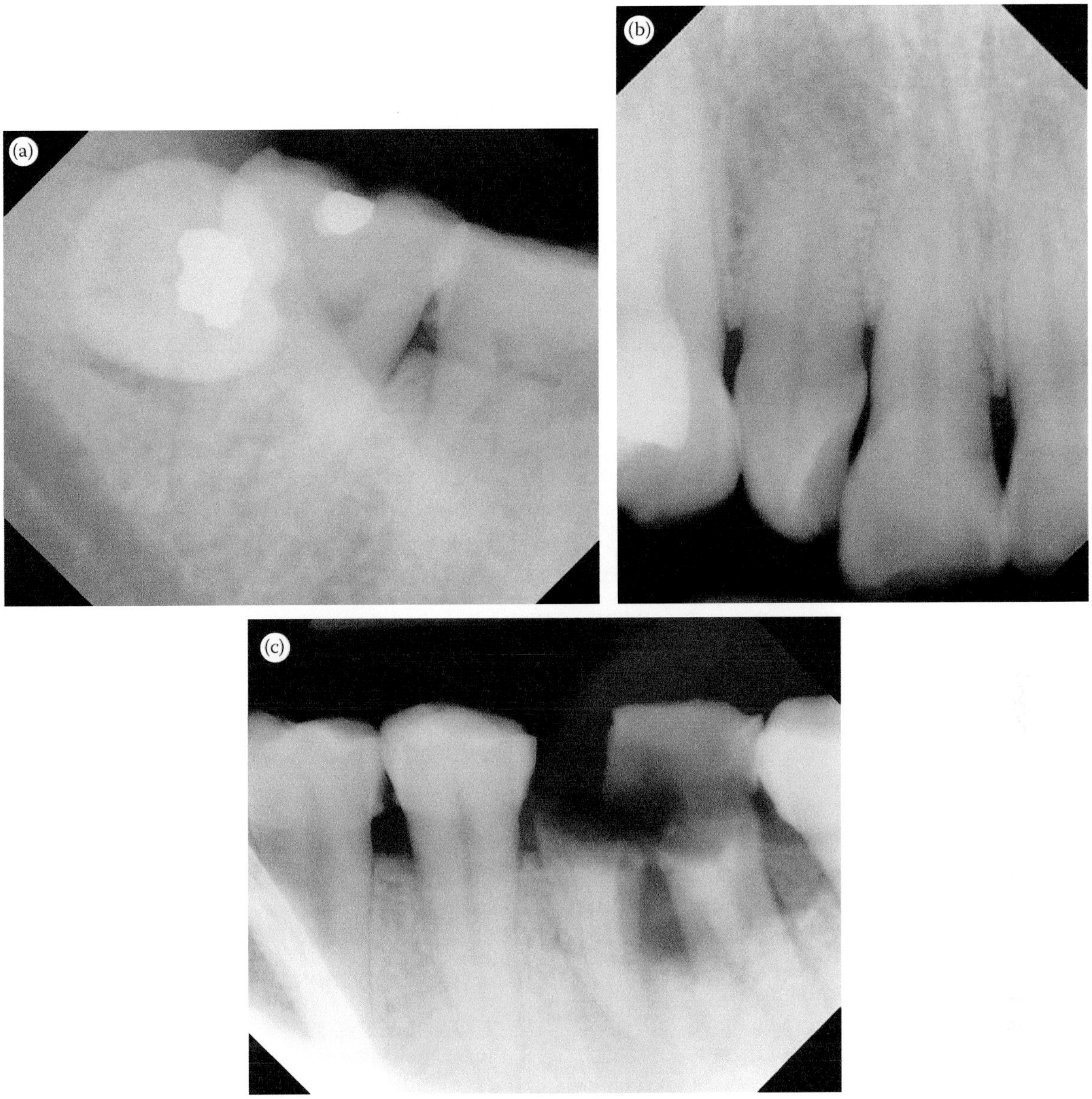

FIGURE 35.7 (a) A postmortem intra-oral image demonstrating impaction of the right lower third molar and restorative amalgams (fillings). (b) A postmortem intra-oral image demonstrating marked rotation of the right upper lateral incisor. (c) A postmortem intra-oral image, extensive destruction of the left lower first molar and surrounding area. (Courtesy of Kenyon International Emergency Services. With permission.)

Intra-oral Imaging Equipment

Intra-oral (or periapical) radiography techniques involve the image receptor (direct or non-direct digital sensors, charged-coupling device or film packet) being placed inside the oral cavity for imaging the dentition.

The following basic kit is recommended for postmortem intra-oral (periapical) imaging of the teeth:

- A portable dental x-ray tube
- Intra-oral x-ray film/digital image receptor or sensor
- Dental processing equipment (if using radiographic film)
- Film/sensor holders for intra-oral and bitewing radiography (for paralleling technique)
- Plastic sheaths for film/sensors or cling film/plastic wrap (to prevent cross contamination)
- Accessory positioning aids; e.g.; sponge wedges, cotton roll, tissue, water bottles
- Surgical tape, plastic-based (Transpore™)
- Small surgical forceps or tweezers (for retrieving dislodged films, positioning aids, and other debris)
- A pen torch
- A supply of water
- Disposable towels.

Other items found around the mortuary make effective positioning aids provided the basic principles of dental radiography techniques are applied. Empty disposable glove boxes, autopsy sponges, and head blocks are helpful in

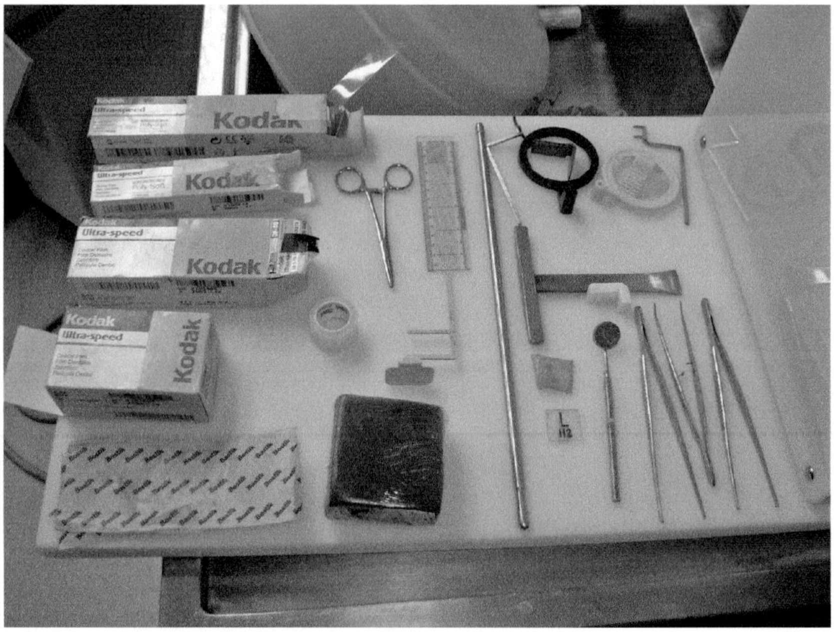

FIGURE 35.8 Postmortem dental radiography kit. (From M. Viner. With permission.)

securing the imaging equipment and the position of the area under examination for the duration of the x-ray exposure Figure 35.8).

Developments in Dental Radiography

The availability of portable x-ray units means that the radiography examination of postmortem dentition can be carried out in a wide range of settings including the mortuary/morgue or at the site of a mass fatality investigation. Small desktop film processors with daylight film-processing capabilities also allow for the provision of dental radiography in limited spaces. The simple technical operation of these types of processors make them straightforward to fix in case of mechanical failure but requirements of water, electrical facilities, and potential hazardous chemicals have to be factored in when planning the mass fatality morgue.

The first intraoral digital sensors for use in dentistry were introduced in the 1980s[7] and their use in mass fatality investigations has significantly reduced examination times in comparison to conventional film radiography.

Direct digital radiography (DR) is the imaging method of choice for postmortem dental radiography as the sensors tend to be fairly robust for the purpose. Sensor-holding equipment is usually included with digital imaging packages and the software can be downloaded onto a laptop computer for added portability.

Teleradiology technology means that x-ray images can be archived in various digital formats and electronically sent to colleagues for review via a secure network. The digital manipulation of images has proved a popular viewing/diagnostic tool and although concerns were initially raised regarding the medico-legal integrity of manipulated images,[8] dental x-ray software now incorporates ways of tracing original images and of changes made subsequently. Digital radiography also has several other advantages, and some disadvantages, compared with conventional x-ray film (Table 35.1).

The advent of highly portable, hand-held x-ray units has contributed to the truly compact nature of dental radiography provision. Images can now be acquired, processed, and shared using equipment no bigger than the size of a small carry-on suitcase.

Radiation Protection

Dental imaging uses ionizing radiation to produce images of the teeth and surrounding areas. The cumulative biological effects of radiation are well documented, and there are international norms that codify occupational dose limits to prevent over-exposure of ancillary staff working in the vicinity of the dental x-ray examination. Those directly involved with the examination, that is, the operator, should never incur a dose higher that 1 mSV/annum.[9]

When providing dental radiography services in the forensic setting, the primary responsibility of the operator/radiographer/radiological technician is to ensure that the risk to other personnel working in the area is reduced to as low as possible by maintaining integrity of the radiation control area (Figure 35.9). The radiation control area is defined as the equidistant boundary around the anode of the dental x-ray tube and the patient which exists during times of x-ray exposure. For intraoral radiography this boundary should be no less than 1.5 m in all directions and all personnel should be clear of this area during the x-ray exposure. Prior to the x-ray exposure the radiographer should ensure that all dental x-ray equipment is in full working order and that it is maintained and routinely tested as appropriate. All mortuary personnel should be briefed on the presence of x-rays in the area and

TABLE 35.1
Table Showing the Advantages and Disadvantages of Film and Digital Radiography Methods

Conventional Film Radiography (Dental)	Digital Radiography
Disadvantages	
Darkroom processing of films is required Health and safety issues (processing chemicals are required) Additional space and plumbing is required Increase in examination time, images take between 3 and 9 minutes to develop	Higher initial costs
Ongoing costs for processing chemicals and film, etc.	Sensor image size is 2–4 mm smaller that that of conventional film, resulting in smaller field of view area
Ongoing cost and additional processing required for duplication of images Higher radiation dose	There may be compatibility issues with various models of digital imaging equipment
Advantages	
Lower initial costs	Lower radiation dose
Imaging area is equal to the dimension of the film resulting in an increased field of view area	Direct (instantaneous) image acquisition resulting in faster examination times
Equipment tends to be less technically complex and easily repairable	Little or no ongoing costs (possible software upgrades required) Lower image duplication costs, images can be archived in DICOM, .JPEG, BMP formats and shared if needed.

precautions to take for personal protection during an exposure. Care should also be taken, as standard practice, to prevent the need for repetition of images of the same area on account of technical error.

During the dental x-ray exposure, in addition to clearing the surrounding area of all personnel as mentioned earlier, the primary x-ray beam should be directed away from personnel and towards a solid wall. Additional lead shielding should be used if this is not possible. Finally, an audible warning should be given by the operator/radiographer/radiography technician to alert nearby mortuary personnel that an x-ray exposure is about to take place to prevent anyone from inadvertently walking into the radiation controlled area.

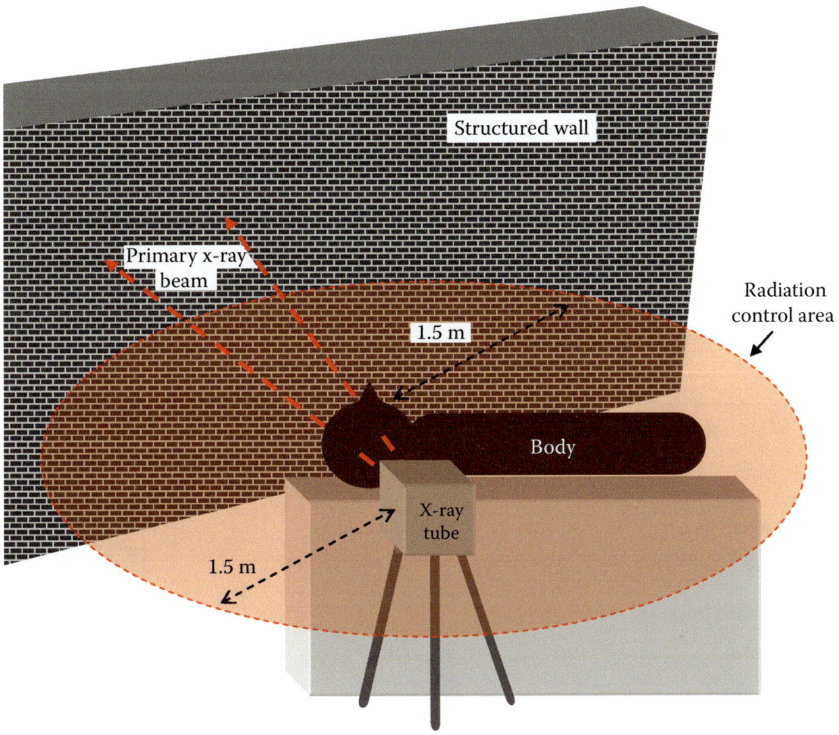

FIGURE 35.9 The radiation controlled area (dental radiography). (From E. Faircloth. With permission.)

Handheld dental x-ray tubes are a new development in x-ray technology and are of value in radiography of the deceased, particularly during mass fatalities. A tripod should be used whenever possible to stabilize the x-ray tube during an exposure avoiding the need for it to be held in position. Where this is not possible, the use of a lead jacket is recommended to prevent the accumulation of unnecessary dose to the operator when multiple x-ray exposure is being carried out.

Imaging Technique

Ideally postmortem dental radiographs should replicate any antemortem films available, using the same technique and x-ray tube angulations. Comparing "like for like" assists the odontologist in the reconciliation process of confirming a positive identification or exclusion. Therefore, postmortem dental images do not always reflect the best technique but should reproduce the appearance of the antemortem image.

Where the antemortem radiographs are not available at the time of the dental examination, it is recommended to obtain a full dental survey; that is a set of intraoral (periapical) radiographs to include all dentition together with bilateral bitewing radiographs. Edentulous areas should also be imaged to demonstrate any retained teeth/tooth roots. This applies particularly to the lower and upper third molars, which can often be impacted and not visible during dental inspection. In mass fatalities, the time available may not allow for full dental surveys, and the survey may have to be adapted to include bilateral bitewing radiographs of the molar regions and intra-oral radiographs, the third molar region being a common site for anatomic variance and restoration.

Provided x-ray facilities are available at the site of the postmortem examination, resection of the mandible and maxillae should not be necessary for the purpose of x-ray examination. Dental radiography on full bodies with the jaws *in situ* is often more straightforward, in comparison with partially or completely disassociated remains. Once oral access has been achieved by the odontologist, image receptors can be inserted and clinical examination undertaken with relative ease. A round water container or similar round-shaped article placed under the shoulders to extend the neck may further assist in providing access to the teeth under examination. The internal soft structures of the mouth (i.e., the tongue) can be both a challenge to maneuver but also an advantage in supporting the position of the holder while the x-ray is being taken. Specific techniques for radiography of the teeth are described in Chapter 39.

Documentation

The importance of documentation cannot be emphasized enough in any forensic investigation, and the same applies to the dental radiography examination. The chain of evidence relates to the audit trail of processes, circumstances, and custody involved with any physical evidence produced as part of a forensic investigation (Figure 35.10). The integrity of the evidence could potentially be questioned if thorough documentary procedures have not been followed. Notes should

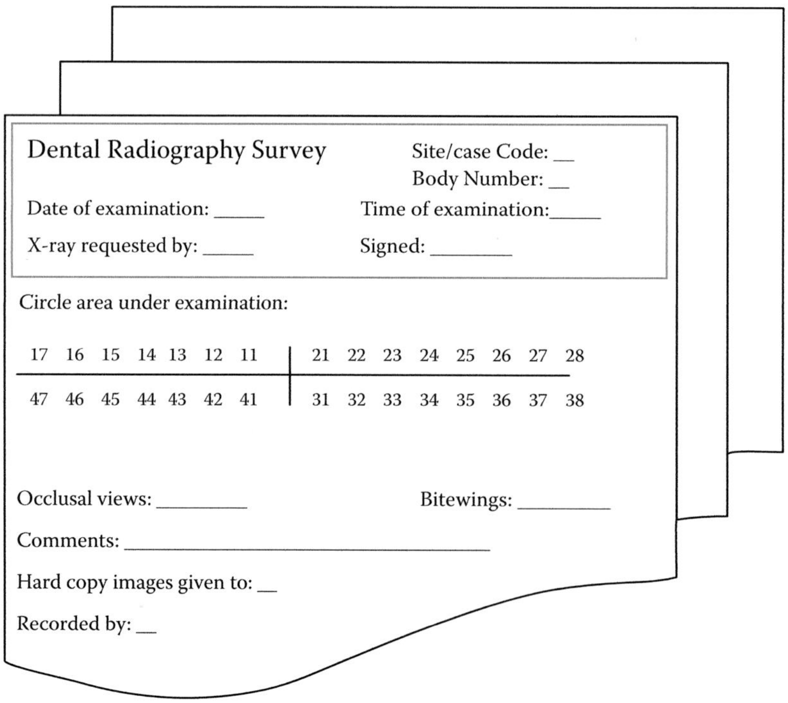

Dental Radiography Survey

Site/case Code: __
Body Number: __

Date of examination: _____
Time of examination:_____

X-ray requested by: _____
Signed: ________

Circle area under examination:

17	16	15	14	13	12	11	21	22	23	24	25	26	27	28
47	46	45	44	43	42	41	31	32	33	34	35	36	37	38

Occlusal views: _________
Bitewings: _________

Comments: ______________________________

Hard copy images given to: __

Recorded by: __

FIGURE 35.10 Recommended notes for the forensic dental radiography examination. (From E. Faircloth. With permission.)

include the date and time of the examination, who it was performed by, and the identity of the odontologist/forensic dentist. Radiographic views should be specified with the number of resultant images produced together with any additional comments about the examination. All notes should be signed and dated, and should clearly state the ongoing chain of custody, that is, who any hard copy images of the examination were given to.

SUMMARY

Dental radiography has evolved significantly over the last 10–15 years with the development of digital imaging and compact, highly portable x-ray tubes. Financial and logistical factors will determine the imaging equipment resources available but with social and political pressure for rapid identification of the deceased, particularly following mass fatality incidents, digital imaging technology is the preferred option for dental radiography of the deceased. Despite these technological changes, the basic principles of dental radiography have remained the same, with techniques being adapted to the forensic setting to provide images comparable to the antemortem odontological assessment for identification purposes.

REFERENCES

1. Leonidas, J. C., Skeletal trauma in the child abuse syndrome, *Pediatric Ann.*, 12, 875–881, 1983.
2. Kleinman, P. K., Marks, S. C., Nimkin, K., Rayder, S. M., and Kessler, S. C., Rib fractures in 31 abused infants: Postmortem radiologic-histopathologic study, *Radiology*, 200, 807–810, 1996.
3. Kleinman, P. K., Marks, S. C., Richmond, J. M., and Blackbourne, B. D., Inflicted skeletal injury: A post-mortem radiologic-histo-pathologic study in 31 infants, *AJR*, 165, 647–650, 1995.
4. Kleinman, P. K., The lower extremity, in Kleinman, P. K., Ed., *Diagnostic Imaging of Child Abuse*, 2nd ed., Mosby, St. Louis, MO, pp. 26–71, 1998.
5. Mason, R. and Bourne, S. A., *Guide to Dental Radiography*, 4th ed., Oxford University Press, Oxford, UK, 1998.
6. Pretty, I. A., Pretty, R. J., Rothwell, B. R., and Sweet, D., The reliability of digitized radiographs for dental identification: A web-based study, *J. Forensic Sci.*, 48 (6), 1325–1330, 2003.
7. Brennan, J., An introduction to digital radiography in dentistry, *J. Orthodont.*, 29, (1), 66–69, 2002.
8. Horner, K., Brettle, D. S., and Rushton, V. E., The potential medico-legal implications of computed radiography, *Br. Dental J.*, 180, 271–273, 1996.
9. Whaites, E., *Essentials of Dental Radiography and Radiology*, 3rd ed., Churchill Livingstone, pp. 53, 2002.

36 Imaging in the Field

Gerald Conlogue, Mark D. Viner, and Nancy Adams

CONTENTS

INTRODUCTION

The incredible advances made over the past 10–15 years in the field of medical radiography are indisputable. Hardware innovations, such as multidetector computed tomography (MDCT), have made it possible to collect data for a volume of tissue and section it to display images of whatever plane or planes are desired. Software has been developed for specialized applications. For example, the "curved reconstruction for dental implants" algorithm can be applied to human remains and generate images that resemble antemortem panoramic radiographs. These marvels of diagnostic imaging would certainly be the first choice for the investigator if they were readily available. However, due to the fixed nature of the equipment, the high capital expenditure and ongoing revenue costs, access is limited. Less sophisticated mobile CT units have become more readily available, but still require a tractor-trailer to haul the equipment to the desired location. Even the basic radiographic units found in most hospitals or private imaging centers are seldom available in the locations where large-scale anthropological or forensic studies take place. It is therefore necessary for the imaging facility to be transported as close to the recovery site as possible. Although the task may appear daunting, prior planning and project management by an experienced team will ensure that a successful outcome can be achieved.

There are a number of advantages to establishing a radiographic facility at, or close to, the site where remains are recovered and/or stored. If the location is remote, equipment can be reduced to the most basic components and literally carried to the site. Two examples illustrate this point. In the spring of 1998, the El Niño rains and resulting landslides made road travel to the village of Lymebamba in the north-central cloud forests of Peru extremely difficult. Although, even when the weather conditions were more conducive to travel, the narrow roads cut into the mountain-sides required at least 12 hours of continuous driving from the coast to reach the village. During the flooding that spring, only small passenger vehicles were capable of making what became a nearly 20 hour trip to reach the destination eliminating any possibility of transporting large pieces of equipment to the site.[1] In December 2004, the Sumatra earthquake and the resulting Indian Ocean tsunami ravaged the coast of Indonesia and the west coast of Thailand (Figure 36.1). The devastation totally eliminated the infrastructure in the affected areas and in this situation, even the most basic of supplies had to be brought to the site.

A second advantage of locating the field radiographic equipment close to the remains pertains to minimizing the disruption of the taphonomic context. A broad definition of taphonomy is the study of all the processes that have taken place before the body was recovered, including the placement of the remains and artefacts in the grave, decomposition, insect activity, dispersal of biological and nonbiological components and any previous disruption of the remains. Reducing the distance the remains and the materials associated with the body, such as clothing or wrapping, are transported to the radiographic examination site will decrease the chances of dislodging or displacing foreign bodies or artifacts. Regardless of how carefully a body is removed from its resting place, travel over even a smooth paved road can change the location or result in the possible loss of a vital clue. An anthropological case to exemplify this point was the disinterment of remains that were thought to be of sixteenth century pirates buried on Isla San Lorenzo off the coast of the port city of Callo, Peru. Once all the equipment had been transported to the island, a tent to house the x-ray unit was erected within 15 m of the excavation site. As each body was lifted out of its coffin, it was placed on a 2 × 0.5 m sheet of corrugated plastic and brought into the tent for radiographs. One of the radiographs revealed a small metallic artifact within a fold of the deteriorated clothing next to the remains. The artifact was removed and determined to be a "miraculous metal" that was first produced in the early 1800s precluding the claim that the individual was one of the Dutch buccaneers buried on the island.[1] A recent case from the Office of the Chief Medical Examiner (OCME) for the State of Connecticut provides a forensic example. All gunshot wound victims are radiographed within the plastic body bag into which they were placed at the scene. The x-rays demonstrated a bullet that was free in the bag outside of the body and clothing (Figure 36.2).

FIGURE 36.1 Following the Indian Ocean tsunami in 2004, much of the local infrastructure was destroyed. Emergency responders were forced to bring in even the most basic of supplies including dry ice to maintain refrigeration for body storage. (Photo: A. Anderson. With permission.)

Although previously not much of a concern in anthropology even fifty years ago, today disruption of remains should take into consideration the ethnic group to which the remains belong. An example of respecting cultural sensitivity is demonstrated by an examination conducted on the mummified and skeletal remains of an ethnic group in the Philippines known as the Eboloi. The remains were located within caves in the remote Kabayan region in the northern section of the island of Luzon. The Eboloi believe that if a mummy or a skeletal component of one of their relatives was removed from the cave it would bring bad luck, even death to the village. Aside from agreeing to the condition that all the radiographs would be taken within each associated cave, the team had to agree to participate in a village ceremony requesting permission from their deceased relatives. Three black pigs were sacrificed and their livers "read" by a shaman. Since no "defects" were noted on the surface of the organs, permission had been granted by the "spirits" and the ceremony was concluded with a feast. A contingent of the villagers accompanied the team to the two caves where the study was conducted to ensure compliance with the agreement. Although working within the caves was a challenge, with the cooperation and assistance of the villagers, satisfactory images were obtained and the project was successful.[1] Similar issues are regularly encountered when previously unidentified Native American burial sites are disturbed during the course of construction work.

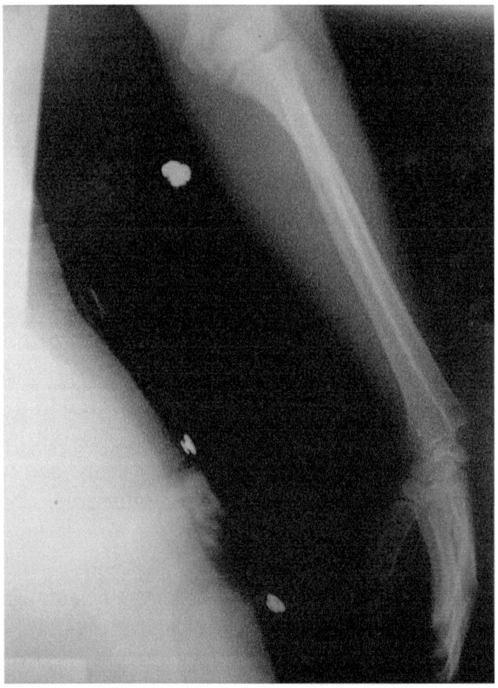

FIGURE 36.2 Radiography of a gunshot victim in the sealed body-bag from which he was recovered from the crime scene shows a bullet lying in the bag. This image enabled both rapid location and retrieval of the ballistic material that might otherwise have been overlooked and provides a record of the location of the bullet in the bag prior to opening, thus maintaining a chain of custody linking it to the crime scene. (Photo: T. Blyth. With permission.)

Even in a forensic situation, the examination of the remains should be conducted with as much cultural and political sensitivity as possible. Following the NATO-led invasion on Kosovo in 1999, it was necessary to exhume and examine the remains of those who had allegedly been killed as part of a systematic process of "ethnic cleansing." Often, the victims had been buried by their relatives in marked graves in religious burial grounds. Partly in response to logistic considerations and also in recognition of the cultural sensitivities of disturbing the remains of those who had been laid to rest, autopsies were carried out in tents at the grave-side or in buildings close to the burial grounds.

Atrocity crimes such as the Yugoslav or Rwandan genocides are often only investigated by the international community following a period of internal military conflict. Such conflict frequently destroys whatever infrastructure may have existed in the region to support forensic investigation, or it may be the case that the scale of the crime to be investigated simply overwhelms local provision. In such cases, it is often necessary to transport all the necessary equipment to the region (which may be very isolated and remote) in order to undertake the investigation. In such international postconflict areas, there may additionally be sound legal reasons why local facilities and personnel should not be used, particularly if the crime is to be tried in an international court. Construction of a local facility by an external international agency may therefore be necessary.

A fourth advantage of the field facility is the function of a triage mechanism. Although the topic will be discussed in more detail later, basically the minimum number of radiographs will be determined by the objectives of the study. Therefore, x-rays can be employed to determine the imaging priorities for a given group of human remains. For example, if a basic series of images reveals a fracture, prosthetic device or osseous pathology, additional more time-consuming radiographic

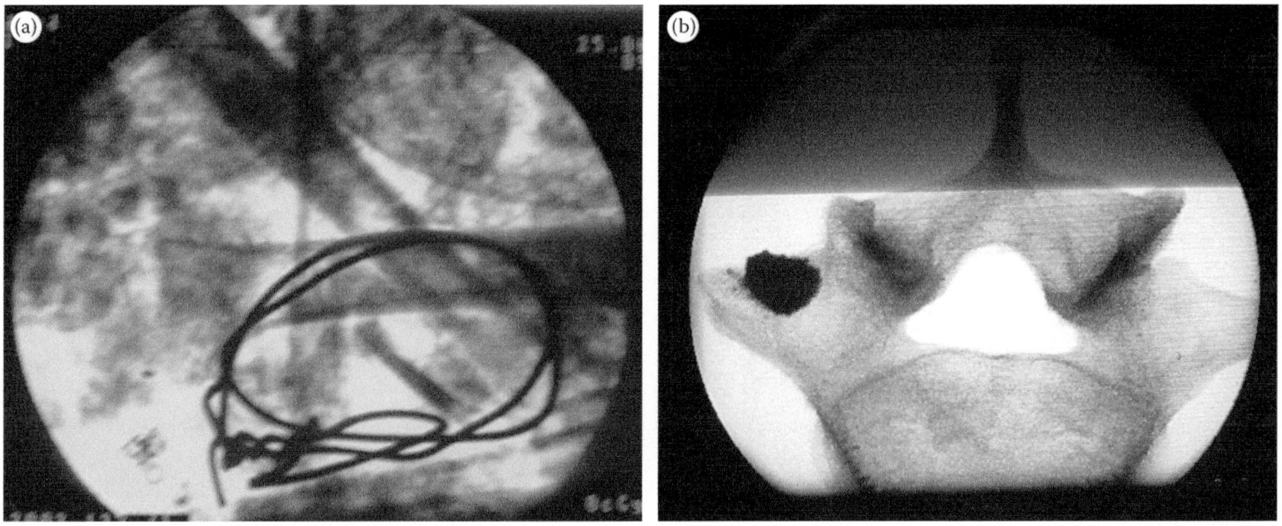

FIGURE 36.3 (a) Fluoroscopy image shows the presence of a wire ligature around the wrists of an exhumed body (b) The location of a bullet lodged in the vertebra is confirmed using fluoroscopy.

projections can be obtained. A rapid screening method, such as fluoroscopy or direct digital radiography, can be used to triage remains in this way. During the United Nations' forensic investigations in the former Yugoslavia all exhumed remains were examined using fluoroscopy to guide the autopsy examination, locate items of forensic significance and to identify areas, such as healed fractures and perimortem trauma, for further imaging[2] (Figure 36.3).

Radiographs can be used to determine the stability of remains prior to being moved. If the images reveal unstable remains, the entire examination may be limited to whatever radiographs can be obtained on site. This situation was the case with the mummified remains of an individual known as the "Soap Lady" at the Mütter Museum at the College of Physicians of Philadelphia in Philadelphia, Pennsylvania. The body had gone through the process of saponification after death and was exhumed in 1875 and placed on exhibit at the museum. Plans were developed in 2006 to update the exhibit and, if possible, move the remains onto another table. A lateral radiograph of the cranio-cervical junction demonstrated that the joint was unstable and moving the mummy could result in irreparable damage (Figure 36.4).

In another situation, a mummy was located in a crypt beneath a church in Popoli, Italy. The x-ray unit was set up in the crypt and the resulting image showed that the remains were stable enough to be removed from the tomb. However, to minimize damage to the clothing and prevent dislodging any artifacts that may have been near the surface, the body was first placed on a wooden board. Next, the entire mummy and the board were tightly covered by wrapping with several roles of plastic film. In this "cocoon-like" state, the mummy was passed vertically out of the crypt into the church.[1]

Similarly, the same procedure was employed with an unwrapped Egyptian mummy located at the Barnum Museum in Bridgeport, Connecticut (Figure 36.5). The triage radiographs can also serve to determine the need for advanced imaging modalities. As previously mentioned, the limited availability of Multi-detector Computed Tomography (MDCT) requires that specific objectives must be defined prior to requesting the study. With a clear target, the amount of time needed to acquire and post-process the images will be greatly reduced. A good example was illustrated by a mummy that had an apparent projectile wound to the distal left femur. Unfortunately, the anterior–posterior, AP, and lateral radiographs did not clearly demonstrate the wound. The remains were transported to an MDCT unit at a private imaging center and a 20 cm volume was collected that included the wound.

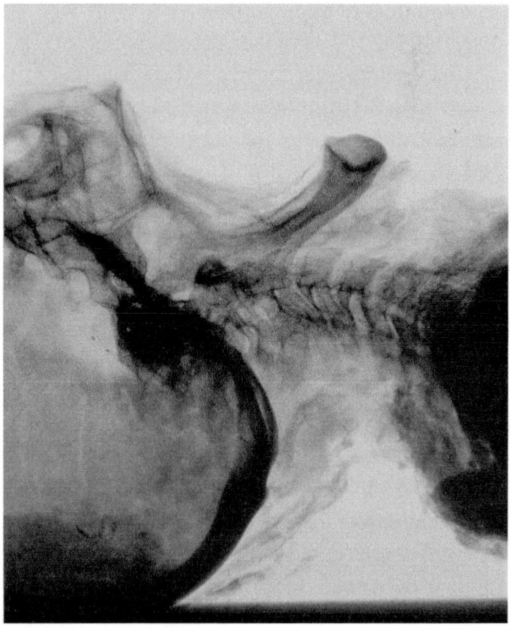

FIGURE 36.4 Lateral Polaroid image of the neck of the mummy known as the Soap Lady. Due to the lack of soft tissue visualized on the radiograph, it was determined that the head would not be stable enough for the mummy to be moved. (From Conlogue, G. With permission.)

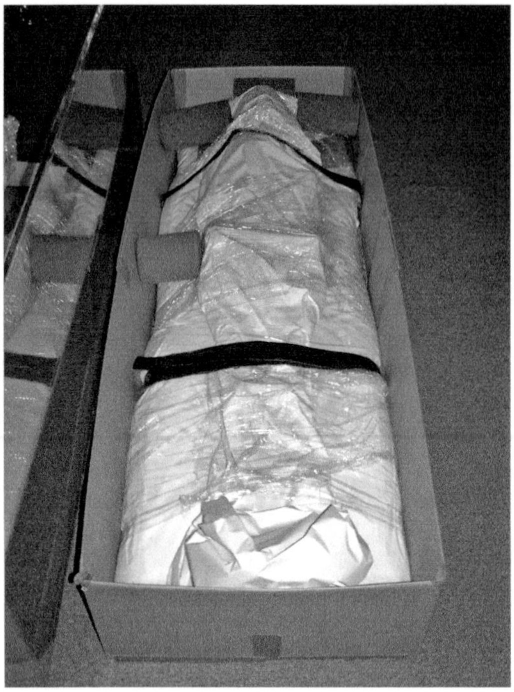

FIGURE 36.5 An Egyptian mummy known as *Pa Ib* from the Barnum Museum in Bridgeport, Connecticut was wrapped in acid-free paper, then "sealed" in plastic wrap, cushioned with foam sponges before being transported in an ambulance to an imaging center for a CT examination. (From Beckett, R.G. and Conlogue, G.J., 2008. *Paleoimaging: Field Applications for Cultural Remains and Artefacts,* p. 160. CRC Press, Boca Raton, FL. With permission.)

Because the tract of the projectile was diagonal, the defect was seen in a number of axial sections. The body was removed from the unit as the radiographer reconstructed off-axis sections to better demonstrate the wound tract. The entire process took less than 30 minutes (Figure 36.6).

A sixth advantage of field radiography is that it can be established in areas where security issues are a concern. Security concerns may relate to security of evidence, security of personnel, or both. In the case of mass fatality incidents (see Chapter 13), it is often necessary to set up an emergency disaster mortuary. This is done for a number of reasons, including the requirement to ensure that the forensic examination can be undertaken in a facility in which staff can work safely without risk of disruption from further disaster events or attack by perpetrators. It is also important that human remains and evidence can be examined without risk of evidence tampering or press intrusion. In these and other instances, particularly in post conflict regions, it may be necessary to transfer remains from a high-security risk area to a lower one in which a safe facility can be established. During the United Nations forensic investigations in post-war Yugoslavia, a central morgue facility was established in an area away from the conflict zones. Human remains were often exhumed from sites within the conflict areas and moved under armed guard to the "safe" facility for examination (Figure 36.7).[2]

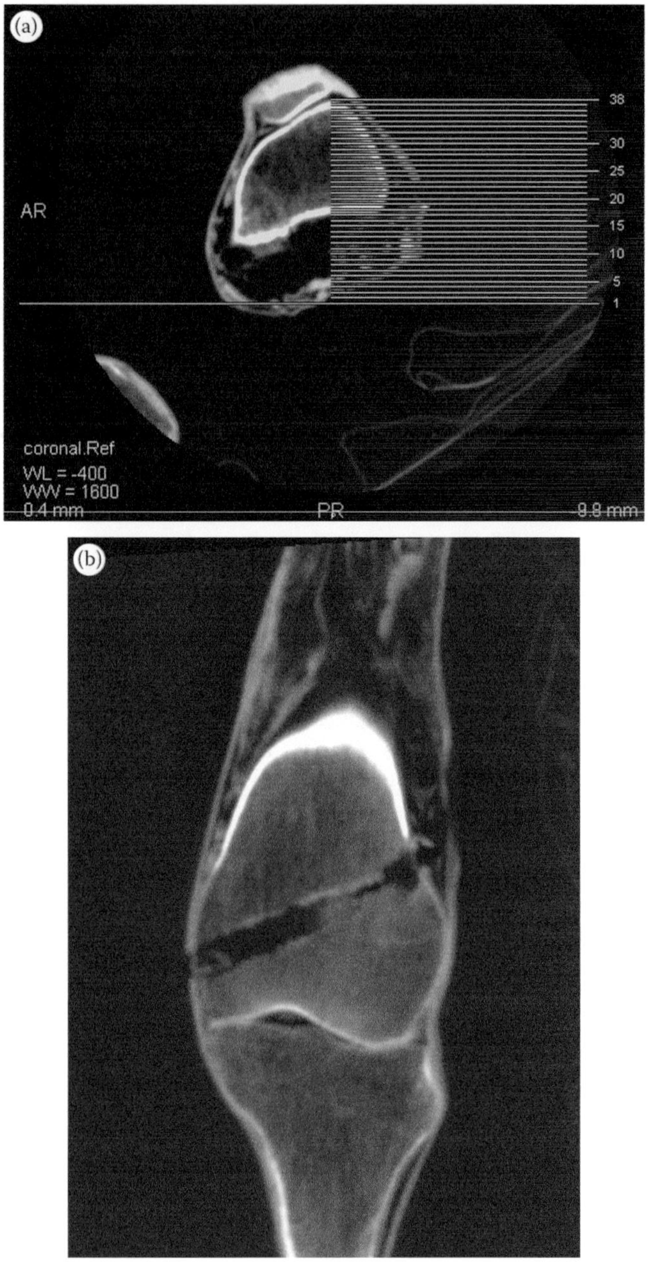

FIGURE 36.6 (a) The axial section of the distal femur that was used to establish the angle and location of the "off-axial" coronal sections. (b) An "off-axis" coronal section through the entire wound track in the distal femur of the mummy known as George-Fred. (From Beckett, R.G. and Conlogue, G.J., 2008. *Paleoimaging: Field Applications for Cultural Remains and Artefacts*, p. 150. CRC Press, Boca Raton, FL. With permission.)

Having discussed the advantages of field radiography, the next step is to consider an approach to establishing the facility. The first consideration is to determine precisely the objectives of the proposed study. Included at this stage is an approximation of the number of individuals who will be examined and the condition of the remains. Three examples will be provided. The first is a collection of 215 mummies in the previously noted village of Lymebamba, Peru. Cultural sensitivity prevented unwrapping of the remains. Therefore,

FIGURE 36.7 (a) During the investigation of atrocity crimes in the former Yugoslavia, human remains from grave-sites located in areas of high security risk were kept under 24-h armed guard by NATO forces. (b) Bodies were transferred under armed guard to secure locations. (Note the refrigerated containers used to transfer the remains.) (c) Communication in postconflict areas is frequently difficult and often only possible using satellite phone to maintain links between the morgue and the grave site. (Figures b and c, from B. Tonello. With permission.)

the imaging objects included the determination of the sex of the individual, the age at the time of death and any pathology or trauma that might be present. As all the remains were preserved, the only time constraint was the period the team could remain at the site. If all the mummies were not radiographed during this first session another session could be scheduled.[3]

The second example was demonstrated by the situation following the 2004 tsunami in Thailand. In this case, best described as a survey of body bags, the number of victims and their bloated condition elevated time as the principal factor. These fatalities, being the result of a natural disaster, were not considered to be part of a crime scene and thus identification of the victims was the primary object. Given the scale of the operation, imaging was restricted to dental surveys and these were the major primary identification method used.

In the third example, skeletal remains from a mass grave in the former Yugoslavia were the focus of the study. In this situation, the objectives are not only to identify the victims but to also document evidence of skeletal trauma due to repeated systematic assaults that were alleged to have taken place in the weeks leading up to the deaths of the victims. In these cases, the imaging protocol was agreed as an initial triage followed by a full dental survey and radiographs of all ribs for signs of healing fractures. This was modified according to the findings of the initial triage.

Following the process of defining the objective, the logistical problems associated with where the study must be conducted needs to be considered. Included in this category is the accessibility of the site. The condition of the roads, if present, will determine the size and quantity of equipment that can be transported. Road access also permits re-supply during long duration studies.

Once access has been discussed, the availability of electricity must be considered. Since the x-ray equipment requires electricity, if it is not present at the location, a generator must be included in the items to be transported.

Generally, a 5 kW generator will be sufficient but may weigh as much as 60 kg. In addition, the quantity of fuel, either gasoline or diesel, necessary to operate the generator must be calculated. If the correct fuel is not available in close proximity to the field site, it must be packed with the rest of the equipment. If power is available, the type, either 110 or 220 V, must be known. X-ray units used in the United States, Canada, Mexico and parts of Europe operate on 110 V. Some of these units can be changed internally to 220 V. If that conversion is not possible, a transformer must be included with the equipment.

After electricity, the availability of water must be considered. Frequently, many of us take the accessibility of water for granted, but it can be a scarce commodity in many places around the world. In addition, the supply can be drastically reduced after natural disasters. Radiographic film processing requires large quantities of water, particularly in the washing phase of the process. Every attempt should be made to minimize the water requirements of the study. Such strategies as collecting and filtering rainwater and using recirculation pumps should be considered. One possibility is to use a table-top automatic processor with a water bottle (Figure 36.8). The tank on this unit requires emptying and refilling with water regularly, depending upon usage. A large drum can be used to refill the bottle and this can be filled remotely and brought to site.

Once these initial factors have been established the next phase in the development of the plan can be determined.

FIGURE 36.8 Agfa table-top processor. This unit has shallow tanks and chemicals are gravity-fed from 2 L bottles; Developer (D), Fixer (F), and Water (W) that can be replenished manually as necessary.

Included in this category would be the selection of the x-ray unit. It is strongly advised that this process should be undertaken with the advice of a radiologic technologist or radiographer who is familiar with portable and mobile x-ray equipment. Although sales and service personnel may be very knowledgeable about the equipment their firm sells, they may have very little to no expertise in operating the equipment outside of the confines of the traditional medical setting.

If the location is remote or maximum flexibility of the equipment is desired, a portable x-ray unit without the support stand would prove the best choice (Figure 36.9). If funding is not available to purchase new equipment, old units no longer in service at hospitals, medical centers, dental practices, or even veterinary practices may serve the purpose. This equipment would certainly prove adequate for mummified and skeletal remains, but insufficient for situations where tissues are hydrated, such as following the tsunami.

Once the x-ray unit has been selected, the image receptor (IR) is the next factor to be considered. Although many hospitals and imaging centers have replaced conventional radiographic film with a digital format, a number of factors must be examined before the final decision is made. Film was the original recording media and dates back to 1895. When film is employed with an intensifying screen, the exposure time required to produce an acceptable image is reduced. The more efficient the screen is at converting x-ray to visible light, the lower the exposure time. However, the increase in screen speed results in a loss of detail on the resulting film (see Chapter 34). In the case of the body bag survey, detailed radiographs are not necessary to achieve the study objective. For the study of the group of mummies in Leymebamba, Peru, detail was a requirement to achieve the desired objectives. In order to increase the recorded detail on the film, instead of a cassette equipped with intensifying screens, a nonscreen film holder (Figure 36.10) would provide the best results. However, because without the intensifying screen the exposure time would have to be increased by a factor of between 100 and 400, and this approach would not be advisable if the study group was more than two mummies. Prolonged exposure times drastically reduce the life of the x-ray tube.

Regardless of the type of device used to hold the film, conventional radiographic film requires a light-tight enclosure to load and unload the film holder, followed by processing with chemicals and washing the processed film with copious quantities of water. Generally in a field situation, manual processing in a darkroom is the most common approach. The chemicals must be transported with all the other equipment. Powdered chemicals can be found in all major cities around the globe and are the easiest to transport. Unfortunately, powders can present a messy situation when combined with water prior to their use. Certainly in remote regions, powdered chemicals are not going to be available; therefore, the quantity of chemical required must be calculated and purchased before leaving for the study site. Chemicals

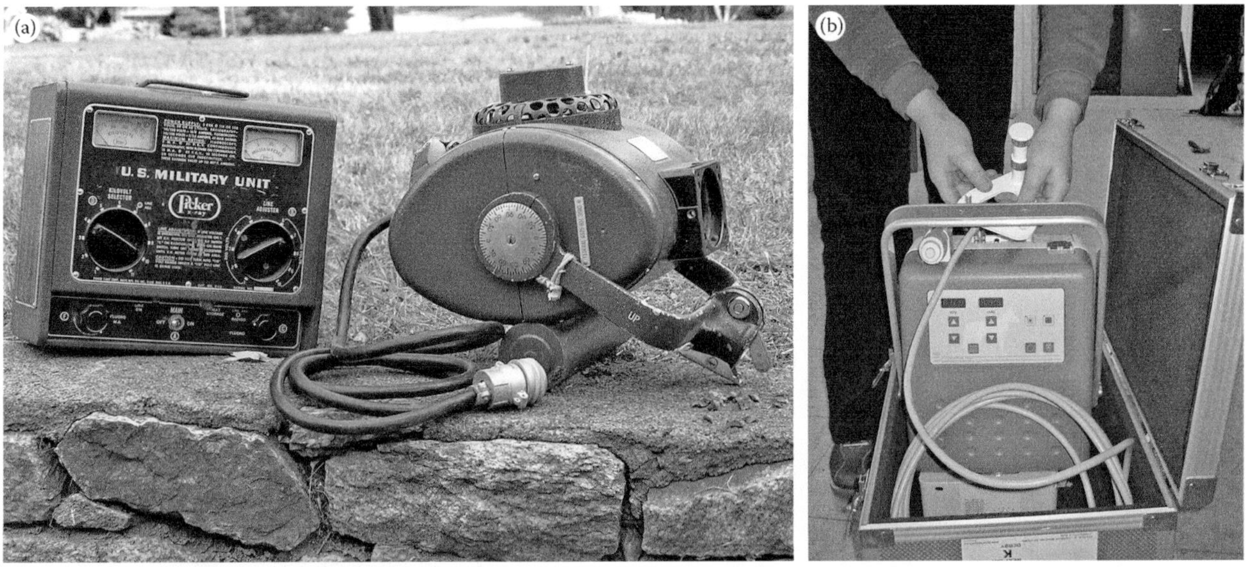

FIGURE 36.9 (a) A 1950s vintage Picker army field unit that was reduced to its minimal components; the x-ray tube and the control unit. (From Conlogue, G. With permission.) (b) A modern portable veterinary unit offers a flexible solution.

will need to be disposed of in accordance with prevailing environmental legislation and good practice. Ideally, they should be stored in appropriate containers and then removed for disposal by a specialist contractor. Local photo-processing labs or pharmacists may be able to assist with this, but in remote situations this will not be possible. Dilution of chemicals with large quantities of water prior to disposal and flushing with water during disposal may be the only viable alternative.

Light-tight enclosures will not normally be found at the study site; however, spaces with few or no windows, such as bathrooms, can be easily adapted to create a suitable work place. Duct- taping four layers of black gardening or a single layer black pool liner over windows and around door frames will render an area sufficiently light-tight to permit loading and unloading of film holders (Figure 36.11). If the space is also going to serve as a film processing area, a few more modifications are necessary. A bath tube or shower enclosure is the most suitable location to place the tanks for processing the film. The porcelain or tile surface can be easily cleaned when the project is completed.

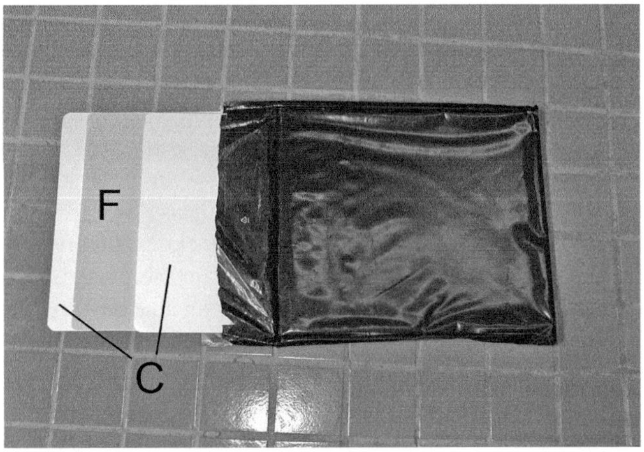

FIGURE 36.10 A simple nonscreen film holder composed of the black plastic liner from a box of 14 × 17 in. (35.6 × 43 cm) film. To protect the film (F), it was placed between two sheets of cardboard (C) before sliding it into the plastic bag. The open end of the bag was folded over and taped closed.

A portable darkroom can be easily constructed using a PVC pipe frame covered with black felt and black pool liner. Bringing the darkroom eliminates the need to find a suitable room at the study site. In addition, because the frame is constructed with widely available PVC pipe, it may be found at a major city close to where the operation will be located. Therefore, only the less readily available covering material needs to be included with the x-rays, film, and cassettes. One consideration when working with film-processing chemistry is the requirement for adequate ventilation for the darkroom. Increased ventilation can be achieved by means of a small electrically-powered fan introducing air at a high level and a corresponding vent for air extraction at a lower level to prevent chemical fumes rising. Care should be taken to install these in such a way that the light-tight seal of the darkroom is not compromised. In extreme situations, respirators may offer an alternative (Figure 36.12).

Attempts have been made to find a replacement for conventional x-ray film. Because visible light and x-rays are both forms of electromagnetic radiation, photographic film can be employed to record x-ray images. Between 1992 and 2008 Polaroid photographic film was used very successfully by the team from the Bioanthropology Research Institute at Quinnipiac University (BRIQ) in Hamden, Connecticut. Despite the largest size film being only 8 × 10 in. (20 × 25 cm), it eliminated the need for wet processing and the "dry" process produced an image approximately 60 s. Unfortunately, since 2008, Polaroid no longer make the film. As a possible

FIGURE 36.11 A bathroom can be easily converted into a darkroom by covering the window with black gardening plastic or pool liner. Note the nonscreen film holder resting on the closed toilet bowl. (From Beckett, R.G. and Conlogue, G.J., 2008. *Paleoimaging: Field Applications for Cultural Remains and Artefacts*, p. 44. CRC Press, Boca Raton, FL. With permission.)

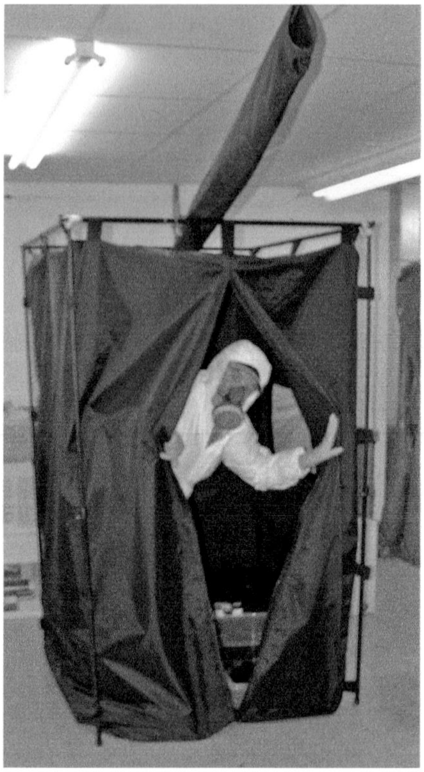

FIGURE 36.12 A commercially available portable darkroom provides a rapid solution to film processing in the field. Note the upper air-vent providing ingress of fresh air drawn in by an electrically operated fan at ground level. Respirators should be used when adequate air changes cannot be maintained.

replacement, the team has been experimenting with Ilford photographic paper as a recording media. Although the paper still needs to be wet processed, it can be done in a small cylinder tank in approximately two minutes. The advantage of using the paper is in the relative ease of establishing the radiographic technical factors and then using those factors for conventional x-ray film.

As previously mentioned, many medical facilities have abandoned conventional film for a digital alternative. The digital format can be divided into two categories: direct digital radiography (DR) and computed radiography (CR). DR has a direct link between the image receptor and the computer (Figure 36.13). Within a second or so after the exposure, the image will appear on the monitor. The longest cord between the imaging plate and the computer is approximately 25 ft, although recently wireless remote systems have been introduced. The extremely short turn-around time between the exposure and the appearance of the image on the monitor is the principal advantage of the system. There are three primary disadvantages. First, if the image receptor is damaged, for example dropped, imaging cannot resume until another plate can be acquired. These can be very expensive components—in the region of $80,000—and damage to the plate may mean the suspension of the project pending replacement, or at worst, cancellation due to cost. However, this consideration must be balanced with the fact that it is not necessary to move the receptor anywhere near as often as for CR, as all the image processing happens with the plate "*in situ*." The second drawback is that multiple exposures cannot be made on the plate. (Multiple exposures are only necessary when the remains are too thick to be penetrated by a single exposure; for example, a lateral abdomen on an individual that weighs over 100 kg.) As soon as the first exposure is made, the computer processes the data. The third drawback is the weight, size, and thickness of the detector, which may make it difficult, to position in certain circumstances. Some of the newer wireless systems now offer a range of smaller and lighter detectors but this will add to the cost.

The other system, CR, has an image receptor that resembles the conventional film cassette. Within the cassette is a photostimulable phosphor plate. Following an exposure, the plate is placed into a reader. Within the reader, the plate is removed and scanned by a laser that assists in interpreting the radiation dose that reaches the plate. Once the plate has been read, the image appears on the monitor, the plate is "cleared" for another exposure and placed back into the cassette. The advantage of the CR system is that, if damaged, the plate can be replaced at a much lower cost (around $3000–4000). In addition, multiple plates are provided with each reader. Since the plate is not "processed" until it is

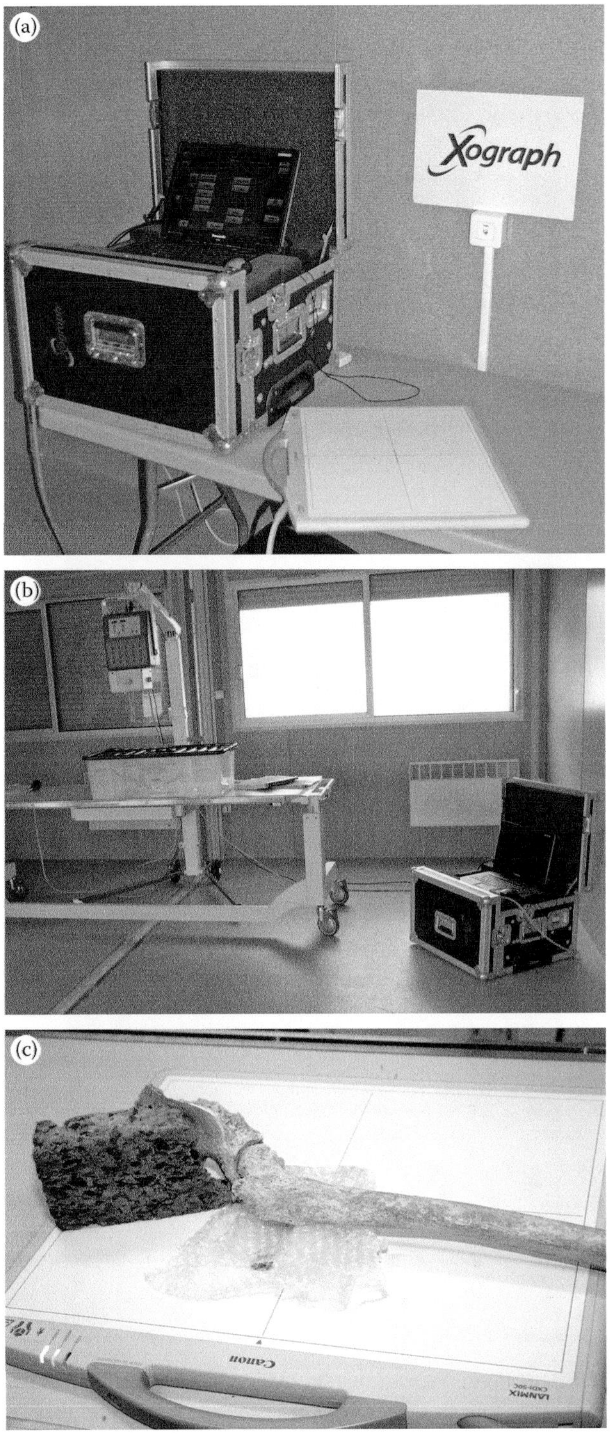

FIGURE 36.13 (a and b) A 17 in. × 14 in. DR detector is linked to a high-frequency portable x-ray tube and rugged laptop computer to provide a truly portable DR solution. (c) Images can be evaluated without the need to remove the detector from underneath the subject. This can be considerably beneficial with subjects that are difficult to position.

placed into the reader, multiple exposures can be made prior to processing. The principle disadvantages are the time that it takes before the image appears on the monitor and the fact that the plate has to be removed from its position under the object for processing, thus disturbing the positioning before the image can be assessed for quality. Generally, it takes approximately 30 s or more from the time the plate is placed into the reader before an image appears. In clinical situations, such as an emergency room setting, time is an extremely important factor. However, in a field radiographic facility the several minutes that would be required before the images appears on a monitor would allow for the x-ray tube to cool between exposures. In a field setting there may be an increased probability of damage to the image receptor and the possibility of introducing dust or moisture into the CR plate mechanism. Since several plates are provided with the CR reader, the project would continue as long as sufficient plates remained operational.

The next item to be considered would be the ancillary equipment required to support the x-ray tube and image receptor and material that could be utilized as positioning aids. This is a category that should be kept to a minimum in order to decrease the total amount of material that must be transported. Although resources may be available at, or close to the site it would be advisable to devise at least a light-weight back-up system. There are basically two approaches to the x-ray tube support system depending on the intended direction of the x-ray beam. The x-ray tube can be supported above the remains directing the beam toward the floor. Two factors must be taken into consideration: the weight of the x-ray tube and the source-to-image receptor distance (SID). The tube will generally weigh between 10 and 20 kg, therefore the support must be stable enough to safely secure that weight. As the SID increases, the stability of the system becomes more of a concern. A 40 in. (100 cm) SID is required for the x-ray beam to cover a 14 × 17 in. (35 × 43 cm) cassette. If a smaller size image receptor (IR) is being used, the SID can be reduced. Although the BRIQ team has duct-taped the x-ray tube to a thick branch, it has equally successfully used aluminum electromechanical tubing (EMT) cut into 20 in. (50 cm) sections to construct an x-ray tube support system (Figure 31.14). Aluminum is relatively light-weight and the sections can be assembled at multiple heights. At the lowest height, 20 in. (50 cm), the support permits the tube to be directed horizontally for lateral projections. The system has been used up to 80 in. (200 cm) to cover a 14 × 36 in. (35.6 × 91 cm) cassette. However, it was discovered that the radiation dose to the equipment operator could be reduced if the x-ray was directed up from the floor. In addition, keeping the x-ray tube on or close to the floor eliminates the possible problem of the tube being knocked over or failure of the support system with subsequent disastrous consequences. A wooden frame was constructed with casters to allow the x-ray tube to roll around on the floor (Figure 36.15).

Once the tube support system has been determined, the approach to supporting the IR needs to be considered. If the x-ray beam is directed from above the remains, the approach is less complex. The remains can be placed on any radiolucent material, such as plastic sheeting. However, a solid wooden plank will serve the purpose. The flat surface should be placed on boards or other material that will elevate the

FIGURE 36.14 A frame constructed of EMT used to support the x-ray tube.

body at least 2 in. (5 cm) above the floor to the permit the IR to be slid under the body.

When the x-ray tube is placed on the floor, the IR support system becomes a little more complex. The EMT system described above to support the x-ray tube can be modified to serve the same function for the IR. In this case, the tubing frame straddles the remains that have been placed on a table. A wooden table can serve the purpose as long as the supports under the table top permit at least a 17 in. (43 cm) clear space to radiograph through. If the table is low, concrete blocks can be placed under the legs to elevate it. A sheet of plastic or even cloth, such as canvas, is stretched across the EMT frame above the body region to be radiographed. The plastic is held taught with two clamps on each side (Figure 36.16).

Lateral projections should be obtained using a horizontally directed beam. As previously described, the EMT tubing frame certainly works very well. However, a stack of chairs or boxes can serve the same purpose. The important point to remember is that the direction of the x-ray beam should be away from where people are standing and preferably towards an exterior concrete wall.

The final items to be considered under ancillary equipment are positioning aids. These are radiolucent objects that are employed to assist in placing the remains into positions that will render conventional medical imaging projection. In a forensic situation, the images may be required to compare with premortem radiographs taken at a medical or dental facility. With skeletal remains, positions that best demonstrate trauma may be required. Commercially available foam positioning aids in a variety of shapes are commonly used in the medical setting; however, they are relatively expensive. In addition, although light-weight they take up a great deal of space. The BRIQ team has purchased foam padding to pack various pieces of equipment and in the field utilized the foam as a positioning aid.

Up to this point everything that has been discussed should have been done prior to arriving at the study site. In addition, a radiographer or radiologic technologist and a radiation physicist should have been included in the team

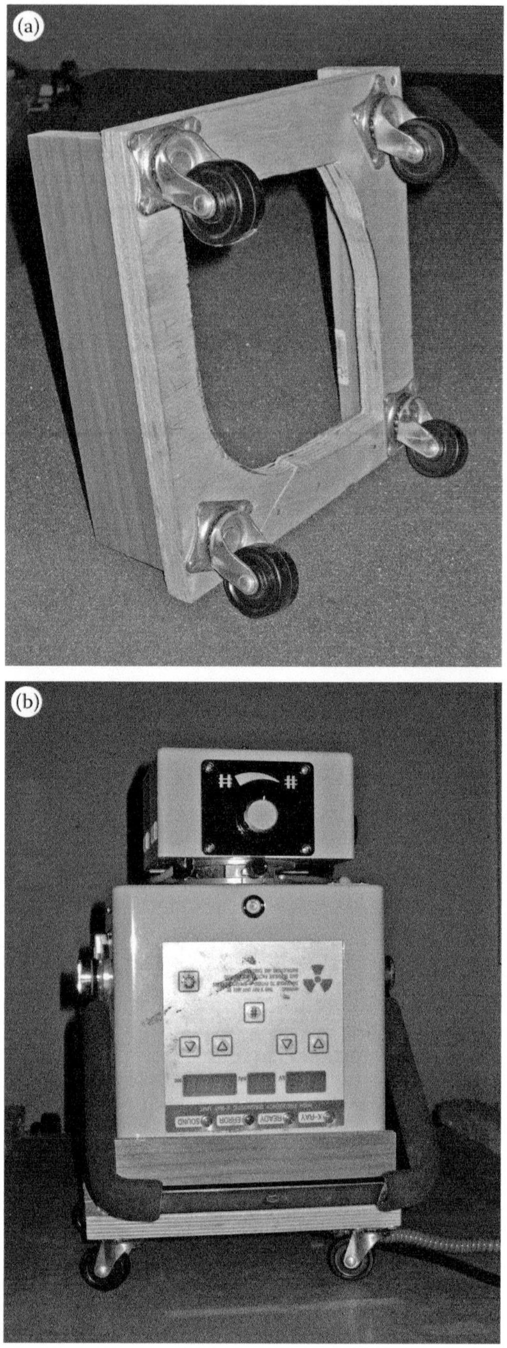

FIGURE 36.15 (a) An x-ray tube support constructed with casters. (b) The x-ray tube is mounted on the movable support so the x-ray beam can be directed from below the object being radiographed.

designing the plan. Their expertise in equipment and image production and radiation protection will ensure that the best selections are made. However, due to the inevitable changes that will likely occur at the proposed study site, it is mandatory that a radiographer be included in the team to implement the study. Regardless of how meticulously planned, the team must be prepared to adapt to the actual conditions faced upon arrival. For example, only half of the film arrives, the remainder has been either lost or damaged. It must be decided how to compensate for the shortfall and

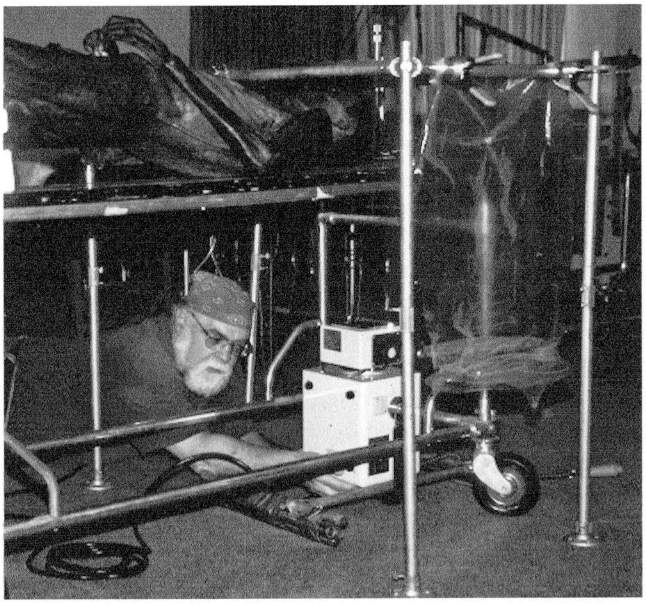

FIGURE 36.16 The x-ray tube was placed beneath the mummy known as James Penn. With the x-ray beam directed vertically, the x-ray cassette was placed on a sheet of plastic stretched across the EMT frame. The plastic sheet was held taut by clamps. (From Beckett, R.G. and Conlogue G.J., 2008. *Paleoimaging: Field Applications for Cultural Remains and Artefacts*, p. 39. CRC Press, Boca Raton, FL. With permission.)

still achieve the study objectives. Similarly, the anticipated location for the placement of the x-ray unit and examination of the remains has been changed. Without a radiographer to compensate for the new conditions, the entire project could be jeopardized.

Included in the on-site preparations is a radiation protection plan for individuals who might be exposed to the x-ray. Working within the agreed procedures established in conjunction with the radiation physicist, the radiographer will employ the well-known rules of time, distance, and shielding (Chapter 34) and take into consideration the location selected for the study. The first factor, exposure time, can be minimized by limiting the number of repeat radiographs. When necessary, a radiographer should be able to make minor adjustments in the exposure factors to ensure the image quality is maintained. Proper placement of the examination table and the x-ray unit are important in achieving the other two factors to reduce radiation exposure. Maximizing distance can be accomplished in three ways. Firstly, the x-ray facility should be established away from areas where others may be working or frequently walking. Secondly, if possible, the x-ray unit should be placed on the floor with the beam directed up toward the ceiling. In this position, less scatter radiation is generated toward the equipment operator. Thirdly, the exposure switch cord should be extended at least five meters from the unit.

Shielding is an excellent method to ensure radiation protection. Mobile lead shields are available; however, they are heavy and may be difficult to transport to isolated areas. In many locations around the globe, concrete and brick may be common building materials. Extending the exposure cord to the maximum length and getting around a doorway, particularly if it is constructed of concrete, will provide maximum protection. If a wall is not present, but concrete blocks are available, a three-foot (one meter) high wall can be easily assembled to crouch behind during exposures. A small handheld digital dosimeter will be invaluable in establishing a safe working area.

Once the x-ray unit and examination table have been put into place, the image receptor is the next consideration. If film is the recording media, at the very least a light-tight enclosure must be established to load and unload the cassettes. If manual or automatic film processing facilities are not available, the light- tight enclosure must be large enough to accommodate the tanks necessary to process the exposed film. Since large quantities of water are required to wash the film, this component can be located away from the "darkroom" and close to where the wet films would be hung to dry.

With everything in place to take the exposure and process the film, it is time to establish the imaging protocols for the study. Even though this facet of the project had been discussed prior to arrival on site, modifications may be necessary. Perhaps the number of individuals to be examined has changed or the conditions of all or some of the remains have been altered. It is at this stage that the minimum number and projections required to achieve the objectives of the study must be established.

Before the first films can be taken, everyone involved in the project should now be involved in the final informational phase. The work health and safety concerns should be thoroughly discussed with everyone on site. Rules should be established concerning ensuring the health and safety of all the personnel, and protocols for all aspects of the work should be established. The operational policies and procedures applicable to field situations are essentially the same (with minor modifications) as those for operating in the medical examiner's office or the osteology laboratory, and are discussed at length in Chapter 34.

The following case studies detail strategies that have been developed for imaging in field conditions to suit particular circumstances.

INVESTIGATION OF WAR CRIMES AND HUMAN RIGHTS ABUSES

As discussed above, radiography is increasingly deployed in the forensic investigation of war crimes and human rights abuses. United Nations forensic teams have operated in many regions and forensic pathology teams deployed in Rwanda, Bosnia, Croatia, Kosovo, East Timor, Sierra Leone, and other postconflict zones have utilized field radiography units. The scope of these investigations has varied with the alleged crimes under investigation, the scale of the conflict, the state of preservation of the human remains and the prevailing security conditions.

In the case of the former Yugoslavia, investigations took place over a number of years. Because of the climatic conditions it was not possible to excavate during the winter months, and excavation and morgue teams and equipment were deployed on an annual basis every summer from 1996 to 2002. Each year, it was necessary to establish a secure morgue site, define the forensic and imaging protocols for each investigation, select and deploy appropriate equipment, and organize staffing to cover the five-month period of the investigation. Where possible, the same secure site was used in consecutive years and the exhumed remains were transferred from different crime scenes within the region. There were three main sites used:

- An existing forensic pathology unit on the campus of a major teaching hospital
- A purpose-built morgue on the site of a city burial ground
- A converted factory.

In each case, different problems were encountered and had to be overcome. The forensic pathology unit had obviously not been designed to cope with the scale of the investigation and the number of bodies examined. Makeshift body holding areas and evidence stores had to be established and workflow adapted to include three main phases of radiology: initial triage, secondary dental and skeletal examination, and further examinations on request. Whilst an imaging room, darkroom, and mobile x-ray machine were available, the volume of work required the deployment of a mobile C-Arm fluoroscope and dedicated dental film processor (Figure 36.17).[2]

The "purpose-built" facility had been designed locally with no input from a radiographer or radiation physicist, and revised workflow and radiation protection barriers had to be introduced to overcome these. The electricity supply in the first months after the war was variable and required the adaptation of existing fluoroscopy equipment to cope with voltage fluctuations. In the absence of a darkroom and suitable water supply, make-shift darkrooms were constructed to house military-specification processing equipment that was able to recirculate the wash water (Figure 36.18). The converted factory was planned with specialist input from radiographers, radiation physicists, anthropologists, and pathologists who had been involved in previous deployments. Adaptations and additions to the building facilitated an excellent workflow, fully lead-lined x-ray facility, state-of-the-art mobile fluoroscopy and film processing equipment, and a purpose-built darkroom. These facilities were constructed in a matter of weeks prior to deployment by Dutch and German military personnel according the specifications provided by the advisory team[2] (Figure 36.19).

In other situations, rapid deployment minimized the amount of prior planning that could be done. Investigations during a 4-day mission to Sierra Leone were severely hampered by an unstable security condition and lack of equipment. However, supplies of film and cassettes, dental film, chemicals, film hangers, and duct tape carried in the team's luggage, combined with an x-ray machine borrowed from a military field hospital, some buckets purchased in the local

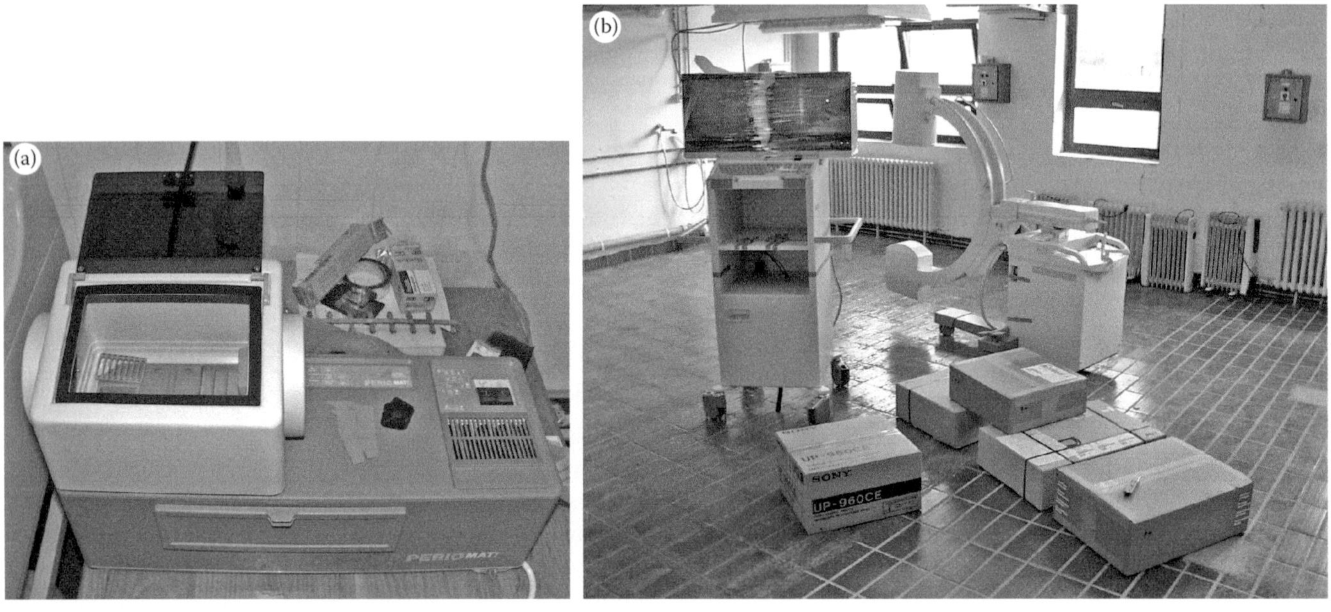

FIGURE 36.17 (a) Desktop portable dental processor used by UN forensic teams in the former Yugoslavia. This unit is bottle-fed (requiring no running water supply) and can be used to develop both intraoral and occlusal films. Note the daylight loading hatch enabling the unit to be used in the absence of a darkroom. (b) Digital C-Arm fluoroscope used by UN forensic teams in Kosovo. This unit had a high-capacity digital disk drive as well as a thermal printer for rapid image recording for reference during autopsy. (From B. Torello. With permission.)

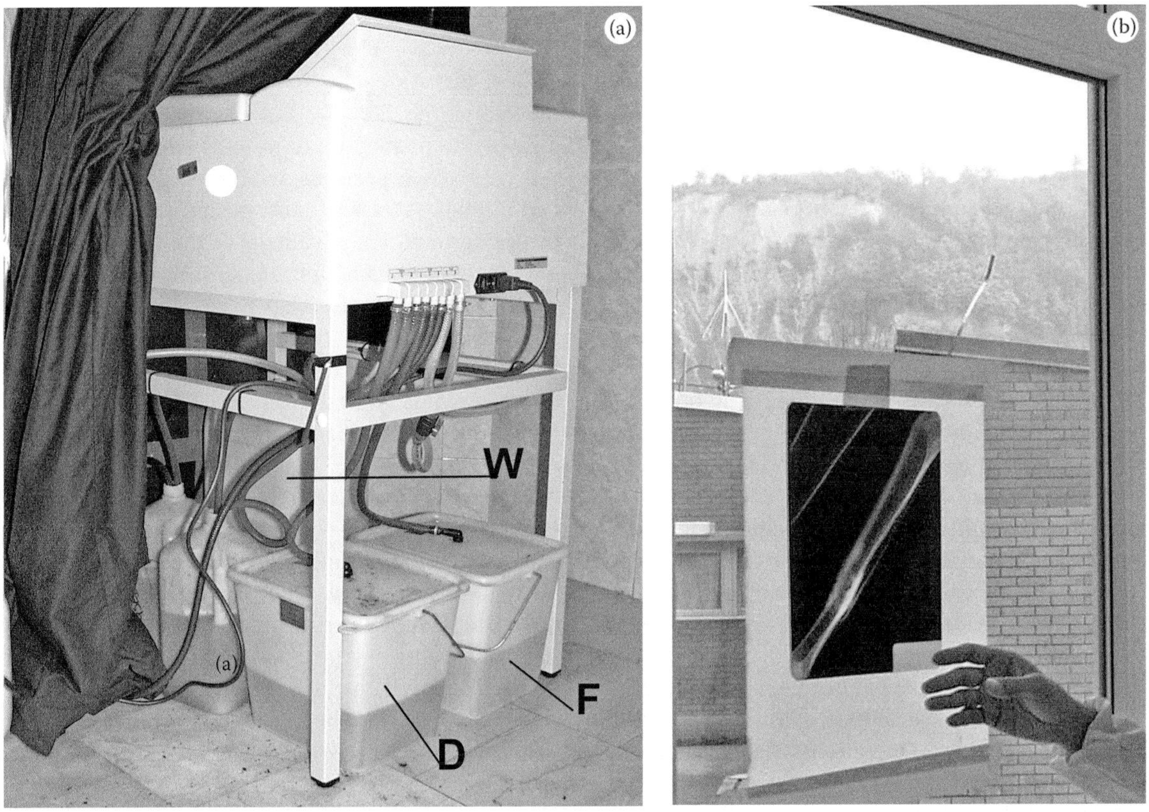

FIGURE 36.18 (a) Makeshift darkroom in Bosnia constructed using light-tight curtains. The automatic processor uses a water recirculation pump, allowing it to be used in the absence of running water supply. The water in the tank (W) requires changing 2 or 3 times per week and this can be done off-site. Developer (D) and Fixer (F) tanks supply the processing chemicals. (b) A simple x-ray viewer can be constructed by taping a diffusion screen to an outside window.

FIGURE 36.19 (a) With forward planning, lead sheeting was fixed to the walls of an existing room in a disused Kosovo factory to give adequate radiation protection. (b) An existing washroom in the factory was converted into a darkroom. The lack of windows, an existing water supply, white-tiled walls and tiled floor with drain make this an ideal location. (From B. Tonello. With permission.)

market and an equal measure of determination and ingenuity quickly established a workable imaging facility (Figure 36.20).

MASS FATALITY INCIDENTS

Mass fatality incidents are an ever-present challenge for the deployment of imaging facilities in field conditions. The process for the investigation of such incidents has been discussed in detail in Chapters 13 and 14 and the arrangements that are in place to respond to such incidents will differ from country to country and state to state.

In the United States, prior to 1990, there was no organized civilian team capable of responding to a national disaster with mass casualties. Either state and local officials or the U.S. military assumed the responsibility for identifying the deceased in mass disasters. As the National Disaster Medical System developed to take care of survivors in these situations, it became apparent that a civilian organization was needed to take care of the casualties. The National Association of Funeral Directors proposed the idea of a team of forensic and mortuary specialists dedicated to carrying out morgue operations in catastrophic events. As a result, the Disaster Mortuary Operational Response Team (DMORT) was established. Today, there are ten regional DMORT teams with approximately 1200 members available to respond to any mass casualty event in the United States if the body count is so high that local and state authorities are overwhelmed. In addition to medical specialists, the teams include a variety of skilled individuals ranging from computer specialists to heavy equipment operators to medico-legal investigators, mortuary officers, mental health specialists, and chaplains. Because of the nature of the work and the confidentiality and legal issues involved, each team member undergoes a comprehensive background check with fingerprinting; provides a meticulous personal, educational, and work history, along with copies of professional licenses and certifications; supplies documentation of all required immunizations; and obtains a valid U.S. passport. In addition, each team member must take numerous educational courses either on-line or in the classroom. The teams are similar to the National Guard in that they go about their daily routines until something happens and local authorities request assistance. Once a team is activated and deployed, they become employees of the federal government until they demobilize. The composition of a team is determined by the event and the needs of the local authorities. Some deployments have only required certain members with specific skills, whereas others have required every available member, with many making multiple deployments throughout the event.

Since its inception, each deployment for DMORT has provided learning opportunities and the experience to build on and improve its operations. Although some deployments may be carried out in a local or state medical examiner's facility, others may require the Deployable Portable Morgue Unit (DPMU). These are fully equipped morgues with the equipment for each section palletized and ready to go (Figure 36.21). The DPMU may be transported by truck, rail, or air. As each event is different, so are the requirements for the DPMU. For example, only the dental and x-ray sections may be required to enhance the local facilities and those sections would be depalletized and made operational. At other times, the entire morgue is required to be operational. An advantage of the DPMU is that it can go anywhere and function under any circumstances. Even when the infrastructure has been destroyed, the morgue can go active utilizing tents and truck-mounted generators. When the complete morgue is set up, it requires a footprint of approximately 10,000 sq. ft.

Following Hurricane Katrina, two DPMUs were set up, one in Baton Rouge, LA, and the other in Gulfport, MS. Both were initially set up in warehouses (Figure 36.22) and needed tents as well because the warehouses had suffered extensive damage and additional protection from the elements was necessary. Due to the extent of the devastation after Hurricanes Katrina and Rita and the number of fatalities and cemetery disruptions, DMORT team members rotated deployments for nearly nine months. During the Katrina event, work began on a semi-permanent facility that would become the world's largest morgue. Contained in a protected 37-acre compound, an 18,000 square-foot morgue, dormitory accommodations for 300 personnel, and a 52,000 square-foot building housing dining, laundry, meeting, workout, storage, administration, and information resource facilities was constructed to carry out the majority of morgue operations (Figure 36.23).

Today the facility is in "stand-by" mode but is utilized for training purposes in preparation for the next disaster. This facility, as well as the DPMUs, utilizes portable digital x-ray systems, specifically the type of equipment used by the military in combat field hospitals: compact, very portable, rugged, and dependable (Figure 36.24). In the morgue configuration, each section of the morgue has its own bay containing its equipment. Radiology is a significant element in the mass disaster morgue operations, and its laptop computer is networked with the laptop computers in the pathology and anthropology sections to allow for fast access to images. Usually, the three sections are placed together so that consultation between the individuals in each section is easily accomplished. Although the morgue configuration will of necessity vary according to the type of disaster, there are admitting, photography, decontamination, fingerprinting, DNA sampling, pathology, anthropology, dental, radiology, and discharge sections, as well as areas deemed "clean" and "dirty" for personnel to put on and remove Personal Protective Equipment (PPE) gear. If the event is man made and involves explosive materials, shrapnel, unexploded ordnance or other unknown materials, then x-ray is going to be the first stop, examining the bodies in containers to search for potentially harmful objects to personnel before proceeding. Regardless of the placement of the radiology equipment, distance from the source is of primary consideration as well as adequate shielding. The x-ray unit

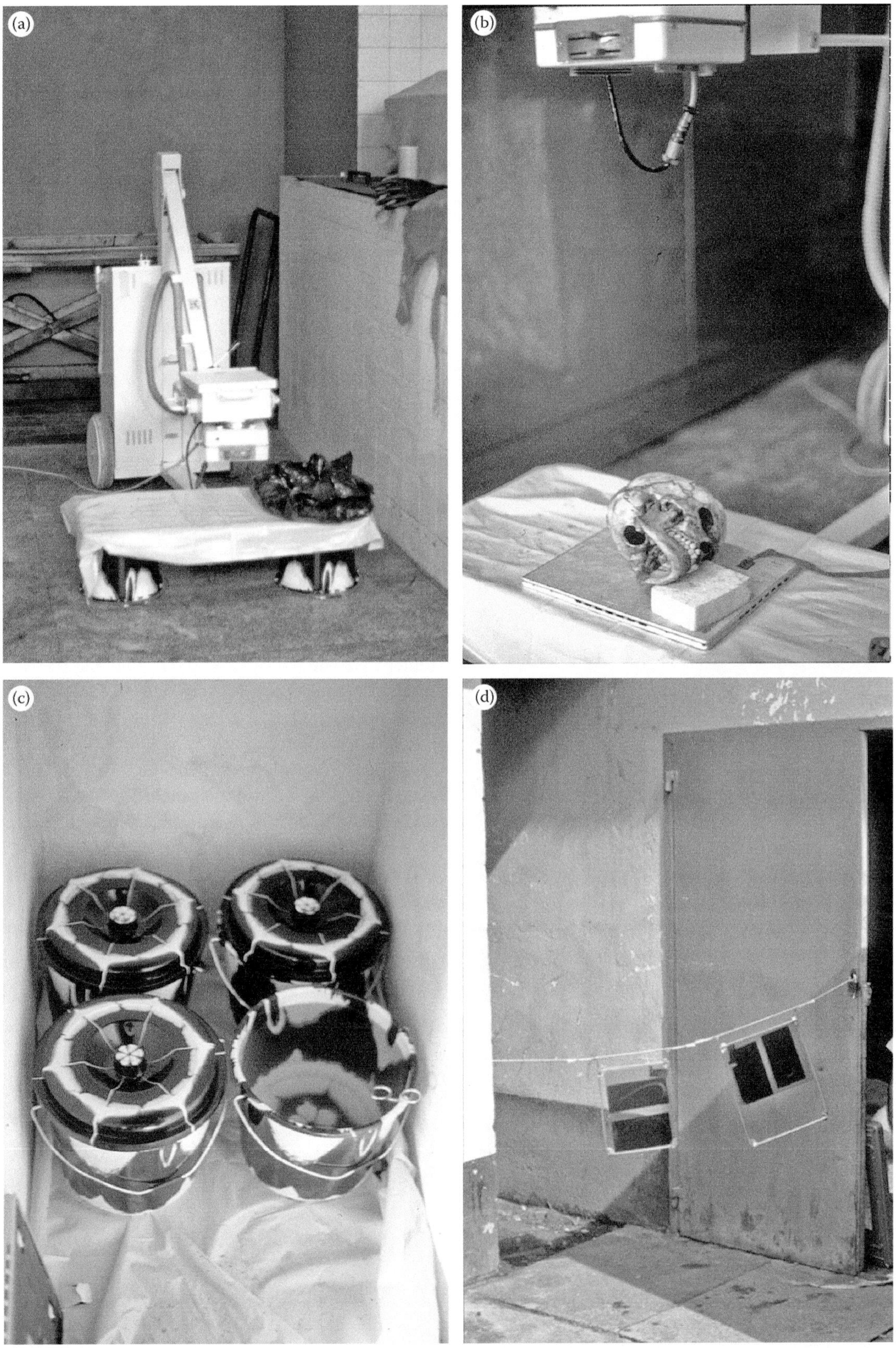

FIGURE 36.20 (a and b) A table constructed from four buckets and a piece of wood, and covered with plastic sheeting made an ideal support for skeletal remains in this morgue in Sierra Leone. (c) Films were manually developed in buckets of chemicals in a toilet cubicle that was made light-tight using duct tape and thick black plastic sheeting. (d) Films dried quickly on the washing line in the African sun.

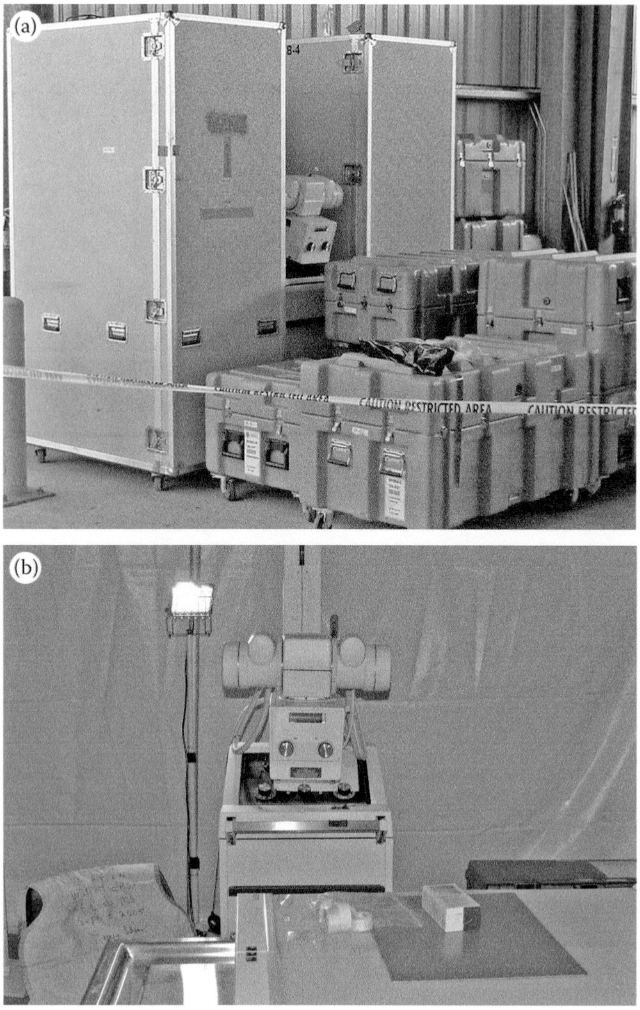

FIGURE 36.21 (a) A DMORT portable morgue, including x-ray units and supplies, is kept palletized and ready to go. (b) The x-ray station can be deployed and set up rapidly.

must be surrounded by a 6-foot radius and lead shielding devices are supplied.

In morgue operations, as postmortem data are accumulated in the morgue, antemortem data is being collected from a number of sources, including family and friends, local medical facilities, physicians, and law enforcement authorities (Figure 36.25). Both types of information are entered into specialized databases that eventually match the information for a positive identification. Once DMORT operations stand down, all records are turned over to the local medical examiner or coroner.

Radiology is considered to be a vital component in mass casualty morgue operations and, even if the entire team is not activated, usually the radiographer is. It is important for any radiographer to consider all aspects related to a mass disaster response before applying for a responder team, whether federal or state. One factor is financial: although a member's job is protected by law, the employer does not have to pay the individual while deployed, and the government pay scale may not be equitable. Usually a deployment is two weeks in length; however, the scale of the disaster may necessitate more than one deployment. It is especially difficult for someone with very young children, or who is the primary caregiver, to participate in this type of organization. The radiographer must be prepared to function and live under extreme conditions. After Katrina, the first teams out lived in vehicles, dug latrines, ate "Meals Ready to Eat" and used bottled water for everything. On other deployments, the teams had adequate lodging and restaurants, but regardless, it is physically demanding. The radiographer must remember that on a deployment, he/she may be tasked to perform other duties besides radiography. Perhaps the greatest consideration must be given to the mental and emotional effects in a mass disaster. Dealing with the condition of the remains on such a vast scale, survivors and relatives, the news media, physical and environmental circumstances, all

FIGURE 36.22 (a) DMORT morgue unit set up in a damaged warehouse following hurricane Katrina. (b) The interior of the morgue.

FIGURE 36.23 An aerial view of the morgue facility at Carville, LA, set up following hurricane Katrina. The morgue itself is in the upper left corner.

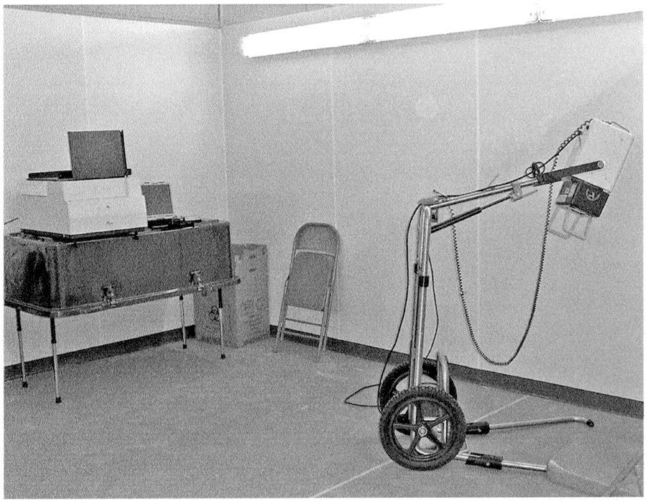

FIGURE 36.24 High-frequency portable x-ray system using CR deployed at the Katrina morgue.

FIGURE 36.25 Postmortem data, including radiological findings are entered into a database for comparison with antemortem data.

will take a toll on the individuals. These are very stressful conditions, and radiographers must be prepared to deal with situations that they may never have imagined. It is not a job for the faint of heart.

REFERENCES

1. Beckett, R.G. and Conologue, G.J., 2008. *Paleoimagin: Field Applications for Cultural Remains and Artefacts*, CRC Press, Boca Raton, FL.
2. Viner, M.D., 2008. The use of Radiology in Mass Fatality Events in Adams, B.J. and Byrd, J.E., Eds., *Recovery Analysis and Identification of Commingled Human Remains*, pp. 145–183. Humana Press, Totowa, NJ.
3. Conlogue, G., Guillen, S., Salazar, J., 1999. Field Radiography of Mummified Peruvian Remains. 26th Annual Meeting of the Paleopathology Association. Columbus, Ohio.

37 Radiology of Special Objects, Antiquities, and Mummies

Gerald Conlogue

CONTENTS

Imaging anthropological material, such as skeletal and mummified remains, or archaeological artifacts, such as pottery, follows all the basic imaging principles previously considered. A more thorough exploration of the topic, more appropriately termed paleoimaging, is presented by Beckett and Conlogue;[1] however, several important points will be discussed here.

There are several distinct differences between forensic and anthropological or archaeological imaging. Although there are occasions when only a single mummy, complete or partial skeleton, or an individual artifact may be the extent of an examination, generally these studies include multiple subjects. The forensic exception would be the examination of remains recovered from a mass grave as discussed in Chapter 36. In this situation, the procedures outlined for mass casualty incidents in Chapter 13 would be modified and the entire process from excavation to imaging would follow an anthropological model similar to the procedures employed with ancient remains. (Unlike a forensic situation where the objective is to collect evidence that can be used in a criminal procedure, paleoimaging focuses on collecting data that will hopefully provide a more complete understanding of a particular culture or segment of the population.) Evidence of a violent demise can be detected with imaging (Figure 37.1); however, more commonly paleoimaging documents less dramatic events, such as skeletal changes related to pathologic or degenerative processes.

Individuals involved in forensic imaging are usually working under time constraints. Radiography of the victim is just another step in the procedure prior to autopsy and the ultimate destruction of the integrity of the tissues. Anthropological and archaeological materials are not in danger of decomposition and the imaging can either be repeated or another modality, such as computed tomography, can be undertaken even years later.

CHOICE OF IMAGE RECORDING MEDIA AND IMPLICATIONS FOR SELECTION OF EXPOSURE FACTORS

The first factor that must be determined in a paleoimaging study is the recording media that will be used. If film has been selected and related considerations, such as film holders and means of processing determined, the most important technical factor is the selection of the penetrating power of the x-ray beam or peak kilovoltage (kVp). It is important to remember that sufficient penetration is necessary for the image to be recorded on film. However, over penetration, even at reduced milliamps (mA) will give the processed film a very "gray" overall appearance. The bone will not be "white." If the kVp is too low, there will be insufficient penetration and the bone will not be visible on the processed image. Therefore, for every object, when film is the recording media, there is an optimal kVp setting. If skeletal and mummified remains are the subjects of the investigation, the optimal kVp is 55. Consequently, the mAs will be the variable technical setting depending upon factors such as the relative speed of the intensifying screens within the cassette and the distance between the x-ray source and the imaging receptor.

Generally, an increased kVp is required for archaeological materials. As the objectives of most of these studies are related to construction techniques or the condition of the object, the principal factor is the part-thickness or density. For example, 80 kVp provides satisfactory penetration of most pottery (Figure 37.2), but 100 kVp is required to penetrate objects such as life- size plaster replicas of ancient and Renaissance sculptures to determine the location of internal metallic rods (Figure 37.3).

If instead of film, a digital image receptor (IR), such as used in computed radiography (CR) or direct digital radiography (DR) system has been selected, the role of kVp is very different. With CR and DR, kVp no longer controls image contrast as it did with film. Rather, the role of kVp is to provide sufficient penetration to excite the majority of electrons within the IR plate. Depending on the system, the captured x-ray energy is converted to an electrical signal or data that can be manipulated by the computer. The resulting image contrast is therefore controlled by preset mathematic manipulations or algorithms applied to the data. The 55 kVp that was optimal for skeletal material is too low for the some IR systems and results in a "washed-out" gray image, termed

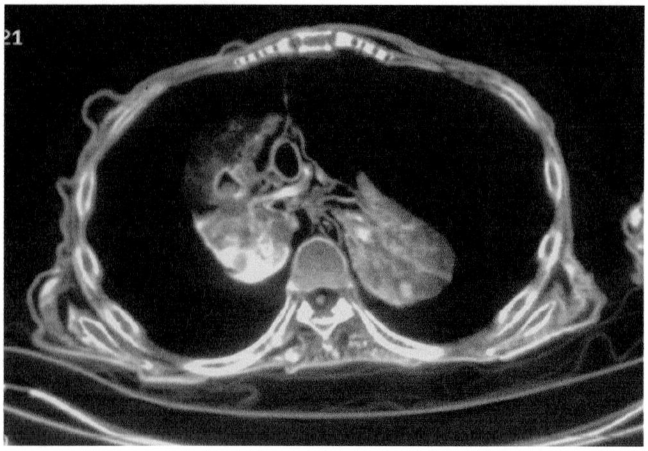

FIGURE 37.1 An axial section from a CT study of a mummy known as Andy, demonstrating rib fractures and a shift of mediastinal structures consistent with a tension pneumothorax and probable mechanism of death in this case. (Adapted from *Paleoimaging: Field Applications for Cultural Remains and Artifacts*, Beckett, R.G., Conlogue, G.J., p44, CRC Press. With permission.)

quantum mottle, because there was insufficient "signal" produced. A lack of signal in an electronic system is considered noise and can be stated as a low signal to noise ratio (S/N). If the kVp is increased to at least 80, the S/N is sufficient to produce an acceptable image. Although digital systems permit the processed image to be manipulated, a procedure known as postprocessing, there are limits to the possible improvement of image quality once the data has been captured by IR.

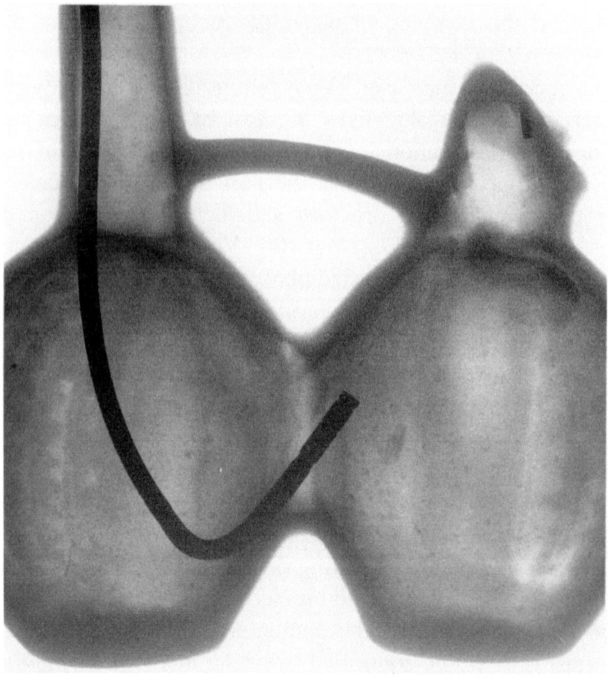

FIGURE 37.2 A radiograph of an example of Peruvian pottery documenting the position of the endoscope used to examine the interior of the artefact. The image was recorded on Polaroid photographic film.

Similarly, for archaeological materials, the kVp must be increased to a minimum of 100 to insure a sufficient S/N that results in a quality image. As the algorithms incorporated into the medical digital imaging systems were developed for hydrated living tissues, the resulting images on these systems will not be optimal. Industrial CR has proven more satisfactory for both anthropological and archaeological materials. The algorithms in the nonmedical system were developed for a wide range of materials ranging from plastics and composites to metals. The principal difference between the medical and nonmedical systems is that the latter has higher a resolution. For example, the Fuji medical CR plates permit 100 μm resolution, while their industrial plates can resolve down to 50 μm.[2] However, due to the increased resolution, the nonmedical CR plate requires 10 times the radiation dose to provide a sufficient S/N.

LOCATION OF EXAMINATION AND IMPLICATIONS FOR RADIOGRAPHIC TECHNIQUE

Once the recording media has been determined and the logistics of the study have been discussed, the remaining key features of a paleoimaging project must be considered on site where the study is to be conducted. There are several reasons for the delay. Unlike forensic cases, the only time-limiting factor is dependent on the length of time that the team can remain at the study site.

Careful positioning of skeletal or mummified remains is critical in order to ensure that the resulting images can be used for comparison purposes. With disarticulated skeletal material, although labor intensive, positioning specimens to obtain anterior-posterior (AP) and lateral projections can usually be easily achieved with only a few bits of radiolucent foam rubber. However, mummified remains present a number of problems, including a body in an immovable flexed position (Figure 37.4), inside a large bundle of textiles where the exact body position is unknown (Figure 37.5), or within a display and the body cannot be moved (Figure 37.6). Therefore, a greater number and shape of radiolucent positioning aid may be required to obtain the necessary images.

In addition to acquiring images for comparison purposes, mummified remains may be radiographed for more esthetic or educational reasons. One of the study objectives may be to produce an image of the entire mummy or mummies that can be displayed. Since the x-ray beam diverges from its source, multiple images taken of each region, for example of the skull, chest, abdomen, pelvis, and extremities cannot be seamlessly assembled to produce a single image of a mummy. The source-to-image receptor distance (SID) must be great enough to cover the entire mummy. For example, in order to cover an object 36 in. (91 cm), the x-ray tube must be positioned 72 in. (182 cm) from the IR. If the x-ray tube and the mummy are not moved between exposures, multiple images of the various regions of the body can be taken (Figure 37.7)

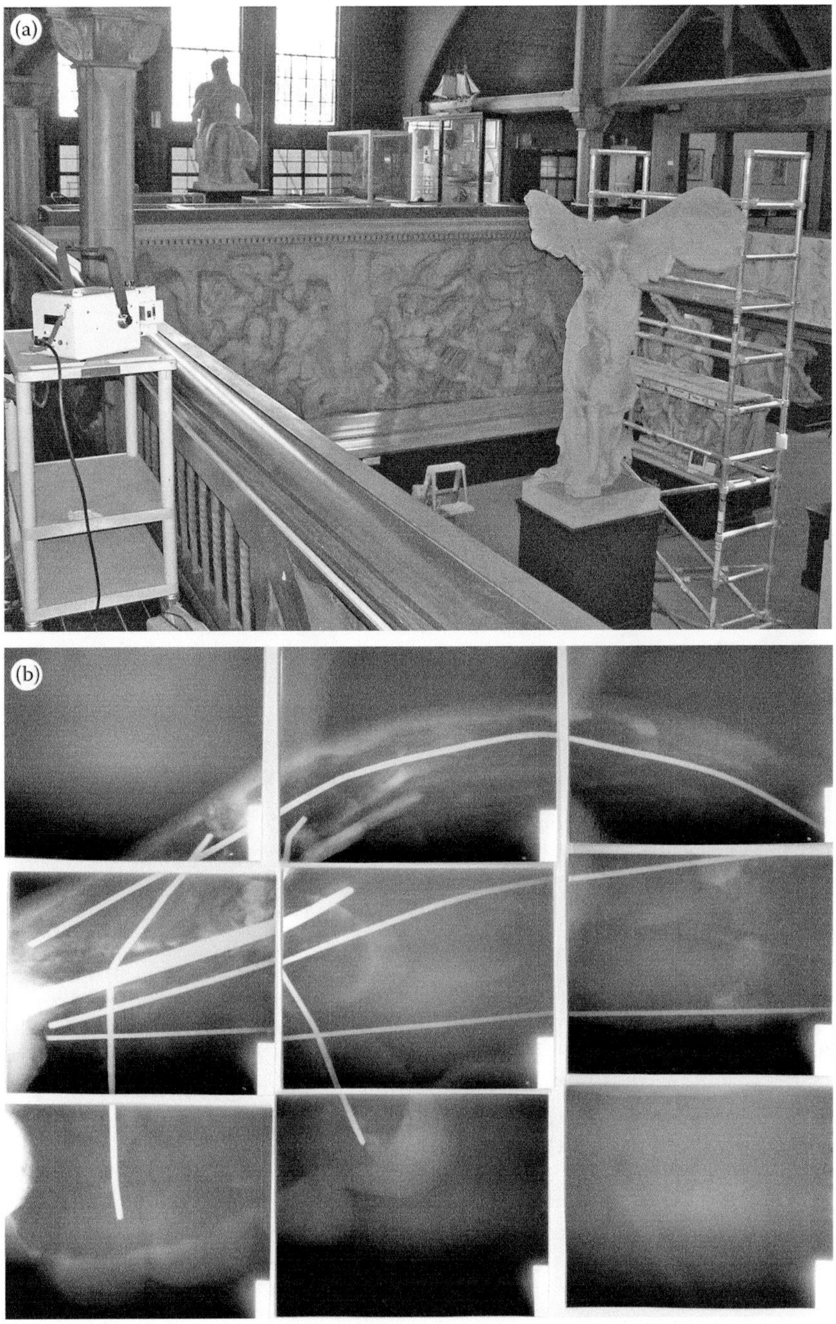

FIGURE 37.3 (a) Position of the x-ray tube for a series of radiographs of the right wing of a plaster reproduction of the Winged Victory of Samothrace at the Slatter Museum on Norwich, Connecticut. (b) The nine radiographs of the right wing assembled to demonstrate the placement of the medal support rods.

and later assembled into a single image that can be used for display purposes (Figure 37.8).

Standard positions are not quite the same for archaeological materials and vary according to the objective of the study. If pottery shards are the subject, usually the objective is simply to document the degree of complexity or composition of the material, and a single projection with the piece parallel to the IR plane is all that is required (Figure 37.9). For intact objects, in addition to the composition, visualization of the construction or assembly technique might be required. Here an arbitrary AP projection is the starting position. The object is then rotated 90° for a lateral view. After examining those two views, additional projections may be required. From the AP position, it may be necessary to rotate the object 45° in either direction for oblique projections or to place the object upright on the IR and direct the x-ray beam through the object for a superior-inferior (SI) projection.

CASE STUDY

Probably the most significant difference between forensic imaging and paleoimaging is that the latter can continue over

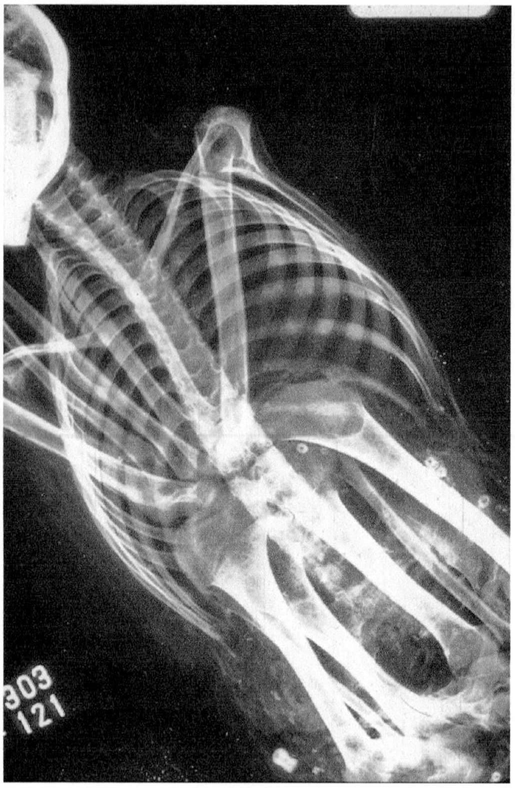

FIGURE 37.4 An AP radiograph of a Chachapoya mummy from north central Peru. The superimposition of the legs makes it difficult to assess the abdominal, lumbar, pelvic, and lower extremity structures.

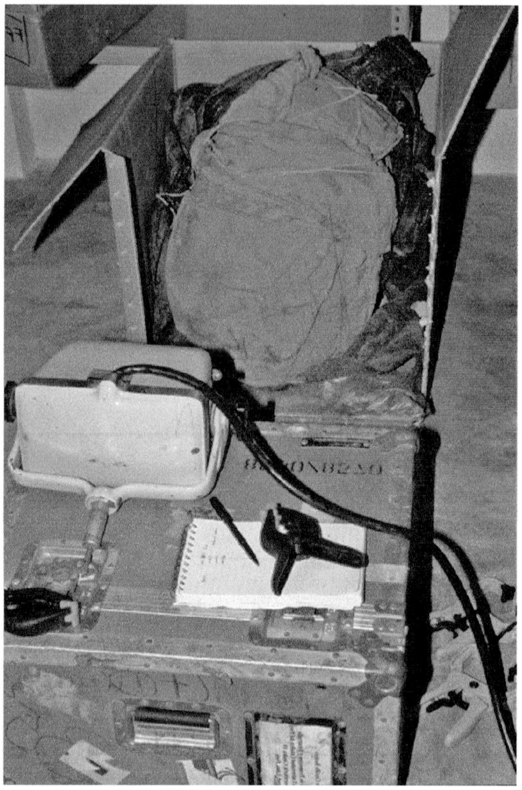

FIGURE 37.5 A large Chiribaya mummy bundle from the south coastal region of Peru lying within a cardboard box with the side of the box opened. The x-ray tube is resting on a transport case and positioned so the x-ray beam will be directed horizontally.

an extended period of time and utilize several imaging modalities. In order to illustrate this and the previously described points a specific case will be presented. In 1987, the curator of the Mütter Museum of the College of Physicians of Philadelphia, Pennsylvania, wanted to assess the condition of a mummy known as the "Soap Lady" (Figure 37.10). The mummified remains, that had gone through a process known as saponification, had been donated to the Museum in 1875 by Dr. Joseph Leidy, Due to the extremely fragile condition of the remains, pieces of waxy adipocere were continuously falling off the body; the mummy, therefore, could not be moved to an imaging center for an examination with state-of-the-art

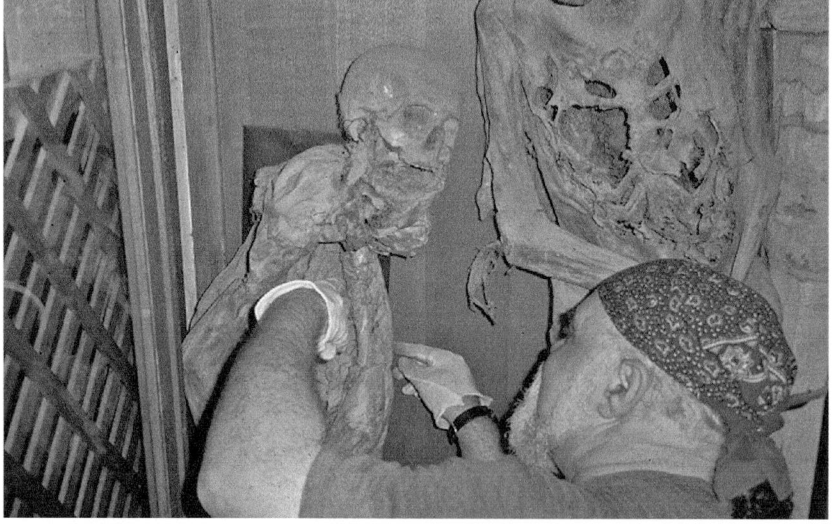

FIGURE 37.6 Placing a nonscreen film behind a mummy in the exhibit in the Church of the Dead in Urbania, Italy.

FIGURE 37.7 The x-ray tube was attached to a tine of the forklift truck to produce a 144″ (3.66 m) SID. The mummy was positioned on a sheet of Plexiglas that was elevated 2″ (5 cm) off the floor the permit the 14″ × 17″ cassette to be placed under the mummy. To provide a safe place to take the exposure, a lead shielding area was created at the foot of the mummy.

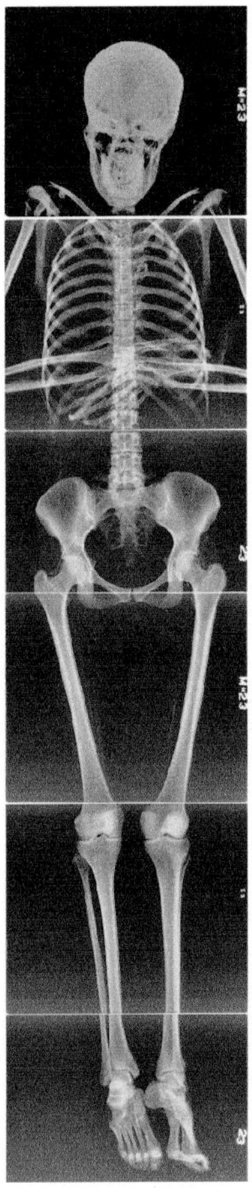

FIGURE 37.8 A composite image of one of the mummies from Guanajuato, Mexico, that was on exhibit at the Detroit Science Museum in Detroit, Michigan.

imaging equipment. The solution was to bring equipment to the museum and conduct a basic radiographic study on site. A 1952 model Profex portable x-ray unit was donated for the project. Since a darkroom was not located on the premises, a 1960s vintage Polaroid Radiographic System was located and used to determine the radiographic exposure setting required to produce optimal images with conventional radiographic film.

The resulting radiographs revealed a number of previously unknown facts regarding the remains.[3] Straight pins with rounded, solid heads, and cuff buttons hidden from a visual examination under the adipocere covering the corpse were easily detected on the radiographs (Figure 37.11). A pin and button were removed and analysis indicated that they were manufactured in the early 1800s, refuting the 1792 burial date provided by Leidy. In addition, the skeleton demonstrated little evidence of degenerative changes, suggesting a much younger individual than suggested by the renowned anatomist and scientist from the University of Pennsylvania. He probably provided the misinformation so the actual identity of the individual could not be determined. Other than a bit of residual brain tissue noted in the posterior portion of the calvarium, it appeared that no organs within the body survived the decomposition process (Figure 37.12). From a technical standpoint, the study demonstrated that mobile radiography could provide diagnostic quality images at

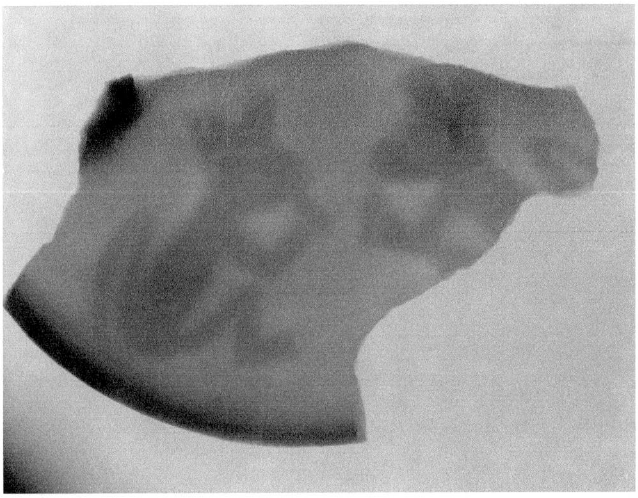

FIGURE 37.9 A radiograph of a pottery shard recovered from the north coastal region of Peru. The homogeneous appearance of the material suggests the intact piece had been finely crafted.

FIGURE 37.10 The "Soap Lady" mummy at the Mütter Museum of the College of Physicians of Philadelphia in Philadelphia, Pennsylvania, after the display case cover had been removed. (From Mary Sonntag. With permission)

locations away from imaging centers. In addition, a trained radiographer employing skills acquired in a clinical setting could obtain the necessary radiographic projections without moving fragile remains (Figure 37.13).

In 2001, as part of a National Geographic Channel series, *The Mummy Road Show*, the Soap Lady was reexamined. Along with conventional radiography, a mobile computed tomography (CT) unit was brought into the museum.[4] Philips Medical Systems arranged for the unit, a Philips Tomoscan® model M, to be transported from the Georgetown University Medical Center in Washington, DC to Philadelphia. The axial CT images demonstrated the presence of internal organs such as the heart, lungs, and kidneys (Figure 37.14). Additionally, radiopaque masses were seen in the right upper and left lower quadrants. As mobile CT was not a possibility 14 years earlier, the 2001 project emphasized the advances in imaging technology. It also emphasized the advantages of CT over conventional radiography, including the elimination of superimposition and the ability to visualize structures with very low densities, like dehydrated lung tissue. Although it was not reported by the radiologist who reviewed the study, both the radiopaque masses could have been precisely measured. Due to magnification, accurate measurements are not possible with conventional radiography. Finally, it must be emphasized that the availability of the mobile CT unit was only due to the cooperation and generosity of the manufacturer and medical facilities. Without such gestures, the ability to conduct future studies may not be possible.

In 2007 the Soap Lady was revisited once again. The new curator was planning to update the exhibit that was constructed

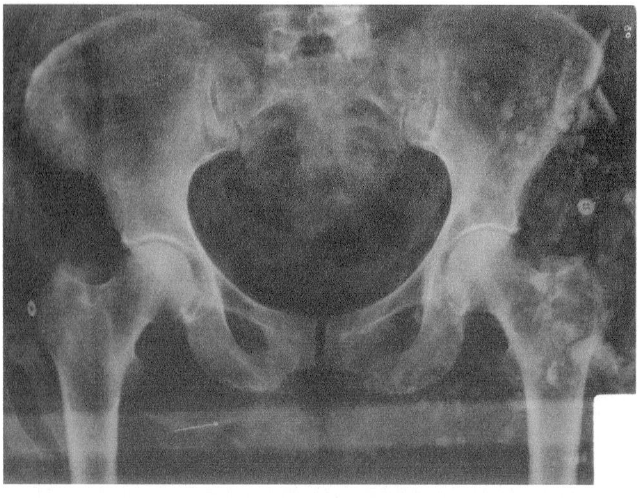

FIGURE 37.11 The straight pin with a rounded, solid head and two cuff buttons were revealed on the radiograph of the pelvis. (Adapted from *Paleoimaging: Field Applications for Cultural Remains and Artifacts*, Beckett, R. G., Conologue, G. J., CRC Press. With permission.)

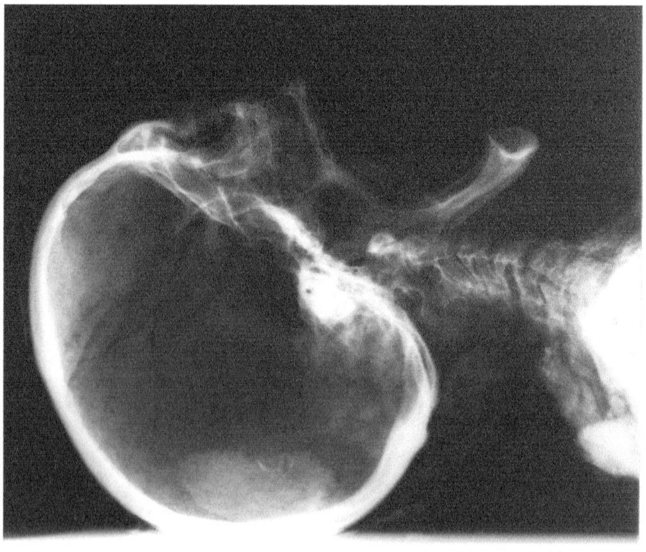

FIGURE 37.12 The lateral radiograph of the skull demonstrating residual brain tissue in the posterior region of the calvarium.

after the 1987 study. This provided the opportunity to apply another advance in imaging technology. Instead of the conventional radiographic film used in the original study, photostimulable phosphor plates employed in computed radiography would serve as the IR. Since the mummified tissues were dehydrated, industrial instead of medical algorithms were used to process the data collected from the plates. Finally, because the images were going to be incorporated into the exhibit, a single AP and lateral projection of the entire mummy would produce the most esthetic images. The maximum height that could be obtained with the SkyJack® provided a possible 124″ (315 cm) SID. The x-ray tube was fastened on to the base of the bucket before the platform was elevated to the maximum distance (Figure 37.15). With a 124″ SID, the x-ray beam would cover a maximum length of 62″ (157 cm). Since the mummified remains could not be moved off the table, the sides of the table precluded getting the IR closer than about 12″ (30 cm) from the posterior surface of the mummy. Because of this increased magnification, the 62″ IR length would exclude the distal lower legs and feet from the image. In order to obtain a continuous 62″ imaging surface, two IR plates were positioned lengthwise and two crosswise. A 1/8″ (3 mm) particleboard was cut to 64″ (163 cm) × 18″ (46 cm). The particleboard served as a "stiff"

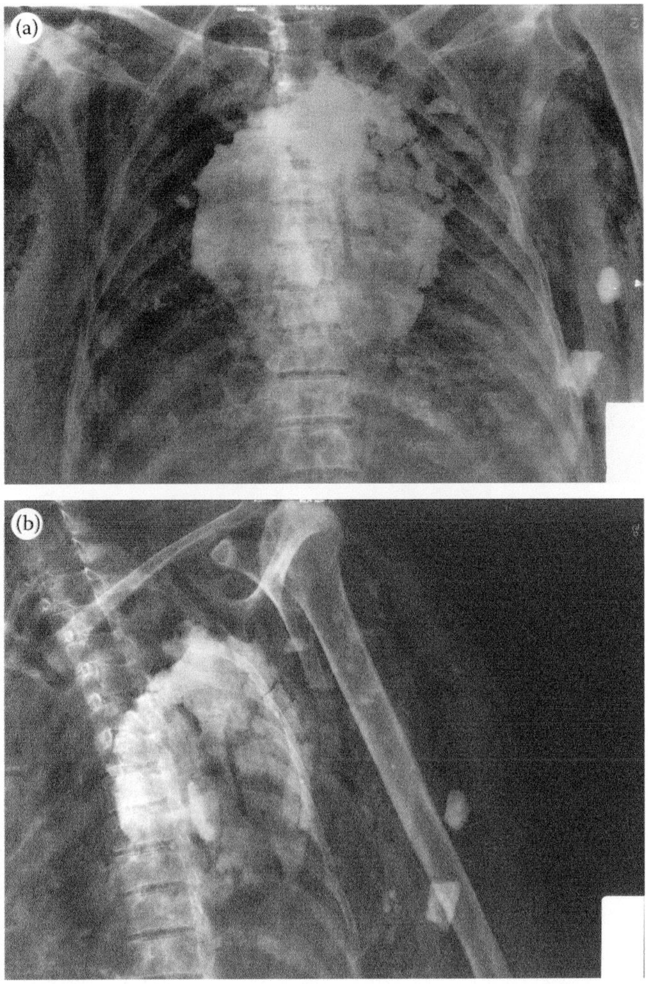

FIGURE 37.13 (a) AP radiograph of the chest demonstrating a large radiopaque mass in the center of the chest. (b) Oblique projection of the chest clearly indicating the mass was located outside of the body. (Adapted from *Paleoimaging: Field Applications for Cultural Remains and Artifacts*, Beckett, R. G., Conlogue G. J., CRC Press. With permission.)

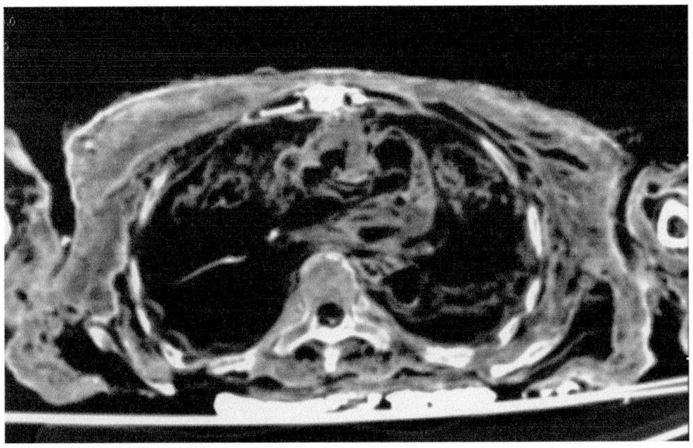

FIGURE 37.14 Axial image through the chest demonstrating the location of the mass on the posterior surface of the thorax and the remnants of the heart and lungs in the interior.

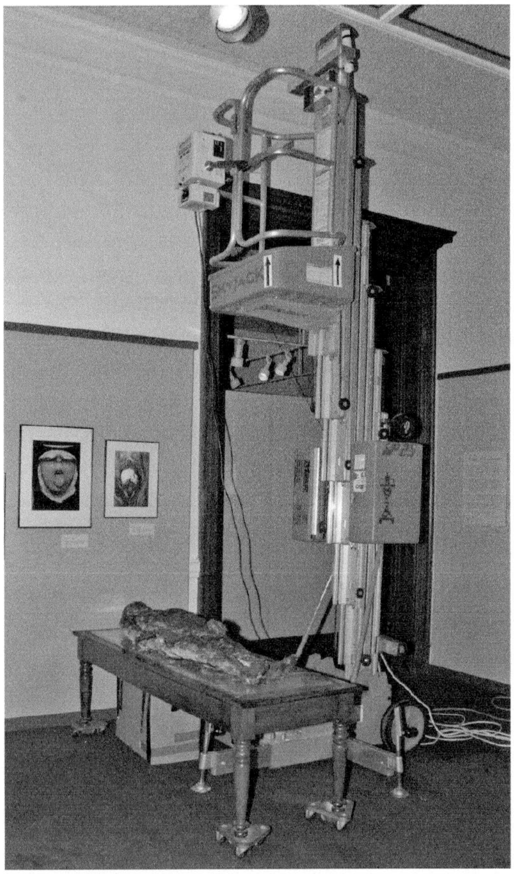

FIGURE 37.15 The x-ray tube fastened to the base of the bucket of the hydraulic elevated platform device, the SkyJack®. (From Mary Sonntag. With permission.)

FIGURE 37.16 IR plates taped onto the particleboard support. (From Mary Sonntag. With permission.)

surface to place the IR plates on. The particleboard was placed upon a sheet of black pool liner measuring approximately 84″ (213 cm) × 54″ (137 cm). The latter provided a light-tight cover for the IR plates. In order to "clear" the IR plates for the x-ray exposure, they were exposed to the fluorescent lights in the Museum corridor for 30 min. Each plate was placed upon the particleboard support and held in place with strips of duct tape (Figure 37.16). After all the plates were fixed to the support, the black pool liner was wrapped around the support and all edges were tapped. The wrapped plate holder was positioned as close as possible to the bottom of the table supporting the mummy. A total of 12 exposures at 86 kVp at 34 mAs were taken. The plate holder was removed from under the table and brought to a temporary darkroom created in a basement office. Each IR plate was removed from the support and placed into a light-tight transport box. The next day the plates were brought to FUJIFILM NDT Systems in Stamford, Connecticut for processing. Using a rubber algorithm, the resulting images demonstrated a higher resolution that could have been achieved with conventional radiographic film. In addition, since there was no space or gap between the plates they could be seamlessly reassembled to produce one continuous image of the mummy (Figure 37.17).

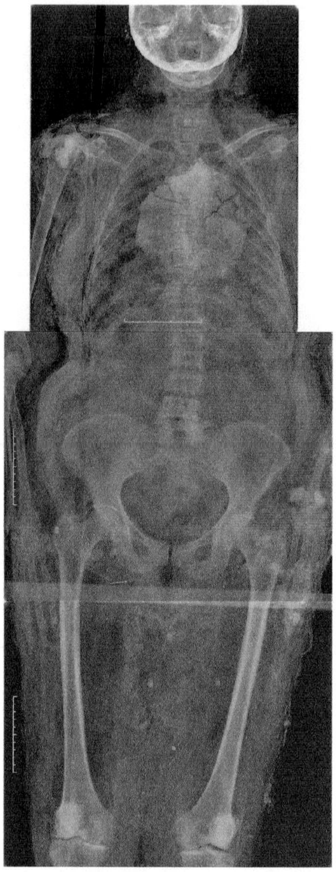

FIGURE 37.17 The three individual CR images of the Soap Lady assembled into a single image that can be displayed.

REFERENCES

1. Beckett, R., and Conlogue, G., *Paleoimaging: Field Applications for Cultural Remains and Artifacts*, CRC Press, Boca Raton, FL, 2009.
2. Lombardo, R., Personal communication, August 28, 2009. R.T.R. Technical Specialist. FUJIFILM NDT. Stamford, CT.
3. Conlogue, G. J., Schlenk, M., Cerrone, F., and Ogden, J. A., Dr. Liedy's Soap Lady: Imaging the past, *Radiol. Technol.*, 60, 411–415, 1989.
4. *Soap Lady. The Mummy Road Show*. The National Geographic Channel. Engel Brothers Media. New York, NY, 2001.
5. *Paleoimaging: Field Applications for Cultural Remains and Artifacts*, Beckett, R.G., Conlogue, G.J., p44, CRC Press.

Section X

Essentials of Diagnostic Imaging

38 Production of the Radiographic Image

Charles W. Newell, Chucri M. Jalkh, and Mark D. Viner

CONTENTS

INTRODUCTION

There are two key elements in producing a quality radiograph: (1) proper execution of the positioning methods for the subject under investigation, and (2) application of the correct exposure factors necessary to yield a radiograph of proper density and contrast. Positioning methods for deceased human subjects are described in Chapter 39, but without proper density and contrast, the usefulness of correct positioning methods is markedly diminished.

While considerable attention is paid to the use of film-based systems employing cassettes with screens in this chapter, the reader should note at the outset that digital imaging systems are steadily replacing film-based systems. Therefore, for the sake of clarity, the use of the words film or cassette refers to film-based systems and the word detector refers to digital systems. Moreover, the function of film and detector are identical as they both serve as image receptors. Additional information regarding digital systems is found later in this chapter. Regardless of the imaging system employed, this chapter will address the fundamental factors affecting radiographic quality and provide suggestions for establishing exposure factors for postmortem radiography.

FACTORS CONTROLLING IMAGE QUALITY

The factors that control image quality are density, contrast, detail, shape distortion, and magnification distortion. All of these factors are essential to produce a quality image. It is therefore important to understand the meaning of each factor and its role in the formation of the radiographic image.

DENSITY

Density refers to the overall blackness of the image. If there is too much density, the image will be overexposed. If there is too little density, the image will be underexposed. In either case, the quality of the image will be diminished. A radiographic image must therefore exhibit a proper amount of density to demonstrate anatomical structures.

There are numerous factors that produce changes in density and the reason for such changes depends on the factor involved. For instance, an increase in some factors can lead to a decrease in density and conversely, when some factors decrease, density increases. Some factors that affect density include:

- Exposure factors (mA, kVp, and exposure time)
- Distance from the x-ray tube to the detector or film
- Size of collimator setting
- Tissue thickness
- Developing temperature (for film-based systems)
- Scatter radiation
- Pathologic conditions

Contrast

Contrast is the difference between adjacent densities on a radiograph. In short, it is the difference between the black, white, and gray shades as seen in a radiographic image. The significance of contrast is that it enables the difference between bone, soft tissue and other body structures to be visualized. As contrast is essentially the difference between shades of density, the same factors that affect density also affect contrast. Therefore, any change in density will usually affect contrast.

Detail

Detail is the sharpness of an image as recorded in a radiographic image. Detail determines how well a small structure or minute fracture line will be visualized. Detail is often referred to as definition or radiographic sharpness. These terms are synonymous. Factors affecting detail are as follows:

- Focal spot size
- Distance
- Object detector/film distance
- Motion
- Imaging system speed (film/screen speed or digital detector speed)

Of the factors listed above, two require further comment because they are variables that have to be set whenever postmortem radiography is performed. As previously indicated, distance literally means the space or distance between the x-ray tube and the detector/film. Since the x-ray tube is the source of x-rays and the detector/film is the image receptor, distance is referred to as *source image-receptor distance*. The acronym is SID, and it is used throughout this chapter to refer to distance.

It is also common to find instances where space or distance exists between the patient (object) and the detector/film. This form of distance is referred to as *object image receptor distance* (OID).

The important point to be made regarding SID and OID is that both control magnification which, in turn, affects detail. The OID, therefore, should always be maintained at the least possible distance and SID at the greatest possible distance. It should be noted, however, that all images possess some degree of magnification, even when they are produced under the most favorable conditions. In recognition of this fact, distance parameters have been established to standardize the degree of magnification for all radiographic examinations. Thus, absent equipment limitations or other obstacles, chest radiographs are typically radiographed at a 72-in. (180-cm) SID and other body parts are radiographed at a 40-in (100-cm) SID. As for OID, the practice of maintaining OID at an absolute minimum still holds true.

Shape Distortion

Shape distortion refers to instances where the shape of an anatomical structure, as projected on a radiographic image, differs from its normal shape. In other words, the image is a misrepresentation of the true shape of a body part. Factors that control shape distortion are[1]:

- Angulation of the x-ray tube to the detector/film cassette
- Angulation of the body part to the detector/film cassette
- Angulation of the detector/film cassette to the body part

Shape distortion arising from angulation of the body to the detector/film is less apt to occur because oblique projections are not routinely performed in postmortem radiography. However, AP projections are routinely performed and therefore careful attention should be paid to ensure that the central ray is perpendicular to the detector/film. Otherwise the image will be elongated. In addition, it is important to ensure that the central ray is centered to both the longitudinal and transverse axes of the part being radiographed. Centering problems can be reduced by centering the body part to the cassette and then centering the central ray to the center of the cassette/detector. Off-centering not only produces shape distortion, but it can also cause grid cut-off when grids are employed.

Since lateral projections are employed in postmortem radiography, attention must also be paid to proper alignment of the body part to the detector/film. This can be easily achieved by making sure the body part and the cassette are always parallel to one another. Failure to maintain a parallel relationship between body part and cassette often yields an image that is foreshortened.

Shape distortion can also be used to better visualize a body part. Angling the central ray and oblique projections are two examples.

Magnification Distortion

As the term suggests, magnification refers to the enlargement of an anatomic part as viewed on a radiographic image. Factors controlling magnification are SID and OID.

All radiographic images possess some degree of magnification. To avoid greater magnification, the established distance parameters should be adhered to: 72 in./180 cm for chest exams and 40 in/100 cm. for all others while keeping the OID to the least possible distance. It should be noted that the effect of OID on magnification is much greater than that of SID. Therefore, while there is some margin of error for SID, there is very little for OID.

Exposure Factors

Essential Exposure Factors

The fundamental components of an x-ray unit have been already described in Chapter 34. One of the components not discussed was the x-ray control panel (Figure 38.1). The control panel enables exposure factors to be selected or changed by manipulating various control devices. The exposure factors regulated via the control panel are milliamperage (mA), exposure time (s), and kilovoltage (kVp). These factors are discussed below.

Milliamperage

Milliamperage (mA) determines the quantity of electrical current flowing through an x-ray tube during exposure. Since mA is a quantitative factor, it controls the quantity of x-rays produced which, in turn, affects radiographic density. As the mA is increased, density is increased. When mA is decreased, density will decrease.

Fixed radiographic units have a wide range of mA settings, which are adjusted to meet the needs of a particular examination or exposure technique. Typical mA settings include 50, 100, 200, 300, 400, 500, 600, 700, 800, and 1000. In contrast, mobile x-ray units have fewer mA settings. In fact, many contemporary mobile units have a fixed or nonvariable mA. In these units, density is usually controlled or varied by a timer selector or by a milliampere-second selector.

Timer Control

The timer control determines the length or duration of the exposure. Similar to mA, as exposure time is lengthened density is increased and vice versa. Like mA settings, the timer control has a wide range of settings. For example, timer selections may range from a short exposure time of 0.016 s up to 1 or more seconds. For some x-ray units, including modern mobile units, the manual timer selector has been omitted. In these units, the exposure time is automatically set as the milliampere-second selector is changed. Timer selections may be expressed in one of three modes on the control panel: fractions, decimals, or milliseconds. However, if a timer selector is present on the control panel of a mobile unit, it is unlikely to be a millisecond timer.

Of the three modes listed above, millisecond settings are not useful when calculating milliampere-seconds, which are described later in this section. When confronted with this type of timer, it will be necessary to convert milliseconds to a decimal or fraction equivalent. This is accomplished as follows:

Examples of milliseconds (ms) include 1, 2.5, 3, 4, 5, 6, 10, 20, 50, 100, etc.

Since a millisecond (ms) is 1/1000 of a second, placement of the ms number over 1000 converts the millisecond to a fraction. For example, 10 ms can be expressed as 10/1000 of a second. Thus, 50 ms is the same as 50/1000 of a second, and 100 ms is the same as 100/1000 of a second, and so forth.

As can be seen, conversion of a millisecond to a fraction often yields an unwieldy fraction. To simplify matters, milliseconds can be converted into decimal equivalents by dividing it by 1000. Using the examples above, 10/1000 becomes 0.01s, 50/1000 is 0.05s and 100/1000 is 0.10s. As these examples suggest, placing a decimal point three places

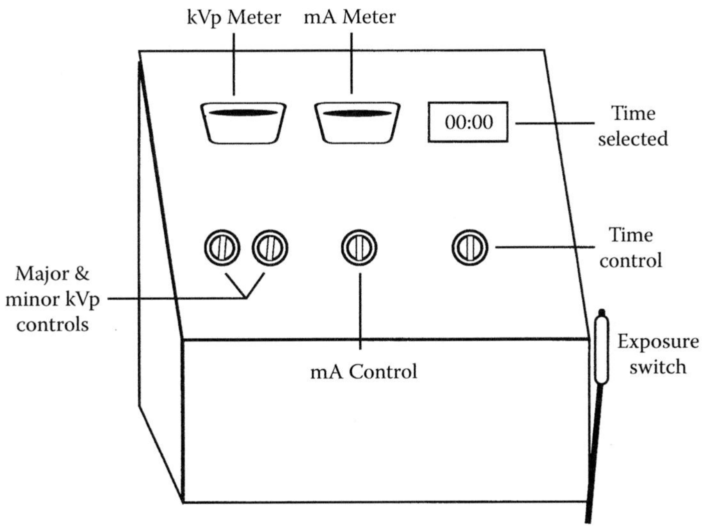

FIGURE 38.1 Schematic drawing of a control panel.

to the left of the last millisecond number produces the same effect as dividing the millisecond number by 1000.

Kilovoltage Control

Kilovoltage controls the penetrating ability of the x-ray beam. The penetrating ability of x-rays is often referred to as the quality of the x-ray beam. In general, a higher kVp is applied to thicker and/or more dense structures, whereas a lower kVp is used for thinner and/or less dense structures. Although the function of kVp is to control penetration, it also plays a secondary role in the production of radiographic density. Kilovoltage also controls the amount of scatter radiation produced in a given exposure, which can have a degrading effect on radiographic contrast.

Kilovoltage settings on the control panel typically range from 40 to 120 kVp. There are usually two kVp control selectors. One selector is designated as a major control and allows changes to be made in increments of 10 kVp. The other selector is referred to as a minor control and it provides changes in increments of 1 to 2 kVp. Some mobile units have a single kVp control which allows changes in increments of 5 kVp, for example, 50, 55, 60, 65, etc. Despite fewer selection alternatives, these units provide sufficient settings to achieve the proper level of kVp.

Exposure Switch

While the exposure switch is not a true factor affecting density and contrast, it is included in this discussion because it is an essential part of the control panel. The purpose of the exposure switch is to initiate the x-ray exposure. The exposure switch may be located directly on the surface of the control panel or wired directly to a detachable cord suspended from the control panel. Some x-ray units employ two switches to initiate the exposure, while others employ a single switch. Regardless of the number of switches employed, they require a two-step process to initiate the exposure. The first step requires the preparation ("prep") or rotor component of the switch to be depressed. When the prep switch is depressed, the tube filament begins to heat up and in the case of rotating anode tubes a whirring or high-pitched sound will be heard from the x-ray tube as the anode begins to rotate. The next step requires the exposure component of the switch to be depressed, at which time x-rays are produced. Care should be taken to avoid depressing the rotor switch unnecessarily, as this will reduce the useful life of the tube.

EXPOSURE AND DISTANCE

Although SID plays a significant role in affecting image magnification, it can also affect radiographic density. Briefly stated, when all of the other exposure factors (kVp, mA, s) remain constant, density can be increased when the SID is reduced and decreased when the SID is increased. While the preceding statement is accurate, small changes in distance, of say, 6–in., will not significantly affect density, for example. However, since distance parameters have been established over the years for routine radiographic examinations, the established distances should be adhered to in order to avoid magnification-related problems. Moreover, exposure factor recommendations in this chapter are based on specific distances.

CALCULATING MILLIAMPERE-SECONDS

A milliampere-second (mAs) is the product of mA and exposure time. Collectively, mA and exposure time control the quantity of x-rays produced. As a result, mA and exposure time directly control radiographic density. The acronym mAs will be used throughout this chapter when reference is made to milliampere-seconds.

Since mAs is the product of mA and exposure time, mAs is determined by the formula mA × s = mAs. This formula is applied in the following examples.

Time in fractions

200 mA × 3/20 s = 30 mAs
300 mA × 1/60 s = 5 mAs
100 mA × 1/10 s = 10 mAs
200 mA × 1 s = 200 mAs

Time in decimals

300 mA × 0.3 s = 90 mAs
100 mA × 0.1 s = 10 mAs
200 mA × 0.5 s = 100 mAs
400 mA × 0.3 s = 120 mAs

Time in milliseconds

200 mA × 100 ms (0.1 s) = 20 mAs
100 mA × 50 ms (0.05 s) = 5 mAs
400 mA × 20 ms (0.02 s) = 8 mAs
600 mA × 4 ms (0.004 s) = 2.4 mAs

Note: milliseconds must be converted to fractional or decimal equivalents to calculate mAs.

As previously indicated, some x-ray units employ an mAs selector in place of separate mA and time controls. In these units, the mAs selector is simply adjusted to the proper level of mAs. For other units, the mA and exposure time should be selected to produce the correct mAs. Regardless of the system employed, the correct mAs must be selected to achieve the proper radiographic density.

Digital Radiography

As noted previously, digital radiography systems are steadily replacing the film-based system. Similar to digital cameras, digital radiography systems do not require film, nor the attending requirement for film development. While the absence of film (filmless) and the development process are clearly advantageous, other advantages of digital systems over the film-based system include the ability to:

- Acquire images in a digital format
- Store a large number of digital images with substantially reduced space requirements via CDs

- Transmit images through electronic means, for example, global/local internet
- Change image density/brightness following initial exposure while the image is viewed on a video monitor. This capability is often referred to as post processing
- Produce quality diagnostic images employing a wide range of exposure factors. A two-word definition of this capability is "exposure latitude," which essentially means a wide margin of error is allowed when applying exposure factors to yield a diagnostic image, that is, mAs, kVp. However, this capability does not replace the necessity of applying appropriate technical factors through the use of technique charts, nor does it replace or reduce the importance of proper positioning methods described in Chapter 39.

As discussed in Chapters 34 and 26, there are currently two digital radiography systems. These systems are known as Computed Radiography (CR) and Direct Digital Radiography (DR). The fundamental difference in these systems is the type of image receptor/detector employed and the manner in which x-rays are converted into electronic signals. Other differences include their physical construction, size, cost, etc. As mentioned in Chapter 35, postmortem radiography is often more difficult to perform than the same examination performed on living subjects and therefore, a brief section of this chapter will focus on CR. At the present time the flexibility offered by CR systems tends to better accommodate the needs of postmortem radiography when compared to the DR systems that are widely available. Current CR systems are also readily adaptable to existing radiographic equipment, which is cost-effective. It should be noted, however, that the latest generation wireless DR systems would appear to offer the same flexibility as CR with the advantage of almost "instant" image processing. However, such units are only just beginning to enter clinical practice and have not been fully evaluated in the forensic setting.

A CR system is depicted in Figure 38.2. The basic hardware components of a CR system consist of a Computer (PC) and Keyboard, Monitor, CR Cassette and the Reader. Among other functions, the computer converts an analog signal to a digital image and is generally responsible for the overall function of the CR system. The CR cassette is similar to a film-based cassette in its outward appearance and functions as the image receptor/detector. However, the CR cassette contains an x-ray sensitive plate instead of film. The plate captures the initial latent/invisible image. Because of its procedural flexibility, the use of the CR cassette readily allows the user to perform the various positions described in Chapter 39. The final component of the CR system is the reader. The CR cassette is inserted into the reader and the reader scans the exposed CR plate with a laser beam. As a result of the laser scan, the plate emits light, which is converted into electrons by a photomultiplier tube assembly. The electrons constitute an analog electrical signal. The computer can now digitize the electrical signal and convert the signal to a visible image, which is displayed on the monitor. The final step in the CR process occurs in the reader wherein the CR plate is scanned with bright white light, which removes all of the previous information recorded on the plate. This final step is known as erasure and is necessary to avoid superimposing images acquired from previous exposures. Once erasure is performed, the plate can be reused. A summary of the basic steps in producing a CR image is as follows:

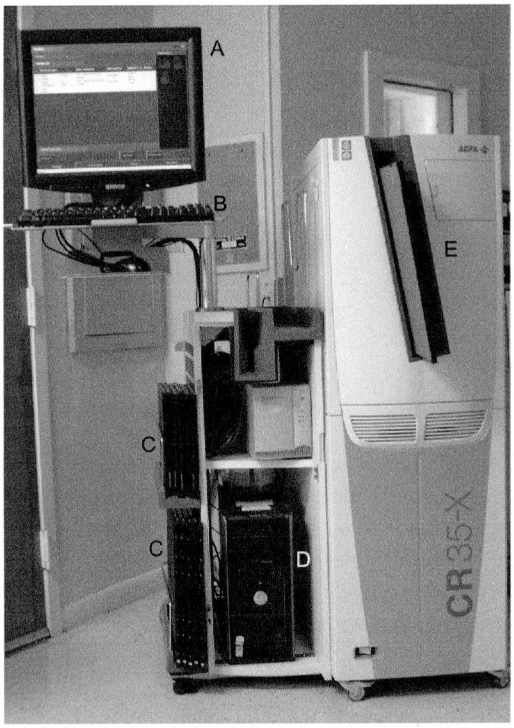

FIGURE 38.2 CR Unit: (A) Monitor, (B) Keyboard, (C) CR Cassettes, (D) Computer, (E) Reader with CR cassette inserted.

- X-rays expose CR cassette
- CR plate is scanned by a laser beam in the reader with light emission
- Electrical signal is generated
- Signal is digitized via analog–digital conversion
- X-ray image is viewed on the monitor
- CR plate is erased.

The CR system requires a few final considerations, which will be useful to the operator:

- The dose measurements listed in Table 38.1 are often referred to as sensitivity numbers or exposure indices. Sensitivity numbers are established by the vendor in consultation with those individuals employed within a given medical facility. In Table 38.1, sensitivity numbers are provided in a range of numbers. A sensitivity number is permanently recorded on each image and is based on the amount of x-ray radiation reaching the CR plate. To produce a technically correct image, the technical exposure factors employed

TABLE 38.1
Relative Dose Measurement for CR Systems by Vendor

	Vendor (units)	AGFA (lgM Value)	KODAK (Exposure Index)	FUJI (S number)	Detector Exposure Estimate (mR)
Under-exposed	Repeat	< 1.45	< 1250	> 1000	< 0.20
	Review	1.45–2.04	1250–1849	301–1000	0.7–0.2
Correctly exposed	Acceptable range	2.05–2.35	1850–2150	150–300	1.3–0.7
Over-exposed	Review	2.36–2.95	2151–2750	50–149	4.0–1.3
	Repeat if necessary	> 2.95	> 2750	< 50	> 4.0

Source: "An Exposure Indicator for Digital Radiography", American Association of Physicists in Medicine, (2009). www.aapm.org/pubs/reports/rpt_116.pdf

must yield an image possessing a sensitivity number that lies *within* the established range. Thus, a number above the range indicates overexposure and numbers below the established range indicate underexposure. Although exposures above the range have no effect upon the postmortem patient, they do indicate that additional scatter radiation was produced, which is dangerous to the user. Overexposures also place additional heat loads on the x-ray tube, which will shorten the life expectancy of the tube. Under exposure produces a common artifact known as "noise." Noise appears as a mottled effect (small black and white dots/grainy appearance) on the image, which obscures image contrast and reduces visibility of detail. To resolve this problem, the mAs must be increased, which will raise the sensitivity number to the correct index range. It should also be noted that a given x-ray exposure produces an electrical signal that is proportional to the amount of x-rays delivered to the CR plate. The degree of proportionality is referred to as signal- to-noise ratio (SNR). A low SNR will yield a mottled/grainy image

- All digital systems are sensitive to scatter radiation. Therefore accurate collimation is recommended. Additionally, grid covered CR cassettes must be employed for the same reason grids are employed in film-based cassettes. Otherwise image contrast will decline
- While the computer in a digital system plays an important role in image density/brightness, the use of appropriate kVp is absolutely essential
- CR cassettes are expensive. Care should therefore be exercised when handling them
- It is a good practice to erase CR cassettes at the beginning of the day. This will avoid the appearance of a ghosting effect upon images taken later in the day.

Digital detectors, either Direct Digital (DR) or CR, also differ in speed and sensitivity between systems. Due to the very wide latitude of these systems, it is not necessary to change detector types in order to image different body parts, to decrease radiation dose, or to increase speed. However, whilst this provides a significant advantage over film-based systems, the ability of digital systems to produce an image of apparently acceptable density and contrast in most situations can lead the operator to believe that an acceptable image has been obtained. At the extremes, the acceptable density and contrast is achieved at the expense of detail and increased "noise" or at greatly increased radiation doses. It is thus important that digital systems are set up and calibrated appropriately for the imaging examinations to be undertaken. Exposure factors for each and every examination (a technique chart) should be set for each digital system in order to achieve the optimum detail and SNR ratio consistent with the needs of the imaging service. This is discussed later in this chapter. The principles of adapting exposure factors to adjust for grid conversions and the inter-relationship between milliampere-seconds and kilovoltage described in the following sections, should be applied when making adjustments to exposures in digital systems.

ADAPTING EXPOSURE FACTORS

While kVp, mA, time, and SID are the primary factors affecting density and contrast, there are secondary factors which affect density and contrast as well. Two secondary factors which often vary in terms of their effect upon radiographic density are grids and the intensifying screens used in film-based systems.

Intensifying screens for use with film differ in accordance with their speed and mAs must be adapted from one screen speed to another to produce the same density. In the same way, there are different types of grids and therefore the mAs must be changed when different grids are employed to maintain equivalent density. Understanding the process of changing exposure factors to accommodate differences in screen speed and grids will enable exposure factors to be adapted whenever changes such as these occur.

In addition to adapting exposure factors to accommodate differences in screens and grids, it is also important to understand how and when to adjust mAs and kVp. Differences in part thickness and density of an anatomical structure are two of the more common reasons why kVp and/or mAs must be adjusted. Changing the kVp or mAs is also the way to correct over- or underexposed films when radiographs must be repeated. Methods used to adapt exposure factors to various circumstances are described below.

TABLE 38.2
Speed Values and Conversion Factors

Speed Values	Conversion Factors
50	0.5
100	1
200	2
400	4
600	6
800	8

Source: Modified from Kodak Health Sciences Publication, No. M3-138, 1984 and No. M5-15, 1993. Eastman Kodak Co., Rochester, NY.

Screen Conversions

The speed of intensifying screens was typically designated as slow, medium, and high speed. This designation no longer holds true since there are more than three speeds available. As a result, the speed of an intensifying screen is designated by a specific number. For example, the medium-speed screen is now identified by the number 100. The number 100 is now the standard against which all intensifying screens are compared. Thus, a 200-speed screen is twice as fast, as the 100-speed screen, and a 50-speed screen is one-half the speed of a 100-speed screen. Some of the screen speeds available today are shown in Table 38.2.

The speed values listed in Table 38.2 can be used to match radiographic density when changing from one screen speed to another. This is accomplished by employing the conversion factors listed in Table 38.2. The conversion factors were obtained by converting the 50 speed value to 0.5 and deleting the zeros in the remaining speed values. The conversion factors are placed in a formula to calculate screen speed changes as shown below.

$$\frac{\text{Original mAs} \times \text{Original Screen Value}}{\text{New Screen Value}}$$

Examples of screen conversions are shown below:

1. *Original factors:* 20 mAs employing a 100-speed screen.
   New factor: 50-speed screen

   $$\frac{20\,\text{mAs} \times 1}{0.5} = 40\text{ mAs}$$

   The mAs is *doubled* to maintain the same density.
2. *Original factors:* 10 mAs employing a 200-speed screen.
   New factor: 400 speed screen

   $$\frac{10\,\text{mAs} \times 2}{4} = 5\text{ mAs}$$

   The mAs is *reduced* by one-half to maintain the same density.
3. *Original factors:* 30 mAs employing a 50 speed screen.
   New factor: 200-speed screen

   $$\frac{30\,\text{mAs} \times 0.\ 5}{2} = 7.5\,\text{mAs}$$

   The mAs is *reduced* by a factor of 4 to maintain the same density.

As can be seen, if the screen speed value increases, the mAs must be reduced to maintain the same density and vice versa. If the speed of the screen being used is not known, the advice of a representative from a commercial film/screen company should be sought

The Use of Grids and Grid Conversions

The principles of secondary radiation grids (referred to simply as "grids") in minimizing unwanted scattered radiation falling upon the image receptor were discussed in Chapter 34. Grids are not necessary for all examinations. For example, grids need not be used in radiographic examinations of the hands, forearm, feet, or lower legs. In general, a grid should be employed whenever the body part measures 12 cm or greater.[2] Although grids improve the quality of a radiograph, they can produce problems if used incorrectly. Improper grid usage typically results in a loss of film blackening. The loss of film blackening is caused by grid cut-off. Following is a list of methods employed to avoid grid cut-off.[3]

1. As previously mentioned, lead strips in a grid are generally arranged parallel to one another and extend in a straight line along the longitudinal axis of the grid. For example, in a grid 14 in. × 17 in., the lead strips extend in the direction of the greatest dimension, that is, the 17-in. direction. With this arrangement in mind, the x-ray tube should never be angled against the long axis of the lead strips. It is, however, permissible to angle the tube longitudinally with the lead strips. In fact, there will be several examples in the chapter on positioning that require angulation of the CR with the long axis of the grid.
2. The grid and the x-ray tube must be perpendicular to one another. If the grid is angled against the CR, significant grid cut-off will occur.
3. Some grids require a specific distance to be maintained between the x-ray tube and the grid. If the distance requirement is violated, loss of film blackening will be seen along the longitudinal edges of the radiograph. Distance requirements are indicated on the front of the grid.

If the central ray is not centered to the longitudinal center of the grid, loss of film blackening will be seen on one side of

TABLE 38.3
Grid Ratios and Conversion Factors

Grid Rations	Conversion Factors
No grid	1
5:1	2.5
6:1	3
8:1	4
10:1	4.5
12:1	5
16:1	6

Source: Modified from *Characteristics and Applications of X-Ray Grids*, Liebel-Flarsheim Co., Cincinnati, OH. 1989.

the radiograph. Grids have a labeled front and back side. If the front side does not face the x-ray tube, a distinct loss of film blackening will be seen on both sides of the radiograph.

Grids differ from one another in accordance with their grid ratio. Grid ratio is essentially a ratio of the height of the lead strips to the distance between the lead strips. The distance between the lead strips is actually the interspace, which is filled with radiolucent material. The relationship between lead lines and interspaces is expressed as a ratio, such as 6:1, 12:1, etc. The ratio of a grid is printed on the front of the grid. An important point to remember is that higher-ratio grids absorb more scatter radiation than lower-ratio grids. Common grid ratios are shown in Table 38.3.

Although higher-ratio grids absorb more scatter radiation, they also require more x-ray exposure (mAs) to yield the same density when compared to lower-ratio grids. In fact, grids of different ratios require different amounts of mAs to produce the same density. In addition, whenever the exposure factors applied to a grid must, for whatever reason, be applied directly to a screen (no grid), the mAs must always be reduced. When variables such as these are encountered, the mAs must be changed to maintain the same density. This can be achieved by applying the conversion factors as in Table 38.3. The formula for grid conversions is similar to that applied to screen conversions. The formula is expressed as follows:

$$\frac{\text{Original mAs} \times \text{New Grid}}{\text{Original Grid}}$$

Examples of grid conversions are shown below:

1. *Original factors:* 30 mAs employing a 6:1 grid.
   New factor: 12:1 grid

   $$\frac{30\,\text{mAs} \times 5}{3} = 50\,\text{mAs}$$

   The mAs is *increased* by 20 to maintain the same density.
2. *Original factors:* 10 mAs employing a 5:1 grid.
   New factor: no grid

   $$\frac{10\,\text{mAs} \times 1}{2.5} = 4\,\text{mAs}$$

   The mAs is *reduced* by 6 to maintain the same density.
3. *Original factors:* 50 mAs employing an 8:1 grid.
   New factor: 16:1 grid

   $$\frac{50\,\text{mAs} \times 6}{4} = 75\,\text{mAs}$$

   The mAs is *increased* by 25 to maintain the same density.

The examples above illustrate that grids with different ratios cannot be interchanged without making appropriate changes in mAs. However, a variety of both screens and grids can be utilized confidently by applying the mAs conversion factors for screens and grids.

Milliampere-Second and Kilovoltage Adjustments

Both mAs and kVp can be adjusted to accommodate a variety of exposure factor needs. Examples include correcting for over- or underexposed radiographs and for the numerous variations that exist amongst patients. Since there is a direct relationship between mAs and density, there are a few fundamental considerations to keep in mind:

1. If the mAs is doubled, the density will be doubled. If the mAs is tripled, the density will be tripled, etc. For example, if 50 mAs is changed to 100 mAs, the density will be doubled. In the same way, if 50 mAs is changed to 150 mAs, the density will be three times greater.
2. If the mAs is cut in half, the density will be reduced by one-half. If the mAs is reduced by a factor of 3, the density will be reduced to one-third of the original, etc. For example, if 50 mAs is changed to 25 mAs, the density will be reduced by one-half. Similarly, if 50 mAs is reduced to 16 mAs, the density will be decreased by two-thirds.

The question of how much to increase or decrease the density in a given situation can only be answered once sufficient experience has been acquired in making mAs adjustments. However, it is important to note that it requires a *minimum* change of 30% in mAs to see a visible change in density. This is referred to as the 30% rule.[4] Examples of the 30% rule are shown below. (*Note: Only the mAs has been changed.*)

1. If an exposure of 50 mAs produces a radiograph that is slightly underexposed, what is the *minimum* amount of mAs necessary to correct the image?

   $$30\% \times 50 = 15,$$

$$50 + 15 = 65 \text{ mAs (the new mAs)}.$$

2. If an exposure of 100 mAs produces a radiograph that is slightly overexposed, what is the *minimum* amount of mAs necessary to correct the image?

$$30\% \times 100 = 30,$$

$$100 - 30 = 70 \text{ mAs (the new mAs)}.$$

In actual practice, the majority of mAs changes will be greater than 30% (50%, 100%, etc.). Thus, 30% is simply the *minimum* starting point for making exposure changes using mAs.

The kVp can also be used to make changes in density. While there are other ways to effect changes in density using kVp, the authors have found the 20% kVp rule to be an effective method. The 20% rule is based on the premise that a 20% (0.20) increase in kVp will produce a doubling effect on density. Likewise, if the kVp is reduced by 20%, the density will be reduced by one-half. Other variations in the 20% rule include increasing or decreasing the kVp by 10% (0.10) to yield a 50% increase/decrease in density. While a 5% (0.05) change in kVp will produce a 25% change in density, such a change has little significance in the clinical setting. Examples of the 20% rule are shown below. (*Note: only the kVp has been changed.*)

1. A radiograph employing 10 mAs at 70 kVp is judged to be 100% overexposed.
   Solution:

$$70 \text{ kVp} \times 0.20 = 14,$$

$$70 - 14 = 56.$$

   The new exposure factors are 10 mAs and 56 kVp.
2. A radiograph employing 20 mAs at 80 kVp is judged to be 50% underexposed.
   Solution:

$$80 \text{ kVp} \times 0.10 = 8,$$

$$80 + 8 = 88.$$

   The new exposure factors are 20 mAs and 88 kVp.

While the 20% rule can be used to effect changes in density, it should be understood that changing the kVp affects image contrast. Significant changes in image contrast may be considered objectionable under certain circumstances. In addition, anatomical structures require a minimum or effective kVp. Effective kVp refers to the minimum kVp required to penetrate a given structure.[5] Thus, the 20% rule must be applied with effective kVp in mind. As with mAs changes, the amount of kVp change applied to a given situation can only be determined following sufficient practical experience.

Structure Thickness and Density as Factors in Radiographic Density

As the term suggests, structure thickness refers to the thickness of a body structure. Variations in body thickness are particularly evident in the chest and abdomen. In general, thicker structures require greater exposure factors. For exposure factor determinations, thickness is measured in centimeters by employing a measuring device referred to as a caliper (Figure 38.3). The technique chart presented later in this chapter provides exposure factors that are based, in part, on the measurements of anatomic structures. Some areas will list specific measurements while others will list measurements in ranges of centimeters. To utilize the chart effectively, the structure must be correctly measured. The correct procedure to follow is to measure the area of the part through which the central ray will pass. This area is typically placed in the center of the cassette.

With experience, it will not be necessary to measure parts such as the hands, wrists, feet, etc. unless there is an obvious decrease or increase in a body part. The chest and abdomen should always be measured, including thoracic and lumbar spine examinations. The density of a structure refers to how close the atoms or molecules of a structure are bound together, that is, the compactness of a structure. Hence, bone has a higher/greater density than fat. Depending on the quantity of each, muscle, blood, and water often exhibit similar densities. Air has the lowest density. Higher-density structures absorb more x-rays and require more exposure

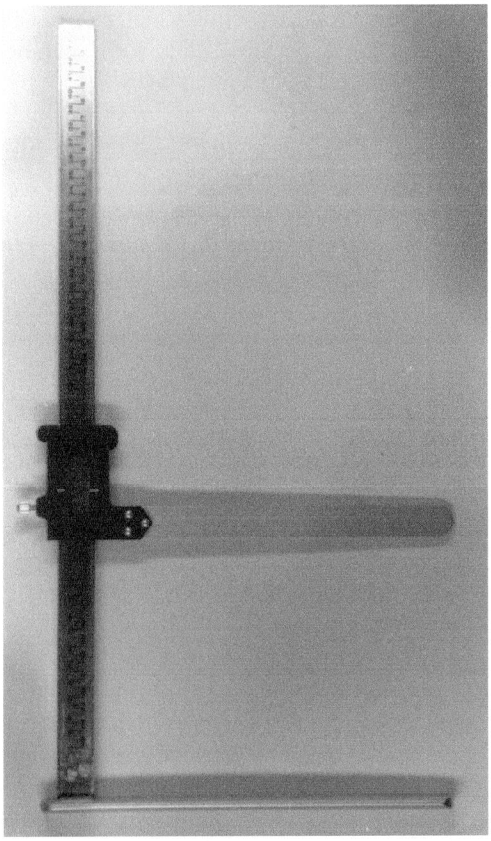

FIGURE 38.3 Radiographic calliper.

(kVp and mAs) to yield satisfactory radiographic densities. The opposite is true of less dense structures. The thickness and density of a structure are major considerations in the selection of exposure factors. This is particularly true in postmortem radiography where the lungs are not inflated with air, or instances where the tissues are gas-filled due to decomposition or dehydrated from incineration. Variables such as these require special consideration. A few suggestions are offered below.

1. Despite the other suggestions that follow, it should be understood that it is virtually impossible to determine or quantify many of the variables that may arise based on a visual inspection of a cadaver. With this in mind, it is strongly recommended that a single projection be taken and then the radiograph reviewed to evaluate what changes, if any, should be applied to the succeeding radiographs. The previously described mAs or kVp rules can then be applied to make the appropriate corrections. An image taken for this purpose is referred to as a scout image. Providing that the corrections are employed properly, the scout image is the best method for ensuring that the remaining radiographs are correctly exposed. The time spent in reviewing the results of the scout image will, in the long run, reduce the time for the overall procedure by decreasing the number of repeat exposures.
2. With less certainty than the scout image method, there are some rules of thumb which can be helpful in establishing exposure factors for a number of physical variables as described below.
   a. For a lung containing water or other fluid, the original mAs can be increased by 34–40%, or the original kVp increased by 6–8.
   b. Freezing greatly increases tissue density and causes "ice artifacts."[6] If possible, bodies should be thawed before radiography. Trial exposure on frozen tissue should begin with a 40–50% increase in mAs or an increase of 8–10 kVp.
   c. For gas-filled tissues, original mAs can be decreased by 30%, or the kVp decreased by 5–6.
   d. Incinerated and/or dehydrated body parts require a decrease in the original mAs of 30–40%, or a decrease in the original kVp of 6–8.
   e. Denuded bones or skeletal remains require a substantial reduction in mAs. Two suggestions are offered below:
      i. All extremities require the mAs to be reduced by one-half of the usual mAs applied to extremities. A nongrid cassette is employed for all exposures.
      ii. The remaining skeletal parts can be radiographed employing a mAs ranging from 2 to 4. A nongrid cassette is employed for all exposures.
      iii. The suggested kVp range for all exposures is 60–70.

Single-Phase and Three-Phase Generators

In the event a three-phase generator is used for postmortem radiography, a simple conversion method can be employed: to convert single-phase exposure factors to a three-phase x-ray generator, simply reduce the original mAs by one-half. Thus, the mAs for a three-phase generator will be one-half that of a single-phase generator.

Technique Chart

The exposure factors listed on the technique chart below match the positions/projections described in Chapter 39. The exposure factors were developed to meet the needs of a particular radiographic unit. It is doubtful that this chart, or any other technique chart, will completely satisfy the exposure requirements for all conditions or all x-ray units. This is particularly true in light of the previous discussion regarding structure thickness and density. Thus, the technique chart should be considered only as a guide. In short, it is the beginning point from which the medical examiner assistant can formulate his/her own exposure factors. Hence, the suggestion of taking a scout image is equally applicable to the technique chart.

Prior to using the technique chart, the type of generator used, screen speed, and grid ratio should be noted. Any differences between these factors and those employed at another site can be equalized by applying those conversion factors discussed in this chapter—for example, screen values, grid conversions, etc. Differences in equipment performance can also be equalized by applying mAs and kVp adjustments. However, the suggested SID should not be changed. Once all of the conversions are made, any further changes should be made by changing either the mAs or kVp, but not both.

Finally, it should be noted that the chart is divided into anatomic regions. Adjacent to each region is a listing of projections appropriate for a specific region. The mAs and kVp listings are based on measurements of the thickness of an anatomic structure. The measurements are listed as a range of centimeters, which correspond to a specific projection. The measurements listed for each anatomic region usually vary when the projection changes. Should the thickness of a part to be radiographed measure above or below the centimeter ranges listed, the kVp can be increased or decreased by 2 for each centimeter of variation. For example, if a structure measures 12 cm and the listed centimeter range is 8–10, the kVp can be increased by 4 to accommodate the difference in thickness. Four kVp would be subtracted if the structure measured 6 cm. Hence, kVp should be varied by 2 kVp/cm to accommodate changes in thickness.

The technique chart in Table 38.4 was prepared using the following equipment:

- Single-phase x-ray unit
- 400 Screen/film system
- 8:1 Grid with 103 lines per inch

TABLE 38.4
Technique Chart For Film-Screen Systems

Region	Projection	Cm. Range	mAs	kVp	Grid	SID
Hand and wrist	AP/PA	3–5	3	55–58	No	40″
	Lateral	3–6	3	60–62	No	40″
Forearm	AP	6–8	3	62–64	No	40″
	Lateral	7–9	3	64–68	No	40″
Humerus	AP	6–10	20	62–66	No	40″
	Lateral	6–10	20	66–70	No	40″
Shoulder	AP	10–14	20	64–66	No	40″
Foot	AP	5–8	3	60–62	No	40″
	CTL	6–9	3	64–66	No	40″
Tibia-Fibula	AP	9–13	3	75–78	No	48″
	CTL	8–12		72–76	No	48″
Knee	AP	10–13	20	66–70	Yes	40″
	CTL	9–12	20	62–66	Yes	40″
Femur	AP	15–18	25	72	Yes	40″
	CTL	14–17	25	70	Yes	40″
Hip	AT	15–20	40	76	Yes	40″
		21–25	60	76		
		26–31	100	80		
	CTL	23–28	125	85	Yes	40″
Pelvis	AP	15–20	40	76	Yes	40″
		21–25	70	76		
		26–31	125	80		
Lumbar spine	AP	15–20	50	72	Yes	40″
		21–25	100	76		
		26–31	200	80		
	CTL	20–25	100	80	Yes	40″
		26–30	200	85		
		31–36	300	85		
Thoracic spine	AP	15–20	50	75	Yes	40″
		21–25	100	78		
		26–31	150	80		
	CTL	19–24	75	80	Yes	40″
		25–31	150	80		
		32–37	225	85		
Cervical spine	AP	10–13	20	68	Yes	40″
	CTL	10–13	20	90	Yes	72″
			10	90		48″
Chest	AP	15–20	4	100	Yes	48″
		21–25	8	100		
		26–31	12	100		
	CTL	19–24	12	100	Yes	48″
		25–31	24	100		
		32–37	36	100		
Abdomen	AP	15–20	40	72	Yes	40″
		21–25	80	76		
		26–31	150	80		
	CTL	20–25	80	80	Yes	40″
		26–30	160	85		
		31–36	240	85		
Skull	AP	18–21	30	80	Yes	40″
	CTL	14–16	30	68	Yes	40″
	AP axial	18–21	30	85	Yes	40″
	CR 30° caudal					

continued

TABLE 38.4 (continued)
Technique Chart For Film-Screen Systems

Region	Projection	Cm. Range	mAs	kVp	Grid	SID
Facial bones	AP	18–21	30	78	Yes	40″
	CR 15° cephalad					
	AP axial	18–21	30	80	Yes	40″
	CR 30° cephalic					
	CTL	14–16	25	70	Yes	40″
Mandible	AP	17–20	25	76	Yes	40″
	CR 15° caudal					
	CTL oblique	8–12	8	66	No	40″
	CTL	14–16	25	70	Yes	40″

Note: Should a grid be deleted from the above suggested exposure factors, one must apply the grid conversion factors discussed previously in this chapter.

Exposure Control for Digital Imaging

Technique charts are an essential quality assurance component of any diagnostic x-ray department. They are especially important in CR and DR, where the wide exposure latitude makes it possible to obtain clinical images that are severely underexposed, resulting in noisy or grossly overexposed images, resulting in possible data saturation and high radiation doses. Improper exposures in DR are more likely to go unnoticed since image processing can correct these deficiencies to a much greater degree than film-screen based systems due to the wide latitude of the image receptor. Exposure charts are especially critical when using CR in the absence of automatic exposure devices, such as is the case with portable equipment or when undertaking "out of bucky" work, such as extremity examinations and cross-table laterals—precisely the type of situations that are encountered in the forensic setting.

It is important to understand what constitutes a properly exposed radiograph. In film-screen systems, the limited dynamic range of the film determines the exposure latitude. In the case of DR, wider exposure latitude is permitted due to the differing energy responses of the systems and the variability introduced by dose estimation systems. Most CR and DR systems incorporate a "relative dose" measurement indicator, which is displayed as a numerical value for each exposure made (Table 38.1). Each vendor has also defined a recommended range of relative exposure that defines an appropriately exposed radiograph and thus relative exposure values falling outside this range can be identified as over- or under exposures and evaluated accordingly. Overexposures should not be repeated unless information has been lost, but their evaluation is important in order to revise technique charts and reduce the overall exposure to the operators. It is important for the technologist, radiologist, and medical examiner to review underexposed radiographs to determine a threshold of acceptability below which radiographs will need to be repeated.[7]

Exposure charts should thus be devised for each digital x-ray system and detector combination. As digital systems allow for a choice of speed class, separate technique factors may need to be derived for each speed class that will be used. An example technique chart for a CR system is shown in Table 38.5.

The technique chart (Table 38.5) was prepared using the following equipment:

- Three-phase 12 pulse generator
- 8:1 Grid with 100 lines per inch

IMAGE IDENTIFICATION

Image identification refers to the placement of information on an image, which serves to distinguish one patient from another. It is also used to determine which extremity (right or left) was radiographed, or to indicate a patient's right or left side. Proper identification plays an important role when older studies must be compared to more recent examinations. Accurate image identification is essential in forensic imaging, both to identify the subject part under examination and to maintain continuity of evidence. In this regard, film identification is an inherent part of the record-keeping system. Examples of data ideally included in image identification are listed below.

- Date of examination
- Time of Examination
- Place of Examination/Name of institution
- File, case, or accession number
- Name of subject (if known)
- Gender (if known)
- Date of birth (if known)
- Technologist/Radiographer Identification (Initials or numbers)
- Referring Medical Examiner Identification
- Anatomic Markers (Left, Right etc.)

The radiograph may eventually be admitted as evidence and the entered information is part of the chain of evidence. It is recommended that as much data as possible and practicable is recorded on the image *at the time of exposure and/or prior to processing.*

TABLE 38.5
Example Technique Chart For CR Systems

Region	Projection	Speed Class	Cm. Range	mAs	kVp	Grid	SID
Hand and Wrist	AP/PA	100	3–5	2.2–4.4	54	No	40″
	Lateral	100	6–8	4.5–9.0	60	No	40″
Forearm	AP	100	6–8	2.9	64	No	40″
	Lateral	100	7–9	6	64	No	40
Humerus	AP	100	6–10	2.9	72	No	40″
	Lateral	100	6–10	4.8	72	No	40
Shoulder	AP	400	16–20	6–12	76	Yes	40″
Foot	AP	100	5–8	4.4	64	No	40″
	CTL	100	6–9	9.3	64	No	40
Tibia-fibula	AP	100	9–13	6.6	68	No	48″
	CTL	100	8–12	6.6	68	No	48″
Knee	AP	100	10–13	6.6	70	Yes	40″
	CTL	100	9–12	6.6	70	Yes	40
Femur	AP	400	15–18	16	76	Yes	40″
	CTL	400	14–17	16	76	Yes	40″
Hip	AP	400	15–20	10	80	Yes	40″
			21–25	20	80		
			26–31	60	80		
	CTL	400	23–28	80	85	Yes	40″
Pelvis	AP	400	15–20	10	80	Yes	40″
			21–25–26–31	20	80		
				60	80		
Lumbar Spine	AP	400	15–20	10	80	Yes	40″
			21–25	20	80		
			26–31	60	80		
	CTL	400	20–25	30	80	Yes	40″
			26–30	80	80		
			31–36	160	90		
Thoracic Spine	AP	400	15–20	8.6	76	Yes	40″
			21–25	18	76		
			26–31	52	76		
	CTL	400	19–24	15	76	Yes	40″
			25–31	40	76		
			32–37	120	76		
Cervical Spine	AP	400	10–13	6.6	76	Yes	40″
	CTL	400	10–13	10	76	No	72″
				10	76	Yes	48″
Chest	AP	200	15–20	4	80	Yes	48″
			21–25	6.6	80		
			26–31	10	80		
	CTL	200	19–24	10	90	Yes	48″
			25–31	20	90		
			32–37	26	90		
Abdomen	AP	400	15–20	10	80	Yes	40″
			21–25	20	80		
			26–31	60	80		
	CTL	400	20–25	30	80	Yes	40″
			26–30	80	80		
			31–36	160	90		
Skull	AP		18–21	13.3	80	Yes	40″
	CTL		14–16	6.6	76	Yes	40″
	AP Axial CR 30° Caudal		18–21	20	80	Yes	40″

continued

TABLE 38.5 (continued)
Example Technique Chart For CR Systems *(New Table)*

Region	Projection	Speed Class	Cm. Range	mAs	kVp	Grid	SID
Facial bones	AP CR 15° Cephalad		18–21	13.3	76	Yes	40″
	AP Axial CR 30° Cephalad		18–21	13.3	80	Yes	40″
	CTL		14–16	5	76	Yes	40″
Mandible	AP CR 15° Caudal		8	8	74	Yes	40″
	CTL Oblique		6.6	6.6	64	No	40″
	CTL		8	8	74	Yes	40″

Source: From Viner, M. With permission.
Note: This chart is provided as an example. X-ray generator output characteristics and CR plate energy sensitivities vary widely and a separate chart should be devised for each system in use.

Since lead is radiopaque, lead letters are placed on the front of a cassette to indicate a variety of information: for example, right or left side, right or left leg, etc. Lead numbers are used whenever it is necessary to record information in numerical form. Lead may also be fashioned into arrows. Lead arrows are used to indicate when a patient is upright or, in conjunction with an R or L, to indicate which side of the patient's body is directed upward.

There are several ways to place information onto the radiograph. In the case of a digital radiograph, image data can be recorded digitally by entering the desired information for each examination into the imaging system program. This information is then displayed on the image header (Figure 38.4).

Informational data can be printed or written (legibly) directly onto a film with a permanent-ink ball-point or felt-tip pen or felt marker immediately after film processing.

Both these methods are considered "post processing" methods, achieved by an additional step in the process. They, therefore, introduce the possibility of incorrect labeling through error and are open to the possibility of tampering.

Neither will prevent errors in side-to-side location of the body or part. That must be done before the detector/film is exposed using lead markers.

Lead letters and numbers can be used to enter informational data as the detector/film is exposed if "R" or "L" markers are included, then the problem of lateralization also is solved (Figure 38.5).

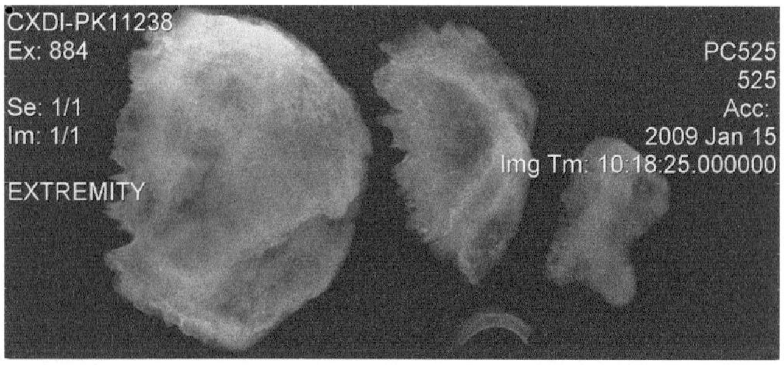

FIGURE 38.4 Digital image showing identification information displayed.

Data such as those listed above may be placed on a film by a device referred to as an identification printer or camera. The printer requires the use of cassettes with small metal windows located in the corner of the cassette. The window is referred to as an ID blocker. The ID blocker has two functions: to block or prevent x-rays from reaching the area of the film directly beneath the blocker and to automatically open when placed in the printer. When the ID blocker is placed in the printer and opened, light passes through an index card and information previously typed on the card is printed or flashed on the film. Figure 38.6 illustrates the ID printer system described above. (*Note*: Some digital CR systems also have an indicating mark on the cassette that defines the position of the "identification window." This position is merely indicative as the image identification information can be moved to a different location on the image during postprocessing.

1. While ID printers and lead letters serve an important function, they must be properly utilized to achieve their intended purpose. A few suggestions are offered below regarding the proper use of image identification devices.
2. Care must be exercised to prevent the placement of anatomic structures over identification markers or the placement of ID devices beneath anatomic areas of interest. Should either of these occur, anatomic structures may be hidden.
   It is important to identify the correct side of a patient's body. In the supine or prone recumbent position, either the right or left side of the body can be identified. For instance, if it is decided to mark the right side, then a right lead marker must be placed on the right side of the patient's body. Examinations that allow discretionary placement of markers to indicate a patient's right or left side when the patient is in the prone or supine position are listed below. In addition, the examinations listed

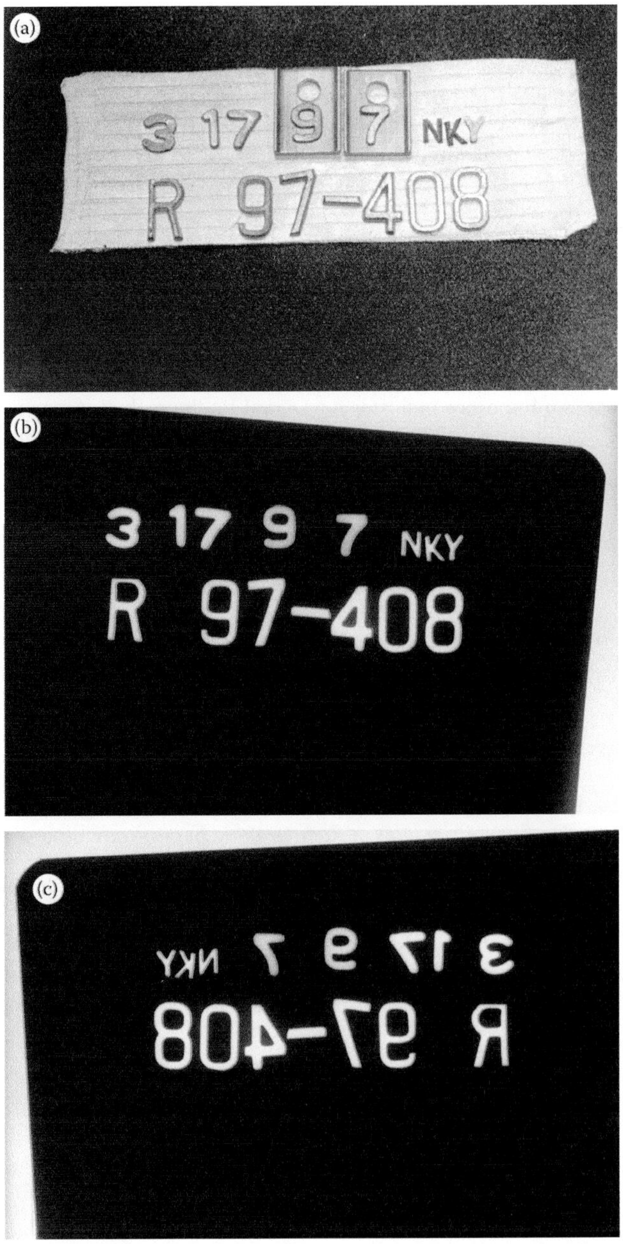

FIGURE 38.5 (a) Lead numbers placed on adhesive tape providing data (left to right, top to bottom): month, day, year, initials of person making the exposure, right side of body, accession number. The tape is placed face-down (sticky-side down) on top of the cassette before exposure. Exposed and processed films are viewed as if facing the patient or body. Therefore, if the lead data images are readable, (b) the x-ray beam direction was posteroanterior; if the numbers and letters are backward, (c) the beam was directed anteroposterior.

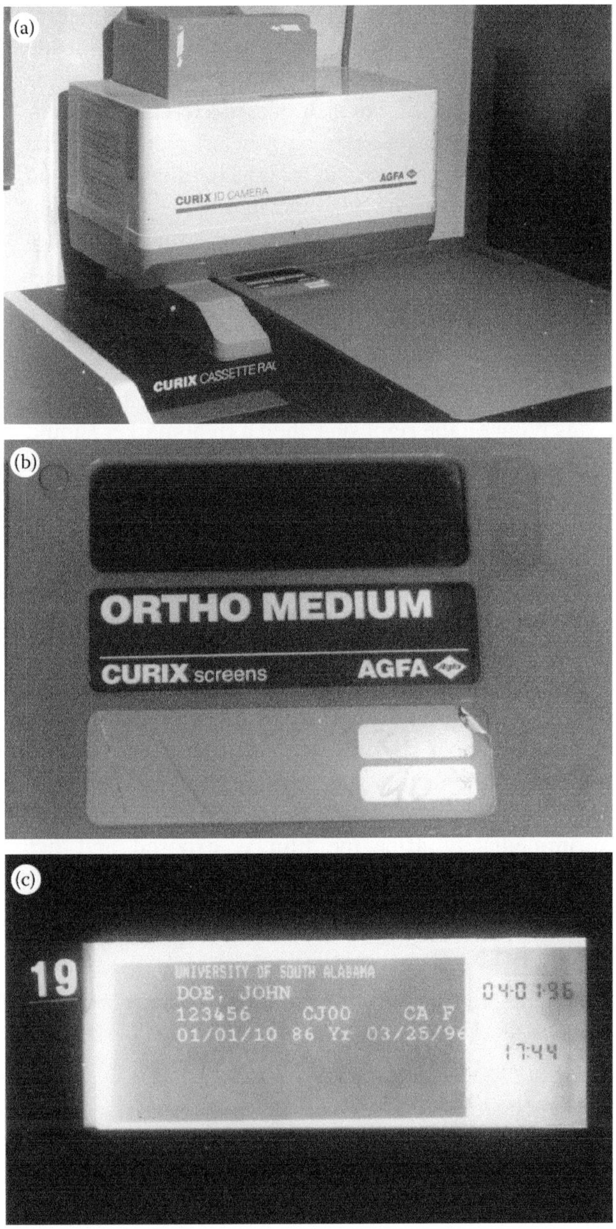

FIGURE 38.6 (a) ID printer with cassettes in place. (b) ID blocker. (c) Exposed and processed film showing imprinted data.

below limit the discretionary placement of right and left markers to AP and PA projections.

- Cervical spine
- Chest
- Thoracic spine
- Abdomen
- Lumbar spine
- Pelvis

3. It is important to remember that right and left markers are also used to differentiate between body parts that are often identical in their radiographic appearance. Examples include the right or left foot, leg, thigh (femur), hand, and forearm. Other body parts possessing a right and left side or segment include the skull and mandible. As can be seen, proper use of right and left markers is essential to correctly identify body parts.
4. Previous mention was made of the use of lead arrows to indicate direction. The use of lead arrows is typically restricted to upright and lateral decubitus positions. For upright positions, the arrow is placed on the same side as the R or L marker and it

is directed cephalad. When used for the lateral decubitus position, the arrow is placed on the side opposite to the side the patient is resting on, and is directed upward. For example, in a left lateral decubitus, the patient rests on the left side. The lead arrow is placed next to a right marker located on the patient's right side, and is directed away from the patient's left side. The arrow simply indicates which side of the patient is up.

5. For lateral and cross-table lateral projections, the correct procedure is to identify the side of the patient placed against the cassette. For example, if the cassette is next to the patient's right side or if the patient is lying on the right side, then a right marker should be placed on the cassette.
6. The placement of lead markers in oblique projections is the same for those examinations previously listed under supine or prone positions: for example, chest, lumbar spine, abdomen, etc. However, when a body part, such as a hand, wrist, or leg, is placed in the oblique position, the part must be identified with the appropriate right or left lead marker.

Lead markers of all kinds should be taped to the cassette before exposure factors are selected and the exposure is made. The ID printer is used to flash information onto the film immediately after the film is exposed and before processing.

REFERENCES

1. Thompson, T., *Cahoon's Formulating X-Ray Techniques*, 9th ed., Duke University Press, Durham, NC, 1979, p. 130.
2. Thompson, T. T., *Cahoon's Formulating X-Ray Techniques*, 9th ed., Duke University Press, Durham, NC, 1979, p. 100.
3. Thompson, T. T., *Cahoon's Formulating X-Ray Techniques*, 9th ed., Duke University Press, Durham, NC, 1979, p. 102.
4. Thompson, T., *Cahoon's Formulating X-Ray Techniques*, 9th ed., Duke University Press, Durham, NC, 1979, 170.
5. Thompson, T., *Cahoon's Formulating X-Ray Techniques*, 9th ed., Duke University Press, Durham, NC, 1979, 229.
6. Notman, D. N. H., Anderson, L., Beattie, O. B., and Amy, R., Arctic paleoradiology: Portable radiographic examination of two frozen sailors from the Franklin expedition (1845–1848), *Am. J. Roentgenol.*, 149, 347, 1987.
7. American College of Radiology, *Practice Guideline for Digital Radiography*, 2007 (Res 42), ACR, 2007, p. 52.
8. "An Exposure Indicator for Digital Radiography", American Association of Physicists in Medicine, (2009). www.aapm.org/pubs/reports/rpt_116.pdf

BIBLIOGRAPHY

Bushong, S. C., *Radiologic Science for Technologists*, 4th ed., C. V. Mosby, St. Louis, 1988.

Carlton, R. R., and Adler, A. M., *Principles of Radiographic Imaging*, Delmar Publishers, Albany, NY, 1992.

Characteristics and Applications of X-Ray Grids, Liebel-Flarsheim Co., Cincinnati, OH. 1989.

Curry, T. S., Dowdey, J. E., and Murry, R. C., *Christensen's Introduction to the Physics of Diagnostic Radiology*, 3rd ed., Lea & Febiger, Philadelphia, 1984.

Devos, D., *Basic Principles of Radiographic Exposure*, Williams & Wilkins, Philadelphia, 1995.

Kodak Health Sciences Publication, No. M3-138, 1984 and No. M5-15, 1993, Eastman Kodak Co., Rochester, NY.

Thompson, T., *Cahoon's Formulating X-Ray Techniques*, 9th ed., Duke University Press, Durham, NC, 1979.

Wallace, J. E., *Radiographic Exposure Principles and Practice*, Davis, F. A., Ed., Philadelphia, 1995.

39 Radiographic Positioning

Charles W. Newell, Chucri M. Jalkh, and Emily Faircloth

CONTENTS

INTRODUCTION

Consistent and accurate positioning is a key element in the overall quality of a radiographic image. This is especially true when attempting to match antemortem and postmortem radiological images of an individual. The task of matching antemortem and postmortem radiological images becomes more complex when the remains of an individual are degraded due to a myriad of environmental factors. The task, however, can be rendered less complex by producing radiographic images that closely approximate those obtained in the antemortem state.

Since it is common practice in many centers to perform postmortem radiography in the absence of a trained radiographer, the focus of this chapter is to provide the medical examiner's assistant with the fundamental skills necessary to produce quality radiological images. To achieve this objective, this chapter will include terminology used in radiographic positioning, essential technical considerations, and radiographic positions to demonstrate specific anatomic structures. Careful attention to the sections on radiographic positioning terminology and technical considerations will enhance understanding of radiographic positioning.

RADIOGRAPHIC POSITIONING TERMINOLOGY

Positioning terminology provides a foundation for understanding and applying the fundamental concepts of radiographic positioning. Terms commonly used in radiographic positioning are assembled under various subheadings in accordance with their general usage, and are defined below.

Anatomic Aspects

Anterior—located before or in front of; generally refers to the ventral or abdominal side of the body (Figure 39.1).

Caudal or caudad—toward the feet; situated beneath the head; inferior in position (Figure 39.1).

Cephalic or cephalad—toward the head; superior in position (Figure 39.1).

Distal—situated far from the point of attachment or origin (Figure 39.1).

External—refers to a structure outside of the body.

Internal—refers to a structure within the body; opposite of external.

Lateral—pertaining to the side; structures situated away from the middle of the body or away from the median plane (Figure 39.1).

Medial—refers to the middle; structures located toward the middle of the body or toward the median plane (Figure 39.1).

Posterior—situated toward the back; generally refers to the back or dorsal portion of a structure; opposite of anterior (Figure 39.1).

Proximal—nearest the point of attachment or origin. Opposite of distal (Figure 39.1).

Anatomic Planes

Anatomic planes are imaginary lines that divide the body into various sections. Three of the more useful planes are defined below (Figure 39.2).

Coronal plane—a vertical plane at right angles to the midsagittal plane that divides the body into anterior and posterior sections.

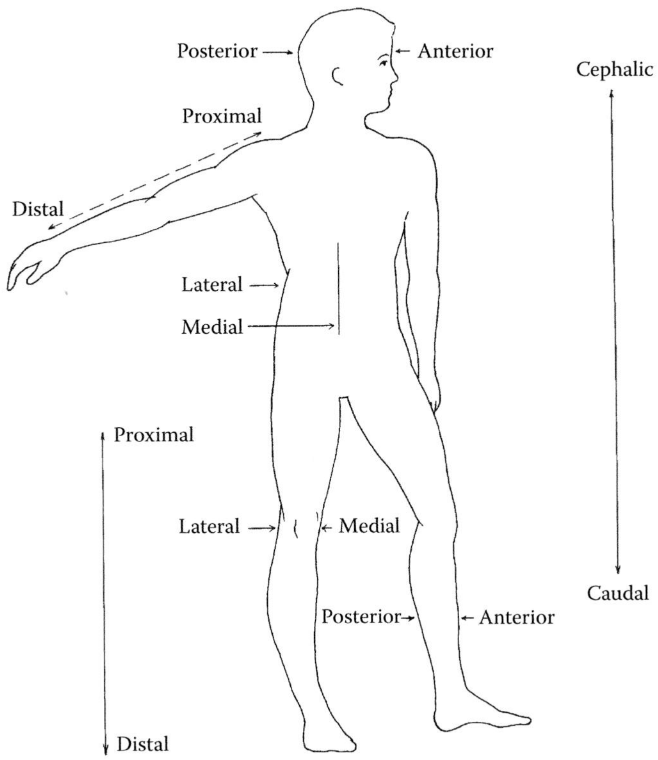

FIGURE 39.1 Anatomic aspects.

Midsagittal or median plane—a vertical plane that divides the body into right and left sections.

Transverse or horizontal plane—a horizontal plane that divides the body into upper and lower sections.

Anatomic Movement

Abduction—moving a body part away from the median plane of the body (Figure 39.3a).

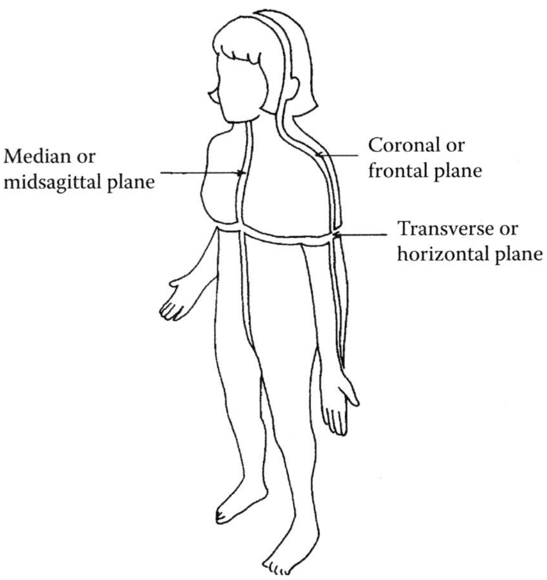

FIGURE 39.2 Anatomic planes.

Adduction—moving a body part toward the median plane of the body; opposite of abduction (Figure 39.3b).

Flexion—the act of bending a body part, such as the knee or elbow joint, which has the effect of reducing the angle of the joint (Figure 39.4a).

Extension—the act of moving a body part, such as the knee or elbow joint, which produces a straightening effect of the joint (Figure 39.4b).

Eversion—the act of turning a body part in an outward direction (Figure 39.5a).

Inversion—the act of turning a body part in an inward direction (Figure 39.5b).

Pronation—refers to placing the abdomen and chest against a flat surface, or the placement of the hand so that the palm faces in a downward direction (Figure 39.6a).

Supination—the act of placing the back of a patient against a flat surface, or the placement of the hand so that the palm faces in an upward direction (Figure 39.6b).

Rotate—the act of turning a body part around an axis; rotating the head or the arm.

Radiographic Positions

Radiographic positioning terminology is employed to describe the actual position of the body while it is being radiographed. These terms, along with their respective acronyms, are listed and defined below.

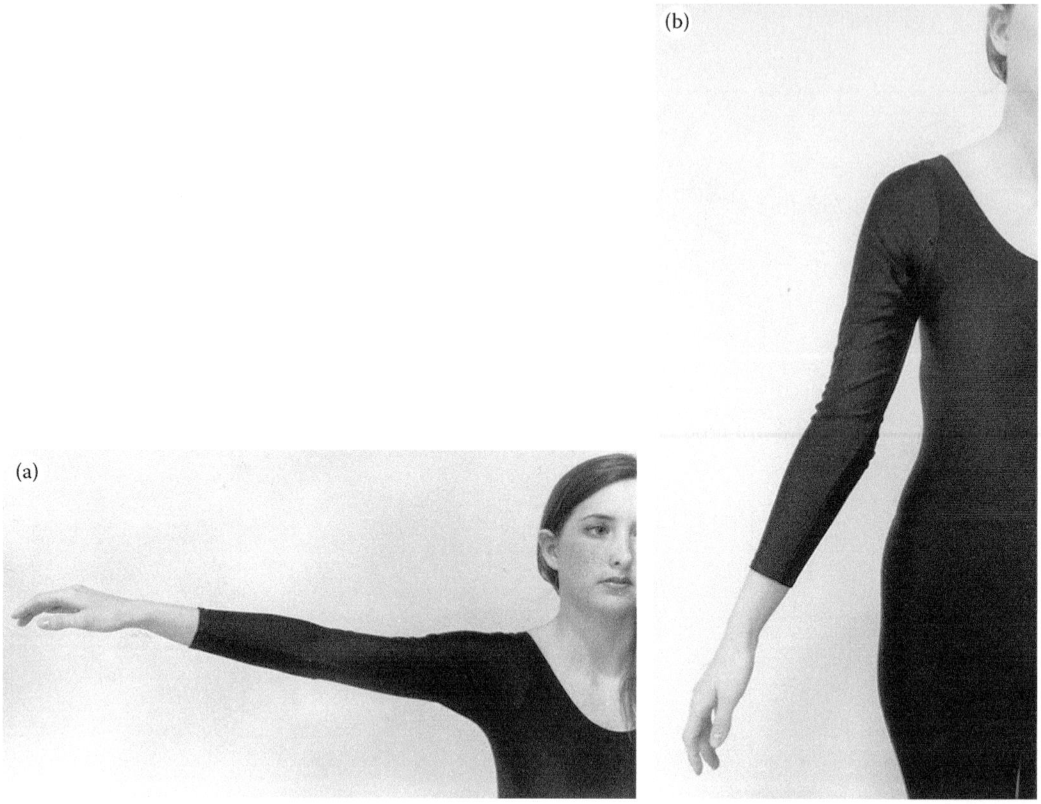

FIGURE 39.3 (a) Abduction and (b) adduction.

Supine—lying on the back with the face upward (Figure 39.7).

Prone—lying on the abdomen with the face downward (Figure 39.8).

Lateral—lying on the left or right side. It is common practice to identify lateral positions as either right or left lateral via the placement of a lead marker. The designation of right or left lateral is determined by which side (right or left) is placed against the film/detector (Figure 39.9).

Oblique—a position whereby a body part is rotated to a predetermined degree between the prone and supine position. Oblique positions may be performed either upright or recumbent, i.e., lying down.

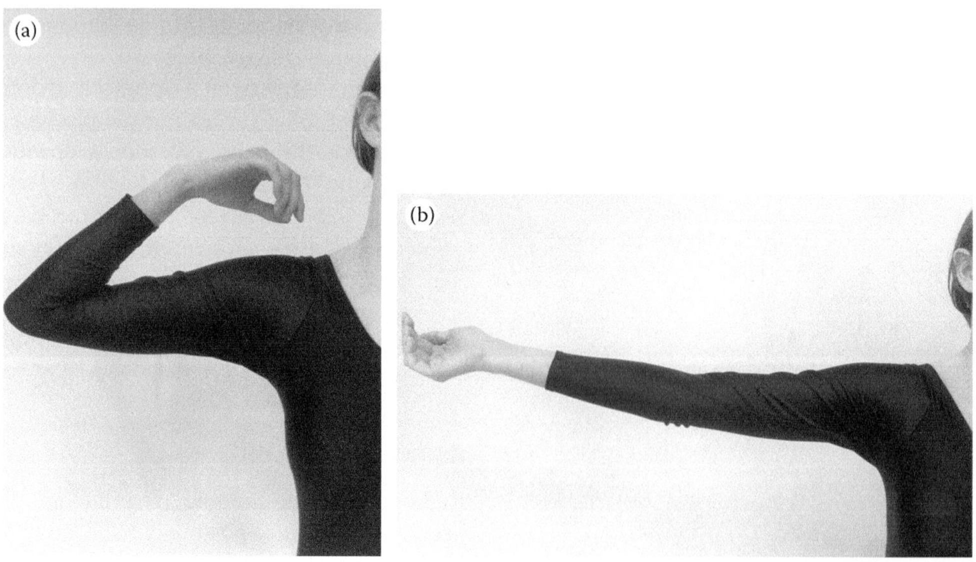

FIGURE 39.4 (a) Flexion and (b) extension.

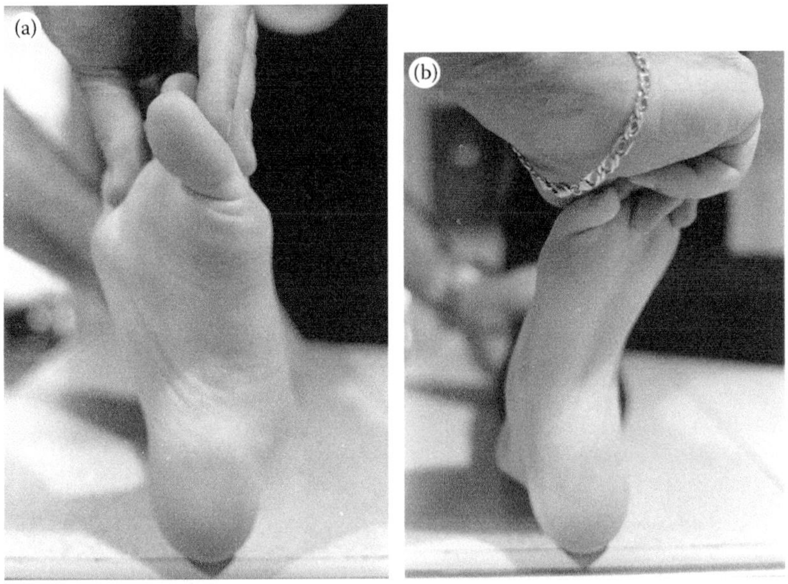

FIGURE 39.5 (a) Eversion and (b) inversion.

Right posterior oblique (RPO)—the model is rotated to place the right posterior side next to the film/detector (Figure 39.10).

Left posterior oblique (LPO)—the model is rotated to place the left posterior side next to the film/detector (Figure 39.11).

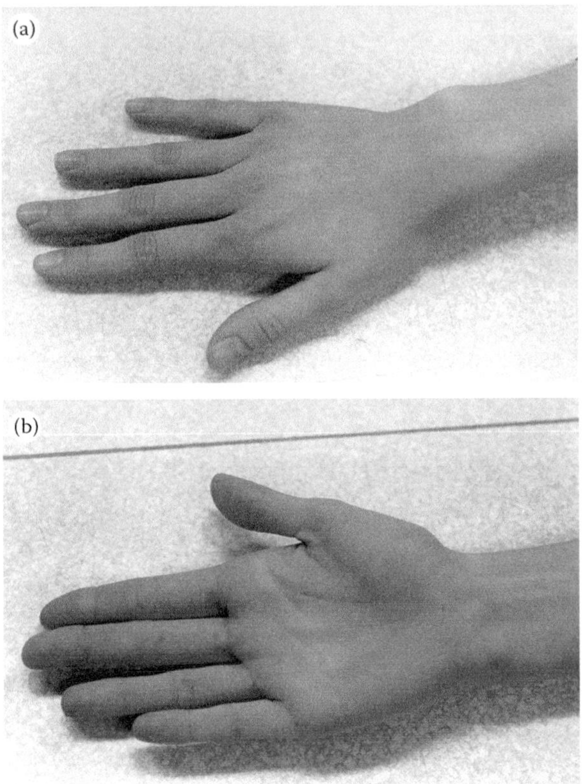

FIGURE 39.6 (a) Pronation and (b) supination.

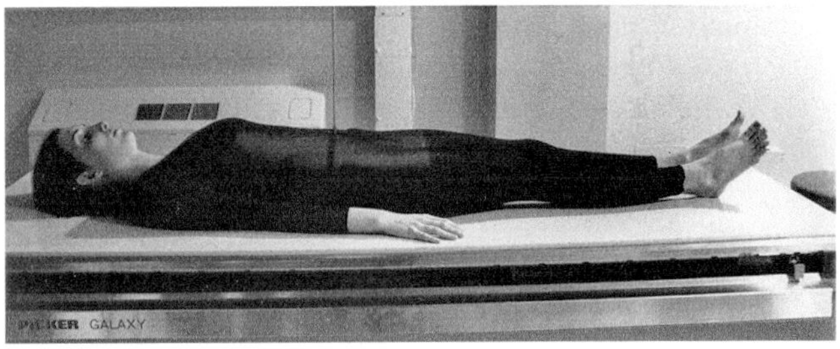

FIGURE 39.7 Supine (anteroposterior).

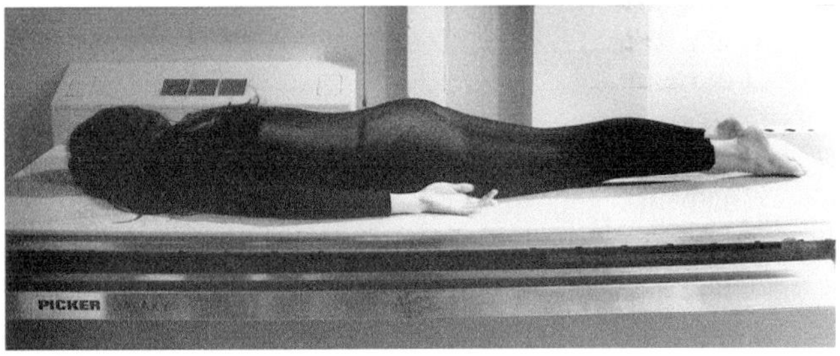

FIGURE 39.8 Prone (posteroanterior).

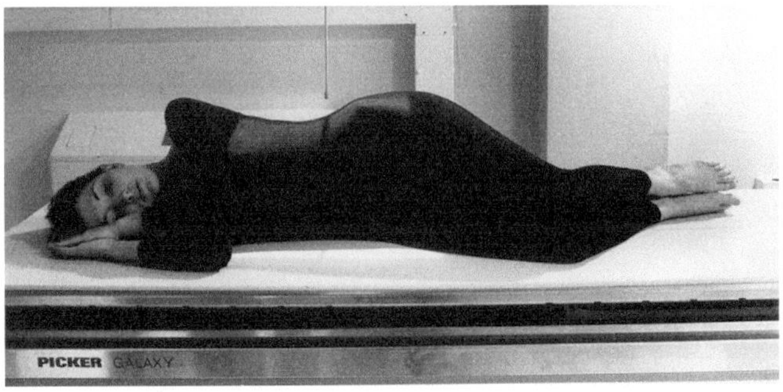

FIGURE 39.9 Lateral (lateral projection).

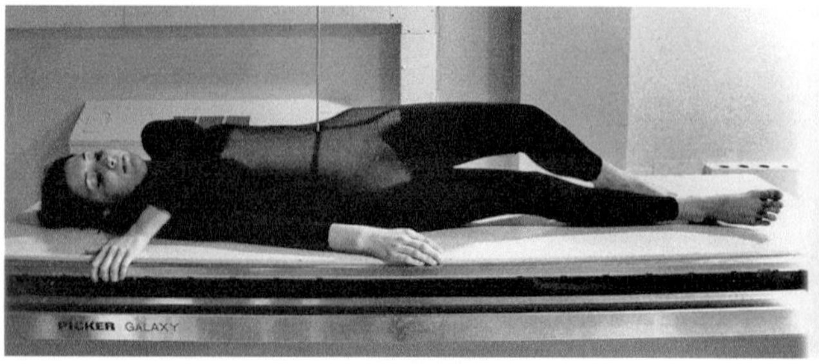

FIGURE 39.10 Right posterior oblique (RPO).

Right anterior oblique (RAO)—rotation of the body to place the right anterior side next to the film/detector.

Left anterior oblique (LAO)—rotation of the body to place the left anterior side next to the film/detector.

Decubitus— indicates that the patient is lying down. It also denotes that the x-ray beam is horizontal or parallel to the floor. Thus, the x-ray beam may enter the patient from either side (right or left), or from the anterior or posterior direction. There are also several ways to perform the decubitus position. In addition, the name of a specific decubitus position is determined by which side (right, left, anterior, or posterior) of the patient's body is placed against the x-ray film/detector.

Dorsal decubitus—the model is supine, and the x-ray beam may enter from either the right or left side. Future reference to this and other similar positions may be referred to as a cross-table lateral projection (Figure 39.12).

Ventral decubitus—the model is prone, and the x-ray beam may enter from either the right or left side (Figure 39.13).

Lateral decubitus—the patient is placed on either the right or left side, and the x-ray beam enters from either the anterior or posterior direction. Once again, the name of a specific lateral decubitus is determined by which side (right or left) of the patient is resting on the x-ray film. For example, Figure 39.14 shows a model lying on her left side. Therefore, the position is referred to as a left lateral decubitus.

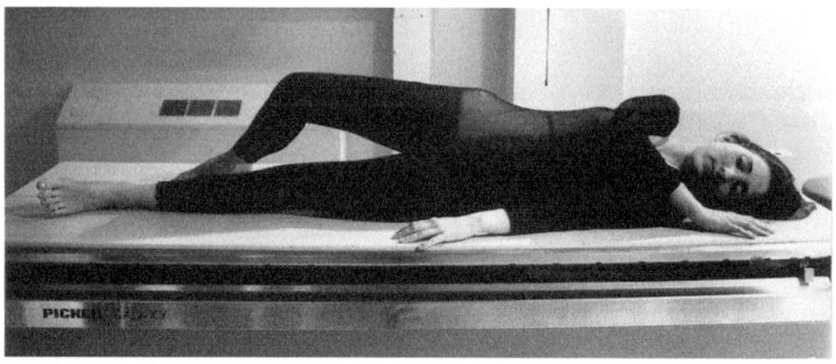

FIGURE 39.11 Left posterior oblique (LPO).

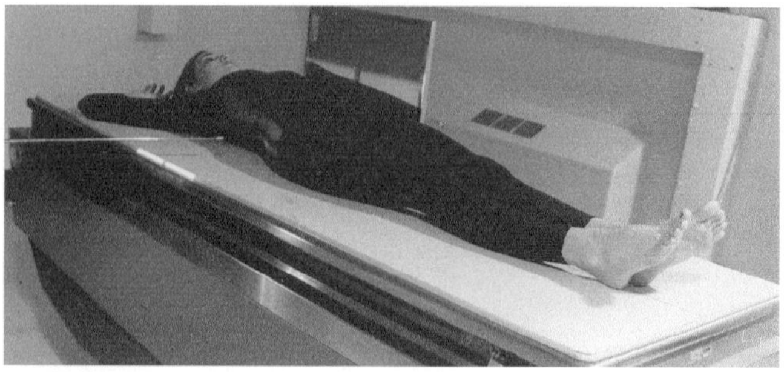

FIGURE 39.12 Dorsal decubitus.

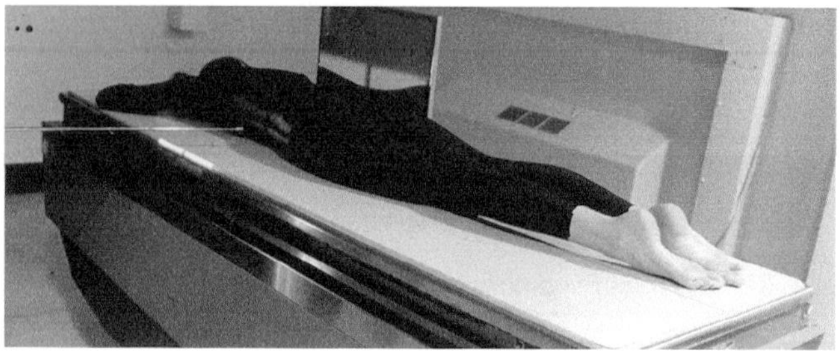

FIGURE 39.13 Ventral decubitus.

Radiographic Projections

The term radiographic position was previously defined as the actual position of the body while it is being radiographed. In contrast, a radiographic projection simply indicates the path of the x-ray beam as it passes through the body. In other words, the term projection describes the route of the x-ray beam as it leaves the radiographic tube, enters the patient, and exits the patient to strike the film/detector. Projection is an important term because all radiographic examinations are identified by a particular projection. Radiographic projections may be performed either upright or recumbent. Despite the distinct difference between the terms position and projection, several illustrations previously used to depict positions are utilized again to further explain various projections in the definitions below.

Anteroposterior (AP)—in Figure 39.7 the x-ray beam enters the anterior surface and exits the posterior surface.

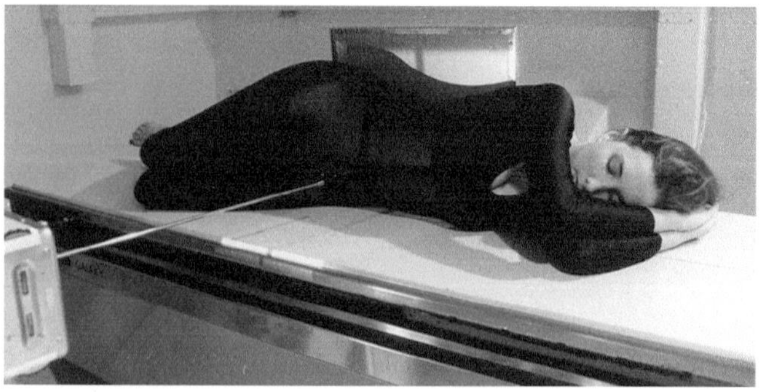

FIGURE 39.14 Left lateral decubitus.

FIGURE 39.15 Axial projection.

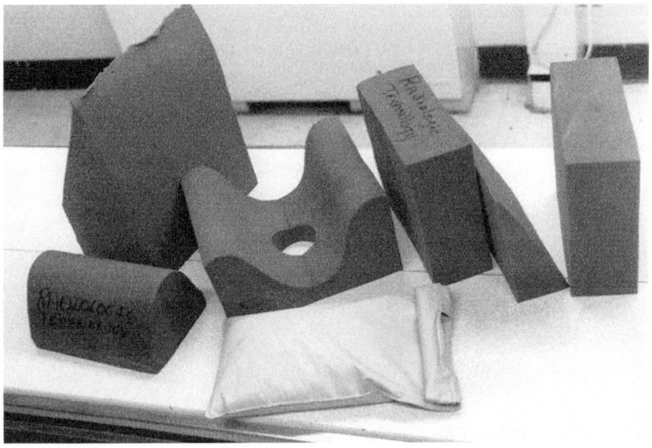

FIGURE 39.16 Sandbag and radiolucent sponge pads.

Posteroanterior (PA)—in Figure 39.8 the x-ray beam enters the posterior surface and exits the anterior surface.

Lateral—in Figure 39.9 the x-ray beam enters the left side of the body and exits the right side. Figure 39.9 is referred to as a right lateral projection.

Oblique—in Figure 39.10, the model has been placed in the right posterior oblique position (RPO). Since the x-ray beam enters the anterior surface, the projection is referred to as a right AP oblique projection. In a similar manner, the position demonstrated in Figure 39.11 would be identified as a left AP oblique projection.

Axial—a longitudinal angulation of the x-ray beam along the long axis of the body (Figure 39.15).

RADIOGRAPHIC POSITIONING

Radiographic positioning entails the proper placement of an anatomic part on an x-ray cassette (or digital x-ray detector) in a manner that yields an image that mirrors the anatomic area of interest. Much of what was addressed in the previous sections focused on positioning terminology and certain technical aspects of radiography. The reader is recommended to refer to those sections when reviewing the radiographic positions in this section.

There are certain principles that must be observed to produce anatomically correct radiographs. Some of the more important principles are discussed below.

Most radiographic examinations require two projections taken at right angles to one another. This requirement is typically fulfiled by taking an AP and lateral, or a PA and lateral projection. The AP/PA and lateral projections yield two images without superimposition of body parts (not lying on top of each other). The AP/PA and lateral projections also clarify the position and depth of a foreign body: a bullet or ingested object. Thus, the AP/PA and lateral projection provides a different perspective of an anatomic part, and pathology demonstrated in one projection may not be demonstrated in the other.

When long bones such as the lower leg or the humerus are radiographed, the distal and proximal joint spaces of the long bone should be included. The inclusion of both joint spaces of a long bone ensures the entire bone is projected on the radiograph.

Regardless of whether an anatomic part is placed in a supine, prone, or lateral position, care must be taken to avoid over- or under rotation of the part. In short, the part must be placed at right angles to the cassette. To ensure proper alignment is achieved, it is recommended that radiolucent sponge pads are used as support devices (Figure 39.16). Radiolucent sponge pads are available with washable surfaces, or can be covered in a washable or disposable material and can be used in place of personnel to support anatomic parts while examinations are performed. Sponge pads can be combined with a sandbag to provide extra support to hold body parts in position, although care should be taken as sandbags are radiopaque and care should be taken not to obscure any relevant areas. In some cases when examining small bones or bone fragments, a soft malleable material, such as Florist's clay or modeling compound, can be used to hold bone or bone fragments in precise position. However, some materials may be radiopaque and care must be taken not to obscure an important area. In addition, some materials may be adhesive and extreme care should be taken when examining dry fragile bony remains.

In order to obtain an optimum examination achieving accurate positioning of structures and avoiding artifacts, all clothing and personal effects should be removed from the area of interest. However, it is sometimes necessary to undertake an initial examination (or "radiographic survey") with the body *in situ* in the unopened body-bag prior to external examination by the pathologist. In such cases, a further examination will be necessary if correctly positioned and optimally exposed radiographs are to be obtained. In all cases, the medical examiner should be consulted before clothing is removed.

SURVEY APPROACH TO RADIOGRAPHIC POSITIONING

Radiographs produced in the radiology department tend to focus on specific anatomic structures, such as the knee, humerus, shoulder, kidney, etc. This method may be impractical in

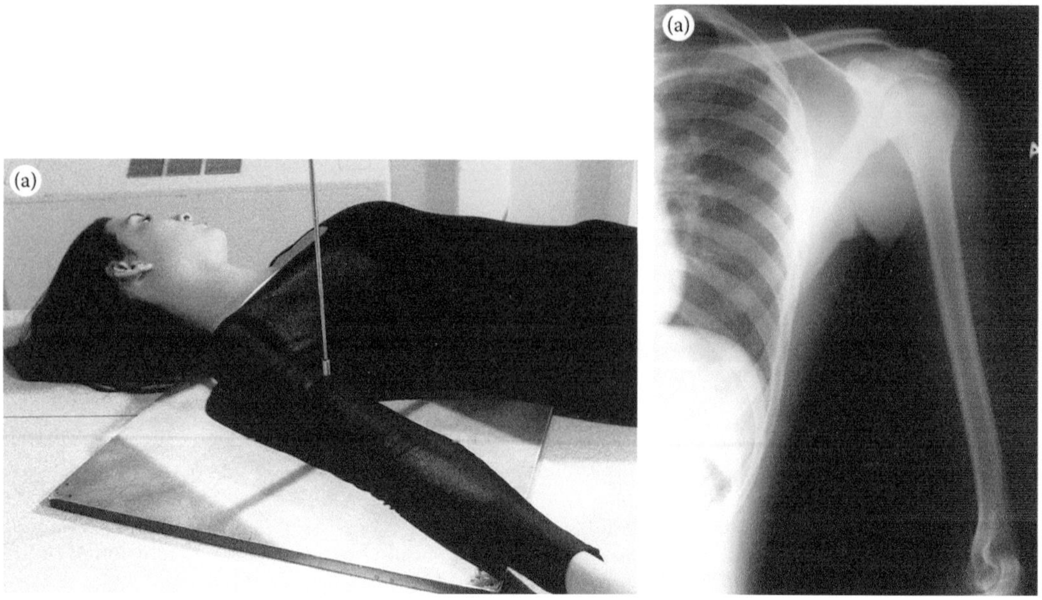

FIGURE 39.17 (a) AP shoulder and humerus position. (b) AP shoulder and humerus projection.

postmortem radiography where, on account of time and other considerations, it is frequently necessary to demonstrate several anatomic structures during a single examination.

In some centers, radiographic surveys may be undertaken using a mobile fluoroscope, enabling the entire body to be examined from head to toe in a matter of a few minutes. The fluoroscope displays a real-time video x-ray image on a computer screen, enabling rapid visualization of pathology and location of projectiles. As the potential for increased radiation dose to operators always exists, such examinations should be undertaken by a trained operator, working in conjunction with a medical examiner. In other centers, radiographic surveys may be undertaken using computed tomography (CAT scanning) or slit-beam scanography. Such procedures have been described in Chapter 29 and are thus outside the scope of this chapter.

In most centers however, radiographic surveys are undertaken using conventional radiographic techniques. The initial part of this section will thus describe radiographic positions using a survey approach. The survey approach endeavors to include several related anatomic structures on a single radiograph. It should also be noted that the needs of postmortem radiography may often be satisfied with a single AP projection. However, in keeping with the principle of providing two projections at right angles, and the importance of matching postmortem and antemortem images, lateral projections are also included.

The projections described throughout this section are intended to accommodate the radiographic needs of a body in the supine recumbent position. Thus, with few exceptions, the focus of this section will be directed to AP and cross-table lateral projections. Moreover, except where indicated, the SID for all projections will be 40 in. (100 cm). In addition, the important aspects of each projection will be discussed under the following headings: cassette size, or image-area in the case of Direct Digital radiography (DR), position of part, central ray, and structures demonstrated. A fifth heading, additional comments, will be added when deemed necessary. Finally, it should also be noted that the important aspects of each projection described below pertain to radiographic positions performed under standard conditions. Obviously, this may not always be the case in postmortem radiography, where anatomic structures may vary greatly in terms of their position, form, or composition. In such cases, the standard positions described below should be referred to, and adapted to the given situation.

Survey Approach: Upper Extremity

AP Projection of the Shoulder to Include the Humerus

Cassette size—14 in. × 17 in. grid cassette.*

Position of part (Figure 39.17a)

- The grid cassette is placed lengthwise beneath the patient, with the top of the cassette approximately 1½ in. above the shoulder. Some adjustment of the grid cassette may be necessary to include the distal humerus.
- The grid cassette must extend to the median plane of the body to include the shoulder girdle.
- Supinate the hand to place the humerus in a true AP position.

* In this chapter, cassette sizes are given in inches. Standard equivalent sizes in centimeters are as follows:
17 in × 14 in—43 cm × 35 cm
14 in × 14 in—35 cm × 35 cm
15 in × 12 in—40 cm × 30 cm
12 in × 10 in—30 cm × 24 cm
10 in × 8 in—24 cm × 18 cm

Central ray (CR)

- Direct the CR perpendicular to the midpoint of the longitudinal and horizontal axis of the grid cassette. Collimate to the size of the cassette.

Structures demonstrated (Figure 39.17b)—An AP projection of the shoulder girdle to include the proximal, mid, and distal portions of the humerus.

Additional comments—In cases where the desired anatomy extends beyond the long axis of the grid cassette, a separate AP projection will be needed to include the distal portion of the humerus.

Lateral Projection of the Humerus

Two methods are described to demonstrate a lateral projection of the humerus.

AP Internal Rotation

Cassette size—14 in. × 17 in. grid cassette.

Position of part (Figure 39.18a)

- Adjust the cassette beneath the humerus to include the proximal and distal joints.
- The humerus is centered to the long axis of the cassette.
- Internally rotate the humerus to a point where the distal joint (epicondyle) is superimposed. Extreme pronation of the hand will usually provide sufficient internal rotation of the humerus. However, the hand must be held in extreme rotation to achieve this projection.

Central ray—Direct the CR perpendicular to the longitudinal and horizontal axis of the grid cassette. Adjust the collimator to an 8 in. × 17 in. setting.

Structures demonstrated (Figure 39.18b)—The entire humerus in the lateral position is achieved through internal rotation.

Additional comments—The cassette may be turned diagonally if the humerus is longer than the cassette. In addition, an alternate projection of the entire humerus can be obtained by moving the arm (humerus) away from the body. This is achieved by placing the hand in the supine position, and abducting the humerus approximately 90° from the midsagittal plane. A cassette is placed adjacent to the lateral aspect of the humerus, and the CR is directed horizontally and perpendicular to the center of the cassette.

Cross-Table Lateral of Mid and Distal Humerus

This projection excludes the proximal portion of the humerus.

Cassette size: 14 in. × 17 in. nongrid cassette.

Position of part (Figure 39.19a)

- Place the cassette between the thorax and the humerus.
- Push the long axis of the cassette as far as permitted toward the axilla.
- The humerus should be supported by a radiolucent sponge or sandbag to ensure inclusion of the posterior aspect.
- The hand is maintained in the supine position.

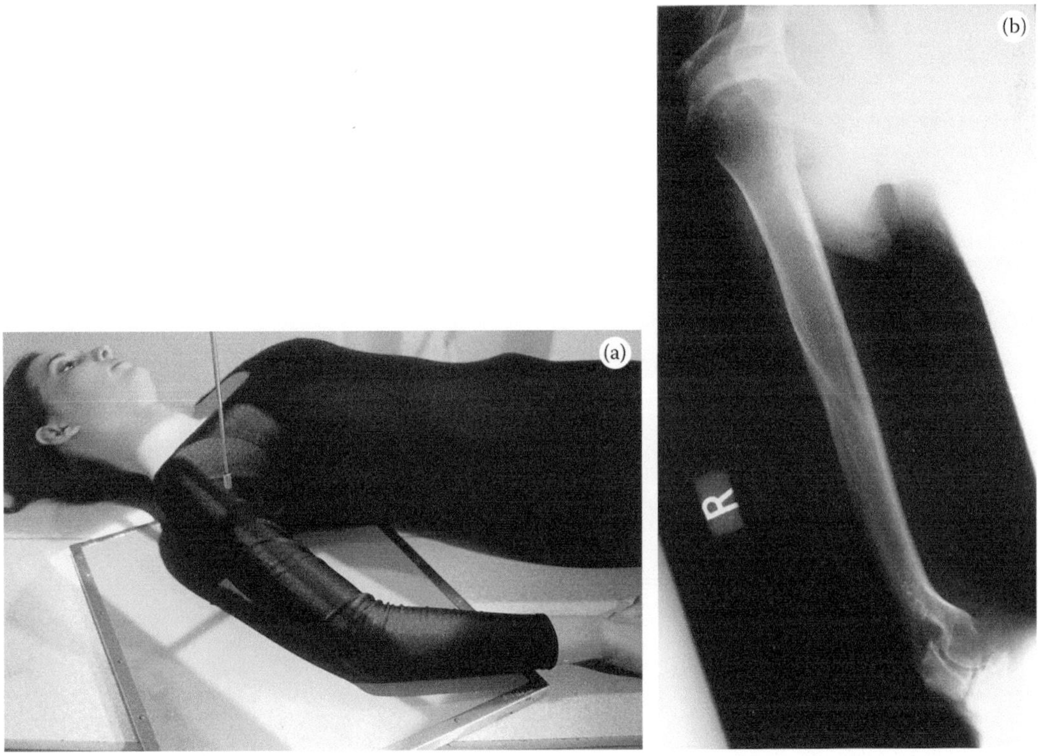

FIGURE 39.18 (a) Lateral humerus position. (b) Lateral humerus projection.

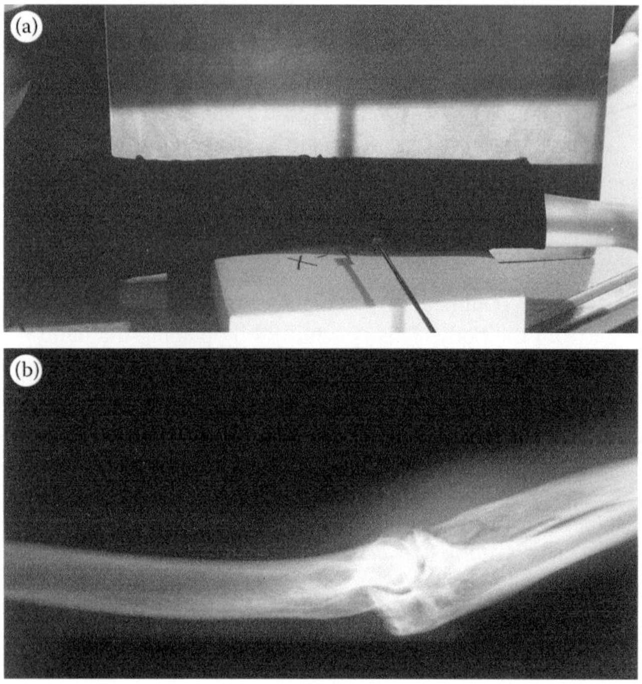

FIGURE 39.19 (a) Cross-table lateral mid and distal humerus position. (b) Cross-table lateral mid and distal humerus projection. Lateral humerus projection.

Central ray

- Direct the horizontal beam perpendicular to the midpoint of the cassette. Collimate to an 8 in. × 17 in. setting.
- *Structures demonstrated* (Figure 39.19b)—A cross-table lateral projection of the mid and distal portions of the humerus.

AP Projection of the Elbow to Include the Hand

Cassette size—14 in. × 17 in. nongrid cassette.

Position of part (Figure 39.20a)

- Adjust the cassette beneath the forearm to include the elbow joint and the hand. This may require the cassette to be turned diagonally. If so, the elbow is placed in one corner of the cassette and the hand in the other corner.
- The hand and forearm must be maintained in a supine position to achieve a true AP projection.

Central ray

- The CR is directed perpendicular to the midpoint of the cassette. Collimate to an 8 in. × 17 in. size.

Structures demonstrated (Figure 39.20b)—An AP projection of the forearm including the elbow and hand.

Lateral Projection of the Elbow to Include the Hand

Cassette size—14 in. × 17 in. nongrid cassette.

Position of part (Figure 39.21a)

- The elbow, forearm, and hand are elevated on radiolucent sponges or sandbags.

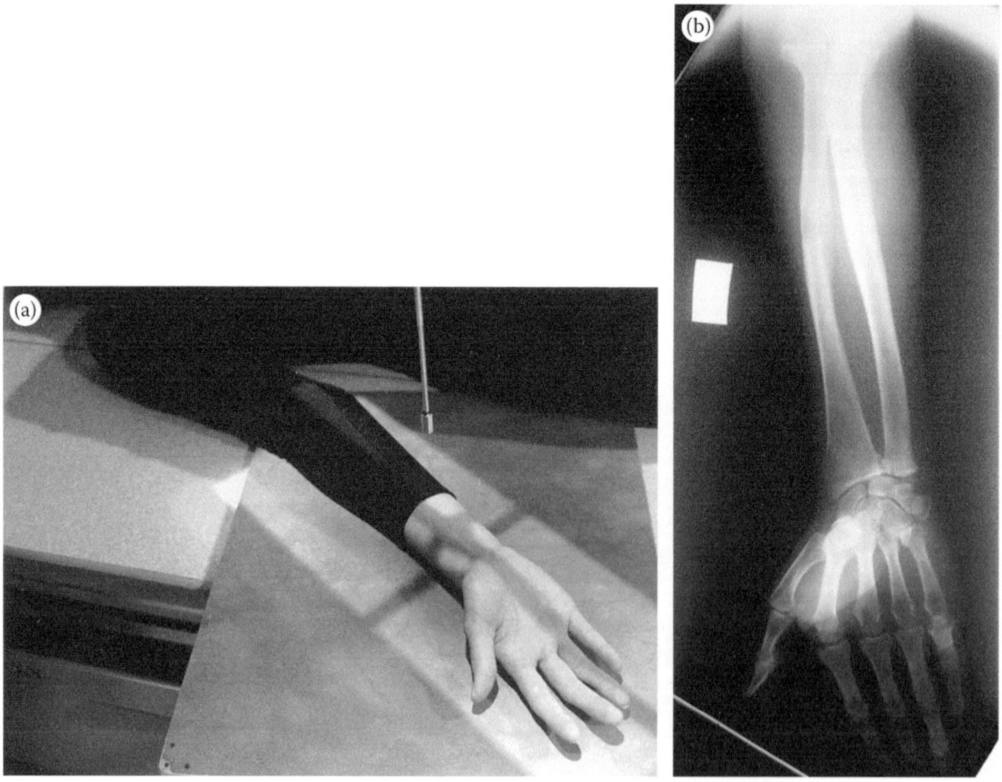

FIGURE 39.20 (a) AP forearm position to include hand. (b) AP forearm projection including hand.

- The long axis of the cassette is placed between the forearm and the side of the patient's abdomen and pelvis.
- The hand and forearm must be maintained in the supine position.

Central ray

- Direct the horizontal beam perpendicular to the midpoint of the cassette. Collimate to an 8 in. × 17 in. size.

Structures demonstrated (Figure 39.21b)—A cross-table lateral projection of the forearm including the elbow and hand.

Survey Approach: Pelvis and Lower Extremity

AP Projection of the Pelvis to Include Both Hips

Cassette size: 14 in. × 17 in. grid cassette.

Position of part (Figure 39.22a)

- A 14 in. × 17-in. grid cassette is placed crosswise (17 in. × 14 in.) beneath the patient.
- The top edge of the grid cassette is positioned 1 1/2 in. above the iliac crest.
- The median plane of the pelvis is centered to the grid cassette.
- The feet, along with the lower legs, are rotated internally. When the feet are properly rotated, they will resemble a pyramid.
- The hips are parallel to the plane of the table/floor to avoid rotation of the pelvis.

Central ray

- The CR is directed perpendicular to the center of the grid cassette. Collimate to 17 in. × 14 in. size.

Structures demonstrated (Figure 39.22b)—An AP projection of the pelvis, including both hips.

Cross-Table Lateral Projection of the Hip (Danelius-Miller Method)[1]

Cassette size—10 in. × 12 in. or larger grid cassette.

Position of part (Figure 39.23a)

- The foot of the side of interest is internally rotated 15°.
- The opposite leg is raised and either placed on a support. The elevated femur should approach an angle of 90° to the plane of the table/floor.
- The horizontal end of the grid cassette is placed just above the iliac crest, and the long axis of the cassette runs parallel with the proximal end of the femur.
- The grid cassette must be held in place. A cassette holder or sand bags can be used for this purpose.
- The cassette must be maintained perpendicular to the table/floor.

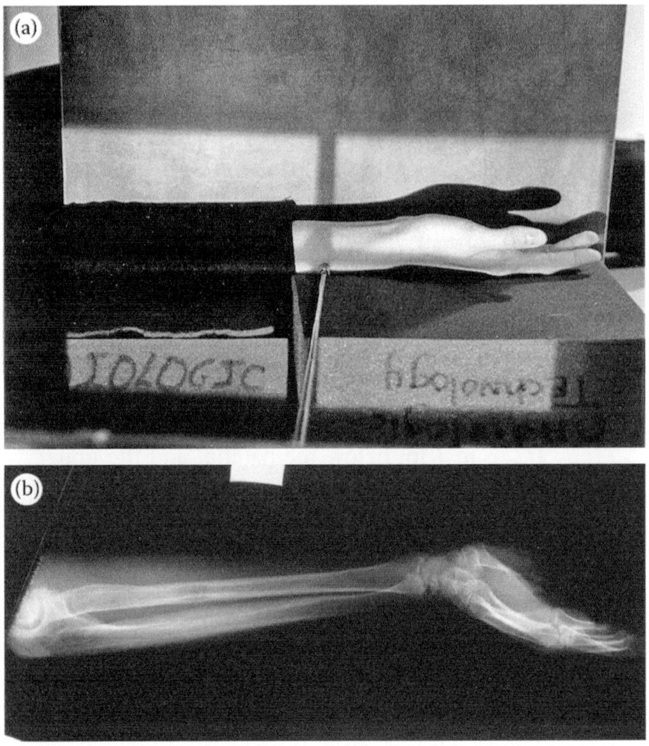

FIGURE 39.21 (a) Cross-table lateral position of forearm and hand. (b) Cross-table lateral projection of forearm and hand.

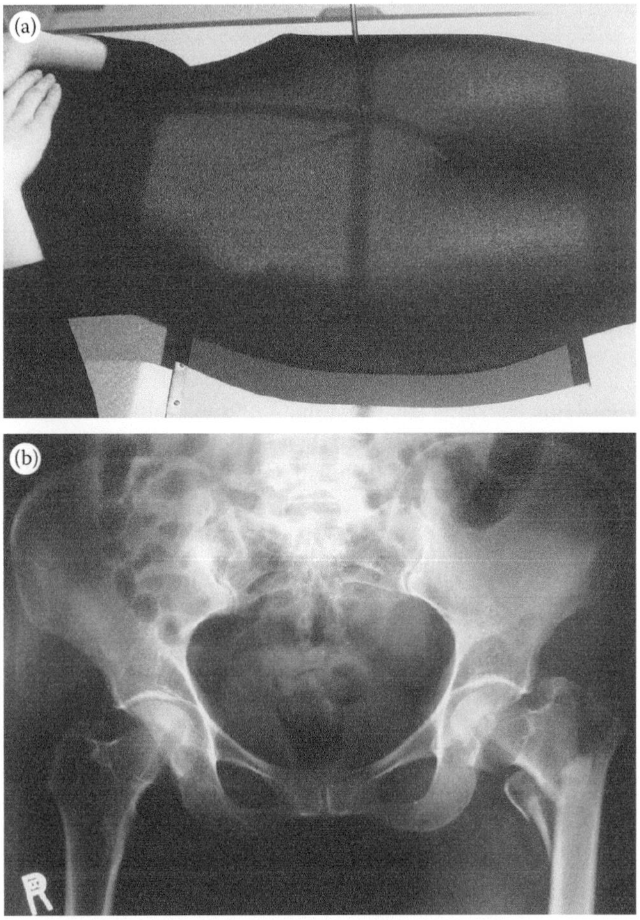

FIGURE 39.22 (a) AP pelvis position. (b) AP pelvis projection.

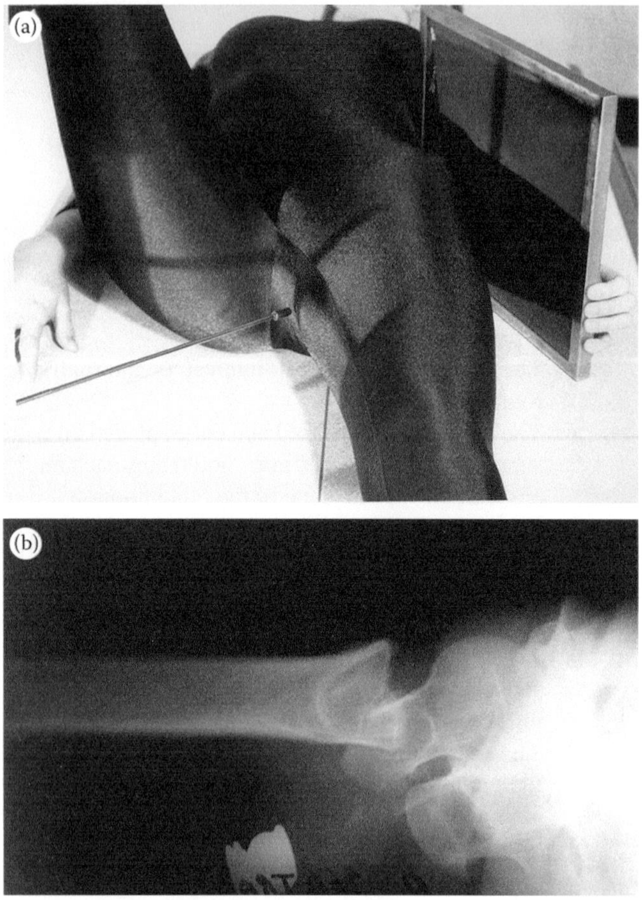

FIGURE 39.23 (a) Cross-table lateral hip position. (b) Cross-table lateral hip projection.

Central ray

- The horizontal beam is directed cephalically to the center of the cassette. As a check, when the CR is angled in this manner it should be perpendicular to the femoral neck and the grid cassette. Collimate to 10 × 12-in. size.

Structures demonstrated (Figure 39.23b)—A cross-table lateral of the proximal femur and hip.

Additional comments—To avoid grid cut-off, the grid cassette must not be angled.

AP Projection of Femur

Cassette size—14 in. × 17 in. and 10 in. × 12 in. grid cassettes.

Position of part (Figure 39.24a)

- Placement of a cassette for an AP femur is frequently influenced by the length of the femur. Since the length of an adult femur often exceeds the length of a cassette/detector, two images may be required to demonstrate both joints. Two images would not be necessary if the pelvis were radiographed, as described above, and included the proximal femur. However, two AP projections are described. For the proximal femur, a 14 in. × 17 in. grid cassette is placed lengthwise beneath the patient.
- The top of the grid cassette is placed at the level of the anterior superior iliac spine (ASIS). The ASIS is a large bony prominence on either side of the pelvic bone. The ASIS lies inferior to the iliac crest.
- The long axis of the femur is centered to the cassette.
- The feet are rotated internally 15°.
- For the distal femur, a 10 in. × 12 in. grid cassette is placed lengthwise beneath the femur.
- The lower border of the grid cassette is placed 1/2 in to 1 in. below the patella. The upper border of the cassette extends cephalad.
- Once again, the long axis of the femur is centered to the cassette.
- The foot should be directed vertically to avoid rotation.

Central ray

- For AP projections of the proximal and distal femur, the CR is directed perpendicular to the center of the grid cassette. Collimator settings for the proximal and distal portions of the femur are 10 in. × 17 in. and 10 × 12 in.

Structures demonstrated (Figure 39.24b)—An AP projection of the femur, which includes the proximal and distal joints.

Cross-Table Lateral Projection of the Femur

Cassette size—14 in. × 17 in. and 10 in. × 12 in. grid cassettes.

Position of part (Figure 39.25a,b)

- Differences between the length of the femur and the cassette are addressed in a manner similar to the AP projection.
- The lateral projection of the proximal femur is performed in the same way as the cross-table lateral of the hip. The cross-table lateral hip projection has been described earlier.
- The cross-table lateral of the mid and distal shaft of the femur requires the long axis of the grid cassette be placed on the lateral aspect of the femur.
- The lower border of the grid cassette is placed 1/2 in to 1 in. below the patella, and the upper border of the cassette extends cephalad.
- A radiolucent sponge may be placed under the femur to elevate it above the table.
- The opposite femur is elevated high enough to avoid superimposition of both femurs (Figure 39.25a).
- When it is not feasible to elevate the leg, the grid cassette may be placed between both femurs with the front of the cassette facing the femur of interest. This method, however, will

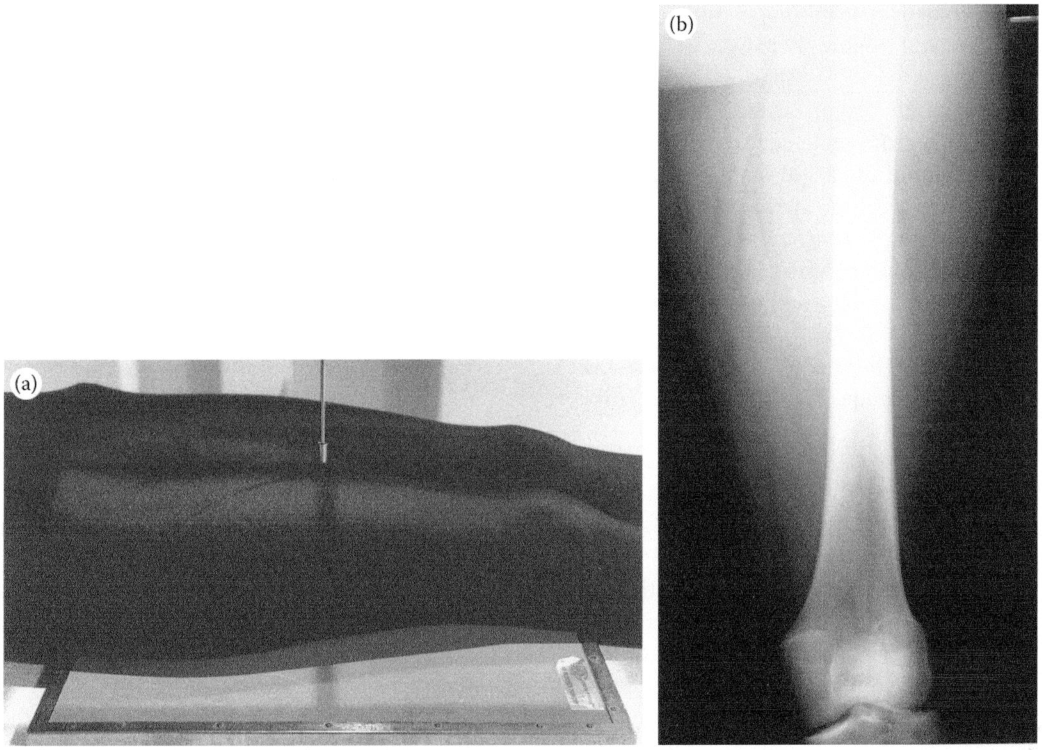

FIGURE 39.24 (a) AP femur position. (b) AP femur projection.

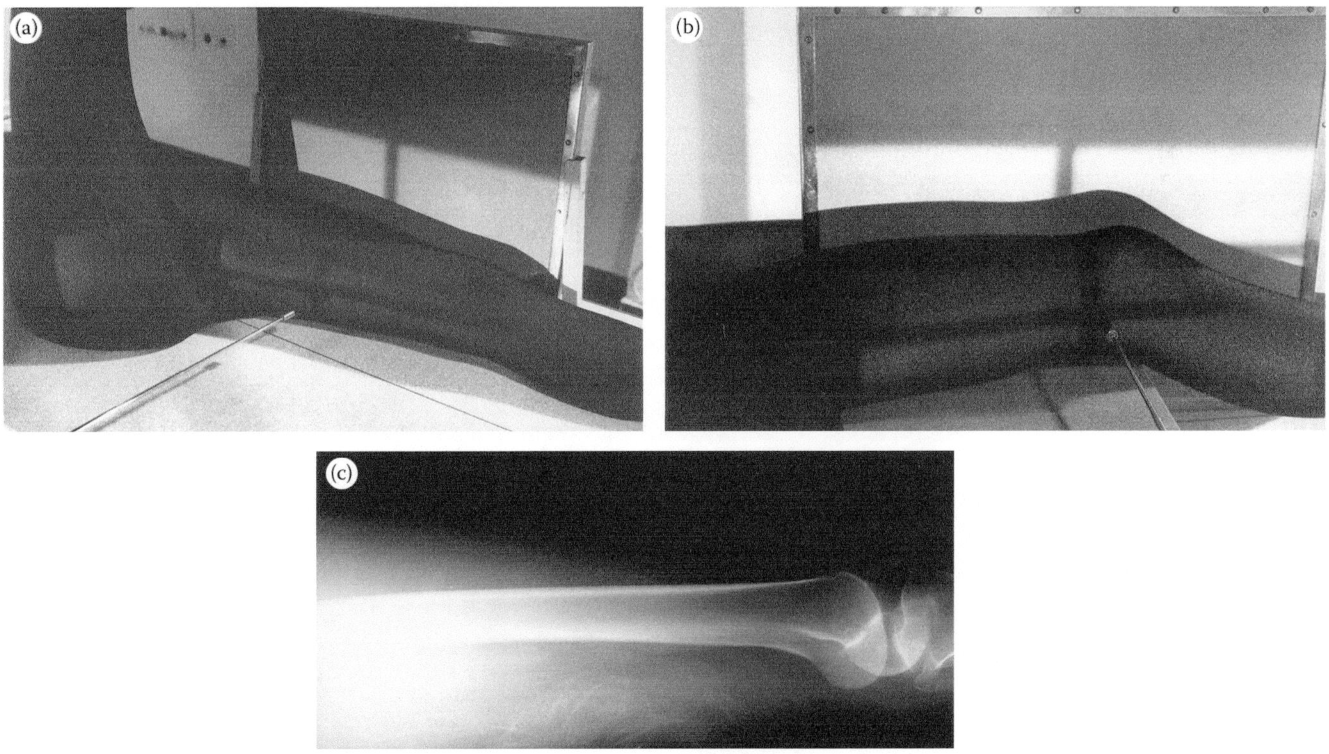

FIGURE 39.25 (a) Cross-table lateral position of mid and distal femur. (b) Cross-table lateral position of femur. (c) Cross-table lateral projection of femur.

not demonstrate as much femoral shaft as when the leg is elevated (Figure 39.25b).

Central ray

- For the proximal femur, direct the CR as described in the cross-table lateral projection of the hip. For the mid and distal shaft of the femur, direct the horizontal beam perpendicular to the midpoint of the grid cassette. The collimator setting for the proximal femur is 10 in. × 12 in., while the setting for the mid and distal shaft is 10 in. × 17 in.

Structures demonstrated (Figure 39.25c)—A cross-table lateral projection demonstrating the mid and distal shaft of the femur.

AP Projection of the Leg (Tibia-Fibula)

Cassette size—14 in. × 17 in. nongrid cassette.

Position of part (Figure 39.26a)

- The foot is maintained in a vertical position to avoid part rotation.
- The cassette is placed lengthwise beneath the leg and is aligned with the long axis of the leg.
- The placement of the cassette is adjusted to include the proximal and distal joints. The cassette may be turned diagonally to accommodate longer legs. Otherwise a separate smaller cassette may be used to demonstrate one of the joint spaces.

Central ray

- Direct the CR perpendicular to the midpoint of the cassette. Collimate to a 10 in. × 17 in. setting.

Structures demonstrated (Figure 39.26b)—An AP projection of the entire tibia and fibula.

Cross-Table Lateral Projection of the Leg

Cassette size—14 in. × 17 in. nongrid cassette.

Position of part (Figure 39.27a)

- The cassette is placed lengthwise between the lower legs. The cassette is adjusted to include the proximal and distal joints. Since the adult tibia and fibula are often longer than the cassette, a second cassette may be needed to include both joint spaces. It may be necessary to elevate the leg of interest on a radiolucent sponge or sandbag to ensure the posterior aspect of the leg is projected on the film.

Central ray

- Direct the horizontal beam perpendicular to the midpoint of the cassette. Collimate to the part.

Structures demonstrated (Figure 39.27b)—A lateral projection of the entire tibia and fibula.

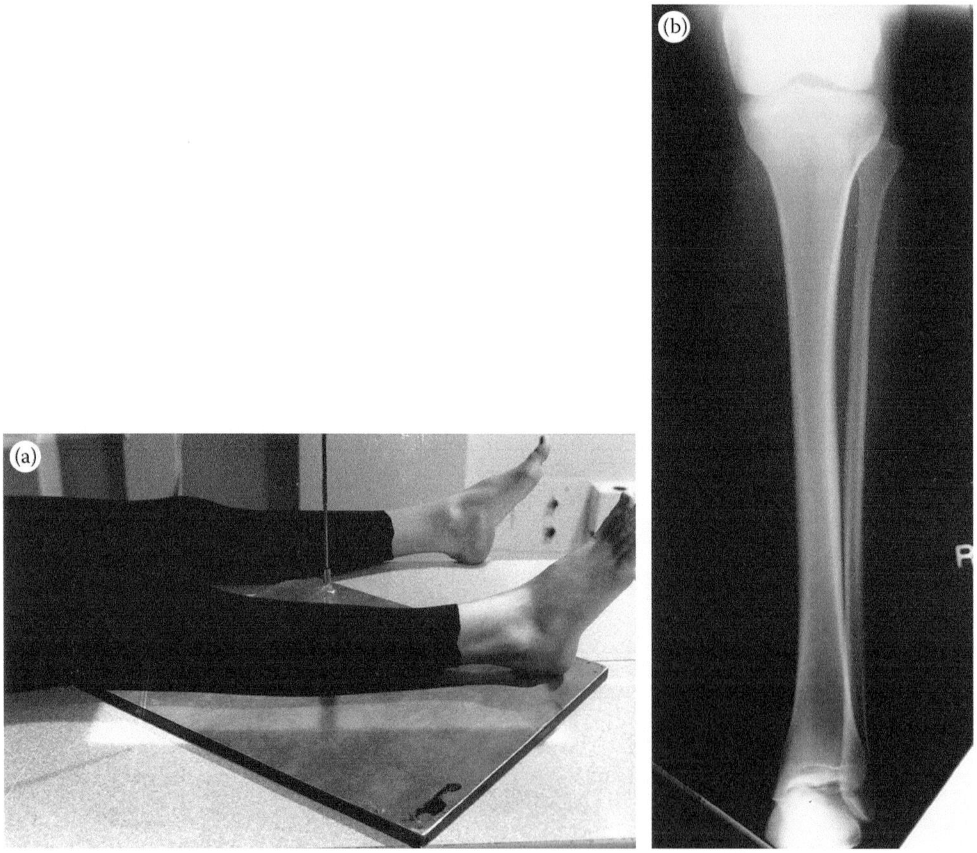

FIGURE 39.26 (a) AP position of tibia-fibula. (b) AP projection of tibia-fibula.

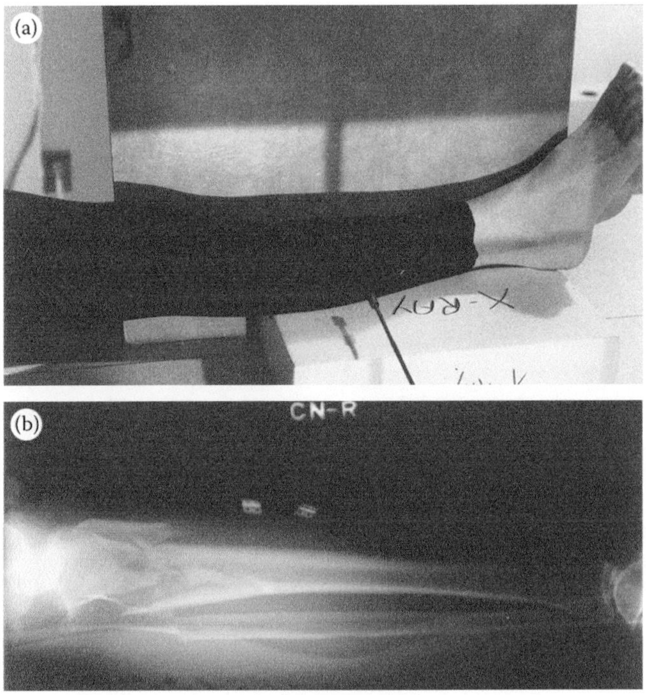

FIGURE 39.27 (a) Cross-table lateral position of tibia-fibula. (b) Cross-table lateral projection of tibia-fibula.

Survey Approach: Chest and Thoracic Spine

Other than technical differences, the positioning requirements for the chest and thoracic spine are essentially the same. Therefore, the chest and thoracic spine will be considered together in the discussion below.

AP Projection of the Chest and Thoracic Spine

Cassette size—14 in. × 17 in. grid cassette.

Position of part (Figure 39.28a)

- The grid cassette is placed lengthwise beneath the thorax. The top of the cassette is about 1 1/2 in. above the shoulders.
- Center the cassette to the midsagittal plane of the patient.
- The thorax should be checked to ensure that it is not rotated.
- If feasible for chest exams, both arms should be raised above the head in order to remove the scapulae from the lung field.

Central ray

- As noted previously, it is important to match antemortem and postmortem radiological images. This is particularly true when an AP postmortem radiograph of the thoracic cavity is compared to the usual PA antemortem chest radiograph. To achieve comparable radiographic studies, the authors recommend the CR be angled 10–15° caudally to the midpoint of the longitudinal and horizontal axis of the grid

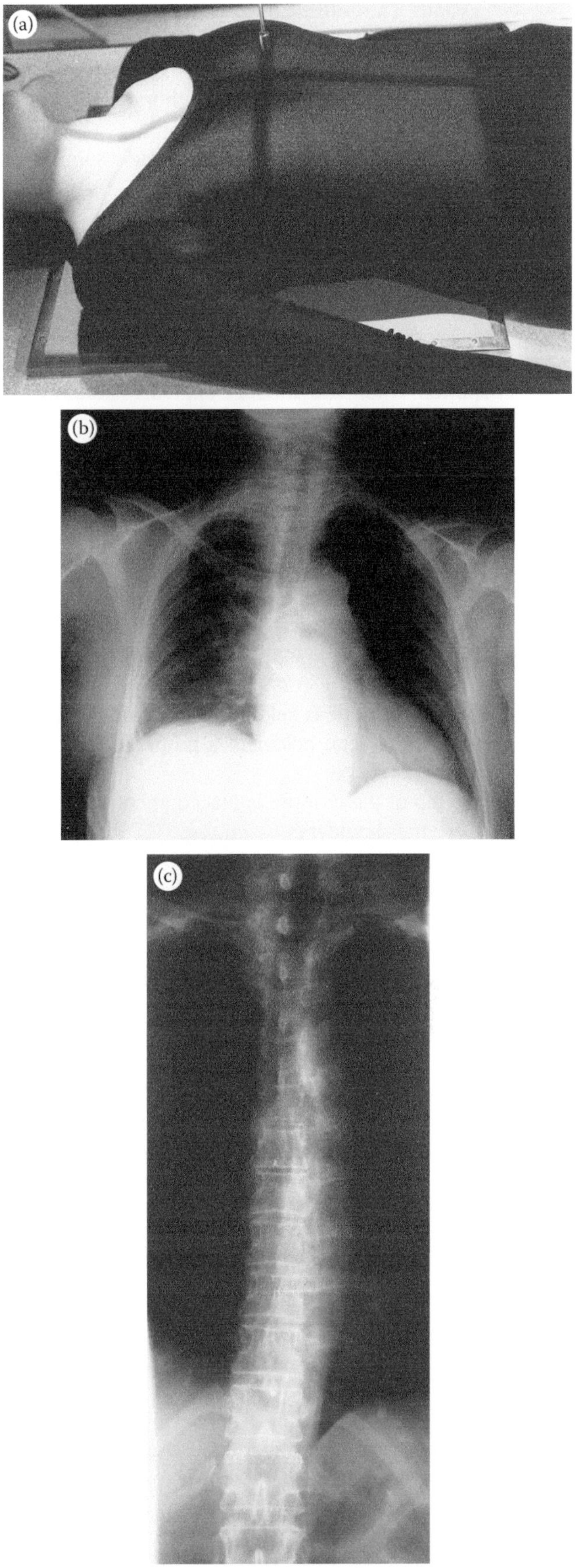

FIGURE 39.28 (a) AP position of chest and thoracic spine. (b) AP Chest projection. (c) AP thoracic spine projection.

cassette. The collimator should also be adjusted to the size of the cassette.

Structures demonstrated (Figure 39.28b,c)

- An AP projection of the entire thoracic viscera should be seen without rotation.
- An AP projection of the entire thoracic spine.

Additional comments

- In cases where the thoracic viscera is larger than the cassette, the grid cassette should be placed crosswise such that the long border (17 in.) of the cassette coincides with the level of the shoulders. However, due to the occurrence of grid cut-off, this approach cannot be applied when the tube is angled. Therefore, a nongrid cassette must be employed in this instance.
- Raise the x-ray tube as much as possible for AP projection of the chest.

Cross-Table Lateral Projection of the Chest and Thoracic Spine

Cassette size—14 in. × 17 in. grid cassette.

Position of part (Figure 39.29a)

- The grid cassette is held in place beside the thorax by a cassette holder or a properly shielded assistant.
- The top of the cassette is placed about 1 1/2 in. above the shoulders.
- The arms should be extended above the head.
- Center the midcoronal plane of the patient to the grid cassette.
- The patient should be elevated on radiolucent sponges to ensure proper visualization of the entire anatomy of the thoracic spine and viscera.

Central ray

- Direct the horizontal beam perpendicular to the midpoint of the longitudinal and horizontal axis of the grid cassette. Collimate to the size of the cassette.

Structures demonstrated (Figure 39.29b)

- A lateral projection of the entire thoracic viscera.
- A lateral projection of the entire thoracic spine.

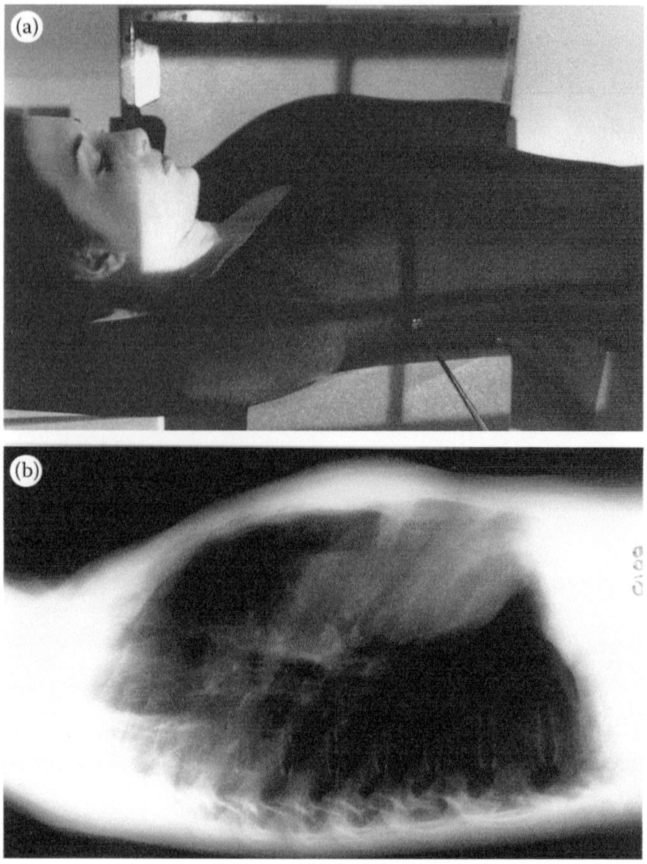

FIGURE 39.29 (a) Cross-table lateral position of chest and thoracic spine. (b) Cross-table lateral projection of chest and thoracic spine.

Survey Approach: Abdomen and Lumbar Spine

Other than technical differences, the positioning requirements for the abdomen and lumbar spine are essentially the same. Therefore, the abdomen and lumbar spine will be considered together in the discussion below.

AP Projection of the Abdomen and Lumbar Spine

Cassette size—14 in. × 17 in. grid cassette

Position of part (Figure 39.30a)

- The grid cassette is placed lengthwise beneath the body.
- The horizontal axis of the cassette is centered to the iliac crest.
- The shoulders and pelvis must lie in the same transverse plane to avoid rotation.
- The midsagittal plane of the body is centered to the longitudinal axis of the cassette.

Central ray

- Direct the CR perpendicular to the midpoint of the longitudinal and horizontal axis of the grid cassette. Collimate to the size of the cassette (14 in. × 17 in.).

Structures demonstrated (Figures 34.30b,c)

- An AP projection demonstrating those abdominal structures situated between the diaphragm and the symphysis pubis.
- An AP projection of the entire lumbar spine.

Additional comments—If the patient is extremely large, it may be necessary to take two separate exposures for an AP projection of the abdomen. The grid cassette will be placed crosswise and one film will show the lower abdomen and the second film will show the upper abdomen.

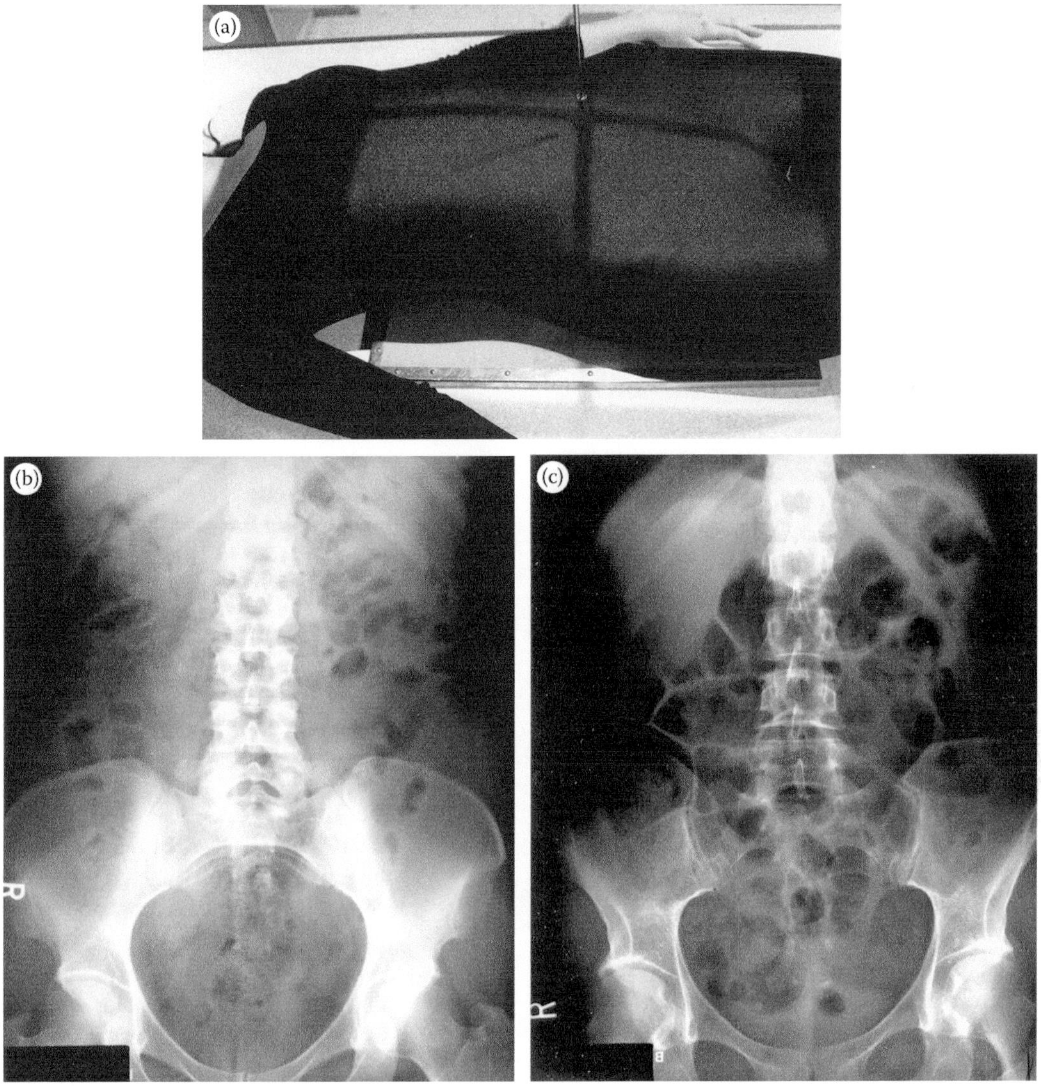

FIGURE 39.30 (a) AP position of abdomen and lumbar spine. (b) AP abdomen projection. (c) AP lumbar spine projection.

Dorsal Decubitus Projection of the Abdomen and Cross-Table Lateral Projection of the Lumbar Spine

Cassette size—14 in. × 17 in. grid cassette.

Position of part (Figure 39.31a)

- The grid cassette is parallel to the midsagittal plane of the body. The grid cassette may be held beside the abdomen by a cassette holder held in place using a pad and sandbags or other positioning aids.
- The top of the iliac crest is centered to the horizontal axis of the grid cassette.
- The midcoronal plane of the body is aligned with the long axis of the cassette.
- The patient should be elevated on radiolucent sponges to ensure proper visualization of the posterior aspect of the abdominal viscera and lumbar spine.

Central ray

- Direct the horizontal beam to the midpoint of the longitudinal and horizontal axis of the grid cassette. Collimate to the size of the cassette.

Structures demonstrated (Figure 39.31b)

- A lateral projection of the abdominal cavity and abdominal viscera.
- A lateral projection of the entire lumbar spine.

Additional comments—If the sacrum and coccyx are not demonstrated, the grid cassette should be adjusted caudally.

Survey Approach: Skull (Cranium)

Proper positioning of the skull requires the application of imaginary positioning lines (Figure 39.32). The positioning

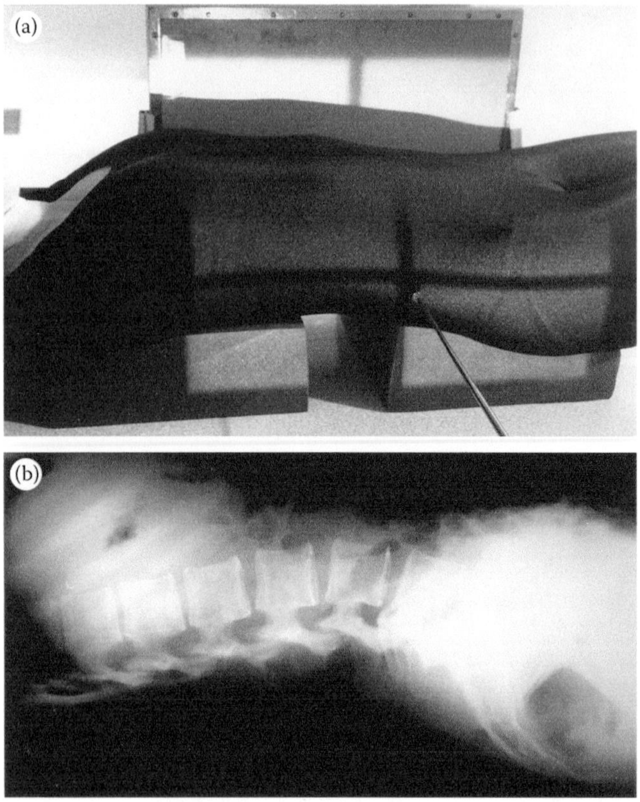

FIGURE 39.31 (a) Cross-table lateral position of abdomen and lumbar spine. (b) Cross-table lateral projection of abdomen and lumbar spine.

lines considered most useful for postmortem radiography of the skull, and their respective acronyms, are listed below:

Interpupillary line (IPL)
Midsagittal plane (MSP)
Orbitomeatal line (OML)
Infraorbitomeatal line (IOML)

AP Projection of the Skull (Cranium)

Cassette size—10 in. × 12 in. grid cassette.

Position of part (Figure 39.33a)

- The grid cassette is placed lengthwise beneath the skull.
- The upper border of the cassette is positioned 1 in. above the top of the skull. Note: the palm of the hand may be used to approximate 1 in.
- The MSP is centered to the long axis of the grid cassette.
- The MSP must be perpendicular to the cassette to avoid rotation.
- The OML is aligned perpendicular to the cassette. Note: it may be necessary to elevate the head on a thin radiolucent sponge to achieve the desired OML alignment.

Central ray

- The CR is directed perpendicular to the midpoint of the grid cassette. Collimate to the size of the cassette.

Structures demonstrated (Figure 39.33b)—An AP projection of the major bony structures in the cranium.

Cross-Table Lateral Projection of the Skull (Cranium)

Cassette size—10 in. × 12 in. grid cassette.

Position of part (Figure 39.34a)

- The grid cassette is placed vertically (12 in. × 10 in.) against the side of the skull.
- The upper border of the grid cassette is placed 3/4 to 1 in. above the top (vertex) of the skull.
- A thin radiolucent sponge is placed beneath the skull to ensure the posterior aspect of the skull is projected on the image.
- The MSP is aligned parallel to the grid cassette.
- The IPL is aligned perpendicular to the grid cassette.
- The IOML is aligned parallel with the long axis of the grid cassette.

Central ray

- Direct the horizontal beam perpendicular to the center of the grid cassette. Collimate to the size of the grid cassette.

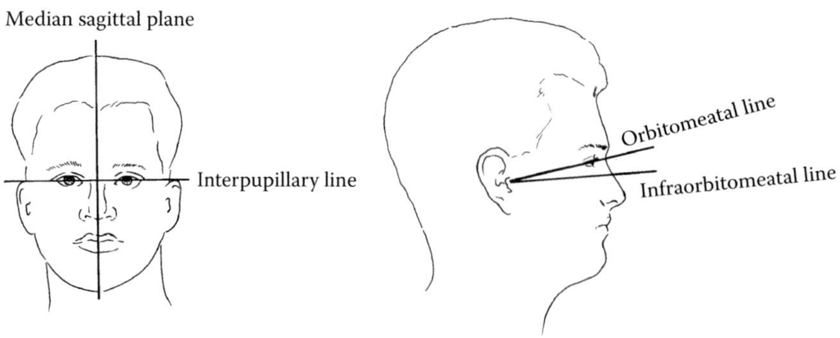

FIGURE 39.32 Cranial positioning lines.

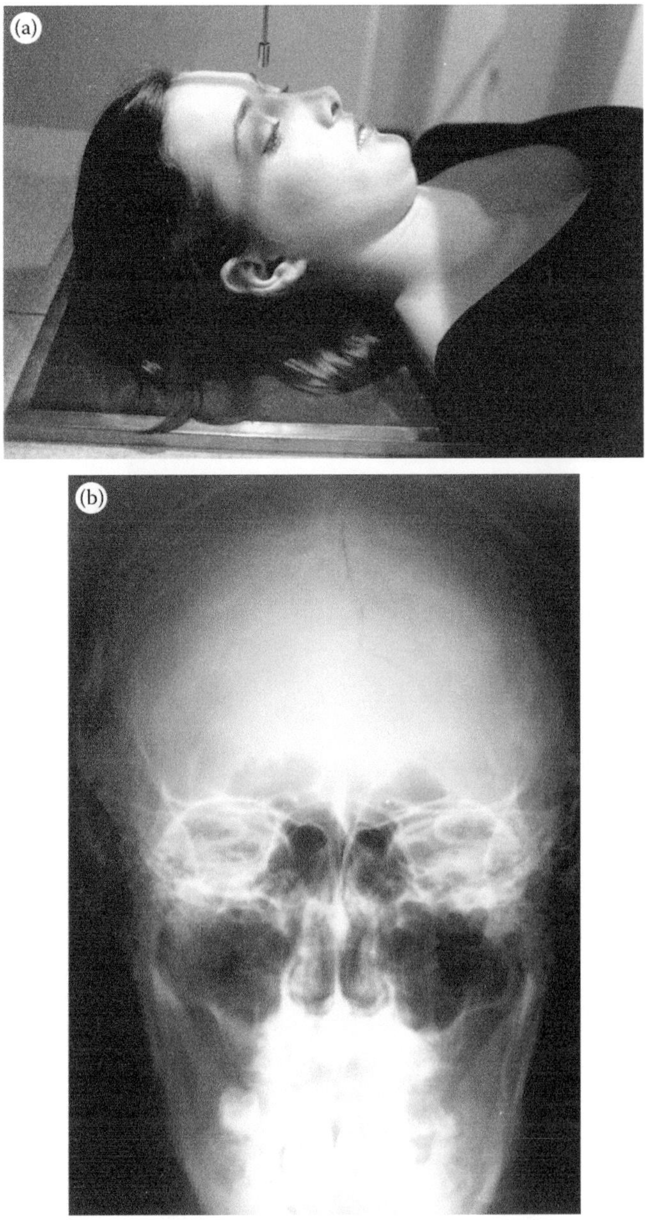

FIGURE 39.33 (a) AP skull position. (b) AP skull projection.

Structures demonstrated (Figure 39.34b)—A lateral projection of the major bones of the skull (cranium).

AP Axial Projection of the Skull (Towne Position)

Cassette size—10 in. × 12 in. grid cassette.

Position of part (Figure 39.35a)

- Place the grid cassette lengthwise beneath the skull.
- The horizontal axis of the grid cassette is centered to the CR (see CR below). Centered in this manner usually requires the upper end of the cassette to be placed approximately 1 in. below the top of the skull.

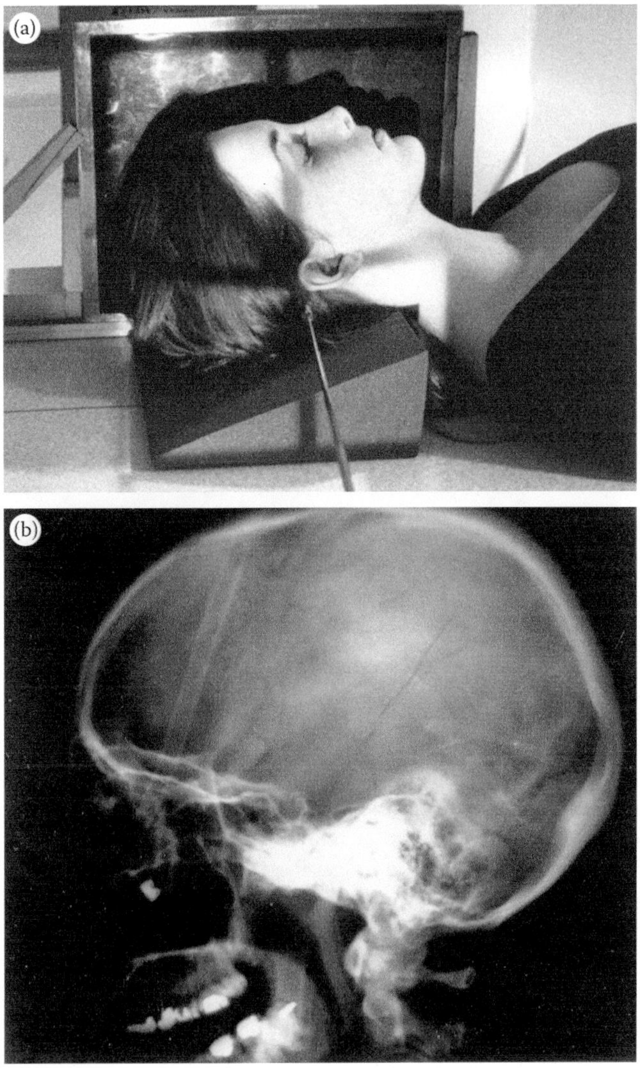

FIGURE 39.34 (a) Cross-table lateral skull position. (b) Cross-table lateral skull projection.

- The MSP is centered to the long axis of the cassette.
- The MSP is perpendicular to the cassette to avoid rotation.
- The OML is aligned perpendicular to the cassette. Note: the IOML may be used. If so, see CR below.

Central ray

- With the OML perpendicular, angle the CR caudally at an angle of 30°. Note: if the IOML is used, the CR is angled 37° caudally.[2]
- The CR enters the skull 1/2 in. above the eyebrows. It exits in the area of the mastoid process.
- The grid cassette is centered to the point where the CR exits the skull. Collimate to the size of the grid cassette.

Structures demonstrated (Figure 39.35b)—An AP projection of the occipital bone and surrounding structures.

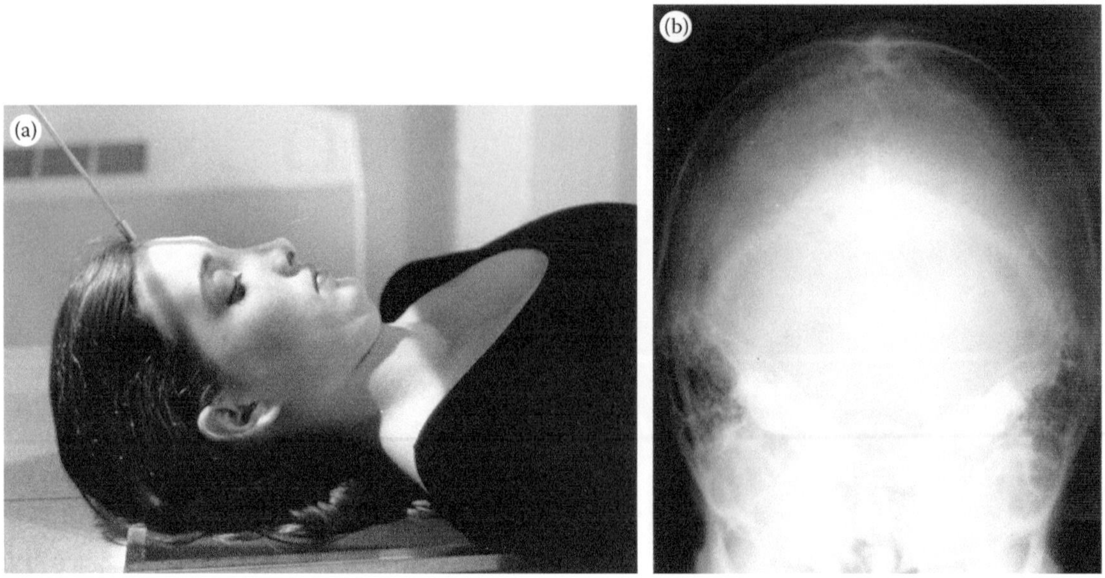

FIGURE 39.35 (a) AP axial skull position (Towne). (b) AP axial skill projection.

RADIOGRAPHIC POSITIONS FOR SPECIFIC STRUCTURES

In contrast to the survey approach, there may be occasions when projections of specific anatomic structures are desired. The positions/projections described below are intended to meet this need. However, several projections previously discussed under the survey approach may also serve to demonstrate specific anatomic structures, for example, forearm, humerus, leg, femur, pelvis, etc. Therefore, these anatomic structures will not be discussed again in this section.

Upper Extremity

AP/PA Projection of Hand and Wrist

Cassette size—10 in. × 12 in. nongrid cassette.

Position of part (Figure 39.36a)

- Place the cassette lengthwise under the wrist and hand.
- Include distal part of forearm.
- Fingers and hand may be pronated or supinated and placed flat on the cassette.

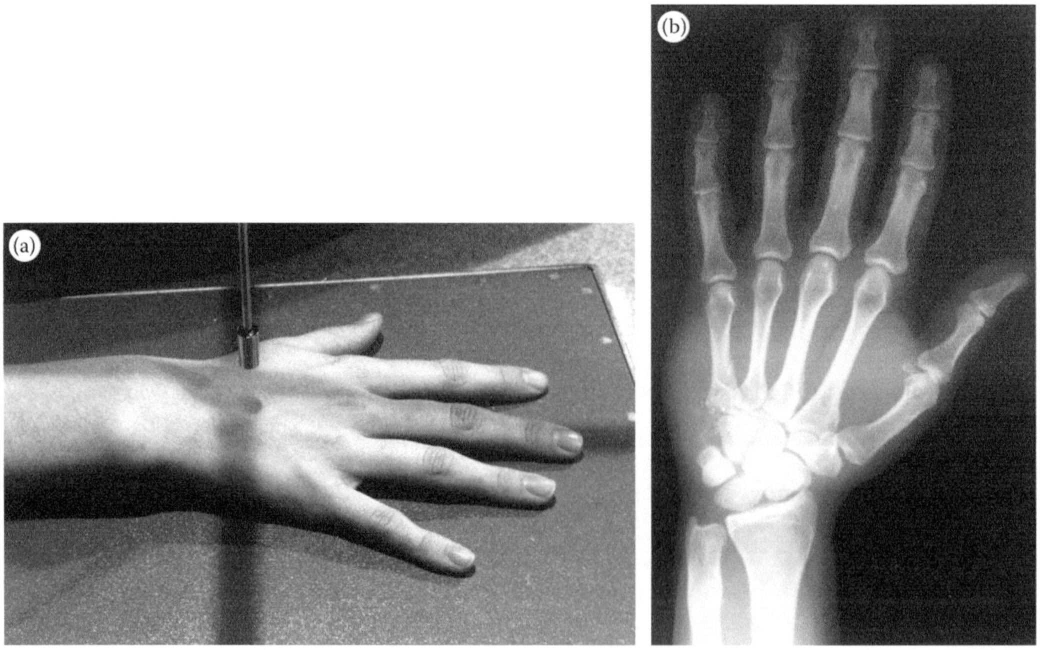

FIGURE 39.36 (a) AP/PA hand and wrist position. (b) PA hand and wrist projection.

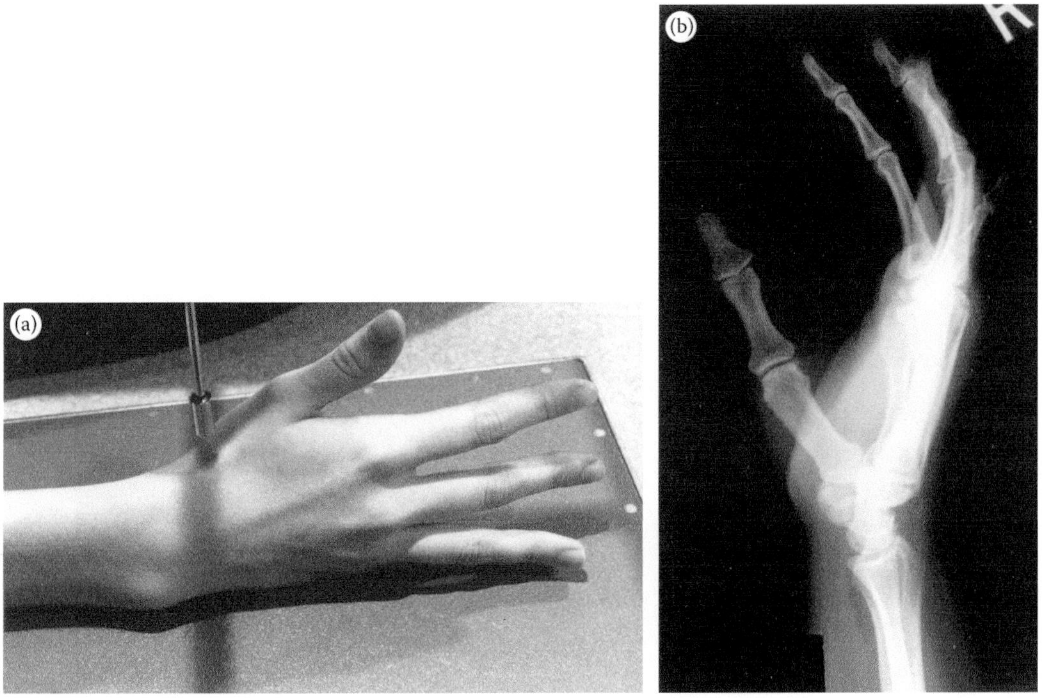

FIGURE 39.37 (a) Lateral hand and wrist position. (b) Lateral hand and wrist projection.

Central ray

- Direct the CR perpendicular to the midpoint of the longitudinal and horizontal axis of the cassette. Collimate to the size of the cassette.

Structures demonstrated (Figure 39.36b)—PA or AP projection of the hand, fingers, and wrist.

Lateral Projection of Hand and Wrist

Cassette size—10 in. × 12 in. nongrid cassette.

Position of part (Figure 39.37a)

- Place the cassette lengthwise under the wrist and hand to include distal part of forearm.
- Rotate the hand and wrist to the lateral position. In this position, the fingers will be lying on top of one another.

Central ray

- Direct the CR perpendicular to the midpoint of the longitudinal and horizontal axis of the cassette. Collimate to the size of the cassette.

Structures demonstrated (Figure 39.37b)—Lateral projection of the wrist, hand, and fingers.

Additional comments—The lateral projection of hand and wrist may also be performed as a cross-table lateral. If done in this manner, the hand should be elevated on a radiolucent sponge. The CR is directed horizontal and perpendicular to the cassette.

AP Projection of the Elbow Joint

Cassette size—10 in. × 12 in. nongrid cassette.

Position of part (Figure 39.38a)

- Place the cassette lengthwise under the fully extended elbow joint.
- Place the elbow joint in the middle of the cassette and supinate the hand to achieve a true AP position.

Central ray

- Direct the CR perpendicular to the midpoint of the longitudinal and horizontal axis of the cassette. Collimate to the size of the cassette.

Structures demonstrated (Figure 39.38b)—AP projection of the elbow joint and distal humerus and proximal forearm.

Cross-Table Lateral Projection of Elbow Joint

Cassette size—10 in. × 12 in. nongrid cassette.

Position of part (Figure 39.39a)

- Place the cassette lengthwise against the medial aspect of the elbow.
- The elbow joint should correspond to the middle of the cassette.
- The fully extended elbow joint should be elevated on a radiolucent sponge.

Central ray

- Direct the horizontal beam toward the cassette.
- CR is perpendicular to the midpoint of the longitudinal and horizontal axis of the cassette. Collimate to the size of the cassette.

Structures demonstrated (Figure 39.39b)—Lateral projection of the elbow joint to include distal humerus and proximal forearm.

Additional comments—A comparable projection can be obtained by extreme internal rotation of the forearm. In this position, the hand will be pronated and the CR is vertical and centered to the cassette.

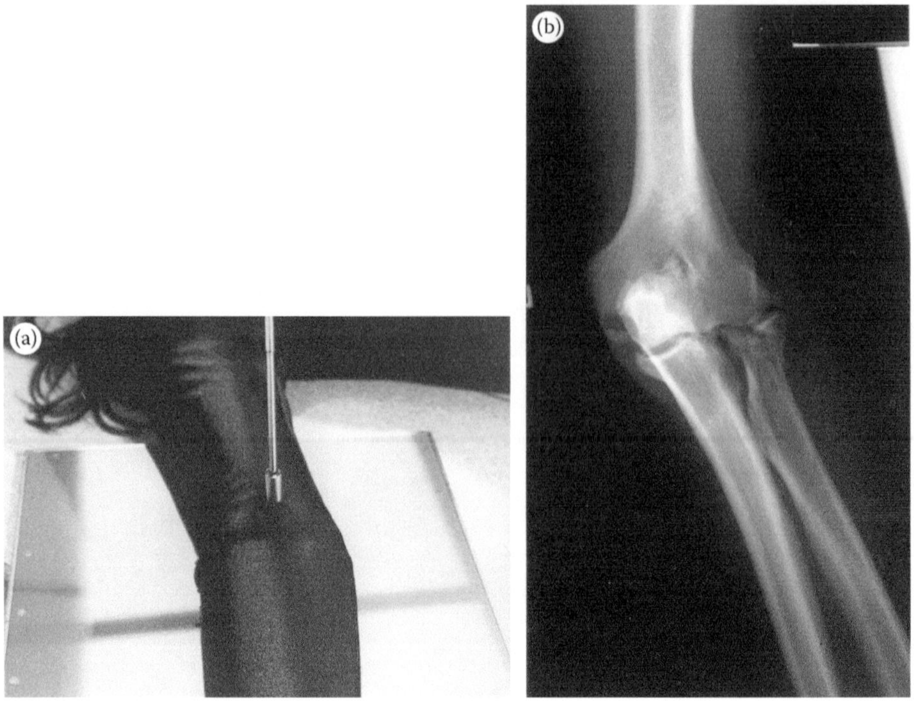

FIGURE 39.38 (a) AP elbow position. (b) AP elbow projection.

AP Projection of the Shoulder Girdle

Cassette size—10 in. × 12 in. grid cassette.

Position of part (Figure 39.40a)

- The grid cassette is placed crosswise beneath the patient, with the top of the cassette approximately 1 1/2 in. above the shoulder.
- The hand should be supinated.
- The midline of the cassette is centered midway between sternum and humerus.

Central ray

- Direct the CR perpendicular to the midpoint of the longitudinal and horizontal axis of the grid cassette. Collimate to the size of the cassette.

Structures demonstrated (Figure 39.40b)—An AP projection of the shoulder girdle to include proximal humerus.

Lower Extremity

AP Projection of Foot

Cassette size—10 in. × 12 in. nongrid cassette.

Position of part (Figure 39.41a)

- Elevate the heel on a radiolucent sponge.
- Place the midline of the cassette lengthwise against the sole of the foot.
- The cassette may be supported by sandbags.

Central ray

- X-ray tube should be angled caudally enough to make the CR perpendicular to the cassette.
- Direct the CR to the midpoint of the longitudinal and horizontal axis of the cassette. Collimate to the size of the cassette.

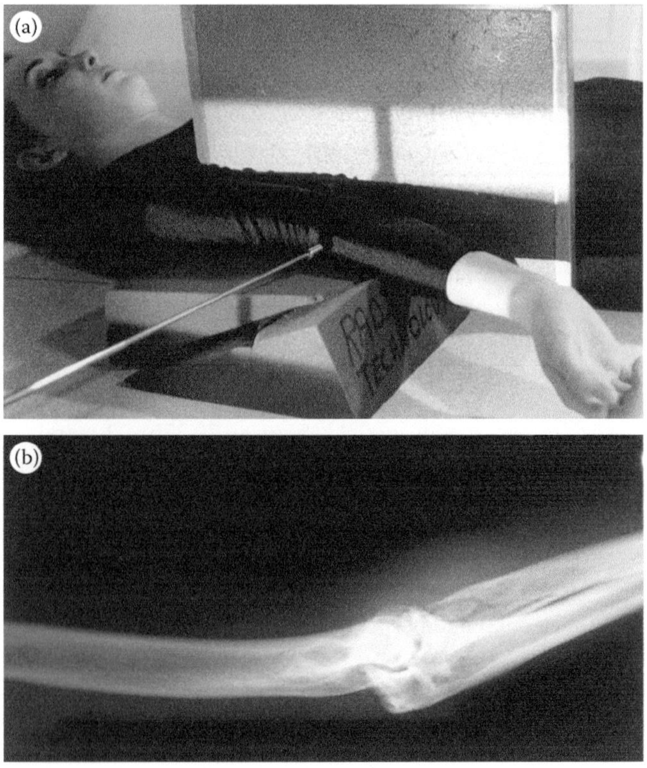

FIGURE 39.39 (a) Cross-table lateral elbow position. (b) Cross-table lateral elbow projection.

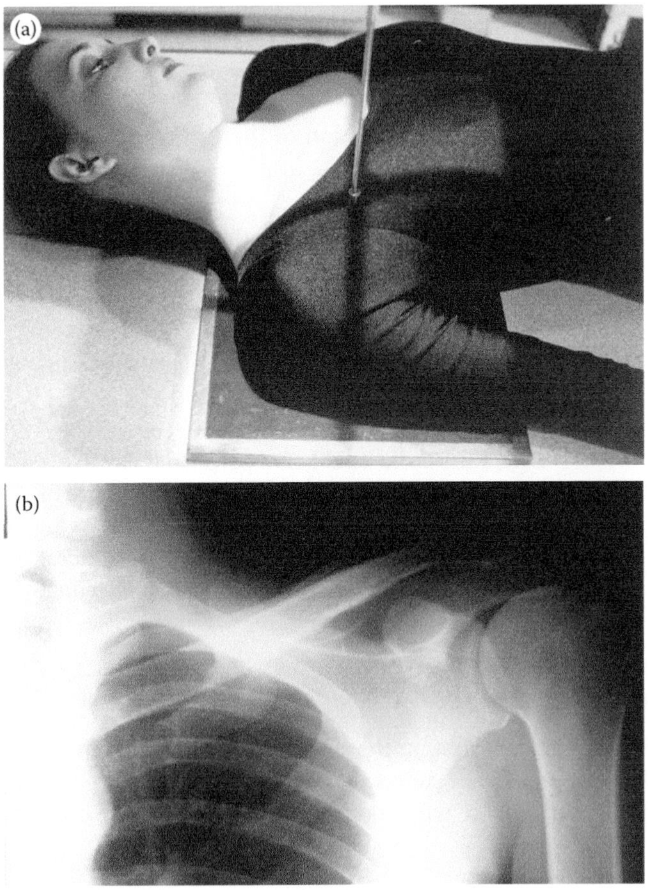

FIGURE 39.40 (a) AP shoulder position. (b) AP shoulder projection.

Structures demonstrated (Figure 39.41b)—AP projection of the foot and toes.

Cross-Table Lateral Projection of the Foot

Cassette size—10 in. × 12 in. nongrid cassette.

Position of part (Figure 39.42a)

- Place the cassette lengthwise against the medial aspect of the foot.
- Heel should be elevated on a radiolucent sponge.
- The long axis of the foot should be centered to the middle of the cassette.

Central ray

- Direct the horizontal beam perpendicular to the midpoint of the longitudinal and horizontal axis of the cassette. Collimate to the size of the cassette.

Structures demonstrated (Figure 39.42b)—Lateral projection of the heel, foot, and toes.

AP Projection of the Knee

Cassette size—10 in. × 12 in. grid cassette.

Position of part (Figure 39.43a)

- The grid cassette is placed lengthwise under the knee joint and is aligned with the long axis of the leg.
- The knee joint should be fully extended.
- The foot is maintained in a vertical position to achieve a true AP position.
- Center the horizontal axis of the grid cassette 1 in. below the patella (kneecap).

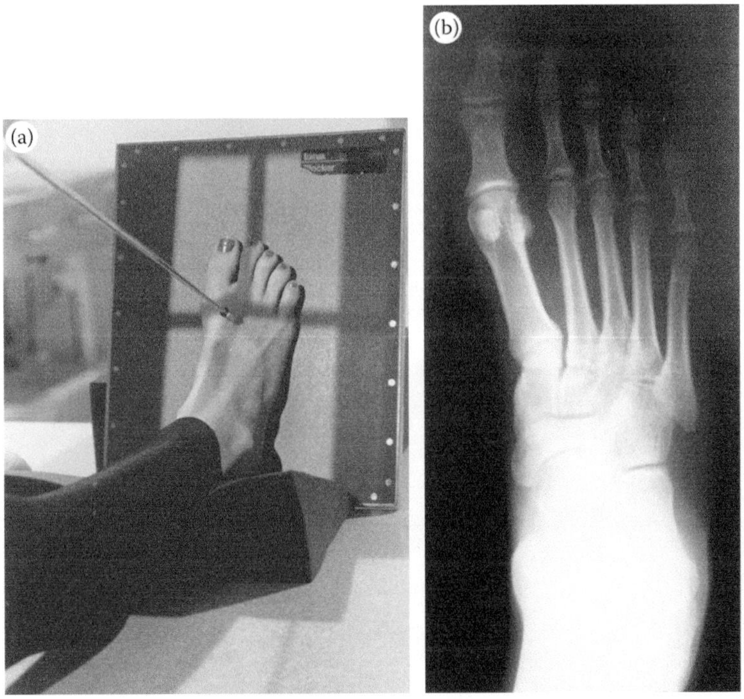

FIGURE 39.41 (a) AP foot position. (b) AP foot projection.

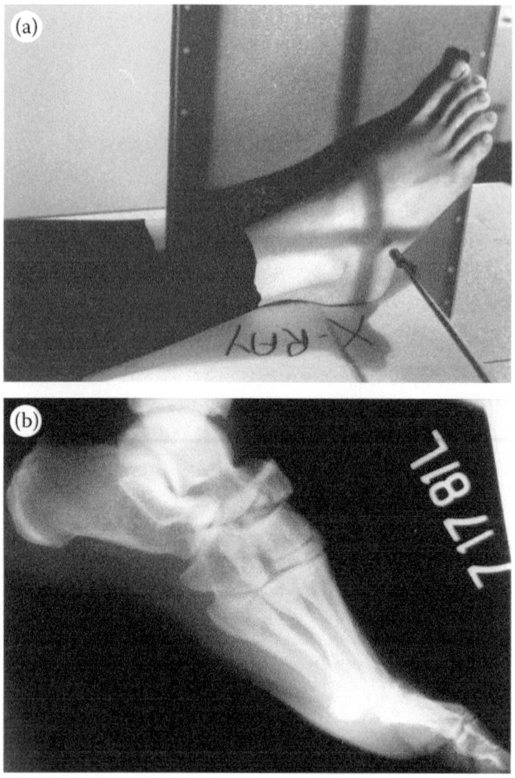

FIGURE 39.42 (a) Cross-table lateral foot position. (b) cross-table lateral foot projection.

Central ray

- Angle the CR 5° cephalad to enter at the midpoint of the longitudinal and horizontal axes of the cassette. Collimate to the size of the cassette.

Structures demonstrated (Figure 39.43b)—AP projection of the knee joint to include distal femur and proximal tibia-fibula.

Cross-Table Lateral Projection of the Knee

Cassette size—10 in. × 12 in. grid cassette.

Position of part (Figure 39.44a)

- The cassette is placed lengthwise between the knees.
- The fully extended knee joint should be elevated on a radiolucent sponge.
- The horizontal axis of the grid cassette is centered 1 in. below the patella.

Central ray

- Direct the horizontal beam perpendicular to the midpoint of the cassette. Collimate to the size of the anatomic part.

Structures demonstrated (Figure 39.44b)—A lateral projection of the knee joint to include proximal tibia–fibula and distal femur.

Vertebral Column

Positions demonstrating the thoracic and lumbar spine were described under the survey approach to radiographic positioning. This section will address positioning of the cervical spine.

AP Projection of the Cervical Spine

Cassette size—10 in. × 12 in. grid cassette.

Position of part (Figure 39.45a)

- Place the grid cassette lengthwise beneath the neck.

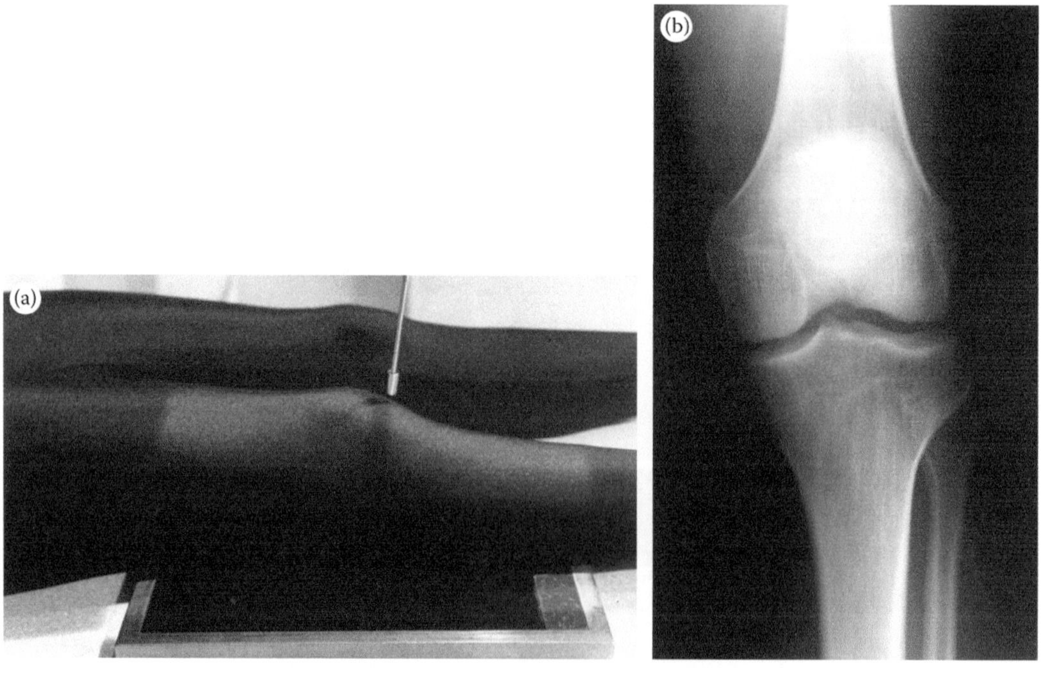

FIGURE 39.43 (a) AP knee position. (b) AP knee projection.

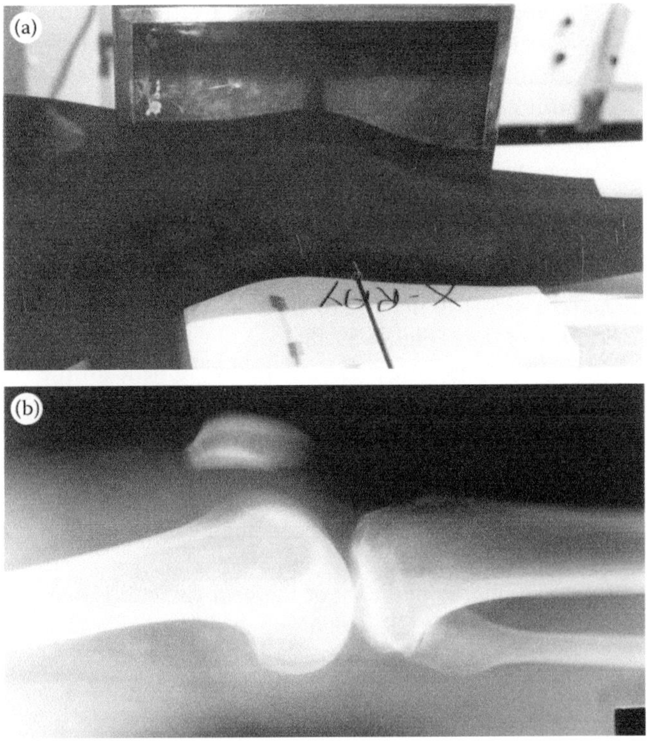

FIGURE 39.44 (a) Cross-table lateral knee position. (b) Cross-table lateral projection.

- Place the upper border of the grid cassette 1 in. above the top of the ear.
- Center the MSP to the long axis of the cassette.
- To avoid rotation, adjust the MSP of the skull perpendicular to the grid cassette.
- Raise the chin as much as possible to avoid superimposition of the chin over the upper cervical vertebrae.

Central ray

- Direct the CR 15–20° cephalic. The CR enters the horizontal axis of the grid cassette. *Note*: the CR must be directed along the long axis of the grid lines.

Structures demonstrated (Figure 39.45b)—An AP projection of the lower cervical vertebrae and the upper thoracic vertebrae.

Cross-Table Lateral of the Cervical Spine

Cassette size—10 in. × 12 in. grid cassette.

Position of part (Figure 39.46a)

- Place the cassette lengthwise next to the shoulder. Note: placement of the cassette next to the shoulder creates a space between the cervical spine and the shoulder (OID). Placement of the cassette in this manner is necessary to demonstrate all of the cervical vertebrae.

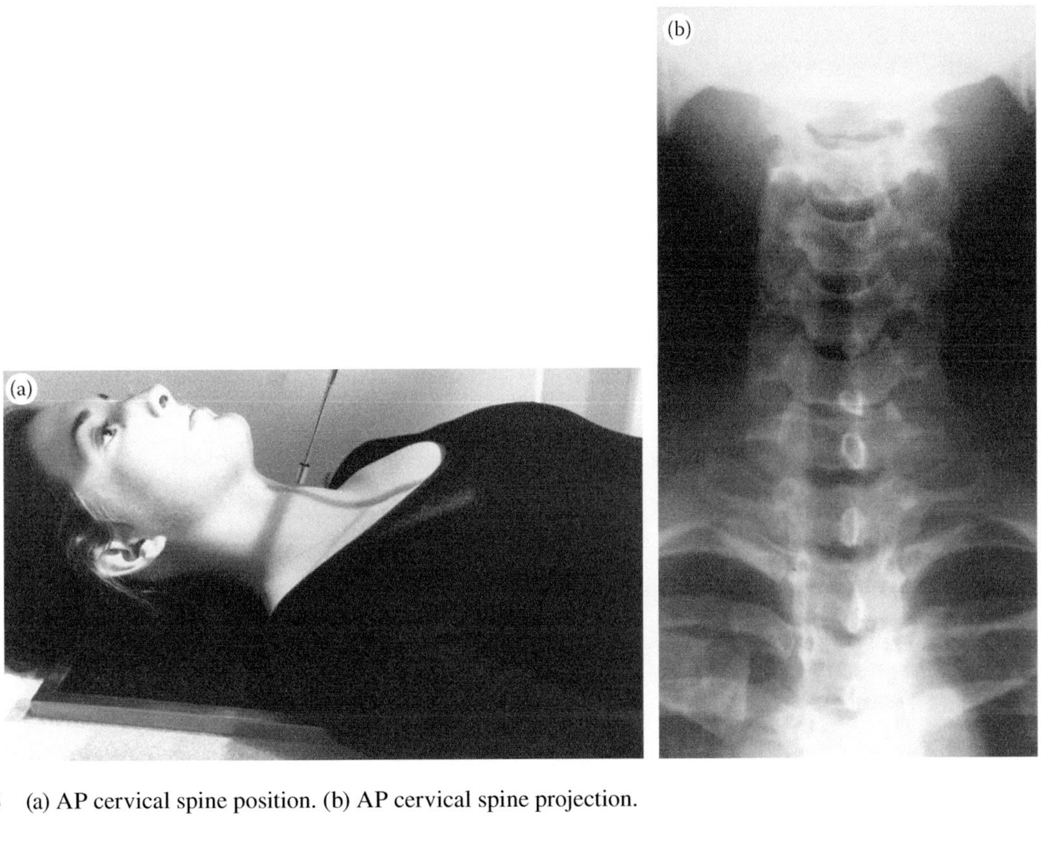

FIGURE 39.45 (a) AP cervical spine position. (b) AP cervical spine projection.

- It is usually necessary to pull the arms downward to avoid superimposing the shoulders over C-6 and C-7.
- Place the upper border of the grid cassette 1 1/2 in. above the top of the ear.
- To avoid rotation, adjust the MSP of the skull perpendicular to the table.
- The IPL should be perpendicular to the grid cassette. Collimate to the size of the grid cassette.

Structures demonstrated (Figure 39.46b)—A lateral projection of the entire cervical spine.

Additional comments—When pulling the arms down, care should be taken to exert equal pressure on both arms. Failure to do so will cause one shoulder to move upward, which will defeat the purpose of the maneuver. A SID of 72 in. (180 cm) is recommended for this projection.

Facial Bones

AP Projection of Facial Bones

Cassette size—10 in. × 12 in. grid cassette.

Position of part (Figure 39.47a)

- Place the grid cassette lengthwise beneath the skull.
- Place the upper border of the cassette 1 in. above the top of the skull.
- The MSP is centered to the long axis of the grid cassette.
- The MSP is perpendicular to the grid cassette.
- The OML is perpendicular to the grid cassette.

Central ray

- The CR is angled 15° cephalad to the midpoint of the cassette.[3] The CR enters the skull 1/2 in. above the bridge of the nose. Collimate to the size of the cassette. Note: the CR must be directed along the long axis of the grid lines.

Structures demonstrated (Figure 23.47b)—An AP projection of the facial bones. The petrous ridges are projected into the lower one-third of the orbit.

AP Axial Projection (Reverse Waters Position)

Cassette size—10 in. × 12 in. grid cassette.

Position of part (Figure 39.48a)

- Place the grid cassette lengthwise beneath the skull.
- The MSP is centered to the long axis of the grid cassette.
- The MSP is perpendicular to the grid cassette.
- The IOML is perpendicular to the grid cassette.
- The horizontal axis of the grid cassette is centered to the CR. Centering in this manner usually requires the upper end of the cassette to be approximately 1/2 in. above the top of the skull.

Central ray

- Angle the CR 30° cephalic.[4] Note: the CR must be directed along the long axis of the grid lines.
- The CR enters the space between the nose and the upper teeth.
- The grid cassette is centered at the point the CR exits the skull. Collimate to the size of the grid cassette.

Structures demonstrated (Figure 39.48b)—An AP survey projection of the facial bones.

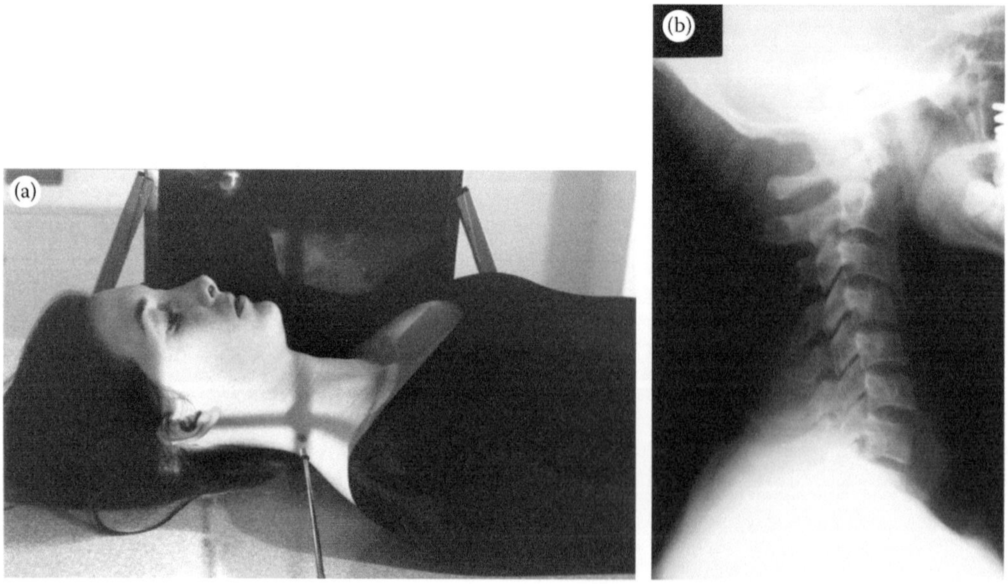

FIGURE 39.46 (a) Cross-table lateral cervical spine position. (b) Cross-table lateral cervical spine projection.

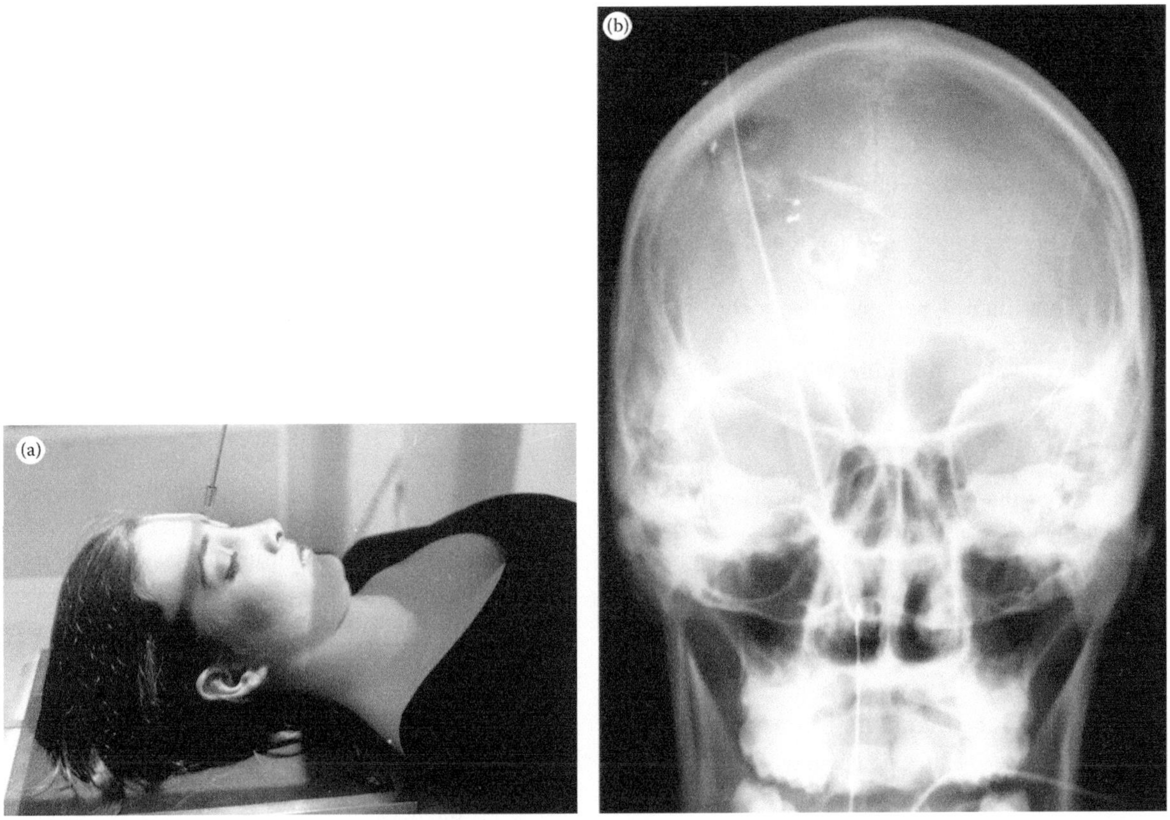

FIGURE 39.47 (a) AP facial bone position. (b) AP facial bone projection.

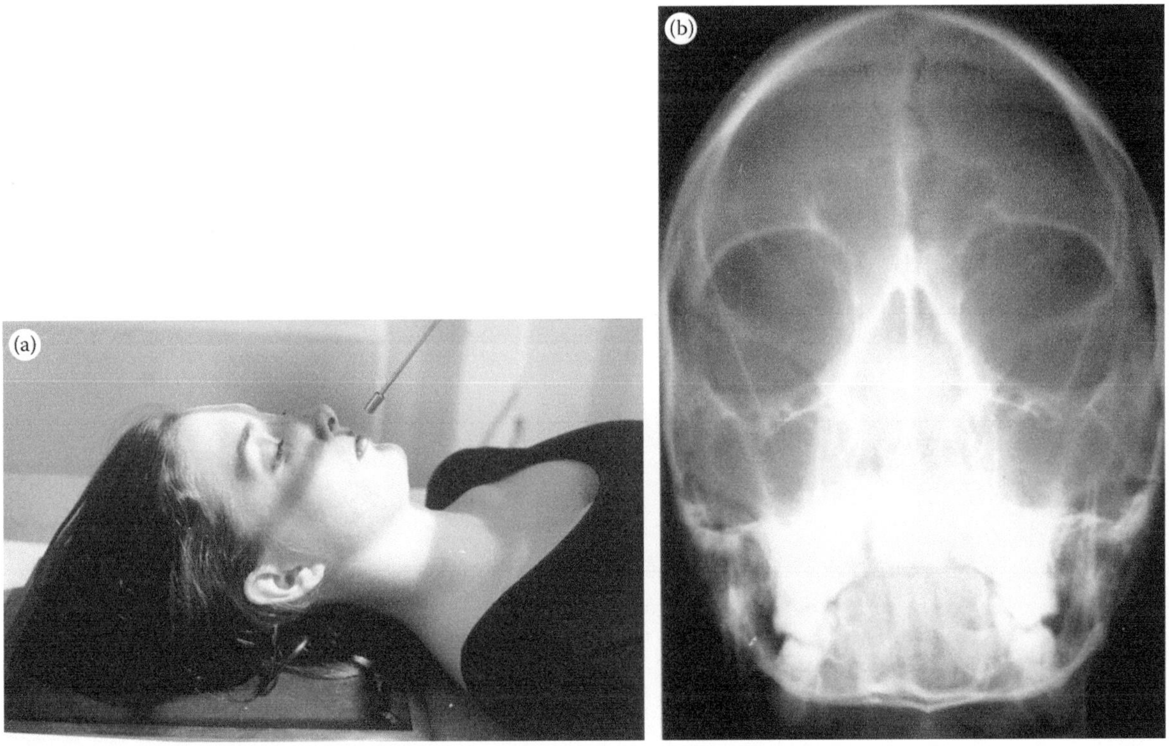

FIGURE 39.48 (a) AP axial position of the facial bones (reverse Waters). (b) AP axial projection of the facial bones.

Cross-Table Lateral Projection of the Facial Bones

Cassette size—10 in. × 12 in. grid cassette.

Position of part (Figure 39.49a)

- The grid cassette is placed lengthwise against the side of the skull.
- The horizontal axis of the grid cassette is centered to the zygoma (cheek bone).
- The MSP is aligned parallel with the grid cassette.
- The IPL is aligned perpendicular to the grid cassette.
- The IOML is aligned parallel with the horizontal axis of the grid cassette.

Central ray

- Direct the horizontal beam perpendicular to the midpoint of the grid cassette. Collimate to the size of the grid cassette.

Structures demonstrated (Figure 39.49b)—A lateral projection of the facial bones.

Mandible

AP Axial Projection of the Mandible

Cassette size—10 in. × 12 in. grid cassette.

Position of part (Figure 39.50a)

- The grid cassette is placed lengthwise beneath the skull.
- Place the upper border of the grid cassette at the top of the ear.
- The MSP is centered to the long axis of the grid cassette.
- The MSP is perpendicular to the grid cassette.
- The OML is aligned perpendicular to the grid cassette.

Central ray—Angle the CR 10 to 15° caudally. The CR should pass through the junction of the lips. Center the CR to the horizontal axis of the grid cassette. Collimate to the size of the grid cassette. *Note*: the CR must be directed along the long axis of the grid lines.

Structures demonstrated (Figure 39.50b)—A survey projection of the mandible including the rami, angle, and body.

Cross-Table Oblique of the Mandible[5]

Cassette size—10 in. × 12 in. nongrid cassette.

Position of part (Figure 39.51a)—The cassette is placed vertically (12 in. × 10 in.) next to the side of the mandible of interest (right or left side). The head is tilted approximately 15°. Tilting the head causes the mandible to move a slight distance from the cassette. Tilting the head assists in reducing superimposition of the mandibular rami.

- Center the cassette to include the structures between the chin and the ear of the side of interest.
- Raise the chin to move the mandible away from the neck.
- The MSP is vertical.

Central ray

- The horizontal beam is directed 15–20° cephalically to the midpoint of the cassette. Note: both obliques are taken for comparison. Collimate to the size of the part.

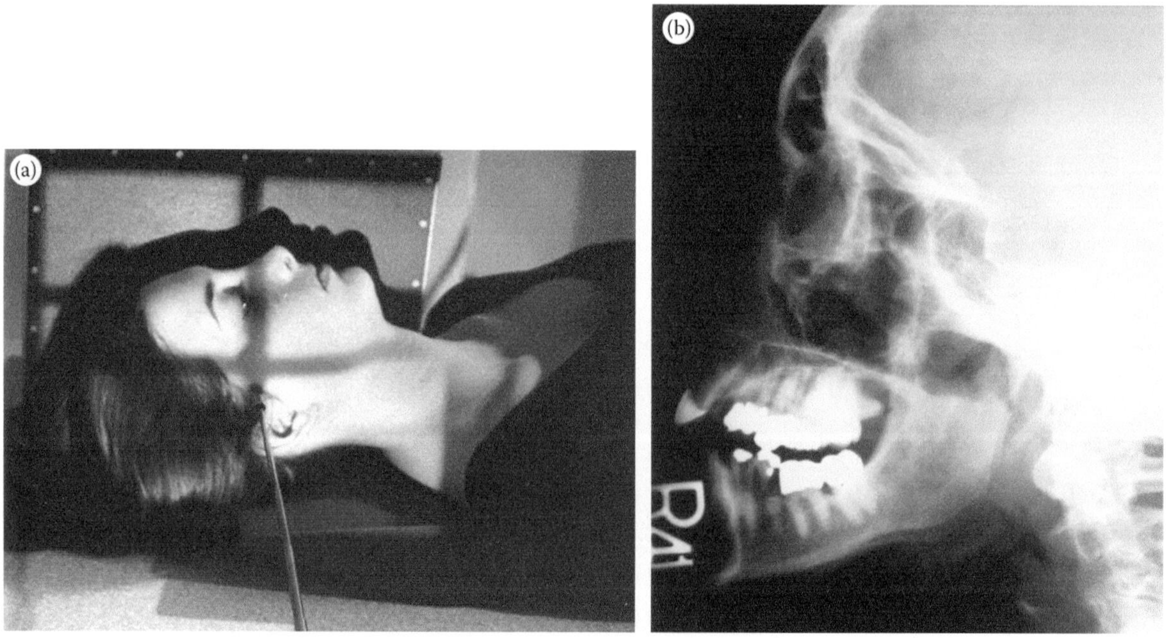

FIGURE 39.49 (a) Cross-table lateral position of facial bones. (b) Cross-table lateral projection of facial bones.

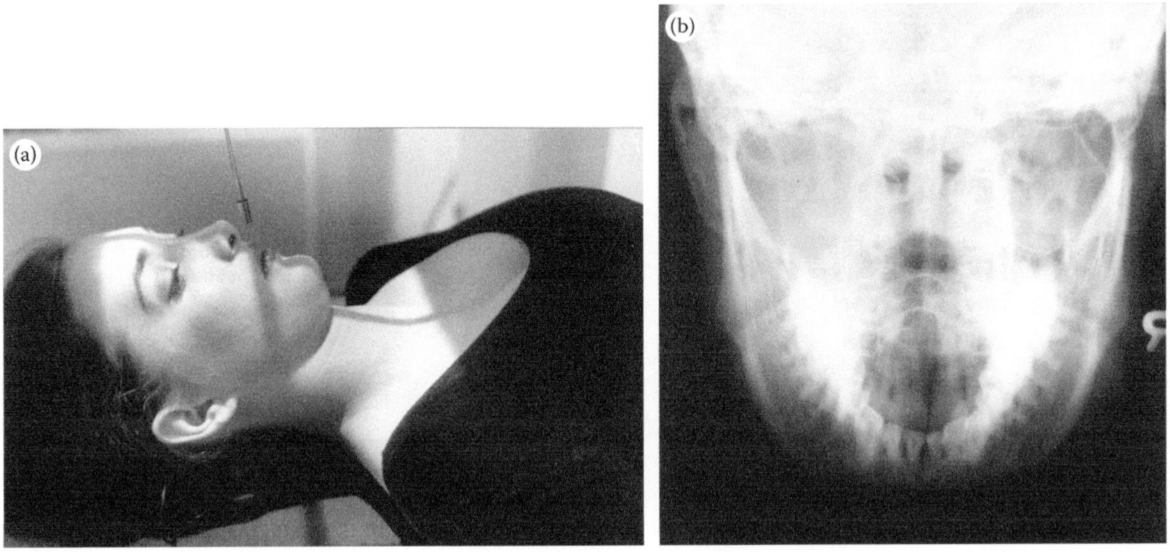

FIGURE 39.50 (a) AP axial position of mandible. (b) AP axial projection of mandible.

Structures demonstrated (Figure 23.51b)—An oblique projection of the mandibular body and ramus of the side next to the cassette.

Cross-Table Lateral of the Mandible

Cassette size—10 in. × 12 in. grid cassette.

Position of part (Figure 39.52a)

- The grid cassette is placed vertically (12 in. to 10 in.) against the side of interest.
- The long axis of the grid cassette is centered to the junction of the lips.
- The MSP is aligned parallel with the grid cassette.
- The IPL is perpendicular to the grid cassette.
- The IOML is aligned parallel with the long axis of the grid cassette.

Central ray

- Direct the horizontal beam perpendicular to the midpolint of the cassette. Collimate to the size of the part.

Structures demonstrated (Figure 39.52b)—A lateral projection of the entire mandible.

RADIOGRAPHIC POSITIONING (DENTAL)

Radiographic Anatomy of the Tooth

Although variable in appearance and function, all teeth have a crown and a root which is secured within the bone of the maxilla or mandible (Figure 39.53).

The adult dentition is made up of 32 teeth: 8 incisors, 4 canine, 8 premolar, and 12 molar teeth. Deciduous or "milk

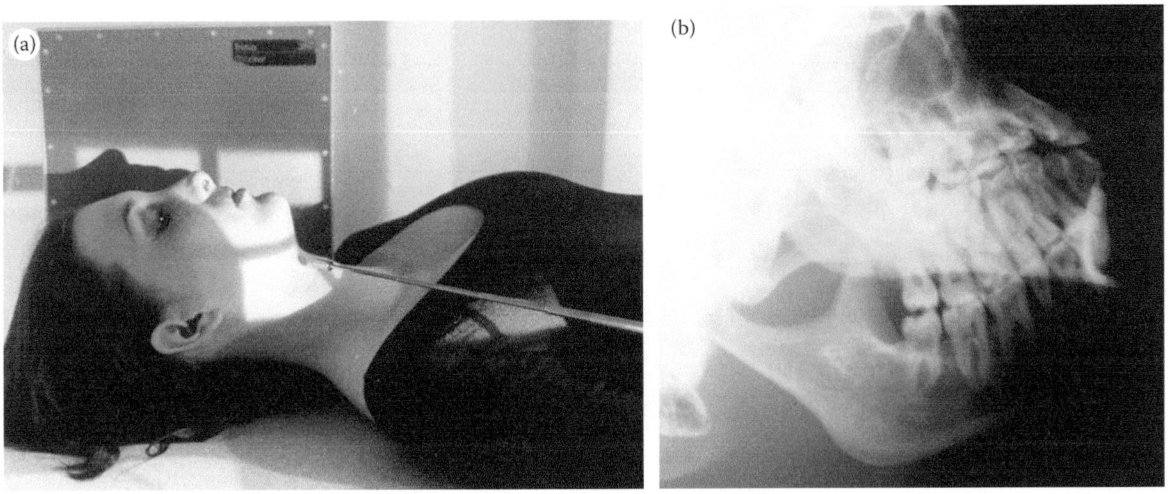

FIGURE 39.51 (a) Cross-table oblique position of mandible. (b) Cross-table oblique projection of mandible.

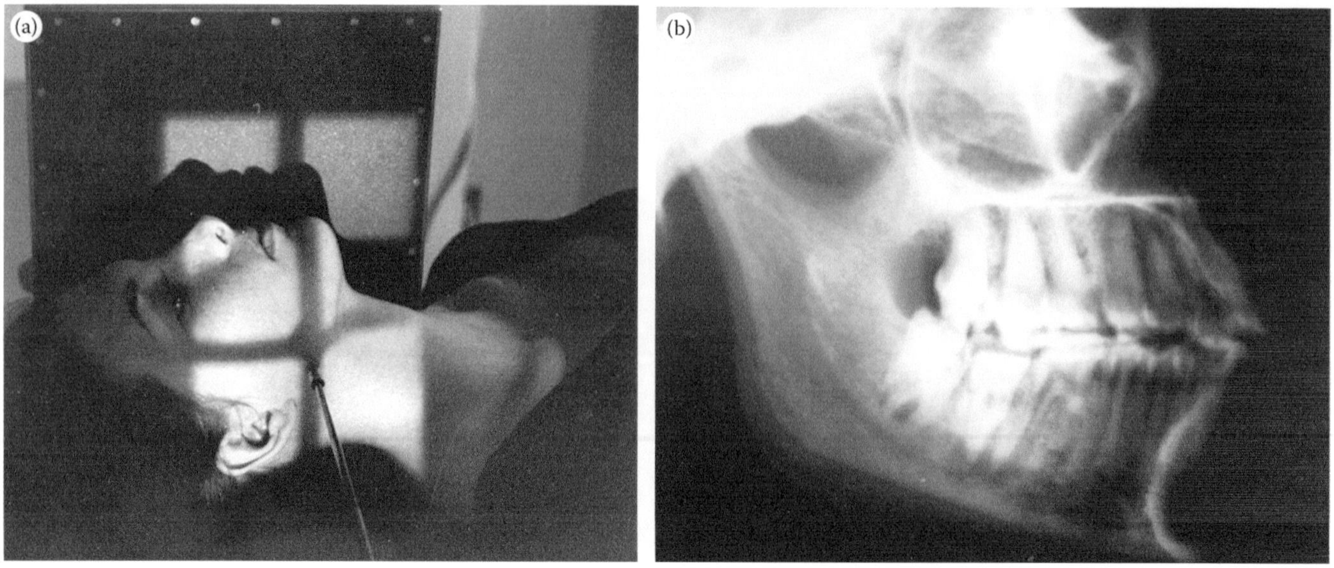

FIGURE 39.52 (a) Cross-table lateral position of mandible. (b) Cross-table lateral projection of mandible.

teeth" normally total 20 with 8 incisors, 4 canines and 8 molars. For dental examination purposes, the dentition is usually divided into four quadrants; left and right upper and left and right lower areas (Figure 39.54a and b).

Dental Notation

In dentistry, the teeth are identified using a variety of coding or notation methods, where a number (or letter) will refer to each tooth individually. Thus, dental notation is the way the teeth are individually identified in writing dental records and referral letters.

A number or letter refers to each tooth, so that each can be individually differentiated. There are many ways in which this is achieved but the three most commonly used methods internationally are the Palmer/Zsigmondy, Universal, and Fédération Dentaire Internationale (FDI) techniques (Figure 39.55a–c).

One of the first notation methods to be developed was by Richard Adolf Zsigmondy in 1861, who divided the teeth into

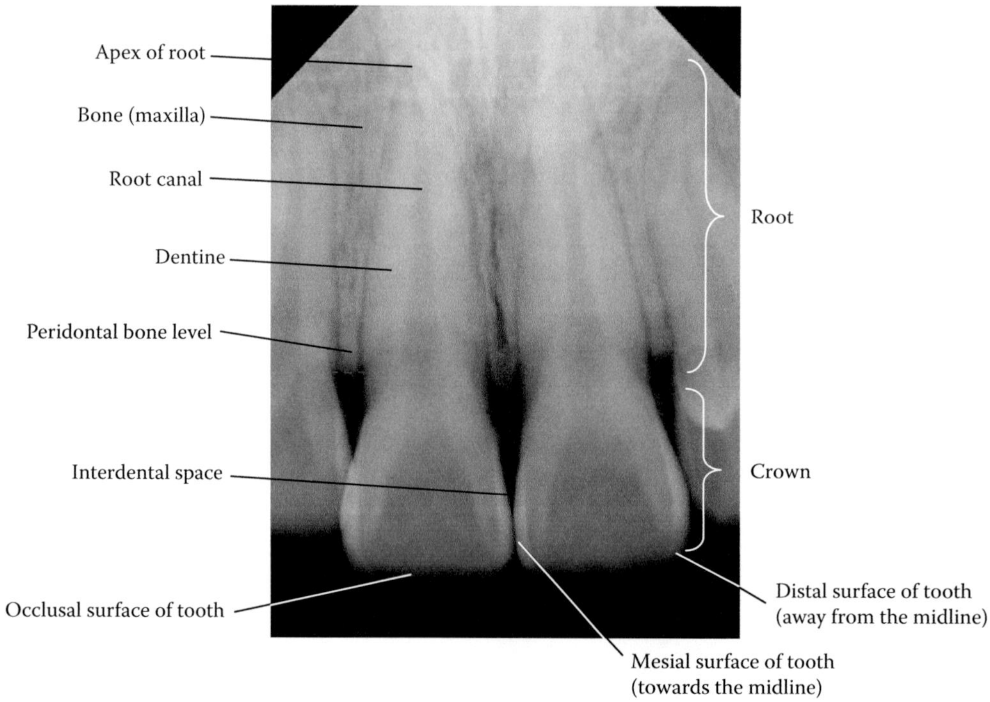

FIGURE 39.53 Radiographic anatomy of the tooth. (From Faircloth, E. With permission.)

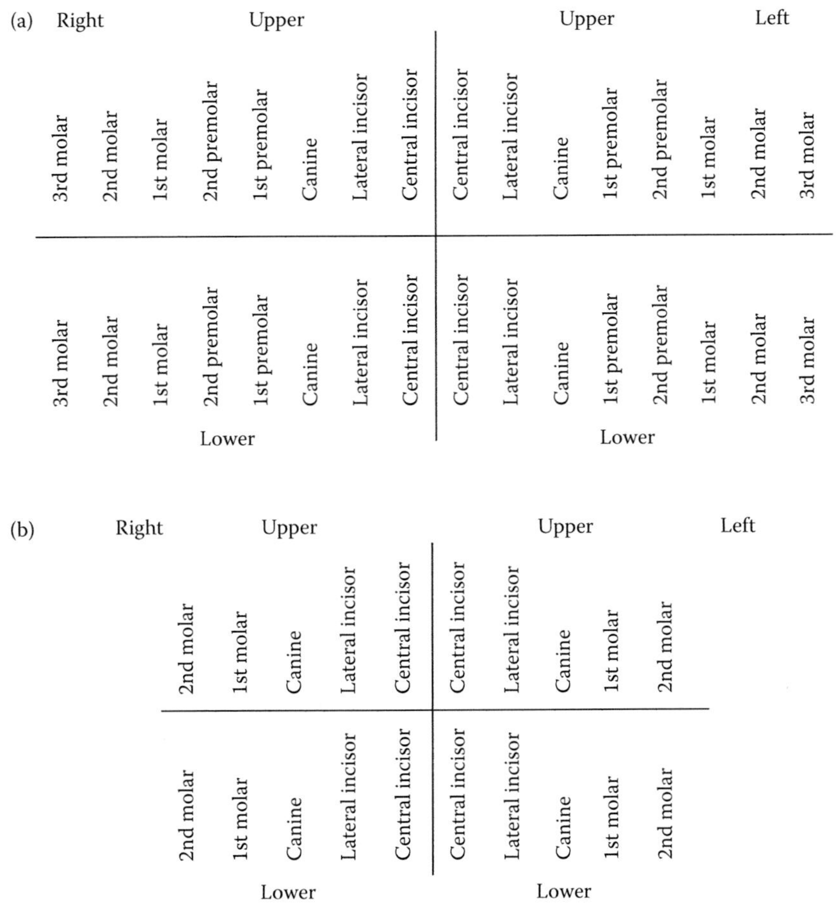

FIGURE 39.54 (a) The four quadrants of the adult dentition. (b) The four quadrants of the pediatric dentition. (Adapted from Faircloth, E.)

(a)

Right																	Left
(upper)	8	7	6	5	4	3	2	1	1	2	3	4	5	6	7	8	(upper)
(lower)	8	7	6	5	4	3	2	1	1	2	3	4	5	6	7	8	(lower)

Deciduous Teeth

(upper)	E	D	C	B	A	A	B	C	D	E	(upper)
(lower)	E	D	C	B	A	A	B	C	D	E	(lower)

(b)

Right																	Left
(upper)	1	2	3	4	5	6	7	8	9	10	11	12	13	14	15	16	(upper)
(lower)	32	31	30	29	28	27	26	25	24	23	22	21	20	19	18	17	(lower)

Deciduous Teeth

(upper)	A	B	C	D	E	F	G	H	I	J	(upper)
(lower)	T	S	R	Q	P	O	N	M	L	K	(lower)

(c)

Right																	Left
(upper)	18	17	16	15	14	13	12	11	21	22	23	24	25	26	27	28	(upper)
(lower)	48	47	46	45	44	43	42	41	31	32	33	34	35	36	37	38	(lower)

Deciduous Teeth

(upper)	55	54	53	52	51	61	62	63	64	65	(upper)
(lower)	85	84	83	82	81	71	72	73	74	75	(lower)

FIGURE 39.55 Internationally applied systems of dental notation (a) Palmer/Zsigmondy (1861). (b) Universal (1968). (c) Fédération Dentaire Internationale (1970). (Adapted from Faircloth, E.)

four quadrants and numbered the teeth in each quadrant from 1 to 8. Deciduous teeth were originally labeled using Roman numerals but these were later updated by Palmer with the alphabetical letters A–E. Using this method, the upper right third molar, for example, was identified as 8|. This method continues to be used by UK dental health care practitioners and all but one of the major UK dental schools.[6] With the advent of typewriters however, the quadrant symbols proved to be difficult to represent using a keyboard and also had a higher margin for error.

The American Dental Association (ADA) originally recommended the use of the Palmer/Zsigmondy notation method for classifying teeth but, as this system used symbols that were difficult to use with keyboards, the ADA approved the use of the Universal notation system in 1968. The Universal system is used widely throughout the United States.

The Fédération Dentaire Internationale (FDI) system comprises two digits, the first of which represents the quadrant in which the tooth is located, where "1" represents right upper quadrant and "3" represents the left lower quadrant. The second digit refers to the position of the tooth in relation to the midline, as in the Palmer/Zigmondy system.

The FDI system is used worldwide, is formally recognized by the World Health Organization and is used in the International Criminal Police Organization (INTERPOL) disaster victim identification (DVI) process (Figure 39.56).

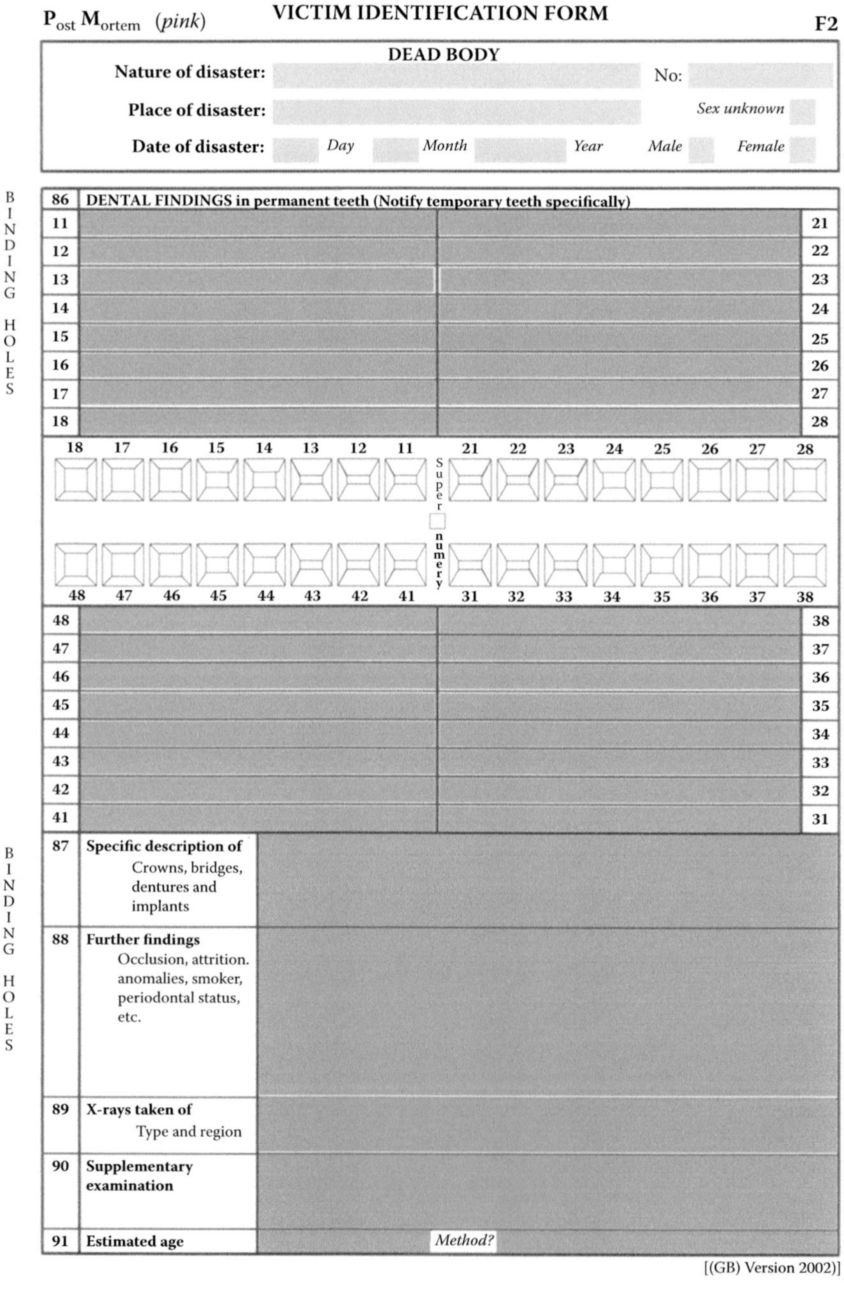
Post Mortem (pink) VICTIM IDENTIFICATION FORM F2

DEAD BODY

Nature of disaster: No:

Place of disaster: Sex unknown

Date of disaster: Day Month Year Male Female

BINDING HOLES

86 DENTAL FINDINGS in permanent teeth (Notify temporary teeth specifically)

11			21
12			22
13			23
14			24
15			25
16			26
17			27
18			28

18 17 16 15 14 13 12 11 Super numerary 21 22 23 24 25 26 27 28

48 47 46 45 44 43 42 41 31 32 33 34 35 36 37 38

48			38
47			37
46			36
45			35
44			34
43			33
42			32
41			31

BINDING HOLES

87	Specific description of Crowns, bridges, dentures and implants	
88	Further findings Occlusion, attrition. anomalies, smoker, periodontal status, etc.	
89	X-rays taken of Type and region	
90	Supplementary examination	
91	Estimated age	Method?

[(GB) Version 2002)]

FIGURE 39.56 The use of the FDI dental notation system for postmortem victim identification records. (Adapted from INTERPOL-PD.)

Intraoral Radiography Techniques

The particular technique for providing dental radiography will depend on the condition of the body or remains.

Intraoral radiographs can be achieved using with the Paralleling technique or the Bisecting Angle technique, both are described in the following sections. The former is considered the superior technique because, although the resultant image is magnified due to the distance of the image receptor from the teeth under examination, there is minimal distortion of the anatomic structure imaged. Images can also be reproduced effectively using this standardized technique. The Bisecting angle technique allows closer proximity of the receptor with the teeth, but this is achieved by angulation of the receptor, resulting in distortion (foreshortening or elongation) of the teeth and surrounding areas. The most appropriate technique should be selected depending on the condition of the remains under examination and the resources available at the time.

The recommended imaging equipment for intraoral radiography are outlined in Chapter 35, p.351.

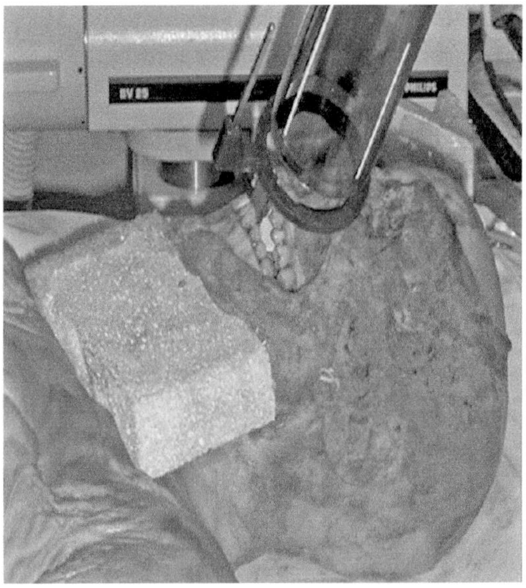

FIGURE 39.57 The use of a dry mortuary sponge to maintain occlusion during x-ray exposure of the canine region. (From Hines, E. With permission.)

Paralleling Technique

Where the maxilla and mandible are *in situ* within the body, the Paralleling technique should be attempted in the first instance. Radiography film/sensor paralleling holders allow for standardized positioning and provide a visual reference outside the oral cavity that can be useful for the accurate positioning and angulation of the x-ray tube.

Minimal oral access is required to achieve good quality images of the incisor and canine regions. The film/sensor holder should be introduced and secured in position with appropriate positioning aids; sponge pads, balloon catheters[7] or other tools.

Gaining access to the oral cavity is one thing, maintaining occlusion with the film/sensor holder in position within the mouth is often quite another. Various methods can be employed to maintain occlusion, avoiding the need for manual support and thus unnecessary exposure of the operator to radiation. Figure 39.57 demonstrates the use of a dry mortuary sponge to keep the mouth closed during the x-ray exposure.

Another innovative tool is a length of rigid plastic, such as a rigid plastic ruler, measuring approximately 1 in. × 2.5 in. (2.5 cm × 6 cm), which makes an effective anchor for the positioning of intraoral films/digital sensors, stabilizing the holder for the duration of the x-ray exposure. This simple, but highly useful tool, is particularly effective for maneuvering the film/sensor into position for imaging of the posterior teeth, the premolars and molars. The film/sensor holder is gently oscillated to obtain the correct positioning, simultaneously using the length of plastic as a winch. This is then left *in situ* acting as an anchor between the occlusal surface and the holder.

Bisecting Angle Technique

Skeletonized or disassociated remains, particularly of the anterior regions of the maxilla and mandible, will not support the use of paralleling film/sensor holders. The bisecting angle technique can be used in this situation as no holders are required and the film/image receptor is positioned next to the tooth under examination (Figure 39.58). To achieve this, the film/receptor must be angled resulting in corresponding angulation of the x-ray tube to compensate for this. To achieve the bisecting angle technique, the central x-ray beam will be perpendicular to the bisecting angle between the long axis of the tooth under examination and the film/receptor. The degree of angulation of the x-ray tube will depend on the area under examination (Table 39.1).

A skull or maxilla may be inverted and supported on an anthropologist's skull ring or sponge pad to make positioning easier remembering, of course, to invert the angulations of the x-ray tube to achieve the required image projection. For completely skeletonized, dry remains, various positioning aids, such as surgical table, sponge pads and modeling clay, may be used to secure the positioning of the film/receptor during the exposure.

Positioning of the head or skull is crucial for this technique; neither should be canted (tilted) or rotated, and the midline should remain perpendicular to the floor. That said, there are many variables that remain with this technique making it difficult to achieve reproducible images of the teeth.

Parallax

The demonstration of three-dimensional objects (i.e., the teeth and jaw) using a two-dimensional imaging medium (i.e., the x-ray image) is a challenge. A technique known as "Parallax imaging" can be used to record three-dimensional information. This imaging technique uses a principle involving the horizontal angulation of the x-ray tube to represent particular areas of anatomy pathology, and other regions of interest.

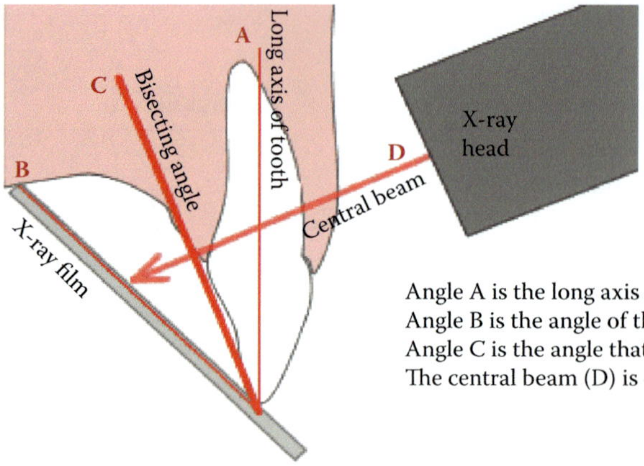

FIGURE 39.58 The bisecting angle technique.

Figure 39.59 demonstrates the principle of the parallax technique, where "A" represents the use of the x-ray tube with no horizontal angulation. The central beam passes through the interdental space avoiding overlapping of the teeth on the resultant image by the superimposition of certain tooth structures (e.g., the roots). Visualization of these areas can be achieved by the horizontal angulation (medial or distal), as shown in "B."

Parallax, an adaptation of the standard intraoral imaging technique, allows the specific localization of particular areas of anatomy in relation to others in the maxilla and mandible. It is particularly useful for the demonstration of molar root morphology, pathology and in the examination of impacted teeth and their relationship to surrounding anatomical areas (Figure 39.60a and b).

Bitewing Radiography

A commonly used diagnostic imaging tool for dental caries (tooth decay) in life, this examination is requested frequently as part of the postmortem dental survey because of the high prevalence of antemortem radiographs with which to compare for identification purposes. The crowns of the maxillary (upper) and mandibular (lower) premolar and molar areas can be demonstrated to show restorations, bone levels, anatomical anomaly, or pathology.

For bitewing radiography, a bitewing film holder or bite tab stickers (attached directly to the film) can be used to hold the film/sensor in position by means of both upper and lower teeth "biting" the holder or tab. Positioning aids will normally be necessary to maintain occlusion during x-ray exposure.

The use of robust bitewing holders is recommended for this examination and these will often require careful manipulation to position accurately. Positioning of the area under examination is crucial: it should not be canted (tilted) or rotated and the midline should remain perpendicular to the floor. Where the maxilla and mandible have become disassociated, for whatever reason, they should be assembled and supported as close to the original anatomical position as achievable to replicate any ante mortem radiographs where possible.

Occlusal Radiography

Occlusal radiographs are rarely indicated in forensic situations but are sometimes useful in the demonstration of

TABLE 39.1
X-ray Tube Angulations for the Bisecting Angle Technique

	Maxillary (Upper) Teeth	
Incisors	Caudal (towards the baseof the skull)	50–60°
Canine	Caudal	40–50°
Premolars	Caudal	30–40°
Molars	Caudal	20–30°
	Mandibular (Lower) Teeth	
Incisors	Cephalic (towards the apex of the skull)	25°
Canine	Cephalic	15–20°
Premolars	Cephalic	10–15°
Molars	Cephalic	0–5°

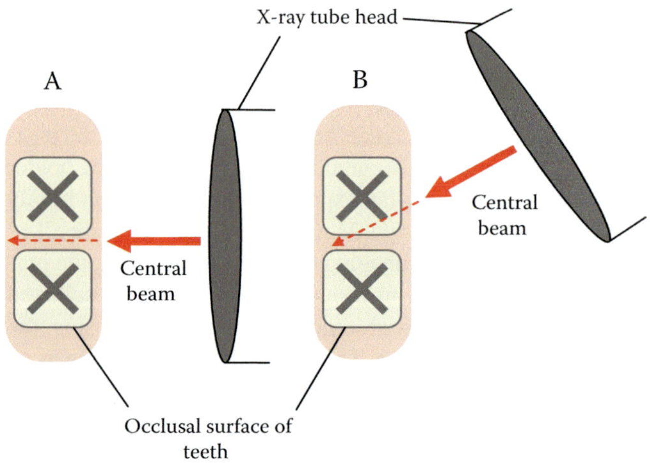

FIGURE 39.59 The principles of Parallax imaging. (Adapted from Faircloth, E.)

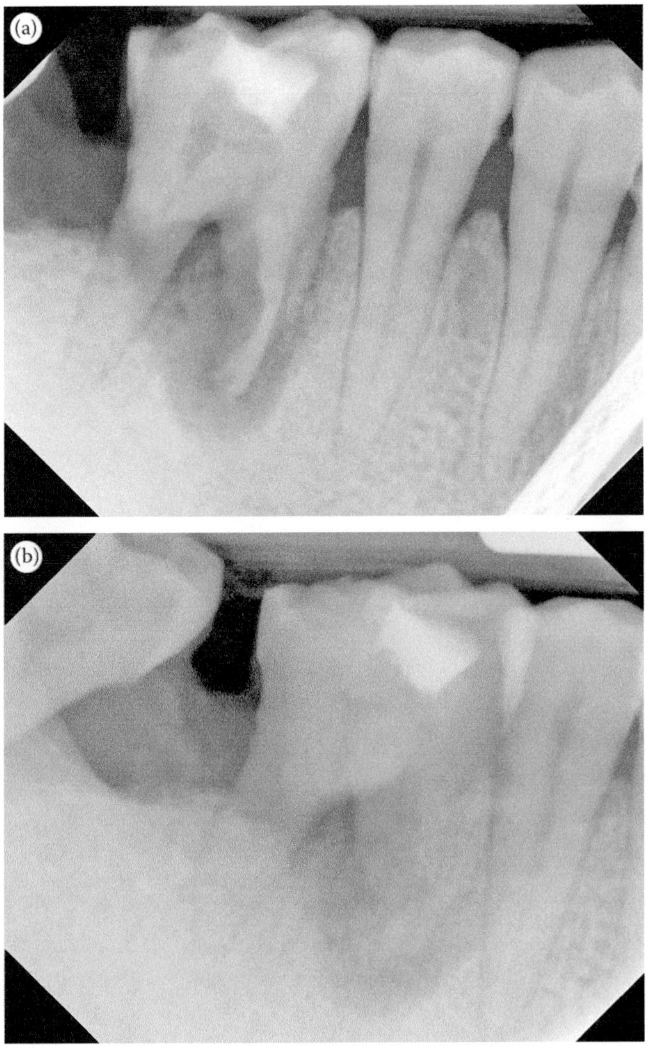

FIGURE 39.60 (a) Intraoral x-ray of the right lower first molar showing extensive pathology involving the mesial root. (b) Intraoral x-ray of the right lower first molar. The parallax technique has been applied to further demonstrate the mesial root and surrounding pathology. (From Kenyon International. With permission.)

anatomical anomalies, restorations (e.g., dental bridge work) and prostheses (dentures). Antemortem orthodontic treatment is also well visualized using occlusal imaging. It is worth noting that direct digital sensors for occlusal radiographs are not generally available in the forensic setting.

The anterior oblique occlusal view is achieved by positioning an appropriately sized (occlusal) film or sensor between the occlusal surfaces of the teeth as far back into the mouth as possible. The dental x-ray tube is positioned with a caudal angle (toward the base of the skull) of 60–65° with the central x-ray beam directed to the center of the film or sensor. Positioning of the head or dentition is important with this technique: the anatomic midline should remain perpendicular to the floor, with the maxillary occlusal plane (a line running along the occlusal surface of the maxillary teeth) parallel to the floor. If this positioning is not achievable, the tube angulation can be adapted to compensate for this, as demonstrated in Figure 39.61.

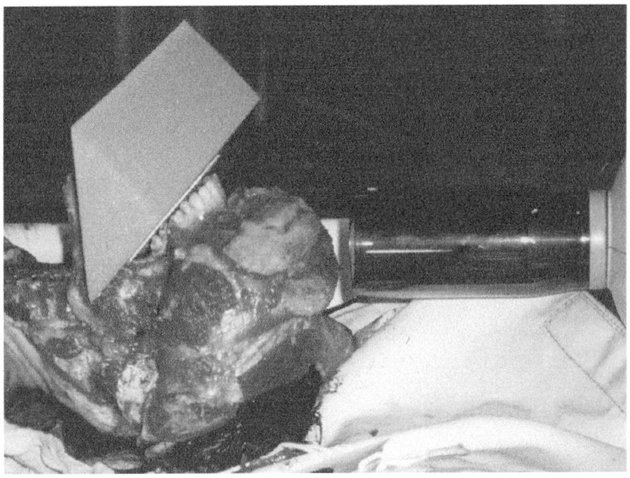

FIGURE 39.61 Positioning of the anterior oblique occlusal. Note the compensatory angulation of the x-ray tube in relation to the anatomic positioning.

REFERENCES

1. Ballinger, P. W., *Merrill's Atlas of Radiographic Positions and Radiological Procedures*, Vol. I, 8th ed., Mosby-Yearbook, St. Louis, 1995, p. 290.
2. Ballinger, P. W., *Merrill's Atlas of Radiographic Positions and Radiological Procedures*, Vol. II, 8th ed., Mosby-Yearbook, St. Louis, 1995, 246.
3. Ballinger, P. W., *Merrill's Atlas of Radiographic Positions and Radiological Procedures*, Vol. II, 8th ed., Mosby-Yearbook, St. Louis, 1995, 244.
4. Ballinger, P. W., *Merrill's Atlas of Radiographic Positions and Radiological Procedures*, Vol. II, 8th ed., Mosby-Yearbook, St. Louis, 1995, 308.
5. Swallow, R. A. and Naylor, E., Eds., *Clark's Positioning in Radiography*, 11th ed., Aspen, Rockville, MD, 1986, p. 243.
6. Blinkhorn, A. S., Choi, C. L., and Paget, H. E., An investigation into the use of the FDI tooth notation system by dental schools in the UK, *European Journal of Dental Education*, 2 (1), 39–41, 1998.
7. du Saucey, M. J. and Brown, K. A., Post-mortem dental radiography: A useful innovation, *Journal of Forensic Odontostomatology*, 9(1), 24–8, 1991.

BIBLIOGRAPHY

Ballinger, P. W., *Merrill's Atlas of Radiographic Positions and Radiological Procedures*, 8th ed., Mosby-Yearbook, St. Louis, 1995.

Bontrager, K. L., *Radiographic Positioning and Related Anatomy*, 3rd ed., Mosby-Yearbook, St. Louis, 1993.

Bushong, S. C., *Radiological Science for Technologists*, 4th ed., C. V. Mosby, St. Louis, 1988.

Carlton, R. R. and Adler, A. M., *Principles of Radiographic Imaging*, Delmar Publishers, Albany, NY, 1992.

Curry, T. S., Dowdey, J. E., and Murry, R. C., *Christensen's Introduction to the Physics of Diagnostic Radiology*, 3rd ed., Lea and Febiger, Philadelphia, 1984.

Dennis, C. A., May, C. R., and Eisenberg, R. L., *Radiographic Positioning Pocket Manual*, Little, Brown, Boston, 1993.

Dowd, S. and Wilson, G., *Encyclopedia of Radiographic Positioning*, W. B. Saunders, Philadelphia, 1995.

Mason, R., and Bourne, S., *A Guide to Dental Radiography*, Oxford University Press, 4th ed., Oxford, UK, 1998.

Meschan, I., *Radiographic Positioning and Related Anatomy*, W. B. Saunders, Philadelphia, 1978.

Mincer, H., Chaudhry, J., Blankenship, *J.* Postmortem Dental Radiography. *Journal of Forensic Sciences*, 53 (2), 405–407, 2008.

Swallow, R. A. and Naylor, E., Eds., *Clark's Positioning in Radiography*, 11th ed., Aspen, Rockville, MD, 1986.

Thomas, C. L., Ed., *Taber's Cyclopedic Medical Dictionary*, 16th edn., F. A. Davis, Philadelphia, 1985.

Thompson, T., *Cahoon's Formulating X-Ray Techniques*, 9th ed., Duke University Press, Durham, NC, 1979.

Wallace, J. E., *Radiographic Exposure Principles and Practice*, F. A. Davis, Philadelphia, 1995.

Whaites, E., *Essentials of Dental Radiography and Radiology*, Churchill Livingstone, 3rd edn., pp. 53, Edinburgh, UK, 2002.

Index

C

G

H

N

O

P

T

U

V

W

X

Y

Z

For Product Safety Concerns and Information please contact our
EU representative GPSR@taylorandfrancis.com Taylor & Francis
Verlag GmbH, Kaufingerstraße 24, 80331 München, Germany